Textbook of Microbiology and Immunology

Subhash Chandra Parija

Textbook of Microbiology and Immunology

Fourth Edition

Springer

Subhash Chandra Parija
Sri Balaji Vidyapeeth (Deemed-to-be-university)
Pondicherry, India

ISBN 978-981-19-3314-1 ISBN 978-981-19-3315-8 (eBook)
https://doi.org/10.1007/978-981-19-3315-8

This Springer imprint is published by the registered company Springer Nature Singapore Pte Ltd.
The registered company address is: 152 Beach Road, #21-01/04 Gateway East, Singapore 189721, Singapore

Lotus Feet of Divine Mother, Maa Sarala,
Almighty Goddess of Learning and Wisdom

My Father
Late Shri Managovinda Parija
Mother
Late Smt Nishamani Parija
Wife Ms Jyotirmayee Parija
and
My Professional Colleagues and Mentors

Preface to the Fourth Edition

It is my great pleasure to author the fourth edition of *Textbook of Microbiology and Immunology*, a book which has received huge support from the readers across India and abroad for the previous editions. A large number of suggestions and feedback were received from the faculty members and students using the previous edition of the book and a sincere attempt has been made to incorporate the appropriate suggestions while preparing this edition.

In the backdrop of the transitions of modern healthcare from generic to personalized, One health replacing the health, medical education is going through transformations to accommodate the new and revise or update the basics. In this landscape, the resurgence of existing pathogens, emerging and re-emerging infections, and the worldwide phenomenon of antimicrobial resistance have made microbiology more relevant where contextual clinical interventions demand more corroboration with accurate diagnosis and syndrome-based investigation rather than classical organismal views. It has been future triggered by our present century's experience with the COVID-19 pandemic where parallel scientific advances in terms of molecular tools and technology and their affordability as a whole have brought whole new revolutions in our understandings and dealings with pathogens. As such, a reflection of these developments in the existing academic book was necessary for the fraternity including the student and teachers.

The recent introduction of "Competency-Based Medical Education" (CBME) by the National Medical Commission, India, has resulted in a sea change in the microbiology syllabus for the undergraduate (MBBS) and postgraduate (MD) students of this country. The approach has shifted from organism-related teaching to system and syndrome-based exposition of infectious diseases which is more relevant for the training of a medical student in Faculty of Medicine.

This fourth edition of *Textbook of Microbiology and Immunology* has, therefore, made the first attempt to be more inclusive and contextual to meet the current demand gap in simplistic approaches keeping stock of the latest development in the field of medical microbiology and immunology. While doing so, this book has blended the traditional organism-based learning and a syndromic approach to infectious disease, together with the introduction of new and modified chapters incorporating the latest information in this field.

This book contains 72 chapters distributed in six specific sections.

Part I deals with the aspects of General Microbiology. Whereas the earlier books in the subject dealt almost exclusively with bacteria in focus, in this edition due importance has been given to other classes of microbes like fungi, viruses, etc. Separate chapters have been assigned to these organisms so that the student can have a baseline knowledge about them. In addition, the chapters on antimicrobial agents and diagnostic methods have been completely rewritten to encompass all pathogenic microbes and parasites. A notable feature is the addition of a new chapter on genomics and proteomics which are gaining increasing importance not only in Microbiology but in almost all specialties of medical sciences.

Part II deals with immunology in the context of medical education and requirements. It has been revised to make it more concise and updated to accommodate the newer developments in this field. Each chapter has detailed components of immune systems, diseases, etc. and has described the underlying processes.

Parts III, IV and V deal with bacteriology, virology and basic mycology, respectively. Wherein the first few chapters of these sections describe the history and current status and approaches involving antimicrobial therapy and laboratory diagnosis including molecular; the rest of the chapters give a specific look into different types of pathogenic bacteria and viruses. This book contains the latest updates on their description, taxonomic status and diseases including the symptoms. It also considers the detection and management roadmaps that have been prescribed by the competent national and international regulatory bodies.

Part VI describes the applied aspects of clinical microbiology. It deals with pathogens in water, milk, air and food, and also collects information on the plethora of healthcare-associated infections. It also updates knowledge on waste handling, management, diagnostic modalities and quality assurance strategies and program in a diagnostic laboratory. As such, this section shall act as a handbook for the practitioner and the laboratory personnel to provide good services by maintaining appropriate measures and regulatory compliances.

Glossary and Further readings are also included, at the end, for the benefit of all.

I hope that the book in its new format will be well appreciated by the readers and can serve not only as a textbook for undergraduate medical students but will also form a foundation base for the postgraduate students on which they can build up their more advanced knowledge. Any suggestion or feedback from the readers will be highly appreciated and can be addressed to me subhashparija@gmail.com for further improvement in subsequent editions.

Pondicherry, India

Subhash Chandra Parija

Preface to the First Edition

The intent of the book is to provide an up-to-date information on microbial diseases which are emerging as an important health problem worldwide. This book has been written to provide a comprehensive coverage of basic and clinical microbiology, including immunology, bacteriology, virology and mycology, in a clear and succinct manner. The book also intends to provide an accurate presentation of clinically relevant information to the learners of medical microbiology.

Textbook of Microbiology and Immunology consists of six sections. Parts I and II deals with general microbiology and immunology, respectively. Parts III, IV and V deals with bacteriology, virology and mycology, respectively. Lastly, Part VI deals with applied microbiology and includes epidemiology and control of community infections, hospital infections, antimicrobial chemotherapy, water analysis and immunisation.

Emphasis, throughout the text, is made on the clinical applications of microbiology to study infectious diseases.

Cultivation and identification of each organism along with pathogenesis of diseases, clinical manifestations, diagnostic laboratory tests, treatment and prevention and control of resulting infections are thoroughly updated to include most recent advances in the field. Details are summarised in the tabular format. Clinical cases are provided in most of the chapters. The book is profusely illustrated with line diagrams and photomicrographs both black and white and colour.

I believe this book will be a useful source of comprehensive information for students mainly the undergraduate students of medicine, allied sciences and others who are interested in medical microbiology.

I welcome reader's views and suggestions for further improvement of the book in the future edition. Suggestions may kindly be e-mailed at subhashparija@yahoo.co.in.

Pondicherry, India

Subhash Chandra Parija

Editorial Board

Acknowledgements

I wish to place on record my sincere thanks to all those who helped us in bringing out this fourth edition *Textbook of Microbiology and Immunology*.

First, I like to thank members of the Editorial Board of the book, my professional colleagues, Prof Abhijit Chaudhury (SVIMS, Tirupati), Prof Shivaprakash M Rudramurthy (PGIMER, Chandigarh) and Prof Tuhina Banerjee (IMS, BHU, Varanasi), for their immense contribution in enriching the contents and also for sharing many images that have found place in the book.

My thanks to my elder brother Shri Kailash Chandra Parija, niece Er Kukumina Ray, son-in-law Er Subhasis Ray, nephew Er Rajkumar Parija, daughter-in-law Ms. Smrithi Parija, daughters Dr Madhuri Parija, son-in-law Dr Ajay Halder, Er Ms. Mayuri Parija and son-in-law Er Shailesh Nandan, and grandchildren Sri Harihar, Ms Shyama and Sri Ram and brother-in-law Dr Biraja Prasanna Das for their support during preparation of the manuscript.

Thanks are due to Prof Ujjala Ghosal (SGPGI, Lucknow), Prof Pramodhini, Dr Namratha Bhosle, Dr Vanathy and Dr Abhijit Poddar of Department of Microbiology, Prof S. Padmavathy, Department of Pharmacology, and Prof Richa Gupta, Department of Physiology, Shri Balaji Vidyapeeth, Pondicherry, for their assistance during preparation of the manuscript.

I like to thank Sri Ram Kumar, IT,Shri Balaji Vidyapeeth and Mr Tamilselvan for preparing images and illustrations of the book.

I am especially thankful to Dr Naren Aggarwal and Dr Bhavik Sawhney of Springer Nature, the Publisher of the book, for their wholehearted support to me in carrying out this project and looking after the countless things which go before the publication of a book.

Pondicherry, India

Subhash Chandra Parija

Contents

About the Author

Subhash Chandra Parija MBBS, MD, PhD, DSc, FRCPath is currently the Vice-Chancellor of Shri Balaji Vidyapeeth, Pondicherry. He was the former Director, also Dean (Research) and Head of the Department of the Jawaharlal Institute of Postgraduate Medical Education & Research (JIPMER), Pondicherry. Dr Parija was the Director of Clinical Services at the BP Koirala Institute of Health Sciences, Dharan, Nepal. He obtained MBBS (1977) from Utkal University, Cuttack; MD (Microbiology) (1981) from Banaras Hindu University (BHU), Varanasi; PhD (Microbiology) (1987) from the University of Madras, Chennai. He is one of the very few Medical Microbiologists of India to be conferred DSc (Microbiology), the highest degree in research, for his contribution in the field of Medical Parasitology by the University of Madras, Chennai. Prof Parija was conferred with the Distinguished BHU Alumnus Award in the field of Medical Sciences.

Prof. Parija has nearly four decades of teaching and research experience, contributing mainly to infectious diseases, especially parasitology, diagnostics and public health. He is actively involved in conducting research programmes on immunology, epidemiology and simple, cost-effective diagnostic tests for parasitic diseases in the community and low resource settings.

Prof. Parija, author of more than 400 research papers, has supervised more than 85 PhD, MD and other postgraduate theses as supervisor and co-supervisor, has 23 copy rights, 1 patent granted and 1 patent published, and transferred one technology in commercialisation of a product. He has authored/edited 17 books including most popular *Textbook of Medical Parasitology*, and the most recent *Effective Medical Communication, The A,B,C,D, E of it*; *Textbook of Parasitic Zoonoses* published by Springer.

Prof Parija is a Fellow of the Royal College of Pathologists, London, and International Academy of Medical Sciences, New Delhi. He is also a Fellow of many professional bodies of eminence such as the National Academy of Medical Sciences, New Delhi; the Indian College of Pathologists, New Delhi; the Indian Academy of Tropical Parasitology, Pondicherry and many others. In recognition of his immense contributions in research in parasitic diseases, Prof Parija has been honoured with more than 26 awards both international and national such as Dr BC Roy National Award of the Medical Council of India, BPKIHS Internal Oration Award, Dr R.V. Rajam Oration Award of the National Academy of Medical Sciences, Distinguished BHU Alumni Award of the Banaras Hindu University, Dr. SC Agarwal Oration Award of the

Indian Association of Medical Microbiologists and Dr BP Pandey Memorial Oration Award of the Indian Association of Medical Microbiologists.

Among others, Prof Parija founded the Indian Academy of Tropical Parasitology, launched a Scientific Journal *Tropical Parasitology*, initiated a quality assurance programme in diagnostic parasitology, also founded Health and Intellectual Property Rights Academy to promote Intellectual Property Rights Activities in Health Sciences, and mentored students both undergraduates and postgraduates, PhD scholars and young faculty to peruse their interest and carrier in parasitic diseases of public health importance.

The current areas of his interest include e-governance, integration of communication technology in medical healthcare and medical education including effective medical communication.

Part I

General Microbiology

Introduction and History of Microbiology

1

Medical Microbiology

Medical microbiology is a branch of microbiology that deals with the study of microorganisms including bacteria, viruses, fungi and parasites of medical importance that are capable of causing diseases in humans. It also includes the study of microbial pathogenesis, disease pathology, immunology and epidemiology of diseases.

Girolamo Fracastoro (1478–1553 AD), a physician, suggested that invisible living creatures caused disease. In his book "*De contagione, contagiosis morbiset curatione* (On Contagion, Contagious Diseases and their Treatment)", published in 1546, he proposed the revolutionary theory that infectious diseases are transmitted from person to person by minute, invisible particles, or seminaria, that are self-replicating and act on the body's humours to cause disease. However, his theories were ahead of time. They were proved only after 200 years when the microscope was invented.

Scientists and Their Contributions to the Development of Microbiology

Anton van Leeuwenhoek: The Microscopist

The first person to observe and describe microorganisms accurately was an amateur microscopist Antony van Leeuwenhoek (1632–1723) of Delft, Holland. Leeuwenhoek earned his living as a draper and haberdasher (a dealer in men's clothing and accessories). However, he spent much of his spare time constructing simple microscopes composed of double convex glass lenses held between two silver plates. His microscopes could magnify around 50–300 times. In 1673, Leeuwenhoek sent detailed letters describing his discoveries to the Royal Society of London. It is clear from his descriptions that he saw both bacteria and protozoa. However, he did not evaluate these organisms as agents of disease.

Louis Pasteur: Father of Microbiology

Louis Pasteur, a French microbiologist, is known as the father of medical microbiology for his immense contributions to the field of medical microbiology (Fig. 1.1). He first coined the term "microbiology" for the study of organisms of microscopic size. Many of his significant contributions are discussed in subsequent sections.

1. **Germ Theory of Disease**: Many scientists have contributed to the theory of spontaneous generation with their experiments, but Louis Pasteur (1822–1895) settled it once for all. Pasteur proved that all life, even microbes, arose only from their like and not de novo

Fig. 1.1 Louis Pasteur (https://en.wikipedia.org/wiki/Louis_Pasteur)

(germ theory of disease). Pasteur had resolved the controversy by 1861 and had shown how to keep solutions sterile.

In the early nineteenth century, support for the germ theory of disease began to accumulate. Agostino Bassi (1773–1856) first showed that a microorganism could cause disease in 1835 by demonstrating that a fungal infection caused the silkworm disease. He also suggested that many diseases were due to microbial infections. In 1845, MJ Berkeley proved that a fungus caused the great potato blight of Ireland.

2. **Pasteurisation**: Pasteur, for the first time, demonstrated that he could kill many microorganisms in wine by heating and then rapidly cooling the wine, a process now called pasteurisation.

   Before pasteurisation, the concept of heating to kill microbes had existed. It was common to protect the beer and vinegar. Nicolas Appert, the inventor of in-container sterilisation, also known as canning, had already shown that treating food with heat could preserve it. However, Pasteur determined the exact time and temperature that would kill the harmful microorganisms in the wine without changing its taste. For this, he obtained a patent for the process by the name of pasteurisation. The process was first used commercially in 1862 to protect milk.
3. **Vaccination**

   In 1877, Pasteur studied anthrax, a disease mainly of cattle and sheep. He developed a vaccine using a weakened strain of the anthrax bacillus, *Bacillus anthracis*. He attenuated the culture of anthrax bacillus by incubation at a high temperature of 42–43 °C and inoculated the attenuated bacilli in the animals. He demonstrated that animals receiving inoculation of such attenuated strains developed specific protection against anthrax. In 1885, he also developed the first vaccine against rabies in humans that saved millions of human lives worldwide. Pasteur coined the term "vaccine" to commemorate Edward Jenner, who used such preparations for protection against smallpox.

Joseph Lister: The Pioneer of Antiseptics

Indirect evidence that microorganisms are the agents of human disease came from English surgeon Joseph Lister (1827–1912). Lister developed an antiseptic surgery system to prevent wound infections from preventing microorganisms from entering the wounds. In this system, the instruments were sterilised by heat and phenol was used in the surgical dressing. Phenol was also used to clean the surgical area. This system was successful, and it transformed the way surgery was done. It also provided strong indirect evidence for the role of microorganisms in disease because phenol, which killed the bacteria, also prevented wound infections.

Robert Koch: The Founder of Koch Postulates

German physician Robert Koch (1843–1910) was the first to directly demonstrate the role of bacteria in causing disease from the study of anthrax.

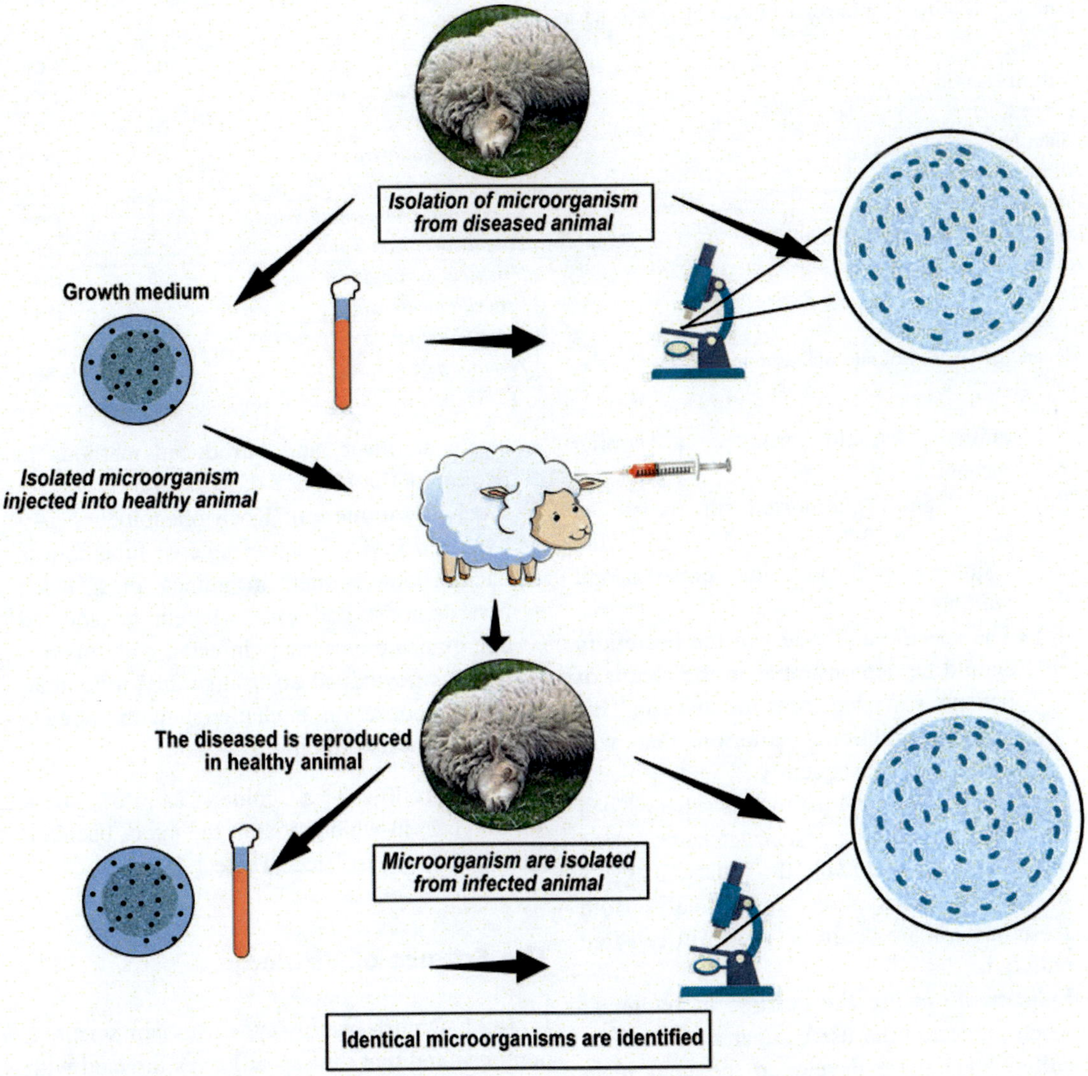

Fig. 1.2 Koch postulates

Koch used the criteria proposed by his former teacher, Jacob Henle (1809–1885), to establish the relationship between *B. anthracis* and anthrax. He published his findings in 1876, briefly describing the scientific method he followed.

1. **Koch Postulates**: Koch postulates (criteria) were useful to prove the claim that a microorganism isolated from disease was indeed causally related to it. A microorganism was accepted as the causative agent of infectious disease only when it satisfied all the following criteria (Fig. 1.2):

(a) The microorganism must be present in every case of the disease but absent from a healthy host.
(b) The suspected microorganism must be isolated and grown in pure culture from lesions of the disease.
(c) The isolated organism, in pure culture, when inoculated in suitable laboratory

Table 1.1 Discovery of important bacterial agents causing human diseases

Scientist	Bacteria	Year
Gerhard Hansen	*Mycobacterium leprae*	1874
Robert Koch	*Bacillus anthracis*	1876
Albert Neisser	*Neisseria gonorrhoeae*	1879
Alexander Ogston	*Staphylococcus aureus*	1880
Friedrich Loeffler	*Corynebacterium diphtheriae*	1884
Louis Pasteur, George Sternberg	*Streptococcus pneumoniae*	1886
Anton Weichselbaum (Vaykselbaum)	*Neisseria meningitidis*	1887
David Bruce	*Brucella melitensis*	1887
Kitasato Shibasaburo	*Clostridium tetani*	1889
Alexandre Yersin, Kitasato Shibasaburo	*Yersinia pestis*	1890

animals, should produce a similar disease.

(d) The same microorganism must be isolated again in pure culture from the lesions produced in experimental animals.

(e) The specific antibodies to the bacterium should be demonstrable in the serum of patients suffering from the disease. This was an additional criterion that was introduced subsequently.

Most human bacterial pathogens satisfy Koch postulates except for *Mycobacterium leprae* and *Treponema pallidum*, the causative agents of leprosy and syphilis, respectively. Both these bacteria are yet to be grown in cell-free culture media.

2. **Solid medium for the culture of bacteria**: Koch pioneered the use of agar as a base for culture media. He developed the pour plate method and was the first to use solid culture media for the culture of bacteria. This development made possible the isolation of pure cultures that contained only one type of bacterium and directly stimulated progress in all areas of bacteriology. Koch also developed media suitable for growing bacteria isolated from the body. He developed the nutrient broth and nutrient agar media that are still widely used worldwide. In 1882, Koch had used these techniques to isolate the bacillus that caused tuberculosis in humans. He also discovered that cholera was caused by *Vibrio cholerae*. He invented the hot air oven and steam steriliser, and introduced methods to determine the efficacy of antiseptics.
3. **Koch Phenomenon**: Koch phenomenon is a hypersensitivity reaction against tuberculosis bacilli demonstrated in guinea pigs. Koch first demonstrated this, wherein he showed that guinea pigs already infected with tubercle bacillus developed an exaggerated inflammatory response on a challenge with tubercle bacillus or its protein.

There followed a golden age of about 40–50 years in which most of the major bacterial pathogens were isolated (Table 1.1).

The Science of Virology

In 1892, Dmitri Ivanovsky, a Russian scientist, demonstrated that the sap of leaves infected with tobacco mosaic disease retains its infectious properties even after filtration through Chamberland filter candles. This provided an operational definition of viruses and an experimental technique by which an agent could be considered a virus.

Beijerinck, a Dutch soil microbiologist, showed that the filtered sap could be diluted and then regain its strength after replication in the living and growing tissue of the plant. The agent could reproduce itself but only in living tissues, not in the cell-free sap of the plant. This explained the failure to culture the pathogen outside its host.

These observations contributed immensely to discovering an organism smaller than bacteria (a filterable agent) that is not observable in the light microscope and can reproduce itself only in living cells or tissues. Beijerinck called this agent a *contagium vivumfluidum* or a contagious living liquid.

The concept of *contagium vivumfluidum* or a contagious living liquid began a 25-year debate about the nature of viruses: whether they were liquids or particles? This conflict was laid to rest when d'Herelle developed the plaque assay in 1917 and the development of electron microscopy by Ruska (1934) when the first electron micrographs of tobacco mosaic virus (TMV) were taken in 1939. After that, viruses were accepted as particles.

Loeffler and Frosch (1898) described and isolated the first filterable agent from animals, the foot-and-mouth disease virus of cattle. Walter Reed and his team in Cuba (1902) recognised the first human filterable virus, the yellow fever virus. Landsteiner and Popper (1909) demonstrated that poliomyelitis was caused by a filterable virus and successfully transmitted the infection to monkeys. Goodpasture (1930) used chick embryos for the cultivation of viruses.

Initially, the term *virus* (taken from the Latin for *slimy liquid* or *poison*) was used interchangeably for any infectious agent and so was applied to TMV and all other agents of the class.

d'Herelle and Twort: Founders of the Principles of Modern Virology

Twort and d'Herelle (1915) independently observed a lytic phenomenon in bacterial cultures, which they attributed to viruses. d'Herelle named these viruses as *bacteriophages*. He developed the use of limiting dilutions with the plaque assay to titre the virus preparation. He suggested that the appearance of plaques in the plaque assay shows the viruses to be particulate or corpuscular.

d'Herelle also demonstrated that the attachment (adsorption) of the virus to the host cell is the first step in the pathogenesis of a virus infection. The attachment of a virus occurred when only bacteria sensitive to the virus were mixed with it, demonstrating the host range specificity of a virus at the adsorption step. He described the process of cell lysis and subsequently the release of infectious virus particles. He developed many other techniques that are still used in virology. d'Herelle was, in many ways, one of the founders of the principles of modern virology.

Immunology as a New Discipline

In the fifth century AD, the earliest smallpox inoculation took place in China. In the 1900s, John Lister, an English merchant, reported the Chinese method to the Royal Society. A Jesuit priest, Father d'Entrecolles, provided the details of this method, and according to him, scabs from pustules are to be collected, and a powder made from them is to be blown into an infant's nose. Pus-coated scabs or thread can be stored, but the operation was usually face-to-face with a patient. In 1747, the same method was used in Japan. A *tika* or dot would be made on the child, usually on the child's foot, by *tikadars* during the pre-colonial period in India.

In 1798, Edward Jenner, an English physician, significantly improved this method. Jenner was fascinated that the milkmaid who had contracted cowpox's mild disease was subsequently immune to smallpox. This made him believe that inoculating cowpox pustule into people may protect them against smallpox. He tested this idea on an 8-year-old boy by inoculating him with fluid from a cowpox pustule, and later, he infected the child with smallpox. As predicted, the child did not develop smallpox. Pasteur followed this up with the development of vaccines for chicken cholera, anthrax and rabies. Although Pasteur proved that vaccination worked, he could not explain how.

The experimental work of Emil von Behring and Shibasaburo Kitasato in 1890 gave the first insight into the mechanism of immunity. They demonstrated that serum contained elements that protected against infections, thus laying the foundation for identifying humoral immunity. In recognition of this work, von Behring received the Nobel Prize in Medicine in 1901.

In 1883, Elie Metchnikoff showed that cells also contribute to the animal's immunity, even before that a serum component could transfer immunity was discovered. He was able to observe that a few white blood cells ingested microorganisms and other foreign material. He named them phagocytes. He also found that these phagocytes were more active in immunised animals. Therefore, he hypothesised that cells other than the serum components were the major effector of immunity. The active phagocytic cells identified by Metchnikoff were most likely blood monocytes and neutrophils.

One of the greatest enigmas facing early immunologists was the specificity of the antibody molecule for foreign material or antigen. The following theories were proposed to explain this mechanism of specificity:

1. The selective theory: The earliest conception of the selective theory dates back to Paul Ehrlich in 1900. In the 1930s and 1940s, the selective theory was challenged by various instructional theories, in which antigen played a central role in determining the specificity of the antibody molecule.
2. The instructional theory: According to the instructional theories, a particular antigen would serve as a template around which the antibody would fold. Friedrich Breinl and Felix Haurowitz first postulated this in the 1930s and redefined it in the 1940s in terms of protein folding by Linus Pauling.
3. The clonal selection theory: The instructional theories were formally disproved in the 1960s, during which information was beginning to appear regarding the structure of Deoxyribonucleic Acid (DNA), Ribonucleic Acid (RNA) and protein. This information offered new insights into the vexing problem of how an individual could make antibodies against almost anything. Then, in the 1950s, selective theories resurfaced due to new experimental data and through the pioneering contributions of Niels Jerne, David Talmadge and F. Macfarlane Burnet. They refined it into a theory that came to be known as the *clonal selection theory*.

The Discovery of Chemotherapeutic Agents

There was no chemical treatment against bacterial infections until the 1930s. Prevention was the only way to protect patients. The Western culture was obsessed with the threat of germs and the responsibility to avoid infection. During this time, there was lots of hope for a wonder drug. In 1889, Paul Vuillemin, a pupil of Louis Pasteur, coined the term "antibiosis".

Paul Ehrlich invented the precursor technique to Gram-staining of bacteria. He demonstrated that dyes react specifically with various components of blood cells and the cells of other tissues. Then he tested the dyes for therapeutic properties. He developed *Salvarsan*, an arsenical compound, in 1909. The compound, known as *the "magic bullet"*, could destroy *T. pallidum*, the causative agent of syphilis. This treatment proved effective against syphilis. This work was of epochal importance, stimulating research that led to the development of sulpha drugs, penicillin and other antibiotics. He, therefore, is known as the father of chemotherapy.

Antibiotics

Antibiotics are a particular class of drugs generally obtained from microorganisms and are used to treat bacterial infections. However, they are ineffective against viral infections and most other infections. Antibiotics are secreted as secondary metabolites by microorganisms to protect them from other microorganisms. The word "*antibiotic*" did not follow immediately, but the drug pyocyanase, a weakly effective antibiotic, was marketed from the late nineteenth century into the 1930s. Sir Alexander Fleming in 1928 accidentally discovered that a substance produced by the fungus destroyed the pyogenic bacteria, staphylococci. This initiated the beginning of the antibiotic era. Other similar antibiotics were discovered in rapid succession. The sulphonamide drugs discovered subsequently offered cures for many bacterial infections.

From their accidental discovery to the present era, several classes of antibiotics have been identified from various microorganisms. It has long served as a tool for humanity for protection against pathogenic bacteria. Unfortunately, this has historically led to the misuse of antibiotics that has offered pathogens to develop resistance against antibiotics. The development of such antibiotic resistance is one of today's leading public health threats. Limited therapeutic options in the absence of antibiotics to treat bacterial infections increase the morbidity and mortality associated with the infectious diseases caused by bacteria. A group of pathogens, collectively called ESKAPE—comprising six bacterial pathogens, *Enterococcus faecium*, *Staphylococcus aureus*, *Klebsiella pneumoniae*, *Acinetobacter baumannii*, *Pseudomonas aeruginosa* and *Enterobacter* spp.—have disturbed health systems globally, leading to life-threatening nosocomial infections due to multidrug resistance. In addition to multidrug-resistant (MDR) bacteria, which are resistant to more than one antimicrobial drug, extensively drug-resistant (XDR) bacteria have been developed, which are resistant to all or almost all approved antimicrobial agents.

Given the limitation of currently circulating antibiotics, a growing need has been felt to find newer classes of antibiotics through extensive research and development (R&D). In this direction, the Global Antibiotic Research and Development Partnership (GARDP) has been developed jointly by the WHO and the Drugs for Neglected Diseases Initiative (DNDi). The GARDP, with an aim to accelerate and deliver new and improved antibiotics to tackle drug-resistant infections, is working with more than 50 public and private sector partners in 20 countries.

Impact of Molecular Biology in Medical Microbiology

In today's world, humans have a greater understanding of the infectious organism due to the ability to look into the gene structure-function level. This approach, which belongs to molecular biology, has greatly escalated efforts towards pathogen detection, surveillance, host-pathogen interaction and finding medical countermeasures that have accumulated more efficiency and accuracy in medical microbiology. In clinical settings, it could detect antimicrobial resistance genes and provide public health information such as strain characterisation by genotyping and also facilitate improved treatment of viral infections by viral resistance and viral load testing against antiviral therapies. **Molecular epidemiology** is far more advanced and newer approaches to epidemiological investigations and has emerged by combining molecular biology with traditional epidemiology. It looks at and investigates diseases by considering genetics and environmental factors at a molecular level. This helps identify important molecules, genes and pathways that influence disease development and enable accurate elucidation of disease aetiology, distribution pattern and penetrance in families and populations. Historically, the phrase "Molecular epidemiology" was introduced by Kilbeurne in 1973 in his article "Molecular Epidemiology of Influenza" and later formalised by Schulte and Perera in their book on Molecular Epidemiology: "principles and practice".

As such, from the traditional approaches of understanding microbes from phenotypes and biochemical properties, the present era is marked with dealing with the DNA, RNA and protein of a microorganism or whole community of microorganisms either culturable or non-culturable approaches. The journey of molecular biology started in 1953 by the discovery of DNA's molecular structure by James Watson and Francis Crick. However, astonishing adoption and transformation in this sector were for discovering the *in-vitro* DNA amplification process, known as the polymerase chain reaction (PCR). This method was discussed by Dr. Kary Banks Mullis in 1983. This landmark discovery and other potential discoveries like finding restriction enzymes and other genome-modifying tools enabled precise manipulation of the genes. It has also contributed to understanding the functions of the genes. The detection of pathogens using PCR has become the hallmark of quality and specificity in the diagnostic laboratory. Specific conserved genes in pathogen (e.g. 16S rRNA for bacteria, structural genes for viruses and 18S rRNA for fungi) are targeted to amplify and detect pathogens in the sample. Further, the advent of highly parallel DNA sequencers and high-throughput mass spectrometers with remarkable mass accuracy and sensitivity is propelling microbiology into a new era, extending its focus from the properties of single organism types in isolation to the operations of whole communities. These new fields, termed **metagenomics**, **metatranscriptomics** and **metabolomics**, look into the microbial taxonomic profile, functional profile and metabolic profiles, respectively.

The Human Genome and Microbiome Projects

Among the most significant scientific achievements of history is the deciphering and mapping of the entire human genome and the associated microbiome through two flagship projects, namely, the Human Genome Project and the Human Microbiome Project.

The health and function of humans are driven by their genes and the resident microorganisms (human microbiome), which interact at the molecular and cellular levels in response to environmental and other factors. The resident microorganisms that outnumber human cells protect an individual from multiple risk factors, including invading pathogens and other diseases. As such, humans are considered "supra-organism", a composite of microbial and human cells wherein human genetic and metabolic features are essentially a blend of human and microbial traits.

From this understanding and with the advent of tools enabling analysing large volume of genes, the **Human Genome Project** was initiated in 1990 by the US Department of Energy (DOE) and the National Institutes of Health, USA. The project was completed in 2013. It completely sequenced approximately 22,300 protein-coding genes that form 99% of the human genome's gene-containing regions with an accuracy of 99.99%. Among many outcomes, the project has led to successful mapping, that is, number, location, size and sequence, of human genes in the entire genome. In addition, it has allowed for the identification of disease-causing alleles and

Table 1.2 Recent Nobel Prize winners

Year	Name/Names(s) of scientists	Research contribution
1993	Kary Mullis	Polymerase chain reaction
1996	Peter C. Doherty and Rolf M. Zinkernagel	Cell-mediated immune defences
1997	Stanley B. Prusiner	Prion discovery
2005	Barry J. Marshall, J. Robin Warren	Discovery of *Helicobacter pylori* and its role in gastritis and peptic ulcer disease
2008	Herald ZurHausen	Discovery of human Papillomaviruses causing cervical cancer
2008	Francoise BarreSinoussi, Luc Montagnier	Discovery of human immunodeficiency virus
2011	Ralph M. Steinman	Discovery of dendritic cell and its role in adaptive immune response
2011	Bruce A. Beutler, Jules A. Hoffmann	Discoveries concerning the activation of innate immunity
2012	Sir John B. Gurdon, Shinya Yamanaka	Mature cells can be reprogrammed to become pluripotent
2015	William C. Campbell and Satoshi Omura	Discoveries concerning a novel therapy (avermectin and ivermectin) against infections caused by roundworm parasites
2015	YouyouTu	Discoveries concerning artemisinin therapy against malaria
2020	Harvey J. Alter, Michael Houghton and Charles M. Rice	Discovery of hepatitis C virus

enabled the production of specific gene probes to detect sufferers and carriers of genetic diseases.

The **Human Microbiome Project** was an experimental extension of the Human Genome Project that aims to generate resources for a comprehensive characterisation of the human microbiome to understand its impact on human health and disease. Funded by the National Institute of Health, USA, the first phase, known as the Human Microbiome Project (HMP) (launched in 2007), studied the human inhabitant microbial communities on nasal, oral, skin, gastrointestinal and urogenital areas. As the second phase of the project, the Integrative HMP (iHMP) (launched in 2014) gathered multiple *-omic data* from both the microbiome and the human host to understand host-microbiome interactions and determine their role in pathogenesis and host immunity. Among many achievements, these projects have enabled capturing the time-lapse "moving pictures" of the human microbiome, and identified healthy and diseased gut microbiota and so on.

The list of recent Nobel laureates in the field of microbiology and infectious diseases is in Table 1.2.

Classification, Nomenclature and Taxonomy of Microbes

2

Classification may be defined as the arrangement of organisms into taxonomic groups (*taxa*) on the basis of their phenotypic (observable) and genotypic (genetic) similarities and differences. It allows for proper and systematic grouping of microorganisms. Organisms are classified into three main kingdoms: animals, plants and Protista. The Protista contain unicellular microorganisms including eukaryotes and prokaryotes.

Classification of Microorganisms

Carl Woese in 1990 put forward the three domains and six kingdom classifications to group all existing living organisms. This is based primarily on variations in the 16S ribosomal RNA structure.

Domain

It is the first and the highest group encompassing all living beings. The three domains are as follows:

1. **Archaea**: These are prokaryotes and members of this group are the most ancient organisms. Though similar to bacteria, these lack nuclear membranes, the cell wall is devoid of peptidoglycan and these can survive in extremes of temperatures or salinity.
2. **Bacteria**: These are "eubacteria" or true bacteria, having lipid-containing cell membranes and their cell wall contains peptidoglycan.
3. **Eukarya**: These have a eukaryote cell structure, with membrane-bound nuclei, and peptidoglycan is absent from the cell wall.

Kingdom

It is the next higher grouping of organisms and in the older classification, there were five kingdoms: Monera, Protista, Fungi, Plantae and Animalia. With the introduction of the domain system, the Monera kingdom has been split into kingdoms Archaebacteria and Eubacteria (Table 2.1). Medically important disease-causing organisms belong to four kingdoms only, with archaebacteria and plantae being non-pathogenic (except for some poisonous substances produced by certain plants).

Medically Important Kingdoms

Kingdom Eubacteria: These are prokaryotic unicellular organisms with a circular DNA and characteristic rRNA. Peptidoglycan is present in the cell wall. There are five main groups of Eubacteria—Proteobacteria, Cyanobacteria, Firmicutes, Chlamydiae and Spirochetes—of which proteobacteria contain the highest number of pathogens.

Kingdom Fungi: For a long time, fungi were classified under the plant kingdom but later a separate kingdom was assigned to these

Table 2.1 Broad classification of living organisms

Archaea	Bacteria	Eukarya
Kingdom: Archaebacteria	Kingdom: Eubacteria	Kingdom: Protista
		Kingdom: Fungi
		Kingdom: Plantae
		Kingdom: Animalia

organisms. The morphology of the fungal structures had been used for a long time in classification. The recent application of genomic studies and sequencing has led to the description of nine phyla, in which members of a few groups are human pathogens.

Kingdom Protista: Protozoan parasites are unicellular organisms which have been classed under the Kingdom of Protista. At present, there are 11 phyla of protozoan parasites of which the following phyla comprise medically important protozoan parasites: Amoebozoa, Trichozoa, Percolozoa, Euglenozoa, Miozoa and Ciliophora.

Kingdom Animalia: In contrast to protozoa, the animalia kingdom contains multicellular organisms, and the helminth parasites and ectoparasites of medical importance are members of this group. Trematodes, cestodes and nematodes are the three major helminth groups studied in medical parasitology, while ectoparasites are largely insects.

In addition to the abovementioned organisms, a distinctly separate group of disease-causing agents are the **viruses**. By their nature, viruses are considered non-living outside a living body, but can replicate inside a live cell or host. The nature of the nucleic acid content of viruses forms the basis of their classification into DNA and RNA viruses. The two groups are further divided into families based on the special features of the nucleic acids, their mode of replication and basic structural features.

Nomenclature and Taxonomy of Microorganism

Nomenclature refers to the naming of microorganisms.

Taxonomy (alternatively "systematics") is the science of classifying organisms leading to their nomenclature. Two kinds of names are usually given to microorganisms—common name and scientific name:

1. The **common or casual name** for a microorganism varies from country to country and is usually known in the local language. For example, tubercle bacillus, typhoid bacillus and gonococcus are common names for communication at the local level.
2. The **scientific name** is the international name that is accepted throughout the world. By accepted taxonomic conventions, the order names end in *ales* (e.g. the order Eubacteriales), the family names end in *aceae* (e.g. the family Enterobacteriaceae) and the tribe names end in *eae* (e.g. the tribe Proteae). The order, family and tribe names begin with capital letters. The genus name also begins with a capital letter, but the species name (e.g. *coli*) begins with a running letter and not a capital letter. Both the genus (e.g. *Escherichia*) and species names are either italicised or underlined when written in the text. The scientific name of the bacterium when written for the first time is written in full (e.g. *Escherichia coli*), but is later mentioned in an abbreviated form (e.g. *E. coli*). When bacteria are referred to as a group, their names are neither capitalised nor italicised or underlined (e.g. streptococci).

The taxonomic processes and nomenclature of bacteria and archaea are governed by the International Committee on Systematics of Prokaryotes and the Bacteriology and Applied Microbiology Division of the International Union of Microbiological Societies. Nomenclature is done as per the International Code of Nomenclature of Prokaryotes (ICNP), formerly the International

Code of Nomenclature of Bacteria (ICNB) or Bacteriological Code (BC). Novel taxa are validly published in the Society's Journal, that is, the International Journal of Systematic and Evolutionary Microbiology (IJSEM; previously International Journal of Systematic Bacteriology). Publication of novel taxa in other scientific journals is also accepted as valid nomenclature subject to the assessment and inclusion in the IJSEM journal as *Approved List of Bacterial Names*.

The nomenclature of eukaryotic microbial groups is provided for by other codes and governed by the organisations mentioned below:

- Fungi and algae by the International Code of Nomenclature for algae, fungi and plants. It is published and managed by the International Association of Plant Taxonomy (IATP),
- Protozoa by the International Code of Zoological Nomenclature. It is published and managed by the International Commission on Zoological Nomenclature.
- Viruses by the International Code of Virus Classification and Nomenclature. It is manged by the International Committee on Taxonomy of Viruses.

Bacterial Taxonomy

Although no universally accepted bacterial classification system is available, three main approaches are usually followed. These include (1) phylogenetic, (2) adansonian and (3) molecular and genetic classifications, which are discussed in the subsequent sections.

1. **Phylogenetic Classification:** Phylogenetic classification is a type of hierarchical classification that presents a branching tree-like arrangement, with one characteristic being employed for divisions at each branch or level. It is called phylogenetic classification because it denotes an evolutionary arrangement of the species.

   This classification groups together the types that are related on an evolutionary basis where several groups are used such as *Divisions*, *Classes*, *Orders*, *Families*, *Tribes*, *Genera* and *Species*. Some characters of special importance, such as Gram staining properties, lactose fermentation, spore formation and so on, are used to differentiate between the major groups. Less important properties (nutritional requirements for growth of bacteria, production of certain enzymes by bacteria and so on) distinguish between the minor groups such as the genera and the species.

   Bergey Manual of Systematic Bacteriology is an authoritative published compilation that describes a phylogenetic classification of bacteria. The manual is a compilation of names and descriptions of bacteria, and is extremely useful for identifying newly isolated bacterial types. It describes their essential characteristics such as the morphology of the bacteria, staining properties, cultural characteristics, biochemical reactions, antigenic structure, the guanine to cytosine ratio of DNA and so on, which are used for identifying and classifying bacteria. It is the definitive reference book accepted worldwide.

2. **Adansonian Classification:** The adansonian classification, the first classification, was proposed by Michael Adanson in the eighteenth century and considers the characteristics expressed at the time of the study. Hence, it is called a **Phenetic system**. This classification gives equal weight to all measurable features and groups of bacteria based on similarities of several characteristics.

   The availability of computer facilities has expanded the scope of phenetic classification. It allows for the comparison of a vast number of properties of several organisms simultaneously. A computer analysis of a large number of characteristics of a bacterium facilitates the identification of several broad sub-groups of bacterial strains that are further sub-divided into species and are represented in a

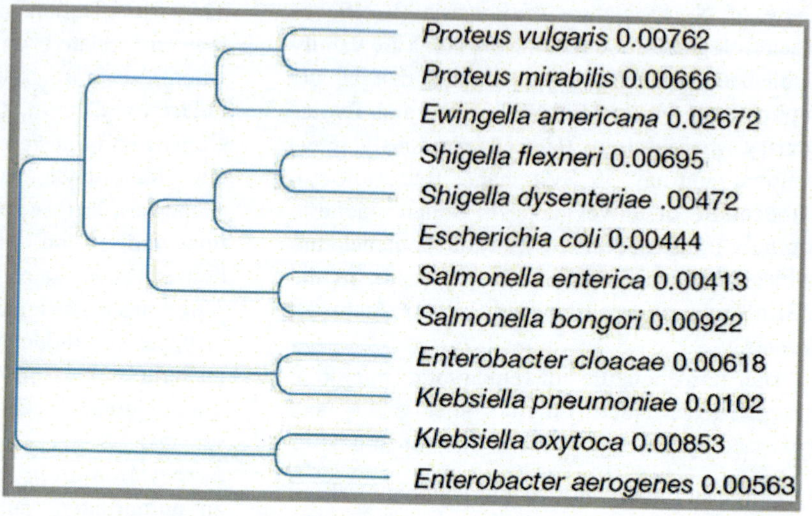

Fig. 2.1 Numerical taxonomy based on the dendrogram

dendrogram (Fig. 2.1). This type of classification, based on a large number of properties, is known as numerical taxonomy.

2. **Molecular and Genetic Classification:** The molecular classification is based on the homology of the DNA base sequences of the microorganisms. First, DNA is extracted from the organism, and the DNA relatedness of the microorganisms is tested, and the nucleotide sequence of the DNA is studied using DNA hybridisation or recombination methods. The degree of hybridisation can be assessed by many methods, such as by using labelled DNA preparations.
   The genetic relatedness can also be assessed by studying the messenger RNA (mRNA). Also, ribosomal RNA (rRNA) analysis is of immense value. Evolutionary relationships among widely divergent organisms have been shown by studying the nucleotide sequence of 16S ribosomal RNA from different biologic sources. It has contributed to the understanding of new groups of bacteria such as the archaebacteria. In recent times, genetic classification is increasingly being used to study viruses.

Intraspecies Classification

Intraspecies classification makes an attempt to sub-classify species of a bacteria based on biochemical properties (bio-types), antigenic properties (serotypes), susceptibility to bacteriophage (phage types) and production of bacteriocins (colicin types). Recently, molecular methods have increasingly been used for intraspecies classification of microorganisms, especially viruses.

Molecular methods of intraspecies classification are of two types—phenotypic and genotypic.

- **Phenotypic methods** are based primarily on the study of expressed characteristics by microorganisms and are carried out by performing electrophoretic typing of bacterial proteins and immunoblotting.
- **Genotypic methods** include direct analysis of genes, and chromosomal and extrachromosomal DNA. These genotypic methods include plasmid profile analysis, restriction endonuclease analysis of chromosomal DNA with Southern blotting, PCR and nucleotide sequence analysis.

Type Cultures

The original cultures of all the reference bacteria, including new species, would be preserved in international reference centres. These culture collections are made available to researchers/ microbiologists for research and comparison.

Table 2.2 Differentiating features of fungi and bacteria

Features	Fungi	Bacteria
Diameter	Approximately 4 μm	Approximately 1 μm
Morphology	Yeast and mould	Cocci, bacilli, spirochete, branching filamentous
Staining property	Gram-positive, non-acid fast, stained with PAS and GMS	Gram-positive, Gram-negative, acid fast
Cell wall content	Chitin	Peptidoglycan
Cell membrane	Sterols present	Sterols absent except mycoplasma
Cytoplasm	Mitochondria and endoplasmic reticulum present	Mitochondria and endoplasmic reticulum absent
Nucleus	Eukaryotic	Prokaryotic
Spores	Sexual and asexual spores for reproduction	Endospores for survival, not for reproduction
Thermal dimorphism	Yes (seen in some fungi)	No

Note: *PAS* periodic acid-Schiff, *GMS* Gomori's methenamine silver

The American Type Culture Collection (ATCC), established in 1925, is now headquartered in Manassas, Virginia, USA.

Nomenclature, Taxonomy and Classification of Viruses

The initial classification of viruses was based on the sites of their isolation or the symptomatology of the disease. Since the formation of the International Committee on the Taxonomy of Viruses (ICTV) in 1966, a systematic classification of viral taxonomy and nomenclature was carried out. The ICTV has been introducing a systematic approach for the classification and nomenclature of the viruses. The ICTV has grouped viruses into families based on (1) the type of nucleic acid they possess, (2) their means of replication and (3) their morphology (e.g. membrane envelope).

The *suffix* virus is used for genus names, viridae for family names and ales for order names. In formal usage, the family and genus names are used in the following manner: for example, family Rhabdoviridae, genus Lyssavirus, human rabies virus. A viral species is a group of viruses that share the same genetic information and ecological niche. These viral species are designated by descriptive common names, such as human herpes virus, with sub-species, if any, designated by a number (e.g. HHV-1).

Depending on the type of nucleic acids that viruses possess, they are classified into two groups: deoxyriboviruses, which contain DNA (DNA virus), and riboviruses, which contain RNA (RNA virus). The DNA and RNA viruses associated with human diseases are divided into six and 13 families, respectively.

Taxonomy and Classification of Fungi

Taxonomy

The members of the fungal kingdom are eukaryotes and although most of them are multicellular, some like the yeast have a single-cell structure. Table 2.2 shows the differentiating features of fungi from bacteria, which is a prokaryotic organism. For a long time, fungi were classified under the plant kingdom, but later a separate kingdom was assigned to these organisms.

The basis of fungal classification largely remains the morphology of the yeast or the mould, together with other features, particularly the mode of reproduction and the type of spores produced. Four phyla were initially described based on these features. Subsequently, the taxonomy of fungi underwent a state of flux, a situation which is being resolved in recent years by the application of genomic studies. Novel relationships between various fungal groups have been revealed by using the tools of molecular biology and through the sequencing of 18S ribosomal RNA. As per the latest understanding, nine phylum-level groups have been described: Opisthosporidia, Chytridiomycota, Neocallimastigomycota, Blastocladiomycota, Zoopagomycota, Mucoromycota, Glomeromycota, Basidiomycota and Ascomycota.

Classification of Fungal Diseases

Mycosis (Pleural: **Mycoses**) is the terminology used for a fungal disease. Fungal infections can be classified as follows:

1. **Superficial mycoses**: These infections are confined to the outer layers of the skin, nail or hair (keratinised layers), and rarely invade the deeper tissue or viscera. Those infections, which are limited to stratum corneum or the outermost layer of the skin, are termed superficial mycoses, while infections which affect

Table 2.3 Classification of fungal infections and causative fungi

Type	Fungal infection	Causative fungi	Tissue/system affected
Superficial mycoses	Tinea versicolor	*Malassezia* spp.	Skin, scalp area
	Tinea nigra	*Hortaea werneckii*	Skin
	White piedra	*Trichosporon* spp.	Hair
	Black piedra	*Piedraia hortae*	Hair
	Dermatophytoses	*Epidermophyton* spp. *Microsporum* spp. *Epidermophyton* spp.	Skin, hair, nail
Subcutaneous mycoses	Sporotrichosis	*Sporothrix schenckii, Sporothrix brasiliensis, Sporothrix globosa*	Skin and subcutaneous tissue
	Chromoblastomycosis	*Fonsecaea* spp. *Phialophora* spp. *Cladophialophora* spp. *Rhinocladiella* spp.	Skin and subcutaneous tissue
	Mycetoma/maduromycosis	*Madurella mycetomatis* *Madurella grisea* *Scedosporium apiospermum* *Exophiala jeanselmei*	Skin and subcutaneous tissue, sometimes bone
	Phaeohyphomycosis	*Curvularia* spp. *Bipolaris* spp. *Exophiala jeanselmei* *Scedosporium prolificans*	Skin and subcutaneous tissue
	Rhinosporidiosis	*Rhinosporidium seeberi*	Nasal mucosa, conjunctiva
	Lobomycosis	*Lacazia (Loboa) loboi*	Skin and subcutaneous tissue
Systemic mycoses	Blastomycosis Histoplasmosis Coccidioidomycosis Paracoccidioidomycosis	*Blastomyces dermatitidis* *Histoplasma capsulatum* *Coccidioides immitis* *Paracoccidioides brasiliensis*	Lungs, other systems may also be affected
Opportunistic mycoses	Cryptococcosis	*Cryptococcus neoformans*	Lungs, CNS
	Candidiasis	*Candida albicans* and other *Candida* spp.	Mucosa, skin, blood-stream and various other systems
	Mucormycosis	*Rhizopus, Mucor*	Nasal and orbital areas, lungs, skin
	Pneumocystosis	*Pneumocystis jirovecii*	Lungs
	Aspergillosis	*Aspergillus flavus, Aspergillus niger, Aspergillus fumigatus* and other Aspergilli	Lungs, allergic disease, Mycotoxicosis

only skin, hair and nails in human are termed dermatophytoses.

2. **Subcutaneous mycoses**: Infections in this category are confined to the subcutaneous tissue and only rarely spread systemically to affect the underlying muscles and bones.
3. **Systemic mycoses**: These infections may involve deep viscera and become widely disseminated. These are frequently caused by dimorphic fungi and affect the lungs.
4. **Opportunistic mycoses**: These are caused by low-virulence fungi infecting primarily the immunocompromised hosts.

Table 2.3 shows the details of fungal infections and the causative agents.

3 Microscopy

The microbial world comprises diverse organisms with a wide range of cellular and molecular organisations. The disease-causing agents may be classified into seven groups: bacteria, fungi, protozoa, helminths, arthropods, viruses and prions. Except for viruses and prions, all have a definite cellular structure. However, only the bacteria have a prokaryotic cell structure, while others (except viruses and prions) have a eukaryotic cellular structure. Among medically important organisms, unicellular morphology is exhibited by bacteria, the protozoan parasites and a few fungi.

The kingdom Protista has been divided into three groups based on cellular organisation and biochemistry differences: prokaryotes, eukaryotes and the most recently described archaebacteria.

Prokaryotes: Bacteria and blue-green algae (Cyanobacteria) are prokaryotes. Bacteria are unicellular free-living organisms having both DNA and RNA. They are capable of performing all essential life processes, for example, growth, reproduction and metabolism. They do not show any true branching except actinomycetales, the higher bacteria. Bacteria lack chlorophyll, unlike blue-green algae, which contain chlorophyll.

Eukaryotes: Fungi, algae other than blue-green, protozoa and slime moulds are eukaryotes.

Archaebacteria: These are more closely related to eukaryotes than to prokaryotes. They, however, do not include any human pathogens.

Differences between prokaryotes and eukaryotes have been summarised in Table 3.1.

Size of Microorganisms

All microbes except the worms are microscopic and very small in size. Their size is measured in units of length called microns or nanometres. A micron (micrometre, μm) is the unit of measurement used in bacteriology.

1 micron (μm) = 1/1000 millimetre (mm)
1 nanometre (nm) = 1/1000 micron (μm)
1 Angstrom unit (A°) = 1/10 nanometre (nm)
Bacteria of medical importance measure 2–5 μm (length) × 0.2–1.5 μm (width)
Size of protozoal parasites: <50 μm. Some may be twice this size
Size of a fungal spore: 2–50 μm in diameter

Microscopy

A microscope is an instrument that uses lenses to produce a magnified image of an object that is invisible to the unaided eye. The naked eye

Table 3.1 Distinguishing features of prokaryotic and eukaryotic cells

Features	Prokaryotes	Eukaryotes
Nucleus		
Nuclear membrane	Absent	Present
Nucleolus	Absent	Present
Chromosome	One, circular	More than one, linear
Location	Free in the cytoplasm, attached to mesosomes	Contained in membrane-bound structure
Replication	Binary fission	Mitotic division
Extrachromosomal DNA	Plasmid	Inside the mitochondria
Cytoplasm		
Cytoplasmic organelles like mitochondria, Golgi apparatus and endoplasmic reticulum	Absent	Present
Cytoplasmic streaming	Absent	Present
Lysosomes	Absent	Present
Ribosomes—protein production site	70S (50S + 30S), free in cytoplasm or bound to cell membrane	80S (60S + 40S), attached to rough endoplasmic reticulum
Chemical composition		
Cell wall	Present except for mycoplasma and some cell-wall-deficient bacteria	Absent, except for fungi that have a chitinous cell wall
Sterols	Absent	Present
Muramic acid	Present	Absent
Energy production site	Electron transport chain located in the cell membrane	Within membrane-bound mitochondria

cannot visualise bacteria because the limit of resolution with the unaided eye is about 200 μm. So, the study of bacteria requires the use of microscopes.

Types of Microscopy

The following types of microscopy are used for the examination of microorganisms:

Optical or Light Microscope

The light microscope uses natural or artificial transmitted light as the light source. The resolving power of a microscope is the ability of the microscope's lens system to distinguish two closely placed objects as distinct and separate entities. Resolving power depends on the wavelength of light used to illuminate the object and the numerical aperture of the microscope. It is about half of the wavelength of light being used. For example, the smallest particle which can be resolved by yellow light with a wavelength of 0.4 μm is about 0.2 μm. The microscope's resolving power can be optimised by properly using a condenser that focuses light on the object's plane. Resolving power is enhanced further by adjusting the medium through which light passes between the object and the objective lens. For example, the use of *immersion oil*, whose refractive index is similar to that of glass, improves resolving power. The *numerical aperture* of the microscope is defined as the light-gathering power of the microscope.

Different types of light microscopy include bright-field microscopy, dark-ground microscopy and phase-contrast microscopy.

1. **Bright-field microscopy**: Bright-field microscopy is the most common form of light microscopy that uses a compound light microscope. A compound light microscope consists of a compound lens system that contains several objective lenses, for example, lenses of low power (×10), high power (×40) and oil immersion (×100). It also contains a fixed ocular (eyepiece) lens of ×10 or ×5. The final magnification is the multiplication of the

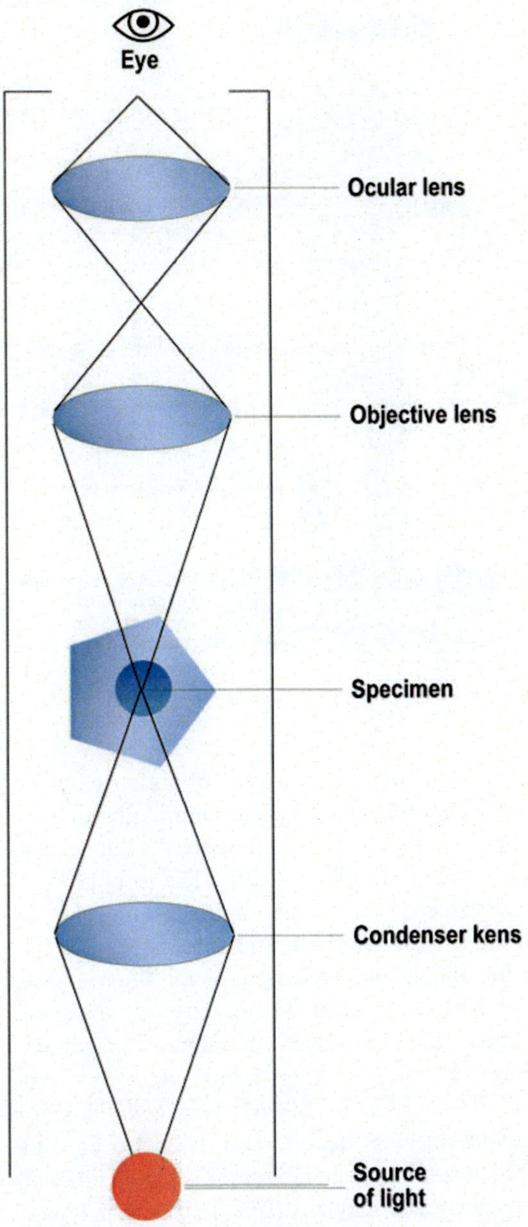

Fig. 3.1 Principle of compound light microscope

power of the objective lens with that of the eyepiece (Fig. 3.1).

Bright-field microscopy has many uses, like examining either wet films or "hanging drop" preparations for demonstrating the motility of flagellated bacteria (e.g. *Escherichia coli*, *Pseudomonas aeruginosa*) and protozoa

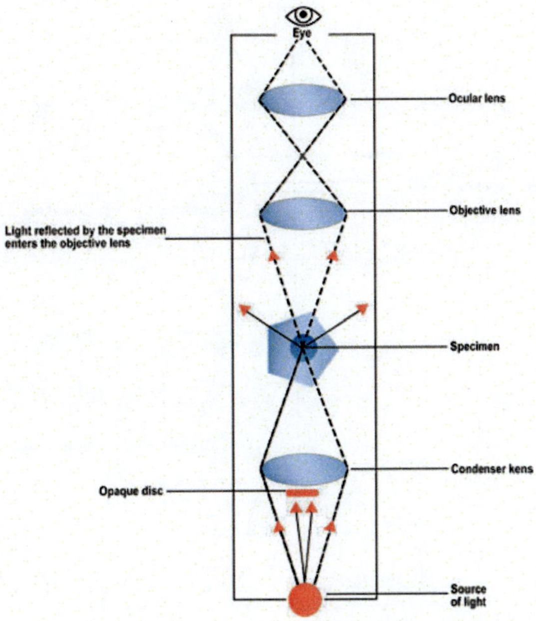

Fig. 3.2 Principle of dark-ground microscopy

(e.g. *Trichomonas vaginalis*, *Giardia intestinalis*). The wet preparation is also helpful in demonstrating microorganisms in urine or faeces and detecting fungi in the skin. In addition, it is helpful for demonstration of the structural details. It is also helpful in measuring the approximate size of the bacteria, fungi and protozoa in stained preparations.

2. **Dark-ground microscopy**: Dark-ground microscopy uses reflected light instead of transmitted light used in the ordinary light microscope (Fig. 3.2). The dark-field condenser with a central circular stop illuminates the object with a cone of light which prevents light from falling directly on the objective lens. Instead, light rays falling on the object are reflected or scattered onto the objective lens. Due to this, the microorganisms appear brightly stained against a dark background.
   Dark-ground microscopy is useful for demonstrating very thin bacteria (e.g. spirochetes) not visible under ordinary illumination. This is a frequently used method for rapid demonstration of *Treponema pallidum* in clinical specimens. It is also

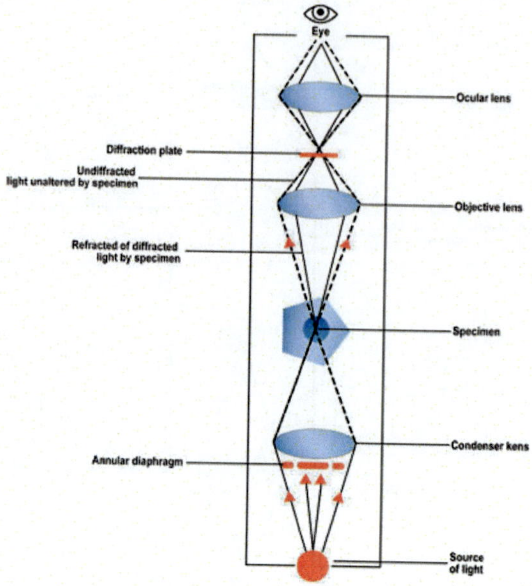

Fig. 3.3 Principle of phase-contrast microscopy

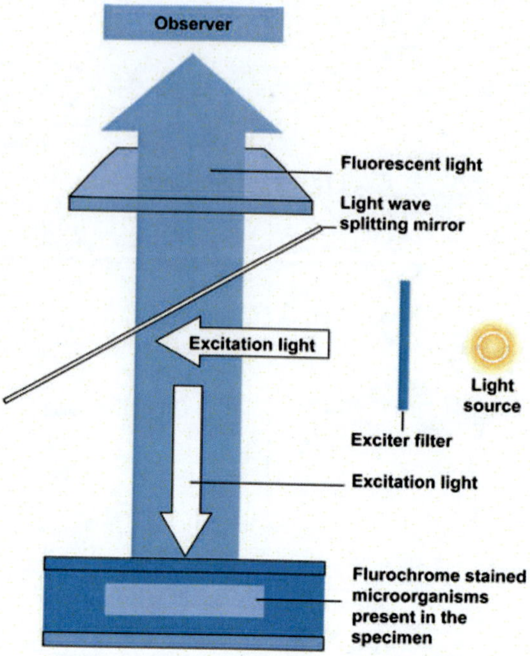

Fig. 3.4 Principle of fluorescence microscopy

helpful for demonstrating the motility of flagellated bacteria and protozoa.

3. **Phase-contrast microscopy**: Phase-contrast microscopy uses a specific optical system that converts differences in phase in an organism into differences in light intensity, thereby producing light and dark contrast in the image (Fig. 3.3). The optical system includes a special condenser and objective lens, which can be fitted to an ordinary light microscope to convert it into a phase-contrast microscope. Phase-contrast microscopy helps examine living microorganisms, particularly protozoa (e.g. *T. vaginalis*, *Entamoeba histolytica*), and the internal structures of a living cell by improving the contrast-differentiating structures within the cell that differ in their thickness and refractive index.

Fluorescence Microscopy

Fluorescence microscopy needs a microscope fitted with an ultraviolet light source. Fluorescence microscopy is based on the principle that specimens stained with fluorescent dye when exposed to ultraviolet light result in the emission of a longer wavelength of light (i.e. visible light) (Fig. 3.4). As a result, the bacteria stained with fluorescent dye appear as brightly glowing objects against a dark background. Auramine O, acridine orange and rhodamine are the fluorescent dyes used to visualise bacteria. The resolution of a fluorescence microscope is much more due to the short wavelength of ultraviolet light.

In diagnostic microbiology, fluorescence microscopy is used to directly demonstrate antigens of a pathogen in clinical specimens using a direct fluorescent antibody test (e.g. direct detection of *Neisseria gonorrhoeae*, *Corynebacterium diphtheriae*). It is also used to estimate antibodies in the serum using an indirect fluorescent antibody test (IFA) (e.g. IFA in leptospirosis, syphilis, brucellosis).

Confocal Laser Scanning Microscope (CLSM)

The first CLSM was developed by Egger and Petran in 1960, almost 10 years after it was conceptualised in 1950s by Minsky. However, the commercial instruments became available in 1987. CLSM helps generate high-resolution optical images with depth selectivity and allows for protein localisation in specific cellular

Table 3.2 Differentiating characteristics of electron microscope and light microscope

Characteristics	Light microscope	Electron microscope
Radiation source	Visible light	Electron beam
Medium of transmission	Air	High vacuum
Nature of lens	Glass	Electromagnet
Focusing mechanism	Lens position is adjusted mechanically	Current to the magnetic lens is adjusted
Changing the magnification	Switch the objective lens	Adjust the current to the magnetic lens
Source of contrast	Differential light absorption	Scattering of electrons
Specimen mount	Glass slide	Metal grid
Best resolution	0.2 μm	0.5 nm
Highest practical magnification	1000–1500	Over 100,000
Affordability	Cheaper	Expensive

compartments. Although it produces lower resolution than an electron microscope, it has become an essential tool in microscopy due to fewer requirements for specimen preparation and its ability to obtain 3D live imaging and access to dynamic cellular and molecular processes. By using fluorescence imaging as the core component, CLSM uses point-by-point examination of samples using laser excitation. It helps obtain images from the focal plane only and removes the background noise resulting from sample thickness. As such, it is not useful in deep tissue imaging.

Electron Microscopy

Electron microscopy, invented by Max Knoll and Ernst Ruska in 1931, uses a beam of electrons instead of light. Electromagnets focus the electron beam, and the object to be examined is kept on the beam's path, thus scattering the electrons and producing an image that is focused on a screen.

Since the resolving power is half of the wavelength and the wavelength of electrons is approximately 0.005 nm compared to the 5000 nm wavelength of visible light, the electron microscope's resolution is very high, theoretically 100,000 times that of a light microscope. However, in practice, the resolving power of the electron microscope is about 0.1 nm.

There have been many developments in electron microscopy that include scanning electron microscopy, immunoelectron microscopy and so on. **Scanning electron microscopy** a technique development that provides a three-dimensional image of the object and high resolution. **Transmission electron microscopy** is a technique where a beam of electrons passes through the specimen, giving higher resolution and ultrastructural details. Finally, **immunoelectron microscopy** is a method to enhance sensitivity and specificity by reacting the specimen with a specific antiviral antibody that results in the clumping of viral particles. In this method, an antibody may also be conjugated with gold to visualise and determine the location of specific antigenic determinants in a specimen.

The electron microscope is widely used for (1) rapid detection of viruses directly in clinical specimens, which is especially useful in detecting non-cultivable viruses, and (2) ultrastructural study of various microorganisms.

Differences between an electron microscope and a light microscope are summarised in Table 3.2.

Sterilisation and Disinfection 4

Microbes are ubiquitous, and many microorganisms are associated with undesirable consequences, such as food spoilage and disease. Therefore, killing microorganisms or inhibiting their growth is essential to minimising their destructive effects. The aim is to (1) destroy pathogens and prevent their transmission, and (2) reduce or eliminate microorganisms that can contaminate water, food and other substances.

Definition of Frequently Used Terms

Sterilisation: The process by which an article, surface or medium is freed of all living microorganisms either in the vegetative or in the spore state is known as sterilisation. Any material subjected to this process is said to be sterile. However, these terms are used only in the absolute sense, that is, an object is sterile or not sterile. It cannot be slightly sterile or almost sterile. Often, sterilisation is done by using a physical agent like heat, but at times it is also done by using chemicals. These are known as **sterilants**, and they have the ability to kill the spores.

Disinfection refers to using a chemical that destroys or removes all pathogenic organisms. It destroys vegetative pathogens but not bacterial endospores. When used in higher concentrations, disinfectants can be toxic to human and other animal tissue. Hence, they are typically used only on inanimate objects. Disinfection also ensures the removal of harmful products of microorganisms such as toxins from the articles. A few examples of disinfection are (1) application of 5% bleach solution to clean the examination table, (2) food utensils used by a diseased individual can be disinfected by boiling and (3) thermometers used for medical purposes can be immersed in an isopropyl alcohol solution between uses.

Sepsis is defined as the microbial growth in the body or microbial toxins in blood and other tissues. **Asepsis** refers to any practice that prevents the entry of infectious agents into sterile tissues and thus prevents infection.

Antiseptics are chemical agents that can be safely used directly on open body surfaces (e.g. skin, mucous membranes), open wounds and surgical cuts to eliminate or prevent the propagation of the vegetative forms of pathogens. Examples of a few antisepsis procedures are (1) preparation of the skin with iodine compounds prior to surgical incisions, (2) treating an open root canal with cotton swabs dipped in hydrogen peroxide, and (3) use of germicidal soap for handwashing.

Bactericidal agents (germicides) are agents used to kill bacteria. **Bacteriostatic agents** are used to stop the multiplication of bacteria. However, they cannot kill the bacteria; **bactericidal agents** may be used as bacteriostatic agents at higher dilutions. **Decontamination** is a process by which an article or area can be made free from contaminants such as microbes, chemicals, radioactive agents and other hazardous materials.

Cleaning before sterilisation or disinfection is always advisable because cleaning reduces the dirt and microbial load, thus making sterilisation more effective.

Sterilisation

Methods of sterilisation can be broadly classified as physical and chemical methods of sterilisation.

Heat Sterilisation

Heat is the most dependable method of sterilisation. It is usually the method of choice for sterilisation unless contraindicated. Higher temperatures that exceed the maximum are microbicidal, whereas lower temperatures below the minimum tend to have inhibitory or micro biostatic effects. Two types of physical heat are used in sterilisation—moist and dry heat.

Many factors influence the sterilisation by heat method. These include (1) nature of heat used for sterilisation—dry or moist, (2) temperature and time of heating, (3) microbial load present on the article or area, (4) ability of the organisms to withstand heat, for example, spores of certain species of bacteria are resistant to normal heating and (5) type of material to be sterilised.

Mode of action In dry heat, protein denaturation, oxidative damage and the toxic effect due to accumulation of high-level electrolytes are the reasons for cell death. On the other hand, moist heat kills microbes by denaturation and coagulation of their proteins. The dry heat technique requires high temperature to kill bacteria or their spores by oxidative damage, and such a high temperature is not recommended for the sterilisation of some of the articles. Most bacterial proteins are resistant in a moisture-free environment and do not undergo denaturation. Hence, the ability of steam to release latent heat when it condenses on a cooler surface is being exploited to kill microbes. In the case of spores, condensed steam on their surface raises the temperature inside spores and increases their water content. It results in the destruction of spores by hydrolysis and the breakdown of the protein.

Thermal death time is the minimum time necessary to kill a suspension of organisms at a fixed temperature in a specified environment. The time required for sterilisation is always inversely proportional to the temperature of exposure and is also related to the number of organisms present in the suspension, their spore-forming ability and strain-specific characteristics. Recommended minimum sterilisation time should be calculated only after an article reaches its optimum sterilisation temperature. The thermal death time is also dependent on the medium in which microbes are suspended. Organic substances, proteins, nucleic acids, starch, gelatine, sugar, fat and oils slow down the sterilisation process by increasing thermal death time, whereas disinfectants and extreme pH conditions accelerate bacterial killing.

Moist Heat Sterilisation

Moist heat can be present in the form of hot water, boiling water or vapourised water (steam). In general, moist heat temperature ranges from 60 to 135 °C. The steam temperature can be regulated by adjusting the pressure in a closed container. Moist heat kills microorganisms by denaturation and coagulation of proteins. Sterilisation by moist heat is of various types as follows:

Sterilisation at a Temperature Below 100 °C

Pasteurisation is an example of sterilisation at a temperature <100 °C. In this method, heat is applied to liquids to destroy potential agents of infection and spoilage without affecting the liquid's flavour and food value. The technique is named after Louis Pasteur, who devised this method. This method is used extensively to sterilise milk and other fresh beverages that are easily contaminated during collection and processing, such as fruit juices, beer and wine.

The flash method and the holder method are the two methods of pasteurisation. In the **flash method**, milk is exposed to heat at 72 °C for

15–20 s. This is followed by a sudden cooling to 13 °C or lower. In the **holder method**, milk is exposed for 30 min to a temperature of 63 °C, followed by cooling to 13 °C or lower, but not less than 6 °C. The flash method is preferable for the sterilisation of milk because it is unlikely to alter the flavour and nutrient content, and it is highly efficient in destroying certain resistant pathogens like *Coxiella* and *Mycobacterium*.

Although pasteurisation inactivates most viruses and destroys the vegetative stages of 97–99% of bacteria and fungi, it does not kill endospores or thermoduric species (yeasts, micrococci and non-pathogenic lactobacilli). Even after pasteurisation, milk is not sterile. It contains about 20,000 or more microbes per millilitre. That is why even a carton of milk that has not been opened also spoils on prolonged storage. Newer techniques in which milk is processed with an *ultrahigh temperature* (UHT) of 134 °C for 1–2 s have now been used to produce sterile milk with a storage life of 3 months.

Vaccine baths are a type of water bath instrument that maintain a low temperature to sterilise heat-labile vaccines. For example, during vaccine production, the vaccine-containing medium is gently heated at 45–55 °C for hours in the vaccine water bath.

Sterilisation at a Temperature of 100 °C

Boiling

Exposing materials to boiling water for 30 min kills most non-spore-forming pathogens, including resistant species, such as the tubercle bacillus and staphylococci. However, it does not kill bacterial spores. Sterilisation by boiling is facilitated by adding 2% sodium bicarbonate to water. Since boiling only once at 100 °C does not kill all the spores, this method cannot be used for sterilisation but only for disinfection. Hence, it is not recommended for sterilising instruments used for surgical procedures. The most significant disadvantage of this method is that the items sterilised by boiling can easily be re-contaminated when removed from the water after boiling.

Free Steam at Atmospheric Pressure (100 °C)

Usually, a Koch or Arnold steam steriliser is used for heat-labile substances that degrade at higher temperatures and pressure, such as during autoclaving. These substances are exposed to steam at atmospheric pressure for 90 min, during which the moist heat kills most vegetative forms of the bacteria except for the thermophiles.

Sterilisation at a Temperature Above 100 °C

This method is otherwise known as **sterilisation by steam under pressure**. Under normal atmospheric pressure at sea level, that is, 1-atm steam can reach a maximum temperature of 100 °C. When pressurised in a closed chamber, the steam temperature can be raised above this point since when a gas is compressed, its temperature rises directly in relation to pressure. So, when the pressure increases by 5 psi above normal atmospheric pressure, the steam temperature rises to 109 °C. When the pressure is increased to 10 psi above normal, its temperature will be 115 °C and at 15 psi (a total of 2 atm), the temperature reaches up to 121 °C. A combined effect of increased pressure and raised temperature kills the microbes. This forms the principle of sterilisation by steam under pressure. A special device is required that can subject pure steam to pressures greater than 1 atm to achieve such pressure-temperature combinations. In health and commercial industries, *autoclaves* are being used for this purpose, which can be comparable with the working principles of home appliance “pressure cooker”.

An **autoclave** is a cylindrical metal chamber with an airtight door at one end and racks to hold materials. A screw clamp fastens the lid, and an asbestos washer makes it airtight. A discharge tap for air and steam is present at the upper side, along with a pressure gauge and a safety valve. Heating is carried out by electricity. The inner chamber is loaded with materials for sterilisation. Steam circulates within the jacket and is supplied under pressure to the inner chamber. The water in the autoclave boils when its vapour pressure equals the surrounding atmosphere. Following the pressure increase inside the closed vessel, the temperature at which the water boils inside the autoclave increases. The saturated steam, which has higher penetrative power, on coming in contact with a cooler surface of the article, condenses into water and releases latent heat to the article surface. For example, when approximately 1600 mL of steam at atmospheric pressure at 100 °C condenses into 1 mL of water at 100 °C, it liberates 518 cal of heat. The gross drop in the steam volume pulls in more steam to that surface area, and this process lasts until the temperature of the surface elevates to that of the steam. Sterilisation is achieved when the steam condenses against the objects in the chamber and gradually raises their temperature. The condensed water facilitates moist conditions that ensure the killing of microbes.

Sterilisation condition: Many studies have shown that the most appropriate temperature–pressure combination for sterilisation by autoclave is 15 psi at 121 °C. The pressure can also be raised to increase temperature (increasing the pressure to 30 psi will raise the temperature by 11 °C), but increasing the temperature will increase the exposure time. This may harm the things which are being sterilised. Over-loading should be avoided because it prevents steam from circulating freely around the contents and impedes the full contact that is necessary. The holding time varies from 10 min for light loads to 40 min for heavy or bulky ones. The average time is 20 min.

Types of autoclaves: From the point of view of industrial design and scope of action, autoclaves belong to three categories:

- Class N autoclaves do not have a vacuum pump. As such, it does not ensure effective steam penetration and allows only sterilisation of instruments with a solid structure.
- Class B autoclaves are the most advanced steam sterilisers. With a fractionated pre-vacuum that removes air from the chamber and ensures effective steam penetration, it can sterilise all types of materials. It is a commonly used autoclave in medical settings and approved medical device.
- Class S autoclaves are an intermediate class between N and B. It has a vacuum pump that removes the air from the chamber before starting the sterilisation process. However, it creates a single-stage pre-vacuum and so is less effective than class B autoclaves.

In most clinical laboratory and hospital settings, class B autoclaves are either vertical or horizontal type. Compared to horizontal autoclaves, the vertical ones can work as a passthrough mechanism and are useful in clean room settings.

Uses of autoclaves: Autoclaves have many applications. It is an excellent method to sterilise heat-resistant materials, such as rubber (gloves), glassware, cloth (surgical dressings), metallic instruments, liquids, paper, some media and some heat-resistant plastics. It is also helpful for sterilising heat-sensitive items, such as plastic Petri plates that need to be discarded. It is also useful for sterilising materials that cannot withstand the higher temperature of the hot-air oven. However, the autoclave is ineffective in sterilising substances that repel moisture (oils, waxes or powders). Types and uses of various moist heat sterilisation methods are summarised in Table 4.1.

Sterilisation controls: Various sterilisation control indicators are used to determine sterilisation efficacy by moist heat. **Thermocouples** are used to record temperatures directly in autoclaves by a potentiometer. The **Browne tube** is the most commonly used chemical indicator of moist heat sterilisation in the autoclave. It contains a red solution that turns green when exposed to a temperature of 121 °C for 15 min in an autoclave. **Biological indicators** such as *Geobacillus stearothermophilus* spores are used as indicators of moist heat sterilisation in autoclaves. This is a thermophilic bacterium with an optimum

Table 4.1 Method and uses of moist heat sterilisation

Method	Uses	Remark
Water bath below 100 °C	Serum, body fluids and vaccines	Only disinfection possible Spores are not killed
Water bath at 100 °C	Glass, metal and rubber items	Some spores still not killed
Arnold steamer: steaming at 100 °C	Culture media containing sugar and gelatin	Preserves properties of media
Autoclave	Culture media and laboratory materials	Kills all the vegetative as well as spore forms of bacteria

temperature of 55–60 °C, and its spores require exposure at 121 °C for 12 min to be destroyed. The efficacy of the autoclave is increased by placing paper strips impregnated with 10^6 spores in envelopes, and these envelopes are kept in different parts of the load inside the autoclave. These strips, after sterilisation, are inoculated into a suitable recovery medium and incubated at 55 °C for 5 days. Spores are destroyed if the sterilising condition of the autoclave is proper.

Intermittent Sterilisation

Certain heat-labile substances (e.g. serum, sugar, egg) that cannot withstand the high temperature of the autoclave can be sterilised by the process of intermittent sterilisation, known as tyndallisation.

Tyndallisation is carried out for 3 days and requires a chamber to hold the materials and a reservoir for boiling water. The items to be sterilised are kept in the chamber and exposed to free-flowing steam at 100 °C for 20 min for three consecutive days. On the first day, all the vegetative forms of the bacteria, yeasts and moulds are killed but not spores. The surviving spores are allowed to germinate to vegetative forms on the second day and are killed on re-exposure to steam. Finally, the third day re-ensures the killing of all the spores by germination to vegetative forms.

Intermittent sterilisation by **inspissation** is used most often to sterilise heat-sensitive culture media, such as those containing sera (e.g. Loeffler serum slope) and egg (e.g. Lowenstein–Jensen medium). This is carried out by heating them at 80–85 °C in an inspissator on three successive days (Fig. 4.1).

Fig. 4.1 Inspissator used to sterilise heat-sensitive culture media, such as Lowenstein–Jensen media. (Courtesy of Prof Tuhina Banerjee)

Dry Heat Sterilisation

Sterilisation by dry heat uses air with a low or no moisture content, where the role of moisture in sterilisation is minimal or nil. In general, the temperature employed for dry heating ranges from 160 °C to more than 1000 °C. The dry heat kills microorganisms by protein denaturation, oxidative damage and the toxic effect of increased electrolytes. The use of dry heat is not so versatile compared to moist heat; however, it has numerous applications in routine sterilisation procedures. Depending on the method used for dry heat sterilisation, the temperature and time employed also vary, but generally dry heat would require more temperature and time than moist heat.

Flaming

Sterilisation of the inoculating loop or wire, the tip of forceps, searing spatulas and so on is done by holding them in the flame of the Bunsen burner till they become red hot. Glass slides, scalpels and mouths of culture tubes are sterilised by passing them through the Bunsen flame without allowing them to become red hot.

Incineration

Incineration safely destroys infective materials by burning them to ashes. Incinerators are regularly used in hospitals and research labs to destroy hospital and laboratory wastes. The method is used to destroy and dispose of infectious material, such as syringes, needles, culture material, dressings, bandages, bedding, animal carcasses and pathology samples in a fast and effective way. However, it is not used for metals and heat-resistant glass materials.

Hot-air Oven

The hot-air oven is the most widely used method and provides another means of dry heat sterilisation. The hot-air oven is electrically heated, which generates hot air. A fan in the oven ensures even distribution of hot air in the chamber. A thermostat ensures hot air circulation at the chamber's desired temperature. The heated and circulated air transfers its heat to the materials inside the chamber. While sterilising by the hot-air oven, it should be ensured that the oven is not overloaded. The materials should be dry and arranged to allow free circulation of air inside the chamber. It is essential to fit the test tubes, flasks and so on with cotton plugs and wrap Petri dishes and pipettes with paper. Sterilisation by hot-air oven requires exposure to 160–180 °C for 2 h and 30 min, ensuring thorough heating of the objects and destruction of spores.

The hot-air oven is used in laboratories and clinics for heat-resistant items that are not sterilised well by moist heat. They are used for sterilisation of (1) glassware (syringes, Petri dishes, flasks, pipettes, test tubes etc.), (2) surgical instruments (scalpels, scissors, forceps etc.), (3) chemicals (liquid paraffin, sulphonamide powders etc.) and (4) oils that are not penetrated well by the steam used in moist heat sterilisation.

Sterilisation Controls

Various sterilisation controls are used. Colour **indicator** methods such as Browne tubes are routinely used for sterilisation control for dry heat sterilisation. After complete sterilisation for 60 min at 160 °C, the Browne tube turns green.

Biological indicators include spores of non-toxigenic *Clostridium tetani* and *Bacillus subtilis* to ensure complete sterilisation by dry heat methods. A paper strip containing approximately 10^6 spores is placed inside an envelope and subjected to sterilisation along with other articles. After the complete sterilisation, paper strips are inoculated into a thioglycollate medium or a cooked meat medium and incubated anaerobically for 37 °C. The medium is incubated for 5 days and observed for any turbidity suggestive of growth. No growth indicates efficient sterilisation. The various parameters of heat sterilization is summarised in Table 4.2.

Cold Sterilisation

This includes methods of sterilisation by filters, radiation and high-frequency sound (sonic) waves carried out at a lower temperature.

Filtration

Filtration is a way to reduce the microbial population in solutions of heat-labile material by using various filters. Filters are used to sterilise these heat-labile solutions. They remove contaminating microorganisms from solutions rather than directly destroying them.

Filters

Candle filters are made up of (1) diatomaceous earth (e.g. Berkefeld filters), or (2) unglazed porcelain (e.g. Chamberland filters). They are available in different grades of porosity and

Table 4.2 Holding period and temperature for sterilisation

Sterilisation process	Holding time	Holding temperature
Dry heat		
Hot air oven	160 °C	2 h
Moist heat		
Temperature below 100 °C		
Pasteurisation		
Holder method	63 °C	30 min
Flash method	72 °C	20 s
Water bath	60 °C	1 h
Inspissation	80–85 °C	30 min for 3 consecutive days
Moist heat at a temperature of 100 °C		
Tyndallisation or intermittent sterilisation	100 °C	20 min for 3 consecutive days
Moist heat at a temperature above 100 °C		
Autoclave	121 °C 126 °C 133 °C	15 min 10 min 3 min

are used widely to purify water for drinking and industrial uses.

Asbestos filters are made up of asbestos, such as magnesium silicate. Seitz and Sterimat filters are examples of such filters. These are disposable and single-use discs available in different grades. They have a high adsorbing capacity and tend to alkalinise the filtered fluid. However, their use is limited by the carcinogenic potential of asbestos.

Sintered glass filters are made up of finely powdered glass particles fused together. They have low absorbing properties and are available in different pore sizes. Although they can be cleaned easily, these filters are brittle and expensive.

Membrane filters are made up of (1) cellulose acetate, (2) cellulose nitrate, (3) polycarbonate, (4) polyvinylidene fluoride, or (5) other synthetic materials. These filters are circular porous membranes and are usually 0.1 mm thick. Membranes with pores about 0.2 μm are used because the pore size is smaller than the size of bacteria, even though a wide variety of pore sizes (0.015–12 μm) are available (Fig. 4.2), Membrane filters remove microorganisms by screening them out in the way a sieve separates large sand particles from small ones.

These filters have many uses. They are used to sterilise pharmaceutical substances, ophthalmic solutions, liquid culture media, oils, antibiotics and other heat-sensitive solutions. They are used to obtain bacteria-free filtrates of clinical specimens for virus isolation. They are also used to separate toxins and bacteriophages from bacteria.

Fig. 4.2 Membrane filter used in filtration of heat-labile serum and other fluids. (Courtesy of Prof Tuhina Banerjee)

Table 4.3 Applications of radiation for sterilisation

Types	Source	Uses	Remarks
Ionising radiation	Cobalt-60–based instruments	For sterilisation of pharmaceuticals like antibiotics, hormones, sutures; and pre-packed disposable items, such as syringes, infusion sets, catheters	Though expensive and fraught with safety risks, it is very effective due to better penetration power
Non-ionising radiation	UV lamps	Only for disinfection of clear surfaces in OTs, laminar flow hoods etc.	Hazardous and not as effective as ionising radiation

Radiations

Ionising and non-ionising radiations are the two types of radiation used for sterilisation (Table 4.3).

Ionising Radiation

Ionising radiation is an excellent sterilising agent with very high penetrating power. These radiations penetrate deep into objects and destroy bacterial endospores and vegetative cells, both prokaryotic and eukaryotic. These are, however, not that effective against viruses. Ionising radiations include (1) X-rays, (2) gamma rays and (3) cosmic rays. Gamma radiation from a cobalt-60 source is used to sterilise antibiotics, hormones, sutures, catheters, animal feeds, metal foils and plastic disposables, such as syringes. This has also been used to sterilise and "pasteurise" meat and other food items.

Irradiation usually kills *Escherichia coli* O157:H7, *Staphylococcus aureus*, *Campylobacter jejuni* and other pathogens. Since there is no detectable temperature increase in this method, this method is commonly referred to as "cold sterilisation". The Food and Drug Administration and the World Health Organization have approved food irradiation and declared it safe.

Gamma radiation from a cobalt-60 source is used to sterilise antibiotics, hormones, sutures, catheters, animal feeds, metal foils and plastic disposables, such as syringes.

Non-ionising Radiations

Non-ionising radiations include infrared and ultraviolet radiations.

Infrared radiations are used for rapid and mass sterilisation of disposable syringes and catheters.

Ultraviolet (UV) radiation, with a 240–280 nm wavelength, is quite lethal and has significant bactericidal activity. It acts by denaturation of bacterial protein and interferes with the replication of bacterial DNA. UV radiation is used primarily for disinfection of closed areas in microbiology laboratories, inoculation hoods, laminar flow and operating theatres. It kills most vegetative bacteria but not spores. Also, it does not penetrate glass, dirt films, water and other substances very effectively.

High-Frequency Sound (Sonic) Waves

High-frequency sound (sonic) waves beyond the human ear's sensitivity are known to disrupt cells. Sonication transmits vibrations through a water-filled chamber (sonicator) to induce pressure changes and create intense points of turbulence that can stress and burst cells in the vicinity. Sonication also forcefully dislodges foreign matter from objects. The heat generated by the sonic waves (up to 80 °C) also contributes to the antimicrobial action.

Gram-negative rods are most sensitive to ultrasonic vibrations, while Gram-positive cocci, fungal spores and bacterial spores are resistant to them. Ultrasonic devices are used in dental and some medical offices to clear debris and saliva from instruments before sterilisation and clean dental restorations. However, most sonic machines are not reliable for regular use in disinfection or sterilisation.

Biological controls used for testing the efficacy of sterilisation techniques are summarised in Table 4.4.

Table 4.4 Biological controls for testing efficacy of sterilisation techniques

Sterilisation technique	Mechanism	Biological control
Autoclave	Moist-heat sterilisation	*Geobacillus stearothermophilus*
Hot-air oven	Dry-heat sterilisation	*Bacillus subtilis*
Ionising radiations	Low-temperature sterilisation	*Bacillus pumilus*
Ethylene oxide	Gaseous sterilant	*Bacillus globigii, Bacillus subtilis*

Sterilisation by Chemical Agents

Several chemical agents are used as antiseptics and disinfectants. All these chemical agents (e.g. alcohols, aldehydes) are described later under disinfection.

Effect of Cold and Desiccation

The main benefit of cold treatment is to slow the growth of cultures and microbes in food during processing and storage. It is essential to know that cold merely retards the activities of most microbes. Although cold temperatures kill some microbes, gradual cooling, long-term refrigeration or deep freezing adversely affect most microorganisms. Cultures of bacteria, viruses and fungi can be preserved for a more extended period in freezing temperatures ranging from −70 to −135 °C. Even at this temperature, some of the psychrophile grow and produce toxic products. *S. aureus*, *Clostridium* species (spore formers), *Streptococcus* species and several types of yeasts, moulds and viruses are the pathogens that can survive for several months in refrigerated food items.

Lyophilisation is a process of freezing and drying. This is one of the most widely used techniques for preserving the microorganisms and cells in a viable condition for years. Cultures are frozen instantaneously and subjected to a vacuum that quickly removes the water. In this method, there is no formation of ice that damages the cells, and not all cells will withstand this process; however, many of them survive after reconstitution of lyophilised culture. Since the effects of chilling, freezing and desiccation are unpredictable and uncertain, they are not regarded as methods of disinfection or sterilisation. Additionally, these procedures will not ensure that pathogens have been killed efficiently.

Disinfection

The process of inactivating microorganisms by direct exposure to chemical or physical agents is known as disinfection. **Disinfectants** are products or biocides that destroy or inhibit the growth of microorganisms on inanimate objects or surfaces. Disinfectants can be sporistatic but are not necessarily sporicidal. **Antiseptics** are products that destroy or inhibit the growth of microorganisms in or on living tissue.

Antiseptics and disinfectants are used extensively in hospitals for various purposes, such as topical and hard surface applications. They play an important role in infection control practices and are helpful in combating nosocomial infections. Differences between sterilisation and disinfection have been summarised in Table 4.5.

Properties of an Ideal Disinfectant

An ideal disinfectant or antiseptic has many characteristics. (1) Ideally, the disinfectant must have a broad range of antimicrobial activity. It should be effective against a wide range of infectious agents (Gram-positive and Gram-negative bacteria, acid-fast bacteria, fungi, viruses and bacterial endospores) at higher dilutions. (2) It should act in the presence of organic matter. (3) It should not be toxic to humans or corrosive. This balance between effectiveness and low toxicity for animals is hard to achieve in practice. Some chemicals are used despite their low effectiveness because they are relatively non-toxic. (4) It should be stable upon storage and not undergo any chemical change. (5) It should be odourless or have a pleasant odour. (6) It should be soluble

Table 4.5 Distinguishing features of sterilisation and disinfection

Features	Sterilisation	Disinfection
Definition	Freeing an article, surface or medium from all living organisms including viruses, bacteria and their spores and fungi and their spores.	Process that reduces the number of contaminating microorganisms, liable to cause infection to a level which is deemed no longer harmful to health. Spores are not killed.
Uses	Objects or instruments coming in direct contact with a break in skin or mucous membrane or entering a sterile body area.	Objects or instruments coming in direct contact with mucous membrane but tissue is intact or via intact skin.
Examples	Surgical instruments, needles, syringes, parenteral fluid, arthroscopes, media, reagents and equipment in laboratory use.	Endotracheal tubes, aspirators, gastroscopes, bed pans, urinals etc.

in water and lipids for penetration into microorganisms. (7) It should be effective in acidic and alkaline media. (8) It should have speedy action. (9) It should be relatively inexpensive if possible.

Action of Disinfectants

Disinfectants act in many ways. They damage the cell wall and alter the permeability of the cell membrane, resulting in exposure, damage or loss of the cellular contents. They alter proteins and form protein salts or cause coagulation of proteins. They inhibit enzyme action and inhibit nucleic acid synthesis or alter nucleic acid molecules. They also cause oxidation or hydrolysis.

Factors Influencing Activity of Disinfectants

Various physical and chemical factors influence the efficiency of disinfectant. These include the following:

1. **Temperature**: Increase in temperature increases the efficiency of disinfectants. However, because disinfectants are chemicals, very high temperatures degrade and weaken their germicidal activity, which results in ineffective disinfection.
2. **pH**: pH influences the antimicrobial activity of disinfectants. For disinfectants like glutaraldehyde and quaternary ammonium compounds, it improves the activity but works opposite for phenols, hypochlorites and iodine.
3. **Relative humidity**: It influences the activity of gaseous disinfectants/sterilants, such as EtO, chlorine dioxide and formaldehyde.
4. **Water hardness**: Divalent cations present in hard water develop insoluble precipitates when in contact with disinfectant. It reduces the efficiency of the disinfectant.
5. **Type of microorganism and duration of exposure**: Vegetative cells are more susceptible than spores. Spores may be resistant to the action of disinfectants. Biofilm-forming microbes are more resistant to disinfectants. Effectiveness also depends on the duration of exposure.
6. **Physiological state of the cell**: Young and metabolically active cells are more sensitive than old dormant cells. Non-growing cells may not be affected.
7. **Environment**: The physical or chemical properties of the medium or substance influence the rate and efficiency of disinfectants. For example, serum, blood, pus or faecal or lubricant material can interfere with the antimicrobial activity of disinfectants.

Types of Disinfectants

Disinfectants include the following:

Phenols

In 1867, Joseph Lister used phenolic compounds to reduce the risk of infection during operations. Phenolic compounds can be bactericidal, bacteriostatic or fungicidal. They act by denaturing proteins and disrupting cell membranes. They are effective in the presence of organic material and remain active on surfaces long after application. Hence, they are the most widely used antiseptics and disinfectants in laboratories and hospitals worldwide. Different phenolic compounds are as follows:

Phenol: It is effective against vegetative forms of bacteria, *Mycobacterium tuberculosis* and certain fungi. It is an excellent disinfectant for faeces, blood, pus, sputum and so on. It has a low degree of activity as compared to other derivatives. It is not suitable for application to the skin or mucous membrane.

Cresols: These are more germicidal and less poisonous than phenol but corrosive to living tissues. They are used for cleaning floors (1% solution), disinfection of surgical instruments and disinfection of contaminated objects. Lysol is a solution of cresols in soap.

Halogenated diphenyl compounds: These include hexachlorophene and chlorhexidine. These are highly effective against both Gram-positive and Gram-negative bacteria. They are used as skin antiseptics and for the cleaning of wound surfaces. Hexachlorophene has been one of the most popular antiseptics because once applied, it persists on the skin and reduces the growth of skin bacteria for longer periods. However, it can cause brain damage and is now used in hospital nurseries only after a staphylococcal outbreak.

Halogens

Halogens are fluorine, bromine, chlorine and iodine—a group of non-metallic elements that commonly occur in minerals, seawater and salts. Although they can occur either in the ionic (halide) or non-ionic state, most halogens exert their antimicrobial activity primarily in their non-ionic state but not in the halide state (e.g. chloride, iodide).

These agents are highly effective disinfectants and antiseptics because they possess microbicidal rather than microbiostatic effects. With longer exposure, they act as a sporicidal agent also. Hence, halogens are used as an active ingredient in approximately one-third of all currently available antimicrobial chemicals. Chlorine and iodine are the only two routinely used halogens because fluorine and bromine are dangerous to handle.

Chlorine and its compounds: For around 200 years, chlorine has been used for antisepsis and disinfection. Liquid and gaseous chlorine and hypochlorite are the major forms of chlorine used to control microbes. These compounds react with water and give hypochlorous acid (HOCl); this oxidises the sulfhydryl (SH) group of cysteine and interferes with the disulphide bond of various enzymes. Thus, the enzymes are denatured permanently. Liquid and gaseous chlorine are used almost exclusively for large-scale disinfection of drinking water, sewage and wastewater from sources such as agriculture and industry. Chlorine kills bacterial cells and endospores, fungi and viruses. Water treatment with chlorine destroys many pathogenic vegetative microorganisms without unduly affecting their taste. Moreover, chlorination at a concentration of 0.6–1.0 part of chlorine per million parts of water makes water potable and safe to use.

Hypochlorites are perhaps the most extensively used of all chlorine compounds. They are used for (1) sanitisation and disinfection of food equipment in dairies, restaurants and canneries, (2) treatment of swimming pools, spas, drinking water and even fresh foods and (3) treatment of wounds; and disinfection of equipment, beddings and instruments. Common household bleach is a weak solution (5%) of sodium hypochlorite used as an all-around disinfectant, deodoriser and stain remover. It is frequently used to treat water supplies as an alternative to pure chlorine.

However, the significant limitations of chlorine compounds are that they are ineffective if used at an alkaline pH and are less effective in the presence of excess organic matter and relatively unstable, especially if exposed to light.

Iodine and its compounds: Iodine is a pungent black chemical that forms brown solutions when dissolved in water or alcohol. Iodine rapidly penetrates the cells of microorganisms, where it disturbs a variety of metabolic functions. It interferes with proteins' hydrogen and disulphide bonds (similar to chlorine). It kills all types of microorganisms if optimum concentrations and exposure times are used. Iodine activity, unlike chlorine, is not as adversely affected by organic matter and pH. The two primary iodine preparations are *free iodine in solution* and *iodophors*.

Aqueous iodine consists of 2% free iodine in solution and 2.4% sodium iodide. It is commonly used as a topical antiseptic prior to surgical procedures and seldom in treating burnt and infected skin. A stronger iodine solution (5% iodine and 10% potassium iodide) is used for disinfecting plastic items, rubber instruments, thermometers and cutting blades.

Iodine tincture comprises a 2% iodine solution and sodium iodide in 70% alcohol. This is used as a skin antiseptic. Iodine is an extreme irritant to the skin and is toxic when absorbed, so at present, these tinctures (5–7%) are not regarded safe routine antiseptics. **Iodine tablets** can be used to disinfect water in emergency conditions or to destroy pathogens in contaminated water supplies.

Iodophors are iodine complexes and neutral polymers, such as polyvinyl alcohol. This formulation allows for an increased degree of penetration and slows the release of free iodine. Free iodine solution, which was used in antisepsis, is replaced by iodophores, which are less prone to staining or irritating tissues. Betadine, povidone and iodine are the common iodophor compounds that contain 2–10% of available iodine.

They are used to prepare the skin and mucous membranes for surgery and surgical hand scrubs. They are also used to treat burns and disinfect equipment. A recent study has shown that betadine solution is an effective means of preventing eye infections in newborns, and it may replace antibiotics and silver nitrate as the method of choice.

Alcohols

Alcohols are among the most widely used disinfectants and antiseptics. They are bactericidal and fungicidal but not sporicidal. They have no action against spores and viruses. Ethyl alcohol and isopropyl alcohol are the two most popular alcohol germicides. They are effective at a concentration of 60–70% in water. They act by denaturing bacterial proteins and possibly by dissolving membrane lipids. They are used as skin antiseptics. Isopropyl alcohol is used for the disinfection of clinical thermometers. A 10–15 min soaking is sufficient to disinfect thermometers. Methyl alcohol is effective against fungal spores.

Aldehydes

Formaldehyde and glutaraldehyde are the two most commonly used aldehydes used as disinfectants. They are highly reactive molecules that combine with nucleic and alkylating molecules. They are sporicidal and can also be used as **chemical sterilants**.

Formaldehyde: It is usually dissolved in water or alcohol before use. In an aqueous solution, it is bactericidal, sporicidal and effective against viruses. Formalin solution is 40% aldehyde in an aqueous solution. It is used to (1) preserve fresh tissue specimens, (2) destroy anthrax spores in hair and wool, (3) prepare toxoids from toxins, (4) sterilise bacterial vaccines, and (5) kill bacterial cultures and suspensions.

Glutaraldehyde: A 2% buffered glutaraldehyde solution can be used as an efficient disinfectant. It is less irritating than formaldehyde and useful in disinfecting hospital and laboratory equipment. Generally, glutaraldehyde disinfects objects within 10 min but may need longer (12 h) to destroy all spores. Glutaraldehyde is especially effective against tubercle bacilli, fungi and viruses. It can be used for cleaning cystoscopes and bronchoscopes, corrugated rubber anaesthetic tubes and face masks, plastic endotracheal tubes, metal instruments and polythene tubing.

Gases

Various gaseous agents are used to sterilise a large volume of heat-sensitive disposable items and instruments. Ethylene oxide, formaldehyde gas and beta-propiolactone are frequently used gaseous agents.

Ethylene oxide: It is a colourless liquid used for gaseous sterilisation. It is active against all kinds of bacteria, spores and viruses. It kills all types of microorganisms by inhibiting proteins and nucleic acids. It is both microbicidal and sporicidal. It is a highly effective sterilising agent because it rapidly penetrates packing materials, including plastic wraps. It is used to sterilise disposable plastic Petri dishes, sutures, syringes, heart-lung machines, respirators and dental equipment. Ethylene oxide is highly inflammable and carcinogenic. Extensive aeration of the sterilised materials is necessary to remove the residual ethylene oxide gas, which is toxic.

Formaldehyde gas: It is used for (1) the fumigation of operation theatres, wards, sick rooms and laboratories, and (2) the sterilisation of instruments and heat-sensitive catheters, clothing and bedding, furniture, books and so on. The formaldehyde gas is produced by adding 150 g of potassium permanganate to 280 mL formalin in 1000 $ft.^3$ of room volume. The room to be sterilised is completely closed and sealed at least for 48 h after fumigation with formalin gas. Sterilisation is achieved by condensation of gas on the exposed surface. The gas is toxic when inhaled and is irritant to the eye; hence its effect should be nullified by exposure to ammonia. It is highly inflammable and carcinogenic.

Beta-propiolactone (BPL): It is a condensation product of ketone and formaldehyde. It is active against all microorganisms and viruses. It is more efficient than formaldehyde for fumigation purposes. It has been used to sterilise vaccines and sera in liquid form. BPL destroys microorganisms more rapidly than ethylene oxide, but it has low penetration capacity and may act carcinogenic. Hence, BPL has not been used as widely as ethylene oxide. Recently, vapour-phase hydrogen peroxide has been used to decontaminate biological wastes.

Hydrogen peroxide fogging: It can be used as a replacement method for fumigation. In this method, fogging machines are filled with hydrogen peroxide as disinfectants and used as per the requirements. Due to the short cycle time and non-toxic nature, this method is advantageous over conventional fumigation techniques.

Surface Active Agents

Surface active agents, such as detergents, alter energy relationships at interfaces producing a reduction in surface tension. Detergents are organic molecules that serve as wetting agents and emulsifiers because they have polar hydrophilic and non-polar hydrophobic ends. In addition, due to their amphipathic nature, detergents solubilise and are very effective cleansing agents. They are different from soaps, which are derived from fats.

Cationic surface-active agents: Cationic detergents are effective disinfectants. Cationic detergents like benzalkonium chloride and cetylpyridinium chloride kill most bacteria but not *M. tuberculosis*, endospores or viruses. They do have the advantages of being stable and non-toxic, but they are inactivated by hard water and soap. These are often used as skin antiseptics and disinfectants for the disinfection of food utensils and small instruments. Quaternary ammonium compounds, such as cetrimide, are the most popular cationic detergents. They act by disrupting microbial membranes and possibly by denaturing proteins.

Anionic surface-active agents: These include soaps prepared either from saturated or unsaturated fatty acids, which act better at acidic pH. The soaps prepared from saturated fatty acids are more effective against Gram-negative organisms, whereas those prepared from unsaturated fatty acids are more active against Gram-positive bacilli and *Neisseria*.

Non-ionic surface active agents: These are non-toxic, and some may even promote bacteria growth.

Amphoteric or ampholytic compounds: These are active against a wide range of Gram-positive and Gram-negative bacteria and also against a few viruses. These are known as **"Tego" compounds**.

Oxidising Agents

This group includes halogens, hydrogen peroxide, potassium permanganate and sodium perborate. They are good disinfectants and antiseptics but are less effective in the presence of organic matter. As a 3% solution, hydrogen peroxide is a weak disinfectant. It is helpful for cleaning wounds and mouthwash or gargle. Potassium permanganate is bactericidal and active against viruses also.

Dyes

The dyes that have been used extensively as skin and wound antiseptics include acridine dyes and aniline dyes.

Acridine dyes include acriflavine, euflavine, proflavine and aminacrine. They show more activity against Gram-positive bacteria than against Gram-negative organisms. They interfere with the synthesis of nucleic acids and proteins in bacterial cells. The yellow acridine dyes, acriflavine and proflavine, are sometimes used for antisepsis and wound treatment in medical and veterinary clinics. **Aniline dyes** such as gentian violet, crystal violet and malachite green are also more active against Gram-positive bacteria than against Gram-negative organisms. They are also effective against various fungi, hence are incorporated into solutions and ointments to treat fungal skin infections such as ringworm.

The dyes, nevertheless, have limited applications because they stain and have a narrow spectrum of antimicrobial activity. They also have no activity against tubercle bacilli. The presence of organic matter also inhibits their actions.

Heavy Metals

Soluble salts of mercury, silver, copper, arsenic and other heavy metals have antibacterial activity, both bactericidal and bacteriostatic. They combine with proteins, often with their sulfhydryl groups, and inactivate them. They may also precipitate cell proteins. Silver compounds are widely used as antiseptics. Silver sulphadiazine is used for burns. Silver nitrate is used as a prophylactic agent in ophthalmia neonatorum in newborns. Copper sulphate is an effective algicide in lakes and swimming pools. Mercuric chloride is used as a disinfectant. These compounds, however, are increasingly replaced by other less toxic and more effective germicides.

Acids and Alkalis

Acids (sulphuric acid, nitric acid, hydrochloric acid and benzoic acid) and alkalis (potassium and sodium hydroxide and ammonium hydroxide) are germicidal. They kill microorganisms by hydrolysis and by altering the pH of the medium. They are rarely used as disinfectants.

Organic acids preserve food by preventing bacterial and fungal growth, and spore germination. They are widely used as they are safe to eat too. Acetic acid in the form of vinegar inhibits bacterial growth. Propionic acid is added to cakes and bread to inhibit mould growth. The addition of lactic acid into sauerkraut and olives prevents the growth of anaerobic bacteria, especially clostridia. Similarly, the addition of benzoic and sorbic acids to beverages, syrups and margarine inhibits the growth of yeasts.

Activities of commonly used disinfectants against various microorganisms are summarised in Table 4.6, and a list of commonly used sterilisation techniques and their utility in some of the important materials are listed in Tables 4.7 and 4.8.

Testing of Disinfectants

The efficiency of disinfectants can be determined by the (1) phenol coefficient (Rideal–Walker) test, (2) Chick–Martin test, (3) capacity (Kelsey–Sykes) test and (4) in-use (Kelsey–Maurer) test.

Phenol Coefficient (Rideal–Walker) Test

Rideal and Walker designed the phenol coefficient test to compare the performance of a disinfectant with that of phenol for the ability to kill

Table 4.6 Actions and uses of antiseptics and disinfectants

Chemical	Mechanism of action	Uses
Ethanol (70%)	Protein denaturation and lipid solubilisation	Skin antiseptics
Isopropanol (50–70%)	Protein denaturation and lipid solubilisation	Skin antiseptics
Silver nitrate	Protein precipitation	Antiseptic for use in eye
Tincture of iodine	Protein inactivation	Skin antiseptics
Detergents	Cell membrane denaturation	Disinfectant
Ethylene oxide gas	Strong alkylation	Disinfectant
Formaldehyde (8%)	Reacts with NH_2, SH and COOH group	Disinfectant
Chlorine	Strong oxidisation	Disinfectant, used in water

Table 4.7 Examples of disinfectants

Disinfectants	Examples
Surface active agents	Cetrimide, sodium stearate, sorbiton, tween 80, benzalkonium chloride, cetylpyridium chloride
Alcohols	Methanol, ethanol, isopropyl alcohol
Aldehydes	Formaldehyde, glutaraldehyde, *ortho*-phlhalaldehyde
Phenol	Cresol, xylenol, chlorphenols, chlorxylenols
Dyes	
Aniline dyes	Crystal violet, gentian violet, brilliant green, malachite green
Acridine dyes	Acriflavine euflavine, proflavine, aminacrine
Gaseous sterilisation	Ethylene oxide, beta-propiolactone (BPL)
Halogens	Chlorine, bromine, iodine
Oxidising agents	Hydrogen peroxide, peracetic acid
Metals	Copper, aluminium, silver, mercury

Table 4.8 Commonly used sterilisation techniques

Sterilisation technique	Articles
Hot-air oven	Glassware, paraffin, glove powder, other oily items
Autoclave	Microbial culture media, decontamination of use culture plates, surgical instruments except sharps, catheters etc.
Incineration	Infective materials other than plastics and animal carcasses
Red heat	Inoculating wires and loops
Alcohol, tincture iodine, Savlon etc.	Skin decontamination, cleaning of personal protective equipment of health care professionals
Filtration	Serum sugars, antibiotic solutions, toxins and other heat sensitive materials etc
Gamma radiations and gas sterilisation	Disposable items such as disposable syringes, gloves, needles, catheters
5% Cresol	Sharps
Glutaraldehyde	Cystoscope, endoscope, rubber items, plastic and polythene tubes
Formaldehyde gas	Sterilisation of operation theatre and rooms
Pasteurisation	Milk

Salmonella Typhi. The phenol coefficient is determined by dividing the dilution of the disinfectant in question, which sterilises the suspension of *S.* Typhi in a given time by the dilution of phenol which sterilises the suspension at the same time.

For the test, a series of dilutions of phenol and the experimental disinfectant is inoculated with the test bacteria *S.* Typhi and *S. aureus*, then placed in a 20 or 37 °C water bath. These inoculated disinfectant tubes are then sub-cultured on a regular fresh medium at 5-min intervals, and the sub-cultures are incubated for two or more days. The highest dilutions that kill the bacteria after a 10-min exposure, but not after 5 min, are used in calculating the phenol coefficient. Here, the reciprocal of the suitable test disinfectant dilution is divided by that for phenol to find the coefficient. For example, suppose that the phenol dilution was 1/90 and the maximum effective dilution for disinfectant X tested was 1/450, then the phenol coefficient of X would be 5.

The higher the phenol coefficient value, the more effective is the disinfectant. Therefore, a phenol coefficient value of more than 1 indicates that the disinfectant is more efficient than phenol. The test, however, does not show the action of the disinfectant in natural conditions, that is, in the presence of organic contaminants.

Chick–Martin Tests

It is a modification of Rideal–Walker test, in which the disinfectant acts in the presence of organic contaminants (e.g. dried yeast, faeces) to simulate natural conditions.

Capacity (Kelsey–Sykes) Test

The capacity (Kelsey–Sykes) test determines the appropriate use of dilutions of the disinfectants. It measures the capacity of a disinfectant to retain its activity when repeatedly used microbiologically. The assessment of a disinfectant is done by its ability to kill bacteria by demonstrating growth or no growth on recovery culture media but not by comparison with phenol. The test is performed under clean and dirty conditions, showing the effectiveness of a disinfectant in the presence of organic material.

In-Use (Kelsey–Maurer) Test

The "in-use" (Kelsey–Maurer) test determines whether the chosen disinfectant is effective, in actual use and in-hospital practice, and also for the period of its use. The disinfectant's effectiveness is determined by its ability to inactivate a known number of standard strains of pathogenic staphylococcus on a given surface within a given time. The in-use test more accurately determines a disinfectant's effectiveness than the phenol coefficient test.

Sterilisation and Disinfection in Hospital Settings

Presently, in all infection-control programmes designed for a health care setup, sterilisation and disinfection play a pivotal role. It was initially conceptualised and practised by Pasteur, Lister and Koch. Acquisition of infection is said to be often associated with the use of contaminated materials or devices in a patient. The classification by Spaulding of the devices used on the patient determines the type of method selected for sterilisation or disinfection.

Spaulding Classification

Critical Items

The items or devices that enter the sterile tissue or the vascular system and pose a high risk of infection if contaminated with microorganisms are classified as critical items. This includes cardiac catheters, implants, surgical instruments and so on. Sterilisation of these articles is done either by autoclaving if heat stable or by using ethylene oxide or hydrogen peroxide gas plasma if heat labile.

Semicritical Items

Semicritical items are those which come in contact with the mucous membranes or non-intact skin. Most of these items like endoscopes and bronchoscopes require a high level of disinfection between patients, which is done by glutaraldehyde, a high-level disinfectant. Semicritical items used for neonates like a thermometer and BP cuffs require only intermediate-level disinfection. As they cannot be treated with glutaraldehyde, alcohol treatment should be followed.

Non-critical Items

Items like bedpans, blood pressure cuffs, bed rails, bedside tables and so on, which do not interfere with the mucous membrane but come in contact only with the intact skin, are non-critical items. As they carry no risk of transmitting microbes directly to the patients, a low-level disinfectant is enough to clean these items.

High-Level Disinfectant

High-level disinfectants are chemicals that can kill all microorganisms except some spores. Some activity against a smaller number of spores can be observed if the contact time increases, such as glutaraldehyde and hydrogen peroxide.

Intermediate-Level Disinfectant

These chemicals can kill all microbes, including non-enveloped viruses, mycobacteria and so on, except spores—for example, alcohol, phenolic compounds and iodophors.

Low-Level Disinfectant

Chemicals that can kill only vegetative bacteria, fungi and lipid-enveloped viruses are called low-level disinfectants, for example, quaternary ammonium compounds.

Newer Methods of Sterilisation of Heat-Sensitive Articles

Hydrogen Peroxide

Low-temperature sterilisation using hydrogen peroxide vapour is useful for sterilising metal and non-metal objects which are heat or moisture sensitive. However, it cannot be used for liquid items or linen materials. The cycle time is about 1 h. It is non-toxic and hence safe for the environment, as well as for workers.

Accelerated hydrogen peroxide is another new disinfectant for high-level disinfection where hydrogen peroxide acts together with low concentrations of surfactants. It can be used for heat-sensitive semicritical items, including flexible endoscopes. It has sporicidal, mycobactericidal and fungicidal activities, apart from acting on bacteria and viruses.

Peracetic Acid

It is a more powerful oxidiser than hydrogen peroxide and has sporicidal activity. It is frequently used in combination with hydrogen peroxide to sterilise hemodialysers and for plasma sterilisation.

Plasma Sterilisation

Plasma has unique properties. It consists of ions, electrons and neutrals. For sterilisation, hydrogen peroxide is passed through a chamber where gas plasma is generated in a vacuum through radiofrequency or microwave (Fig. 4.3). This electrical field causes the gas to ionise; that is, ions and electrons are generated. It owes its sterilising properties to these charged particles, many of which are highly reactive free radicals. As the process is carried out at a low temperature of 37–44 °C, gas plasma can sterilise heat- and moisture-sensitive items. It is used for arthroscopes, urethroscopes, plastic devices,

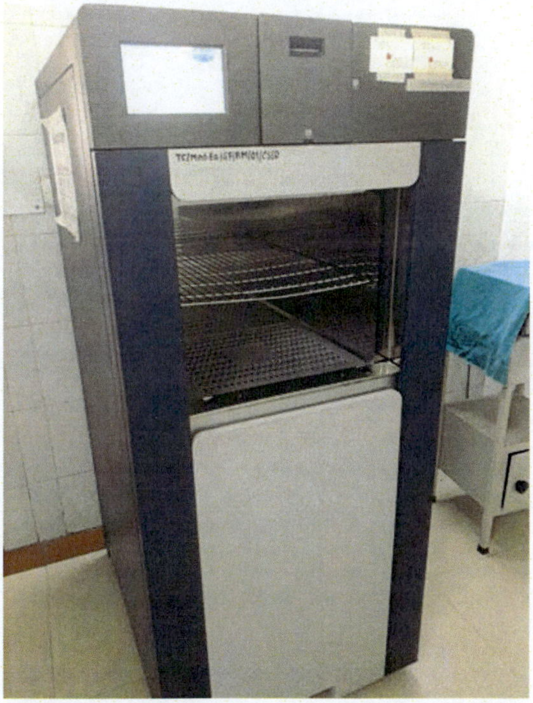

Fig. 4.3 Plasma steriliser. (Courtesy of Prof Tuhina Banerjee)

electrical devices and so on. Liquids or instruments with dead-end lumens cannot be sterilised by this method. This is a safe process, and sterilised items can be used immediately without aeration.

Ozone Sterilisation

Ozone (O_3) is a highly reactive gas and can be generated from molecular oxygen (O_2) by energising it. The extra oxygen atom is loosely bound and is available to oxidise other molecules and hence highly reactive and a powerful microbicidal agent. The instrument is operated electrically 30–35 °C using oxygen and water, and any residual ozone is converted back to molecular oxygen and water vapour at the end of the process. It is used for moisture- and heat-sensitive items like metal and plastic instruments.

5 Microbial Genetics

Genetics is the study of heredity and variation to understand the cause of resemblance and differences between parents and their progeny. The term *genetics* was coined by William Bateson, a British biologist, in 1906.

The unit of heredity is the **gene**, a segment of deoxyribonucleic acid (DNA) that carries in its nucleotide sequence information for a specific biochemical or physiologic property. All hereditary properties are encoded in DNA. Hence, chromosomal DNA plays an essential role in the maintenance of character from generation to generation. Genes carry the information to code for all the necessary components and the actions of life. The genes at each cell division are replicated, and a copy is transmitted to each daughter cell.

Terminologies in Genetics

Codon: A codon is nothing but a sequence of three (**triplet**) adjacent DNA or RNA nucleotides on mRNA that determines the insertion of a specific amino acid or stop signal during protein synthesis. The salient features of the codon are as follows: (1) Codon is **degenerate** means more than one codon may exist for the same amino acid. For example, the triplet codon AUU codes for isoleucine but the other triplets AUC and AUA also code for the same Isoleucine. (2) Codon is **non-overlapping** means three nucleotides in one particular codon do not participate in the other adjacent codon. (3) **Nonsense codons** or **termination codons** or **stop codons** are the three codons (UAA, UAG and UGA) that do not code for any amino acid; instead, they terminate the synthesis of a polypeptide.

Gene: A stretch of DNA that codes for the synthesis of a particular polypeptide is called a **gene**. It is the basic physical and functional unit of heredity. Many genes may exist in a DNA molecule. The bacterial chromosome is made up of a single and circular DNA molecule that is double-stranded.

Introns and exons: **Introns** are the intervening or noncoding regions in DNA that do not code for any protein, whereas the regions of DNA coding for the proteins are called **exons**. The introns and exons are copied during the transcription process, but introns are removed (spliced out) in the RNA molecule before being translated into a protein.

Basic Principles of Molecular Biology

The **central dogma** of molecular biology refers to the flow of genetic information, which is stored and carried by deoxyribonucleic acid (DNA) onto ribonucleic acid (through a process called **transcription**) and then finally to the synthesis of a specific polypeptide (through a process called **translation**). Specific polypeptides such as proteins and enzymes determine the nature and function of a cell. Hence, DNA primarily is an essential component of heredity, the storehouse of all information for protein synthesis in almost all life forms. However, RNA remains the genetic material in some viruses instead of DNA.

DNA Structure

The DNA is the key fundamental component of the gene, which carries the genetic information transcribed onto ribonucleic acid and then translated as the particular polypeptide. The basic structure of the DNA molecule was first described by Watson and Crick, for which they were honoured with the Nobel Prize in Medicine. The DNA molecule is composed of two strands of complementary nucleotides wound together in the form of a double helix (Fig. 5.1). The double helix has a diameter of 2 nm. Each complete turn of the double helix contains 10 nucleotide pairs and is 3.4 nm in length.

Each DNA strand has a backbone of deoxyribose (sugar) and phosphate group residues arranged alternately. To each sugar are attached one of four nitrogenous bases, two purines (adenine and guanine) and two pyrimidines (thymine and cytosine).

The two complementary strands are held together by hydrogen bonds between the nitrogenous bases on the opposite strands. The hydrogen bonding follows a specific binding manner in which hydrogen bonds are formed only between guanine and cytosine and between adenine and thymine. Guanine and cytosine form a complementary base pair, and adenine and thymine form another base pair.

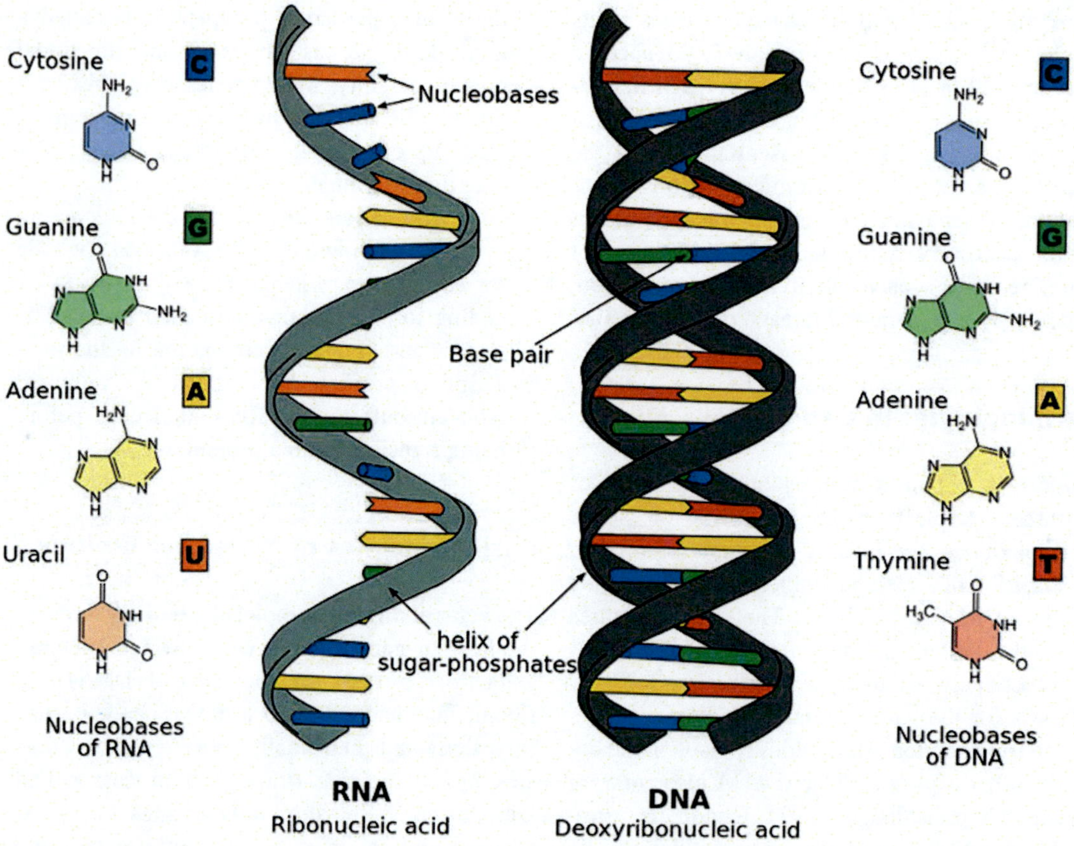

Fig. 5.1 Watson-Cricks structure of DNA: a schematic diagram (https://commons.wikimedia.org/wiki/File:Difference_DNA_RNA-EN.svg)

Therefore, a molecule of DNA contains as many units of adenine as thymine and of guanine as cytosine. The ratio of adenine and thymine to guanine and cytosine is constant for each species but varies widely from one bacterial species to another. During replication, the DNA molecule replicates, first by unwinding at one end to form a fork and then by separating strands at the other end. Each strand then acts as a template for synthesising a complementary strand with which it then forms a double helix.

RNA Structure

The structure of RNA is similar to that of DNA except for two significant differences: (1) In DNA, the sugar is d-2-deoxyribose; in RNA, the sugar is d-ribose. (2) RNA contains the nitrogenous base uracil instead of the thymine present in DNA. Based on structure and function, RNA can be differentiated into three types: (1) Messenger RNA (mRNA), (2) Ribosomal RNA (rRNA) and (3) Transfer RNA (tRNA).

The most common function of RNA is the communication of DNA gene sequences in the form of mRNA to ribosomes. For mRNA synthesis, DNA acts as a template; hence adenine, guanine, cytosine and uracil in mRNA become complementary to the thymine, cytosine, guanine and adenine, respectively, in DNA. The ribosomes, which contain rRNA and proteins, translate this message into the primary structure of proteins via aminoacyl-tRNAs.

Bacterial Genetics

Extrachromosomal DNA Substances

Most bacteria, in addition to chromosomal DNA, possess extrachromosomal genetic elements. Plasmids are such extrachromosomal substances. These are not required for the normal life and functioning of the host bacterium, but their presence in the bacterium confers new properties such as drug resistance, toxigenicity and so on.

Plasmids

Plasmids are extrachromosomal DNA substances. They are replicons maintained as discrete, extra-chromosomal genetic elements in bacteria. They are usually much smaller than the bacterial chromosome (varying from less than 5 to more than several 100 kbp). However, plasmids as large as two million base pairs can occur in some bacteria. Plasmids are circular and double-stranded DNA molecules that encode traits not essential for bacterial viability. They are capable of replicating independently of the bacterial chromosomes. The plasmids can also be present as integrated with bacterial chromosomes, and plasmids integrated with the host chromosome are known as **episomes**. Plasmids are present in both Gram-positive and Gram-negative bacteria.

Types of Plasmids

Plasmids, depending on their transmissibility and nature of the factor, can be of the following types:

1. **Plasmids, depending on transmissibility**, are of two types.
   Transmissible plasmids: They can be transferred from cell to cell by genetic transfer known as conjugation. They are large (mol. wt. 40–100 million) plasmids. They contain more than a dozen genes responsible for the synthesis of the sex pilus and the synthesis of enzymes required for their transfer. Usually, one to three plasmid copies are present in a cell.
   Non-transmissible plasmids: These cannot be transferred from cell to cell because they do not contain the transfer genes. They are small (mol. wt. 3–20 million), usually non-conjugative, and have high copy numbers (typically 10–60 per chromosome). They depend on their bacterial host to provide some functions required for replication and are distributed randomly between daughter cells at the division.
2. **Depending on the nature of factors**, plasmids are of the following types:
   F factor: The F plasmid, also called the F factor, is a transfer factor that contains the

genetic information essential for controlling the mating process of the bacteria during conjugation. The F plasmid of *Escherichia coli* is the prototype for fertility plasmids in Gram-negative bacteria. Strains of *E. coli* with an extrachromosomal F plasmid are called F+ and function as donors, while strains lacking F plasmid are F− and behave as recipients. The conjugative functions of the F plasmid are determined by a cluster of at least 25 transfer (tra) genes. These genes determine (1) expression of pili, (2) synthesis and transfer of DNA during mating, (3) interference with the ability of F+ bacteria to serve as recipients, and (4) other functions.

The F plasmid in *E. coli* can occur as an extrachromosomal genetic element or be integrated into the bacterial chromosome. Both the F plasmid and the bacterial chromosome are circular DNA molecules. Hence, reciprocal recombination produces a larger DNA circle consisting of F-plasmid DNA inserted linearly into the chromosome.

R factor: Resistance factors, also called R factors, are extrachromosomal plasmids. They are circular with double-stranded DNA. R factors occur in two sizes: large plasmids (mol. wt. 60 million) and small plasmids (mol. wt. 10 million). The large plasmids are conjugative "R" factors, containing extra DNA to code for the conjugation process. The small plasmids contain only the "r" genes and are not conjugative. The R factor consists of two components: the resistance transfer factor (RTF) and the resistant determinant (r). The RTF is responsible for the conjugational transfer, while each "r" determinant carries resistance for one of the several antibiotics.

The R factors are responsible for the spread of multiple-drug resistance among bacteria. They carry the genes for various enzymes that can destroy antibiotics and modify the membrane transport system. The R factor may carry one antibiotic resistance gene or may carry two or more of these genes. The R factor carrying more than two genes has many clinical implications: First and foremost, a bacterium carrying such genes will show resistance to more than one type of antibiotic, such as penicillin and aminoglycosides. The second importance is that the use of an antibiotic that selects an organism for a bacterium resistant to one antibiotic (such as penicillin) will select for a bacterium resistant to other antibiotics (such as tetracyclines, aminoglycosides, chloramphenicol, erythromycin). They may also carry genes for resistance to metal ions. For example, the genes code for enzymes that reduce mercuric ions to elementary mercury, thereby making the bacteria resistant to the action of mercuric ions. They also carry resistance to certain bacteriophages by coding for the enzymes, for example, restriction endonucleases that degrade the DNA of the infecting bacteriophages.

Colicinogenic (Col) factor: The Col factor is a plasmid that resembles the F factor in promoting conjugation, leading to self-transfer and, at times, transfer of segments of chromosomes. The Col factor encodes for the production of colicins, which are antibiotics-like substances specifically and selectively lethal to other enteric bacteria. They also encode for the production of diphthericin and pyocyanin produced by *Corynebacterium diphtheriae* and *Pseudomonas pyocyanin*, respectively, which are substances similar to colicins.

Transfer of plasmids during conjugation is responsible for the spread of multiple drug resistance among bacteria. The plasmid responsible for drug resistance consists of two components, namely, RTF and a *resistance determinant* (r) for each of the several drugs. Transferable drug resistance occurs widely among pathogenic and commensal bacteria of humans and animals. The plasmid is also responsible for the production of colicins, the antibiotics-like substances lethal to other Gram-negative bacteria. The plasmid that encodes for the production of colicins is known as the Col factor and is transferred by conjugation.

Bacterial Variations

Bacteria show phenotypic and genotypic variations.

Phenotypic Variations

Phenotype (phaeno = display) is the physical appearance/expression of a particular bacteria. It purely depends on external environmental conditions, and it is a reversible phenomenon. For example, *Salmonella Typhi*, a flagellated bacterium, loses its flagella when grown on phenol agar but regains it when subcultured back on nutrient broth. Similarly, the Lac operon model of *E. coli* is also an example of phenotypic variation. *E. coli* possesses an enzyme called *beta-galactosidase*, which ferments lactose; the gene responsible for the *beta-galactosidase* gets activated only in the presence of lactose. Hence, the presence of lactose in the medium is crucial for switching on the *beta-galactosidase* gene. Some enzymes are produced upon induction with the substrate and are called **inducible enzymes**, whereas many other enzymes are substrate independent and synthesised even in the absence of substrate, and are called **constitutive enzymes**.

Phenotypic variations are mainly dependent on environmental factors, non-heritable, temporary, reversible and limited in number compared to genotypic variations.

Genotypic Variations

Genotype of a particular organism is established by sum total of gene(genome), which are transferable from parents to progeny. Genotype includes complete genome with enormous number of genes, which may or may not be expressed in a given environmental condition. Genotype variations are either due to mutations or by genetic transfer by one of the following methods such as transformation, transduction, lysogenic conversion and conjugation.

Mutations

A mutation is a random, undirected, heritable variation seen in the cell's DNA. It is caused by a change in the base sequence of DNA due to the addition, deletion or substitution of one or more bases in the nucleotide sequence. It can involve any of the genes present in the bacterial chromosome. The mutation results in the insertion of a different amino acid into a protein, resulting in the appearance of an altered phenotype.

Types of Mutations

Mutations are a natural event occurring in dividing cells. The frequency of mutations ranges from 10^{-2} to 10^{-10} per bacterium per division. These occur spontaneously or are enhanced by different mutagens. Mutations are of three types: (1) base substitution, (2) frame-shift mutation and (3) mutations due to transposons or insertion sequences.

Mutation Due to Base Substitution: This type of mutation occurs when one base in the nucleotide sequence is inserted in the place of another. This occurs during replication of DNA either due to an error in the function of the DNA polymerase or due to a mutagen that alters the hydrogen bonding of the base being used as a template in such a manner that the wrong base is inserted. The base substitution mutation may be of two types: *missense mutation* and *nonsense mutation*.

A **missense mutation** is one in which the base substitution results in a codon that specifies a different amino acid to be inserted. A **nonsense mutation** is another type of mutation in which the base substitution produces a terminal codon that stops protein synthesis prematurely. The entire protein function is destroyed during the process of nonsense mutation.

Mutation Due to Frame-Shift Mutation: It is the second type of mutation. This occurs when one or more base pairs are added or deleted in the DNA. Therefore, this leads to shifting the reading frame of the ribosome, resulting in the incorporation of the wrong amino acids downstream from the mutation. Thus, the frame-shift mutation results in the production of an inactive protein.

Mutation Due to Transposons or Insertion Sequence: This is the third type of mutation that occurs when transposons or insertion sequences are integrated into the DNA. These newly inserted pieces of DNA cause profound changes in the gene into which they are inserted and also cause changes in the adjacent genes.

Causative Agents of Induced Mutation

Mutation can be caused by viruses, radiation or chemicals.

Bacterial Viruses

Bacterial viruses (*mutator bacteriophage*) are an example of viruses that cause a high frequency of mutation by inserting their DNA into the bacterial chromosome. Mutations can occur in various genes as viral DNA can insert bacterial chromosomes at many different sites. The mutations caused by these viruses may be either frame-shift mutations or deletions.

Radiations

X-rays and ultraviolet radiations are examples of radiation that can cause mutation in chromosomal DNA. **X-rays** damage DNA in many ways. They cause damage by producing free radicals that can attack the bases or alter them in the strand, thereby changing their hydrogen bonding. They also damage DNA by breaking the covalent bonds that hold the ribose phosphate together. **Ultraviolet radiation** causes damage in DNA by cross-linking the adjacent pyrimidine bases to form dimers. For example, the cross-linking of adjacent thymine to form thymine dimers results in the inability of DNA to replicate correctly.

Chemicals

Various chemicals, such as nitrous acid, alkylating agents and benzopyrene, and base analogues, such as 5-bromouracil, cause mutation in several ways:

Benzopyrene is commonly present in tobacco smoke that binds to existing DNA bases and causes frame-shift mutations. The benzopyrene, a carcinogen and a mutagen, intercalates between the adjacent bases, thereby distorting and offsetting the nucleotide sequence in the DNA.

Nitrous acid and alkylating agents act by altering the existing base in the DNA. This results in the formation of a hydrogen bond with the wrong base. For example, adenine does not form a bond with thymine but makes the wrong pair with cytosine.

Base analogues such as 5-bromouracil have less hydrogen bonding capacity than thymine, so they bind to guanine with better frequency. This results in a mutation due to a transition from an AT base pair to a GC base pair. Iododeoxyuridine, an antiviral drug, also acts as a base-pair analogue.

Effects of Mutations

Mutations in the bacteria cause many changes in their various properties. The mutation alters drug susceptibility, antigenic structure and virulence of mutant bacteria. It also alters the susceptibility of bacteria to bacteriophages, alter their colony morphology and pigment productions, and affect their ability to produce capsules or flagella. The development of drug resistance due to mutations in bacteria is a major health concern.

1. **Conditional Lethal Mutation**

   Lethal mutations occur when some mutations involve vital functions, resulting in the production of nonviable mutants. On the other hand, a conditional lethal mutation is a form of lethal mutation, in which mutation is expressed only under certain conditions, resulting in the production of viable mutants. This is of medical importance because it is used for preparing vaccine strains. Temperature-sensitive strains are the most common example of conditional lethal mutations.

   The temperature-sensitive organisms have the unique property of replicating at a low permissive temperature, such as 32 °C, but cannot grow at a higher restrictive temperature, such as 37 °C. This is due to a mutation that causes changes in an amino acid in an essential protein, allowing these organisms to function at 32 °C but not at 37 °C. The temperature-sensitive influenza virus strain used in the experimental vaccine is an example of conditional lethal mutations. This influenza vaccine contains a virus that can grow at 32 °C and infect the nose, and can replicate and induce immunity. However, the virus cannot grow at 37 °C, hence cannot infect the lungs and does not cause pneumonia.

2. **Spontaneous Mutations**
   In mutations that occur spontaneously in nature, their spontaneous mutation rate varies, that is, large genes mutate more frequently. A fixed frequency of mutation takes place in each gene. Approximately, 10^{-4} to 10^{-10} mutations per bacterium per division occur in each gene in a single bacterium. Spontaneous mutation occurs due to errors in DNA replication while the progeny strand is copied. For example, naturally, thymine forms a bond with adenine, but in case of occurrence of spontaneous mutation, instead of adenine, thymine binds with guanine [T = A (normal); T = G (spontaneous mutation)].
   Bacterial mutations are of prime importance in the field of drug susceptibility testing and live vaccine studies. This is because mutation may alter any characteristic of a bacterium by affecting any gene. The mutants can be distinguished by virulence, type of colony, morphological features, host range, biochemical reactions, nutritional requirements, antigenic structure and so on.

Gene Transfer

Gene or DNA transfer in bacterial cells is broadly of two types: (1) transfer of DNA between bacterial cells; and (2) transfer of DNA within bacterial cells.

Transfer of DNA Between Bacterial Cells

Genetic information can be transferred from one bacterium to another by transformation, transduction and conjugation (Fig. 5.2).

Transformation

It is a process of transferring genetic information through the agency of free DNA from one bacterium to another. This may occur either in nature

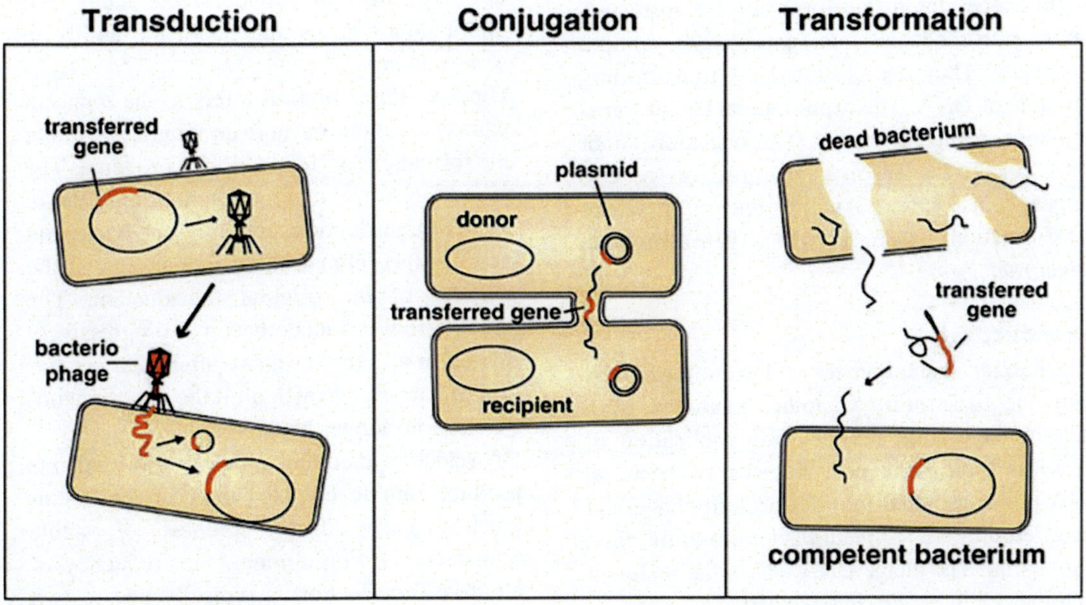

Fig. 5.2 Transmission of genetic material by transduction, conjugation and transformation (https://link.springer.com/protocol/10.1007/8623_2015_46)

or in a laboratory. In nature, DNA is released from a bacterium by lysis, which may be taken up by the recipient bacterium that must be competent. This natural process of transfer of genetic material appears to play no role in disease. However, in laboratory conditions, DNA may be extracted from one type of bacterium and introduced into genetically different bacteria. When purified DNA is injected into the nucleus of a bacterial cell, the process is called *transfection*. Transfection is frequently used in genetic engineering studies.

Griffith (1922), in his classical experiment on mice, demonstrated that neither of the mice died when injected separately with a live, non-capsulated Pneumococcus (nonvirulent) and heat-killed, capsulated Pneumococcus (nonvirulent). However, the mice died when they were injected with a mixture of these strains. He could isolate live, capsulated pneumococci from the dead mice, which were virulent. He demonstrated that some factor in heat-killed, capsulated pneumococci had transferred the material for capsule synthesis in the non-capsulated strains of the bacteria, making them virulent (Fig. 5.3).

McLeod and McCarthy in 1944 demonstrated that DNA extracted from encapsulated, smooth pneumococci could transform non-encapsulated, rough pneumococci into capsulated, smooth organisms. They demonstrated the transforming principle of DNA. The experimental use of transformation revealed important information about DNA and was the first example of genetic exchange in bacteria. Another bacterium where transformation is observed is *Haemophilus influenzae*.

Transduction

Transduction is the transfer of a portion of DNA from one bacterium to another mediated by a bacteriophage (Fig. 5.4). During replication of the virus within the cell, a piece of bacterial DNA is incorporated into the bacteriophage and is carried into the recipient bacterium at the time of infection. The phage DNA within the recipient bacterial cell integrates into the cell DNA during a process called **lysogenic conversion**. Transduction is of two types: generalised transduction and specialised transduction.

1. **Generalised Transduction**

   This occurs when a small fraction of the phage virions produced during the lytic cycle are atypical. It contains an arbitrary section of the bacterial genome rather than phage DNA. Every transducing phage handles a distinct set of closely related genes that portray a small section of the bacterial genome. Transduction carried out by such different populations of phage is called generalised transduction. Each section of the bacterial genome certainly has the same chance of being transferred from the donor bacteria to the recipient bacteria.

   Generalised transduction involves any segment of the donor DNA at random. This occurs because the cell DNA is fragmented after such infection, and pieces of the same DNA, the same size as viral DNA, are incorporated into the bacterial DNA. This occurs at a frequency of about 1 in every 1000 viruses. Generalised transduction may be complete or abortive:

   Complete transduction is characterised by the production of stable recombinants that inherit donor genes and retain the ability to express them.

   Abortive transduction refers to the transient expression of one or more donor genes without the formation of recombinant progeny. The donor DNA fragment does not replicate in abortive transduction, and only one bacterium contains the donor DNA fragment among the progeny of the original transductant. The donor gene products become progressively diluted in all other progeny after each generation of bacterial growth until the donor phenotype can no longer be expressed.

   On selective medium, abortive transductants produce minute colonies that can be distinguished easily from colonies of stable transductants. Furthermore, the frequency of abortive transduction is typically one to two times more than generalised transduction. This indicates that most cells infected by

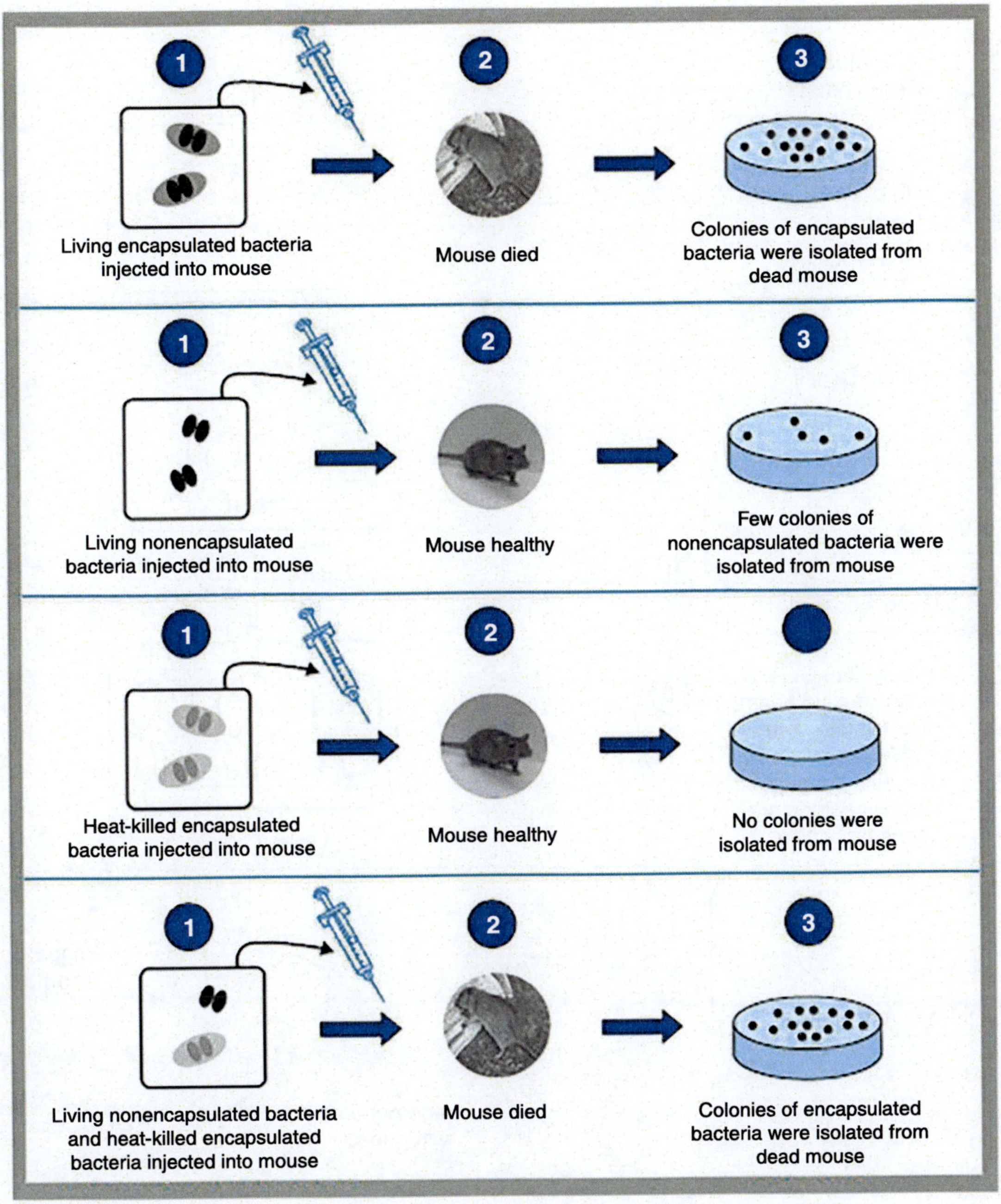

Fig. 5.3 A schematic diagram showing the transmission of DNA by transformation

generalised transducing phages do not produce recombinant progeny.

2. **Specialised Transduction**
   Specialised transduction results from the lysogenisation of the recipient bacterium by specialised transducing phage and expression of the donor genes. Specialised transducing phages are formed only when lysogenic donor bacteria enter the lytic cycle and release phage progeny.
   The specialised transducing phages are rare recombinants that lack part of the normal

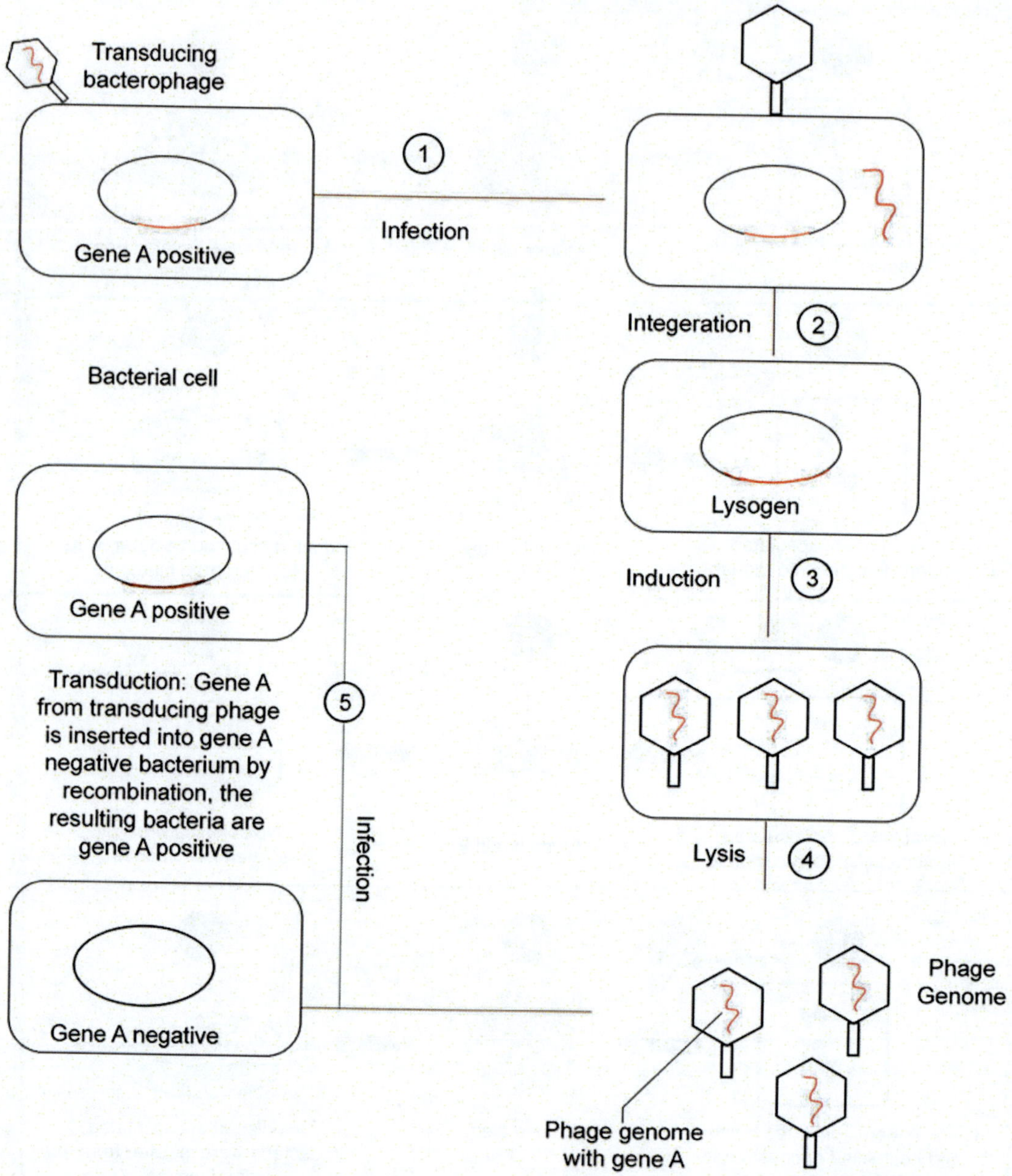

Fig. 5.4 A schematic diagram showing transmission of genetic material by bacteriophage-mediated transduction

phage genome. They contain part of the bacterial chromosome present adjacent to the site of the pro-phage attachment. Many specialised transducing phages are defective. They cannot complete the lytic cycle of phage growth in infected cells unless helper phages are present to provide missing phage functions. Specialised transduction differs from generalised transduction in many ways. The former is mediated only by specific temperate phages, and only a few specific donor genes can be transferred to the recipient bacteria.

Transduction is not only confined to the transfer of chromosomal DNA but also that of plasmids and episomes. Transformation involving the transfer of plasmids from one bacterium to another by transduction is responsible for penicillin resistance in staphylococci. Transduction appears to be the most widespread mechanism of gene transfer among bacteria. This method also provides an excellent tool for the genetic mapping of bacteria. It can occur in many bacteria for which bacteriophages are known. In addition, it may occasionally occur in eukaryotic cells.

The process of lysogenic conversion confers a new property to the bacterial cell; for example, non-pathogenic bacteria can become pathogenic by lysogenic conversion. Bacteriophages encode diphtheria toxin, botulinum toxin, cholera toxin and erythrogenic toxin, and can be transferred from one bacterium to another by transduction.

Conjugation

Bacterial conjugation is a process by which the transfer of genetic material takes between the donor bacterium and the recipient bacterium of related species by direct cell-to-cell contact. Plasmids are the main extrachromosomal elements transmitted by conjugation (Fig. 5.5).

This gene transfer mechanism was first described in *E. coli* K12 by Joshua Lederberg and Edward Tatum in 1946. Bacterial conjugation is also considered sexual reproduction or mating in bacteria. During the conjugation process, the donor or male bacterium possesses a special plasmid called F plasmid (F+) that codes for the projected structure called fimbria or sex pilus or conjugation tube. The recipient or female bacterium does not contain a similar plasmid (F−); hence there is no sex pilus structure. The fimbria of donor connects with the recipient bacterium, and transfer of plasmid DNA and portions of host DNA sometimes occurs. The transferred genetic information is beneficial to the recipient bacterium and may include antibiotic resistance or the ability to use new metabolites.

High-frequency recombination (Hfr): The long length of DNA can be transferred by the process of conjugation. The Hfr strain is a type of F+ cell that has an F plasmid integrated into the bacterial DNA. Hence they acquire the capability of transferring the chromosome to another cell. A

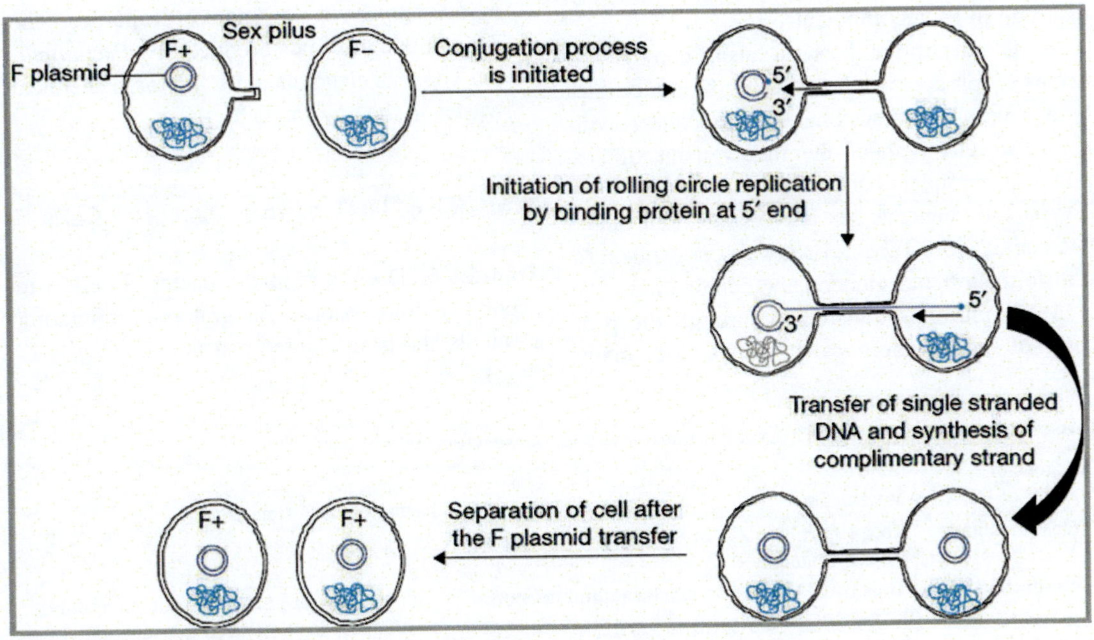

Fig. 5.5 A schematic diagram showing transmission of genetic material by conjugation

whole chromosome can be transferred if integrated with the F plasmid. In this process, the single strand of DNA that enters the recipient F− cell contains a part of the F factor at one end, followed by the bacterial chromosome, and then by the remainder of the F factor. The bacterial genes adjacent to the leading piece of the F factor are the most frequently transferred. The newly acquired DNA recombines with the recipient DNA and becomes an integral component of the genetic material. The complete transfer of the bacterial DNA is usually completed in approximately 100 min.

In the conjugation process between the F+ and F− bacteria, conjugation (i.e. transfer rate) is high in F plasmids, whereas transfer rate is very low in chromosomal genes, which is specifically mediated by the Hfr mutants in F+ bacteria. In mating between the Hfr and F− strains, the tra region of a plasmid is transferred only at the end of the conjugation process after the whole bacterial chromosome has been transferred. In most instances, recombinants produced due to the mating across the Hfr and F− cells fail to acquire the whole set of F− plasmid genes and are characteristically F−, whereas in mating between the F+ and F− strains, the F− plasmid transfers quickly resulting in F+ recombinants.

In Gram-positive bacteria also, conjugation occurs by producing adhesions by the donor bacteria. These adhesions make them cluster with recipient cells without the involvement of sex pili. Some recipient bacteria, for example *Streptococcus* spp., generate sex pheromones that facilitate conjugation. Table 5.1 shows a comparison of transformation, transduction and conjugation.

Many plasmids transmitted through the process of conjugation carry genes for many enzymes, toxins and resistance to one or more antibiotics as follows:

1. Resistance to one or several antibiotics that is mediated by a variety of enzymes.
2. Resistance to ultraviolet light that is mediated by DNA repair enzymes.
3. Resistance to heavy metals such as mercury and silver that is mediated by enzyme reductase.
4. Pili that mediate the adherence of bacteria to epithelial cells. Examples include K88, K99 found in uropathogenic *E. coli*.
5. Exotoxins, including many enterotoxins. Representative toxins encoded by plasmids include heat-labile and heat-stable enterotoxins of *E. coli*, the exfoliative toxin of *Staphylococcus aureus*, hemolysins of *Clostridium perfringens* and tetanus toxin of *Clostridium tetani*.
6. Bacteriocins produced by certain Gram-negative bacteria are lethal for other bacteria. Examples include pyocin produced by *P. pyocyanea* and diphthericin produced by *C. diphtheriae*.
7. A variety of enzymes such as urease of *Helicobacter pylori* and proteases of *Pseudomonas* spp. have degradative properties. The degradative enzymes produced by *Pseudomonas* spp. can clean oil spills and toxic chemical wastes in the environment.

Transfer of DNA Within Bacterial Cells

Transfer of DNA within the bacterial cells can occur by transposons, integrative conjugating elements and programmed rearrangement.

Table 5.1 Differentiating characteristics of transformation, transduction and conjugation

Methods of transfer of DNA	Mechanism	Nature of DNA transferred
Transformation	Recipient cell uptake of free DNA released into the environment	Any gene
Transduction	Transfer of DNA from one bacterium to another by bacteriophage	Any gene in generalised transduction; only select genes in specialised transduction
Conjugation	Transfer of DNA from one living bacterium to another through the sex pilus	Chromosomal or plasmid DNA

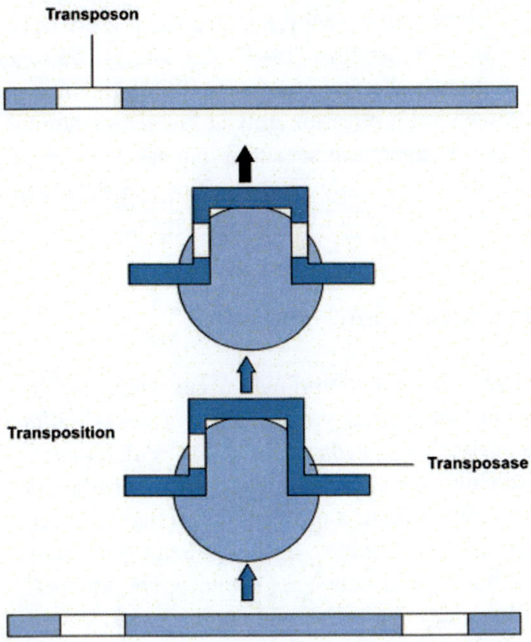

Fig. 5.6 Structure of a transposon

Transposons

Transposons are a type of mobile DNA of 2000–20,000 bp. They can transfer DNA from one site of the bacterial chromosome to another site or a plasmid (Fig. 5.6). Barbara McClintock, a geneticist working in maize genetics, gave the idea of transposons or jumping genes.

The mode of genetic transfer by transposon is called **transposition**. Transposition differs from recombination in that a segment of DNA can be transferred from one molecule to another with no genetic homology with either the transposable element or the donor DNA.

Transposons do not occur independently but have the characteristic of jumping from one part of a chromosome to another or a plasmid. They can also jump from one plasmid to another or back to the chromosome; hence, they are called **jumping genes**. Transposition in prokaryotes usually involves two steps—self-replication and recombination. They jump from one part to another by synthesising a copy of their DNA and inserting it at another site in the bacterial chromosome or the plasmid. Transposons, unlike plasmids, are not self-replicating and depend on chromosomal or plasmid DNA for replication.

The plasmid can contain several transposons carrying drug-resistance genes, which code for drug-resistance enzymes, toxins or metabolic enzymes. These transposons can either cause mutations in the gene into which they insert or alter the expression of nearby genes.

Insertion sequences: The simplest form of transposons are the insertion sequences. These are a type of transposons that have a few bases varying from 800 to 1500 bp. They are found as multiple copies at the end of larger transposon units. These cause mutations by moving from one side to another in the DNA sequence and are believed to control various cellular responses.

Integrative Conjugative Elements

Integrative conjugative elements (ICEs) are ways of horizontal gene transfer and self-transmissible mobile genetic elements. The elements exchange by conjugation but need to integrate into the chromosome to propagate. They cannot replicate autonomously. ICEs integrate into and replicate with the host cell chromosome, whereas plasmids exist as extrachromosomal (usually circular) autonomously replicating DNA molecules.

Programmed Rearrangements

The transfer of DNA within bacteria can also occur by programmed rearrangement. In this programmed rearrangement, the gene moves from a silent site where it is not expressed to an active site where transcription and translation occur.

Many silent genes are present in the DNA that encode variants of the antigens. Presentation of the new gene into the active site occurs sequentially and repeatedly, manifesting in antigenic variations in the bacteria and parasites. This mechanism is responsible for antigenic variations seen in *Neisseria gonorrhoeae*, *Borrelia recurrentis* and *Trypanosome brucei*.

Viral Genetics

Viruses, like other living beings, obey the law of genetics. The viruses show variation in their genomic structure by two principal methods—mutations and recombination.

Mutation

A mutation is the most important mechanism of genetic modification in viruses. Mutations occur spontaneously and readily in viral genomes causing frequent changes in the nucleic acids. This results in new viral strains showing properties different from parental or wild-type viruses. Mutations occur during every viral infection, the frequency of mutations being about 10^{+5}. New variants of strains are identified by their nucleotide sequence, antigenic differences or by differences in their structural or functional properties. Mutations in the essential genes of the virus cause the inactivation of the virus. This is known as lethal mutation. However, mutations in the other genes alter the antigenicity and pathogenicity of the virus and induce drug resistance in viruses. Mutations in viruses may be induced by mutagens, such as irradiation or chemical agents, such as 5-fluorouracil, or even spontaneously.

Lethal Mutations

Mutations in essential genes are known as lethal mutations. Mutations in viruses lead to the formation of various mutants, which show new functional or structural proteins. This leads to the formation of new viral mutants, which are difficult to isolate because the virus cannot replicate. Selective removal or loss of a portion of the viral genome and the subsequent loss of function that it encodes lead to the formation of the deletion mutant. Other mutations may produce:

- **Attenuated mutants**—variant strains that cause less severe infections in humans and animals.
- **Host range mutants**—variant strains showing differences in the tissue type and species of target cells affected by viruses.
- **Plaque mutants**—variant strains showing differences in their size or appearance in infected host cells.
- **Conditional mutants**—for example, temperature-sensitive or cold-sensitive mutants.

Temperature-sensitive mutants can grow only at a low temperature (28–37 °C). This is due to a mutation in the gene for an essential protein that allows viral replication only at low temperatures. These temperature-sensitive mutants are used extensively to study viral genetics and evaluated for possible use in live viral vaccines.

Recombination

Genetic recombination may occur when two different but related viruses infect a cell simultaneously. This leads to extra-molecular genetic exchange between the two viruses, leading to the production of a progeny virion that possesses genes from both the viruses. Genetic recombination may occur between (1) two active infectious viruses, (2) one active and another inactive virus and (3) two inactive viruses.

Intramolecular Recombination

Intramolecular recombination occurs between two closely related DNA viruses. For example, infection of the host cell by two closely related herpes viruses, such as herpes simplex virus types 1 and 2, produces recombinant, new hybrid strains. These new hybrid strains possess genes from both viruses: types 1 and 2. This process has also been observed in closely related RNA viruses. For example, the recombination of Sindbis and eastern equine encephalitis virus has led to the formation of western equine encephalitis virus, another togavirus. Intramolecular recombination may be re-assortment or re-activation.

Re-assortment

It is another process of genetic recombination. It occurs between viruses with segmented genomes, such as influenza virus A and B (8 segments), Reoviridae (10–12 segments), Bunyaviridae (3 segments) and Arenaviridae and Birnaviridae (2 segments). An exchange of segments occurs between these viruses, resulting in new hybrid strains. Synthesis of new strains of influenza A by co-infection with a virus from a different species is an example of genetic recombination by re-assortment that occurs in nature.

Re-activation

It is of two types: (1) cross-reactivation or marker rescue and (2) multiplicity re-activation.

Cross-reactivation (marker rescue): This occurs between an active virus and a related inactive virus, resulting in a progeny possessing one or more genetic traits of the inactivated virus. For example, in influenza virus, when a new epidemic strain (e.g. strain A1) that does not often grow well in eggs as compared to the established strain is grown in eggs along with a standard strain (e.g. strain A2) inactivated by UV irradiation, it results in the production of a new hybrid strain. The hybrid strain may have antigenic properties of strain A2 but genetic characteristics of A1. Such viruses have been evaluated widely in the production of influenza vaccines. Marker rescue is also used to map the genome of the herpes simplex virus.

Multiplicity re-activation: This is a phenomenon in which a live virus may be produced when a cell is infected with two or more (high multiplicity) virus particles of the same virus, each of which has suffered lethal mutation in a different gene due to UV irradiation. Thus, it may be possible to obtain a full complement of undamaged genes from the total genetic pool, resulting in the production of infectious virion progeny. Therefore, there is a potential danger of multiplicity re-activation occurring following the administration of UV-irradiated vaccine. So, UV irradiation is not safe to produce inactivated virus vaccines.

Non-genetic Interaction Between Viral Gene Products

Non-genetic interaction between viral gene products takes place in the following four ways: (1) complementation, (2) phenotypic mixing, (3) genotype mixing and (4) interference.

Complementation

Complementation is a functional interaction between the gene products of two viruses when two viruses (one or both of which may be infective) infect a cell simultaneously. In this process, their gene products, such as proteins and enzymes, complement the action of each other to have an increased yield of one or both viruses. This occurs even between unrelated viruses. In this process, a defective virus strain can be rescued by the replication of another mutant or by a wild-type strain. For example, adenovirus is defective in simian cells, but it can be rescued by SV40. The SV40 virus can grow in monkey cells and produce a gene product that adenovirus needs, thereby complementing adenovirus replication.

Phenotypic Mixing

In this process, virus products from cells infected with different virus strains may be phenotypically mixed and have the properties of one strain but the genome of another. This is not a stable variation; on subsequent passage, the capsid will be found to be of the original type only. In phenotypic mixing, when the nucleic acid of one virus is surrounded by the entire capsid of the other virus, it is called **transcapsidation**. Pseudo types are produced when transcapsidation occurs between different types of viruses, but this is rare.

Genotypic Mixing

All the viruses of vertebrates are haploid except retroviruses, which are diploid, but sometimes several nucleocapsids may be enclosed within a single envelope. This phenomenon of genotypic mixing is known as **polyploidy**. No recombination between the virus's genomes occurs, and it is not a stable change.

Interference

Interference denotes inhibition of the growth of one virus in a host cell when it is simultaneously infected with another virus. Interferon is the most important mediator of interference. Auto interference is another type of interference in which a high multiplicity of infection inhibits the

Table 5.2 Distinguishing characteristics of prokaryotic and eukaryotic transcription

Characteristics	Prokaryotic transcription	Eukaryotic transcription
Transcription and Translation	Occurs simultaneously	Separate
Site	Cytoplasm	Nucleus
RNA polymerase	5 subunits	10–15 subunits
Transcription factor for initiation	Not required	Required
Coverage	Complete genome is transcribed at a time	Different genes are transcribed separately

production of infectious progeny. In contrast, mixed infections sometimes increase the yield of one virus and produce well-marked cytopathic effects in the cell lines. This phenomenon is called **enhancement**.

Transcription

The process of making an mRNA from a specific gene sequence is known as transcription. Although the process is fundamental in prokaryotes and eukaryotes, and is performed by enzymes called RNA polymerases as the first step towards gene expression, there are some differences (Table 5.2). Prokaryotic RNA polymerase is a five-subunit (α, α, β, β′ and σ) holoenzyme. Four of these subunits, that is, α, α, β and β′, compose the polymerase core and the σ is involved only in transcription initiation.

The two promoter consensus sequences, or regions, namely the −10 region, is TATAAT. The −35 sequence, TTGACA, exists in bacteria. During the transcription, the σ subunit of RNA polymerase recognises and binds the −35 region. After this recognition, the core polymerase binds to the entire promoter region, where −10 regions facilitate the unwinding of DNA to initiate RNA synthesis.

For decades, bacterial transcription has remained an excellent target for novel antibacterial development. Actinomycin D and rifampin are the commonly used antibiotics that prevent transcription. The rifamycins bind to bacterial RNA polymerase and prevent the extension of RNA chains beyond a length of 2–3 nucleotides. It is the first-line antituberculosis agent and is among the only antituberculosis agents able to clear the infection and prevent relapse. Some of the reasons that transcription inhibition is the preferred centre for antibacterial discovery are (1) transcription is an essential process for cell viability, (2) bacterial RNA polymerase and other transcription factors are highly conserved. This enables specific action on prokaryotic without harming the eukaryotic transcription, thus resulting in lesser cytotoxicity and (3) numerous high-resolution structures of RNA polymerase and transcription complexes enable structure-based drug design.

Translation

In a prokaryotic organism, transcription and translation occur simultaneously in the cytoplasm. Because the chromosome size is small compared to the eukaryotic genome, all functional genes (the cluster is called an **operon**) for an activity are transcribed from polycistronic mRNA and translated together.

Like the eukaryotes, bacterial translation is divided into initiation, elongation and termination, where 70S ribosomes carry out the elongation process. The **translation** is initiated by the binding of the 30S subunit of the ribosome at a purine-rich region (the Shine-Dalgarno sequence) upstream of the AUG initiation codon of mRNA. The Shine-Dalgarno sequence is complementary to a pyrimidine-rich region on the 16S rRNA of the 30S subunit and plays a crucial role in prokaryotic translation. **Elongation** is a stepwise process where the ribosome slides along the mRNA and stitches the amino acids after it enters the ribosome through the A site, peptide bond formation in the P site and release through the E site. **Termination** is achieved by release factors (RF) proteins where RF1 recognises the UAA and UAG stop codons, and RF2 recognises UGA stop codons. These factors hydrolyse the ester bond in peptidyl-tRNA and release the newly synthesised protein from the ribosome.

Several antibiotics work by preventing specific steps in translation. Some of the antibiotics and their targeted actions are:

1. Tetracyclines like doxycycline prevent the binding of aminoacyl-tRNA by blocking the A (aminoacyl) site of the 30S ribosome.
2. Aminoglycoside antibiotics like streptomycin prevent the formation of the 30S initiation complex. In contrast, kanamycin and tobramycin bind to the 30S ribosome but block the formation of the larger 70S initiation complex. Hygromycin B specifically binds to the 30S subunit in a region that contains the A, P and E sites of tRNA.
3. Erythromycin, a macrolide, binds to the 23S rRNA component of the 50S ribosome and interferes with the assembly of 50S subunits.
4. Lincomycin and clindamycin inhibit peptidyl transferase activity and prevent elongation.
5. Puromycin acts as an analogue of the 3-′-terminal ends of aminoacyl-tRNA. It disrupts synthesis and leads to premature chain termination.

Operons and Regulation of Gene Expression in Prokaryotes

In the bacterial genome, **operons** are a cluster of genes (components of the same biochemical functions) that are transcribed together to give a single messenger RNA (mRNA) molecule (i.e. polycistronic mRNA). It allows for the coordinated regulation, transcription and translation of functionally related genes by encoding multiple proteins.

Because the genes in an operon are functionally related, their expression during transcription is regulated together to determine the fate of a particular biochemical function. This process of expression of gene regulation to switch on or off biological functions is carried out by a set of two major proteins, that is, **repressors** and **activators**. In general, an operon is said to be functional when the genes are transcribed, and proteins are produced in the presence of activators. Likewise, repressors bind to the genes in an operon that prevents their transcription and translation and switches off the biological function.

The most widely studied operon in the bacterial system is the **Lac operon**. The genes in the operon are responsible for the transport and metabolism of lactose by a bacterium. In a 5′ to 3′ direction, the operon consists of a CAP-binding site, promoter, operator and three enzyme-producing genes, namely, Z, Y and A. The Z gene codes for beta-galactosidase, which catalyses the hydrolysis of lactose into glucose and galactose. The Y gene codes for permease, which regulates the cell's lactose permeability, and the A gene codes for transacetylase, which assists the enzyme beta-galactosidase.

The **CAP** is the activator protein that binds to the CAP site to turn the operon on, and the Lac repressor binds to the operator site to turn off the operon. The operon is functional when the genes coding for the enzymes are expressed only when lactose is present and glucose is absent. In the excess presence of glucose, the repressor prevents the transcription of Z, Y and A genes that turn off the operon. A catabolite activator protein activates the transcription of the operon only when glucose levels are low.

Fungal Genetics

Fungal genome contains more noncoding DNA (repetitive elements and introns) than coding DNA (genes). The noncoding DNA, such as introns, telomeres etc., provides the genome plasticity to fungi during environmental/stress adaptation. The fungal genome architecture contains Gene cluster region (gene coding for toxins/secondary metabolites), Gene sparse region (repeat rich region), isochore-like region, subtelomeric region and conditionally dispensable chromosomes.

Fungi show a large diversity in genome size, which ranges from 2 Mb (*Microsporidia*) to 2 Gb

(*Pucciniales*), suggesting multiple genome reduction/expansion events in fungi during host/environmental adaptation. Some obligate parasitic fungi and yeast have small genome sizes, such as *Microsporidia* (2–6 Mb), *Pneumocystis* (8 Mb), *Candida* (12–16 Mb) and *Cryptococcus* (20 Mb). These fungi lost most of the noncoding DNA to acquire a smaller genome and cell size. By contrast, most filamentous fungi have large genome sizes by the expansion of noncoding DNA, such as *Aspergillus oryzae* (37 Mb), *Neurospora crassa* (41 Mb), *Schizophyllum commune* (38.5 Mb) and *Rhizopus* (40–50 Mb).

Genome Evolution

Genomic changes in small-scale variation such as changes in nucleotide level and gene level, changes in alleles within a species, horizontal gene transfer, introgression, and meiotic drive elements and large-scale variation contribute to genome evolution.

Small-Scale Evolution of Fungal Genomes

Small-scale evolution includes mutation (single nucleotide changes), transposition and repetitive sequences expansion/contraction.

- **Hypermutation**
  Fungal isolates with greater mutation rates than reference strains of the same species are called hypermutator phenotypes. The hypermutator strains have been described in human fungal pathogens such as *Cryptococcus neoformans* and *Candida glabrata*. These fungi carry mutations in the DNA mismatch repair pathway, thus displaying hypermutator phenotypes. Hypermutation is beneficial for stress adaptations but disadvantageous in the long term.
- **Fast-evolving regions of the fungal genome**
  Introns, intergenic sequences, centromeres and subtelomeric sites are the fast-evolving regions in the fungal genome. The evolution of introns occurs by ancestral introns retentions and frequent intron loss or gain. Intergenic sequences encode long noncoding RNAs (lncRNAs) to regulate gene expression, morphogenesis, cellular metabolism, mating behaviours, virulence, drug resistance and genomic stability. Centromeres are rapidly evolving sequences among fungi. The centromere length varies between 100 bp and 300 kb among related species. The centromeres architecture and not sequence is critical for its function.
  Other fast evolving genomic regions are the subtelomeres and the repeat-rich regions of fungi that harbour virulence genes. Subtelomeric loci are prone to rapid expansion or contraction and often encode genes associated with virulence and secondary metabolites. The rapid subtelomeric evolution increased the virulence in *Candida auris* and *Aspergillus fumigatus*. Interestingly, *C. auris* lineage with low virulence exhibits high mutation in subtelomeric regions.
- **Allelic transmission**
  The gene transmission between species/kingdoms can grossly impact genomic evolution. Horizontal gene transfer (HGT), genetic introgression and meiotic drive elements are strong forces of genomic evolution in fungi. HGT is more frequently observed in pathogenic fungi such as *Malassezia* species (acquisition of bacterial flavohemoglobin and other genes) and *Microsporidia* (2% genes derived from HGT). Introgression also occurs with high frequency in fungi. For example, the loss and regain of fermentation capability by *Kluyveromyces lactis* by introgression of a subtelomeric locus from *Kluyveromyces marxianus*.

Large-Scale Evolution of Fungal Genomes

Large-scale genomic evolution includes chromosomal structural rearrangements such as deletions, inversions, duplications, and translocations and chromosome copy number variation such as polyploidy and aneuploidy.

- **Chromosomal structural rearrangements**
  Structural variations within chromosomes have a critical role in the evolution of functions in fungal species, including host adaptation, metabolism, pathogenicity and speciation. Chromosome rearrangements facilitated by

centromeres have been observed in *Malassezia*, *Candida* and *Cryptococcus*.

- **Chromosome copy number variation**
  Fungi can undergo both polyploidy (change in chromosome set numbers) and aneuploidy (change in chromosome numbers) for stress adaptation. Polyploidy is common among fungal pathogens. Human fungal pathogen including, *Candida albicans* (haploid, diploid and tetraploid) and *C. neoformans* (haploid and titan cells) induces ploidy variation for rapid stress adaptation whereas aneuploidy is associated with drug resistance in pathogenic fungi like *C. albicans* and *C. neoformans* by isochromosome formation and aneuploidy, respectively.
- **B chromosomes**
  B chromosomes (also called as an accessory or supernumerary chromosomes) are dispensable extra chromosomes found in some fungi. B chromosomes transmit in a non-Mendelian phenomenon and do not interact with the core (A) chromosomes set. Their size ranges from 400 Kb to 3 Mb. The loss or gain of B chromosomes has been observed in stress adaptation, suggesting a critical role in genome plasticity. These chromosomes are essential for imparting virulence and are transmitted horizontally.

Application for Fungal Genetics in Identification

Apart from causing infections among humans, fungi also cause a wide spectrum of diseases in plants and animals. The precise diagnosis of fungi is of utmost importance owing to the morbidity and mortality caused, especially in various susceptible hosts. Among the diagnostic methods though the phenotypic methods are being routinely used among laboratories, they have inherent hindrances of being tedious, time-consuming and entail experience. These roadblocks acting as a major obstacle to the precise identification of fungi underlined the requisite for implementing genotypic methods for routine diagnosis. Due to the multicopy nature (15–20 copies) of the ribosomal DNA gene, this is used as the target for identification. However, single-copy genes such as beta-tubulin, calmodulin and chitin synthase may be necessary to identify some species. The rDNA consists of 18s, internal transcribed spacer 1 (ITS-1) 1, 5.8s, ITS-2 and 28s regions; each copy of rDNA is separated by an intergenic spacer region (IGS). ITS region is more conserved and generally used as the target for identification by sequencing and other methods. Pan fungal primers are available to amplify most fungi, including yeasts and moulds. Multiple modalities like polymerase chain reaction (PCR), multilocus sequence typing (MLST) and DNA sequencing have been exploited for the same. PCR is considered the simplest and most inexpensive modality for which many studies deciphered ITS as the most suitable candidate for fungal barcoding with the highest probability of precise identification of a larger group of fungi. Various studies have also been performed to determine the performance of 18S and 28s rDNA PCR compared to ITS PCR and culture.

Human Normal Microbial Flora and Microbiome

6

Humans get colonised with a microbial population immediately after birth, which persists throughout the individual's lifetime. Normal microbial flora are the microorganisms that inhabit the skin and mucous membranes of a healthy human being and are regularly found at various anatomical sites. The term indigenous microbiota is also used for them. For all practical purposes, they are commensals without any deleterious effect on the host, but some can act as pathogens in certain circumstances.

It has been estimated that humans have approximately 10^{13} cells in their bodies, and about 10^{14} bacteria are associated with them. Most bacteria are present in the large bowel, which constitutes the normal flora. Fundamentally, the organisms are present in parts of the body that are exposed to or communicate with the external environment, namely, the skin, nose, mouth and intestinal and urogenital tracts (Table 6.1). Internal organs and tissues are normally sterile.

In a pregnant mother, the human foetus lives in a sterile environment protected from microbes except when pathogens like cytomegalovirus, rubella virus or *Toxoplasma gondii* cross the placenta for the first 9 months of life. At the time of birth, the newborn is confronted with the mother's vaginal microbes and environmental organisms. The infant's skin surface is initially colonised, followed by the oropharynx, the gastrointestinal tract and mucosal surfaces.

Microbes that can cause serious diseases are normally found in and on the human body. Not the recovery of a specific organism, but the recovery of the organism in a normally sterile site is the hallmark of the pathogenesis of microbial infections. For example, *Escherichia coli* is a normal resident confined to the gastrointestinal tract. If it is demonstrated in the stool, it may be considered normal, but if it is found in the abdominal cavity or the patient's bloodstream, this would be considered abnormal. Similarly, certain organisms are never present as part of the normal microbial flora in humans; hence, their recovery is always associated with clinically significant diseases (e.g. *Bacillus anthracis*, *Brucella* spp., *Francisella tularensis*, *Histoplasma capsulatum*).

Viruses and parasites are not found as commensals in the host and are not considered members of the normal microbial flora.

Normal Microbial Flora

The normal microbial flora is more or less constant for each mammalian species and is broadly divided into transient flora and resident flora.

Transient flora: These organisms may or may not be pathogenic and inhabit a particular site for variable periods from a few hours to a few weeks. Their occurrence is influenced by many factors, including the hospital environment of an admitted patient, where the hospital

Table 6.1 Distribution of normal microbial flora

Body site	Predominant flora	Other organisms
Skin	Staphylococci, diphtheroids, *Propionibacterium*, Micrococci	Anaerobic cocci, non-pathogenic *Neisseria*, *Candida* spp., Streptococci, human Papilloma virus (HPV), adenovirus
Oral cavity	Viridians streptococci, lactobacilli, *Actinomyces*, Spirochaetes	Anaerobic cocci, *Fusobacterium*, *Prevotella*, adenovirus, *Entamoeba gingivalis*
Nose and pharynx	Coagulase negative Staphylococci, Streptococci, diphtheroids, non-pathogenic *Neisseria*, anaerobic non-spore forming bacteria	*Haemophilus*, *Meningococcus*, *Pneumococcus*, *Candida*, *Staphylococcus aureus*, HPV
Conjunctiva	*Staphylococcus epidermidis*, *Propionibacterium acnes*	*Staphylococcus aureus*, *some* streptococci, *Haemophilus* spp., *Neisseria* spp., Moraxella
Stomach	*Helicobacter pylori* (50%), Lactobacilli	
Small and large intestine	Lactobacilli, Entrobacteriales, Enterococci, Bifidobacterium, anaerobic non-spore forming bacteria (Bacteroides, Prevotella, Fusobacterium)	Streptococci, diphtheroids, *Candida* spp. including *Candida albicans*, protozoan parasites (*Entamoeba coli*, *Escherichia hermannii*, *Endolimax nana*, *Iodamoeba butschlii*)
Vagina	Lactobacilli, anaerobic non-spore forming bacteria, *Staphylococcus epidermidis*, diphtheroids, non-pathogenic *Neisseria*	*Staphylococcus aureus*, *Candida* spp., HPV

resident flora can colonise the person. They can assume a pathogenic role if there is a disturbance of the resident flora of the individual.

Resident flora: These are normal flora always present at a particular site. Even if removed by chemicals or drugs, they promptly re-appear again within a short span. Examples include *Staphylococcus epidermidis* on skin or *E. coli* in the intestine.

Diseases result when the interaction between microbes and the human host results in a pathological process. This process is mediated by microbial factors or the host's immune response to the organism's presence. The other microbe and host interaction outcomes are either transient or prolonged colonisation. Infection is another term used for colonisation, which does not imply disease but rather the association of the microbe with the human host for a time.

Factors Determining Colonisation by Microbes

Factors that determine whether exposure to microbes results in transient passage through a human host or prolonged colonisation are complex and involve microbial properties, host characteristics and environmental factors.

The colonisation of skin is accomplished by organisms that can tolerate the dry surface and are resistant to the fatty acids produced by anaerobic bacteria and from the metabolism of sebum triglycerides. The most important factors that determine colonisation in the human body are the properties of the specific organism. For example, the oropharynx provides the organisms with opportunities to colonise saliva, mucosal surface, tongue and gingiva, and the teeth line. Thus, oxygen-sensitive bacteria proliferate in the gingival crevices. *Streptococcus mutans* organisms adhere to the hard surface of teeth by polysaccharides. Bacteria can also bind to cells lining the oropharynx, intestine and vagina via specific receptors for the bacterial pili. This type of adhesion prevents the mechanical elimination of organisms.

Various host factors determine the success of colonisation with the microbe. Nutritional and environmental conditions must favour the survival of microbes. The age of the host also influences microbial colonisation. In addition, the hormonal secretions, alteration of dietary habits, person-to-person interaction, sexual activity and many other factors determine the establishment of the normal microbiota.

Establishment of Normal Flora

The foetus inside the uterus is sterile but gets colonised during its passage through the birth canal. With its first feed, organisms start entering the intestines, and the outside environment helps in the microbial colonisation of skin. It has been found that a stable flora gets established in about 48 h of birth. Numerous factors are responsible for influencing the composition of normal flora, including age, sex, the genetic make-up of the person and dietary habits. Hospitalised patients constitute a special group where the resident flora changes subtly, and many of the hospital pathogens normally found only in health care institutions can cause colonisation. Such organisms pose a serious risk of causing serious infections as the patient is already compromised in some way, either medically or surgically. Bacteria like *Pseudomonas aeruginosa*, *Acinetobacter baumannii* complex, *Staphylococcus aureus*, *Klebsiella*, *Candida albicans* and non-albicans are some of the problem pathogens which can cause even life-threatening infections like pneumonia and sepsis.

Importance of Normal Microbial Flora

The normal microbiota prevents colonisation with potentially pathogenic organisms and maintains a protected environment. For example, when the normal intestinal flora has been reduced or removed by antibiotics, *Clostridium difficile* produces gastrointestinal disease. The production of proteolytic enzymes by microbes augments host factors in the digestion of food. Intestinal bacteria can also synthesise vitamins and other biological products (e.g. biotin, pantothenic acid, pyridoxine, riboflavin, vitamin K). Colicins produced by some bacteria of normal flora prevent harmful effects of the bacteria.

Beneficial Effects of Normal Flora

They confer many benefits to the host. (1) They prevent the colonisation of body sites by pathogenic microbes. They compete with the attachment of the pathogens, and since they outnumber the invading organisms, the latter are eliminated. This has been proved experimentally where germ-free animals can be infected with far fewer pathogenic organisms than normal animals. (2) The intestinal bacteria synthesise vitamins-B12 and -K, and the excess is absorbed by the intestine. (3) Bacteriocins, volatile fatty acids and peroxides generated by the normal flora can effectively kill invading microbes. (4) These organisms act as foreign substances and induce antibody production. Such antibodies cross-react and may act against related but pathogenic bacteria in the body, either in the blood or at mucosal sites.

Harmful Effects of Normal Flora

The organisms which co-inhabit the human body are largely beneficial for the individual, but in certain conditions, they may pose a danger as well. They may act as pathogens if the person's immune status is lowered or there is a breach in the continuity leading to dispersion of these organisms to other body sites. Intestinal microflora may compete with the host and deprive the host of nutrients. Some members of the normal flora may help a pathogen by providing some essential nutrients, and in such cases, a synergism between the pathogenic and the non-pathogenic organisms can occur. Some residents of the upper respiratory tract, like pneumococcus, may get transferred to other individuals, causing infection.

Infections Caused by Resident Flora

There are two situations in which the resident flora may assume a pathogenic role, and such infections are termed **endogenous infections**. First, any major disturbance of the host or host defences like immunosuppression can result in the failure of the resident flora to eliminate or suppress the pathogenic transient flora. In such situations, the transient flora, and sometimes even the resident flora, may invade healthy tissues and cause infection. These infections are also referred to as **opportunistic infections**. Second, any break in the continuity of skin or mucous membrane may cause the resident flora to invade tissues

Table 6.2 Infections caused by normal human flora

Endogenous and opportunistic infections	Causative organism in the normal flora
Infective endocarditis	Viridians streptococci from oral cavity
Dental caries, gingivitis and periodontal disease	Anaerobic organisms from oral cavity
Peritonitis	Aerobic and anaerobic intestinal flora
Urogenital infections	Intestinal and vaginal flora
Pseudomembranous colitis	*Clostridium difficile* from intestinal flora after antibiotic therapy

where they behave as foreign organisms and initiate infection and disease process. Table 6.2 enumerates some of the important infections caused by normal flora.

Microorganisms at Anatomical Sites

The number of resident microorganisms in any individual far outnumbers the human body's total number of cells. The highest number found in the intestines is to the tune of 10^{14}, while the numbers are 10^{10}–10^{12} in the skin and the oral cavity. Some members are predominant at a particular site, while others are present in fewer numbers.

The human body serves as a home for numerous microbes qualitatively and quantitatively. They cover the skin and mucosal surfaces; occasionally, they invade sterile sites to produce disease. These microorganisms are a barrier to more virulent microorganisms, provide vitamins and required growth factors, or exist as commensal inhabitants. The complexity of the microbiota is influenced over time by environmental, host and microbial factors. Thus, the knowledge of the human microbiota forms a fundamental building block of the normal physiological processes in the human body.

Respiratory Tract

Nares are inhabited by a small number of organisms, which include the *Staphylococcus*, *Corynebacterium*, *Peptostreptococcus* and *Fusobacterium* species. The microbial population in the **nasopharynx** is much more complex, with the predominance of the streptococci and *Neisseria* species. *Streptococcus salivarius*, *Streptococcus parasanguinis* and *Streptococcus pneumoniae* can be readily recovered from the nasopharynx. Twelve species of *Neisseria* have been recovered from nasopharyngeal carriers. Colonisation with *Neisseria meningitidis* varies from 10% to 95%, with the highest incidence in young adults. Other *Neisseria* species that colonise the nasopharynx are *Neisseria subflava*, *Neisseria sicca*, *Neisseria mucosa* and *Neisseria lactamica*. Gram-negative coccobacilli that colonise the nasopharynx include the *Moraxella catarrhalis* and *Kingella* species. Non-capsulated strains of *Haemophilus influenzae* are also commonly found in the nasopharynx.

Gastrointestinal Tract

The **oropharynx** is a complex mixture of ecosystems. Gram-positive and Gram-negative cocci predominate in the oropharynx. Overall, anaerobes outnumber aerobes with a ratio of 100:1. The most common anaerobic bacteria present in the oropharynx are *Peptostreptococcus*, *Veillonella*, *Actinomyces* and *Fusobacterium*, whereas the most common aerobic bacteria are *Streptococcus* and *Neisseria*. Gram-negative cocci and coccobacilli also colonise the oropharynx. *Veillonella*, anaerobic Gram-negative cocci, are the most numerous Gram-negative cocci found in the oropharynx.

The predominant anaerobes found in the oropharynx are the *Fusobacterium*, *Bacteroides*, *Porphyromonas*, *Prevotella* and *Selenomonas* species. *Fusobacterium nucleatum* is the most common *Fusobacterium* found in the mouth. Fungal colonisation of the oropharynx is restricted to yeasts; *C. albicans* is present in almost all individuals. *Entamoeba gingivalis* and *Trichomonas tenax* are the only protozoa present in the oropharynx.

The stomach contains hydrochloric acid and pepsinogen, and is an inhospitable organ. For this reason, the normal microbial flora is very sparse in the stomach. The organisms present in the stomach are the acid-tolerant *Lactobacillus* spp., *Streptococcus* spp. and *Helicobacter pylori*. *H. pylori* can become pathogenic and causes gastritis, gastric and duodenal ulcers, and gastric malignancy. In the upper part of the small intestine, the number of microbes is less than 105/mL of fluid. They are predominantly anaerobic, consisting of *Lactobacillus*, *Peptostreptococcus*, *Streptococcus*, *Porphyromonas* and *Prevotella*. The microbial flora shifts to resemble colonic bacteria if upper tract obstruction and stasis occur. This leads to malabsorption syndrome.

The large intestine is the organ most densely populated with microbial flora. It consists of more than 108 aerobic bacteria and 1011 anaerobic bacteria per gram of faeces, various yeasts and non-pathogenic parasites. *E. coli* causes colonisation soon after an infant is born. It is the most common facultative organism responsible for intra-abdominal infections. Other members of Enterobacteriaceae like *Citrobacter* spp., *Klebsiella* spp., *Enterobacter* spp. and *Proteus* spp. can establish residence in the intestine. Various species of *Candida* and protozoa colonise the large intestine.

Genitourinary Tract

The genitourinary tract is generally sterile except for the female urethra and vagina. The ureters, kidneys, prostate and cervix are normally sterile. The female urethra is colonised with a large number of lactobacilli, streptococcal species and coagulase-negative staphylococci. Faecal organisms, such as *E. coli*, *Enterococcus* spp. and *Candida* spp., can also colonise the female urethra, which is transient.

The microbial flora in the vagina is more numerous and diverse. Lactobacilli are the predominant organisms, with the most commonly isolated species being *Lactobacillus acidophilus*, *Lactobacillus fermentum*, *Lactobacillus casei* and *Lactobacillus cellobiose*. Other anaerobes commonly isolated from the vagina are the *Bifidobacterium*, *Peptostreptococcus*, *Porphyromonas* and *Prevotella* species. The aerobic bacteria present in the vagina include coagulase-negative *Staphylococcus*, *Streptococcus* (viridans group and beta-haemolytic strains), *Corynebacterium* spp., *Gardnerella vaginalis*, *Neisseria* spp., *Haemophilus* spp. and members of Enterobacteriaceae. The flagellate *Trichomonas vaginalis* is present in small numbers in healthy women. The *Candida* species and *Torulopsis glabrata* are common members of the microbial flora of the vagina.

Body Surface

Human skin is constantly bombarded by organisms present in the environment. The skin is exposed to extremes of temperature, moisture and chemical disinfectants. There the microbial population is less numerous and complex than at other body sites. The skin is not a homogenous surface. There are relatively dry areas like palms and soles; areas with more apocrine glands, such as axillary, inguinal and perianal areas; and areas rich with sebaceous glands like forehead and nasolabial folds. Most organisms proliferate in the moist area. Thus higher densities of microbes are present in areas rich with sweat glands. The skin of the face, perirectal area or groin has a more complex microbial flora than that at other sites (Fig. 6.1).

Anaerobic bacteria are 10- to 100-fold more numerous on the skin surface than aerobic bacteria. Gram-positive bacteria predominate over Gram-negative bacteria. The bacteria more commonly recovered from skin surfaces are *Staphylococcus* spp., *Micrococcus* spp., *Corynebacterium* spp., *Peptostreptococcus* and *Propionibacterium* spp. *S. epidermidis* is the most commonly isolated bacterium. Gram-negative bacteria are generally not recovered from the skin surface, except during transient colonisation. *Acinetobacter* spp. organisms have adapted to survive in moist areas of the skin.

Fungal colonisation of the skin surface is transient. This includes colonisation with yeasts (*C. albicans*, *Malassezia furfur* and *Rhodotorula*) and dermatophytes. Among these, *M. furfur* is present in most individuals.

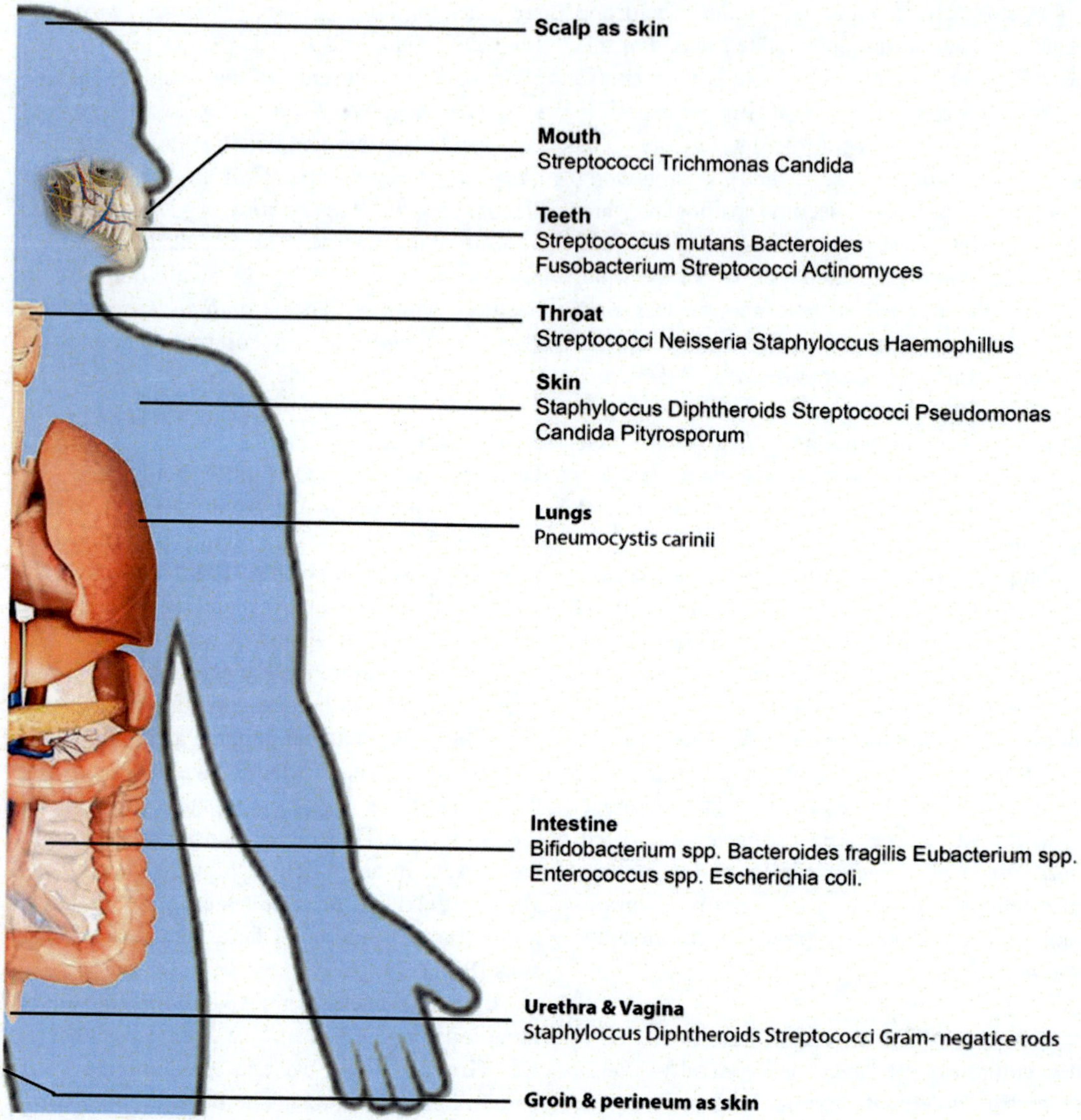

Fig. 6.1 Sites of normal microbial flora of the human

In the conjunctiva, the flushing action of tears and the presence of lysozymes in tears usually make the site free of bacteria. The organisms present in the conjunctiva are *Corynebacterium xerosis*, *Moraxella* spp., *S. aureus* and non-haemolytic streptococci.

Sterile Sites

The body fluids, including blood, cerebrospinal fluid, synovial, pleural, pericardial, peritoneal and other exudates, are normally sterile or transiently infected by microbes. Some microbes from the mouth and gastrointestinal tract can invade the bloodstream in healthy individuals (during tooth brushing or a bowel movement). These organisms are rapidly removed and are of little significance. Thus, the isolation of an organism from body fluids should be considered significant unless the specimen is contaminated during its collection.

Organ tissues are also generally sterile unless they are infected following systemic spread.

Some bacteria like *Mycobacterium tuberculosis* may be disseminated to the lungs, liver and kidneys during initial infection and remain dormant. In this situation, the organism can be recovered from tissue samples. Long-term colonisation with organisms like *Pneumocystis carinii* or latent viruses, such as herpes simplex virus or cytomegalovirus, may commonly occur. The external auditory meatus, being an extension of the skin, is colonised primarily by *S. epidermidis* and diphtheroids.

Effects of Alteration of Normal Microbial Flora

Some conditions could adversely affect the normal microflora of the human body, including skin, gut and so on. It includes diseases such as obesity, inflammatory bowel disease (IBD), diabetes mellitus, metabolic syndrome, atherosclerosis, alcoholic liver disease (ALD), nonalcoholic fatty liver disease (NAFLD), cirrhosis, hepatocellular carcinoma and therapies including immunosuppression, radiation therapy, chemotherapy and antibiotics. Under such conditions, normal flora may be reduced, lost or replaced by other microorganisms, that is, population suppression or replacement. As a result, it allows the pathogens to invade and colonise in the absence of the defence system enacted by the normal flora. Such pathogens, commonly known as opportunistic pathogens, impair the overall health of the susceptible individual and can cause infections depending on the location it has replaced the normal flora. For example, replacement of skin and gut microflora can lead to higher chances of skin diseases (dermatitis) and gastrointestinal complications, respectively.

Pseudomembranous colitis is an acute inflammatory disease of the colon that may cause mild colon inflammation to a severe life-threatening condition. Only after the 1970s, *C. difficile* was found to be associated with pseudomembranous colitis. This condition is mainly caused by the indiscriminate use of antibiotics that disrupts the normal flora's balance, thus providing a suitable environment for the proliferation of the pathogenic microbes. Certain antibiotics like fluoroquinolones, penicillin, clindamycin and cephalosporins may lead to overgrowth of *C. difficile* (which is otherwise present in few numbers as normal flora) by suppressing the growth of the gut flora that normally keeps it in check. The toxins produced by *C. difficile* rise in quantity to cause damage to the colon. This condition is treatable if it is diagnosed on time. Thus to prevent the suppression of gut flora, wherever possible, an accurate dosage of narrow-spectrum antibiotics has to be administered at the right time.

The Human Microbiome

Commensal organisms outnumber body cells by ten to one in the human body. It is estimated that the total number of bacteria in a healthy adult is 3.8×10^{13}. Most of them reside in the gut, particularly the large intestine. **The total genome and gene product of these commensal organisms living in humans is known as the microbiome.** The number of genes in an individual's microbiome is 200 times the number of genes in the human genome. The microbiome's composition is dynamic in infancy, with the first colonisation occurring in the sterile gut and other parts of the body in the newborn. However, by the end of infancy, the core microbiome is fully formed predominantly of anaerobic bacteria such as *Bacteroides*, *Bifidobacterium* and *Clostridium* spp. in the gut.

The microbiome plays a vital role in human development, immunity and nutrition and benefits the individual in several ways. The gut bacteria competitively inhibit the colonisation of pathogenic bacteria and thus protect us from various infections. It also helps in the digestion of food and the production of vitamins K, B12, thiamine and riboflavin. It also plays a vital role in the functioning of the immune system. Disturbance of the normal microbiome is termed dysbiosis and is associated with various autoimmune disorders like diabetes, rheumatoid arthritis, muscular dystrophy and multiple sclerosis. Changes in gene activity and metabolic processes may be responsible for this type of abnormal immune response against self-antigens.

One of the recent procedures being tried is **faecal microbiota transplantation** or **faecal transplantation**, where stool of a normal individual is introduced into the large intestine by instrumentation like colonoscopy to restore the microbiota in a diseased person. It has been tried successfully to cure severe *C. difficile* infections and treat colitis, constipation and irritable bowel syndrome.

Probiotics and Prebiotics

Probiotics are preparations of live beneficial gut microorganisms that maintain the normal gut microflora and benefit the host. There are different groups of probiotic microorganisms. The most common are bacteria belonging to the genus *Lactobacillus* and *Bifidobacterium*. Yeast like *Saccharomyces boulardii* is also used as probiotics. Yoghurt, Sauerkraut, Miso Soup, Soft Cheeses, Kefir, Sourdough Bread, Acidophilus Milk and so on are examples of probiotic foods that contain live preparation of beneficial microorganisms. In case of gastrointestinal complications that lead to loss of gut microflora (e.g. use of broad-spectrum antibiotics kills the gut microflora), probiotics help restore the gut microbiome.

On the other hand, **prebiotics** are the non-digestible dietary ingredient that beneficially affect the host by selectively stimulating the growth or activity of a limited number of bacteria in the colon. Fructooligosaccharides are a primary example of prebiotics. Prebiotics are naturally found in many fruits and vegetables, especially those containing complex carbohydrates, such as fibre and resistant starch.

Microbial Infections

7

The host-parasite relationship is determined by the interaction between host factors and the infecting microorganisms. The outcome of any microbial infection depends on the interaction between the host and the parasite. The relationship between the host and parasite may be symbiosis, commensalism or disease.

Symbiosis describes a situation where both the microorganisms and the host species live together with mutual benefit. There is an element of symbiosis in the relationship between the human host and the gut flora. **Commensalism** is an association in which only the microorganism derives benefit without causing any injury to the host. Most human microbes are commensals. **The disease** is caused by certain microorganisms known as *pathogens*. Microorganisms vary in their ability to cause disease in humans.

Types of Microorganisms

Microorganisms may be of the following types:

Saprophytes: These are free-living microorganisms that live on dead or decaying organic matter. They are usually present in soil and water. They are generally unable to invade the living body.

Parasites: These are microorganisms that live on a living host, derive nutrition from the host without any benefit to the host and cause harm to the infected host.

Commensals: These are microorganisms that live on a living host without causing any injury to the host. Most human microbes are commensals.

Pathogens: A microorganism capable of causing disease, especially if it causes disease in immunocompetent people, is called a *pathogen*. These pathogens, however, represent a very small proportion of the microbial species.

Opportunistic pathogens: A microbe capable of causing disease only in immunocompromised people is an *opportunistic pathogen*. These organisms can cause disease only if one or more of humans' usual defence mechanisms are reduced or altered by accident, by intent (e.g. surgery), an underlying metabolic disorder or an infectious disease (e.g. AIDS).

The words **virulence** and **virulent** are derived from the Latin word *virulentus*, meaning "full of poison". The term *virulentus* is derived from the Latin words *virus* (poison) and *lentus* (fullness), and, in turn, the term *virus* may be related to the Sanskrit word *visham*, meaning "poison". The virulence of a microbe is determined by virulence factors, such as capsules, exotoxins or endotoxins (Table 7.1).

Table 7.1 Important bacterial surface virulence factors

Virulence factors	Bacteria
Capsule	
Polysaccharide capsule	*Streptococcus pneumoniae* *Klebsiella pneumoniae* *Haemophilus pneumonia* *Salmonella* Typhi *Neisseria meningitidis*
Polypeptide capsule	*Bacillus anthracis*
Other factors	
Pili protein	*Escherichia coli*
Protein A	*Staphylococcus aureus*
M protein	*Streptococcus pyogenes*
V and W proteins	*Yersinia pestis*

Commensalism and the Microbiome

Commensal organisms outnumber body cells by ten to one in the human body. It is estimated that the total number of bacteria in a healthy adult is 3.8×10^{13}. The majority of them reside in the gut, particularly the large intestine. The total genome and gene product of these commensal organisms living in humans is the **microbiome**. The number of genes in an individual's microbiome is 200 times the number of genes in the human genome. The microbiome's composition is dynamic in infancy, with the first colonisation occurring in the newborn's sterile gut and other parts of the body. However, by the end of infancy, the core microbiome is fully formed predominantly of anaerobic bacteria such as *Bacteroides*, *Bifidobacterium*, and *Clostridium* spp. in the gut.

The microbiome plays a vital role in human development, immunity and nutrition, and benefits the individual in several ways. The gut bacteria competitively inhibit the colonisation of pathogenic bacteria and thus protect us from various infections. It also helps in the digestion of food and the production of vitamins K, B12, thiamine and riboflavin. It also plays a vital role in the functioning of the immune system. Disturbance of the normal microbiome is termed **dysbiosis**, and it is associated with various autoimmune disorders like diabetes, rheumatoid arthritis, muscular dystrophy and multiple sclerosis. Changes in the gene activity and metabolic processes may be responsible for this type of abnormal immune response against self-antigens.

One of the recent procedures being tried is **faecal microbiota transplantation** or faecal transplantation, where stool of an average individual is introduced into the large intestine by instrumentation like colonoscopy to restore the microbiota in a diseased person. It has been tried successfully to cure severe *Clostridium difficile* infections and treat colitis, constipation and irritable bowel syndrome.

Infection

This is a process when an organism enters the body, increases in number and causes damage to the host. All infections do not invariably result in disease. The term **infection** has more than one meaning: the presence of microbes in the body and the symptoms of the disease. The presence of microbes in the body does not always result in symptoms of the disease.

Infestation indicates scenarios where large numbers of disease-carrying animals or insects occupy a place or host and cause damages including structural, chemical and biological. Infestation increases chances of disease dispersion from one location to another unless controlled otherwise. In most cases, mites, ticks and lice carry the parasitic pathogens and result in infestation, either external or internal, depending on the parasites' location in the host.

Growth of microorganisms in a site on a host is called **colonisation**. In normal conditions, enteric bacteria colonise the intestine, grow and form the normal flora. However, colonisation of pathogens including opportunistic pathogens leads to disease.

Types of Infections

Infections may be of various types:

Primary infection: This condition denotes an initial infection with an organism in a host.

Reinfection: This condition denotes subsequent infection with the same organism in the same host.

Secondary infection: This condition denotes an infection with a new organism in a host whose body resistance is already lowered by a pre-existing infectious disease.

Cross-infection: This condition denotes an infection with a new organism from another host or another external source in a patient already suffering from a disease.

Nosocomial infection: Cross-infections acquired in hospitals are called hospital-acquired, hospital-associated or nosocomial infections.

Iatrogenic infection: This condition denotes a physician-induced infection resulting from therapy with drugs or investigation procedures.

Subclinical infection: Inapparent clinical infections are called sub-clinical infections.

Latent infections: This denotes a condition in which some organisms may remain in a latent or hidden stage in the host, and subsequently, they multiply to produce clinical disease when host resistance is lowered.

Transmission of Infection

Three critical components play an essential role in successfully transmitting microbial diseases. These are (1) sources, (2) mode of transmission and (3) susceptible host.

Sources of Infection

Sources of microbial infections are human, animal, plant, soil or inanimate matter in which organisms usually live, multiply and cause the infections with or without overt clinical manifestations. Humans are usually the common reservoirs of many microbial infections. Animals are reservoirs of zoonotic infections, such as plague (e.g. rats), rabies (e.g. dogs), cysticercosis (e.g. pigs) and so on. The sources of infections may be endogenous or exogenous.

Endogenous sources: The source of infection is the normal bacterial flora present in the human body. These bacteria are usually non-pathogenic but in certain situations become pathogenic and cause infections at different sites in the same host. For example, *Escherichia coli* present as normal flora of the intestine may cause urinary tract infection in the same host. Similarly, viridans streptococci present as part of the normal flora of the mouth may cause infective endocarditis.

Exogenous sources: The source of infection is from outside the host's body. Most of the microbial infections are exogenous. The exogenous sources include the following:

Humans

Humans are the most common sources of infections caused by microorganisms. They may be either patients or carriers. A patient suffering from an active infection is an important source of infection for others.

Carrier: A carrier is a person who harbours pathogenic microorganisms without showing any signs and symptoms of the disease. Carriers are also important sources of infections. A carrier may be a (1) healthy carrier, (2) convalescent carrier, (3) temporary carrier or (4) chronic carrier.

A healthy carrier is a host who harbours the microorganism without ever suffering from the disease caused by that microorganism. **A convalescent carrier** is a host who continues to harbour the microorganism even after recovering from the clinical disease caused by the same pathogen. **A temporary carrier** is a host who harbours the microorganism up to 6 months after recovering from the disease caused by the same pathogen. Finally, **a chronic carrier** is a host who harbours the microorganism for many years after recovering from the clinical disease caused by the same pathogen.

Animals

Animals are also important sources of infection for humans. Symptomatic, as well as asymptomatic, animals can transmit infections to humans. Asymptomatic animals act as a reservoir of human infections. These are called reservoir hosts. Infections transmitted from animals to humans are called **zoonotic infections**. Examples of zoonotic infections include bacterial (e.g. plague, anthrax, bovine tuberculosis), viral (e.g. rabies, Japanese encephalitis), fungal (e.g. dermatophytic infections) and parasitic (e.g. toxoplasmosis, cysticercosis, hydatid disease).

Insects

Insects, such as mosquitoes, ticks, mites, flies, fleas and lice, may transmit a wide variety of microorganisms to humans (Table 7.2). The diseases transmitted by insects are collectively referred to as **arthropod-borne diseases** and the insects transmitting these pathogens are called **vectors**. Insect vectors may transmit the infection in two ways: mechanical transmission and biological transmission.

Mechanical transmission is by vectors such as insects (e.g. domestic flies) carrying enteric

Table 7.2 Microbial infections transmitted by insect vectors

Microbes	Disease	Reservoir	Insect
Bacteria			
Yersinia pestis	Plague	Rodents, especially rats	Rat fleas (*Xenopsylla cheopis*)
Francisella tularensis	Tularaemia	Rabbits and other rodents	Ticks (*Dermacentor variabilis*)
Borrelia recurrentis	Relapsing fever	Humans	Lice (*Pediculus humanus*)
Borrelia burgdorferi	Lyme disease	Rodents and deer	*Ixodes scapularis*
Rickettsia rickettsii	Rocky mountain spotted fever	Dogs, rodents and ticks	Ticks
Rickettsia prowazekii	Epidemic typhus	Humans	Lice (*Pediculus humanus corporis*)
Ehrlichia chaffeensis	Ehrlichiosis	Deer and rodents	Ticks (*Amblyomma americanum*)
Parasites			
Leishmania spp.	Leishmaniasis	Humans, dogs and wild canine	Sandfly (*Phlebotomus*)
Trypanosoma spp.	Trypanosomiasis	Humans, armadillos, antelope and cattle	Reduviid bug, tsetse fly
Plasmodium spp.	Malaria	Humans	*Anopheles* mosquitoes
Wuchereria bancrofti and *Brugia malayi*	Lymphatic filariasis	Humans	Mosquitoes (*Culex, Aedes, Mansonia* spp.)
Viruses			
Japanese encephalitis virus	Japanese encephalitis	Birds	*Culex tritaeniorhynchus* mosquitoes
West Nile virus	West Nile fever	Birds	*Culex* mosquitoes
Chikungunya virus	Chikungunya	Humans and monkeys	*Aedes aegypti* mosquitoes
Dengue virus	Dengue	Humans	*Aedes aegypti* mosquitoes
Kyasanur forest disease virus	Kyasanur forest disease	Forest birds and small mammals	Tick (*Hemaphysalis spinigera*)
Yellow fever virus	Yellow fever	Monkeys and humans	*Aedes aegypti* mosquitoes

Table 7.3 Modes of transmission of bacterial infections

Modes of transmission	Bacterial infections	Causative bacteria
Human to human		
Direct contact	Gonorrhoea	*Neisseria gonorrhoeae*
Indirect contact	Dysentery	*Shigella dysenteriae*
Blood-borne	Syphilis	*Treponema pallidum*
Trans-placental	Congenital syphilis	*Treponema pallidum*
Animal to human		
Direct contact	Cat-scratch disease	*Bartonella henselae*
	Lyme disease	*Borrelia burgdorferi*
Through animal excreta	Dysentery leading to heamolytic uraemic syndrome	*Enterohemorrhagic Escherichia coli* (EHEC)
Handling of fomites	Skin infection	*Staphylococcus aureus*
Soil borne	Tetanus	*Clostridium tetani*
Water borne	Legionnaire disease	*Legionella pneumophila*

bacteria (*Salmonella typhi*, *Shigella* spp. etc.) mechanically on their legs, wings and surface of the body and transfer them to food. **Biological transmission** is carried out by vectors in which the microorganisms multiply or undergo a part of their life cycle before being transmitted to humans. For example, rat flea and female *Anopheles* mosquitoes are biological vectors that transmit plague and malaria, respectively, to humans by biting. Besides acting as vectors, insects also act as reservoir hosts (e.g. ticks in relapsing fever).

Extrinsic incubation period (EIP) is the time taken by a pathogen to grow in an intermediate host before it could transmit by the host. As such, it is a time-dependent journey of a pathogen through several vector tissues and organs. For example, the EIP of malaria (period of sporogony) is the time taken by the parasites to grow in the mosquito and reach the salivary glands to make the mosquito infectious.

Environment

These include food, water and soil and inanimate objects.

Food contaminated with pathogens also acts as a source of infection and causes diarrhoea, dysentery, food poisoning and gastroenteritis. **Water** contaminated with microorganisms also acts as a source of infection and transmits water-borne diseases, such as leptospirosis, cholera, dysentery, hepatitis A infection and so on. **Soil**, a perfect ecosystem for microbial growth, is responsible for the transmission of soil-borne pathogens including *Salmonella enterica*, *Campylobacter* spp., *E. coli*, *Legionella* spp. (pneumonia; Legionnaires' disease), *Mycobacterium leprae* (leprosy), *Shigella* spp. (shigellosis) and so on.

Modes of Transmission

Microbial pathogens causing various infectious diseases are transmitted from one host to another in many ways as follows (Table 7.3).

Contact

Transmission of microorganisms from person to person occurs by direct or indirect contact. **Transmission by direct contact** occurs through touching, kissing, sex and so on. The diseases transmitted by direct contact include common cold, staphylococcal infections and sexually transmitted infections (e.g. gonorrhoea, syphilis, AIDS). The term **contagious disease** was used earlier for the disease acquired by direct contact.

Microorganisms can also be transmitted by **indirect contact** through inanimate objects, such as clothing, handkerchiefs, toys and so on, called fomites. Influenza, tuberculosis and certain superficial fungal infections are diseases transmitted by fomites.

Inoculation

Infections can be transmitted by the inoculation of microorganisms directly into the host's tissues. For example, tetanus is transmitted by direct inoculation of *Clostridium tetani* spores present in the soil to the injured tissues in the host. These spores then germinate to vegetative forms of bacteria and migrate along the neural tissues to cause tetanus. Similarly, through their act of biting, rabid dogs inoculate the rabies virus directly to host tissue and cause rabies in humans.

Iatrogenic infection occurs following the use of unsterile syringes and equipment in a hospital. Hepatitis B and C and HIV infections are examples of iatrogenic infections caused by the use of contaminated syringes and contaminated blood and blood products.

Laboratory-acquired infections (LAI) occurs to the laboratory personnel working with pathogens. It is most common where appropriate laboratory safety measures and personnel training have not been critically considered. A LAI could be obtained through inhalation, percutaneous inoculation (needle and syringe, cuts or abrasions from contaminated items and animal bites), contact between mucous membranes and contaminated material (hands or surfaces) and ingestion (aspiration through a pipette, smoking or eating). Depending on the pathogen being handled, the person may be symptomatic or remain asymptomatic.

Ingestion

Ingestion of water and food contaminated with microorganisms can transmit a wide variety of microbial infections. For example, food poisoning caused by *Bacillus cereus* is transmitted by rice contaminated with bacterial spores that survive boiling. Cholera, typhoid, food poisoning, hepatitis A, poliomyelitis and many parasitic infections are other diseases transmitted by ingestion of contaminated food and water.

Inhalation

Infections are transmitted by inhalation of **droplet nuclei** that are discharged into the air by coughing, sneezing or talking. Respiratory pathogens are shed into the environment by patients in secretions from the nose or throat during coughing, sneezing or talking. Small droplets (less than 0.1 mm in diameter) become airborne as minute particles or droplet nuclei (1–10 μm in diameter), whereas large droplets fall to the ground. Measles, influenza, whooping cough, tuberculosis and aspergillosis are a few examples of infectious diseases acquired by inhalation.

Vectors

Mosquitoes, flies, fleas, ticks, mites and lice are the vectors that transmit many diseases, as mentioned earlier.

Susceptible Host

The infective agent enters the body through four main routes such as genital tract, respiratory tract, gastrointestinal tract and skin. The pathogens can be transmitted either as vertical or horizontal transmission.

Vertical transmission: Certain bacteria (*Treponema pallidum*), viruses (rubella and cytomegalovirus) and parasites (*Toxoplasma gondii*) can be transferred from mother to foetus by a process called vertical transmission. The organisms can be transmitted vertically in three ways: (1) across the placenta, (2) within the birth canal during birth and (3) through breast milk. Table 7.4 summarises a list of diseases transmitted vertically.

Horizontal transmission: Unlike vertical transmission, horizontal transmission occurs from person to person and is not from mother to offspring (Fig. 7.1).

Pathogenesis of Infections

The pathogenesis of infectious diseases is complex. The **outcome of infection** depends on various factors of the microbe and the host. These include (1) the ability of the organism to break host barriers and evade destruction by innate local and tissue host defences, (2) the ability of the organism to replicate, spread, establish infection

Table 7.4 Vertical transmission of microbial pathogens

Modes of transmission	Pathogen	Disease
Breast milk		
Bacteria	*Staphylococcus aureus*	Skin or oral infection
Viruses	Human T-cell leukaemia virus	Asymptomatic
	Cytomegalovirus Human immunodeficiency virus (HIV)	Asymptomatic infection
During passage through birth canal		
Bacteria	Group B streptococcus	Neonatal sepsis and meningitis
	Neisseria gonorrhoeae	Conjunctivitis
	Chlamydia trachomatis	Pneumonia or conjunctivitis
Viruses	HIV	Asymptomatic infection
	Hepatitis B virus	Hepatitis B
	Herpes simplex virus-2	Skin or CNS infection; sepsis
Transplacental		
Bacteria	*Listeria monocytogenes*	Neonatal sepsis and meningitis
	Treponema pallidum	Congenital syphilis
Viruses	Cytomegalovirus	Congenital abnormalities
	Parvovirus B19	Hydrops fetalis
Parasite	*Toxoplasma gondii*	Congenital toxoplasmosis

and cause disease, (3) the ability of the organism to transmit to a new susceptible host and (4) the innate and adaptive immunologic ability of the host to control and eliminate the invading microorganism.

Factors Predisposing to Microbial Pathogenicity

Pathogenicity is the capacity of a pathogen species to cause disease, whereas virulence is used to describe the sum of disease-causing properties of a population (**strain**) within the species. Pathogens can be distinguished from their avirulent counterparts by specific genes or gene clusters in the genome known as pathogenicity islands.

Virulence is a measure of a microbe's ability to cause disease. It is a quantitative measure of pathogenicity and is measured by the number of organisms required to cause disease. It means that a highly virulent microbe requires fewer organisms to cause disease than a less virulent one; hence it is directly dependent on the infectious dose of the organism.

The **virulence of bacteria**, in particular, depends on the following factors:

Adhesions

Adherence of bacteria to the body surface is the most important event in the pathogenesis of the disease. Once bacteria enter the host's body, they must adhere to the cells of a tissue surface. If they do not adhere, they will be swept away by mucus and other fluids that bathe the tissue surface.

Specialised molecules mediate bacterial adherence to the cell surface. The various molecules that mediate adherence to the cell surface are called **adhesins**. These adhesins allow the bacteria to adhere to the surface of human cells, thereby promoting their ability to cause disease. Conversely, microorganisms that lack this mechanism are non-pathogenic. Bacterial adhesins are divided into two major groups: pili (fimbriae) and non-pilus adhesins (fimbrial adhesins).

Pili These fibres extend from the bacterial surfaces and attach bacteria to specific receptors on the host cells. The pili of many Gram-negative bacteria bind directly to sugar residues of glycolipids or glycoproteins present on the membrane of the host cells. They also act as a protein scaffold to which another more specific adhesive protein can attach.

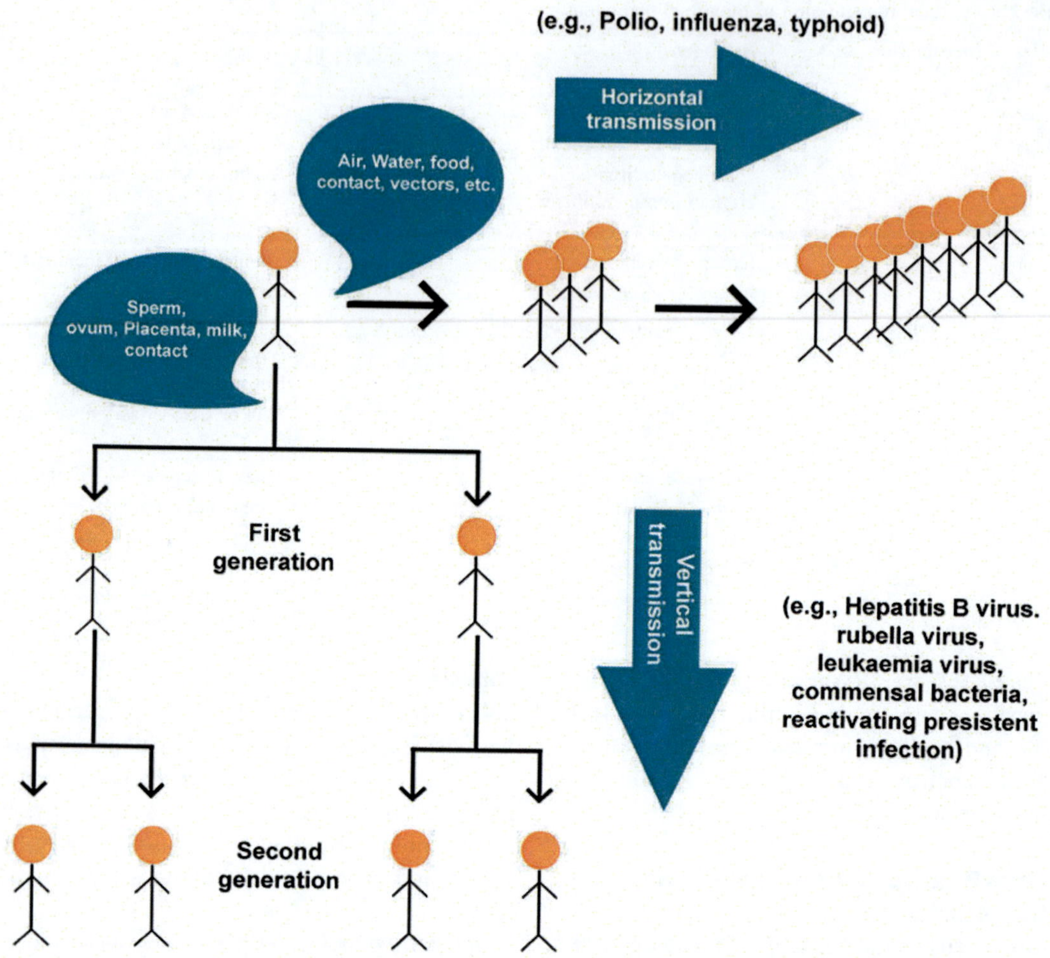

Fig. 7.1 Schematic diagram showing horizontal and vertical transmission of infections

The pili of the Gram-negative bacteria are classified into five types. They are critical in the pathogenesis of infections caused by them. Also, antigenic variation in the actual structural pilin protein can lead to antigenic diversity in the pathogen.

Most *E. coli* strains that cause pyogenic nephritis produce an adhesin protein called P-pili (pyelonephritis-associated pili) encoded by *pap* genes. Many of these adhesin proteins are present at the tip of the pili that bind specific receptors on the surface of the urinary bladder. The binding prevents the bacteria from being washed away from the urinary bladder by the flushing action of the urine.

Similarly, the pili of *Neisseria gonorrhoeae* mediate the organism's attachment by binding to oligosaccharide receptors on the epithelial cells of the urethra. The gonococci use pili as primary adhesins and opacity-associated proteins (Opa) as secondary adhesins to cells. These Opa proteins intercede adherence to polymorphonuclear cells. A part of gonococci survives subsequent phagocytosis by these cells. In organ cultures of the uterine (fallopian) tube, the gonococci affix to the microvilli of non-ciliated cells and emerge to activate engulfment by these cells. These gonococci develop intracellularly and drift to the sub-epithelial space by an unexplained action.

Non-pilus adhesins These include glycocalyx and other adhesins present on the bacterial surfaces. The glycocalyx is a polysaccharide "**slime layer**" secreted by some strains of bacteria that mediates strong adherence to certain structures, such as catheters, prosthetic implants and heart valves. For example, the glycocalyx of *Staphylococcus epidermidis* and certain viridans streptococci allows the bacteria to adhere firmly to the endothelium of the heart valve. The matrix formed by these adhesins-forming proteins is called a biofilm.

Biofilms are important in pathogenesis because they protect the bacteria from host defence and antibiotics. In addition, the biofilms facilitate bacteria colonisation, especially surgical appliances, such as artificial valves or indwelling catheters. *Streptococcus pyogenes* makes use of non-pilus adhesins (such as lipoteichoic acid, protein F and M protein) to bind to epithelial cells. Recently, it has been shown that certain strains of *E. coli* and *Shigella* spp. have surface proteins called **curli**, which help bind bacteria to the host endothelium and extracellular proteins.

Biofilms are important in human infections that are persistent and difficult to treat. A few examples include (1) *S. epidermidis* and *Staphylococcus aureus* infections of central venous catheters; (2) eye infections that occur with contact lenses and intraocular lenses; (3) infections in dental plaque and (4) *Pseudomonas aeruginosa* airway infections in cystic fibrosis patients.

Biofilm also confers an inherent resistance to anti-microbial agents, whether these anti-microbial agents are anti-biotics, disinfectants or germicides. The mechanisms of resistance include (1) delayed penetration of anti-microbial agents through the biofilm matrix, (2) the altered growth rate of biofilm organisms and (3) other physiological changes due to the biofilm mode of growth.

Invasiveness

Invasiveness refers to the ability of an organism to invade the host cells after establishing infection. "**Invasion**" is the term commonly used to describe the entry of bacteria into host cells, implying an active role for the organisms and a passive role for the host cells. For many disease-causing bacteria, invasion of the host's epithelium is central to the infectious process. Some bacteria (e.g. *Salmonella* spp.) invade tissues through the intracellular junctions in the cytoplasm. Some bacteria (e.g. *Shigella* spp.) multiply within host cells, whereas others do not.

Shigella spp. initiate the infection process by adhering to host cells in the small intestine. Multiple proteins, including the *invasion plasmid antigens* (IpA-D), contribute to the process. Once inside the cells, the shigellae are either lysed or escape from the phagocytic vesicle, multiplying in the cytoplasm.

Other bacteria (e.g. *Yersinia* spp.) invade specific types of hosts' epithelial cells and may subsequently enter the tissue. Once inside the host cell, they may remain enclosed in a vacuole composed of the host cell membrane, or the vacuole membrane may dissolve, and the bacteria may disperse within the cell and from one cell to another.

Invasion of tissues followed by inflammation is enhanced by many factors, which include: (1) enzymes, (2) antiphagocytic factors, (3) inflammation and (4) intracellular survival.

1. **Enzymes**: The invasion of bacteria is enhanced by many enzymes. Many species of bacteria produce enzymes that are not intrinsically toxic but do play important roles in the infectious process. Some of these enzymes are discussed below:
   (a) **Hyaluronidases and collagenase**: Hyaluronidases and collagenase are the enzymes that hydrolyse hyaluronic acid and degrade collagen, respectively, thereby allowing the bacteria to spread through subcutaneous tissues.
   Hyaluronidases are produced by many bacteria (e.g. staphylococci, streptococci and anaerobes) and aid in their spread through tissues. For example, hyaluronidase produced by *S. pyogenes* degrades hyaluronic acid in the subcutaneous tissue, thereby facilitating the organism to spread rapidly. *Clostridium perfringens*

produces the proteolytic enzyme **collagenase**, which degrades collagen (the major protein of fibrous connective tissue), and promotes the spread of infection in the tissue.

(b) **Coagulase**: *S. aureus* produces the enzyme coagulase, which coagulates the plasma in association with blood factors. Coagulase contributes to the formation of fibrin walls around staphylococcal lesions, which protects bacteria from phagocytosis by walling off the infected area. The enzyme also causes fibrin deposition on the surfaces of individual staphylococci, which may help protect them from phagocytosis or destruction within phagocytic cells.

(c) **Streptokinase (fibrinolysin)**: Many haemolytic streptococci produce the enzyme streptokinase, which activates plasma proteolytic enzyme. This enzyme can then dissolve coagulated plasma and thereby possibly aids in the rapid spread of streptococci through tissues. Streptokinase has been used to treat acute myocardial infarction to dissolve fibrin clots.

(d) **IgA1 proteases**: Certain pathogenic bacteria produce enzymes IgA1 proteases that split IgA1 at specific proline–threonine or proline–serine bonds in the hinge region and inactivate its antibody activity. IgA1 protease is an important virulence factor of the pathogens, such as *N. gonorrhoeae*, *Neisseria meningitidis*, *Haemophilus influenza* and *Streptococcus pneumoniae*. Production of IgA1 protease allows the pathogens to inactivate the primary antibody found on mucosal surfaces and thereby facilitates the attachment of these bacteria to the mucous membrane.

2. **Antiphagocytic factors**: Many bacterial pathogens are rapidly killed once they are ingested by polymorphonuclear cells or macrophages. However, some pathogens evade phagocytosis or leukocyte microbicidal mechanisms by several antiphagocytic factors.

    (a) **Capsule**: The capsule surrounding bacteria such as *S. pneumoniae* (Fig. 7.2) and *N. meningitidis* is the most important antiphagocytic factor. It retards the phagocytosis of bacteria by preventing

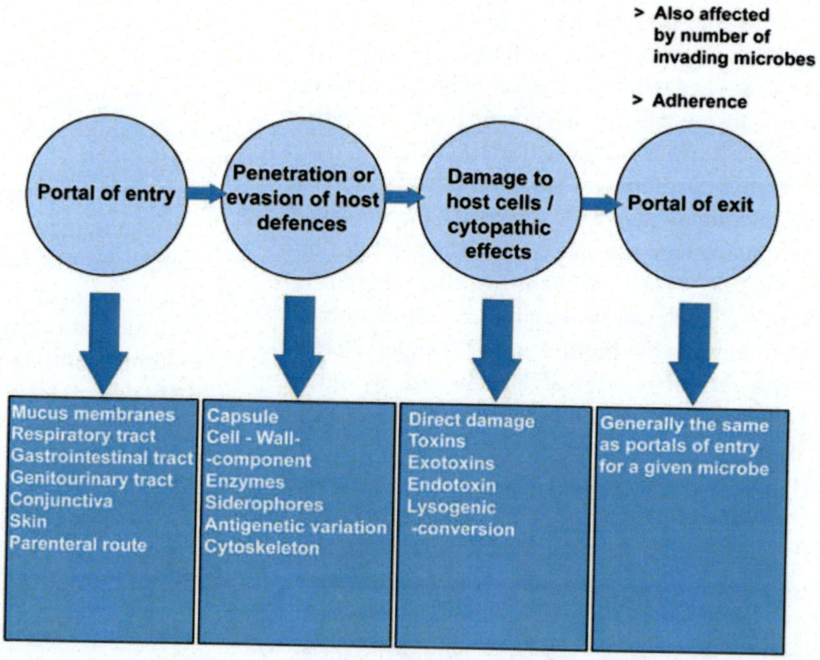

Fig. 7.2 Schematic diagram showing various stages of the infection process

the phagocytes from adhering to the bacteria.

(b) **Cell wall proteins**: Cell wall proteins, such as the protein A and protein M, of *S. aureus* and *S. pyogenes*, especially are antiphagocytic. For example, protein A of *S. aureus* binds to IgG and prevents complement activation. M protein of *S. pyogenes* is antiphagocytic.

(c) **Cytotoxins**: Certain bacteria produce cytotoxins that interfere with chemotaxis or phagocyte killing. For example, *S. aureus* produces haemolysins and leukocidins that lyse and damage RBCs and WBCs.

(d) **Surface antigens**: Surface antigens of bacteria such as Vi antigen of *S. typhi* and K antigen of *E. coli* make the bacteria resistant to phagocytosis and lytic activity of complement.

3. **Inflammation**: Inflammation is an important host defence induced by the presence of bacteria in the body. It is of two types: pyogenic and granulomatous. **Pyogenic inflammation** is the host defence seen primarily against pyogenic or pus-producing bacteria, such as *S. pyogenes*. It typically consists of neutrophils and the production of specific antibodies and elevated levels of complement. **Granulomatous inflammation** is the host defence seen primarily against intracellular granuloma-producing bacteria, such as *Mycobacterium tuberculosis*, *M. leprae* and so on. The response consists of the production of macrophages and CD4+ T cells.
4. **Intracellular Survival**: A few mechanisms suggested for intracellular survival of bacteria include (1) inhibition of phagolysosome fusion, (2) resistance to the action of lysosomal enzymes and (3) adaptation to cytoplasmic replication.

   1. Bacteria (such as *Chlamydia*, *M. tuberculosis*) interfering with the formation of phagolysosomes in a phagocyte can evade the host defence process and survive intracellularly. Thus, these bacteria live within cells and are protected from attack by macrophages and neutrophils. On the other hand, bacteria that do not interfere with the formation of phagolysosomes are killed by the phagocytes.
   2. Capsular polysaccharide in *Mycobacterium lepraemurium* and mycoside in *M. tuberculosis* makes these bacteria resistant to the action of lysosomal enzymes.
   3. Certain bacteria, such as rickettsiae, escape from the phagosome into the host cell's cytoplasm before the phagosome fuses with a lysosome and hence they remain intracellularly.

A list of intracellular pathogens is given in Table 7.5.

Toxigenicity

Toxins produced by bacteria are generally classified into two groups: exotoxins and endotoxins.

Exotoxins

Exotoxins are heat-labile proteins that are produced by several Gram-positive and Gram-negative bacteria. These are bacterial products secreted into tissues and directly harm tissues or trigger destructive biological activities (Fig. 7.3). The genes coding for these proteins is frequently encoded on a plasmid or bacteriophage DNA. Some important toxins encoded by plasmids are the tetanus toxin of *C. tetani* and heat-labile and

Table 7.5 Intracellular microbial pathogens

Pathogens	Examples
Bacteria	*Mycobacterium* spp., *Listeria monocytogenes*, *Brucella* spp., *Legionella pneumophila*, *Francisella* spp., *Yersinia pestis*, *Salmonella* Typhi, *Shigella dysenteriae*, Rickettsia, Chlamydia
Viruses	All viruses
Parasites	*Leishmania* spp., *Trypanosomacruzi*, *Plasmodium* spp., *Babesia* spp., *Toxoplasma gondii*, *Cryptosporidium parvum*, *Microsporidium* spp.
Fungus	*Histoplasma capsulatum*

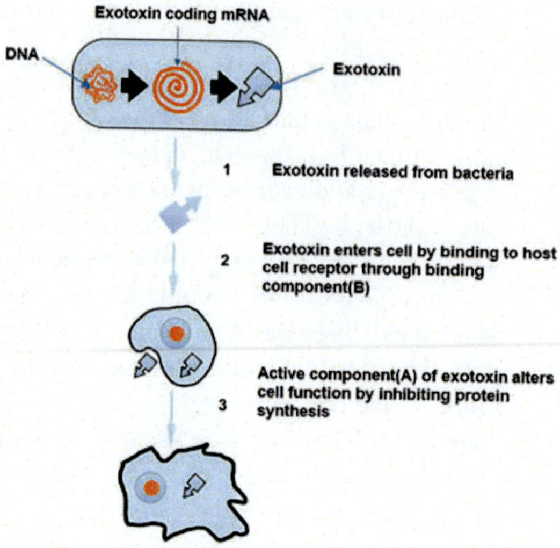

Fig. 7.3 Schematic diagram showing mode of action of exotoxin

heat-stable toxins of enterotoxigenic *E. coli*. Toxins encoded by bacteriophage DNA are the cholera toxins, the diphtheria toxin and the botulinum toxin.

Exotoxins are good antigens; they induce the synthesis of protective antibodies called antitoxins. Some of these antitoxins are useful in the treatment of botulism, tetanus and other diseases. Exotoxins treated with formaldehyde or acid or heat can be converted into toxoids. The toxoids lack toxicity but retain antigenicity. Hence, these are used in protective vaccines.

Exotoxins are some of the most toxic substances known. They are highly potent even in minute amounts. Botulinum toxin is the most potent one, and it has been estimated that 3 kg of botulinum toxin can kill all persons in the world. Similarly, the fatal dose of tetanus toxin for a human is estimated to be less than 1 lULg.

Many toxins have a dimeric A–B subunit structure. The diphtheria toxin, the tetanus toxin, the cholera toxins and the enterotoxin of *E. coli* are examples of exotoxins with an A–B subunit structure. A is the active subunit that possesses the toxic activity, and B is the binding subunit responsible for the binding of exotoxin to specific receptors of the membrane of human cells.

These toxins are very specific in their mechanism of action and act at specific tissue sites. The biochemical targets of A–B toxin include ribosomes, transport mechanisms and intracellular signalling (cyclic adenosine monophosphate, CAMP; G protein production); all these cause diarrhoea, loss of neuronal functions or even death.

Unlike endotoxins, exotoxins have specific pharmacological activities and do not produce fever.

Endotoxins

The term endotoxin was coined in 1893 by Pfeiffer to distinguish the class of toxic substances released after lysis of bacteria from the toxic substances (exotoxins) secreted by Gram-negative bacteria, but not by Gram-positive bacteria. They are lipopolysaccharide (LPS) components of the outer membrane of Gram-negative bacteria. Unlike exotoxins, these form an integral part of the cell wall, which are actively released from the cells. The genes encode the enzymes that produce the LPS on the bacterial chromosome, but not on plasmids or bacteriophage DNA, which usually encodes the exotoxins. They are heat-stable and released from the bacterial cell surface by disintegrating the cell wall. They are weakly antigenic and do not induce or poorly induce protective antibodies. Hence, their action is not neutralised by the protective antibodies. They cannot be toxoided.

Endotoxic shock: Endotoxins at low concentration induce a protective response, such as fever, vasodilation and activation of immunity and inflammatory response. However, endotoxins at very high concentrations, as seen in the blood of patients with Gram-negative bacterial sepsis, cause a syndrome of endotoxic shock. Endotoxic shock is characterised by fever, leucopenia, thrombocytopenia, sudden fall of blood pressure, circulatory collapse and sudden death. This is because a high endotoxin concentration can activate the alternative pathway of complement and cause vasodilation and capillary

Table 7.6 Mechanisms of endotoxin-mediated toxicity in common clinical manifestations

Clinical features	Mechanism
Fever	Interleukin-1
Inflammation	Activation of alternative pathway of complement (C3a, C5a)
Disseminated intra-vascular coagulation (DIC)	Activation of Hageman factor
Shock (hypo-tension)	Bradykinin, nitric oxide
Leucopoenia, thrombocytopaenia, decreased peripheral circulation, and perfusion to organs	Secondary to DIC

leakage, resulting in high fever, hypertension and shock. It also causes the blood coagulation pathway activation, leading to disseminated intravascular coagulation. Endotoxins are not destroyed by autoclaving; hence infusion of a sterile solution containing endotoxins can cause serious illness (Table 7.6).

Table 7.7 summarises the differences between exotoxins and endotoxins.

Superantigens: Superantigens are a special group of toxins. These bind simultaneously with the T cell receptor and major histocompatibility complex class II (MHC II) molecules on another cell. It leads to nonspecific activation of T cells, causing a life-threatening autoimmune-like response due to the production of a large amount of interleukins, such as IL-1 and IL-2. Furthermore, nonspecific stimulation of T cells can also lead to the death of activated T cells, leading to the loss of specific T-cell clones. Some examples of superantigen are Staphylococcal enterotoxin (toxic shock syndrome toxin) of *S. aureus* and erythrogenic toxin of type A or C of *S. pyogenes*.

Genetic Materials

Genes

Made with DNA, a gene is a functional unit that is transcribed and translated into one functional mRNA and protein, respectively. All functions in a body are encoded by different genes that are usually present in chromosomes. In prokaryotes, some of the genes are also present in plasmid, for example, antibiotic resistance genes.

Table 7.7 Distinguishing features of endotoxins and exotoxins

Features	Endotoxin	Exotoxin
Nature	Lipopolysaccharide	Protein (polypeptide)
Source	Gram-negative bacterial cell wall	Gram-positive bacteria and some Gram-negative bacteria
Location of genes	Chromosome	Plasmid or bacteriophage
Nature of secretion	Not secreted by the bacterial cell	Actively secreted by the bacteria
Release of toxin	Cell lysis	Filtration of bacterial cultures
Heat stability	Highly stable (withstand even 100 °C for an hour)	Heat-labile, destroyed mostly at 60 °C
Mode of action	Mediated by interleukins (IL-1) and tumour necrosis factor	Mostly enzyme-like action
Effect	Nonspecific (fever, shock etc.)	Specific pharmacological effect
Tissue affinity	No	Specific affinity for certain tissues
Diseases	Gram-negative bacterial sepsis, meningococcemia	Botulism, diphtheria, and staphylococcal toxic shock syndrome
Fatal dose	Only large doses are fatal	Small doses (even a few micrograms) are fatal
Antigenicity	Poorly antigenic	Highly antigenic
Neutralisation by antibodies	Ineffective	Neutralised by specific antibodies
Vaccine	No effective vaccine	Specific toxoids are available

Plasmids
Plasmids are extra-chromosomal DNA materials and some of them carry specific genes responsible for drug resistance or virulence factors. Plasmids can be transferred from one bacterium to other bacteria through conjugation. Thus, certain bacterial virulence characteristics may be solely dependent on the presence or absence of the plasmid inside the cell. *E. coli* colonisation of intestinal mucosa through surface antigen and enterotoxin production by *E. coli* and *S. aureus* are the examples of plasmid-borne virulence characteristics.

In the modern era, antibiotic resistance is a major concern in treating bacterial infections. The main reason for the raise in antibiotic resistance is plasmids carrying genes for antibiotic resistance. These plasmids can be easily transferred among bacteria.

Bacteriophages
In *Corynebacterium diphtheriae*, virulence is due to the diphtheria toxin. The gene responsible for diphtheria toxin production is not an integral part of the *Corynebacterium* genome but is acquired through lysogenic conversion from a beta or other tox + corynephages.

Communicability
The survival and distribution of the parasite in a community mainly depend on its ability to spread from one host to another, called communicability. Virulence and communicability are two different entities, and they need not be correlating to each other. In some cases, a highly virulent organism may kill its host, thereby hindering its existence in the community.

Generally, intestinal and respiratory infections have high communicability because these pathogens spread among a community through the secretory materials of the infected individuals. However, in certain infections, humans act as a dead-end host, thereby interrupting the communicability of the pathogen. Thus, only a pathogen strain with high degrees of virulence and communicability can cause epidemic and pandemic diseases.

Other Factors
The 50% lethal dose (**LD_{50}**) is the number of organisms required to kill half of the hosts. The 50% infectious dose (ID_{50}) is the number of microbes required to cause infection in half of the hosts. The infectious dose of an organism required to cause disease varies among the pathogenic bacteria.

Infecting dose is the minimum number of organisms required to cause an infection. It is one of the most important factors that determine the transmission and epidemic potential of foodborne pathogens. The lower the infection dose, the higher is the pathogenicity of a pathogen. For example, the *E. coli* O157:H7 has an infecting dose of 100 bacteria.

Pathogenesis of Infectious Disease

The infection process involves the sequential stages of (1) entry of the organisms and evasion of the local defences, (2) adherence to cell surfaces, (3) growth and multiplication of the bacteria at the site of adherence, (4) manifestations of disease and (5) termination of the disease.

Skin, mucus, ciliated epithelium and secretions containing antibacterial substances (e.g. lysozyme) are the natural barriers of the human and animal hosts that prevent microbial entry. However, these barriers are sometimes broken (e.g. a break in the skin, an ulcer in the intestine, a tumour), thereby allowing the entry of microbes into the host (Table 7.8). On entry, the microbes spread through blood circulation to other sites in the body.

Immunopathogenesis
In certain diseases, the symptoms are caused not by the organism itself but due to the immune response to the presence of such organism. For example, immune complexes deposited in the glomerulus of the kidney cause post-streptococcal glomerulonephritis. Antibodies produced against the M proteins of *S. pyogenes* cross-react with joint, heart and brain tissues producing disease

Table 7.8 Routes of entry of microbial pathogens

Routes of entry	Bacteria	Viruses	Fungi
Skin and mucus membrane	*Clostridium tetani*, *Leptospira*	Hepatitis B virus (HBV), human immunodeficiency virus (HIV)	Dermatophytes
Respiratory tract	*Streptococcus pneumoniae*, *Neisseria meningitidis*, *Haemophilus influenzae*, *Mycobacterium tuberculosis*	Rhinovirus, RSV (respiratory syncytial virus), Epstein–Barr virus, influenza virus	*Cryptococcus neoformans*, *Histoplasma capsulatum*, *Pneumocystis jirovecii*
Gastrointestinal tract	*Shigella* spp., *Salmonella* spp., *Vibrio* spp.	Hepatitis A or E virus, poliovirus	*Candida albicans*
Genital tract	*Neisseria gonorrhoeae*, *Treponema pallidum*	HIV, human papilloma virus	*Candida albicans*

manifestations of rheumatic fever. Similarly, the host immune response is an important cause of disease symptoms in patients suffering from syphilis caused by *T. pallidum*, Lyme disease caused by *Borrelia* and other diseases.

Clinical Manifestations

Diseases caused by various pathogenic and opportunistic pathogens cause a variety of clinical manifestations in infected human hosts.

Stages of an Infectious Disease

There are four discrete stages of an infectious disease.

Incubation Period

This denotes the time interval between the entry of the infective agent and the onset of clinical manifestations of the disease. During this period, the infective agent undergoes multiplication after reaching the selective tissue.

Prodrome Period

It is the time during which only nonspecific symptoms of the disease occur.

Specific Illness Period

It is the time during which the characteristic features of the disease occur.

Recovery Period

It is the time during which the symptoms resolve and health is restored.

Not all the cases that recover become free of the organism. Some become chronic carriers and act as sources of infection for others. Though some cases may not develop an infection, they can act as a link in the transmission of infection. This is called **sub-clinical infection.**

Termination of an infectious disease may occur by the resolution or continuation of the disease. This depends on a complex interaction of host immunity with pathogens and host response to treatment with specific anti-microbial agents.

Epidemiology of Infectious Diseases

The diseases that can be spread from one person to another are **communicable diseases**. Most microbial infections are communicable diseases. Three epidemiological terms are often used to describe infection: endemic, epidemic and pandemic. The infection that occurs at a persistent, usually low level in a certain geographical area is called **endemic**. The infection that occurs at a much higher rate than usual is known as an **epidemic**. Infection that spreads rapidly over large areas of the world is known as a **pandemic**.

Geographical Distributions

Geographical distribution of the infections, including bacterial, viral, fungal and parasitic, depends on many factors, including (1) the presence of vectors for the transmission of disease, (2) co-factors facilitating the transmission of disease and (3) the presence of a susceptible population in the particular area. For example, the presence of suitable arthropod vectors for a particular virus in large numbers increases the risk of a high prevalence of that viral disease in that community.

The microbial infections in a community may occur as an (1) outbreak, (2) endemic, (3) epidemic, or (4) pandemic. These terms only denote the spread of infection and are not a measure to determine the virulence or pathogenicity of a pathogen. The **outbreak** of a disease usually occurs from a common source, for example, infected food, and is seen only in clusters of people. The World Health Organization (WHO) in its Disease Outbreak News (DONs) monitors and provides information on outbreaks globally. Some of the recent outbreaks include the Lassa fever in Great Britain, UK, and the Nipah virus in Kerala, India.

Endemics (Greek éndēm(os), where én means "in or within" and dēm(os) means "people") are diseases that are confined to a particular geographical location. An outbreak turns to endemic when it spreads from a cluster to the entire population of a region. Most of the parasitic infections like malaria are considered endemic in tropical regions.

Epidemics of infection, in contrast, occur in large geographical areas. An epidemic usually occurs due to the introduction of a new strain of virus, parasite or bacteria into an immunologically susceptible population. Some of the notable epidemics are the H1N1 influenza epidemic in the USA, MERS in the Middle East, the Lassa fever in Mali and Ebola in West Africa.

Pandemics are usually worldwide epidemics resulting from the introduction of a new virus, for example, the HIV. The COVID-19 pandemic occurred due to the introduction of new strains of coronavirus.

Microbial diseases also show seasonal differences; for example, enteric viruses are more prevalent in the summer due to poor hygiene and the use of contaminated water because of a shortage of water supply. Similarly, respiratory viruses are more prevalent during winter due to overcrowding, which facilitates the spread of viruses.

Events like climate change, urbanisation and deforestation have nearly dissolved the wildlife-human boundaries. It has allowed pathogen spillovers and its spread to newer geographical regions. As a single contributing factor, climate change has triggered a cascade of climate adaptation events in the pathogen that in turn is influencing the host-pathogen-transmission triangle. Vector-borne diseases like malaria and dengue are reportedly spreading to European regions that are becoming warmer and turning into global diseases of concern. Spillovers of new, emerging and re-emerging viruses have become more frequent. For example, the new SARS-CoV-2 virus has brought a pandemic after spillover from bats.

Reservoir of Infection

Infected humans are usual reservoirs of viruses. They serve as a source of infection for other susceptible hosts. It includes sexually transmitted diseases, polio, measles, mumps, streptococcal infection and so on. Because humans are the sole reservoir for these infections, eradication of the disease from the last reservoir can completely eradicate the disease itself. Eradication of smallpox is an example of such an event.

For many other diseases, animals act as reservoirs. Pathogens transmitted from these animal reservoirs to humans are called zoonotic diseases; for example, rabies is a zoonotic viral disease, and the infection is transmitted from infected dogs, bats, foxes, cats and so on to humans. More than 200 pathogens are known to be zoonotic. It is estimated that 60% of emerging infectious diseases are zoonotic. Recent viral outbreaks like MERS, SARS, Ebola, Lassa, Nipah and the recent COVID-19 pandemic have a zoonotic origin.

Mosquitoes, ticks and sand flies are arthropods that act as vectors for transmission of malaria, lymphatic filariasis, togaviruses, flaviviruses, bunyaviruses and reoviruses. Some important vector-borne diseases are presented in Table 7.2.

In 2017, the WHO launched the "Global Vector Control Response (GVCR) 2017–2030" to provide strategic guidance to countries for urgent strengthening of vector control to prevent diseases and respond to outbreaks.

Emerging and Re-emerging Infections

The term "**emerging and re-emerging diseases**" was coined in 2000 by the Nobel Laureate Joshua Lederberg. "Emerging infectious diseases" are diseases of infectious origin whose incidence in humans has increased in the past two decades or threatens to increase in the near future. Emerging infectious diseases were those that, like AIDS and Ebola, were previously unknown to have afflicted humans. Re-emerging diseases, such as cholera and plague, were familiar scourges whose incidence was rising or whose geographical range was expanding. The WHO has added a few more clauses to describe these two categories:

Emerging infectious diseases (EIDs) are diseases that are newly recognised, newly introduced or newly evolved, or that have recently and rapidly changed in incidence or expanded their geographical, host or vector range (WHO/FAO/OIE 2004). They also include diseases that have affected a given region in the past, declined over time or were controlled, but again resurfaced in growing numbers (WHO 2014).

Re-emerging diseases are diseases that reappear after they have been on a significant decline. Re-emergence may happen because of a breakdown in public health measures for diseases that were once under control. They can also happen when new strains of known disease-causing organisms appear. Human behaviour affects re-emergence. For example, overuse of antibiotics has led to disease-causing organisms that are resistant to medicines. It has allowed a return of diseases that once were treatable and controllable. Re-emerging diseases include malaria, tuberculosis, cholera, pertussis, influenza, pneumococcal disease and gonorrhoea.

Table 7.9 provides a list of emerging and re-emerging diseases. A study revealed that 60.3% of EIDs infecting humans between 1940 and 2004 were zoonotic with wildlife playing a great role, and one in five was transmitted by invertebrate vectors such as mosquitoes, ticks

Table 7.9 List of emerging and re-emerging infections

Bacterial diseases	Rickettsial diseases	Viral diseases	Parasitic diseases
Emerging diseases			
Anthrax Diphtheria Cholera Leptospirosis Meningitis *Escherichia coli* O157:H7 (Toxin) Lyme disease	Scrub typhus	SARS Avian Influenza AIDS Dengue/DHF HFMD Japanese encephalitis Measles Nipah Rabies West Nile fever Zika virus	Malaria
Re-emerging diseases			
Tuberculosis Cholera Pertussis Pneumococcal disease Gonorrhoea	Scrub typhus	Influenza COVID 19 Monkey pox	Malaria

Table 7.10 List of Blueprint Priority Diseases (WHO 2018–2019, WHO 2020b)

Viral infections
Middle East respiratory syndrome
Severe acute respiratory syndrome
Marburg
Ebola
Lassa fever
Viral haemorrhagic fevers
Rift Valley fever
Zika
Nipah and Henipavirus diseases
Chikungunya
Covid-19
Bacterial infections
Tuberculosis

and midges. A WHO-endorsed expert committee (WHO 2018–2019) has updated the list (Table 7.10) of current and future priority emerging diseases based on seven criteria. This list has been recently updated again with the addition of the novel coronavirus disease 2019 (COVID-19) that recently hit the world (WHO 2020b).

8 Diagnostic Methods in Microbial Infections

Laboratory diagnosis of an infectious disease is one of the most important functions of Medical/Clinical microbiology and forms an integral part of this speciality. In a patient suspected of suffering from an infection, the specimen is obtained from the individual and sent to the microbiology laboratory to ascertain the causative organism. After various tests and procedures, a positive or negative report is sent to the treating clinician, confirming or disproving the initial diagnosis. In some cases, particularly bacterial infections, an antimicrobial sensitivity profile is also sent along with the identity of the organisms, which helps the clinician in instituting pathogen-directed antimicrobial therapy.

The Diagnostic Principles

Identifying a microbe in a patient's specimen can be done in three broad manners. These include (1) visualisation of the microbe or demonstration of microbial product or isolation of the microbe in pure form (as per Koch's Postulates) followed by identification, (2) detection of patient's immune response by detection of antibodies (most common) or patient's immune defence system and (3) amplification of pathogen's nucleic acid.

In practice, the following procedures are employed and selected as per the requirement and laboratory capacity:

Appropriate specimen collection and transport
Microscopy
Cultivation (culture) of microorganisms
Animal pathogenicity tests
Identification of culture-grown microbes
Antimicrobial susceptibility test
Detection of microbial antigen
Serological tests
Assessment of patient's immune defence system
Recent approaches in diagnostic microbiology

Appropriate Specimen Collection and Transport

The pre-analytical phase of sample collection and its transport plays a crucial role in the test outcome since a sample that is not suitable can result in misleading test results. The specimen should reflect the infection process and is commonly obtained from the lesion's principal site. An adequate amount of sample is collected since the number of microorganisms may be scanty in the sample collected. Normally, any organism in a specimen collected from a normally sterile site is reflective of the infection; the same does not hold when samples are taken from areas such as skin, mouth and throat, where commensal flora are also present. In the latter cases, the area should be appropriately cleaned before sample collection, or some invasive procedure can be used to prevent contamination with unwanted flora. Thus, a good idea about the nature of flora from the collection site is necessary for the correct interpretation of results. All specimens are collected in sterile containers and transported to the laboratory in the quickest possible time. In case of delay in transport or processing, specimens are stored at 4–8 °C in a refrigerator. Transport media in tubes or vials are commercially available, which are used when shipping the specimen to laboratories situated at a distance and may require a few hours for transport.

Microscopy

Light microscopy and **fluorescent microscopy** play an important role as a first step in the initial examination of the specimen. By microscopy, we get a preliminary idea about the nature of the infective organism and the presence of other host-derived cells like leucocytes.

Unstained direct examination of the wet preparation of the sample helps detect bacterial motility, identify protozoan or helminthic parasites and confirm fungal structures.

Stained preparation of the sample is a general procedure primarily used for bacteria and useful for other groups of organisms. A stain is a dye that is applied to a sample for better visualisation of microorganisms. The most common stain used in microbiology is the Gram stain (Fig. 8.1), which helps us to distinguish between Gram-positive and Gram-negative bacteria and gives us an idea about the morphology, whether cocci or bacilli. Acid-fast stain is used for *Mycobacterium*, while other stains such as Giemsa and Trichrome are used for various other specimens to visualise other groups of organisms.

Fluorescent staining (Fig. 8.2) is used for directly staining the sample with mono- or polyclonal antibodies tagged with a fluorescent dye. It provides a rapid and specific means of detecting

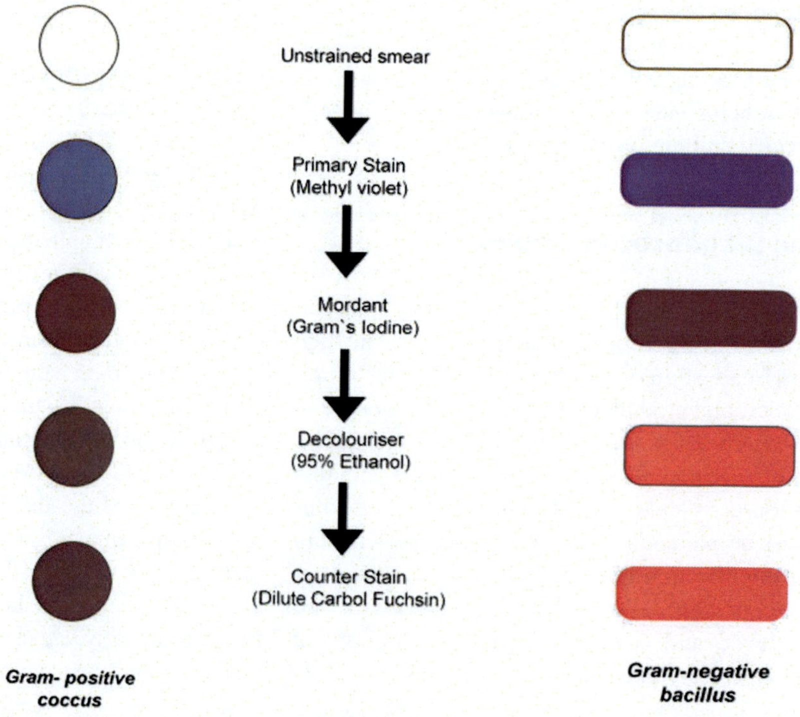

Fig. 8.1 Principle of Gram staining

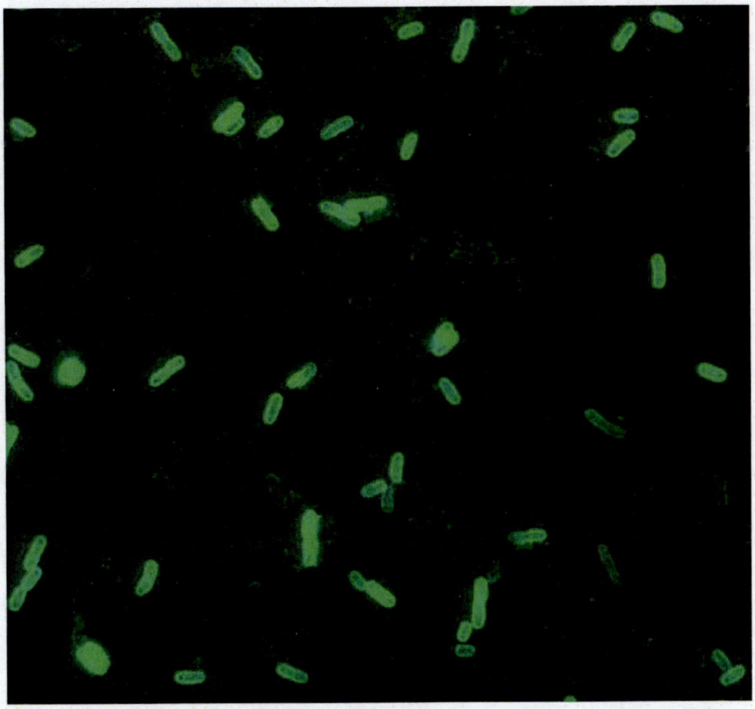

Fig. 8.2 Fluorescent staining of tissue sample showing *Haemophilus influenzae* bacteria (https://phil.cdc.gov/Details.aspx?pid=1946)

Table 8.1 Commonly used staining techniques in clinical microbiology

Staining method	Utility
Gram stain	Can be used for diverse specimens. Identification of Gram-positive and Gram-negative bacteria and their shape
Acid-fast stain (Ziehl-Neelsen Method, Kinyoun Method, Modifications)	Staining of *Mycobacteria*, *Nocardia* Staining of some protozoan parasites
Albert stain	*Corynebacterium diphtheriae*
Giemsa stain	Blood samples for parasites Viral inclusion bodies in cells
Lactophenol cotton blue stain	Fungi from culture material
Gomori methenamine silver stain	Fungi in tissue sections
Trichrome stain	Parasites in stool
Direct fluorescent stain	Detection of an organism or its antigens

the microorganism or its antigen. Microscopy remains the gold standard for identifying most of the protozoan parasites and helminths. The important stains used are given in Table 8.1.

Cultivation (Culture) of Microorganisms

The microorganisms present in a specimen need to be grown by some means so that proper identification can be done. Bacteria and fungi grow on artificial media, while viruses and bacteria such as *Chlamydia* and *Rickettsia* can only be grown inside cells in tissue culture. Both liquids and solid artificial culture media are widely used. While liquid culture is very sensitive when very few organisms are present in a specimen, solid media are necessary for growing discrete colonies in pure form from a specimen and are also helpful for counting the number of bacteria present in a

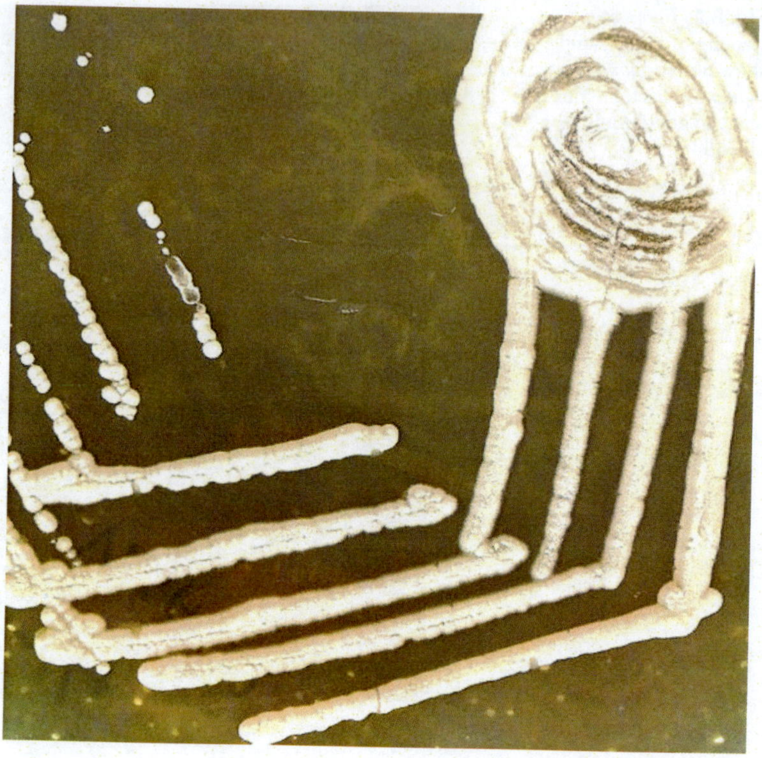

Fig. 8.3 Culture of yeast on Sabourauds Dextrose Agar (SDA) medium (Courtesy of Prof S. Pramodhini)

sample expressed as colony-forming units/ml. This is because a colony of microorganisms contains thousands of cells derived from a single cell that has been cultured. There is no single medium capable of growing all the bacteria, hence judicious use should be made in selecting the media that can support the growth of probable organisms present in a particular specimen. Incubation at 37 °C for 48–72 h is sufficient for most bacteria's visible growth. For fungi, a lower temperature (27 °C) and a more prolonged incubation period are generally needed except for yeasts and yeast-like fungi, which can grow at 37 °C (Fig. 8.3).

Most viruses and a few bacteria can only be grown in viable cells derived from humans or animals. This technique involves laboratory expertise and is time-consuming and hence not frequently used in many laboratories. Cultivation for parasites is not performed in routine laboratories, except occasionally for leishmania, trypanosome and trichomonas. Egg hatching technique to grow the larval form for definitive identification of a helminth parasite can also be undertaken.

Identification of Culture-Grown Microbe

Bacteria or fungi grown in a culture medium are further identified to the genus and, in many instances, to the species level by using a battery of phenotypic characteristics. These include, among others, growth conditions such as temperature, time and atmospheric requirements; colonial appearance; staining properties; biochemical reactions using various substrates; and antigenic properties. The polymerase chain reaction can also be performed on the culture material to amplify the DNA or RNA.

Filamentous fungi are principally identified based on colony characters and microscopic morphological features. Virus identification in tissue culture relies on the changes in the cells brought about by the viral growth (*cytopathic effects*) (Fig. 8.4) along with the detection of viral antigens and/or viral nucleic acid by amplification techniques.

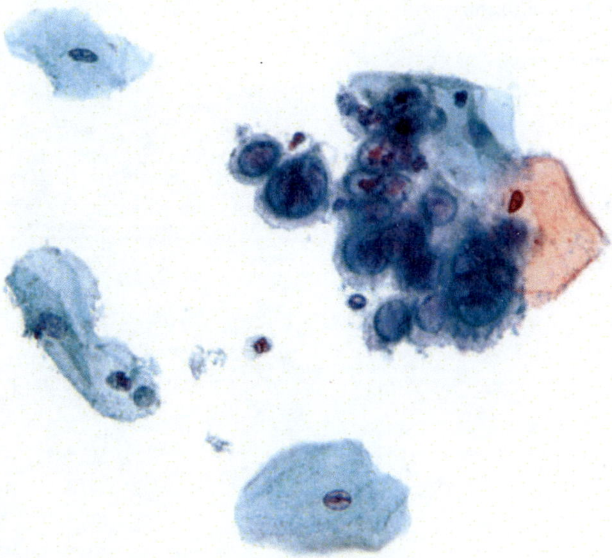

Fig. 8.4 Cytopathic effect—"Cell rounding"—type induced by the HSV-1 on Vero cells (https://commons.wikimedia.org/wiki/File:CPE_rounding.jpg)

Antimicrobial Susceptibility Test

A logical step after isolating the bacterial pathogen from a patient's sample is to find out the sensitivity of the isolate to the antibiotics that can be used for targeted therapy of the infection. Traditionally, this is done with antibiotic-impregnated discs containing a fixed concentration of the drug and applied to a lawn culture of the bacteria. The zone of inhibition of bacterial growth around the disc gives an idea of whether the particular isolate is sensitive or resistant to the antibiotic. This is the principle of the disc diffusion technique (Fig. 8.5).

A more accurate method is to determine the **Minimum Inhibitory Concentration** (MIC) by breakpoint dilution. MIC is carried out either on a solid medium or more often in a liquid medium or by using the Epsilometer (E) test. MIC determination is also an integral part of automated systems where simultaneous identification and sensitivity profile are available after the requisite hours of incubation. Although the above techniques are standardised for many yeast and yeast-like fungi, the same is not available for many filamentous fungi. Antiviral susceptibility testing may be done by phenotypic methods such as plaque reduction, dye uptake or enzyme immune assays, but it is not suitable for most routine laboratories. Similarly, except for a few protozoan parasites like malaria, phenotypic methods for drug sensitivity testing are lacking.

In the last two to three decades, molecular testing to determine the expression of drug resistance gene/s is gaining steady popularity and gradually replacing the phenotypic method. This method can be used for bacteria, viruses, parasites or fungi, and in many instances, it may be done directly with the patient's sample.

Animal Pathogenicity Tests

Laboratory animals have an important role in the fields of infectious and non-infectious diseases. In microbiology, animal inoculation of a patient's sample to isolate the causative agents has been most commonly done for viruses, some bacteria like tubercle bacilli and protozoan parasites like toxoplasma. Apart from diagnostic use, animal studies give invaluable information on the pathogenesis and immune response to microorganisms. They have played a pivotal role in vaccine development. Small animals such as monkeys, rabbits, guinea pigs, hamsters, mice and rats are commonly used for diagnostic purposes. Depending on the likely pathogen, the

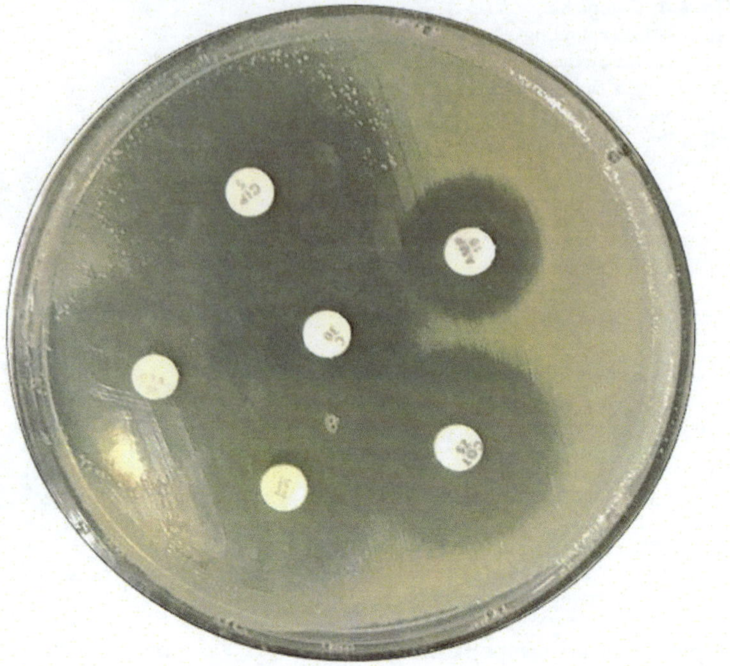

Fig. 8.5 Disc diffusion technique. (Courtesy of Prof. S. Pramodhini)

specimen can be inoculated through various routes such as intra-peritoneal, intra-cerebral and subcutaneous. In most instances, the animal needs to be sacrificed to study the internal organs and detect the pathogen. As a result of stringent ethical restrictions on animal use and the development of newer and more sensitive methods, animal inoculation purely for the diagnosis of infection is no longer in use.

Serological Tests

Detection of Antibodies

Antibody detection in a patient's body fluids, particularly serum, is commonly practised. It gives indirect proof of infection, which may or may not indicate a disease state since even in asymptomatic infection, antibody levels may be increased. Most of the tests available can detect the IgG fraction, which in many cases is retrospective since they appear after about 2–4 weeks of infection. On the other hand, IgM appears after 7–10 days and is a better indicator of acute infection. The dictum of a serological test is to look for rising titre, that is, a fourfold or more rise in antibody titre between acute and convalescent sera collected 3–4 weeks apart. However, since it cannot be followed in actual practice, a single serum sample is commonly tested and either high IgG or the presence of IgM/IgA/IgE is taken as a diagnostic proof of the infection process.

Enzyme immune assay in various formats is the most popular procedure used in the majority of laboratories. ELISA is useful for testing a large number of samples at a time Rapid immunochromatographic tests are also often used as a point-of-care (POC) test to give immediate results. **Chemiluminescence** is another rapid method for antibody testing. Here luminol and derivatives of isoluminol are used as chemiluminescent labels. It is highly specific and considered to be more sensitive compared to ELISA. The **indirect fluorescence antibody** (IFA) test was once quite popular, but EIA has largely replaced it. This is because of the ease of performance of the test, quick turnaround time and freedom from observer's subjective interpretation. However,

IFA remains the standard test in certain instances like chlamydial infections.

Detection of Antigens

In many instances, it is easier to detect the microbial antigen in the patient's sample, and it is both a quick and a specific method. Polyclonal or preferably monoclonal antibodies tagged with a fluorescent dye or an enzyme can be used to detect the antigen by direct fluorescent technique or enzyme immune assay.

Rapid point-of-care immunochromatography tests based on enzyme immune assay are also available. Other methods for antigen detection include the latex agglutination test and co-agglutination test, where the antibody is coated on latex particles or the Cowan strain of *Staphylococcus aureus*, respectively. Monoclonal antibodies used in the test format can identify species and may even be used for strain separation in the same species based on antigenic structure. Antigen testing can provide quick results, but these tests lack sensitivity, particularly if the antigen concentration is low. Antigen tests are widely used for capsular antigen detection for *Streptococcus pneumoniae*, *Haemophilus influenzae* type b and *Cryptococcus neoformans* in CSF; *Rotavirus*, *Crptosporidium*, *Giardia* and *Entamoeba histolytica* in stool; and hepatitis B virus and HIV antigens in the blood.

Assessment of Cell-Mediated Immune System

A few tests can help in understanding the status of the body's immune system and can be of help in diagnosis as well. The intradermal tests for diagnosing infectious diseases detect mainly the Type IV hypersensitivity reaction or Delayed-Type Hypersensitivity (DTH) to the injected antigens (exception: Casoni's test for hydatid disease is Type I hypersensitivity reaction). Intradermal tests for detecting DTH are read at 48–72 h with some exceptions like the Lepromin test. While taking the reading, the induration size is more important than erythema. Although many such tests have been described, only the Tuberculin test (Fig. 8.6) is routinely performed in many countries. *In vitro* immune functional assay for tuberculosis in the form of Interferon-Gamma Release Assays (IGRAs) has replaced the tuberculin test in many countries to diagnose latent tuberculosis. QuantiFERON-TB Gold® and T-SPOT.TB® are examples of two commercially available IGRAs.

Fig. 8.6 Tuberculin test

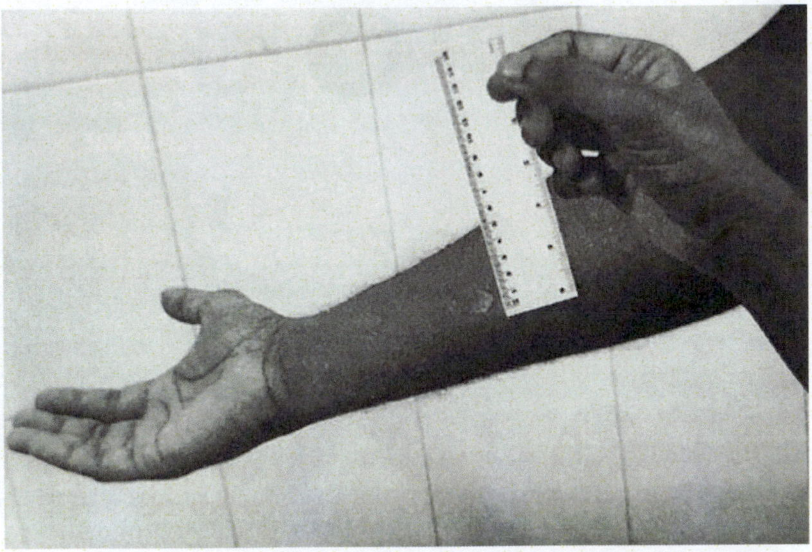

Recent Approaches in Diagnostic Microbiology

Nucleic Acid Amplification Tests (NAAT)

Developed at the turn of the last century, NAAT has revolutionised the field of microbiology with its wide applications. These include polymerase chain reaction (PCR), real-time polymerase chain reaction (real-time PCR), ligase chain reaction (LCR), transcription-mediated amplification (TMA), nucleic acid sequence-based amplification (NASBA), strand displacement amplification (SDA) and so on.

The conventional polymerase chain reaction where the amplified DNA is visualised by gel electrophoresis is being rapidly replaced with real-time PCR in diagnostic microbiology. Real-time PCR gives a picture of the amplification process in real time, and there is no need for post-amplification electrophoresis, which eliminates contamination. The copy number of nucleic acid can also be determined if a set of standards with known copy numbers is included, which gets amplified along with the test sample (quantitative PCR, qPCR).

In multiplex PCR, primers for more than one pathogen can be included in the test system, which allows to group the tests according to the disease syndromes. For example, the commercial system of Biofire Film Array system from bioMerieux company has got respiratory, gastro-intestinal, blood culture and meningitis-encephalitis panels which can detect 20–25 pathogens in each panel. This helps the clinician to take a syndromic approach and order the tests accordingly. For details of these other techniques, refer to Chap. 9.

Most of the NAATs require sample extraction and processing before detection. It requires laboratory set-up, consumables and experts to handle the infectious agent safely and ensure efficient extraction. However, many new tools have recently come up that can give results directly without processing the sample. One such example is GeneXpert® (Fig. 8.7) System (developed by Cephid), which has automated entire molecular diagnostics. The process known as CB-NAAT uses PCR reaction reagents in a cartridge and only requires direct loading of the sample on them. The machine automatically performs the extraction to results without manual intervention.

Fig. 8.7 GeneXpert®. (Courtesy of Prof. Ujalla Ghosal)

It effectively reduces safety concerns that arise from the human handling of infectious agents.

Sequencing

Sequencing of a gene has established itself as a powerful diagnostic tool where other methods have given inconclusive results. Sequencing of ribosomal DNA using panfungal primers is now the gold standard for the identification of fungi. DNA sequencing can also be used for precise species identification and is widely used in bacterial, fungal, viral and parasitic pathogen identification and strain differentiation.

Automated Methods

Automated methods are now available, which take only hours for the characterisation of isolates. These include the detection of specific enzymes, toxins, antigens or metabolic end products. Automated systems for bacterial identification include Phoenix bacterial identification system (BD biosciences), MALDI-TOF (Matrix-assisted laser desorption/ionisation time-of-flight), VITEK-2 bacterial identification, MicroScan WalkAway system and so on. The cartridge-based GeneXpert is also an automated nucleic acid amplification process. These methods have got more added advantages than the conventional methods in the aspect of continuous monitoring, more sensitive, rapid and fewer chances of contamination. The latest platform of Matrix Associated Laser Desorption/Ionisation Time-of-Flight (MALDI-TOF) is increasingly being used for the identification of the bacteria, yeast and also to some extent moulds.

Detection of Metabolites

Microbial metabolites are defined as the intermediate products of the metabolism of microorganisms. Metabolites can be grouped into primary and secondary metabolites. Primary metabolites are essential for the proper growth of microorganisms. Examples of primary metabolites are ethanol, lactic acid and certain amino acids. Secondary metabolites have no role in development, growth and reproduction. They are produced during the stationary phase of growth. Examples of secondary metabolites include antibiotics, mycotoxins and so on. Nuclear magnetic resonance (NMR) spectroscopy, Mass spectrometry (MS), Gas chromatography (GC) and Liquid chromatography (LC) coupled to mass spectrometry (MS) are the frequently used methods for the detection of metabolites.

Immunochromatographic Tests

Lateral flow immunochromatographic tests (ICTs) are the commonly used rapid diagnostic test (RDT) for the detection of infectious diseases. In contrast to complex molecular tools and technologies that require clinical establishments for diagnosis, such ICTs offer portability and are perfect as a **point-of-care** (POC) **test**. For the detection of pathogens, it contains biomarkers (antigen) for the pathogens incorporated on a paper strip. When a sample containing the pathogen is placed on the paper, it develops lines corresponding to the binding of the antibody (present in the sample) and the biomarker indicating the presence of the particular antigen or antibody. Such tests are easy to interpret and do not require any technical expertise.

9 Genomics, Proteomics and Molecular Biology in Microbiology

The science of microbiology had relied for a long time on phenotypic markers such as culture, morphology, biochemical and serological properties of the microbes to characterise them. But it was realised early that gaps existed in the knowledge about the biology of these organisms and many unresolved issues remained in the taxonomy, identification, pathogenesis and so on. The discovery of the structure and finer details of DNA in the 1950s heralded the application of DNA-based studies of microorganisms. Phenotypic studies started getting supplemented by genotypic studies, which brought about a revolution in our knowledge about microbes. In recent times, the study of the proteins encoded by the genes has opened up a new field of proteomics with diverse applications.

From discovering DNA structure and the genetic engineering strategies, science has never looked back and opened a new avenue where the gene structure and function played a significant role in technological advancements. The recent development and applications of genomics, proteomics and molecular biology have contributed immensely to Medical Microbiology, especially in diagnostic microbiology.

Genomics

DNA is the building block of the genes in a living being. The set of genes constitutes the chromosome and the whole framework forms the genome. Thus the genome is the whole of the genetic material of the organism, or, in other words, the complete set of genes is designated as the genome. The term was first coined by Hans Winkler in 1920 from the words *ge*ne and chromos*ome*. The introduction of the chain termination method to sequence DNA in 1977 by Frederick Sanger followed by whole genome sequencing technologies paved the way for the modern era of genomics.

Technologies

Sanger chain termination method and Maxam-Gilbert Sequencing methods were developed in the 1970s but could sequence only small fragments and were manual methods. Thus Sanger method has a read length of only 400–900 base pairs (bp). The Shotgun approach which came soon after was useful to study the sequence of a larger piece of DNA (more than 1000 bp) by breaking up the DNA into small fragments, followed by sequencing each piece separately. Pyrosequencing technology was described in 1993 and has a capacity to sequence up to 700 bp. Newer automated methods of sequencing started appearing in 1990s and these commercial sequencers have established their places in the twenty-first century. They are not only automated but also allow the entire genome to be sequenced (Whole Genome Sequencing, WGS) in one go and popularly they are known as "next-generation" or "second generation" sequencers. Some of these technologies such as Single molecule real-time sequencing and Nanopore DNA sequencing are referred to as long-read sequencers and have read lengths of 20,000 nucleotides or more, with average read lengths of 5 kb. They are also known as "third generation" sequencing methods. In contrast, a large number of short-read commercial sequencers based on different technologies are available. These include, among others, Illumina dye sequencing, SOLiD sequencing by ligation, DNA nanoball sequencing, ion semiconductor sequencing and so on. Efforts are continuing to develop faster, more high-throughput technologies.

Applications

The era of microbe sequencing started with bacteriophages in 1976. *Haemophilus influenzae* was the first pathogenic microorganism which was fully sequenced in 1995. Since then, a large number of bacteria, fungi, viruses and parasites have been studied and their whole genome sequence is known. However, although all the genes of these organisms have been identified, the functions of many of them still remain unknown or not annotated. The applications of genomic studies can be grouped as follows:

Identification of the Pathogen

Whole genome sequencing (WGS) is one of the definitive methods for strain identification. Although unlikely to replace the conventional methods or MALDI-TOF (Matrix-assisted Laser Desorption/Ionisation Time-of-Flight) in this aspect, WGS is most useful for those organisms which are difficult to diagnose by routine methods. At present, the identification of organisms by PCR amplification of gene/s or 16S rDNA sequencing is nothing but the product of genomic studies of organisms.

Identification Directly from Sample and Metagenomic

Metagenomics is a methodology where all the DNA in a given sample is sequenced. Using bioinformatics tools, the non-pathogenic organisms and human DNA are excluded and the pathogenic DNA is identified. Samples like urine or CSF taken from normally sterile sites will give better results than from a sample with large microbiota like stool. Identification is highly accurate and results can be obtained even if few organisms are present.

Detection of Resistance to Antibiotics

Resistance to certain antibiotics can be predicted by WGS by analysis of the data. Mutations which are responsible for rifampicin resistance can also be detected. The currently used methods may be more of practical use for detecting resistance to common bacteria. Nevertheless WGS may help in detecting resistance to slowly growing organisms or for cases where phenotypic methods are not standardised.

Diagnostic Test Development

New diagnostic tests for existing and novel pathogens have been developed using sequencing tools. The genome sequences are also useful for the identification of targets for the development of diagnostic PCR and serological tests.

Detection of Virulence Factors

Apart from being able to detect antimicrobial resistance mentioned above, genome data may also be useful for detecting toxins of organisms and also new ones which may not be detectable by existing methods.

Epidemiological Typing of Strains

Restriction fragment length polymorphism, pulsed-field gel electrophoresis and so on are the investigative methods commonly used for epidemiological tracking of strains during outbreaks or epidemics. But all these methods have varying resolution power and often do not give reproducible results. WGS has much higher discriminatory power and has the potential to replace the traditional methods in the not too distant future. It can be considered the ideal typing method since it can identify a single genome change between isolates. Many studies have employed single nucleotide polymorphism for strain differentiation. WGS can also help in mapping the route of transmission and how the pathogen spread during an epidemic.

Proteomics

The term proteome was first coined in 1994 by Wilkins and his co-workers to describe the "PROTein complement expressed by a genOME". It is the set of all proteins expressed by the genome and the study of proteome is known as proteomics.

Scopes of Proteomics

It has wide-ranging scope encompassing everything related to proteins. These include annotation of genomes, protein expression, protein function, modifications of proteins after translation, localisation of proteins and interaction between proteins. Towards this end, proteomics is divided into the following three subdivisions:

1. **Protein expression proteomics**: Here protein expression of the whole proteome between samples can be compared which can identify new proteins or pathogen-specific proteins.
2. **Structural proteomics**: It is also known as the "cell map" where all the proteins of a cell are identified, along with their compartmentalisation or location and their interaction with each other.
3. **Functional proteomics**: It involves the study and characterisation of a selected group of proteins which can give information about the pathogenesis of a disease, signalling of proteins and protein-drug interactions.

Tools for Proteomics

The present-day proteomic analysis essentially uses two methods. In the "bottom-up" method, short peptides generated by protein fragmentation are analysed for proteome identification, while in the "top-down" method whole protein is analysed. Essentially, in the first approach, the protein is first digested into peptides with enzymes, followed by their separation by liquid chromatography. Alternatively, in gel-based technology, the proteins are first separated by gel electrophoresis, followed by enzymatic digestion. The final step involves the use of mass spectrometry which analyses the peptide fragments and produces the mass spectroscopy spectrum. The data is then studied using various database search engines which give the qualitative identity of the proteome. In the "top down" approach, proteolytic methods are not used and it is mainly useful for purified proteins. The third approach is a quantitative method where amino acids or peptides are either labelled or non-labelled to find out the relative abundance of the proteins in a sample.

Applications in Microbiology

Although still in its early stages, proteomics has made its inroads in microbiology. The following are some of its important applications.

Elucidation of Host–Pathogen Interaction

Proteomic analysis can reveal important information regarding host and pathogen interaction and the microbe expressed proteins involved in the pathogenesis can be clearly identified. It has been found that microbial proteins interact with host proteins and alter the normal cellular process.

Pathogen Identification

Proteomic analysis has arrived in a big way into the identification of a pathogen from culture material or even a clinical sample by the use of Matrix-assisted Laser Desorption/Ionisation Time-of-Flight Mass Spectroscopy (MALDI-TOF MS). It is now available in many diagnostic laboratories around the world, including in India. Apart from identifying organisms at the species level, it can also help in strain differentiation. In this method, the protein is extracted from the sample and mixed with the matrix (an energy absorbent organic compound) on a target plate. The sample within the matrix is subjected to short laser pulses and desorption and ionisation produces an ionic state and the ions get separated on the basis of their mass-to-charge (*m*/*z*) ratio. These ions reach the detector at the time-of-flight tube. The *m*/*z* ratio of an ion is measured by determining the time required for it to travel the length of the flight tube. As a result, a characteristic spectrum is produced called peptide mass fingerprint. Final identification of the organism is done by comparing the fingerprint with the existing databases. It has found application in bacterial, fungal and viral identification directly from a sample. Apart from identification, this technique has also been employed to detect antimicrobial resistance.

Development of Vaccines

Protective antigens can be identified using proteome analysis and bioinformatics tools which can be used for future vaccine development.

Drug Development

Another area of potential use is designing of appropriate antibiotics for resistant organisms. The resistance mechanism can be studied and the targets can be identified and a response library can be created for future drug development.

In essence, proteomics and genomics have entered medical microbiology in a big way and are likely to make rapid progress in the future. These approaches may ultimately pave the way for personalised medicine or individual patient-centric treatment of infectious diseases.

Molecular Biology

The use of molecular tools and techniques in the diagnostic laboratory has been remarkably adopted with applications ranging from genetic testing, epidemiologic studies, tissue typing, in situ hybridisation, forensics, parentage testing and detection, identification and classification of microorganisms for disease diagnosis.

Molecular techniques used in microbiological diagnosis include nucleic acid amplification, DNA sequencing and typing, direct molecular probe techniques (in situ hybridisation) and nucleic acid quantification. Across the platforms, several commercially available IVD certified reagents, equipment and test kits are now available that have helped clinicians in giving more evidence-based patient care. It has almost swept away previous generation methodologies that were primarily manual and dependent on biochemical and morphological observations. Rapid growth in precision medicine, cell and gene therapy and its translation into modern healthcare have necessitated functional molecular testing where devices have been developed to analyse large genomes, analyse, interpret and maintain data for accurate and timely assessment. Routine clinical testing, disease diagnostics and Specialty tests such as genetics, immunology, oncology and endocrinology are now reported based on genetic markers and gene functions. The rapid development of AI has introduced machine dependencies capable of running the service to reporting with minimal manual interventions.

Molecular Biology Techniques in Microbiology

From discovering DNA structure and the genetic engineering strategies, science has never looked back and opened a new avenue where the gene structure-function played a significant role in technological advancements. The benefits were also extended to medical microbiology, with most rewards enjoyed in molecular pathology and diagnostic microbiology. The use of molecular tools and techniques in the diagnostic laboratory has been remarkably adopted with applications ranging from genetic testing, epidemiologic studies, tissue typing, in situ hybridisation, forensics, parentage testing and detection, identification and classification of microorganisms for disease diagnosis.

Molecular techniques used in microbiological diagnosis include nucleic acid amplification, DNA sequencing and typing, direct molecular probe techniques (in situ hybridisation) and nucleic acid quantification. Across the platforms, several commercially available IVD certified reagents, equipment and test kits are now available that have helped clinicians in giving more evidence-based patient care. It has almost swept away previous generation methodologies that were primarily manual and dependent on biochemical and morphological observations. Rapid growth in precision medicine, cell and gene therapy and its translation into modern healthcare has necessitated functional molecular testing where devices have been developed to analyse large genomes, analyse, interpret and maintain data for accurate and timely assessment. Routine clinical testing, disease diagnostics and Specialty tests like genetics, immunology, oncology and endocrinology are now reported based on genetic markers and gene functions. The rapid development of AI has introduced machine dependencies capable of running the service to reporting with minimal manual interventions.

Nucleic Acid Probes

Nucleic acid probes, otherwise known as DNA probes, are segments of DNA and RNA labelled with radioisotopes or enzymes that can hybridise to complementary nucleic acids with high specificity.

DNA probes have practical value because they can detect specific nucleotide sequences in unknown samples. The probes carry reporter particles, such as radioactive labels, which are isotopes that emit radiation, or luminescent labels, which give off visible light. Reactions can be exposed by setting the photographic film in contact with the test reaction. Fluorescent probes consist of dyes that can be seen with ultraviolet light. Probes can be used in a broad range of analytic procedures.

Two diverse nucleic acids can ***infuse*** by coalescing at their integral sites. All distinct combinations are probable: single-stranded DNA can coalesce with other single-stranded DNA or RNA, and RNA can incorporate with other RNA. This property forms the basis of specially formulated oligonucleotide tracers called **gene probes.**

Applications of DNA probes

DNA probes are non-amplified methods; they only detect DNA in specimens without amplifying the same (Fig. 9.1). Many DNA probes have been developed and used in clinical microbiology.

They are used to directly detect microbes in clinical specimens (e.g. *Mycobacterium tuberculosis*, *Streptococcus pyogenes*, *Candida albicans*). They are also used for (1) identification of culture isolates, (2) strain identification for epidemiological typing, (3) identification of virulence factors of microbial agents and (4) identification of toxins (e.g. diphtheria toxin).

Some regions of human DNA exhibit substantial variability in the distribution of restriction sites. This variability is termed **restriction fragment length polymorphism (FLP)**. Oligonucleotide probes that hybridise with RFLP DNA fragments can be used to trace DNA from a small sample to its human donor. Thus, the technique is valuable to forensic science. RFLP applications in medicine include identifying genetic regions closely linked to human genes with dysfunctions coupled to genetic disease. This information is valuable in genetic counselling.

Blotting Techniques

Blotting is a well-established technique in molecular biology to detect specific biomolecules (DNA, RNA or protein) from a complex mixture after separating it in a gel matrix following gel electrophoresis. The separated biomolecules are

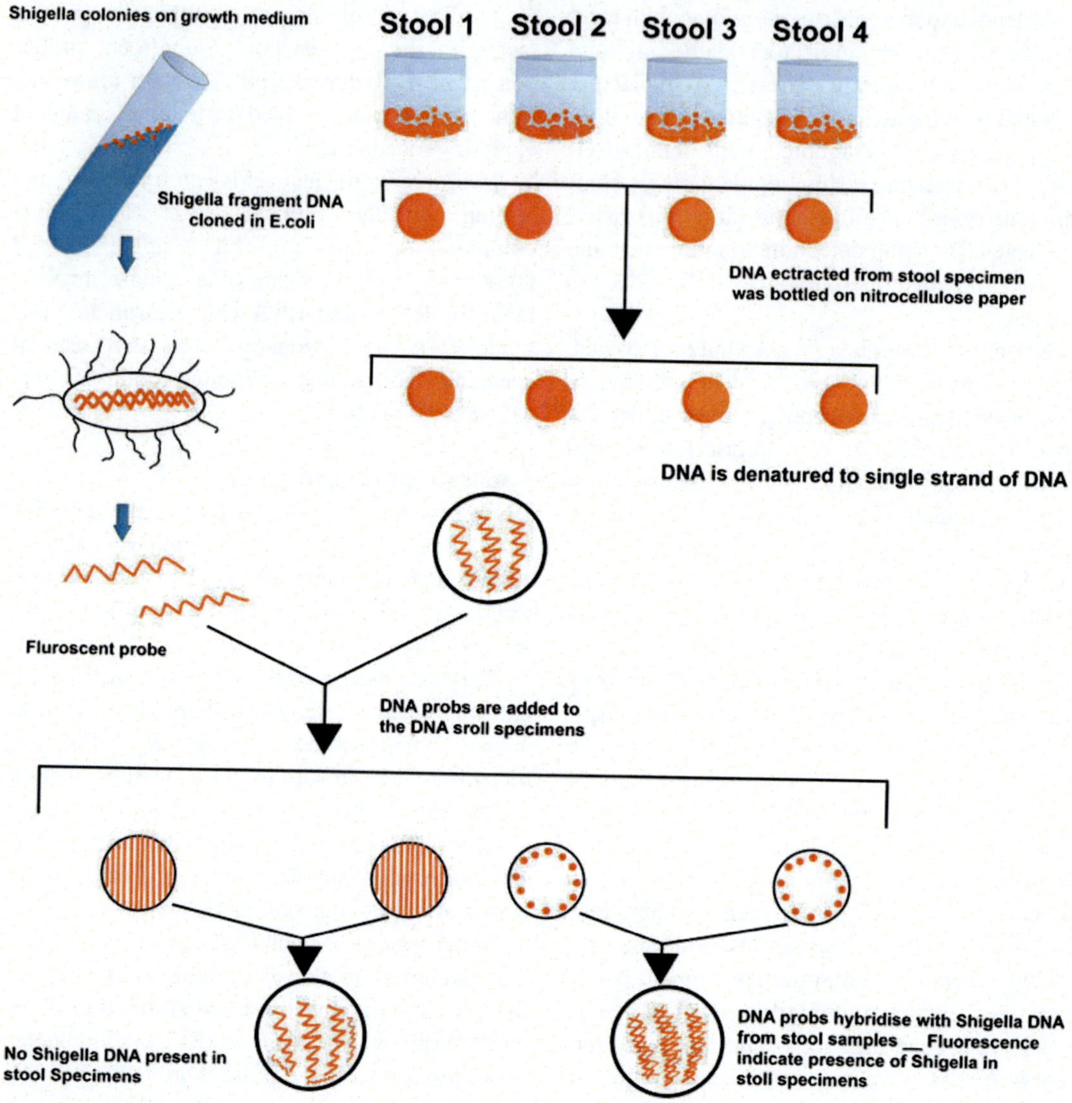

Fig. 9.1 DNA probes

transferred on an inert membrane, the process called blotting, and then a biomolecule is located by using a labelled probe specific to the biomolecule. Initially, the radiolabelling of probes was followed that later mostly shifted to fluorescent and chemiluminescent technologies. As such, blotting is adjunct to gel electrophoresis, and the result is dependent on the resolution ability of the gel electrophoresis.

Depending on the biomolecules to be blotted and identified, blotting methods are classified into the following three types:

Southern Blotting

Developed by Edwin M. Southern in 1975, it refers to detecting specific DNA in a mixture of DNA. DNA mixtures separated in agarose gel electrophoresis are transferred on the nitrocellulose membrane and then detected using a DNA probe. The probe being specific to the interest hybridises with the DNA of interest and produces a signal that could be visualised. The application of Southern blotting includes the identification of one or more restriction fragments that contain a gene or other DNA sequence of interest and in the detection of RFLPs used in the construction of genomic maps.

Northern Blotting

It applies to RNA detection wherein the process remains the same as Southern blotting. It has found potential application in determining expression patterns of genes between tissues, organs, developmental stages, environmental stress levels and pathogen infection.

Western Blotting

It refers to the detection of protein. In this, a mixture of proteins separated in Polyacrylamide Gel Electrophoresis is transferred on a nitrocellulose membrane. The desired protein is detected by using labelled antibodies specific to the protein (Fig. 9.2).

DNA Amplification Techniques

The chain reaction was the first technique which became popular in a short time. Apart from PCR, there are similar techniques which are used in various settings.

Polymerase Chain Reaction (PCR)

In 1983, Kary Mullis developed a new technique that made it possible to synthesise large quantities of a DNA fragment without cloning it. This technique is called polymerase chain reaction (PCR) and has great practical importance and impact on biotechnology. This technique can prepare large quantities of a particular DNA sequence.

In PCR, oligonucleotide sequences identical to those flanking the targeted sequence are first synthesised. These synthetic oligonucleotides are usually about 20 nucleotides long and serve as primers for DNA synthesis. Pieces ranging in size from less than 100 bp to several 1000 bp in length can be amplified, and only 10–100 pmol primer is required. The concentration of target DNA can be as low as 10–15 mL.

The reaction mix for PCR contains (1) the target DNA, (2) a substantial excess of the desired primers, (3) a thermostable DNA polymerase and (4) four deoxyribonucleoside triphosphates. Only DNA polymerases that can function at high temperatures can be employed in the PCR technique. *The two popular enzymes used in the PCR are Taq polymerase* from the thermophilic bacterium *Thermus aquaticus* and the *Vent polymerase* from *Thermococcus litoralis*.

The PCR cycle takes place in three steps as follows (Fig. 9.3):

- **Step 1**: The target DNA containing the sequence to be amplified is heat-denatured to

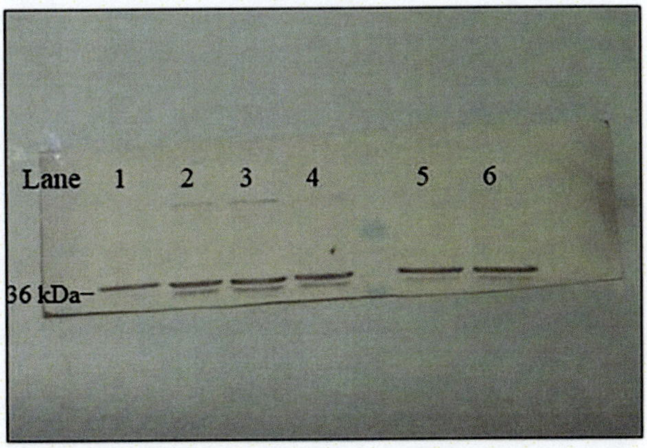

Fig. 9.2 Western blotting

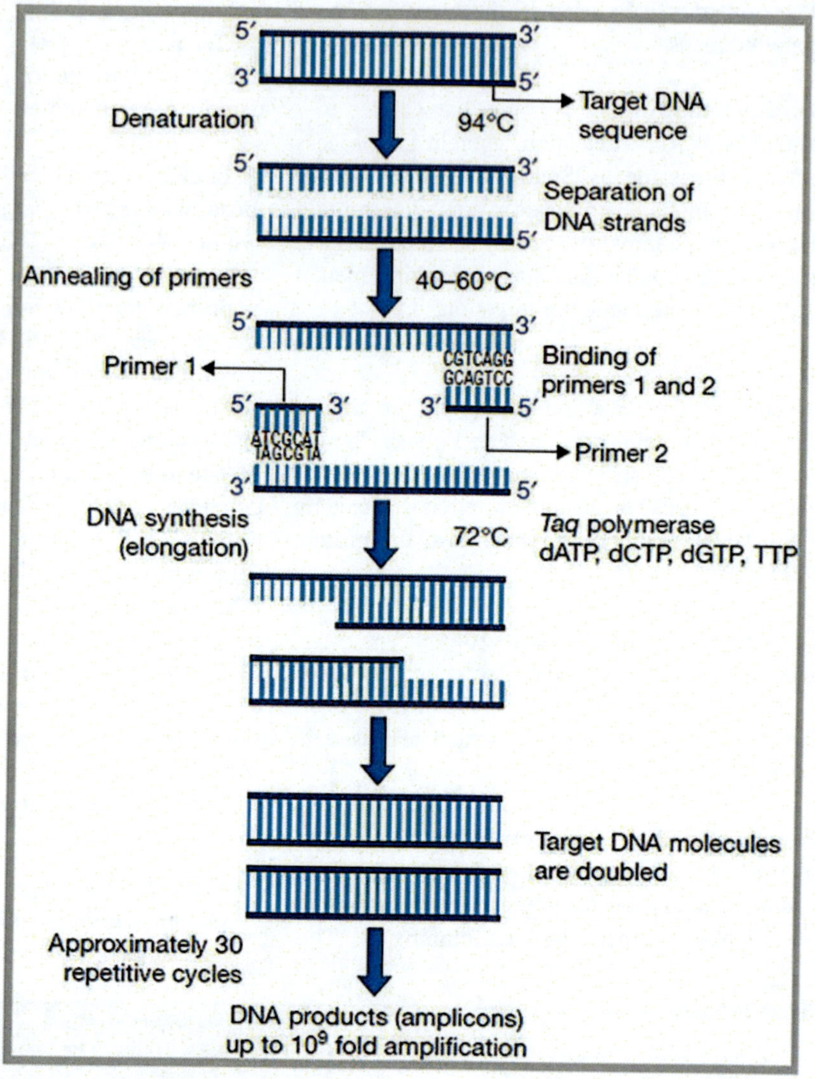

Fig. 9.3 Steps in the PCR cycle

separate its complementary strands. Normally the target DNA is between 100 and 5000 bp in length.

- **Step 2**: The temperature is lowered so that the primers can anneal to the DNA on both sides of the target sequence. Because the primers are present in excess, the targeted DNA strands normally anneal to the primers rather than each other.
- **Step 3**: DNA polymerase extends the primers and synthesises copies of the target DNA sequence using the deoxyribonucleoside triphosphates.

At the end of one cycle, the targeted sequences on both strands are copied. The four strands from the first cycle are copied to produce eight fragments when the three-step cycle is repeated. The third cycle yields 16 products. Theoretically, 20 cycles will produce about one million copies of the target DNA sequence, and 30 cycles yield around one billion copies.

Multiplex PCR: It is another modification of PCR in which two or more target sequences can be demonstrated simultaneously in a single specimen at the same time (Fig. 9.4). This

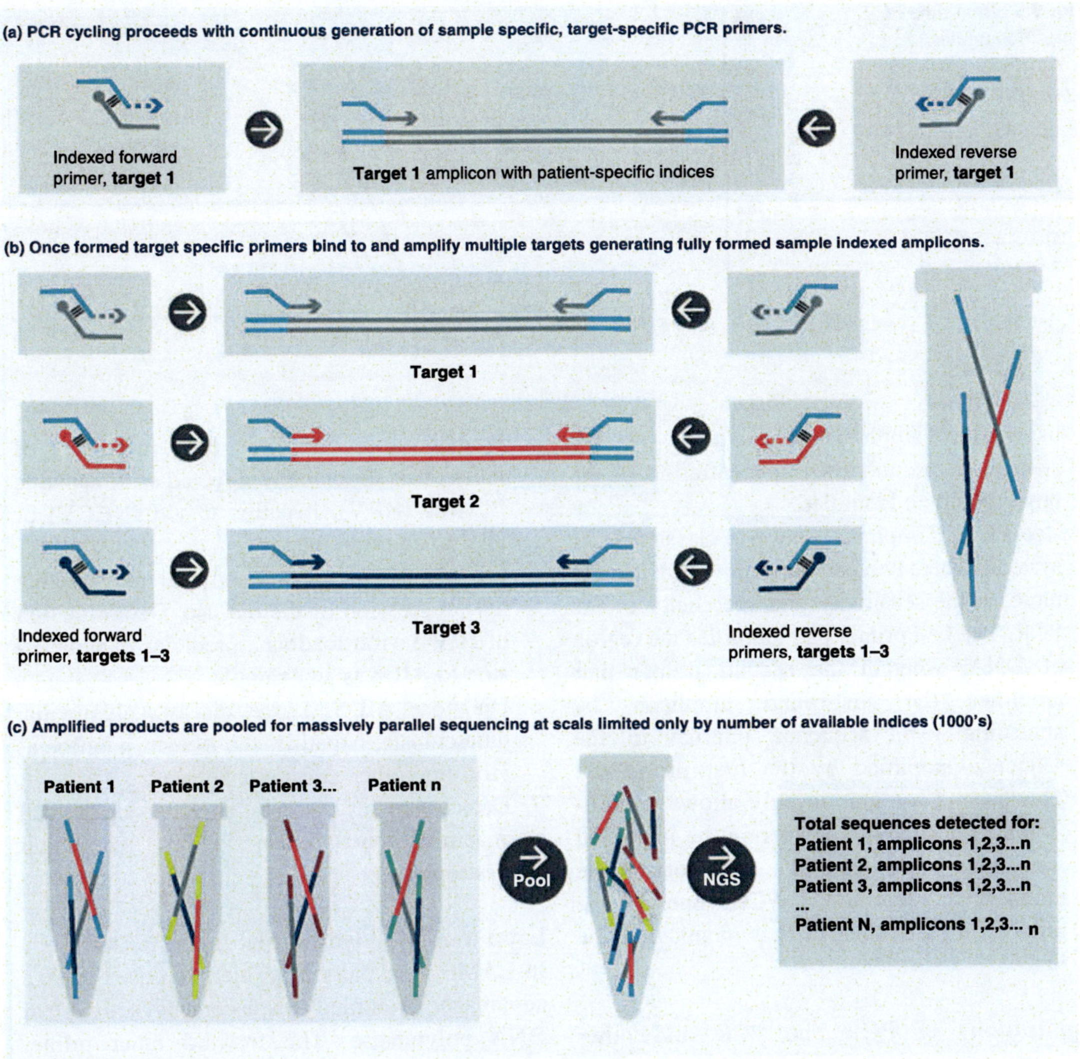

Fig. 9.4 Multiplex PCR (https://commons.wikimedia.org/wiki/File:BASIC_MULTIPLEXD_NGS_RC-PCR.jpg)

method uses two or more primer sets to amplify different targets. Multiplex is now increasingly evaluated for simultaneous demonstration of two or more pathogens in a clinical specimen.

Real-time PCR (RT PCR): It is the most recent development. It is so named because the PCR amplicons can be detected in real time (Fig. 9.5). In fact, "real time" refers to the detection of amplicons after each PCR cycle. Results can be obtained in considerably less time than it takes to perform a conventional PCR assay. Several commercial instruments are available that combine PCR amplification of target DNA with the detection of amplicons in the same closed vessel. Probe detection formats involve detecting fluorophores. Results could be either quantitative or qualitative. For the quantification, called quantitative RT PCR (qRT PCR), standards in different concentrations are also amplified along with the target. From a standard curve developed

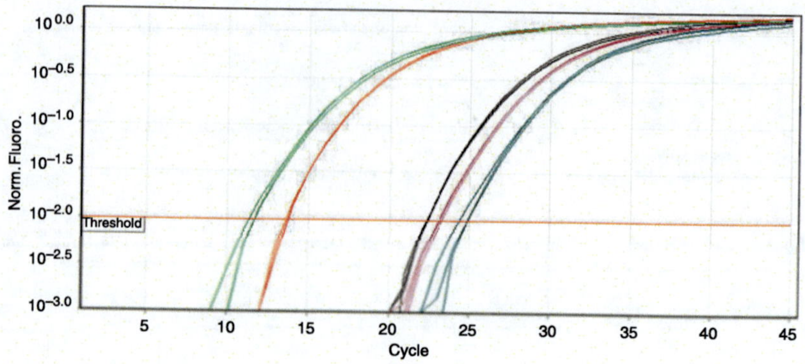

Fig. 9.5 Real-time PCR (https://commons.wikimedia.org/wiki/File:Qpcr-cycling.png)

using the amplification patterns of the standards, the unknown concentration of the target could be identified.

Nested PCR: It is a modification of classical PCR systems where two pairs of primers are used to increase the sensitivity and specificity of the PCR. The first primer pair amplifies the region of DNA, wherein the second primer pair amplifies the first-round amplicon by annealing to a sequence internal to the sequence amplified by the first prime set. This second recognition only allows amplification of desired DNA wherein the increased sensitivity arises from the high total cycle number. In semi-nested PCR, one of the primers in the second PCR is identical to the first.

Applications of PCR The PCR and other molecular techniques have already proven exceptionally valuable in many areas of molecular biology, medicine and biotechnology. These methods are useful for the following reasons:

1. Amplify very small quantities of a specific DNA and provide sufficient material for accurately sequencing the fragment or cloning it by standard technique.
2. Detect previously unrecognised or uncultivable microorganisms.
3. Detect the genes responsible for drug resistance. They supplement conventional antimicrobial susceptibility testing to detect methicillin resistance in staphylococci, rifampicin resistance in *M. tuberculosis* and so on.
4. Predict and monitor the response of individuals chronically infected with hepatitis B virus (HBV), hepatitis C virus (HCV) or HIV to antiviral therapy.
5. Provide useful information that may predict the progression of the disease. Determination of HIV-1 viral load as a predictor of progression to AIDS is an example.
6. Diagnose AIDS, Lyme disease, chlamydia, tuberculosis, hepatitis, the human papillomavirus and other infectious agents.
7. Detect genetic diseases such as sickle cell anaemia, phenylketonuria and muscular dystrophy.

Loop-Mediated Isothermal Assay (LAMP)

In LAMP, four pairs of primers are used to take advantage of strand displacement activity of DNA polymerase. The forward inner primer hybridise with a target to initiate LAMP. The DNA synthesis is initiated upon the annealing of the outer primer to its complementary region. Its extension causes strand displacement that releases a ssDNA molecule. This molecule serves as template for second inner and outer primers that produces a stem-loop DNA structure. Next, LAMP enters in the cycling phase in which one inner primer hybridises to the stem-loop and initiates displacement DNA synthesis. This yields the original stem–loop molecule and a new stem–loop DNA with a stem twice as long (Fig. 9.6)

Compared to PCR, LAMP is 10- to 100-fold more sensitive. It also does not require an initial template denaturation step at 95 °C and allows results to be visualised by the naked eye

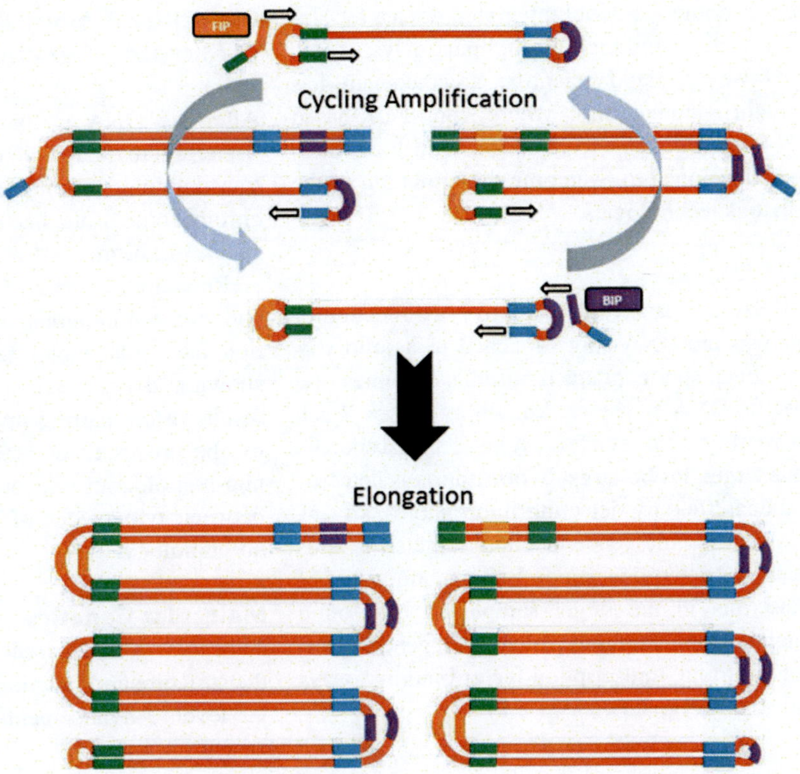

Fig. 9.6 LAMP ((Cycling amplification and elongation steps of reverse transcription loop-mediated isothermal amplification method)-https://commons.wikimedia.org/wiki/File:Cycling_amplification_and_elongation_steps_of_RT-LAMP_method.png)

(e.g. turbidimetry). This makes it ideal for PoC applications.

Ligase Chain Reaction

It is a type of probe amplification method where two oligonucleotide probes labelled with capture haptens are used. They hybridise side by side on each of the denatured target DNA with the formation of a nick. The DNA ligase molecule is used to seal the nick and such ligated molecule is allowed to amplify. The amplification product is detected by microparticle enzyme immunoassay.

Transcription-Mediated Amplification (TMA) and Nucleic Acid Sequence-Based Amplification (NASBA)

Like LAMP, these are also isothermal processes where there is no need for a thermal cycler. In principle, both are very similar and are essentially RNA amplification techniques. The RNA is converted to cDNA by a process of reverse transcription and then the RNA copies are synthesised by using RNA polymerase. In TMA, reverse transcriptase with inherent RNAse H property is used while in NASBA Avian Myeloblastosis Virus Reverse Transcriptase (AMV RT) is used. The final products are detected by hybridisation.

Strand Displacement Amplification (SDA)

It is a technically easy-to-use isothermal technique which can detect minute amounts of nucleic acids. Real-time detection of amplified products is possible using fluorescent probes, similar to real-time PCR.

Non-amplification DNA Techniques

In molecular diagnosis, the first-generation technologies were completely dependent on nucleic acid amplification technologies (NAATs) using PCRs. Despite PCRs being highly sensitive, they require large, sophisticated equipment to maintain different working temperatures, which has largely prevented them from becoming a novel integrated device for single workflow and high-throughput analysis.

However, rapid scientific advancement has brought innovations where non-PCR-based DNA amplification technologies and tools have been developed to overcome the limitation of the previous technologies.

DNA Microarray

A DNA microarray (DNA chip) is a collection of microscopic DNA spots attached to a solid surface. Each spot contains thousands of copies of a specific DNA sequence, known as probes. Each microarray includes a few probe sets for a specific gene target. Probe-target hybridisation is detected and quantified by detecting fluorophore-, silver-, or chemiluminescence-labelled targets. It thus helps to determine the relative abundance of transcripts in the target sample. It has found potential applications in genotyping, gene expression profiling, transcription factor binding assays and so on (Fig. 9.7).

Molecular Biology Applications in Microbial Genetics

After the DNA is transferred from one donor bacterium to the recipient through transformation, transduction or conjugation, it combines with the chromosome of the bacterium by a process called **recombination**.

Recombination is of two types: homologous and non-homologous. Homologous recombination takes place between two pieces of DNA, showing extensive homologous reunions. This results in the pairing up and exchange of pieces by the processes of breakage and reunion. The non-homologous recombination takes place between two pieces of DNA, showing little or no homology.

Molecular Genetics

Molecular genetics basically deals with studying the structure and function of genes at the molecular level. Advancements and methodologies in

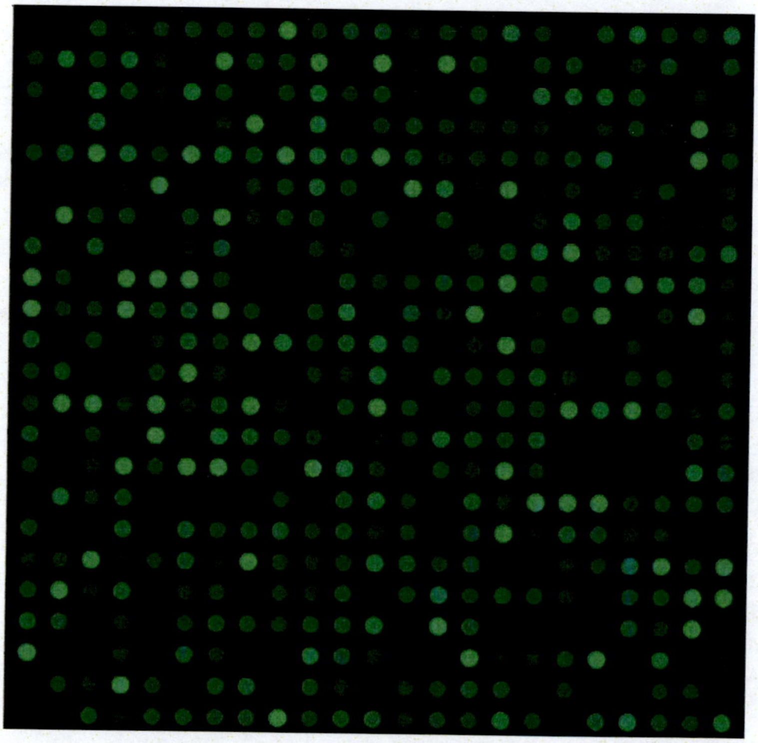

Fig. 9.7 DNA microarray (https://commons.wikimedia.org/wiki/File:Microarray_or_sequencing_chip_(green).svg)

molecular genetics in analysing and manipulating DNA can help understand heredity, genetic diseases and genetic variation. In this section, a few techniques and their applications are discussed later.

Genetic Engineering

Genetic engineering is the process of modifying the genome of a particular organism. In recent years, with the help of engineering, bacterial genetics has modified biology. At a sensible cost, desired DNA can be detected and amplified, and their genes can be expressed at desired content in other expression systems at a reasonable cost. It directly impacts society, especially agriculture and the healthcare system.

Restriction Endonucleases

Restriction endonucleases are enzymes produced by bacteria and are used to split large fragments of DNA into smaller pieces which can be visualised under UV light when run on an agarose gel. Many restriction enzymes are available in nature, and they specifically cleave double-stranded DNA at a particular site. In nature, bacteria use these restriction endonucleases to destroy foreign DNA, which may enter the bacterial cytoplasm. Restriction enzymes are generally used in recombinant DNA technology and molecular epidemiological studies. *EcoRI*, *HindIII* and so on are examples of frequently used restriction endonucleases.

Hybridisation Technique

Hybridisation probes are used routinely in the cloning of DNA. The amino acid sequence of a protein is used to deduce the DNA sequence from which a probe may be constructed and employed to detect a bacterial colony containing the cloned gene. cDNA, encoded by mRNA, is used to detect the gene that encodes the mRNA.

Hybridisation is of different types. The hybridisation of DNA to RNA is known as **Northern blot**, which provides quantitative information about RNA synthesis. The hybridisation of DNA to DNA is known as the **Southern blot**. This method is useful to detect specific DNA sequences in restriction fragments separated on gels. These blots can be used to detect overlapping restriction fragments. **Western blot** is a technique used to detect genes, in which antibodies are used to detect cloned genes by binding to their protein products.

Cloning these fragments makes it possible to isolate flanking regions of DNA by a technique known as **chromosomal walking**.

DNA Sequencing

DNA sequencing shows gene structure that helps research workers to find out the structure of gene products.

DNA sequencing has many applications. DNA sequence analysis demonstrates regulatory regions that control gene expression and genetic "hot spots" particularly susceptible to mutation. Information obtained by DNA sequencing makes it possible to understand or alter the function of genes. A comparison of DNA sequences shows evolutionary relationships that provide a framework for the definite classification of microorganisms, including viruses. In addition, the comparison of DNA sequences facilitates the identification of conserved regions, which are useful for developing specific hybridisation probes to detect microorganisms, including viruses in clinical samples.

Maxam-Gilbert technique and **Sanger (dideoxy termination) methods** are used routinely for DNA sequence determination. Maxam-Gilbert technique depends on the relative chemical liability of different nucleotide bonds, whereas the Sanger method interrupts the elongation of DNA sequences by incorporating dideoxynucleotides into the sequences.

Genetic Mapping

Genetic mapping is mapping the entire chromosome to develop methods to identify gene locations and new genes and understand their function. In 1913, Alfred Sturtevant created the first genetic chromosome map from the fruit fly (*Drosophila melanogaster*). It has been used to find genes responsible for inherited disorders such as cystic fibrosis and Duchenne muscular dystrophy. It also helps find many genes that are believed to play a role in developing more common disorders such as asthma, heart disease, diabetes, cancer and psychiatric conditions.

Recombinant DNA Technology

In recombinant DNA technology, a particular phenotype for DNA bounded is recognised and combined. Once purified, the gene or genes are combined with other pieces of DNA to recombinant DNA molecules. These are inseminated (gene cloning) by infusion into an organism that is not necessarily the initial gene donor or in the same division.

Cloning Vectors and Hosts

A good recombinant vector has two indispensable qualities: it must be capable of carrying a significant piece of the donor DNA and the cloning host must readily accept it.

Cloning Vectors

Cloned vectors include plasmids, bacteriophages and hybrid vectors.

1. **Plasmids** are excellent vectors because they are small, well characterised, easy to manipulate and can be transferred into appropriate host cells through transformation. *Escherichia coli* plasmid carries genetic markers for resistance to antibiotics, although it is restricted by the relatively small amount of foreign DNA it can accept.
2. **Bacteriophages** are also good vectors because they can naturally inject DNA into bacterial hosts through transduction. The *Charon*2 phage is a modified phage vector that lacks a large part of its genome; hence, it can carry a relatively large segment of foreign DNA.
3. **Hybrid vectors** have been developed by splicing two different vectors together. A *cosmid* is an example of a hybrid vector that combines a plasmid and a phage and can carry relatively large genomic sequences. A hybrid *E. coli*-yeast vector can be inserted in bacterial and yeast cloning hosts.

Cloning Hosts

E. coli is the traditional cloning host still used in most experiments. This is because this bacterium was the initial recombinant host used and the protocols using these are well approved, comparatively easy and authenticated. There are hundreds of functional cloning vectors which have been refined for the organism. The main drawback with *E. coli* is its lack of flexibility incorrectly expressing eukaryotic genes.

The yeast *Saccharomyces cerevisiae* is another alternative host used for research and certain industrial processes. This host, being eukaryotic, acquired mechanisms for processing and adapting eukaryotic gene products. Many known approaches may also assign different bacteria (*Bacillus subtilis*), plant and animal cell cultures and even live animals to be attributed as cloning hosts.

Biological Products of Recombinant DNA Technology

Pharmaceutical companies use recombinant DNA technology to manufacture medicines that other means cannot produce. Diseases caused by a lack of essential hormones (e.g. diabetes and dwarfism) are now being treated by replacing the genes of the missing hormone. Porcine and bovine insulin were once the only forms available to treat diabetes, even though such animal products were used to cause allergic reactions in certain sensitive individuals. In contrast, dwarfism that cannot be treated with animal growth hormones was treated only with human growth hormone (HGH) obtained from the pituitaries of cadavers. At one time, not enough HGH was available to treat thousands of children in need. However, now the scenario is changed by the advent of recombinant HGH. Recombinant technology has changed the outcome of these and many other conditions by enabling the large-scale manufacture of lifesaving hormones and enzymes of human origin.

Genetically Modified Organisms

The process of artificially introducing foreign genes into organisms is termed **transfection**, and the recombinant organisms produced in this way are called **transgenic** or genetically modified organisms.

Gene Therapy

Gene therapy is a technique for replacing a faulty gene with a normal one in individuals with fatal or extremely debilitating genetic diseases. The inherent benefit of this therapy is to permanently cure the physiological dysfunction by repairing the genetic defect.

The first gene therapy experiment in humans was started in 1990 by researchers at the National Institutes of Health USA. A 4-year-old girl was the subject who was suffering from a severe immunodeficiency disease produced by the lack of enzyme adenosine deaminase (ADA). She was given her own blood cells known to contain a functional ADA gene that had been modified. Later, other children suffering were administered a similar type of therapy. So far, the children have shown exceptional advancement and continue to be healthy, but the treatment is not permanent and must be repeated. Proper scientific controlled trials are required before its induction as a routine clinical practice.

Part II

Immunology

10 Immunity

The term immunity is derived from the Latin word "immunis" (exempt), which originally referred to the protection from legal prosecution offered to the Roman senators during their tenures in office. Subsequently, this term was adopted to designate the naturally acquired protection against diseases, such as measles or smallpox. It indicated that an individual could develop life-long resistance to a certain disease after having contracted it only once. The cells and molecules responsible for immunity constitute the **immune system**, and their collective and coordinated response to foreign substances is called the **immune response.**

The concept of immunity has existed since ancient times. An example is a Chinese custom of making children resistant to smallpox by making them inhale powders made from the skin lesions of patients recovering from the disease. Thucydides records the first European mention of immunity in Athens during the fifth century BC. In describing the plague in Athens, he wrote in 430 BC that only those who had recovered from it could nurse the sick because they would not contract the disease a second time.

Von Behring and Shibasaburo Kitasato's experiments in 1890 gave an idea about the mechanism of immunity. This earned Von Behring a Nobel Prize in medicine in 1901. Kitasato showed that serum from animals immunised to diphtheria could transfer the immune state to non-immunised animals. Since then, immunology has come a long way as a field of study. It has been and remains one of the hottest research fields, as shown by the statistic that about 17 Nobel Prizes have been awarded to scientists involved in immunological research.

The immune system's primary function is to prevent or limit infections by pathogenic microorganisms, such as bacteria, viruses, parasites and fungi. Recognising microorganisms and foreign substances is the first event in the host's immune responses. The body's defence mechanisms can be broadly divided into **innate (native)** and **acquired (adaptive) immunity**.

Innate or Native Immunity

Innate or native immunity is the immunity present by birth in an individual. It is due to an individual's genetic and constitutional make-up; hence, prior contact with microorganisms or their products is not essential. It acts as the first line of defence of the host immune system. The mechanisms involved in innate immunity are present in place even before exposure to the foreign agent. They are not specific to any infectious agent and do not seem to improve response on repeated exposures. Phagocytic cells (e.g. macrophages and neutrophils), barriers (e.g. skin and mucous membrane) and various antimicrobial compounds synthesised by the host play important roles in innate immunity. Innate immunity may be classified as individual immunity, racial immunity and species immunity.

Individual Immunity

It denotes resistance to infection, which varies among individuals of the same race and species and is genetically determined. For example, if one homozygous twin develops tuberculosis, there is very likely that the other twin will also develop tuberculosis. However, there is a very low possibility of the other twin suffering from tuberculosis in heterozygous twins.

Racial Immunity

It denotes a difference in susceptibility or resistance to infection among different races within the same species. For example, races with sickle cell anaemia prevalent on the Mediterranean coast are immune to infection caused by the malaria parasite *Plasmodium falciparum*. This is due to a genetic abnormality of erythrocytes, resulting in sickle-shaped erythrocytes that prevent parasitisation by *P. falciparum*. Similarly, individuals with a hereditary deficiency of glucose-6-phosphatase dehydrogenase are also less susceptible to infection by *P. falciparum*.

Species Immunity

It denotes a total or relative resistance to a pathogen shown by all members of a particular species. For example, chickens are resistant to *Bacillus anthracis*, rats are resistant to *Corynebacterium diphtheriae* and humans are susceptible to these bacteria. The exact reason for such a type of immunity is not known.

Factors Influencing Innate Immunity

The factors that may influence the host's innate immunity include the age and nutritional status of the host.

Age

Extremes of age make an individual highly susceptible to various infections. The immature immune system partly explains this in very young children and waning immunity in older individuals. The foetus-in-utero is usually protected from maternal infections by the placental barrier. However, human immunodeficiency virus (HIV), rubella virus, cytomegalovirus and *Toxoplasma gondii* cross the placental barrier and cause congenital infections.

Very old people are susceptible to suffering more than young people from a disease (e.g. pneumonia) and have high mortality. Measles, mumps, poliomyelitis and chickenpox are a few examples of the diseases that cause more severe clinical illness in adults than in young children.

Nutrition

Nutritional status of the host plays a vital role in innate immunity. Both humoral and cell-mediated immunities are lowered in malnutrition. For example, neutrophil activity is reduced, interferon response is decreased and C3 and factor B of the complement are decreased in protein-calorie malnutrition. The deficiency of vitamin A, vitamin C and folic acid makes an individual highly susceptible to infection by many microbial pathogens.

Hormones

Individuals with certain hormonal disorders become increasingly susceptible to infection. For example, individuals suffering from diabetes mellitus, hypothyroidism and adrenal dysfunction are increasingly susceptible to staphylococcal infection, streptococcal infection, candidiasis, aspergillosis, zygomycosis and many other microbial infections. Similarly, pregnant women are more susceptible to many infections due to higher levels of steroids during pregnancy.

Mechanisms of Innate Immunity

Innate immunity of the host performs the two most important functions. It kills invading microbes and activates acquired (adaptive) immune processes. Innate immunity, unlike adaptive immunity, does not have any memory and does not improve after re-exposure to the same microorganism. The innate immunity acts by the following mechanisms:

Epithelial Surfaces

They act as mechanical barriers and prevent the entry of microorganisms into the body.

Skin

The intact skin prevents the entry of microorganisms. When there are breaks in the skin due to wounds, scratches or abrasion, microorganisms can easily enter and cause infection. That is why pathogens can enter the body when insects with pathogenic organisms bite the skin surface (e.g. mosquitoes, mites, ticks, fleas and sandflies). Also, glands in the skin secrete sebum, which consists of lactic acid and fatty acids that maintain the skin pH between 3 and 5. This inhibits the growth of most microorganisms.

Mucous Membrane

The mucous membrane forms a large part of the outer covering of the human host's gastrointestinal, respiratory, genitourinary and many other tracts. Several non-specific defence mechanisms act to prevent the entry of microorganisms.

Epithelial cells of mucous membranes secrete mucus that entraps invading microorganisms. Saliva, tears and other body secretions wash away invading microorganisms, preventing their attachment to the initial site of infections. These secretions also contain antibacterial or antiviral substances that can kill invading pathogens. In the lower respiratory tract, the mucous membrane is covered by cilia, the hair-like protrusions of the epithelial cell membranes. The synchronous movement of cilia moves mucus-entrapped microorganisms from the lower respiratory tract towards the nose. Gastric acidity also destroys microorganisms since few ingested microorganisms can survive the low pH of stomach contents.

The **vaginal microbiota** is predominantly made up of lactobacilli. They not only block the entry of pathogens, but the lactic acid produced produces an acidic environment with a pH 3.0–4.5. These also contribute to innate defence mechanisms by inhibiting or killing invading pathogens. In contrast to the female reproductive tract, the male reproductive tract is sterile. The flushing action of urine can remove many microbes which might have gained access to the urethra. In addition, the seminal plasma has got many antibacterial substances, among which semenogelin plays an important role. Major bactericidal activity is shown by peptides derived by proteolysis of seminogelin I and II, which is dependent on zinc ions present in the seminal fluid in high concentration.

Several enzymes are of importance in innate defence. Lysozyme is most important, found in saliva, tears, respiratory tract and reproductive tracts, which can disrupt the peptidoglycan layer of bacteria, causing its death.

Microbial Antagonisms

Non-pathogenic organisms tend to colonise the epithelial cells of mucosal surfaces. These normal florae generally compete with pathogens for attachment sites on the epithelial cell surface and necessary nutrients. For example, in antibiotic-associated diarrhoea, extraneous organisms such as *Staphylococcus* may replace normal gut flora and cause enterocolitis. Similarly, germ-free animal models have been found to be highly susceptible to multiple infections. Hence, normal flora is an important constituent of innate immunity.

Antibacterial Constituents of Blood and Tissues

The **complement system** plays a vital role in destroying invading organisms from blood and tissues. In complementary to this, several antibacterial substances have been described to play a critical role in innate immunity. Anthrax groups of bacilli are sensitive to a relatively thermostable substance called beta-lysine. Lukins and plakins are the basic polypeptides extracted from leukocytes and platelets, respectively, found to have an antibacterial effect. Acidic substances, such as lactic acid produced in muscle and inflammatory sites, are lethal to invading microbes. Interferons are naturally secreted signalling proteins of host cells in response to pathogen invasion and are commonly used in acute viral infection therapy. Lactoferrin, calprotectin and psoriasin are proteins found in mucosal secretions at multiple body sites. These can cause sequestration of metal ions such as iron, manganese or zinc and deprive the bacteria or the fungi of these metals, causing inhibition of their growth. In addition, they can also disrupt the cell membrane. The antimicrobial peptides (AMPs), made up of less than 100 amino acids, are a group of defence molecules present in many living beings, including the skin and epithelium of humans. The important AMPs are α- and β-defensins and cathelicidin, which can kill many pathogens by causing membrane damage of bacteria, fungi, parasites and even viruses. They can also enter the microbial body and cause inhibition of DNA or RNA synthesis.

Cellular Factors

Although the skin and mucosa defence system provides a formidable barrier to invading pathogens, these microbes can cross the epithelial barrier and gain entry into tissue spaces in many instances. In such circumstances, a non-specific immune response is mediated by various cells that can detect foreign organisms.

The phagocytic cells make up the **second defence barrier**, and such cells include monocyte/macrophages, neutrophils and dendritic cells. Many tissues harbour resident macrophages, such as the Kupffer cells in the liver or the Langerhans cells in the skin, which can take care of pathogens. Lymphocytes associated with certain mucosal surfaces like the intestine (**Mucosa Associated Lymphoid System or MALT**) provide both innate and adaptive immune functions at these sites. **Natural Killer (NK)** cells are particularly important in destroying intracellular pathogens like viruses and malignant cells by causing the death of the host cell, harbouring the pathogen and activating adaptive immune response by cytokine signal. Several new cell subsets have been described in recent years, which play a dual role in innate and adaptive immunity. These include γ-δ T Cells, NK T cells, Innate Lymphoid cells (ILCs) and alternatively activated macrophages (AAM).

Phagocytosis

It is another important defence mechanism. It is a process of ingestion of extracellular particulate material by cells, such as neutrophils, blood monocytes and tissue macrophages. It is a type of endocytosis in which invading microorganisms are ingested by the phagocytic cells. During phagocytosis, the cell's plasma membrane expands around the particulate material, which may include whole pathogenic microorganisms. The engulfment of the matter ultimately forms large vesicles called phagosomes.

Interestingly, certain microbes such as *Mycobacterium* and *Brucella* have been known to possess immune evasion mechanisms by multiplying inside phagosomes. In such cases, phagocytosis eventually helps disseminate and progress the disease. However, in agranulocytosis (depleted phagocytic cells) and chronic granulomatous diseases (presence of functionally deficient phagocytes), patients are more prone to infection, which indicates the importance of phagocytic cells in innate immunity. The natural killer (NK) cells, a subset of lymphocytes, also play an important role in non-specific immunity. Interferons activate NK cells and specifically kill virally infected cells and tumour cells.

Toll-Like Receptors (TLRs)

These are proteins present on the host immune cells, such as dendritic cells and macrophages, with unique pattern recognition capabilities. They recognise unique, conserved proteins of microbes and activate immune systems. Thirteen different TLRs can recognise different pathogens and enhance innate immune mechanisms by phagocytosis and inflammation.

Fever

Fever results in increased host metabolism and, in some cases, the destruction of the invading pathogen. For example, fever stimulates interferon production in viral infections and, in turn, kills the virus and causes recovery from infection. Before the penicillin discovery, therapeutic induction of fever was one of the treatment modalities used to kill *Treponema pallidum* in syphilis patients.

Inflammation

Tissue damage caused by a wound or an invading pathogenic microorganism induces a complex sequence of events, collectively known as the inflammatory responses. The end result of inflammation may be the activation of a specific immune response to the invasion or clearance of the invader by components of the innate immune system. The four cardinal features of inflammatory responses are *rubor* (redness), *calor* (temperature rise), *dolor* (pain) and *tumor* (swelling). Histamine, kinins, acute-phase proteins and defensin are the important mediators of inflammatory reactions.

Histamine

It is a chemical substance produced by a variety of cells in response to tissue injury. It is one of the principal mediators of the inflammatory response. It binds to receptors on nearby capillaries and venules, causing vasodilatation and increased permeability.

Kinins

These are other important mediators of inflammatory response. They are normally present in blood plasma in an inactive form. Tissue injury activates these small peptides, which then cause vasodilatation and increased permeability of capillaries. Bradykinin also stimulates pain receptors in the skin. This effect probably serves a protective role because pain typically causes an individual to protect the injured area.

Acute-phase Reactant Proteins

These include C-reactive proteins and mannose-binding proteins that form part of the innate immunity. These proteins are produced at an increased concentration in plasma during an acute-phase reaction as a non-specific response to microorganisms and other forms of tissue injury. They are synthesised in the liver in response to cytokines called **pro-inflammatory cytokines**, namely, interleukin-1 (IL-1), interleukin-6 (IL-6) and tissue necrosis factor (TNF). They are called pro-inflammatory cytokines because they enhance the inflammatory responses.

Defensins

These are another important component of innate immunity. They are cationic peptides that produce pores in the membrane of the bacteria and thereby kill them. These are present mainly in the lower respiratory tract and gastrointestinal tract. The respiratory tract contains β-defensins, whereas the gastrointestinal tract contains α-defensins. The α-defensins also exhibit antiviral activity. They bind to the CXCR4 receptors and block the entry of the HIV virus into the cell. How these defensins differentiate microbes from some cells is not known.

Table 10.1 summarises host defence barriers of innate immunity.

Table 10.1 Non-specific host defence barriers of innate immunity

Factors involved	Active component	Effector mechanism
Anatomic barriers		
Skin	Epithelial cells joined by tight junctions, salt of sweat glands (long-chain fatty acids and lactic acid)	Skin peeling, organic acids, pH (3–5) dislodge bacterial growth, longitudinal flow of fluid
Mouth	Saliva	Flushing effect
Gastrointestinal tract	Indigenous intestinal flora, gastric secretions, wall, antibacterial substances	Peristalsis, movement of mucus by cilia, acidic pH, flushing, lysozyme; bile salts, natural antibodies, liver filtration
Respiratory tract	Tracheal cilia, coughing reflex	Mucociliary elevator, pulmonary surfactant, forcible removal
Nasopharynx	Nasal hair, mucus	Mechanical barrier, entrap and swallow
Eye	Tears	Flushing, lysozyme
Circulation and lymphoid organs	Phagocytic cells, NK cells	Phagocytosis, intracellular killing and antibody-dependent cytolysis
Breast milk	Antibodies (secretory IgA, IgM, IgG), complement proteins, bifidus factor, fibronectin, interferon, lactoferrin, lysozyme, mucin, oligosaccharides, cellular components	Protection of newborn
Phagocytic/endocytic barriers		
Cell-associated receptors	TLRs	Pattern recognition of pathogen and elimination
Acute phase proteins	CRP mannose binding protein, alpha-1 acid glycoprotein	Inflammation
Semen	Spermine and zinc	Antibacterial activity
Serum	Lactoferrin and transferrin, interferons, TNF-α, lysozyme, fibronectin, complement proteins	Iron binding, antiviral proteins, phagocyte activation, peptidoglycan hydrolysis, opsonisation and phagocytosis, enhance intracellular killing
Physiologic barriers		
Adult vagina	Acidic pH	Retard bacterial growth
Urinary system	Acidic pH and flow	Retard bacterial growth, flushing
Fever	High temperature	Destroys

Acquired (Adaptive) Immunity

Adaptive immunity is also called acquired immunity since the potency of immune response is acquired by experience only. It is the resistance acquired by an individual during life. In other words, it refers to antigen-specific immune response. Differences between innate and acquired immunity are summarised in Table 10.2. Characteristic features of acquired immunity include the following:

1. **Antigenic specificity**: The immune system can recognise and distinguish antigens. This specificity property is initiated and undertaken by antigen-presenting cells (APCs) such as activated T cells, B cells, macrophages, dendritic cells and thymic epithelial cells.
2. **Diversity**: This property enables the immune system to respond to a large variety of antigens. Many genes form the basis of such diversity.
3. **Immunologic memory**: The development of immunological memory is the important feature of acquired immunity in which pathogens are "remembered" through memory B cells and memory T cells. Repeated exposure to the same antigen leads to a secondary response that is more specific, heightened and quick.

Table 10.2 Distinguishing features of innate and acquired immunity

Features	Innate immunity	Acquired immunity
Definition	The resistance to infection that an individual possesses by virtue of genetic and constitutional make-up	The resistance that an individual acquires during life
Cells	Dendritic leucocytes, NK cells, mast cells, granulocytes/ macrophages, basophils etc.	Killer CD8 T cells, helper CD4 cells, B cells, antigen presenting cells
Molecules involved	Cytokines, complement components, interferons, acute phase reactants	Antibodies, cytokines
Types	Non-specific and specific	Active and passive
Activation time	Immediate	Delayed
Specificity	For structures shared by groups of related microbes	For antigens of microbes and non-microbial antigens
Memory	None; repeated exposure brings response like primary response	Yes; secondary response much faster than primary response
Physical and chemical barriers	Skin, mucosal epithelia and antimicrobial chemicals	Lymphocytes in epithelia and antibodies secreted at epithelial surfaces
Blood and tissue antimicrobial substances	Complement; leukins from leukocytes, plakins from platelets, lactic acid found in muscle tissue, lactoperoxidase in milk and interferons (antiviral)	Antibodies

4. **Self/non-self-recognition**: Recognition and distinguishing specific "non-self" (foreign) antigens in the presence of "self" during antigen presentation is another feature. This self-tolerance property prevents injury to the host during responses to foreign antigens. Multiple mechanisms regulate and ensure this self-tolerance property; any defect in these mechanisms may lead to autoimmunity.

Types of Acquired Immunity

Acquired immunity against a microbe may be induced by the host's response to the microbe or by transferring antibodies or lymphocytes specific to the microbes. It is of two types: active immunity and passive immunity.

Active Immunity

The immunity induced by exposure to a foreign antigen is called active immunity. It is also called adaptive immunity, as it represents the adaptive response of an individual after contact with a specific pathogen or other antigens. This contact may be in the form of (1) clinical or sub-clinical infection, (2) immunisation with live or killed infectious agents or their antigens or (3) exposure to microbial products, such as toxins and toxoids. In all these circumstances, the host's immune system is stimulated to elicit an immune response consisting of antibodies and activated helper T (Th) cells and cytotoxic T lymphocytes/cells (CTLs). Active immunity is mediated by humoral immunity and cell-mediated immunity. Different immune system components and functions mediate these two types of immunities in different ways to kill different types of pathogens (Table 10.3).

Active immunity develops after a latent period, during which the host's immunity is geared up to act against the microorganism. Hence, it is slow in onset, especially during this primary response. However, once the active immunity develops, it is long-lasting, which is the primary advantage of active immunity.

Types of Active Immunity

Active immunity is of two types: natural active immunity and artificial active immunity.

1. **Natural active immunity**: It is acquired by natural clinical or sub-clinical infections. Such natural immunity is long-lasting. For example, individuals suffering from smallpox become immune to a second disease attack.

Table 10.3 Differentiating characteristics of cell-mediated and humoral immunity

Characteristics	Cell-mediated immunity	Humoral immunity
Definition	Immunity that identifies and destroys infected cells	Immune response mediated by macromolecules found in the extracellular fluids
Pathogen target	Protects against fungi, viruses and facultative intracellular bacterial pathogens	Protects against extracellular bacterial pathogens and viruses infecting respiratory or intestinal tract and prevents recurrence of viral infections
Type of hypersensitivity	Mediated by delayed (type IV) hypersensitivity	Mediated by immediate (types I, II and III) hypersensitivity
T cell/B cell	Only T-cell-dependent antigens lead to cell-mediated immunity	B cells directly bind soluble antigens resulting in the production of antibodies
Cells involved	Both CD4+ and CD8+ T cells are involved	Only T_H cells are involved
Action	Provides immunological surveillance and immunity against cancer	No major role in immunological surveillance
	Participates in rejection of homografts and graft-versus-host reaction	May be involved in early graft rejection due to preformed antibodies

Similarly, a person with infection by Hepatitis A virus develops immunity, which may be lifelong. Temporary immunity is induced by the majority of wild infections where the individual remains protected for a few months to a few years.

A special kind of naturally acquired protective immunity is **premunition**, which is commonly seen with some parasitic infections in hyperendemic areas. It is also known as **infection-immunity**, where previous infections protect an individual against subsequent attacks by a larger number of parasites. It is seen in malaria-endemic zones, and the reason behind it is not exactly clear. Either a low level of parasitaemia in the host or the presence of antibodies may be responsible for this phenomenon.

2. **Artificial active immunity**: It is induced in individuals by vaccines. There is a wide range of vaccines available against many microbial pathogens. These may be live vaccines, killed vaccines or vaccines containing bacterial products.

Live vaccines are made by attenuation (reduced virulence) of pathogens, which can be administered orally or parenterally. These vaccines initiate infection without harming the host (without disease), but it activates the immune system as natural infection. Thus, immunity acquired by live vaccines lasts for a long time and may require booster doses.

Killed vaccines are less immunogenic than live vaccines and need repeated doses of administration. In general, a minimum of two doses are required to generate immunity. The first dose is called **the primary dose**, and the subsequent doses are known as **booster doses**. The route of administration is usually parenteral, and vaccine potency may be enhanced by combining with an adjuvant such as aluminium phosphate.

Examples of **bacterial vaccines** include live (BCG for tuberculosis), killed (cholera vaccine), sub-unit (typhoid VI antigen) and bacterial products (tetanus toxoid). The bacterial polysaccharide vaccines are made up of capsular polysaccharides (Pneumococcal, Meningococcal and Haemophilus influenza type b). **Viral vaccines** include live (Sabin—oral polio vaccine), killed (Salk—injectable polio vaccine) and sub-unit (hepatitis B vaccine).

Passive Immunity

When immunity is conferred by the transfer of serum or lymphocytes from a specifically immunised individual, it is known as passive immunity. This is a useful method for rapidly conferring resistance without waiting for the development of an active immune response. Passive immunity may be natural or artificial.

Table 10.4 Distinguishing characteristics of active and passive immunity

Characteristics	Active immunity	Passive immunity
Host participation	Actively done by the host's immune system	No active participation
Mode of induction	Through antibodies induced by infection or by immunogens	Antibodies transferred directly
CMI/humoral immunity	Involves both	By ready-made antibodies
Types	*Natural*—clinical or inapparent infection *Artificial*—induced by vaccines	*Natural*—transfer of maternal antibodies through the placenta *Artificial*—injection of immunoglobulins
Action	After lag period	Immediate
Durability and effectiveness	Durable; effective protection	Transient; less effective
Immunological memory	Present	Absent
Booster dose	Increases the protective effect	Less effective due to immune elimination
Negative phase	Present	Absent
Immunodeficient states	Not applicable	Applicable

Natural Passive Immunity

It is observed when IgG is passed from mother to foetus during pregnancy. This forms the basis of preventing neonatal tetanus in neonates by active immunisation of pregnant mothers. It is achieved by administering tetanus toxoid to pregnant mothers during the last trimester of pregnancy. This induces a high level of antibodies in the mother against tetanus toxin, which is subsequently transmitted from mother to foetus through the placenta. The antibodies subsequently protect neonates after birth against the risk of tetanus. Natural passive immunity is also observed by the passage of IgA from mother to newborn during breastfeeding.

Artificial Passive Immunity

It is induced in an individual by administering preformed antibodies, generally in the form of antiserum, raised against an infecting agent. Administration of these antisera makes large amounts of antibodies available in the recipient host to neutralise the action of toxins.

The preformed antibodies against rabies and hepatitis A and B viruses and so on given during the incubation period prevent virus replication and alter the course of infection. The immediate availability of a large number of antibodies is the main advantage of passive immunity. However, the short lifespan of these antibodies and the possibility of hypersensitivity reaction if antibodies prepared in other animal species are given to hypersensitive individuals to these animal globulins (e.g. serum sickness) are the two noted disadvantages of passive immunity. Differences between passive and active immunity are summarised in Table 10.4.

Combined Passive-Active Immunisation

It is carried out by giving both preformed antibodies (antiserum) and a vaccine to provide immediate protection and long-term protection, respectively, against a disease. This approach is followed to prevent certain infectious conditions, namely, tetanus, rabies and hepatitis B.

Local Immunity

The immunity at the site of invasion and multiplication of a pathogen is called local immunity. Secretory IgA antibodies confer local immunity in various body secretions. These antibodies are produced locally by plasma cells present on mucosal surfaces or in secretory glands. Additionally, attenuated live viral vaccines are given orally or intranasally to induce local immunity in gut mucosa and nasal mucosa.

Herd Immunity

Herd immunity refers to an overall *immunity level in a community*. Eradication of an infectious disease from a population depends on developing a high level of herd immunity against the pathogen. Conversely, when herd immunity against disease is very low, indicating a larger number of susceptible people in the community, an epidemic of disease is likely to occur.

Measurement of Immunity

The resistance offered by an individual's immune system to the invading pathogen is called immunity. Since the pathogen alters immunity, it cannot be calculated accurately. In general, an estimate of immunity is calculated by collecting data from many individuals and applying appropriate statistical tools.

The type of immune response in immunity can be calculated qualitatively or quantitatively depending on the pathogen. A simple method of immunity measurement is the demonstration of a specific antibody in an individual to a particular pathogen. However, the demonstration of a single antibody is not always an indicator of immunity because the host can produce several antibodies and cell-mediated responses towards a particular pathogen. The antibody detection may be done by several methods such as ELISA, agglutination, precipitation, western blotting, hemagglutination inhibition, neutralisation and many others. If the knowledge about the **protective antigen** of the pathogen is unclear, then the serological methods employed to measure immunity yield an incomprehensive result. However, in certain organisms like *Corynebacterium*, where pathogenicity is due to a well-defined antigen, the immunity of an individual towards diphtheria can be evaluated accurately by *in vivo* (Schick test) and *in vitro* assays. Similarly, if protection against disease is cell mediated, skin tests for delayed hypersensitivity and *in vitro* cell-mediated immunity tests may be employed to evaluate individual immunity.

11 Structure and Function of Immune System

The lymphoreticular system consists of intricate networking of cells with diverse morphology, which spreads across various organs and tissues of the human body, and its cumulative effect is responsible for the immune response. It is made of lymphoid and reticuloendothelial components. The lymphoid cells consisting of lymphocytes and plasma cells confer specific immunity. Similarly, the reticuloendothelial system consisting of phagocytic cells and plasma cells confers non-specific immunity. The important function of these immune cells is to clear invading foreign antigens or microbes from blood and tissues.

The specific immune response to an antigen is of two types: humoral or antibody-mediated immunity, mediated by antibodies produced by plasma cells, and cell-mediated immunity (CMI), mediated by sensitised lymphocytes. The immune system is classified into various tissues, hence broadly called **lymphoid** or **immune tissues**. These tissues, which developed into an organ of highly specific functions, are described as **lymphoid organs**.

Lymphoid System

The lymphoid system includes lymphoid organs and lymphoid tissues. The lymphoid organs are classified into central (primary) and peripheral (secondary) lymphoid organs based on their function.

Central (Primary) Lymphoid Organs

Central or primary lymphoid organs are the major sites for lymphopoiesis. These organs have the ability to produce progenitor cells of the lymphocytic lineage. These are the organs in which precursor lymphocytes proliferate, develop and differentiate from lymphoid stem cells to become immunologically competent cells (ICCs). The primary lymphoid organs include the thymus and bone marrow. In mammals, T cells mature in the thymus and B cells in the foetal liver and bone marrow. After acquiring immunological competency, the lymphocytes migrate to secondary lymphoid organs to induce an appropriate immune response on exposure to antigens.

Thymus

The thymus is the first lymphoid organ to develop. It is a bi-lobed flat organ located above the heart and surrounded by a capsule. Each lobe is separated into lobules and is connected to each other by trabeculae, connective tissue. The lobules are separated into two sections: cortex and medulla. The stroma of the organ is composed of dendritic cells, epithelial cells and macrophages (Fig. 11.1). The thymus attains its maximal size during puberty and atrophies, with an increase in total fat content of the organ and a decrease in the cortical and medullary cells.

The **cortex** consists mainly of (1) cortical thymocytes, the T lymphocytes, which are

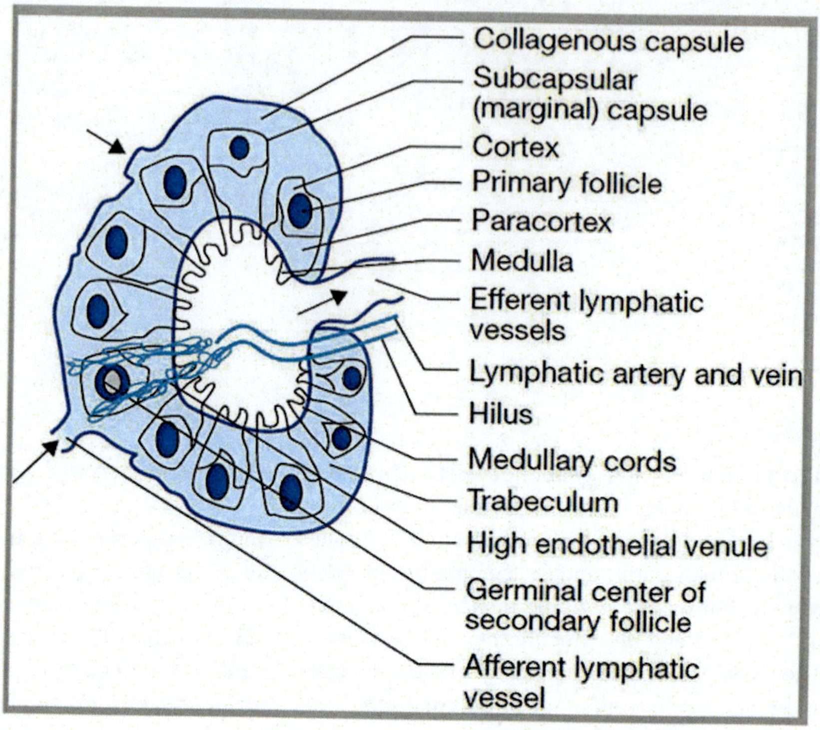

Fig. 11.1 A schematic diagram of the thymus

immunologically immature cells, and (2) a small fraction of plasma cells and macrophages. Additionally, the cortex has two cells, the nurse cells and the cortical cells, which are subpopulations of epithelial cells, forming a network in the cortex. The **medulla** mainly constitutes mature T lymphocytes. The epithelial cell-to-lymphocyte ratio is more in the medulla when compared to the cortex. One of the conspicuous and unique features of the medulla is the presence of concentric rings of squamous epithelial cells known as Hassall corpuscles.

A vast diversity of T cells is produced in the thymus, which distinguishes and acts against a multitude number of major histocompatibility complex (MHC) antigens. The thymus promotes the demise of those T cells that can no longer identify MHC antigen and which react with self-antigen MHC that may evoke autoimmune disease. Almost 95% of the thymocytes die in the thymus by apoptosis, without even developing into mature cells.

Functions

The thymus is the only clearly individualised primary lymphoid organ in mammals. It has many functions. The production of thymic lymphocytes is the primary function of the thymus. It is a major organ for the proliferation of lymphocytes in the body. The thymus is believed to play a crucial role in determining the differentiation of T lymphocytes. New surface antigens acquire through the maturation of lymphocytes which are called T lymphocytes (thymus dependent). During the development of the organ, the thymus holds control of the immunological competence of the cells. Lymphocyte proliferation in the thymus, unlike in the peripheral lymphoid organs, is not dependent on antigenic stimulus.

The T lymphocytes play a major role in CMI. The neonatally thymectomised mice are associated with gross deficiency of CMI, which leads to lymphopenia, deficient graft rejection and running disease. Another classical example is the DiGeorge syndrome, where congenital aplasia has been associated with deficiency of CMI due to the absence of thymus.

Bone Marrow

Some lymphoid cells develop and mature within the bone marrow and are called **B cells** (B for **bursa of Fabricius** or bone marrow). The function of the bursa of Fabricius in birds is played by the bone marrow in humans. Bone marrow is the site for stem cell proliferation, the origin of pre-B cells and their maturation to immunoglobulin-producing lymphocytes. For this, stromal cells combine directly with the B cells to secrete different cytokines essential for the maturation of B cells. A selection process occurs within the bone marrow to remove antibody receptors of self-reacting B cells. This process is similar to the thymic selection that occurs during the maturation of T cells.

Bone marrow dysfunction: The bone marrow is the most important site for the development and maturation of B cells. Nevertheless, in addition to B cells, the bone marrow contains various immune cells, including $CD4^+$ and $CD8^+$ T cells, regulatory T cells, NK-T cells, dendritic cells and myeloid-derived suppressor cells. Any condition that compromises bone marrow function adversely affects the individual's immune response. Bone marrow dysfunction or failure may be congenital or acquired. Fanconi anaemia, Schwachman-Diamond syndrome, aplastic anaemia, sideroblastic anaemia, myelodysplastic syndrome, leukaemias and so on are conditions where the patients become immunocompromised and susceptible to various infections.

Peripheral (Secondary) Lymphoid Organs

Peripheral or secondary lymphoid organs consist of lymph nodes, spleen and non-encapsulated structures, such as mucosa-associated lymphoid tissues (MALT). These organs serve as the sites for the interaction of mature lymphocytes with antigens.

Lymph Nodes

The lymph nodes are extremely numerous and disseminated all over the body. They play a dynamic role in the initial or inductive states of the immune response. Lymph nodes measure 1–25 mm in diameter and are surrounded by a connective tissue capsule. The lymph node has two main parts: cortex and medulla. The reticulum or framework of the lymph node is composed of phagocytes and specialised types of reticular or dendritic cells.

Cortex: The cortex and the deep cortex, also known as the paracortical area, are densely populated by lymphocytes. Roughly spherical areas containing densely packed lymphocytes, termed primary lymphoid follicles or nodules, are found in the cortex. B and T lymphocytes are found in different areas in the cortex.

The primary lymphoid follicles predominately contain B lymphocytes. They also contain macrophages, dendritic cells and some T lymphocytes. The primary follicles are very densely packed with small lymphocytes, not actively involved in an immune response. The larger, less dense follicles, termed secondary follicles, are found in the cortex of a lymph node draining an area where an infection has occurred. The secondary follicles contain clear germinal centres where B lymphocytes actively divide due to antigenic stimulation.

T lymphocytes are found predominantly in the deep cortex or paracortical area; hence, the paracortical area is termed T dependent. Also, interdigitating cells exist in the paracortical area, where the antigen is presented to the T lymphocytes.

Medulla: Medulla is predominately packed with medullary cords and is less densely populated. These cords are elongated branching bands of the lymphocytes, plasma cells and macrophages. They drain into the hilar efferent lymphatic vessels. Plasma cells are also found in the medullary cords.

After cell division, B cells undergo a rigorous selection process where more than 90% of cells die through apoptosis. When the lymph takes an antigen into a regional node, it is trapped and presented by interdigitating dendritic cells present in the paracortex region, along with class II MHC molecules, resulting in the T_H cells activation. The primary activation of the B cells takes place in the paracortex region, rich in T cells. After the activation of Th and B cells, they form small foci, consisting of proliferating B cells mainly near the margins of the paracortex. Some of the B cells in the foci distinguish into plasma cells that secrete IgM and IgG.

Functions

Lymph nodes act as filters for the lymph, the fluid and cellular content of the lymphocytic circulatory system. They also provide sites forming ring-shaped structures composed of lymphocytes, monocytes and dendritic cells for initiating immune responses. In lymphoid cells, plasma cells producing antibodies were generated after the activation of B cells by the antigen. However, only a small fraction of activated B cells find their way to enter the germinal centres. These B cells undergo a few processes of cell division during which the genes mutate at a high level that encodes for antibodies.

Spleen

The spleen is the largest lymphoid organ. It is a large, ovoid secondary lymphoid organ located high in the left abdominal cavity. The spleen parenchyma is heterogeneous and is composed of white and red pulp. It is surrounded by a capsule made of connective tissue. Unlike the lymph nodes, the spleen is not supplied by lymphatic vessels. Instead, blood-borne antigens and lymphocytes are carried into the spleen through the splenic artery. The narrow central arterioles, derived from the splenic artery after multiple branching, are surrounded by lymphoid tissue (periarteriolar lymphatic sheath). T lymphocytes are present in the lymphatic sheath of the white pulp surrounding the arteriole. B lymphocytes are distributed primarily in the germinal centre, perifollicular area and mantle layer, which lies close to the arterioles peripherally.

Functions

The spleen plays a significant role in mounting immune responses to antigens in the bloodstream. The circulating antigens are trapped by the macrophages present in the marginal zone. These macrophages then process the antigen, migrate deeper into the white pulp and initiate the immune response by interacting with T and B lymphocytes.

The spleen also facilitates in filtering or clearing of (1) infectious organisms, (2) aged or defectively formed elements (e.g. spherocytes, ovalocytes) and (3) particulate matter from the peripheral blood. In addition, the spleen traps blood-borne antigens and microbes. The macrophages lining up the splenic cords perform the primary filtering function.

Clinical significance of splenectomy: In children, splenectomy often leads to an increased incidence of bacterial sepsis caused primarily by *Streptococcus pneumoniae*, *Neisseria meningitidis* and *Haemophilus influenzae*. In adults, the adverse effects are less, although, in some adults, it makes the host more susceptible to blood-borne bacterial infections.

Mucosa-Associated Lymphoid Tissues (MALTs)

MALTs consist of the lymphoid tissues of the intestinal tract, genitourinary tract, tracheobronchial tree and mammary glands. The MALT is non-capsulated and contains both T and B lymphocytes, and the latter predominate. Structurally, these tissues include clusters of lymphoid cells in the lamina propria of intestinal villi, tonsils, appendix and Peyer patches.

MALTs play a significant role in the defence system of humans. This is evidenced by the large population of plasma cells producing antibodies in MALT, which exceeds the number of plasma cells in the spleen, bone marrow and lymph nodes combined. In addition to spleen and lymph nodes, MALTs facilitate interaction among circulating leucocytes. Such lymphoid tissues present in the gut from the adenoids and **tonsils** to the follicles in the colon are called the gut-associated lymphoid tissue (GALT) and those in the respiratory tract are called the bronchus-associated lymphoid tissue (BALT). The follicles present in the Peyer patch are highly rich in B cells that differentiate to produce IgA antibodies in plasma cells. M cells, specialised epithelial cells, are found predominantly in the dome epithelia of **Peyer patches**, mainly in the ileum. These cells engulf the bacteria, virus, small particles and so on and transport them to the sub-mucosal macrophages, where these materials are presented to T and B lymphocytes after processing.

Cells of the Lymphoreticular System

The followings are the cells of the lymphoreticular system (Fig. 11.2).

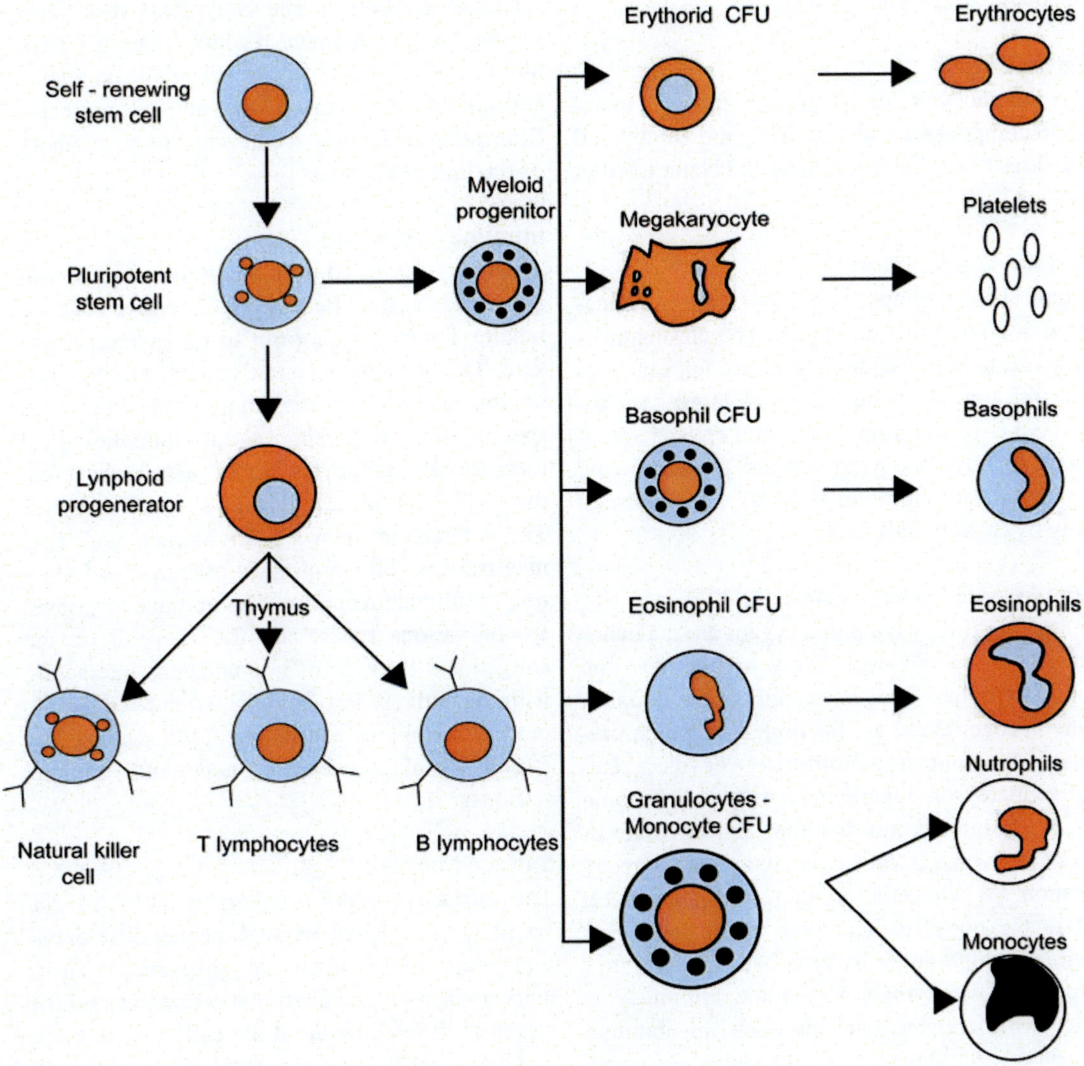

Fig. 11.2 Proliferation and development of cells of lymphoreticular system

Lymphocytes

The lymphocytes occupy a very special place among the leucocytes. They make up 20–40% of total blood leucocytes and 99% of the cells found in the lymph. The lymphocytes consist of heterogeneous populations of cells that differ significantly in terms of origin, lifespan, preferred areas of settlement within the lymphoid organs, surface structure and function. They differentiate from stem cells in the foetal liver, bone marrow and thymus into two main functional classes: B cells and T cells. They are found in the peripheral blood and all lymphoid tissues.

Classification

The lymphocytes, depending on the cell membrane components and functions, are subdivided into three types: T cells, B cells and Natural Killer (NK) cells.

Lymphocyte Structure

A typical small lymphocyte is 6–10 μm in diameter with a large round nucleus. The chromatin is dark and dense, and the cytoplasm appears as a thin rim around the nucleus. The large lymphocyte is 14–16 μm from which null cells arise. Of about 10^{12} lymphocytes that are present in the human body, approximately 10^9 are replaced daily by fresh lymphocytes.

Lymphocytic Recirculation

Central or primary lymphoid organs are the major sites for lymphopoiesis. These organs have the ability to produce progenitor cells of the lymphocytic lineage. These are the organs in which precursor lymphocytes proliferate, develop and differentiate from lymphoid stem cells to become ICCs. These cells mix together and circulate in the body, which is known as “lymphocytic recirculation”. Whenever a foreign particulate antigen enters the body, it is carried to the local and regional lymph nodes by lymphatics, where it is imprisoned. In lymph nodes, the lymphocytes react with antigens and produce an immune response. Lymphocytes are the only leucocytes that continuously migrate from tissues back into the blood using the lymphatic system and create complete blood–tissues–lymph–blood circulation. Generally, the majority of the lymphocytes complete this circulation several times during their lifespan. A cycle of circulation requires approximately 1 or 2 days. These recirculating lymphocytes will be picked by lymphoid tissues whenever necessary. T cells mainly constitute the circulating lymphocytes, and B cells are more sessile in nature.

Functions

The major cellular components of the immune response are both T and B lymphocytes. They participate in immune reactions due to their ability to interact specifically with antigenic substances and react to non-self-antigenic determinants. They also contribute to the memory of the immune system.

Immune Responses

The lymphocytes educate themselves in a central lymphoid organ to become **ICCs**. **Naïve cells** are mature T and B cells prior to the antigen exposure. The naïve cells can be activated upon stimulation with a specific antigen to produce a specific immune response and immunological memory. Lymphocytes carry surface receptors that will help in recognising specific antigens. The interaction between an antigen and ICC may result in immunological tolerance or induction of immune response. The immune responses are of various types, and they depend on the antigen and the type of immune cells interacting with the antigen. For example, stimulated T cells produce cytokines and induce CMI, whereas the B cell stimulation leads to a humoral immune response.

Surface Markers

The surface antigen markers present on the lymphocytes help understand the cell differentiation stage and functional properties. These markers denote the functional properties and the stages of differentiation of the cells.

Owing to the advancement in recombinant DNA technology, it became easy to identify vast numbers of surface antigens present on various lymphocytes through monoclonal antibodies. Hence, depending on the cluster of monoclonal antibodies interacting with a particular antigen, cluster differentiation (CD) numbers were designated. The CD number helps identify a specific subset of lymphocytes, which is important in understanding an individual's immunological status in certain infections such as HIV. Thus, far more than 150 CDs have been identified. Table 11.1 includes a list of a few CDs and their association with different types of immune cells.

Differences Between T and B Lymphocyte Cells

The CD markers can be used as a sole, clear marker to differentiate between T and B lymphocytes; T cell possesses CD3 receptor, whereas B cells have Ig molecule on their surface. Similarly, many other methods are also available.

Rosette formation with sheep RBCs (SRBC or E rosette) can be seen in T cells but not with B cells. Here, CD2 antigen on T lymphocyte interacts with CD2 receptors present on the RBC surface to form a rosette. The B cell surface has a C3 (CR2) receptor, which binds to sheep RBCs coated with antibodies and complement (EAC rosettes). T cells lack this receptor, but Epstein–Barr virus possesses a CR2 receptor. Mature B cells have immunoglobulin molecules on their surface, which will help interact with specific antigens, but T cells do not have Ig on their surface; instead, they have T-cell receptors (TCR). The B cells have distinct, separate Ig surface molecules for a particular antigen. Approximately 10^5 Ig molecules are present on each B cell. The IgM is the first Ig to appear on B cells, followed by IgG, IgA, IgE or IgD. Thymus-specific antigens are present only in T cells but not in B cells.

Blast formation in T cells can be seen on treatment with mitogens such as concanavalin (Con A), phytohemagglutinin (PHA) and so on. In contrast, blast formation in B cells can be seen after interaction with bacterial endotoxins, Cowan 1 strain of *Staphylococcus aureus* and Epstein–Barr virus. Under electron microscopy, T cells do not show any cytoplasmic projection, but B cells possess an enormous number of microvilli and a widespread filamentous surface.

Differences between T cells, B cells and macrophages are summarised in Table 11.2.

T Cell

T-Cell Maturation

Thymus-derived cells, T lymphocytes or T cells, are so designated because the thymus plays a crucial role in their differentiation. They are the key players in adaptive immunity. They participate directly in immune responses and orchestrate and regulate the activities of other cells. T-cell precursors migrate from the yolk sac, bone marrow and liver to primary lymphoid organs such as the thymus during the early embryonic and post-natal stages of life. In the thymus, progenitor T cells undergo differentiation, and the initial identifiable cell in the T-cell lineage possesses a CD7 molecule on its surface; subsequently, they

Table 11.1 CD markers and cell type association

CD markers	Cell type association
CD1	Thymocytes, Langerhans cells
CD2	T-cell SRBC receptor
CD3	T-cell antigen receptor complex
CD4	Helper T cell (receptor for HIV)
CD8	Suppressor/cytotoxic T cells
CD19	B cells
CD20	B cells
CD25	Activated T cells
CD28	T cells
CD69	Activated T cells

Table 11.2 Differentiating features of T cells, B cells and macrophages

Features	T cell	B cell	Macrophages
Antigen recognition receptors	T-cell receptors	Membrane-bound Ig	Membrane-bound receptors
Surface glycoprotein antigens	CD3	CD19	CD40
IgM on surface	–	+	–
Immunoglobulin synthesis	–	+	–
IL-2, IL-4, IL-5 and gamma interferon synthesis	+	–	–
Receptor for Fc fragment of IgG	–	+	–
EAC rosette (receptors for C3 component of complement)	–	+	–
Sheep red blood cells (SRBC) rosette (E rosette)	+	–	–
Effect of cell-mediated immunity	+	–	+
Thymus-specific antigens	+	–	–
Maturation in the bursa or its equivalent	–	+	–
Antigen receptor recognises processed peptides with MHC	+	–	–
Innumerable microvilli on the surface	–	+	–
Glass surface adherence	–	–	+
Blast transformation with:			
(a) Anti CD3	+	–	–
(b) Anti-Ig	–	+	–
(c) PHA	+	–	–
(d) Concanavalin A	+	–	–
(e) Endotoxin	–	+	–

acquire all the characteristics of mature T cells, including synthesis of the T-cell receptor.

T lymphocytes show the following features: (1) They constitute 65–80% of the small lymphocytes present in the circulation, (2) are usually present in the inner sub-cortical regions but not seen in the germinal centres of the lymph nodes, (3) have a longer lifespan (months to years) than B lymphocytes, (4) are stimulated to divide after exposing to certain mitogens, such as phytohemagglutinin or concanavalin A and (5) most human T cells have receptors for sheep erythrocytes on their surface which help them in rosette formation. This property is used to identify T cells in a mixed population of cells.

T-Cell Receptors (TCR)

These are heterodimer glycoproteins expressed on the T-cell surface. TCR for antigen is made of two polypeptides, namely, alpha and beta. These peptides are associated with CD3 proteins (Fig. 11.3). Single T cell has unique TCR on their surface; hence, an individual possesses hundreds of millions of T cells. Activated T cells and activated B cells produce many cells specific to those antigens.

T-cell alpha and beta polypeptides show many similarities to immunoglobulin heavy chain in the following ways: (1) the genes coding for T-cell polypeptides are formed by rearrangement of multiple regions of DNA. (2) there are V (variable), D (diversity), J (joining) and C (constant) segments. These segments can rearrange in various combinations to yield more than 10^7 diversified receptor proteins. (3) RAG-1 and RAG-2 genes catalyse the gene rearrangement process. These two genes are similar in T and B cells and encode the recombinase enzymes. However, they differ from immunoglobulins by having two chains rather than four chains in immunoglobulins. They also recognise antigens only in conjunction with MHC proteins, whereas immunoglobulins recognise free antigens.

Immature T cells: These cells carrying TCR for self-antigens are destroyed in the thymus during the early course of life. This is the basis of

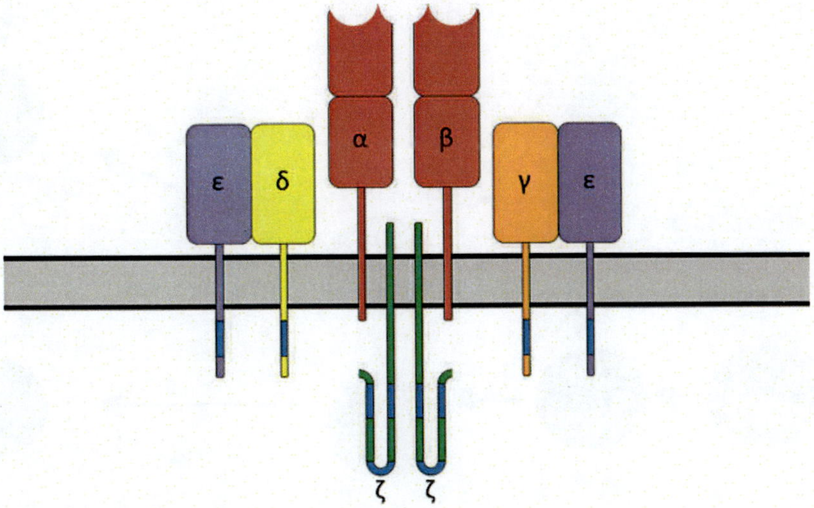

Fig. 11.3 Structure of T-cell receptor (https://en.wikipedia.org/wiki/T_cell_receptor)

self-tolerance, that is, no immune response to autoantigens. However, T cells capable of interacting with self-antigens are produced during their entire life, but these T-cell clones are regularly deleted by the antigen-specific suppressor cells to check on autoimmunity.

Mature T cells: These cells consist of many CD molecules on them (CD 1, 2, 3, 4, 8 etc.) but upon functional maturity, they acquire a specific set of CDs. Based on the presence of CD molecules, T lymphocytes can be divided into two major subsets, namely, **CD 8^- 4^+ (T helper cell)** and **CD 8^+ 4^- (T cytotoxic cells)**. These mature cells then develop MHC restriction; that is, T helper cells respond only when antigens are presented along with human leucocyte antigen (HLA) class II molecules, and T cytotoxic cells respond only when antigens are presented along with HLA class I molecules. Both CD4 and CD8 T cells have TCRαβ on their surface, which facilitates T-cell interaction with processed antigens.

Certain T cells possess TCR γδ; their actual role in the immune response is unclear. However, it is believed that these cells participate in innate immune response, particularly against intracellular bacterial pathogens, epithelial lining defence and also in homeostasis. Overall, cells bearing TCR γδ are considered immune surveillance cells.

During T-cell maturation, surface antigens progressively change, and these changes can be identified by analysing their surface markers. The identification of CD markers has got a diagnostic implication. For example, T-cell malignancies can be classified into acute or chronic based on the presence of premature or mature T cells, respectively. Similarly, in HIV infection, CD4 count plays a significant role in understanding antiviral therapy efficacy.

Types of T Cells

Based on their functions, target cells and surface markers, the T cells can be classified into **effector T cells** and **regulatory T cells** (Fig. 11.4). Depending on whether they have CD4 or CD8 proteins on their surface, T cells are subdivided into two major groups: $CD4^+$ T cells and $CD8^+$ T cells. Mature T cells have either CD4 or CD8 proteins, but never both. The regulation of immune responses is mediated primarily by helper ($CD4^+$) T cells, which produce interleukins. Various effector functions are mediated primarily by cytotoxic ($CD8^+$) T cells, which kill allografts, tumour cells and virus-infected cells.

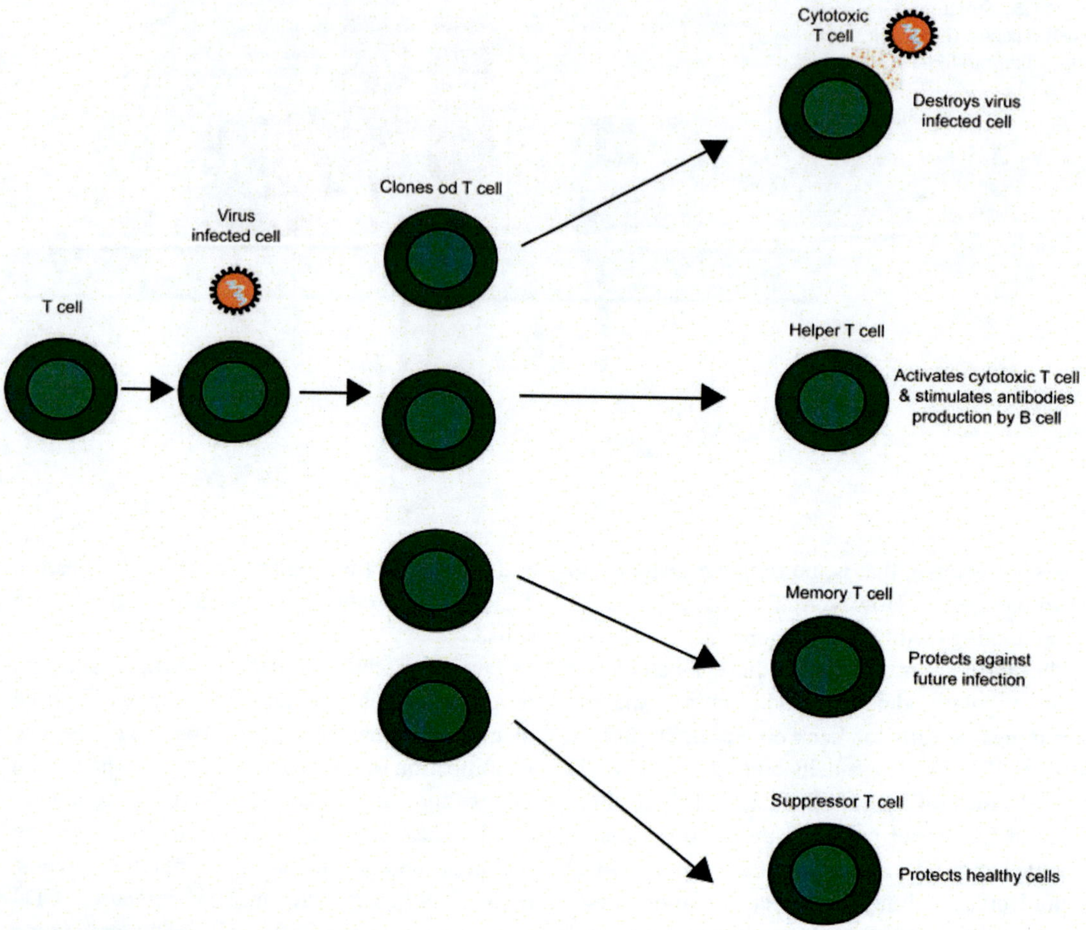

Fig. 11.4 Different T cells and their functions

Effector T Cells

Effector T cells are the most important cells which perform immune response functions. Current knowledge has revealed that they can modulate both innate and adaptive immune responses. When the naïve T cells differentiate into effector T cells, there is a complete change in the cytokine types produced by these cells. The new set of cytokine production is driven mainly by activated cells of the innate and adaptive immune systems.

T Helper or $CD4^+$ T Cells

CD4 cells are also known as helper T (Th) cells. They constitute about 65% of peripheral T cells and are found mainly in the thymic medulla, tonsils and blood. CD4 displayed on the surfaces of these T cells recognises a non-peptide-binding portion of MHC class II molecules. Therefore, $CD4^+$ T cells are only limited to recognising MHC class II complexes. The helper T lymphocytes are involved in inducing and regulating immune responses.

$CD4^+$ T cells perform the following helper functions: (1) The CD4 cell helps transform B cells to plasma cells, (2) helps $CD8^+$ T cells convert into activated cytotoxic T cells and (3) helps mediate delayed-type hypersensitivity reactions through macrophages. The main functions of helper T cells are summarised as follows: (4) help in the antigen-specific activation of B cells and effector T cells, (5) h-1 cytokines activate cytotoxic inflammatory and delayed hypersensitivity reactions, (6) Th-2 cells help in the production of interleukins which encourage the production of antibodies, especially IgE and (7) Th-2 cytokines are associated with regulation of strong antibody and allergic responses.

All these functions are mediated by Th-1 and Th-2 cells—the two subpopulations of CD4$^+$ T cells. The Th-1 cells activate cytotoxic T cells by producing IL-2. They help develop hypersensitivity responses by producing IL-2 and gamma interferon primarily. The Th-2 cells perform B-cell helper functions by producing IL-4 and IL-5 primarily.

The balance between Th-1 and Th-2 cells is regulated by gamma interferon and IL-12. Gamma interferon inhibits the production of Th-2 cells, whereas IL-12 increases the number of Th-1 cells, thereby increasing host defence against microorganisms that are controlled by a delayed hypersensitivity reaction. Table 11.3 shows a comparison of Th-1 and Th-2 cells.

T Cytotoxic or CD8$^+$ T Cells

CD8$^+$ T cells are also known as cytotoxic T (Tc) and suppressor T (Ts) cells. They account for approximately one-third of all mature CD3$^+$ cells. They are found mainly in the human bone marrow and gut lymphoid tissue. CD8$^+$ T glycoprotein expressed on the surfaces of these T cells recognises a non-peptide-binding portion of MHC class I molecules. Hence, CD8$^+$ T cells are restricted to recognising MHC class I complexes.

CD8$^+$ T cells perform mainly cytotoxic functions. They kill (1) virus-infected cells, (2) allograft cells and (3) tumour cells. Two pathways help in the mediation of T-cell-mediated cytotoxicity. One pathway releases perforin proteins which form channels in the targeted cell membranes. These channels facilitate the diffusion of enzymes such as granzymes (serine esterases) into the cytoplasm. The exact mechanism of granzymes in inducing apoptosis has not been well established, but granzyme-induced apoptosis is a Ca^{2+}-dependent process. Another pathway depends on signals delivered by the cytotoxic cell to the target cell, which requires cell-to-cell contact. This pathway is Ca^{2+} independent.

The ratio of CD4$^+$ and CD8$^+$ T cells is approximately 2:1 in normal human peripheral blood. This may be significantly altered in immunodeficiency, autoimmune and other disorders. Differences between CD4 and CD8 T cells are summarised in Table 11.4.

Table 11.3 Distinguishing features of Th-1 cells and Th-2 cells

Features	Th-1 cells	Th-2 cells
Enhances cell-mediated immunity and delayed hypersensitivity	Yes	No
Enhances antibody production	No	Yes
Activation of cytotoxic T lymphocytes	Yes	No
Stimulated by IL-12	Yes	No
Stimulated by IL-4	No	Yes
Produces IL-2 and gamma interferon	Yes	No
Produces IL-4, IL-5, IL-6 and IL-10	No	Yes

Table 11.4 Differentiating characteristics of helper T cells (CD4) and cytotoxic T (CD8) cells

Helper T cells (CD4)	Cytotoxic T cells (CD8)
Carries CD4 marker on the cell surface	Carries CD8 marker on the cell surface
Helps or induces immune responses	The main function is the cytotoxic effect
Class II MHC present on CD4 cell helps in recognising the antigen	Class I MHC present on CD8 cell helps in recognising antigen
It activates macrophages to kill intracellular microbes by cytokine secretion	It directly destroys virus-infected and tumour cells

Memory T Cells

As the name suggests, Memory T cells confer host immunity with the ability to respond rapidly and vigorously for many years after the initial exposure to a microbe or other foreign substances.

The memory produced against a specific antigen shows the following characteristics: (1) memory cells survive for many years and have the capacity to multiply, (2) a large number of memory cells are produced following a primary infection, and consequently during secondary response is enhanced and is greater than the primary response, (3) a small quantity of antigens is sufficient to activate memory cells and requires less co-stimulation than that of naïve and unactivated T cells and (4) greater amounts of interleukins are produced by activated memory cells than that of naïve T cells when they are first activated.

Regulatory T Cells

Treg cells have $CD4^+$ molecules on their surface. The main functions of these cells are regulation of immune response and deletion of cell clones which are active against autoantigens. TGF β is the major cytokine produced by the suppressor T cells.

Activation of T Cells

Recognition of complex on the surface of antigen-presenting cells (APCs), such as macrophages and dendritic cells, consisting of both the antigen and a class II MHC protein by TCR present on T cells, is most important for activating helper T cells. Two signals are required to activate T cells. The first is the interaction of the antigen and the MHC protein with the T-cell-receptor-specific antigen. IL-1 secreted by the macrophages is also necessary for efficient helper T-cell activation. The second is a co-stimulatory signal required for activation of T cells.

Functions of T Cells

T cells perform two crucial functions: cytotoxicity and delayed hypersensitivity.

Cytotoxicity activity of T cells is required primarily to destroy virus-infected cells and tumour cells. It also plays an important role in graft rejection. Delayed hypersensitivity reactions against antigens of many intracellular pathogens are mediated by the CD4 cells, particularly the Th-1 subset cells and macrophages. The CD4 cells produce interleukins, such as gamma interferon, macrophage activation factor and macrophage inhibition factor, which mediate delayed hypersensitivity reactions.

B Cells

B Cell Maturation

The bone-marrow-derived lymphocytes are known as B lymphocytes or B cells. **Plasma cells** are derived from mature B cells. Both B cells and plasma cells synthesise and secrete immunoglobulin. The bursa of Fabricius, a lymphoid organ present near the caudal end of the gut, plays a significant role in differentiating B cells in birds. However, a mammalian equivalent to the bursa is yet to be found. Here the early stages of maturation of these lymphocytes occur in the bone marrow.

Nearly 30% of the recirculating small lymphocytes are composed of B cells. The B cells have a short lifespan of days or weeks. Nearly 10^9 B cells are produced every day. B cells are found in the germinal centres of the lymph nodes, in the white pulp of the spleen and in the MALT. B cells perform two important functions. First, they differentiate into **plasma cells** and produce antibodies. Second, they can **present antigens** to helper T cells.

The **clonal selection theory** explains the origin of antibody formation. According to this postulation, each immunologically competent B cell possesses receptors for either IgM or IgD that can combine with one antigen or closely related antigens. After binding the antigen, the B cell is activated to proliferate and form a clone of cells. Selected B cells are transformed into plasma cells that secrete antibodies specific to the antigen. Plasma cells synthesise the immunoglobulin with the same antigenic specificity as those carried by activated B cells. The same clonal selection also occurs with T cells.

B-cell precursors, during embryogenesis, first proliferate and develop in the foetal liver. From there, they migrate to the bone marrow, the main site of B-cell maturation in adults. Unlike T cells, they do not require the thymus for maturation. Pre-B cells exhibit only M heavy chains in the cytoplasm but do not have surface immunoglobulins and light chains. Pre-B cells are found in the bone marrow, whereas B cells are found in the circulation. Owing to allelic exclusion, individual B cells are committed to producing a single class of immunoglobulin molecules consisting of either **kappa or lambda light chain**. These B cells produce antibodies to a specific antigenic epitope. B cells develop immunological tolerance towards self-antigens by **clonal deletion or anergy** during the initial developmental stage.

B cells mature in two phases: **antigen-independent phase**, which consists of stem cells and pre-B cells, and **antigen-dependent phase**, which consists of the cells such as activated B cells and plasma cells that proliferate on interactions of antigen with B cells (Fig. 11.5).

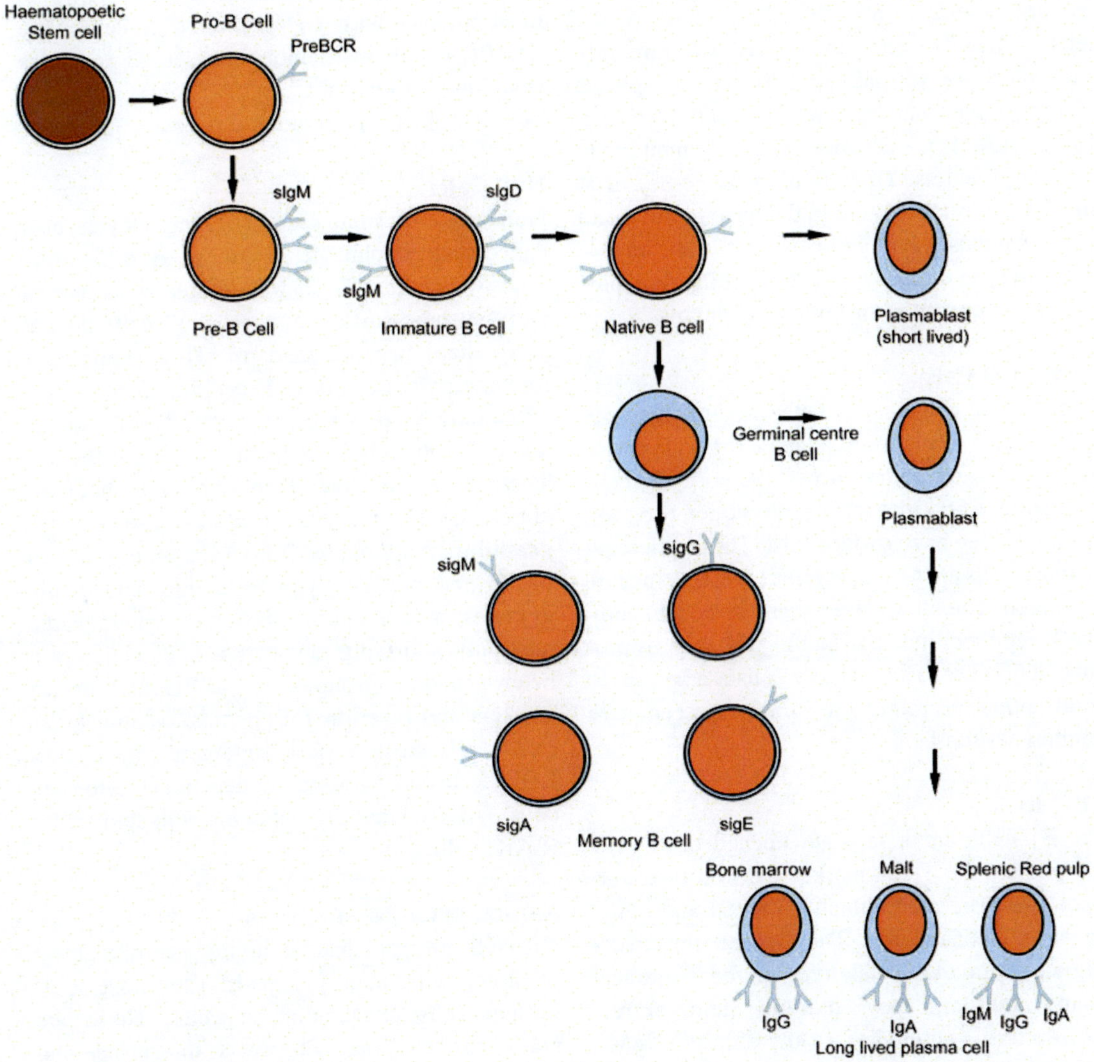

Fig. 11.5 B-cell maturation process

Types of B Cells

Plasma Cells

Activation of B cells leads to the production of many plasma cells. Subsequently, the plasma cells produce a large number of specific immunoglobulins complementary to the epitope of the antigen. Plasma cells originate from terminally differentiated B cells. Plasma cells are oval or egg-shaped structures characterised by a stellate (star-like pattern) nucleus, non-staining Golgi and basophilic cytoplasm.

The main function of the plasma cells is to produce and secrete all the classes of immunoglobulins into the fluids around the cells. They secrete thousands of antibody molecules per second, specific to the epitope of the antigen, for a few days and then die. They, however, do not express membrane immunoglobulins. They divide very poorly, if at all, and are usually found in the bone marrow and the perimucosal lymphoid tissues. They have a short lifespan of 30 days, during which they produce large quantities of immunoglobulins.

Memory B Cells

Some of the activated B cells produce memory cells, which remain in a stage of quiescence for months or years. Most memory B cells have surface IgG that acts as the antigen receptor, but some even have surface IgM. These quiescent memory cells are activated rapidly on re-exposure to antigen. The interleukins secreted by the memory T cells facilitate the memory B cells to produce antibodies. The memory cells are helpful in rapid antibody production during the secondary immune responses.

B1 Cells

The B1 cells are a separate lineage of B cells and possess the T-cell marker CD5 on their surface. These are predominant in foetal and initial stages of neonatal life. The B1 cell precursors migrate from the foetal liver to the peritoneal cavity. Further, they multiply and secrete polyreactive antibodies (low-affinity IgM), including autoantibodies at higher concentrations. These T- independent natural IgM antibodies are produced without any antigen stimuli and may contribute to the initiation of autoimmune disorders.

Activation of B Cells

Activation of B cells to produce the full range of antibodies first requires recognition of the epitope by the T-cell-antigen receptor and the production of IL-4 and IL-5 by the helper T cells. In addition, it also requires other co-stimulatory interactions of CD28 on the T cells with B7 on the B cells. The CD28–B7 interaction is essential to produce IL-2. It also includes CD40L on the T cells, which must interact with CD40 on the B cells. The CD40L–CD40 interaction is essential for class switching from IgM to IgG and switching between other immunoglobulin classes.

Null Cells

Null cells are a small section of lymphocytes, also called null lymphocytes, synthesised in the bone marrow. Unlike T- and B-lymphocytes which carry specific surface markers, null cells do not carry these specific markers. These cells lack immunologic memory and specificity.

Among lymphocytes, about 5–10% of cells are null cells which lack the features of T and B cells. These cells are larger than small lymphocytes, almost double the size, and also called **large granular lymphocytes (LGL)** containing sunken nuclei and profuse cytoplasm with azurophilic granules, mitochondria, ribosomes, endoplasmic reticulum and Golgi apparatus. LGL cells are a heterogeneous group with different surface markers and functional properties. Natural killer (NK) cells are the most important type among LGL, followed by antibody-dependent cytotoxic cells (ADCC) and lymphokine-activated killer (LAK) cells.

Natural Killer (NK) Cells

NK cells are specialised in killing a certain group of target cells, mainly virus-infected host cells, foreign cells and malignant cells. These are non-phagocytic, non-adherent to the surface and provide the first line of defence against infection and act spontaneously without prior sensitisation;

therefore, it is considered an innate immune response. The thymus is not required for the development. The NK cells are also involved in the regulatory mechanisms such as reciprocal interaction with dendritic cells, T cells, macrophages and so on. However, their activities are not enhanced by prior exposure and are not specific to any virus.

NK cells comprise approximately 5–10% of peripheral lymphocytes and are found in the spleen and peripheral blood. The number remains normal even in severe combined immunodeficiency disease. They lack both T-cell (CD3) and surface IgM and IgD (surface immunoglobulin) markers. They also lack immunologic memory and, unlike cytotoxic T cells, do not have any TCRs, and the killing of target cells does not require recognition of MHC proteins. These cells do not carry antigen receptors of any kind but can recognise antibody molecules bound to target cells and destroy those cells using the exact general mechanisms involved in T-lymphocyte cytotoxicity. They also have a cognition mechanism that allows them to destroy tumour cells and virus-infected cells.

NK cells develop within the bone marrow and lack TCR but possess another set of receptors called killer activation receptors and killer inhibition receptors. They also possess NK T cells, another subset of T cells, which share some functional characteristics with the NK cells. However, unlike NK cells, these NK T cells are stimulated by lipids, glycolipids and hydrophobic peptides presented by a non-classical class I molecule CD1D and secrete large amounts of cytokines, especially IL-4.

Functions of the NK Cells

The main functions of NK cells are to kill virus-infected cells and tumours. They do so by secreting cytotoxins such as perforins and granzymes similar to those of cytotoxic T lymphocytes and also by FasL-mediated apoptosis. They kill the viruses without specific antibodies by a mechanism called ADCC. Both IL-12 and gamma interferons are potent activators of NK cells.

Antibody-Dependent Cytotoxic Cells (ADCC)

These cells were called killer (K) cells initially but are now classified as NK cells. They have surface receptors for the Fc part of Ig and are a subpopulation of LGLs. They can lyse cells sensitised with IgG antibodies in a distinct mechanism different from Tc cells.

Lymphokine-Activated Killer (LAK) Cells

These are NK lymphocytes that are treated with interleukin-2 (IL-2) and are specifically cytotoxic to a wide range of tumour cells. IL-2 is also demonstrated to be a growth factor for NK cells. These LAK cells can be used in treating some tumours (e.g. renal cell carcinoma).

Phagocytic Cells

Phagocytosis is a crucial defence mechanism in animals by which a cell engulfs and destroys invading particles such as bacteria and fungi. Phagocytosis is carried out by a phagosome that may be certain types of WBCs or their derivatives like mononuclear macrophages/polymorphonuclear macrophages. The process begins with a macrophage or neutrophil recognising the invading organism and engulfing it to form a phagocytic vesicle or phagosome.

Macrophages

Macrophages measure about 20–80 μm and are the largest lymphoid cells in peripheral blood. Monocytes undergo a series of changes or differentiate to become macrophages. They are found in all tissues, more commonly in the liver, spleen and connective tissues. Macrophages are involved in different functions and are named according to their location, for example, adipose tissue macrophages, Kupffer cells, Langerhans cells, osteoclasts and so on. Mononuclear macrophage cells originate from precursor cells in the bone marrow and later become monocytes. Multinucleated cells and epithelioid cells seen in granulomatous inflammatory lesions such as tuberculosis originate from mononuclear macrophage cells.

Clinical Significance of Macrophages

These are essential in eliminating cell debris and dead cells. In addition, macrophages, along with dendritic cells, play a crucial role in initiating immune response by presenting antigens. The antigen is presented to the corresponding helper T cells after digestion by the macrophages.

Macrophages express many surface receptors, including Ia proteins, Fc part of IgG, activated complement components and various lymphokines. **Mac I** (macrophage-1 antigen or macrophage integrin) protein antigens are found on mouse macrophages; similarly, MI markers are found on human macrophages closely related to CR3, a cell receptor for C3 components.

Macrophage Participation in Immune Response

Macrophage participates in the induction and execution of the specific immune response. The antigen is trapped and presented in optimal concentration to the lymphocytes. If the concentration of antigen presented is too high, it may produce immunological tolerance, and if it is too low, it may not be immunogenic. For certain antigens, antigen processing and presentation by macrophages are required to generate antibodies.

Both T cells and macrophages require surface determinants coded by the same MHC genes for the processing and presentation of antigens. After the antigen is presented to the surface of the infected macrophage, MHC class II stimulates the T cell. Conversely, if a different MHC antigen is presented, T cells are not stimulated; this condition is called **MHC restriction**.

The phagocytic activity is greater in macrophages, particularly after activation by soluble mediators released during immune responses, than in monocytes. Differentiation of monocyte into a tissue macrophage involves several changes. The macrophage enlarges 5–10 folds, with intracellular organelles of the cells increasing and exhibiting complexity. Further, the cell attains high phagocytic ability and produces hydrolytic enzymes at higher levels. The cell also starts the process of secreting a variety of soluble factors.

Macrophage-like cells have a diverse set of functions in different tissues and hence are named based on the location of the tissue. For example, (1) histiocytes in connective tissues, (2) osteoclasts in the bone, (3) microglial cells in the brain, (4) alveolar macrophages in the lung, (5) Kupffer cells in the liver and (6) mesangial cells in the kidney.

Functions of Macrophages

Macrophages perform three main functions: phagocytosis, antigen presentation and cytokine production (Table 11.5).

1. **Phagocytosis**: Phagocytosis of bacteria, viruses and other foreign particles is the most important function of macrophages. The macrophages on their cell surfaces have Fc receptors that interact with the Fc component of the IgG, thereby facilitating ingestion of the opsonised organisms. They also have receptors for C3b, another important opsonin. After ingestion, the phagosome containing the microbe fuses with a lysosome. The microbe within the phagolysosome is killed by reactive oxygen, reactive nitrogen compounds and lysosomal enzymes.
2. **Antigen presentation**: After ingestion and degradation of foreign materials, the antigen fragments are presented on the macrophage cell surface in conjunction with class II MHC proteins for interaction with the TCR of $CD4^+$ helper T cells. Degradation of the foreign

Table 11.5 Important features of macrophages and their mechanism of actions

Features	Mechanism of action
Phagocytosis	Ingestion and killing of microbes in phagolysosome
Antimicrobial and cytotoxic activities	*Oxygen-dependent killing*: by superoxides, nitric oxide and hydrogen peroxide *Oxygen-independent killing*: by tumour necrosis factor, lysozyme and hydrolytic enzymes
Antigen processing	Phagocytic antigen

protein is stopped following the association of antigen with the class II MHC proteins in the cytoplasm. This is followed by the transportation of the complex to the cell surface by transporter proteins.

3. **Cytokine production**: Macrophages produce several cytokines, including IL-1, TNF and IL-8. IL-1 plays a vital role in activating helper T cells, whereas TNF plays an important mediator in inflammatory reactions. IL-8 attracts neutrophils and T cells to the site of infection.

Microphages

These are the polymorphonuclear leucocytes; these include neutrophils, eosinophils and basophils. Both neutrophils and eosinophils are phagocytic, whereas basophils are not.

Neutrophils

These are the predominant phagocytic cells observed during acute infections. Phagocytic activity increases in the presence of opsonin, whereas the phagocytic activity of neutrophils is non-specific in nature.

Eosinophils

These play an important role in defence against parasitic infections, though their phagocytic role is significantly lower than neutrophils. They are commonly found in tissues rather than blood; the important features of eosinophil are the presence of small, round homogeneous and large ovoid granules. The granules consist of various hydrolytic enzymes, which help in the extracellular killing of large parasitic cells.

Basophils

These cells are found in blood and tissues (mast cells). They are non-phagocytic granulocytes that function by releasing pharmacologically active substances (histamine, serotonin and other hydrolytic enzymes) from their cytoplasmic granules. These substances play a significant role in certain allergic responses.

Mast cells are derived from the myeloid stem cells and differ from basophils. They are present widely in various tissues, such as the mucosal epithelial tissue of the respiratory, digestive, and genitourinary tracts, connective tissues of many organs and the skin. Mast cells possess cytoplasmic granules in large numbers, similar to circulating basophils, which include histamine and other pharmacologically active compounds. These cells play a significant function in the development of allergies in combination with blood basophils.

Antigen-Presenting Cells (APCs)

Antigen-presenting cells (APCs) include macrophages, dendritic cells and B lymphocytes. These APCs possess MHC class II molecules on their surface. T helper cell only recognises if the processed antigens are presented with MHC II molecules of APCs. In turn, T helper cells activate both humoral and cell-mediated immunity.

Dendritic Cells

Dendritic cells resemble neuronal dendrites with many long, narrow processes; hence, they are called so. These long processes are beneficial in making efficient contacts with foreign materials. They are mainly found in the skin (e.g. Langerhans cells) and the mucosa, from where they migrate to local lymph nodes for the presentation of antigen to helper T cells.

Dendritic cells play a vital role in primary immune response where they present the antigen into T cells. They are derived from bone marrow cells and express class II MHC proteins on their surface and also present antigen to $CD4^+$ T cells. They have little or no role to play in phagocytosis. They also serve as professional APCs, although macrophages and B cells are the major APCs.

Four types of dendritic cells are known: (1) Langerhans cells, (2) interstitial dendritic cells, (3) myeloid cells and (4) lymphoid dendritic cells. All these cells constitutively express high levels of both class II MHC molecules and members of the co-stimulatory B7 family. The mature and immature forms of Langerhans cells and interstitial dendritic cells travel to drain in the lymph nodes, after a microbial invasion or during the inflammation process. These cells present the antigen to the Th cells, which helps initiate immune responses.

Follicular Dendritic Cells

These cells are similar to the dendritic cells except for their sites of presence and functions. These cells are present in B-cell-containing germinal centres of the follicles in the spleen and lymph nodes. These cells do not present antigen to helper T cells but combine with antigen-antibody complexes by Fc receptors found on their surfaces.

Abnormalities of Immune Cells

These are generally inherited defects in specific or non-specific immune systems. Affected individuals are susceptible to different infections depending upon the nature of the abnormality. Some examples due to abnormalities of immune cells include chronic granulomatous disease caused by the defective intracellular killing mechanism of phagocytes, DiGeorge syndrome where congenital aplasia of the thymus occurs and so on.

Major Histocompatibility Complex (MHC)

The major histocompatibility complex (MHC) was first detected as the genetic locus encoding the glycoprotein molecules (transplantation antigens) responsible for the rapid rejection of tissue grafts transplanted between genetically non-identical individuals. Gorer, in 1930, was the first to identify the antigens responsible for allograft rejection that led to the discovery of a major histocompatible complex. In mice, he demonstrated two blood group antigens (antigen 1 and antigen 2). Antigen 1 was found in all strains of mice, whereas antigen 2 was found in certain strains of mice and was responsible for allograft rejection. This was named H2 antigen and was found to be the major histocompatible antigen. This antigen was coded by a closely linked multiallelic cluster of genes called the MHC, named as H2 complex.

The name histocompatibility was coined because this discovery was made based on the transplantation experiments. The alloantigens present on the surface of the human leucocytes determine the histocompatibility during transplantation. Hence, human MHC molecules are synonymously called HLA and MHC complexes are termed HLA complex (Fig. 11.6).

Histocompatible antigen denotes the cell surface antigens that induce immune responses to an incompatible host, resulting in allograft rejection. The MHC in humans is known as the **HLA complex**. In humans, these alloantigens are present on the surface of leucocytes and are called HLA and the set of genes encoding for them is named the HLA complex. Carbohydrate antigens of erythrocytes (blood groups) and glycoprotein antigens of cell membranes are the two major transplantation antigens of humans. Snell, Dauseset and Benacerraf (1980) were awarded the Nobel Prize for their work on MHC and genetic control of immune responses.

Functions of MHC

The MHC performs many functions. It kills virus-infected cells and tumour cells. In addition, it takes part in the non-specific killing of virus-infected cells and tumour cells. The killing of cells is independent of antigen presentation by MHC proteins. The killing of cells is carried out by perforins and granzymes, which cause apoptosis of the target cell. Killing is activated by the failure of a cell to present antigen with class I MHC or by reducing class I MHC proteins on the cell surface.

Classes of Proteins

Almost all animal species, including human beings, possess these genes on a segment of a chromosome pair, coding for three different classes of proteins. **Class I proteins** are responsible for histocompatibility, allograft acceptance and rejection. **Class II proteins** play an important

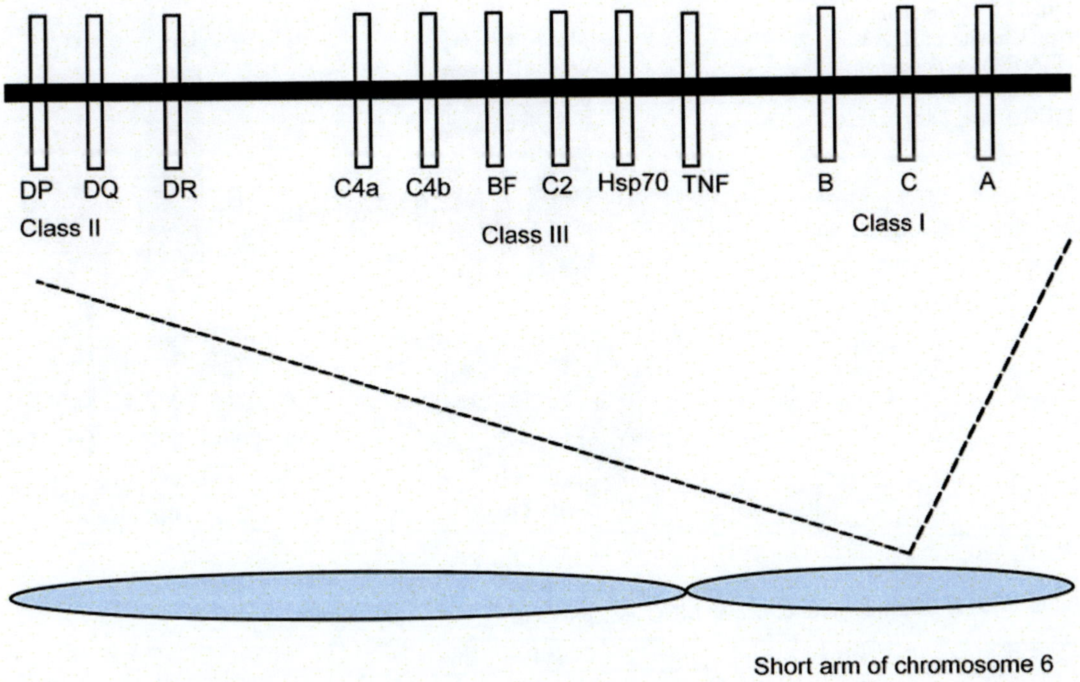

Fig. 11.6 A schematic diagram of HLA complex

role in the regulation of immune response. Finally, **class III molecules** involve components of the complement system such as the C2, C4 and B factor.

HLA Complex

In humans, the HLA complex of genes is located on the short arm of chromosome 6, containing several genes that are critical to immune function (Figs. 11.6 and 11.7). The HLA complex of genes is classified as **Class I** (HLA-A, HLA-B and HLA-C), **Class II** (HLA-DP, HLA-DQ and HLA-DR.) found within the HLA-D region of the HLA complex, and **Class III** (TNFs alpha and beta cells) complement loci which encode for C2, C4 and factor B of the complement system, respectively.

A **locus** means the position where a particular gene is located on the chromosome. HLA loci are usually multiallelic, meaning that genes present on the locus can be any one of several alternative forms. There are 24 alleles at the HLA-A locus and 50 at the HLA-B locus. Each allele expresses a distinct antigen. The HLA system is pleomorphic, and every person inherits one set of HLA genes from each parent. The WHO official Committee has recommended a set of guidelines for the nomenclature of the HLA system. Recognised alleles and their corresponding antigens are indicated by alphabet(s) and a number (e.g. HLA-A1, HLA-DR7).

HLA Molecules

The HLA genes encode various enzymes and structural molecules essential for the activation and function of B and T cells. The genes encode MHC proteins classified into three groups or classes known as the class I, class II and class III molecules.

HLA Class I Molecules

These are glycoproteins found on the surface of virtually all nucleated cells in the body. Class I proteins are encoded by the HLA-A, HLA-B and HLA-C loci. Allelic genes encode approximately 20 different proteins at the A locus, 40 at the B locus, and 8 at the C locus (Table 11.6). The entire class I is composed of a transmembrane

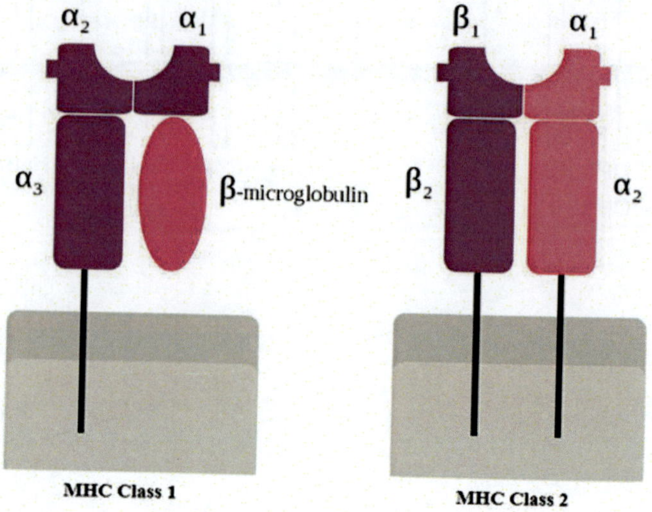

Fig. 11.7 Diagrammatic representation of HLA I and HLA II (https://commons.wikimedia.org/wiki/File:MHC_Class_2.svg)

Table 11.6 Important features of human MHC gene products

Features	Class I	Class II
Genetic loci (partial list)	HLA-A, -B and -C	HLA-DP, -DQ and -DR
Polypeptide composition	MW 45,000 Da + MW 12,000 Da	Light chain (MW 33,000 Da), light chain (MW 29,000 Da), light chain (MW 30,000 Da)
Cell distribution	All nucleated somatic cells and platelets	Antigen-presenting cells (macrophages, B cells etc.) and activated human T cells
Present peptide antigens to	CD8 T cells	CD4 T cells
Size of peptide bond	8–11 residues	10–30 or more residues

glycoprotein of 45,000 Da, heavy chain, non-covalently associated with a β_2-microglobulin (a non-HC-encoded polypeptide of MW 12,000 Da). Like that of an immunoglobulin molecule, the heavy chain has a variable and constant region. The variable region is highly pleomorphic. The polymorphism of these molecules is important in the recognition of self and non-self. The constant region of the heavy chain binds with the CD8 proteins of the cytotoxic T cells. Class I proteins are involved in graft rejection and cell-mediated cytolysis.

HLA Class II Molecules

These are glycoproteins, and unlike class I proteins, they have a restricted tissue distribution. They are chiefly found on macrophages, B cells and other APCs, such as dendritic cells of the spleen and Langerhans cells of the skin. Their expression on other cells (e.g. endothelial cells) can be induced by gamma interferon. They are highly polymorphic glycoproteins composed of two non-covalently associated transmembrane glycoproteins of MW of about 33,000 and 29,000 Da.

The genes on the HLA-D locus encode class II proteins. This locus retains control of immune responsiveness, and different allelic forms of these genes confer striking differences in the ability to mount an immune response against a given antigen. The class II MHC locus also includes genes encoding proteins involved in antigen processing, for example, transporter associated with antigen processing (TAP). Class II proteins are primarily responsible for the graft-versus-host and mixed leucocyte responses.

HLA Class III Molecules

The class III MHC locus encodes complement proteins (C2, C4) of the classical pathways, properdin and factor B of the alternative pathway and several cytokines. HLA class III molecules are polymorphic.

HLA Typing

HLA typing or tissue typing is usually performed to determine the closest MCH match between donors and recipients before transplantation surgery. The methods commonly used in the laboratory include (1) molecular methods using DNA sequence, (2) serological assays and (3) mixed lymphocyte culture (MLC) techniques. All these methods are used to determine the haplotype, that is, the class I and class II alleles on both donor and recipient chromosomes.

DNA probe and PCR are highly specific and sensitive methods used to detect the different alleles. Serological assays using a battery of antibodies specific for different class I and class II proteins are also used to demonstrate the alleles. If these two methods fail to provide sufficient data, additional information can be obtained by performing the MLC technique, also known as mixed lymphocyte reaction.

Mixed Lymphocyte Reaction

This test is performed by using stimulator lymphocytes from a potential donor, which are first killed by irradiation. Then it is mixed with live responder lymphocytes from the recipient. The resultant mixture is then incubated in cell culture to allow DNA synthesis, which is detected by adding tritiated thymidine. The more the DNA synthesis in the responder cells, the more foreign are the class II MHC proteins of donor cells. Therefore, a large amount of DNA synthesis indicates that class II (HLA-D) MHC proteins of donors and recipients are not similar. This shows an unsatisfactory match between donor and recipients, and thereby graft is likely to be rejected. On the other hand, the least production of DNA suggests the best donor and a good match between the donor and recipients.

Applications of HLA Typing

Its highly pleomorphic nature is responsible for the distinct difference between the HLA molecules of one person to another person.

HLA typing has vast applications. HLA typing is a required test before **transplantation** to know the compatibility between donor and recipient HLA molecules. It is also used in resolving **paternity disputes**. It is also used in **anthropological studies**. The HLA molecules of humans are distinct and large variations can be seen based on the different ethnicities and races. The different population studies based on HLA typing suggest that humans originated from Africa and then migrated to other parts of the world. It is also used as a **genetic marker** for finding the association of HLA with diseases, such as the association of HLA-B27 with ankylosing spondylitis and HLA-DR4 with rheumatoid arthritis.

It is now known that MHC molecules bind peptide antigens and present them to T cells. Thus, these transplantation antigens are responsible for antigen recognition by the TCR. In this respect, the TCR is different from the antibody.

Antibody molecules interact with antigen directly; the TCR only recognises antigen presented by MHC molecules on another cell, the APC. The TCR is specific for the antigen, but the antigen must be presented on a self-MHC molecule. The TCR is also specific for the MHC molecule. If the antigen is presented by another allelic form of the MHC molecule *in vitro* (usually in an experimental situation), there is no recognition by the TCR. This phenomenon is known as **MHC restriction**. Both classes I and II MHC molecules are involved in this phenomenon. Peptide antigens associated with class I MHC molecules are recognised by $CD8^+$ cytotoxic T lymphocytes, whereas class II-associated peptide antigens are recognised by $CD4^+$ helper T cells.

12 Antigen

An antigen is a molecule which, when introduced parenterally into the body, initiates the production of an antibody with which it reacts specifically in an observable manner.

Molecules that can be recognised by the immunoglobulin receptor of B cells or by the T-cell receptor when complexed with major histocompatibility complex (MHC) are called **antigens**. The word antigen is a shortened form of the words "antibody generator". Antigens are substances that react with antibodies, whereas **immunogens** are molecules that induce an immune response. In most cases, antigens are immunogens, and the terms are used interchangeably. The antigens that are not immunogenic but can participate in immune reactions are termed **haptens**. The term **immunogenicity** means the ability of an antigen to elicit an immune reaction in the form of a B-cell or T-cell response, whereas the term **antigenicity** means just the ability to combine specifically with the products of the above responses. All immunogenic molecules are antigenic too, but all antigenic molecules cannot be considered immunogenic. Thus, haptens can be said to lack immunogenicity.

Determinants of Antigenicity

Several factors have been identified that make a substance immunogenic.

Molecular Size

Antigenicity depends on the molecular size. In general, protein molecules with large molecular weight are highly antigenic. Substances with molecular weights of about 100,000 Da and more are highly immunogenic, whereas those with less than 5000 Da are generally not immunogenic. This property has been exploited in experimental studies by using high molecular weight proteins such as bovine gamma globulin (MW 150,000 Da) to induce an immune reaction. Substances with low molecular weight may be made antigenic by adsorbing these on carrier particles, such as bentonite, kaolin and other inert particles.

Foreignness

To be immunogenic, a molecule must be recognised as non-self, that is, foreign. The molecule is considered self or non-self by the immune system depending on whether or not the molecule was exposed to the immune system during foetal development.

Foreignness implies the ability of the host to tolerate self-antigens. Tolerance to self-antigens develops by contact with them in the initial phases of the development of the immune system, particularly during the development of lymphocytes. In general, the more distantly related two species are, the greater the immunogenicity of a molecule from one species will be when exposed to the other. For example, the bovine serum albumin is more immunogenic in a chicken than in a goat. A graft from an unrelated human will be rejected within about 2 weeks unless immunosuppressive drugs are used, but a graft from a chimpanzee will be rejected within hours even if drugs are used. In contrast, a kidney graft from an identical twin will be accepted readily.

Chemical-Structural Complexity

Proteins are the most potent immunogens, followed by polysaccharides. Nucleic acids and lipids are inefficient in eliciting a good immune reaction, although they may act as haptens. The structural complexity of a protein contributes to its immunogenicity. Chains of single amino acids or single sugars are poorly immunogenic, but if different amino acids or sugars are combined in the same molecule, the immunogenicity is greatly enhanced. In cell-mediated immunity, the response of T cells to the peptide component of the proteins depends on how the peptide is recognised and presented by the MHC cells. Therefore, protein structure plays an essential role in its immunogenicity, especially in inducing cellular immunity.

The lipid-specific antibodies are not easily produced; hence, they do not play a significant role in immunity. However, these antibodies have a role in measuring certain lipid-based molecules and drugs. These antibodies are produced first by treating lipids with haptens and then conjugating with suitable carrier molecules, such as the proteins (e.g. haemocyanin or bovine serum albumin).

Stability

Highly stable and non-degradable substances (e.g. some plastics, metals, or chains of δ-amino acids) are not immunogenic. This is because internalisation, processing and presentation by antigen-presenting cells (APCs) are always essential for mounting an immune response. Therefore, very stable substances (such as silicon) have been successful as non-immunogenic materials for reconstructive surgeries, such as breast implants.

On the other hand, if a substance is very unstable, it may break up before an APC can internalise it and hence become immunogenic. In addition, large, insoluble complexes are more immunogenic than smaller, soluble ones. This is because macrophages find it easier to phagocytose, degrade and present the insoluble complexes than the soluble complexes.

Other Factors

The biological system also plays an essential role in determining the immunological efficiency of an antigen. For example, some substances are immunogenic in one individual but not in others (i.e. responders and non-responders). This is because individuals may lack or have altered genes that code for the receptors for antigen on B cells and T cells, or they may not have the appropriate genes needed for the APC to present antigen to the helper T (Th) cells.

Dosage and route of the antigen by which it comes into contact with the immune system also influence the immunogenicity of the antigen. Very low doses of antigen do not stimulate an immune response, either because too few lymphocytes are contacted or because a non-responsive state is elicited. Conversely, an extremely high dose also fails to elicit tolerance.

Repeated administration of antigens (booster doses) may be required to enhance the host's immune response to certain antigens. This is particularly important in the case of vaccines, where a prerequisite immune level needs to be attained. Hence the booster doses of vaccines, such as DPT (diphtheria, pertussis, tetanus), DT (diphtheria, tetanus) and so on, are given to ensure good protective levels of antibodies. Generally, antigens are administered by the parenteral route to produce a good level of antibodies. The antigens can be given by (1) intravenous,

(2) subcutaneous, (3) intradermal, (4) intramuscular, (5) intra-peritoneal and (6) mucosal routes. Usually, the subcutaneous route of administration proves to be better than intravenous routes at eliciting an immune response.

Adjuvants These are the substances that, when mixed with an antigen and injected with it, boost the immunogenicity of the antigen. Adjuvants increase both the strength and the duration of the immune response. Adjuvants boost the immunogenicity of antigens in several ways. Adjuvants such as aluminium potassium sulphate (alum) and Freund's water-in-oil adjuvant prolong antigen persistence by forming a depot at the injection site. Alum precipitates the antigen and releases it a little at a time. The water-in-oil emulsion forms small droplets with the antigen and also releases these slowly over time.

Freund's complete adjuvant contains, in addition to the emulsifying factors, heat-killed mycobacteria. The bacterial components activate macrophages and increase both the production of IL-1 and the level of B7 membrane molecules, which enhances the immune response. The increased expression of Class II MHC increases the ability of APC to present antigen to Th cells. B7 molecules on the APC bind to CD28, a cell surface protein on Th cells, triggering costimulation, an enhancement of the T-cell immune response.

Some adjuvants, such as synthetic polyribonucleotides and bacterial lipopolysaccharides (LPS), stimulate non-specific lymphocyte proliferation and bring about their action.

Antigenic Specificity

Antigenic specificity of the antigen depends on antigenic determinants or epitopes.

Epitopes

An *epitope* is defined as the immunologically active region of an immunogen that binds to antigen-specific membrane receptors on lymphocytes or secreted antibodies. The interaction between cells of the immune system and antigens occurs at many levels, and the complexity of any antigen is mirrored by its epitope (Fig. 12.1). There are two types of epitopes: B-cell epitopes and T-cell epitopes.

B-Cell Epitopes

These are antigenic determinants recognised by B cells. The B-cell epitope can combine with its receptor only if the antigen molecule is in its native state. The complementary surfaces of the antibody and the antigen molecules appear to be relatively flat. Smaller molecules often fit nicely within a particular depression or groove in the antigen-binding site of the antibody molecule. The B-cell epitope is about six or seven sugar residues or amino acids long. They are also often found in regions of proteins with higher mobility; this may make it possible for an epitope to shift just a bit to fit into an almost-right site.

T-Cell Epitopes

T-cell epitopes recognise amino acids in proteins but do not recognise polysaccharides or nucleic acid antigens. This is why polysaccharides are considered T cell-independent antigens and proteins as T-cell-dependent antigens. The primary sequence of amino acids in proteins determines the antigenic determinants recognised by T cells. Free peptides are not recognised by T cells, whereas the complex of MHC molecules and peptides are recognised by T cells. Thus, for a T-cell response, it should recognise both the antigenic determinant and the MHC, and therefore it is said to be MHC restricted. In general, T-cell epitopes or antigenic determinants are small and are only 8–15 amino acids long. The antigenic determinants are limited to those parts of the antigen that can bind to MHC molecules.

There are two types of epitopes of protein antigens depending on their interaction with the paratope (antigen-binding site on the antibody molecule) and its structural configuration:

1. **Conformational epitope**: This constitutes the most common type of epitope, which has a

Fig. 12.1 Epitopes (linear and conformational)

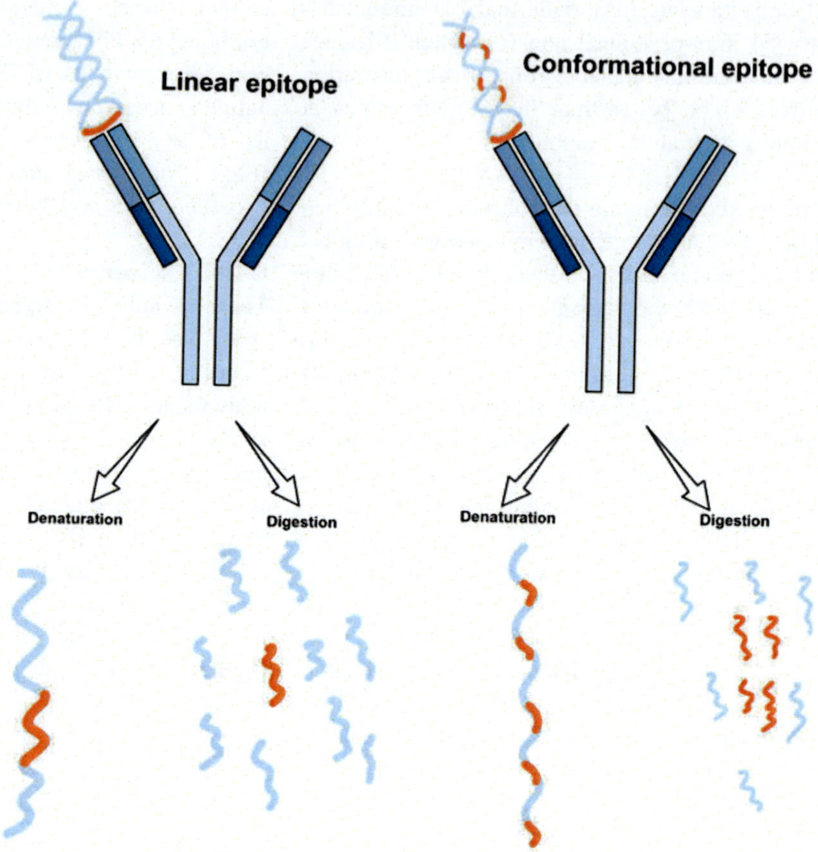

non-continuous amino acid sequence. As a result, their interaction with the corresponding paratope is determined by the tertiary structure of the antigen.

2. **Linear epitope**: Here, the epitope is made up of a continuous sequence of amino acids and the interaction with the paratope is determined by the primary structure of the antigen. T lymphocytes recognise only the linear epitopes, while B cells and antibodies can recognise both linear and conformational types of epitopes.
3. **Paratope**: A paratope is the region of an antibody that recognises and binds to the epitope of an antigen. Paratopes are produced by the complementary binding of light and heavy chains resulting in a three-dimensional structure. In theory, 10^4 heavy chains can combine with 10^4 light chains to generate 10^8 different paratopes. The paratope region is present in the variable region of the antibody and consists of 5–10 amino acids.

Types of Specificity

The specificity of antigens is of various types.

Species Specificity

Tissues of all individuals in a species possess certain species-specific antigens. However, some degree of cross-reaction occurs between antigens from related species. The species specificity shows a phylogenetic relationship. The phylogenetic relationship is useful (1) in tracing evolutionary relationships between species and (2) in identifying species from blood and seminal stains in forensic medicine.

Isospecificity

It is determined by the presence of isoantigens or histocompatibility antigens.

Isoantigens are antigens are found in some, but not all, species members. A species may be grouped depending on the presence of different isoantigens in its members. These are genetically determined. Human erythrocyte antigens, based on which individuals are classified into different blood groups, are the best examples of isoantigens in humans. The blood groups are of primary importance in (1) transfusion of blood and blood products, (2) isoimmunisation during pregnancy and (3) providing valuable evidence in paternity disputes, the results of which are supplemented by more recent DNA fingerprinting tests.

Histocompatibility antigens are the cellular determinants specific for each individual of a species. These antigens are associated with the plasma membrane of tissue cells. Human leukocyte antigen (HLA) is the major histocompatibility antigen determining homograft rejection. Therefore, HLA typing is essential before carrying out tissue or organ transplantation from one individual to another.

Autospecificity

Self-antigens are generally non-antigenic. However, sequestrated antigens (such as eye lens protein and sperm) are exceptions because these are not recognised as self-antigens. This is because the immune system never encounters corneal tissue and sperm during the development of tolerance to self-antigens. Therefore, these tissues become immunogenic if accidentally or experimentally released into the blood or tissues.

Organ Specificity

Antigen characteristics of an organ or tissue are called organ-specific antigens. These antigens found in the brain, kidney and lens tissues share the same antigen specificity even from different animal species. Organ-specific antigens, such as brain-specific antigens, shared by human and sheep brains, are one example. The anti-rabies vaccines prepared from sheep brain, when given, may induce an immune response in some humans, causing damage to the neural tissues of the recipient. This may result in neuroparalytic complications in some individuals.

Heterophile Specificity

It is determined by the presence of **heterophile antigens**. The same or closely related antigens, sometimes present in tissues of different biological species, classes or kingdoms, are known as heterophile antigens. Antibodies against the heterophile antigens produced by one species cross-react with the antigens of other species. This property is exploited for the diagnosis of many infectious diseases. Weil–Felix reaction, Paul–Bunnell test and cold agglutination tests are examples of serological tests that use such heterophile antigens.

Weil–Felix reaction is a test used for the diagnosis of rickettsial infections, in which the strains of *Proteus* species (such as OX 19, OX 2 and OX K) are used to detect heterophile antibodies produced against rickettsial pathogens. **Paul–Bunnel test** is used to diagnose infectious mononucleosis caused by Epstein–Barr virus infection by demonstrating heterophile antibodies that agglutinate sheep erythrocytes. A **cold agglutinin test** is performed to diagnose primary atypical pneumonia caused by *Mycoplasma pneumoniae* by demonstrating heterophilic antibodies.

Biological Classes of Antigens

Based on their ability to induce antibody production, antigens are classified as (1) T-cell-dependent antigens, (2) T-cell-independent antigens and (3) superantigens.

T-Cell-Dependent Antigens (TD)

The majority of naturally occurring proteins are T-cell-dependent antigens, and these antigens cannot induce antibody production upon direct contact with B cells. T-dependent antigens are

Table 12.1 Distinguishing features of T-cell-dependent antigens and T-cell-independent antigens

Characteristics	T-cell-dependent antigen	T-cell-independent antigen	
		Type 1	Type 2
Structural complexity	Complex	Simple	Simple
Chemical nature	Soluble proteins—protein, glycoprotein, lipoprotein	Bacterial cell wall components (e.g. LPS)	Capsular polysaccharides
Antibody class involved	Ig—G, M, A, D and E	IgM	IgM
Isotype switching	Yes	No	Yes
Polyclonal response	No	Yes	No
Affinity maturation	Yes	No	No
Memory response	Yes	No	No
Present in most pathogen	Yes	No	No

made up of a few copies of several antigenic determinants. These antigens bind to the Ig receptor present on the B cells, which are then internalised and processed into smaller peptides and then expressed on the B cell surface. Further, B cells form a complex with MHC II and present themselves to T helper cells. Upon recognising processed antigenic determinants, Th cell secretes cytokines and expresses CD40 ligands on it. After that, the Th cell interacts with CD 40 markers present on the B cells in the presence of cytokines to produce antibodies against the specific antigens.

T-Cell-Independent Antigens (TI)

These antigens stimulate B cells to produce antibodies without any direct involvement of T cells. Most microbial antigens such as carbohydrates, lipids and certain nucleic acids are T-independent antigens. Further, these antigens are of two types.

Type-1 antigens include endotoxins and lipopolysaccharides (LPS) of microbes. They are mitogenic, thus helping B cell multiplication and polyclonal B cell activation.

Type-2 antigens include polymeric compounds such as polysaccharides (pneumococcal capsular polysaccharides) or proteins (flagellar proteins). These antigens, along with cytokines, activate B cells to produce specific antibodies and augment other immune cell functions such as dendritic cells, mast cells, macrophages and NK cells. Differences between T-cell-dependent antigens and T-cell-independent antigens are listed in Table 12.1.

Boivin antigens are lipopolysaccharide analogues which are endotoxins released from Gram-negative bacteria cell wall.

Superantigens

Superantigens are a class of molecules that can interact with APCs and T lymphocytes in a non-specific way. The superantigens act differently by interacting with both MHC class II molecules of the APC and the Vβ domain of the T-lymphocyte receptor in a sequential manner (Fig. 12.2). They do not require processing and hence can bind to MHC molecules outside the peptide or antigen-binding groove. Superantigens are considered the most powerful mitogen, and a

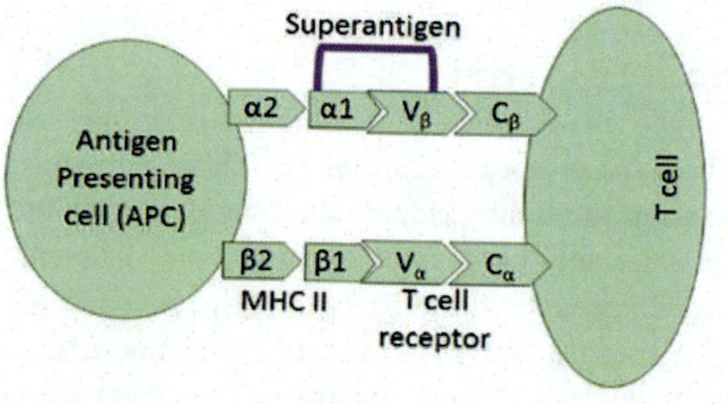

Fig. 12.2 Superantigen (https://link.springer.com/article/10.1007/s00203-020-01932-w)

dose of less than 0.1 pg/mL can be lethal. This interaction results in the activation of a larger number of T cells (up to 25%) than conventional antigens (1%), leading to massive cytokine expression and immunomodulation. As a result, there is polyclonal activation of T cells with the production of large amounts of cytokines like IFN-γ. IFN-γ, in turn, causes the activation of macrophages to produce IL-1, IL-6 and TNF-α. The net result is the development of a "cytokine storm". Examples of superantigens are staphylococcal enterotoxins, toxic shock syndrome toxins, exfoliative toxins and viral proteins. Most recently, the SARS-Coronavirus-2 has been shown to produce a cytokine storm leading to death in many instances. The diseases associated with superantigens in humans are listed in Table 12.2. Distinguishing features between antigen and superantigen are listed in the Table 12.3.

Haptens

A hapten is a low molecular weight substance that is antigenic but not immunogenic. It cannot elicit an immune response by itself but can combine with a product of a specific immune response (**Incomplete antigen**). A carrier molecule which is commonly a protein, combined with a hapten, makes the complex immunogenic. A hapten-carrier conjugate (**Complete antigen**) can produce three different types of antibodies: against the hapten, against the protein and new epitopes formed due to conjugation. Urushiol toxin is found in poison ivy, and when it enters the body, it is converted to quinone, the hapten. It reacts with skin protein and acts as an antigen causing contact dermatitis. Penicillin and halothane are other examples of haptens that can cause human disease. Haptens can be of two types. **Simple hapten** contains only one epitope or antibody-binding site, and the hapten-antibody complex cannot be visualised in a precipitation reaction. **Complex hapten** has two or more antibody-binding sites, and the complex can be visualised in a precipitation reaction.

Haptens fail to activate helper T cells due to their inability to bind to MHC proteins since haptens are not proteins, and only proteins can be presented by MHC proteins. Moreover, haptens cannot activate B cells since they are

Table 12.2 Association of superantigens with human diseases

Pathogen	Superantigen	Human disease
Staphylococcus aureus	Toxic shock syndrome toxin	Toxic shock syndrome
	Enterotoxins (18 types)	Food poisoning
	Exfoliative/epidermolytic toxins A and B	Scalded skin syndrome (Ritter disease)
Streptococcus pyogenes	Streptococcal pyrogenic exotoxins (SPE) A, C	Toxic shock syndrome
	Streptococcal pyrogenic exotoxins (SPE) E, M	Acute rheumatic fever
Yersinia pseudotuberculosis	*Yersinia pseudotuberculosis* mitogenic factor (YPM-A,B)	Kawasaki disease
Mycoplasma	Mycoplasma arthritidis mitogen	Acute and chronic proliferative arthritis
Epstein–Barr Virus activated by a human endogenous retrovirus (HERV)	K18	Insulin-dependent diabetes mellitus
Human Herpesvirus-8 (Kaposi's Sarcoma Herpesvirus)	Unknown	Multicentric Castleman's disease
SARS-CoV-2	Unknown	Coronavirus disease 2019 (COVID-19)
HIV	HIV-gp 120 (B-cell SAg)	Induces the release of proinflammatory, angiogenic, and lymphangiogenic factors from human lung mast cells

Table 12.3 Distinguishing characteristics of antigens and superantigens

Characteristics	Antigens	Superantigens
Type	Almost all proteins	Some of the proteins
Examples	Tumor antigen (Cancer) Staphylococcus cell wall antigen (Streptococcus nephritis) Nuclear antigen (Systemic Lupus Erythematosus)	Staphylococcus enterotoxin (food poisoning) Staphylococcus pyogenic exotoxins (Psoriasis) Toxic shock syndrome toxin I (Toxic shock syndrome) B cell SAg (HIV)
Processing	Requires processing by Antigen Presenting Cells (APCs)	Does not require processing by Antigen Presenting Cells (APCs)
Antigen Presentation and recognition	Recognition is MHC I or II restricted	Presented by MHC II class
Recognition by	TcRα and β chains are involved	Only TcRα chains are involved

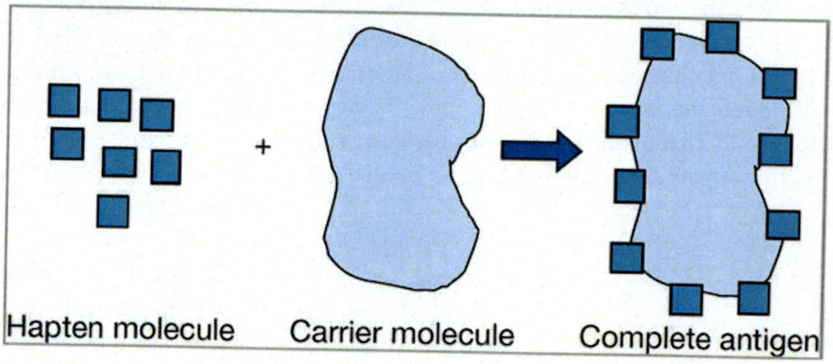

Fig. 12.3 Hapten-carrier conjugate

univalent. On the contrary, they can activate B cells when covalently bound to a "carrier" protein; since then, they form an immunogenic hapten-carrier conjugate (Fig. 12.3). Then, they bind to the IgM receptor on the B cells, and the hapten-carrier protein complex is internalised. B cells present a peptide of the carrier protein to the helper T cells in association with Class II MHC protein. The activated helper T cells then produce interleukins, which stimulate the B cells to produce antibodies to the hapten.

Animals immunised with such a conjugate produce antibodies specific for (1) the hapten determinant, (2) unaltered epitopes on the carrier protein and (3) new epitopes formed by combined parts of the hapten and the carrier. The hapten-carrier molecule is bound to surface immunoglobulins on B cells via the hapten epitopes. The hapten-carrier molecule is then taken in, processed and pieces of the carrier are presented by these B cells and T_H cells. In the body, the formation of hapten-carrier conjugates is the basis for developing allergic responses to drugs, such as penicillin.

Pathogen Determinants Recognition by Innate Immune System

Usually, the **adaptive immune system** reacts specifically to a particular pathogen by recognising its discrete antigenic determinants with the help of specific receptors present on the immune cells. In contrast, the components of the innate immune system can recognise the unique molecular patterns shared by many pathogens. These unique molecular patterns shared by multiple pathogens are termed pathogen-associated molecular patterns (PAMPs), and the innate immune receptors that recognise PAMPs are called pattern recognition receptors (PRRs). A single PRR can recognise a particular PAMP present on different pathogens.

Broadly, PRRs are categorised into three classes.

Toll-Like Receptors (TLRs)

TLRs are transmembrane receptors on the host immune cells such as dendritic cells and macrophages with unique pattern recognition capabilities. There are 13 different types of TLRs that can recognise different pathogens and enhance innate immune mechanisms by phagocytosis and inflammation at the site of infection.

Scavenger Receptors (SR)

SRs are expressed by macrophages, dendritic cells and certain endothelial cells. They can bind to the bacterial cell wall components such as LPS, peptidoglycans and damaged or apoptotic cells. SRs are involved in lipid metabolism and play an important role in clearing damaged host cells and other molecules. Examples of scavenger receptors include CD-36, CD-68, SRB-1 and so on.

Mannose Receptors

These are found on the surface of phagocytes and can bind to the mannose-rich glycans, the components of the microbial glycoproteins and glycolipids. Mannose-rich glycans are short carbohydrate chains with mannose or fucose as the terminal sugar, and they are usually present in microbes but not in humans.

13 Antibodies

Antibodies are globulinproteins (immunoglobulins) that are synthesised in serum and tissue fluids and react specifically with the antigen that stimulates their production. Three types of globulins are present in the blood: alpha, beta and gamma.

The antibodies are the gamma globulins. Antibodies are one of the major plasma proteins, and against infection, they are often referred to as the "first line of defence". Von Behring and Kitasato performed the first experiments that proved the physical existence of antibodies in 1890. They demonstrated that serum obtained from rabbits immunised with tetanus or diphtheria toxins could prevent disease in mice infected with such pathogens. The unknown substance that was present in serum and protected on transfer was named "*antitoxin*" by Tizzoni and Cattani in 1891. Subsequently, experimental works by Paul Ehrlich and Jules Bordet demonstrated that a protective response could be generated even against whole cells (erythrocytes). The more inclusive term *antibody* subsequently replaced the term *antitoxin*.

Tiselius and Kabat accomplished the first successful attempt to identify antibody molecules in 1939 by electrophoretic mobility (Fig. 13.1). They demonstrated that hyperimmunisation increased the concentration of γ-globulins in serum and that this fraction contained antibody activity. Since most antibodies are found in these Gamma globulin fractions, these are named immunoglobulins (Ig). Porter in 1959 succeeded in digesting rabbit immunoglobulin G (IgG) with the proteolytic enzyme papain. Edelman and Poulik using a similar method split myeloma globulins into two distinct components, which subsequently were termed heavy (H) and light (L) chains.

The World Health Organization (WHO) in 1964 coined the term immunoglobulin (Ig) for the term antibody. The immunoglobulins synthesised primarily by plasma cells and lymphocytes are structural and chemical concepts. The immunoglobulin includes antibody globulins and the cryoglobulins, macroglobulins, and abnormal myeloma proteins. Thus, all antibodies, which are biological and functional models, are immunoglobulins, but not all immunoglobulins may be antibodies.

Immunoglobulins or antibodies constitute 20–25% of the total serum protein. Depending on physiochemical and antigenic differences, they are divided into five classes of immunoglobulins: (1) immunoglobulin G (IgG), (2) immunoglobulin M (IgM), (3) immunoglobulin A (IgA), (4) immunoglobulin E (IgE) and (5) immunoglobulin D (IgD).

The most crucial function of antibodies is to confer protection against microbial pathogens. Antibodies confer protection in many ways: (1) they prevent attachment of microbes to mucosal surfaces of the host, (2) they reduce the virulence of microbes by neutralising toxins and viruses, (3) they facilitate phagocytosis by opsonisation of microbes and (4) they activate

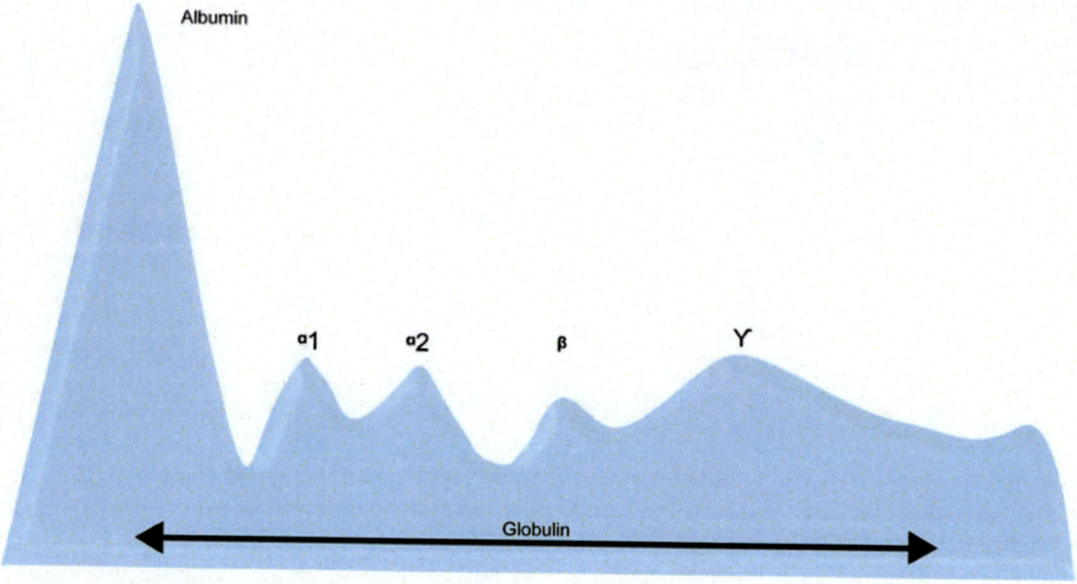

Fig. 13.1 Normal serum protein electrophoresis pattern

complement, leading to complement-mediated activities against microbes.

Structure of Immunoglobulins

Immunoglobulins are glycoproteins. A single antibody molecule has two identical heavy chains and two identical light chains, H2L2, or a multiple (H2L2) *n* of this basic four-chain structure. This gives immunoglobulin an overall "Y" or "T" shape, which is the most widely recognised feature of immunoglobulin structure.

Digestion of Immunoglobulins by Enzymes

The immunoglobulin molecule can be broken into many "sections" or "fragments" by the action of proteolytic enzymes.

The enzyme **papain** cleaves just above the interchain disulphide bonds linking the heavy chains. Peptide bonds in the "hinge" region are broken on the treatment of antibody molecule with papain, producing two identical Fab fragments and one Fc fragment. The Fab fragments produced during cleavage noncovalently bind to the antigen.

The **enzyme pepsin** cleaves just below these bonds, generating different digestion products (Fig. 13.2). Treatment with pepsin cleaves immunoglobulin but at a different site, producing an Fc fragment and two Fab fragments, F $(ab)_2$, which upon exposure to reducing conditions are separated into Fab monomeric units.

Each immunoglobulin molecule, in essence, contains the following components.

Immunoglobulin Chains

They are a complex structure of four polypeptide chains: two identical heavy (typically 55 kDa each) chains and two identical light chains (25 kDa each). The terms "heavy" and "light" refer to the molecular weights of the chains, whereas light chains have a molecular weight of 25,000 Da. The heavy chains are longer, and the light chains are shorter (Fig. 13.3). Sub-isotypes exist for α and γ chains, and this leads to the existence of subclasses of the respective immunoglobulins.

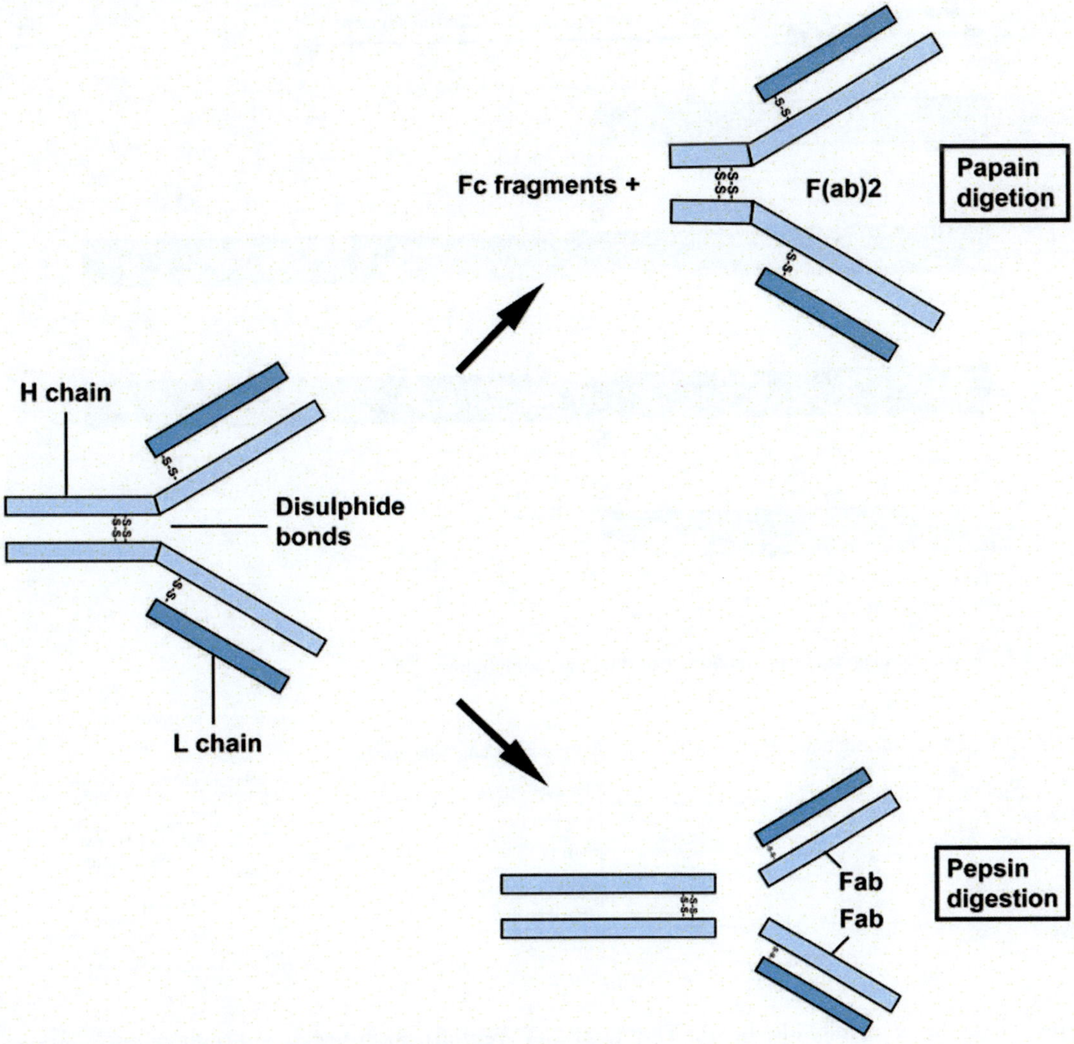

Fig. 13.2 Proteolytic cleavage of antibody by papain and pepsin

Heavy Chains

An immunoglobulin molecule has two heavy chains. Each heavy chain is made of 420–440 amino acids. The two heavy chains are held together by one to five disulphide (S–S) bonds. Each heavy chain is bound to a light chain by a disulphide bond and non-covalent bonds such as salt linkages, hydrogen bonds and hydrophobic bonds to form a heterodimer of heavy and light chains (H-L). Similar non-covalent interactions and disulphide bridges link the two identical heavy and light (H-L) chains to each other to form the basic four-chain $(H\text{-}L)_2$ antibody structure.

The heavy chains of a given antibody molecule determine the class of that antibody. For example, IgM contains mu (μ), IgG contains gamma (γ), IgA contains alpha (α), IgD contains delta (δ) and IgE contains epsilon (ε) heavy chains (Table 13.1). These heavy chains are structurally and antigenically distinct for each class of immunoglobulin. They differ in size, carbohydrate content and antigens.

Light Chains

An immunoglobulin molecule has two light chains. Each light chain is made of 220–240 amino acids. A light chain is attached to the

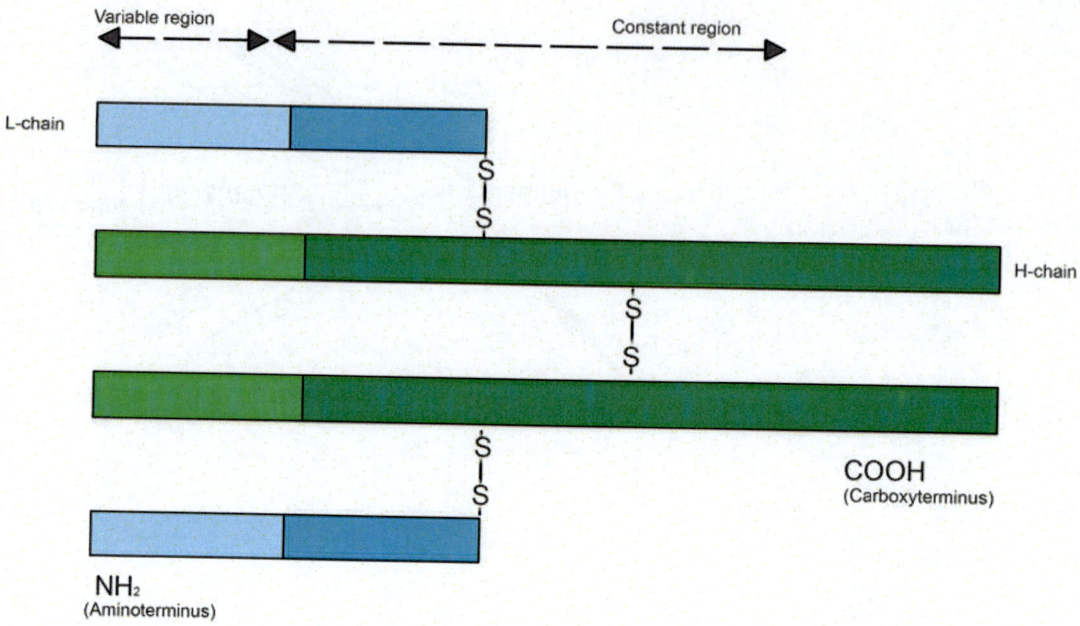

Fig. 13.3 Schematic diagram of the monomer of the immunoglobulin

Table 13.1 Classes of immunoglobulins with their heavy chains and subclasses

Immunoglobulin	Heavy chain	Subclasses
IgG	Gamma (γ)	γ_1, γ_2, γ_3, γ_4
IgM	Mu (μ)	None
IgA	Alpha (α)	α_1, α_2
IgE	Epsilon (ε)	None
IgD	Delta (δ)	None

heavy chain by a disulphide bond. The light chains are structurally and chemically similar in all classes of immunoglobulins. They are of two types: kappa (κ) and lambda (λ). These two differ in their amino acids present in constant regions. Each immunoglobulin has either two κ or two λ chains but never both. The κ and λ chains are present in human serum in a ratio of 2:1.

Variable and Constant Regions

Each polypeptide chain of an immunoglobulin molecule contains an amino-terminal part and a carboxy-terminal part. The amino-terminal part is called the variable region (**V region**) and the carboxy-terminal part is called the constant region (**C region**). Both heavy and light chains contain variable and constant regions.

Variable region: The amino-terminal half of the light or heavy chain, consisting of 100–110 amino acids, is known as variable or *V regions* (VL in light chains and VH in heavy chains). V region is different for each class of immunoglobulin. The variable regions of both light and heavy chains consist of three highly variable regions known as **hypervariable regions**. The antigen combining sites Fab of the antibody molecule that consists of only 5–10 amino acids each are present in the hypervariable region of both light and heavy chains. These antigen-binding sites are responsible for the specific binding of antibodies with antigens. The high specificity of antibodies is primarily due to these hypervariable regions.

Constant region: The carboxyl-terminal half of the molecule is called the constant (C) region. It consists of two basic amino acid sequences. The Fc fragment, found to crystallise under low ionic conditions, is present in the constant region of the heavy chain. The constant region of the heavy chain has many biological functions. It is responsible for activating the complement, binding to cell surface receptors, placental transfer and many other biological activities. The constant region of the light chain has no biological function.

Hinge Region: It is a flexible amino acid stretch present in between C_H1 and C_H2 heavy chains of immunoglobulin molecules which links the two heavy chains by disulphide bonds. It is rich in cysteine and proline amino acids and has an extremely variable amino acid sequence. Different classes and subclasses of immunoglobulins have a varying number of disulphide bonds. The hinge region is present in γ, α and δ heavy chains but ε and μ heavy chains are devoid of it. However, ε and μ heavy chains have an additional hinge-like domain with approximately 110 amino acids.

Immunoglobulin Domains

These regions are composed of three-dimensional folded structures with repeating segments called **domains**. Each heavy chain consists of one variable (VH) and three constant (CH) domains. IgG and IgA have three CH domains (CH1, CH2 and CH3), whereas IgM and IgE have four domains (CH1, CH2, CH3 and CH4). Each light chain consists of one variable (VL) and one constant domain (CL).

Biosynthesis of Immunoglobulins

B lymphocytes and plasma cells take part in the synthesis of immunoglobulins. Resting B cells synthesise only small amounts of immunoglobulins that mainly get incorporated into cell membranes. Plasma cells, the most differentiated B cells, are specialised to produce and secrete large amounts of immunoglobulins.

Heavy and light chains are synthesised in separate polyribosomes of the plasma cell. Plasma cells secrete free light chains; free heavy chains are not secreted by plasma cells. Heavy chains require light chains to form a complete immunoglobulin molecule. Synthesised heavy chains are transported to the endoplasmic reticulum for glycosylation.

Both IgM and IgA are polymeric antibodies, which have one additional polypeptide chain, that is, the J chain. The J chain is synthesised by all plasma cells, including those that produce IgG. However, it is only incorporated into polymeric forms of IgM and IgA. IgM proteins are assembled in two steps. First, the monomeric units are assembled. Then, five monomers and one J chain combine via covalent bonds to produce a pentameric molecule.

Metabolism of Immunoglobulins

The half-life (T 1/2) of immunoglobulin is one of the most commonly used parameters to assess the catabolic rate of immunoglobulins. The half-life is the time taken to reduce circulating immunoglobulins to half after reaching equilibrium. The half-life is found by injecting a radiolabelled isotope (^{131}I) with immunoglobulin.

The IgG is the immunoglobulin class with the longest half-life (average of 21 days) except IgG3. The IgG3 has a considerably shorter half-life (average of 7 days) that is nearer to that of IgA (5–6 days) and IgM (5 days). IgE and IgD have a half-life of 1–5 days and 2–8 days, respectively.

Immunoglobulin Classes

The structure and biological functions of five classes of immunoglobulins (IgG, IgM, IgA, IgE and IgD) are described.

Immunoglobulin G (IgG)

IgG is the most abundant class of immunoglobulins in the serum, forming about

Table 13.2 Properties of immunoglobulins

Characteristics	IgG	IgA	IgM	IgD	IgE
Structure	Monomer	Dimer	Pentamer	Monomer	Monomer
Percentage of total serum	80%	10–13%	5–8%	0.2%	0.002%
Location	Blood, lymph and intestine	Blood, lymph and B cell surface	Secretions	B cell surface, blood and lymph	Bound to mast and basophil cell
Sedimentation coefficient	7	7	19	7	8
Molecular weight (kDa)	150	160	900	180	190
Carbohydrate (%)	3	8	12	13	12
Serum concentration (mg/mL)	12	2	1.2	0.03	0.00004
Half-life (days)	23	6–8	5	2–8	1–5
Heavy chain	$\gamma_1, \gamma_2, \gamma_3, \gamma_4$	α_1, α_2	μ	Δ	ε
Light chain	κ or δ	κ or δ	κ or δ	κ or δ	κ or δ
Complement binding	Classical pathway	Alternate pathway	Classical pathway	None	None
Placental transport	+	–	–	–	–
Present in milk	+	+	–	–	–
Seromucous secretion	–	+	–	–	–
Heat stability (56 °C)	+	+	+	+	–
Binding to tissue	Heterologous	None	None	None	Homologous
Opsonisation	+	–	–	–	–

80% of the total serum immunoglobulin. The average serum concentration of IgG is about 8–16 mg/mL. It contains less carbohydrate than other immunoglobulins. IgG is a 7S immunoglobulin with a molecular weight of 150,000 Da. It has a half-life of 23 days—the longest among all the immunoglobulins. Other properties of the IgG are given in Table 13.2.

IgG structure

The IgG molecule contains two γ chains and two κ, or γ chains joined together by disulphide bonds. The γ chain is a 51-kDa, 450-amino acid residue heavy polypeptide chain. It consists of one variable VH domain and a constant (C) region with three domains designated CH1, CH2 and CH3. The hinge region is situated between CH1 and CH2. Proteolytic enzymes, such as papain and pepsin, cleave an IgG molecule in the hinge region to produce Fab and F (ab′) 2 and Fc fragments.

Biological Activities

In response to infection, IgG antibodies appear late after the appearance of IgM antibodies but persist for a longer period. It confers protection against the microorganisms present in the blood and tissues. It is distributed equally in the intra- and extravascular compartments. It is the only immunoglobulin that crosses the placenta; hence, it confers natural passive immunity to newborns. It takes part in precipitation, complement fixation and neutralisation of toxins and viruses. In addition, it binds to microorganisms and facilitates the phagocytosis of microorganisms.

Subclasses of IgG

There are four subclasses of IgG (IgG1, IgG2, IgG3 and IgG4) in humans with four corresponding γ chain isotypes designated γ-1, γ-2, γ-3 and γ-4. IgG1, IgG2, IgG3 and IgG4 have a difference in their hinge regions. The number and position of the disulphide bond which links two γ chains in IgG are also different. Human γ chain isotypes show only a 5% difference in amino acid sequence in the hinge region. Inter-heavy γ chain disulphide bonds are formed in the hinge region by cysteine residues. There are two inter-heavy chain disulphide bonds in IgG1 and IgG4, whereas IgG2 has four and IgG3 has 11.

Functions of IgG: The functions of four subclasses of IgG vary. IgG1, IgG3 and IgG4 are special because these are the only immunoglobulins with the ability to cross the placental barrier. Therefore, they play an important role in protecting the developing foetus against infections. IgG3, IgG1 and IgG2, in order of their efficiency, are effective in the activation of the complement. IgG1 and IgG3 bind with high affinity to Fc receptors on phagocytic cells and thus mediate opsonisation. On the other hand, IgG4 has an intermediate affinity for Fc receptors, while IgG2 has an extremely low affinity.

The functions of four subclasses of IgG vary. IgG1, IgG3 and IgG4 are special because these are the only immunoglobulins with the ability to cross the placental barrier. Therefore, they play an important role in protecting the developing foetus against infections. IgG3, IgG1 and IgG2, in order of their efficiency, are effective in the activation of the complement. IgG1 and IgG3 bind with high affinity to Fc receptors on phagocytic cells and thus mediate opsonisation. On the other hand, IgG4 has an intermediate affinity for Fc receptors, while IgG2 has an extremely low affinity.

Immunoglobulin M (IgM)

IgM constitutes about 5–8% of total serum immunoglobulins. It is distributed mainly intravascularly. It is a heavy molecule (19S) with a molecular weight varying from 900,000 to 1,000,000 Da (**millionaire molecule**). It has a half-life of 5 days (Table 13.2).

It is also the first immunoglobulin synthesised by a neonate at about 20 weeks of age. IgM is not transported across the placenta; hence, the presence of IgM in the foetus or newborn indicates intrauterine infection. Therefore, the detection of IgM antibodies in serum is helpful for the diagnosis of congenital infections, such as syphilis, rubella and toxoplasmosis.

IgM Structure

IgM is basically a pentamer, composed of five immunoglobulin subunits (monomeric subunits, IgMs) and one molecule of the J chain. Each monomeric IgM comprises two light chains (κ or γ light chains) and two heavy chains (μ). The heavy chains are larger than those of IgG by about 20,000 Da, corresponding to an extra domain on the constant region (CH4). Two subclasses of IgM (IgM1 and IgM2) are described, which differ in their μ chains. IgM1 consists of μ_1 chain, and IgM2 consists of μ_2 chain (Fig. 13.4). Serum IgM has a higher valency when compared to other isotypes, which is due to its pentameric structure with 10 binding sites. Ten small hapten molecules can bind to one IgM molecule, but only fewer molecules can bind due to steric hindrance.

Monomeric IgM, with a molecular weight of 180,000 Da, is expressed as a membrane-bound antibody on B cells. As mentioned earlier, the J chain found in the IgM molecule was believed to play a significant role in the secretion of its polymerised form. Being present on the membrane of B cells, IgM acts as the antigen-binding molecule in the antigen-antibody complex.

Treatment of serum with 2-mercaptoethanol destroys IgM without affecting IgG antibodies. This forms the basis for differential estimation of IgM and IgG antibodies in serum pre-treated with 2-mercaptoethanol.

Biological Activities

IgM is the first immunoglobulin produced in a primary response to an antigen. The immunoglobulin confers protection against the invasion of blood by microbial pathogens. The deficiency of IgM antibodies is associated with septicaemia.

Because of its high valency, Pentameric IgM is more efficient than other isotypes in binding antigens with many repeating epitopes, such as viral particles and red blood cells. It is more efficient than IgG in activating complement. Complement activation requires two Fc regions

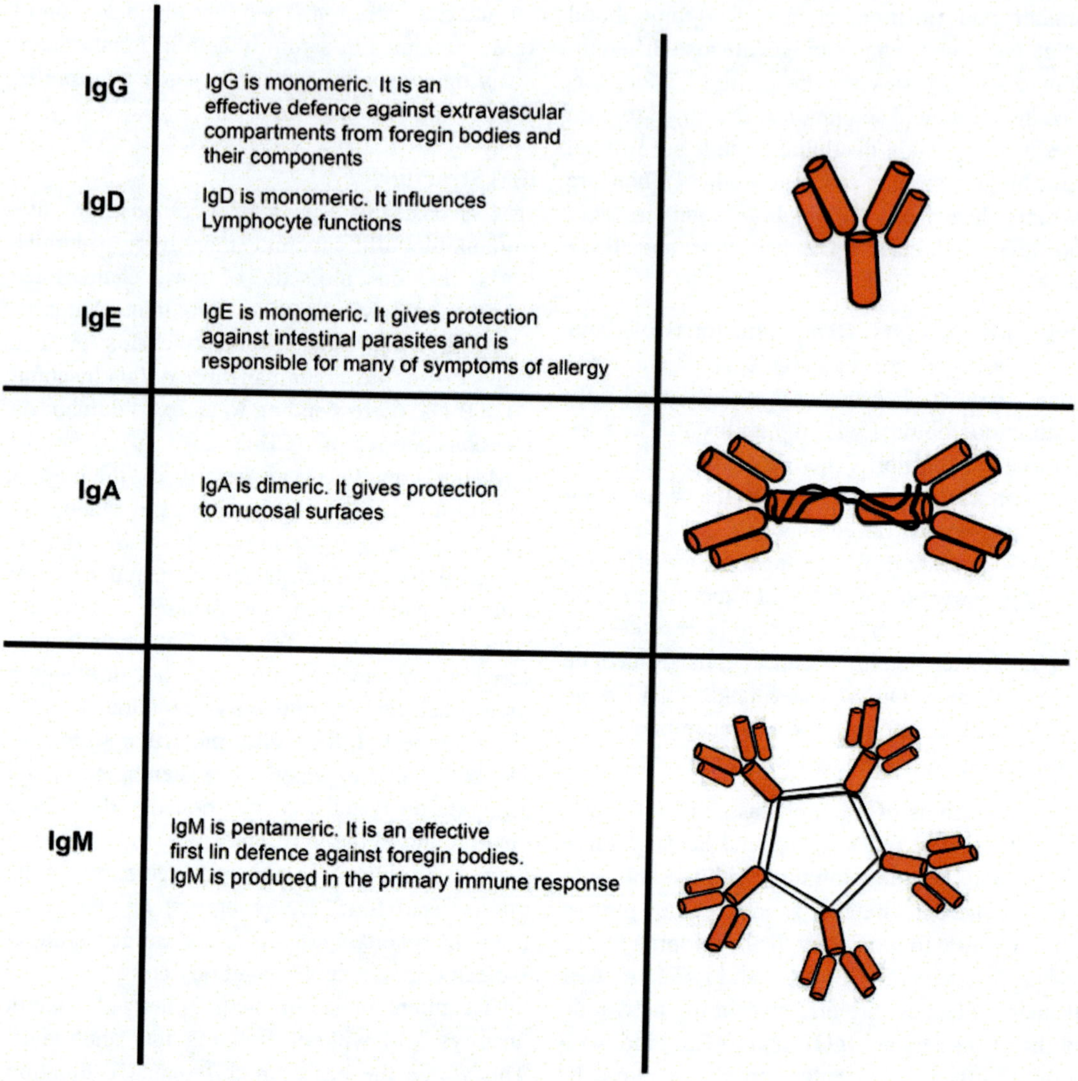

Fig. 13.4 Schematic diagram of immunoglobulins and their functions

nearby, and this requirement is fulfilled by the pentameric structure of a single molecule of IgM.

IgM antibodies are short-lived and disappear early compared to IgG. The presence of IgM antibody in serum, therefore, indicates recent infection.

Immunoglobulin A (IgA)

IgA is the second major serum immunoglobulin, comprising nearly 10–15% of serum immunoglobulin. It has a half-life of 6–8 days. It is the major immunoglobulin in saliva, tears and colostrum (Table 13.2).

IgA Structure

IgA occurs in two forms: **serum IgA** and **secretory IgA**.

Serum IgA is present in the serum and is a monomeric 7S molecule with a molecular weight of 60,000 Da. It has a half-life of 6–8 days. It has two subclasses, IgA1 and IgA2, which are two α-chain isotypes α-1 and α-2, respectively. The α-2 chain has two allotypes, A2m (1) and A2m (2), and does not have disulphide bonds linking heavy to light chains. Differences in the two chains are found

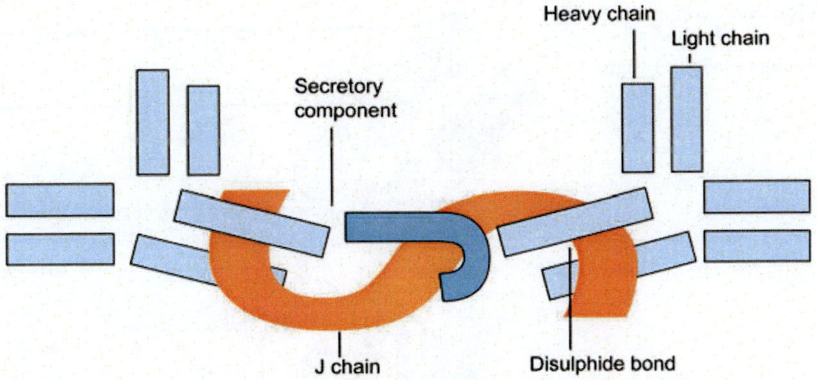

Fig. 13.5 Schematic diagram of secretory immunoglobulin A (IgA)

in two CH1 and five CH3 positions. Thus, there are three varieties of a-heavy chains in humans.

Secretory IgA is a dimer or tetramer 11S with a molecular weight of 400,000 Da. It consists of a J-chain polypeptide and a polypeptide chain called secretory component (SC) or secretory piece (Fig. 13.5). An additional segment of 18-amino acid residues at the penultimate position of the chain contains a cysteine residue where the J chain can be attached through a disulphide bond. The SC is a polypeptide with a molecular weight of 70,000 Da produced by epithelial cells of mucous membranes. The secretory piece protects IgA from denaturation by enzymes secreted by bacteria colonising the intestinal mucosa. It consists of five immunoglobulin-like domains that bind to the Fc region of the IgA dimer. This interaction is stabilised by a disulphide bond between the fifth domain of the SC and one of the chains of the dimeric IgA. IgA-secreting plasma cells are concentrated along mucous membrane surfaces. The daily production of secretory IgA is greater than that of any other immunoglobulin.

Secretory IgA is the major immunoglobulin present in external secretions, such as breast milk, saliva, tears and mucus of the bronchial, genitourinary and digestive tracts. IgA inhibits the adherence of microorganisms to the intestinal mucosa surface, thereby preventing their entry into host tissues. IgA activates the complement not by the classical pathway but by the alternative pathway.

Biological Activities

Secretory IgA protects the mucous membranes against microbial pathogens. As IgA is polymeric, it can cross-link multiple epitopes of a large antigen. It also performs effector functions at the mucous membrane site through which most pathogenic organisms enter. Secretory IgA binds to the viral and bacterial surface antigens to form a complex and inhibits the pathogens from attaching to the mucosal cells. This prevents bacterial colonisation and viral infection. These complexes are eliminated by ciliated epithelial cells in the respiratory tract or gut peristalsis.

Breast milk contains secretory IgA and many other molecules that protect newborns against infection during the first month of life. Breastfeeding plays a vital role in maintaining the health of newborns because the immune system of infants is not fully functional. Defence against certain bacteria such as *Neisseria gonorrhoeae*, *Vibrio cholerae* and *Salmonella* spp. and viruses such as reovirus, polio and influenza are provided by secretory IgA.

Immunoglobulin E (IgE)

IgE constitutes less than 1% of the total immunoglobulin pool. It is present in serum at a very low concentration (0.3 μg/mL). It is mainly found extravascularly inlining of the respiratory and intestinal tracts. IgE is an 8S molecule with a molecular weight of 190,000 Da and a half-life of 2–3 days. Unlike other immunoglobulins that are heat stable, IgE is a heat-labile protein—easily inactivated at 56 °C in 1 h. IgE does not cross the placenta or does not fix the complement. It is susceptible to mercaptoethanol (Table 13.2).

Fig. 13.6 Schematic diagram of immunoglobulin E (IgE)

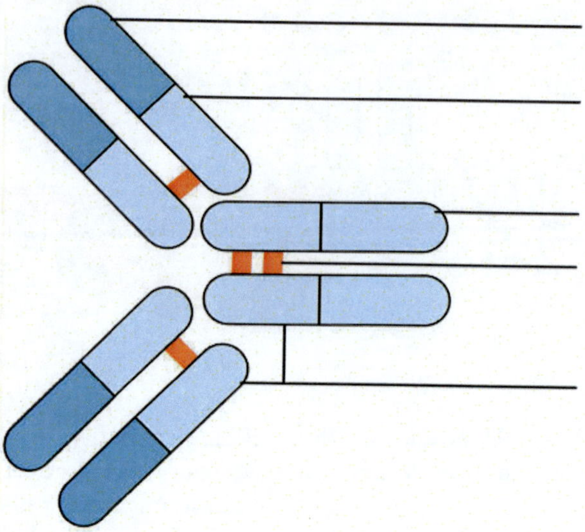

Structure

Two heavy polypeptide chains, along with two κ or two λ light chains, fastened together by disulphide bonds, comprise an IgE molecule. This chain is a 72-kDa, 550-amino acid residue polypeptide chain. It consists of one variable region, designated VH and a four-domain constant region, designated CH1, CH2, CH3 and CH4. This heavy chain does not possess a hinge region. In humans, the **e** heavy chain has 428 amino acid residues in the constant region (Fig. 13.6).

Biological Activities

IgE, also known as **reaginic antibody**, mediates the type I immediate hypersensitivity (atopy) reactions. It is responsible for the symptoms of hay fever, asthma and anaphylactic shock. There are FC receptors in the membrane of blood basophils and tissue mast to which IgE binds. When there is cross-linkage of receptor-bound IgE molecule with antigen, the basophils and mast cells are induced to translocate their granules to the plasma membrane. This then releases their content to the extracellular environment. This process is known as degranulation. Owing to this, many active mediators are released, which induces allergy.

Localised mast cell degranulation induced by IgE may also release mediators that facilitate a build-up of various cells necessary for antiparasitic defence. It plays an essential role in defence against helminthic infections in humans.

Immunoglobulin D (IgD)

IgD comprises less than 1% of serum immunoglobulins. It is a 7S monomer with a molecular weight of 180,000 Da. The half-life of IgD is only 2–3 days (Table 13.2). IgD has the basic four-chain monomeric structure with two δ heavy chains (molecular weight 63,000 Da each) and either two κ or two λ light chains (molecular weight 22,000 Da each) (Table 13.2). IgD is present on the surface of B lymphocytes, and both IgD and IgM serve as recognition receptors for antigens. However, the role of IgD in immunity continues to remain elusive.

Table 13.3 summarises the roles of various immunoglobulins in human defence.

Abnormal Immunoglobulins

Abnormal immunoglobulins are other structurally similar proteins found in the serum in certain

Table 13.3 Role of immunoglobulins in human defence

Immunoglobulins	Functions
IgG	Highest opsonisation and neutralisation Protects foetus and newborn
IgM	First antibody produced in response to the initial infection Especially effective against microorganisms and agglutinating antigens
IgA	Localised protection on mucosal surfaces Found in secretion; higher expression in breast milk
IgD	Serum function not known Found on the surface of mature B cells
IgE	Allergic reaction Possibly lysis of parasitic worms

pathological conditions, such as multiple myeloma, heavy chain disease and cryoglobulinemia and sometimes in healthy individuals.

Multiple Myeloma

Bence–Jones (BJ) proteins were the earliest abnormal proteins described in 1847 and are found in patients with multiple myeloma. These proteins are the light chains of immunoglobulins and hence occur as either κ or λ form. In a patient, it may occur as either κ or λ but never in both forms. BJ proteins have a peculiar property of coagulating at 40–60 °C and redissolving again at a higher temperature of 80–100 °C.

In multiple myeloma, plasma cells synthesising IgG, IgA, IgD or IgE are affected. Myeloma involving IgM-producing plasma cells is known as **Waldenstrom macroglobulinemia**. This condition is characterised by excessive production of the respective myeloma proteins (M proteins) and that of their light chains (BJ proteins). BJ protein can also be detected in amyloidosis cases. The urine dipstick test for proteins is positive if BJ protein is present. It can be detected using the conventional electrophoresis or capillary zone electrophoresis, while immunofixation electrophoresis and immunoelectrophoresis are used for confirmation. In recent years, the free light chain assay has become popular, which measures the concentrations of free κ or λ light chains. A ratio of κ:λ of 0.26–1.65 is considered the reference range. Any increase or decrease from the above lower and upper limits denotes the presence of excess κ and λ chain proteins, respectively.

The advancement in understanding the immunoglobulin function happened only after studying myeloma proteins. For example, the immune sera were absorbed with other myeloma proteins and were used to identify allotypic, idiotypic and isotypic.

Heavy Chain Disease

Heavy chain disease (HCD) is a different disorder, which is a lymphoid neoplasia, characterised by an excess production of heavy chains of the immunoglobulins. This condition is characterised by the production of truncated monoclonal isolated heavy chains without the light chains. There are three varieties, of which the α-HCD is the commonest form with well-defined features, while the other two varieties, namely γ-HCD and μ-HCD, have variable presentations. They are all rare types of B-cell lymphoplasma-cell proliferative disorders. α-HCD presents as lymphoma of mucosa-associated lymphoid tissue (MALT), γ-HCD as lymphoplasmacytoid non-Hodgkin lymphoma, while μ-HCD can present as non-Hodgkin's lymphoma or chronic lymphocytic leukaemia.

Cryoglobulinemia

Cryoglobulinemia is a condition characterised by the presence of cryoglobulins in blood. The condition may not always be associated with disease but is often found in patients with macroglobulinemia, systemic lupus erythematosus and myelomas. Most cryoglobulins consist of either IgG or IgM or their mixed precipitates. In cryoglobulinemia, serum from the patient precipitates on cooling and redissolves on warming.

Antigen Determinants (Paratope) of Immunoglobulin

There are three major types of immunoglobulin antigen determinants or paratopes: isotypes, allotypes and idiotypes. These contribute to immunoglobulin specificity.

Isotypes

The *isotype* of an immunoglobulin refers to the distinct differences in the constant region of the immunoglobulin's light or heavy chain. They are antigenic determinants that characterise the various classes and subclasses of heavy chains and those of light chains. Immunoglobulins are classified based on various heavy chain isotypes. Heavy chains are distinguished by the presence of heavy chain markers, such as μ, γ, *α*, δ and ε in the immunoglobulins IgM, IgG, IgA, IgD and IgE, respectively. The light chains are also distinguished by isotype markers, such as κ and γ. Isotypes are present in all members of a species and all normal human beings. However, some individuals with immunodeficiencies may lack one or more isotypes. Anti-isotype antibodies are used to quantify the immunoglobulin classes and subclasses and diagnose certain immunodeficiency disorders characterised by the absence or decreased production of antibodies.

Allotypes

The *allotype* refers to allelic differences in both variable and constant regions of immunoglobulin, and the prefix "allo" means different in individuals of a species. The allotype markers are present in the constant regions of light and heavy chains. They are Am on *α* heavy chains, Gm on γ heavy chains and Km on κ light chains. Allotype markers are absent on μ, δ and ε heavy chains and λ light chains. More than 25 Gm types, 3 Km allotypes and 2 Am on IgA have been described. Allotypes are present in some but not in all members of a species and are inherited in a simple Mendelian fashion. Individual allotypes are present in individual members of a species but not in all members of species. The allotypes have importance in monitoring bone marrow grafts, paternity tests, and forensic medicine for identifying an individual from blood stains or semen.

Idiotypes

The *idiotype* refers to a specificity that is associated with the hypervariable region. These are nothing but antigenic determinants created by combining sites of an immunoglobulin molecule. Idiotype markers are found on the Fab fragment of an antibody molecule in the hypervariable region of the heavy and light chains. Idiotypes are specific for each antibody molecule. Anti-idiotypic antibodies produced against Fab fragments prevent antigen-antibody interaction. It has been postulated that anti-idiotypic antibodies have a role in regulating immune responses. These antibodies have been found useful in the therapy of B-cell malignancies and vaccine development by stimulating B cells to produce antibodies.

Antibody Diversity

Antibodies have the diversity to respond to a variety of antigens. Antibody diversity means the presence of a large number of antibodies that can bind to a variety of antigens. It is one of the important structure–function relationships in antigen recognition. Antibody repertoire accounts for the overall antibodies with different specificities. The immune system can create billions of different antibodies with a limited number of genes, possibly by recombining different DNA segments during B cell development prior to antigen exposure. The gene fragments used during recombination determine the specificity of generated antibodies.

Clinical Significance

The development of engineered antibodies could be possible for therapeutic purposes based on the knowledge of immunoglobulin genes to treat various lymphomas and autoimmune diseases.

Meanwhile, the genetic variation and mutability in certain viruses cause constantly changing antigens which could, in turn, generate diverse immune responses in the host. This poses a significant challenge in developing vaccines. For instance, the human immunodeficiency virus (HIV) constantly changes its antigen expression pattern, creating significant challenges in developing a vaccine against it.

Class Switching

During an immune response in the host, B lymphocytes can switch expression of immunoglobulin (antibodies) from isotype IgM to other isotypes IgG, IgE or IgA. This biological mechanism is called immunoglobulin class switching, also termed **class-switch recombination (CSR)** or **isotype switching**. During this phenomenon, in the constant region of the antibody, the heavy chain is changed (switched), but the variable region of the heavy chain remains the same. Since the variable region remains unaltered, the class switching process does not affect the antigen specificity and affinity.

Antibody class switching occurs in mature B cells via its membrane-bound antibody molecule (termed B-cell receptor) to generate different antibody classes. It occurs by a unique type of recombination within special G-rich tandem repeated DNA sequences (GAGCT and TGGGG) called switch sites (or SS) located upstream of each heavy chain. Specific recombinases bind to these switch sites and undertake recombination to happen. Class switching is determined by three crucial factors: (1) switch regions/sites, (2) specific switch recombinases and (3) cytokine signals (e.g. IL-4).

Clinical Significance

Failure in immunoglobulin class switching could be responsible for various disease complications. Primary immunodeficiency diseases that include recombination deficiencies (Ig CSR deficiencies) or hyper IgM syndromes (HIGM) are characterised by defective CD40 signalling of B cells. Affected patients' serum shows low levels of IgG and IgA, but in contrast normal or elevated levels of IgM, leading to the increased susceptibility of an individual to frequent bacterial infections of the skin and respiratory tract.

Genetic Regulations of Immunoglobulin Production

The human body can produce an innumerable number of different antibodies with different specificities. This type of diversity is possible because of three reasons: the presence of a large number of genes for the variable regions of the heavy and light chains in the embryonic life; DNA deletions and rearrangements and RNA splicing resulting in the generation of a large number of B cell lineages encoding different heavy and light chains; and high frequency of mutations in the heavy and light chains. It is to be noted that different chromosomes encode the heavy and light chains: Chromosome 14 encodes for heavy chains, while chromosomes 2 and 22 encode for the κ and λ light chains.

Antibody diversity is generated in two stages: somatic recombination and somatic hypermutation.

Somatic Recombination of V(D)J Recombination

V(D)J stands for Variable, Diverse and Joining regions recombinations. B cell maturation results in splicing out the DNA of all genes from each region except one and combining the three remaining genes to form one VDJ segment. Each of these three regions contains multiple gene segments. The heavy chain gene contains gene segments of all three regions, but the light chain gene lacks the D region. V(D)J segment, along with a constant region gene, is the background for subsequent antibody production. For the heavy chains of antibodies and the β and δ genes of the T-cell receptor, one D and one J segments are chosen and brought together to form the D-J rearrangement. Next, a V region is selected and joined with the D-J segment to form a complete VDJ exon. Rearrangements for the immunoglobulin light chain genes and the α and

γ genes of T-cell receptors occur in a single V-J recombination step due to the absence of a D segment from both loci. As a result of DNA rearrangement, one copy of each type of gene segment can go to any given lymphocyte, generating an enormous antibody diversity to the tune of 3×10^{11} combinations. The whole process occurs in the bone marrow for B cells and the thymus for T cells. There is a near-random manner in which V, D and J segments combine and this random nature results in the vast diversity of proteins that can match the innumerable antigens in nature.

Somatic Hypermutation

B cell activation results in rapid cell division, and, as a consequence, the genes encoding the variable regions of both heavy and light chains undergo a high rate of point mutation. This causes diversification of B-cell receptors so that the immune system adapts itself to recognise the plethora of antigens encountered by it during an individual's lifetime. This process is highly programmed so that only one nucleotide change per variable gene per cell division occurs. As a result, both high-affinity and low-affinity antibodies can be generated, but normally the high-affinity antibodies out-compete the low-affinity antibodies for better function and survival. This phenomenon is termed affinity maturation.

Monoclonal Antibodies

Antibodies that arise from a single clone of cells (e.g. myeloma) are homogeneous and are called monoclonal antibodies. For example, in multiple myeloma, antibodies are produced by a single clone of plasma cells against a single antigenic determinant, and hence antibodies are monoclonal. The monoclonal antibodies differ from polyclonal antibodies, which are heterologous and formed by several different clones of plasma cells in response to antigen.

Kohler and Milstein, in 1975, were the first to describe a method for producing monoclonal antibodies against the desired antigen, for which they were awarded the Nobel Prize in 1984. Monoclonal antibodies are produced by the fusion of myeloma cells with antibody-producing cells, resulting in the production of hybridomas. Such hybridomas produce virtually unlimited antibodies that are useful in research and diagnostics.

Preparation of Monoclonal Antibodies

Hybridoma cells are prepared in the following ways (Fig. 13.7):

- First, an animal (e.g. mouse) is immunised with the antigen of interest.
- Spleen cells (lymphocytes) are then fused with mouse myeloma cells and grown in culture, which is deficient in the enzyme hypoxanthine phosphoribosyltransferase (HPRT).
- Fusion of the cells is facilitated by adding certain chemicals, such as polyethylene glycol. The fused cells are grown in a special culture medium (HAT medium) that supports the growth of the fused hybrid cells but not of the parent cells.
- Finally, the resulting clones of cells are screened to produce antibodies to the antigen of interest.
- These clones are then selected for continuous cultivation. The hybridomas can be maintained indefinitely and will continue to produce monoclonal antibodies. Myeloma cell provides hybrid cell immortality, whereas splenic plasma cell provides the antibody-producing capacity.

Humanised (Chimeric) Monoclonal Antibodies

Since mouse monoclonal antibodies are not suitable for human therapeutic use, humanised monoclonal antibodies, called chimeric antibodies, have been produced with modification of the original technique. The chimeric antibodies consisting of human constant regions and mouse variable regions are being prepared for use in the treatment of leukaemia. Chimeric antibodies are also used to kill tumour cells by delivering toxins, such as diphtheria, to tumour cells or by killing tumour cells through complement-mediated cytotoxicity.

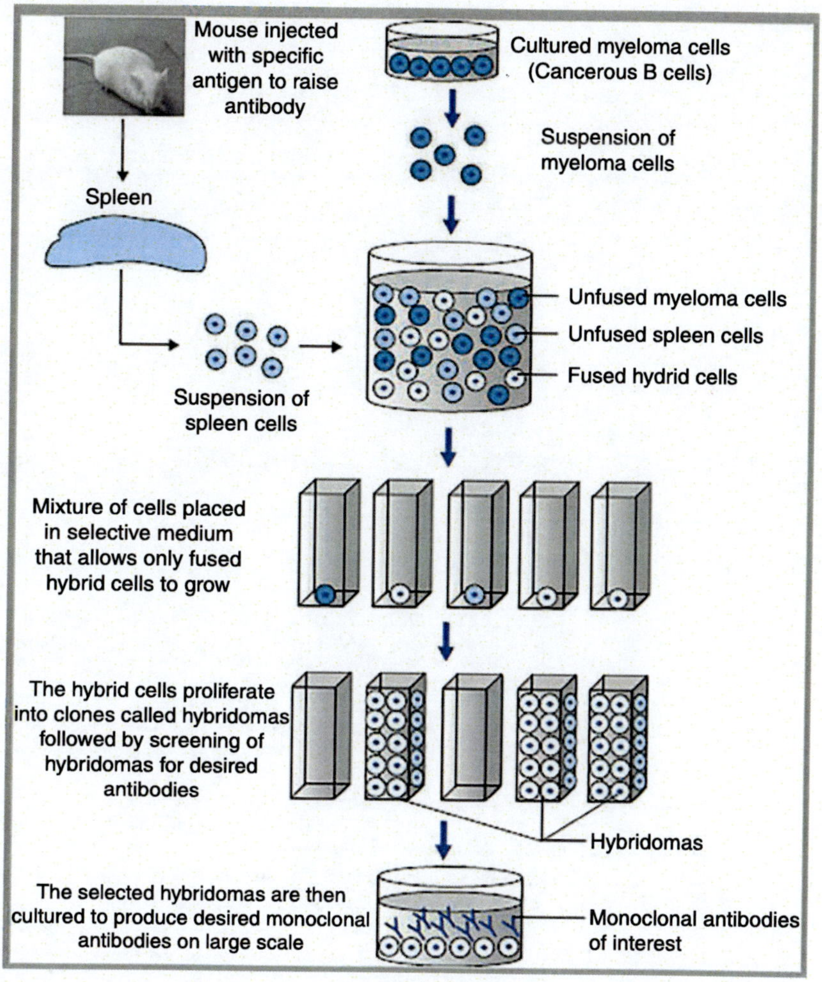

Fig. 13.7 A schematic diagram showing the production of monoclonal antibodies

Application of Monoclonal Antibodies

Monoclonal antibodies are now used widely in research and therapeutics. They are used in various clinical situations, such as to treat cancer and autoimmune diseases. They are used to induce immunosuppression in transplant surgery and prevent infectious diseases. They are extensively used in diagnostic applications in bacterial, viral, parasitic and other infections. Monoclonal antibodies are used widely as diagnostic reagents in ELISA, immunofluorescence and other techniques in diagnostic microbiology.

Antibody Engineering

Recent advances in biotechnology have made possible the development of human monoclonal antibodies. This was made possible by fusing human genes (of a particular antibody fragment) and bacteriophage genes. Bacteria were then infected with these bacteriophages to produce large quantities of desired antibodies. This approach has been successfully employed to build whole libraries of antibodies. This approach has made revolutionary changes in the field of immunotherapy.

Complement System 14

The term **complement** refers to the ability of some non-specific proteins in human serum to complement, that is, augment the effects of other immune system components. The complement system, an important component of the human innate host defence system, consists of approximately 20 proteins present in normal human serum.

The Complement

The complement system is an extremely powerful system comprising rapidly acting glycoproteins, several pro-enzymes and components, called triggered enzyme cascades. Each pro-enzyme in the enzyme cascade activates molecules of the succeeding component culminating in amplifying the response at each step. The complement system also includes kinin and fibrinolytic and the coagulation system. All normal individuals always have complement components in their blood.

Properties of Complement

Complement is present in sera of all mammals, including humans, and in lower animals, including birds, amphibians and fishes. These are heat-labile substances that are inactivated by heating serum at 56 °C for 30 min. These are glycoproteins and are synthesised primarily by liver cells and to a very less extent by macrophages and many other cell types. The synthesis rate of the various complement glycoproteins increases when complement is activated and consumed.

The complement usually does not bind to the antigen or antibody but only to the antigen–antibody complex. The importance of the complement lies in the fact that it contributes to both the acquired and innate immunities of an individual.

Nomenclature of Complement

Complement components are designated by numerals, namely, C1–9. These components circulate in the plasma as pro-enzymes that are functionally inactive. Activation involves cleavage by proteolysis into peptide fragments. The fragments are designated with lowercase suffixes—for example, C3 is cleaved into two fragments, C3a and C3b. Normally, the large fragment is designated "b", and the small fragment "a". However, for historical reasons, concerning the fragments of C2, the large fragment is designated C2a, and the small one is designated C2b.

Components of Complement

The main components of the complement system are all protein molecules with different functions. These include the complement protein C; factors P, B and D; mannose-binding lectin (MBL); and MBL serine proteases (MASP).

C1 Complex

This complex consists of a single C1q molecule bound to two molecules, each of C1r and C1s. C1q is a calcium-dependent sugar-binding protein that belongs to the collectin family of proteins. When C1q binds to a pathogen, it causes a conformational change in the C1r-C1s component. This leads to activation of C1r, which has enzymatic activity and causes cleavage of C1s, an active serine protease. C1q forms a bridge between innate and adaptive systems since it can bind directly to a pathogen and antigen–antibody complex. The activated C1 enzymes act on C4 and C2, causing their cleavage and initiating the classical pathway.

Factor P (Properdin)

It is a positive regulatory factor capable of binding to microbial surface and stabilising the C3b. Bb complex in the alternative pathway. Its deficiency predisposes the individual to meningococcal infections.

Factors B, D

These are components of the alternative pathway. Factor D is a protease that can cleave Factor B to Ba and Bb.

MBL and MASP

Mannose- or mannan-binding lectin is a serum protein that binds to the mannose residue on bacterial and viral surfaces. It is similar to C1q and belongs to the same collectin family. MASP-1 and MASP-2 are two enzymes, and MBL forms a complex with them. The binding of the MBL complex to a pathogen surface causes activation of MASP-1 and MASP-2, resulting in cleavage of C4 and C2. In this way, the lectin pathway causes complement activation analogous to the classical pathway.

Functions of Complement

There are four main effects of complement.

1. It causes **lysis of cells** (such as bacteria, viruses, allografts and tumour cells).
2. It generates mediators of **chemotaxis** that participate in triggering specific cell functions, inflammation and secretion of immunoregulatory molecules.
3. It facilitates **opsonisation**, the process by which bacteria are more readily and more efficiently engulfed by phagocytes.
4. It causes **immune clearance**, in which immune complexes from the circulation are removed and are transported to the spleen and liver.

Complement Activation

Complement activation occurs through any of the following three pathways: (1) **classical pathway**, (2) **alternative or properdin pathway** and (3) **lectin pathway**. Of these, alternative and lectin pathways are important in the host's innate immunity since they activate the complement system when the human host is infected by a microorganism for the first time because the antibody required to trigger the classical pathway is not present.

All three activation pathways lead to the activation of C3, resulting in the production of C3b. Hence, C3b is the central molecule in activating the complement cascade. The C3b has two critical functions to perform. First, it combines with other components of the complement to produce C5 convertase, the enzyme that leads to the production of **membrane attack complex (MAC)**; and second, it opsonises bacteria due to the presence of receptors for C3b on the surface of the phagocytes.

The final steps leading to the formation of an **MAC** are the same in all the pathways. A sequential, rapid cascading pattern ensues when these complement components are activated, which can break away or bind to a previously activated complement component complex. Usually, once each component is activated, it amplifies the successive steps in the series by activating many other molecules involved in the cascading process.

Classical Pathway of Complement Activation

The classical pathway is a chain of events in which complement components react in specific sequences as a cascade resulting in cell lysis. It is activated by an antibody bound to antigen but never by the native or free antibody.

Activators of the Classical Pathway

The classical pathway of complement is activated by immunoglobulins IgM and IgG. Usually, the most efficient immunoglobulins are IgG3, followed by IgG1 and IgG2. IgG4 do not show their efficiency in activating the classical pathway. Native, free IgG or IgM do not activate the complement system. A single, native IgG molecule will not bind and activate the complement pathway. However, if antibodies of the IgG class are aggregated by antigen binding, this will result in complement fixation and activation. Antigen–antibody complex formation helps induce the Fc portion of the IgM molecule, which exposes a binding site for the C1 component of the complement system to undergo conformational changes.

The classical pathway is also activated by Staphylococcal protein A, C-reactive protein and deoxyribonucleic acid (DNA).

Steps in Activation of the Classical Pathway

Activation of the classical pathway of the complement system is usually very simple; it begins with the reaction of soluble antigen–antibody complex (immune complexes) formation or in the presence of a suitable target (a bacterial cell) the antibody binds to the antigen (Fig. 14.1). Following are the sequential steps in the activation of the classical pathway:

1. **Activation of C1 is the first step** in the cascade of classical pathway activation. The C1 is a complex of three different types of molecules: **C1q, C1r** and **C1s**. C1q first combines with the Fc portion of the bound antibody, IgM or IgG. C1q has six combining sites. Effective activation occurs after C1q is attached to IgM or IgG by at least two of its binding sites. Two IgG molecules or one molecule of IgM can therefore initiate this process.
2. **Binding of C1 to the antigen–antibody complex** is the next step. C1q binding in the presence of calcium ions leads to activation of **C1r** and **C1s**, leading to C1qrs complex. Activated C1s is an esterase that splits C4 into two fragments: a small soluble fragment (C4a) and a larger fragment (C4b). C4a has anaphylatoxin activity, and C4b binds to the cell membrane along with C1.
3. C4b in the presence of Mg^{2+} splits C2 into C2a and C2b. The smaller fragment (C2b) diffuses away, while the larger fragment (C2a) remains attached to C4b. The resulting C4b2a complex possesses enzymatic activity and is called **C3 convertase**, which converts C3 into an active form.
4. The C3 convertase activates thousands of C3 molecules and splits these molecules into C3a and C3b. One single molecule of C3 can generate more than 200 molecules of C3b, which results in massive amplification of the cascading reaction at this step. At this juncture, C3b binds with C4b2a to give rise to a trimolecular complex called C4b2a3b, which is also known as C5 convertase. Then this C5 convertase enzyme splits the C5 molecule into two separate molecules, C5a and C5b, in which C5a diffuses away, while the C5b molecule attaches to another new molecule that is C6 to initiate the formation of C5b-9 complex, otherwise known as MAC.

Formation of the MAC: This involves the participation of the complement components C5b, C6, C7, C8 and C9. The initial step in the MAC formation is the cleavage of C5 with the formation of C5b. Then, C5b binds to C6, and the

Fig. 14.1 Classical pathway of activation of the complement

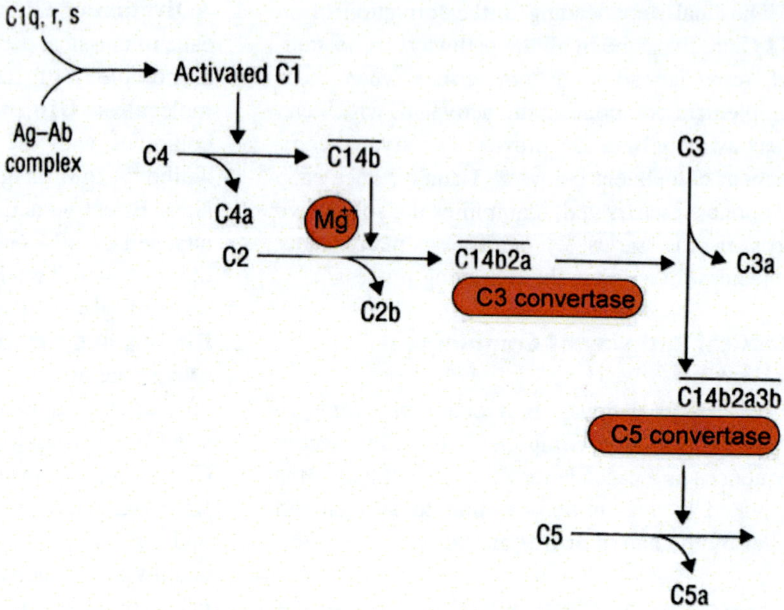

resulting C5b,6 complex binds to C7. As a result of these reactions, a conformational change occurs, and a hydrophobic site is exposed on C7, which then inserts into the cell membrane through its lipid layer. In the subsequent stages, C8 and C9 also get bound to C5b,6,7 with the exposure of their hydrophobic sites and insertion into the lipid bilayer of the cell membrane. The polymerisation of 10–16 molecules of C9 induced by C8 results in the pore-forming complex or MAC, represented as C5b,6,7,8,9. Electron microscopic studies have shown that the MAC has a hydrophobic external end that helps in binding and a hydrophilic internal channel. The channel has a diameter of about 100 Å, which allows the cell's internal contents to leak out. It also causes disturbance of proton gradient of the cell membrane, attack by lysozyme and eventual destruction of the cell (Fig. 14.2).

Alternative Pathway of Complement Activation

Pillemer first described the alternative pathway in 1954. It differs from the classical pathway in (1) the nature of activating substances and (2) the sequence of events itself. The alternative pathway is unique in not requiring antigen–antibody complexes to activate the complement. This pathway does not depend on antibodies and does not involve the early complement components (C1, C2 and C4) for activation of the complement. It, therefore, can be activated before the establishment of an immune response to the infecting pathogen (Fig. 14.3).

Activators: of the alternative pathway of complement activation include (1) IgA, (2) IgD,

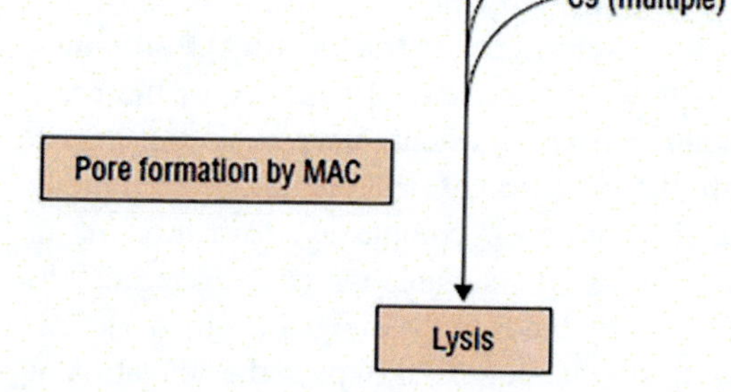

Fig. 14.2 Formation of membrane attack complex (MAC)

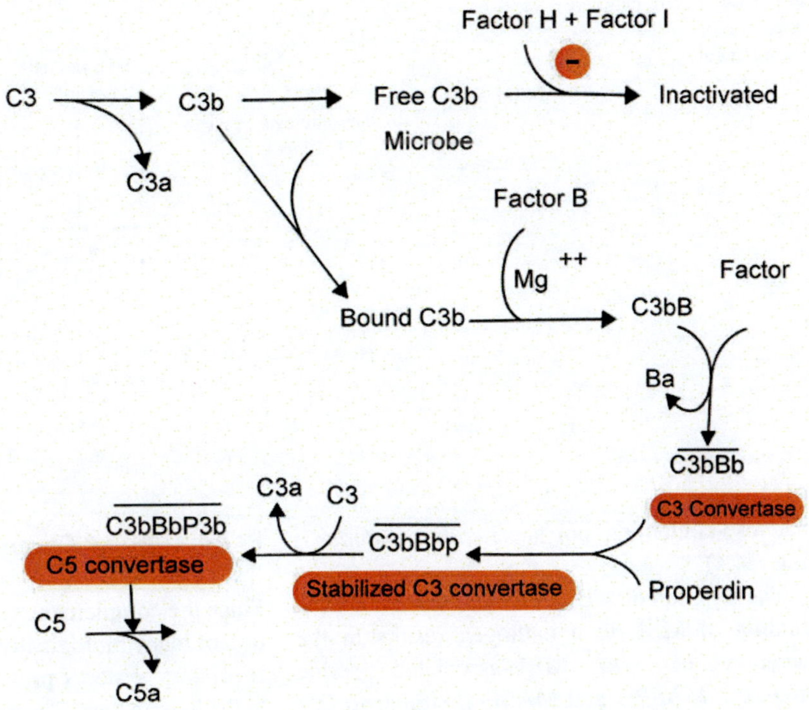

Fig. 14.3 Alternative pathway of activation of the complement

(3) bacterial endotoxin, (4) cobra venom factor and (5) nephritic factor.

Steps of Activation of Alternative Pathway

The initial component of the alternative pathway involves four serum proteins: C3b, factor B, factor D and properdin.

1. In the first step, a serum protein called factor B (**C3 pro-activator**) binds with C3b to form the C3bB complex. The Mg^{2+}, the only ion required to activate the alternative pathway, stabilises the interaction between C3b and factor B.
2. The C3bB is split into two fragments, Ba and Bb, by another serum protein called factor D **(C3 pro-activator convertase).** The Ba is released into the medium, and the Bb binds to C3b, forming the C3bBb complex, which possesses the **C3 convertase** activity.
3. The C3bBb complex is extremely labile and has a half-life of only 5 min, but by binding with properdin (also called factor P), it stabilises C3 convertase, which becomes heat stable.
4. The C3 convertase then hydrolyses C3 to produce the MAC, finally, leading to further steps in the cascade, in the same way as occurs in the classical pathway.

Lectin Pathway of Complement Activation

Lectin pathway utilises all the components of classical pathway activation, except the C1 proteins. As the name suggests, the lectin pathway is triggered by the **mannose-binding lectins**. The mannose-binding lectin (MBL) is one such acute-phase protein produced during the inflammatory response that takes part in this pathway of complement activation. MBL is a large serum protein that binds to non-reduced mannose, fructose and glucosamine on *Neisseria*, *Salmonella*, *Listeria*, *Candida albicans*, *Cryptococcus neoformans* and other cell surfaces with

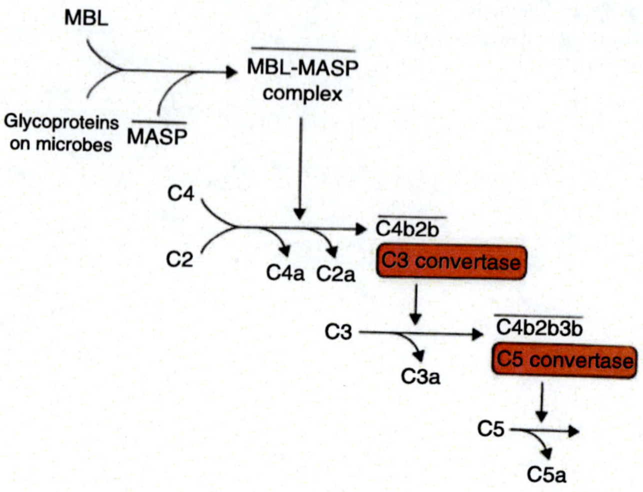

Fig. 14.4 Lectin pathway of activation of the complement

mannose-containing polysaccharides (*mannans*) (Fig. 14.4).

The function of MBL is similar to that of C1q. Binding of MBL to a pathogen results in the secretion of two MBL-associated serine proteases: MASP-1 and MASP-2, similar to C1r and C1s, respectively. For activation of MASPs, a trimolecular complex formation of MBL/MASP-1/MASP-2 is required and forms a C1-like active complex. Subsequently, it causes cleavage and activation of C4 and C2 to produce C5 convertase without any need for specific antibody binding. Subsequently, it proceeds to produce MAC in the same way that occurs in the classical and alternative pathways.

The formation of the MAC is the terminal sequence of all three different pathways, including the classical pathway. All three pathways converge at the step involving the formation of the MAC. The MAC forms a transmembrane channel of 100 Å diameter in the cell. That allows the free exchange of ions between the cell and its surrounding medium. In turn, this results in a rapid increase in osmotic pressure of the cell due to the rapid influx of ions into the cell and their association with cytoplasmic proteins. Ultimately, this would lead to a rapid influx of water molecules inside the cell, swelling of the cell and some cell types may also rupture, resulting in cell **lysis**.

Differences between classical, alternative and lectin pathways are summarised in Table 14.1.

Regulation of Complement Activation

Since the complement system involves the formation of many biologically active substances, many regulatory systems prevent unwarranted damage to the human host. The following are regulators of the complement activation.

Inhibitors

C1 esterase inhibitors (C1sINH) are present in normal serum. These prevent the formation and function of the C1qrs complex by causing C1s to dissociate from C1qrs. The C1sINH inhibitor is a heat-labile alpha serum. It also inhibits multiple esterases found in blood, including plasmin, kininogen and the Hageman factor. This inhibitor will not prevent normal complement cascade but would check autolytic prolongation of complement pathways. **S protein** is another example of a complement pathway inhibitor. The S proteins are present in normal serum and bind to C67, thereby controlling the cytolytic action of the MAC (Fig. 14.5).

Inactivators

Multiple substances have inhibitory effects over different steps of the activation sequence of the classical pathway. These are considered to be host cell protective mechanisms. These include:

1. **Factor Ia** serum beta-globulin, earlier called C3b, C4bINAC or conglutinogen-activating

Table 14.1 Comparison of classical, alternative and lectin pathways of complement activation

Characteristic	Classical pathway	Alternative pathway	Lectin pathway
Mode of activation	Chain of events occur in specific sequence following C1 activation	Activation of C3 without involvement of C1, C4, C2	Binding between mannose-binding lectin and mannose residues on microorganisms
Activators	Requires binding of C1 to antigen–antibody complex	Bacterial endotoxins, IgA and IgD, cobra venom factor and nephritic factor	No role for antibodies; similar to alternate pathway
Innate immunity	Cannot be considered as a component of innate immunity	Component of the innate immunity	Component of innate immunity

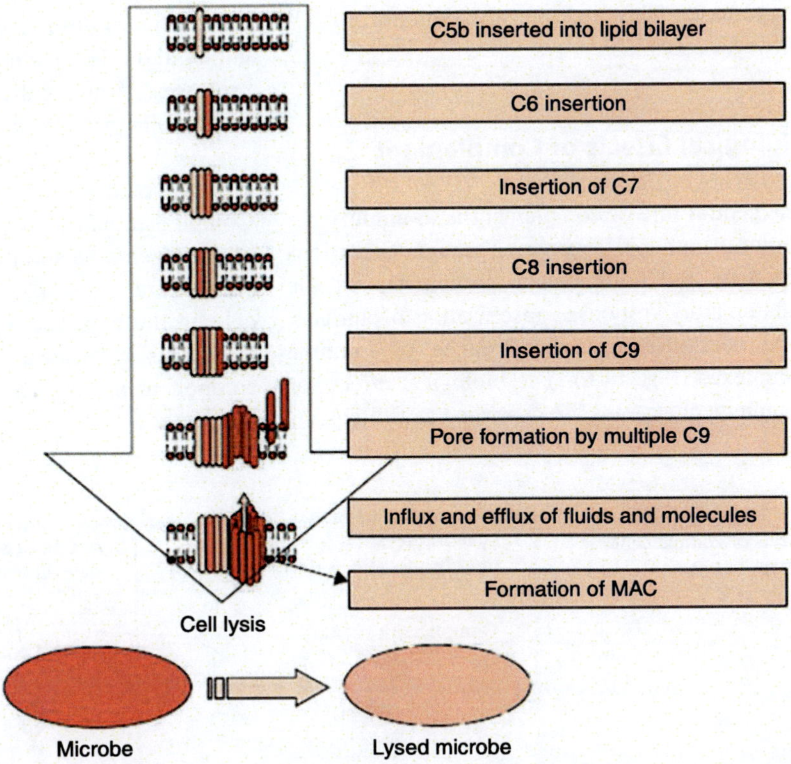

Fig. 14.5 Action of membrane attack complex (MAC)

factor (KAF), by means of the alternative pathway, provides homeostatic control of C3 activation.

2. **Factor H** is another serum beta-globulin that augments Factor I in modulating C3 activation.
3. **Anaphylatoxin activator**, an alpha-globulin, has the enzymatic ability to break down the anaphylatoxins (C3a, C4a and C5a) released during complement pathways.
4. **C4-binding** protein checks the activity of cell-bound C4b.
5. **Decay-accelerating factor (DAF)** is another inhibitory substance located in a large variety of host cell membranes. This factor is termed so because it can accelerate the dissociation of active C4b2a complexes by stopping their ability to activate native C. Additionally, DAF can also bind to membrane-bound C4b and C3b to prevent its subsequent interaction with C2a and factor B, respectively, because of

which the formation of two types of C3 convertases (C4b2a and C3bBb) will be stopped. Due to this, C3 breakdown would decrease, and the host cells tend to undergo sparing from complement-mediated membrane damage.

Regulation of alternative pathway

The alternative pathway has its own set of regulatory proteins and mechanisms. It is mediated by the binding of factor H to C3b and cleavage of this complex by specific plasma inhibitor factor I, a protease. This reduces the amount of C5 convertase available.

Biological Effects of Complement

The central role of complement is to amplify the humoral immune response. Through its various products, the complement participates in the inflammatory response, opsonisation of antigen, viral neutralisation and clearance of immune complexes (Fig. 14.6). The biological effects of complement systems are described as follows.

Phagocytosis

C5a is a chemotactic molecule specifically recognised by polymorphonuclear leukocytes or phagocytic cells. This chemical substance is responsible for the migration of leukocytes to a tissue, where an antigen–antibody reaction is occurring. At that specific reaction site, phagocytic cells start to recognise the opsonised particles, and finally, the cell engulfs them inside and neutralises.

The C5a has a role in the chemotaxis of neutrophils. Similarly, it also activates these cells, which in turn cause their reversible aggregation and release some stored enzymes such as proteases. It also enhances the adhesiveness of neutrophils to the endothelium.

Opsonisation

Complement plays a vital role in the opsonisation of pathogenic bacteria and viruses. Bacteria and viruses are easily phagocytosed by phagocytic cells in the presence of complement component C3b. This is because the receptors for the C3b component are present on the surface of many phagocytes.

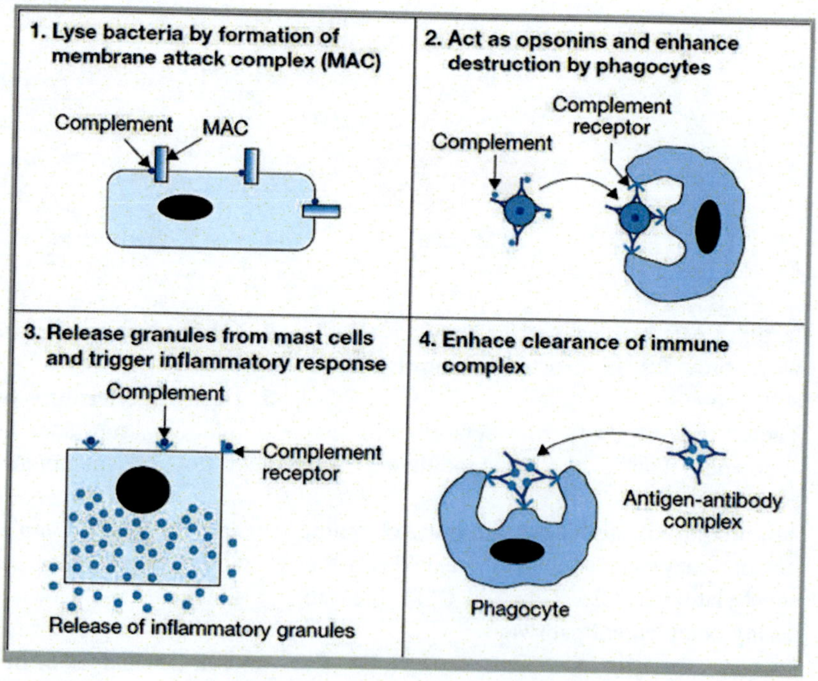

Fig. 14.6 Biological effects of complement system

Hypersensitivity Reactions

Complement participates in type II (cytotoxic) and type III (immune complex) hypersensitivity reactions. The C3a, C4a and C5a components stimulate degranulation of mast cells with the release of mediators, such as histamine. Upon binding with basophils and mast cell receptors, the C3a fragments induce them to release vasoactive molecules such as heparin and histamine. Once histamine is released into the tissues, it causes an increased blood capillary permeability and contraction of smooth muscles. Further, fluids are released into the surrounding tissue, causing tissue swelling and oedema.

These two molecules, C3a and C5a, have been reported to possess direct activity on endothelial cells resulting in increased vascular permeability. The end consequences are more or less similar to a classical anaphylactic reaction where IgE antibodies bound to the mast cells and basophil membranes react with the corresponding antigen. Hence, C3a and C5a are also called anaphylatoxins.

Cytolysis

Complement mediates cytolysis. Insertion of C5b–9 complex (MAC) into the cell membrane leads to killing or lysis of erythrocytes, bacteria and tumour cells. The insertion of the MAC complex results in disruption of the membrane, leading to the entry of water and electrolytes into the cell.

Autoimmune Diseases

In autoimmune disorders such as rheumatoid arthritis and systemic lupus erythematous, the levels of the complement system components are reduced in serum. The complement system also plays a pivotal role in the pathogenesis and disease progression of autoimmune haemolytic anaemia.

Endotoxic Shock

Endotoxin elicits a robust immune response; it is also an efficient activator of the alternative complement pathway. Therefore, there will be a massive C3 fixation and platelet adherence during endotoxic shock. Because of the huge destruction of platelets, the platelet factors are released into the blood, resulting in disseminated intravascular coagulation and thrombocytopenia. Similar pathogenesis mechanisms may be observed in Gram-negative bacterial septicaemia and dengue haemorrhagic fever. Shwartzman reaction is a rare reaction of a body to an endotoxin, which results in thrombosis in the affected tissue. Protection against Shwartzman reaction is conferred in individuals with depleted complement levels in their serum.

Generally, the complement system gets activated on Gram-negative bacterial invasion, but some Gram-negative and most Gram-positive bacteria have evolved immune-evasion strategies to overcome the complement-mediated damage and are listed in Table 14.2.

Conglutination

In bovine serum, there is an unusual protein present called conglutinin (K). This protein is responsible for clumping particles or cells coated with C. This type of agglutination reaction is called conglutination. Conglutinin always binds with the C3 molecule only. Therefore, conglutinin acts as an antibody to C. However, it is not an immunoglobulin molecule, but it necessitates the presence of Ca^{2+} for its activity. Immunisation with complement-coated materials can be useful in producing antibodies with conglutinin-like activity such as immunoconglutinin (IK).

Deficiency of Complement

Complement plays a vital role in the well-being of humans. The deficiency of various components may result in many diseases. **Angioedema** is caused by an inherited deficiency of C1 esterase inhibitors. The low level of C1 esterase inhibitors leads to overproduction of esterase. This leads to an increase in the release of anaphylatoxins, which cause capillary permeability and oedema. **Complement-mediated haemolysis** is caused by acquired deficiency of DAF. The condition manifests clinically as paroxysmal nocturnal haemoglobinuria. Inherited or acquired deficiency of C5–8 components greatly enhance **susceptibility to *Neisseria* bacteraemia and other**

Table 14.2 Microbial evasion of complement-mediated damage

Pathogen	Microbial components	Mechanism of evasion	Examples
Gram-positive bacteria	Peptidoglycan layer	Insertion of membrane attack complex (MAC) is prevented by a thick layer of peptidoglycan	*Streptococcus*
	Bacterial capsules	Physical barrier between C3b deposited on bacterial membrane and CR1 on phagocytic cells	*Streptococcus pneumoniae*
Gram-negative bacteria	Lipopolysaccharides (LPS)	Sidechains of LPS (or antigen) prevent the insertion of MAC into bacterial membrane, preventing cell damage	Resistance strains of *Escherichia coli* and *Salmonella typhimurium*
	Other membrane proteins	MAC specifically acts with membrane proteins and fails to insert into bacterial membrane	Resistance strains of *Neisseria gonorrhoea*
	Elastase	Chemotactic and anaphylatoxic properties of C3a and C5a are inactivated by microbial elastase	Elastase of *Pseudomonas aeruginosa*
Other pathogens	Proteins that mimic complement regulatory proteins	Various proteins present in bacteria, viruses, fungi and protozoa inhibit the complement cascade	Vaccinia virus, herpes simplex, Epstein–Barr virus, *Toxoplasma gondii* and *Candida albicans*

infections. Deficiency of C3 leads to severe recurrent **pyogenic sinusitis** and **respiratory infections**. The synthesis of sufficient quantities of complement is reduced in patients with severe liver disease, such as chronic hepatitis or alcoholic cirrhosis. These patients, therefore, are highly susceptible to infections caused by pyogenic bacteria. Diseases associated with deficiencies of complement components are listed in the Table 14.3.

Biosynthesis of Complement

Various components of the complement are synthesised at various sites of the body. For example, C1 is synthesised in the intestinal epithelium, C2 and C4 in the macrophages, C5 and C8 in the spleen and C3, C6 and C9 in the liver. There is an increase in C3, C4, C5 and C6 in the acute phase of inflammation. Complement and some other plasma proteins known as acute-phase reactants show a rise in acute inflammation.

Quantitation of Complement

Estimation of complement proteins is useful in various conditions where there are hereditary deficiencies of a particular component or disturbance in complement levels due to an ongoing

Table 14.3 Diseases associated with deficiencies of complement components

Deficiency	Disorder	Salient features
C1 esterase inhibitor	Hereditary angioedema	Transient but recurrent localised oedema in the skin, gastrointestinal tract and respiratory tract
C1q	Associated with hypogammaglobulinemia and severe combined immunodeficiency disease	Repeated infections
C2 and C4	Similar to systemic lupus erythematosus (SLE)	Due to failure in clearance of immune-mediated complexes
C3	Severe recurrent pyogenic infections	*Streptococcus pneumoniae* infections
C5	Impaired chemotaxis	Increased susceptibility to bacterial infection
C5–C8	Bacteraemia	Gram-negative diplococci and toxoplasmosis
C5–C9	Multiorgan disease syndrome	Multiorgan involvement and Neisserial infections

disease process. Recurrent infections, mainly by *Neisseria* spp. due to complement defects, autoimmune diseases like systemic lupus erythematosus associated with inherent complement deficiency, membrano-proliferative glomerulonephritis, haemolytic uraemic syndrome, hereditary angioedema and paroxysmal nocturnal haemoglobinuria are some of the conditions where complement quantitation is indicated.

Various methods are available for complement estimation differing in sensitivity and specificity. They can be grouped as follows:

Functional Assays

Total Complement Activity of the Classical, Lectin and Alternative Pathways

The total complement activity can be assessed by the haemolytic tests, which were the first test developed for this purpose. The assay can be performed using antibody-sensitised sheep erythrocytes in tubes or agarose gel. The test is reported as reciprocal of the sample dilution causing 50% (CH50) or 100% (CH100) haemolysis. Guinea pig, rabbit or chicken red blood cell (RBC) is employed for detecting alternate pathway activity (AH50). Enzyme-linked immunosorbent assay (ELISA) tests have also been developed where the functional activity of the three pathways can be detected using immobilised IgM, lipopolysaccharide or mannose, which act as activators of the classical, alternative and lectin pathways, respectively.

Activity of Individual Components

Here, plasma from which complement has been removed is taken and purified specific complement protein is added to it so that the haemolytic activity for that pathway is regained.

Immunochemical Assays

Radial immunodiffusion, ELISA or nephelometry can detect and quantify individual complement components. C3 and C4 components are commonly measured in many laboratories by these methods.

Analysis of Complement Activation Products

In contrast to the haemolytic assays or quantification of individual components, the analysis of the increase in complement activation in disease conditions gives a more realistic picture. Newer assays using monoclonal antibodies in a capture ELISA format can detect the activation products at different stages and individual pathways. C1rs-C1 inhibitor, C4d and C4bc for the classical pathway; Ba, Bb and C3bBbP in the alternative pathway; C3 and its activation products; and finally, the C5b-9 terminal product all can be measured by this technique and are commercially available.

Assays for Complement-Binding Autoantibodies

C1q or C1 inhibitor autoantibodies can be measured in a patient's serum using purified complement as antigen by ELISA.

Antigen–Antibody Reactions 15

The interactions between antigens and antibodies are known as antigen–antibody reactions. The reactions are highly specific, and an antigen reacts only with antibodies produced by itself or with closely related antigens.

The antigen–antibody reactions have many uses. Since these reactions are essentially specific, they have been used in the laboratory for the diagnosis of infectious diseases by detection of either the antigen or the antibody *in vitro*. The antigen and antibody reactions also form the basis of immunity against microbial diseases in vivo. It may cause tissue injury in hypersensitivity reactions and autoimmune diseases in the host.

General Features of Antigen–Antibody Reactions

Antigen and antibody reactions show the following features:

1. Antigen–antibody reactions usually show **a high degree of specificity**. This is because an antigen combines only with a homologous antibody and vice versa. The specificity, however, is not absolute. However, cross-reactions between antigens and antibodies are sometimes responsible for causing diseases in hosts and for showing false and positive laboratory tests.
2. During the antigen–antibody reaction, not fragments but **entire molecules react**, and there is no irreversible chemical alteration in either antigens or antibodies.
3. Electrostatic, hydrogen, van der Waals and hydrophobic interactions are the intermolecular forces involved in antigen–antibody reactions.
4. **The antigen–antibody combination occurs at the surface**. For that reason, the "good fit" between an antigenic determinant and an antibody-combining site determines the stability of the antigen–antibody reaction. Multiple bonding between the antigen and the antibody ensures that the antigen will be bound tightly to the antibodies.
5. Both antigens and antibodies participate in the formation of **precipitates** and **agglutinates**.
6. The combination of the antigen and antibody molecules is **firm but reversible**. The firmness of the reaction is influenced by affinity and avidity.

Affinity denotes the intensity of attraction between antigen and antibody. Low-affinity antibodies bind antigen weakly and tend to dissociate readily, whereas high-affinity antibodies bind antigen more tightly and remain bound longer. High-affinity binding is believed to result from a very close fit between the antigen-binding sites and the corresponding antigenic determinants that facilitate the development of strong non-covalent interactions between antigen and antibody.

Avidity measures the overall strength of binding of an antigen with many antigenic determinants and multivalent antibodies. Avidity is a better indicator of the strength of interactions in real biological systems than affinity. Therefore, the avidity of an antigen–antibody reaction is dependent on the valencies of both antigens and antibodies and is greater than the sum total of individual affinities.

Stages in Antigen–Antibody Reactions

The antigen–antibody reaction occurs in three stages: primary, secondary and tertiary:

Primary Stage

The primary stage is the initial interaction between antigen and antibody. It is rapid and reversible but without any visible effects. The ionic bonds, hydrogen bonds, van der Waals forces and hydrophobic interactions are the weaker intermolecular forces that bind antigen and antibodies together in this primary stage. However, covalent binding, a stronger intermolecular force between antigen and antibody, does not occur in this stage. The primary reaction can be measured by estimating both free and bound antigens or antibodies separately in the reaction mixture. Several methods, including radioactive isotopes, fluorescent dyes, ferritin or other markers, are used for this purpose.

Secondary Stage

The secondary stage is an irreversible interaction between antigen and antibody, with visible effects, such as agglutination, precipitation, neutralisation, complement fixation and immobilisation of motile organisms. The binding between antigen and antibody during this stage occurs by covalent binding. A single antibody is capable of causing different types of antigen–antibody reactions, and a single antigen is capable of inducing the production of different classes of immunoglobulins, which differ in their biological properties.

Table 15.1 Effectiveness of the immunoglobulin classes in different serological reactions

Serological reaction	IgG	IgA	IgM
Precipitation	Strong	Variable	Weak
Agglutination	Weak	Moderate	Strong
Compliment fixation	Strong	Negative	Weak
Lysis	Weak	Negative	Strong

The results of agglutination, precipitation, neutralisation and other tests are usually expressed as a titre. **Titre** is the highest dilution of serum that gives a positive reaction in the test. A higher titre means a greater level of antibodies in serum. For example, a serum with a titre of 1/128 contains more antibodies than a serum with a titre of 1/8.

Tertiary Stage

Some in vivo antigen–antibody reaction turns into a chain reaction, leading to neutralisation or destruction of injurious antigen. Tissue damage can be seen in clinical allergies and other immunological diseases. Humoral immune response against infectious diseases involves the tertiary stage of antigen–antibody reaction.

Depending on the type of serological reaction, various immunoglobulin classes react differently (Table 15.1).

Measurement of Antigen and Antibody

The antigens and antibodies involved in primary, secondary and tertiary stages of antigen–antibody reaction can be measured by various methods. The unit of measurement may be in terms of mass, units and titre value. Usually, antibodies are titrated against sera. On the other hand, antigens may also be measured using titration. The serum antibody titre is the highest dilution of the serum in which observable reaction against the specific antigen can be seen. Sensitivity and specificity are the two critical parameters of all serological tests.

Sensitivity (true-positive rate) is the ability of a particular test to detect antibodies or antigens in very small quantities. A highly sensitive test does not miss any positive cases. That is, false-negative results will be absent or minimal.

Specificity (true-negative rate) is the ability of a particular test to correctly identify specific antigens or antibodies present in a homologous mixture. Highly specific tests do not show any false-positive reactions. As the number of false-positive cases increases, the specificity of that particular test decreases.

Generally, sensitivity and specificity are inversely proportional.

Quantitative Assays

These are used to get the actual concentration of an antigen or, more commonly, an antibody in a sample. In the older immunoassays, different dilutions of the sample were used to find out the endpoint, and the reciprocal of the dilution or the titre was taken as an indirect measure of the quantity of the analyte in the sample. Widal agglutination test or the classical Venereal Diseases Research Laboratory (VDRL) tests are examples of such types of immunoassays. In recent years, technological advances have helped in accurate determination of the analyte in the sample to micro- or pico-gram levels. Enzyme-linked immunosorbent assay (ELISA), Fluorescent immunoassay, Chemiluminescence immunoassay, Microarray immunoassay and some Point-of-care assays can accurately quantify an antibody or antigen in a sample. Detection of protective antibody levels by ELISA after vaccination by hepatitis B or tetanus toxoid are two such areas of application.

Qualitative Assays

In contrast to quantitative assays, in many situations, the mere presence or absence of an antigen or antibody in a specimen is enough to conclude a patient's disease condition. In such cases, any classical tests like agglutination or precipitation test or modern techniques like ELISA or Fluorescent antibody test can be performed.

Antigen–Antibody Reactions Used in the Laboratory

Antigen-antibody reaction-based tests are widely used in the laboratory to detect either serum antibodies or antigens to diagnose a wide variety of infectious diseases. These tests are also used to diagnose autoimmune diseases and in typing blood and tissues before transplantation (Table 15.2).

Precipitation Reaction

Precipitation is a type of antigen–antibody reaction in which the antigen occurs in a soluble form. A precipitation test is a test in which an antibody interacts with the soluble antigen in the presence of an electrolyte at a specified pH and temperature to produce a precipitate. A **lattice** is formed between the antigens and antibodies; it is visible as an insoluble precipitate in certain cases. Lattice formation depends on the affinity of both the antigen and antibody. Antibodies that aggregate soluble antigens are called **precipitins**. Instead of sedimenting, the reaction is known as flocculation when the precipitate remains suspended as **floccules**. For a precipitate to occur, antibody must have a bivalent structure of the binding site; a precipitate will not form with monovalent Fab fragments. In addition, the antigen must have at least two copies of the same epitope or have different epitopes that react with different antibodies present in polyclonal antisera; the antigen must be either bivalent or polyvalent in nature.

Mechanism of Precipitation

Antigen and antibody reaction occurs optimally only when the proportion of the antigen and antibody in the reaction mixture is equivalent (zone of equivalence; Fig. 15.1). On either side of the equivalence zone, precipitation is prevented

Table 15.2 Commonly used tests in clinical microbiology

Test	Uses
Flocculation test	Detection of reaginic antibodies in syphilis by VDRL test
Radial immunodiffusion	Detection of fungal antigen and antibodies
Counter-current immunoelectrophoresis	Detection of both antigen and antibodies in bacterial, viral, fungal and parasitic diseases
Slide agglutination test	Identification of bacterial isolates, such as *Salmonella, Shigella, Vibrio*
Tube agglutination test	Detection of antibodies in bacterial infections, for example, Widal test for enteric fever
Latex agglutination test	Quantitation and detection of antigen and antibodies
Haemagglutination test	Detection of both antigens and antibodies in viral and parasitic infections
Coagglutination test	Detection of microbial antigens
Complement fixation test	Quantitation and detection of antibodies
Direct immunofluorescence test	Detection and localisation of antigen in a cell or tissue
Indirect immunofluorescence test	Detection of specific antibodies in the serum
Sandwich ELISA	Detection of antigens and antibodies
Indirect ELISA	Quantitation and detection of antibodies
Radioimmunoassay	Quantitation of hormones, drugs, etc.
Western blot	Detection of antigen-specific antibody

because of an excess of either antigen or antibody. The zone of antibody excess is known as the pro-zone phenomenon, and the zone of antigen excess is known as the post-zone phenomenon.

Marrack, in 1934, proposed the ***lattice hypothesis*** to explain the precipitation mechanism. The Marrack hypothesis assumes that each antibody molecule must have at least two binding sites, and an antigen must be multivalent.

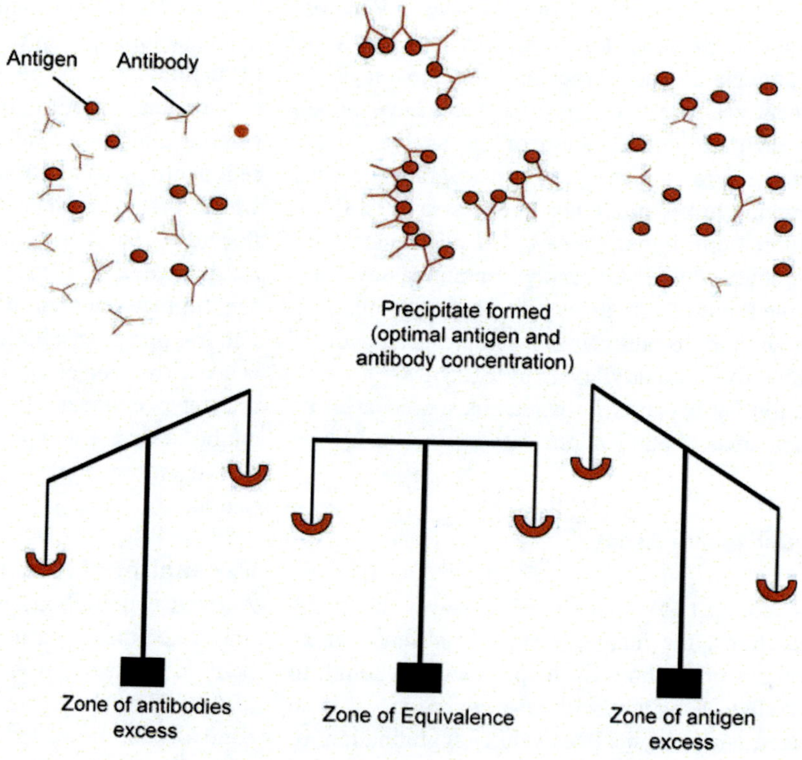

Fig. 15.1 Prozone phenomenon

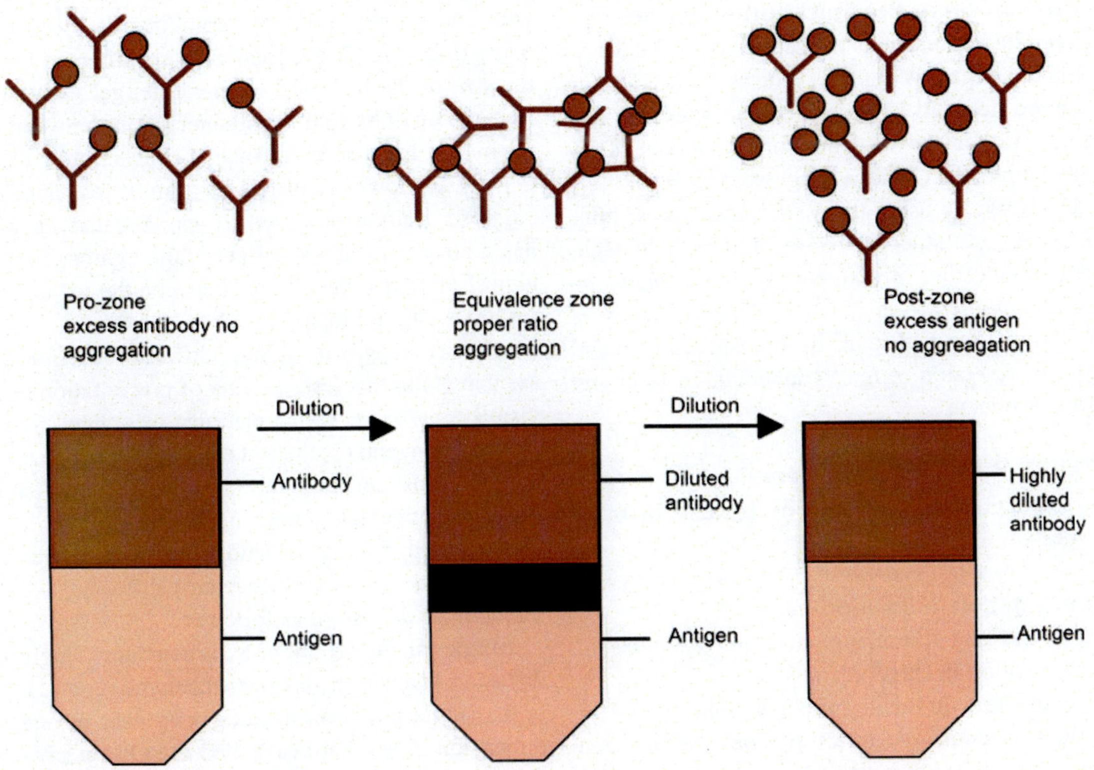

Fig. 15.2 Marrack lattice hypothesis

In the **zone of equivalence** where optimum precipitation occurs, the number of multivalent sites of antigen and antibody is approximately equal. In this zone, precipitation occurs due to random, reversible reactions whereby each antibody binds to more than one antigen and vice versa, forming a stable network or lattice—combining the results in a multimolecular lattice that increases in size until it precipitates out of solution.

There is too much antibody in the **pro-zone phenomenon** for efficient lattice formation. This is because the antigen combines with only a few antibodies, and no cross-linkage is formed.

In the **post-zone phenomenon**, small aggregates are surrounded by excess antigen, and again no lattice network is formed. Thus, for precipitation reactions to be detectable, they must be run in the zone of equivalence (Fig. 15.2).

The pro-zone and post-zone phenomena are considered in the interpretation of serological tests because false-negative reactions can occur in either of these conditions. A false-negative reaction suspected to be due to the pro-zone phenomenon can be rectified by diluting the antibody and performing the test. In the post-zone phenomenon, excess antigen presence might lead to obscure results due to fewer antibodies. Typically, such a test should be repeated with an additional patient sample taken about a week later. The delay would allow the immune system for the further production of antibodies. If the test results are negative, it is unlikely that the patient has that particular antibody.

Application of Precipitation Reactions

Precipitation reactions are sensitive for the detection of antigen. Antigen as little as 1 μg of protein can be detected by this method. They are less sensitive for the detection of antibodies. They are employed (1) in a diagnostic microbiology laboratory, by using the VDRL test for syphilis, grouping streptococci by the Lance field technique, detection of toxigenicity of *Corynebacterium diphtheriae* etc.; (2) for identification of blood and seminal stains in forensic science; and (3) for detection of food adulterants in food and food products.

Types of Precipitation Reactions

Precipitation reactions can be broadly of three types:

Precipitation in Solution

Ring test and flocculation test are examples of precipitation in solution.

Ring test: In this test, antigen solution is layered over antiserum in a test tube. Precipitation between antigen and antibodies in antiserum solution is marked by the appearance of a ring of precipitation at the junction of two liquid layers. C-reactive protein (CRP) and streptococcal grouping by the Lancefield methods are examples of the ring test.

Flocculation test: Flocculation test may be performed in a slide or tube. VDRL test for detection of reaginic antibodies in syphilis is an example of a *slide flocculation test.* In this test, a drop of VDRL antigen suspension is added to a drop of patients' serum on a cavity slide, and the result is recorded after shaking the slide on a VDRL shaker. In a positive test, the floccules appear, which can be demonstrated well under a microscope. Kahn's test for syphilis is an example of the *tube flocculation test.* The tube flocculation test for standardisation of toxins and toxoids is another example.

Immunodiffusion (Precipitation in Gel)

The precipitation test in soft (1%) agar gel or agarose gel is termed as **the immunodiffusion** test. In this test, reactants are added to the gel, and an antigen–antibody combination occurs using diffusion. The diffusion rate is affected by the size of the particles, temperature, gel viscosity, amount of hydration and interactions between the matrix and reactants. An agar concentration of 0.3–1.5% allows for diffusion of most reactants. Agarose is often preferred to agar because agar has a strong negative charge, while agarose has almost none, so interactions between the gel and reactants are minimised.

The advantages of immunodiffusion reactions are that (1) in this test, the line of precipitation is visible as a band, which can also be stained for preservation; and (2) the test can be used to detect identity, cross-reaction and non-identity between different antigens in a reacting mixture.

Immunodiffusion reactions are classified based on the number of reactants diffusing and direction of diffusion, as follows:

Single diffusion in one dimension: Single diffusion in one dimension, as the name suggests, is the single diffusion of antigen in agar in one dimension. It is otherwise called **the Oudin procedure** because it was pioneered by Oudin, who used gels for precipitation reactions for the first time. In this method, the antibody is incorporated into agar gel in a test tube, and the antigen solution is poured over it. Over time, the antigen diffuses downward towards the antibody in agar gel, and a line of precipitation is formed. The number of precipitate bands shows the number of different antigens present in the antigen solution.

Single diffusion in two dimensions: Single diffusion in two dimensions is also called **radial immunodiffusion**. In this method, antiserum solution containing antibody is incorporated in agar gel on a slide or Petri dish. The wells are cut on the surface of the gel. The antigen is then applied to a well-cut into the gel. When antibody already present in the gel reacts with the antigen, which diffuses out of the well, a precipitation ring is formed around the wells. The diameter of the ring is directly proportional to the concentration of the antigen. Therefore, the greater the amount of antigen in the well, the farther the ring will be from the well.

Radial immunodiffusion has been used quantitatively to estimate antibodies and antigens in the serum. It is used to measure IgG, IgM, IgA and complement components in the serum, antibodies to influenza virus in sera, and serum transferrin and α-fetoprotein. However, currently, the test has recently been replaced by more sensitive and automated methods, such as nephelometry and enzyme-linked immunosorbent assays (ELISAs).

Double diffusion in one dimension: This method is also called **Oakley– Fulthorpe procedure**. In this method, the antibody is incorporated in agar gel in a test tube, above which a layer of plain agar is placed. The antigen is then layered on top of this plain agar. Over time, the antigen and antibody move towards each other through the intervening layer of plain agar. In this zone of plain agar, both antigen and antibody react to form a band of precipitation at their optimum concentration.

Double diffusion in two dimensions: This method is also called the **Ouchterlony procedure.** In this method, both the antigen and antibody diffuse independently through agar gel in two dimensions, horizontally and vertically. The test is performed by putting wells in the agar gel poured on a glass slide or in a Petri dish. The antiserum consisting of antibodies is placed in the central well, and different antigens are added to the wells surrounding the centre well. After an incubation period of 12–48 h in a moist chamber, the lines of precipitins are formed at the sites of a combination of antigens and antibodies.

Three types of reactions can be demonstrated: (1) line of precipitation at their junction forming an arc represents serologic identity or the presence of a common epitope in antigens, (2) a pattern of crossed lines demonstrates two separate reactions and indicates that the compared antigens are unrelated and share no common epitopes, and (3) fusion of two lines with a spur indicates cross-reaction or partial identity. In this last case, the two antigens share a common epitope, but the antigen and traverse do not capture some antibody molecules through the initial precipitin line to combine with additional epitopes found in the more complex antigen.

Double diffusion in two dimensions has been used for (1) demonstration of antibodies in serodiagnosis of smallpox, (2) identification of fungal antigens and (3) detection of antibodies to extractable nuclear antigens. Elek precipitation test in the gel is a special test used for demonstration of toxigenicity of *C. diphtheriae*.

Precipitation in Agar with an Electric Field (Electroimmunodiffusion)

The immunodiffusion methods described above are too slow, and the results can only be obtained after several hours. On the other hand, using an electric current can cause quick movement of the molecules resulting in much faster antigen–antibody precipitation, which is the underlying principle for all electrodiffusion techniques described below.

1. **Immunoelectrophoresis:** It is a process of combination of immunodiffusion and electrophoresis. It is a method in which different antigens in serum are separated according to their charge under an electric field. In this method, a drop of antigen is placed into a well in agar on a glass slide. An electric current is then passed through the agar. According to their charge and size, antigens move in the electric field during electrophoresis. Following electrophoresis, a trough is cut into the agar and is filled with the antibody and diffusion is allowed to occur. As the antigen and antibody diffuse towards each other, they form a series of lines of precipitation. The main advantage of immunoelectrophoresis is that several antigens can be identified in serum. The method detects normal and abnormal proteins in human serum, such as myeloma proteins.
2. **Counter-current immunoelectrophoresis (CIE):** It depends on the movement of antigen towards the anode and antibody towards the cathode through the agar under an electric field. The test is performed on a glass slide with agarose in which a pair of wells is punched out. One well is filled with antigen and the other with antibodies. Electric current is then passed through the gel. The migration of antigen and antibody is greatly facilitated

under an electric field, and the line of precipitation is made visible in 30–60 min.

The test has many uses. It is a rapid and highly specific method for detecting antigens and antibodies in the serum, cerebrospinal fluid (CSF) and other body fluids in diagnosing many infectious diseases, including bacterial, viral, fungal and parasitic. It is commonly used for hepatitis B surface antigen (HBsAg), α-fetoprotein, hydatid and amoebic antigens in the serum and cryptococcal antigen in the CSF.

3. **Rocket electrophoresis:** This technique is an adaptation of radial immunodiffusion developed by Laurell. In this method, an antibody is incorporated in the gel, and the antigen is placed in wells cut in the gel. Electric current is then passed through the gel, facilitating the migration of antigen into the agar. This results in forming a precipitin line that is conical in shape, resembling a rocket. That is why it is known as rocket electrophoresis. The rocket's height, measured from the well to the apex, is directly proportional to the amount of antigen in the sample. Rocket electrophoresis is used mainly for quantitative estimation of antigen in the serum.
4. **Two-dimensional electrophoresis:** This technique is a variant of rocket electrophoresis. This is a double-diffusion technique used for qualitative and quantitative analyses of sera for a wide range of antigens. This test is a two-stage procedure. In the first stage, antigens in the solution are separated by electrophoresis. Then, in the second stage, electrophoresis is carried out again but perpendicular to the first stage to obtain rocket-like precipitation.

A small trough is first cut in agar gel on a glass plate and filled with the antigen solution in this test. Electric current is then passed through the gel, and the antigens migrate into the gel at a rate proportional to their net electric charge. In the second stage, after electrophoresis, the gel piece containing the separated antigens is placed on a second glass plate, and the agar containing antibody is poured around the gel piece. A second electric potential is applied at right angles to the first direction of migration. The pre-separated antigens then migrate into the gel containing antibodies at a rate proportional to their net charge and precipitate with antibodies in the gel, forming precipitates.

This method is qualitative in that it identifies different antigens present in the serum solution and quantitative in that it detects the amount of different antigens present in the solution.

Agglutination Reaction

Agglutination is an antigen–antibody reaction in which a **particulate antigen** combines with its antibody in the presence of electrolytes at a specified temperature and pH, resulting in the formation of visible clumping of particles. Agglutination occurs optimally when antigens and antibodies react in equivalent proportions. In agglutination reactions, serial dilutions of the antibody solution are made, and a constant amount of particulate antigen is added to serially diluted antibody solutions. After several hours of incubation at 37 °C, clumping is recorded by visual inspection. The titre of the antiserum is recorded as the reciprocal of the highest dilution that causes clumping. Since the cells have many antigenic determinants on their surface, the phenomenon of antibody excess is rarely encountered.

Occasionally, antibodies are formed that react with a cell's antigenic determinants but do not cause any agglutination. They inhibit the agglutination by the complete antibodies added subsequently. Such antibodies are called **blocking antibodies**. Anti-Rh antibodies and anti-brucella antibodies are a few examples of such blocking antibodies. Agglutination reactions have various applications in detecting antigens and antibodies in serum and other body fluids. They are very sensitive, and the test result can be read visually with ease.

Agglutination reactions are mostly similar to precipitation reactions in their fundamentals and share similar features. This reaction is analogous to the precipitation reaction in that antibodies act as a bridge to form a lattice network of antibodies and the cells that carry the antigen on their surface. Because cells are so much larger than a soluble antigen, the result is more visible when the cells aggregate into clumps.

Agglutination differs from precipitation reaction in that since the former reaction takes place at the surface of the particle involved, the antigen must be exposed and bind with the antibody to produce visible clumps.

Types of Agglutination Reactions

Agglutination reactions are broadly of two types: direct agglutination and passive agglutination. The agglutination reactions that employ the antigens found naturally on a particle are known as direct agglutination. This is different from passive agglutination, which employs particles coated with antigens not normally found on their surfaces.

Direct Agglutination

Direct agglutination reactions can broadly be of the following types.

Slide agglutination test: It is a basic agglutination reaction performed on a slide. Identification of bacterial types represents a classic example of a direct slide agglutination that is still used today. In this test, a suspension of bacteria is prepared and is added to a drop of standardised antiserum. A positive reaction is indicated by the clumping of bacteria and clearing of the background solution. Clumping occurs instantly or within seconds in a positive test. A control consisting of antigen suspension in saline without adding antiserum is included on the same slide. It is used to validate the results and detect possible false-positives due to autoagglutination of the antigen.

Slide agglutination is used as a routine procedure (1) to identify bacterial strains, such as *Salmonella*, *Shigella*, *Vibrio*, etc., isolated from clinical specimens; and (2) for blood grouping and cross-matching.

Tube agglutination test: Tube agglutination test, as the name suggests, is performed in glass tubes. Typically, in these tests, the patient's serum is diluted in a series of tubes and bacterial antigens specific for the suspected disease are added to it. Antigen and antibody reactions are demonstrated by demonstrating visible clumps of agglutination. It is a standard method used for quantitative estimation of antibodies in the serum. Tube agglutination tests are routinely used for demonstration of antibodies in the serum for serodiagnosis of enteric fever and brucellosis, such as:

Heterophile agglutination test: This test depends on the demonstration of heterophilic antibodies in serum present in certain bacterial infections:

1. **Weil–Felix test** is an example of heterophile agglutination reaction for serodiagnosis of rickettsial infections. In this test, the cross-reacting antibodies produced against the rickettsial pathogen are detected using cross-reacting related antigens (e.g. *Proteus* strains OXK, OX19 and OX2). Although the antibodies are produced against rickettsial organisms, they cross-react with antigens of *Proteus* strains OXK, OX19 and OX2. **Paul–Bunnell test** is another heterophile agglutination test used to detect antibodies in infectious mononucleosis by using sheep erythrocytes as antigens. Finally, the ***Streptococcus* MG agglutination test** is a similar test used to detect antibodies to *Mycoplasma pneumoniae* causing primary atypical pneumonia.
2. **Antiglobulin (Coombs) test:** The Coombs test was originally devised by Coombs, Mourant and Race to detect incomplete anti-Rh antibodies that do not agglutinate Rh + erythrocytes in saline. When serum-containing incomplete anti-Rh antibodies are mixed with Rh + erythrocytes in saline, incomplete antibody antiglobulin coats the surface of erythrocytes but does not cause any agglutination. When such erythrocytes are treated with antiglobulin or Coombs serum (rabbit antiserum against human seven globulin), then the cells are agglutinated.

   Coombs test is of two types: (a) direct Coombs test and (b) indirect Coombs test.

   (a) **Direct Coombs test:** In this test, the sensitisation of red blood cells (RBCs)

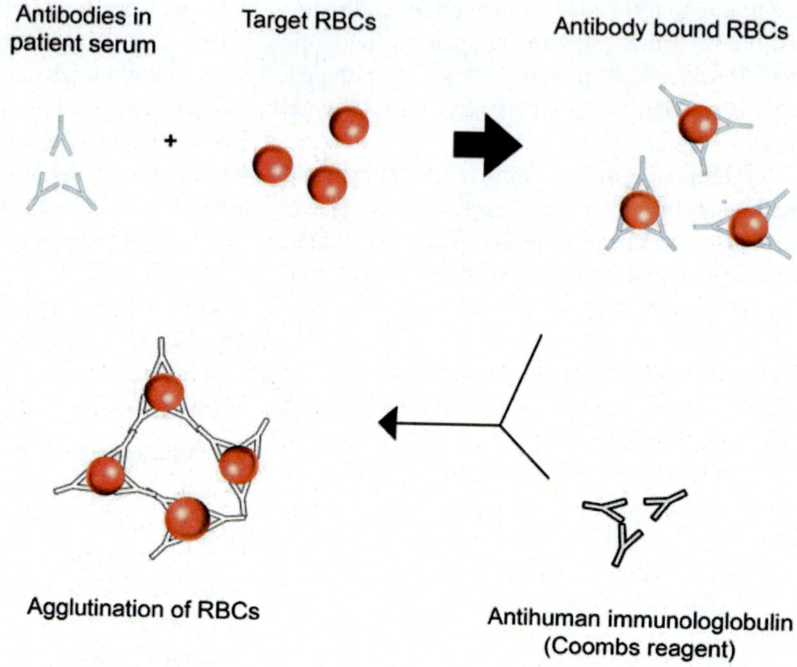

Fig. 15.3 Principle of the indirect Coombs test

with incomplete antibodies occurs in vivo. The cell-bound antibodies can be detected by this test in which antiserum against human immunoglobulin is used to agglutinate a patient's red blood cells.

(b) **Indirect Coombs test:** In this test, RBCs' sensitisation with incomplete antibodies occurs *in vitro*. In this test, the patient's serum is mixed with normal red blood cells and antiserum to human immunoglobulin is added. Agglutination occurs if antibodies are present in the patient's serum (Fig. 15.3).

Coombs tests are used to detect anti-Rh antibodies and incomplete antibodies in brucellosis and other diseases.

Passive Agglutination

Passive agglutination employs carrier particles that are coated with soluble antigens. This is usually done to convert precipitation reactions into agglutination reactions since the latter are easier to perform and interpret and are more sensitive than precipitation reactions to detect antibodies. When the antibody instead of antigens is adsorbed on the carrier particle for detection of antigens, it is called **reverse passive agglutination**.

Until the 1970s, erythrocytes were the major carrier particles used to coat antigens. Recently, however, a variety of other particles, including polystyrene latex, bentonite and charcoal, have been used for this purpose. Particle size varies from 7 μ for RBCs to 0.05 μ for very fine latex particles. The use of synthetic beads or particles provides the advantage of consistency, uniformity and stability. Reactions are also easy to read visually. The passive agglutination reaction can be of the following types depending on the carrier particles used.

1. **Latex agglutination test:** It is a test that employs latex particles as a carrier of antigen or antibodies. In 1955, Singer and Plotz accidentally found that IgG was naturally adsorbed to the surface of polystyrene latex particles. Latex particles are inexpensive, relatively stable and do no cross-reactivity with other antibodies. Therefore, these particles can be coated with antibodies to detect the antigen in the serum and other body fluids. The use of monoclonal antibodies has reduced the

cross-reactions resulting in a reduction of false-positive reactions. Additionally, the large particle size of the latex facilitates better visualisation of antigen–antibody reactions by the naked eye observation. The tests are usually performed on cardboard cards or glass slides, and positive reactions are graded on a scale of 1+ to 4 + .

The latex agglutination tests are used to rapidly identify antigens of group B *Streptococcus*, *Staphylococcus aureus*, *Neisseria meningitidis*, *Cryptococcus neoformans* etc. The tests have also been useful for detecting soluble microbial antigens in urine, spinal fluid and serum to diagnose various infectious diseases. These tests are being used to detect RA factor, anti-streptolysin O (ASLO), CRP etc., in serum specimens.

2. **Haemagglutination test:** RBCs are used as carrier particles in haemagglutination tests. RBCs of sheep, humans, chicks etc. are commonly used in the test. When RBCs are coated with antigen to detect antibodies in the serum, the test is called **the indirect haemagglutination (IHA) test**. The IHA is the most commonly used test for serodiagnosis of many parasitic diseases, including amoebiasis, hydatid disease and toxoplasmosis.

   When antibodies are attached to the RBCs to detect microbial antigens, it is known as **reverse passive haemagglutination (RPHA)**. The RPHA has been used extensively in the past to detect viral antigens, such as in HBsAg in the serum for diagnosis of hepatitis B infection. The test has also been used to detect antigens in many other viral and parasitic infections.

   (a) **Viral haemagglutination:** Many viruses, including influenza, mumps and measles, can agglutinate RBCs without antigen–antibody reactions. This process is called viral haemagglutination. This haemagglutination can be inhibited by an antibody specifically directed against the virus, and this phenomenon is called **haemagglutination inhibition**. This forms the basis of the viral haemagglutination inhibition test, which detects antibodies in patients' sera that neutralise the agglutinating viruses. The patient's serum is first incubated with a viral preparation to perform this test. Then the RBCs that the virus is known to agglutinate are added to the mixture. If the antibody is present, this will combine with viral particles and prevent agglutination, and a lack of or reduction in agglutination indicates the presence of antibody in the patient's serum.

3. **Coagglutination test:** Coagglutination is a type of agglutination reaction in which Cowan I strain of *S. aureus* is used as a carrier particle to coat antibodies. Cowan I strain of *S. aureus* contains protein A, an anti-antibody, that combines with the Fc portion of immunoglobulin, IgG, leaving the Fab region free to react with the antigen present in the specimens (Fig. 15.4). In a positive test, protein A bearing *S. aureus* coated with antibodies will be agglutinated if mixed with a specific antigen. The advantage of the test is that these particles

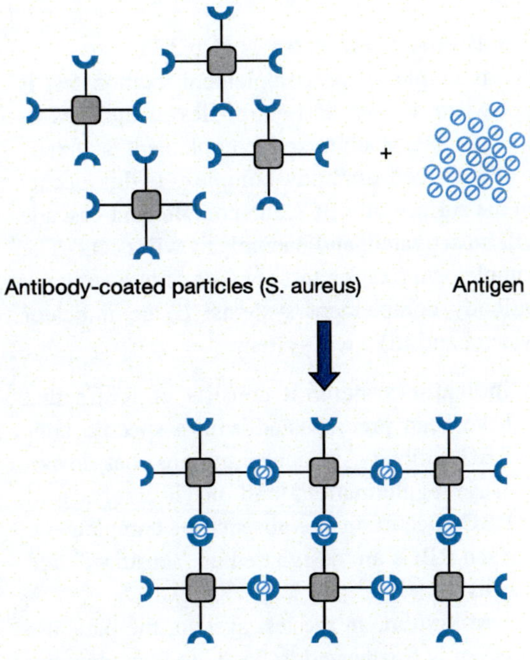

Fig. 15.4 Principle of the coagglutination

show greater stability than latex particles and are more refractory to changes in ionic strength.

Coagglutination test has been used for (1) detection of cryptococcal antigen in the CSF for diagnosis of cryptococcal meningitis; (2) detection of amoebic and hydatid antigens in the serum for diagnosis of amoebiasis and cystic echinococcosis, respectively; and (3) grouping of streptococci and mycobacteria and for typing of *Neisseria gonorrhoeae.*

Complement-Dependent Serological Tests

A complement system is a group of serum proteins present in normal serum. The system consists of 20 or more serum proteins interacting with each other and cell membranes. It is a biochemical cascade that helps to clear pathogens from the body. It aids the antibodies in lysing bacteria and promotes phagocytosis and immune adherence. The complement-dependent serological tests may be of the following types.

Complement Fixation Test (CFT)

The principle of the complement fixation test is that when antigen and antibodies of the IgM or the IgG classes are mixed, complement is "fixed" to the antigen–antibody complex. If this occurs on the surface of RBCs, the complement cascade will be activated, and haemolysis will occur. The complement fixation test consists of two antigen–antibody complement systems: (i) an indicator system, and (ii) a test system.

1. **Indicator system:** It consists of RBCs that have been preincubated with a specific anti-RBC antibody, in concentrations that do not cause agglutination, and no haemolysis of RBCs occurs in the absence of complement. Such RBCs are designated as "sensitised" red cells.
2. **Test system:** In the test system, the patient's serum is first heated to 56 °C to inactivate the native complement. The inactivated serum is washed to avoid cross-reactivity because these anti-RBC antibodies may interfere with the assay. The purified serum is then mixed with the serum and diluted with fresh pig serum, which is a complement. Then the mixture is incubated for 30 min at 37 °C to allow antibody–antigen to form complexes (Fig. 15.5).

In the complement fixation test, "sensitised" red cells are added to the mixture. If the red cells are lysed, it indicates that there were no antibodies specific to the antigen in the patient's serum. The complement, therefore, was not consumed in the test system and was available to be used by the anti-RBC antibodies, resulting in haemolysis. This reaction is considered negative. Conversely, the test is considered positive if the red cells are not lysed. This is because no lysis of the cells indicates that the patient's serum had antibodies specific to the antigen, which have a "fixed" complement. Hence, no complement was available to be activated by the indicator system.

The complement fixation reactions were used earlier to diagnose many infections, such as (1) Wassermann test for syphilis, and (2) tests for demonstration of antibodies to *M. pneumoniae*, *Bordetella pertussis*, many different viruses and to fungi (such as *Cryptococcus* spp., *Histoplasma* and *Coccidioides immitis*). However, since this test is technically very cumbersome and often difficult, it is no longer used nowadays.

Types of Complement Fixation Test

1. **Indirect complement fixation test:** This test is carried out to test the sera that cannot fix guinea pig complement. These include avian sera (e.g. parrot, duck) and mammalian sera (e.g. cat, horse). The test is carried out in duplicate, and after the first test, the standard antiserum known to fix the complement is added to one set. Haemolysis indicates a positive test. If the serum contains an antibody in a positive test, the antigen would have been used up in the first test. Standard antiserum added subsequently would fail to fix the complement, therefore causing haemolysis.

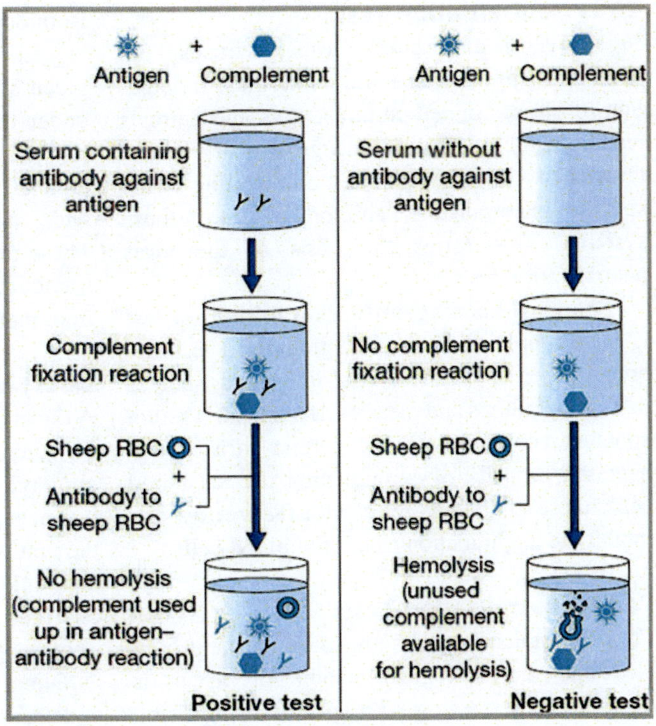

Fig. 15.5 Complement fixation test

2. **Conglutinating complement adsorption test:** It is an alternative method for systems that do not fix guinea pig complement. Sheep erythrocytes, sensitised with bovine serum, are used as the indicator system. The bovine serum contains conglutinin, a β-globulin that acts as an antibody to the complement. Therefore, conglutinin causes agglutination of sensitised sheep erythrocytes if combined with complement, known as **conglutination**. If the horse complement had been used up by the antigen–antibody reaction in the first step, the agglutination of the sensitised cells does not occur.
3. **Immune adherence test**: Immune adherence test is a test in which certain pathogens (e.g. *Vibrio cholerae*, *Treponema pallidum*) react with specific antibodies in the presence of complement and adhere to erythrocytes or platelets. The adherence of cells to bacteria is known as immune adherence. This facilitates phagocytosis of the bacteria.
4. **Immobilisation test:** The immobilisation test is a complement-dependent test in which certain live bacteria, such as *T. pallidum*, are immobilised when mixed with the patient's serum in the presence of complement. This forms the basis of the *T. pallidum* immobilisation test. A positive test shows serum to contain treponemal antibodies.
5. **Cytolytic or cytocidal reactions:** When a live bacterium, such as *V. cholerae*, is mixed with its specific antibody in the presence of complement, the bacterium is killed and lysed. This forms the basis of the test used to measure anti-cholera antibodies in the serum.

Neutralisation Tests

Neutralisation is an antigen–antibody reaction in which the biological effects of viruses and toxins are neutralised by homologous antibodies known as neutralising antibodies. These tests are broadly of two types: virus neutralisation and toxin neutralisation.

Virus Neutralisation Tests

In virus neutralisation tests, viruses are neutralised by their specific antibodies. Inoculation of viruses in cell cultures, eggs and animals results in the replication and growth of viruses. Growth of viruses is inhibited on injecting virus-specific neutralising antibodies into these systems. This forms the basis of the virus neutralisation test.

The viral haemagglutination inhibition test is an example of a virus neutralisation test frequently used to diagnose viral infections, such as influenza, mumps and measles. If a patient's serum contains antibodies against certain viruses that have the property of agglutinating the red blood cells, these antibodies react with the viruses and inhibit the agglutination of the red blood cells.

Toxin Neutralisation Tests

Toxin neutralisation tests are based on the principle that the biological action of toxin is neutralised by reacting with specific neutralising antibodies called antitoxins. Examples of neutralisation tests include:

Schick test, an in-vivo test, is used to demonstrate immunity or susceptibility against diphtheria. This test depends on the ability of circulating antitoxin to neutralize diphtheria toxin administered intradermally.

Anti-streptolysin O (ASLO) test, an *in-vitro* test, is used to detect the ASLO antibody in patients with past Group A streptococci infection. Raised or rising levels of ASLO antibodies indicate past or present infection, and Nagler reaction is used for rapid detection of *C. welchii*. This test is used to detect alpha-toxin of *Clostridium perfringens* in clinical specimens.

Immunofluorescence

The property of certain dyes absorbing light rays at one particular wavelength (ultraviolet [UV] light) and emitting them at a different wavelength (visible light) is known as fluorescence. Fluorescent dyes, such as fluorescein isothiocyanate and lissamine rhodamine, can be tagged with antibody molecules. They emit blue-green and orange-red fluorescence, respectively, under the fluorescence microscope's ultraviolet (UV) rays. This forms the basis of the immunological test. Immunofluorescence tests have wide applications in research and diagnostics. These tests are broadly of two types: (1) direct immunofluorescence test and (2) indirect immunofluorescence test.

Direct Immunofluorescence Test

A direct immunofluorescence test is used to detect the unknown antigen in a cell or tissue by employing a known labelled antibody that interacts directly with the unknown antigen. If the antigen is present, it reacts with labelled antibodies, and the antibody-coated antigen is observed under the UV light of the fluorescence microscope (Fig. 15.6).

Direct immunofluorescence test is widely used to detect bacteria, parasites, viruses, fungi or other antigens in CSF, blood, stool, urine, tissues and other specimens. The test is used for antemortem diagnosis of rabies by detecting rabies virus antigen in the skin smear collected from the nape of the neck in humans and the saliva of dogs. The test is also used to detect *N. gonorrhoea*, *C. diphtheriae*, *T. pallidum* etc., directly in appropriate clinical specimens.

The need to prepare separately labelled antibodies for each pathogen is the principal disadvantage of the direct immunofluorescence test.

Indirect Immunofluorescence Test

The test detects specific antibodies in the serum and other body fluids for the serodiagnosis of many infectious diseases.

Indirect immunofluorescence is a two-stage process. In the first stage, a known antigen is fixed on a slide. Then the patient's serum to be tested is applied to the slide, followed by careful washing. If the patient's serum contains antibodies against the antigen, it will combine with the antigen on the slide. In the second stage, the combination of antibody with antigen

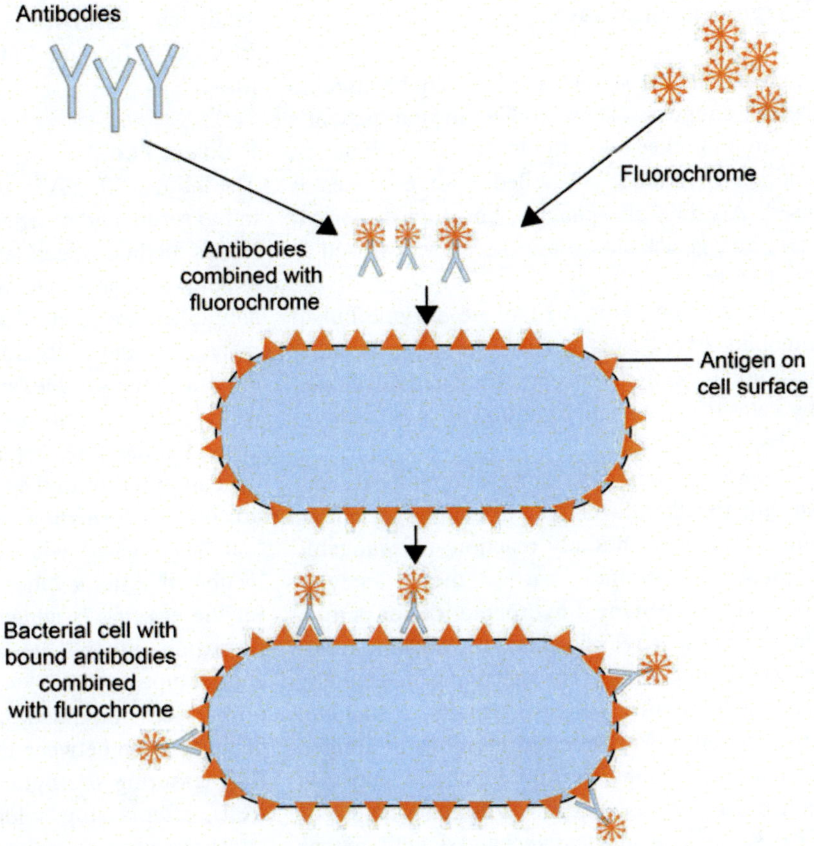

Fig. 15.6 Direct fluorescent antibody test

can be detected by adding a fluorescent dye-labelled antibody to human IgG, which is examined by a fluorescence microscope.

The first step in the indirect immunofluorescence test is incubating a fixed antigen (e.g. in a cell or tissue) with an unlabelled antibody, which becomes associated with the antigen. Next, after careful washing, a fluorescent antibody (e.g. fluorescently labelled anti-IgG) is added to the smear. This second antibody will become associated with the first, and the antigen–antibody complex can be visualised on the fluorescence microscope.

The indirect method has the advantage of using a single labelled antiglobulin (antibody to IgG) as a "universal reagent" to detect many different specific antigen–antibody reactions. The test is often more sensitive than the direct immunofluorescence test.

Indirect immunofluorescence test is used widely to (1) detect specific antibodies for serodiagnosis of syphilis, leptospirosis, amoebiasis, toxoplasmosis and many other infectious diseases; (2) identify the class of a given antibody by using fluorescent antibodies specific for different immunoglobulin isotypes; (3) identify and enumerate lymphocyte subpopulations by employing monoclonal antibodies and cytofluorographs; and (4) detect autoantibodies, such as antinuclear antibodies in autoimmune diseases.

The major limitation of immunofluorescence is that the technique requires (1) expensive fluorescence microscope and reagents, (2) trained personnel and (3) has a factor of subjectivity that may result in erroneous results.

Enzyme Immunoassays

Enzyme immunoassays (EIAs) can be used to detect antigens or antibodies in the patient's serum and other body fluids. In EIA techniques, antigen or antibody labelled with enzymes is used. Alkaline phosphatase, horseradish peroxidase and galactosidase are the enzymes used in the EIA tests.

The commonly used EIAs are enzyme-linked immunosorbent assays (ELISAs). The ELISA technique was first conceptualised and developed by Peter Perlmann and Eva Engvall at Stockholm University, Sweden.

These assays involve using an immunosorbent specific to either the antigen or antibody. Following the antigen–antibody reaction, chromogenic substrate specific to the enzyme (*o*-phenylenediamine dihydrochloride for peroxidase, *p*-nitrophenyl phosphate for alkaline phosphatase etc.) is added. The reaction is detected by reading the optical density. Usually, a standard curve based on known concentrations of antigen or antibody is prepared from which the unknown quantities are calculated. Different types of ELISAs are available to detect and quantify either the antigen or antibodies in serum and other body fluids. These include indirect ELISA, sandwich ELISA, competitive ELISA and ELISPOT (Enzyme-linked immunosorbent spot) assay.

Indirect ELISA

The indirect ELISA is used to estimate antibodies in the serum and other body fluids quantitatively. In this method, specimens are added to microtiter plate wells coated with antigens to which specific antibodies are to be detected. After a period of incubation, the wells are washed. If antibody was present in the sample, antigen–antibody complex would have been formed and will not get washed away. On the other hand, if the specific antibody were not present in the specimen, there would be no complex formation. Next, an anti-isotype antibody conjugated with an enzyme is added and incubated. After another washing step, a substrate for the enzyme is added. If there were complex formation in the initial step, the secondary anti-isotype antibody would have bound to the primary antibody, and there would be a chromogenic reaction between the enzyme and substrate. By measuring the optical density values of the wells, after a stop solution has been added to arrest the chromogenic reaction, one can determine the amount of antigen–antibody complex formed in the first step (Fig. 15.7).

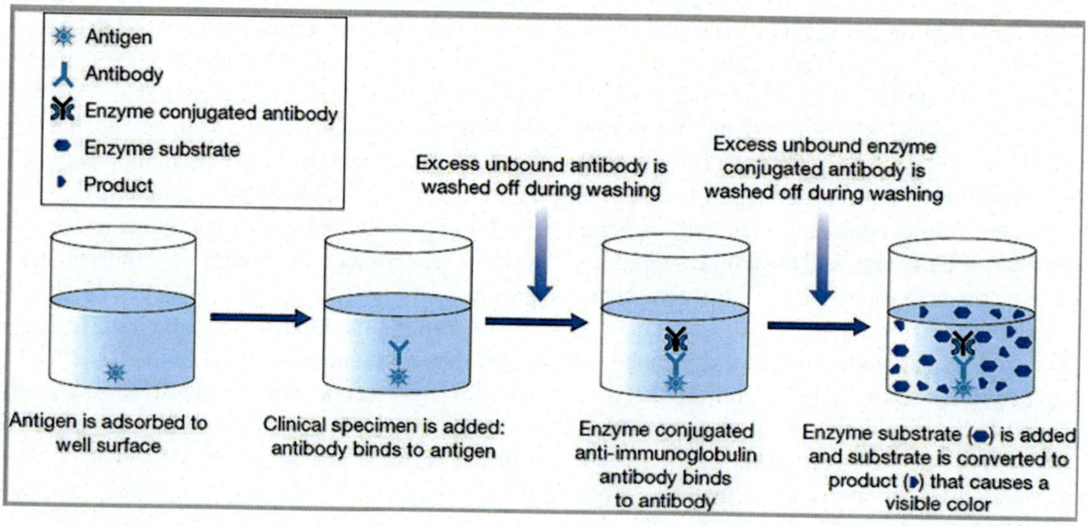

Fig. 15.7 Indirect ELISA test

The test is extensively used to determine serum antibodies for diagnosing human immunodeficiency virus (HIV) infection, Japanese encephalitis, dengue and other viral infections.

Sandwich ELISA

The sandwich ELISA is used for the detection of antigen. In this test, the known antibody is coated and immobilised onto the wells of microtiter plates. The test sample containing the suspected antigen is added to the wells and is allowed to react with the antibodies in the wells. After washing the well, a second enzyme-conjugated antibody specific for a different epitope of the antigen is added and allowed to incubate. After removing any free secondary antibody by rewashing, the specific substrate is added, and the ensuing chromogenic reaction is measured. The chromogenic reaction is then compared with a standard curve to determine the exact amount of the antigen present in the test sample. In a positive test, an enzyme acts on the substrate to produce a colour, and its intensity can be measured by a spectrophotometer or ELISA reader. The change in colour can also be observed by the naked eye.

The sandwich ELISA detects rotavirus and enterotoxin of *Escherichia coli* in faeces.

Competitive ELISA

Competitive ELISA is another technique used to estimate antibodies present in a specimen, such as serum. The principle of the test is that two specific antibodies, one conjugated with enzyme and the other present in test serum (if the serum is positive for antibodies), are used. Competition occurs between the two antibodies for the same antigen. The appearance of colour indicates a negative test (absence of antibodies), while the absence of colour indicates a positive test (presence of antibodies).

In this test, the microtiter wells are coated with HIV antigens. The sera to be tested are added to these wells and incubated at 37 °C, and then washed. If antibodies are present in the test serum, an antigen–antibody reaction occurs. The antigen–antibody reaction is detected by adding enzyme-labelled-specific HIV antibodies. In a positive test, no antigen is left for these antibodies to act. Hence, the antibodies remain free and are washed away during washing. When the substrate is added, no enzyme is available to act on it. Therefore, a positive result indicates no colour reaction. In a negative test, in which no antibodies are present in the serum, antigen in the coated wells is available to combine with enzyme-conjugated antibodies, and the enzyme acts on the substrate to produce colour.

Competitive ELISA is the most commonly used test for detecting HIV antibodies in serum in patients with HIV.

Capture ELISA

This works on ELISA principles, and these methods are much simpler, rapid and reliable in the diagnostic scenario. For example, card and dipstick methods are widely used to diagnose various infectious diseases nowadays. Because of its rapid and reliable nature, capturing ELISA is very useful in the initial diagnosis and early treatment of many diseases.

Cassette ELISA (Cylinder ELISA)

Cassette ELISA is a simple modification of the conventional ELISA technique; it is one of the highly useful diagnostic methods, and even a very few samples can be processed. The whole test procedure finishes within 10–20 min. Each sample can be processed individually, as cassette ELISA test units are packed separately in a disposable pouch. There is no need to use any instruments to carry out the assay procedure, and the assay results can be read through naked eyes. Another advantage of this test is that the test unit includes in-built control, which helps in instant validation of the assay. The HIV tri-dot test used to detect HIV type 1 and 2 antibodies is an example of cassette ELISA. In this test, specific HIV type 1 and 2 antibodies are immobilised on the nitrocellulose membrane at two different fixed spots along with another spot for the control.

ELISPOT Assay

ELISPOT assay is a modification of ELISA. It allows the quantitative determination of the number of cells in a population producing antibodies specific for a given antigen or an antigen for which one has a specific antibody. These tests have found application widely in the measurement of cytokines.

Applications of ELISA

ELISA is extensively used in the diagnosis of a wide variety of infectious as well as non-infectious diseases. These are widely employed to diagnose HIV, hepatitis virus, rotavirus, *Helicobacter pylori*, *Salmonella*, etc. A few examples of such uses are as follows:

1. Detection of viral infections such as HIV, Japanese encephalitis, dengue, hepatitis, influenza.
2. Detection of rotavirus, *Entamoeba histolytica*, *Giardia intestinalis*, *Taenia*, etc. in stool samples.
3. Detection of bacterial infections including tuberculosis, syphilis, enteric fever, *H. pylori* infections, etc.
4. Detection of parasitic infections like neurocysticercosis, amoebiasis, lymphatic filariasis, etc.
5. Detection of *E. coli*, *Campylobacter*, *Salmonella*, etc. in food and food products.

Radioimmunoassay

When radioisotopes instead of enzymes are used as labels to be conjugated with antigens or antibodies, the technique of detection of the antigen–antibody complex is called radioimmunoassay (RIA). In 1960, Rosalyn Yalow and Solomon Berson of the Veterans Administration Hospital in New York described RIA for the first time. They were awarded the Nobel Prize in 1977 for their work on developing RIA for peptide hormones.

The classical RIA methods are based on the principle of competitive binding. Labelled and unlabelled antigens compete to bind with the antibody when radiolabelled and unlabelled antigens are incubated together with the corresponding antibody. The quantity of unlabelled antigen is directly proportional to the amount of radiolabelled antigen, which is not bound.

In the first step, mixtures of fixed amounts of labelled antigen, a variable amount of cold antigen and samples with unknown concentrations are prepared—following which similar amounts of antibodies are added to the mixtures. Second antibody or reagents that promote precipitation of antigen–antibody complexes are added to precipitate out antigen–antibody complexes. The amount of radiolabelled antigen precipitated with the antibody is determined by counting radioactivity. Finally, a standard curve is constructed by plotting the percentage of antibody-bound radiolabelled antigen against known concentrations of a standardised unlabelled antigen. The antigen concentrations in patient samples are extrapolated from that curve. The major advantage of RIA is its extremely high sensitivity.

Applications of RIA

The test can be used to determine very small quantities (e.g. nanogram) of antigens and antibodies in the serum. The test is used to quantify hormones, drugs, HBsAg and other viral antigens. The main drawbacks of the RIA include (1) the cost of equipment and reagents, (2) the short shelf-life of radiolabelled compounds and (3) the problems associated with the disposal of radioactive waste.

Western Blot Techniques

Western blotting is called so because the procedure is similar to Southern blotting, which Edwin Southern developed to detect deoxyribonucleic acid (DNA). However, while Southern blotting is done to detect DNA, Western blotting is done to detect proteins.

Western blotting is usually done on a tissue homogenate or extract. It uses sodium dodecyl sulphate-polyacrylamide gel electrophoresis (SDS-PAGE), a type of gel electrophoresis, to first separate various proteins in a mixture based on their shape and size. The protein bands thus obtained are transferred or blotted onto a

Fig. 15.8 Western blot test

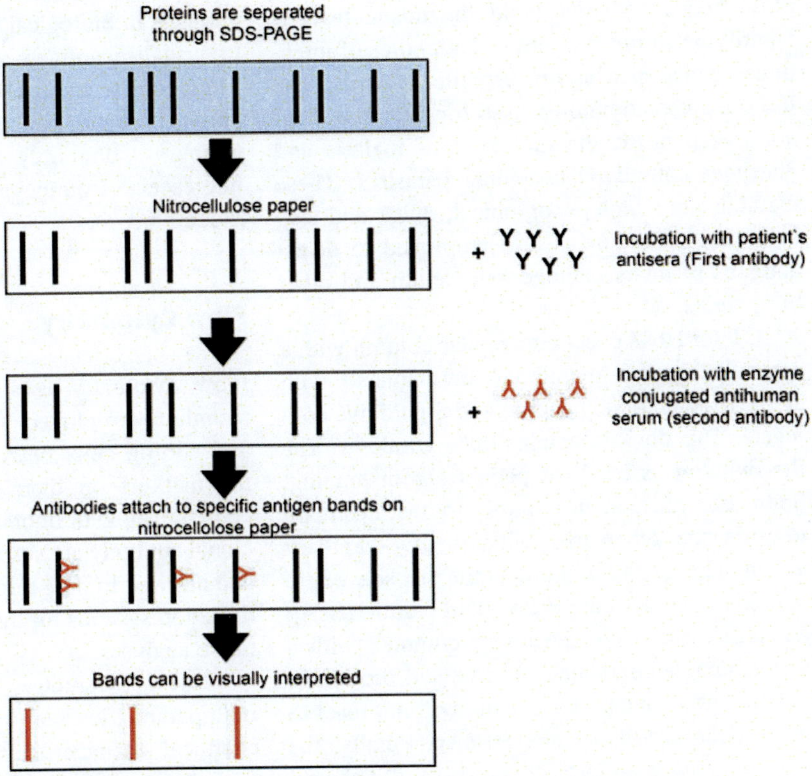

nitrocellulose or nylon membrane where they are "probed" with antibodies specific to the protein to be detected. The antigen–antibody complexes that form on the band containing the protein recognised by the antibody can be visualised in various ways, including enzyme immunoassays or radioimmunoassays. The enzyme-linked antibodies against the protein are generally used detection procedures. After binding of the enzyme–antibody conjugate, the addition of substrate, which will produce coloured and insoluble products, causes the appearance of a coloured band at the site of the target antigen.

The higher sensitivity can be improved to determine the site of protein interest if a chemiluminescent compound and suitable enhancing agents are used to produce light at the antigen site (Fig. 15.8).

Applications of Western Blot

Western blot identifies a specific protein in a complex mixture of proteins. In this method, antigen with known molecular weight can be separated by SDS-PAGE. Then the same can be blotted onto a nitrocellulose membrane. The sample suspected to contain antibodies is used to probe any specific antigens. Antibody–antigen reaction is detected using an enzyme-linked secondary antibody specific to the test species or a radiolabelled antibody. It is also used to estimate the size of the protein and the amount of protein present in the mixture.

The Western blot test is most widely used as a confirmatory test for the diagnosis of HIV, where this procedure is used to determine whether the patient has antibodies that react with one or more viral proteins or not. Western blotting is also used to demonstrate specific antibodies in the serum to diagnose neurocysticercosis and tubercular meningitis.

Immunochromatographic Tests (ICTs)

ICT is a qualitative, single-step test technique used to diagnose various infections. One of the most frequently used rapid diagnostic tests is identifying serum antigens or antibodies of infectious agents such as the hepatitis B virus. The ICTs are also commonly used for the rapid diagnosis of parasitic diseases such as malaria and diarrhoea caused by coccidian parasites. These methods are simple, economical, quick and reliable. The procedure is also employed to detect antigens of many pathogens in serum and other body fluids.

ICT for HbsAg is a rectangular column with a centrally situated membrane impregnated with anti-HBsAg antibody and colloidal gold dye conjugate. This device contains three windows. The first window is used for patient serum loading. Soon after that, serum starts moving upstream towards the other two windows by capillary action. The second window is the test site where colloidal gold conjugate and anti-HBsAg antibodies are fixed, forming a complex with a serum sample containing HBsAg and producing red coloured bands. The test site will not react to normal serum without any HBsAg. Finally, the third window is an in-built control, and it should be positive for all the samples. If the third window bands are not visible, the test should be considered invalid, and serum should be retested with fresh ICT.

Chemiluminescence Assay

The chemiluminescence assay uses chemiluminescent compounds that emit energy in the form of light during the antigen–antibody reactions. The emitted lights are measured, and the concentration of the analyte is calculated. The assay is a fully automated method used for drug sensitivity testing of *Mycobacterium tuberculosis*.

Immunohistochemical Technique

Immunohistochemistry (IHC) or immunohistochemical technique is the process of detecting surface or intracellular antigens in cells of a tissue section based on the principle of fluorescent-tagged antibodies that bind specifically to antigens in biological tissues. This method helps to visualise antigen–antibody reactions in situ. Immunohistochemistry is widely used to diagnose abnormal cell populations in cancerous tumours. Commonly used fluorescent dyes are fluorescein isothiocyanate, phycoerythrin and rhodamine.

Flow Cytometry

Flow cytometry is a laser-based fluorescence technique employed for identifying, counting and sorting cells bearing particular antigens or the markers on their surface. In this method, cells stained with fluorescence dye-tagged-monoclonal antibody are suspended in a stream of fluid and allowed to pass through lasers and electronic detection systems for their qualitative and quantitative analyses.

Flow cytometry allows simultaneous multiparametric analysis of the physical and chemical characteristics of up to thousands of particles per second. Determination of different molecules on the cell surface (antigenic markers) and differentiation of cell size and shape can be done by using forward and side scattering options available in the flow cytometry machine settings.

Flow cytometry is routinely used in the diagnosis of health disorders. It is used to count blood cells, including differential leukocyte count (DLC), for diagnosis, treatment and prognosis of leukaemia and lymphomas; to count and sort T-cell subsets (CD4 and CD8 counts in HIV patients); to study the cell cycle and programmed cell death (apoptosis); etc. It has many other applications in basic research clinical trials.

Turbidimetry and Nephelometry

Turbidimetry and nephelometry are the two methods used to detect and quantitate precipitation reactions in serum and are based on the phenomenon of light scattering by precipitates in a solution.

Turbidimetry is a measurement of turbidity or cloudiness of precipitate in a solution.

Nephelometry is an improvement on this technique because it measures the light scattered at a particular angle from the incident beam as it passes through a suspension containing the antigen–antibody precipitate. The amount of light scattered is an index of the concentration of the solution. Beginning with a constant amount of antibodies, an increasing amount of antigen would increase antigen–antibody complexes. Thus, the relationship between antigen concentrations, as indicated by the antigen–antibody complex formation, and light scattering approaches linearity. The exact values of the antigen or antibody in the serum can be estimated using a computer through this system. Laser beams have been used as the source of incident light to improve the sensitivity of this system. Nephelometry is now becoming the method of choice for use in various laboratories to measure plasma proteins, including IgG, IgM and IgA, complement components, RA (rheumatoid arthritis) factor, and ASLO (anti-streptolysin O).

Immunoelectron Microscopic Assays

Immunoelectron microscopy, first developed in 1941, is a method for visualising and characterising antigen–antibody reactions in tissues and cell cultures and identifying viruses in tissues or cultured cells. Gold is the most preferred probe since it is available in various sizes from 5 to 150 nm and is highly electron opaque. Normally, transmission electron microscopy is used in this technique. The antigen–antibody complex can be seen using colloidal gold of known diameter conjugated to the antibody molecule. The immunoferritin technique is an older method where electron-dense ferritin molecules are conjugated to the immunoglobulin and used as a probe. Gold-labelled enzymes as probes have also been used to localise biomolecules like nucleic acids.

Immune Response 16

The specific reactivity induced by an antigenic stimulus in a host is known as an immune response. The immune system is developed in a host primarily to protect the host from the harmful effects of invading pathogens or infectious organisms by stimulating various immune responses. The host immunological system can recognise foreign substances, and the immune response that follows is associated with macrophages and B-cells or T-cells. There are two main sites where pathogens may reside in an infected host—extracellularly in tissue spaces or intracellularly within a host cell; the immune system has different ways of dealing with pathogens at these sites.

Types of Immunity

An encounter with a microbial or viral agent usually elicits a complex variety of responses. The immune response can be antibody-mediated (humoral), cell-mediated (cellular) or both. Humoral immunity acts mainly against extracellular pathogens, whereas cell-mediated immunity (CMI) acts against intracellular pathogens (Table 16.1).

Humoral Immunity (Antibody-Mediated Immunity)

Humoral immunity is mediated by soluble host proteins called antibodies or immunoglobulins. It is based on the action of antibodies and complements. It provides the first line of defence against a wide range of invading foreign substances. It is directed primarily against (1) extracellular bacteria, in particular exotoxin-producing bacteria, such as *Corynebacterium diphtheriae*, *Clostridium tetani* etc., (2) bacteria whose virulence is due to polysaccharide capsules (e.g. *Haemophilus influenzae*, *Neisseria meningitidis*, *Streptococcus pneumoniae*) and (3) certain viruses that cause infection through the respiratory or intestinal tract. It helps to prevent recurrence of viral infections. The humoral immunity also participates in the pathogenesis of hypersensitivity reactions (types I, II and III) and certain autoimmune diseases.

Production of Antibodies

It is the main feature of humoral immune responses. The antibody production involves three steps: (1) the afferent limb that consists of the antigen entry, its location and outcome in the tissues along with its contact with suitable immune

Table 16.1 Differentiating characteristics of cell-mediated and humoral immunity

Characteristics	Cell-mediated immunity	Humoral immunity
Definition	Immunity that identifies and destroys infected cells	Immune response mediated by macromolecules found in the extracellular fluids
Pathogen target	Protects against fungi, viruses and facultative intracellular bacterial pathogens	Protects against extracellular bacterial pathogens and viruses infecting respiratory or intestinal tract; and prevents recurrence of viral infections
Type of hypersensitivity	Mediates delayed (type IV) hypersensitivity	Mediates immediate (types I, II and III) hypersensitivity
T-cell/B-cell	Only T-cell-dependent antigens lead to cell-mediated immunity	B-cells directly bind soluble antigens resulting in production of antibodies
Cells involved	Both $CD4^+$ and $CD8^+$ T-cells are involved	Only T_H cells are involved
Action	Provides immunological surveillance and immunity against cancer	No major role in immunological surveillance
	Participates in rejection of homografts and graft-versus-host reaction	May be involved in early graft rejection due to preformed antibodies

cells; (2) the central function that involves antigen processing and control of antibody production; and (3) the efferent limb that consists of antibody secretion, its distribution in tissues, blood and other body fluids and its manifestations in combating antigen of a pathogen.

The production of antibodies follows a characteristic pattern as follows:

1. **Lag phase**: This is the immediate phase following exposure to antigen. During this phase, no antibodies are detected in circulation.
2. **Log phase**: This is the next phase characterised by a steady rise in antibody titres in the circulation.
3. **Plateau**: This is a phase of equilibrium between antibody synthesis and catabolism.
4. **Phase of decline**: This phase is characterised by an increase in antibody catabolism compared to antibody production, leading to a fall in antibody titre in the circulation.

Humoral Immune Response

It is of two types: primary and secondary.

Primary Response

During the primary response, when an individual encounters an antigen for the first time, antibody response to that antigen is detectable in the serum after a longer lag period (**lag phase**) than occurs in the secondary response. The serum antibody concentration continues to rise (log phase) for several weeks, and then it sustains (plateau phase) for a while and finally, antibody level declines (phase of decline); it may drop to very low levels. A small clone of B-cells and plasma cells specific for the antigen are formed during this primary response.

The lag period is typically of 7–10 days duration but can be longer, even for weeks, depending on the nature of the antigen. For example, the lag phase may be as long as 2–3 weeks with some antigens, such as diphtheria toxoid, whereas it may be as short as a few hours with pneumococcal polysaccharides. The lag period also depends on the dose of the antigen and the route of administration, whether oral or parenteral. IgM is the first antibody to be formed, followed by IgG, IgA or both. IgM levels tend to decline sooner than IgG levels (Fig. 16.1).

Secondary Response

The antibody response is typically more rapid in the secondary response due to a second encounter with the same antigen or a closely related "cross-reacting" antigen, months or years after the primary response. The lag period is typically very short (only 3–5 days). The antibody level is also much higher than that during the primary response.

These changes in secondary response are attributed to the persistence of antigen-specific

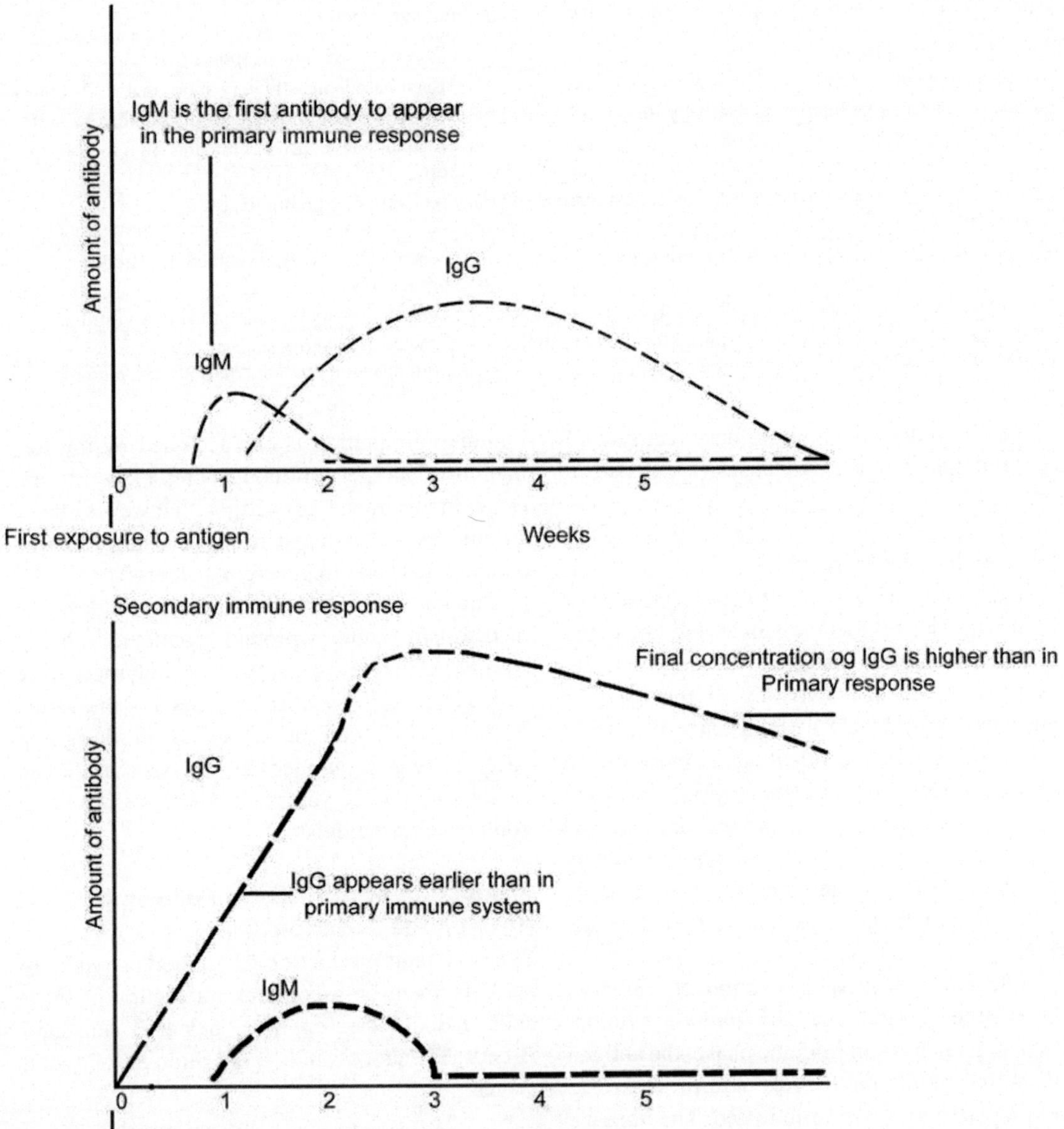

Fig. 16.1 A schematic diagram showing primary and secondary response

"**memory cells**" following the first contact with the antigen. These memory cells proliferate in large numbers to produce large clones of specific B-cells and plasma cells that mediate the secondary response. The amount of IgM produced in the secondary response is qualitatively similar to that produced after the first contact with the antigen; however, much more IgG is produced, and the IgG level tends to persist much longer than in the primary response. Furthermore, such an antibody tends to bind antigen more firmly (i.e. to have higher affinity) and thus to dissociate less easily. Improved antibody binding is due to mutations in the deoxyribonucleic acid (DNA) that encodes the antigen-binding site. This process is called **somatic hypermutation**.

Table 16.2 Differences between primary and secondary humoral immune response

Primary immune response	Secondary immune response
Lag phase most of the times as long as weeks or months	Lag phase is usually very short, that is, 3–4 days
Amount of antibody production is relatively low	High levels of antibody production can be seen
Antibody level reaches phase of decline relatively early	Antibody level usually remains high for a longer time
Induces IgM antibody production (very low concentrations of IgG production)	IgG antibody predominates
Naïve B-cells are involved in antibody production	Memory B-cells are involved in antibody production
Both thymus-dependent and independent antigens are involved	Only thymus-dependent antigen involved
Relatively high dose of antigens is required for immunisation; incorporation of adjuvants favours immunisation	Low dose of antigens is sufficient for immunisation and adjuvants are not required

The differences between primary and secondary humoral immune responses are listed in Table 16.2.

Elimination of Antigens

Route of administration of antigen affects the site of localisation of these antigens in the body. For example, most of the antigens introduced subcutaneously are localised mainly in the draining lymph nodes, and only a small amount is there in the spleen. On the other hand, most of the antigens introduced intravenously are localised in the spleen, liver, bone marrow, kidney and lungs but not in lymph nodes. Approximately three-fourths of these antigens are broken down by reticuloendothelial cells and excreted in the urine.

Particulate antigens are removed in two phases—the non-immune and immune elimination phases. In the non-immune phase, the antigen is engulfed by the phagocytes, and processed to make it inactive and then eliminated. The immune phase begins with the appearance of the first antibody to that antigen. This antibody forms complexes with the antigen, which are then phagocytosed and eliminated. This immune phase removes antigen at an accelerated rate.

Soluble antigens are eliminated in three stages or phases—equilibration, metabolism and immune elimination. In the phase of equilibration, the antigen diffuses into the extravascular spaces. The antigen levels fall owing to catabolic decay in the metabolism phase. The phase of immune elimination is characterised by the formation of antigen–antibody complexes, leading to rapid clearance of antigen. However, these complexes can damage tissue, causing immune complex diseases (e.g. serum sickness).

The rate of antigen elimination depends on its metabolism rate, protein antigens being eliminated in a few days, whereas polysaccharide antigens require months or years for the same. This principle can be exploited in designing vaccines (e.g. pneumococcal polysaccharide can persist for up to 20 years in humans, so it can be a good vaccine candidate).

Production of Antibodies Following Antigen Presentation

Synthesis and production of antibodies typically are dependent on a complex interaction of three cells, namely macrophages, helper T-cells and B-cells. The process undergoes through following stages:

Antigen Processing

This is the first step in which antigens are presented to immunocompetent cells by antigen-presenting cells (APCs), such as macrophages and dendritic cells. Processing by macrophages appears to be a prerequisite for forming antibodies against many T-cell-dependent antigens (Fig. 16.2), such as proteins and erythrocytes. However, antibody production does not require macrophage participation in the case of T-cell-independent antigens. Both the

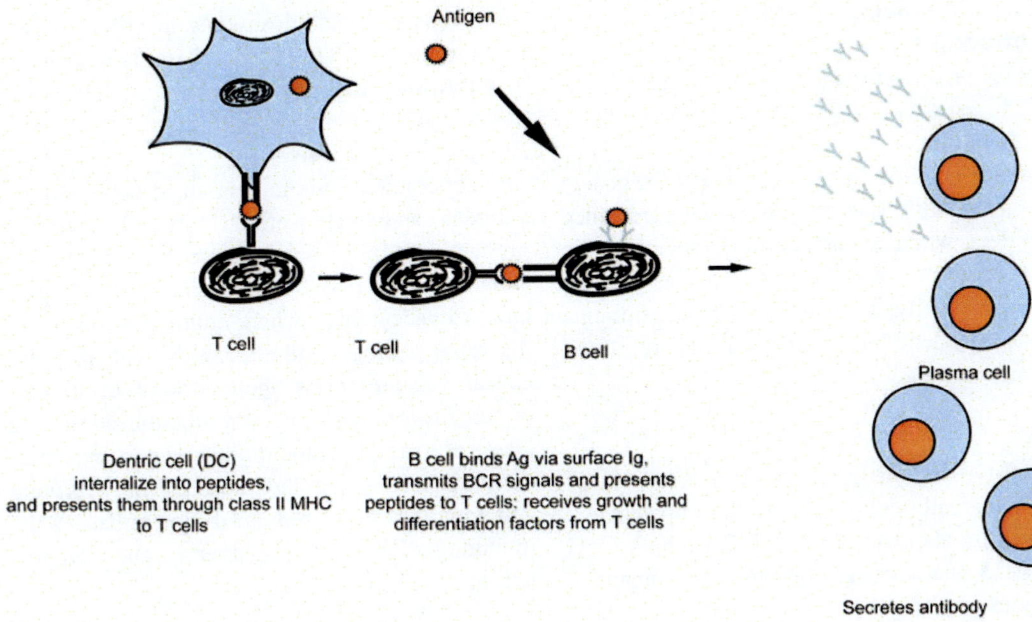

Fig. 16.2 T-cell-dependent immune response

macrophages and dendritic cells present either native or processed antigen at the cell surface. Macrophages play a key role by modulating the optimum dose of antigen presented to lymphocytes to induce immune responses.

T- and B-Cell Activation

After processing antigens by a macrophage, antigen fragments appear on surfaces of macrophages in association with Class-II MHC (Major Histocompatibility Complex) proteins. The antigen-Class-II MHC protein complex binds to specific receptors present on the surface of helper T-cells. Subsequently, these helper T-cells produce cytokines that activate B-cells, producing antibodies specific to that antigen. The activated cytokines are interleukin-2 (IL-2: T-cell growth factor), IL-4 (B-cell growth factor) and interleukin-5 (B-cell differentiation factor). The activated B-cells undergo clonal proliferation and differentiate to form plasma cells, producing specific immunoglobulins (antibodies). Major host defence functions of antibodies include neutralisation of toxins and viruses and opsonisation (coating) of the pathogen, which aids its uptake by phagocytic cells.

Although helper T-cells play a key role in the formation of antibodies, certain substances (e.g. polysaccharides) can activate B-cells directly without the help of T-cells. Such substances are called T-cell-independent antigens. These antigens, however, induce only the production of IgM antibodies but no other antibodies by B-cells. This is because B-cells require IL-4 and IL-5 to switch classes to produce IgG, IgA and IgE. These IL-4 and IL-5 are produced by T-helper-cells only.

B-cells perform two important functions: (1) They recognise antigens with their surface IgM that acts as an antigen receptor; besides, they present epitopes to helper T-cells in association with Class-II MHC proteins. IgM antigen receptor on the B-cells recognises foreign proteins and lipids, carbohydrates, DNA, ribonucleic acid (RNA) etc. (2) On the other hand, Class-II MHC proteins present protein fragments to the helper T-cells. The IgM antigen receptor binds with this wide variety of molecules that stimulate B-cells to produce antibodies against all the molecules possible.

Factors Influencing the Production of Antibodies

Many factors affect the production of antibodies.

Genetic Factors

These factors influence the response of the host to antigen. Persons responding to antigens are called responders, whereas persons not responding are called non-responders. These differences are controlled genetically and controlled by the immune response (Ir) gene located in the short arm of the sixth chromosome.

Age

The embryo and the infant, at birth, are not fully immunologically competent. Full competence is achieved by about the age of 5–7 years for IgG and 10–15 years for IgA by the development of lymphoid organs.

Nutritional Status

Malnutrition affects both the humoral and cell-mediated immunities. Deficiencies of amino acids and vitamins have been shown to decrease the production of antibodies.

Route of Antigen Administration

Induction of immune response in a host depends on the route of administration of the antigen. Parenteral administration of the antigen induces a better immune response than the oral or nasal routes.

Multiple Antigens Administration

Antibody responses vary when two or more antigens are administered simultaneously. Antibody responses to one or more of them may be diminished due to antigenic competition or enhanced as seen after vaccination with the triple vaccine (diphtheria, pertussis and tetanus), or may be similar. Hence, different antigens' nature and relative proportions should be carefully adjusted for optimal effect.

Antigen Doses

A minimum critical antigen dose is essential to elicit an optimum immunological response. A very high or small dose fails to stimulate the immune system. This phenomenon is referred to as **immunological paralysis**.

Antigen Interactions

The concomitant administration of two or more antigens can have two effects. Either the antibody production against one or more antigens will be enhanced, or it may be diminished. It has importance in vaccinology where multiple serotypes of the same pathogen (like a multivalent pneumococcal vaccine) or antigen cocktail of diverse organisms are used for active immunisation. The classical example is the diphtheria-pertussis-tetanus vaccine, where the presence of a protein component of the pertussis fraction enhances the immunogenicity of diphtheria and tetanus toxoids.

Adjuvants

Adjuvants are substances that enhance the immunogenicity of an antigen. The adjuvants delay the release of an antigen from the injection site and prolong the antigenic stimulus. The substances that are used as adjuvants include: (1) Freund's incomplete adjuvant (protein antigen incorporated in the water phase of water in oil emulsion); (2) Freund's complete adjuvant (incomplete adjuvant along with the suspension of killed tubercle bacilli); (3) aluminium salts, both phosphate and hydroxide and (4) others, such as silica particles, beryllium sulphate, endotoxin etc.

An adjuvant functions in many ways. It (1) causes sustained release of antigen from the depot, (2) enhances the immunogenicity of non-antigenic substances, (3) increases the concentration and persistence of antibodies, (4) induces and enhances CMI, (5) activates macrophages, and (6) stimulates lymphocytes non-specifically.

Immunosuppressive Agents

Immunosuppressive agents are those that suppress the immune response. They are used in transplantation surgery and in situations that require suppression of host immunity. The agents are as follows:

X-irradiation: Sub-lethal dose of irradiation is toxic to replicating cells and is used to suppress antibody formation. Antibody production ceases after 24 h of receiving irradiation.

Radiometric drugs: These include alkylating agents (such as cyclophosphamide, nitrogen mustard), which suppress antibody production. Cyclophosphamide, given for 3 days, completely suppresses the antibody response. It selectively prevents the replication of B-cells.

Corticosteroids: Corticosteroids are anti-inflammatory drugs that diminish the responsiveness of both B- and T-cells. They alter the maturation of activated cells by suppressing the production of interleukins. They suppress delayed hypersensitivity, but in therapeutic doses for a short period, they have little effect on the production of antibodies.

Antimetabolites: These include folic acid antagonists (such as methotrexate); analogues of purine (6-mercaptopurine and azathioprine); and analogues of cytosine (cytosine arabinose) and uracil (5-fluorouracil). These substances inhibit DNA and RNA syntheses, inhibiting cell division and differentiation, essential for cellular and humoral immune responses. These are usually used for the prevention of graft rejection.

Antilymphocyte serum (ALS): It is a heterogeneous antiserum raised against T-lymphocytes. The ALS acts mainly against circulating lymphocytes but not against lymphocytes in lymphoid organs. It is mainly used to prevent graft rejection in transplantation surgery.

Effect of Antibodies

The primary defence mechanism of the host is to produce antibodies against infectious pathogens and their products. Antibodies act by the following mechanism:

Neutralisation

Antibodies bind to the pathogen or foreign substance, thus blocking the pathogen's binding to their targets. For example, antibodies produced against the bacterial toxins prevent their binding to the host cells, thus making the toxin ineffective. In the same way, antibodies prevent infection or colonisation of virus or bacterial pathogen by binding to them and blocking the binding of the pathogen to target cells.

Opsonisation

Antibody opsonises the pathogen or foreign substance by binding to them and assisting in phagocytes' uptake and destruction of substances. The Fc receptors of the phagocytes interact with the Fc region of another antibody, thus facilitating the pathogen for phagocytosis.

Complement Activation

Antibody activates the complement cascade that effectively kills certain bacteria and viruses. Additionally, pathogens can be opsonised by certain components of the complement cascade (e.g. C3b), thus rendering the uptake of the pathogen via complement receptors by phagocytes.

Antibodies can be induced in the host actively by using vaccines or acquired passively to confer immediate protection against the pathogen. For example, hyperimmunised sera containing ready-made antitoxins against toxins of tetanus, botulism or diphtheria are given to neutralise the actions of these toxins immediately in the body. Also, hyperimmune sera containing a high titre of specific antibodies are given to inhibit attachment and replication of rabies and hepatitis A and B viruses early during the period of incubation.

Tests for Detection of Humoral Immunity

The measurement of IgG, IgM and IgA in the patient's serum is the primary method for detecting humoral immunity. Radial immunodiffusion and immunoelectrophoresis are the methods frequently employed for the measurement of antibodies.

Cell-Mediated Immunity (CMI)

Cell-mediated immunity (CMI) is a specific type of acquired immune response not mediated by antibodies but by sensitised T-cells. This form of immunity is transferred from donor to recipient, not with antisera but with intact lymphocytes;

hence it is called a cell-mediated immune reaction. It is stimulated by T-helper type I (Th1) lymphocytes and the cytokines released in response to antigens. This type of immunity kills the infected cells by recognising antigens found on them. CMI mainly targets the microbes that survive the phagocytosis and also that infect non-phagocytes. CMI removes the virus-infected cells actively and protects against protozoans, intracellular bacteria and fungi. It plays an important function in transplant rejection and graft–host reaction.

Functions of CMI

It performs the following immunological functions: (1) It confers immunity in diseases caused by obligate intracellular pathogens such as bacteria (*Mycobacterium tuberculosis*, *Mycobacterium leprae*, *Brucella* etc.), viruses (smallpox, measles, mumps etc.), fungi (*Histoplasma capsulatum*, *Blastomyces dermatitidis*, *Coccidioides immitis* etc.) and parasites (*Toxoplasma gondii*, *Leishmania donovani* etc.). (2) It participates in immunological surveillance and immunity against cancer. (3) It plays an important role in the pathogenesis of delayed hypersensitivity reactions and certain autoimmune diseases, such as autoimmune thyroiditis, encephalitis.

Induction of CMI

The induction of the CMI depends on the nature of the antigenic stimulus. It is best developed following infection with intracellular parasites. Only T-cell-dependent antigens induce CMI. Killed vaccines or other non-living antigens do not induce CMI unless administered with Freund's adjuvants. Induction of CMI involves sequence stages of (1) antigen processing and presentation to T-cells, (2) recognition of antigen by T-cells, and finally (3) release of cytokines by Tc lymphocytes.

1. **Antigen processing and presentation to T-cells**: Induction of CMI begins with the presentation of foreign antigen by APCs to T-lymphocytes. Proteins from exogenous antigens, such as bacteria, are internalised via endocytic vesicles into APCs, such as macrophages (Fig. 16.3). Then, they are exposed to cellular proteases in intracellular vesicles. Peptides, approximately 10–30 amino acid residues in length, are generated in endosomal vesicles. The endosomal vesicles can then fuse with exocytic vesicles containing Class-II MHC molecules.
2. **Recognition of antigen by T-cells**: T-cells recognise antigens only when presented with MHC molecules. The combination of foreign antigen and Class-I MHC molecule is recognised by $CD8^+$ cells. These $CD8^+$ cells, after recognition, differentiate into Tc and Ts lymphocytes. On the other hand, $CD4^+$ cells recognise the combination of antigen and Class-II MHC antigen, after which they are differentiated into Th and Td cells. Subsequently, the sensitised T-lymphocytes undergo blast transformation, clonal proliferation and differentiation into memory cells and effector cells, such as Th, Tc, Td and Ts.
3. **Release of cytokines by Tc lymphocytes**: Finally, the lymphokines, which are biologically active products responsible for various manifestations of CMI, are released by the activated lymphocytes resulting in the lysis of the target cells.

Cytokines

The interleukins, lymphokines, chemokines, interferons (IFNs) and tumour necrosis factors (TNFs) altogether are called cytokines. Usually, hormones or growth factors are not considered cytokines. However, some overlaps in the terminology still exist. These small stretches of peptide cell signalling molecules regulate the host's immunological, inflammatory and reparative responses.

Cytokines are highly active and potent hormone-like substances that mediate biological effects even at femtomolar concentrations. Cytokines actively participate in a wide range of biological activities varying from chemotaxis to

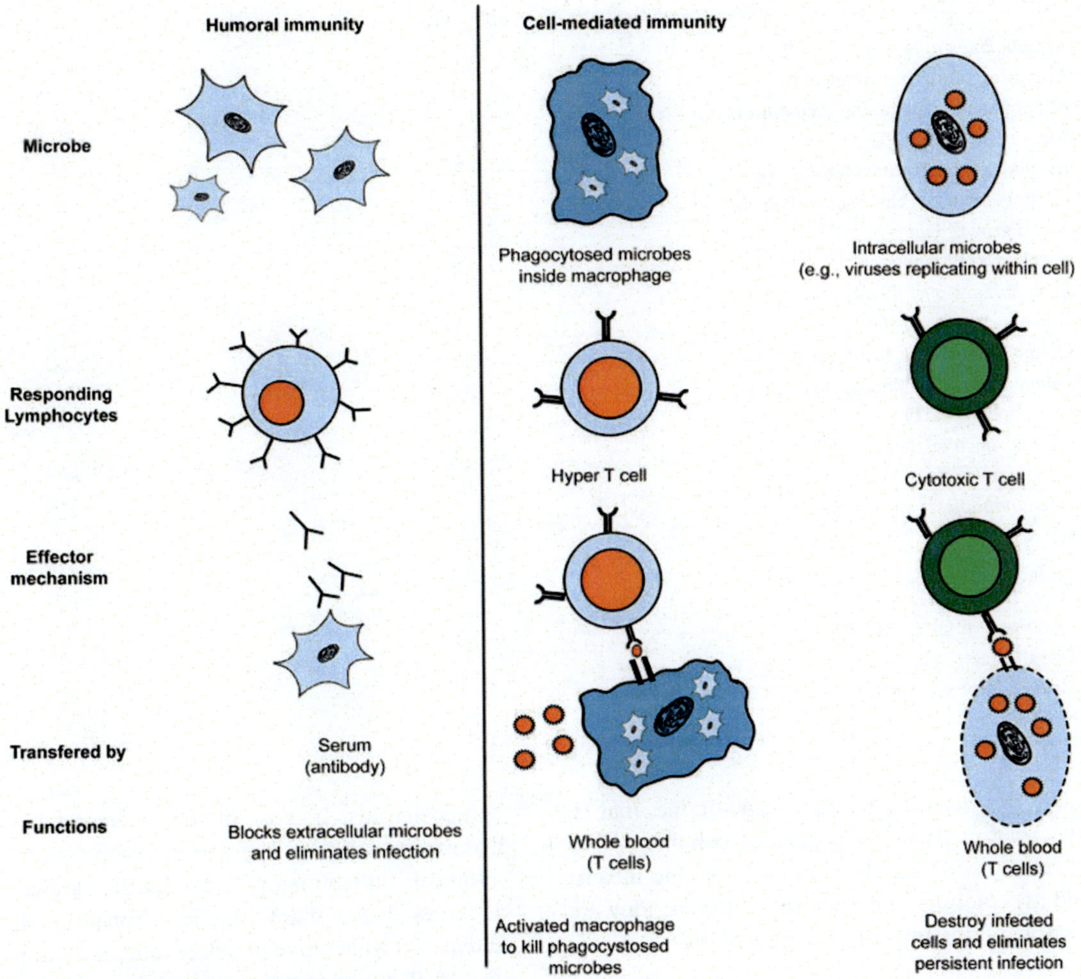

Fig. 16.3 Humoral and cell-mediated immune responses

activation of specific cells. They are produced in response to specific antigenic stimuli, and they are not stored as preformed proteins. They are synthesised by the initiation of gene transcription, and their messenger-RNAs (mRNAs) also have a very short period.

Cytokines are biologically active substances secreted by monocytes, lymphocytes and other cells and are actively involved in innate immunity, adaptive immunity and inflammation. The initial terminologies used to identify cytokines are based on their established biological effects (Table 16.3). The substances that are released from T-lymphocytes are called **lymphokines**. On the other hand, substances released from macrophages or monocytes are termed **monokines**. The secretory products of leucocytes are called **interleukin**, and they have regulatory effects on other immune cells. Similarly, interferons (IFNs), tumour necrosis factor (TNF) and other secretory products have such regulatory effects. These small proteins with specific biological activities (important in cell signalling) have been termed cytokines.

The cytokines differ from endocrine hormones because: (1) cytokines are not the products of specialised cells, but they are being produced by immune system components, which are widely distributed in the body such as lymphocytes, macrophages, fibroblasts and platelets; (2) the

Table 16.3 Classification of lymphokines on the basis of effector cells

Lymphokines affecting macrophages
Migration inhibiting factor (MIF)
Macrophage activation/aggregation factor (MAF)
Macrophage chemotactic factor (MCF)
Lymphokines affecting lymphocytes
Blastogenic/mitogenic factor (BF/MF)
T-cell growth factor
B-cell growth factor
Lymphokines affecting granulocytes
Chemotactic factor (CF)
Colony-stimulating factor (GM-CSF)
Lymphokines affecting cultured cells
Lymphotoxin (LT)
Tumour necrosis factor (TNF)
Interferon (IFN)
Haematophoiesis stimulators
GM-CSF
IL-3, IL-7
Other lymphokines
Skin reactive factor (SRF)
Transfer factor (TF)
Platelet derived growth factor (PDGF)
Neuroleukin

mode of action is paracrine or autocrine, that is, they act directly on the producing cell themselves or nearby cells but not through systemic means; and (3) cytokines are pleiotropic, that is, they are involved in various biological activities.

Important Cytokines

Due to the introduction of recombinant DNA technology, many cytokine genes were cloned and characterised. Few cytokines synthesised by recombinant DNA technology are being used as protein therapeutics. The list of important cytokines, and their sources and functions are highlighted in Table 16.4.

Interleukins

Different interleukins have regulatory effects on other immune cells.

Interleukin-1 (IL-1): It is a protein produced mainly by activated macrophages and monocytes. Its production is stimulated by antigens, toxins and inflammatory processes but inhibited by cyclosporine and corticosteroids. It is an important interleukin that mediates a wide range of metabolic, physiological, inflammatory and haematological activities. It has many important functions. It activates a wide range of target cells, including T- and B-lymphocytes, neutrophils, epithelial cells and fibroblasts to proliferate, differentiate or synthesise specific products. For example, it stimulates helper T-cells to produce IL-2 and stimulates B-cells to proliferate and synthesise antibodies etc. It acts on the hypothalamus to cause fever associated with infections and other inflammatory reactions.

Interleukin-2 (IL-2): IL-2 is mainly produced by helper T-cells. It is a major T-cell growth factor. It stimulates both helper and cytotoxic T-cells to grow. IL-2 stimulates the monocytes and natural killer (NK) cells and activates the growth of B-cells. IL-2 activates the T-cells in an autocrine manner resulting in the IL-2R expression and IL-2 production. IL-2 binds to IL-R, stimulating cell division. If an antigen no longer activates the T-cells, the IL-2R will

Table 16.4 Important functions of cytokines

Cytokines	Source	Functions
Interleukins		
IL-1 (α and β)	Macrophages	Activates helper T-cells; causes fever
IL-2	Th-1 cells	Activates helper, cytotoxic T-cells and B-cells
IL-3	Th cells, NK and mast cells	Supports growth and differentiation; stimulates histamine release
IL-4	Th-2 cells	Stimulates B-cell growth; increases isotype switching and IgE; up-regulates Class-II MHC expression
IL-5	Th-2 cells	Stimulates B-cell differentiation; increases eosinophils and IgA
IL-6	Th, macrophages and fibroblasts	Affects numerous processes, which includes differentiation of B-cells, neutrophil mobilisation, reproduction, bone metabolism and ageing
IL-7	Stromal cells of bone marrow and thymus, dendritic cells, hepatocytes etc.	Proliferation of all lymphoid lineage cells; growth factor in B- and T-cell maturation
IL-8 (CXCL8)	Macrophages and other cells	Neutrophil chemotactic factor; helps in the induction of chemotaxis in target cells
IL-9	Th cells	T-cell growth and maturation; prevents apoptosis
IL-10	Th-1 cells	Inhibitor of IFN-γ synthesis; immune-regulatory in function
IL-11	Bone marrow stromal cells	Induces acute-phase proteins; activates B-cells, hepatocytes and megakaryocytes
IL-12	Dendritic cells, macrophages, neutrophils etc.	T-cell-stimulating factor; stimulates growth and functions of T-cells
IL-13	T-cells	Inhibitory to mononuclear functions
IL-17	Th-17 cells	It is a potent mediator of delayed hypersensitivity reactions and also a proinflammatory marker
IL-18	Macrophages, dendritic cells, intestinal epithelial cells etc.	It promotes the production of different proinflammatory mediators
Colony-stimulating factors (CSF)		
GM-CSF	Macrophages, T-cells, mast cells, fibroblasts and endothelial cells	Stimulation of stem cells to produce neutrophils, monocytes, eosinophils and basophils
G-CSF	Macrophages, endothelium and various leukocytes	Induces bone marrow to produce granulocytes and stem cells and also helps in migration of these cells into the peripheral circulation
M-CSF	Fibroblasts and endothelium	Stimulation of macrophage growth
Tumour necrosis factors (TNF)		
TNF-α	Macrophages, monocytes	Activates neutrophils and increases their adhesion to endothelial cells; mediates septic shock; causes necrosis of tumours, lipolysis, wasting, and antiviral and antiparasitic effects
TNF-β	T-cells	Induction of other cytokines
Interferons (IFNs)		
IFN-α	Leukocytes	Inhibits viral replication
IFN-β	Fibroblasts	Inhibits viral replication
IFN-γ	Th-1, Tc and NK cells	Inhibits viral replication; increases expression of Class-I and Class-II MHC; stimulates phagocytosis and killing by macrophages and NK cells
Others		
TGF-β	Platelets, mast cells and lymphocytes	Induces increased IL-1 production; induces class switch to IgA; limits inflammatory response and promotes wound healing
LIF	T-cells	Stem cell proliferation and eosinophil chemotaxis

eventually decay and end the proliferative phase.

Interleukin-3 (IL-3): IL-3, upon interaction with the IL-3 receptor, facilitates its function. It is involved in the proliferation of multipotent haematopoietic stem cells into myeloid progenitor cells. IL-3 acts as a bone marrow stem cell growth factor and is also called a multi-colony-stimulating factor (CSF).

Interleukin-4 (IL-4): It is a protein produced mainly by helper T-cells and macrophages. It stimulates the development of Th-2 cells, the subset of helper T-cells that produces IL-4 and IL-5 and enhances humoral immunity by producing antibodies. It is also required for class (isotype) switching from one class of antibodies to another within antibody-producing cells.

Interleukin-5 (IL-5): It is a protein produced by helper T-cells. It promotes the growth and differentiation of B-cells and eosinophils. It enhances IgA synthesis and stimulates the production and activation of eosinophils.

Interleukin-6 (IL-6): It is a protein produced by helper T-cells and macrophages. It stimulates the production of acute-phase proteins by the liver. It also acts on the hypothalamus to cause fever.

Other interleukins (ILs): IL-10, IL-12 and IL-13 are the other interleukins that affect lymphocytes. Activated macrophages and Th-2 cells produce IL-10. It is predominantly an inhibitory cytokine. It inhibits the production of type-I IFN. It inhibits the production of IFN-γ by Th-1 cells, which changes the immune system's responses to Th-2 cells. IL-10 inhibits cytokine production by activated macrophages and inhibits the Class-II MHC expression and co-stimulatory molecules in the macrophages, which results in immune response depression.

IL-12 is produced by the activation of dendritic cells and macrophages. IL-12 induces the IFN-γ production and stimulates the differentiation of Th cells to become Th-1 cells. In addition, it enhances the cytolytic functions of Tc and NK cells.

IL-13 is produced by Th-2 cells. It is associated with the pathogenesis of allergic airway disease (asthma). It is involved in the occurrence of hyper-responsiveness seen in asthma.

Colony-Stimulating Factors (CSFs)

Colony-stimulating factors (CSFs) are secreted glycoproteins that stimulate stem cells in the bone marrow to form specialised white blood cells. Haemopoietic stem cells are cultured on a semi-solid matrix such as soft agar, and then exogenous factors (stimulants) can be administered to achieve proliferation and differentiation of the cells into specific kinds of blood cells. For example, factors capable of forming macrophage colonies are called **macrophage colony-stimulating factors**, for granulocytes, granulocyte colony-stimulating factor etc. Certain factors induce the growth of all haemopoietic cell types and are called multi-CSFs, for example, IL-3. CSFs have clinical applications in patients undergoing chemotherapy. Neutropenia and bone marrow are the known causes of chemotherapy. CSFs tend to reduce the duration of a patient being neutropenic and thus help to survive the damage to bone marrow. It is also useful in treating haemopoietic dysfunction in infections.

Interferons (IFNs)

Typically, a cell infected with a virus will release IFNs, which helps adjacent cells to gain antiviral properties (**interfere with viral replication**), hence the name IFN was coined for this cytokine. So far, more than 20 IFNs have been identified in humans and other animals. IFNs can be divided into three major classes. Leucocytes produce IFN-α, fibroblasts produce IFN-β and T-cells activated by antigens, mitogens or IL-2 exposure produce IFN-γ.

Other than antiviral activity, IFNs also have other functions such as activation of NK cells and macrophages, up-regulation of antigen presentation by elevated expression of MHC molecules present on the APCs etc. Flu-like symptoms, fever and myalgia commonly found in infections are mainly due to the production of IFNs and other cytokines. IFN-γ is responsible for many immunological effects and antitumour activity.

Transforming Growth Factor-Beta (TGF-β)

TGF-β is an inhibitory cytokine produced by T-cells and other types of cells. It helps in the activation of macrophages and inhibits T-cell proliferation. TGF-β acts on endothelial cells and polymorphonuclear leukocytes, preventing the proinflammatory cytokine effects. In essence, it suppresses the immune response when it is not required after an infection, promoting the healing process.

Tissue Necrosis Factor-α (TNF-α)

The activated macrophages produce TNF-α against the microbes, mainly to the lipopolysaccharide of Gram-negative bacteria. It mediates acute inflammation and produces chemokines that are chemotactic cytokines. TNF-α stimulates endothelial cells by producing adhesion molecules, which mediates the arrival of the macrophages and neutrophils to the site of infection. It also induces the hypothalamus to stimulate fever and promotes the acute-phase protein production.

Chemokines

Chemokines are a subtype of low-molecular-weight cytokines with a characteristic structural pattern. More than 50 chemokines varying in size from 68 to 120 amino acids have been identified. The alpha-chemokines, such as IL-8, are produced by activated mononuclear cells, which attract neutrophils. The beta-chemokines such as RANTES—"regulated upon activation, normal T-cell expressed and secreted"—and MCAF—"monocyte chemotactic and activating factor"—are produced by activated T-cells and attract macrophages and monocytes.

Chemokines are produced by endothelial cells, resident macrophages and various cells present at the site of infection. They attract either macrophages or neutrophils to the site of infection; hence they are involved in chemical-induced migration of leukocytes—a process called chemotaxis. Specific receptors for chemokines are present on the surface of monocytes and neutrophils. They also facilitate the migration of white cells into the tissue to reach the infected area. They do so by activating integrins on the surface of neutrophils and macrophages that bind to the intercellular adhesion molecule (ICAM) proteins on the surface of the endothelium.

Tests for Detection of CMI

The CMI can be detected by *in vivo* and *in vitro* tests as follows:

CMI *In-Vivo* Tests

Skin tests are useful to detect delayed hypersensitivity reactions to common antigens that come in contact with the body. Purified protein derivatives in the tuberculin test, dinitrochlorobenzene or dinitrofluorobenzene, are the antigens used for skin testing. Most normal people respond with delayed-type reactions to these skin antigens. The absence of reactions to these skin tests suggests impairment of the CMI.

CMI *In-Vitro* Tests

Many *in vitro* tests are available for the detection of CMI.

1. **Migration inhibiting factor (MIF) test**: This test is performed to determine the CMI by making a semi-quantitative assessment of the migration inhibition of leukocytes. This test depends on the principle that cultured T-cells produce macrophage migration inhibiting factor on exposure to the antigen to which they are sensitised. In this test, human peripheral leukocytes are incubated in capillary tubes in a culture chamber containing culture fluid. In the absence of antigen, the leukocytes migrate out to the open end of the tube to form a fan-like pattern. In the presence of antigen, the leukocytes are prevented from migrating.
2. **Lymphocyte blast transformation**: A large number of T-cells undergo blast transformation when exposed to certain mitogens, such as phytohemagglutinin and concanavalin. Sensitised T-lymphocytes are transformed into large blast cells with a great increase in DNA synthesis on exposure to the specific antigen. An increase in DNA synthesis is

then measured by incorporating tritiated thymidine.

3. **Enumeration of T-cells, B-cells and subpopulation**: Fluorescence-activated cell sorter is used to count the number of each type of cells, whether T- or B-cells. In this method, cells are labelled with monoclonal antibodies tagged with a fluorescent dye, such as fluorescent or rhodamine. Then, the number of cells that fluoresce is registered bypassing those single cells through a laser light beam.
   The total number of B-cells can be counted using fluorescein labelled antibodies against all immunoglobulin classes. Specific monoclonal antibodies directed against T-cell marker allows the counting of T-cells, CD4 helper cells, CD8 suppressor cells and other cells. The ratio of CD4 to CD8 in a normal person is 1.5 or more but becomes less than one in a person with acquired immunodeficiency syndrome (AIDS).
4. **Rosette formation**: Rosette is a lymphocyte attached to three or more sheep erythrocytes. Most T-cells form rosettes when mixed with sheep erythrocytes. T-cell rosette is referred to as E-rosette. T-cells can be determined by counting E-rosettes. Rosette formation is useful to detect T-cells, hence the CMI of the host.
5. **Molecular methods**: Some other assays have been described, including the finding of alteration of surface markers like CD25 and CD71 after activation of lymphocytes and measurement of T-cell cytokine by ELISPOT (Enzyme-linked immunosorbent spot) assay. For tuberculosis, the interferon-gamma (IFN-γ)-released assays measure T-lymphocyte production of IFN-γ in response to antigen stimulation, a measure of cell-mediated immunity.

Transfer Factor (TF)

Lawrence, in 1954, first reported the transfer of CMI in humans by injection of an extract from the leukocytes from the immunised host. The extract from leukocytes contained a soluble substance known as **transfer factor (TF)**. The transferred immunity is specific in that CMI can be transferred only to those antigens to which the donor is specific. Transferred immunity is systemic and not local. TF does not transfer humoral immunity.

TF is a nucleotide with a low molecular weight of 2000–4000 Da. It is resistant to trypsin but gets inactivated by heating at 56 °C for 30 min. It is non-antigenic. The exact mode of action of TF is not known. It is believed to stimulate the release of lymphokines from sensitised T-lymphocytes. TF has many clinical applications. It is used to treat disseminated infections associated with deficient CMI, such as tuberculosis and lepromatous leprosy. It is also used to treat malignant melanoma and other types of cancer. It is employed to treat Wiskott–Aldrich syndrome and other T-cell-deficient syndromes.

Immunological Tolerance

Immunological tolerance is a state of specific immunologic unresponsiveness to a particular antigen to which a person has been exposed earlier. The immune tolerance prevents the body to mount an immune response against the self-antigen.

Types of Immune Tolerance

The immune tolerance may be of two types: natural or acquired.

Natural tolerance: Natural tolerance is non-responsiveness to self-antigens. It develops during the embryonic life, and any antigen that comes in contact with the immune system during its embryonic life is recognised as self-antigen. The self-antigen would not induce any immune response. Burnet and Fenner also postulated that foreign antigens would not induce immune responses if administered during embryonic life.

Acquired tolerance: Acquired tolerance develops when a potential immunogen induces a state of unresponsiveness to itself. The antigen must be repeatedly or persistently administered to maintain the acquired tolerance. This is probably necessary because of the continuous production of new B- and T-cells that must be rendered tolerant.

Specific immunological tolerance: It is the absence of immune response against a specific or targeted antigen and maintenance of such non-response in the individual. This has been explored for the treatment and prevention of autoimmune diseases. This kind of therapy will allow the individual to tolerate autoreactive immune cells but permit the immune system to react to other antigens. Many such strategies are in experimental phases, and a few, like soluble-peptide-induced tolerance and DNA-vaccination-induced tolerance, and oral- or nasal-induced tolerance, have entered clinical trial stages.

Factors Influencing Immunological Tolerance

The induction of immune tolerance depends on many factors.

1. **Species and immune competence of the host**: Tolerance depends on the immunological maturity of the host. Embryos and neonatal animals are immunologically immature, hence are more susceptible to induction of tolerance. Rabbits and mice can be made tolerant more rapidly than guinea pigs and chickens.
2. **Physical nature, dose and route of administration of antigens**: These depend on the **nature of the antigen**. Soluble antigens and haptens can induce more immune tolerance than the aggregated antigens. For example, heat-aggregated human gamma globulin is more tolerogenic than de-aggregated gamma globulin in mice. It may be due to enhanced phagocytosis of aggregated proteins than soluble antigens by macrophages, in which they can be presented to antibody-forming cells, thus inducing antibody synthesis.

The induction of tolerance is **dose-dependent** also. For example, a very simple molecule induces tolerance more readily than a complex one. Repeated minute doses and high doses of antigen induce B-cell tolerance, whereas a moderate degree of the same antigen may be immunogenic. The **route of administration** is also important. In guinea pigs, intravenous or oral administration of certain haptens causes tolerance, whereas intradermal administration causes induction of immunity.

T-cells become tolerant more readily than B-cells and remain tolerant longer. Tolerance is overcome spontaneously or by injection of cross-reacting antigens. For example, in rabbits, tolerance to bovine serum albumin can be abolished by immunisation with cross-reacting human serum albumin. The administration of immunosuppressive drugs can enhance tolerance. For example, tolerance is enhanced in patients who have received organ transplants.

Theories of Immune Response

There are two sets of theories of an immune response. These are instructive theories and selective theories.

Instructive Theory

Instructive theory suggests that an immunocompetent cell can synthesise antibodies of all specificity. The antigen directs the immunocompetent cell to produce complementary antibodies.

Two instructive theories are postulated. **Direct template theory** was first postulated by Breinl and Haurowitz. They suggested that a particular antigen or antigenic determinant could serve as a template model for folding the antibody produced. Thus, the antibody was assumed to configure complementarily to the antigen template. **Burnett and Fenner** postulated **indirect template theory** in 1949. They suggested that the entry of antigenic determinants into the antibody-producing cells induced a heritable change in these cells. A genocopy of the antigenic determinant was incorporated into the genome and transmitted to the progeny cells.

However, both instructive theories that tried to explain specificity and secondary responses are no longer accepted.

Selective Theories

Three selective theories were postulated as follows:

Sidechain Theory

This theory was proposed by Ehrlich. According to this theory, immunocompetent cells have surface receptors capable of reacting with antigens, which have complementary side chains. When antigens are introduced into the host, they combine with those cell receptors with a complementary fit. This inactivates the receptors. There is an overproduction of the same type of receptors that circulate as antibodies as a compensatory mechanism.

Natural Selection Theory

This theory was proposed by Jerne. According to this theory, millions of globulin molecules were formed against all possible range of antigens during embryonic life. When introduced to the host, the antigen combines selectively with the globulin molecule that has the nearest complementary fit. The globulin with the combined antigen stimulates antibody-forming cells to produce the same type of antibody.

Both Side chain theory and Natural selection theory, like instructive theories, are abandoned.

Clonal Selection Theory

The clonal selection theory is most widely accepted and provides a framework for a better understanding of the specificity, immunological memory and the property of recognition of self and non-self by adaptive immunity. Burnet suggested that immunological specificity existed in the cell but not in the serum and proposed the most acceptable clonal selection theory. According to this theory, many clones of immunologically competent cells (ICCs) bearing specific antibody patterns are produced during foetal development by the somatic mutations of ICCs against all possible antigens.

This theory suggests that an individual ICC exhibits membrane receptors, which are highly specific against a particular antigen. ICC has a unique receptor specificity, which is decided long before the antigen is presented to lymphocytes. Cells are activated upon the antigen binding to its specific receptor and cause cellular proliferation to form clones, synthesising the antibody.

Immune Network Hypothesis

Niels Jerne postulated the **immune network hypothesis** of an adaptive immune system in 1974. According to this theory, the immune system consists of lymphocytes and immunoglobulin molecules bearing variable regions (V), which form an interacting network. The V regions are highly variable, and they bind to various antigens and even to other V regions of the same immune system. Thus, the immune system is a complex network of V–V interactions.

The **idiotypes** are the shared characteristic between a group of immunoglobulins, and they have distinct amino acid sequences at the antigen-binding site and adjacent parts of the V region. At times, the idiotype acts as an antigenic determinant and may induce anti-idiotypic antibodies. Further, these anti-idiotypic antibodies can induce antibodies to themselves, continuing this process. This, in turn, forms an idiotype networking, which is responsible for regulating the amount of antibodies produced and the number of antibody-forming cells in action. This postulate by Niels Jerne about antibody production and regulation of the immune system fetched a Nobel Prize for Physiology and Medicine in 1984.

It is estimated that an individual's immune system produces approximately 10^8 various antibody molecules. According to the longstanding dogma of "one gene-one protein", producing such a huge number of antibody molecules requires millions of genes. However, the discovery of split genes responsible for immunoglobulin production has resolved this problem and clarified the genetic basis of antibody diversity. For this outstanding discovery, Japanese scientist Susumu Tonegawa was awarded a Nobel Prize for Physiology and Medicine in 1987.

Hypersensitivity

17

Hypersensitivity reaction denotes an immune response resulting in exaggerated or inappropriate reactions harmful to the host, following contact with specific antigens. It is a harmful immune response in which tissue damage is induced by exaggerated or inappropriate immune responses in a sensitised individual on re-exposure to the same antigen. Both the humoral and cell-mediated arms of the immune response may participate in hypersensitivity reactions. Prince of Monaco first observed the deleterious effects of jellyfish on bathers. Subsequently, Portier and Richet suggested a toxin to be responsible for these effects and coined the term "anaphylaxis".

Hypersensitivity essentially has two components. First **sensitising** or **priming dose** (first dose) of antigen is essential, which is required to prime the immune system, followed by a **shocking dose** (second dose) of the same antigen that results in injurious consequences.

Classification of Hypersensitivity Reactions

Hypersensitivity has been broadly classified into immediate and delayed types depending on the time taken for the reactions and the mechanisms that cause the tissue damage. In the former, the response is seen within minutes or hours after exposure to the antigen, and in the latter, the process takes days together to manifest as symptoms.

Gell and Coombs classified hypersensitivity reactions into four categories based on the time elapsed from exposure of antigen to the reaction and the arm of the immune system involved. Types I, II and III are antibody-mediated and are known as immediate hypersensitivity reactions; type-IV is cell-mediated (i.e. mediated by cell-mediated immunity) and is known as delayed hypersensitivity reactions. Type-V hypersensitivity reaction has been described later. It is called stimulatory-type reaction and is a modification of type-II hypersensitivity reaction.

Type-I (anaphylactic, IgE or reagin-dependent) hypersensitivity: It is also known as IgE-mediated or reagin-dependent hypersensitivity reactions. IgE is responsible for sensitising mast cells and providing recognition of antigen for immediate hypersensitivity reactions. Individuals sensitised with antigen will have IgE molecules fixed on the mast cells and basophils. An anaphylactic antigen reacts with IgE-fixed cells and releases pharmacologically active substances from the basophils and mast cells during subsequent exposure. The short time lag between exposure to antigen and onset of clinical symptoms is due to the presence of pre-formed mediators in the mast cells. Thus, the time taken for these reactions to initiate is minimal, so the onset of symptoms seems to be immediate.

Type-II (cytotoxic) hypersensitivity: It is mainly mediated by IgG antibodies, which react with the cell surface or tissue antigens. Further, in the presence of complements or mononuclear cells, destruction of the cell takes place.

Type-III (immune complex) hypersensitivity: Here, the antigen–antibody complexes mediate hypersensitivity reaction. These immune complexes may deposit in small blood vessels and other tissues to interfere with their functions.

Type-IV delayed cell-mediated hypersensitivity: This is due to the cell-mediated response and does not involve antibodies. The antigen specifically stimulates T-lymphocytes, and in turn, secretion of lymphokines and phagocyte accumulation occurs.

Type-V stimulatory hypersensitivity: This type of hypersensitivity reaction is a modification of type-II reaction. Here, antibodies react with host cell surface receptors, leading to cell proliferation and differentiation.

Differences between immediate and delayed hypersensitivities have been summarised in Table 17.1.

Type-I Reactions (Immediate Anaphylactic) Hypersensitivity

Type-I hypersensitivity reaction is commonly called an allergic or immediate hypersensitivity reaction. This reaction is always rapid, occurring within minutes of exposure to an antigen, and always involves IgE-mediated degranulation of basophils or mast cells. The type-I reaction can occur in two forms: anaphylaxis and atopy.

Anaphylaxis

Anaphylaxis is an acute, potentially fatal and systemic manifestation of immediate hypersensitivity reaction. Individuals, once sensitised, remain so for a long period.

Mechanism of Anaphylaxis

Both antigens and haptens can cause anaphylaxis. The sensitising antigen and shocking antigen are immunologically related. The initiator of type-I reaction is otherwise known as an allergen. Typical allergens include plant pollen, proteins (e.g. foreign serum and vaccines), certain food items (e.g. eggs, milk, seafood and nuts), drugs (e.g. penicillin and local anaesthetics), insect products (venom from bees, wasps and ants), dust mites, mould spores and animal hair.

The exact reason for these substances to act as allergens is not known, although they show some common characteristics. Because these reactions are T-cell dependent, T-cell-independent antigens such as polysaccharides cannot elicit type-I reactions.

Cytotoxic IgE antibody is the key immunoglobulin molecule. The binding of IgE to the mast cells is also known as sensitisation because IgE-coated mast cells are ready to be activated on repeat antigen encounters. It occurs when an antigen (**allergen**) binds to IgE on the surface of mast cells. On exposure to the antigen, Th2 cells specific to the antigen are activated, leading to the stimulation of B-cells to produce IgE antibodies (Fig. 17.1). The IgE then binds to the Fc portion of mast cells and basophils with high affinity. On re-exposure to the antigen, the allergen cross-links the bound IgE, followed by activation of IgE and degranulation of basophils and mast cells to release pharmacologically active mediators within minutes.

In anaphylaxis, the **effector cells** include (1) mast cells, (2) basophils and (3) eosinophils. All these three cells contain cytoplasmic granules whose contents are the major mediators of allergic reactions. Also, all these three cell types produce lipid mediators and cytokines that induce inflammation. Mast cells are the prime mediators of anaphylaxis. Secretion of pre-formed contents of their granules occurs by a regulated process of exocytosis—synthesis and secretion of lipid mediators; and synthesis and secretion of cytokines.

Pharmacological Mediators of Anaphylaxis

Instead of a single substance, many substances are responsible for all manifestations of anaphylaxis. These include primary mediators and secondary mediators.

Table 17.1 Differentiating features of immediate and delayed hypersensitivity reactions

Properties	Immediate hypersensitivity	Delayed hypersensitivity
Type	Type-I, type-II, type-III	Type-IV
	Manifests within minutes to hours	Manifests in days
Mediators	Antibodies are the mediators	T-cells are involved
Route of sensitisation	Sensitisation possible through any route	Sensitisation possible through intradermal route
Passive transfer with serum	Possible	Not possible
Desensitisation	Easy but short lived	Difficult but long lasting

Primary Mediators

These are the **pre-existing compounds** present inside basophils and mast cells and include the followings:

1. **Histamine:** It is the most important mediator of anaphylaxis. It is found in a pre-formed state in granules of mast cells and basophils. It causes vasodilatation, increased capillary permeability and smooth muscle contraction. It is the principal mediator of allergic rhinitis (hay fever), urticaria and angioedema. Antihistamines that block histamine receptors are relatively effective against allergic rhinitis but not against asthma.
2. **Slow-reacting substances of anaphylaxis (SRS-A):** These are produced by leukocytes. These consist of several leukotrienes, which do not occur in a pre-formed state but are produced during anaphylaxis reactions. Leukotrienes are principal mediators of bronchoconstriction in asthma and are not inhibited by antihistamines. They cause increased vascular permeability and smooth muscle contraction.

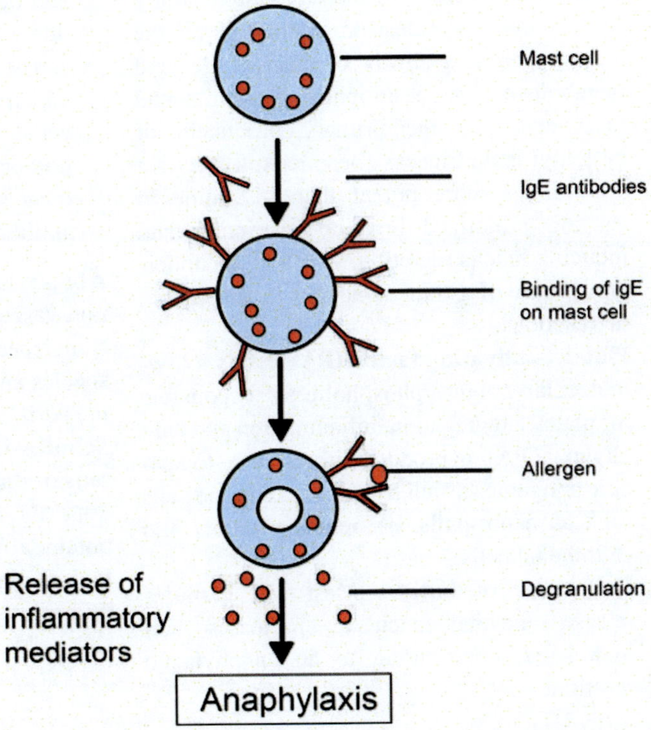

Fig. 17.1 A schematic diagram showing type-I hypersensitivity reaction

3. **Serotonin:** Serotonin or 5-hydroxytryptamine is found in a pre-formed state in mast cells and platelets. It causes vasoconstriction, increased capillary permeability and smooth muscle contraction.
4. **Eosinophilic chemotactic factors of anaphylaxis (ECF-A):** It is found in a pre-formed state in granules of mast cells. It attracts eosinophils to the site of action. However, the role of eosinophils is not clear in type-I hypersensitivity reactions. Nevertheless, it is believed to reduce the severity of type-I hypersensitivity by releasing histaminase and arylsulphatase enzymes that degrade histamine and SRS-A, respectively.
5. **Secondary mediators** are produced upon stimulation by mast cells, basophils and other leukocytes.
6. **Prostaglandins and leukotrienes:** The prostaglandins and leukotrienes are generated after the degranulation of mast cells. Prostaglandins are lipid autacoids derived from arachidonic acid and cause bronchoconstriction and dilatation, platelet aggregation and increased permeability of capillaries. Leukotrienes belong to the eicosanoid family and are synthesised through the 5-lipoxygenase pathway of arachidonic acid metabolism. The eosinophils, basophils and mast cells are the primary producers of cysteinyl leukotrienes. These leukotrienes are 1000 times more potent than histamine in causing prolonged bronchoconstriction and inducing mucus secretion, which is the underlying cause of asthmatic attacks in anaphylactic reactions.
7. **Platelet-activating factor (PAF):** It is a low-molecular-weight phospholipid responsible for platelet aggregation, inflammation and anaphylaxis. PAF is produced in response to specific antigenic stimuli by various immune cells such as neutrophils, basophils, platelets and endothelial cells.
8. **Enzymatic mediators:** Mast cell granules release enzymes such as proteases and hydrolases, contributing to an anaphylactic reaction.

Types of Anaphylaxis

1. **Cutaneous Anaphylaxis** It is a wheal and flare reaction followed by intradermal inoculation of a small shocking dose of antigens to a previously sensitised host. The wheal is a central, irregular, blanched, elevated, pale area caused by oedema, surrounded by a flare in an area of erythema. A cutaneous anaphylactic skin test helps identify allergens responsible for the atopic disease (type-I hypersensitivity disorder). In highly sensitised individuals, introducing a small quantity of antigen (like in skin tests) may lead to serious complications and death. Adrenaline injections are used for emergency treatment of allergic reactions (type-1).
2. **Passive Cutaneous Anaphylaxis (PCA)** It is an extremely sensitive test to detect antibodies *in vivo*. Ovary first described this method in 1958. In PCA, a small quantity of antibodies is passively transferred to a normal animal by intradermal injection. After a latent period (4–24 h), a homologous antigen is injected intravenously along with Evans blue dye, following which immediate bluing of skin at the site of intradermal injection can be observed due to the wheal and flare reaction. PCA test can be utilised to detect human IgG antibodies, which is heterocytotropic and able to bind with cells of other species, whereas IgE is homocytotropic and cannot bind to cells of other species, so PCA test cannot detect IgE antibody.

Anaphylaxis *in vitro*: The vascular and non-vascular smooth muscle tissues obtained from specifically sensitised animals of various species and humans contract to a specific antigen *in vitro*, and this phenomenon is called the **Schultz–Dale phenomenon**. Passive sensitisation of normal animal tissue can also be done by appropriate treatment with the serum obtained from sensitised animals. This immunological phenomenon is useful for demonstrating immediate-type hypersensitivity in actively or passively sensitised tissues.

Clinical Manifestations of Anaphylaxis

Anaphylaxis is a severe, life-threatening condition that damages multiple organs usually. The time for the symptoms to start exhibiting is based on the hypersensitivity level and the amount, exposure region and diffusibility of the antigen.

Multiple organs are affected usually, including the skin (angioedema, urticaria, pruritus and flushing), cardiovascular system (cardiac arrhythmias and hypotension) and respiratory tract (laryngeal oedema and bronchospasm). When the death occurs, it is usually due to laryngeal oedema, intractable bronchospasm, hypotensive shock or cardiac arrhythmias developing within the first 2 h (Table 17.2).

Anaphylactoid reaction: This appears to be clinically similar to anaphylactic reaction but differs in many ways. First, it is not IgE-mediated. Second, the inciting agents (such as drugs or iodinated contrast media) stimulate basophils and mast cells directly to release mediators without any involvement of the IgE.

Management and Prevention of Anaphylaxis

Desensitisation is an effective way to prevent systemic anaphylaxis. It is of two types: acute desensitisation and chronic desensitisation.

Acute desensitisation involves administering small amounts of antigen to which the person is sensitive at an interval of 15 min. The complex of antigen–IgE is produced in small quantities; hence enough mediators are not released to produce a major reaction. However, this action is short-lived because of the return of hypersensitivity reaction due to the continued production of IgE.

Chronic desensitisation involves the long-term administration of antigen to which the person is sensitive at an interval of weeks. This stimulates the production of IgA- and IgG-blocking antibodies that prevent the subsequent antigen from binding to mast cells, preventing the reaction.

Administration of drugs to inhibit the action of mediators, maintenance of airways and support of respiratory and cardiac functions form the mainstay of treating anaphylactic reactions.

Atopy

The term atopy was first coined by Coca to denote a condition of familial hypersensitivities that occur spontaneously in humans. Atopy is a recurrent, non-fatal and local manifestation of immediate hypersensitivity reaction. Atopy is associated with mutations in certain genes encoding the alpha chain of the interleukin-4 (IL-4) receptor. These mutations facilitate the effectiveness of IL-4, resulting in an increased production of IgE synthesis by B-cells.

The atopy shows a high degree of familial predisposition and is associated with a high level of IgE. It is localised to a specific tissue, often involving epithelial surfaces at the site of antigen entry. It is mediated by IgE antibodies, which are **homocytotropic** (i.e. species-specific). Only human IgE can fix to surface of mast cells.

Common manifestations of atopy are asthma, rhinitis, urticaria and atopic dermatitis. The commonest atopic reaction is bronchial asthma. Atopic individuals produce high levels of IgE in response to allergens as against the normal individuals who do not. This depends on an individual's propensity to mount a Th2 response because it is only Th2 cell-derived cytokines that stimulate the heavy-chain isotype switching to the IgE class in B-cells. Stimulation of heavy-chain isotype switching to the IgE class may be influenced by various factors, including inherited genes, the nature of the antigens and the history of antigen exposure.

Atopic hypersensitivity is non-transferable by lymphoid cells but transferable by serum. This observation was used in the past to diagnose PCA reaction by Prausnitz–Kustner reaction.

Prausnitz–Kustner Reaction

This is based on the special affinity of IgE antibody for skin cells. In this experiment, serum was collected from Kustner, who suffered from gastrointestinal allergy to certain cooked fish. The same serum was given intradermally to Prausnitz, who was then given another intradermal injection of a small quantity of cooked fish into the same site 24 h later. This resulted in a wheal and flare at the injection site within minutes. The test,

Table 17.2 Distinguishing features of mechanisms of hypersensitivity reactions

Type	Syndromes caused	Immune reactant	Antigen	Mechanism
Type-I	*Anaphylaxis and atopy*: Hay fever, asthma, hives and eczema	IgE	Soluble antigen	IgE-mediated (histamine, tryptase, kininogenase, ECF-A); allergen-specific IgE antibodies bind to mast cells; cross-linking of IgE induces degranulation of mast cells
Type-II	*Cytotoxic*: Blood transfusion reactions, erythroblastosis fetalis and autoimmune haemolytic anaemias	IgG or IgM	Cell-bound antigen	Antibodies against cell surface antigen, causing damage by ADCC or complement activation; IgG or IgM ab binds to cellular antigen, leading to complement activation and cell lysis; IgG can also mediate ADCC with cytotoxic T-cell, NK cells, macrophages and neutrophils
Type-III	*Immune complex*: Arthus reaction, farmer lung, serum sickness, rheumatoid arthritis, necrotising vasculitis, glomerulonephritis, SLE, immune complex in hepatitis B and malaria	IgG or IgM	Soluble antigen	Mediated by immune complexes containing complement-fixing antibodies; antigen–antibody complexes are deposited in the tissues; complement activation provides inflammatory mediators and enzymes released from neutrophils damage tissues
Type-IV	*Cell mediated*: Contact dermatitis, tubercular lesions and graft rejection	T-cells	Soluble or cell-bound antigen	T_H cells cause the release of cytokines, which stimulate macrophages or cytotoxic T-cells to mediate direct cellular damage

ADCC: antibody-dependent cell-mediated cytotoxicity; *ECF-A*: eosinophilic chemotactic factors of anaphylaxis; *NK*: natural killer; *SLE*: systemic lupus erythematosus

however, is not done nowadays due to the risk of transmission of certain blood-borne viral infections, such as hepatitis B, hepatitis C and human immunodeficiency virus (HIV).

Atopy Sensitivity

Overproduction of IgE antibodies leads to atopic sensitivity, frequently corresponding with decreasing levels of IgA. IgA deficiency may be a predisposing factor for atopy. In the case of normal individuals, IgA, which lines the respiratory and intestinal mucosa, deals with inhalant and ingestion antigens, respectively; therefore there is no participation of IgE-forming cells. In IgA deficiency, antigens cause overproduction of IgE by massive stimulation of IgE-producing cells.

When an antigen combines with IgE-fixed cells, pharmacologically active substances are released, which leads to atopic symptoms. Depending on the site of entry of the antigen, clinical symptoms vary—conjunctivitis (following exposure through eyes), rhinitis (following exposure through the nose), gastrointestinal symptoms (intestinal route) and dermatitis (through the skin). Occasionally, the symptoms may be away from the portal of entry. For example, ingestion of antigens leads to urticaria. Usually, the treatment of atopy is the specific desensitisation of the host.

Radioallergosorbent test (RAST), enzyme-linked immunosorbent assay (ELISA) and passive agglutination tests are the frequently used tests for detecting IgE in the serum for diagnosis of atopy.

Type-II Reactions (Cytotoxic Hypersensitivity)

Type-II reaction is mediated by antibodies directed against antigens on the cell membrane that activates complement, thereby causing antibody-mediated destruction of cells (Fig. 17.2). A membrane attack complex damages the cell membrane during activation of the complement.

The reactions involve a combination of IgG or IgM antibodies with the cell-fixed antigens or alternately circulating antigens absorbed onto cells. The antigen–antibody reaction leads to complement activation, resulting in forming a membrane attack complex. This complex then acts on the cells, causing damage to the cells, as seen in complement-mediated lysis in Rh haemolytic disease, transfusion reaction or haemolytic anaemia. Similarly, the antibodies combining with tissue antigens contribute to the pathogenesis of Goodpasture syndrome, pemphigus and myasthenia gravis.

Free antigens or haptens may bind to the cell surface in some instances. Subsequently, the combination of specific antibodies with this antigen or hapten leads to a reaction that damages the cell. Certain drugs (such as penicillin, quinidine, phenacetin) may induce haemolysis of red blood cells (RBCs) in this way. They attach to the surface of red blood cells and induce the formation of IgG antibodies. These autoantibodies then react with the red blood cell surface, causing haemolysis. Similarly, quinacrines attach to the surface of platelets and induce autoantibodies that lyse the platelets, causing thrombocytopaenia.

Antibody-Dependent Cell-Mediated Cytotoxicity (ADCC)

It is another mechanism that involves the binding of cytotoxic cells with Fc receptors in the Fc-binding part of the antibodies coating the target cells. The antibody coating the target cell can also cause its destruction by acting as an opsonin. This mechanism is important in immunity against large-sized pathogens, such as the helminths.

Type-III Reactions (Immune-Complex-Mediated Hypersensitivity)

Type-III reaction is mediated by antigen–antibody immune complexes, which induce an inflammatory reaction in tissues.

In many situations, reactions between the various antigens and antibodies in the body lead to the formation of immune complexes (Fig. 17.3). In the normal course, these immune complexes are normally removed by the mononuclear-phagocyte system through the participation of RBCs. However, the body may be exposed to an excess of antigens in many conditions, such as persistent infection with a microbial organism, autoimmunity to self-components and repeated contact with environmental agents. When the clearance capacity of this system is exceeded, deposition of the complexes takes place in various tissues.

Immune complexes are deposited (1) on blood vessel walls, (2) in the synovial membrane of joints, (3) on the glomerular basement membrane of the kidneys, and (4) on the choroid plexus of the brain. Sometimes, immune complexes are formed at the site of inflammation itself. These in situ immune complexes are potent activators of the complement system and cause disease.

Arthus reactions and serum sickness reactions are two typical manifestations of type-III hypersensitivity.

Arthus Reactions

Arthus reaction is an inflammatory reaction caused by the deposition of immune complexes at a localised site. This reaction is named after Dr. Arthus, who first described this reaction. The time taken for the reaction from the antigen exposure is 5 h generally. Tissue damage is caused by the deposition of antigen–antibody immune complexes and complements. Through its product of activation, the complement causes vascular occlusion and necrosis.

Hypersensitivity pneumonitis is the clinical manifestation. Farmer lung, cheese-washer lung, wood-worker lung and wheat-miller lung are examples of hypersensitivity pneumonitis associated with different occupations. All these conditions are caused by the inhalation of fungi or bacteria present in different products handled by infected people. Arthus reaction can also occur locally at the site of tetanus immunisation if toxoids are given at the same site within a very short period of 5 years.

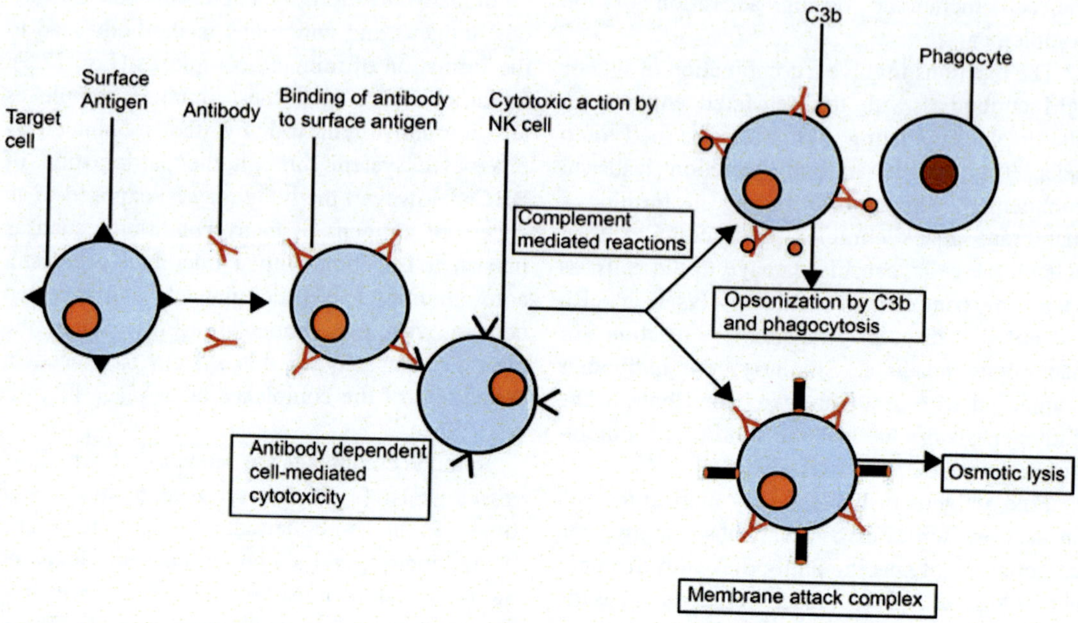

Fig. 17.2 A schematic diagram showing type-II hypersensitivity reaction

Serum Sickness

Serum sickness is a systemic inflammatory reaction caused by the deposition of immune complexes at many body sites. The condition manifests after a single injection of a high concentration of foreign serum. It appears a few days to 2 weeks after injection of foreign serum or certain drugs, such as penicillin. However,

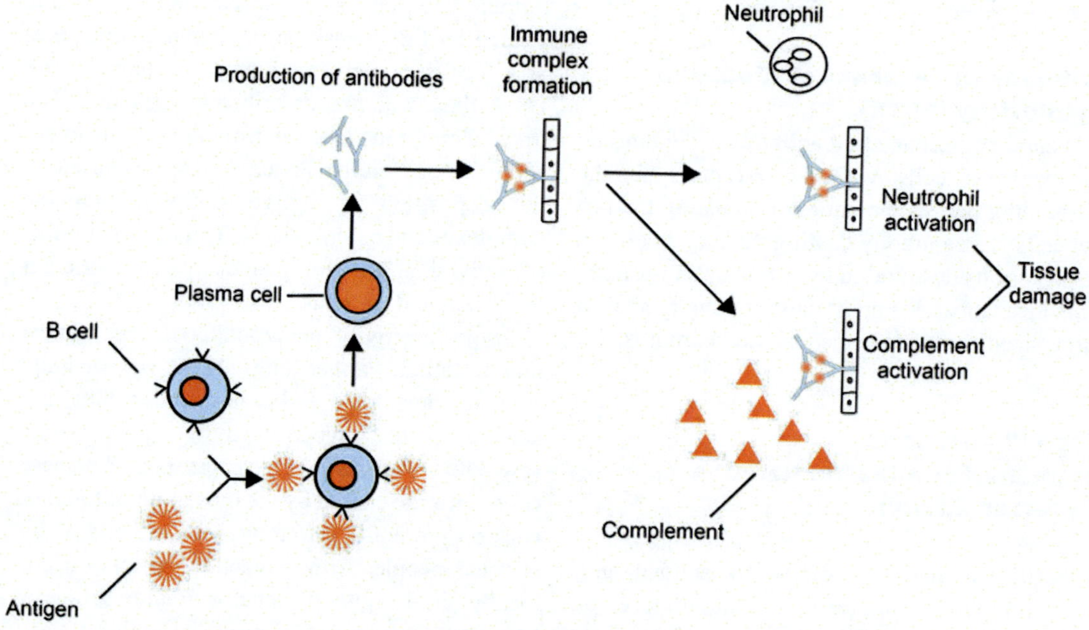

Fig. 17.3 A schematic diagram showing type-III hypersensitivity reaction

serum sickness is considered an immediate hypersensitivity reaction because symptoms appear immediately after the immune complex formation.

Unlike type-I hypersensitivity reaction, a single injection acts as priming and shocking doses. The typical manifestations are fever, lymphadenopathy, rashes, arthritis, splenomegaly and eosinophilia. The disease is self-limited and clears without sequelae.

Type-IV Delayed (Cell-Mediated) Hypersensitivity

Type-IV hypersensitivity reaction is called delayed-type hypersensitivity (DTH) because the response is delayed. It starts hours or days after primary contact with the antigen and often lasts for days. It differs from the other types of hypersensitivity by being mediated through cell-mediated immunity. This reaction occurs due to the activation of specifically sensitised T-lymphocytes rather than the antibodies (Fig. 17.4).

Initially described by Robert Koch in tuberculosis as a localised reaction, this form of hypersensitivity was known as tuberculin reaction. Later, it was renamed as DTH on the realisation that the reaction can be elicited in various pathologic conditions. DTH reactions are mainly of two types: (1) contact dermatitis type and (2) tuberculin (infection) type (another type, cutaneous basophil hypersensitivity, is discussed later).

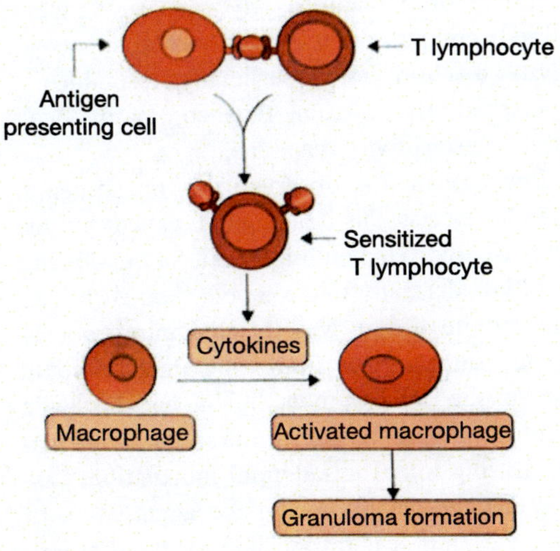

Fig. 17.4 A schematic diagram showing type-IV hypersensitivity reaction

Contact Dermatitis

Contact dermatitis is a manifestation of DTH occurring after sensitisation to certain substances. These include drugs, such as sulphonamides and neomycin; plant products, such as poison ivy and poison oak; chemicals, such as formaldehyde and nickel and cosmetics, soaps and other substances. This reaction manifests when these substances acting as haptens enter the skin and combine with body proteins to become complete antigens to which a person becomes sensitised. On second exposure to the same antigen, the immune system responds by the attack of cytotoxic T-cells that cause damage, mostly in the skin.

The condition manifests as itching, erythema, vesicle, eczema or skin necrosis within 12–48 h of the second exposure.

Tuberculin (Infection) Type of Delayed Hypersensitivity

Tuberculin reaction is a typical example of delayed hypersensitivity to antigens of microorganisms, which is being used to diagnose the disease.

Tuberculin Skin Test

This test is carried out to determine whether an individual has been exposed previously to *Mycobacterium tuberculosis* or not. In this test, a small amount of tuberculin (purified protein derivative, PPD), a protein obtained from the *M. tuberculosis* cell wall, is intradermally injected. The exhibition of a slightly swollen, red, firm lesion after 48–72 h in the place of injection indicates a positive test. A positive test indicates that the person has been infected with the bacteria but does not confirm the presence of the disease tuberculosis. However, if a person with a tuberculin-negative skin test becomes positive, it indicates that the patient has been recently infected.

The skin test, however, can even become negative in (1) infected persons receiving therapy with immunosuppressive drugs (such as corticosteroids and anticancer drugs) and (2) in those suffering from the diseases associated with suppressed cell-mediated immunity (such as acquired immunodeficiency syndrome [AIDS], sarcoidosis, lymphoma, post-measles vaccination).

Various Other Skin Tests are used to Detect DTH

These include many skin tests in bacterial, fungal, viral and helminthic infections. Positive skin tests in coccidioidomycosis, paracoccidioidomycosis and other fungal infections suggest exposure to the fungi. In both viral and parasitic infections, skin tests are less specific and less useful than serological tests for diagnosis.

The lepromin test is a useful test for leprosy. A positive lepromin test indicates the presence of tuberculoid leprosy with intact cell-mediated immunity. On the other hand, a negative lepromin test indicates the presence of lepromatous leprosy with impaired cell-mediated immunity.

Differences between contact dermatitis and tuberculin-type hypersensitivity reaction are summarised in Table 17.3.

Type-V Reactions (Stimulatory-Type Hypersensitivity)

In this type of hypersensitivity reaction, antibodies combine with antigens on the cell surface, which induces cells to proliferate and differentiate and enhances the activity of effector cells. Type-V hypersensitivity reaction plays an important role in the pathogenesis of Graves' disease, in which thyroid hormones are produced in excess quantity. It is postulated that a long-acting thyroid-stimulating antibody, an autoantibody to thyroid membrane antigen, combines with thyroid-stimulating hormone (TSH) receptors on a thyroid cell surface. Interaction with the TSH receptor produces an effect similar to the TSH, resulting in excess production and secretion of thyroid hormone, responsible for Graves' disease.

Table 17.4 summarises important features of various types of hypersensitivity.

Cutaneous Basophil Hypersensitivity

It is a distinct form of lymphocyte-mediated hypersensitivity reaction (resembling the tuberculin response) with a delayed time-course different from both the classic delayed-type and immediate hypersensitivity reactions. Basophils on arrival at the local site (basophil infiltration) may cause degranulation, and the released vasoamine may have protective or regulatory effects. This phenomenon was formerly called Jones–Mote reaction. It occurs in humans, guinea pigs and other animals. Cutaneous basophil hypersensitivity may be induced by sensitisation with various antigenic immunogens (viruses, allografts, parasites, fungal antigens etc.). This reaction is mediated by both T- and B-lymphocytes, and its clinical significance is yet to be known.

Shwartzman Reaction

Shwartzman reaction is also termed the **Shwartzman phenomenon**. It is a rare reaction of a body or a local tissue characterised by haemorrhagic necrosis due to antigen–antibody reaction to particular types of toxins, called endotoxins and other certain bacterial substances (a variety of polyanions, glycogen or antigen/antibody complexes). It is not an actual hypersensitivity reaction but superficially resembles a hypersensitivity reaction, thus commonly dealt with hypersensitivity reactions.

Shwartzman first observed this phenomenon in 1928; hence this reaction is called by his name. He injected a culture filtrate of *Salmonella* Typhi intradermally into a rabbit, then 24 h later same culture filtrate was injected intravenously into the same rabbit. Following this, he could observe the development of a haemorrhagic necrotic lesion at the site of intradermal inoculation. During initial intradermal inoculation, vascular damage occurs due to the aggregation of leukocytes, and it is responsible for the development of intravascular clotting, necrosis and

Table 17.3 Distinguishing features of contact and tuberculin-type hypersensitivity

Characteristics	Contact hypersensitivity	Tuberculin-type hypersensitivity
Site	Epidermal	Intradermal
Antigen	Organic chemicals, poison ivy, metals etc.	Tuberculin, lepromin, leishmanin skin tests etc.
Reaction time	48–72 h	48–72 h

Table 17.4 Comparison of various types of hypersensitivity reactions

Features	Type-I	Type-II	Type-III	Type-IV
Antigen	Exogenous	Cell surface	Soluble	Tissue and organ
Antibody	IgE	IgG, IgM	IgG, IgM	None
Reaction time	15–30 min	Minutes to hours	3–8 h	48–72 h
Transfer	Antibody	Antibody	Antibody	T-cells
Conditions	Hay fever, allergy and asthma	Erythroblastosis fetalis and Goodpasture syndrome	Systemic lupus erythematosus; serum sickness	Tuberculin test, poison ivy etc.

haemorrhagic lesion formation following intravenous inoculation of antigens.

It is a **two-stage reaction** where intravascular fibrin thrombi are produced by the first encounter with endotoxin, which results in the reticuloendothelial blockade. Upon a second encounter with endotoxin, the reticuloendothelial blockade prevents the clearance of thrombi, resulting in tissue necrosis and/or haemorrhage.

The initial intradermal dose (preparatory) and the intravenous dose (provocative) need not be of the same or related endotoxins to develop a Shwartzman reaction. Because of this non-specificity and short interval between two doses, the Shwartzman reaction cannot be considered a hypersensitivity reaction. If both preparatory and provocative doses are given intravenously, the animal dies within 12–24 h, and this reaction was first described by Sanarelli in the experimental cholera model. Hence, this reaction is called as **Sanarelli–Shwartzman reaction**.

Shwartzman phenomenon is usually observed during delivery or abortion when foreign bodies are introduced into the tissues of the female reproductive system. It can also be observed in some clinical conditions, such as fulminant meningococcal septicaemia (**Waterhouse–Friderichsen syndrome**).

Autoimmunity

18

Autoimmunity is a condition when the body produces autoantibodies and immunologically competent T-lymphocytes against its tissues. Conditions where the mechanisms of self-tolerance fail are termed autoimmune disorders, and the phenomenon is called autoimmunity. Autoimmunity means protection against self; however, it leads to injury to self in practice. Autoimmunity is involved in different and distinct diseases at the clinical level, such as rheumatoid arthritis, insulin-dependent diabetes mellitus, haemolytic anaemias, systemic lupus erythematosus (SLE), multiple sclerosis and myasthenia gravis.

Ehrlich, in 1901, first postulated the existence of tolerance to self-antigens as also those situations where this mechanism would fail, leading to *horror autotoxicus*. More recently, understanding the various immunological mechanisms and disorders has led to the same conclusions.

Autoimmunity diseases are chronic and non-reversible. They are characterised by (1) any levels of immunoglobulins; (2) presence of demonstrable level of autoantibodies; (3) deposition of lymphocytes and plasma cells; and (4) immunoglobulins or their derivatives at the site of the lesion. Treatment with corticosteroids or other immunosuppressive therapy benefits the patient.

Immunological Tolerance

Tolerance is a state of specific immunological unresponsiveness to a certain antigen or epitope, although the immune system is otherwise functioning normally. The antigens present during embryonic life are usually considered self and do not stimulate an immunologic response; hence the host remains tolerant to those antigens. The absence of an immune response in the foetus is due to the deletion of self-reactive T-cell precursors in the thymus. On the other hand, the antigens not present during the maturation process are considered non-self and usually elicit an immunologic response against those antigens.

Mechanisms of Tolerance

Both T-cells and B-cells participate in intolerance, but T-cell tolerance plays a major role in immunosuppressive cell tolerance, which is explained by theories of clonal deletion, clonal anergy and clonal ignorance.

Clonal deletion

The theory of clonal deletion described by Burnet, Fenner and Medawar based on their studies on mice was the first theory of tolerance. Recent studies suggest that T-lymphocytes acquire the ability to distinguish self from non-self by the process of clonal deletion during the early phases of life. This process involves killing T-cells (*negative selection*) that act against antigens, mainly self-MHC – major histocompatibility complex – molecules present in the foetus during that time. The self-reactive cells die by apoptosis, a process of programmed cell death.

Clonal anergy

Clonal energy is a process that leads to the incapacitation of the self-reactive T-cells. These cells become incapable of mounting an immune response due to lack of proper co-stimulation and are called anergic.

Clonal ignorance

This term describes self-reactive T-cells that ignore self-antigens. These self-reactive T-cells ignore self-antigens because the antigens are present in very small quantities. Also, these self-reactive cells are kept ignorant by physical separation from the target antigens, such as the blood-brain barrier.

B-cells' tolerance to self-antigens is also due to (1) clonal deletion of B-cell precursors while they are in the bone marrow, and (2) clonal anergy of B-cells in the periphery.

Mechanisms of Autoimmunity

The following mechanisms have been proposed for the pathogenesis of autoimmunity:

Release of Sequestrated Antigens

Certain tissues, such as sperm, the central nervous system, the lens and the uveal system of the eye, are sequestrated or hidden. These sites are normally never exposed to the immune system for various reasons. These are called immunologically privileged sites. When these hidden or sequestrated antigens are released/exposed as a result of injury, the host immune system—both cellular and humoral—does not consider them as self but as foreign, and hence attacks them. For example, lens protein is enclosed within its capsule and has no contact with circulation. Therefore, immunological tolerance against lens protein is not developed during foetal life. Following injury or cataract surgery, when this antigen is leaked into circulation, it elicits an immune response, resulting in damage to another eye's lens.

Similarly, developing sperms are found within the lumen of the testicular tubules, which are sealed off early in embryonic development, prior to the development of the immune system. These developing sperms are enclosed within a sheath of tightly joined Sertoli cells, hence are never accessed to immune cells. If these are exposed by surgery or vasectomy and injury, an immune response occurs against the sperm, producing spermatogenesis that may lead to male sterility.

Deoxyribonucleic acid (DNA), histones and mitochondrial enzymes are the intracellular antigens normally sequestrated from the immune system. However, certain viral or bacterial infections and exposure to radiation and chemicals can damage these cells and release sequestrated intracellular antigens into circulation. These antigens then elicit a strong immune response. The autoantibodies are produced against these antigens, which combine with subsequently released sequestrated antigens. This results in the formation of immune complexes, which causes damage to tissues. For example, following infection by mumps, the virus causes damage to the basement membrane of seminiferous tubules, thereby eliciting an immune response and resulting in orchitis.

Antigenic Alteration

Certain physical, chemical or biological factors may alter tissue antigens, resulting in the formation of new cell surface antigens called **neoantigens**. Intracellular pathogens, including viruses, may induce alteration of cell antigens. These neoantigens are no longer recognised as self. Therefore, they appear foreign to the immune system, eliciting an immune response. Procainamide-induced SLE, myxovirus (viral neuraminidase acting on erythrocytes inducing altered antigen), infectious mononucleosis etc. are examples of autoimmune diseases caused by this mechanism.

Epitope Spreading

Epitope spreading is the term used to describe the new exposure of sequestrated autoantigens due to cell damage caused by viral infections. It is another process that is believed to contribute to the pathogenesis of autoimmunity. These newly exposed autoantigens or epitopes stimulate autoreactive T-cells, resulting in autoimmune diseases. For example, in experimental animal infection, multiple sclerosis-like diseases are caused by an infection with an encephalomyelitis virus. In this condition, the self-reactive T-cells are directed against cellular antigens but not against the virus that causes the sclerosis-like disease.

Molecular Mimicry

Molecular mimicry is a process in which infection by a particular microbial pathogen is associated with the subsequent development of specific autoimmune diseases. Sharing of M protein of *Streptococcus pyogenes* and myosin of cardiac muscle is one of many examples of molecular mimicry. Repeated infections with *S. pyogenes* induce the production of antibodies against certain M proteins that cross-react with myosin of cardiac muscle, resulting in damage to cardiac tissue, leading to rheumatic fever (Fig. 18.1). Similarly, infection with *Shigella* spp. or *Chlamydia* spp. may result in Reiter's syndrome and infection with *Campylobacter* spp. may lead to Guillain–Barre syndrome. Development of encephalitis in certain cases following vaccination with rabies Semple vaccine is due to molecular mimicry of sheep brain antigens, used in the vaccine, with neural tissues in the brain.

Cross-Reacting Foreign Antigen

In rare instances, infectious agents process antigens similar to those on the host tissue cells. Thus, the antibodies generated by the immune system to that particular antigen show cross-reactivity with self-antigens, and it makes the basis of autoimmunity. These cross-reactive autoantibodies produced by the external antigen stimulus can cause severe damage to host organs. Several species have organ-specific antigens; these heterologous antigens, when injected, may elicit an immune response, causing damage to the specific organ of the host.

An example of organ-specific cross-reactive antigen is the neural vaccine developed by rabies injected sheep brain tissue, which was used for the vaccination against rabies in humans. The sheep brain tissue antigen present in the vaccine stimulates antibody production, and these antibodies are cross-reactive to human brain tissues, causing neurological injury.

Other cross-reacting foreign antigens, such as Streptococcal M protein, and antibodies developed against these antigens are cross-reactive to the heart muscle membranes, resulting in rheumatic fever. Thus, the repeated streptococcal infection may result in heart muscle damage. Some nephritogenic strains of streptococci possess similar antigens present on the human renal

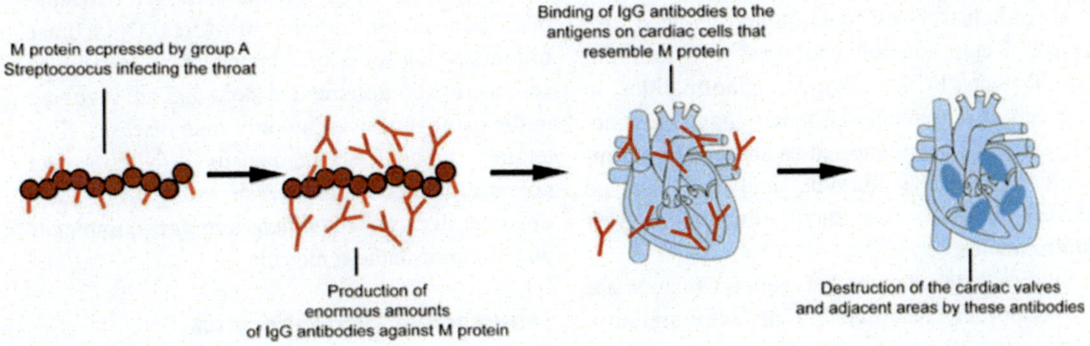

Fig. 18.1 Pathogenesis of acute rheumatic fever

glomeruli; thus, infection with such strains may lead to glomerulonephritis. A similar autoimmune disorder can be seen in Chagas disease caused by trypanosomes in which some of the trypanosoma antigens resemble antigens present on the surface of the specialised nerve cells involved in proper bowel movements. The cross-reacting antibody reacts with these nerve cell antigens, disrupting the bowel movement pattern.

Polyclonal B-Cell Activation

Polyclonal B-cell activation has a crucial role in the pathogenesis of the systemic autoimmune disease. The possible mechanism of polyclonal B-cell activation involves activation of multiple B-cell clones by certain stimuli such as chemicals (cyclophosphazenic compounds with ethylene-amino groups), enzymes (trypsin), antibiotics (nystatin) and infections with some of the microbes. In some infectious diseases, non-specific antibodies are formed; for example, in mycoplasma infection, anti-human erythrocyte cold antibodies react with red blood cells (RBCs) of humans, and in infectious mononucleosis, anti-sheep erythrocyte antibodies are produced. The IgM polyclonal antibodies produced during polyclonal B-cell activation resemble the natural antibodies of CD5+ B-cells.

Altered T- or B-Cell Function

Different experimental models suggest that CD4+ TH cells are involved in the development of autoimmunity. The two T-cell subsets, Th1 and Th2 cells, are antagonists to each other. Elevated levels of helper T-cell function and reduced suppressor T-cell function could cause autoimmunity. Defects in the thymus, abnormalities in stem cell development and macrophage dysfunction have also been reported as causes of autoimmunity. Similarly, idiotype and anti-idiotype network defects also have implications for autoimmunity.

Studies have shown that genetic factors are also important determinants of autoimmunity. There is a familial clustering; monozygotic twins are more prone to autoimmune disorders than dizygotic twins. Mutations in single genes have also been postulated as reasons for autoimmunity. Though there are multiple proposed mechanisms to alter T- or B-cell function, so far, no single mechanism has shown its actual role in autoimmunity.

Pathogenesis of Autoimmunity

Each of the mentioned immune mechanisms, as mentioned subsequently, contribute a significant role in specific diseases or may be concurrently involved specifically in multi-organ associated and systemic autoimmune disorders.

Autoantibodies and Associated Diseases

Autoantibody-associated disorders are characterised by the presence of autoantibodies in the individual's serum and by the deposition of autoantibodies in tissues. Autoantibodies may be directly involved in the pathogenesis of certain diseases, while in others they may serve simply as disease markers without a known pathogenic role. They may also be instrumental in triggering various pathogenic mechanisms leading to tissue injury and cell death. Autoantibodies play a key role in the pathogenesis of (1) myasthenia gravis, (2) pemphigus vulgaris and (3) various autoimmune cytopaenias.

Immune Complexes Carrying Autoantigens

The formation of immune complexes between self-antigens and autoantibodies, leading to end-organ damage, is another pathogenic mechanism seen in autoimmune disorders. Only those immune complexes of adequate size manage to activate the complement system and are involved in the pathogenesis of autoimmune diseases. Systemic lupus erythematosus (SLE) and polyarteritis nodosa are classic examples of autoimmune diseases in which immune complexes play a major pathogenic role.

Self-Reactive T-Lymphocytes

These antigens that are sequestrated or concealed from the circulation, which are not exposed to the developing T-cells in the thymus, do not develop self-tolerance. Mature T-cells' exposure to such concealed antigens at a later stage results in their stimulation. Induction of autoantibodies to sperms after vasectomy, sympathetic ophthalmitis and the presence of antibodies to myocardial cells after myocardial infarction are examples.

Inappropriate Expression of Class-II MHC Molecules

This can also sensitise self-reactive T-cells in certain other situations. This is supported by the clinical observations showing increased frequency of autoimmune diseases in families and increased clinical concordance rates in monozygotic twins. Polyclonal B-cell activation may also lead to the initiation of the autoimmune disease process.

Animal models of autoimmunity: Better understanding of the autoimmune disease has been facilitated by many experimental studies in animal models. The usual models used are mice and rats. Experimentally induced autoimmune diseases in these animals include (1) myasthenia gravis, (2) multiple sclerosis, (3) rheumatoid arthritis and (4) Hashimoto thyroiditis.

Classification of Autoimmune Disorders

Different molecules, cells and tissues are affected in autoimmune diseases. Table 18.1 summarises affected tissue, target antigens and resultant autoimmune diseases. Autoimmune diseases can be broadly classified as organ-specific autoimmune diseases and systemic autoimmune diseases.

Organ-Specific Autoimmune Diseases

These are diseases in which autoantibodies are produced targeting only the tissue of a single organ, thus affecting it solely. These diseases can be further sub-grouped based on tissue damage as: (1) diseases triggered by a cell-mediated autoimmune reaction, and (2) diseases mediated by the action of autoantibodies.

Diseases Triggered by a Cell-Mediated Autoimmune Reaction

Some of the diseases where the main mechanism of cell damage is directly mediated by lymphocytes are discussed below.

Hashimoto thyroiditis: Primarily, it is a sub-clinical disease where no thyroid dysfunction is apparent, and therapy is not required until the later phases of the disease. The disease is developed by a variety of unknown factors that are triggered by a cell-mediated autoimmune reaction. The disease occurs commonly during the third to fifth decades, with a female-to-male ratio of 10:1. It produces both autoantibodies and Th1 cells specific for thyroid antigens. It is the most common form of thyroiditis, and it usually has a chronic evolution.

A slow progression functionally characterises the disease to hypothyroidism with an insidious onset of symptoms. Most patients become hypothyroid with symptoms of malaise, fatigue, cold intolerance and constipation. The detection of antithyroglobulin antibodies usually confirms the diagnosis.

Addison disease (chronic primary hypoadrenalism): The condition, also called chronic primary hypoadrenalism, is caused in two ways, either exposure to exogenous agents or idiopathic. The idiopathic manifestation was found to be involved by certain immune-related mechanisms because, in more than half of the patients, autoantibodies have been found against the microsomes of adrenal cells. So, these autoantibodies are found to have a major role in the pathogenesis of the disease. The diagnosis is confirmed by demonstration of antiadrenal antibodies by indirect immunofluorescence test. Addison disease is frequently associated with other autoimmune diseases, such as thyroiditis, pernicious anaemia and diabetes mellitus.

Diseases Mediated by Autoantibodies

In a few autoimmune disorders, antibodies act as agonists that bind to hormone receptors instead of normal ligands and trigger improper activity. This

Table 18.1 Autoimmune disorders

Affected tissue	Disease	Target-antigen tissue affected	Immune response
Organ-specific autoimmune diseases			
Adrenal cortex	Addison disease	Adrenal cells	Autoantibodies
Erythrocytes	Autoimmune haemolytic anaemia	Membrane protein of RBC	Autoantibodies
Kidneys and lungs	Goodpasture syndrome	Lung basement, renal and membrane	Autoantibodies
Thyroid gland	Graves' disease	Thyroid-stimulation hormone receptor	Autoantibodies
Thyroid gland	Hashimoto thyroiditis	Thyroid-stimulation hormone receptor antibodies to thyroglobulin, "long-acting thyroid stimulator"; IgG antibody to thyroid membrane	Stimulating T-cells and autoantibodies
Platelets	Idiopathic thrombocytopenic purpura	Platelet membrane protein	Autoantibodies
Liver	Autoimmune hepatitis	Hepatocyte antigens, cytochrome p450-IA2	Autoantibodies
Pancreas	Insulin-dependent diabetes mellitus	Pancreatic beta cells	T_H cells and autoantibodies
Skeleton muscle	Myasthenia gravis	Acetylcholine receptors	Autoantibodies
Stomach	Pernicious anaemia	Gastric parietal cells, intrinsic factor	Autoantibodies
Kidney	Poststreptococcal glomerulonephritis	Basement membrane	Antigen–antibody complex
Sperm	Male sterility	Spermatogonia	Autoantibodies
Skin	Pemphigus	Desmogleins	Autoantibodies
Systemic autoimmune diseases			
Lower spine	Ankylosing spondylitis	Vertebrae	Immune complex
Myelin of CNS	Multiple sclerosis	Brain or white matter	T_H, T_C cells and autoantibodies
Synovial membranes and joints	Rheumatoid arthritis	Connective tissue, IgG	Autoantibodies and immune complex
Connective tissue	Scleroderma	Nuclei, heart, lungs, GI tract and kidney	Autoantibodies
Tear ducts	Sjögren syndrome	Salivary gland, liver, kidney and thyroid	Autoantibodies
Bile ducts	Primary biliary cirrhosis	Oxoacid dehydrogenase complexes	Autoantibodies
Skin, joints, vasculature, muscle and kidney	SLE	DNA, nuclear protein, RBC and platelets	Autoantibodies and immune complex
Melanocytes	Vitiligo	Tyrosinase-related protein 2	Autoantibodies

CNS: central nervous system; *DNA*: deoxyribonucleic acid; *GI*: gastrointestinal; *SLE*: systemic lupus erythematosus; *RBC*: red blood cell

leads to the high output of mediators or elevation in cell growth. On the other hand, some autoantibodies act as antagonists and block receptor function by binding to hormone receptors. This usually leads to impaired or decreased production of mediators and moderate atrophy of the diseased organ.

Myasthenia gravis: Myasthenia gravis is the prototype autoimmune disease mediated by blocking antibodies. It is a disorder of neuromuscular transmission. Autoantibodies produced in the patient with these diseases attach to acetylcholine receptors lying on the motor end plates of muscles. Furthermore, these antibodies prevent the usual binding of acetylcholine and lead to complement-mediated lysis of the cells.

Increased muscular fatigue and weakness, especially becoming more pronounced following exercise, are the usual symptoms of myasthenia gravis. The face, tongue and upper extremities are also frequently involved. Skeletal muscle involvement is usually proximal. The disease is usually marked by spontaneous remission periods. Thymic abnormalities are frequent in myasthenia gravis. About 10% of the patients develop malignant tumours of the thymus (thymomas). The finding of anti-acetylcholine receptor antibodies confirms the diagnosis.

Graves' disease: It occurs due to the production of autoantibodies against the thyrotrophic hormone. The thyroid-stimulating hormone (TSH) receptor antibodies (also known as long-acting thyroid stimulators, thyroid-stimulating immunoglobulin and thyroid-stimulating antibodies) stimulate the activity of the thyroid gland in Graves' disease. These antibodies can stimulate the production of thyroid hormones by activating the adenylate cyclase system after binding to the TSH receptor. These antibodies are detected in 80–90% of patients with Graves' disease and are usually of the IgG isotype.

Protruding eyeballs, also known as exophthalmos, is the classical presentation of the condition. Other symptoms of hyperthyroidism include increased metabolic rate with weight loss, nervousness, weakness, sweating, heat intolerance and loose stools. This disease is more prevalent in women in their 30s. Thyroid gland biopsy shows scattered lymphoplasmacytic interstitial infiltration. Diagnostic tests show elevated levels of thyroid hormones T3 and T4 (triiodothyronine and thyroxine) and elevated uptake of T3 and antithyroid receptor antibodies.

Autoimmune orchitis: Primary autoimmune orchitis is defined by infertility and asymptomatic orchitis associated with anti-sperm antibodies (ASA: 100%) attached to the basement membrane of seminiferous tubules in infertile men, without any systemic disease. It is characterised by inflammation of the testis and the presence of specific anti-sperm antibodies (ASA). Secondary autoimmune orchitis is indicated by symptomatic orchitis and/or testicular vasculitis corresponding with a systemic autoimmune disease, particularly vasculitis.

Systemic Autoimmune Diseases

The autoantibodies are produced against a wide range of host tissues in systemic autoimmune diseases. These diseases reflect a general defect in immune regulation that results in hyperactive T-cells and B-cells. Tissue damage is widespread, more generalised and leads to multi-system disorders. Tissue damage is caused by cell-mediated immune responses, autoantibodies or the accumulation of immune complexes.

Systemic lupus erythematosus (SLE)

SLE is a generalised autoimmune disease associated with humoral and cellular immunity abnormalities with various clinical presentations. It is most common in females of childbearing age. The clinical manifestations of SLE vary among different patients. The kind of organ (vital versus non-vital) that becomes involved determines the seriousness and the overall prognosis of the disease. A wide range of autoantibodies to a vast range of tissue antigens, such as DNA, histones, RBCs, platelets, leukocytes and clotting factors, are produced in patients with SLE. When these autoantibodies combine with their specific antigens, it produces a variety of symptoms. Demonstration of lupus erythematosus (LE) cells by incubating normal neutrophils with damaged leukocytes pre-incubated with sera obtained from SLE patients is typical of SLE. The cell is a peculiar-looking polymorphonuclear leukocyte that has ingested nuclear material.

Sunlight exposure is the first environmental factor to be identified that influences the clinical evolution of SLE in a patient. Exposure to sunlight may precede the clinical expression of the disease or disease relapse. This is due to the Langerhans cells of the skin and keratinocytes that release significant amounts of interleukin-1 upon exposure to ultraviolet (UV) light.

Infections and also drugs are responsible for setting off the disease process in certain susceptible individuals.

Laboratory diagnosis of SLE is made by demonstrating antinuclear antibodies (ANAs) in various tissues and cell lines as substrates by an indirect immunofluorescence test. Demonstration of nuclear fluorescence after incubating the cells with the patient's serum indicates a positive test. Four patterns of fluorescence can be seen, indicating different types of ANAs. ANA tests have high sensitivity but low specificity. To rule out the diagnosis of SLE, a negative result should be obtained. A high value of titre does not confirm SLE but strongly suggests SLE. A false-positive test can be seen in persistent infection, and collagen and autoimmune disorders. Anti dsDNA are found almost exclusively in SLE (60–70% of the patients). Levels of serum anti-DNA antibodies may vary with disease activity, but they are poor predictors of disease activity. There are several techniques to detect significantly elevated circulating immune complex levels in SLE patients during the acute phase of the disease.

Treatment of the condition is carried out by administering corticosteroids and other anti-inflammatory drugs. Other examples of systemic autoimmune diseases include multiple sclerosis and rheumatoid arthritis.

Rheumatoid arthritis

It is a chronic progressive autoimmune disease-causing inflammation primarily affecting joints and resulting in painful deformity and immobility. It is the most common type of autoimmune arthritis and is associated with serositis, myocarditis and vasculitis. Women are more prone to suffer from this autoimmune disorder. It is a systemic whole-body disorder principally affecting synovial tissues, ultimately causing swollen and oedematous joints due to the infiltration of lymphocytes and plasma cells. **Rheumatoid factor (RF)** is the circulating autoantibody (antibody produced against self-tissues) found in rheumatoid arthritis affected individuals. It is raised against the Fc portion of IgG (an antibody against an antibody), and altogether RF and IgG form immune complexes that contribute to the disease process.

RF in the serum of suspected rheumatoid arthritis patients can be detected by using agglutination tests such as the Rose–Waaler test, which is often modified from time to time.

Polyarteritis nodosa

It is a necrotising angiitis of small- or medium-sized muscular arteries. It is commonly associated with renal and visceral vessels. It may sometimes end fatally with coronary artery thrombosis, cerebral haemorrhage or even gastrointestinal tract bleeding. Though it is considered an autoimmune disorder, the specific antibodies responsible for the polyarteritis nodosa have not yet been recognised. Infantile polyarteritis nodosa is a condition found in infants. In polyarteritis nodosa, the **rosary sign** (localised swellings appear like strings of the rosary) is considered as one of the diagnostic features of the vasculitis.

Sjögren Syndrome

It is a chronic autoimmune disorder, usually involving a triad of conjunctivitis sicca (dry eyes), xerostomia (dryness of mouth with or without salivary gland enlargement) and rheumatoid arthritis. The autoimmunity of this disorder is mainly due to the destruction of the exocrine glands, particularly the salivary and lacrimal glands, by host white blood cells. This syndrome may be classified into primary or secondary types. The primary Sjögren syndrome occurs by itself, but secondary Sjögren syndrome occurs in patients with collagen diseases. Individuals suffering from Sjögren syndrome possess antinuclear antibodies and rheumatoid factors in their sera, which can be used for diagnostic purposes.

Laboratory Diagnosis of Autoimmune Diseases

Detection of autoantibodies remains the cornerstone for the diagnosis of autoimmune conditions. However, it has to be kept in mind that the mere presence of such antibodies is not indicative of disease since these autoantibodies may be present in healthy individuals and patients suffering from non-autoimmune diseases. Thus, the patient's clinical condition should be co-related with the laboratory findings to interpret test results accurately. The tests for autoantibodies can be divided into screening tests and additional confirmatory tests.

Screening Test (Detection of Antinuclear Antibody, ANA)

This is the initial test to rule in or rule out a suspected autoimmune disease. Indirect immunofluorescent antibody test (IFAT) using human HEp2 cells as the antigen is considered the gold standard, although this test is also available in ELISA (Enzyme Linked Immunosorbent Assay) or immunoblot formats. This test screens approximately 150 autoantibodies directed against nuclear antigens and cell cytoplasm. If positive, the titre and the pattern of immunofluorescence should be noted. Titres above 1:80 are considered positive. The fluorescent patterns like homogenous, diffuse, speckled, peripheral suggest some specific autoimmune diseases. In contrast, ELISA provides high sensitivity but poor specificity. As mentioned earlier, the mere presence of ANA is not conclusive of autoimmune disease because it is present in up to 15% of normal individuals without any disease and 10–40% of the elderly population. Moreover, it may be positive in unrelated infectious diseases like hepatitis C, tuberculosis and certain parasitic diseases.

Tests for Specific Autoantibodies

If ANA is positive in a patient strongly suspected of having an autoimmune disease, further testing should be done to diagnose specific conditions like systemic lupus erythematosus, mixed connective tissue disease, Sjögren's syndrome, systemic sclerosis, polymyositis and CREST (calcinosis, Raynaud phenomenon, esophageal dysmotility, sclerodactyly, and telangiectasia) syndrome. Various autoantibodies like dsDNA, chromatin, Smith, ribonuclear protein, centromere-b can be detected by ELISA or Western blot methods.

Other Tests

In addition to the above autoantibodies, other parameters like antineutrophil cytoplasmic antibody (ANCA), rheumatoid factor, complement levels, immunoglobulin fractions, HLA B27 may be useful in the diagnosis of some autoimmune diseases.

Treatment of Autoimmune Diseases

Treatment of autoimmune diseases has undergone a revolutionary change in the last two decades with the introduction of biological drugs that can modulate or block various immune system pathways. Traditionally, anti-inflammatory agents, analgesics and immunosuppressants like corticosteroids were the mainstay of therapy. However, in recent years, with a better understanding of the immune process and pathogenesis of autoimmune diseases, various inhibitors and monoclonal antibodies directed against immune response molecules are increasingly being licensed and used. Some of these target cytokines like TNF-α, IL-17, IL-1 receptor, while others target T-cells (CD28) or B-cells (CD20). More recent agents are directed against Janus kinases (JAKs) and integrins.

HLA Association with Autoimmune Diseases

The risk for many autoimmune diseases appears to be associated with the presence of particular human leukocyte antigen (HLA) genes. Some autoimmune diseases have increased frequencies in persons carrying certain HLA genes. Table 18.2 summarises MHC association with some important autoimmune diseases.

Table 18.2 MHC associations with autoimmune diseases

HLA gene	Disease
A3	Haemochromatosis
B8	Myasthenia gravis
B27	Psoriatic arthritis
	Acute uveitis
	Juvenile rheumatoid arthritis
	Inflammatory bowel disease
	Ankylosing spondylitis
	Reiter's disease
Cw6	Psoriasis vulgaris
DR2	Goodpasture syndrome
	Multiple sclerosis
DR3	Graves' disease
	Multiple sclerosis
	Myasthenia gravis
	SLE
DR4	Pemphigus vulgaris
	Rheumatoid arthritis
DR3/DR4 heterozygote	Type-I insulin-dependent diabetes mellitus
DR5	Hashimoto thyroiditis

HLA: human leukocyte antigen; *MHC*: major histocompatibility complex; *SLE*: systemic lupus erythematosus

Immunodeficiency Diseases

19

Immunodeficiency disorders are conditions where the defence mechanism of the body is impaired. They show enhanced susceptibility to infection with common or opportunistic microorganisms and increased susceptibility to malignancies. Immunodeficiency can occur in T-cells, B-cells, complement and phagocytes—the major components of the immune system. Immunodeficiency disorders can be classified as primary immunodeficiencies or secondary immunodeficiencies.

Primary Immunodeficiencies

Primary immunodeficiency is a condition that arises in the immune system due to a genetic or developmental defect. In such cases, the genetic defect present at birth sometimes may not reveal until later in life. Most primary immunodeficiencies are inherited from parents to offspring and may affect adaptive or innate immune functions. Primary immunodeficiency diseases can be classified as: (1) disorders of specific immunity, (2) disorders of complement and (3) disorders of phagocyte deficiencies.

Disorders of Specific Immunity

Humoral Immunodeficiencies

Humoral deficiencies include the following conditions.

X-Linked Hypogammaglobulinemia

X-linked hypogammaglobulinemia, infantile agammaglobulinemia or X-linked agammaglobulinemia (XLA) is the prototype of "pure" B-cell deficiency. This disease is transmitted in the majority of cases as a sex-linked trait. The gene responsible for this defect codes for the B-cell progenitor kinase or Bruton tyrosine kinase (Btk) protein located on the locus Xq21.2–22. Btk plays an important role in B-cell differentiation and maturation, and is also part of the group of tyrosine kinases involved in B-cell signaling in adult life.

In most cases, mutations in Btk are associated with infantile agammaglobulinemia. X-linked hypogammaglobulinemia is associated with extremely low levels of IgG, and lack of other classes of immunoglobulin is the characteristic feature of this disease. Individuals with XLA disease lack peripheral B-cells and are often affected by recurrent bacterial infections, starting from 9 months of age.

The most common infections are caused by the pyogenic organisms such as *Staphylococcus aureus*, *Haemophilus influenzae*, *Neisseria meningitidis*, *Streptococcus pneumoniae* with the following condition such as pharyngitis, sinusitis, bronchitis, pneumonia, meningitis, pyoderma, purulent conjunctivitis, otitis media, empyema, purulent arthritis and septicaemia. The recurrent bronchopulmonary infections cause bronchiectasis and chronic obstructive lung diseases. These patients are diagnosed with a high frequency of *Giardia lamblia* infections that cause chronic diarrhoea and malabsorption.

Patients with agammaglobulinemia condition have a higher chance to develop paralytic polio upon attenuated virus vaccination. They also have a high probability of developing chronic viral meningoencephalitis caused by echovirus. About 30–35% of the cases develop arthritis in the large joints caused by *Ureaplasma urealyticum* infections. This condition is best treated with replacement therapy using gamma globulin (a plasma fraction containing predominantly IgG, obtained from normal healthy donors) administered intravenously.

Selective Immunoglobulin Deficiencies

In this condition, only one or more of the immunoglobulins are deficient in serum, while the others remain normal or elevated. IgA deficiency is the most common example of selective immunoglobulin deficiency. Indication of IgA deficiency is the mere absence of serum and secretory IgA. In such conditions, the level of IgA is <5 ng/dL. However, the levels of other immunoglobulin classes are normal or increased.

IgA plays a significant role in the defence of mucosal surface and is the principal immunoglobulin in secretions. Thus, individuals with deficiency of IgA have a high incidence rate of respiratory, gastrointestinal and urogenital infections. These patients also have an increased incidence of autoimmune diseases, such as systemic lupus erythematosus and rheumatoid arthritis. Some atopic individuals have a high incidence of this IgA disorder. Certain patients with selective deficiency of IgA have significant titres of antibody against IgA and develop anaphylactic reactions when receiving IgA-containing blood transfusions.

Selective IgA deficiency is diagnosed by demonstrating less than 5 mg/dL of IgA in serum. They, however, have normal levels of IgG and IgM antibodies. Some individuals develop antibodies against IgG, IgM and IgA.

Hyper-IgM Syndrome

The condition is characterised by a high concentration of serum IgM but a very low concentration of serum IgG, IgA and IgE. They have normal numbers of T-cells and B-cells. Some of these immunodeficiencies are X-linked, and some are inherited as autosomal recessives. Patients with this condition are susceptible to recurrent microbial infections and many autoimmune disorders, such as thrombocytopenia, neutropenia and haemolytic anaemia.

Transient Hypogammaglobulinemia of Infancy

The condition is characterised by hypogammaglobulinemia in infants due to progressive catabolism of maternal IgG during the second and third months of life. This condition may persist until 2–3 years of age and become progressively more accentuated. In many cases, the delay in the synthesis of immunoglobulin is due to the deficiency in the function of T_H cells. Intravenous gamma globulin can be detected till the immunoglobulin levels are normalised in children. However, normal functions of the immune system develop with time in most children.

Interleukin-12 (IL-12) Receptor Deficiency

Patients with interleukin-12 receptor deficiency are highly susceptible to disseminated mycobacterial infections. Lack of interleukin-12 receptor prevents IL-12 from initiating a Th-1 response, which is essential to prevent mycobacterial infections.

Transcobalamin II Deficiency

It is an autosomal recessive trait associated with vitamin B_{12} deficiency, megaloblastic anaemia and intestinal villous atrophy. This, in turn, results in low levels of plasma cells, decreased immunoglobulin levels and impairment of phagocytosis. These disorders can be restored by intake of vitamin B_{12}. However, phagocytic functions cannot be restored.

Cellular Immunodeficiencies

Cellular immunodeficiencies include the followings:

Thymic Aplasia (DiGeorge Syndrome)

DiGeorge syndrome or thymic aplasia is a classic example of a pure T-cell deficiency. This syndrome is not hereditary, even though it is a congenital immunodeficiency. It is caused by the viral infection in the intrauterine prior to the eighth week of life and involves microdeletions in the chromosome region 22q11. From an immunological perspective, this syndrome arises in the third and fourth pharyngeal clefts due to embryogenesis defects during 6–8 weeks of foetal life, causing the thymus and parathyroid development deficiency.

Tetany and hypocalcaemia, both characteristics of hypoparathyroidism, are observed in DiGeorge syndrome, in addition to the defects in T-cell immunity. Defective cell-mediated immunity (CMI) may increase the patient's susceptibility to opportunistic infections and render the individual vulnerable to a graft-versus-host reaction in blood transfusion recipients. However, serum immunoglobulin levels are within normal limits, and there is a normal immune response following immunisation with commonly employed immunogens.

Considerable success in treating DiGeorge syndrome has been achieved with foetal thymic transplants and the passive administration of thymic humoral factors.

Chronic Mucocutaneous Candidiasis

Some patients with chronic infection of skin and mucosa with *Candida albicans* have exhibited a selective deficiency of cell-mediated immunity. Affected individuals develop severe and widespread forms of candidal infections. There will be a selective deficiency in reactivity to *Candida* antigens in skin tests and lowered *in vitro* proliferation of lymphocytes' response to *C. albicans*. When tested with other antigens and mitogens, the functions of T-lymphocytes remain normal. Furthermore, a humoral immune response is found to be normal for *C. albicans*. The success rate with antimycotic agents for symptomatic therapy is very minimal.

Purine Nucleoside Phosphorylase (PNP) Deficiency

This is an autosomal recessive inherited trait resulting in the deficiency of PNP enzymes responsible for the sequential breakdown of purines into hypoxanthine and ultimately uric acid. The patients with PNP deficiency show decreased CMI and recurrent chronic infection. The usual clinical presentation of these PNP deficient individuals involves hypoplastic anaemia, candidiasis, diarrhoea and recurrent pneumonia. The serum's low uric acid levels due to defected purine metabolism may be used for diagnostic purposes. Treatment options are quite less.

Combined Immunodeficiencies

Combined B-cell and T-cell deficiencies include the following conditions.

Severe Combined Immunodeficiency (SCID)

The condition includes many syndromes with severe deficiency of both humoral and cell-mediated immune responses. All these are inherited diseases with a defect in the differentiation of early stem cells. These are of two types: X-linked and autosomal.

X-linked SCID is linked with a defective gene coding for a polypeptide chain common to most interleukin receptors like IL-2, IL-4, IL-7, IL-11 and IL-15. It results in a decreased level of IL-2 production and T-cell and B-cell lymphopenia. After immunogen challenges, there will be a lack of cellular immunity, delayed-type hypersensitivity and the normal synthesis of antibodies.

Autosomal SCID is due to the mutation in the gene encoding a tyrosine kinase called ZAP-70, which plays an important role in signal transduction in T-cells. Other SCID patients show mutations in other genes, such as RAG-1 or RAG-2.

SCID is an infant disease; the affected individuals die within the initial 2 years of life. The infants develop a measles-like rash exhibiting hyperpigmentation and develop severe, repeated infections, mostly pulmonary infection. These infants exhibit severe susceptibility to *Pneumocystis carinii, C. albicans* and other pathogens. The bone marrow graft from HLA-DR (Human Leukocyte Antigen–DR isotype) matched siblings can help to normalise all these forms of SCID. This grafting method is usually successful; however, there is a higher chance for graft-versus-host-disease (GVHD) development.

Graft-Versus-Host Disease (GVHD) is observed in SCID patients receiving unirradiated blood transfusions. It can also develop due to the contamination of fresh blood with viable T-lymphocytes after transfusion. It is usually followed by fever, splenomegaly, thrombocytopenia, Coombs positive haemolytic anaemia, maculopapular rash in the volar surfaces, diarrhoea and protein-losing enteropathy. Poor outcome is seen in full-blown conditions, and usually patient dies within 10–14 days after the onset of symptomatology. This reaction can be averted with the help of frozen or irradiated blood products in transfusion. An effort in producing a bone marrow with only stem cells appears to be a promising treatment option.

Wiskott–Aldrich Syndrome

Wiskott–Aldrich syndrome is an X-linked recessive immunodeficiency disease of infants. It is characterised by thrombocytopenia, eczema and increased IgA and IgE levels. There is decreased cell-mediated immunity. The inability to mount an IgM response to the capsular polysaccharide of bacteria is the most important defect. There will be an increase in the IgA and IgE levels. On the other hand, the IgM levels were found to be lower, although IgG levels remained normal. The T-cells lack the markedly fimbriated surface of normal T-cells under electron microscopy. The defect appears to be caused by the inability of T-cells to provide help to B-cells. Bone marrow transplantation corrects the deficiency.

Ataxia-Telangiectasia

It is an autosomal recessive disease caused by gene mutations that encode deoxyribonucleic acid (DNA) repair enzymes. This condition is characterised by ataxia, telangiectasia and recurrent infections in babies of 2 years of age. IgA deficiency and lymphopenia commonly occur.

Nezelof Syndrome

Nezlelof syndrome denotes a group of disorders of various origins, characterised by depressed CMI and abnormal immunoglobulin synthesis. In this disorder, patients are susceptible to a wide range of recurrent microbial infections. Thymic dysplasia occurs with lymphoid depletion and plasma cells in abundance in the spleen, lymph node and other parts of the body. Antigenic stimuli do not induce antibody production by B-cells. Haemolytic anaemia is common. Treatment options are bone marrow transplantation, transfer factor and thymus transplantation. Constant antimicrobial therapy is very much needed to combat recurrent microbial infections.

Major Histocompatibility Complex (MHC) Class II Deficiency

It is an autosomal recessive disease failing to express MHC molecules on the surface of antigen-presenting cells, such as macrophages and B-cells. This results in a deficiency of CD4 T-cells. The lack of these helper T-cells results in the production of deficient antibodies.

Immunodeficiency with Thymoma

It is a benign thymic tumour, usually seen in adults. The condition is associated with impaired CMI, agammaglobulinemia and also aplastic anaemia. Surgical removal is the mainstay of treatment.

Short-Limbed Dwarfism is an autosomal recessive trait characterised by short-limbed dwarfism, ectodermal dysplasia, thymic defects and susceptibility to recurrent infections.

Episodic lymphopenia with lymphocytotoxin is a familial disorder in which the patient suffers from profound and episodic depression of T-cell functions. This is due to the action of circulating complement-dependent lymphotoxin, which acts as an antilymphocyte antibody. This leads to a lack of immunological memory, abolishing the secondary immune response

Complement Deficiencies

Most of the complement deficiency disorders in humans are genetic mediated and usually autosomal recessive traits. Complement deficiencies include the following conditions:

Recurrent Infections

This condition is caused by a deficiency of C1, C3 or C5, or even C6, C7 or C8 components of the complement. Patients with C3 deficiency are highly susceptible to *S. aureus* and other pyogenic bacteria infections. Similarly, patients with C6, C7 or C8 deficiency are more susceptible to bacteraemia with *N. meningitidis* or *Neisseria gonorrhoeae*.

Autoimmune Diseases

Patients with deficiencies in C2 and C4 components have diseases resembling systemic lupus erythematosus or other autoimmune diseases. Patients with C2 deficiency are usually asymptomatic, and C2 deficiency is the most common complement defect.

Paroxysmal Nocturnal Haemoglobinuria

It is a disease characterised by haemoglobinuria during the night when the patient is asleep. The haemoglobinuria occurs due to complement-mediated haemolysis, especially at night. This is because the lower concentration of oxygen in the blood during sleep increases the susceptibility of the red blood cells to lysis. This occurs in patients with a defect in the gene for the molecules that attach decay-accelerating factor (DAF) and other proteins to the cell membrane. This results in a deficiency of DAF on the surface of blood cell precursors, leading to increased activation of complement and increased haemolysis.

Hereditary Angioedema

Hereditary angioedema is a disease caused by a deficiency of C1 inhibitor, a component of the complement. This deficiency results in an increased production of vasoactive components, such as C3a and C5a, which leads to increased capillary permeability and oedema in the larynx and several other organs. Steroids (such as oxymetholone and danazol) are used to treat the condition because they increase the concentration of C1 inhibitors, thereby preventing the increased production of more C3a and C5a.

Phagocyte Deficiencies

Phagocyte deficiencies include the following conditions (Table 19.1).

Chronic Granulomatous Disease (CGD)

It is a disorder inherited in two-thirds of the cases as an X-linked trait and the remaining as an autosomal recessive trait.

The clinical condition is evident prior to the second year of life and is linked with an enzyme NADPH—(reduced) nicotinamide adenine dinucleotide phosphate—oxidase deficiency. The deficiency of the enzyme causes a reduced intake of oxygen and glucose utilisation in the neutrophils and monocytes using the hexose monophosphate shunt pathway. Though neutrophils engulf the microorganism by phagocytosis, it does not produce superoxide and other oxygen intermediates, which involve the respiratory burst. These conditions combined result in reduced intracellular removal of bacteria and

Table 19.1 Disorders of phagocytosis

Disease	Specific deficiency
Chronic granulomatous disease	Defective bactericidal activity due to deficient NADPH oxidase activity
Myeloperoxidase deficiency	Leucocytes have reduced myeloperoxidase
Chediak–Higashi syndrome	Inclusions in leucocytes; diminished phagocytic activity
Job's syndrome	Chemotaxis
Lazy leucocyte syndrome	All phagocytic cell locomotion depressed
Leucocyte G6PD deficiency	Deficient G6PD in leucocytes

NADPH: (reduced) nicotinamide adenine dinucleotide phosphate fungi. Hence, the individual with these manifestations shows high susceptibility to microbial infection caused by *Aspergillus*, *Serratia marcescens* and *Staphylococcus epidermidis*, which have less virulence comparatively.

Individuals with CGD exhibit pneumonia, hepatosplenomegaly, osteomyelitis, abscesses and draining lymph nodes. The diagnosis of CGD is carried out by using the quantitative nitro blue tetrazolium (NBT) test and the quantitative killing curve. Therapy includes interferon-gamma, antibiotics and surgical drainage of abscesses.

Chediak–Higashi Syndrome

This syndrome is a disorder in children caused by an autosomal recessive mode of inheritance. It is detected by the presence of large, stable lysosomal granules in leucocytes and the slow degranulation process. In blood platelets, the large cytoplasmic granular inclusions exist similar to their appearance in white blood cells and can be visualised in peripheral blood smears by light microscopy. It is identified by defective neutrophil chemotaxis mechanism and changes in the cells' ability to destroy the ingested microorganisms. The majority of affected individuals die during childhood, although occasional cases may live longer.

There is no effective therapy other than antibiotics to treat the infecting microorganisms. High doses of ascorbic acid have been shown to restore normal chemotaxis, bactericidal activity and degranulation.

Job Syndrome

Job syndrome is caused by the failure of helper T-cells to produce interferon- gamma, which reduces macrophages' ability to kill bacteria. This results in an increased production of Th-2 and consequently an increased production of IgE. All these, in turn, cause more histamine production that prevents certain components of inflammatory reaction and inhibits chemotaxis. Therefore, the patient with this syndrome suffers repeatedly from staphylococcal abscesses and eczema with a high level of IgE.

Leucocyte Adhesion Deficiency

It is an autosomal recessive disease caused by a mutation in the gene encoding the B-chain of an integrin that mediates the adhesion of leucocytes to microbes. This causes poor adhesion of neutrophils to endothelial surfaces, hence phagocytosis of bacteria.

Myeloperoxidase Deficiency

It is a disease associated with the deficiency of the enzyme myeloperoxidase, which is essential for producing hypochlorite, a microbicidal agent. The deficiency of this enzyme is frequently seen but has little clinical importance. This is because other intracellular killing mechanisms of leucocytes are intact.

Tuftsin Deficiency

Tuftsin is a tetrapeptide primarily produced in the spleen. Tuftsin specifically binds to the receptors on phagocytic cells and modulates their migration, and phagocytic and bactericidal activities. Tuftsin deficiency can occur as hereditary or after splenectomy. Patients with tuftsin deficiency suffer from frequent bacterial infections, especially by capsulated bacteria.

Lazy Leucocyte Syndrome

It is caused due to the abnormality of leucocyte function, that is, a deficiency of chemotaxis. These patients are more susceptible to bacterial infection, with regular stomatitis, gingivitis and otitis. The patient's bone marrow shows a normal number of neutrophils, but neutropenia exists in peripheral blood. Further, leucocytes exhibit diminished and slower responses to chemical and inflammatory stimuli.

Hyper-IgE Syndrome

It is a rare syndrome characterised by early onset of eczema, recurrent bacterial infections such as staphylococcal skin abscesses, frequent lung infections, eosinophilia, ten times increased IgE levels in serum. CMI and humoral responses, however, are unaltered.

Shwachman Disease

The condition is mainly due to decreased neutrophil mobility, malfunctioning of the pancreas and abnormalities in the bone. Because of this deficiency, patients suffer from recurrent bacterial infections.

Actin-Binding Protein Deficiency

In this patient, defective actin-binding protein is responsible for the slow mobility of leucocytes and frequent infection.

Secondary Immunodeficiencies

Secondary immunodeficiencies occur secondary to numerous diseases or conditions or as a consequence of therapeutic measures that causes the immune system depression. Secondary immunodeficiency occurs in most immunodeficient individuals, which are due to either pathological conditions altering the immune mechanism or the introduction of therapeutic compounds causing immunosuppressive effects. By far, the most common secondary immunodeficiency is acquired immunodeficiency syndrome (AIDS), which results from infection with the human immunodeficiency virus (HIV).

Secondary immunodeficiencies are more common than primary immunodeficiencies and may be categorised as: (1) humoral deficiencies, (2) cellular deficiencies, (3) complement deficiencies and (4) phagocytic deficiencies.

Humoral Deficiencies

Common Variable Hypogammaglobulinemia

This occurs in persons aged between 13 and 35 years. In this condition, the number of B-cells is normal, but the ability to produce IgG and other immunoglobulins is greatly reduced. The cell-mediated immunity is normal. Patients with this condition are highly susceptible to infections caused by *S. pneumoniae*, *H. influenzae* and other pyogenic bacteria. The administration of intravenous gamma globulin reduces the infections caused by these bacteria.

Malnutrition: In malnutrition, the synthesis of IgG is reduced due to a low supply of amino acids. People with malnutrition, hence, are susceptible to infection by pyogenic bacteria.

Cellular Deficiencies

Acquired immunodeficiency syndrome (AIDS): Patients with AIDS caused by HIV are highly susceptible to infection by many opportunistic pathogens, including bacteria, viruses, fungi and parasites. This is attributed to the loss of helper T-cell activity. The virus specifically infects and kills the cells bearing CD4 surface receptors. The immunity is highly suppressed, and failure of immune surveillance leads to a high incidence of tumours. For detail, refer to Chap. 68.

Measles

T-cell function is altered, but immunoglobulins are normal in patients suffering from measles. Patients show a temporary suppression of delayed hypersensitivity.

Complement Deficiencies

Liver Failure

The synthesis of complement proteins is reduced in chronic hepatitis B or C and liver failure caused by alcoholic cirrhosis. Hence, these patients are highly susceptible to infection by pyogenic bacteria.

Malnutrition

In severe malnutrition, the synthesis of complement proteins by the liver is reduced due to a low supply of amino acids. Therefore, people with malnutrition are susceptible to infection by pyogenic bacteria.

Phagocyte Deficiencies

Neutropenia

The condition is characterised by a low neutrophil count (less than 500 μL^{-1}), caused commonly by

Table 19.2 Immunodeficiency syndromes

Disease	Specific deficiency	Molecular defect
B-cell defects		
X-linked agammaglobulinemia	Absence of B-cells; very low IgG levels	Mutant tyrosine kinase
Selective IgA deficiency	Very low IgA levels	Failure of heavy-chain gene switching
Transient hypogammaglobulinemia of infants	Delay in initiation of IgG synthesis	
Common variable immunodeficiency	Total IgG is less; inability of B-cells to differentiate	
Immunodeficiency with hyper-IgM	Low IgA and IgG; elevated IgM	
Transcobalamin II deficiency	Metabolic defects of vitamin B_{12} deficiency	
T-cell defects		
Thymic aplasia (DiGeorge syndrome)	Absence of T-cells	Defective development of pharyngeal pouches
Chronic mucocutaneous candidiasis	Deficient T-cell response to *Candida*	Unknown
Purine nucleoside phosphorylase (PNP) deficiency	PNP deficiency	Autosomal recessive
Both B- and T-cell defects		
Nezelof syndrome	Deficient T-cell and B-cell immunity	
Ataxia telangiectasia	Lack of serum and secretory IgA, IgE	Autosomal recessive
Wiskott–Aldrich syndrome	Depressed cell-mediated immunity; serum IgM	X-linked disease
Severe combined immunodeficiency	Deficiency of both T-cells and B-cells	Defective IL-2 receptor, kinases, recombinases
Immunodeficiency with thymoma	Impaired cell-mediated immunity; thymic tumour; agammaglobulinemia	
Omenn's syndrome	Both T-cells and B-cells develop but are oligoclonal	Autosomal recessive
Complement disorders		
Hereditary angioedema	Deficiency of C1 protease inhibitor	Excess C3a, C4a and C5a generated
Complement component deficiencies	Insufficient C3, C6, C7, C8	Unknown
Recurrent Neisserial infections	Defective MAC assembly	C5, C6, C7, C8

cytotoxic drugs, such as those used in cancer therapy. The patients are susceptible to severe bacterial infections caused by pyogenic bacteria, such as *S. aureus* and *S. pneumoniae*. Immunodeficiency diseases have been summarised in Table 19.2.

Laboratory Diagnosis of Immunodeficiency Disorders

Laboratory studies are necessary to arrive at a diagnosis and the clinical findings in primary immunodeficiency disorders. The laboratory tests can be divided into identifying antibody deficiencies, cellular (T-cell) defects, neutrophil disorders and complement deficiencies. Basic screening with a complete blood count is essential in all such cases.

Tests for Antibody Deficiency

The screening test includes measuring IgG, IgM and IgA in the patient's serum. Specific antibody measurement can also be done by the response to protein or polysaccharide vaccines, and values are taken before and after vaccine administration. In addition, a flow cytometry study should be undertaken to identify and enumerate the number of B-cells in circulation. Detection of gene defect is an advanced test where the sequencing of a gene or whole-genome analysis can confirm a particular diagnosis.

Tests for Cell-Mediated Immune Deficiency

The blood count gives a preliminary idea about these deficiencies where we find decreased lymphocyte count. This can be confirmed by using flow cytometry using markers specific for different T-cell subsets. Functional assays for T-cell activity are possible by cell culture studies. The T-cells are cultured under the influence of mitogens like phytohaemagglutinin or antigens like tetanus toxoid to look for proliferation or by their ability to produce specific cytokines like interferons. Genetic defects can be detected by mutation analysis following sequencing, which is highly specific to diagnose some of the primary immunodeficiency disorders.

Tests for Neutrophil Function

There is normal or elevated neutrophil count in conditions like chronic granulomatous disease and leukocyte adhesion deficiency. In such cases, functional testing can be done by a dye reduction test or the nitro blue tetrazolium test where the cells are challenged with a bacterium or candida, and the absence of the oxidative burst can be detected. Another test using flow cytometry can measure the oxidative burst of such activated neutrophils using the dye dihydroergotamine-123. This test is highly sensitive. The diagnosis of leukocyte adhesion deficiency (LAD) type 1 also relies on flow cytometry. However, in all such cases, the most definitive method of diagnosis is gene sequencing.

Tests for Complement Defects

The total haemolytic complement assay or CH50 remains the best screening test for complement deficiencies. For defects in the alternative pathway, AH50 can be done. A negative CH50 or AH50 assay is an indication of complement deficiency.

A basic algorithm can be followed for suspected primary immunodeficiency disorders, which can be summarised as follows:

(a) Basic screening: Blood count and serum IgM, IgG and IgA levels.
↓
(b) Flow cytometry analysis: Differentiation of B- and T-lymphocytes, followed by detection of their subsets.
↓
(c) Functional testing.
↓
(d) Mutation analysis by Sanger's or Whole-genome sequencing.

Immunology of Transplantation and Malignancy

20

Successful transplantation requires a multitude of important steps: surgical asepsis, development of surgical techniques of vascular anastomosis, genetic matching of donors with hosts, use of agents that could suppress the immune system and prevention of infection in both recipients and donor. The development of strict antiseptic techniques contributes immensely to controlling the infection, while the proper use of immunosuppressive drugs and tissue typing increases the success rate of transplantation.

Transplant Immunology

The transfer of cells, tissues or organs from one site of one individual to another or in between two individuals is known as transplantation. In the transplantation between two individuals, the **donor** is an individual who provides the transplanted organ, and the **recipient** is an individual who receives the transplant.

Classification of Transplants

Transplants could be classified based on many criteria:

1. Based on the nature of organ or tissue transplanted—for example, kidney, heart, skin transplant.
2. Based on the anatomical site of origin of the transplant and the site of its placement, they can be classified as (i) Orthotopic: these grafts are placed in natural positions, for example, in skin grafts; and (ii) Heterotopic: grafts placed in abnormal sites, for example, transplantation of thyroid tissue in a subcutaneous pocket.
3. Based on vitality, it may be of living or dead materials. These include (i) Vital grafts: live grafts, such as the kidney or heart, which are expected to survive and function physiologically in the recipient; and (ii) Structural or static grafts: non-living transplants like bone or artery that merely provide scaffolding on which new tissue is laid by the recipient.

Types of Transplants/Grafts

There are four different basic types of transplants. These reflect the genetic relationship of the donor with the recipient. The degree of immune response to a graft varies with the type of graft in transplantation (Fig. 20.1).

Autograft: An autograft is the transfer of an individual's tissue or organ from one site to another site in the body. In simpler terms, in this context, the recipient is also the donor. Examples of autograft transplants include skin transplants in burn patients and patients suffering from coronary heart disease who undergo bypass surgery.

Syngraft: A syngraft is a transfer of tissue between two genetically identical individuals,

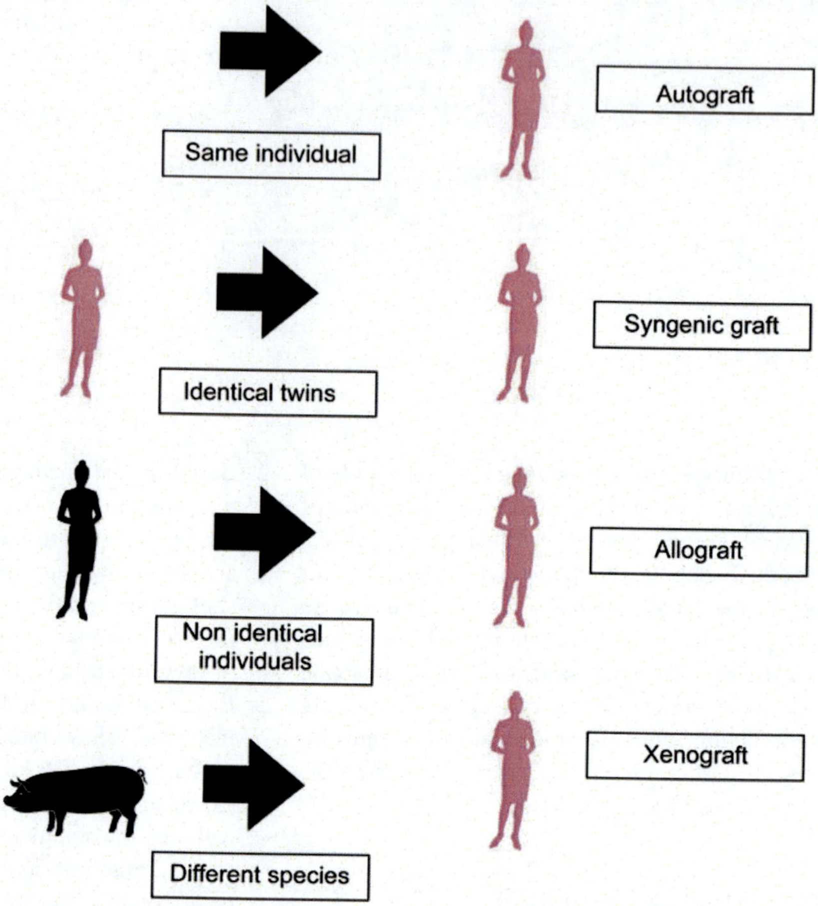

Fig. 20.1 Grafts in transplantation

that is, identical twins. In 1954, a syngraft, the first successful human kidney transplant, was performed between identical twins.

Allograft: The transfer of an organ or tissue between genetically diverse members of a similar species is referred to as an allograft, that is, from one human to another. This is the predominant form of transplantation today, and allografts have dominated transplant research for many years.

Xenograft: A xenograft is the transfer of tissues or organs between members of different species. It represents the most disparate genetic relationships and is always rejected by an immunocompetent recipient.

Transplantation immunology is the study of the events that occur after an allograft, or a xenograft, is removed from a donor and then transplanted into a recipient.

The Allograft Reaction

Allografts are rejected by a process called allograft reaction. The immune response provoked by the recipient against the graft due to the incompatibility between tissue antigens of the donor and recipient can be defined as graft rejection. The problem of rejection was first recognised when attempts to replace damaged skin on burn patients with skin from unrelated donors were found to be relatively unsuccessful. During a period of 1–2 weeks, the skin would undergo necrosis and peel off. The failure of such grafts led scientists like Peter Medawar and many others to study skin transplantation in animal models. These experiments established that the failure of skin grafting was caused by an inflammatory reaction, now called rejection. Results of several

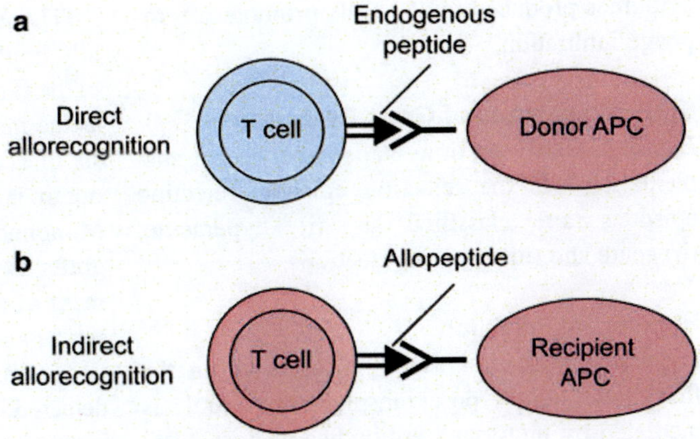

Fig. 20.2 Allograft rejection. (**a**) Direct and indirect allorecognition. (**b**) Acute and chronic allograft rejection mechanisms

experimental studies imply that adaptive immune response is responsible for rejection.

Mechanisms of Allograft Rejection

The recognition by the T-cells of a graft recipient by allogeneic MHC (Major Incompatibility Complex) molecules can be presented in two ways: (i) direct presentation and (ii) indirect presentation (Fig. 20.2).

Direct Presentation

Direct presentation includes the recognition of an intact MHC molecule displayed by donor antigen-presenting cells (APCs) in the graft. It depends on the similarity in the structure of an intact foreign (allogeneic) molecule and self-MHC molecules. As many as 2% of an individual's T-cells can recognise and respond to a single foreign MHC molecule, and this high frequency of T-cells reactive with allogeneic MHC molecules is one reason that allografts elicit strong immune responses *in vivo*.

Indirect Presentation

The "indirect presentation" involves the recognition of processed allogeneic MHC molecules but not an intact MHC molecule. The indirect presentation may result in recognition by CD4+ T-cells itself. This is due to the antigen initially acquired by the endosomal vesicular pathway and presented by class II MHC molecules. Few antigens of graft cells that are phagocytosed tend to access the class I MHC pathway of antigen presentation, indirectly recognised by CD8+ T-cells.

Stages of Graft Rejection

Graft rejection might occur either in the sensitisation phase or in the effector mechanisms phase.

The Phase of Sensitisation

In this phase, lymphocytes of the recipient, which are antigen-reactive, multiply in the presence of alloantigens on the graft. During this phase, CD4+ and CD8+ T-cells detect antigens expressed on the foreign graft cells and multiply. The proliferating CD4+ T-cell is the major cell that can directly identify class II alloantigens or hosts APCs presenting alloantigen peptides. The proliferated population of activated Th cells plays an important role in provoking the different effector mechanisms involving allograft rejection.

The Phase of Effector Mechanisms

Destruction of the graft by the immune cells occurs in this phase. Many effector mechanisms are involved in allograft rejection. The commonest are cell-mediated reactions involving delayed-type hypersensitivity and cytotoxic T-lymphocyte (CTL)-mediated cytotoxicity. The least common mechanisms involved are antibodies with complement lysis and destruction through antibody-dependent cell-mediated cytotoxicity (ADCC).

An influx of T-cells and macrophages into the graft is the hallmark of graft rejection involving cell-mediated reactions. Histologically, the infiltration in many cases resembles that seen during a delayed-type hypersensitive response, in which cytokines produced by T_H cells promote macrophage infiltration.

Clinical Features of Graft Rejection

Based primarily on time elapsed between transplantation and the rejection episode, rejection episodes are classified as (i) hyperacute, (ii) acute and (iii) chronic rejections.

Hyperacute Rejection

Hyperacute rejection usually occurs within the first few hours post-transplantation and is mediated by preformed antibodies against ABO or MHC antigens of the graft. The key role in this rejection can be due to the antibodies directed against other alloantigens, such as vascular endothelial antigens. If the antibodies bind to the tissues that are transplanted, rejection can be due to either (i) by complement system activation, which results in chemotactic migration of granulocytes and turning on the inflammatory cascades, or (ii) by ADCC.

Acute Rejection

Acute rejection occurs mostly in the first few days or weeks after transplantation. When acute rejection occurs in the first few days after grafting, it may imply a secondary immune response (**second-set response**). This signifies that the patient was earlier sensitised to the human leukocyte antigens (HLAs) already present in the organ donor (this may be due to an earlier transplant, pregnancy or blood transfusions). When graft rejection occurs in the first week after grafting, it usually refers to a first-set primary immune response (**first-set response**). About 70% of graft recipient patients account for one or more rejection episodes.

T-lymphocytes predominantly mediate acute rejection. CD4+ helper T-lymphocytes are believed to play a key role in acute rejection of the graft. In a rejected organ, the cellular infiltrates usually contain a majority of monocytes and T-lymphocytes of either helper or cytotoxic phenotypes, and to a lesser extent of B-lymphocytes, natural killer (NK) cells, eosinophils and neutrophils. All these cells can play an important role in the process of rejection.

The initial diagnosis of acute rejection is usually implied to suspicion clinically. Deteriorating of the functional capacity of the grafted organ acts as the main basis for the suspicion of the diagnosis of acute rejection. The biopsy of the grafted organ is a basic requirement for confirmation. Mononuclear cell infiltration in tissues of rejected grafts is a characteristic finding. The measurement of cytokines (such as interleukin [IL]-2) in serum and urine (in the case of renal transplants) is another diagnostic approach. In most cases, if detected early, acute rejection can be reversed by increasing the dose of immunosuppressive agents or by briefly administering additional immunosuppressants.

Delayed or Chronic Rejection

This is characterised by an insidiously progressive loss of function of the grafted organ. The functional deterioration associated with chronic rejection appears to be due to both immune and non-immune processes. Vascular endothelial injury is the most common feature. Granulocytes, monocytes and platelets are found to increasingly adhere to injured vascular endothelium. The damaged endothelium is usually enveloped by a layer of platelets and fibrin, which helps in the multiplication of fibroblasts and smooth muscle cells. It eventually leads to enlarging of lesions in the blood vessels, which advances towards fibrosis and occlusion.

The greater the agreement between donor and recipient, the more likely the transplant will survive. For instance, when a six-antigen match involves the allocation of two HLA-A antigens, two HLA-B antigens and two HLA-DR antigens among donor and recipient. Although antigenically different grafts may persist, while a strong immunosuppressive drug, such as cyclosporine, has been used, the persistence of the graft can be still enhanced by having as many antigenic matches feasible.

Histocompatibility Antigens

The recipient tissue does not possess antigens that are present on the grafted tissue. Hence, host tissue mounts an immune response against the transplant. If the antigens present on the grafted tissue are similar to that of the recipient, then there will not be any immune response even though donor and recipient are not syngeneic, which results in successful transplantation. The two parent strains with genotypes AA and BB upon breading produce first-generation (F_1) hybrid AB. The F_1 hybrid comprises representative antigens from both the parents, and hence, F_1 hybrids can accept grafts from either of the parents. However, transplantation in the reverse direction from F_1 hybrid to parents will not be successful because parent strain AA reacts with antigen B of the F_1 hybrid, and parent strain BB reacts with antigen A of F_1 hybrid.

Eichwald–Silmser effect: Most of the time, the transplants between the members of highly inbred animals are found to be successful, but the grafts are rejected only when the donor is male and the recipient is female. The reason for the rejection is, grafted male tissue (XY) possesses Y chromosome determined antigens, which are absent in the female recipient tissue (XX). However, a male can receive graft successfully from a female. This kind of unilateral sex-linked histocompatibility is called the Eichwald–Silmser effect.

The transplantation or histocompatibility antigens are cell-surface glycoproteins expressed on nucleated cells participating in the graft rejection process. The system of cell antigens that plays a vital role in allograft transplantation is called the "**major histocompatibility system**". The major histocompatibility system of humans is called the human leukocyte antigen (HLA) system.

Histocompatibility Testing

Blood Group Analysis

Basic ABO blood grouping determines the presence or absence of antigens on donor red blood cells (RBCs) and the presence or absence of specific antibodies to the same antigens in the recipient's serum or blood cells. Transplantation with ABO-incompatible graft results ends up with rejection.

HLA Compatibility Test

HLA is a protein (or marker) located on the surface of white blood cells and other tissues in the body, and its compatibility with a donor is a foremost important factor for the survival of allograft. The best transplant outcome depends upon the close match between the patient's HLA and the donor's HLA. HLA typing and tissue matching are used to test HLA compatibility.

HLA Typing Methods

HLA typing is not the same as ABO blood typing. HLA typing is performed in laboratories by separating blood into components such as red blood cells, white blood cells, platelets and plasma or serum. HLA typing is performed on the white blood cells known as lymphocytes.

Microlympho-Cytotoxicity Test

In this test, purified lymphocyte suspensions from donor and recipient are added to microwells containing a standard panel of typing antisera (containing monoclonal antibodies), complement and trypan blue stain. Cells carrying antigens corresponding to the HLA antiserum are killed by complement-mediated membrane damage. These cells can be detected using vital stains such as eosin or trypan blue. Cell death indicates that the test cells (recipient and donor cells) possess the antigens being tested for, namely HLA-A, HLA-B and HLA-DR antigens. Although this method of HLA typing is preliminary, it does not provide direct information about sequence variation in alleles.

Molecular Methods

Molecular methods are highly specific and help distinguish subtle variations in the HLA genes. In this method, deoxyribonucleic acid (DNA) is amplified by polymerase chain reaction (PCR) using sequence-specific primers and digested with a specific restriction enzyme (a technique called PCR-RFLP or polymerase chain reaction-restriction fragment length polymorphism) to generate a specific restriction pattern. Alleles are identified according to the pattern.

Tissue Matching or Tissue Typing

It is a procedure in which the tissues of a prospective donor and recipient are tested for compatibility prior to transplantation. This is done by the functional assay called mixed lymphocyte reaction or culture (MLR or MLC). This assay measures the proliferative response of lymphocytes on exposure to HLA-incompatible antigens from one individual (recipient) to lymphocytes from another individual (donor). The intensity of the reaction (proliferative response) is a measure of the antigenic disparity between the donor and recipient lymphocytes. This is a one-way test in which donor lymphocytes are killed, and only recipient lymphocytes are permitted to be transformed in response to the incompatible antigens on the donor cells. The MLR can be executed on lymphocytes from the alike individuals (AUTO-MLR), from dissimilar individuals of the same species (ALLO-MLR) or individuals from diverse species (XENO-MLR).

Prevention of Graft Rejection

Immunosuppression of the host prevents graft rejection (Table 20.1). It is achieved by treatment with radiation, corticosteroids and antilymphocyte serum. Cyclosporin A and rapamycin are also used, which cause immunosuppression by specific inhibition of T-cells.

Graft-Versus-Host (GVH) Reaction

When a patient with a known immunodeficiency disease such as primary, secondary or iatrogenic acquires a graft of an organ rich in immunocompetent cells, there is a reasonable risk in which a graft-versus-host (GVH) reaction can be evoked. This is in contrast to graft rejection due to the reaction of the host to the grafted tissue (host-versus-graft response).

The GVH reactions occur when the following conditions are present: (i) the donor graft must contain immunocompetent T-cells, (ii) the host must be immunocompromised and must not reject the graft, and (iii) the recipient should express transplant antigens, such as MHC proteins, which are absent in the graft.

GVH reactions are a significant problem in transplantation of (i) bone marrow or thymus in infants and children with primary immunodeficiencies, (ii) bone marrow in adults and (iii) small bowel, lung and even liver—the organs that have a substantial amount of lymphoid tissues, and reactions occur because the donor T-lymphocytes become activated, proliferate and differentiate into helper and effector cells in the irradiated, immunocompromised host. These activated T-cells attack the host cells and tissues, producing the signs and symptoms of GVH disease. The donor's cytotoxic T-cells play a key role in destroying the recipient's cells. The crucial role of the donor T-cells is demonstrated by the fact that removing these T-cells from a bone marrow graft prevents GVH reactions.

Table 20.1 Therapeutic uses of immunosuppressive agents in transplantation

Immunosuppressive agent	Mode of action
Azathioprine	Inhibition of nucleotide synthesis of multiple cells
Cyclophosphamide	Inhibition of nucleotide synthesis of multiple cells
Cyclosporine	Inhibition of transcription of cytokines in lymphocytes
Corticosteroids	Inhibition of transcription for cytokines and products involved in inflammation in multiple cells
Antimetabolites, mycophenolate mofetil	Inhibitors of de novo purine synthesis
Sirolimus	Inhibition of transduction induced by cytokines in T-cells
Irradiation	DNA damage in all rapidly proliferating cells
Anti-CD4 and CD8 antibodies	Interference with T-cell receptor binding of CD4 and CD8 T-cells
Tacrolimus	Inhibitors of phosphatase required for interleukin 2 production

Transplantation of organs such as the heart and kidneys—poor in endogenous lymphoid tissues—rarely results in a GVH reaction.

All the known immunosuppressive drugs utilised in the prevention and treatment of rejection have also been used to treat the GVH reaction. The tranquiliser drug such as thalidomide was known for its teratogenic effects and has been successfully used in treating chronic GVH non-responder to conventional immunosuppressants.

Privileged Sites

In the human body, there are anatomical regions that are minimally exposed to the immune system, and even foreign antigens entering these regions are protected and do not elicit any immune response. These areas include the brain and central nervous system (CNS), the eyes and the testes. It seems to be nature's mechanism to protect these highly vulnerable tissues from the collateral damage, which is part of any immune response in the body. Previously, it was thought that the self-tolerance at these sites is due to the physical barrier that blocks lymphocytes' entry and other immune cells. However, cell-tracking studies have shown a constant turnover of bone-marrow-derived cells to the brain and retina, and circulating T-cells can enter the CNS tissues before the onset of autoimmune diseases. The previous idea about the absence of lymphatics and antigen-presenting cells at these sites has also been disproved by the fact that the CNS (not the brain parenchyma) has got lymphatic drainage system and by the presence of antigen-presenting cells in the retina.

According to recent studies, it has been proposed that Fas death receptors are responsible for immune evasion at the privilege sites. FasL is expressed by very few cells, including the cells lining the inner chambers of the eyes and Sertoli cells in the testes. FasL interacts with Fas on activated T-cells and causes apoptosis of these T-cells before they can differentiate into effector cells and can attack the tissues. In addition to that, immunosuppressive cytokines and regulatory T-cells also have some role to play in immune privilege.

Immunology of Malignancy

Tumour immunology has been defined as part of immunology that deals with the antigens on tumour cells and the immune response to them. Tumour cells express developmental antigens that are usually seen only in the prenatal period due to their loss of differentiation. The alpha-fetoprotein and carcinoembryonic antigen are examples of such antigens. Tumours or neoplasia are said to develop when the balance between cell death and renewal is disturbed such that numerous clones of a single cell group are produced in an uncontrolled fashion.

A tumour that is not capable of indefinite growth and does not invade the healthy surrounding tissue extensively is **benign**. A tumour that continues to grow and becomes progressively invasive is **malignant**; the term cancer refers specifically to a malignant tumour. In addition to uncontrolled growth, malignant tumours exhibit **metastasis**. In this process, a small group of cancerous cells move from a tumour, attack the blood or lymphatic vessels and drift to other tissues, where they continue to multiply. By this process, a primary tumour at one site advances to a secondary tumour at another site.

Medical Evidence of Immune Response in Malignancy

Many clinical observations show the immune system response in preventing, arresting and rarely curing malignancies.

Spontaneous Regression

In neuroblastoma and malignant melanoma cases, it has been reported that, in rare instances, there is a chance of spontaneous regression of full-blown tumours. The reasons for the spontaneous regression of tumours could be an analogy of the immune response occurring during infections.

Chemotherapy

In rare cases, chemotherapy to choriocarcinoma and Burkitt lymphoma patients may lead to dramatic recovery and cure. At times, a single dose may result in a complete cure. Interestingly, in a few cancer types, including hypernephroma with pulmonary metastases, surgical removal of primary cancer often slows down the metastases. The reason could be that removing a large mass of cancer cells facilitates the immune system to act effectively on the remaining cancer cells and regress metastases. Usually, the immune response is more effective if the tumour size is below the critical mass.

Overcoming Defence Mechanisms

During an autopsy, certain cancers are unexpectedly observed without their clinical prevalence. This indicates the suppression of the growth of malignant cells by the host immune system, but some of them are successful in overcoming the defence mechanisms to develop into clinical cancers.

Cellular Response

The presence of lymphocytes, plasma cells and macrophages infiltrating tumours provides evidence of immune response against malignancy. Tumours with such indications of cellular filtration have a better disease prognosis than those without them.

Immunodeficiency States

The immune system plays a natural role in preventing tumour formation. That indicates that if the immune system of an individual is deficient, he/she should be more prone to cancer. For example, the high prevalence of cancer, particularly lymphoreticular malignancies, can be seen in conditions such as congenital immunodeficiency states, acquired immunodeficiency syndrome (AIDS) and individuals going through immunosuppressive therapy.

Features of Malignant Cells

Malignant cells exhibit many features: (i) once cells become malignant, they stop functioning normally and add to the burden on the body by competing with the normal cells for space and nutrition, (ii) the more "undifferentiated" a cell is, the lesser its functionality and more its malignant nature, (iii) they undergo rapid and uncontrolled division and (iv) they lose their homing instinct and start invading the basement membrane and enter the vasculature to spread to dissimilar tissues, leading to metastasis and spread of cancer.

It has been postulated that the immune system is partly responsible for protecting the body against the development of malignancies. At the same time, the prevalence of numerous cancers in immunocompetent individuals indicates that the immune system has only a partial role in protecting against malignancies and is not very efficient at it.

Tumour Antigens

Tumour cells also express unique molecules that can be classified into two groups as follows:

Tumour-Specific Antigens

Tumour-specific antigens, also called **tumour-specific transplantation antigens (TSTAs)**, are unique to tumours. They are not found in other cells of the body. They are usually the products of mutated genes seen in cancer cells. Cytosolic processing of the abnormal proteins yields unique peptides and, when presented by the appropriate MHC class I molecules, elicits a cell-mediated immune response.

Various physical and chemical carcinogens cause malignancies by inducing mutation in key genes that modulate cell growth. ***Ras's proto-oncogene*** products, including the p21Ras proteins and other related gene products, are an example of TSTA. These Ras proteins affix to guanine nucleotides such as GTP (Guanosine triphosphate) and GDP (Guanosine diphosphate), which exhibit intrinsic GTPase activity. Some mutations are associated with Ras's genes in malignant cells. They turn out to end in single amino acid substitutions at specific positions such as 12, 13 or 61, which produces elevated enzymatic activity of the related gene product. As a result, these cells achieve transforming capacity. Moreover, these products are also recognised as foreign antigens by the cellular immune response.

Another mode through which the tumour cells may express unique and novel antigens is by integrating with proviral genomes. These virus-induced tumours usually have their genome integrated with the proviral genome; hence the proteins encoded and expressed are sometimes novel and recognised by the cellular immune response. Viruses that have been implicated in tumorigenesis include Epstein–Barr virus (EBV), hepatitis B virus (HBV), hepatitis C virus (HCV).

Tumour-Associated Transplantation Antigens (TATAs)

Tumour-associated transplantation antigens (TATAs) are the other class of tumour antigens. These antigens are expressed by (i) tumour cells and also by (ii) normal cells at low levels or only during the process of differentiation. The expression of these antigens is considerably depressed or enhanced after malignant transformation. They can be of the following types:

1. **Tumour-associated carbohydrate antigens:** They represent an abnormal form of mucin-associated antigen detected in breast and pancreatic cancers.
2. **Differential antigens:** These include CD10 and prostate-specific antigens (PSAs). The latter is used as a diagnostic indicator in prostatic cancer.
3. **Oncofetal antigens:** These antigens are found in embryonic and malignant cells but are absent in normal adult cells. Alpha-fetoprotein and carcinoembryonic antigens are examples of this antigen found in hepatomas and colonic cancers, respectively. Silent tumour-associated genes are not expressed in normal cells but are actively transcribed in tumour cells. The tissue-specific genes or differentiation genes are usually seen on the surface of normal cells, or they may shed to the circulation; however, in normal cells, their usual levels of expression are very low. This finds practical application in the diagnosis of malignancies, as illustrated by the PSA assay for prostate carcinoma diagnosis.
4. **Prostate-specific antigen (PSA): PSA** is a kallikrein-like serine protease formed solely from the epithelial cells in the prostate gland. This antigen is measurable at comparatively higher levels in seminal plasma and very low levels in the serum of healthy men. The evaluation of serum PSA levels proves to be a very helpful marker in diagnosing prostate carcinoma. It is also confirmed to be the most purposeful serum marker for neoplasia. In healthy men, at the age group of 21–30, the levels of PSA range between 0.65 and 0.66 ng/mL, and at the age group 61–70, between 1.15 and 0.68 ng/mL. Considerably, the increased levels of PSA can be demonstrated in patients (63–86%) with prostatic carcinoma, and it directly corresponds to the stage of carcinoma.

In essence, tumour antigens capable of stimulating an immune reaction may be one of the different types. First, these antigens are uniquely expressed by tumour cells alone. Also, there are the products of genes that have been mutated during the transformation process, leading to the expression of abnormal products. Second, certain antigens expressed by tumours are present only when normal cells are undergoing the process of differentiation, and these are also readily recognised by the immune system. Finally, the over-expressed antigens by the tumour cells elicit a good immune response.

Immune Reactions against Tumours

Tumour antigens can elicit a comprehensive immune response involving both the cellular and humoral immune responses.

Cellular Immune Responses

T-lymphocytes play an important role in tumour immunity. They act both as cytotoxic effector cells and as central modulating cells. Through these effector cells, they control the specific cell-mediated antitumour immune responses and upregulate non-specific killing mechanisms. The T-lymphocytes' activation by tumour cell substances results in antigen recognition, which leads to the secretion of non-specific immunoregulatory factors. These factors can upregulate the tumour-killing function of monocytes, leukocytes and products. Second, these enhance the ability of NK cells and monocytes to participate in ADCC against tumour cells. Macrophages also play an important role in tumour response. Clustering of macrophages around tumour cells is associated with tumour regression and is seen in the case of numerous cancers.

Humoral Immune Responses

B-lymphocytes produce tumour-specific antibodies, which may induce complement-dependent cytotoxicity of tumour cells or may mediate ADCC. ADCC can be mediated by various cells expressing Fc7 receptors (NK cells, monocytes or macrophages and granulocytes) by recognising and destroying IgG-coated tumour cells.

Immunosurveillance

The formation of cancer cells within our body is a usual phenomenon. Trillions of normal cells present in the body, thousands in a day, would be subjected to malignant degeneration in response to the carcinogenic stimuli. Our immune system plays an important role in stopping the growth of these cancerous cells and halting the development of definite malignancy.

Ehrlich initially put forward the concept of surveillance in immune response, and later on, it was modified by Thomas and Burnet. Ehrlich first suggested that though cancer cells frequently arise in the body, they are recognised as foreign and eliminated. Later, Burnet postulated the surveillance in immune response theory. He proposed that the immune system usually be vigilant over the cells in our body. Additionally, he suggested that on recognising a cell or a cluster of cells that have become cancerous, efforts will be taken to destroy them, hence inhibiting the growth of the tumour.

Mechanisms

Immunological surveillance is very important to check cancer cell growth and proliferation. Immunological surveillance mechanisms mainly work based on the following:

1. The ability of the immune system to recognise and target the cells (cancerous) that express antigens that are not found on normal cells. Failure of this would lead to an increased incidence of cancer. For instance, cells infected with the following oncogenic viruses express different antigens from normal cells. Oncogenic viruses such as human papillomavirus pose a risk factor for cervical cancer, Kaposi's sarcoma is caused by the virus KSHV (Kaposi's sarcoma-associated herpesvirus), Burkitt lymphoma is caused by Epstein–Barr virus and hepatitis B develops to liver cancer.
2. A few gene products of the host are over-expressed or under-expressed in cancer cells. For example, a protein named NY-ESO1 is over-expressed in tumour cells (e.g. melanoma, and lung and breast cancers), but it is not being expressed by normal cells. Another example is increased HER2 – "Human Epidermal growth factor Receptor 2" – a growth factor receptor found on some kind of tumour cells (some breast cancers, lymphomas).

Despite the surveillance mechanisms mentioned above, if there is the development of a tumour, then it represents a lapse in surveillance. This failure may be linked to the very quick rate of proliferation of cancer cells, low immunogenicity of tumour antigens or the production of certain cytokines like transforming growth factor-β (TGF-β) that suppresses cell-mediated immunity (CMI). Tumour cells often express low class I MHC molecules to escape from the destruction caused by CD8+ CTLs (cytotoxic lymphocytes).

Immunotherapy of Cancer

Immunotherapy of cancer involves various approaches, as mentioned in the subsequent sections.

Non-Specific Active Immunotherapy

This includes treatment with various non-specific immune modulators. Bacillus Calmette–Guerin (BCG) vaccine has been shown to possess antitumour activity. When injected directly into certain solid tumours, the vaccine may cause regression of the tumour. The antitumour effect is believed to be due to the activation of macrophages and NK cells. The BCG therapy has been reported to be beneficial in treating bladder cancer, malignant melanomas, stage I lung cancer and certain leukaemias. *Corynebacterium parvum* also possesses antitumour activities. Its antitumour effect is due to its ability to stimulate macrophages and B-cells. It shows a synergistic effect when used in conjunction with cyclophosphamide. It is useful in the treatment of metastatic breast cancer and various types of lung cancer.

Other non-specific immunomodulators include: (1) dinitrochlorobenzene (DNCB), evaluated in squamous and basal carcinoma, (2) levamisole for stimulating CMI and macrophage function, (3) interferon to stimulate NK cell function, (4) cytokine IL-2 to stimulate killing of cancer cells by cytotoxic T-cells, (5) NK cells, macrophages and thymic hormones to restore T-cell function and (6) tuftsin to stimulate phagocytic cells.

Antigen-Specific Therapy

Antigen-specific therapy includes (i) vaccination with tumour antigen, (ii) treatment with transfer factor, (iii) treatment with immune ribonucleic acid (RNA), (iv) treatment with monoclonal antibodies raised against tumour-associated antigens (TAAs) given alone or in conjunction with the cytotoxic drug and (v) modification of tumour antigenicity by treatment with neuraminidase.

Passive Immunotherapy

Passive immunotherapy is the earliest method of special therapy comprising antibodies or other immune system components made *in vitro* and administered to patients to provide immunity against the growing tumour and neutralise the circulating tumour antigens. However, passive immunotherapies do not stimulate a patient's immune system the way vaccines do. Monoclonal antibody (mAb) therapy is the most widely used cancer immunotherapy today among passive immunotherapies. Monoclonal antibodies target and bind only to cancer cell-surface specific antigens.

Adoptive Immunotherapy

Adoptive immunotherapy is a promising approach for the treatment of cancer. This immunotherapy is based on the infusion of naturally occurring lymphocytes or gene-engineered T-cells, which could mediate tumour regression in patients with metastatic cancer. However, the primary challenge in this adoptive immunotherapy is to identify tumour-specific targets to avoid off-tumour, on-target toxicity and, additionally, this immunotherapy might be ineffective if there is a large mass of tumour cells.

Immunomodulation

It is defined as the stimulation or suppression of the immune system with the help of drugs or biological agents to fight against cancer, infection or other diseases, and such agents are termed immunomodulators. Stimulation or augmentation of the immune response is useful in the therapy of immunodeficiency to fight infections and malignancies. On the other hand, weakening or suppressing the immune system is desirable in autoimmune disorders and organ transplantations.

The agents that help augment the immune response include cytokines (IL-2 and α-interferon), IL-11, erythropoietin granulocyte-macrophage colony-stimulating factor (GM-CSF) and granulocyte colony-stimulating factor (G-CSF). BCG vaccine boosts cell-mediated immunity and is used in the therapy of bladder cancers. Drugs like thalidomide, lenalidomide and pomalidomide cause release of IL-2 and are used in malignancies.

The agents that can blunt the immune response can act at various targets. Some of them interfere with antigen presentation (anti-CD 154, CTLA4-Ig), while others can interfere with T-cell activation (calcineurin inhibitors, including cyclosporine A and tacrolimus) or T-cell proliferation (sirolimus, mycophenolate mofetil, leflunomide). These agents find their main use in preventing organ rejection after transplantation.

Immunohaematology and Immunoprophylaxis

21

Immunohaematology is the study of blood group antigens and antibodies and their interactions in health and disease. Immunoprophylaxis is the prevention of disease either by active or passive immunity.

Immunohematology

Ehrlich and Morgenroth first described blood groups in goats based on antigens of their red cells in an article published in the *Berliner klinische Wochenschrift* in 1900. Subsequently, Karl Landsteiner, a Viennese pathologist, successfully identified the human ABO blood groups for which he was awarded the Nobel Prize 30 years later. After this initial discovery, blood grouping was developed as a science, and many different systems of grouping were designed based on many isoantigens on the surface of erythrocytes. The **ABO** and **Rh blood systems** are among the well-known human blood groups described in the literature.

ABO Blood Group System

ABO blood group system was the first human red cell antigen system to be characterised. The ABO blood group substances are glycopeptides with oligosaccharide side chains (Fig. 21.1).

The ABO system contains four blood groups, and the blood group of an individual is determined by the presence or absence of two distinct antigens, A and B, on the surface of the red cell membrane. Red cells of blood group A carry antigen A, cells of blood group B carry antigen B and cells of blood group AB have both A and B antigens (Table 21.1). On the other hand, blood group O cells have neither A nor B antigens. Four blood groups are also differentiated by the presence or absence of two distinct isoantibodies in the serum. Serum of blood group A individuals have anti-B antibodies, blood group B have anti-A antibodies and blood group O have both anti-A and anti-B antibodies. The blood group AB does not contain any anti-A and anti-B antibodies in the serum.

The frequency of ABO blood groups differs across the world. The most prevalent blood group is O, and less abundantly seen is AB. In the Indian context, the highest to the lowest distribution of blood group is as follows: O (40%), B (33%), A (22%) and AB (5%). The naturally occurring IgM anti-A and anti-B are identified in the serum in the absence of the corresponding red cell antigen in the ABO blood group system. Even though alloantibodies are described as naturally occurring, the antibodies anti-A and anti-B are not measurable in the blood of newborn infants. These antibodies are detectable at about 3 months of age. Even when the person gets older, the naturally occurring anti-A and anti-B antibodies' concentration in the blood becomes less detectable, and hence in the serum of some elderly patients, these antibodies are undetectable.

Fig. 21.1 ABO blood group system

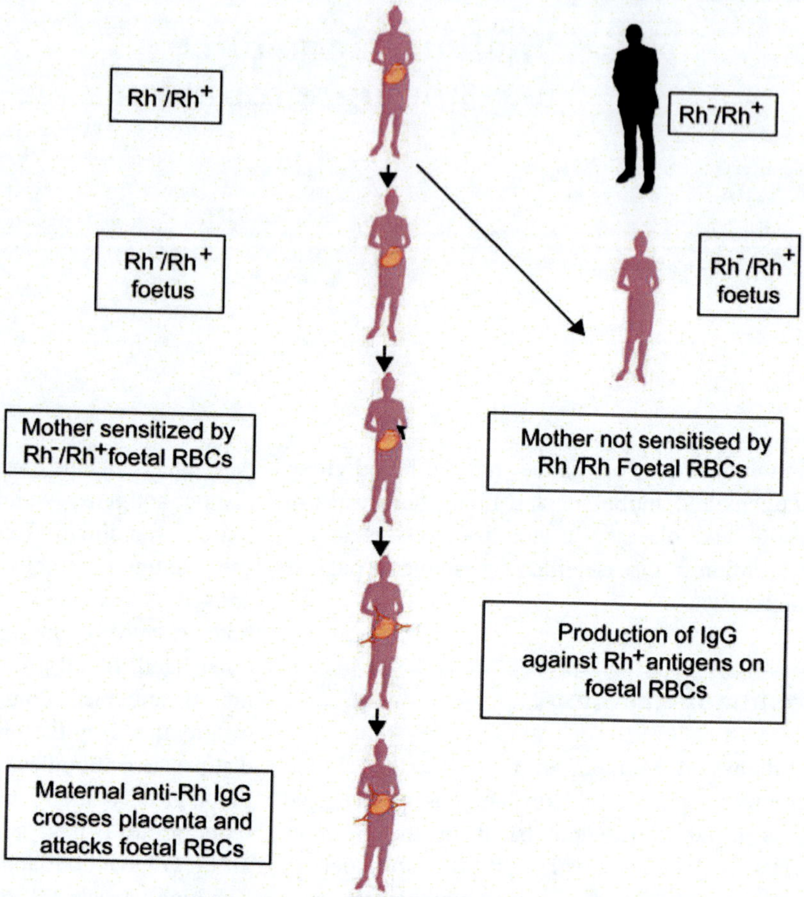

Table 21.1 Distribution of ABO antigens on red cells and antibodies in the serum

Blood group	Antigens on red blood cells	Antibodies in serum	Genotype	Occurrence (%) in India
A	A	Anti-B	AA or AO	22
B	B	Anti-A	BB or BO	33
AB	A and B	None	AB	5
O	None	Anti-A and Anti-B	OO	40

Naturally occurring anti-A and anti-B antibodies are IgM in nature and react optimally between 4 and 18 °C. However, these saline agglutinating antibodies are less reactive at 37 °C. ABO-incompatible pregnancy or transfusion may lead to the production of immune antibodies, which are IgG molecules capable of agglutinating albumin and reacting optimally at 37 °C. Isoantibodies in the presence of complement act as haemolysins, and these antibodies are clinically more significant than natural antibodies because of their involvement in severe transfusion reactions.

H Antigen

Red cells of all ABO groups possess a common antigen, the H antigen, or H substance. H antigen is a glycoprotein and structurally is an l-fucose. It is a precursor for the production of A and B antigens. A and B antigens are formed by adding *N*-acetylgalactosamine and galactose, respectively, to l-fucose of H antigen.

Due to its universal distribution, the H antigen is not that important in blood grouping or transfusion. However, in rare instances, such as in "Bombay", or OH blood, both A and B antigens and H antigens are absent in the blood. Individuals with the “Bombay” blood group have anti-A, anti-B and anti-H antibodies, hence are not compatible with most of the red cells.

RH Blood Group System

In 1939, Philip Levine discovered that most women who gave birth to newborns with a haemolytic disease whose sera contained an antibody that reacted with the infant red cells, most (85%) of them were Caucasians. Landsteiner and Wiener, in 1940, infused blood into rabbits and guinea pigs from the monkey *Macacus rhesus*; this led to the discovery that the antibody that agglutinated rhesus (Rh) red cells points to have a similar specificity as compared to the neonatal antibody. Donors whose cells showed agglutination with the antibody to Rh red cells have an Rh-positive blood group, while those donors whose cells were not showing agglutination have an Rh-negative blood group. A recent discovery reveals that the antibody discovered by Landsteiner and Wiener shows reaction with an antigen (LW), which is distinct but is similar to the one that is detected in haemolytic disease, even though the Rh nomenclature remains.

Rh Blood Group Antigens

Rh blood group system refers to the five major Rh antigens, namely C, c, D, E and e, and there are many other less frequent Rh antigens present. Of all the known Rh antigens, RhD, antigen D is most significant. The individuals usually either have or do not have the **antigen D**, which is present on the surface of the red blood cells (RBCs). These individuals are usually termed as “RhD-positive” (who have RhD antigen) or “RhD-negative” (who do not have antigen) following their ABO blood type. There are several alloantigenic determinants within the Rh system.

Rh Typing

The Rhesus (Rh) blood group system is next in importance to the ABO system in blood transfusion practice, but it is not of equal significance in every country because the frequency of the most important Rhesus antigens, i.e. **antigen D**, varies in different populations. Clinically, the antigen D has a lot of medical importance. This is because RhD-negative individuals who receive RhD-positive erythrocytes by transfusion can develop alloantibodies that may cause haemolytic reactions, called Rh incompatibility reactions. A person is grouped as Rhesus (Rh) positive or negative based on the presence or absence of antigen D (Rho) on red cells and hence can be accomplished by testing with **anti-D (anti-Rh factor)** serum. The distribution of the Rh-positives varies across the globe and between the races. Indians show 93% Rh positivity and the remaining 7% are Rh-negative, whereas European descendants carry 85% Rh antigen.

Du antigen is a variant of antigen D, and individuals belonging to Du subtypes react to some but not all anti-D sera. Du subtypes that are not capable of reacting with anti-D sera would have the ability to absorb anti-D antibodies. Du subtype detection of an individual can be done by direct Coombs test, in which the Du subtype RBCs treated with anti-D sera is used. In blood donation, Du subtype individuals are regarded as Rh-positive. However, during blood transfusion to Du individuals, Rh-negative blood transfusion is recommended to avoid the complications that might develop due to prior sensitisation of that individual with Rh-positive blood.

Antigen D even creates a problem in RhD-negative mothers bearing a child with RhD-positive red cells obtained from the father. An entry of foetal erythrocytes into the maternal circulation at either parturition or trauma caused during the pregnancy, such as in amniocentesis, will ultimately lead to alloimmunisation recorded against the RhD antigen. This may continue to cause haemolytic disease of the newborn (HDN) in future pregnancies. This now can be prevented by administering the Rh (D) immunoglobulin to the women within 72 h of delivery.

Unlike the ABO system, there are no natural antibodies against Rh antigens. Antibodies against Rh antigens develop only in certain situations, such as Rh-incompatible pregnancy or transfusion. Most of these antibodies are IgG antibodies and a few IgM antibodies. These are incomplete antibodies and can be detected in newborn blood by direct Coombs test and in mother blood by indirect Coombs test.

According to the International Society for Blood Transfusion, more than 30 human blood grouping systems are present, but ABO and Rh blood group systems have clinical importance as they cause haemolytic disease. ABO and Rh blood grouping have a wide application in genetics, forensic medicine, anthropology and tissue typing.

Medical Importance of Blood Transfusion

The process of transferring blood or blood-based products from one person into another can be defined as blood transfusion. Blood transfusions often are life-saving in certain situations, such as extensive blood loss due to trauma or during major surgery. It may also be used to treat severe anaemia, leucopenia or thrombocytopenia caused by a systemic blood disease. Individuals with haemophilia, sickle-cell disease or other haemoglobinopathies may require frequent blood transfusions.

Prior to blood transfusion, there is a list of procedures that need to be accomplished to ensure the correct blood group selection for the patient. These are carried out to ensure the ABO and Rh compatibility between donor and recipient in which both the blood groups are typed, and to exclude the presence of antibodies in the recipient's serum, as this could react with the transfused red cells.

Choice of Donor

Many conditions must be satisfied in choosing a donor for safety in blood transfusion. It should be ensured that (1) plasma of the recipient should not contain any antibodies that will haemolyse the donor's red blood cells, (2) plasma of the donor should not have any antibodies that will haemolyse the recipient's red blood cells and (3) red blood cells of the donor should not have any antigen that is absent in the recipient. In that case, transfused cells are considered "foreign antigens" by the recipient's immune system, which will mount an immune response.

The O group has been designated as the universal donor. In the transfused O blood group, the anti-A and anti-B antibodies do not ordinarily cause any damage to the red cells of A or B group recipients because they will be rendered ineffective by dilution in the recipient's plasma. However, some isoantibodies in high titres (1:200 or above) may be present in O group plasma, which can damage recipient red cells. This is known as the dangerous O group.

Individuals with an ABO blood group having red blood cells expressing antigens A and B, but the serum not consisting of antibodies anti-A and anti-B are termed as universal recipients. Hence, red blood cells that contain any of the ABO antigens, that is, from an individual with type A, B, AB or O, may be transfused to the universal recipient without inducing a haemolytic transfusion reaction. A universal recipient is Rh-positive who has RhD antigen on the surface of erythrocytes, hereby to prevent causing a haemolytic transfusion reaction.

Cross-Matching

The easiest way to identify antibodies in the recipient's serum that can cause haemolysis of the red cells transfused is by testing the patient's serum with the donor's cells (i.e. major cross-match).

Coombs cross-match: Most of the time, the saline slide test does not detect Rh and other minor incompatibilities. Therefore, the Coombs cross-match was used to discriminate the other minor incompatibility during cross-matching. In Coombs cross-match, washed donor cells and recipient serum are incubated in a water bath at 37 °C for 2 h. Following this, a direct Coombs test is performed. This detects all types of incompatibilities along with the detection of incomplete antibodies.

Risks Related to the Blood Transfusion of Infectious Origin

Transmission of infectious agents through blood is the most important complication. Due to this, there are constant developments in screening and testing; because of that, in present days, the blood products obtained are much safer than before. However, the risk cannot be ruled out completely because many infections undergo window periods during which they are not measurable by diagnostic assays. Many infectious agents could be transmitted if the blood is transfused from infected individuals. These include (1) viruses, such as human immunodeficiency viruses I and II (HIV I and II), hepatitis B virus (HBV), hepatitis C virus (HCV), hepatitis D virus (HDV) and cytomegaloviruses (CMV); (2) bacteria, such as *Treponema pallidum* and *Leptospira interrogans*; and (3) protozoa, such as *Toxoplasma gondii*, *Leishmania donovani* and *Plasmodium* species. Of these pathogens, the transmission of HIV I and II, HBV and HCV is a major concern.

Haemolytic Disease of Newborn (Erythroblastosis Foetalis)

Haemolytic disease of the newborn (HDN) is an alloimmune condition. This condition develops in the foetus, where the foetus will have IgG antibodies transferred from the mother via the placenta. These antibodies subsequently attack and lyse the red blood cells in the foetal circulation, resulting in reticulocytosis and anaemia. This condition in the foetus varies from mild to very severe, and foetal death may occur due to heart failure, that is, hydrops foetalis. When the disease progresses to a moderate or severe stage, erythroblasts can be seen in the foetal blood, and this condition of the disease is termed as erythroblastosis foetalis.

Immune-mediated destruction of foetal and/or newborn erythrocytes occurs in the presence of IgG antibodies of maternal blood directed against the antigens on the foetal blood cells. This is because IgG is the only antibody that can cross the placenta and reach the foetal circulation. Anti-D and anti-A or anti-B are the two types of antibodies most usually involved in haemolytic disease of the newborn. The Anti-A or anti-B antibodies are usually IgM in nature, but in certain conditions, IgG antibodies can also develop as seen in the case of group O mothers. This can be seen as an alternative to stimulation of the immune system (certain vaccines are known to contain blood group substances or polysaccharides that are cross-reactive) or may occur without any known cause.

When the body is exposed to a foreign antigen, antibodies are produced in the body. In a mother, there is exposure to a foreign antigen that in turn produces IgG (as opposite to IgM, which cannot cross the placenta); the IgG will react with the antigen, and if persistent in the foetus, may influence in utero and shall be present after delivery.

Detection of Rh Antibodies

Anti-Rh antibodies are mainly of IgG class, and they cannot agglutinate Rh-positive cells in saline. This is because they are "incomplete". A few anti-Rh antibodies are "complete" and can agglutinate Rh-positive cells in saline. These "complete" antibodies are of IgM class and are irrelevant in the pathogenesis of the haemolytic disease as they are unable to cross the placenta.

Detection of IgG anti-D antibodies can be carried out by (1) using a colloid medium such as 20% bovine serum albumin; (2) using red cells treated with enzymes such as trypsin, pepsin, ficin or bromelin; and (3) the indirect Coombs test (the most sensitive method).

Identification of Rh Incompatibility

Rh typing should be a part of the routine antenatal examination. Foetal complications should be expected when the woman is Rh-negative, and the father is Rh-positive. Women with Rh-incompatible pregnancies should be screened for Rh antibodies using the indirect Coombs test at 32–34 weeks of pregnancy and once a month afterwards. Intrauterine transfusion with Rh-negative blood is indicated if the haemolytic disease is diagnosed antepartum. The red cells introduced will circulate and will survive normally. It may be necessary for premature delivery followed by transfusion in some cases. If a baby is born with haemolytic disease, exchange transfusion with Rh-negative, ABO-compatible blood must be the treatment of choice.

ABO Haemolytic Diseases

ABO haemolytic diseases occur in a very less number of cases, even though materno-foetal ABO incompatibility is very common. The condition is usually seen in O group mothers bearing blood group A or B foetus. It occurs largely in O group mothers because the isoantibodies are largely IgG in nature and can cross the placenta. It does not occur in mothers with blood group A or B because natural antibodies are mainly IgM in nature, which does not cross the placenta and sensitise the foetus.

Unlike haemolytic disease of the newborn, the ABO haemolytic diseases can occur even in a first child even without prior immunisation. This is because the ABO disease is caused by naturally occurring maternal isoantibodies. ABO haemolytic disease is a much milder condition than that of Rh disease. The diagnosis of ABO incompatibility is made by a positive indirect Coombs test but a negative direct Coombs test. Peripheral blood smear characteristically shows spherocytosis.

Immunoprophylaxis

Immunoprophylaxis is the prevention of disease by the production of active or passive immunity. Owing to the regular vaccination programmes, the frequency of occurrence of various diseases such as poliomyelitis, diphtheria, measles, mumps, tetanus, pertussis (whooping cough), rubella (German measles) has declined drastically.

Disease prevention is possible by vaccination, a cost-effective method. Vaccines have largely contributed to the fight to eradicate the deadly global scourges, like smallpox. The last case of naturally acquired smallpox was reported in October 1977; since then, there have been no reports from around the globe. Other diseases like diphtheria, pertussis, tetanus, measles, mumps, rubella and poliomyelitis, also known as **"vaccine-preventable diseases"**, have been successfully brought down to negligible levels in most developed nations and in some cases in the developing nations as well.

Active or passive immunisation can be effective in developing immunity to infectious organisms. Immunity, in either case, can be acquired by two means: (1) natural processes, usually by transfer from mother to foetus or by the previous infection by the organism; and (2) artificial processes, by injection of antibodies or administration of vaccines etc. Passive immunity may be induced by agents, including antibodies from human or animal sources. However, active immunity is attained by inoculation of the pathogen that may induce immunity but is not capable of causing a disease or with antigenic elements of the pathogen.

Immunoprophylaxis is broadly of two types: First is routine immunisation, which is used for all children. Second is immunisation of individuals or selected groups, used for individuals or groups exposed to the risk of specific diseases.

Active Immunisation

Active immunisation can be acquired either by natural or artificial means. Natural immunity may be due to infection with a microorganism, whereas artificial immunity may be artificially acquired by administering a vaccine. As the name implies, active immunisation involves the active role played by the host immune system. The antigen-reactive B- and T-cells proliferate, resulting in the formation of memory cells. Various vaccines based on active immunisation have dramatically reduced the number of deaths (especially among children) from infectious agents. Vaccines may be (1) live attenuated, (2) killed or (3) in the form of toxoids (Table 21.2).

Live-Attenuated Vaccines

As the name suggests, these vaccines contain live-attenuated organisms that have lost their pathogenicity but have antigenicity. The attenuated organisms are the suspensions of live organisms with reduced virulence. These organisms multiply in the body and provide a continuous antigenic stimulus, resulting in the production of protective antibodies without producing the disease. They are also capable of generation of memory cells. For viruses and intracellular bacteria, these vaccines can provide a cell-mediated immune response since they can grow inside the host cell.

Table 21.2 Types of vaccines

Types of vaccines	Characteristics	Vaccines licensed for use in infectious diseases
Inactivated	Contain copies of whole viruses that are killed	Polio, influenza, hepatitis A, rabies
Live attenuated	Contain copies of whole organisms that are live but attenuated	Polio, mumps, measles, rubella, rotavirus, BCG
Toxoid	Inactivated toxin	Diphtheria, tetanus
Subunit	Isolated and purified viral proteins	Pertussis, influenza, hepatitis A and B
Virus-like particles	Viral proteins that mimic the structure of virus but do not contain genetic material	Human papillomavirus
Viral vector	Viral genetic material packed inside other harmless virus	Ebola
Nucleic acid vaccine	Viral DNA or mRNA	SARS-CoV-2

BCG bacillus Calmette–Guérin, *DNA* deoxyribonucleic acid, *mRNA* messenger ribonucleic acid, *SARS-CoV-2* severe acute respiratory syndrome-Coronavirus-2

Bacille Calmette–Guérin (BCG), smallpox vaccine, oral polio vaccine (OPV), measles, mumps and rubella (MMR) vaccine and yellow fever vaccine are some of the examples of live vaccines. These live vaccines are usually contraindicated for use in immunocompromised patients, such as patients with HIV (human immunodeficiency virus), leukaemia, malignancies. Some of the important features of live-attenuated vaccines include:

1. The live-attenuated vaccines have the rare potential to revert to the virulent form. Vaccine-derived paralytic poliomyelitis when using the oral polio vaccine has been reported.
2. The organism in the live vaccine may be able to cross the placenta and cause foetal infection. So, similar to immunocompromised patients, these vaccines should be avoided during pregnancy.
3. The vaccines need a cold chain to maintain the viability of the organisms. High temperatures can inactivate these vaccines, particularly the viral vaccines, leading to vaccine failure.

Killed Inactivated Vaccines

These vaccines contain killed pathogens, hence do not replicate in the body. At least three doses of the killed vaccine followed by a booster dose are essential to confer protective immunity. These vaccines may not produce an adequate immune response, and the response may be short-lived. However, in contrast to live vaccines, they have an excellent safety profile and can be given to immunocompromised patients and pregnant women. Typhoid, cholera, pertussis, pneumococcal, rabies, hepatitis B and influenza vaccines are examples of killed-attenuated vaccines.

Toxoids

Toxoids are modified toxins that have retained their antigenicity but have lost their toxicity. These are given to confer protection against diseases caused by the production of toxins. Two toxoids most widely used for immunisation are tetanus toxoid and diphtheria toxoid. DPT—diphtheria, pertussis, tetanus—vaccine is a triple vaccine consisting of tetanus and diphtheria toxoids and pertussis killed vaccine.

Adsorption of toxoids to a mineral carrier, such as aluminium hydroxide or aluminium phosphate, enhances the antigenicity of toxoids. This is because these adsorbed toxoids remain longer in a depot after injection and continue to stimulate the host's immune system for a longer time. These are usually prepared by treating toxins with formalin.

Subunit Vaccines

Subunit vaccines are specialised vaccines prepared by purifying the fragments of major immunogenic components of microorganisms and are produced by recombinant deoxyribonucleic acid (DNA) technology (hepatitis B subunit vaccine). The immune response occurs, but there is no certainty that memory cells will be produced.

Cellular Fractions

Cellular fractions or bacterial polysaccharides are polymers with repetitive sugar units linked by glycosidic bonds. Certain vaccines are prepared from cellular fractions like extracellular polysaccharides, capsular polysaccharides (CPS) and lipopolysaccharides of microorganisms. Purified cell-wall components and CPS from *Neisseria meningitidis* (meningococcal vaccine) and *Streptococcus pneumoniae* (pneumococcal vaccine) are used to produce vaccines against these pathogens. The safety and efficacy of these vaccines are high, but they are of limited use.

Recombinant Vaccines

The concept behind the recombinant vaccine is that a more targeted immune response is produced when the most immunogenic protein or peptide (epitope) from the infectious agent is administered. These proteins are produced in an expression system (bacteria, yeast or mammalian system) by recombinant technology. The expressed proteins are then purified, combined with an adjuvant, and administered to boost the immune response (e.g. hepatitis B vaccine and pertussis vaccine). Since 1981, recombinant protein vaccine against hepatitis B infection has been available. This vaccine uses the surface antigen of hepatitis B, and it is the first recombinant vaccine to be cloned in yeast.

Conjugate Vaccine

It is a subtype of subunit vaccine where the bacterial polysaccharide, which is poorly immunogenic by itself, is bound to a carrier protein that helps in increasing the long-term immune response. Various protein carriers are used for conjugation, including diphtheria and tetanus toxoids. Thus, it can be used in infants where plain polysaccharide vaccines are ineffective. The three conjugate vaccines commonly used are *Haemophilus influenza* type b (Hib), pneumococcal 7 valent, 10 valent and 13 valent vaccines and meningococcal A vaccine.

DNA Vaccines

These are novel potential vaccine candidates, sometimes referred to as third-generation vaccines, and involve the direct introduction into the tissue of a DNA sequence encoding an antigen on a plasmid vector. Since plasmids are non-living entities, there is no risk for reversion to a pathogenic phase. The advantages include stimulation of both T- and B-lymphocytes, vaccine stability, absence of any organism and ease of large-scale production. These vaccines were first tried for HIV and, later, for influenza virus, human papillomavirus, hepatitis viruses and malaria parasite in animal models. Phase 1 clinical trials, however, have shown disappointing results with poor immunogenicity for both cell-mediated and antibody-mediated immunity, although they were found to be safe without many side effects. No DNA vaccine is licensed for human use yet but has been approved for West Nile Virus in horses and canine melanoma. A DNA-based Coronavirus disease-2019 (COVID-19) vaccine was recently approved in 2021 for emergency use in India.

mRNA Vaccines

This type of vaccine uses a copy of the messenger ribonucleic acid (mRNA). The mRNA encodes for the antigen of a pathogen that is delivered to the immune cells, which in turn produces the designated protein (antigen) of the organisms resulting in an immune response. This type of approach has been used to develop a widely used vaccine against severe acute respiratory syndrome-Coronavirus-2 (SARS-CoV-2) to combat the COVID-19 pandemic.

Edible Vaccines

It is another new approach that is being explored, where antigens will be delivered through edible plant products. They can be produced by introducing the desired gene into a plant to manufacture the encoded protein. The envelope protein of a specific virus or bacteria that has no pathogenicity is used for this purpose. Advantages include affordability, ease of mass production and the ease of administration by consuming the plant product like its fruit, and very importantly, their ability to induce mucosal immunity at the gut level. Potato, tomato, banana, tobacco, rice and edible algae are some of the plants that are potential candidates, and the diseases that have been targeted in various studies include, among others, cholera, hepatitis A and B, Norwalk disease, influenza and malaria.

Immunisation Schedule

World Health Organisation (WHO) Immunisation Programmes

Globally, in 1975, smallpox was reported to be successfully eradicated through effective vaccination programmes and strengthened surveillance. Following this, to control vaccine-preventable diseases, in 1978, the Expanded Programme on Immunisation (EPI) was launched in India. Diphtheria, pertussis, tetanus, poliomyelitis, typhoid and childhood tuberculosis were the diseases initially selected to cover 80% of all the infants. After a while, in 1985, this programme was universalised and named as Universal Immunisation Programme (UIP). In this programme, the typhoid vaccine was withdrawn, and the measles vaccine was included. UIP was implemented in a phased manner with the primary immunisation schedule for all infants and tetanus toxoid immunisation for all pregnant women, and intended to cover all the districts around the country by 1990.

Universal Immunisation Programme (UIP)

Its vision is to achieve and sustain universal immunisation with three doses of DPT and OPV, one dose of both measles and BCG in infants, and two primary doses or one booster dose of tetanus toxoid in pregnant women. This also includes storing and transporting vaccines by a reliable cold-chain system and attaining self-sufficiency in the production of all required vaccines.

The UIP became a component of the Child Survival and Safe Motherhood Programme (CSSM) in 1992; it also became an important part of the Reproductive and Child Health Programme (RCH) in 1997. The cold-chain system was strengthened throughout the country, and extensive training programmes were launched. Polio Eradication Programme with the National Immunisation Days (NIDs) and active surveillance for acute flaccid paralysis (AFP) for intensified polio eradication was started in 1995–1996. This programme was laid down with the assistance of the National Polio Surveillance Project.

National Immunisation Schedule

The immunisation programme in India was introduced in 1978 as the Expanded Programme of Immunisation (EPI), which gained momentum in 1985 and was expanded as the Universal Immunisation Programme (UIP) to be implemented in a phased manner to cover all districts in the country by 1989–1990. UIP became a part of the Child Survival and Safe Motherhood Programme in 1992. Since 1997, immunisation activities have been an important component of the National Reproductive and Child Health Programme and are currently one of the key areas under the National Rural Health Mission (NRHM) since 2005. Under the Universal Immunisation Programme, the Government of India is providing vaccination to prevent the following vaccine-preventable diseases, i.e. diphtheria, pertussis, tetanus, polio, measles, a severe form of childhood tuberculosis (BCG), hepatitis B, *H. influenza* type b (Hib) and *Rotavirus* diarrhoea. One significant achievement is that from 2011, no wild poliovirus case was reported from the country and India was certified as a "polio-free country" along with other South-East Asia Region countries of the WHO.

National Technical Advisory Group on Immunisation recommended introduction of injectable polio vaccine (IPV) as an additional dose along with a third dose of DPT in the entire country from the first quarter of 2016. In 2013, India, along with other South-East Asia Region, declared commitment towards measles elimination and rubella/congenital rubella syndrome (CRS) control by 2020. Towards this direction, the rubella vaccine is to be introduced as MR—measles, rubella—vaccine replacing the measles-containing vaccine, with first dose at 9 months and the second dose at 16–24 months. In addition, the Japanese encephalitis vaccine has been introduced in endemic areas in 20 states. The introduction of a pentavalent vaccine incorporating Hib, DPT and hepatitis B is another new step that has been taken as part of the immunisation schedule. National Immunisation Schedule is mentioned in Table 21.3.

Passive Immunisation

Passive immunisation is carried out by administering human and animal sera, which serve as the readymade source of prepared antibodies against a particular pathogen. These are given to an individual to confer immediate protection against a particular pathogen. The immunity, however, is of short duration.

Human Sera

Normal human immunoglobulins include (1) pooled human immunoglobulin and (2) specific hyperimmune immunoglobulin.

Table 21.3 Universal immunisation schedule in India

Age	Vaccine	Route of administration
Primary vaccination		
At birth	BCG, Hepatitis B, OPV (zero dose)	Intradermal, intramuscular, oral
6 Weeks	OPV (first dose), Pentavalent[a] (first dose), IPV (fractional dose)	oral, intramuscular
10 Weeks	OPV (second dose), Pentavalent (second dose)	Oral, intramuscular
14 Weeks	OPV (third dose), Pentavalent (third dose), IPV (fractional dose)	Oral, intramuscular
9–12 Months	Measles and rubella vaccine (MR; first dose), vitamin A (first dose), JE[b]	Subcutaneous, oral
Booster doses		
16–24 Months	OPV, DPT and MR (second dose), vitamin A (16–18 months, then one dose every 6 months up to the age of 5 years), JE	Oral, intramuscular, subcutaneous
5–6 Years	DPT	Intramuscular
10 Years	TT	Intramuscular
16 Years	TT	Intramuscular
Pregnant women		
Gestational period		
As early as possible during pregnancy (first contact)	TT (first dose)	Intramuscular
One month after the first dose	TT (second dose)	Intramuscular
If previously vaccinated, within 3 years	TT (booster)	Intramuscular

BCG bacillus Calmette–Guérin, *DPT* diphtheria, pertussis, tetanus, *IPV* injectable polio vaccine, *OPV* oral polio vaccine, *TT* tetanus toxoid

Note: [a]The pentavalent vaccine contains five antigens: diphtheria, pertussis, tetanus, hepatitis B and *Haemophilus influenza* type b

[b]JE vaccine is introduced in select endemic districts (first dose: 9–12 months; second dose: 16–24 months). Route of administration is subcutaneous

Pooled Human Immunoglobulin

It is prepared from normal human serum containing a high concentration of a particular antibody. It is usually used for short-term prophylaxis of hepatitis A, measles.

Specific Hyperimmune Immunoglobulin

Specific hyperimmune immunoglobulin is prepared from patients convalescing from the disease or individuals actively immunised against a specific disease. These hyperimmune sera are available against a variety of diseases. Examples are human rabies immunoglobulin (HRIG) against rabies, human tetanus immunoglobulin (HTIG) against tetanus, hepatitis B immunoglobulin (HBIG) against hepatitis B and varicella-zoster immunoglobulin (ZIG) against varicella-zoster. These hyperimmune sera are usually given intramuscularly. They are usually not given shortly before or after active immunisation to avoid inhibition of immune responses.

Animal Sera

Animal sera are prepared in animals. These sera rich in antibodies are produced by active immunisation of animals, such as horses. Antitetanus serum and antivenom sera are examples of such sera used against tetanus and snake bites, respectively. The disadvantages of these sera are that in some recipients, they may cause serum sickness and anaphylaxis; moreover, they are eliminated much more rapidly from humans.

Combined Active and Passive Immunisation

Combined active and passive immunisation is often carried out to confer slowly developing immunity and immediate passive immunity, respectively, against certain diseases, such as diphtheria, tetanus and rabies.

Individual Immunisation

The vaccines mentioned in the universal immunisation schedule are selected based on economic considerations and the epidemiology of infectious diseases in the region. But there are many vaccines available, which can supplement the schedule depending on the individual cases and affordability.

Hepatitis B Vaccine

Hepatitis B virus (HBV) is transmitted from one person to another through blood and body fluids and primarily infects the liver. Healthcare personnel and people constantly exposed to infected blood or other body fluids are at a high risk of infection. Infected mothers are the commonest mode of spread to infants globally. Chronic HBV carriers are generally 90% of the infants infected at birth through their mothers or approximately 30–50% of those before the age of 5 years. Adults who are newly infected contribute to only a 6–10% risk of chronic infection. Therefore, routine administration of hepatitis B immunisations at birth is highly recommended.

Vaccination with at least three doses of the hepatitis B vaccine is recommended. Infants are given a dose of the hepatitis B vaccine at birth. HBV in combination with the DPT and inactivated polio vaccines are recommended for children between the ages of 6 weeks and 7 years. HBV and Hib (*H. influenza* type b) conjugate vaccine may be given to children between the ages of 6 weeks and 15 months. Dialysis patients and immunocompromised people may require additional doses.

MR Vaccine

MR (measles, rubella) vaccine is given for preventing both measles and rubella disease in the child. MR vaccine is being introduced instead of MMR vaccine as there is not enough evidence to suggest that mumps is a disease of public health importance. MR vaccine gives lifelong protection in most children.

MR campaign: MR campaign is a special campaign to vaccinate all children of 9 months to <15 years of age group with one dose of MR vaccine. The MR campaign dose is given to all targeted children, both immunised and unimmunised, irrespective of prior measles/ rubella infection. The goal of the MR campaign is to accelerate population immunity by reaching

100% target children with the MR vaccine that will reduce cases and deaths from measles and disabilities from CRS (congenital rubella syndrome).

Typhoid Vaccine

Recently, two newer vaccines have replaced the older vaccine used in typhoid fever. These vaccines are oral live-attenuated vaccines using a weakened strain of *Salmonella* Typhi that is given orally and a parenteral capsular polysaccharide vaccine that is given by injection.

The **oral typhoid vaccine** can be administered to children 6 years of age or older and adults. It is given in a total dosage of four capsules, one capsule taken orally every other day. A booster dose is needed every 5 years for people who continue to remain at risk. The **polysaccharide typhoid vaccine** is given by injection to children older than 2 years. One shot is enough to provide protection. A booster dose is needed every 2 years for people who remain at risk for exposure. Efficacy of 50–80% was seen from the two licensed vaccines. The oral vaccine has shown a protective efficacy of 62% for at least 7 years after the last dose.

Vaccination During Outbreaks

There has been controversy regarding the usefulness of administering a vaccine during the outbreak of the particular disease against which the vaccine is being used. This problem has been studied in detail regarding measles vaccination to children during a measles outbreak. The weight of evidence suggests that because of the high transmissibility of the virus, mass vaccination will not stop the epidemic but will reduce the death in settings where most infections occur in children below 5 years of age. The lessons learnt from the measles study have been applied to the recent COVID-19 pandemic, where it has also been found that although the available vaccines cannot stop infection, they definitely reduce the severity of illness and mortality.

Newer Vaccines in the Experimental or Clinical Trial Stages

The traditional vaccine development approach relied on the isolation of the organisms and then employing whole or part of the organisms as antigen to develop an immune response. This approach has been largely superseded in recent years due to the knowledge gained from genomic studies and the associated proteomics. It allows the prediction of all antigens without growing the organisms. The potential vaccine candidates are identified by proteomic analysis of the organism in silico with the help of computers, and this approach has been termed **Reverse Vaccinology**. With the modern approach, new vaccines are being explored not only for diseases for which no vaccines are available but also to improve the existing vaccines in terms of immunogenicity and duration of effectiveness. Studies are underway to develop new vaccine candidates for various bacterial, viral, protozoan and helminthic diseases. Many of them have entered clinical trials, like those of malaria, acquired immunodeficiency syndrome (AIDS), tuberculosis, kala-azar etc.

Part III

Bacteriology

Introduction to Bacteriology 22

Microbes are tiny little creatures, invisible to naked eyes. Initial taxonomical classification broadly categorised bacteria under the animal and plant kingdoms. However, currently, bacteria are placed under a separate kingdom called **Protista**. The kingdom Protista mainly comprises **Eukaryotes** and **Prokaryotes.** Organisms such as fungi, protozoa, algae (other than blue-green algae) and slime moulds are included in Eukaryotes, whereas bacteria and blue-green algae belong to Prokaryotes.

Classification

Traditionally, all life forms are classified into kingdoms. Such kingdom bases classification started with two kingdom classifications of Linnaeus in 1735 that slowly evolved and expanded by different individuals with the addition of more kingdoms. In the latest, seven kingdoms distributed in two superkingdoms (empire) based classification were introduced in 2015.

Superkingdom (Empire)	Kingdom
Prokaryota (or Prokaryotes)	Bacteria
	Archaea
Eukaryota (or Eukaryotes)	Protozoa
	Chromista
	Plantae
	Fungi
	Animalia

In contrast, classification was also done to determine life form evolution based on genetic information. Such classification made by Carl Woese has introduced the concept of domains instead of kingdoms. Life has been three distinct domains, namely Eubacteria, Archaebacteria and Eukarya. Phylogenetically it was noted that Eubacteria (we know as bacteria) were separated first in evolution, whereas Archaebacteria and Eukarya (including plants, fungi, animals, humans) diverged later in the timeline.

Viruses, a non-living entity, were not considered in any classification systems and hence are neither prokaryotes nor eukaryotes.

The Linnaean system of nomenclature is followed for naming a bacterium, like *Escherichia coli*, where *Escherichia* is the genus and *coli* is the species. The name should be italicised or underlined.

The basis of the classification of bacteria is not completely perfect, and a combination of phenotypic and genetic characters is commonly used. The phylogenetic basis of classification considers various morphological, biological and antigenic properties of the bacterium. A phylogenetic tree is constructed based on weightage given to individual characteristics (which is different for various attributes), reflecting the organism's evolutionary position. On the other hand, the phenetic or Adansonian classification or its further refinement, the Numerical Taxonomy method, is similar but gives equal weightage to each characteristic. The genetic relatedness of the organisms is the basis for molecular classification where traditionally the guanine+cytosine content has been used. The G=C content for a particular species is remarkably constant. The classification system has been undergoing tremendous changes by using genome sequencing in recent years. The classification relies on average nucleotide identity (ANI), where whole genomes of organisms are compared in contrast to the previous generation approaches on GC content analysis. Most of the important pathogenic bacteria have undergone whole-genome sequencing leading to precise identification and degree of relatedness between closely associated members in a group.

In addition to the species level identification, many bacteria can be subdivided below the species level into subspecies by either antigenic properties of cell components or a set of biochemical and related tests. This has helped describe the pathogenesis and the variation in the disease spectrum of a particular species. The use of terms like subspecies, serogroups, serovars or biotypes refers to this type of sub-classification.

Shape of Bacteria

Depending on their shape, bacteria are classified into several types.

1. **Cocci**: The cocci (*kokkos*, berry) are oval or spherical cells. These may be arranged in pairs (e.g. pneumococci, meningococci and gonococci), tetrads (micrococci), chains (e.g. streptococci) and clusters (e.g. staphylococci).
2. **Bacilli**: The bacilli (*bacillus,* rod) are rod shaped. These bacilli may show either of the following arrangement. (1) *Coccobacilli*: The length of the bacteria is approximately the same as its width, for example, *Brucella*. (2) *Streptobacilli*: These are arranged in chains, for example, *Streptobacillus*. (3) *Comma shaped*: They exhibit a curved appearance, for example, *Vibrio;* and (4) *Spirilla:* They exhibit rigid spiral forms, for example, *Spirillum.*
3. **Spirochetes**: Spirochetes (*Spira,* coil; *chaite,* hair) are slender, flexuous spiral forms, for example, *Treponema.*
4. **Actinomycetes**: Actinomycetes (*actin,* ray; *mykes,* fungus) are branching filamentous bacteria resembling fungi. They possess a rigid cell wall.
5. **Mycoplasmas**: These are cell wall deficient bacteria without stable morphology. They occur as round or oval bodies and as interlacing filaments in nature.

At times cell wall synthesis of bacteria becomes defective either spontaneously or due to antibiotics like penicillin. As a result, bacteria lose their distinctive shape, and such cells are called protoplast, spheroplasts or L forms.

An arrangement of the bacteria depends on its plane of cell division (binary fission) and the tendency of daughter cells to remain attached even after complete cell division. For example, corynebacteria arrange themselves in cuneiform or Chinese letter pattern. Similarly, most cocci arrange themselves in various patterns, as described earlier.

Bacterial Anatomy

Bacterial cell shows the following structures:

Cell envelope: It provides a structural and physiological barrier between the protoplasm of the cell and the external environment. It gives bacteria rigidity and shape and protects bacteria against osmotic lysis. The cell envelope primarily consists of two components: (1) rigid cell wall and (2) cytoplasmic or plasma membrane beneath the cell wall.

Cell interior: It encloses the protoplasm, which consists of (1) cytoplasm, (2) cytoplasmic inclusions (mesosomes, ribosomes, inclusion granules and vacuoles) and (3) nucleus, a single circular DNA (Fig. 22.1).

Additional structures: Some bacterial cells may have additional appendages such as fimbriae for adhesion, and flagella for locomotion. Similarly, some bacteria are covered with a slime

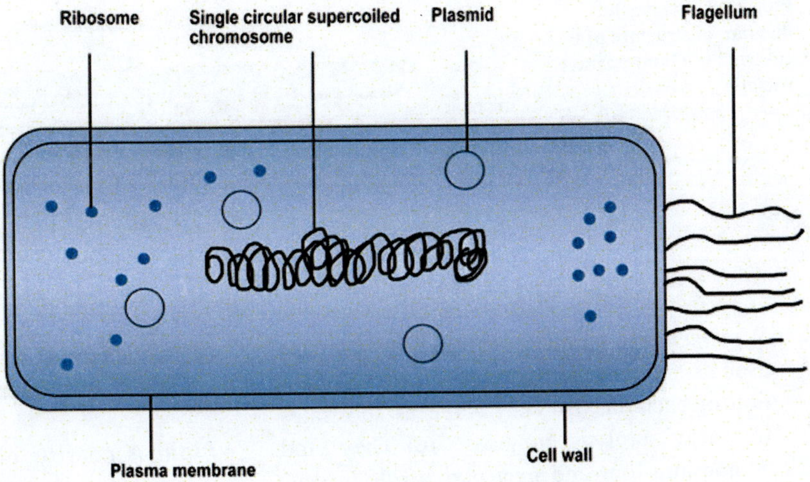

Fig. 22.1 Schematic diagram of structure of bacteria

layer or well-defined capsule, which helps in immune evasion.

Cell Envelope

This includes cell wall and cytoplasmic membrane.

Cell Wall

Prokaryotic cells are almost always bounded by a fairly rigid and chemically complex structure between the cell membrane and capsule/slime layer called the cell wall. The cell wall provides shape to the cell and protects bacteria from changes in osmotic pressure, which measures 5–20 atmospheres within the bacteria cell. The cell wall cannot be visualised by direct light microscopy and does not stain with simple stains.

Structure: Generally, cell walls of bacteria range from 10 to 25 nm in thickness and makeup to 20–30% of the dry weight of the cell. Peptidoglycan is the main component of the cell wall and is responsible for the shape and strength of the cell. It is a disaccharide and contains two sugar derivatives—*N*-acetyl glucosamine and *N*-acetyl muramic acid—joined together by short peptide chains. The spaces formed during peptidoglycan scaffolding contain other chemicals, and their composition varies from species to species.

Bacterial cells can be classified into Gram-positive or Gram-negative based on the structural differences in the cell walls. In general, the cell walls of Gram-positive bacteria have simpler chemical structures than Gram-negative bacteria. The cell wall bears bacterial antigens that are important in virulence and immunity.

Gram-Positive Cell Wall

The Gram-negative cell wall is thick (15–80 nm) and more homogeneous than the thin (2 nm) Gram-negative cell wall. The Gram-positive cell wall contains a large amount of peptidoglycan present in several layers that constitute about 40–80% of the dry weight of the cell wall (Fig. 22.2). The Gram-positive cell wall consists primarily of teichoic and teichuronic acids.

1. **Teichoic acids**: Teichoic acids are polymers of polyribitol phosphate or polyglycerol phosphate-containing ribitol and glycerol. Teichoic acids are of two types—wall teichoic acid (WTA) and lipoteichoic acid (LTA).
   Teichoic acids have many functions. (1) They constitute major surface antigens of those Gram-positive species that possess them. In *Streptococcus pneumoniae*, the teichoic acids bear the antigenic determinants called Forssman antigen. In *Streptococcus pyogenes*, LTA is associated with the M protein that protrudes from the cell membrane through the peptidoglycan layer. The long M protein molecules together with the LTA form microfibrils that facilitate the attachment of

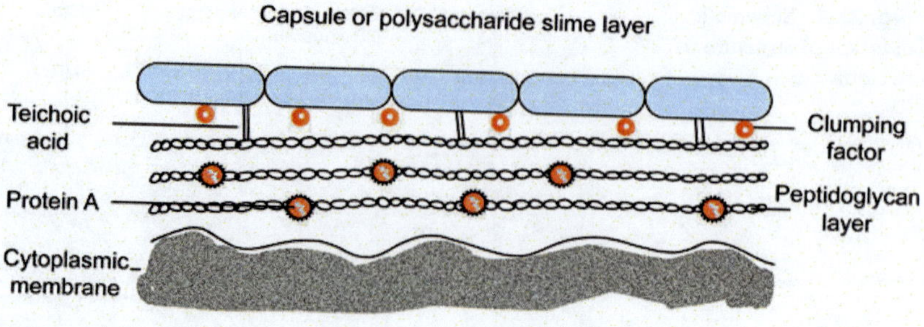

Fig. 22.2 Schematic diagram of structure of the cell wall of Gram-positive bacteria

S. pyogenes to animal cells. (a) They are also used as antigens for the serological classification of bacteria. (b) They serve as substrates for many autolytic enzymes'. (c) They bind magnesium ions and may play a role in supplying this ion to the cell. (d) They play a role in the normal functioning of the cell wall and provide an external permeability barrier to Gram-positive bacteria. (e) Membrane teichoic acid serves to anchor the underlying cell membrane.

2. **Teichuronic acid**: Teichuronic acid consists of repeat units of sugar acids (such as *N*-acetyl mannuronic or D-glucuronic acid). They are synthesised in place of teichoic acids when phosphate supply to the cell is limited.

   The Gram-positive cell wall also contains neutral sugars (such as mannose, arabinose, rhamnose and glucosamine) and acidic sugars (such as glucuronic acid and mannuronic acid) which occur as sub-units of polysaccharides in the cell wall.

Gram-Negative Cell Wall

The Gram-negative cell wall is much more complex than the Gram-positive cell wall. Peptidoglycan content in the Gram-negative cell wall is significantly less than in the Gram-positive cell wall. The Gram-negative cell wall contains (1) lipopolysaccharides, (2) outer membrane, (3) lipoprotein layer and (4) peptidoglycan (Fig. 22.3).

1. **Lipopolysaccharides**: Lipopolysaccharides (LPS) are complex molecules present in the outer membrane of Gram-negative bacteria. Structurally, the LPS consist of three main components—lipid A (Region III), the core oligosaccharide (Region II) and the O polysaccharide or O-antigen (Region I) (Fig. 22.4).

   Lipid A consists of hydroxymyristic acid, a unique fatty acid associated with the endotoxic activity of the LPS. The endotoxic activities include pyrogenicity, lethal effect, tissue necrosis, anticomplementary activity, B-cell mitogenicity, immunoadjuvant property and antitumour activity.

   Core oligosaccharide: This includes two characteristic sugars—keto-deoxy octanoic acid (KDO) and a heptose—joined together by lipid A. This is genus specific and similar to all Gram-negative bacteria. Lipooligosaccharides (LOS) are smaller glycolipids. They are relatively short, multi-antennary (i.e. branched) glycans present in bacteria (e.g. *Neisseria meningitidis, Neisseria gonorrhoea, Haemophilus influenzae* and *Haemophilus ducreyi*) that colonise mucosal surfaces. They exhibit extensive antigenic and structural diversity even within a single strain. LOS are an important virulence factor. Epitopes on LOS have a terminal *N*-acetyllactosamine (Gal((3)l-4-GlcNAc) residue, which is immunochemically similar to the precursor of the human erythrocyte antigen. Sialylation of the *N*-acetyllactosamine residue *in vivo* provides the organism with the environmental advantages of molecular mimicry of a host antigen and the biologic masking thought to be provided by sialic acids.

   O polysaccharide or O-antigen: It is the portion extending outwards from the core. It

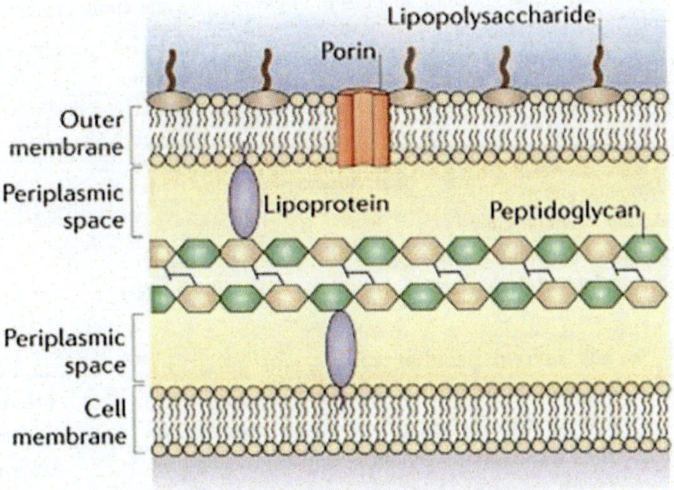

Fig. 22.3 Schematic diagram of structure of the cell wall of Gram-negative bacteria (Source: https://www.nature.com/articles/nrmicro3480)

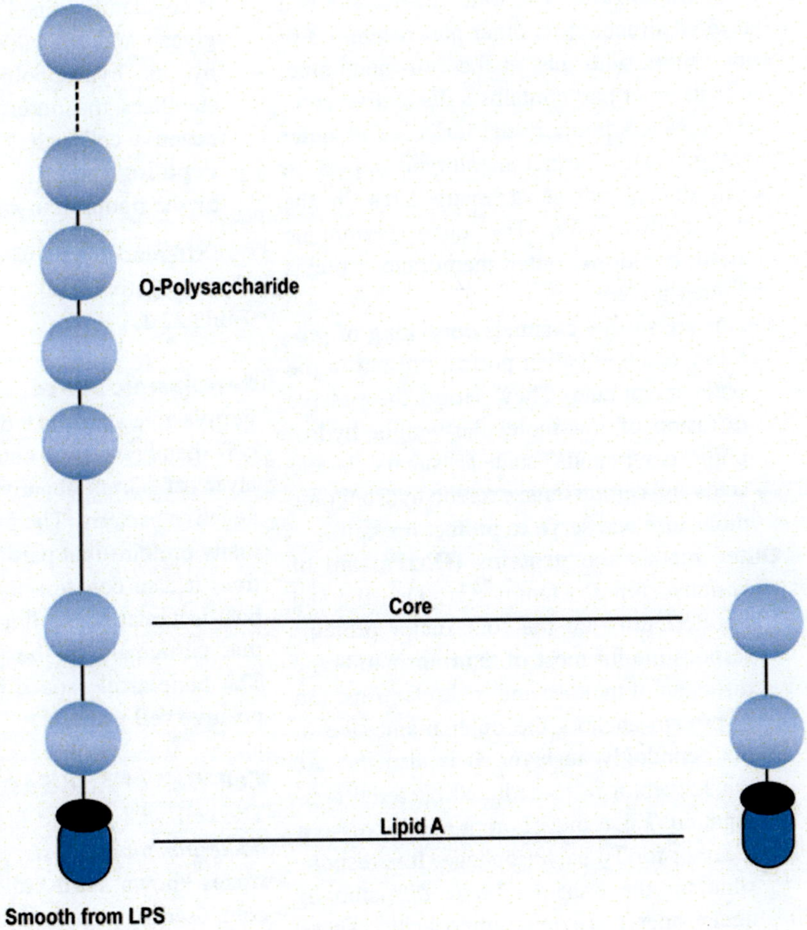

Fig. 22.4 Schematic diagram of structure of a lipopolysaccharide

Table 22.1 Distinguishing features of Gram-positive and Gram-negative bacteria cell wall

Features	Gram-positive cell wall	Gram-negative cell wall
Thickness	15–80 nm	2 nm
Lipid content	2–5%	15–20%
Teichoic acid	Present	Absent
Variety of amino acid	Few	Several
Aromatic amino acid	Absent	Present
Action as endotoxin	No	Yes
Sulphur-containing amino acid	Absent	Present
Treatment with lysozyme	Protoplast	Spheroplast

has several peculiar sugars and varies in composition between bacterial strains, conferring species-specific antigen specificity. It is exposed to the host immune system. Gram-negative bacteria may thwart host defences by rapidly changing the nature of their O side chains to avoid detection.

2. **Outer membrane**: The outer membrane is a bilayered structure; its inner part resembles in composition with that of the cell membrane, while its outer part contains a distinctive component called lipopolysaccharide. The outer membrane and plasma membrane appear to be in direct contact at many sites in the Gram-negative wall. The outer membrane consists of porins, outer membrane proteins and other proteins.

   Porins are special channels consisting of protein molecules called porins, present in the outer membrane. They permit the passive diffusion of low-molecular-weight hydrophilic compounds, such as sugars, amino acids and certain ions; exclude hydrophobic molecules and serve to protect the cell.

   Outer membrane proteins (OMPs) are of various types. Omp C, D, F and PhoE&LamB are the four major proteins responsible for most of the transmembrane diffusion of maltose and maltodextrins. **On A** protein anchors the outer membrane to the peptidoglycan layer. It is also the sex pilus receptor in F-mediated bacterial conjugation. **Tsx** is one of omps and serves as a receptor for T6 bacteriophage. It is responsible for the transmembrane diffusion of nucleosides and some amino acids. These are involved in the transport of specific molecules, such as vitamin B_{12} and iron-siderophore complexes; it also contains a limited number of minor proteins, such as enzymes, phospholipases and proteases.

3. **Lipoprotein layer**: The lipoprotein layer is mainly composed of Braun lipoprotein. Braun lipoprotein is a small lipoprotein that is covalently joined to the underlying peptidoglycan and embedded in the outer membrane by its hydrophobic end. The lipoprotein stabilises the outer membrane of the Gram-negative cell wall.
4. **Peptidoglycan**: It is 2–3 nm thin and is bound by the lipoprotein and plasma membrane.

Differences between Gram-positive and Gram-negative cell walls are summarised in Table 22.1.

Periplasmic Space

Periplasmic space is a distinct space between the cell membrane and outer membrane (innermost layer of Gram-negative cell wall) in Gram-negative bacteria. The periplasmic space contains many proteins that participate in nutrient acquisition. It also contains many hydrolytic enzymes, beta-lactamases binding proteins and enzymes that participate in the peptidoglycan synthesis. The periplasmic space is less distinct in Gram-positive cell walls.

Cell Wall of Acid-Fast Bacilli

The cell wall of acid-fast bacilli, such as *Mycobacterium tuberculosis*, contains large amounts of waxes known as mycolic acids. The cell wall is composed of peptidoglycan and lipid. The inner lipid bilayer contains mycolic acids linked to an arabinose glycan protein, and the outer layer contains other extractable lipids. Its structure makes these bacteria resistant to harsh chemicals, including detergents and strong acids. During staining, if the dye is introduced into these cells by brief heating or treatment with detergents, they resist decolourisation by sulphuric acid or acid alcohol and are therefore called acid-fast organisms.

Inhibition of Cell Wall Synthesis

Many agents, such as antibiotics, lysozyme and bacteriophages, interfere or inhibit the synthesis of bacterial cell wall components, resulting in the formation of defective bacteria or atypical forms of bacteria. These defective forms include i) cell wall deficient forms, ii) pleomorphic bacteria and iii) involution forms.

Cell Wall Deficient Forms

These may be of various types, such as—protoplasts, spheroplasts, mycoplasma and L-forms.

Protoplasts are defective unstable forms of bacteria with an intact cytoplasmic membrane but without any cell wall. In hypertonic media, these are produced from Gram-positive cells on treatment with lysozyme.

Spheroplasts are defective forms derived from Gram-negative bacteria in the presence of EDTA (ethylene diaminetetraacetic acid). The EDTA disrupts the outer membrane allowing access to lysozyme and resulting in the formation of spheroplasts. Spheroplasts are osmotically fragile and still retain the outer membrane and entrapped peptidoglycan.

Mycoplasma is naturally occurring bacteria without a cell wall. They do not possess any definite shape. They are very minute in size, measuring 50–300 nm in diameter.

L-forms is named after Lister Institute, London, where the abnormal form of *Streptobacillus moniliformis* was first demonstrated. The L-forms do not exhibit any regular size and shape. They may be spherical or disc-shaped, about 0.1–20 μm in diameter. They are difficult to cultivate and usually require a medium that is solidified with agar and has the right osmotic strength. They are produced more readily with exposure to penicillin than with lysozyme. Some bacterial species produce L-forms spontaneously. L-forms in the host may produce chronic infections and are relatively resistant to antibiotic treatment.

Pleomorphic Bacteria

Pleomorphic bacteria (e.g. *Yersinia pestis*) may show considerable variation in size and shape called pleomorphism.

Involution Forms

The involution forms are those that, on the ageing of culture, show swollen and aberrant forms, especially in high salt concentration.

Demonstration of Cell Wall

The cell walls can be demonstrated by (1) differential staining procedure, (2) electron microscopy, (3) plasmolysis, (4) microdissection, (5) mechanical rupture of the cell and (6) serological test by exposure to specific antibodies.

Cytoplasmic Membrane

Cytoplasmic membrane or plasma membrane is a thin (5–10 nm) semipermeable membrane that acts as an osmotic barrier. It lies beneath the cell wall separating it from the cell cytoplasm. Bacterial plasma membranes usually have a higher proportion of protein than eukaryotic membranes. Cell membrane primarily consists of phospholipids and proteins. It also contains enzymes associated with DNA biosynthesis, cell wall polymers and membrane lipids. They usually differ from eukaryotic membranes in lacking sterols, such as cholesterol, except in *Mycoplasma.* The cytoplasmic membrane acts as a semipermeable membrane regulating the inflow and outflow of metabolites to and from the protoplasm. It also helps in electron transport and oxidative phosphorylation.

Cell Interior

This encloses the protoplasm, which comprises many components such as cytoplasm, cytoplasmic inclusions and nucleus.

Cytoplasm

Bacterial cytoplasm is a colloidal suspension of various organic and inorganic solutes in a viscous watery solution. It consists of nearly 70% water. The cytoplasm contains all the biosynthetic components required by a bacterium for growth and cell division, together with genetic material.

Cytoplasmic Inclusions

These include ribosomes, mesosomes and intracytoplasmic inclusions bodies.

1. Ribosomes

The cytoplasmic matrix often is packed with ribosomes. Ribosomes look like small, featureless particles at low magnification in electron micrographs. They consist of two subunits of the 30S and 50S, giving a net 70S. Ribosomes are important because they serve as the sites of protein synthesis. Matrix ribosomes synthesise proteins destined to remain within the cell, whereas plasma membrane ribosomes make proteins for transport to the outside. They are also the sites of actions of several antibiotics, such as aminoglycosides, macrolides and tetracyclines.

2. Mesosomes

These are vesicular convoluted or multilaminated structures formed by invagination of the plasma membrane into the cytoplasm. Mesosomes are of two types—septal and lateral. The septal mesosome attached to the bacterial DNA coordinates nuclear and cytoplasmic divisions during binary fission. The function of lateral mesosomes remains to be known. Mesosomes are analogous to the mitochondria of eukaryotes and are the principal sites of respiratory enzymes in bacteria.

3. Intracytoplasmic Inclusion Bodies

Intracytoplasmic inclusion bodies are present in the protoplasm of bacteria. They may be of two types: (1) organic inclusion bodies, which usually contain either glycogen or polyhydroxy butyrate, and (2) inorganic inclusion bodies, which may be of polyphosphate granules or sulphur granules. Their main function is believed to be storage. They are also the sources of carbon, inorganic substances and energy. Some inclusion bodies also function to reduce osmotic pressure.

Examples of intracytoplasmic inclusion bodies include metachromatic granules or **volutin granules, starch inclusions** and **lipid inclusions.** Volutin granules, typically present in *Corynebacterium diphtheriae*, can be demonstrated by Albert stain. Similarly, starch granules present in the bacteria can be demonstrated on staining with iodine. Lipid inclusion found in mycobacteria is demonstrated by Sudan black dye. Bacterial cytoplasm, unlike that of eukaryotes, lacks endoplasmic reticulum and mitochondria. It also does not show any protoplasmic streaming.

Nucleoid

The **bacterial nucleoid is unique** because it is neither enclosed in a nuclear membrane nor associated with any nucleolus. It is a haploid and replicates by simple fission. The nucleus of the bacteria consists of a single circle of double-stranded deoxyribonucleic acid (DNA) arranged in a supercoiled circular structure. It measures about 1000 μm when straightened.

The chromosome is located in an irregularly shaped region called nucleoid but is often referred to as a bacterial chromosome because of the analogy with the eukaryotic structure. The nucleoid is visible through the light microscope after staining with the Feulgen stain, specifically reacting with DNA. In actively growing bacteria, the bacterial DNA can account for up to 20% of the volume of the bacterium and has projections that extend into the cytoplasmic matrix. Careful electron microscopic studies often have shown the nucleoid to be in contact with either the mesosome or the plasma membrane.

Many bacteria also possess smaller circles of extrachromosomal DNA called **plasmids**. The plasmids are double-stranded DNA molecules, usually circular, that can exist and replicate independently. Plasmids are not required for host growth and reproduction, although they may carry genes that confer the bacterium with properties such as antibiotic resistance or the capacity to produce toxins or enzymes.

Additional Structures

Many Gram-positive and Gram-negative bacteria possess a gel-like layer outside the envelope when growing in their natural environments. When a gel-like layer forms a well-defined condensed layer around the bacterial envelope, it is called a **capsule.** The surface appendages of some bacteria include surface appendages such as flagella and fimbriae or pili.

Capsule

The **capsule** surrounding the bacteria is demonstrable by a light microscope. When this gel-like layer is narrower, detectable only by indirect serological methods or by electron microscope but not by light microscope, it is called a **microcapsule**. The capsule is mostly made up of polysaccharides, often referred to collectively as exopolysaccharides. Exopolysaccharides are sometimes neutral homopolysaccharides (e.g. the glucans and fructans of many oral streptococci) or negatively charged (Table 22.2). However, *Bacillus anthracis* has a capsule comprising of polyamine acids, such as D-glutamic acid. The D-glutamic acid is probably analogous to the negatively charged polysaccharide capsule.

Functions

The capsule has various functions. (1) It contributes to the invasiveness of bacteria by protecting the bacteria from phagocytosis. (2) It also prevents bacteria from generating an immune response in infected hosts. (3) It facilitates adherence of bacteria to surfaces. *Streptococcus mutans*, for example, owes its capacity to the glycocalyx to adhere tightly to tooth enamel to its glycocalyx. (4) It plays a role in forming biofilms.; and (5) the glycocalyx layer of the capsule may also play a role in resistance to desiccation.

Demonstration of Capsule

The capsule is fully hydrated and can be demonstrated by light microscopy in either living or stained bacteria as follows:

Special Capsular Staining Methods

These include the Welch method and M'Faydean capsule stain.

Welch's method uses copper as mordant. This involves treating fixed smear with hot crystal violet solution followed by rinsing with copper sulphate solution. The latter is used to remove the excess stain because conventional washing with water would dissolve the capsule. The copper salt also gives colour to the background, with the result that the cell and background appear dark blue and the capsule of the bacteria a much paler blue. Using polychrome methylene blue stain, McFadyen capsule stain is a frequently used method for demonstrating the capsule of *B. anthracis*.

Table 22.2 Chemical composition of capsules of various bacteria

Bacteria	Polymer	Chemical sub-units
Bacillus anthracis	Polypeptide	D-glutamic acid
Enterobacter aerogenes	Complex polysaccharide	Glucose, fructose, glucuronic acid
Neisseria meningitidis	Homopolymers and heteropolymers	Partially O-acetylated *N*-acetylmannosamine phosphate
Streptococcus pneumoniae	Complex polysaccharides (many types)	Rhamnose, glucose, glucuronic acid
Streptococcus pyogenes (group A)	Hyaluronic acid	*N*-acetylglucosamine, glucuronic acid
Streptococcus salivarius	Levan	Fructose

Negative staining with India ink is also known as the wet India ink method. It is the simplest way to demonstrate a capsule. It is carried out by mixing a suspension of bacteria with an equal volume of Indian ink on a slide, covering it with a coverslip, and examining it under a microscope. The capsule appears as a clear zone around the cell. This method is useful for improving the visualisation of encapsulated bacteria in clinical samples, such as blood or cerebrospinal fluid.

Serologic methods are also used to demonstrate capsules since they are antigenic. **Quellung's reaction** is such a serological method for demonstrating the capsule. When a suspension of bacterium is mixed with its specific anticapsular serum and methylene blue and examined under a microscope, the capsule becomes very prominent and appears swollen due to increased refractoriness. This method is useful for rapidly identifying capsular serotypes of *S. pneumoniae*, *N. meningitidis*, *H. influenzae*, *Yersinia*, *Bacillus* etc.

Slime Layer

An amorphous viscid colloidal material secreted by some bacteria extracellularly is termed as loose or free **slime** or **glycocalyx**. They are resistant to proteolytic enzymes and protein-denaturing agents. The slime layer protein protects the cell from wall-degrading enzymes and bacteriophages. It plays an important role in maintaining cell shape, and it may also be involved in cell adhesion to host epidermal surfaces.

Surface Appendages

The surface appendages of the bacteria include flagella and fimbriae or pili.

Flagella

Bacterial flagella are thread-like appendages intricately embedded in the cell envelope. These structures are responsible for conferring motility to the bacteria. The arrangement of flagella varies between different bacterial species.

Depending on the arrangement, flagella can be of the following types: **Monotrichous** (single polar flagellum), for example, *Vibrio cholerae*; **Lophotrichous** (multiple polar flagella), for example, Spirilla; **Peritrichous** (flagella distributed over the entire cell), for example, *Salmonella typhi*, *E. coli* and **Amphitrichous** (single flagellum at both the ends), for example, *Spirillum minus* (Fig. 22.5).

Structure: The flagella are 3–20 μm in length and 0.01–0.03 μm in diameter. The main part of the filament is made up of protein subunits called flagellin arranged in several helices around a central hollow core. The flagellum is attached to the bacterial cell body by a complex structure consisting of a hook and a basal body. The basal body bears a set of rings, one pair in Gram-positive bacteria and two pairs in Gram-negative bacteria, through which the bacteria rotate either in a clockwise or an anticlockwise direction. Above the filament base is the hook, a short curved structure between the external filament and basal body. This part produces a propeller-like repulsion from the revolving flagellum (Fig. 22.6).

Spirochetes are motile bacteria but without any external flagella. They are motile due to the presence of an axial filament. Axial filament consists of a bundle of flagellum-like structures between the cell surface and an outer sheath and connects one end of the cell to the other. They are sometimes called the **endoflagellates**.

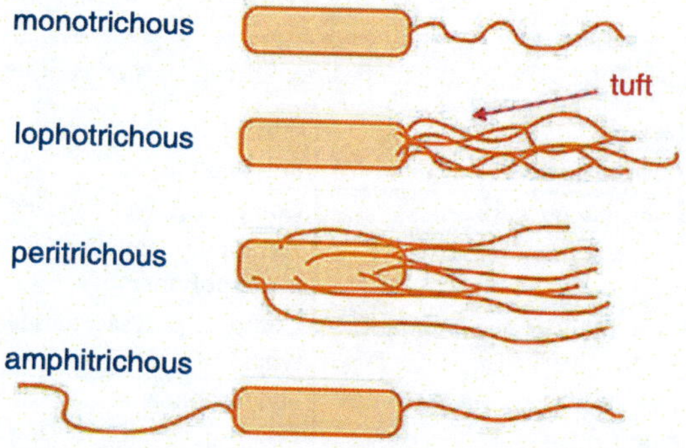

Fig. 22.5 Arrangement of the bacterial flagella

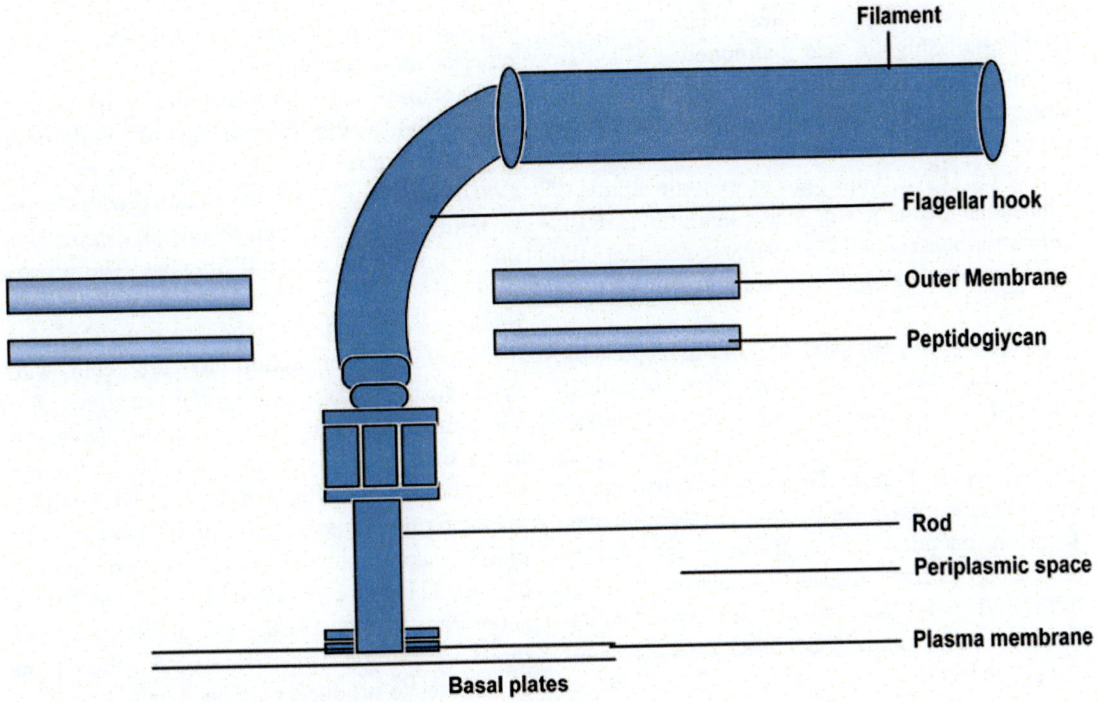

Fig. 22.6 Structure of a bacterial flagellum

Functions

Flagella are primarily responsible for the motility of bacteria by chemotaxis. They may play a role in bacterial survival and pathogenesis. They are highly antigenic, possess H antigens, and some of the immune responses to infection are directed against these proteins. The flagella of different bacteria differ antigenically. Flagellar antibodies are not protective but help in serodiagnosis.

Demonstration of Flagella

Direct and indirect methods can demonstrate the flagella.

Direct methods include a direct demonstration of the capsule by electron microscope. These also include a demonstration of the capsule after staining by special staining methods, such as the Ryu method and Hugh–Leifson method. Since flagella are very thin structures, these staining methods demonstrate flagella by increasing their thickness by mordanting with tannic acid.

Indirect methods of flagella demonstration include the bacteria's motility by (1) dark-ground microscopy, (2) hanging drop method or (3) observing spreading type growth on semisolid media, such as mannitol motility medium.

Pili (Fimbriae)

Pili or fimbriae are synonymous for most purposes. They are hair-like filaments that extend from the cell surface and are found almost exclusively on Gram-negative bacteria. They are composed of structural protein sub-units termed pilins.

Structure: The pili are shorter and straighter than flagella, although the basic structure is similar. Like flagella, it consists of helices of a protein called *pilins*. Minor proteins termed adhesins are located at the tips of pili and are responsible for the attachment properties. Pilins are arranged around a hollow core but without a motor. They are 0.5 μm long and 10 nm thick. They are antigenic. Pili haemagglutinate.

RBCs of guinea pigs are specifically inhibited by mannose, based on which they are classified into four types. **Type 1**, these occur in *E. coli*, Klebsiella, Shigella and Salmonella. They are mannose sensitive. **Type 2** are present in *Salmonella gallinarum* and *Salmonella pullorum*, devoid of any haemagglutinating or adhesive properties. **Type 3** are present in some strains of Klebsiella, Serratia etc. They agglutinate RBC only after heating and are mannose resistant, and **Type 4** are mannose resistant and occur in Proteus.

Sex pili is a specialised kind of pili responsible for the attachment of donor and recipient cells in bacterial conjugation. These pili are longer (10–20 μm) and vary 1–4 in number. These are of two types: **F pili**: They specifically adsorb male-specific RNA and DNA bacteriophages. They are encoded by sex factor F and fertility inhibition-positive resistance factors (fi+ R factors). **I pili**: They adsorb male-specific filamentous DNA phages, encoded by col. factor and fi—R factor.

Functions of Pili

Pili play a major role in the adherence of symbiotic and pathogenic bacteria to host cells, which is a necessary step in initiating infection. Transfer of bacterial DNA takes place through sex pili during conjugation.

Demonstration of Pili

The pili can be detected directly by electron microscope; and by agglutination of guinea pigs, fowl, horses and pigs RBCs. They agglutinate human and sheep RBCs weakly. The haemagglutination can be specifically inhibited by D-mannose.

Some Gram-positive bacteria do not possess typical pili but instead possess a fine fibrillar arrangement of proteins on their surfaces known as **fibrils**. These fibrils bind to the host surfaces. M-protein of *S. pyogenes* is an example of Gram-positive bacteria possessing fibrils.

Bacterial Spores

Spores are highly resistant resting phase of some of the bacteria. The organism survives in spores, a dormant state, for longer periods of starvation and other adverse conditions.

Sporulation is a primitive process of differentiation with the formation of endospores (e.g. spores of aerobic *Bacillus* spp. and anaerobic *Clostridium* spp.). The sporulation process begins in nutrition-deprived conditions. It begins with the formation of an axial filament. The process continues with the infolding of the membrane to produce a double membrane structure whose facing surfaces correspond to the cell wall-synthesising surface of the cell envelope. Two spore membranes then engage in the active synthesis of special layers that form the cell envelope: the spore wall and the cortex, lying outside the facing membranes. In the newly isolated cytoplasm or core, many vegetative cell enzymes are degraded and are replaced by a set of unique spore constituents. During sporulation, each cell forms a single internal spore; the spore germinates to produce a single vegetative cell.

Morphology

The bacterial endospores are a highly resistant resting phase of bacteria. The spore shows the following structures:

1. **Core**: The core contains a complete nucleus (chromosome), all of the components of the protein-synthesising apparatus and an energy-generating system based on glycolysis. The heat resistance of spores is partly due to their dehydrated state and the presence of large amounts (5–15% of the spore dry weight) of calcium dipicolinate in the core.
2. **Spore wall**: This is the innermost layer surrounding the inner spore membrane. It contains normal peptidoglycan and becomes the cell wall of the germinating vegetative cell.
3. **Cortex**: It is the thickest layer of the spore envelope containing unusual peptidoglycan. It is extremely sensitive to lysozyme, and its autolysis plays a role in spore germination.
4. **Protein coat**: It is composed of a keratin-like protein containing many intramolecular disulphide bonds; this layer confers relative resistance to antibacterial chemical agents due to its impermeability (Fig. 22.7).

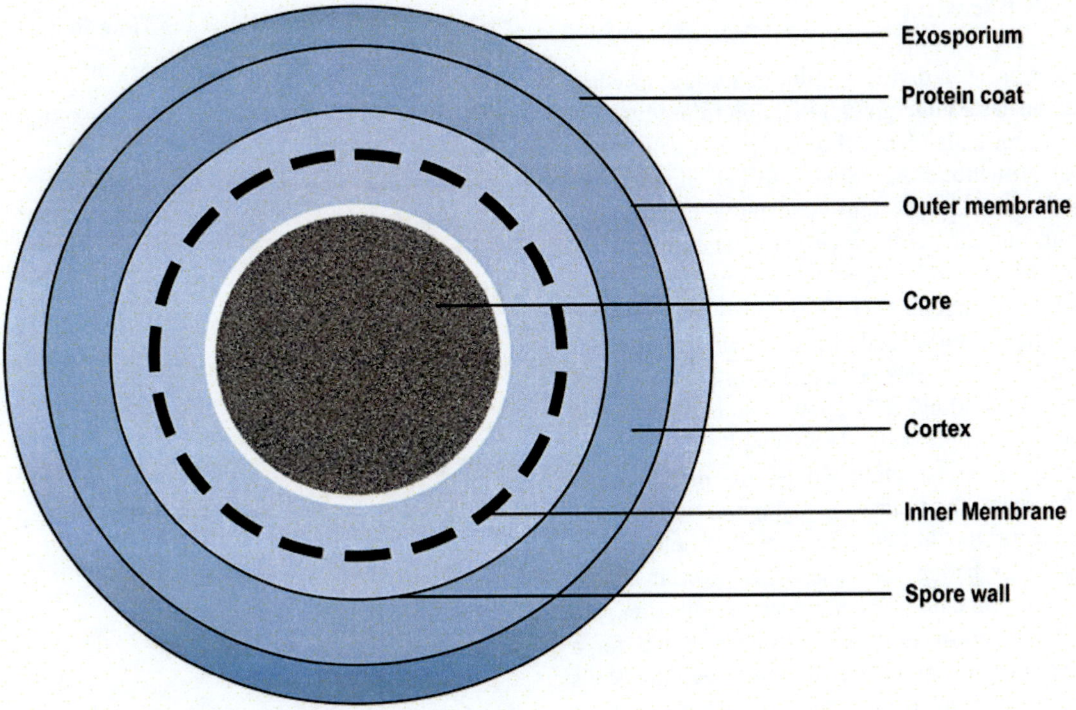

Fig. 22.7 Diagrammatic representation of bacterial spore

The spores may vary among different species depending on the spores' position, shape and relative size. For example, spores may be central (*Bacillus*), sub-terminal (*Cl. perfringens*) or terminal (*Cl. tetani*); maybe oval (*Cl. tertium*) or spherical (*Cl. tetani*) in shape; and maybe bulging (*Clostridium*) or non-bulging (*Bacillus*).

Properties of Spores

Bacterial spores are resistant to ordinary boiling, disinfectants and heating. Spores of all medically important bacteria are destroyed by autoclaving at 121 °C for 15 min. The conversion of a spore into a vegetative cell under suitable conditions is known as germination. The germination process occurs in three stages: activation, initiation and outgrowth.

Demonstration of Spores

Spores may be seen as refractile bodies in unstained preparations. Spores are usually demonstrated by staining with malachite green or carbol fuchsin. In Gram staining, spores appear as an unstained refractile body inside the cell. The spores appear as red acid-fast bodies on staining by modified ZN stain (using 0.25–0.5% sulphuric acid instead of 20% sulphuric acid).

The spores have certain uses in clinical microbiology. The spores of *Bacillus stearothermophilus* are used to indicate proper sterilisation by autoclaving. These spores are destroyed at a temperature of 121 °C for 10–20 min, the time required for autoclaving. The spores of certain bacteria, such as *B. anthracis*, are misused as bioterrorism agents.

Physiology (Growth and Multiplication) of Bacteria

Bacterial growth can be defined as an orderly increase of all the cell's chemical components. Cell multiplication is a consequence of growth that increases the number of bacteria making up a population or culture.

Multiplication

Most bacteria divide by **binary fission**, in which the bacteria undergo cell division to produce two daughter cells identical to the parent cell. Bacterial growth can be equated to cell number: one bacterium divides into two, these two produce four, and then eight and so on. The growth rate of a bacterium is therefore measured by measuring the change in bacterial number per unit time. During bacterial cell division, nuclear division takes place before cytoplasmic separation. Hence, dividing cells show two nuclear bodies inside a single cell. The cytoplasmic division occurs by constricting the plasma membrane or developing the transverse septum across the cell. Some of the bacteria arrange themselves in a particular pattern because of the partial attachment of the daughter cells after cell division.

Generation time is the time required for a bacterium to give rise to two daughter cells under optimum conditions. The generation time for most pathogenic bacteria, such as *E. coli*, is about 20 min. The generation time is longer (i.e. 20 h) for *M. tuberculosis* and the longest (i.e. 20 days) for *Mycobacterium leprae*. A bacterium replicates and multiplies rapidly, producing millions of cells within 24 h. For example, *E. coli* in about 7 h can undergo 20 generations and produce 1 million cells. In about 10 h, it can undergo 30 generations and produce 1 billion cells and in 24 h produces 10^{21} cells (Fig. 22.8). However, in actual practice, the multiplication of bacteria is arrested after a few cell divisions due to exhaustion of nutrients and accumulation of toxic products.

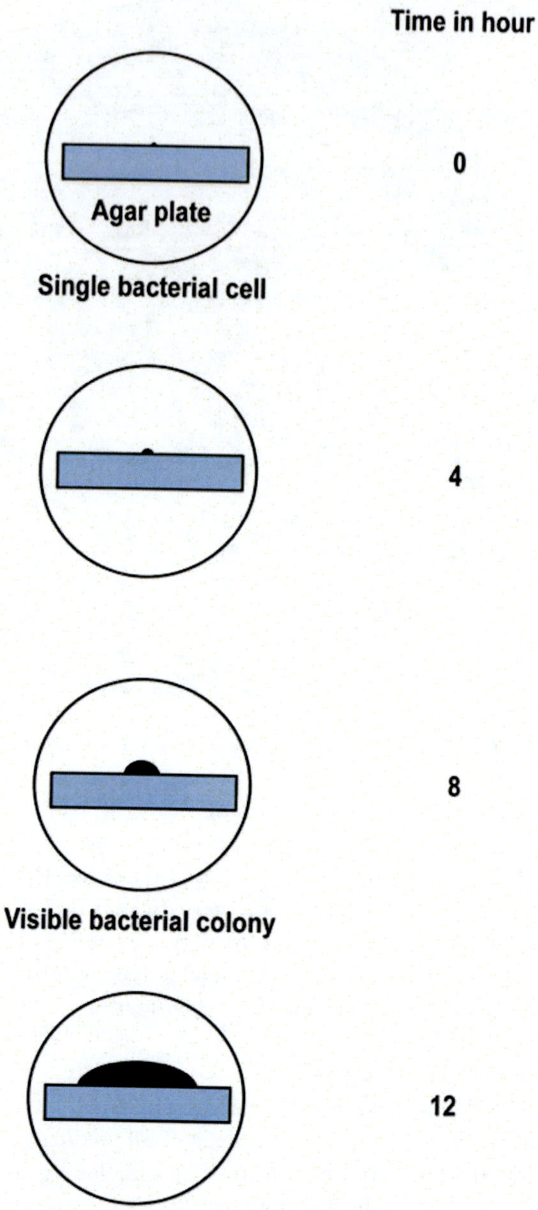

Fig. 22.8 Generation time of the bacteria

Bacterial Growth

When grown in a small vessel containing a liquid culture medium (**batch culture**), bacterial growth is arrested after a few cycles of cell divisions due to depletion of nutrient supplies or the accumulation of toxic by-products of bacterial metabolism. In certain research and industrial setups, maintenance of a **continuous culture** of bacteria is necessary, which can be accomplished with the help of a **chemostat** or **turbidistat**. These are special devices used to replenish depleted nutrients of the culture medium, and they also remove toxic metabolic by-products along with the dead bacterial cells.

In the case of pathogenic bacteria, the condition for multiplication inside the host tissue may be intermediate between batch culture and continuous culture; because the host may be a source of inexhaustible nutrients, but its immune system combats infection. In liquid culture, bacterial growth is diffused, but each bacterial colony represents a clone of cells derived from a single parent cell in solid media.

Bacterial Growth Curve

When a broth culture is inoculated with a small bacterial inoculum, the population size of the bacteria increases showing a classical pattern. The bacterial growth curve shows four distinct phases (Fig. 22.9).

1. **Lag phase**: After a liquid culture broth is inoculated, the multiplication of bacteria does not start immediately. It takes some time to multiply. The time between inoculation and the beginning of multiplication is known as **the lag phase**. In this phase, the inoculated bacteria become acclimatised to the environment, switch on various enzymes and adjust to the environmental temperature and atmospheric conditions. During this phase, there is an increase in the size of bacteria but no appreciable increase in the number of bacterial cells. The cells are active metabolically.
   The duration of the lag phase varies with the bacterial species, nature of culture medium, incubation temperature etc. It may vary from 1 h to several days.
2. **Log phase**: This phase is characterised by rapid exponential cell growth (i.e. 1 to 2 to 4 to 8). The bacterial population doubles during every generation. They multiply at their maximum rate. The bacterial cells are small and uniformly stained. The microbes are sensitive to adverse conditions, such as antibiotics and other antimicrobial agents.
3. **Stationary phase**: After the log phase, the bacterial growth almost stops completely due to lack of essential nutrients, lack of water oxygen, changes in pH of the medium etc. and accumulation of their toxic metabolic wastes. The death rate of bacteria exceeds the rate of replication of bacteria. Endospores start forming during this stage. Bacteria become Gram variable and show irregular staining. Many bacteria start producing exotoxins.
4. **Decline phase**: During this phase, the bacterial population declines due to the death of cells. The decline phase starts due to the accumulation of toxic products and autolytic enzymes and the exhaustion of nutrients. Involution forms are common in this stage.

The total and viable counts plotted on the bacterial growth curve run parallel up to the stationary phase. After that, viable count enters the

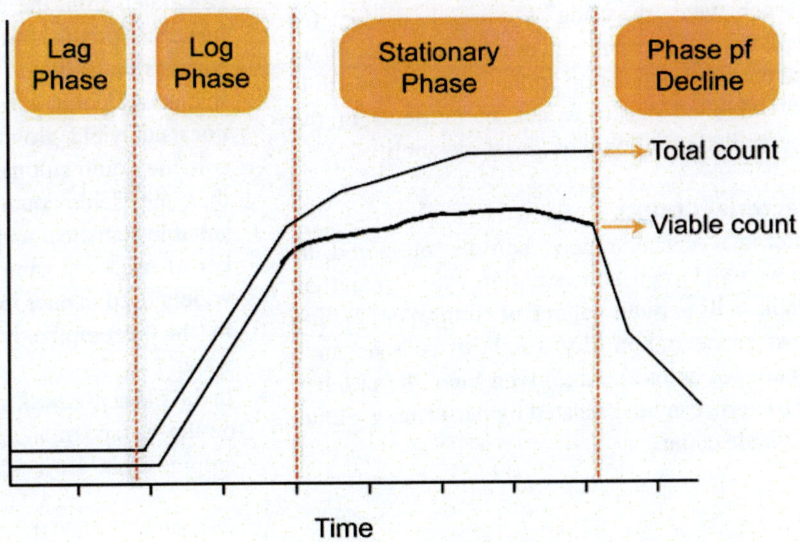

Fig. 22.9 Bacterial growth curve

Table 22.3 Bacterial growth rate during different phases of bacterial growth curve

Phase	Growth rate
Lag	Zero
Log or exponential	Constant
Stationary	Zero
Decline	Negative (death)

phase of decline, whereas total count continues steadily without any phase of decline. However, in the case of autolytic bacteria, even the total count shows the phase of decline.

Bacteria undergo various **morphological** and **physiological** changes during various phases of the growth curve. Towards the end of **the Lag phase**, bacterial cell attains maximum size. In **the Log phase**, bacterial cells appear smaller and stain uniformly. In **the Stationary phase**, bacterial cells are Gram variable. In this phase, bacteria possess plenty of intracellular storage granules responsible for uneven gram staining characteristics. Sporulation is common in this phase. Most of the bacteria produce secondary metabolites (e.g. exotoxins) and antibiotics. **Involution forms** are quite common in **the phase of decline.**

The growth rate during different phases of the bacterial growth curve is summarised in Table 22.3.

Continuous culture: It is a method of culture useful for industrial and research purposes. This is achieved by using a special device for replenishing nutrients and continuously removing the bacterial population so that bacteria growth is not inhibited due to a lack of nutrients or the accumulation of toxic bacterial metabolites.

Bacterial Counts

Microbial concentrations can be measured in terms of (1) cell concentration (the number of viable cells per unit volume of culture) or (2) biomass concentration (dry weight of cells per unit volume of culture). At a given time, the number of bacteria can be estimated by performing a total or viable count.

Total Count

This denotes the total number of bacteria in the sample, irrespective of whether they are living or dead. This can be obtained by (a) Counting chambers can be used for direct counting under the microscope. (b) Electronic devices such as Coulter counter can be used for total count determination. (c) Stained smears prepared by spreading the known volume of the culture over a measured area on the slide are also used for direct counting. (d) Absorptiometer or nephelometer uses opacity measurement for total count calculation. (e) Relative numbers of bacterial cells in a smear of a culture mixed with known numbers of other cells can also be used to compare and determine the total cont. (f) Wet weight or dry weight calculation: here, bacterial cells are separated by either centrifugation or filtration, and then wet or dry weight is calculated to determine total count by using appropriate formulae and (g) Chemical assay of cell components such as nitrogen is also being used for total count determination.

Viable Count

This usually indicates the number of living or viable bacteria. This count can be obtained by dilution or plating method.

- In the **dilution method**, several tubes with liquid culture media are incubated with varying dilutions of sample and the viable count is calculated from the number of tubes showing bacterial growth. In this method, the bacterial suspension is diluted by using several tubes so that at one point, a unit quantity of suspension does not yield growth upon inoculation into suitable liquid culture media (extinction). Further, the viable count can be calculated using suitable statistical methods based on the number of tubes showing growth. This method is widely used in microbiological testing of water for the presumptive coliform count in drinking water.
- In the **plate method**, a diluted sample is spread on the surface of an agar plate. The number of colonies that grow after a suitable incubation

timc indicates the viable count of the bacteria. The method described by Miles and Misra (1938) is in common use, in which serially diluted bacterial suspensions are poured on the Petri plate containing solid medium and colonies are counted to obtain the viable count.

Bacterial Nutrition

The minimum requirements for the growth of bacteria include water, a source of carbon, a source of nitrogen and certain inorganic salts. These are required for the synthesis of proteins, enzymes etc. For example, nitrogen is required to synthesise proteins, DNA, RNA and adenosine triphosphate (ATP). Sulphur is required for certain amino acids and vitamins, and phosphorus is required for nucleic acids, ATP and phospholipids. In addition, inorganic ions, such as potassium, sodium, iron, magnesium, calcium and chloride, are required to facilitate enzymatic catalysis and maintain chemical gradients across the cell membrane.

Some bacteria grow in a variety of simple media. *E. coli* and other members of the family Enterobacteriaceae are examples of bacteria that can grow in a variety of simple media containing the inorganic salts and with a source of energy, the simplest being glucose. The inorganic salts in the media provide major essential elements of carbon, hydrogen, oxygen, nitrogen, phosphate and sulphur. These chemicals are usually present in the media and are not added specifically.

On the other hand, *H. influenzae* and other related bacteria are very fastidious and have certain growth requirements. They require certain amino acids, vitamins and other growth factors that are supplied by adding yeast extract and meat digests to the media. They also require the addition of blood or serum for their growth.

Factors Affecting Growth of Bacteria

A variety of factors affect the growth of bacteria. These are discussed in the subsequent sections:

Oxygen

Bacteria can be broadly classified into aerobic and anaerobic based on their oxygen requirements.

Aerobic Bacteria

Aerobic bacteria are those which grow in the presence of oxygen. They may be (1) **Obligate aerobes**—which can grow only in the presence of oxygen (e.g. *Pseudomonas aeruginosa*); (2) **Facultative aerobes**—which are ordinary aerobes but can also grow without oxygen (e.g. *E. coli*). Most of the pathogenic bacteria are facultative aerobes; and (3) **Microaerophilic bacteria**—those bacteria that can grow in the presence of low oxygen and the presence of low (4%) concentration of carbon dioxide (e.g. *Campylobacter jejuni*).

Some fermentative organisms (e.g. *Lactobacillus plantarum*) are aerotolerant but do not contain the enzyme catalase or superoxide dismutase. Oxygen is not reduced, and therefore hydrogen peroxide (H_2O_2) and nascent oxygen (O_2^-) are not produced.

Anaerobic Bacteria

Obligate anaerobes are bacteria that can grow only in the absence of oxygen (e.g. *Clostridium botulinum, Clostridium tetani*). These bacteria lack the enzymes superoxide dismutase and catalase; hence oxygen is lethal to these organisms.

The apparent oxygen toxicity of anaerobic bacteria is a less understood mechanism. However, it is attributable to the accumulation of hydrogen peroxide and other toxic peroxides inside bacterial cells in the presence of oxygen. Aerobic bacteria possess an enzyme called **catalase** which converts harmful hydrogen peroxide into harmless products, but anaerobes are devoid of this enzyme. Another reason for apparent oxygen toxicity is that obligate anaerobes possess key enzymes of metabolism which work only in the reduced state, that is, in the absence of oxygen.

Bacterial metabolism plays a key role in understanding a bacterium's oxygen-dependent or independent nature. In aerobic bacteria, metabolic pathway intermediates and energy are obtained only by oxidation, and oxygen serves as the final hydrogen acceptor molecule, while in anaerobes final hydrogen acceptor molecule will be other than oxygen. Facultative anaerobes may act vice versa. Aerobes use atmospheric oxygen as the ultimate electron acceptor (aerobic respiration); the complex carbon and energy source may be completely oxidised to carbon and water, the simpler molecules. During this process, energy is generated by the production of energy-rich phosphate bonds and the conversion of adenosine diphosphate (ADP) to ATP; this process is called **oxidative phosphorylation**. In the case of anaerobic bacteria, nitrates or sulphates act as the final electron acceptor molecule instead of oxygen (anaerobic respiration).

Anaerobic metabolism involves a series of oxidoreduction reactions; here, carbon and energy sources act as electron acceptors and donors. This process is known as **fermentation**, and it yields various organic end products such as acids alcohols along with carbon dioxide and hydrogen gas. Fermentation facilitates energy production by forming energy-rich phosphate bonds by adding organic phosphate into intermediate metabolites. This process is called **substrate-level phosphorylation.** Further, these energy-rich phosphate bonds are utilised to convert ADP into ATP.

Oxidation-Reduction (Redox) Potential

It is a system that is very important in determining bacterial growth. The redox potential of the system indicates the readiness of all the components of a system to take part with electrons. The redox potential of a medium can be estimated by measuring the electrical potential difference set up between the medium and an unattackable electrode immersed in it; this electrical potential is measured in millivolts. On the other hand, oxidation-reduction indicators such as methylene blue can be used to measure the redox potential of a medium by noting the colour change in the medium. This is a simpler but less accurate method of estimating the redox potential.

Carbon Dioxide

The organisms that require higher amounts of carbon dioxide (CO_2) for their growth are called **capnophilic bacteria**. They grow well in the presence of 5–10% CO_2 and 15% O_2. In a candle jar, 3% CO_2 can be achieved. Examples of such bacteria include *H. influenzae, Brucella abortus.*

Temperature

The optimum temperature for most of the pathogenic bacteria is 37 °C. The optimal temperature, however, is variable. Depending on their temperature range, the growth of bacteria is grouped as follows:

Psychrophiles

These bacteria are cold-loving microbes that grow within a temperature range of 0–20 °C. Most soil and water saprophytes belong to this group.

Mesophiles

These are moderate temperature loving microbes that grow between 25 and 40 °C. Most pathogenic bacteria belong to this group.

Thermophiles

These are heat-loving microbes. They can grow at a high-temperature range of 55–80 °C. *B. stearothermophilus* is an example.

Bacterial viability to temperature may vary from one particular group of bacteria to other. In sterilisation, heat is being used as one of the important modalities of killing bacteria. Dry heat kills bacteria by oxidation and charring, whereas moist heat kills bacteria by coagulation and denaturation. **Thermal death point**: It is the lowest temperature that kills a bacterium under standard conditions in a given time. Under moist conditions, the thermal death point of most of the vegetative, mesophilic bacteria will be between 50 and 65 °C, but spores have a higher range of 100–120 °C. A very few bacterial species do not survive in low temperatures; otherwise, most survive well. Bacterial specimens stored in a refrigerator or deep freezer can be easily revived. Rapid freezing by dry ice (solid carbon dioxide) or storage and stabilisation with glycerol reduces the risk of death of bacteria during long-term storage.

pH

Most pathogenic bacteria grow between pH 7.2 and 7.6. Very few bacteria, such as lactobacilli, can grow at acidic pH below 4.0. Many food items, such as pickles and cheese, are prevented from spoilage by acids produced during fermentation. *V. cholerae* is an example of a bacteria that can grow at an alkaline (8.2–8.9) pH.

Light

Depending on the source of energy bacteria make use of, they may be classified as **phototrophs** (bacteria deriving energy from sunlight) or **chemotrophs** (bacteria deriving energy from chemical sources).

Osmotic Pressure

Microbes obtain almost all their nutrients in solution from the surrounding water. Hence factors such as osmotic pressure and salt concentration of the solution affect the growth of bacteria. Bacteria, by virtue of the mechanical strength of their cell wall, can withstand a wide range of external osmotic variations. Organisms requiring high osmotic pressures are called osmophilic bacteria. Sudden exposure of bacteria to a hypertonic solution may cause the withdrawal of water due to osmosis, leading to shrinkage of the protoplasm (**plasmolysis**). On the other hand, sudden transfer of bacteria from the concentrated solution to distilled water may cause excessive imbibition of water leading to swelling and bursting of the cell (**plasmoptysis**).

Certain lower forms of bacteria fail to grow in cell-free culture media and require living cells to grow. *Treponema pallidum* and *M. leprae* are two pathogenic bacteria that cannot grow in an artificial culture media but can only be cultured when inoculated into living animals.

Bacteriocins

In 1925, Gratia found the production of highly specific antibiotic compounds produced from a strain of *E. coli* that was active against another strain of the same species. Later on, such compounds produced by *E. coli* and other members of Enterobacteriaceae were called by the name **colicin**. Interestingly, the same phenomenon was observed in many other species of bacteria; hence these colicin-like substances produced by various bacteria were called by a generic name **bacteriocin**. Bacteriocin is a highly specific antibiotic-like substance produced by certain strains of bacteria which are active against other strains of the same or different species. Based on the species of origin, bacteriocins are called by specific names such as **colicins** from *E. coli*, **pyocins** from *P. aeruginosa*, **megacins** from *Bacillus megaterium* and **diphthericins** from *C. diphtheriae*.

Bacteriocins are made of proteins, but few of them have associated lipopolysaccharide contents derived from the bacteria's cell wall producing them. The bacteriocins resemble phages in many aspects; they adsorb specific bacterial surfaces targeting specific receptors, even some bacteriocins and phages have similar receptor sites on the bacterial cell wall. Under an electron microscope, bacteriocins like pyocins appear like the tail structures of phages. Hence, they may be reflected as a product of the defective genome of phages that had integrated with the bacteria's genome. Bacteriocins specifically kill certain strains of the same species of bacteria; hence these are used in bacterial typing of certain bacterial species such as *P. aeruginosa, Proteus* spp., *Shigella sonnei*.

The presence of **Col factors** (colicinogenic factor) determines the ability of bacteria to synthesise bacteriocins. Col factors are episomes, and they can be transferred from one bacterium to another through transduction and conjugation. Bacteria having col. factor can be induced for bacteriocin production by certain physical and chemical agents such as UV rays, nitrogen mustard.

Bacteriocin typing scheme mainly depends on the ability of the bacteriocin produced by the test strain of bacteria to kill the standard indicator strains of bacteria. On the other hand, in **phage typing**, the sensitivity of the test strains to the lytic action of the phage plays a major role. Generally, the plate diffusion technique is used for bacteriocin typing. In this method, test strain is inoculated as a broad streak at the centre of the solid culture medium. After incubation, the growth of the test bacterium should be scraped off completely, and the remaining bacterial cells are killed by exposing them to chloroform vapour. Further, standard indicator bacterial strains are streaked right angles to the original inoculum and plates are incubated. The pattern of inhibition of standard strain indicates the bacteriocin type of the test bacterium.

Antimicrobial Therapy 23

Any chemical used in treatment, relief or prophylaxis of disease is defined as a **chemotherapeutic drug** or agent. When chemotherapeutic drugs are given to control infection, the practice is termed **antimicrobial chemotherapy**. Antimicrobial drugs (also termed anti-infective drugs) are a special class of compounds capable even in high dilutions of destroying or inhibiting microorganisms. The origin of modern antimicrobial drugs is varied.

Antibiotics are substances produced by the natural metabolic processes of some microorganisms that can inhibit or destroy other microorganisms. **Synthetic antimicrobial drugs** are derived from dyes or other organic compounds through chemical reactions. Although division into these two categories has been traditional, they tend to overlap because most antibiotics, termed **semi-synthetic antibiotics**, are now chemically altered in the laboratory. The current trend is to use the term **antimicrobic** for all antimicrobial drugs, regardless of origin.

Antimicrobial drugs vary in their spectrum of activities. They may be broad-spectrum or narrow-spectrum antibiotics. **Broad-spectrum or extended-spectrum** antibiotics are active against a wider range of different microbes. For example, tetracyclines are active against various Gram-positive and Gram-negative bacteria, rickettsiae, mycoplasmas and even protozoa. **Narrow-spectrum** antibiotics are effective against one or very few microbes. For example, vancomycin is active against certain Gram-positive bacteria (such as staphylococci and enterococci) or griseofulvin, which is used only against fungal skin infections.

Table 23.1 provides a list of cephalosporins used in clinical practice.

Mechanisms of Action of Antimicrobials

Antimicrobial drugs may be bactericidal or bacteriostatic. Whereas a bacteriostatic drug and bactericidal drugs kill bacteria. They are very much useful in (1) life-threatening situations, (2) endocarditis, (3) patients with low polymorphonuclear count (below 500 μL^{-1}) and (4) conditions in which bacteriostatic drugs do not cause a cure.

Bacteriostatic drugs inhibit the growth of bacteria but do not kill them. They depend on the host defence mechanisms such as phagocytes to kill the bacteria. Hence these drugs are not used when the patient has too few neutrophils.

The mode of action of antimicrobial drugs varies depending on its site of inhibition. Some of the antimicrobials and their mode of action are classified as follows:

Inhibition of Protein Synthesis

The majority of antibiotics inhibit translation and prevent the formation of functional proteins

Table 23.1 List of cephalosporins used in clinical practice

	First generation	Second generation	Third generation	Fourth generation
Drugs	Cephalexin, cephalothin, cephradine, cephaloridine	Cefoxitin, cefaclor, cefamandole, cefuroxime, cefprozil, cefmetazole	Cefoperazone, cefotaxime, ceftazidime, ceftizoxime, ceftriaxone, cefixime	Cefpirome, cefepime
Antibacterial activity	**Gram-positive bacteria** *Staphylococcus aureus*, *Streptococcus* spp. except for enterococci **Gram-negative bacteria** *Escherichia coli*, *Klebsiella*, *Haemophilus influenzae*, *Proteus mirabilis*	First-generation spectrum + *Enterobacter* spp., *Serratia* spp., *Proteus vulgaris*, *Citrobacter* spp. + Gram-negative anaerobes	Second-generation spectrum + *Pseudomonas aeruginosa*, *Neisseria gonorrhoeae* including β-lactamase-producing strains	Third-generation spectrum + extended Gram-negative coverage including *Citrobacter* spp. and *Enterobacter* spp. resistant to third-generation cephalosporins

essential for cellular activities. These drugs target different steps in protein synthesis. Some examples are:

(a) Tetracyclines, doxycycline, prevents the binding of aminoacyl-tRNA by blocking the A (aminoacyl) site of the 30S ribosome.
(b) Aminoglycoside antibiotics like streptomycin prevent the formation of the 30S initiation complex. In contrast, Kanamycin and tobramycin bind to the 30S ribosome but block the formation of the larger 70S initiation complex. Hygromycin B specifically binds to the 30S subunit in a region that contains the A, P and E sites of tRNA.
(c) Erythromycin, a macrolide, binds to the 23S rRNA component of the 50S ribosome and interferes with the assembly of 50S subunits.
(d) Lincomycin and clindamycin inhibit peptidyl transferase activity and prevent elongation.
(e) Puromycin acts as an analogue of the 3-′-terminal ends of aminoacyl-tRNA. It disrupts synthesis and leads to premature chain termination.

Inhibition of Cell Wall Synthesis

Beta-lactam antibiotics like penicillin and cephalosporin inhibit bacterial cell wall synthesis. Likewise, glycopeptide antibiotics like vancomycin also inhibit cell wall synthesis. These cell wall synthesis inhibitors have a long history of use; however, resistance to these groups has emerged very rapidly due to misuse. Table 23.1 provides a list of cephalosporins used in clinical practice.

Due to its characteristic four-membered beta-lactam ring, the beta-lactam antibiotics inhibit cross-linking of bacterial cell wall components that essentially prevent the formation of a new cell wall. In the absence of a cell wall, the cell bursts due to osmotic pressures. The enzyme beta-lactamase degrades the antibiotic and protects the agent in some resistant bacteria.

In contrast, glycopeptide antibiotics like vancomycin bind the acyl-D-ala-D-ala residues that interlink two peptidoglycan layers of the cell wall. This binding prevents layers from cross-linking with each other and effectively reduces the strength of the cell wall. It results in dislodging of the cell wall and makes the cell vulnerable to osmotic pressure.

Table 23.2 Cellular target sites of antibiotics

Drug	Target site	Mechanism of action
β-lactams	Cell wall	Bactericidal, interfere with cross-linking of cell wall peptidoglycan molecules
Erythromycin, fusidic acid, tetracycline	Ribosomes	Bacteriostatic or bactericidal, interfere with translocation and attachment of t-RNA, thus inhibiting protein synthesis
Polyenes	Cytoplasmic membrane	Bacteriostatic, disrupt the yeast cell membrane
Metronidazole, idoxuridine, acyclovir	Nucleic acid replication	Bactericidal, interfere with DNA replication

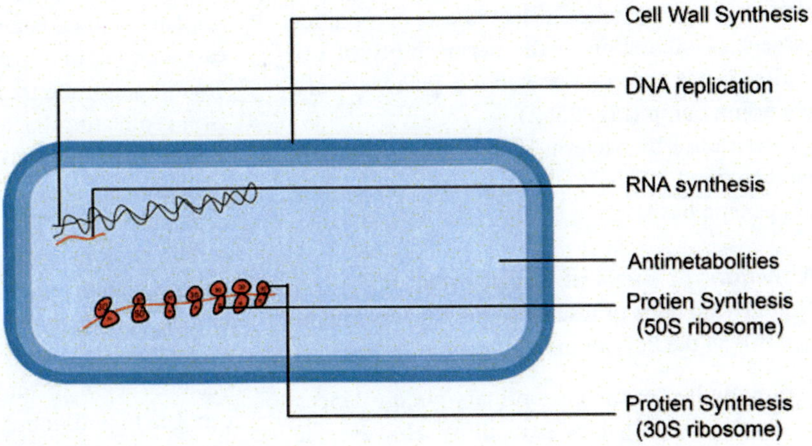

Fig. 23.1 Schematic diagram showing various mechanisms of actions of antibiotics

Inhibition of Nucleic Acid Synthesis

Nalidixic acid and fluoroquinolones are examples of antibiotics that inhibit nucleic acid synthesis. They belong to the group of antimicrobials known as quinolones. These antibiotics bind to the subunits of DNA gyrase and topoisomerase, thus preventing DNA coiling and inducing the formation of cleavage complexes. This results in the termination of the DNA synthesis process that essentially stops the downstream mechanisms, i.e. transcription and translations.

Action on the Bacterial Cell Membrane

Antibiotics (e.g. Nisin) and peptide antibiotics (like Polymyxin B or E) bind to the Lipid II and Lipid A of the cell membrane. Such binding alters the membrane's permeability by developing open pores in the membrane (for Nisin) or preventing the respiratory processes occurring on the membrane (for polymyxin B). It results in the loss of cytoplasmic components through the pore or prevents cellular respiration, which can kill the cell.

Cellular target sites of antimicrobial drugs are listed in Table 23.2 and Fig. 23.1.

Resistance to Antimicrobial Drugs

Antibiotic resistance is seen more commonly in hospital-acquired infections than in community-acquired infections. Antibiotic-resistant bacteria are more commonly seen in hospital environments due to the widespread use of antibiotics in hospitals that are selected for these bacteria. These hospital strains of bacteria are characterised by developing resistance to multiple antibiotics simultaneously. Common examples of such strains of bacteria showing drug resistance include hospital strains of *Staphylococcus aureus* and Gram-negative enteric bacteria, such as

Escherichia coli and *Pseudomonas aeruginosa*. Resistance to multiple antibiotics is mediated by plasmid-carrying several genes that encode enzymes responsible for the resistance.

Mechanisms of Antibiotic Resistance

There are many different mechanisms by which microorganisms might exhibit resistance to drugs. These are (1) production of enzymes, (2) production of altered enzymes, (3) synthesis of modified targets, (4) alteration of the permeability of cell wall, (5) alteration of metabolic pathways and (6) efflux pump (Fig. 23..2).

Resistance by bacteria against antibiotics may be classified as of (1) non-genetic basis and (2) genetic basis.

Non-genetic Basis of Resistance

Non-genetic basis of resistance plays a less important role in the development of drug resistance:

1. Certain bacteria under ordinary circumstances are usually killed by penicillins. However, if they lose their cell wall and become protoplast, these bacteria become non-susceptible to the action of cell-wall-acting drugs such as penicillins.
2. In certain conditions, such as in the abscess cavity, bacteria can be walled off, which prevents drugs from penetrating effectively into bacteria. However, surgical drainage of pus makes these bacteria again susceptible to antibiotics.
3. Presence of foreign bodies (such as surgical implants and catheters) and penetration injury caused by splinters and sharpeners make successful antibiotic treatment more difficult.
4. Non-replicating bacteria in their resting stage are less sensitive to the action of cell wall inhibitors such as penicillin and cephalosporins. This is particularly true for certain bacteria such as *Mycobacterium tuberculosis* which remains in the resting stage in tissues for many years, during which it is insensitive to drugs. However, when these bacteria begin to multiply, they become susceptible to antibiotics.

Fig. 23..2 Schematic diagram showing the mechanisms of drug resistance in antibiotics

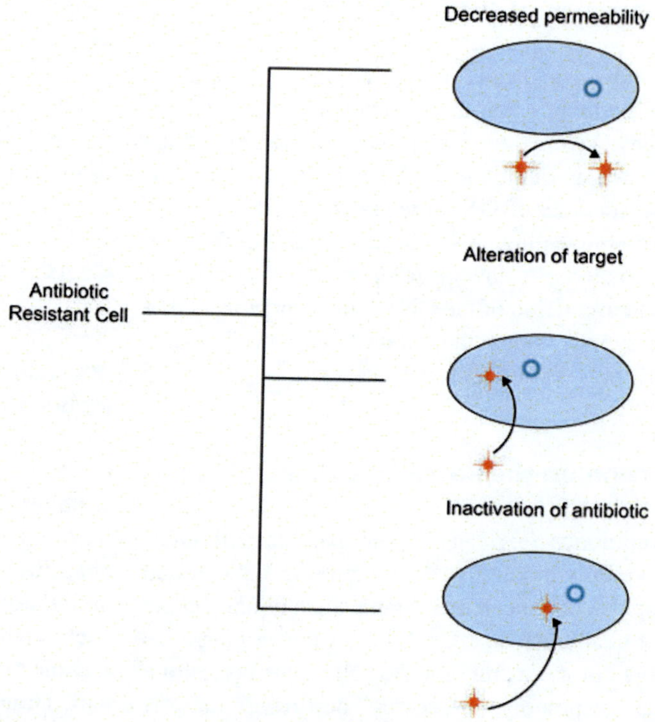

Genetic Basis of Drug Resistance

The genetic basis of drug resistance, mediated by a genetic change in bacteria, is most important in developing drug resistance in bacteria. This is broadly of three types: (1) chromosome-mediated resistance, (2) plasmid-mediated resistance and (3) transposon-mediated resistance.

Chromosome-Mediated Resistance

Chromosome-mediated resistance occurs as a result of **spontaneous mutation**. This is caused by a mutation in the gene that codes for either the target of the drug or the transport system in the membrane of the cell wall, which controls the entry of drugs into cells. The frequency of chromosomal mutation is much less than the plasmid-mediated resistance. It varies between 10^{-7} and 10^{-9}.

Mutational resistance in *M. tuberculosis* is of much importance in treating tuberculosis. Treatment of tuberculosis with two or more antitubercular drugs is carried out to prevent the emergence of multi-drug resistance in tuberculosis. This is based on the principle that it is most unlikely that the bacterium, *M. tuberculosis*, becomes concurrently resistant to all the antibiotics given simultaneously. There is a less likely resistant mutation because on treatment with antitubercular drugs, a mutant of *M. tuberculosis* resistant to one drug, and it appears, will be destroyed by the other drug.

Plasmid-Mediated Resistance

Plasmid-mediated drug resistance in bacteria occurs by the transfer of plasmid and genetic materials. It is mediated by a resistance plasmid, otherwise known as the R factor.

R Factors

These are circular, double-stranded DNA molecules that carry the genes responsible for resistance against various antibiotics. These factors may carry one or even two or more resistant genes. The genes encode various enzymes that destroy the antibiotics by degrading antibiotics or modifying the membrane transport systems. For example, the genes code for enzymes like β-lactamases that destroy the β-lactam ring (responsible for the anti-bactericidal action of β-lactam antibiotics, such as penicillins and cephalosporins).

Plasmid-mediated resistance plays a very important role in antibiotic usage in clinical practice. This is because a high rate of transfer of plasmids from one bacterium to another occurs by conjugation, plasmids mediate resistance to multiple antibiotics, and plasmid-mediated resistance occurs mostly in Gram-negative bacteria.

Transposons-Mediated Drug Resistance

Drug resistance is also mediated by transposons that often carry the drug resistance genes. Transposons are small pieces of DNA that move from one site of the bacterial chromosome to another and from the bacterial chromosome to plasmid DNA. Many R factors carry one or more transposons.

Specific Mechanisms of Resistance in Antibiotics

Vancomycin

Resistance to vancomycin is mediated by change in the D-ALA-D-ALA part of peptide in the peptidoglycan to D-ALA-D-lactate. This results in the inability of vancomycin to bind to the bacteria. Vancomycin resistance in *Enterococcus* spp. has been increasingly documented in different clinical conditions.

Vancomycin-Resistant Enterococci (VRE)

The first VRE was detected in 1986 in the European countries (the United Kingdom and France), and later in the United States, an *Enterococcus faecalis* clinical isolate was reported for VanB resistance. VRE has been reported and documented in many other countries such as Australia, Belgium, Denmark, Germany, Italy, Spain, Sweden, the Netherlands, Canada and Malaysia. However, the human VRE infections incidence rate is comparatively lower (1–3%) in European countries than in the United States, where the rate is high and alarmingly increasing.

In enterococci, seven types of glycopeptide resistance have been documented, namely, the

VanA, VanB, VanC, VanD, VanE, VanG and VanL, which are described based on their specific ligase genes (e.g. *vanA*, *vanB*). Related gene clusters have been found in non-pathogenic organisms. The common end point of these phenotypes is the formation of a peptidoglycan precursor with decreased affinity for glycopeptides, resulting in decreased inhibition of peptidoglycan synthesis. Peptidoglycan precursors ending in the depsipeptide D-alanyl-D-lactate are produced in VanA, VanB and VanD strains, whereas VanC, VanE and VanL (recently described in an *E. faecalis* strain) isolates produce precursors terminating in D-alanyl-D-serine, instead of the normally occurring D-alanyl-D-alanine.

The *vanA* gene cluster was originally detected in the Tn*1546* transposon. These or related genetic elements are usually carried by plasmids and occasionally by host chromosomes; this plasmid-borne transposon has also been found in clinical trials isolates of *S. aureus* (vancomycin-resistant *S. aureus* strains).

In enterococci, the glycopeptide resistance is categorised into the intrinsic class (characteristic of species) and the acquired class. The intrinsic class is the characteristic of the motile species such as *Enterococcus casseliflavus/flavescens* and *Enterococcus gallinarum*, members of which carry the naturally occurring *vanC-1* and *vanC-2/vanC-3* genes, respectively. These enterococci show variable minimum inhibitory concentrations (MICs) of vancomycin, with many falling in the susceptible range, and clinical failures have been reported following vancomycin use. In general, the isolation of these species does not require strict infection control isolation procedures unless they are highly resistant, suggesting the added presence of potentially transferable *vanA* or *vanB* genes.

Vancomycin-Resistant *Staphylococcus aureus* (VRSA)

In 1997, the first clinical isolate of *S. aureus* with diminished susceptibility to vancomycin (strain Mu50) was described in Japan. This strain displayed a vancomycin MIC of 8 μg/mL, which is in the range of intermediate susceptibility (4–8 μg/mL) as per the current Clinical and Laboratory Standards Institute (CLSI) breakpoints. It thus was referred to as vancomycin-intermediate *S. aureus* (VISA) or glycopeptide-intermediate *S. aureus* (GISA).

Others from various countries followed this initial report. The precursors of these VISA isolates are the heteroresistant-VISA (hVISA or hGISA) with sub-populations of cells that can grow at a vancomycin concentration of 8 μg/mL. As of late 2008, nine clinical isolates of VRSA harbouring the enterococcal *vanA* gene have been described in the United States, with MICs ranging from 32 to 1024 μg/mL.

Aminoglycosides

Three important mechanisms mediate resistance to aminoglycosides.

1. Plasmid-dependent resistance to aminoglycosides enzymes is the most important mechanism. It depends on the production of plasmid-mediated phosphorylating, adenylating and acetylating enzymes that destroy the drugs.
2. Chromosomal resistance of microbes to aminoglycosides is the second mechanism. Chromosomal mutation in genes results in the lack of a specific protein receptor on the 30S subunit of the ribosome, essential for binding the drug.
3. A "permeability defect" is the third mechanism of resistance. This leads to an outer membrane change that reduces the active transport of the aminoglycoside into the cell so that the drug cannot reach the ribosome. Often this is plasmid-mediated.

Tetracyclines

Resistance to tetracyclines occurs by three mechanisms: (1) efflux, (2) ribosomal protection and (3) chemical modification. The first two are the most important. Efflux pumps, located in the bacterial cell cytoplasmic membrane, pump the drug out of the cell. *Tet* gene products protect the ribosome, possibly through mechanisms that induce conformational changes. These conformational changes either prevent the binding of the tetracyclines or cause their dissociation from the ribosome. This is often plasmid controlled.

Macrolides

Resistance to macrolides, such as erythromycin, is caused by a plasmid-encoded enzyme that methylates the 23S ribosomal RNA, thereby blocking the binding of the drug.

Sulphonamides and Trimethoprim

Resistance to sulphonamide is caused by a plasmid-mediated transport system that actively exports the drug out of bacteria. It is also caused by a chromosomal mutation in the gene that codes for the target enzyme dihydrofolate synthetase, resulting in a reduced binding affinity of the drug.

Resistance to trimethoprim is caused by a chromosomal mutation in the gene coding for dihydrofolate reductase, the enzyme that reduces dihydrofolate to tetrahydrofolate.

Quinolones

Quinolone resistance occurs mainly due to chromosomal mutations that modify the bacterial DNA gyrase. Resistance is also caused by changes in the outer membrane proteins of the bacteria, which results in reduced uptake of the drugs into bacteria.

Penicillins and Cephalosporins

Three mechanisms mainly mediate resistance to penicillin: (1) production of penicillin-destroying enzymes, (2) mutation genes coding for PBP and (3) reduced permeability to the drug.

1. Production of Penicillin-Destroying Enzymes (β-Lactamases)

Resistance to penicillins may be determined by the organism's production of penicillin-destroying enzymes (β-lactamases). β-lactamases, such as penicillinases and cephalosporinases, open the β-lactam ring of penicillins and cephalosporins and abolish their antimicrobial activity. β-lactamases have been described for many species of Gram-positive and Gram-negative bacteria. Some β-lactamases are plasmid-mediated (e.g. penicillinase of *S. aureus*), while others are chromosomally mediated (e.g. many species of Gram-negative bacteria such as *Enterobacter* spp., *Citrobacter* spp., *Pseudomonas* spp.).

Certain species of Gram-negative bacilli, such as *Klebsiella pneumoniae* and *E. coli*, produce one particular group of β-lactamases. These enzymes confer bacteria the ability to hydrolyse the β-lactam rings of cefotaxime, ceftazidime or aztreonam and hence described as **extended-spectrum-lactamases.**

2. Mutation in Genes Coding for PBP

This form of resistance occurs due to the absence of some penicillin receptors (PBP) and occurs due to chromosomal mutation. Both low-level and high-level resistance that is seen in *Streptococcus pneumoniae* to penicillin G and *S. aureus* to nafcillin is due to this mechanism.

3. Reduced Permeability to Drug

Low-level resistance of *Neisseria gonorrhoea* to penicillin is caused by poor permeability of the drug. However, high-level resistance is mediated by a plasmid coding for penicillinase.

Cephalosporins are resistant to β-lactamases in varying degrees.

Metronidazole

Metronidazole is a bactericidal drug that acts by inhibiting DNA synthesis. It is effective against anaerobes and protozoa.

Chloramphenicol

Microorganisms resistant to chloramphenicol produce the enzyme chloramphenicol acetyltransferase that destroys drug activity. A plasmid usually mediates the production of this enzyme.

Antitubercular Drugs

Antitubercular drugs include isoniazid, ethambutol, rifampicin, pyrazinamide and streptomycin. Isoniazid is a bacteriostatic agent. It penetrates well into the tissue fluid and also acts on intracellular organisms. Resistance to isoniazid is mainly due to the loss of enzyme catalase that activates isoniazid to active metabolites that inhibit mycolic acid synthesis.

Ethambutol acts by interfering with RNA metabolism. The bacteria develop resistance due to mutation in the gene coding for arabinosyltransferase, which synthesises arabinogalactan in the mycobacterial cell wall. Rifampin inhibits RNA synthesis. The bacteria develop resistance due to mutation in the gene coding for DNA-dependent RNA polymerases. Pyrazinamide resistance is due to mutation of the gene coding for bacterial amidase, which converts it to its active form, pyrazinoic acid.

Therefore, combination therapy in tuberculosis is essential to prevent the emergence of drug resistance.

Carbapenems

Production of Carbapenemases, Including NDM Metallo-Beta-Lactamase

New Delhi metallo-beta-lactamase-1 (NDM-1) is an enzyme produced by bacteria and causes resistance to a broad range of bacteria range of beta-lactam antibiotics. They also comprise the carbapenem family of antibiotics, the mainstay of antibiotics used to treat antibiotic-resistant bacterial infections. The gene coding for the NDM-1 enzyme is one of the constituents of a large gene family coding for the carbapenemases, a beta-lactamase enzyme. The "superbugs" are those bacteria producing carbapenemases whose infections are difficult to treat, and generally, they are sensitive to polymyxins and tigecycline.

The first NDM-1 was reported in a *K. pneumoniae* clinical isolate in 2008 from an Indian-origin Swedish patient. Later, it was found in bacteria isolated from clinical samples of various countries such as Japan, India, Pakistan, United Kingdom, United States, Canada and Brazil. The Gram-negative bacilli, such as *E. coli* and *K. pneumoniae*, are the most common bacteria known to produce the NDM-1 enzyme; however, the gene responsible for this enzyme can transfer from one bacterial strain to another via horizontal gene transfer.

The NDM-1 was named after New Delhi, the capital city of India, as it was first described by a Swedish national who was sick with an antibiotic-resistant bacterial infection acquired in India in December 2009. This patient was not treated with antibiotics in a New Delhi hospital, but his infection was not cured. However, soon after his return to Sweden, the *K. pneumoniae* strain was identified with the novel gene for carbapenem resistant, following which a new mechanism of resistance was established stating that "clearly arose in India, but there are few data arising from India to suggest how widespread it is", though, the exact geographical region for the resistance has not been confirmed. On the other hand, a study conducted during March 2010 from a Mumbai hospital detected carbapenem-resistant bacterial isolates in patients mostly of *bla*NDM-1 gene.

In May 2010 in the United Kingdom, an individual from Coventry was reported with *E. coli* infection expressing the NDM-1 gene. The patient was of Indian origin and had undergone dialysis treatment in India approximately 18 months ago. The bacterium was highly resistant to all antibiotics in the initial tests, but later, it was found to be susceptible to colistin and tigecycline. Hence, the utilisation of many nations' healthcare systems by the international traveller and patients is believed to be the reason for the "rapid spread of NDM-1 gene with potentially serious consequences."

Antibiotic Sensitivity Testing

Antibiotic sensitivity testing is carried out to determine the appropriate antibiotic agent for a particular bacterial strain isolated from clinical specimens. Antibiotic sensitivity testing can be carried out by diffusion or dilution methods, as follows:

Disc Diffusion Tests

Disc diffusion tests are the most commonly used methods in a laboratory to determine the susceptibility of bacteria isolates to antibiotics. In this method, as the name suggests, discs impregnated with known concentrations of antibiotics are placed on an agar plate that has been inoculated with a culture of the bacterium to be tested. The plate is incubated at 37 °C for 18–24 h. After diffusion, the concentration of antibiotics usually remains higher near the site of the antibiotic disc but decreases with distance. Susceptibility to the particular antibiotic is determined by measuring the zone of inhibition of bacterial growth around the disc (Fig. 23.3).

Selection of media: The medium that supports both test and control strains is selected for antibiotic susceptibility testing of the bacteria. For example, Mueller–Hinton agar is used for testing Gram-negative bacilli and *Staphylococcus* spp., blood agar for *Streptococcus* spp. and *Enterococcus* spp. species, chocolate agar for *Haemophilus influenzae* and Wellcotest medium for sulphonamides and cotrimoxazole. The medium is prepared by pouring onto the flat horizontal surface of Petri dishes of 100 mm to a depth of 4 mm. The pH of the medium is maintained at 7.2–7.4. More alkaline pH increases the activity of tetracyclines, novobiocin and fusidic acid, whereas an acidic pH reduces the activity of aminoglycosides and macrolides, such as erythromycin. The plates after preparation may be stored at 4 °C for up to 1 week.

Preparation of the inoculum: For testing the antibiotic sensitivity, the bacteria are first isolated in pure culture on a solid medium. At least three to four morphologically similar bacteria colonies to be tested are touched and inoculated into the appropriate broth and incubated at 37 °C for 4–6 h. The bacterial suspension density in the broth is adjusted to 1.5×10^8 cfu/mL by comparing its turbidity with that of 0.5 McFarland opacity standard tube. The broth is inoculated on the medium by streaking with sterile swabs. A sterile cotton swab is dipped into the broth, and excess broth is removed by rotation of the swab against the sides of the tube above the fluid level.

Antibiotic discs: Only clinically relevant antibiotics are tested in antibiotic susceptibility tests. Antibiotic discs (6-mm filter paper discs) can be prepared from pure antimicrobial agents in laboratories or obtained commercially. The discs are applied with sterile forceps, a sharp needle or a dispenser onto the surface of the medium, streaked with test strains, and the reading is reported after incubating the plate for 18–24 h at 37 °C aerobically.

Types of Disc Diffusion Tests

Disc diffusion tests are of the following types:

Kirby–Bauer Disc Diffusion Method

Kirby–Bauer disc diffusion method is the most common method used routinely to determine the antibiotic sensitivity of bacteria isolated from clinical specimens. Both test and control strains are tested in separate plates in this method.

The test is performed by inoculating the test organism in a suitable broth solution, followed by incubation at 37 °C for 2–4 h. Then 0.1 mL of the broth is inoculated on the surface of the agar medium by streaking with a sterile swab. This method uses either nutrient agar or Mueller–Hinton agar in Petri dishes. The inoculated medium is incubated overnight at 37 °C. The susceptibility of the drug is determined from the zones of inhibition of bacterial growth surrounding the antibiotic discs. The diameters of the zone of inhibition are calculated with vernier callipers or a thin transparent millimetre scale to the nearest millimetre. The point of abrupt diminution of the zone is considered the zone edge. A maximum of six antibiotic discs are tested in a Petri dish of 85 mm size (Fig. 23.4).

Interpretation of the zone size is made as per the interpretation chart. Bacteria can be considered sensitive, intermediate or resistant to antibiotics depending on the zone size.

Stokes Method

This is a disc diffusion method, which uses built-in controls against many variables. In this method, the Petri dish containing the Mueller–Hinton agar is partitioned into three equal parts horizontally. The test strain is inoculated in the central portion of the plate, whereas the control strain is inoculated in the upper and lower third

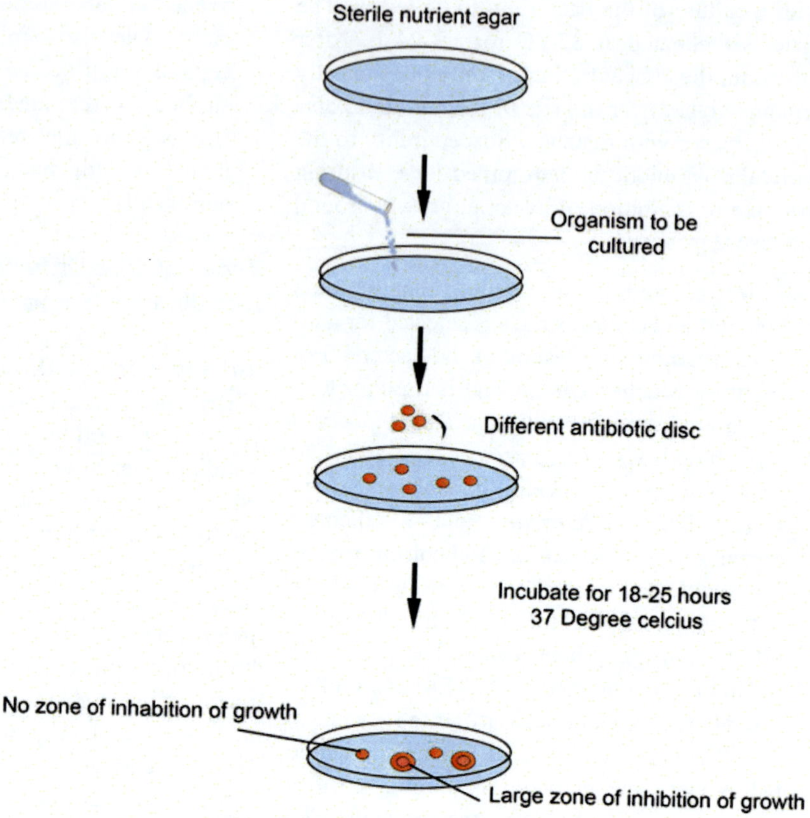

Fig. 23.3 Schematic diagram showing the performance of antibiotics sensitivity testing by disc diffusion method

portion of the plate. However, in the modified Stokes method, the control strain is inoculated in the central part, but test strains are inoculated on the upper and lower third of the plate. The plates are incubated at 37 °C and observed for zones of bacterial inhibition around the discs (Fig. 23.5). A maximum of six antibiotic discs can be applied on a 100 mm Petri dish.

The result is reported by comparing the zones of inhibition of test and control bacteria. The zone

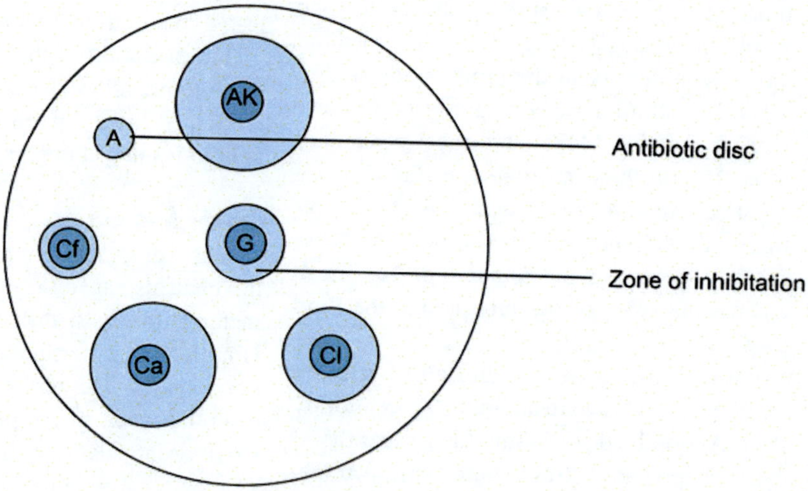

Fig. 23.4 Schematic diagram showing Kirby–Bauer disc diffusion method of antibiotic sensitivity

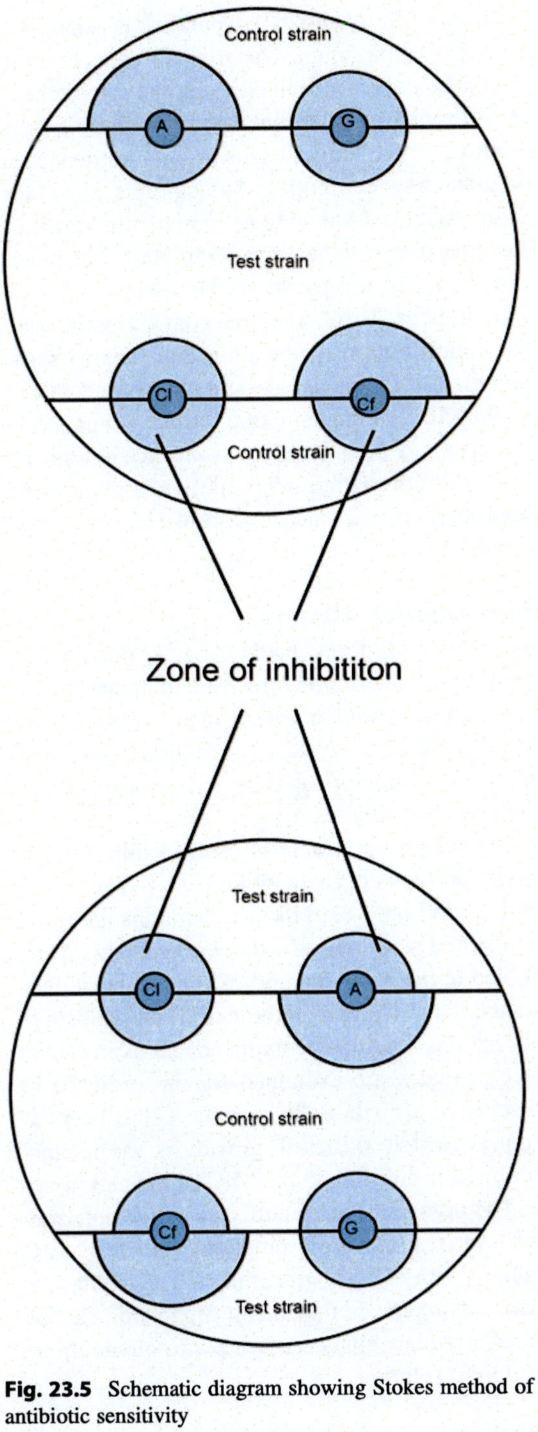

Fig. 23.5 Schematic diagram showing Stokes method of antibiotic sensitivity

sizes are measured from the edge of the disc to the edge of the zone. It is interpreted as follows:

Sensitive (S): The zone of the test bacterium is equal to or more than that of the control strain. The differences between the zone sizes of control and test strains should not be more than 3 mm if the zone size of the test bacterium is smaller than that of control.

Intermediate (I): The zone size of the test bacterium should be at least 2 mm, and the differences between the zone of test and control strain should be at least 3 mm.

Resistant (R): The zone size of the test bacterium is 2 mm or less.

Interpretation of Disc Diffusion Tests

Results of disc diffusion tests, such as Kirby–Bauer and Stokes method, are interpreted as follows:

Sensitive (S): Infection treatable by the normal dosage of the antibiotic.

Intermediate (I): Infection may respond to a higher dosage.

Resistant (R): Unlikely to respond to the usual dosage of the antibiotics.

However, for certain bacteria and antibiotic discs, the following may be kept in mind while interpreting results of disc diffusion tests:

1. *Proteus mirabilis*, *Proteus vulgaris* and other bacteria producing swarming produce a thin film on the agar surface, often extending into inhibition zones. In such situations, the zones of swarming should be ignored, and the clear outer margin should be measured to determine the zones of inhibition.
2. Many strains of MRSA grow very slowly in the presence of methicillin. They produce growth within the zone of inhibition on incubation for more than 48 h. This problem can be overcome by incubating the bacteria at 30 °C or by using 5% salt agar and incubating at 37 °C.
3. Penicillinase-producing *Staphylococcus strains* often fail to secrete enough enzymes to neutralise penicillin close to the antibiotic

disc. In such a situation, it may show a zone of inhibition, but with the presence of large colonies at the edge of the zone and without any gradual fading away of the growth of the bacteria towards the disc. In such conditions, the zones of inhibition should not be considered, and they should be reported resistant irrespective of the zone size.

4. Trimethoprim and sulfamethoxazole should be tested separately to know whether the bacterium is sensitive to both or only to one of these. These should never be tested in combination, the way these two drugs are used in combination in clinical practice.

Dilution Tests

Dilution tests are performed to determine the MIC of an antimicrobial agent. **MIC** is defined as the lowest concentration of an antimicrobial agent that inhibits the growth of organisms. Estimation of the MIC is useful to a) regulate the therapeutic dose of the antibiotic accurately in the treatment of many life-threatening situations, such as bacterial endocarditis and b) test antimicrobial sensitivity patterns of slow-growing bacteria, such as *M. tuberculosis*. Following methods are carried out to determine the MIC.

Broth Dilution Method

The broth dilution method is a quantitative method for determining the MIC of an antimicrobial agent that inhibits the growth of organisms *in vitro*. In this method, the antimicrobial agent has serially diluted in Mueller–Hinton broth by doubling dilution in tubes and then a standard suspension of the broth culture of the test organism is added to each of the antibiotic dilutions and control tube. This is mixed gently and incubated at 37 °C for 16–18 h. An organism of known susceptibility is included as a control. The MIC is recorded by noting the lowest concentration of the drug at which there is no visible growth, as demonstrated by the lack of turbidity in the tube. The main advantage of this method is that this is a simple procedure for testing a small number of isolates. The added advantage is that the bacteria's minimum bactericidal concentration (MBC) can be determined using the same tube.

The MBC is determined by sub-culturing from each tube, showing no growth on a nutrient agar without any antibiotics. Sub-cultures are made from each tube, showing no growth into the nutrient agar plates without any antibiotics. The plates are examined for growth, if any, after incubation overnight at 37 °C. The tube containing the lowest concentration of the drug that fails to show any growth on the sub-culture plate is considered as the MBC of the antibiotic for that strain. Broth dilution may be of two types—macrodilution and microdilution. Broth microdilution is done using microtiter plates and is considered the "gold standard."

Agar Dilution Method

The Agar dilution method is a quantitative method for determining the MIC of an antimicrobial agent against the test organism. It is useful to test organisms from serious infections like bacterial endocarditis or to verify equivocal results of the disc diffusion test.

Mueller–Hinton agar is used in this method. Serial dilutions of the antibiotic are made in agar and poured onto Petri dishes. Dilutions are made in distilled water and added to the agar melted and cooled to not more than 60 °C. One control plate is inoculated without antibiotics. The organism to be tested is inoculated and incubated overnight at 37 °C. Plates are examined for the presence or absence of growth of the bacteria. The concentration at which bacterial growth is completely inhibited is considered the MIC of the antibiotic.

The organisms are reported sensitive, intermediate or resistant by comparing the test MIC values with CLSI guidelines. The method's main advantage is that many organisms can be tested simultaneously on each plate containing an antibiotic solution.

Epsilometer Test (E-test)

Epsilometer test (E-test), based on the principle of disc diffusion, is an automated system for measuring the MIC of a bacterial isolate. In this method, a plastic absorbent strip is immobilised

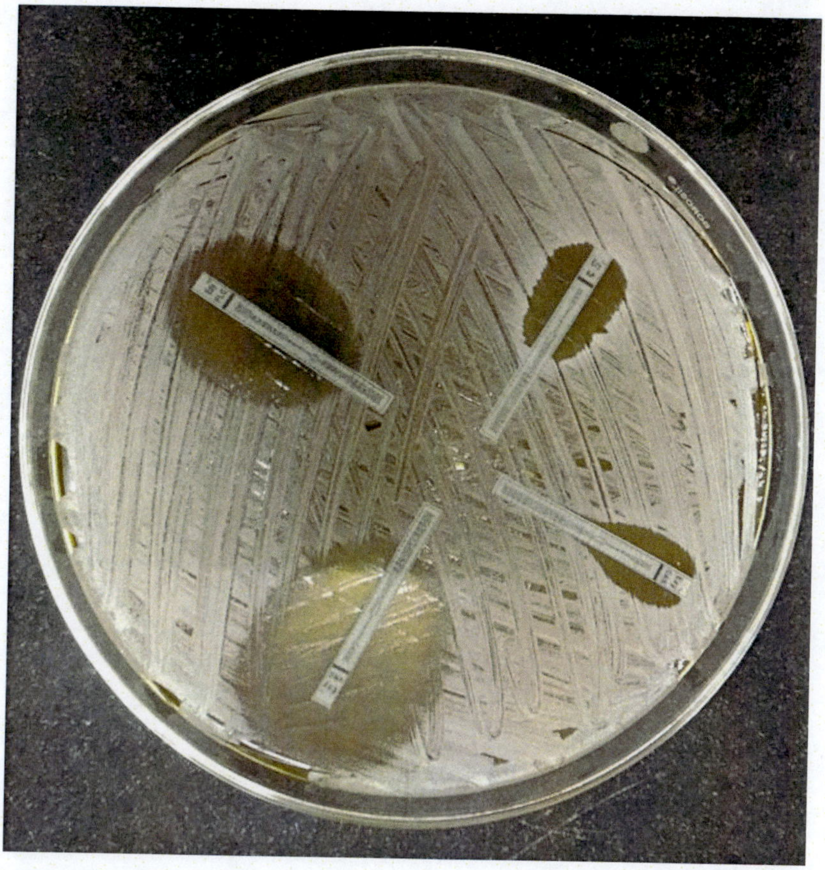

Fig. 23.6 Epsilometer test (E-test) (Courtesy of Prof S. Pramodhini)

with a continuous gradient of antibiotics on one side. On the other side, MIC interpretative scale is used corresponding to 15 twofold MIC dilutions. This strip is then placed on an agar plate inoculated with the test organism with the MIC scale facing towards the opening side of the plate. An elliptical zone of growth inhibition is seen around the strip after incubation at 37 °C overnight. The MIC is read from the scale at the intersection of the zone with the strip. The end point is always read at complete inhibition of all growth, including hazes and isolated colonies. E-test is a very useful test for easy interpretation of the MIC of an antibiotic (Fig. 23.6).

Antibiotics Assays in Body Fluids

Antibiotics assays are carried out to demonstrate antibiotics' toxic and therapeutic levels in the blood and other body fluids. In this method, the MIC of the antibiotic for the test organism is done first by the standard broth dilution method. The test with the patient's serum is done concurrently. Serial dilutions of the patients' serum in nutrient broth are carried out, and to each, a standard drop of culture is added and incubated for 18 h. The highest dilution of serum which has inhibited the growth of the assay organism is noted. This is equivalent to the MIC of the antibiotic in question. Multiplication of the MIC by the serum dilution gives the antibiotic concentration that was present in the patient's undiluted serum.

Laboratory Diagnosis of Bacterial Diseases

24

The clinical manifestations of bacterial diseases may be non-specific and protean. Nevertheless, they provide a clue in the diagnosis of bacterial diseases. Laboratory diagnosis, therefore, plays an important role in confirming the bacterial aetiology of suspected bacterial infections.

The procedures for the laboratory diagnosis, including identification of bacteria, involve the steps of (1) microscopic demonstration of bacteria in clinical specimens, (2) culture and isolation of bacteria in the culture media or laboratory animals, (3) identification of bacteria based on their growth characteristics in medium, and biochemical characteristics followed by antibiotic sensitivity, (4) detection of specific bacterial antigens and antibodies in serum and various other body fluids, (5) detection of nucleic acids of bacteria by molecular techniques, (6) typing of bacterial strains and (7) rapid identification methods. The identification of bacteria is supplemented by automation methods wherever applicable.

Conventional Tests in Diagnostic Bacteriology

Microscopy

The **wet mount preparation** is useful for demonstrating the motility of bacteria by light microscopy or demonstration of spirochetes by dark-ground microscopy. **Stained preparations** are used to demonstrate structural details of the bacteria by producing colour contrast.

Staining Methods

Various staining techniques are used in the microscopy of bacteria.

Simple Stains

Methylene blue and basic fuchsin are simple stains that provide colour contrast but impart the same colour to all the bacteria in a smear.

Negative Stains

Bacteria are mixed with dyes, such as Indian ink, that produce a uniform dark-coloured background against which the unstained organisms stand out in contrast. This is used to demonstrate the bacterial capsule that is usually not stained by simple stains. India Ink method for demonstration of the capsule is a common example.

Impregnation Stains

Cells and structures that are too thin to be visualised by the light microscope can be rendered visible by the impregnation of silver on their surface. The silver impregnation method is common for staining spirochetes, such as *Treponema pallidum*, *Leptospira*, *Borrelia*.

Differential Stains

Differential stains impart different colours to different bacteria or different bacterial structures.

The commonly used differential stains include Gram stain, acid-fast stain and Albert stain.

1. Gram Stain

Gram stain was devised by Christian Gram, a Danish microbiologist, in 1884, as a convenient method for classifying bacteria. Gram stain also provides useful information on the structure of bacterial cell envelopes.

Principle

The exact mechanism of Gram reaction is unclear, but various theories have been put forward. Gram-positive bacteria's protoplasm is more acidic than Gram-negative bacteria so it may be the reason for the strong binding of the basic primary dye to Gram-positive bacterial protoplasm. The cell wall of Gram-positive bacteria is made up of a thick peptidoglycan layer, which enables Gram-positive bacteria to retain the dye–iodine complex. The Gram-negative bacterial cell wall is rich in lipid content. Thus, decolourisation with organic solvents like acetone makes way for secondary dye contents.

Decolourisation is an important step in Gram staining because prolonged decolourisation may damage Gram-positive bacterial cell walls and make them appear Gram-negative. On the other hand, inadequate decolourisation may cause all cells to appear Gram-positive.

Procedure

1. Heat fix the slide, followed by *Primary staining* with basic dyes, such as methyl violet, crystal violet, for about 1 min.
2. *Application of mordant* in the form of a dilute solution of iodine for a minute.
3. *Decolourisation* with ethanol, acetone or aniline (10–30 s).
4. *Counterstaining* with acidic dyes, such as carbol fuchsin, safranine or neutral red for 30 s.

Gram staining is an essential procedure that is used in the identification of bacteria. The stain differentiates bacteria into two broad groups. **Gram-positive bacteria** are those that resist decolourisation and retain the primary dye–iodine complex, appearing violet. They have a relatively thick amorphous wall and more acidic protoplasm, which retain the basic violet dye and iodine complex within the cell. **Gram-negative bacteria** are decolourised by organic solvents and take counterstain, appearing red. The decolourising agent, such as acetone or ethanol, used during staining disrupts this membranous envelope, and the dye and iodine complex is washed out of Gram-negative bacteria.

Application

It is an essential and routine protocol used in diagnostic microbiology. Since Gram-positive and Gram-negative bacteria have different growth requirements, antibiotic susceptibility and pathogenicity, proper Gram staining and interpretation play a vital role in the diagnosis and treatment.

There are certain groups of bacteria, such as *Mycobacterium tuberculosis* and *Mycobacterium leprae* that cannot be considered typical Gram-negative or Gram-positive bacteria. This is because these bacteria either do not take up the Gram stain or have a different envelope type. These mycobacteria possess a waxy envelope containing complex glycolipids that make them impervious to the Gram stain.

2. Acid-Fast Stain

The acid-fast stain was discovered by Ehrlich, who found that after staining with aniline dyes, tubercle bacilli resist decolourisation with acid. The method was modified by Ziehl and Neelsen subsequently; hence it is widely known as Ziehl–Neelsen (ZN) stain.

Principle

Acid fastness is due to the (1) high content of lipids, fatty acids, components of mycolic acid, and (2) higher alcohols found in the cell wall of the *Mycobacterium*. Acid fastness also depends on the integrity of the cell wall.

Procedure

1. Fixed smears are first stained by a strong carbol fuchsin with heat application. Heating facilitates the entry of phenolic carbol fuchsin stain into the bacteria.

2. It is then decolourised with 5–20% (depending on the bacteria to be stained) sulphuric acid.
3. It is then counterstained with a contrasting dye, such as methylene blue. The acid-fast bacilli (AFB) retain the red colour of carbol fuchsin and appear bright red in stained smears. On the other hand, pus cells and epithelial cells present in the smear take up the blue colour of the counterstain and appear blue.

Application
The ZN smear is best used for staining bacteria such as *M. tuberculosis, M. leprae*, Nocardia and Actinomyces. It is also used for staining of oocysts of intestinal coccidian parasites (such as *Cryptosporidium, Cyclospora, Isospora*).

3. Albert Stain
Albert stain is used for staining the volutin granules of *Corynebacterium diphtheriae*.

Principle
Volutin granules of *C. diphtheria* have an affinity for basic dyes and are called metachromatic granules. These granules are stained bluish-black against green protoplasm on staining by Albert stain.

Procedure
1. Fixed smears are stained with Albert I stain; excesses stain is drained and washed.
2. Pour excess Albert II stain on the slide to cover the smear entirely.
3. Remove the excess stain and observe under microscopy.

Application
They are used to identify *C. diphtheriae*, the causative agent of diphtheria. These bacilli arrange themselves in V or L form. At the poles of these bacteria, **metachromatic granules** are present, which takes up Albert stain and helps in easy identification. Metachromatic granules are also called **polar bodies**, **volutin** or **Babes-Ernst granules**. These granules can also be stained by Neisser and Ponder staining technique.

Culture

Laboratory diagnosis of an infection is usually confirmed by isolating and culturing microorganisms in artificial media. Bacteria are cultured in either liquid (**broth**) or solid (**agar**) artificial media.

Culture Media
Koch pioneered the use of agar as a base for culture media. He developed the pour plate method and was the first to use solid culture media for the culture of bacteria. At first, Koch cultured bacteria on the sterile surfaces of cut, boiled potatoes. This was unsatisfactory because bacteria would not always grow well on potatoes. He then tried to solidify regular liquid media by adding gelatine. Separate bacterial colonies developed after the surface had been streaked with a bacterial sample. The sample could also be mixed with liquefied gelatine medium. After hardening the medium, separate colonies were produced by individual bacteria present in the sample. Although gelatine has several advantages, it has few drawbacks to be used as a solidifying agent, as gelatin was digested by most bacteria and melted if the temperature rises above 28 °C. One of Koch's assistants, Fannie Eilshemius Hesse, the wife of Walther Hesse, provided an easy alternative material for the gelatin problem. She recommended agar instead of gelatin as a solidifying agent, which she had been using to make jellies. Agar does not melt until the temperature reaches 100 °C and is not digested by bacteria. Also, the Petri dish (plate), a container for solid culture media, was developed by Richard Petri, one of Koch's assistants.

Ingredients of Culture Media
The basic constituents of culture media include the following.

1. Agar
Agar is the main component used universally for the preparation of solid media. It is obtained from various seaweeds and, after necessary processing, is usually available as a powder or as long shreds.

Agar has a unique property of melting at 98 °C and solidifying at 42 °C. The jellifying property of agar varies depending on the type of agar used (e.g. New Zealand agar is more jellifying than Japanese agar). Agar does not have any nutritive value and is also not affected by the growth of bacteria.

Agar is usually used in a concentration of 2–3%. Agar is hydrolysed at high temperatures and high acid or alkaline pH.

2. Peptone

Peptone is another important ingredient of cultural media. It is a complex mixture of partially digested proteins. It is obtained by digestion of lean meat or other protein materials (such as heart muscle, casein, fibrin or soy flour) with proteolytic enzymes (e.g. pepsin, trypsin or papain).

Peptone is an important source of nutrition for bacteria to grow. It contains peptones, proteoses, amino acids, inorganic salts (phosphates, potassium and magnesium) and certain accessory factors such as nicotinic acid and riboflavin. **Neopeptone** and **protease peptone** are special types of peptones with high nutritive value.

3. Other Ingredients

Other common media ingredients include water, sodium chloride and other electrolytes, meat extract, yeast extract, malt extract, blood and serum. Meat extract is available commercially as *Lab-Lemco*. It contains inorganic salts, carbohydrates, certain growth factors and protein degradation products. Blood or serum is usually used for enriching the culture of bacteria. Usually, 5–10% defibrinated sheep, or human blood is used.

Types of Culture Media

Culture media can be classified into several ways as follows:

1. Liquid Media, Semi-Solid Media and Solid Media

The media depending on their consistency can be liquid media, semi-solid media and solid media.

Liquid Media

Liquid media provide greater sensitivity for the isolation of small numbers of microorganisms. Examples of liquid media include peptone, nutrient broth, sugar media and enrichment media. Liquid media offer a few other advantages. Mixed cultures identification in liquid media need further sub-culturing on solid media; hence, individual, isolated colonies can be processed and identified. Quantification of bacterial growth cannot be done ordinarily in liquid media. Bacteria grown in liquid cultures often form colloidal suspensions. Uses of some common liquid media are summarised in Table 24.1.

Solid Media

Agar is used as a solidifying agent in most culture media. By varying the agar concentration, it is possible to make the medium solid or semi-solid. However, somewhat less sensitive than liquid media, solid media provide isolated colonies that can be quantified and identified. Some genera and species of bacteria can be recognised based on their colony morphologies.

Nutrient agar (Fig. 24.1) prepared by adding 2% agar to nutrient broth is the simplest and most common medium used routinely in diagnostic laboratories. The composition and uses of some common solid media are summarised in Table 24.2.

Semi-Solid Media

The addition of a reduced concentration of agar (0.2–0.4%) makes the medium semi-solid, which facilitates the spread of the bacteria in the medium. Examples of semi-solid media are mannitol motility medium and other motility testing mediums.

2. Simple, Complex, Synthetic and Special Media

Simple, complex, synthetic and special media are the different types of media used for bacteria culture.

Table 24.1 Composition and uses of some common liquid media

Medium	Composition	Uses
Peptone water	Peptone, sodium chloride, water	Routine culture, a base for sugar fermentation test, indole test
Nutrient broth	Peptone water, meat extract	Routine culture
Glucose broth	Nutrient broth, glucose	Blood culture, a culture of fastidious organisms such as streptococci
Brain heart infusion broth	Sodium citrate, sodium chloride, sodium phosphate, dextrose peptone, brain and heart infusion broth (ox), sodium polyanethol sulphonate (SPS)	Whole blood, bone marrow, body fluid culture
Alkaline peptone water	Peptone water (pH 8.6)	Enrichment medium for *Vibrio*
Selenite-F broth	Peptone water, sodium selenite	Enrichment medium for faeces for *Salmonella* and *Shigella*
Tetrathionate broth	Nutrient broth, sodium thiosulphate, calcium carbonate, iodine solution	Culture of faeces for *Salmonella*
Robertson cooked meat (RCM) broth	Nutrient broth, pre-digested cooked meat of ox heart	Anaerobic bacterial culture

Simple Media

The simple or basal media include nutrient broth and peptone water, which form the basis of other media.

The nutrient broth is an example of a simple liquid medium that consists of peptone, meat extract, sodium chloride and water. Adding 0.5% glucose to a nutrient broth makes it glucose broth.

Nutrient agar is an example of a simple solid medium. The medium is used routinely to isolate many bacteria from clinical specimens.

Complex Media

Most of the media other than basal media are usually known as complex media [e.g. chocolate agar, MacConkeyagar, Robertson cooked meat (RCM) medium, Lowenstein–Jensen (LJ) medium]. Complex media have some complex ingredients, consisting of many chemicals in unknown proportions. This is an undefined medium because the amino acid source contains various compounds with the exact composition unknown.

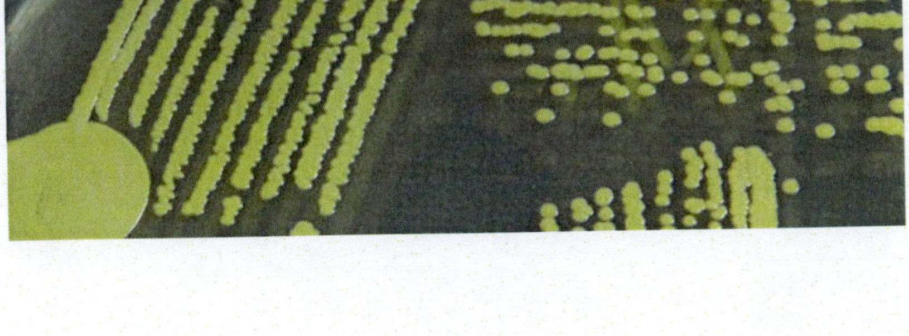

Fig. 24.1 Nutrient agar showing colonies of *Staphylococcus* spp. (Courtesy of Prof S. Pramodhini)

Table 24.2 Composition and uses of some common solid media

Medium	Composition	Uses
Nutrient agar	Nutrient broth, agar 2%	Routine culture
MacConkey medium	Peptone, lactose, sodium taurocholate, agar, neutral red	Culture of Gram-negative bacteria such as *Escherichia coli*
Blood agar	Nutrient agar, 5% sheep or human blood	Routine culture, the culture of fastidious organisms such as *Streptococcus* spp.
Chocolate agar	Heated blood agar	Culture of *Haemophilus influenzae* and *Neisseria*
Deoxycholate citrate agar	Nutrient agar, sodium deoxycholate, sodium citrate, lactose, neutral red etc.	Culture of *Shigella* spp. and *Salmonella* spp.
Thiosulphate citrate bile salt sucrose agar	Thiosulphate, citrate, bile salt, sucrose, bromothymol blue, thymol blue	Culture of *Vibrio cholerae*
Loeffler serum slope	Nutrient broth, glucose, horse serum	Culture of *Corynebacterium diphtheriae*
Lowenstein–Jensen medium	Coagulated hen's egg, mineral salt solution, asparagine, malachite green	Culture of *Mycobacterium tuberculosis*

Synthetic Media

A defined medium, also known as synthetic medium, contains known quantities of all ingredients. All the chemicals used are known, and they do not contain any animal, yeast or plant tissue. These media consist of trace elements, vitamins, and defined carbon and nitrogen sources required by certain microbes. Glucose or glycerol is often used as carbon sources and ammonium salts or nitrates as inorganic nitrogen sources.

Dubos medium with Tween 80 is an example of this medium.

Special Media

These include (1) enriched media, (2) enrichment media, (3) selective media, (4) indicator or differential media, (5) transport media and (6) sugar media.

Enriched Media

The enriched media are invariably solid media that facilitate the growth of certain fastidious bacteria. These media are prepared by adding substances like blood, serum and egg to the basal media to meet the nutritional requirements of more exacting and more fastidious bacteria. Blood agar, chocolate agar, Loeffler serum slope (LSS) and LJ medium are examples of enriched media. Blood agar is an enriched medium in which nutritionally rich whole blood supplements constitute the basic nutrients. The blood agar media heated slowly to 75 °C, turned brown colour, and hence is named chocolate agar.

1. **Brain–heart infusion broth (BHIB):** It is a highly nutritive media composed of nonenzymatic infusion from calf brain and cow heart along with peptones and dextrose. It can be used as a regular media and is also useful in isolating more fastidious bacteria. Brain–heart infusion agar can be obtained by adding agar to the broth.
2. **Blood agar:** As the name suggests, blood is an integral part of blood agar, and the remaining composition is the same as nutrient agar. Sheep, horse, cow and pig blood are the source of blood. In India, sheep blood is commonly used for blood agar preparation. Blood agar is a solid, enriched and non-selective medium. It is widely used for routine diagnostic purposes and supports the growth of aerobic and anaerobic bacteria and certain fungi. In the case of certain Gram-positive cocci, blood agar is used to assess haemolysis; thus, blood agar acts as a differential media. Haemolysis on blood agar can be categorised into three types (Fig. 24.2):

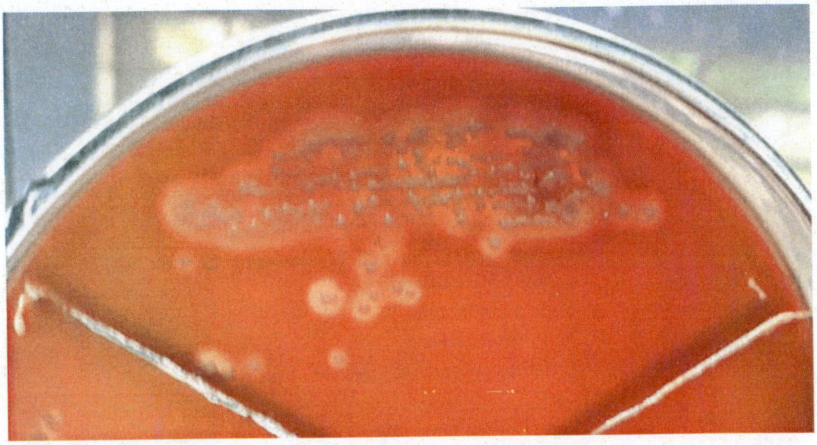

Fig. 24.2 Blood agar showing beta haemolytic colonies of group A *Streptococci*

(a) **Beta haemolysis:** Certain bacteria such as *Staphylococcus aureus* show complete lysis of red blood cells and haemoglobin around their colonies on blood agar. This kind of clear haemolysis is called beta haemolysis.
(b) **Alpha haemolysis:** Certain bacteria such as viridans streptococci show partial lysis of red blood cells and haemoglobin around their colonies on blood agar, resulting in greenish discolouration. This kind of partial, greenish discolouration is regarded as alpha haemolysis.
(c) **Gamma haemolysis:** In gamma haemolysis, there will not be any haemolysis around the bacterial colonies grown on a blood agar medium.

3. **Chocolate agar**: This solid medium is similar in composition to blood agar, but because of heating, contents of RBCs leak out and give chocolate colour to the medium; hence it is called chocolate agar. During heating, RBCs release substances like haemoglobin, haemin (**X factor**) and nicotinamide adenine dinucleotide (**V factor**) into the medium. These contents are necessary for isolating fastidious organisms such as *Haemophilus influenzae, Neisseria meningitides, Neisseria gonorrhoeae, Moraxella* spp. etc.

Enrichment Media

Enrichment media are liquid media that stimulate the growth of certain bacteria or suppress the growth of others for isolation of desired pathogenic bacteria. Commensal bacteria, such as *Escherichia coli* present in faeces tend to overgrow pathogenic ones in stool specimens. In such situations, enrichment media (such as selenite-F broth or tetrathionate broth) are used to isolate *Salmonella* Typhi and *Shigella* spp. from faeces. Enrichment media are useful for isolating wanted bacteria from the stool and other specimens containing more than one species of bacteria.

Selective Media

These are solid media that contain substances that inhibit the growth of all but a few bacteria but at the same time facilitate isolation of certain bacteria:

Thayer-Martin medium: This medium is used to isolate *N. gonorrhoeae* from clinical samples selectively. It contains 3.0 mg vancomycin, 7.5 mg colistin and 12.5 units of nystatin per millilitre of agar. These antimicrobials at this concentration suppress the growth of microorganisms other than *N. gonorrhoeae*.

LJ medium: It is used for the Mycobacterial culture. LJ medium is an egg-based medium consisting of hen's eggs, mineral salts, asparagine, glycerol and malachite green. Malachite green inhibits most of the microorganisms present in the sputum, thus selectively supporting the growth of the *Mycobacterium* on the LJ medium. LJ medium is the primary isolation medium for *Mycobacterium*, and

Table 24.3 Some selective media used in routine microbiology laboratories

Medium	Colony characteristics		Bacteria inhibited
Mannitol salt agar	Big yellow colonies of *Staphylococcus aureus*	Small pink colonies of *Staphylococcus epidermidis*	*Streptococcus* spp.
Thayer–Martin medium	Grey colonies of *Neisseria meningitidis* and *Neisseria gonorrhoeae*		Gram-positive cocci
MacConkey agar medium	Lactose fermenters: red colonies, for example, *Escherichia coli*	Lactose non-fermenters: colourless colonies of *Salmonella* spp.	Gram-positive cocci
Thiosulphate citrate bile salt sucrose agar	Sucrose fermenter: yellow colonies of *Vibrio cholerae*	Sucrose non-fermenters: green colonies of *Vibrio parahaemolyticus*	Enteric bacilli
Charcoal yeast extract	Cut glass colonies of *Legionella* spp.		Gram-positive cocci
Lowenstein–Jensen medium	Rough, tough and buff colonies of *Mycobacterium tuberculosis*	Smooth and pigmented colonies of atypical *Mycobacterium* spp.	Cocci
Sabouraud dextrose agar	Most fungi		Most bacteria

sterilisation of LJ medium can be done by inspissation.

Some selective media used in routine microbiology laboratories are summarised in Table 24.3.

Indicator or Differential Media

Indicator or differential media distinguish one microorganism from another growing on the same media by their growth characteristics. These media depend on the biochemical properties of a microorganism growing in the presence of specific nutrients or indicators such as neutral red, phenol red, eosin or methylene blue. The medium changes colour when a bacterium grows in them. For example, *S. typhi* grows as black colonies on Wilson and Blair's medium containing sulphite. With respect to certain media, terms used for media can be interchangeable. For example, blood agar is an enriched media, but it can also differentiate haemolysis produced by the Gram positive. Thus, blood agar can be called enriched and/or differential media. Eosin methylene blue (EMB), the medium used for differential for lactose and sucrose fermentation, and MacConkey, differential for lactose fermentation are a few examples.

MacConkey agar: Primary use of MacConkey agar is to differentiate Gram-negative bacteria based on their ability to ferment lactose present in the medium. It also inhibits the growth of Gram-positive bacteria. Lactose fermenting bacteria such as *E. coli* produce pink colonies, whereas non-lactose fermenting bacteria such as *Salmonella* spp. form pale or colourless colonies on MacConkey agar (Fig. 24.3). Fermentation of lactose in the medium makes it acidic and leads to the formation of pink colonies in the presence of neutral red.

Transport Media

Transport media are used to maintain the viability of certain delicate organisms in clinical specimens during their transport to the laboratory. They typically contain only buffers and salt. They lack carbon, nitrogen and organic growth factors, and hence do not facilitate microbial multiplication. Stuart transport medium for *N. gonorrhoeae* is an example.

Sugar Media

Sugar media contain 1% "sugar", which in microbiology denotes any fermentable substance such as glucose, sucrose, lactose and mannitol routinely used for fermentation tests. It consists of 1% sugar in peptone. The indicator used in sugar media is the Andrade indicator that consists of 0.005% acid fuchsin in 1 N NaOH. The production of acid after fermentation of sugar is

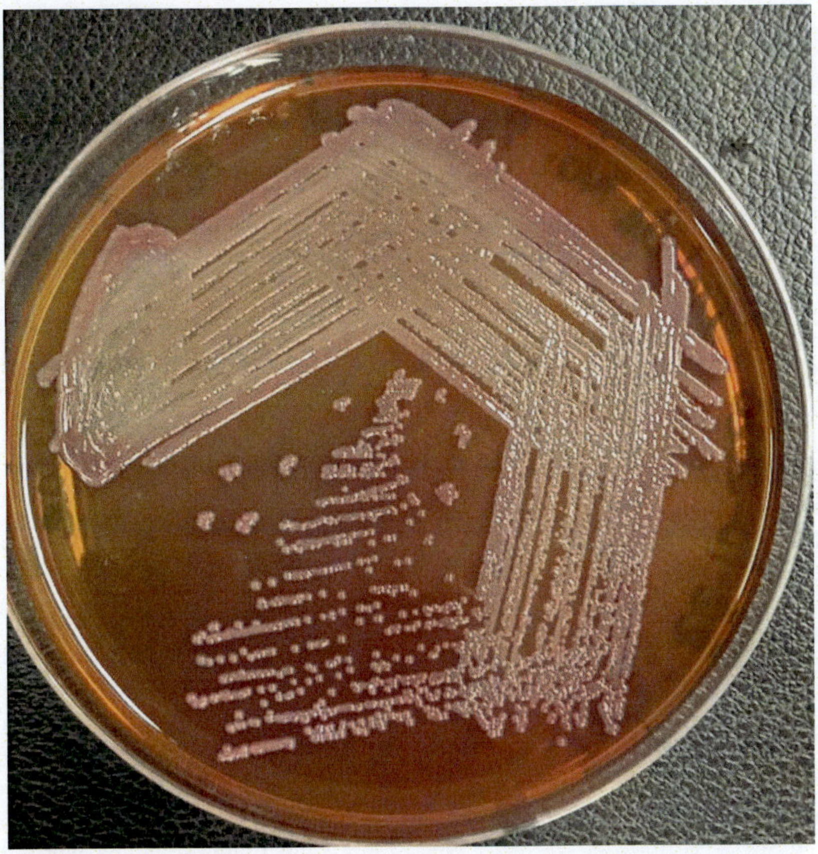

Fig. 24.3 MacConkey agar showing growth of lactose fermjenting colonies

indicated by the change of colour of the medium to pink due to the presence of an indicator. Durham tube is kept inverted inside the sugar tube to demonstrate gas production. Production of gas is indicated by the demonstration of gas bubbles in the Durham tube.

Anaerobic Media

These media support the growth of anaerobic/facultative anaerobic bacteria.

RCM medium is routinely used for anaerobic bacterial culture. Further, the identification of anaerobic bacteria requires special media to demonstrate their biochemical properties. Neomycin blood agar is also commonly used in anaerobic culture; it helps to isolate the bacteria grown in RCM.

Thioglycolate medium is semi-solid. It supports the growth of aerobes, facultative anaerobes and anaerobes at the top, middle and bottom of the medium, respectively. This medium forms an oxygen gradient from top to bottom, thus facilitating the growth of organisms of all oxygen requirements. This medium is boiled and cooled before use to eliminate its oxygen. It contains an oxygen indicator that turns pink or blue if exposed to oxygen. The stab culture method is used to inoculate this medium.

Incubation of Culture Media

For bacteria, two agar plates should be inoculated and kept both in aerobic and anaerobic conditions at 37 °C for a minimum of 48 h. Aerobic culture can usually be regarded as negative if no growth is present after 48 h of incubation. In the case of anaerobic culture, 7 days of incubation is sufficient to declare no growth. If *Propionibacterium acnes* is suspected, anaerobic culture plates should be incubated for 14 days. Nowadays, dehydrated media are available for wide use in diagnostic laboratories because of the simplicity of procedure and are less cumbersome to prepare. These dehydrated media are prepared by simply reconstituting in distilled water and sterilising it before use.

Culture Methods

Culture methods are very crucial in a microbiology laboratory. Various culture methods are carried out to (1) isolate bacteria in pure culture and identify the same by performing various tests, (2) demonstrate biochemical, antigenic and other phenotypic and genomic properties of the isolated colonies. (3) demonstrate the susceptibility of the isolated bacteria to antibiotics, bacteriophages, bacteriocins etc. (4) prepare antigens for various uses; and (5) maintain stock culture.

Methods of Culture for Aerobic Bacteria

Various methods are used for culturing of bacteria. These include (1) streak culture, (2) lawn culture, (3) pour-plate culture, (4) stroke culture, (5) stab culture and (6) liquid culture.

1. Streak Culture

Streak culture is the most useful method for obtaining discrete bacteria colonies. It is carried out by streaking on the surface of a solid media plate using a platinum or nichrome loop of 2–4 mm diameter.

In this method, a loopful of the inoculum is placed near the peripheral area of the plate. The inoculum is then spread with the loop to about one-fourth of the plate with close parallel strokes. From the primary inoculum, it is spread thinly over the plate by streaking with the loop in parallel lines. The loop is flamed and cooled between the streaks to obtain isolated colonies. The inoculated culture plate is incubated at 37 °C overnight to demonstrate colonies. Confluent growth occurs at the primary inoculum but becomes progressively thinner, and well-separated colonies are demonstrated on the final streaks of the inoculum. Single isolated colonies obtained by this method are very useful for studying various bacteria properties.

2. Lawn Culture

The lawn culture provides a uniform layer of bacterial growth on a solid medium. It is carried out by flooding the surface of the solid media plate with a liquid culture or suspension of bacteria, pipetting off the excess inoculum and finally incubating the plate overnight at 37 °C. Alternatively, the culture plate may be inoculated by a sterile swab soaked in liquid bacterial culture or suspension and incubated overnight to demonstrate the bacterial colonies.

The lawn culture method is useful (1) to carry out antibiotic sensitivity testing by disc diffusion method, (2) to carry out bacteriophage typing, and (3) to produce a large amount of bacterial growth required to prepare bacterial antigens and vaccines.

3. Pour-Plate Culture

The pour-plate culture is used to determine an approximate number of viable organisms in liquids, such as water or urine. It quantifies bacteria in urine cultures and estimates the viable bacterial count in a suspension.

This method is carried out in tubes containing 15 mL of molten agar. The molten agar in tubes is left to cool in a water bath at 45 °C. The inoculum to be tested is diluted in serial dilution. Then 1 mL each of diluted inoculum is added to each tube of molten agar and mixed well. The contents of tubes are poured into sterile Petri dishes and allowed to set. After overnight incubation of these Petri dishes at 37 °C, colonies are distributed throughout the medium's depth, which can be counted using a colony counter.

4. Stroke Culture

Stroke culture provides a pure growth of bacteria for carrying out slide agglutination and other diagnostic tests for bacterial pathogenicity. It is carried out in tubes usually containing nutrient agar slopes.

5. Stab Culture

Stab culture is prepared by stabbing the medium in tubes with a long, straight wire and incubating at 37 °C. Stab culture is frequently used for

(1) maintaining stock cultures, (2) demonstration of oxygen requirement of bacteria and (3) demonstration of gelatin liquefaction of bacteria.

6. Liquid Culture

Liquid culture is prepared in liquid media enclosed in tubes, flasks or bottles. The medium is inoculated by touching with a charged loop or adding the inoculum with pipettes or syringes and incubating at 37 °C, then subculturing onto solid media for final identification.

Liquid culture is specifically used (1) for blood culture and for sterility tests, where the concentration of bacteria is expected to be mallei (2) for the culture of specimens containing antibiotics and other antibacterial substances, as these are rendered ineffective by dilutions in the medium, and (3) when large yields of bacteria are required. A major disadvantage of liquid culture is that it does not provide a pure culture of the bacteria, and also, the bacterial growth does not exhibit special characteristic appearances.

Methods of Culture for Anaerobic Bacteria

Obligate anaerobes are bacteria that can live only in the absence of oxygen. These anaerobes are killed when exposed to the atmosphere for as briefly as 10 min. Some anaerobes are tolerant to small amounts of oxygen. On the other hand, facultative anaerobes are those anaerobes that grow with or without oxygen.

Anaerobic bacterial culture is used to grow anaerobes from a clinical specimen. Culture and identification of anaerobes are essential for initiating appropriate treatment. The failure to do so may have serious consequences, such as amputation, organ failure, sepsis, meningitis and even death.

1. Specimen Collection

Specimens frequently used for anaerobic culture include (1) blood, bile, bone marrow, cerebrospinal fluid, direct lung aspirate and tissue biopsy from a normally sterile site; (2) fluid aspirated from a normally sterile site, such as a joint; (3) pus specimens from a dental abscess, burn wound, abdominal or pelvic abscess; and (4) specimens from the knife, gunshot or surgical wounds.

Collecting a contamination-free specimen and protecting it from oxygen exposure during collection forms the mainstay of anaerobic culture. The specimens must be obtained from an appropriate site without contaminating the sample with bacteria from the adjacent skin, mucous membrane, or tissue.

Abscesses or fluids are usually collected using a sterile syringe and then tightly capped to prevent air entry. Tissue samples are placed into a degassed bag and sealed or into a gassed-out screw-top vial that may contain oxygen-free pre-reduced culture medium and tightly capped. The specimens need to be plated as rapidly as possible onto culture media to isolate bacteria.

Swabs are always avoided when collecting specimens for anaerobic culture because cotton fibres may be detrimental to anaerobes. Coughed throat discharge (sputum), rectal swab, nasal or throat swab, urethral swab and voided urine are some of the specimens that are not suitable for processing anaerobic cultures.

2. Culture Media

The commonly used media for anaerobic culture include Robertson cooked meat broth, thioglycolate broth, Willis and Hobbs media and neomycin blood agar. Robertson cooked meat (RCM) broth is the most widely used medium in an anaerobic culture. It consists of nutrient broth and pieces of fat-free minced cooked meat of ox heart with a layer of sterile liquid paraffin over it. Unsaturated fatty acids and even glutathione and cysteine in the meat extract utilise oxygen for auto-oxidation. The medium before inoculation is usually boiled at 80 °C in a water bath to make the medium free of oxygen. The media after inoculation and incubation allows the growth of even strict anaerobes and indicates their saccharolytic or proteolytic activities as meat is turned red or black, respectively.

Methods of Anaerobiasis

Anaerobic cultures are carried out in an environment free of oxygen, followed by incubation at 95 °F (35 °C) for at least 48 h before the plates are examined for growth. The cultures of anaerobic bacteria are carried out as follows:

(a) **McIntosh and Fildes anaerobic jar:** It is the most widely used and dependable method of anaerobiosis. It consists of a glass or metal jar with a metal lid that can be clamped airtight with the help of a screw. The lid has one inlet tube and another outlet tube. The outlet tube is connected to a vacuum pump by which the air is evacuated out of the jar. The inlet tube is connected to a source of hydrogen supply. The lid has two electric terminals also that can be connected to an electric supply. The underside of the lid contains a catalyst (e.g. alumina pellets coated with palladium) that catalyses the combination of hydrogen with residual oxygen present in the air. This method ensures complete anaerobiosis.

The culture media are inoculated with the specimens suspected to contain anaerobic bacteria. The inoculated media are then kept inside the jar, and the lid is closed airtight. The anaerobiosis in the jar is carried out by first evacuating the air from the jar through an outlet tube with the help of a vacuum pump. The outlet tube is closed, and then the sealed jar containing the culture plates is replaced with hydrogen gas and passed through the inlet tube till reduced atmospheric pressure is brought to normal atmospheric pressure, which is monitored on the vacuum gauge as zero. The electrical terminals are then switched on to heat the catalyst that catalyses a combination of hydrogen with residual oxygen and ensures complete anaerobiosis in the jar.

Reduced methylene blue is used as the indicator of anaerobiosis in the jar. If anaerobiosis is complete, it remains colourless; if anaerobiosis is not complete, it turns blue on exposure to oxygen.

(b) **Gaspack system:** It is a simple and effective method of production of hydrogen gas for anaerobiosis. It does not require the cumbersome evacuation method and gas filing after evacuation. Carbon dioxide, which is also generated, is required for growth by some anaerobes. Water activates the gas pack system, resulting in hydrogen and carbon dioxide production. Hydrogen combines with oxygen in the air in the presence of a catalyst and maintains anaerobiosis. In this method, the inoculated plates are kept along with the gas pack envelope with water added in the airtight jar.

It is the method of choice for preparing anaerobic jars. It is commercially available as a disposable envelope containing chemicals that produce hydrogen and carbon dioxide by adding water.

(c) **Anaerobic glove box:** This innovation is developed to isolate anaerobic bacteria. It uses a clear-vinyl large chamber with gloves attached to it and is supplied with a gas mixture of 80% nitrogen, 10% carbon dioxide and 10% hydrogen. A lock is fitted at one end of the chamber with two hatches, one for the outlet and the other for the inlet to the chamber. After placing specimens in the lock, the outlet hatch is closed, and the air is exchanged with the gas mixture in the lock. The inlet hatch is opened then to bring the specimen into the chamber.

(d) **Anoxomat:** This is a fully automated system that evacuates a portion of the jar contents and refills the jar with an anaerobic gas mixture. During this procedure, the oxygen concentration in the air is rarefied. For an anaerobic atmosphere, this procedure is repeated three times, after which the oxygen concentration is rarefied to 0.16%. A small catalyst removes this very small percentage of oxygen content. Anoxomat is capable of producing micro-aerophilic conditions also. The method is increasingly used to process clinical specimens to isolate anaerobic bacteria.

Other Methods of Anaerobiosis

These include the following methods.

(a) **Reducing agents:** Inclusion of reducing agents such as 1% glucose, 0.1% thioglycolate, 0.1% ascorbic acid and

0.05% cysteine into the culture medium would reduce the oxygen concentration of the medium, in turn favouring an anaerobic environment.

(b) **Broth:** The addition of pieces of red-hot metallic iron into the broth and layering it with sterile vaseline creates an anaerobic environment inside the broth. Culture medium like Smith-Noguchi medium contains fresh animal tissues, which facilitate reduced environment within the medium and supports the growth of many anaerobes.

(c) **Agar concentration:** Slightly increased agar concentration reduces the oxygen diffusion into the medium, facilitating anaerobic bacterial growth.

Methods of Isolating Pure Cultures

The methods used to isolate pure cultures are listed below.

1. **Surface plating** technique is routinely used for bacterial isolation in diagnostic bacteriology. Isolated colonies of interest are further used for biochemical or molecular identification.
2. **Enrichment, selective and indicator media** play an important role in isolating pathogenic organisms from clinical samples such as faeces, which consist of numerous normal florae.
3. **Pre-treatment** of specimens with specific bactericidal substances destroys unwanted normal flora without harming the integrity of the organism of our interest. For example, in *Mycobacterium* culture, sputum samples need alkali treatment to kill the abundant commensal flora present in them. However, this pre-treatment does not affect *Mycobacterium* cell integrity.
4. **Incubation at different temperatures** can also be used to isolate certain bacteria in pure culture. For example, *Neisseria catarrhalis* can be separated from *N. meningitides* by incubating the mixed culture at 22 °C, as this temperature is suitable only for the growth of *N. catarrhalis*.
5. **Heating** is used to isolate *Clostridium tetani* from dust or similar sources. Heating kills all vegetative forms of bacteria, but only spores of spore-forming bacteria survive, which can be isolated by culture.
6. **Craigie tube** is used to separate motile bacteria from non-motile bacteria.
7. **Animal inoculation** is also a good old method for isolating pathogenic bacteria from a clinical sample or mixed culture. For example, Anthrax bacilli can be identified from other sporulating, non-pathogenic, aerobic bacteria by inoculating into a suitable animal model. Only anthrax bacilli can multiply and cause fatal septicaemia. Further, it can be cultured from heart blood.
8. **Selective filters** are used to separate different sizes of bacteria. Usually, bacterial filters are used to separate viruses from bacteria.

Laboratory Identification of Bacteria

It is not always possible to identify microorganisms by microscopic methods alone due to the similarity of bacteria in their morphology and staining characters. Hence, further study of microorganisms is essential for their identification. For this purpose, organisms have been grown in artificial media in the laboratory.

Conventional Tests in Diagnostic Bacteriology

The identification of bacteria involves the following steps: (1) morphology of bacterial colony on solid medium, (2) growth in a liquid medium, (3) Staining reactions, (4) Hanging drop preparations, (5) biochemical reactions, (6) animal pathogenicity testing in laboratory animals, (7) antibiotic sensitivity, (8) typing of bacterial strains, (9) serology and (10) rapid bedside tests.

Morphology of Bacterial Colony on Solid Medium

Morphology of the bacterial colony on solid medium depends on many factors, such as nature of culture medium, temperature and time of

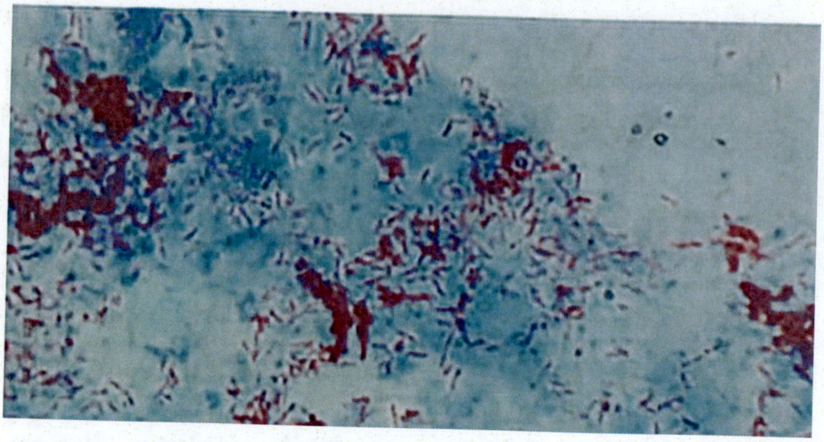

Fig. 24.4 Ziehl–Neelsen staining to differentiate acid-fast bacilli from non-acid-fast bacilli

incubation, age of culture and number of subcultures done.

The characteristics noted include (1) **Size:** The colonies may be a few millimetres in size: pinhead size (*S. aureus*) or pinpoint (streptococci). (2) **Shape:** The shape may be circular or irregular. (3) **Surface:** Surface of the colonies may be smooth, rough or granular. (4) **Elevation:** The colonies on the medium may be flat, raised, low convex, convex or umbonate. (5) **Edge:** The edge may be entire, lobate, crenated, undulate or ciliate's. (6) **Opacity:** The colonies may be transparent, translucent or opaquing. (7) **Pigment and haemolysis:** Certain bacterial colonies are associated with the production of pigments and haemolysis around them. (8) **Consistency:** Bacterial colonies may be mucoid, friable firm and butyrous.

Growth in Liquid Medium

Nutrient broth and peptone water are frequently used as liquid media for the growth of bacteria. The characteristics of bacterial growth in liquid media provide some clues in presumptive identification. For example, streptococci produce granular deposits at the bottom of the liquid medium, whereas most Gram-negative bacteria produce uniform turbidity. *Pseudomonas* spp. and other aerobic bacteria produce surface pellicles in liquid media. The pigment production of some bacteria, such as *Pseudomonas aeruginosa* can be appreciated in liquid media.

Staining Reactions

The study of morphology and staining characteristics is useful in identifying the isolate. Smears prepared on glass slides from bacterial colonies grown on either solid or liquid media are examined for bacteria by appropriate staining. Gram staining is most widely used to differentiate between Gram-positive and Gram-negative bacteria. Ziehl–Neelsen stain differentiates acid-fast bacilli (e.g. *M. tuberculosis*, *M. leprae*) from non-acid-fast bacilli (Fig. 24.4). Albert stain is used to demonstrate metachromatic granules in *C. diphtheriae*.

Hanging Drop Preparation

Hanging drop preparation helps to distinguish motile bacteria from those of non-motile bacteria.

Biochemical Reactions

Biochemical reactions are very important in identifying bacterial isolates and different bacterial species. These tests depend on the presence of certain enzymes, such as catalase, oxidase, urease, gelatinase, produced by the bacteria. Commonly used biochemical tests are described below.

Catalase Test

It is used to detect the presence of enzyme catalase in a bacterium. The enzyme catalase catalyses the breakdown of hydrogen peroxide with the release of free oxygen. It is found in most aerobic and facultative anaerobic bacteria. The main exception is *Streptococcus* spp. It is not found in anaerobes. Red blood cells contain catalase and their presence, and therefore, give a false-positive result.

The catalase test is primarily used to differentiate *Staphylococcus* and *Streptococcus*. In this test, a small amount of culture to be tested is picked from a nutrient agar plate with a sterile platinum loop or glass rod, and this is inserted into hydrogen peroxide solution (3%) held on a slide or in a tube. Immediate production of air bubbles in the solution denotes a positive test, and no bubbles indicate a negative test.

Oxidase Test

This test determines the presence of enzyme oxidase in many bacteria. The enzyme oxidase catalyses the oxidation of reduced cytochrome by molecular oxygen. Kovac oxidase reagent that contains tetramethyl-*p*-phenylenediaminedihydrochloride is the main reagent used in the oxidase test. The dye serves as an alternate substrate for the cytochrome oxidase reaction. In the reduced state, the reagent is colourless, but when oxidised, it becomes purple.

The dry filter paper method is performed by impregnating filter paper strips with 1% Kovac oxidase reagent. The paper is smeared with the bacterial colonies tested by a glass rod. In a positive test, the smeared area on the filter paper turns deep purple within 10 s. No colour change indicates a negative test.

Oxidase-positive bacteria include *Neisseria* spp., *Vibrio* spp., *Aeromonas* spp., *Plesiomonas* spp., *Pseudomonas* spp. and *Moraxella* spp. Oxidase-negative bacteria include all members of the family Enterobacteriaceae.

Indole Test

The test is used to detect the ability of bacteria to decompose amino acid tryptophan into indole, which accumulates in the medium. Tryptophan or peptone broth is the medium used for the indole test. The test is performed by inoculating the medium with bacteria, incubating at 37 °C for 24–48 h. Then, five drops of Kovac reagent containing amyl or isoamyl alcohol, *p*-dimethyl amino benzaldehyde and concentrated hydrochloric acid are added to the inoculated medium. A positive test is indicated by the formation of a red ring at the surface of the medium. No colour change indicates a negative test (Fig. 24.5).

Indole-positive bacteria are *E. coli*, *Vibrio cholerae*, *Proteus vulgaris* and *Edwardsiella* spp. Indole-negative bacteria are *S. typhi*, *Klebsiella* spp. and *Proteus mirabilis*.

Sugar Fermentation Tests

A carbohydrate fermentation test is done to determine the ability of a bacterium to ferment a specific carbohydrate incorporated in a basal medium, producing acid or acid with visible gas. The carbohydrates include glucose, lactose, sucrose, maltose. The sugar medium contains 1% sugar. Andrade indicator is a solution of acid fuchsin to which sodium hydroxide is added. It is a pH indicator that is added to the basal medium, which indicates the formation of organic acids.

The test is performed by inoculating the sugar media with bacteria, incubating at 37 °C for 18–24 h. The change of the colour to pinkish-red (acidic) is considered a positive test result, whereas yellow to colourless (alkaline) sugar indicates a negative test result. Production of gas is indicated by the appearance of gas bubbles in the Durham tube.

All members of the family Enterobacteriaceae ferment glucose, whereas *Salmonella* spp. ferments glucose and mannitol and *E. coli* and *Klebsiella* spp. ferment glucose and lactose.

Oxidation–Fermentation Test

Oxidation–fermentation test (OF test) determines the oxidative or fermentative metabolism of carbohydrates by a bacterium. OF test is used to know whether a bacterium has the enzymes necessary for the aerobic breakdown of glucose (i.e. oxidation) and/or for the fermentation of glucose. The test differentiates Enterobacteriaceae (fermenters) from the members of the family Pseudomonadaceae (the non-fermenters).

Fig. 24.5 Indole test

Kligler Iron Agar/Triple Sugar Iron Agar Test

Kligler iron agar (KIA) and triple sugar iron agar (TSI) tests are used to determine the ability of an organism to attack a specific carbohydrate incorporated into a basal growth medium, with or without the production of gas, along with the production of hydrogen sulphide (Fig. 24.6). KIA medium contains two carbohydrates: lactose and glucose, in a ratio of 10:1. TSI contains a third carbohydrate, sucrose, in addition to lactose and glucose.

The test is performed by inoculating KIA or TSI with an inoculating needle by stabbing the butt, streaking the slant, and incubating at 37 °C for 18–24 h. During incubation, the bacterium first utilises glucose and then lactose and sucrose. After 18–24 h, the glucose concentration is depleted in the slant and the butt. The bacteria start oxidative degradation of the peptone present in the slant, resulting in the production of alkaline by-products, thereby changing the indicator to a red colour. Anaerobic fermentation of glucose in the butt produces a large volume of acid, which neutralises the alkalinity caused by peptone degradation; hence, the butt remains yellow. Yellow colour (acidic), therefore, indicates fermentation of carbohydrates, and red colour (alkaline) indicates no fermentation. The presence of bubbles in the butt indicates gas production during the fermentation of carbohydrates.

Certain bacteria produce H_2S, which is detected as a black precipitate that blackens the slant and butt of the medium. The medium is turned black due to a combination of H_2S with ferric ions from ferric salts present in the medium to form ferrous sulphide as black precipitates. Three basic fermentation patterns are observed on KIA medium after incubation at 37 °C for 24 h:

1. **Fermentation of glucose only (alkali/acid or K/A reaction; K denotes alkaline reaction and A denotes acidic reaction):** After 18–24 h, the low glucose concentration is completely used up, and the organism starts to utilise the peptones present in the medium, resulting in an alkaline pH in slant (red). In the

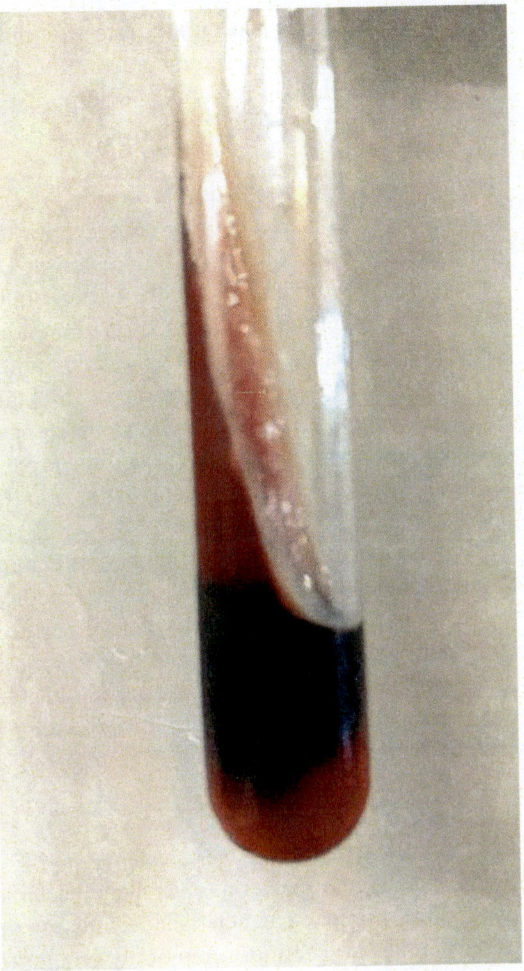

Fig. 24.6 Kligler iron agar showing production of hydrogen sulphide

butt, a yellow colour exists due to the anaerobic degradation of glucose.

2. **Fermentation of lactose and glucose (acid/acid or A/A reaction):** Lactose is present in 10 times the amount of glucose. In 18–24 h, the lactose is not depleted and therefore, acidic conditions exist in the butt and slant.
3. **Neither lactose nor glucose fermented (alkali/alkali or K/K reaction):** Many Gram-negative, non-enteric bacilli cannot ferment glucose or lactose. A reaction of K/K results from the aerobic catabolism of peptone by the organism.

H_2S production and carbohydrate fermentation patterns are generally characteristic for specific bacterial groups, especially the Enterobacteriaceae.

Urease Test

The test is used to determine the ability of an organism to split urea into ammonia by the enzyme urease. Production of ammonia makes the medium alkaline; thus, the indicator phenol red changes to red or pink in colour. The test is performed in Christensen urease medium. Urease-positive bacteria produce a pink colour.

The urease test is useful for screening urease production by *Proteus* spp., *Morganella* spp. and some *Providencia* spp. *Proteus* spp. is rapidly urease positive—usually within 6 h. Another example of urease producing organism is *Helicobacter pylori*, the production of which is essential for its survival in the acidic pH of the stomach.

Citrate Test

The citrate test is used to demonstrate the ability of an organism to utilise citrate as the sole source of carbon for metabolism. Koser citrate medium (a liquid medium) and Simmon citrate medium (a solid medium) are used for the test. Growth in the Simmon medium is accompanied by a rise in pH to change the medium from its initial green colour to deep blue. Hence, growth with blue colour on the slant indicates a positive test, and no growth without any colour change indicates a negative test. In Koser liquid medium, turbidity indicates a positive test and absence of turbidity indicates a negative test.

Nitrate Reduction Test

The test is used to determine the presence of enzyme nitrate reductase in the bacteria. The enzyme reduces nitrate to nitrites or free nitrogen gas. The test is carried out by inoculating the broth containing 1% potassium nitrate (KNO_3) and incubating it at 37 °C for up to 5 days. Then 1–2 drops of a reagent consisting of a mixture of 1 mL of naphthylamine and 1 mL of sulphanilic acid are added. The red colour developed within a few minutes signifies a positive reaction, whereas the absence of colour indicates a negative reaction.

Nitrate reduction test is used mainly to identify members of the family Enterobacteriaceae, which are usually positive for nitrate reduction test.

Methyl Red Test

The test detects the ability of an organism to produce and maintain stable acid end products during the fermentation of glucose, thereby maintaining a sustained pH below 4.5. The test is performed by inoculating a glucose phosphate broth and incubating it at 37 °C for 2–5 days. Five drops of 0.04% of MR solution are then added to the inoculated medium, mixed, and the result is read immediately. Development of red colour denotes a positive test, and yellow colour indicates a negative test. *E. coli*, *Yersinia* spp. and *Listeria monocytogenes* are examples of MR-test-positive bacteria.

Methyl red test is commonly used to differentiate *E. coli* (MR positive) from *Enterobacter* (MR negative) and *Yersinia* spp. (MR positive) from other Gram-negative, non-enteric bacteria (MR negative).

Voges–Proskauer Test

Voges–Proskauer (VP) test detects the production of acetyl methyl carbinol (acetoin), a natural product formed from pyruvic acid in the course of glucose fermentation. In the presence of alkali and atmospheric oxygen, the acetoin is oxidised to diacetyl that reacts with alpha-naphthol to produce a red colour. This test is performed by inoculating the glucose phosphate broth with the organism and incubating it at 37 °C for 48 h. Then approximately three drops of alpha-naphthol are added, followed by the addition of 1 drop of 40% potassium hydroxide. The reagents are mixed well and are allowed to stand for 30 min. In the positive test, the pink colour appears in 2–5 min, deepening to magenta in 30 min. The solution remains colourless for 30 min in the negative test. *Staphylococcus* spp., *V. cholerae* biotype eltor, *Klebsiella* spp. and *Enterobacter* spp. are the common examples of VP-test-positive bacteria.

Indole, MR, VP and citrate tests, commonly referred to as "IMViC tests", are the most frequently used tests to identify enteric, Gram-negative bacteria.

Hydrogen Sulphide Production

These tests are carried out to demonstrate H_2S produced by certain bacteria. Production of H_2S is demonstrated by inoculating the bacteria in the media containing lead acetate, ferric ammonium citrate or ferrous acetate and incubating overnight at 37 °C. H_2S-producing bacteria by their enzymatic actions produce H_2S from sulphur-containing amino acids in the medium. Production of sulphur produces black colour in the media, visible to the naked eye.

Filter paper strip is another method of demonstration of the production of H_2S. In this method, a filter paper strip impregnated with lead acetate is kept between the cotton plug and the medium in the tube. The blackening of the paper indicates the production of H2S. *P. mirabilis*, *P. vulgaris* and *Salmonella* spp. are some examples of the H_2S-producing bacteria.

Laboratory Animals

Some pathogenic bacteria and their bacterial metabolites produce characteristic lesions in laboratory animals. The most commonly employed animals are rats, guinea pigs, mice and rabbits. Depending on the organisms to be tested, these animals may be inoculated by subcutaneous, intramuscular, intravenous, intraperitoneal or intracerebral routes. The identification of the bacteria is made depending on the post-mortem findings and cultural properties of the bacteria. For example, guinea pigs are commonly used for performing animal pathogenicity of *C. diphtheriae*, *Clostridium perfringens* and *M. tuberculosis*.

Antibiotic Sensitivity

Determination of antibiotic sensitivity of an isolate from a patient is essential for the choice of drug therapy. In some cases, the sensitivity of an organism to a particular agent helps in the identification. For example, *Streptococcus pyogenes* is sensitive to bacitracin and *Streptococcus pneumoniae* to optochin. The sensitivity to the antibiotic can be determined by disc diffusion test, serial dilution tests and E-test, as described in detail in Chap. 23.

Typing of Bacterial Strains

The ability to discriminate between similar strains of bacteria may be important in tracing sources or modes of spread of infection in a community. Typing methods are widely used for epidemiological studies. These include phenotypic techniques and genotypic techniques.

Phenotypic Techniques

Phenotypic techniques include (1) **Bio typing:** It relies on a set of biochemical reactions to distinguish different strains within a given species. Antimicrobial susceptibility testing is an example of this type. (2) **Serotyping:** Different strains of organisms of the same species can be differentiated based on the difference in the expression of antigenic determinants on the cell surface. (3) **Bacteriocin typing:** This is used in the case of bacterial species for which many lytic bacteriophages have been identified, and (4) **Phage typing:** This has been the mainstay of strain discrimination and is widely used in epidemiological studies.

Gas-Liquid Chromatography

Gas-liquid chromatography, commonly referred to as gas chromatography, is a highly sensitive tool for detecting fatty acids in microorganisms. It has long been used for the chemotaxonomic analysis of microorganisms wherein cellular fatty acids are processed through transesterification, saponification and extraction. Finally, it is derivatised to form fatty acid methyl esters (FAME) and passed through the gas chromatography's silica columns. Individual FAME present in the sample produces a specific peak in a gas chromatogram with respect to the time of retention in the column. A novel species develops a pattern in gas chromatogram that is different from closely related species of the same genus.

Genotypic Techniques

Genotypic techniques depend on differences related to the genome of bacteria. Genotypic techniques employed to differentiate strains of bacteria include plasmid profile analysis and restriction endonuclease analysis of chromosomal DNA.

Serology

1. Agglutination of biochemically confirmed bacteria with specific antisera facilitates further identifying the isolated bacteria. Agglutination test identifies presumptive isolates of pathogens (e.g. *Salmonella*) from clinical samples. It is also used for the grouping of beta-haemolytic streptococci.
2. A group of serological tests like flocculation tests, neutralisation tests, hemagglutinin-inhibition tests, enzyme-linked immunosorbent assays (ELISAs) and chemiluminescence immunoassays are available for detection of antigen or antibodies that develop in response to certain diseases, including infections and autoimmune disorders.

Flocculation tests result in precipitation or flocculation when antigen–antibody reacts. Flocculant precipitate confirms a positive indication of a targeted disease. For example, the venereal disease research laboratory (VDRL) test for syphilis is a very popular nontreponemal, qualitative test for screening syphilis where antibodies, if present in the body in response to the syphilis is, precipitated using specific antigens.

ELISA

ELISA uses enzyme-tagged antibodies that, when bound to the target, could produce colour by forming a chemiluminescent product after breaking a substrate. ELISA is performed using serum, plasma, cell culture, cell lysates, saliva, tissue lysates and urine.

1. **Direct ELISA** uses one enzyme-linked antibody (primary) that binds to the target antigen. Such antigen–antibody binding could be detected when a substrate is added to the

solution, and the enzyme in the primary antibody breaks it to produce a coloured product.

2. **Indirect ELISA** uses two antibodies. At first, an unconjugated primary detection antibody binds to the specific antigen. Then, the enzyme-conjugated secondary antibody is added that binds to the Fc region of the primary antibody.
3. **Sandwich ELISA** also uses two antibodies, both of which bind to the antigen in a manner that sandwich an antigen molecule in between two antibodies. This type of ELISA uses an antibody-coated plate on which the sample is added. Antigen specific to the primary antigen is present in the sample binds to the antibody. A secondary enzyme conjugate antibody is then added that also binds to the other end of the antigen and produces colour when a substrate is added to the complex.

Rapid (Bed Side) Tests

Latex particle agglutination (LAT), agglutination (Co-A) are the most frequently used techniques in the clinical laboratory to detect microbial antigens directly in clinical specimens rapidly. Antibody to a specific antigen is bound to latex particles in LAT or a heat-killed and treated protein A-rich strain of *S. aureus* of Co-A to produce agglutination.

Immunochromatographic Tests

Lateral flow immunochromatographic tests (ICTs) are the commonly used rapid diagnostic test (RDT) to detect infectious diseases. In contrast to complex molecular tools and technologies that require clinical establishments for diagnosis, such ICTs offer portability and are perfect as a point-of-care (POC) test. For the detection of pathogens, it contains biomarkers (antigen) for the pathogens incorporated on a paper strip. When placed on the paper, the sample containing the pathogen develops lines corresponding to the binding of the antibody (present in the sample) and the biomarker. Such tests are easy to interpret and do not require any technical expertise.

Newer Tests in Diagnostic Bacteriology

Nucleic Acid-Based Tests

Molecular methods, such as nucleic acid probes, polymerase chain reaction (PCR), and other amplification procedures, are increasingly used to identify microorganisms.

Genetic probes are based on detecting unique nucleotide sequences with the DNA or RNA of a microorganism. The hybridisation of the sequence with a complementary sequence of DNA or RNA follows the cleavage of the double-stranded DNA of the microorganism in the specimen. The genetic probe-based technique is also used for the detection of *M.tuberculosis* infection.

PCR has major applications in detecting infections caused by difficult to culture microorganisms (e.g. HIV) or that have not yet been successfully cultured (e.g. Whipple disease bacillus).

Automated Methods

Automated methods are now available, which take only hours to characterise isolates. These include detecting specific enzymes, toxins, antigens or metabolic end products. Several instruments like GeneXpert have been developed that detects pathogens from samples in a cartridge-based nucleic acid amplification process.

MALDI-TOF

Like DNA, pathogens have unique proteome signatures and could be identified using MALDI-TOF. Very recent, it is gaining attention and emerging as an alternative tool for early identification of bacteria in blood cultures, urinary tract infections (UTIs), cerebrospinal fluids, respiratory tract infections, stool samples etc. The technology produces results within 30 min and reduces the need for microbiological processing for diagnosis.

Staphylococcus

25

The genus *Staphylococcus* consists of 61 recognised species, most of which are animal pathogens or commensals. The bacteria belonging to this genus are aerobic and facultative anaerobic, catalase-positive, oxidase negative and are arranged in clusters, pairs or tetrads.

Staphylococcus aureus is the most important human pathogen. The other important human pathogens are coagulase-negative staphylococci (CONS), including *Staphylococcus epidermidis*, *Staphylococcus saprophyticus*, *Staphylococcus haemolyticus*, *Staphylococcus hominis*, *Staphylococcus warneri*, *Staphylococcus saccharolyticus*, *Staphylococcus schleiferi* and *Staphylococcus lugdunensis*.

Staphylococci can acquire resistance to many antibiotics and therefore can cause major clinical and epidemiological problems in hospitals. Human infections caused by *Staphylococcus*, *Micrococcus* and *Stomatococcus* are summarised in Table 25.1.

Staphylococcus aureus

S. aureus is an important human pathogen that causes a spectrum of clinical diseases. These range from superficial skin lesions like folliculitis to deep-seated abscesses and various pyogenic infections like endocarditis, osteomyelitis. The bacterium also causes toxin-mediated diseases, such as food poisoning, toxic shock syndrome (TSS) and staphylococcal scalded skin syndrome (SSSS).

Properties of the Bacteria

Morphology

They are Gram-positive cocci, measuring around 1 μm in diameter. They are non-motile, non-sporing. They are non-capsulated. They, however, contain a microcapsule that can be visualised by electron microscope only, but not by a light microscope.

The cocci are typically arranged in irregular grape-like clusters. This appearance is due to the incomplete separation of daughter cells during successive divisions of bacteria, which takes place in perpendicular planes. The grape-like clustering is seen when the bacteria are grown in solid media, but usually, short chains are seen when grown in liquid media. In smears taken from pus, the cocci are present either singly or in pairs, clusters, or short chains of three or four cells.

Culture and Biochemical Properties

Staphylococci are aerobes and facultative anaerobes but can also grow in the absence of oxygen. They grow at a temperature range of 10–42 °C (optimum temperature 37 °C) and a pH range of 7.4–7.6 (optimum pH 7).

Table 25.1 Human infections caused by *Staphylococcus*, *Micrococcus* and *Rothia* spp.

Bacteria	Diseases	
Staphylococcus aureus	Skin infections	Impetigo, folliculitis, furuncle, carbuncle, paronychia, wound infection
	Systemic infections	Bacteraemia, osteomyelitis, septic arthritis, endocarditis, pneumonia, meningitis, deep-seated abscess
	Toxin-mediated infections	Food poisoning, toxic shock syndrome (TSS), staphylococcal scalded skin syndrome (SSSS)
Staphylococcus epidermidis	Opportunistic infections	Intravenous catheter infections, CSF shunt infections and catheter-associated peritonitis endocarditis in "immunocompromised" patients
Staphylococcus saprophyticus	Urinary tract infection	Sexually active young women
Rothia spp.	Opportunistic infections	Bacteraemia, Enterocolitis, Endocarditis

Nutrient Agar

S. aureus produces round, convex, well-defined colonies measuring 2–4 mm in diameter. The colonies show a butyrous consistency with a smooth glistening surface.

S. aureus produces characteristic golden-yellow colonies due to the production of a non-diffusible golden-yellow pigment. The pigment is believed to be a lipoprotein allied to carotene. The pigment production is enhanced by incubation at 22 °C in the presence of oxygen. **Milk agar** and **1% glycerol monoacetate agar** are other media that facilitate pigment production. The growth gives a characteristic "oil paint" appearance on nutrient agar slopes.

Blood Agar

S. aureus produces a clear zone of haemolysis (beta-haemolysis) surrounding the colonies (Fig. 25.1). Haemolysis is well marked on sheep

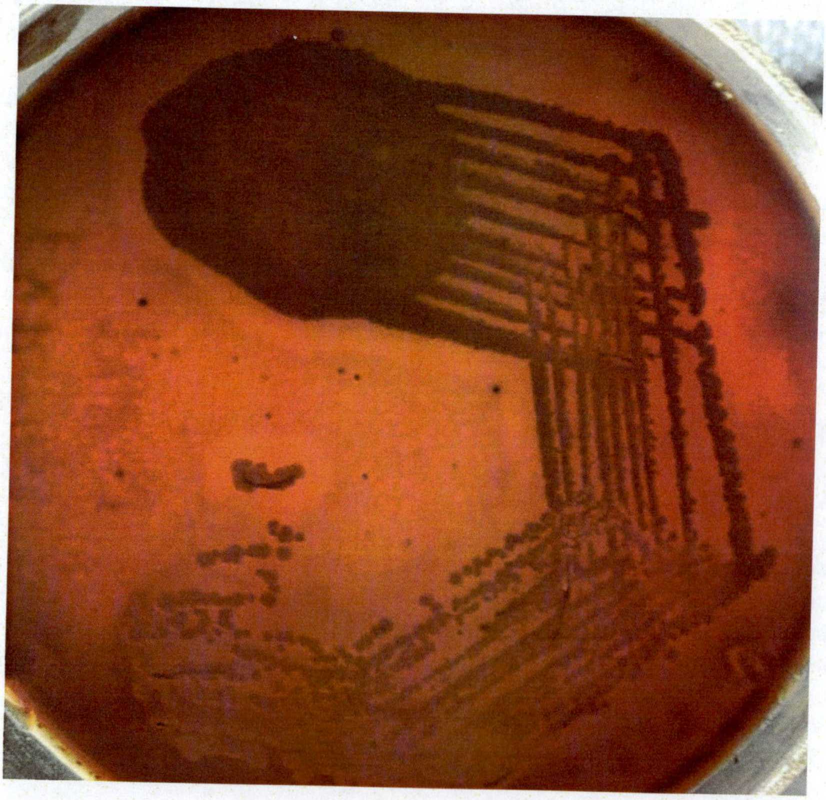

Fig. 25.1 Blood agar plate showing beta-haemolysis surrounding the colonies of *Staphylococcus aureus* (Courtesy of Prof S. Pramodhini)

or rabbit blood agar, especially when incubated in an atmosphere of 20–25% CO_2. Sheep blood agar is used for primary isolation. Haemolysis is weak on horse blood agar. Human blood is not used, as it may contain antibiotics or other inhibitors. Other species of *Staphylococcus* do not produce haemolysis.

MacConkey Agar

S. aureus produces small pink colonies due to lactose fermentation.

Mannitol salt agar, milk agar and glycerol monoacetate agar are the commonly used selective media for isolating *S. aureus* from clinical specimens containing normal bacterial flora (e.g. stools). Mannitol salt agar contains 1% mannitol, 7.5% sodium chloride and 0.0025% phenol red indicator. Most strains of *S. aureus* ferment mannitol with acid production, which gives rise to yellow zone formation around the colonies.

Other Properties

S. aureus is coagulase positive. Coagulase production is used to differentiate *S. aureus* from *S. epidermidis* and other CONS. It is phosphatase positive. Phosphatase production can also differentiate *S. aureus* from *S. epidermidis*, as the latter either does not produce or has very weak phosphatase activity. It is oxidase negative and catalase-positive. It produces enzyme catalase (unlike *Streptococcus*), which degrades H_2O_2 into nascent oxygen and water.

S. aureus ferments mannitol, sucrose, maltose and trehalose under aerobic conditions, producing acid but no gas. Fermentation of mannitol is of diagnostic importance because most strains of *S. aureus* ferment mannitol, while *S. epidermidis* and *S. saprophyticus* do not ferment mannitol.

Susceptibility to physical and chemical agents: The cocci withstand moist heat at 60 °C for 30 min but are killed after 30 min. They are also killed rapidly by disinfectants, such as phenol, chlorhexidine and hexachlorophene. The cocci are very sensitive to aniline dyes, such as crystal violet. The dye at a concentration of 1: 500,000 inhibits the growth of the cocci on the blood agar medium but permits the growth of streptococci.

Pathogenesis

S. aureus causes broadly two types of diseases: (1) **infection** by multiplying in tissues causing inflammation and (2) **toxin-mediated diseases** also by liberating toxins.

Virulence Factors

S. aureus produces several virulence factors (Table 25.2), which include (1) cell-surface proteins, (2) cell-associated factors, (3) extracellular enzymes and (4) toxins.

Cell-Surface Proteins

Protein A

Protein A is the major protein in the cell wall and has a molecular weight of 13,000 Da. It is present in large quantities in the cell wall of certain strains of *S. aureus*, such as the Cowan's strain of *S. aureus* (SAPA). This is a group-specific antigen. The antigen is present in more than 90% of strains of *S. aureus*. Protein A is absent in both the CONS and micrococci.

It is an important virulence factor since it has non-specific interaction with the Fc portion of the immunoglobulin G (IgG), leaving the Fab portions free to combine with a specific antigen. It is chemotactic, anticomplementary and antiphagocytic. It causes platelet injury and elicits hypersensitivity reactions.

Coagulase

S. aureus has a unique ability to clot various mammalian plasma. Clotting of plasma is brought about by the action of the enzyme coagulase secreted by the pathogenic strains of *S. aureus*. The enzyme coagulase is of two types: (1) free coagulase and (2) bound coagulase.

Free Coagulase

It is a heat-labile and filterable enzyme. It has eight antigenic types (A, B, C, D, E, F, G and H). Most human *S. aureus* strains produce

Table 25.2 Virulence factors of *Staphylococcus aureus*

Virulence factors	Biological functions
Cell wall–associated polymers	
Polysaccharide peptidoglycan	Inhibits chemotaxis of inflammatory cells
Capsular polysaccharide	Inhibits phagocytosis and chemotaxis
Teichoic acid	Mediates attachment of staphylococci to mucosal cell
Cell surface proteins	
Protein A	Chemotactic, anticomplementary and antiphagocytic; causes platelet injury; and elicits hypersensitivity reactions
Clumping factor	Bound coagulase
Extracellular enzymes	
Coagulase	The enzyme coats the bacterial cells with fibrin, rendering them resistant to opsonisation and phagocytosis
Catalase	Produces nascent oxygen which causes oxidative damage to host tissue
Hyaluronidase	Hydrolyses hyaluronic acids present in the matrix of the connective tissues, thereby facilitating the spread of bacteria in the tissues
Penicillinase	Inactivates penicillins
Nuclease	Hydrolyses DNA
Lipid hydrolases	Hydrolyses lipids
Toxins	
Toxic shock syndrome toxin (TSST)	Super-antigen, stimulates the release of large number of interleukins (IL-1 and IL-2)
Enterotoxin	Super-antigen, acts by producing large amounts of interleukins (IL-1 and IL-2)
Exfoliative toxin	Splits intercellular bridges in the stratum granulosum of epidermis of the skin
Leukocidin toxin Panton-valentine leucocidin	Leukocidin is thermostable and causes lysis of leukocytes
Cytolytic toxins (Alpha, Beta, Delta haemolysis	Causes lysis of erythrocytes

antigenic type A. The enzyme coagulase, in association with coagulase-reacting factor (CRF) present in plasma, converts fibrinogen to fibrin. In the absence of CRF, coagulase cannot bring about clotting like in the case of the guinea pig plasma. This fibrin coats the bacterial cells, rendering them resistant to opsonisation and phagocytosis, making bacteria more virulent. All coagulase-producing staphylococci are, by definition, *S. aureus*. Coagulase production is demonstrated by the tube coagulase test, which is important for identifying *S. aureus*.

Bound Coagulase

It is otherwise known as the clumping factor. It is a heat-stable protein and is present in the cell wall. This enzyme brings about clumping of the staphylococci when mixed with plasma by directly acting on fibrinogen. Lysis of the cell releases the enzyme. Unlike free coagulase, the clumping factor does not need CRF for its action; only one type has been identified to date. Bound coagulase is not a virulence factor.

Differences between coagulase and clumping factors are summarised in Table 25.3.

Cell Wall-Associated Factors

These include capsular polysaccharide, protein A, peptidoglycan and teichoic acid (Fig. 25.2).

Capsular Polysaccharide

Few strains of *S. aureus* are capsulated. These strains are more virulent than the non-capsulated ones. The capsule protects the bacteria from phagocytosis. The capsule also facilitates adherence of the cocci to host cells and to prosthetic implants.

Cell Wall Polysaccharide Peptidoglycan

It activates the complement, stimulates the production of the antibodies and inhibits chemotaxis by inflammatory cells. *S. aureus* cell wall is rich

Table 25.3 Differentiating features of coagulase and clumping factors

Coagulase	Clumping factor
Produced extracellularly	Present on the surface
Detected by tube test	Detected by slide test
Heat labile	Heat stable
Eight serotypes	One serotype
Needs CRF	Does not need CRF
A virulence factor	Not a virulence factor

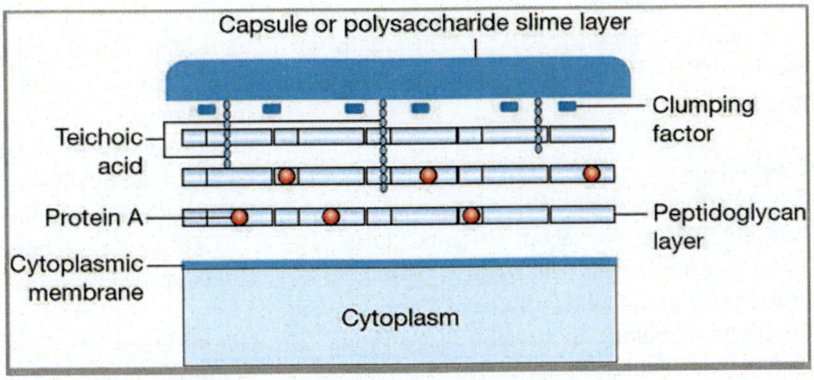

Fig. 25.2 Cell wall structure of *Staphylococcus aureus*

in peptidoglycans. Peptidoglycan is a polysaccharide polymer that provides rigidity to the bacteria's cell wall. It has the characteristic pentaglycine bridges that link tetrapeptides to the muramic acid residues.

Teichoic Acid

It mediates the attachment of staphylococci to mucosal cells. It is the major antigenic determinant of the cell wall of *S. aureus*. It is a polymer of ribitol phosphate. Antibodies to teichoic acids develop in endocarditis and certain other staphylococcal infections.

Extracellular Enzymes

The enzymes include coagulase, catalase, hyaluronidase, penicillinase and other enzymes.

Catalase

The enzyme catalase reduces H_2O_2 to nascent oxygen and water. This nascent oxygen causes oxidative damage to host tissue. This enzyme is produced after phagocytosis or during the metabolism of the bacteria. All strains of staphylococci produce catalase, unlike streptococci.

Hyaluronidase

The enzyme hyaluronidase hydrolyses the acidic mucopolysaccharides present in the matrix of the connective tissues, thereby facilitating the spread of bacteria in tissues.

Penicillinase

More than 90% of *S. aureus* produce the enzyme penicillinase. The enzyme inactivates the penicillin group of antibiotics, hence is responsible for the widespread occurrence of penicillin-resistant staphylococci. The gene for this enzyme is acquired through plasmids.

Other Enzymes

These include phosphatase, deoxyribonucleases, nucleases, proteases, phospholipase and lipases.

Toxins

Toxins include toxic shock syndrome toxin (TSST), enterotoxin, exfoliative toxin, leucocidins and haemolysins.

Toxic Shock Syndrome Toxin

TSST is an antigenic protein with a molecular weight of 22,000 Da and resembles enterotoxin F, and exotoxin C. Production of toxin is pH-dependent and occurs at pH 7–8. The toxin causes toxic shock syndrome (TSS). TSST is a super-antigen and hence a potent stimulant of T lymphocytes, resulting in the release of a large number of interleukins (IL-1 and IL-2) and tumour necrosis factor, ultimately manifesting in TSS. *S. aureus* strains are responsible for menstruation-associated TSS and half of the strains responsible for non-menstruation associated TSS produce TSST-1. The strains producing TSST-1 belong to the bacteriophage group I.

Enterotoxin

Enterotoxin is a heat-stable protein capable of resisting boiling for about 30 min. It is also gut-enzyme resistant. The toxin is produced by nearly one-third of all the strains of *S. aureus*, and these strains belong to bacteriophage group III (6/47). Nine antigenic types (A, B, $C_{1,\ 2,\ and\ 3}$, D, E, G, H, I and J) of enterotoxins have been described, out of which types A and B are most important. These proteins are of molecular weights ranging from 26,000 to 30,000 Da.

The toxins are super-antigens. These, like other super-antigens, act by producing large amounts of interleukins, IL-1 and IL-2. They are responsible for causing staphylococcal food poisoning and pseudomembranous enterocolitis post-antibiotic therapy.

Exfoliative Toxin

Exfoliative toxin is of two types: (1) toxin A (molecular weight of 30,000 Da) and (2) toxin B (molecular weight of 29,500 Da). Toxin A is heat stable, while toxin B is heat labile. The toxin is antigenic, and specific antibodies against the toxin are protective. The strains producing this toxin belong to bacteriophage group II.

The toxin breaks intercellular bridges in the stratum granulosum of the epidermis and causes its separation from the underlying tissue, resulting in blistering and exfoliating skin disease. The toxin in localised form causes bullous impetigo and, in generalised form, causes SSSS in children below 4 years of age.

Leucocidins

Leukocidins include alpha-lysin, Panton-Valentine leukocidin (PV-leukocidin) and leukolysin.

Alpha-lysin is the most important leucocidin. It causes marked necrosis of the skin and haemolysis by damaging the cell membrane, releasing low-molecular-weight substances from the damaged cells.

Panton-Valentine leukocidin (PVL) is six in number, consisting of two components. The molecular weight is around 32 kDa. These toxins cause the death of human leukocytes and macrophages without causing any lysis. The toxin has been associated with community-acquired methicillin-resistant *S. aureus* **methicillin-resistant staphylococci** (MRSA).

Leucolysin is thermostable and causes lysis of leukocytes and necrosis of tissues *in vivo*.

Haemolysins

S. aureus produces four haemolysins: alpha (α), beta (β), gamma (γ) and delta (δ) haemolysins.

Alpha-haemolysin is a protein with a molecular weight of 33 kDa. It has lethal effects on various cell types and lyses erythrocytes of several animal species.

Beta-haemolysin is sphingomyelinase that is active on a variety of cells. It is a protein with a molecular weight of 35 kDa. It is a hot-cold haemolysin; that is, its haemolytic properties are increased by exposure of the RBCs to cold temperatures.

Delta-haemolysin is a protein with a molecular weight of 8 kDa. It acts primarily as a surfactant.

Gamma-haemolysin consists of three proteins. The three delta-haemolysin proteins interact with one of the two PV-leukocidin proteins.

Clinical Syndromes

S. aureus is pyogenic bacteria that cause localised lesions in contrast to streptococcal lesions that are spreading in nature. Staphylococci adhere to the damaged skin, mucosa or tissue surfaces. At these sites, they evade defence mechanisms of the host, colonise and cause tissue damage. They produce disease by (1) multiplying in tissues, (2) liberating toxins and (3) stimulating inflammation.

The diseases caused by *S. aureus* can be divided into two groups: (1) inflammatory staphylococcal diseases and (2) toxin-mediated staphylococcal diseases.

Inflammatory Staphylococcal Diseases

These include the following conditions:

Staphylococcal skin infections include impetigo, folliculitis, furuncles, carbuncles, paronychia, surgical wound infection, blepharitis and post-partum breast infection. *S. aureus* is the most common cause of boils. The infection is acquired either by self-inoculation from a carrier site, such as the nose or through contact with another person harbouring the bacteria.

Bacteraemia and septicaemia may occur from any localised lesion, especially wound infection or due to intravenous drug abuse. *S. aureus* is an important cause of acute bacterial endocarditis of normal or prosthetic heart valves which is associated with high mortality. *S. aureus* is the most common cause of osteomyelitis in children. The bacteria reach bone through the bloodstream or by direct implantation following trauma.

Staphylococcal pneumonia may occur in post-operative patients following a viral respiratory infection, leading to empyema; it also leads to chronic sinusitis. It causes deep-seated abscesses in any organ after bacteraemia.

Toxin-Mediated Staphylococcal Diseases

These include staphylococcal food poisoning, staphylococcal toxic shock syndrome and staphylococcal scalded skin syndrome.

Staphylococcal food poisoning is caused by enterotoxin. The enterotoxin is a preformed toxin already present in the contaminated food before consumption. Milk and milk products and animal products like fish and meat kept at room temperature after cooking are mainly incriminated. When kept at room temperature, the contaminating staphylococci multiply and produce toxin adequate (as little as 25 μg of toxin B can lead to illness) to cause food poisoning. The toxin acts by stimulating the release of large amounts of interleukins IL-1 and IL-2. It is fairly heat resistant and so is not inactivated by brief cooking.

Often a food handler, who is a carrier of *S. aureus* (nose, skin) or is suffering from staphylococcal skin infection, is the source of infection. The onset of symptoms is sudden, appearing within 2–6 h of ingestion of food. It is a self-limiting condition characterised by nausea, vomiting, abdominal cramps and watery, non-bloody diarrhoea.

Staphylococcal Toxic Shock Syndrome (STSS) is caused by TSST. The toxin is a superantigen, which causes STSS by stimulating the release of large amounts of interleukins IL-1 and IL-2 in the body.

The STSS is an acute and potentially life-threatening condition similar to Gram-negative sepsis and septic shock. It is a multi-system disease characterised by fever, hypotension, myalgia, vomiting, diarrhoea, mucosal hyperaemia and an erythematous rash followed by desquamation of the skin, particularly on palms and soles. This condition was first documented in 1980 in the United States among the menstruating women who used highly absorbent vaginal tampons; the vaginal swab from these women showed a heavy growth of *S. aureus*. This condition can also occur in other individuals who have a local site of staphylococcal infection on skin or mucosa or any other extra-genital site.

Staphylococcal Scalded Skin Syndrome (SSSS) is caused by the exfoliative toxin exfoliatin. The condition is seen commonly in infants and children. It is associated with extensive exfoliation of the skin, in which the outer layer of the epidermis is separated from the underlying tissue. The appearance of extensive bullae characterises it. These bullae, when ruptured, may leave behind scalded, red, tender skin. The lesion typically starts peri-orificially or in skin folds. It usually resolves within 10 days.

Pemphigus neonatorum and bullous impetigo are the milder forms, whereas Ritter's diseases in

Table 25.4 Various specimens collected in staphylococcal infections

Specimen	Condition
Pus	Suppurative lesions and osteomyelitis
Sputum	Respiratory infections
Blood	Bacteraemia
Faeces and vomitus	Food poisoning
Urine	Urinary tract infections
Nasal and perineal swab	Suspected carriers

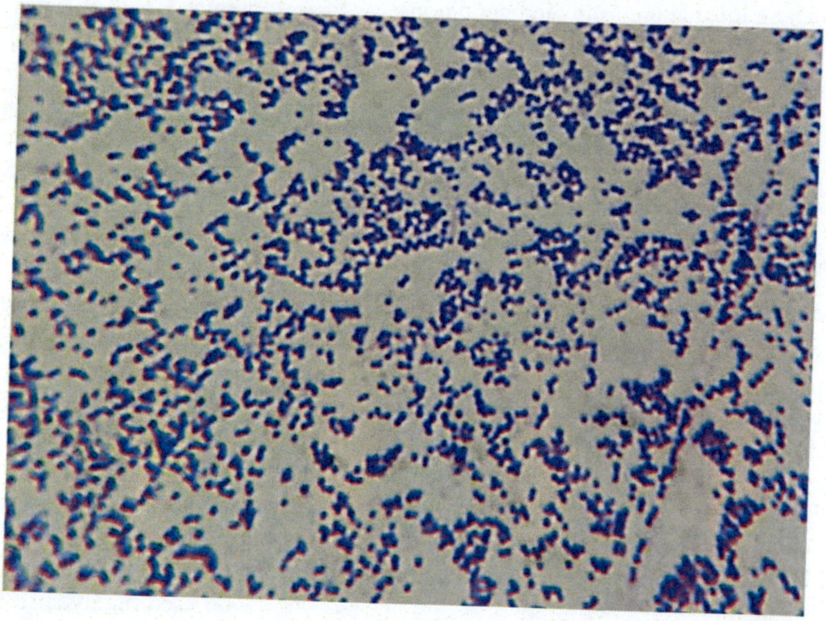

Fig. 25.3 Gram-stained smear showing staphylococci (×1000) (Courtesy of Prof S. Pramodhini)

the newborn and toxic epidermal necrolysis in the older persons are the severe forms of the SSSS.

Laboratory Diagnosis

Specimens

Specimens to be collected for demonstration of staphylococci depend on the nature of the lesion (Table 25.4).

Microscopy

Demonstration of Gram-positive cocci arranged in clusters and pus cells in the Gram-stained smears of pus (Fig. 25.3), wound exudate etc., are the characteristic features of pyogenic infection caused by *S. aureus*. It is noteworthy that microscopy alone is not adequate to differentiate various species of staphylococci or micrococci from one another. It is also of no value for sputum and other specimens where mixed bacterial flora is present.

Culture and Identification

The specimens are inoculated onto nutrient agar and blood agar and incubated at 37 °C for 24 h. Large, circular, smooth, convex and glistening colonies showing golden-yellow pigments can be observed on nutrient agar. On blood agar, the colonies show a zone of beta-haemolysis; any other species of staphylococci do not show this.

Specimens from heavily contaminated sources, such as vomitus and faeces, are inoculated on selective media (e.g. mannitol salt agar or salt milk agar). These media inhibit the growth of Gram-negative bacteria but allow the growth of staphylococci and certain other Gram-positive cocci.

Coagulase Test

Coagulase test is an important test to detect *S. aureus*. The test is done in two ways: tube coagulase test and slide coagulase test.

A tube coagulase test is carried out to detect free coagulase. In this test, 0.1 mL of an overnight broth culture is mixed with 0.5 mL of a 1:10 dilution of human or rabbit plasma. The plasma-broth culture mixture is incubated in a water bath at 37 °C for 3–6 h. In a positive test, the plasma is coagulated and does not flow (Fig. 25.4). Human or rabbit plasma, rich in CRF, is used in the test. The plasma is collected in vials containing anticoagulants, such as oxalate, heparin or Ethylene diamine tetraacetic acid (EDTA). Citrated plasma is not used because if the specimen is contaminated with Gram-negative bacilli, the latter may utilise the citrate and produce a false-positive reaction.

Slide coagulase test detects the bound coagulase or the clumping factor. The test is performed by mixing a dense suspension of the staphylococci with a loopful of undiluted rabbit plasma on a slide. In a positive test, clumping takes place within 10 s.

Antibiotic Sensitivity Testing

Antibiotic sensitivity testing plays an important role in treatment because often staphylococci develop resistance to routine antibiotics. Hence, an antibiotic susceptibility test can be used as a guide to treat patients with appropriate drugs.

Detection of MRSA in a Laboratory

The laboratory's screening for MRSA is performed with a 30 μg disc of cefoxitin as per the present guidelines. Cefoxitin is a surrogate marker for detection of the MRSA strains. Besides, oxacillin screen agar containing 6 μg/ml of oxacillin in Mueller-Hinton agar supplemented with 4% NaCl is one of the alternative methods of

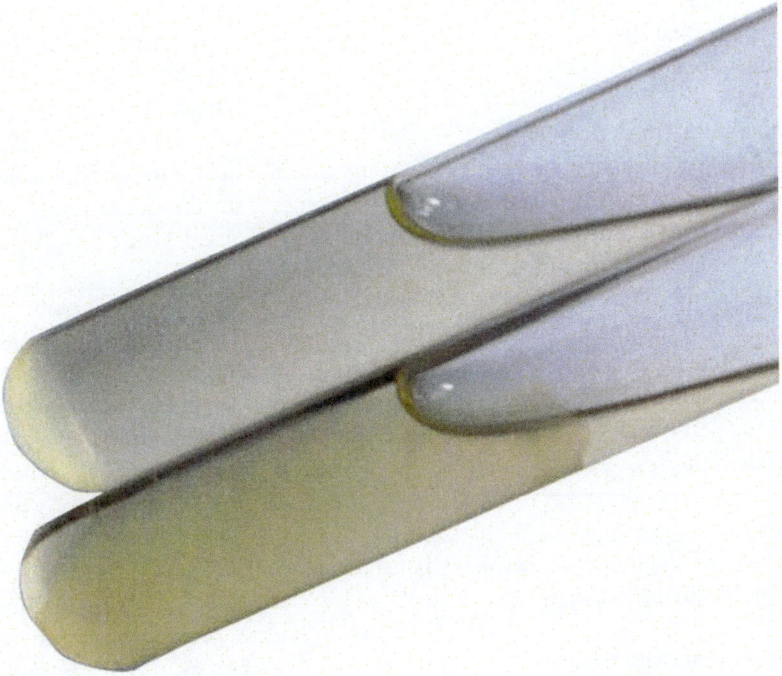

Fig. 25.4 Tube coagulase test

Table 25.5 Phage typing of human isolates of *Staphylococcus aureus*

Group	Phage
I	29, 52, 52A, 79, 80
II	3A, 3C, 55, 71
III	6, 42E, 47, 53, 54, 75, 77, 83A, 84, 85
IV	–
V	94, 96
Not allocated	81, 95

testing for MRSA. In addition, there are FDA-approved molecular assays for detecting the *mecA* gene and commercially available chromogenic agars for MRSA detection. Finally, latex agglutination or immunochromatographic membrane tests for PBP2a can be used to detect MRSA.

Serology

These may be helpful in the diagnosis of hidden deep infections. The body produces anti-staphylolysin antibodies in response to a staphylococcal infection, which may be useful in diagnosing deep-seated staphylococcal infections. If these antibody titres rise to more than 2 units/mL, it may indicate staphylococcal infections. The anti-staphylolysin antibody test has mainly been used to diagnose chronic osteomyelitis, and elevated results are also evident in joint infections.

Molecular Detection

For correct detection of species of *S. aureus* in clinical samples, several antibiotic-resistant genes like *MecA*, *SCCMec* as well as coagulase genes (*CoA*) and those for the cell wall components (PBP2a) are often detected by polymerase chain reaction (PCR). Classical PCR, RT PCR and multiplex PCR are the common tools used for detection. Commercial kits and platforms like GeneXpert (Cephid) are available for faster diagnostic purposes.

Typing

Typing of staphylococcal isolates is useful in an epidemiological survey.

1. Bacteriophage Typing

Strains of staphylococci can be typed by bacteriophage typing, which is useful in epidemiological studies. Bacteriophage typing is based on the susceptibility of cocci to bacteriophages. This is carried out by the pattern method, where a set of 23 standard typing phages of *S. aureus* is used to type staphylococcal isolates and distinguish them from one another by their patterns of susceptibility to lysis. In this method, the strain of *S. aureus* to be typed is inoculated on a nutrient agar plate to produce a lawn culture. After drying the plate, various phages at their routine test dilution (RTD) are applied over marked squares on a plate. Such plates are then incubated overnight at 30 °C and observed for the presence or absence of lysis of the colonies by the phages.

The phage-type of a strain is known by the designation of the phages that lyse it. Thus if a strain is lysed by phages 83A, 84 and 85, it is called phage-type 83A/84/85. By this method, most of the strains of staphylococci can be classified and divided into five lytic groups, while a few cannot be classified and constitute the unclassified group (Table 25.5). The National reference centre for staphylococcal phage typing in India is in the Department of Microbiology, Maulana Azad Medical College, New Delhi.

2. Molecular Typing

These include plasmid profile, DNA fingerprinting using RFLP, random amplification of polymorphic DNA(RAPD) ribotyping and PCR-based analysis of genetic pleomorphism. Sequence-based typing methods (MLST) and PFGE have recently been used in epidemiological investigations.

3. Other Typing Methods

S. aureus has been classified into six biotypes (A, B, C, D, E and F) by serotyping. Most human pathogenic strains belong to biotype A.

Treatment

Systemic antibiotics are necessary for deep-seated and systemic infections. Benzylpenicillin is the drug of choice for penicillin-sensitive strains of *S. aureus*. Erythromycin, vancomycin or first-generation cephalosporins are recommended for patients with an allergy to penicillin.

Antibiotics Resistance in Staphylococci

The emergence of bacterial resistance of different types among the staphylococcal strains word wide is a matter of serious concern.

Penicillin Resistance in Staphylococci

Penicillin resistance in bacteria has been increasingly recognised since 1945. Nearly 80% or more strains of *S. aureus* are resistant to penicillin. It is of three types:

1. Beta-lactamase-mediated resistance

This type of resistance may be due to the production of the enzyme penicillinase (beta-lactamase), which is plasmid mediated. This enzyme inactivates penicillin by splitting the beta-lactam rings. Staphylococci produce four types of penicillinases (A, B, C and D). Penicillinase plasmids are transmitted to the staphylococci by both transduction and conjugation. The plasmids also carry markers of resistance to heavy metals, such as arsenic, cadmium, mercury, lead and bismuth, as well as to other antibiotics, such as erythromycin and fusidic acid. Beta-lactamase producing trains are sensitive to beta-lactamase inhibitor formulations such as amoxicillin-clavulanic acid.

2. Chromosomal-mediated resistance

This type of resistance is characterised by a reduction in the affinity of the penicillin-binding proteins (PBPs) present on the cell wall of the staphylococci to the beta-lactam antibiotics. This also contributes to the resistance of staphylococci to penicillins and other beta-lactam antibiotics.

3. Tolerance to penicillin

Staphylococci developing tolerance to penicillin are only inhibited but not killed.

Penicillin-resistant strains can be treated with beta-lactamase-resistant penicillins such as cloxacillin, methicillin or vancomycin.

Methicillin-Resistant *Staphylococcus aureus* (MRSA)

This occurs due to a reduction in the affinity of the penicillin-binding proteins (PB2A) present on the cell wall of the staphylococci to the beta-lactam antibiotics. This is regulated by a set of staphylococcal cassette chromosomal mec genes (*SCCmec*), especially *the mecA* gene. The chromosomal *mecA* gene is responsible for producing PB2A proteins, thereby conferring a high degree of intrinsic resistance to all the B-lactams. MRSA denotes the resistance of *S. aureus* to penicillin and all other beta-lactam antibiotics, including the third-generation cephalosporins and carbapenems.

MRSA strains are classified as community-acquired MRSA (CA MRSA) or hospital-acquired (HA MRSA). The HA-MRSA and CA-MRSA strains can be differentiated in certain aspects, though not with complete certainty. The *SCCmec* types I, II and III are usually associated with HA-MRSA, while *SCCmec* types IV and V are seen in CA-MRSA. Drug resistance in CA-MRSA is less as compared to HA-MRSA. The gene encoding the bicomponent toxin named Panton-Valentine leucocidin *(luk-PV)* is commonly seen in CA-MRSA. HA-MRSA infections usually are associated with invasive procedures or devices and can spread by health care workers due to improper infection control practices. CA-MRSA usually spreads by skin-to-skin contact in overcrowded communities with inadequate hygiene. However, with an extensive circulation of both types of MRSA, this distinction is gradually losing its importance.

For invasive infections due to MRSA strains, vancomycin or daptomycin is the drug of choice. Alternative agents used for second-line therapy include telavancin, ceftaroline and linezolid.

Vancomycin resistance *Staphylococcus aureus* (VRSA)

These are usually the strains of MRSA that have also acquired resistance to vancomycin. VRSA strains have been recognised in some places around the globe. This type of resistance is due to the acquisition of *vanA* gene clusters. There are only limited drugs available for the treatment of VRSA. These include daptomycin, linezolid, telavancin, ceftaroline and quinupristin–dalfopristin.

While VRSA is not commonly encountered, MRSA strains with intermediate or reduced susceptibility to vancomycin are common entities in clinical settings. These strains with intermediate susceptibility to vancomycin are called **Vancomycin intermediate resistance (VISA)**. Any strain with a heterogeneously resistant population of *S. aureus* is called h-VISA, which ultimately progresses to VISA. Infections due to VISA strains are treated similarly to those for VRSA. The use of appropriate infection control practices, including contact precautions and adherence to hand hygiene by healthcare personnel, can reduce the spread of VISA and VRSA.

Epidemic Methicillin-resistant (EMRSA)

has been observed in some strains of MRSA. These strains have been noted to get easily disseminated in hospital environments, thus causing many healthcare-associated infections (HAIs). Epidemic strains of MRSA are defined as those identified in two or more patients in two or more hospitals. Among others, EMRSA-15 has emerged as a dominant clone in hospitals across the globe. Infections due to these strains are treated as those with MRSA. The colonisation of healthcare workers with these strains is treated with applications of nasal mupirocin (2%,) and whole-body antisepsis (3% hexachlorophene emulsion or 20% cetrimide shampoo).

Epidemiology

Geographical Distribution

Staphylococcal infections are found throughout the world. Nearly one-third of the adult population is an asymptomatic carrier of staphylococci. Hospital infections caused by *S. aureus* are worldwide in distribution.

Habitat

Staphylococci are the primary pathogens of humans and animals. They are present as commensals on the skin, in the glands of the skin and on mucous membranes. The cocci are commonly found in the intertriginous skin folds, perineum, axillae and vagina. Approximately 35–50% of normal adults carry *S. aureus* in the anterior nares, 10% in the perineum and 5–10% in the vagina.

Reservoir, Source and Transmission of Infection

Human cases and carriers are important reservoirs of infection.

Human cases of cutaneous and respiratory infections shed large numbers of staphylococci into the environment for a prolonged period. Staphylococci colonise the skin very early in life (in neonates on the umbilical stump). Staphylococci shed by the patients and carriers contaminate handkerchiefs, bed linens, blankets and other inanimate fomites and persist in them for weeks. *S. aureus* found in the nose and sometimes on the skin, especially in hospital staff and patients, is the main source of infection in hospitals.

Staphylococcal carriers play a major role in transmitting the infections. Staphylococcal carriage is very common in the babies born in the hospitals; their umbilical stump is usually colonised with Staphylococci. On the other hand, healthy adults are persistent or intermittent nasal carriers, and they play a key role in the epidemiology and pathogenesis of staphylococcal infections. However, the persistent nasal carriers are at high risk of self-infection, especially during surgical site infections, other infections and bacteraemias. Few persistent carriers are called shredders, and they release a large number of staphylococci into their surroundings, which can be persistent for many days to weeks. Hence, proper handwashing and topical agents used on the skin and nose, such as mupirocin and chlorhexidine to eradicate the agents, effectively prevent and control nosocomial infections caused by MRSA.

Domestic animals, such as cows, can also be reservoirs of staphylococcal infection.

Staphylococcal infections may be acquired through self-inoculation from the nose or other sites in patients who harbour staphylococci (endogenous infection) or direct contact with infected humans, carriers, and less frequently, animals (exogenous infection). Exogenous infection can also be acquired by close contact with infected fomites or inhalation of air droplets in a heavily contaminated environment.

Hospital-Acquired *S. aureus* Infections

Infections caused by MRSA have been increasingly reported worldwide since 1980. The infection is also being increasingly reported now from different hospitals. MRSA usually colonises the broken skin and can cause a wide range of local and systemic staphylococcal infections.

Hospital staffs harbouring MRSA are the chief source of infection for the patients. These strains can cause a wide range of infections, including bacteraemia, endocarditis and pneumonia. These strains are increasingly recognised as important agents of hospital-acquired infection in hospitalised patients undergoing prosthetic heart valve surgery.

S. aureus is the most common cause of hospital-acquired infections. Certain strains of *S. aureus* causing hospital infections are known as hospital strains. They exhibit certain properties: (1) they are usually resistant to penicillin, methicillin and other routinely used antibiotics, (2) they belong to certain bacteriophage types and (3) some of the strains (e.g. phage-type 80/81) are known to cause hospital infections throughout the world. Such strains are known as "epidemic strains."

Prevention and Control

S. aureus infection does not cause any life-long immunity. It causes repeated infections in a susceptible host.

There is no effective immunisation with toxoids or bacterial vaccines against staphylococcal infection. Cleanliness, frequent handwashing and aseptic management of lesions help control *S. aureus* infection. Treating with nasal creams containing neomycin or bacitracin prevents recurrent infections in nasal carriers of *S. aureus*. The topical application of antimicrobial agents prevents the dissemination of infection from the abscesses.

Other Coagulase-Positive Staphylococci

Other than *S. aureus*, few other *Staphylococcus* species such as *Staphylococcus intermedius* and *Staphylococcus hyicus* show coagulase positivity upon slide or tube coagulation tests. These species are involved in animal infections but do not cause human infections.

Coagulase-Negative Staphylococci (CONS)

Coagulase-negative staphylococci (CONS) are the normal flora of the skin. CONS are opportunistic bacteria. They cause infections in debilitated or immunocompromised patients and patients fitted with urinary catheters, cardiac valves, pacemakers and artificial joints.

Staphylococcus epidermidis

S. epidermidis forms white colonies on blood agar. It is catalase positive, coagulase negative and does not ferment mannitol. It tolerates salt, survives drying and is highly antibiotic resistant. It is a normal skin commensal. The carriage rate is as high as 100%. This bacterium is transmitted by self-inoculation or contact with infected patients and hospital personnel.

The ability to produce slime (Biofilm) is an important virulence factor of the bacterium. *S. epidermidis* causes infection by adhering itself to the surface of the intravenous plastic catheters and prosthetic devices. The adherence is believed to be facilitated by polysaccharide glycocalyx known as slime, produced in large quantities by the bacteria. Slime also inhibits the action of lymphocytes and neutrophils. *S. epidermidis* is an important agent of hospital-acquired infection. The bacteria are usually multi-drug resistant. It causes infection in patients with intravenous catheters and other prosthetic devices, such as heart valves leading to bacteriaemia. It causes central-line-associated bloodstream infection (BSI). It also causes endocarditis in patients with prosthetic valves, especially among drug addicts. Vancomycin is the drug of choice to treat infection caused by *S. epidermidis*.

Staphylococcus saprophyticus

S. saprophyticus forms white colonies on blood agar. It is catalase positive, coagulase negative and does not ferment mannitol. It normally inhabits the skin and genital mucosa. The bacterium causes
urinary tract infection by endogenous spreading colonised women. It adheres to the epithelial cells lining the urogenital tract. It causes dysuria, pyuria and haematuria. It also causes urethritis, catheter-associated urinary tract infections, prostatitis in elderly men and rarely, sepsis and endocarditis. Urinary tract infection caused by *S. saprophyticus* can be treated with quinolones such as norfloxacin or with trimethoprim sulfamethoxazole.

S. epidermidis and *S. saprophyticus* are distinguished from each other by their reaction to antibiotic novobiocin—*S. epidermidis* is sensitive to novobiocin, while *S. saprophyticus* is resistant. Polymyxin B sensitivity is carried out by using a 300-U polymyxin B disk on an overnight culture of staphylococci on Mueller–Hinton agar. Polymyxin resistance is shown by an inhibition zone of <10 mm.

The differences between *S. aureus*, *S. epidermidis* and *S. saprophyticus* are summarised in Table 25.6.

Other Coagulase-Negative Staphylococci

Many other CONS have been reported recently to cause human infections. These include the following:

S. haemolyticus causes bacteraemia, endocarditis, urinary tract infection and wound infection. *S. saccharolyticus* causes endocarditis. *S. hominis* causes bacteraemia in cancer patients. *S. schleiferi* causes wound infections, bacteraemia and in-dwelling catheter infections. *S. lugdunensis* causes endocarditis, peritonitis, osteomyelitis and breast abscesses. *Staphylococcus simulans* causes septicaemia, osteomyelitis and septic arthritis.

Table 25.6 Differentiating features of *Staphylococcus aureus*, *Staphylococcus epidermidis* and *Staphylococcus saprophyticus*

Test	*Staphylococcus aureus*	*Staphylococcus epidermidis*	*Staphylococcus saprophyticus*
Coagulase	Positive	Negative	Negative
Clumping factor	Positive	Negative	Negative
Heat-stable nuclease	Positive	Negative	Negative
Urease	Variable	Negative	Positive
β-galactosidase	Negative	Negative	Positive
Alkaline phosphatase	Positive	Positive	Negative
Polymyxin B	Resistant	Resistant	Sensitive
Novobiocin	Sensitive	Sensitive	Resistant
Acid from mannitol	Positive	Negative	Negative
Acid from trehalose	Positive	Negative	Negative
Acid from mannose	Positive	Negative	Negative
PYR test	Negative	Negative	Positive

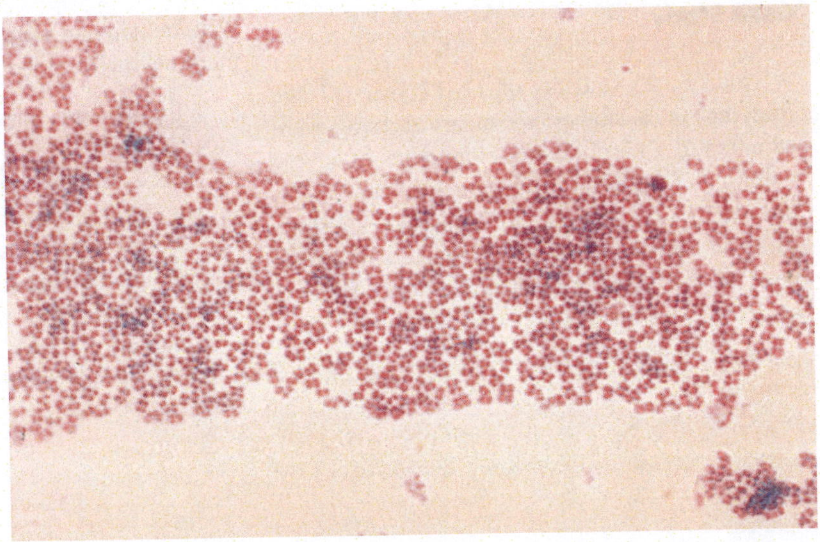

Fig. 25.5 Gram-stained picture of *Micrococcus* (×1000)

Micrococcus

Micrococci, compared to staphylococci, are larger and measure up to 2 μm in diameter. In smears, they appear as Gram-positive cocci arranged in tetrads (Fig. 25.5). In culture, they produce colonies with yellow, pink or red pigments. It is doubtful that they are human pathogens. Hugh and Leifson's oxidative fermentation test can differentiate micrococci from staphylococci. In this biochemical test, micrococci show an oxidative pattern, whereas staphylococci produce a fermentative pattern. The distinguishing characteristic features of *Staphylococcus* and Micrococci are summarised in Table 25.7.

Table 25.7 Distinguishing characteristics of *Staphylococcus aureus* and *Micrococcus* spp.

Character	*Staphylococcus aureus*	*Micrococcus* spp.
Gram staining	Gram positive cocci usually in grape like clusters Uniform staining	Gram positive cocci in tetrads Frequently staining is irregular Usually larger than *Staphylococcus* species
Colony characters	Medium to large (0.5–1.5 μm); smooth, slightly raised, low convex, opaque; most colonies pigmented creamy yellow (golden yellow); beta-haemolytic colonies can be seen	Small to medium (1–2 μm); opaque, convex; non-haemolytic; wide variety of pigments (white, tan, yellow, orange, pink)
Modified oxidase	Negative for cytochrome C	Positive for cytochrome C
Coagulase test	Positive	Negative
Oxidative–fermentative (OF) test	Breakdown carbohydrates by fermentation	Breakdown carbohydrates by oxidative mode
Aerobic acid production of glycerol in the presence of erythromycin	Positive	Negative
Aerotolerance	Facultative anaerobes	Usually grow aerobically
Lysostaphin Sensitivity Test (200 μg/μL)	Lysed with lysostaphin (sensitive)	Not lysed with lysostaphin (resistant)
Furazolidone susceptibility (100 μg of furazolidone disc)	Sensitive	Resistant
Bacitracin sensitivity test	Resistant to 0.04 U of bacitracin	Susceptible to 0.04 U of bacitracin

Case Study

A group of 25 students of 11–12 years of age studying in a higher secondary school were admitted to a hospital with complaints of severe vomiting and diarrhoea within 3 h of consuming the food prepared in their school. History revealed that a new cook appointed a few days back in the school prepared the food. All the students consumed the same food. Children were treated and were discharged after observing them overnight in the hospital.

- What is the possible cause of this food poisoning?
- What is the possible reservoir for this organism that was responsible for this outbreak?
- What tests will you perform to establish the etiological diagnosis of the condition?
- What steps you will take to prevent this infection?

Streptococcus, Enterococcus and *Pneumococcus* 26

Streptococci are Gram-positive cocci arranged in pairs or chains and require complex nutrients for their growth. Streptococci are part of the normal flora in humans and animals. They are catalase negative, by which they are distinguished from staphylococci. They are relatively fastidious bacteria requiring enriched medium, such as blood agar, for their growth.

Streptococcus

The streptococci are classified into aerobes and obligate anaerobes based on their oxygen requirement. Obligate anaerobes are designated as peptostreptococcus, described in detail in Chap. 30. The aerobes and facultative anaerobes are further classified as follows.

1. Streptococci based on oxygen requirement are classified into aerobes and obligate anaerobes
   Streptococci are further classified based **on haemolysis on blood agar**. Brown (1919) classified these aerobic streptococci into three groups based on their growth in 5% horse blood agar:
   (a) **Alpha-haemolytic streptococci** produce a narrow, greenish zone of partial haemolysis of RBCs around the colonies. The greenish discolouration is due to the formation of a reduced product of haemoglobin. Most alpha-haemolytic streptococci are found as commensals in the upper respiratory tract of humans and may cause opportunistic infections. Viridans streptococci and pneumococci are examples.
   (b) **Beta-haemolytic streptococci** produce a well-defined, clear and colourless zone of haemolysis (2–4 mm wide) around the colonies. RBCs in the zone of haemolysis are completely lysed. This lysis is due to the liberation of enzymes streptolysin O (SLO) and streptolysin S (SLS). The term haemolytic streptococci apply only to beta-haemolytic streptococci. Most of the pathogenic streptococci belong to this group, and amongst them, *Streptococcus pyogenes* is the most important one.
   (c) **Gamma-haemolytic streptococci** do not produce any haemolysis or discolouration on blood agar. These non-haemolytic streptococci are generally found as commensals. These cocci are now placed under a separate genus, *Enterococcus*.
2. Streptococci are also classified based on the cell wall carbohydrate antigen(C)
   Lancefield classification, suggested by Lancefield (1933), is a serological classification of the beta-haemolytic streptococci. The classification is based on the detection of group-specific carbohydrate antigen (C antigen) on the cell wall of the streptococci. In this classification, the beta-haemolytic streptococci are classified into 21 serological

Table 26.1 Key features of medically important streptococci

Species	Lancefield serogroup	Haemolysis	Habitat in humans	Diseases
Streptococcus pyogenes	A	Beta	Skin and throat	Acute group A streptococcal respiratory infection (pharyngitis or tonsillitis), skin infection (pyoderma), rheumatic fever following respiratory infection and glomerulonephritis following respiratory or skin infection
Streptococcus agalactiae	B	Beta	Female genital tract and rectum	Neonatal meningitis and sepsis and infections in adults
Streptococcus dysgalactiae sub spp. *equisimilis*	C	Beta	Throat	Endocarditis, bacteraemia, pneumonia, meningitis and mild respiratory tract infections
Streptococcus anginosus	A, C, F and G untypable	Beta (alpha and gamma)	Throat, female genital tract and colon	Sub-cutaneous or organ abscesses, endocarditis and mild upper respiratory tract infections
Enterococcus sp.	D	Gamma (alpha and beta)	Colon	Endocarditis, suppurative infections and urinary tract infections
Streptococcus bovis	D	Gamma	Colon	Endocarditis
Viridans streptococci	Not typed	Alpha	Female genital tract, mouth and colon	Dental caries, sub-acute endocarditis and intraabdominal suppurative infections

groups known as **Lancefield groups**, designated from A to W (except for I and J). Most haemolytic streptococci that cause human infections belong to group A. Group A streptococci are also known as *S. pyogenes*, while group B streptococci are known as *Streptococcus agalactiae* (Table 26.1).

Based on the M, T and R protein antigens present on the cell surface, *S. pyogenes* has been further classified into 80 serotypes. This classification is known as **Griffith typing**. M protein is the most important type-specific antigen. This serotyping is important for epidemiological studies (Fig. 26.1).

Streptococcus pyogenes (Group A Streptococci)

S. pyogenes is the most important human pathogen. It causes pyogenic infections (bacterial pharyngitis and cellulitis), toxin-mediated diseases (scarlet fever and toxic shock syndrome) and immunological diseases (Acute glomerulonephritis and rheumatic fever).

Properties of the Bacteria

Morphology

S. pyogenes is Gram-positive *cocci* measuring 0.6–1.0 μm in diameter. It is arranged in long chains. Some strains of *S. pyogenes* and group C streptococci produce capsules during the first 2–4 h of growth. The capsule is composed of hyaluronic acid containing repeating glucuronic acid and *N*-acetyl glucosamine molecules. It is chemically similar to host connective tissue and is therefore non-antigenic. Capsulated strains produce mucoid colonies on the blood agar. The bacteria are catalase negative; by this property, they are distinguished from staphylococci.

Culture

S. pyogenes grows on enriched media, such as blood agar. It produces small white to grey colonies with a clear zone of beta-haemolysis. The presence of 10% CO_2 enhances the growth and haemolysis of colonies.

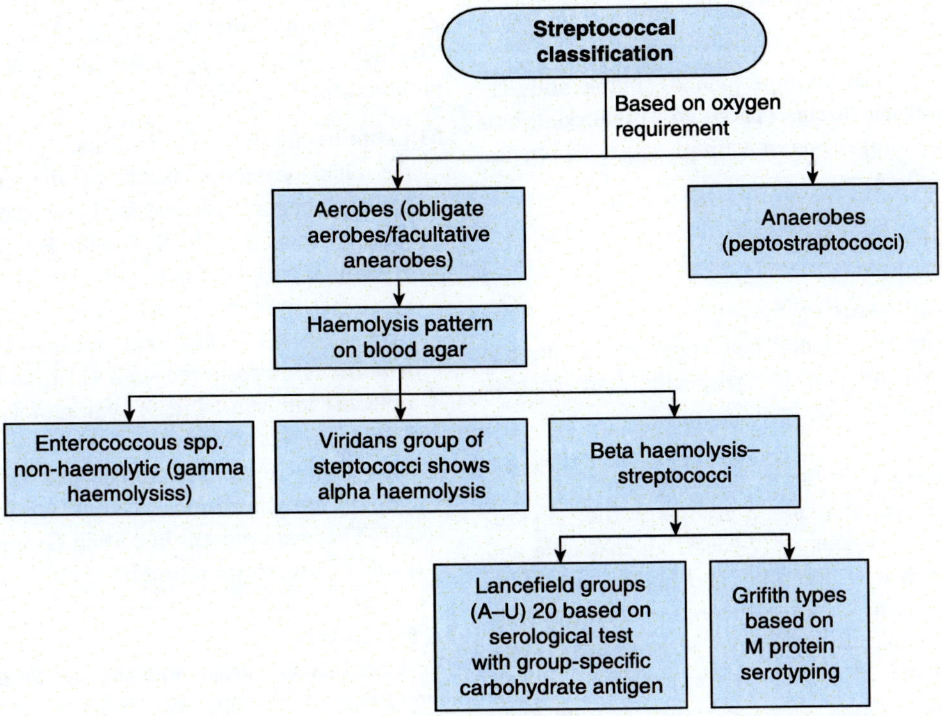

Fig. 26.1 Flow chart showing the classification of streptococci

Table 26.2 Virulence factors of *Streptococcus pyogenes*

Virulence factors	Biological functions
Cell wall associated polymers and proteins	
Capsule	Prevents phagocytosis
Teichoic acid	Binds to epithelial cells
M protein	Adhesin and antiphagocytic; inactivates C3b—an important complement factor responsible for phagocytosis; interferes with opsonisation via the alternative complement pathway
F protein	Mediates attachment to epithelial cells
Enzymes	
Streptokinase	Breaks down the fibrin barrier around the infected site, thereby facilitating spread of the infection
Deoxyribonucleases	Depolymerises free DNA present in the pus
Hyaluronidase	Hydrolyses hyaluronic acids in the matrix of the connective tissues
Toxins	
Streptococcal pyrogenic exotoxins (SPEs)	Dissolves the clot, thrombi and emboli; thereby facilitates spread of the bacteria in tissues
SLO and SLS	Lyse erythrocytes, leucocytes and platelets; stimulate production of lysosomal enzymes
Pyrogenic exotoxins	Release large amounts of cytokines from helper T cells and macrophages; rapidly destroy tissues

Pathogenesis

Streptococci have more than 20 soluble antigens, enzymes and toxins (Table 26.2) that contribute to the pathogenesis of various stages of streptococcal diseases.

Virulence Factors

Cell Wall-Associated Factors

These include several factors present on the cell wall and include group-specific carbohydrate antigen, protein antigen and pili.

1. Group-Specific Carbohydrate Antigen

The cell wall contains a group-specific polysaccharide, a polymer of *N*-acetyl glucosamine and rhamnose. It is non-toxic and hapten in rabbits. Based on group-specific carbohydrate (C) antigen, *S. pyogenes* strains have been divided into 21 groups (A–W) except I and J by Lancefield; hence they are known as Lancefield groups. The "C" antigen can be extracted by acid extraction with hydrochloric acid (Lancefield method), formamide extraction at 150 °C (Fuller method), autoclaving (Rantz and Randall method) and enzyme produced by *Streptomyces albus* extraction (Maxted method). After the extraction, the carbohydrate component is treated with type-specific antiserum by a precipitation reaction or by immunofluorescence to group the isolates of *S. pyogenes* (Fig. 26.2).

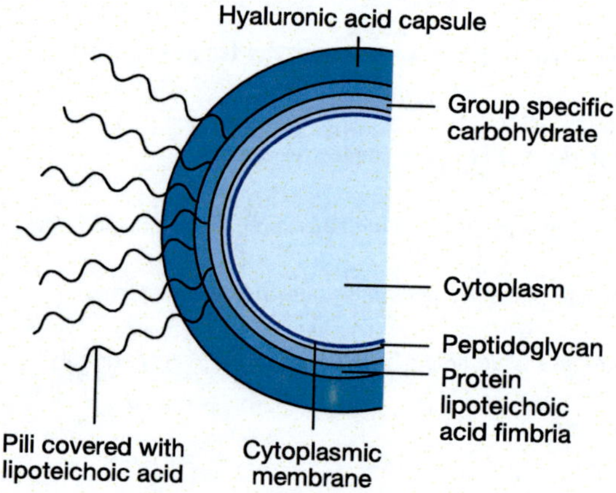

Fig. 26.2 Antigenic structure of *Streptococcus pyogenes*

2. Type-Specific Proteins

The cell wall of *S. pyogenes* has three major proteins: M, T and R.

M proteins are the most important protein. It is the chief virulence factor of the cocci. It inhibits phagocytosis and facilitates the attachment of cocci to epithelial cells. Based on M protein, *S. pyogenes* is divided into more than 80 (1–60) serotypes. *S. pyogenes* of M types 1,3,5,6,14,18,19 and 24 are rheumatogenic.

T proteins are trypsin-resistant (T) proteins and are acid- and heat-labile. T typing of strains of *S. pyogenes* is useful in the epidemiological surveillance of the infection caused by cocci.

R proteins are pepsin sensitive but trypsin resistant. These are present in certain Group A, B, C and G streptococci strains.

3. Pili (Fimbria)

These are hair-like projections that appear outside the Group A streptococci capsule. It partly consists of M protein and has a lipoteichoic acid which helps in the adhesion of streptococci into epithelial cells.

Toxins

1. Haemolysins:

Streptolysin O and streptolysin S are two types of **haemolysins** produced by Group A Streptococci.

Streptolysin O is an oxygen-labile protein. It causes beta-haemolysis only when colonies are grown under the surface of the blood agar plate. It is antigenic, and anti-streptolysin antibodies (ASO) against it develop in streptococcal infection. Demonstration of ASLO is important for determining a recent group A streptococcal infection and the late complications of streptococcal infections after the organisms have been eliminated from the host.

The SLO cross-reacts with similar haemolysins produced by streptococci of groups C and G, pneumolysins of *Streptococcus pneumoniae*, tetanolysin of *Clostridium tetani*, theta toxin of *Clostridium perfringens*, cereolysin of *Bacillus cereus* and listeriolysin of *Listeria monocytogenes*.

Streptolysin S is a serum-soluble (hence named as S) and oxygen-stable protein. It is non-antigenic. Hence, no antibodies against this toxin are demonstrated in serum. This toxin is responsible for haemolysis around the colonies grown on the surface of the blood agar. It inhibits chemotaxis and is antiphagocytic.

2. Streptococcal pyrogenic exotoxins (SPEs):
The toxin, otherwise called erythrogenic toxins, is of three types: A, B and C. Types A and B production is regulated by a bacteriophage phage in their genome. The chromosome mediates type C. These toxins act as superantigens and are responsible for scarlet fever and streptococcal toxic shock syndrome.

Enzymes

1. Streptokinase:
This enzyme is produced by *S. pyogenes* and group C and G streptococci. The enzyme is of two types: streptokinase A and B. The streptokinase activates plasminogen to form plasmin which breaks down the fibrin barrier around the infected site, thereby facilitating the spread of the infection. This thrombolytic property of the enzyme is made good use of in the medical management of myocardial infarction. Treatment is done by giving streptokinase intravenously.

2. Deoxyribonucleases (DNAase or streptodornase):
Most strains of *S. pyogenes* produce these enzymes. The enzymes are four types: deoxyribonucleases A, B, C and D. The enzyme depolymerises free DNA present in the pus, thereby reducing the viscosity of pus and helping in the spread of the infection. The enzymes are antigenic. Demonstration of anti-DNase B antibody in serum is diagnostic of *S. pyogenes* infections, particularly in skin infections.

3. Nicotinamide adenine dinucleotide (NADase):
NADase specifically acts on the coenzyme NAD to release nicotinamide. The actual biological importance of The NADase is not clear. However, it is believed to have leucotoxic activity.

4. Hyaluronidase:
The enzyme is produced by *S. pyogenes* and other groups, such as B, C and G streptococci, *Streptococcus suis*, *S. anginosus* and *S. pneumoniae*. The enzyme splits hyaluronic acid present in host connective tissue, thereby facilitating the spread of the bacteria through tissues. The enzyme is antigenic.

5. Serum opacity factor:
This is a lipoproteinase produced by some M types of *S. pyogenes*. This enzyme produces opacity when applied to agar gel containing swine or horse serum; hence it is known as serum opacity factor (SOP).

6. Other enzymes:
These include neuraminidase, amylase, esterase, lipase and beta-glucuronidase. However, their role in the pathogenesis of the disease remains unclear.

Clinical Syndromes

S. pyogenes produces a variety of clinical manifestations. These infections can be classified broadly as suppurative streptococcal infections, toxin-mediated infections and non-suppurative infections.

Suppurative Streptococcal Infections

Adherence of *S. pyogenes* to the surface of host cells is the first stage in the pathogenesis of the disease. The cocci adhere to the epithelium of the pharynx with the help of pili, lipoteichoic acid, F proteins and M proteins. Subsequently, *S. pyogenes* invades into epithelial cells mediated by M protein, F protein and other antigens of the cocci. Invasion of the cocci is suggested to be responsible for the persistence of infection, such as streptococcal pharyngitis and invasion into deep tissues.

S. pyogenes is the most common bacterium causing **pharyngitis** or **sore throat**. The incubation period is 1–4 days. Pharyngitis is characterised by inflammation of pharyngeal mucosa with exudate formation, tender enlarged cervical lymph nodes, fever and leucocytosis. The condition is commonly seen in children, and it spreads by droplet nuclei. Uncomplicated pharyngitis resolves within 3–5 days.

Pyoderma or **impetigo** is a localised skin infection, primarily affecting the face, arms, legs and other exposed parts of the body. The infection is acquired by direct contact with an infected person or fomites. The condition is caused by a limited number of *S. pyogenes* (49, 53–55, 59–61 etc.) of *S. pyogenes*. The condition is seen mainly in young children. In the tropics, impetigo is one of the important causes of acute glomerular nephritis in children.

Erysipelas is an acute and diffused skin infection affecting superficial lymphatics. It is characterised by red, swollen, and indurated skin with well-marked and raised borders. Commonly, the face and legs are affected. Erysipelas occurs most commonly in young children or older adults.

Cellulitis is the infection of the skin and subcutaneous tissues. It is characterised by local inflammation like oedema, erythema, tenderness, fever, headache, malaise and other systemic manifestations. It is spreading in nature, often without any apparent focus of infection.

Necrotising fasciitis is a rapidly spreading streptococcal infection of superficial and deep fascia. The infection is caused by certain M strains of *S. pyogenes* (M types 1 and 3), which produce pyrogenic exotoxins. These strains are also called "flesh-eating strains" due to the extensive destruction of muscle and fascia caused by them. The condition is also associated with a toxic shock-like syndrome, leading to disseminated intravascular coagulation and multi-system failure.

Toxin-Mediated Infections

Scarlet fever is a complication of streptococcal pharyngitis caused by certain strains of *S. pyogenes* producing pyrogenic exotoxins. It manifests as fever, pharyngitis and a characteristic rash. The rash is followed by desquamation. However, with penicillin and other antibiotics, the condition has become rare.

Toxic shock syndrome (TSS) is caused by certain strains of *S. pyogenes* (M serotypes 1 or 3) with a prominent hyaluronic acid capsule. This is a condition similar to staphylococcal toxic shock syndrome. The condition manifests initially as pain at the site of inflammation and non-specific systemic complaints, such as nausea, vomiting, diarrhoea, fever and chills. The condition progresses subsequently to multi-organ failure and shock. The condition occurs in people of all ages.

Non-Suppurative Streptococcal Infections

Many *S. pyogenes* cell components have antigenic similarities with various human-tissue antigens. The antigenic cross-reactions may be the reasons for rheumatic fever and other streptococcal diseases. Due to antigenic cross-reactions, the immune system produces autoantibodies, which damage human tissue or organs, leading to various clinical manifestations.

Acute glomerulonephritis (AGN) is a non-suppurative complication of *S. pyogenes* infection. The onset of infection typically occurs 2–3 weeks following skin infection or pharyngitis. It is caused by certain nephritogenic types such as pharyngeal (M types 1 and 12) and pyodermal strains (M types 49, 53–55 and 59–61). M protein-type 49 skin infection is most frequently implicated. The disease occurs due to the deposition of antigen–antibody complexes on the glomerular basement membrane, initiating inflammation. This leads to hypertension, generalised oedema, haematuria and proteinuria.

Rheumatic fever is an autoimmune disease exacerbated by the recurrence of streptococcal infection. It is caused by certain rheumatogenic strains of *S. pyogenes*. Rheumatic fever occurs due to antigenic cross-reaction between streptococcal proteins and the connective tissue antigens of the heart and joints. It has a latent period of 2–4 weeks. It is characterised by fever, migrating polyarthritis, presence of Aschoff nodules and carditis. Damage to heart valves may occur during infection (Table 26.3).

Table 26.3 Differentiating features of acute rheumatic fever and post-streptococcal glomerulonephritis

Feature	Acute rheumatic fever	Post-streptococcal glomerulonephritis
Hereditary factors	Contribute	Not known
Initial site of infection	Throat	Skin and throat
Prior exposure to *Streptococcus pyogenes*	Essential	Not required
Serotypes involved	All serotypes	Pyodermal strains (M types 42, 49, 53–55, 59–61); throat infection strains(1, 2 and 4)
Immune response	Prominent	Lower
Complement level	Unaffected	Decreased
Repeated attacks	Common	Absent
Penicillin prophylaxis	Essential	Not required
Course of the disease	Static/ progressive	Complete recovery
Prognosis	Variable	Excellent

Acquired immunity to streptococcal infection is based on developing specific antibodies against the antiphagocytic epitopes of M protein. The acquired immunity against a particular M type of streptococci lasts longer in untreated than treated persons.

Laboratory Diagnosis

Specimens

The nature of specimens to be collected depends upon the disease manifestations. Throat swabs, nasal swabs, pus, cerebrospinal fluid, blood, sterile body fluid etc. are collected for culture. Pikes transport medium is used in case of delay in receipt of specimens. The serum is used for serology.

Microscopy

Gram staining of pus or exudates is a rapid and presumptive diagnostic procedure. The presence of Gram-positive cocci in pairs and chains (Fig. 26.3) in association with leucocytes is suggestive of streptococcal infection. However, Gram staining is of no value in pharyngitis or genital infections because the streptococci are found as part of the normal flora in the oropharynx.

Culture and Identification

Throat swab culture is the most specific for the diagnosis of streptococcal pharyngitis. Ideally, the throat swab specimen should be collected from the tonsils and posterior part of the oropharynx because more bacteria are present at this site than in the anterior part of the mouth. After collecting the specimen, they are plated immediately on the blood agar plate; in case of delay, they are sent to the laboratory in Pikes transport medium.

In the laboratory, the specimens are inoculated on a **5% sheep blood agar** and incubated at 37 °C aerobically in the presence of 5–10% CO_2 for 2–3 days. The culture of the swabs on blood agar shows a clear zone of beta-haemolysis surrounding the small translucent to opaque colonies (Fig. 26.4).

Crystal violet blood agar is a selective medium for the culture of *S. pyogenes*. The addition of 0.0001% crystal violet to blood agar makes the medium highly selective for *S. pyogenes*. This inhibits all other Gram-positive cocci while allowing selective growth of *S. pyogenes*. **PNF medium** (horse blood agar containing polymyxin B sulphate, neomycin sulphate and fusidic acid) is another selective medium used to isolate *S. pyogenes*. These two

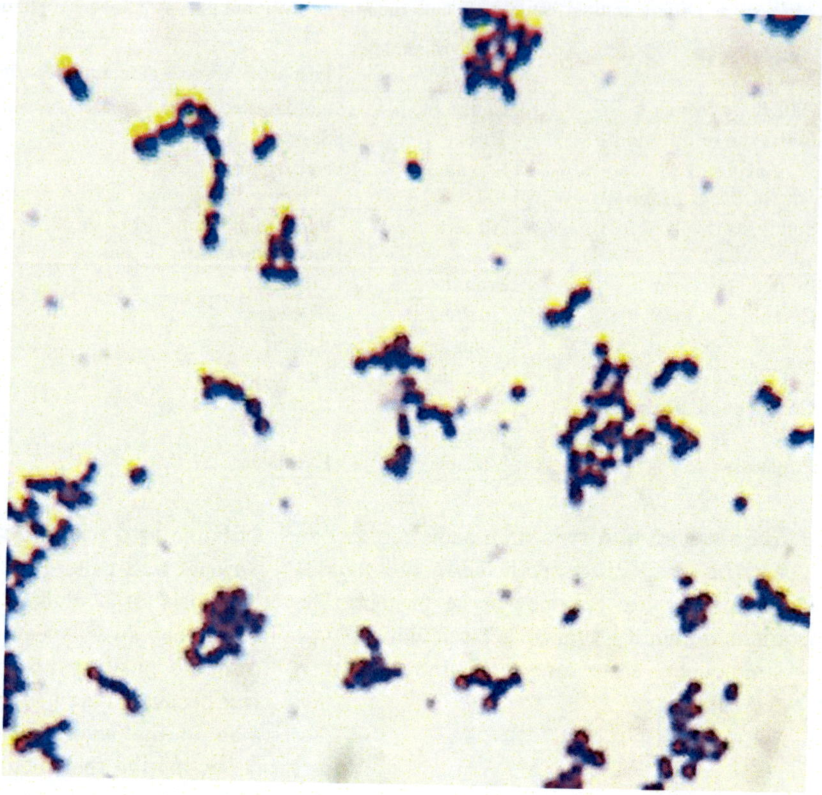

Fig. 26.3 Gram-stained pus smear showing streptococci in chains (×1000)

media are used to isolate streptococci from specimens with mixed flora.

Biochemical reactions are of little diagnostic value in the identification of cocci.

Rapid Direct Antigen Detection Test

Detection of group-specific carbohydrate antigen A directly in the throat swabs by direct fluorescent antibody test is very rapid and specific. The result for this test is obtained within 4 h. The test is as specific as a culture but is less sensitive.

Bacitracin Sensitivity Test

S. pyogenes can be distinguished from other streptococcal groups by their sensitivity to bacitracin. Bacitracin test is simple to perform and is useful for presumptive identification of *S. pyogenes*. It is positive in more than 95% of *S. pyogenes* strains and negative in non-group A streptococci (Fig. 26.5).

Serology

Serology is of value in the diagnosis of acute glomerulonephritis and rheumatic fever.

Antistreptolysin O (ASLO) antibodies appear in serum 3–4 weeks after initial infection by *S. pyogenes*. A titre greater than 200 indicates streptococcal infections. Higher antibody titres are found in patients with acute rheumatic but not in patients with glomerulonephritis and streptococcal pyoderma.

Antideoxyribonuclease B (anti-DNAse B) and other streptococcal enzyme antibodies are also demonstrated in *S. pyogenes* infections. The demonstration of antibodies against these antigens (titre > 300) may prove useful in diagnosing streptococcal pharyngitis and pyoderma.

Streptomyces test is a simple and convenient passive haemagglutination test. It is used to detect the streptococcal-specific antibodies in the patient

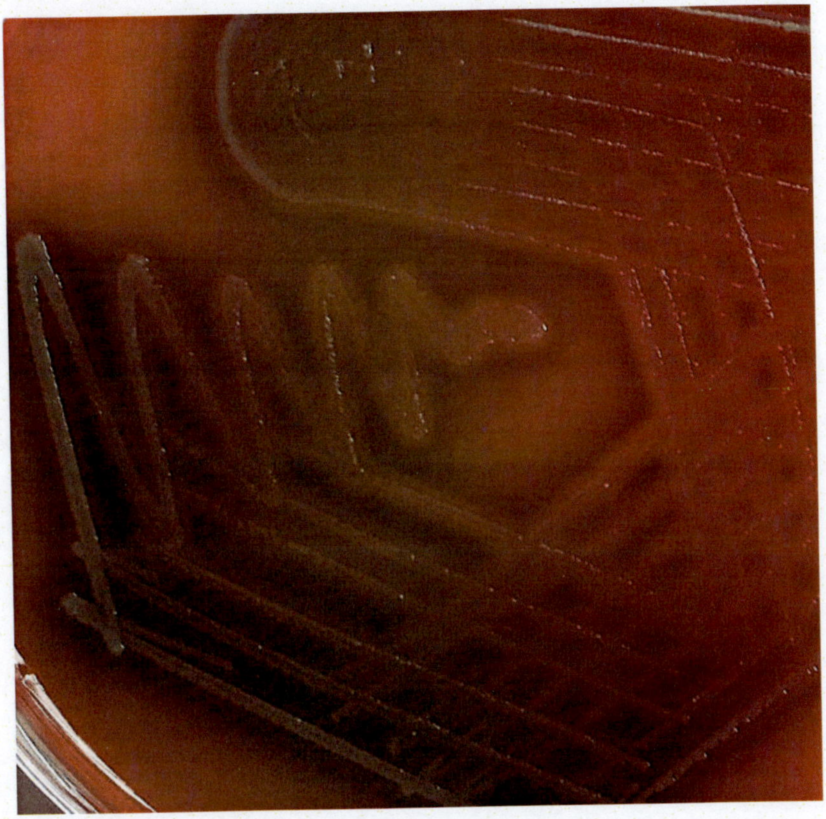

Fig. 26.4 Blood agar plate showing beta-haemolysis surrounding the colonies of *Streptococcus pyogenes* (Courtesy of Prof S. Pramodhini)

sera. The test is for all types of human streptococcal infections.

Typing

S. pyogens typing is based on MTR protein, commonly used for epidemiological studies. This typing method employs various precipitation or agglutination techniques by using specific antisera.

Treatment

Treatment of *S. pyogenes* infections by antibiotics varies depending upon the clinical conditions. Penicillin is highly effective against *S. pyogenes*. As of now, no penicillin-resistant strains of *S. pyogenes* have been documented in clinical practice. Penicillin, therefore, remains the drug of choice, except in penicillin-allergic individuals and complicated or invasive diseases. Erythromycin and clindamycin are given to patients allergic to penicillins. Recently, strains resistant to erythromycin have been reported. Sulphonamides and tetracycline are not used for streptococcal infections.

Epidemiology

Geographical Distribution

S. pyogenes infections are worldwide in distribution. Prevalence of streptococcal *pyoderma* is higher in tropics with no seasonal variation, whereas it is more common in winter months in temperate countries.

Rheumatic fever is most frequently observed in children aged 515 years, the age group most susceptible to *S. pyogenes* infections. The attack rate of rheumatic fever following upper respiratory tract infection is approximately 3% for

Fig. 26.5 Bacitracin sensitivity of *Streptococcus pyogenes*

persons with untreated or inadequately treated infections.

Habitat

Streptococci are normal flora of the oral cavity, nasopharynx, skin, fingernails, perianal region and intestine of humans.

Reservoir, Source and Transmission of Infection

Respiratory and salivary secretions in the form of droplets and contaminated fomites are the sources of *S. pyogenes* infection. Person-to-person transmission is the main route of transmission. The infection is transmitted through respiratory droplets by direct contact with an infected patient, fomites or arthropod vectors. Children with untreated acute infections spread organisms by their salivary droplet and nasal discharge. Overcrowding (crowded homes and classrooms) is an important factor in transmission. Both impetigo and pharyngitis are more likely to occur in children living in crowded homes and under poor hygienic conditions.

Prevention and Control

Chemoprophylaxis is most important in the prevention of acute glomerulonephritis or rheumatic fever.

Long-term chemoprophylaxis using penicillins to prevent streptococcal infection is recommended for children with a history of acute rheumatic fever. Antibiotic prophylaxis prevents streptococcal re-infection and further damage to the heart. Antibiotic prophylaxis is not recommended for patients with glomerular nephritis.

A multivalent streptococcal vaccine containing multiple M protein epitopes has been evaluated. Its efficacy is being proved in animal models. The vaccine is still in the experimental stage and is yet to be used in clinical practice.

Other Haemolytic Streptococci

Streptococci belonging to groups B, C, F and G and rarely H, K, O and R can also be associated with human infections.

Group B Streptococci (GBS)

S. agalactiae is the only species belonging to group B streptococci. This is a pathogen of the cattle causing bovine mastitis, hence named "agalactiae". *S. agalactiae* is Gram-positive cocci arranged in pairs and short chains in clinical specimens and is morphologically similar to *S. pyogenes*.

S. agalactiae is a commensal in the genitourinary tract and lower gastrointestinal tract. A vaginal carriage rate as high as 40–50% has been observed in some pregnant women. More than 50% of infants born to these mothers through vaginal delivery are colonised with *S. agalactiae*.

S. agalactiae in neonates can cause either early-onset or late-onset infections. Early-onset infection is acquired either in utero or from the mother's vagina during delivery. The clinical symptoms develop during the first week of life. The condition is characterised by septicaemia, meningitis or pneumonia. Late-onset infection is acquired from the mother or from another infant (environment) during 2–12 weeks of life. The condition manifests as septicaemia and meningitis. *S. agalactiae* in pregnant women causes urinary tract infection, particularly immediately after delivery. In non-pregnant women and men, *S. agalactiae* can cause osteomyelitis, arthritis, peritonitis and skin infections.

After overnight incubation, the cocci grow readily on blood agar and produce large colonies with beta-haemolysis. The **CAMP test** was first described in 1944 by Christie, Atkins and Munch–Peterson. The basis for the test is that *S. agalactiae* produces a diffusible and heat-stable protein, known as a CAMP factor that accentuates RBCs' haemolysis. The staphylococci produce an enzyme sphingomyelinase C which binds to the RBCs present in the blood agar. RBCs undergo haemolysis on exposure to CAMP factor liberated by *S. agalactiae*, producing a butterfly appearance (Fig. 26.6). Haemolysis showing a

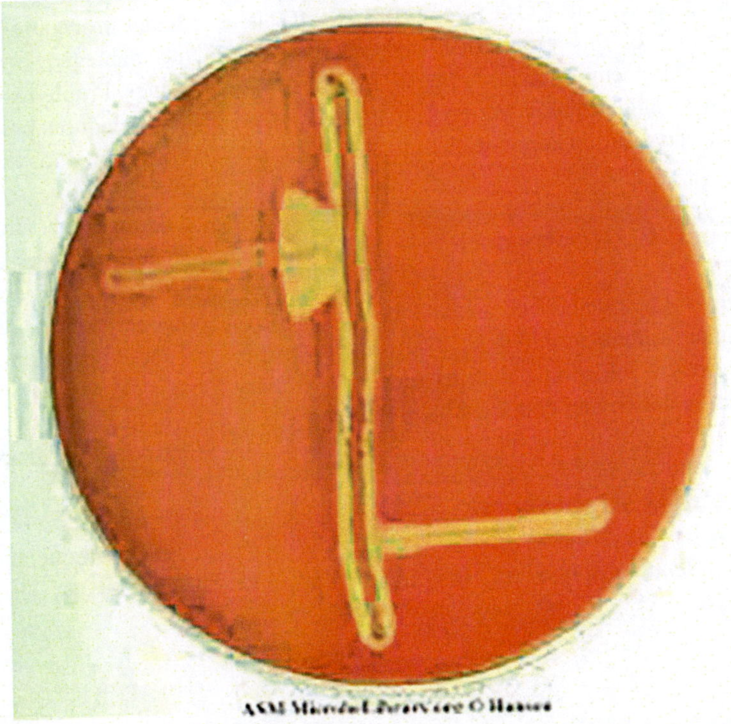

Fig. 26.6 CAMP test for *Streptococcus agalactiae* (https://commons.wikimedia.org/wiki/File:CAMP_test.JPG)

typical butterfly appearance indicates a positive test. *S. agalactiae* is CAMP positives. *Agalactiae* is hippurate hydrolysis positive. It hydrolyses the hippurate to produce hippuric acid.

Penicillin is the drug of choice. Vancomycin is given to persons allergic to penicillins. Recently, strains resistant to erythromycin and tetracycline have been documented.

Group C Streptococci

Group C streptococci are usually pathogens of animals. *Streptococcus equisimilis* is a group C *Streptococcus* which can cause occasional infections in humans. It causes upper respiratory tract infections, pneumonia, osteomyelitis, endocarditis, brain abscess and puerperal sepsis. *S. equisimilis* shows tolerance to treatment with penicillin; therefore, patients may not respond to treatment with penicillin.

Group F Streptococci

Group F Streptococci are called "minute streptococci". They grow well on blood agar in the presence of CO_2. Streptococcus MG is a member of this group that can cause primary atypical pneumonia in humans. Demonstration of cold agglutinins in serum is diagnostic of primary atypical pneumonia.

Group G Streptococci

Group G streptococci are the commensals of humans, monkeys and dogs. They may occasionally cause infections, such as tonsillitis, urinary tract infection and endocarditis in humans.

Group D Streptococci

Streptococcus bovis is important group D streptococci. It is associated with human infections, such as urinary tract infections and rarely endocarditis. It is susceptible to penicillins.

Viridans Streptococci

Viridans streptococci are a heterogeneous group of alpha-haemolytic and non-haemolytic streptococci. These bacteria produce a green pigment on the blood agar and are called *viridans* (Latin for "green"). Most isolates of viridans streptococci do not possess a group-specific carbohydrate; hence, they cannot be classified under the Lancefield classification of streptococci. A few examples are *Streptococcus mitis, Streptococcus mutans, Streptococcus salivarius* and *Streptococcus sanguinis*. These are commensal flora in the oral cavity, oropharynx, gastrointestinal tract and genitourinary tract.

Viridans streptococci can cause a variety of infections. They are commonly implicated in dental caries, sub-acute bacterial endocarditis and intra-abdominal suppurative infections. *S. sanguinis* is the most common causative agent of bacterial endocarditis in individuals with pre-existing heart lesions. There is transient bacteraemia following tooth extraction or other dental procedures, after which bacteria adhere to the damaged heart valves or prosthetic heart valves. Prophylactic use of antibiotics before dental procedures prevents such complications.

Earlier, most strains of Viridans streptococci were sensitive to penicillins. However, moderately resistant and highly resistant strains have been reported recently, particularly in the *S. mitis* group. Broad-spectrum cephalosporins or vancomycin are recommended for the treatment of these penicillin-resistant strains.

Other Streptococci

A few other streptococci have pathogenic potential and infrequently may cause human infections. It includes:

Table 26.4 Distinguishing features of *Streptococcus* and *Enterococcus*

Characteristics	*Streptococcus*	*Enterococcus*
Arrangement of bacteria	Pairs or in short chains	Pairs of oval cocci
Growth in the presence of 40% bile	Negative	Positive
Growth in the presence of 6.5% sodium chloride	Negative	Positive
Growth at 45 °C	Negative	Positive
Growth at pH 9.6	Negative	Positive

Zoonotic Streptococci

Streptococcus equi and *Streptococcus equinus* cause infections in farm animals and sometimes cause human infections. On the other hand, in recent years, *S. suis* has gained prominence as a causative agent of meningitis and sepsis in humans with up to 15% mortality. It is a normal commensal of the upper respiratory tract of pigs, and infections are confined to workers in pig farms or those working in abattoirs. *Streptococcus iniae* is a pathogen of freshwater fish like tilapia and may cause cellulitis, meningitis or sepsis among fish handlers.

Nutritionally Variant Streptococci (NVS)

These are normal commensals of the upper respiratory tract and oral cavity in humans and are so named because they require pyridoxine or L-cysteine as growth factors and grow as satellite colonies around helper bacteria which provide them with these factors. Originally designated the species of *Streptococcus adjacen*t and *Streptococcus defectives,* they were later reclassified into two new genera of *Abiotrophia* and *Granulicatella*. They are implicated in several bacteraemia cases progressing to endocarditis. A significant resistance of these organisms to penicillins and other β-lactams has been noted.

Enterococcus

The enterococci were classified earlier as group D streptococci because they possess the group D-cell wall antigen. These enterococci, however, showed several distinctive features (Table 26.4) by which they were separated from the streptococci. The genus *Enterococcus* has 16 species. *Enterococcus faecalis* and *Enterococcus faecium* are two important species known to cause human infections (Table 26.5).

Enterococci are commonly found in humans' and animals' gastrointestinal and genital tracts. Enterococci now are emerging as important agents of nosocomial infections. They cause urinary tract infections, especially in hospitalised patients. Indwelling catheters and urinary tract instrumentation are important predisposing factors. They are frequently isolated from cases of wound infections, particularly intra-abdominal. They also cause bacteraemia, infection of the bile duct and endocarditis.

The antimicrobial therapy includes a combination of aminoglycosides with penicillin, ampicillin or vancomycin. Enterococci are less sensitive

Table 26.5 Classification of *Enterococcus*

Groups	Species
Group I	*Enterococcus avium*
Group II	*Enterococcus faecalis*, *Enterococcus faecium* and *Enterococcus gallinarum*
Group III	*Enterococcus durans*
Group IV	*Enterococcus sulfureus* and *Enterococcus cecorum*
Group V	*Variants of Enterococcus faecalis* and *Enterococcus gallinarum*

to penicillin and gentamicin and are resistant to cephalosporins. Plasmid-mediated resistance is a major cause of resistance amongst drug-resistant strains of *Enterococcus*. Vancomycin-resistant *Enterococcus* has been emerging in recent years.

Pneumococcus

S. pneumoniae is a normal inhabitant of nasopharynx. They are the common cause of pneumonia worldwide.

Properties of Bacteria

Morphology

S. pneumoniae is Gram-positive cocci. In stained clinical specimens, they appear typically "lanceolate -shaped" with one end pointed and the other end round. They are arranged in pairs (diplococci) with the broad ends in apposition to each other. In cultures, they usually appear more rounded and are arranged in short chains. A polysaccharide capsule completely envelops each pair of cocci. The capsule is visualised by staining it directly with specific stains or by India ink negative staining or by Quellung reaction. The capsules are present in strains isolated from clinical specimens but are lost on repeated cultivation, which is called smooth-to-rough variation.

Culture and Biochemical Properties

S. pneumoniae is an aerobe and a facultative anaerobe. It grows at an optimum temperature of 37 °C (range 25–42 °C) and pH 7.8 (range 6.5–8.3). The growth is enhanced by the presence of 5–10% CO. Pneumococci are fastidious. They grow only on an enriched media such as blood agar or chocolate agar (supplemented with blood products) which will supply nutrients pH buffers etc.

On **blood agar** morphology of the colonies varies depending on the nature of the strain (whether capsulated or non-capsulated), the type of incubation (whether aerobic or anaerobic) and the time of incubation. Colonies on blood agar in anaerobic incubation show beta-haemolysis (greenish discolouration), but show alpha-haemolysis in aerobic incubation (Fig. 26.7).

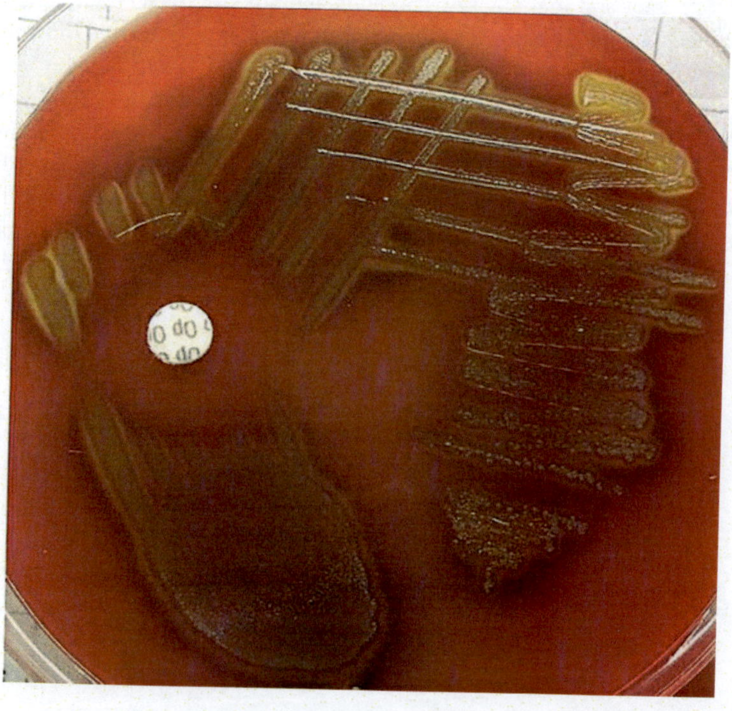

Fig. 26.7 Blood agar showing alpha-haemolysis and optochin sensitivity on blood agar produced by *Streptococcus pneumoniae*

Capsulated strains after overnight incubation produce round and mucoid colonies measuring 1–3 mm in diameter. Some strains, for example, type 3 *S. pneumoniae* (most virulent), produce copious quantities of capsular material and hence produce large mucoid colonies. Non-capsulated strains produce small and flat colonies.

On prolonged incubation, the colonies undergo autolysis and the centres become flattened or depressed (umbonation) and edges become raised, giving the colonies a typical draughtsman appearance (Fig. 26.8). The central flattening or depression is due to the production of intracellular enzymes, such as amidase, which lyses the bacteria. Bile salts, sodium lauryl sulphate and other surface-active agents enhance the process of autolysis of the bacterial colonies.

The capsules are present in strains isolated from clinical specimens but are lost on repeated cultivation. This is called **smooth-to-rough variation**. Non-capsulated rough (R) strains are avirulent; these forms arise as spontaneous mutants and outgrow the parental smooth (S) strains in artificial culture. In tissues, R forms are eliminated by phagocytosis. Transformation of a rough strain to a smooth one is also possible by treatment with smooth capsular substance containing DNA.

In the **liquid media** such as glucose broth, pneumococcal growth shows uniform turbidity. Due to the action of intracellular enzymes, pneumococci undergo autolysis in liquid culture, which can be enhanced by the sodium lauryl sulphate, bile salts and other surface-active agents. However, autolysis will not occur in heat-killed cultures.

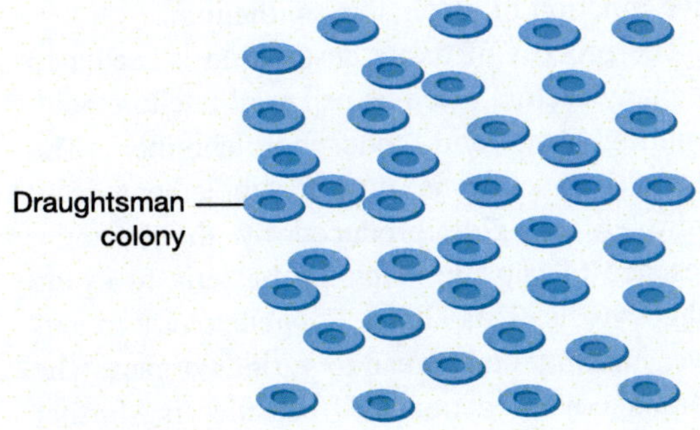

Fig. 26.8 Draughtsman colonies of *Streptococcus pneumoniae*

S. pneumoniae ferments many sugars, producing acid only but no gas. Fermentation of sugars is carried out in Hiss serum water or in serum agar slopes. *S. pneumoniae* ferments inulin, and this is an important test to differentiate it from those streptococci that do not ferment inulin. Bile solubility is a constant feature of pneumococci and is positive in all the capsulated and some non-capsulated variants. Pneumococci are catalase and oxidase negative.

Other Properties

Susceptibility to Physical and Chemical Agents

Pneumococci are delicate bacteria. They are killed by heating at 52 °C for 15 min and by usual strengths of disinfectants. Pneumococcal colonies die on prolonged incubation.

Optochin Sensitivity Test

Pneumococci are sensitive to optochin (ethylhydrocupreine)—a useful property to differentiate these from streptococci. Optochin sensitivity is usually performed by a paper disc containing 5 μg of the compound; an inhibition zone of 14 mm or more indicates sensitivity to optochin. In recent times, occasional optochin-resistant pneumococci have also been documented. The optochin targets an H + ATPase in pneumococci and resistance is thought to be due to a point mutation in one of sub-unit *ar* or c of the H + ATPase. This resistance to optochin is not related to the virulence of the bacteria.

Antigenic Structure

S. pneumoniae has three important antigenic components: C polysaccharide, F antigen and M protein.

1. **C polysaccharide** is present on the surface of cell wall. The exposed part of teichoic acid, which is linked to peptidoglycan layer and extends through the capsule, is known as C

polysaccharide or C antigen. This is a species-specific antigen and in no way related to the group-specific C carbohydrate antigen found in beta-haemolytic streptococci. The C polysaccharide present in the cell wall of pneumococci precipitates with C-reactive protein (CRP), a serum globulin.

The **CRP** is an acute-phase substance synthesised in the liver. It is not produced specifically against C antigen of pneumococci. It is present in low concentrations in healthy individuals, but the concentration increases in inflammation, malignancies and bacterial infections. CRP increases during pneumonia and disappears during convalescence, hence can be used as a prognostic tool. CRP is used as an index of the treatment in rheumatic fever and certain other conditions.

2. **F antigen** is the other type of teichoic acid, which is covalently bound to the lipids in cytoplasmic membrane. It is so called because it cross-reacts with the Forssman surface antigen of the mammalian cells. F proteins mediate the binding of pneumococci to the host cell surface.
3. **M protein** is a type-specific protein similar to the M protein of *S. pyogenes*, but is immunologically distinct. Antibodies against M proteins are not protective, as they do not inhibit phagocytosis of pneumococci.

Pathogenesis

Pneumococci cause disease primarily by their capacity to multiply in host tissues by avoiding the host defence mechanisms.

Virulence Factors

The polysaccharide **capsule** surrounds the cell wall. The capsule is essential for virulence, its role being to protect the cocci from phagocytosis. Virulent strains of *S. pneumoniae* have a complex polysaccharide capsule. The acidic and hydrophilic nature of the capsule allows the bacteria to escape phagocytosis by macrophages (Fig. 26.9).

The antibody to Pneumolysin is a membrane-damaging toxin produced by pneumococci. The toxin alters the mucociliary clearance function of respiratory epithelium and inhibits phagocytic cell oxidative burst essential for intracellular killing of the bacteria. It activates the classical complement pathway resulting in the production of C3a and C5a, thereby contributing to the disease process.

IgA protease, the enzyme produced by pneumococci disrupts secretory IgA-mediated clearance of the bacteria and thereby enhances the ability of the bacteria to colonise mucosa of the upper respiratory tract.

Cell wall associated polymers and proteins include teichoic acid, peptidoglycan, protein adhesin, phosphorylcholine, F protein that contribute to pathogenesis of pneumococcal diseases in various ways (Table 26.6).

Pathogenesis of Pneumococcal Disease

S. pneumoniae causes disease through the phages of (1) colonisation and invasion, (2) tissue destruction and (3) avoidance of opsonisation and phagocytosis.

S. pneumoniae colonises the oropharynx by adhering to the epithelial cells of pharynx. The cocci also release an enzyme, secretory IgA protease, which destroys the secretory IgA and thereby enhances the ability of the cocci to colonise mucosa of the upper respiratory tract. Pneumolysin liberated by the cocci destroys the ciliated epithelial cells and phagocytic cells by binding to cholesterol in the epithelial cell membrane. This adversely affects mucociliary clearance function of respiratory epithelium.

The process of tissue destruction is mediated by many factors, such as cell wall teichoic acid, peptidoglycan and phosphorylcholine. The destructive action is further supplemented by hydrogen peroxide produced by the bacteria. Presence of capsule helps the bacteria to evade phagocytosis. This allows *S. pneumoniae* to survive, multiply and spread to various organs. The antiphagocytic action of capsule is further supplemented by pneumolysin, which causes

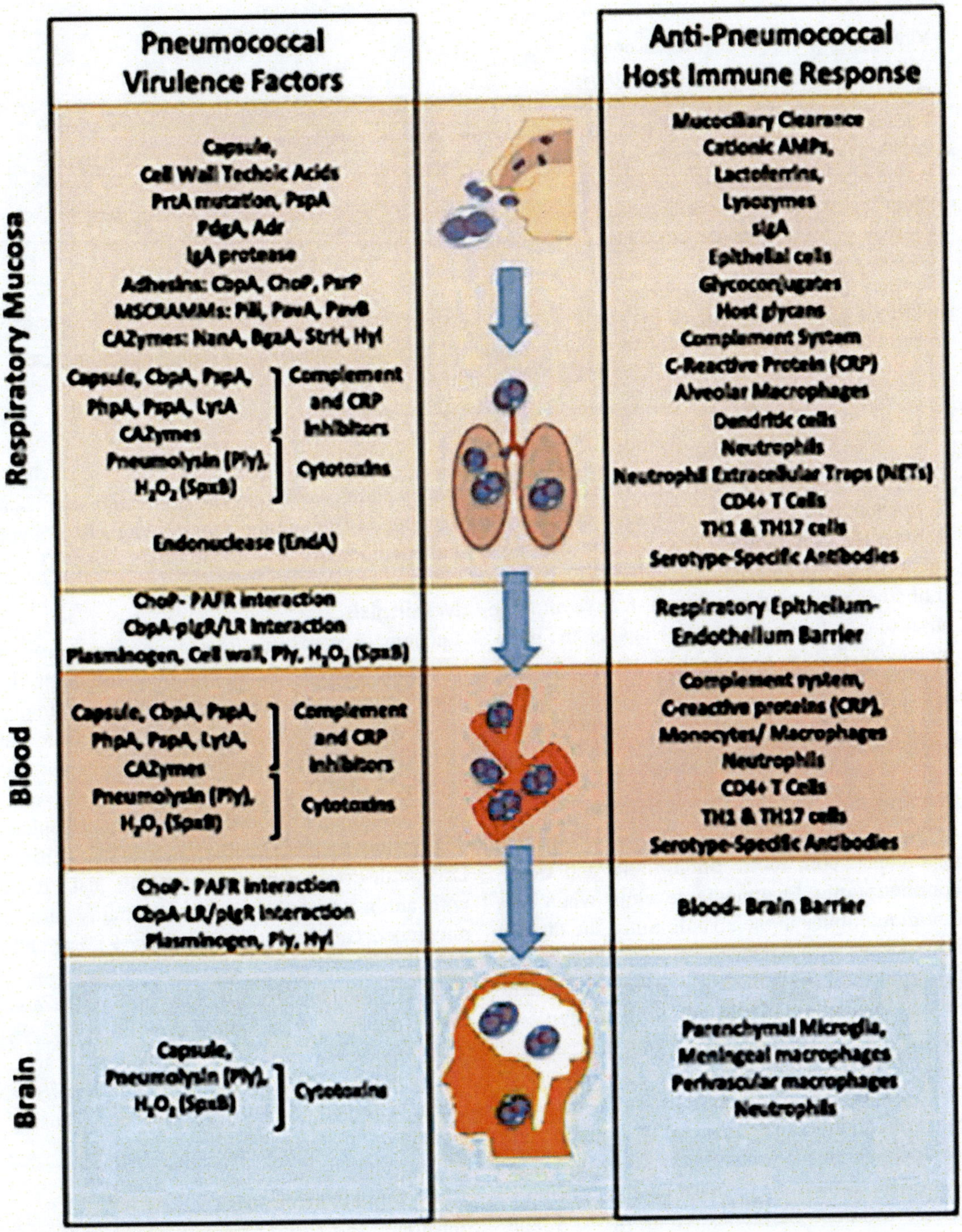

Fig. 26.9 *Streptococcus pneumoniae* virulence determinants and host immune responses associated with pneumococcal infection in an anatomical site-specific manner (https://pneumonia.biomedcentral.com/articles/10.1186/s41479-016-0007-9)

inhibition of phagocytic cell oxidative burst, required for intracellular killing of bacteria.

Twenty-three serotypes of *S. pneumoniae* are known to cause disease in humans. Out of these,

Table 26.6 Virulence factors of *Streptococcus pneumoniae*

Virulence factors	Biological functions
Cell wall associated polymers and proteins	
Capsule	Prevents phagocytosis
Teichoic acid	Binds to epithelial cells and activates alternative complement pathway
Peptidoglycan	Activates alternative complement pathway
Protein adhesin	Binds to epithelial cells
Phosphorylcholine	Mediates invasion of host cell by cocci
F protein	Mediates attachment to epithelial cells
Enzymes and toxins	
Secretory IgA protease	Destroys secretory IgA
Pneumolysin	Alters mucociliary clearance function of respiratory epithelium and inhibits phagocytic cell oxidative burst essential for intracellular killing of the bacteria
Autolysin	Kills the organism after exponential growth

serotypes 3, 4, 6B, 9V, 14, 18C, 19F and 23F of *S. pneumoniae* causes the majority of invasive disease. Host immunity is type specific with production of anticapsular antibodies. These antibodies appearing in serum 5–8 days after the onset of infection are protective against the pneumococcal serotype causing the infection.

Clinical Syndromes

Pneumonia

S. pneumoniae is the leading cause of bacterial pneumonia, both **lobar pneumonia** and **bronchopneumonia**. Pneumonia develops when bacterium multiplies in the alveoli. Since the disease is associated with aspiration and is localised in the lower lobes of the lungs, it is called lobar pneumonia. Pneumonia is common at the extreme of ages, in children and in elderly, who have a more generalised bronchopneumonia.

Pneumococci are the most common bacterial cause of childhood pneumonia, especially in children younger than 5 years. Serotypes 6, 14, 18, 19 and 23 are responsible for most cases of pneumonia in children. Serotypes 1, 3, 4, 7, 8 and 12 cause pneumonia in adults. *Haemophilus influenzae* and *Moraxella catarrhalis* are the other causes of acute pneumonia. *Mycoplasma pneumoniae*, *Chlamydia pneumoniae* and *Legionella* spp. are the causative agents of atypical pneumonia.

Meningitis

Pneumococcus is the most common cause of pyogenic meningitis in children, although the condition can occur in all age groups. Meningitis is always secondary to other pneumococcal infections, such as pneumonia, bacteraemia and infections of the ear or sinuses. The bacteria reach the brain through blood stream or from nasopharynx, following head trauma or dural tear particularly with cerebrospinal fluid (CSF) leak. Even with antibacterial therapy, the mortality due to pneumococcal meningitis is nearly 25%in children and in adults.

S. agalactiae, *Escherichia coli*, *Neisseria meningitidis*, *H. influenzae* type B, *L. monocytogenes*, *Pseudomonas* spp., *Flavobacterium meningosepticum* and *Staphylococcus aureus* are the other bacteria causing meningitis.

Other Suppurative Infections

Sinusitis and otitis media caused by *S. pneumoniae* occur in patients with prior viral infections. Other conditions include endocarditis, prosthetic or natural joint septic arthritis, myositis conjunctivitis, and brain and epidural abscesses. All these infections result from an infection of tissues during bacteraemia.

Laboratory Diagnosis

Specimens

Sputum, endotracheal aspirate, bronchoalveolar lavage fluid, CSF, pleural fluid, joint fluid, abscess fluid, bones and other biopsy material are the specimens collected for Gram staining and culture.

Microscopy

Gram staining of sputum is a rapid method for diagnosis of acute pneumonia. Stained smears showing lanceolate-shaped, Gram-positive cocci in pairs surrounded by a capsule are good evidence of pneumococcal infection. The morphology of the pneumococci may be altered in the patient receiving antibacterial therapy.

Gram staining of CSF is used for demonstration of Gram-positive diplococci—inside the polymorphs as well as outside in a CSF smear. Gram staining is positive in 90% of these cases (Fig. 26.10).

Gram staining of a buffy coat or blood smear is frequently positive in cases of overwhelming pneumococcal sepsis and is useful for rapid presumptive diagnosis of this condition. In acute pneumococcal otitis media, Gram stain of an aspirated fluid smear from the middle ear is useful to demonstrate the bacteria.

Culture and Identification

Sputum is plated on blood agar and incubated in the presence of 5–10% carbon dioxide. Grey colonies with alpha-haemolysis are observed after overnight incubation. Sputum culture may be negative due to normal flora outgrowing pneumococci or due to rapid autolysis.

Diagnosis of pneumococcal meningitis is confirmed by CSF culture and is positive in most of untreated cases. However, the culture is negative in more than 50% of cases who have received treatment even with a single dose of antibiotics. In the acute phase of pneumonia, the blood can be cultured in glucose broth. Demonstration of the pneumococci in the blood shows bad prognosis. Culture of aspirated fluid from the middle ear or from the sinus is a definitive method for diagnosis of otitis media or sinusitis. However, culture is not recommended for specimens collected from the nasopharynx or from the external ear.

S. pneumoniae colonies are identified by many tests.

S. pneumoniae is identified by its sensitivity to optochin **(Optochin sensitivity test)**. In this method, a filter paper disc containing optochin

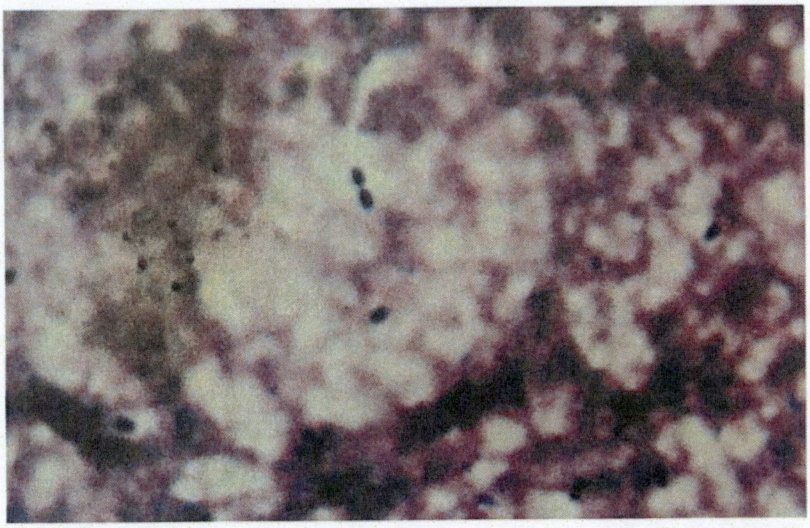

Fig. 26.10 Gram-stained smear of CSF showing *Streptococcus pneumoniae* in pairs (×1000)

(*ethylhydrocupreine dihydrochloride*) is applied on the middle of blood agar plate streaked with pneumococci and is incubated overnight. In a positive test, a zone of inhibition of 14 mm or more is observed around the disc after overnight incubation.

Bile solubility test is a very useful test to identify *S. pneumoniae*. It detects an autolytic enzyme, amidase, present in pneumococci, which breaks the bond between alanine and muramic acid of the peptidoglycan of the pneumococcal cell wall. The enzyme amidase is activated by bile salts present in bile, resulting in lysis of pneumococci. The test is carried out by applying a loopful of 10% sodium deoxycholate solution on the young colonies in the blood agar. Most colonies of pneumococci are dissolved within a few minutes.

Pneumococci ferment inulin; hence **inulin fermentation test** is a useful test to differentiate pneumococci from streptococci as the latter do not ferment it.

S. pneumoniae can be isolated from clinical specimens containing few pneumococci by **intraperitoneal inoculation in mice**. Pneumococci are demonstrated in the peritoneal exudate and heart blood of the mice, which die 1–3 days after inoculation.

Table 26.7 summarises important biochemical tests used to differentiate pneumococci from viridans streptococci.

Serology

The indirect hemagglutination, indirect fluorescent antibody test and ELISA are used to demonstrate specific **pneumococcal antibodies** in invasive pneumococcal diseases.

Latex agglutination test using the latex particles coated with anti-CRP antibody is employed to detect CRP. The CRP is used as a prognostic marker in acute cases of acute pneumococcal pneumonia, acute rheumatic fever and other infectious diseases. CRP is found in sera from cases of acute pneumonia but is absent during the convalescent phase of the disease. **Procalcitonin** level is found to be elevated in the invasive pneumococcal diseases, and this biomarker can be used as a prognostic marker to determine antibiotic treatment response.

SSS can be detected in the CSF by precipitation or by latex agglutination test. CIEP is also used to detect capsular polysaccharide antigen in the blood, urine, CSF etc., by CIEP. Recently an immunochromatographic test (ICT) has been used or the detection of the antigen in the urine.

Molecular Detection

PCR-based methods are very much useful in the identification of pneumococcus in patients undergoing antibiotic therapy. Further, molecular methods can be used for epidemiological purposes. Real-time PCR targeting autolysin gene, lytA and permease gene of the pia ABC transporter, piaB are the gold standard for PCR-based detection of *S. pneumoniae*. Other target includes surface adhesion gene (psaA) and the spn9802 gene fragment.

Typing of Pneumococci

Serotyping

S. pneumoniae is classified into more than 90 different serotypes (1–90) based on the antigenic structure of the capsular polysaccharide. Of these, only 23 serotypes are associated with pneumococcal diseases. Serotyping of *S. pneumoniae* is not carried out routinely and is done only for epidemiological studies. Serotyping is carried out by the following methods:

1. **Quellung reaction** is due to swelling (Latin *quellung*: swelling) of the capsule in the presence of the type-specific antiserum against SSS. This reaction was first described by

Table 26.7 Important biochemical properties of streptococci

Characteristics	*Streptococcus pyogenes*	*Streptococcus agalactiae*	Viridans streptococci
CAMP test	Negative	Positive	Negative
Hippurate hydrolysis	Negative	Positive	Negative
Bacitracin susceptibility	Sensitive	Resistant	Resistant

Fig 26.11 Quellung reaction

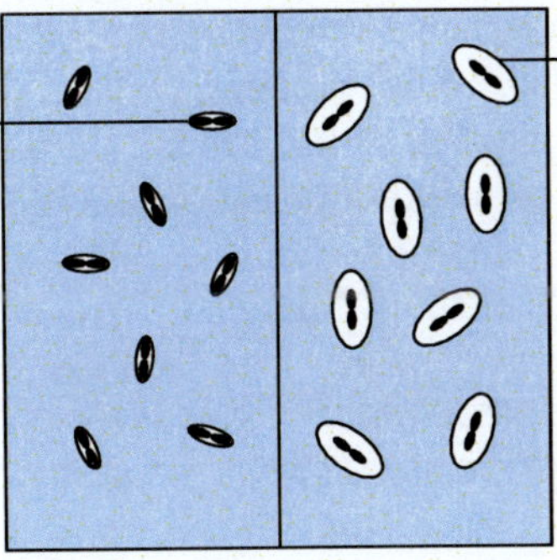

Neufeld in the year 1902. In this test, a drop of type-specific antiserum is added to a drop of suspension of pneumococci on a glass slide along with a drop of methylene blue solution. The capsule, in the presence of the specific homologous antiserum, becomes apparently swollen, clearly delineated and refractile (Fig. 26.11).

2. **Counter-current immunoelectrophoretic (CIEP)** demonstrates precipitation of the SSSs with specific serum. This is also a rapid test used to serotype pneumococci.
3. **Latex agglutination test** is another rapid test used for serotyping of pneumococci.

Molecular Typing

Pulsed-field gel electrophoresis (PFGE) and amplified fragment length polymorphism are the DNA finger printing methods, which are being used currently for typing of pneumococci. Other praised methods include multilogues sequence typing (MLST), multilocus enzyme electrophoresis.

Distinguishing features of Pneumococci and viridans streptococci are summarised in the Table 26.8.

Treatment

Most pneumococci are susceptible to penicillin, amoxicillin and erythromycin.

Most pneumococci are susceptible to penicillin. Strains that are susceptible to penicillin are also susceptible to nearly all other antibiotics. But since 1977, **penicillin-resistant pneumococci** are being increasingly documented. Penicillin-resistant strains may be moderately resistant (minimum inhibitory concentration or MIC, >0.1–1 (μg/mL) or highly resistant (MIC > 2 μg/mL). The resistance to penicillin and other beta-lactam antibiotics is not mediated by

Table 26.8 Distinguishing features of pneumococci and viridans streptococci

Characters	*Streptococcus pneumoniae*	*Streptococcus viridans*
Morphological characteristics		
Bacterial arrangement	Pairs of lanceolate diplococci	Pairs of oval cocci, short chains
Polysaccharide capsule	Present	Absent
Cultural characteristics		
Colonies	Draughtsman colonies	Dome shaped colonies
Growth in liquid medium	Uniform turbidity	Granular turbidity
Bile solubility	Positive	Negative
Optochin sensitivity	Positive	Negative
Inulin fermentation	Positive	Negative
Quellung reaction	Positive	Negative
Intraperitoneal mice inoculation	Fatal	Non-pathogenic

production of beta-lactamase enzymes, but is due to the modifications of proteins, such as penicillin-binding proteins (PBPs) found on the cell wall. These penicillin-resistant strains are also resistant to multiple drugs, such as cefotaxime, ceftriaxone, erythromycin, tetracycline, macrolides and trimethoprim-sulfamethoxazole (TMP-SMX). Resistance is seen most often in *S. pneumoniae* serotypes 6, 9, 14, 19 and 23.

Amoxicillin is the drug of choice for treatment of otitis media, sinusitis and pneumonia caused by penicillin-resistant pneumococci with intermediate resistance. Ceftriaxone is the drug of choice for non-CNS invasive pneumococcal diseases caused by penicillin- and ceftriaxone-resistant pneumococci. Vancomycin is a reserved antibiotic, used only if the pneumococcus is resistant to ceftriaxone.

Epidemiology

S. pneumoniae is worldwide in distribution. *S. pneumoniae* is a normal inhabitant of throat and nasopharynx. Nasopharyngeal colonisation occurs in approximately 5–75% of the population. Colonisation is more common in children than in adults.

Pneumococci are strict human pathogens. They cause infection exclusively in humans. Horizontal transmission requires close person-to-person contact, hence overcrowding facilitates spread of infection. Respiratory and pharyngeal secretions of carriers and patients are the sources of infection. Infection is acquired by inhalation of droplets nuclei and by coming in contact with contaminated fomites. Pneumococcal infection accounts for more deaths than any other vaccine-preventable disease. Children between 6 months and 4 years of age and adults over 60 years of age are most commonly at risk for pneumococcal infection.

Prevention and Control

Pneumococcal vaccines play an important role in the prevention of pneumococcal diseases.

23-valent pneumococcal polysaccharide vaccine is available for use against pneumococcal infections. The vaccine confers 80–90% protection. The vaccine contains capsular antigens from each of 23 serotypes of *S. pneumoniae* most commonly involved in human infections. The vaccine is effective and safe in children older than 5 years. It is not recommended for use in children younger than 2 years.

7-valent pneumococcal conjugate vaccine (PCV7) is made available recently for the immunisation of infants and toddlers up to 2 years age. It is targeted against serotypes 4, 6B, 9V, 14, 18C, 19F and 23F, conjugated to CRM 197 protein of *Corynebacterium diphtheriae*.

Distinguishing characteristics of pneumococcal polysachride vaccine and pneumococcal conjugate vaccine is summarised in the Table 26.9.

Table 26.9 Distinguishing characteristics of pneumococcal polysaccharide and conjugate vaccine

Characteristics	Polysaccharide vaccine (23 valentPPV)	Conjugate vaccine (13 valentPCV)
Capsular polysaccharide antigen	25 mcg from each of 23 serotypes	2.2 mcg from 12 serotype and 4.4 from serotype 6B
Preservative	0.25% phenol	Nil
Carrier protein	Nil	34 mcg of CRM
Adjuvant	Nil	Aluminium phosphate
Immunogenic memory	No	Yes
Stimulation of T cell immunity	No	Yes
Stimulation of mucosal immunity	No	Yes
Immunogenic in children <2 years	No	Yes

Table 26.10 Composition and uses of pneumococcal conjugate vaccine and polysaccharide vaccines

Type	Composition	CDC recommendations
Pneumococcal conjugate vaccines		
PCV13	Purified capsular polysaccharide of 13 serotypes of *Streptococcus pneumoniae* (1, 3, 4, 5, 6A, 6B, 7F, 9V, 14, 19A, 19F, 18C and 23F) conjugated to a nontoxic variant of diphtheria toxin known as CRM197	All babies and children younger than 2 years old Children 2 through 18 years old with certain medical conditions .
PCV15	Purified capsular polysaccharides from 15 serotypes of *S. pneumoniae* (1, 3, 4, 5, 6A, 6B, 7F, 9V, 14, 18C, 19A, 19F, 22F, 23F and 33F) individually conjugated to a nontoxic variant of diphtheria toxin known as CRM197	Adults 65 years or older Adults 19 through 64 years old with certain medical conditions or risk factor
PCV20	Purified capsular polysaccharides from 20 serotypes of *S. pneumoniae* (1, 3, 4, 5, 6A, 6B, 7F, 8, 9V, 10A, 11A, 12F, 14, 15B, 18C, 19A, 19F, 22F, 23F and 33F) individually linked to a nontoxic variant of diphtheria toxin known as CRM197	Adults 65 years or older Adults 19 through 64 years old with certain medical conditions or risk factor
Pneumococcal polysaccharide vaccine		
Pneumococcal polysaccharide vaccine or PPSV23	Purified preparations of pneumococcal capsular polysaccharide. PPSV23 contains polysaccharide antigen from 23 types of pneumococcal bacteria	Children 2 through 18 years old with certain medical conditions Adults 19 years or older who get PCV1

Since 2009, two novel PCVs, **PCV10** and **PCV13** were introduced that give protection against 10 and 13 serotypes, respectively. Both single- or double-dose vaccine formulations are available. These vaccines offer better coverage for multiple serotypes that commonly cause disease in low- and middle-income countries. It also results in reduction of CAP hospitalisations in children, mostly in developed countries. PCV10 and PCV13 have been introduced in national immunisation programs by several nations by replacing PCV7 (Table 26.10).

Case Study

A 6-year-old child was admitted to hospital with high-grade fever (>103 °C), headache, stiff neck, vomiting, lethargy and altered sensorium for 3 days. Gram staining of the CSF smear showed Gram-positive cocci. The same bacteria were isolated from CSF by culture. The patient died despite prompt treatment with ceftriaxone.

- Which are the most likely bacteria to cause this fulminant condition?
- What other diseases are caused by this bacterium?
- What are the vaccines available against the infection caused by the bacteria in children?
- Do the bacteria show any resistance to antibiotics?

Neisseria and *Moraxella* 27

The genus *Neisseria* consists of Gram-negative, aerobic cocci, typically arranged in pairs (diplococci) with adjacent sides flattened together.

Genus *Neisseria*

The genus contains 29 recognised species. They are oxidase-positive and mostly catalase-positive. They ferment sugars with the production of acid but no gas. *Neisseria gonorrhoeae* and *Neisseria meningitidis* are the two important species that cause human infections. These two species are strictly pathogens for humans, whereas the other *Neisseria* species are commensals of the mouth and upper respiratory tract and hence cause opportunistic infections (Table 27.1).

Neisseria gonorrhoeae

In 1879, Neisser described the gonococcus in gonorrhoeal pus for the first time. In 1985, Bumm cultured this organism and inoculated it into healthy volunteers to prove its pathogenicity. *N. gonorrhoeae* is a strict human pathogen. It is the causative agent of gonorrhoea, one of the most common sexually transmitted diseases worldwide.

Properties of the Bacteria

Morphology

N. gonorrhoeae is a Gram-negative and aerobic diplococcus. It is mostly intracellular—found within the polymorphonuclear (PMN) leucocytes—and some cells contain as many as a hundred cocci. Smears from the pus sample show the intracellular kidney-shaped cocci, typically arranged in pairs with concave sides facing each other.

Culture and Biochemical Properties

N. gonorrhoeae is a fastidious coccus. It is aerobic and requires complex media for growth. The cocci grow on enriched media, such as blood or chocolate agar. These cannot grow on ordinary media, such as nutrient agar or Mueller–Hinton agar. The growth of bacteria is enhanced by incubation in a humid atmosphere supplemented with 5–10% CO_2.

Thayer–Martin Medium

Thayer–Martin medium (chocolate agar medium containing antibiotics, such as colistin, nystatin and vancomycin) and modified New York City medium (a translucent medium containing vancomycin, colistin trimethoprim and either nystatin or amphotericin B) are selective media used for isolation of gonococci from the clinical specimens containing mixed microbial flora. In these media,

Table 27.1 Human infections caused by *Neisseria* species

Bacteria	Diseases
Neisseria gonorrhoeae	Gonorrhoea, disseminated gonococcal infections (DGI), ophthalmia neonatorum and gonococcal diseases, such as, anorectal gonorrhoea, gonococcal pharyngitis and acute perihepatitis
Neisseria meningitidis	Meningitis, meningococcaemia and other meningococcal diseases, such as, meningococcal pneumonia, septic arthritis, purulent pericarditis and endophthalmitis
Other *Neisseria* species	Opportunistic infections

the growth of contaminating bacteria is suppressed, including commensal *Neisseria.* On these media, *N. gonorrhoeae* produces small, translucent and convex colonies which are soft and easily friable. Four types of colonies (T1, T2, T3 and T4) of gonococci have been recognised. Types 1 and 2 are small and are brown-pigmented colonies. The strains producing these colonies that possess pili are virulent and cause acute cases of gonorrhoea. Types 3 and 4 are large and are non-pigmented colonies. The cocci producing these colonies do not possess pili and are avirulent.

N. gonorrhoeae ferments glucose with the production of acid but no gas. It does not ferment maltose, lactose, sucrose or fructose. This is an important feature to differentiate *N. gonorrhoeae* from *N. meningitidis*. *N. gonorrhoeae* utilises glucose only, whereas *N. meningitidis* utilises both glucose and maltose. It is oxidase- and catalase-positive.

Rapid carbohydrate utilisation test (RCUT) is based on the ability of a bacterium to ferment a sugar by pre-existing enzymes present in it. The test result is not dependent on the growth of the bacterium on a sugar-containing medium. It is a rapid and sensitive test; hence, the utility of this test is recommended.

Other Properties

Susceptibility to Physical and Chemical Agents

The gonococci are highly delicate bacteria. They die rapidly on drying. They are also killed by soap and many other disinfectants, such as phenol, chlorhexidine, hexachlorophene and antiseptics. They are killed at a temperature as low as 25 °C. Freeze drying (−70 °C) or storing in liquid nitrogen is the most effective method for storing gonococci for a longer period.

Antigenic Structure

The cell wall of *N. gonorrhoeae* is similar to any other Gram-negative bacteria. These contain the following proteins (Fig. 27.1).

1. **The outer membrane proteins (OMP)** are present in the outer membrane. They mediate the uptake of iron essential for the growth and

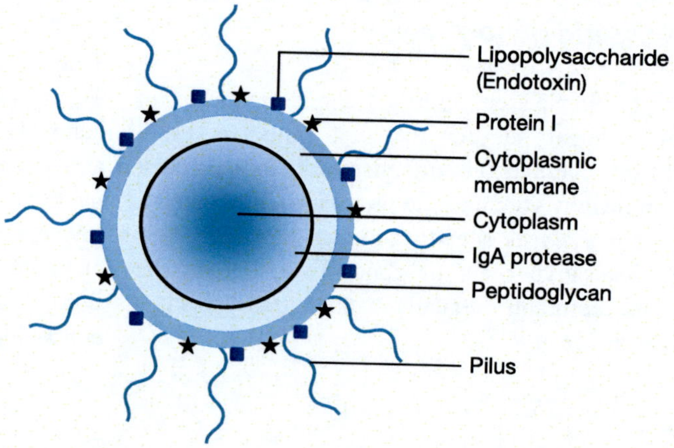

Fig. 27.1 Schematic diagram of *Neisseria gonorrhoeae*

metabolism of the cocci. They promote iron intake by binding to haemoglobin, transferrin and lactoferrin. These proteins are of three types: (1) Por proteins, (2) Opa proteins and (3) Rmp proteins

Por proteins, earlier known as protein I, are porin proteins that form pores or channels in the outer membrane. Por proteins are of two types: Por-A and Por-B, each with various antigenic variations. Strains producing Por-A proteins are commonly associated with disseminated disease because these proteins prevent the killing of gonococci in the serum by the serum-complement components. The antigenic variations observed in Por proteins form the basis for the serotype classification of *N. gonorrhoeae.*

Opa proteins, also known as opacity proteins, were formerly known as protein II. These proteins are found in the membrane and mediate adherence of the bacteria to each other and the eukaryotic cells. Strains producing Opa proteins produce opaque colonies in culture.

The Rmp proteins, also known as reduction-modifiable proteins, were formerly known as protein III. These are proteins found in the outer membrane of gonococci and lead to the production of antibodies that block serum bactericidal activity against gonococci.

2. **Lipo-oligosaccharide (LOS)** is another major antigen present in the bacteria's cell wall. This antigen consists of lipid A and oligosaccharides similar to that of lipopolysaccharide (LPS) of Gram-negative bacteria. However, LOS does not show antigenic variation as found in LPS. LOS possesses endotoxic activity. IgA1 protease and beta-lactamase are the other important proteins. IgA1 protease degrades secretory IgA1, whereas beta-lactamase degrades beta-lactam rings in penicillin.

Pathogenesis

N. gonorrhoeae causes disease both by multiplying in tissues and by causing inflammation.

Virulence Factors

N. gonorrhoeae does not form a true carbohydrate **capsule**, unlike *N. meningitidis*. Instead, it forms a polyphosphate capsule loosely associated with its cell surface. The capsule is most evident in freshly isolated gonococci, and it prevents phagocytosis of the gonococci.

Pili are hair-like structures that extend from the cytoplasmic membrane through the outer membrane. The pili are composed of the proteins known as *pilins*, which are repeating protein subunits. There is a marked antigenic variation in gonococcal pili. More than 100 serotypes are known. The pili mediate attachment of gonococci to non-ciliated epithelial cells. They also contribute to virulence by preventing ingestion and killing of gonococci by neutrophils (Table 27.2).

Pathogenesis of Gonorrhoea

It is a sexually transmitted disease. *N. gonorrhoeae* infection occurs following

Table 27.2 Virulence factors of *Neisseria gonorrhoeae*

Virulence factors	Biological functions
Capsule	Prevents phagocytosis
Pili	Mediate attachment of gonococci to non-ciliated epithelial cell Prevent ingestion and killing of gonococci by neutrophils
Por proteins	Confer resistance to serum killing of gonococci by preventing fusion of phagolysome in neutrophils
Opa proteins	Mediate bacterial adherence to each other and to the eukaryotic cells
Rmp proteins	Produce antibodies that block serum bactericidal activity against gonococci
LOS	Possesses endotoxic activity of the bacteria
IgA protease	Destroys IgA immunoglobulin
Beta-lactamase	Degrades beta-lactam rings of penicillin
Plasmids	Plasmid-borne virulence determinants contribute to antimicrobial resistance

mucosal inoculation during vaginal, anal or oral sexual contact. Ophthalmia neonatorum is acquired non-sexually. This infection occurs following a conjunctival inoculation during vaginal delivery.

Gonorrhoea is primarily a genital infection restricted to the urethra in men and cervix in women. The gonococci attach to the mucosal cells and cause disease. Subsequently, they enter the cells, multiply inside and pass through them into the sub-epithelial space, establishing the infection. The pili and Opa proteins aid in the attachment of gonococci to the mucosa. The Opa proteins also facilitate the subsequent migration of gonococci into the epithelial cells. The Por proteins inhibit phagolysosome fusion in the phagocytes, protecting the phagocytosed bacteria from intracellular killing. The LOS of the gonococcal cell wall stimulates the production of tumour necrosis factor-alpha (TNF-α) and other inflammatory responses. All these contribute to most of the symptoms associated with gonococcal infection.

Clinical Syndromes

Gonorrhoea

The incubation period varies from 2 to 8 days.

Gonorrhoea in Men

It is characterised by urethritis with burning micturition and serious urethral discharge. Subsequently, the discharge becomes more profuse, purulent and even blood-tinged. Occasionally, the infections progress towards periurethral tissues causing abscesses, and multiple discharging sinuses are known as water can perineum. Anorectal gonorrhoea (proctitis) and gonococcal pharyngitis occur in homosexual men following rectal intercourse or by orogenital contact, respectively. Pharyngitis is most commonly acquired during orogenital contact.

Gonorrhoea in Women

Endocervix is the primary site of **gonorrhoea in women** because gonococci invade only the endocervical columnar epithelial cells. The bacteria cannot infect the squamous epithelial cells in the vagina of post-pubescent women. The urethra, rectum and pharynx are the other sites of infection in women. The infection is mostly asymptomatic in women. The presence of vaginal discharge, dysuria, dyspareunia and mild lower abdominal pain are the common symptoms.

Pelvic inflammatory disease (PID) is the most important complication in females following gonococcal infection. Increased vaginal discharge or purulent urethral discharge, dysuria, lower abdominal pain and intermenstrual bleeding are the common symptoms of PID. Tubal scarring, ectopic pregnancy and infertility are the major complications in women following PID.

Disseminated gonococcal infections (DGI) occurs because of the haematogenous spread of gonococci from the primary site of infection. The DGI is mostly seen in untreated asymptomatic women and persons with complement deficiency. The strains that cause DGI are characterised by their resistance to the bactericidal action of serum, marked sensitivity to penicillin and autotropism for arginine, hypoxanthine and uracil (for growth, they require these substances in the medium).

Gonococcal vulvovaginitis occurs in pre-pubertal girls through sexual contact.

Ophthalmia Neonatorum

It is a non-sexually transmitted infection caused by *N. gonorrhoeae*. This is a condition of bilateral conjunctivitis of a neonate born by vaginal delivery to an infected mother. Gonococci can cause permanent injury to the eye in a very short time. Of late, the condition has become rare following the instillation of 1% silver nitrate solution in newborn babies' eyes.

Other Gonococcal Infections

Anorectal gonorrhoea (proctitis) and gonococcal pharyngitis occur in homosexual men following rectal intercourse or by orogenital contact, respectively. Pharyngitis is most commonly acquired during orogenital contact. Pharyngitis is often asymptomatic; however, it may present as exudative pharyngitis with cervical lymphadenopathy.

Purulent gonococcal conjunctivitis occurs in adults following autoinoculation of gonococci into the conjunctival sac from a primary site of infection, such as the genitals. Conjunctivitis may rapidly progress to panophthalmitis and loss of the eye unless promptly treated.

Acute perihepatitis (Fitz–Hugh and Curtis syndrome) is caused by the spread of *N. gonorrhoeae* or *Chlamydia trachomatis* to the liver capsule and above the peritoneum from the fallopian tubes directly.

Repeated gonococcal infections occur due to the antigenic changes of the pili and OMPs. The various body sites affected by gonococci is shown in Fig. 27.2. Persons with a deficiency of the late-acting complement components (C6–C9) are at risk of disseminated infections.

Laboratory Diagnosis

Specimens

Urethral discharge in males and cervical discharge in females are the specimens of choice in **acute gonococcal infection**. In females, the cervix is first cleaned of the exudate while collecting endocervical discharge. A swab is then placed into the external os and rotated for several seconds. The high vaginal swab is not satisfactory. In males, discharge present at the meatus is collected. The meatus is first cleaned with gauze soaked in saline. The urethral discharge is then collected with the help of a platinum loop. If no discharge is present at the meatus, urethral specimens are collected by inserting and rotating a small swab 2–3 cm into the urethra. Calcium

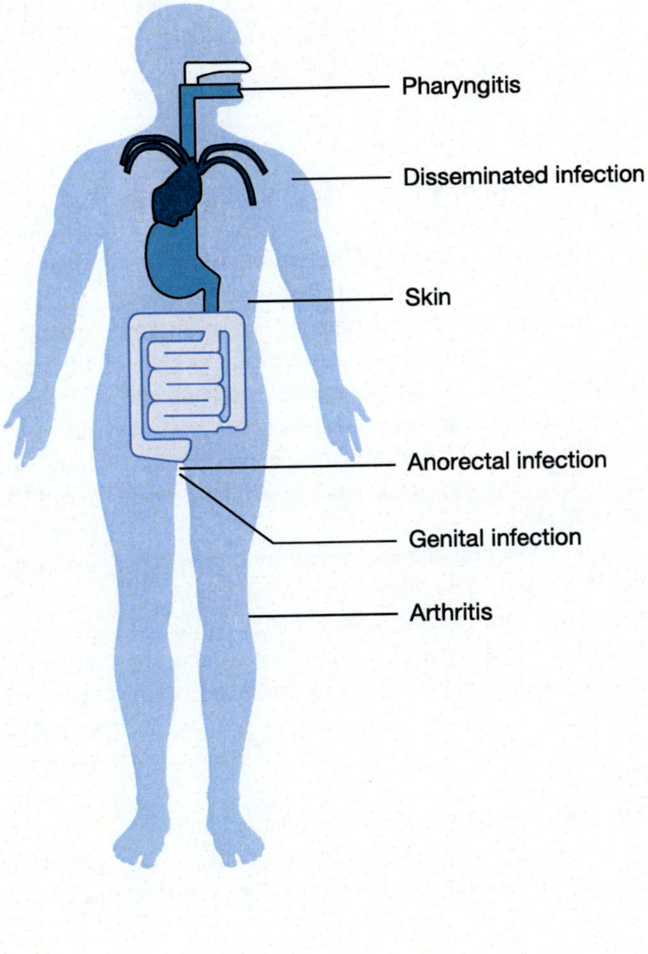

Fig. 27.2 Sites of infection caused by *Neisseria gonorrhoeae*

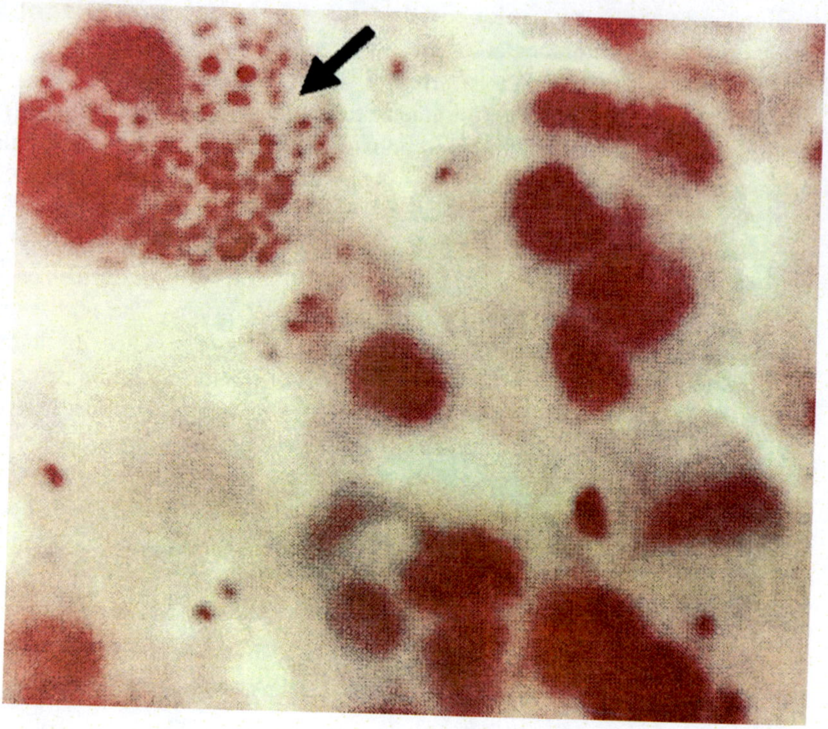

Fig. 27.3 Gram-negative intracellular *Neisseria gonorrhoeae* in Gram-stained smear of pus exudate (×1000)

alginate or rayon swab on a metal shaft is usually used for this purpose.

In chronic gonococcal infection, since urethral discharge is less, the exudate after the prostatic massage or morning drops of secretion and urine is collected. Rectal specimens are frequently useful in asymptomatic women and homosexual and bisexual men. Samples are collected from the pharynx, urethra, cervix, rectum and blood and synovial fluid in patients with possible DGI.

After collection, the specimens are transported and processed immediately in the laboratory. If the delay is unavoidable, specimens are collected in charcoal-impregnated swabs. The swabs are transported to the laboratory in Stuart transport medium.

Microscopy

Demonstration of typical Gram-negative intracellular diplococci is characteristic of *N. gonorrhoeae* (Fig. 27.3). In men, the presence of four or more PMN leucocytes per oil-immersion field in Gram-stained urethral exudate smear is diagnostic of urethritis. Gram stain helps in the presumptive diagnosis of the gonococcal infection. It is more than 90% sensitive and 98% specific for diagnosing gonococcal infection in symptomatic males. However, in asymptomatic males, the sensitivity of the Gram stain is only 60% or less. The demonstration of 10 or more PMN in the wet mount examination of the centrifuged early morning (the first) voided urine under high power suggests urethritis in men.

In women, the presence of more than 10 PMN per high-power field on an endocervical smear is suggestive of cervicitis. Gram stain of endocervical smears is less sensitive (50–60%) and 82–90% specific in both symptomatic and asymptomatic women.

Gram stain is not a sensitive method for detecting gonococci in patients with anorectal gonorrhoea, pharyngitis and skin lesions. Specificity is also less because commensal *Neisseria* species in the oropharynx and gastrointestinal tract can be confused with those of *N. gonorrhoeae*. However, it is not a sensitive method for detecting gonococci in patients with anorectal gonorrhoea, pharyngitis and skin lesions.

Detection of Gonococcal Antigen

Direct fluorescent antibody (DFA) test and direct enzyme immunoassays (EIA) are increasingly being used. These two tests detect gonococcal antigens in urethral discharge, endocervical discharge and other clinical specimens for diagnosis.

Culture and Identification

Isolation of *N. gonorrhoeae* from specimens by culture confirms the diagnosis of gonorrhoea. Genital, rectal and pharyngeal specimens are inoculated immediately on a non-selective medium (e.g. blood agar or chocolate agar) and on a selective medium (e.g. Modified Thayer–Martin medium).

The colonies of gonococci are examined on chocolate agar after 48 h of incubation at 35–36 °C in the presence of 5–10% CO_2. The colonies are small, round, translucent and convex with a finely granular surface. On Thayer–Martin medium, the colonies show similar morphology as that on chocolate agar. The mixed microbial flora present in the clinical specimens is suppressed by the selective media. However, the vancomycin present in the selective media inhibits some strains of gonococci.

N. gonorrhoeae ferments glucose with the production of acid but no gas. It does not ferment maltose, lactose, sucrose or fructose. This is an important feature to differentiate *N. gonorrhoeae* from *N. meningitidis*. *N. gonorrhoeae* utilises glucose only, whereas *N. meningitidis* utilises both glucose and maltose.

Rapid carbohydrate utilisation test (RCUT) is a rapid sugar fermentation test. It is based on the ability of a bacterium to ferment sugar by pre-existing enzymes present in the gonococci. It is a rapid and sensitive test.

Strain Typing

N. gonorrhoeae strains can be typed by auxotyping and serotyping.

Auxotyping is based on the addition of specific nutrients and co-factors in the medium for the growth of gonococci. There are over 30 auxotypes. The most common auxotypes are prototrophic or wild type (proto), praline-requiring type (pro) and the strains requiring arginine, hypoxanthine and uracil (AHU).

Serotyping is based on the OMP "porin", which is further divided into serovars (e.g. 1A-4, 1B-12) based on agglutination with a panel of monoclonal antibodies. This helps in contact tracing and has epidemiological uses in discriminating strains with antibiotic resistance and severity of infections.

Serology

The serological tests are not used in routine diagnoses.

Molecular Detection

Nucleic acid hybridisation, signal amplification, polymerase chain reaction, strand displacement amplification and transcription-mediated amplification are the FDA-approved methods for molecular detection of *N. gonorrhoea*. DNA probes (*Gen probe*) are commercially available to detect bacteria in the genital and other clinical specimens directly. These probes are specific, sensitive and rapid for nucleic acid detection of *N. gonorrhoeae*. The results become available within 2–4 h.

Treatment

Penicillin is the drug of choice for penicillin-sensitive strains of *N. gonorrhoeae*.

Penicillin-resistant Strains of *N. gonorrhoeae*

Initially, gonococci were highly sensitive to penicillin. However, since 1957, strains of gonococci with decreased sensitivity to penicillin have been documented. Most of them are beta-lactamase (penicillinase) produced by virtue of plasmid transmission. These strains show a high level of resistance to penicillin. Some strains of *N. gonorrhoeae* not producing beta-lactamase yet showing resistance to penicillin have also been reported. This resistance is mediated chromosomally and is of a low level. Chromosomal-mediated resistance to other antibiotics, such as tetracycline, erythromycin and aminoglycosides, has also been reported.

Ceftriaxone, cefixime, ciprofloxacin or ofloxacin are the alternative drugs in cases of penicillin resistance or penicillin-allergic individuals. A single dose of ceftriaxone 125 mg intramuscularly or cefixime (400 mg), ciprofloxacin (500 mg) or ofloxacin (400 mg) as a single dose orally is also effective.

WHO Guidelines for Treatment

For genital and anorectal infections, WHO recommends dual therapy with ceftriaxone 250 mg intramuscular (IM) along with azithromycin 1 g per oral (PO) as a single dose followed by cefixime 400 mg PO as a single dose along with azithromycin 1 g PO as a single dose. Treatment failures are managed by either of the following regimens as ceftriaxone 500 mg IM as a single dose with azithromycin 2 g PO as a single dose or cefixime 800 mg orally as a single dose with azithromycin 2 g PO as a single dose or gentamicin 240 mg IM as a single dose with azithromycin 2 g PO as a single dose or spectinomycin 2 g IM as a single dose along with azithromycin 2 g PO as a single dose.

Epidemiology

Geographical Distribution

Gonococcal infection is reported throughout the world. WHO estimated ~87 million gonococcal infections globally in 2016. However, the trend is increasing where emerging antimicrobial resistance is worsening the global situation. It disproportionately affects marginalised populations. The prevalence is high in sexually active young adults living in low-income countries, among sexual and gender minorities, racial/ethnic minorities and indigenous communities and sex workers compared to the general population in high-income countries. However, this condition has a low incidence rate in European countries and is nearly eliminated in Sweden. In pregnant women, the prevalence of gonorrhoea infection is 10% in Africa, 5% in Latin and 4% in Asia.

Habitat

N. gonorrhoeae is exclusively a human pathogen. The gonococci are only found in infected conditions. In infected women, the gonococci are most commonly found in the endocervix and, in infected men, found in the urethra. Gonococci can also be found in the pharynx, rectum and eyes in both men and women. The gonococci are not found as normal human flora in the mucosa of the urethra, cervix or vagina.

Reservoir, Source and Transmission of Infection

Only humans, especially asymptomatic infected men and women, are reservoirs of infections. Asymptomatic carriage is more common in women than in men. Purulent urethra or cervical discharge is the most common source of infection. The infection is transmitted primarily by sexual contact. *N. gonorrhoeae* infection occurs following mucosal inoculation during vaginal, anal or oral sexual contact. Increased sexual contact with infected partners increases the risk of acquiring the infection. Less frequently, ophthalmia neonatorum is acquired non-sexually. This infection occurs following a conjunctival inoculation during vaginal delivery. Less frequently, the disease is transmitted through the rectum, oropharynx or through the birth canal.

The gonococci are not found as normal human flora in the mucosa of the urethra, cervix or vagina. Fomites do not play any role in transmitting the disease because gonococci die rapidly outside the human body.

Prevention and Control

Currently, there is no effective vaccine available against *N. gonorrhoeae*. Chemoprophylaxis by the prophylactic use of penicillin is also ineffective and may promote the development of resistant strains. Therefore, health education, early detection of cases, tracing of contacts and follow-up and screening of sexual contacts are important in preventing gonorrhoeal infections. Gonococcal conjunctivitis in newborns is prevented by using erythromycin ointment.

Non-Gonococcal or Non-Specific Urethritis

Non-gonococcal urethritis (NGU) or non-specific urethritis (NSU) refers to a case of chronic urethritis where gonococci cannot be demonstrated. Some of these cases may be due to **Reiter's syndrome.** These include the triad of NGU, conjunctivitis and arthritis. Some of them may be gonococcal infections that cannot be diagnosed by routine tests, as they exist in L-forms, which is undetectable. It is caused by *C. trachomatis*, *Ureoplasma urealyticum* and *Mycoplasma hominis*. Occasionally, Herpes virus, cytomegalovirus, *Candida albicans*, *Trichomonas vaginalis* and bacteria, such as *Acinetobacter lwoffii* and *Gardnerella vaginalis*. Since the aetiology of the disease is not so clear and well defined, the management of NGU remains a difficult task.

Neisseria meningitidis

N. meningitidis causes a spectrum of diseases ranging from meningococcaemia (rapidly fatal) to transient bacteraemia (relatively benign). It is also the second most common cause of community-acquired meningitis in adults.

Properties of the Bacteria

Morphology

N. meningitidis is Gram-negative, spherical or oval cocci arranged in pairs with the adjacent sides flattened. The cocci are usually found inside the leucocytes in smears from pus cells and other specimens. They measure 0.6–0.8 μm in diameter. Freshly isolated bacteria are usually capsulated. Meningococci are oxidase- and catalase-positive.

Culture and Biochemical Properties

Meningococci are strict aerobes. They grow optimally at a temperature between 36 and 39 °C and at an optimum pH of 7.4–7.6. Their growth is enhanced by incubation in a moist atmosphere in the presence of 5% CO_2.

Meningococci are fastidious bacteria with complex nutritional requirements. They do not grow on ordinary media but grow well on the medium enriched with blood or serum, such as blood agar, chocolate agar and Mueller–Hinton agar. The blood or serum promotes the growth of bacteria by neutralising inhibitory substances found in the media rather than by providing additional requirements.

On **blood agar**, *N. meningitidis* produces small, round (1–2 mm in diameter), convex, grey and translucent non-pigmented colonies with entire edges after 24 h of incubation. At 48 h, the colonies become larger with an opaque raised centre and crenated with a transparent margin. It does not produce any haemolysis on blood agar. Strains of meningococci with large polysaccharide capsules appear as mucoid colonies. Meningococci produce large colonies on chocolate agar.

Thayer–Martin medium with antibiotics (vancomycin, colistin, nystatin and trimethoprim) and **New York City medium** are the selective media commonly used to isolate the bacteria from clinical specimens containing mixed bacterial flora.

N. meningitidis is oxidase- and catalase-positive. These two tests are important biochemical markers for the preliminary identification of this organism. *Alcaligenes* spp., *Aeromonas* spp., *Vibrio* spp., *Campylobacter* spp. and *Pseudomonas* spp. are the other bacteria that are oxidase-positive.

Oxidase tests can be performed in two ways. In the first method, 1% solution of oxidase reagent (test can—diamine—dihydrochloride) is poured on the culture media; the *Neisseria* colonies turn deep purple. In the second method, a few colonies of *Neisseria* are rubbed with a glass rod on a strip of filter paper moistened with oxidase reagent. A deep purple colour develops immediately.

N. meningitidis ferments glucose and maltose with acid but no gas. It does not ferment sucrose or lactose. Fermentation tests are required for the final identification of *Neisseria* species. It does not produce hydrogen sulphide and does not reduce nitrates.

Other Properties

Susceptibility to Physical and Chemical Agents

N. meningitidis is a highly delicate organism. It is highly sensitive to heat, desiccation and disinfectants.

Antigenic Structure

Depending on group-specific capsular polysaccharide antigens, meningococci are sub-divided into 13 **serogroups**. Serogroups A, B, C, D, X, Y, Z, W135, 29E, H, I, K and L are of human importance. Group A is responsible for most of the epidemics of meningitis. Group C is associated with localised outbreaks of meningitis. Serogroups Y and W135 meningococci cause disease more commonly than the serogroups X and Z. Meningococci that lack group-specific antigens are considered non-pathogenic.

Each serogroup includes many **serotypes**. The classification of meningococci serogroups into serotypes is based on the differences in their outer membrane proteins and the oligosaccharide part of LOS. For example, group A meningococci have a single serotype, whereas group B and C meningococci consist of many serotypes.

Pathogenesis

N. meningitidis colonises the human nasopharynx. Under specific conditions, the meningococci invade the bloodstream and then reaches the brain, causing meningitis.

Virulence Factors

N. meningitidis is surrounded by a prominent **polysaccharide capsule**. The capsule contributes to virulence by inhibiting phagocytosis. It protects meningococci from destruction by the leucocytes.

LOS endotoxin is present in the outer membrane of *N. meningitidis*. It is responsible for damage to the blood vessels associated with meningococcal infections. The continuous production and release of endotoxin by meningococci causes severe endotoxin reaction in patients with meningococcal disease.

IgA protease acts by clearing the secretory IgA. It, thereby, helps the meningococci attach to the epithelial cells of the upper respiratory tract. The virulence factor of meningococci are summarised in Table 27.3.

Pathogeneses of Meningitis

Meningococci are acquired by airborne droplets of nasopharyngeal secretions of infected hosts. Initially, *N. meningitidis* causes a localised infection by colonising the nasopharynx. From this site, the meningococci invade the sub-mucosa and reach the central nervous system (CNS) by the invasion of the bloodstream. Meningococci can also reach the CNS by direct contiguous spread from the nasopharynx. They travel along the olfactory nerve and through the cribriform plate reach the subarachnoid space of the

Table 27.3 Virulence factors of *Neisseria meningitidis*

Virulence factors	Biological functions
Capsule	Prevents phagocytosis
LOS endotoxin	Causes damage of the blood vessels Stimulates release of TNF-α which results in host cell damage
IgA protease	Destroys IgA immunoglobulin, thereby helps gonococci to attach to the epithelial cells of the upper respiratory tract

unionised CNS, the bacteria multiply. Uncontrolled multiplication of bacteria in the cerebrospinal fluid (CSF) causes a cascade of meningeal inflammation resulting in meningitis.

Clinical Syndromes

Meningococcal infection of the nasopharynx is usually sub-clinical. Asymptomatic nasopharyngeal carriage of meningococci is of short duration and resolves within several weeks.

Meningitis

Meningitis is a febrile illness of short duration and is characterised by headaches and a stiff neck. The noted complications are mental obtundation, stupor and coma due to increased intracranial pressure. The condition is most common in children and young adults. Meningococcal meningitis is rapidly fatal, requiring early diagnosis and treatment. Serious complications like cranial nerve dysfunctions, hearing loss, paralysis, seizures. The prognosis of untreated infections is mostly fatal. With treatment, the prognosis in children is better than in adults. In children, chances of post-meningitis developmental delays always remain.

Meningococcaemia

Meningococcaemia with or without meningitis is a life-threatening condition. The condition presents as an acute fever with petechial rash. Small petechial rashes are found on the trunk and lower extremities. The rashes subsequently may coalesce to form large haemorrhagic lesions.

Waterhouse–Frederickson Syndrome

It is an overwhelming systemic infection caused by meningococci. This condition is characterised by severe disseminated intravascular coagulation, shock and multi-organ failure, including the destruction of the adrenal glands. The condition is invariably fatal. It is most commonly seen in persons suffering from deficiency of C5–C9 components of the complement.

Recurrent meningococcal meningitis is seen in patients with hereditary deficiency of various complement system components.

Laboratory Diagnosis

Specimens

Blood and CSF are the specimens of choice for demonstration of meningococci in the early stage of meningitis. **Blood** is collected in blood culture media containing glucose broth or sodium taurocholate broth. Nasopharyngeal swabs are useful to detect carriers. They are collected using sterile swabs and are transported in **Stuart transport medium** to the laboratory.

CSF is collected by lumbar puncture and blood by venepuncture in strict aseptic conditions. Meningococcal meningitis produces various inflammatory changes in the blood, and CSF in bacterial meningitis is more turbid. It contains more than 1000 WBC/μL, and the cells are predominantly PMN cells. The total protein content is increased. The total glucose level, which is normally 60% of simultaneous blood glucose level, is lowered (hypoglycorrhachia). The intracranial pressure may be elevated.

CSF is never refrigerated as *Haemophilus influenzae*, another agent of meningitis, may die at the cold temperature. It is transported immediately to the laboratory for processing. The CSF received in the laboratory is processed in three parts: The first part is centrifuged, and a smear is prepared from the deposit for Gram staining. The supernatant is tested for meningococcal antigens. The second part of CSF is used for direct culture. Finally, the third part is incubated overnight with an equal volume of glucose broth and then sub-cultured onto the blood agar and chocolate agar.

Microscopy

Gram staining of the CSF is a very useful method for detecting meningococci. They are seen as Gram-negative diplococci present mainly inside the leucocytes. Some may even be present extracellularly. In fulminant meningococcaemia, Gram staining of the peripheral blood buffy coat may show Gram-negative diplococci.

Culture and Identification

Isolation of *N. meningitidis* from the CSF, blood and other clinical specimens by culture confirms the diagnosis of meningococcal infection. The CSF is inoculated immediately on blood agar or chocolate agar and incubated at 35–36 °C in the presence of 5% CO_2. The colonies on blood agar after 24 h of incubation are small, round (1–2 mm in diameter) and convex. They are grey and translucent non-pigmented colonies with entire edges. At 48 h, the colonies become larger with an opaque raised centre and crenated with a transparent margin. It does not produce any haemolysis on blood agar. **Strains of meningococci with large polysaccharide capsules appear as mucoid colonies**. Meningococci produce large colonies on chocolate agar.

Blood is inoculated immediately into blood culture bottles containing glucose broth or sodium taurocholate broth and incubated at 35–36 °C. Sub-cultures are made on blood agar or chocolate agar from these broths and are pre-incubated overnight at 35–36 °C in the presence of 5% CO_2. The cultures should be incubated for 4–7 days with a daily sub-culture. Blood culture is often positive during the early stage of meningitis and in meningococcaemia.

Other specimens, such as nasopharyngeal swabs and petechial exudates, are processed similarly as described earlier for CSF. **Thayer–Martin medium with antibiotics** (vancomycin, colistin, nystatin and trimethoprim) and **New York City medium** are the selective media used to isolate the bacteria from nasopharyngeal swabs containing mixed bacterial flora.

N. meningitidis ferments glucose and maltose with acid but no gas. It does not ferment sucrose or lactose.

Serogrouping of meningococci grown on culture is carried out by slide agglutination with specific hyperimmune serum. Serotyping of strains is useful for identifying meningococci strains for epidemiological studies.

Antigen Detection

Detection of soluble polysaccharide antigen in the CSF is useful for diagnosing meningitis. Latex agglutination test and ELISA using specific antibodies are useful to detect the soluble antigen in the CSF. Antigen detection is useful when bacteria are scanty in the CSF. However, antigen detection is not useful in meningitis caused by Group B meningococci. This is because *N. meningitidis* serogroup B is relatively non-immunogenic and does not react with specific antibodies.

Serology

ELISA is useful for demonstrating antibodies against specific polysaccharide antigens in the serum. Serology is useful in chronic meningococcal infections where cultures have proved negative for meningococci.

Molecular Detection

PCR is a highly sensitive technique than the culture, even in the patients' receiving antibiotics. Unlike culture, the results are available within 24 h. PCR has been used to detect *N. meningitidis* DNA in clinical specimens. The test is useful in detecting meningococcal DNA in CSF even in small quantities and is a highly sensitive technique than the culturing method. The drawback of this test is its high cost and the need for expertise to operate the PCR. The test, therefore, is only used in a large-scale outbreak when many specimens are to be analysed and in a tertiary healthcare centre.

Treatment

Prompt and specific antimicrobial therapy of meningococcaemia or meningococcal meningitis is most crucial. Intravenous penicillin G is the drug of choice for the treatment of meningitis. The MIC of penicillin usually ranges from 0.01–0.05 μg/mL against meningococcal isolates.

Chloramphenicol, rifampicin, erythromycin, tetracycline and cephalosporins (ceftriaxone, cefotaxime and cefuroxime) are useful in the treatment of bacterial meningitis. Ceftriaxone has an additional advantage of eradicating the nasopharyngeal carriage of meningococci. Chloramphenicol is useful for patients who are allergic to penicillin. Meningococci are not susceptible to vancomycin and polymyxin. Recently, meningococci resistance to sulphadiazine (MIC ≥ 0.128 μg/mL) has been documented.

Epidemiology

Geographical Distribution

Meningococcal disease occurs worldwide. *N. meningitidis* serogroup A usually causes epidemics; serogroup B causes both epidemics and outbreaks, while serogroup C mostly causes localised outbreaks. Endemic meningitis is more common in children below the age of 5 and elderly people. Large outbreaks of meningococcal disease have occurred in central African countries, with attack rates as high as 400–500 cases per 100,000 people. Epidemics of meningococcal disease have occurred in many parts of the world.

Meningococci of group A are associated with diseases in underdeveloped countries; meningococci of group B, C or Y are responsible for most (90%) of the cases of meningococcal diseases in the developed countries (Table 27.4).

Habitat

N. meningitidis is primarily a pathogen of humans. Meningococci are found in the nasopharynx and oral cavity. Asymptomatic carriage of *N. meningitidis* varies from as low as 1% to as high as 40% in the population. The carriage rates are highest in school-going children, in young adults and the population with low economic status.

Reservoir, Source and Transmission of Infection

Human is the only reservoir of meningococcal infection. Nasopharyngeal secretion is the most common source of infection. Meningococci are transmitted by airborne droplets of infected nasopharyngeal secretions (the most common source of infection). Family members living in crowded conditions or the people who live in close populations (such as military barracks and prisons) and older people are more susceptible to infection.

The exact prevalence of meningococcal disease in India is largely unknown. However, large outbreaks and sporadic cases have been reported, which indicate a low prevalence of around 2.7% among all confirmed cases of meningitis. Meningococcal disease in India is caused predominantly by serogroup A though other serogroups, namely B, C, W and Y, have also been reported.

Prevention and Control

This includes antimicrobial chemoprophylaxis and vaccines.

Antimicrobial Chemoprophylaxis

Antimicrobial chemoprophylaxis of close contacts is the key component in preventing secondary cases of sporadic meningococcal disease. Person-to-person transmission can be interrupted by the administration of antibiotics which eradicate the asymptomatic nasopharyngeal carrier state. Sulphonamides, rifampin, minocycline,

Table 27.4 Epidemiology of *Neisseria meningitides* serogroups

Serogroups	Epidemiological features
A	Associated with epidemics in under-developed countries
B	Associated with epidemics and outbreaks Associated with most cases of meningitis in developed countries
C	Associated with outbreaks Associated with most cases of meningitis in developed countries
X	Meningococcal pneumonia
Y	
W135	

ciprofloxacin and ceftriaxone are frequently used to eradicate meningococci from the nasopharynx.

Vaccines

Four types of meningococcal vaccines, including two conjugate vaccines and recombinant protein vaccines, are available against meningococci. The conjugate vaccines contain meningococcal A, C, W and Y polysaccharides conjugated to diphtheria toxoid protein carrier. It does not contain any preservative or an adjuvant. The other vaccines contain recombinant proteins like Neisserial adhesin A (NadA), Neisserial Heparin Binding Antigen (NHBA) and factor H binding protein (fHbp) along with Outer Membrane Vesicles (OMV) and aluminium hydroxide as adjuvants. The latter protects against infections caused by the B serogroup.

Quadrivalent meningococcal polysaccharide vaccine (MPSV4) is effective against the meningococci infections caused by serogroups A, C, W135– and Y. This vaccine is injected intramuscularly. These vaccines are not effective in children under 2 years of age.

Tetravalent meningococcal polysaccharide-protein conjugate vaccine (MCV4) is being used for the people aged 11–55 years for vaccination against meningococci in the United States.

The conjugate vaccines that protect against the four serogroups are administered to adolescents in two doses. The first dose is given at 11 or 12 years of age, followed by the second dose at 16 years of age. In addition, this is also administered to the at-risk population, including immunocompromised people, travellers to endemic regions and laboratory personnel. The recombinant vaccine against serogroup B is recommended for people 10 years or older who are at increased risk for serogroup B meningococcal diseases like patients with splenectomy, complement deficiencies and outbreaks. Usually, more than a single dose is required for effective protection as protective antibodies decrease quickly after vaccination with the latter type.

No vaccine is available against the group B meningococci due to their non-immunogenic property of capsular antigen. However, lately, progress has been achieved in developing vaccine preparation against group B meningococci. At present, the vaccine is in the experimental stage and constitutes an OMP that helps stimulate group-specific bactericidal antibodies.

Commensal *Neisseria*

Commensal *Neisseria* is found as part of normal bacterial flora, mostly of the respiratory tract. These include *Neisseria flavescens*, *Neisseria sicca*, *Neisseria lactamica* and *Neisseria subflava*. The commensal *Neisseria* differs from pathogenic *Neisseria* species. They can grow on ordinary agar not enriched with blood and serum and can also grow at 22 °C. They do not require a high percentage of CO_2 for their growth. They produce greenish-yellow or yellow colonies on the media.

N. flavescens and *N. sicca* have been associated with isolated cases of meningitis, osteomyelitis, acute otitis media and acute sinusitis. Most of these strains are susceptible to penicillins. *Lactamica* is frequently isolated from the nasopharynx.

Genus *Moraxella*

Neisseria catarrhalis—which was later designated as *Branhamella catarrhalis* and is now renamed *Moraxella catarrhalis*—is a commensal of the upper respiratory tract and is occasionally found in the female genital tract. It is a recognised respiratory opportunistic pathogen in the immunocompromised host and hospitalised people.

Genus *Moraxella* consists of at least seven species that can cause infections in humans. These are *M. catarrhalis*, *Moraxella lacunata*, *Moraxella nonliquefaciens*, *Moraxella osloensis*, *Moraxella phenylpyruvica* and *Moraxella atlantae*. These are Gram-negative diplococci or coccobacilli.

M. catarrhalis is the most important human pathogen. It is found as a normal commensal in the respiratory tract of humans. It is a Gram-negative coccus measuring about 0.8 μm in diameter. It is arranged singly or in pairs with adjacent sides flattened. The bacteria are also found in

Table 27.5 Differential characteristics of *Neisseria* Species

Neisseria species	*Neisseria gonorrhoeae*	*Neisseria meningitidis*	*Neisseria lactamica*	*Neisseria sicca*	*Neisseria flavescens*	*Neisseria catarrhalis* (*Branhamella catarrhalis*)
Growth in medium						
BA at 22 °C	–	–	V	+	V	+
CHA	–	–	V	+	+	+
NA at 35 °C	–	V	+	+	V	V
Thayer–Martin	+	+	+	–	+	+
Fermentation of						
Glucose	+	+	+	+	–	–
Maltose	–	+	+	+	–	–
Sucrose	–	–	–	+	–	–
Lactose	–	–	+	–	–	–

V variable

groups of tetrads. Most strains grow on nutrient agar, blood agar, chocolate agar and MacConkey agar. After 24 h of incubation on blood agar, they produce non-haemolytic, white or greyish, convex colonies with entire margins. On prolonged incubation at 4 h, the colonies become large with elevated margins and a raised opaque centre. The bacteria do not grow on the media selective for gonococci due to the presence of colistin, to which it is sensitive.

M. catarrhalis is oxidase- and catalase-positive. It is differentiated from *Neisseria* species by a positive DNAse test and a positive tributyrin hydrolysis test. It is an opportunistic pathogen, a common cause of bronchitis and bronchopneumonia seen in patients with chronic pulmonary disease and elderly patients. The bacterium is also associated with sinusitis and otitis, most commonly in healthy people. In immunocompromised patients and children, it causes infections (e.g. otitis media, maxillary sinusitis, meningitis, septic arthritis, endocarditis, sepsis) in immunocompromised patients and children. It is multi-drug resistant and is resistant to penicillins due to the production of beta-lactamases but is susceptible to other antibiotics, such as trimethoprim-sulphamethoxazole, erythromycin, chloramphenicol, tetracyclines and cephalosporins.

Differential characteristics of Neisseria species are listed in the Table 27.5.

Case Study

A 22-year-old female complained of lower abdominal pain on and off for the last 3 months. She complained of a feeling of heaviness in the pelvis and pain during sexual intercourse. A tender mass was found on the right side during the examination. Gram staining of cervical swab showed plenty of pus cells and a few Gram-negative cocci. She gave a history of allergy to penicillins.

- Which is the most likely genital infection the patient is suffering from?
- Which is the most likely bacterium to cause this genital condition?
- What other diseases are caused by this bacterium?
- How you will confirm the diagnosis of this condition in the laboratory?
- What antibiotics can you use in this patient to treat the condition?

Corynebacterium and Coryneform Bacteria

28

The genus *Corynebacterium* consists of 136 recognised species, of which at least 30 species are known to be associated with human diseases.

Genus *Corynebacterium*

Corynebacterium diphtheriae, the causal agent of diphtheria, is the most widely studied species. Non-diphtherial corynebacteria—collectively referred to as **diphtheroids**—originally were believed to be mainly contaminants. These diphtheroids have recently been recognised as pathogenic, especially in immunocompromised hosts (Table 28.1).

Corynebacterium diphtheriae

C. diphtheriae is the most important species causing diphtheria. Diphtheria is an acute upper respiratory tract illness characterised by sore throat, low-grade fever and an adherent membrane on the tonsil(s), pharynx and/or nose.

Properties of the Bacteria

Morphology

C. diphtheriae is a Gram-positive bacillus showing maximum pleomorphism on Gram staining. The bacteria characteristically appear in palisades or as individual cells lying at sharp angles to each other in V and L formation. This Chinese letter pattern formation or cuneiform arrangement is caused by the incomplete separation of the daughter cells during division when the organism is grown on nutritionally inadequate media, such as coagulated egg medium or Loeffler coagulated serum.

The bacterium measures 3–6 × 0.6–0.8 μm, slender and sometimes has swollen ends. Most bacteria have two to three granules at the swollen ends, which give reddish-purple colour when stained with Loeffler alkaline methylene blue. The rest of the bacterium is unevenly stained with the dye. The granules are also known as metachromatic granules, Babe–Ernest granules or volutin granules. These granules, which are the accumulation of polymerised phosphates, are responsible for the beaded appearance of the bacteria. The granules can be demonstrated by special staining techniques, such as Albert stain, Neisser stain and Ponder stain.

Culture and Biochemical Properties

C. diphtheriae grows best under aerobic conditions. It grows on media enriched with blood, serum or egg.

Loeffler serum slope is an enriched medium frequently used for the culture of *C. diphtheriae*. The bacteria, on this medium, produce a luxuriant growth in 6–8 h at a temperature of 37 °C. Initially, the colonies are small, circular, opaque and white, but on prolonged incubation, the

Table 28.1 Human infections caused by *Corynebacterium* species

Corynebacterium species	Diseases
Corynebacterium diphtheriae	Diphtheria (respiratory and cutaneous) Diphtheria of other sites (external ear, the eye) and the genital mucosa; pharyngitis and endocarditis
Corynebacterium ulcerans	Cutaneous infections Diphtheria-like lesions, such as pharyngitis and respiratory disease
Corynebacterium jeikeium (group JK)	Wound infections Septicaemia, foreign body (catheter, prosthesis) infections and endocarditis
Corynebacterium urealyticum (group D2)	Urinary tract infection
Corynebacterium pseudotuberculosis	Native and prosthetic valve endocarditis Pneumonia, lung abscesses, tracheobronchitis and suppurative lymphadenitis
Corynebacterium haemolyticum	Pharyngitis
Corynebacterium striatum	Respiratory tract infections Foreign body infections
Corynebacterium pseudodiphtheriticum	Pneumonia, lung abscesses, tracheobronchitis, Endocarditis and lymphadenitis

colonies become larger and show a distinctly yellow tint. Loeffler serum slope does not support the growth of streptococci and pneumococci.

MacLeod or Hoyle tellurite blood agar media are examples of selective media used for the culture of *C. diphtheriae*. Tellurite (0.04%) present in the media inhibits the growth of other contaminant bacteria. On the tellurite agar, *C. diphtheriae* produces characteristic grey or black coloured colonies after 48 h of incubation (Fig. 28.1). *C. diphtheriae* reduces tellurite to metallic tellurium, which is incorporated in the colonies, giving them a characteristic grey or black colour. *C. diphtheriae* is classified into three distinct biotypes (mitis, intermedius and gravis) based on the colony morphologies on cysteine tellurite agar (Table 28.2).

C. diphtheriae ferments glucose, galactose, maltose and dextrin, producing acid but no gas. The bacteria do not ferment lactose, mannitol and sucrose. Hiss serum water is always used for testing the fermentation of sugars.

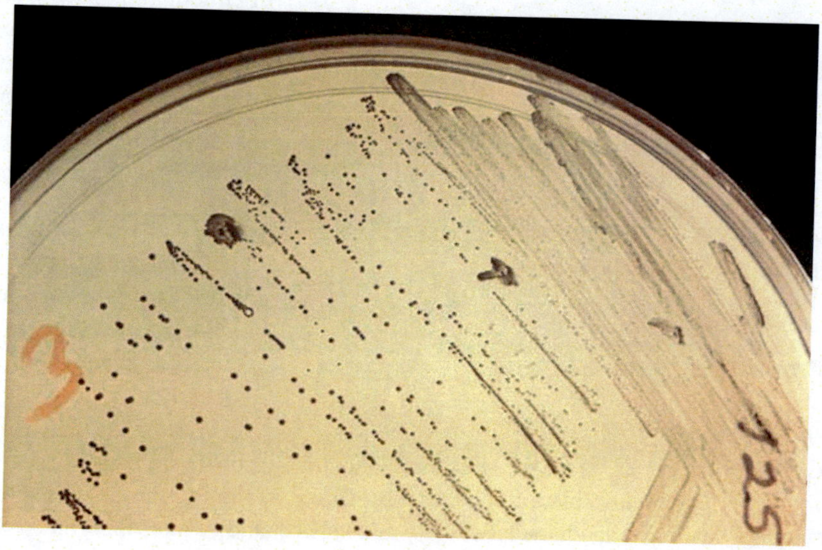

Fig. 28.1 Tinsdale medium agar showing black colonies of *Corynebacterium diphtheriae* (https://phil.cdc.gov/Details.aspx?pid=16564)

Table 28.2 Salient features of various biotypes of *Corynebacterium diphtheriae*

Features	Gravis	Intermedius	Mitis
Size	Short rods	Long rods	Long curved rods
Pleomorphism	+	+++	++++
Snow shoe and Tear drop forms	+	−	−
Granules	Few or no	Few	Prominent
Staining reaction	Uniform	Irregular	Irregular
Growth on Tellurite Agar			
Colony	18 h: 1–2 mm; Greyish black centre, paler 2–3 days: Flat with raised dark centre	18 h: 1 mm, Misty. 48 h: Dull granular centre with smoother glistening edges, with a lighter ring near the edge	Size variable. 48 h: Flat with a central elevation
Colony morphology	Daisy head	Frog's egg	Poached egg
Surface	Matte	Shining	Glossy
Consistency	Brittle	Weak buttery	Buttery
Haemolysis	Variable	Non-haemolytic	Usually, haemolytic
Starch fermentation	+	−	+
Phage types	14	3	4
Toxigenicity of strains	95%	99%	85%
Antigenic types	13	4	40
In broth medium	Pellicle formation and granular deposits	No pellicle, granular deposit	Diffuse turbidity

Other Properties

Susceptibility to Physical and Chemical Agents

Diphtheria bacilli are readily killed by heating at 58 °C for 10 min and 100 °C for 1 min. The usual strengths of antiseptics destroy them. They are resistant to the action of light, desiccation and freezing. They remain fully virulent and viable in-floor dust and blankets even for up to 5 weeks.

Pathogenesis

Diphtheria is a classic example of toxin-mediated bacterial disease.

Virulence Factors

Diphtheria toxin produced by *C. diphtheriae* is the key virulence factor of the bacteria.

Diphtheria Toxin

Diphtheria toxin is an exotoxin. This is produced only by *C. diphtheriae* (tox+ strains) containing intracellular structural *tox* genes. The gene responsible for toxin production is present in the prophage chromosome (corynephages (tox+)). However, the *tox* gene expression is regulated by a bacterial repressor protein. This repressor protein is activated by iron molecules, and thus, iron influences the production of toxins. The presence of iron at 0.1 mg/L in the medium is optimum for producing the toxin. Non-toxigenic strains can become toxigenic by acquiring beta bacteriophages. This process is known as phage conversion or lysogenic conversion. Such lysogenic bacteria produce high levels of toxin only under iron-deficient conditions.

Diphtheria toxin is a powerful exotoxin. As little as 0.0001 mg can cause death in a 250 g guinea pig in 96 h. It is a protein with a molecular weight of 58,000 Da. It consists of two functionally distinct polypeptide chain fragments A (24,000-Da protein) and B (38,000-Da protein). A disulphide bridge links both fragments. Fragment A is the active component of the toxin. Fragment B is the binding component that binds toxins to the cells. Neither of the fragments is toxigenic on their own, but act together to cause toxigenicity. The toxin is activated by the enzymes such as proteases present in the infected cells. The toxin binds to a specific receptor (now known as the HB-EGF receptor) present on susceptible cells and enters by a process known as receptor-mediated endocytosis.

Pathogenesis of Diphtheria

C. diphtheriae is transmitted by nasopharyngeal secretions or infected skin lesions. It usually enters the body through the upper respiratory tract but can also enter through the skin, genital tract or eye. Infection begins by adherence of toxigenic strains in the nasopharyngeal cavity. Non-toxigenic *C. diphtheria* can often indirectly obtain the tox gene by releasing toxigenic corynebacteriophage and in situ lysogenic conversion.

Growth and multiplication of *C. diphtheriae* are restricted to the nasopharynx, the site of infection. Toxaemia and the systemic manifestation of diphtheria occur due to the absorption of toxin from the membrane site in the nasopharyngeal cavity.

The toxin binds to the HB-EGF receptor is present on cells and enters the cell by receptor-mediated endocytosis. A fragment of the toxin is cleaved and released from the B subunit. Fragment A then gains entry into the cell and catalyses ADP ribosylation. This leads to the inhibition of the NAD in protein synthesis. Ultimately, inactivation of all the host–cell EF-2 (elongation factor) molecules causes cell death, which clinically manifests as the necrotic lesion of diphtheria.

Clinical Syndromes

The clinical manifestations of diphtheria depend upon the following: (1) immune status of the patient, (2) virulence of the bacteria and (3) the site of the infection.

Respiratory Diphtheria

It is usually caused by toxigenic strains (tox+) of *C. diphtheriae.* The incubation period varies from 2 to 5 days. Sore throat is the usual initial symptom in the absence of systemic complaints.

The initial lesion usually occurs on the tonsils and oropharynx, and from this site, it may spread to the nasopharynx, larynx and trachea. The organisms multiply rapidly in the epithelial cells, resulting in a local lesion and secreting exotoxins that cause necrosis of the cells in that area. The combination of cell debris and exudative inflammatory response (leading to accumulation of red blood cells, necrosed cells, bacteria, fibrin and lymphocytes) results in the formation of the characteristic pseudomembrane of diphtheria (Fig. 28.2). This is formed on the palate, pharynx, epiglottis, larynx or trachea and may extend to the tracheobronchial tree. It is thick, leathery, greyish-blue or white. The membrane adheres very tenaciously to the underlying mucosa, and if attempts are made to remove it forcibly, the raw bleeding surface is exposed. Spreading of the membrane down the bronchial tree can occur, causing respiratory tract obstruction and dyspnoea. The condition is associated with marked oedema of the tonsils, uvula, submandibular region and anterior neck (bull neck).

Complications occur following the effect of toxins on various sites. These include myocarditis, neurotoxicity, kidney failure, pneumonia.

Cutaneous Diphtheria

Cutaneous diphtheria is generally caused by non-toxigenic strains (tox− strains) of *C. diphtheriae.* The condition is an indolent non-progressive infection of the skin. It is characterised by a superficial, non-healing ulcer with a grey-brown membrane. Extremities are affected more often than the trunk or head. Cutaneous diphtheria often causes no toxicity while producing natural immunity. However, it may cause epidemics in poorly immunised populations.

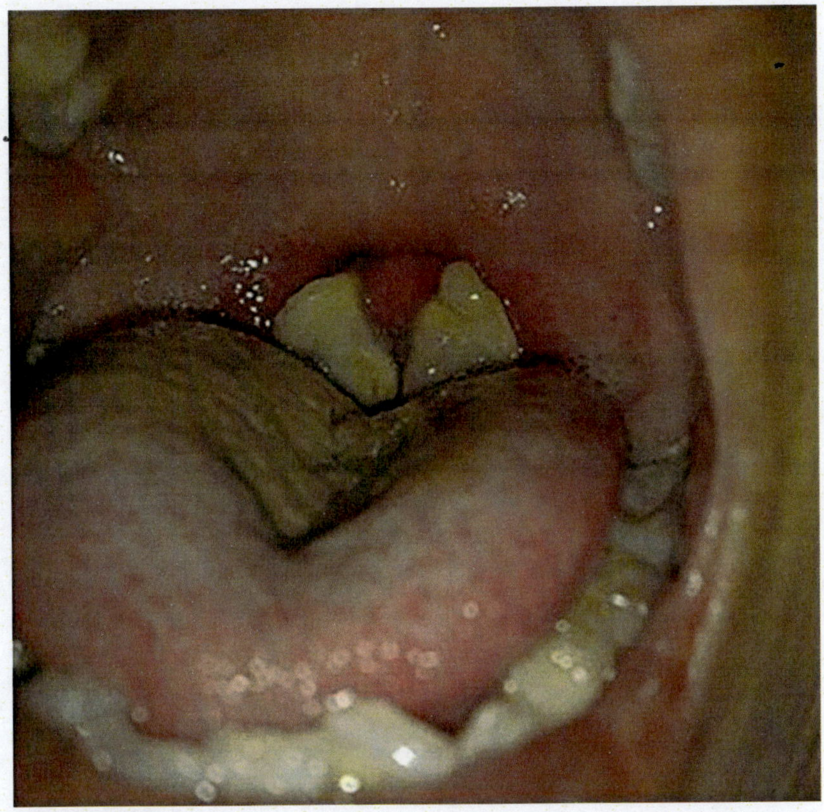

Fig. 28.2 Pseudo membrane in oropharynx of an infected child (https://link.springer.com/chapter/10.1007/978-3-319-74835-1_19)

Diphtheria of Other Sites

The external ear, eye (usually the palpebral conjunctivae) and genital mucosa are the other sites of diphtheria. Septicaemia caused by *C. diphtheriae* is rare but is invariably fatal.

Laboratory Diagnosis

Laboratory diagnosis is carried out not for the treatment of individual cases. It is carried out only for epidemiological purposes and for the initiation of control measures of diphtheria.

Specimens

These include swabs from the nose, throat, pieces of pseudomembrane, if possible, even from beneath the membrane, biopsy tissue etc. At least two swabs from the infected site are obtained. The first swab is used to make a direct smear, and the other is used for the culture.

The specimens are collected as soon as possible when diphtheria is suspected, even if treatment with antibiotics has been started. These are then transported to the laboratory in a sterile empty container or silica gel sachets for immediate processing.

Microscopy

Gram stain of the smear shows Gram-positive bacilli with swollen ends. The bacteria characteristically appear in palisades or as individual cells lying at sharp angles to each other in V and L formation. This Chinese letter pattern formation or cuneiform arrangement is caused by the incomplete separation of the daughter cells during division.

Albert, Neisser or Ponder stain of direct smears shows metachromatic granules in the bacilli (Fig. 28.3). These granules are also known as Babe–Ernest granules or volutin granules. These granules are rich in polymerised

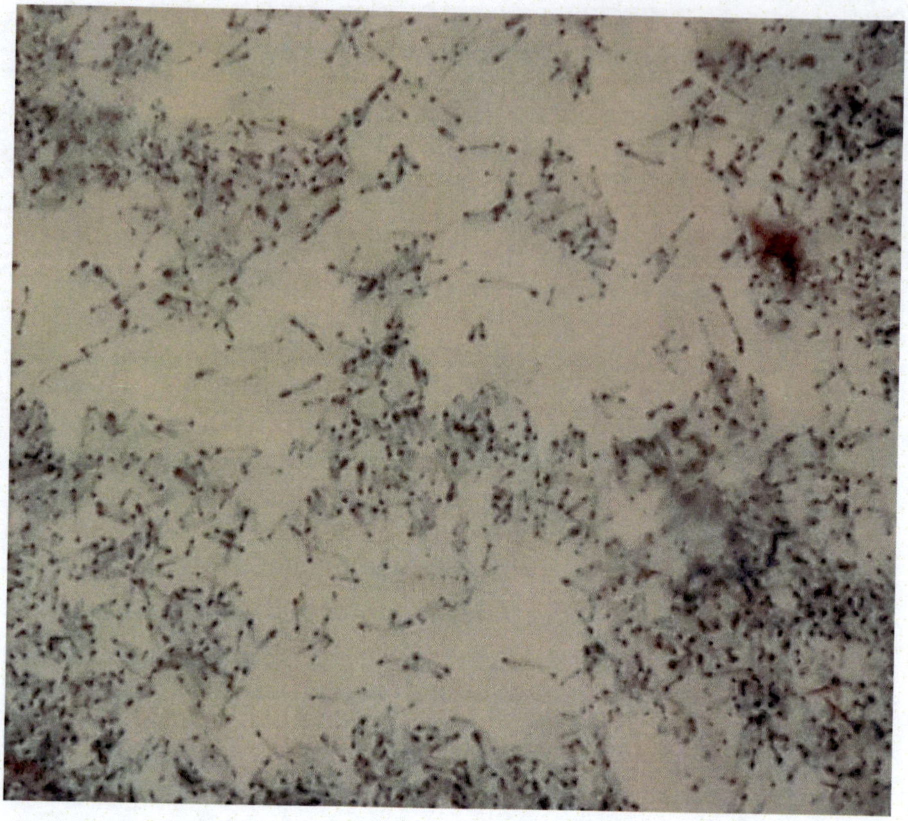

Fig. 28.3 Albert-stained smear showing granules of *Corynebacterium diphtheriae* (×1000)

phosphates, which are responsible for the beaded appearance of the bacteria. Most bacteria have two to three granules at the swollen ends, which give reddish-purple colour when stained with Loeffler alkaline methylene blue.

Examination of smears, however, has many limitations. First, diphtheria bacilli may not always be demonstrated in the smears. Second, it may be difficult to differentiate the bacilli from commensal corynebacteria frequently found in the throat.

Direct Immunofluorescence

It is useful for detecting and identifying *C. diphtheriae* in smears collected from the pseudomembrane using a specific antibody tagged with fluorescent dye.

Culture and Identification

The culture of specimens for *C. diphtheriae* is essential to confirm the diagnosis of diphtheria. The specimens are inoculated on non-selective media (e.g. sheep blood agar) as well as on selective media (e.g. tellurite agar) or enriched media (e.g. Loeffler, Hoyle, Mueller or Tinsdale medium).

Loeffler serum slope is an enriched medium frequently used for the culture of *C. diphtheriae*. After inoculation of the specimens, the bacteria produce a luxuriant growth in 6–8 h at a temperature of 37 °C. Initially, the colonies are small, circular, opaque and white, but on prolonged incubation, the colonies become larger and show a distinctly yellow tint. Loeffler serum slope does not support the growth of streptococci and pneumococci. Hence culture on sheep blood agar is required to differentiate *C. diphtheria* from these bacteria causing pharyngitis, mimicking diphtheria.

MacLeod or Hoyle tellurite blood agar is very useful for isolating *C. diphtheriae* from contacts, carriers and convalescents. This is because clinical specimens obtained contain several other bacteria. Tellurite (0.04%) present in the media inhibits the growth of other contaminant bacteria. On the tellurite agar, *C. diphtheriae* produces characteristic grey or black coloured colonies after 48 h of incubation. The bacteria reduce tellurite to metallic tellurium, which is incorporated in the colonies, giving them a characteristic grey or black colour. *C. diphtheriae* is classified into three distinct biotypes, mitis, intermedius and gravis, based on the colony morphologies on cysteine tellurite agar (Table 28.2).

Grams staining of colony isolate shows Gram-positive bacilli showing a *Chinese letter pattern* formation or *cuneiform arrangement.* Identified by biochemical tests and production of toxins as follows:

C. diphtheriae ferments many sugars (glucose, galactose, maltose and dextrin), producing acid but no gas. Hiss serum water is always used for testing the fermentation of sugars. The bacteria do not ferment lactose, mannitol and sucrose. Some strains of virulent *C. diphtheriae* ferment sucrose.

Toxigenicity Testing

All strains of *C. diphtheriae* are tested for the production of toxins. *In vivo* tests and *in vitro* tests can demonstrate the production of toxins. Currently, *in vivo* tests such as subcutaneous and intradermal tests in guinea pigs and rabbits are not being used because of restrictions imposed by animal protection acts. *In vitro* tests are increasingly used.

In Vitro Tests for Detection of Diphtheria Toxin

Elek gel precipitation test is an immunoprecipitation test for demonstration of biological activity of the toxin, initially described in 1949. It is an *in vitro* neutralisation reaction between toxin and antitoxin. A modified test recommended by Diphtheria reference Unit, WHO, is currently used. In this method, a rectangular strip of filter paper impregnated with diphtheria antitoxin (10 IU/mL) is placed on horse serum agar. *C. diphtheriae* to be tested for toxicity is streaked on the medium at right angles to the filter paper strip. A known toxigenic strain of *C. diphtheriae* (positive control) and non-toxigenic strain of *C. diphtheriae* (negative control) are also inoculated along with the test strain at right angles to the strip. The plate is incubated for 48 h. The toxin produced by the growth of the test strain diffuses into the agar and reacts with the antitoxin, thereby forming the line of precipitation. In strains that are negative, no precipitin lines are seen (Fig. 28.4).

Tissue Culture Test

Chinese hamster ovary cell lines and African green monkey kidney cell lines are used to detect the production of toxins. The toxigenicity of diphtheria bacilli is demonstrated by incorporating strains in an agar overlay of cell culture monolayers. The toxin produced by the bacteria diffuses into the cells below and kills the cells in the monolayer.

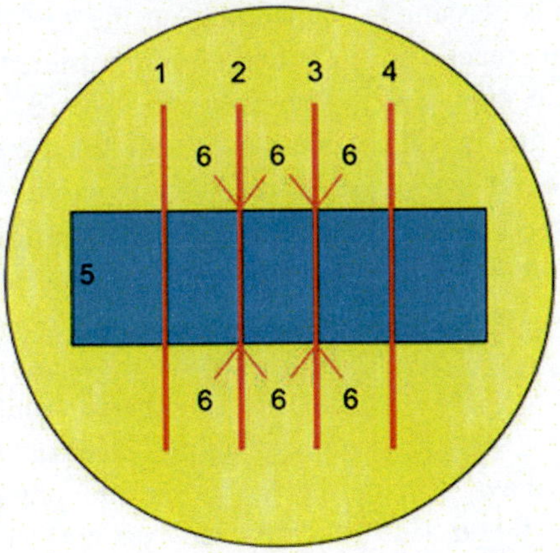

1. Non-toxigenic *Corynebacterium diphtheriae (*Negative control)
2. Toxigenic *Corynebacterium diphtheriae* (Positive control)
3. Toxigenic *Corynebacterium diphtheriae* (Test organism)
4. Non-toxigenic *Corynebacterium diphtheriae* (Test organism)
5. Strip with antitoxin
6. Lines of Precipitation

Fig. 28.4 Diagrammatic representation of Elek gel precipitation test

Sero

Immunochromatography and ELISA tests are being used to detect the toxin of isolates of *C. diphtheriae* from clinical specimens.

Molecular Detection

A polymerase chain reaction is a useful method for directly detecting the toxigenic strain of *C. diphtheriae* in a clinical specimen. The assay allows detection of the diphtherial toxin gene (TOX) in the bacteria. The test can also detect non-viable *C. diphtheriae* organisms in specimens collected even after starting antibiotic therapy.

Among all the tools, real-time PCR is the preferred choice for detecting diphtheria toxin-encoding gene *tox* in an isolate or specimen. It shows accuracy to differentiate *C. diphtheriae* and *Corynebacterium ulcerans* despite the tox genes of both organisms differ from each other in about 5% of their base pair. TaqMan hydrolysis and LightCycler hybridisation probes are the two most widely applied real-time PCR detection methods used in diagnosis.

Typing of *Corynebactrium diphtheriae*

The use of "Typing" of *C. diphtheriae* strains by various methods helps in global monitoring of the spread of this important pathogen. Further, it helps in outbreak containment, identifying the index case in any outbreak, and understanding the spread in the community, country and inter-continental spread.

Serotyping

Serotyping of *C. diphtheriae* strains is classified into various serotypes by agglutination reactions. Mitis strains have been classified into 40 types, intermedius into four types, and gravis into 13 types. Gravis type-II strains are found world-wide, whereas types I and III are commonly found in Great Britain. Type IV is mainly found in Egypt, whereas type V is found in the United States.

Bacteriophage Typing

So far, about 15 bacteriophage types have been identified. The mitis includes types I and III, intermedius includes types IV and VI, avirulent gravis includes type VII and remaining bacteriophage types come under virulent gravis type of *C. diphtheriae.*

Biotyping

Bio typing by automated systems using biochemical tests is frequently used. Four biotypes, namely mitis, gravis, intermedius and belfanti, are noted, of which most of the outbreaks reported are due to gravis.

Molecular Typing

Molecular typing of *C. diphtheriae* strains can also be done using molecular biology techniques. These include restriction endonuclease digestion patterns of *C. diphtheriae* chromosomal DNA and genetic probe for cloned corynebacterial insertion sequences. In ribotyping, strains are typed with the BstEII enzyme, and individual patterns are compared and analysed. Besides, pulsed-field gel electrophoresis (PFGE), random amplification of polymorphic DNA (RAPD) and amplified fragment length polymorphism (AFLP) have also been used in typing.

Treatment

Treatment of diphtheria is started immediately on clinical suspicion and diagnosis of diphtheria.

Diphtheria antitoxin is the mainstay of therapy in diphtheria. Diphtheria antitoxin is a hyper-immune antiserum produced in horses. The dosage recommended is 20,000–100,000 U depending on the severity of the infection. The antitoxin is administered by the intravenous route with infusion over 30–60 min. The antitoxin neutralises only free toxin before the toxin enters the cells but is ineffective after the toxin has entered the cell. Antitoxin appears to be of no value in treating cutaneous diphtheria and in the treatment of asymptomatic carriers.

Antimicrobial therapy is useful in the treatment of diphtheria. Antibiotics limit the production of toxins, eradicate diphtheria bacteria from infected hosts and prevent transmission of the bacteria to patient contacts. Penicillin and erythromycin are the only antibiotics recommended for treatment. Both antibiotics are equally effective in resolving fever and local symptoms. Erythromycin is marginally superior in eradicating the carrier state. Elimination of the bacteria is demonstrated by at least two successive negative nose and throat cultures or skin cultures obtained 24 h' apart after the completion of the therapy. Treatment with erythromycin is repeated if culture results remain positive. Antibiotic therapy, however, is not a substitute for antitoxin therapy.

Epidemiology

Geogrphical Distribution

The largest epidemic of diphtheria occurred from 1990 to 1996 throughout the states of the former Soviet Union. Most cases were reported among adolescents and adults rather than children. More than 110,000 cases and 2900 fatalities from diphtheria were reported during the epidemic. In the United States, Europe and Eastern Europe, recent diphtheria outbreaks have occurred largely among alcohol and/or drug abusers.

Habitat

The upper respiratory tract of an infected host is the primary habitat of *C. diphtheriae*. The bacteria also inhbit the superficial layer of the skin lesions.

Reservoir, Source and Transmission of Infections

Humans are the only natural host and thus are the only significant reservoirs of infection of *C. diphtheriae*. Common sources of infection include infective droplets or nasopharyngeal secretions. Direct human contact facilitates the transmission of the disease. Patients with active infection are more likely to spread the infection.

When diphtheria was endemic, the disease was most commonly seen in children younger than 15 years. Recently, the epidemiology has shifted to adults. This is because adults lacked natural exposure to toxigenic *C. diphtheriae* in the vaccine era and also received fewer booster doses.

Both symptomatic and asymptomatic individuals transmit diphtheria and contribute to global epidemiology. However, the asymptomatic carriers result in ~70% fewer infections cases than symptomatic individuals. Such asymptomatic cases are identified through contact tracing.

Prevention and Control

In diphtheria, immunity against clinical diseases depends on the presence of antitoxin in the bloodstream in response to clinical or sub-clinical disease or active immunisation. Infants below 6 months of age carry IgG antibodies derived from the immune mother either transplacentally or through breastfeeding. The antibodies later developed are IgG and IgA types. Most young children are highly susceptible to infection in areas where diphtheria is endemic, and mass immunisation is not practised.

Prevention of diphtheria can be achieved by either active, passive or combined immunisation.

Active Immunisation

Vaccines

Active immunisation by vaccination with **diphtheria toxoid** is the key to preventing diphtheria. Two types of toxoids are available. Fool toxoid is prepared by treating the exotoxin with formaldehyde. This treatment inactivates the toxic effects but maintains the antigenicity. Adsorbed toxoid is prepared by adsorbing purified toxoid, usually with aluminium phosphate as an adsorbent. This also acts as an adjuvant and increases the immunogenicity of the vaccines. Adsorbed toxoid, therefore, is more immunogenic than the formol toxoid. Administration of the vaccine stimulates the production of antibodies with specific protective properties. Serum antitoxin concentration of 0.01 IU/mL is usually accepted as the minimum protective level, and 0.1 IU/mL provides a definitely protective level.

DTP or triple vaccines are typically combined with tetanus toxoid and acellular pertussis (triple DTaP vaccine). It is the vaccine of choice for children aged 6 weeks to 6 years, as recommended by the Universal Immunisation Programme. Diphtheria toxoid is given in a dosage of 10–25 Lf (limit of flocculation) to all the children.

The primary immunisation schedule for infants and children consists of the first three doses of DTP, separated by a minimum of 4 weeks. The fourth dose is given 1 year after the third dose. A booster dose is given at the age of 4–6 years. A further booster dose is given at school entry. For adults and children older than 7 years, a lower dose of diphtheria toxoid (Td) is the vaccine of choice, and the primary series comprises three doses. For pregnant mothers, the National Immunization schedule advocated the administration of the adult dose of diphtheria vaccine, Td, as mentioned above, early in pregnancy followed by a second dose following 4 weeks after Td-1. In case of vaccination as per the above schedule in a previous pregnancy within 3 years, a single booster dose of Td is required.

The vaccine decreases local tissue spread of *C. diphtheria*. It also prevents toxic complications and diminishes transmission of *diphtheria*. The vaccine provides herd immunity when at least 70–80% of a population is immunised. However, it does not prevent the subsequent respiratory or cutaneous carriage of toxigenic *C. diphtheriae.*

Passive Immunisation

Passive immunisation is carried out by administering 500–1000 U of **antidiphtheritic serum (ADS)** injected subcutaneously. The effective neutralisation of diphtheria toxin is mediated by antitoxin with a long half-life period present in the ADS. ADS is usually recommended as an emergency measure for treating susceptible persons. It helps to reduce the fatality rate most effectively.

Combined Immunisation

Combined immunisation is carried out by simultaneous administration of ADS and diphtheria toxoid. The ADS is given in one arm, whereas the toxoid is given in the other arm, followed by a complete schedule of vaccination with toxoid.

Other Pathogenic *Corynebacterium* Species

C. ulcerans usually causes skin infections but is occasionally associated with diphtheria-like lesions, such as pharyngitis and respiratory disease. It resembles *C. diphtheriae* gravis type but differs from it by (1) reducing nitrate to nitrite, (2) liquefying gelatin and (3) fermenting trehalose. *C. ulcerans* produces two types of toxins: one resembles that of *C. diphtheriae,* and the other resembles that of *Corynebacterium pseudotuberculosis*. It is also pathogenic for animals. The bacteria can cause infection in cows and can be transmitted to humans through milk. Antidiphtheria toxin confers protection against the infection caused by *C. ulcerans*.

C. pseudotuberculosis causes infection in immunocompromised hosts. It causes native and prosthetic valve endocarditis, pneumonia, lung abscesses, tracheobronchitis and suppurative lymphadenitis. *Corynebacterium jeikeium* can is another species that colonise the skin of healthy people. It is an opportunistic pathogen in immunocompromised patients with haematological disorders or intravascular catheters.

Corynebacterium urealyticum is an important agent of urinary tract infection. The bacterium has been reported to cause chronic or recurrent cystitis, bladder stones and pyelonephritis. People with prior urinary tract abnormalities or recent urologic procedures are at the highest risk for this disease. *Corynebacterium striatum* is found on catheters in patients with neutropenia and malignancies. The bacterium causes respiratory tract infections and foreign body infections. *Corynebacterium haemolyticum* causes as many as 10% of all pharyngitis cases in patients between 10 and 30 years. The bacteria produce an extracellular toxin that causes an erythrogenic rash associated with pharyngitis.

Diphtheroids

Other *Corynebacterium* species resembling *C. diphtheriae* may sometimes be mistaken for

Table 28.3 Distinguishing features of *Corynebacterium diphtheriae* and diphtheroids

Characteristics	*Corynebacterium diphtheriae*	Diphtheroids
Metachromatic granules	Present	Few or absent
Staining property	Uniformly stained	More uniformly stained
Arrangement	Cuneiform arrangement	Parallel rows
Culture	Grows on enriched media	Grows on ordinary medium
Sugar fermentation	Negative	Positive
Toxins	Produce toxins	Do not produce toxins with few exceptions (e.g. *Corynebacterium ulcerans*, *Corynebacterium pseudotuberculosis*)

diphtheria bacilli and are called diphtheroids or non-diphtherial corynebacteria.

Diphtheroids are present widely in nature. They are found as commensals in the skin, throat, conjunctiva and other areas. Only recently the role of these organisms in causing human infections has been recognised. The most common diphtheroids include *Corynebacterium xerosis* found in the human conjunctival sac and *Corynebacterium pseudodiphtheriticum* found in the human throat; also, these are found to be associated with human infections. Differences between *C. diphtheriae* and diphtheroids are summarised in Table 28.3.

Coryneform Bacteria

Besides genus *Corynebacterium*, other genera of irregularly shaped, Gram-positive bacteria have been identified to colonise and cause disease in humans. The other coryneform genera include *Arcanobacterium*, *Oerskovia*, *Brevibacterium* and *Turicella*.

Arcanobacterium is associated with polymicrobial wound infection, pharyngitis and less commonly septicaemia and endocarditis. The use of penicillin or erythromycin has proved effective.

Oerskovia causes septicaemia, endocarditis, meningitis and soft tissue infections. It shows a high degree of resistance to commonly used antibiotics.

Brevibacterium has been found to colonise the skin of humans. The organism has been associated with malodorous feet in some colonised people, septicaemia, osteomyelitis etc.

Turicella colonises the ears of the persons infected with *Turicella* and also of healthy persons.

Case Study

An 8-year-old child was admitted to the hospital with low-grade fever (<100 °F), malaise, dysphagia and headache. Examination revealed a fibrinous membrane on the pharynx. A pair of swabs collected from the membrane was sent to the microbiology laboratory to diagnose *C. diphtheriae*. The child was administered antitoxin by the intravenous route and treated with antibiotics. Subsequently, the laboratory report confirmed isolation of a toxigenic *C. diphtheriae* from the specimen 3 days later.

- Why treatment for diphtheria was started immediately without waiting for the confirmatory report?
- What tests are performed in a microbiology laboratory to demonstrate *C. diphtheriae* in clinical specimens?
- Are non-toxigenic strains of *C. diphtheriae* capable of causing disease in humans?

Bacillus

29

The family Bacillaceae consists of rod-shaped Gram-positive bacteria that form endospores. The family includes two main groups of spore-forming bacteria: The anaerobic spore-forming bacteria of the genus *Clostridium* and the aerobic or facultatively anaerobic spore-forming bacteria of the genus *Bacillus*.

The Genus *Bacillus*

The genus *Bacillus* is frequently known as aerobic spore bearers. They are ubiquitous and are present in soil, dust, air and water. These bacteria are also frequently isolated as contaminants in bacteriological culture media. The genus *Bacillus* consists of more than 50 species. *Bacillus anthracis* and *Bacillus cereus* are the two most important species that cause infections in humans and animals. *B. anthracis* causes anthrax, while *B. cereus* causes food poisoning. The human infections caused by *Bacillus* species are summarised in Table 29.1.

Bacillus anthracis

B. anthracis, the causative agent of anthrax, has a worldwide distribution. Anthrax is caused by inhalation, skin exposure or gastrointestinal (GI) absorption. The use of anthrax as a biological warfare agent is an additional concern now.

B. anthracis is of considerable historical interest due to many reasons. It was the first pathogenic bacterium to be seen under a microscope. It was the first bacterium shown to be the cause of a disease. From studies on anthrax, Koch established his famous postulates in 1876. Robert Koch cultured the *B. anthracis* using pure culture in 1877 and proved the anthrax bacilli's ability to form endospores. He also produced anthrax in experimental animals by administering *B. anthracis* spores. It was the first bacterium to be isolated in pure culture and shown to possess spores. For the first time, spores were discovered from *B. anthracis* and *Bacillus subtilis* by Koch and Cohn, and they established the germ theory of disease. It was the first bacterium used to prepare attenuated vaccines by Louis Pasteur.

Properties of the Bacteria

Morphology

B. anthracis is a Gram-positive spore-forming bacillus. It is a very large bacillus measuring 1–1.2 am in width and 3–5 μm in length. In smears from infected tissues, the bacteria are found as single, in pairs and short chains, the entire chain being surrounded by a capsule. In culture, *B. anthracis* grows as long chains and may appear similar to streptobacilli. In these chains, the bacilli are arranged end-to-end, and the ends of the bacilli are truncated, not rounded, or often concave and somewhat swollen. This gives the chain of bacilli a **bamboo-stick appearance**.

Table 29.1 Human infections caused by *Bacillus* species

Bacteria	Diseases
Bacillus anthracis	Anthrax (cutaneous, gastrointestinal and inhalational) and anthrax meningitis
Bacillus cereus	Gastroenteritis, intravenous catheter septicemia and endocarditis
Bacillus licheniformis	Gastroenteritis
Bacillus thuringiensis	Gastroenteritis, Opportunistic infections
Other *Bacillus* species	Opportunistic infections

It is capsulated. The bacterium forms the capsule only when grown on nutrient agar containing 0.7% sodium bicarbonate in the presence of 5–20% carbon dioxide. The capsule is a polypeptide (*polymer of d-glutamic acid*) in nature. It protects the bacteria against leucocytic phagocytosis and lysis. It is usually formed in the tissues but is absent when grown in ordinary culture conditions. It is easily visualised using a methylene blue or India ink stain. A 60-megadalton plasmid, pX02, mediate the production of the capsule. Therefore, the transfer of this plasmid to non-capsulated *B. anthracis* by transduction makes the recipient strain capsulated.

Spores

Gram staining of *B. anthracis* shows the characteristic squared ends with spores as unstained spaces. Anthrax spores are oval, central in position and are refractile. They are of the same width as the bacillary body, so they do not cause bulging of vegetative cells. The anthrax spore consists of the central protoplast, cortex and spore coat.

The central protoplastor germ cell of the spore carries the constituents of the future vegetative cell. It also consists of dipicolinic acid, which is responsible for the heat resistance property of the spore. The cortex surrounding the protoplast consists largely of peptidoglycan (murein), which protects the spore from radiation and heat. As the spores germinate, the cortical membrane or protoplast wall acts as the cell wall for the new vegetative cell. The sport coat comprises 50% of the spore's volume and protects the spores from enzymes, chemicals etc.

The spores are highly refractile and are resistant to staining, heat, cold, radiation, desiccation and disinfectants. Spores are never found in host tissues unless the infected body fluids are exposed to ambient air.

Culture and Biochemical Properties

B. anthracis is an aerobe and facultative anaerobe. The bacteria grow at a temperature range of 12–45 °C, the optimum temperature being 37 °C. They grow on a wide range of media, including ordinary nutrient media and several selective media.

On **nutrient agar** after 24 h of incubation, *B. anthracis* produces greyish and granular colonies measuring 2–3 mm in diameter. Under low-power microscopy, the edges of the colony appear as long, interlacing chains of bacilli, resembling locks of matted hair, which gives them a medusa head appearance with an uneven surface and wavy margin.

B. anthracis on a solid medium containing 0.05–0.5 U of penicillin/mL produces large, spherical colonies within 3–6 h and occurs in chains on the surface of the agar, resembling a string of pearls. This property is known as the string of pearls reaction and is useful in differentiating *B. anthracis* from *B. cereus* and other aerobic spore bearers.

In a **gelatin stab**, there is growth down the stab line with lateral spikes, longer near the surface, giving an *inverted fir tree* appearance. The liquefaction process is slow and late, which occurs after 7 days at 20 °C and starts at the surface (Fig. 29.1).

Knisely Polymyxin B-lysozyme-EDTA-thallous acetate (PLET medium) agar medium is a selective medium used to isolate *B. anthracis* from mixtures containing other spore-bearing bacilli. The medium comprises heart infusion agar, polymyxin, lysozyme, ethylene diamine tetraacetic acid (EDTA) and thallous acetate.

Production of capsular material is associated with forming a characteristic **mucoid** or **"smooth" (S)** colony type. Capsulated bacteria on serum or bicarbonate medium produce smooth or mucoid

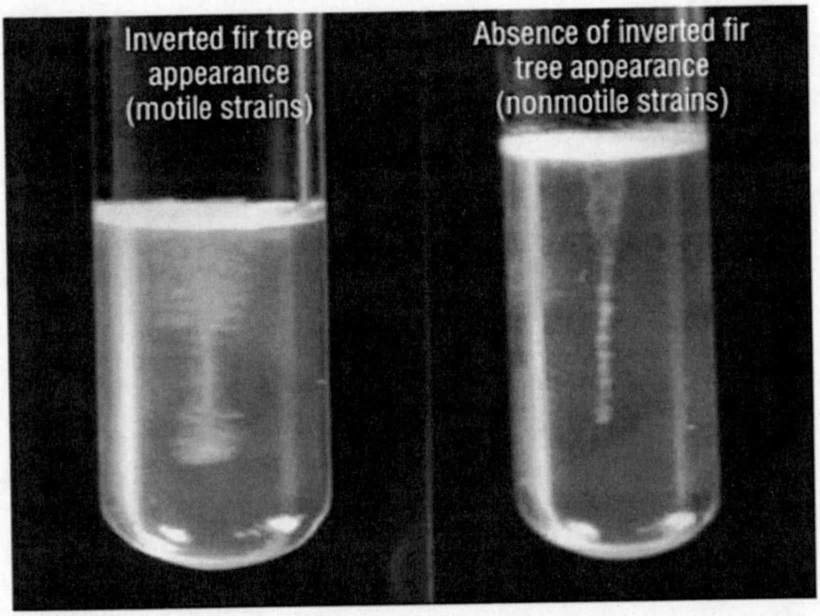

Fig. 29.1 Inverted fir tree appearance of colony of *Bacillus anthracis* in a gelatin stab culture

colonies. They are virulent strains. Rough (R) variants that lack capsules are relatively avirulent.

B. anthracis are catalase positive. They demonstrate a weak lecithinase reaction on egg-yolk agar, which gives a narrow zone of opalescence around the colonies. They reduce nitrate to nitrite and produces acid from glucose, maltose, sucrose, trehalose and dextrin, but not from lactose, arabinose, d-xylose or d-mannitol.

Other Properties

Susceptibility to Physical and Chemical Agents:

Bacillus vegetative forms are susceptible to moist heat; they are killed at 60 °C in 30 min. *Bacillus* spores are resistant to hostile physical and chemical conditions, ranging from desert sands and hot springs to Arctic soils and freshwater to marine sediments. They remain viable at temperatures, pH values and salt concentrations at which few other organisms could survive. The spores may survive in the dry state and certain soils for 50 years or more. They survive in 5% phenol for weeks. The spores are killed by moist heat at 100 °C in 60 min. They are also killed by 4% (w/v) formaldehyde or 4% (w/v) potassium permanganate in a few minutes.

Destruction of the spores in animal products is achieved by **duckering**. Duckering is a procedure that was used to reduce anthrax spores in wool below critical infection levels without destroying the animal hairs and bristles. In this procedure, 2% formaldehyde is used at 30–40 °C for 20 min for the disinfection of wool, while 0.25% formaldehyde solution is used at 60 °C for 6 h for disinfection of animal hair and bristles.

Antigenic Properties

Three types of antigens are present in anthrax bacilli.

The capsular antigen is present in virulent capsulated strains of *B. anthracis*. Antibodies against the capsular antigen are not protective. **Cell wall antigen** is present in the cell wall and is made up of *N*-acetylglucosamine and d-galactose. It cross-reacts with blood group A antigen and capsular polysaccharide of type 14 *Pneumococcus*. The antibody against the cell wall antigen is not protective. The **somatic antigen** is a heat-labile protein present in the bacterial body. It stimulates the immune system to produce protective antibodies.

Table 29.2 Virulence factors of *Bacillus anthracis*

Virulence factors	Biological functions
Bacterial capsule	Protects anthrax bacilli against leucocytic phagocytosis and lysis; if engulfed, resist killing and digestion.
Anthrax toxin complex	Anthrax toxin complex is plasmid-encoded and comprises of the following three proteins.
Protective antigen (PA)	Entry of the bacilli into the host cell. Binds the complex to receptors on macrophage surface.
Oedema factor (EF)	Causes cellular oedema within the target tissue and also inhibits neutrophil function. Blocks adenyl cyclase pathway within cells.
Lethal factor (LF)	Release of tumour necrosis factor-α and interleukin-1 by macrophages.

Pathogenesis

Virulence Factors

Virulence of *B. anthracis* (Table 29.2) depends on the bacterial capsule and anthrax toxin complex.

Bacterial Capsule

The poly-D-glutamyl capsule is non-toxic. It protects the bacteria phagocytes. It has a significant function in the pathogenesis of anthrax and has the main role in the development of infection.

Anthrax Toxin Complex

Local and fatal effects are due to the toxins produced by *B. anthracis*: Anthrax toxin is an exotoxin that is a protein in nature. Anthrax toxin complex comprises three components: protective antigen, oedema factor and lethal factor.

Protective antigen (PA) is an 82.7 kDa protein. It is called PA because antibodies against this antigen are protective against the action of anthrax toxin. The PA is the binding (B) domain of anthrax toxin and is necessary for the entry of the bacteria into the host cell. It binds to the cell receptors in the target tissue. After the attachment, an additional binding site is exposed due to the cleavage of a fragment that combines with EF to form an oedema toxin or LF to form a lethal toxin.

Oedema factor is an 88.9 kDa protein. It is a component of the oedema toxin. It is a calmodulin-dependent adenylate cyclase that converts adenosine triphosphate (ATP) to cyclic adenosine monophosphate (cAMP). This causes an increase in the cellular cAMP levels, leading to cellular oedema within the target tissues. The oedema factor also causes inhibition of neutrophil function and lowers the production of tumour necrosis factor (TNF-alpha) and interleukin-6 (IL-6) by monocytes.

The lethal factor is a 90.2 kDa protein. It is a zinc metalloprotease that inactivates mitogen-activated protein kinase, leading to the inhibition of intracellular signalling. It stimulates the release of TNF-alpha and interleukin-1 (IL-1) by macrophages. This mechanism is thought to contribute to sudden death from toxic effects in animals with high degrees of bacteraemia. The local and fatal effects of the infection are due to the toxins produced by *B. anthracis*.

Pathogenesis of Cutaneous Anthrax

Cutaneous anthrax is the most common form of anthrax and constitutes more than 95% of the naturally occurring anthrax. The condition is caused by the entry of spores through the skin lacerations, abrasions or through fly bites, usually on the face, neck or arms.

After ingestion by macrophages at the entry site, the spores germinate to form the vegetative forms of the bacteria. Virulent anthrax bacilli multiply at the site of infection with the production of capsules and toxins. Phagocytes migrate to the area, but the encapsulated anthrax bacilli resist phagocytic engulfment; or, if engulfed, resist killing and digestion. The anthrax toxin causes further impairment of phagocytic activity and its lethal effect on leucocytes, including phagocytes, at the site. The lethal toxin or oedema toxin is internalised by endocytosis. Oedema toxin produces the characteristic oedema of anthrax. Subsequently, the bacteria and their toxins enter the circulation, causing systemic morbidity.

Clinical Syndromes

B. anthracis in humans causes anthrax of various types, such as cutaneous anthrax, gastrointestinal anthrax, inhalational anthrax and anthrax meningitis.

Cutaneous Anthrax

This is commonly found in farmers and persons handling infected carcasses. People carrying hides of infected animals may also develop cutaneous anthrax, hence called "Hide Porters disease". Cutaneous anthrax may progress through the stages of papule, vesicle and eschar.

The lesion begins as a painless, pruritic **papule** at the inoculation site of spores. The papule, which is the primary lesion, becomes a 1–2-cm vesicle within 2 days. The **vesicle** is filled with clear or serosanguineous fluid containing numerous large, Gram-positive bacilli and very rare leucocytes. A characteristic non-pitting, gelatinous oedema surrounds the lesion. The vesicle enlarges, and satellite vesicles may develop. Subsequently, the vesicle ruptures undergo necrosis and enlarge, forming an ulcer covered by a characteristic **black eschar** (Fig. 29.2). The skin in the surrounding areas may become oedematous and necrotic but not purulent. The name anthrax, meaning coal, comes from the eschar, which is black coloured. The lesion is called **malignant pustule** due to its characteristic appearance; however, these lesions are neither malignant nor pustular. Lesions are painless but, on occasion, are slightly pruritic. Occasionally, **multiple bullae** develop along with marked toxic effects, and the lesions, especially on the face or neck, become massively oedematous.

Cutaneous anthrax usually remains localised. Without treatment, the condition spreads to the blood and disseminates to cause systemic infection in nearly one-fifth of the cases. Prompt antibiotic therapy prevents the dissemination of the infection but does not affect the natural history of the lesion. With treatment, the mortality due to cutaneous anthrax is approximately 1%.

Gastrointestinal Anthrax

The condition is caused by ingestion of undercooked meat of infected carcasses containing spores. The anthrax spores invade the mucosa of the GI tract and reach mesenteric lymph nodes. The spores in lymph nodes germinate to vegetative forms and begin multiplying, causing the occlusion of the lymphatic system.

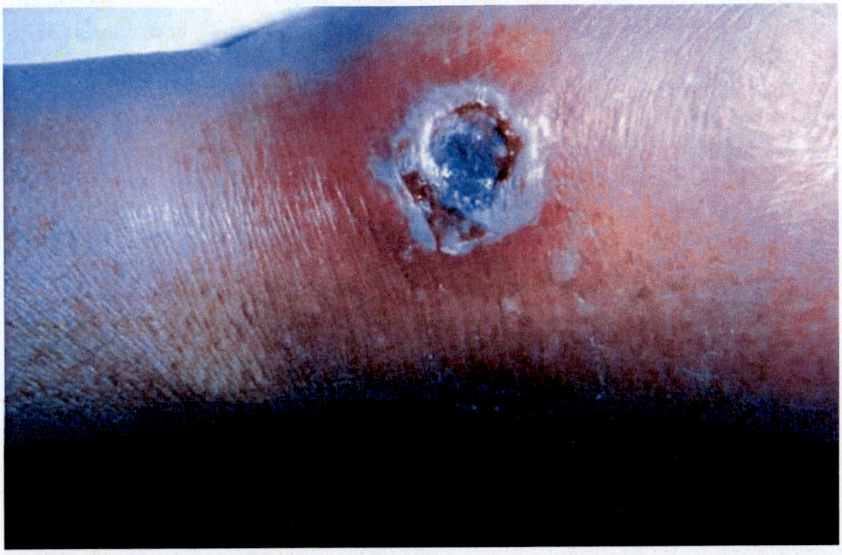

Fig. 29.2 Black eschar with a rim of erythema and oedema: cutaneous anthrax (https://phil.cdc.gov/Details.aspx?pid=2033)

This leads to ascites, haemorrhagic adenitis and oedematous stomach and intestine. Abdominal pain and fever are the first symptoms to appear 2–5 days after ingesting the food. These symptoms are followed by nausea, vomiting and diarrhoea. Without antibiotic therapy, death is rapid. Mortality is very high and is always more than 50%.

Inhalational or Pulmonary Anthrax

Inhalational or pulmonary anthrax is also known as *wool sorters disease*. The condition occurs after inhaling spores into the lungs. The median lethal inhalational dose for humans—extrapolated from data on experimental infection in monkeys—has been estimated to be 2500–55,000 spores. The spores are present in the dust or the filaments of wool from infected animals, particularly in wool factories. Alveolar macrophages engulf the spores and present them to the mediastinal lymph nodes. Anthrax in the lungs causes pulmonary oedema and haemorrhagic mediastinitis, but not pneumonia. Death occurs within 24–36 h. The disease has a high fatality rate.

Anthrax Meningitis

Anthrax meningitis may occur commonly due to bacteraemia from inhalational anthrax but is a less common manifestation of anthrax. The meninges are characteristically haemorrhagic and oedematous. Cerebrospinal fluid (CSF) is typically haemorrhagic and exhibits a polymorphonuclear pleocytosis. Numerous large, encapsulated, Gram-positive bacilli are demonstrated in the CSF. Mortality is as high as 100%, but occasionally, patients treated with antibiotics have survived.

Laboratory Diagnosis

It is essential to follow biosafety (level II) precautions while handling clinical specimens suspected of anthrax in a microbiology laboratory.

Specimens

These include vesicular fluid, fluid from under the eschar (in cutaneous anthrax); blood, lymph node or splenic aspirates (in septicaemic and inhalational anthrax); CSF (in anthrax meningitis); and sputum and blood (in inhalational anthrax).

Microscopy

The smears are stained with Gram stain, polychrome methylene blue (McFadyean stain) and Giemsa stain.

Gram-stained smear of a skin lesion (vesicular fluid or eschar), CSF or blood show encapsulated, broad, large, Gram-positive bacilli. The bacteria are found as single, in pairs and short chains. The bacilli do not show any spores. By Gram staining, presumptive identification of anthrax can be made (Fig. 29.3).

Giemsa-stained smears show purple bacilli surrounded by a red capsule.

Blood smears containing anthrax bacilli stained with **polychrome methylene blue** show an amorphous purple material around the bacilli. This amorphous purple material represents the disintegrated capsule and is a characteristic of anthrax bacillus. This reaction is known as **the McFadyean reaction** and is used for presumptive diagnosis of anthrax (Fig. 29.4). In smears from infected tissues, the bacteria are found as single, in pairs, and short chains, with the entire chain being surrounded by a capsule.

Sudan black B stain is used to demonstrate fat globules within the bacilli.

Direct fluorescent antibody (DFA) test is used for capsule and polysaccharide cell wall antigen of the bacteria.

Culture and Identification

Clinical specimens are inoculated on blood or nutrient agar or specialised selective media to isolate *B. anthracis*. Knisely PLET medium is used to isolate *B. anthracis* from soil and clinical materials containing numerous other spore-forming bacteria.

The absence of haemolysis on blood agar and the formation of medusa head colonies are the characteristic features of *B. anthracis*. Gram-

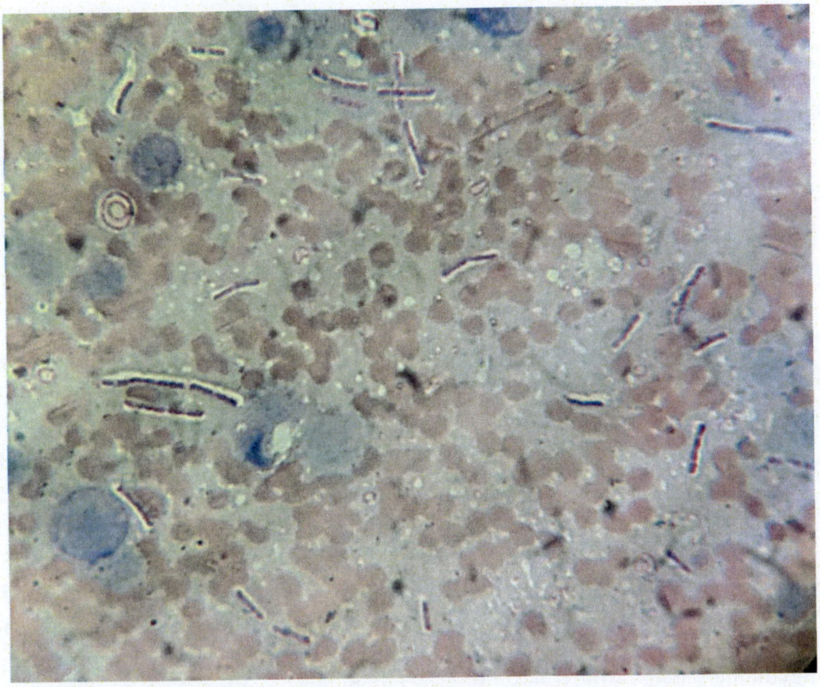

Fig. 29.3 Gram-stained smear shows broad large Gram-positive bacilli. *Note*: The bacilli do not show any spores (×1000)

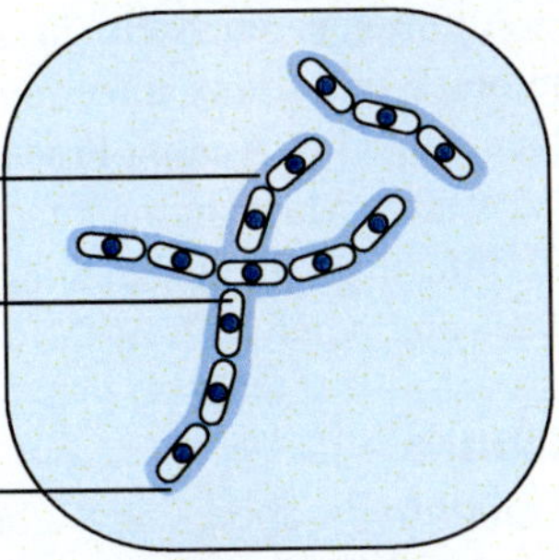

Fig. 29.4 Polychrome methylene blue-stained smears show an amorphous purplish material, remnant of the capsular material around the bacillus (McFadyean's reaction) (×1000)

positive, spore-forming bacilli in the Gram-stained smear of the colony, non-motile and capsulated bacteria, positive string-of-pearls test and lysis by gamma-phage are the identifying features of *B. anthracis*.

Animal Inoculation

B. anthracis can be isolated by collecting specimens from the contaminated tissue and applying them over the shaven skin sites of guinea pigs. The bacteria present in the specimen penetrate the skin through minute abrasions. Later, the bacilli are demonstrated in the sputum and heart blood of the guinea pigs, which die 2–3 days after the inoculation.

Serology

Serodiagnosis of anthrax is based on demonstrating antibodies against a protective antigen in the patient's serum.

ELISA test has been used extensively to demonstrate serum IgG against the protective antigen. The test is considered positive if a single acute-phase serum shows a high titre of antibodies or a fourfold greater rise in the antibody titre is observed between acute and convalescent serum specimens. The test is highly sensitive and specific. Specific IgG anti-PA antibodies are detected as early as 10 days after the onset of symptoms, but a high IgG level is observed only 40 days after the onset of symptoms.

Ascoli thermo-precipitation test is used in animals, mainly for rapid diagnosis when the tissue received is putrid and viable bacilli are unlikely to be found. The tissues are ground up in saline, boiled for 5 min and filtered. When this extract is layered over the anti-anthrax serum in a narrow tube, a ring of precipitate appears at the junction of two liquids within 5 min in a positive case.

Molecular Detection

B. anthracis-specific primers can be used for further molecular analysis. MLVA (multiple-locus and strain characterisation) and other techniques can be used for epidemiological studies and strain characterisation.

Treatment

B. anthracis is generally sensitive to penicillin; therefore, penicillin G is the first-line treatment. Penicillin, such as amoxicillin or amoxicillin/clavulanic acid, effectively treats cutaneous anthrax. However, after October 2001, ciprofloxacin or doxycycline is usually recommended due to the possibility of genetically engineered penicillin-resistant anthrax strains.

Patients with inhalational anthrax are treated with a multidrug regimen of either ciprofloxacin or doxycycline along with at least one more antibiotic, such as vancomycin, imipenem, meropenem and aminoglycosides. A recombinant human monoclonal antibody (Raxibacumab) has been used in prophylaxis since 2012 for inhalational anthrax.

Ciprofloxacin or doxycycline for 60 days is effective against cases of GI anthrax. Doxycycline is not suitable for use in suspected cases of anthrax meningitis because it has poor penetration of the central nervous system.

Anthrax immune globulin intravenous (AIGIV), a human polyclonal antiserum, has also been used with antibiotics to manage serious manifestations of anthrax.

Post-Exposure Prophylaxis

CDC recommends oral antibiotic therapy in persons exposed to *B. anthracis*. The recommended drugs are ciprofloxacin or doxycycline. Alternate agents include levofloxacin, clindamycin, amoxicillin or penicillin for penicillin-susceptible strains. Antimicrobial drug prophylaxis for 60 days for immediate protection and a 3-dose series of Anthrax Vaccine Adsorbed for long-term protection is recommended.

Epidemiology

Geographical Distribution

Anthrax is distributed worldwide and is present in the soil as extremely resistant spores. They cause infection in humans, farms and wild animals fed with contaminated food or grazed on contaminated ground. The extent of disease is approximately 20,000–100,000 cases annually throughout the world. Despite vaccination programs, the condition is frequently documented in the Middle East, Africa, Asia, Latin America and the Indian subcontinent. Epidemics of human anthrax have been reported in Russia and Zimbabwe.

Anthrax is **enzootic** in India. An epizootic outbreak of anthrax infections in sheep has been documented in Andhra Pradesh, Tamil Nadu and Karnataka. Many human cases of cutaneous and GI infections have been documented in these states and other parts of the country.

Habitat

B. anthracis spores are ubiquitous. They are found in the soil, water and air. Spores are formed in culture or soil under unfavourable conditions for growth. Anthrax spores need oxygen for sporulation. These germinate when exposed to a nutrient-rich condition, such as the tissues or blood of an animal or human host. Rainfall stimulates anthrax spore germination while flies and vultures spread the spores.

Reservoir, Source and Transmission of Infection

Infected animals are reservoirs of infection. Anthrax in humans is primarily zoonotic (Fig. 29.5). Infected animals or contaminated animal products (such as hides, wool, hair and ivory tusks and poorly cooked infected meat) are the important sources of infection for humans.

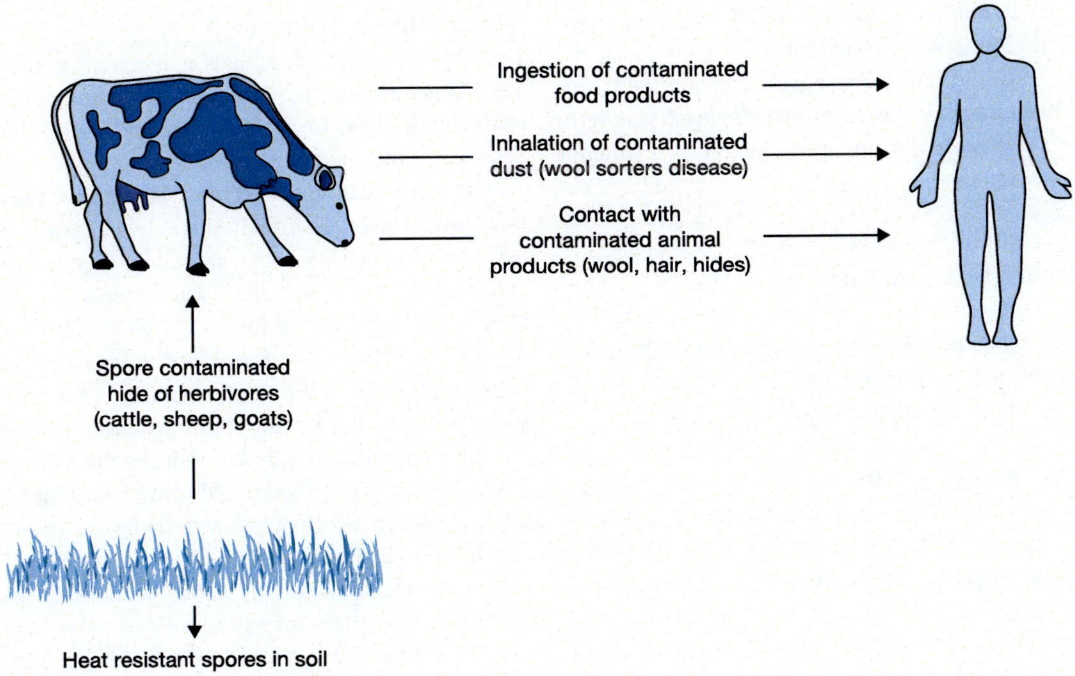

Fig. 29.5 Transmission of anthrax

Humans acquire anthrax infection by close contact with infected animals or contaminated animal products, such as hides, wool, hair and ivory tusks (cutaneous anthrax). Ingestion of poorly cooked infected meat contaminated with anthrax spores (GI anthrax). Inhalation of aerosolised anthrax spores associated with the large-scale processing of hides and wool in enclosed factory spaces (**inhalational anthrax**).

Direct spread of the disease from humans to humans is rare. Infected animals discharge many bacilli from their mouth, nose and rectum. These bacilli sporulate in the soil and remain viable over a long period. The contaminated soil is an important reservoir of infection for animals. Infection occurs in susceptible animals by ingesting the spores present in the soil.

Anthrax and Biological Warfare

Inhalation of spores disseminated by terrorists is an important transmission mode in **biological warfare**. The possible use of anthrax spores in bioterrorism was a major concern. *B. anthracis*, a biological weapon used in bioterrorism, was created by aerosols containing anthrax spores. Weapon-grade anthrax spores in bioterrorism may be dispersed as an aerosol for mass effect or by restricted spore dissemination and contamination via letters or packages.

CDC Guidelines for Identification After the Threat of Bioterrorism

CDC has established a Laboratory Response Network (LRN). It is a three-tier system with the following components:

1. **Sentinel Labs**: These are local clinical laboratories that can conduct the preliminary tests of a specimen to rule out the presence of *B. anthracis*. If it is unable to rule out, and anthrax is strongly suspected, the specimen is sent to the next level or reference lab. Morphology of the bacterium, growth characteristics on sheep blood agar and haemolysis, catalase test and motility are done to rule out anthrax. If all the characteristics are present, it is suspected to be *B. anthracis*.
2. **Reference Labs**: These public health laboratories can verify whether the organism

is *B. anthracis*. Once verified, it is sent to the highest level national lab
3. **National Labs**: These are CDC or armed force laboratories that carry on advanced testing of the organism's properties, including testing for antibiotics.

Prevention and Control

These include chemoprophylaxis with antibiotics, vaccination in humans and animals, and decontamination of animal products.

Chemoprophylaxis

Chemoprophylaxis with antibiotics is indicated for people who have been exposed to anthrax but do not have symptoms of the disease. This is useful to reduce the risk or progression of disease due to inhaled anthrax spores. Ciprofloxacin, tetracyclines including doxycycline, or penicillin administered for 60 days is effective.

Vaccination in Humans

Humans show intermediate susceptibility to *B. anthracis*. A non-cellular human anthrax vaccine called anthrax vaccine adsorbed (AVA) is available for individuals in high-risk occupations such as veterinarians and animal handlers. The vaccine is developed using the protective antigen (PA) harvested from the *B. anthracis* culture filtrate of an avirulent, non-encapsulated strain that produces PA during their growth.

Pre-Exposure Immunisation

Pre-exposure immunisation for anthrax is recommended for high-risk individuals. It consists of three subcutaneous injections at 2 weeks apart followed by three more subcutaneous injections administered at 6, 12 and 18 months. Annual booster injections for the anthrax vaccine are necessarily required to retain a protective level of immunity. The anthrax vaccine is recommended to use **as part of the post-exposure treatment** of healthy individuals from 18 to 65 years of age who are at risk of exposure to anthrax, along with antibiotics.

Vaccination in Animals

Control of anthrax in animals is essential to prevent transmission of infection to humans. A non-encapsulated toxigenic strain has been used effectively for vaccination in livestock.

Pasteur's anthrax vaccine was the first vaccine invented by Louise Pasteur, which was an active immunisation of animals to prevent anthrax. Pasteur vaccine was prepared by attenuating anthrax bacilli by growing them at 42–43 °C. **Sterne vaccine** contains spores of a non-encapsulated avirulent mutant strain of *B. anthracis*. The mutant strain produces sublethal amounts of the toxin that induces the formation of protective antibodies. **Maulucci vaccine** is prepared using stable attenuated Carbozoo strain in 2% saponin. The animal is protected for a year with a single injection of spore vaccine. It is extensively used in animals; however, it is not safe for human use, but in Russia, these vaccines have been used for vaccination in humans.

The vaccination of animals has also been used to protect people living in an area endemic for the disease and protect persons dealing with animal and animal products.

Decontamination of Animal Products

These include proper sterilisation of animal products, such as wool, hides, and improvement of personal hygiene. Animal carcasses suspected of having died of anthrax are buried deep in burnt lime (calcium oxide) or cremated to prevent soil contamination.

Anthracoid Bacilli

Bacillus species resembling *B. anthracis* are collectively called **anthrocoid** or **pseudo anthrax bacilli**. These are opportunistic pathogens with low virulence. They are ubiquitous organisms, present virtually in all environments. Some of them are frequent laboratory contaminants. *B. cereus* is the most important pathogen causing GI, ocular and catheter-related infections. *B. subtilis* may act as an opportunistic pathogen, causing eye infections and septicaemia. *Bacillus* *licheniformis* has also been incriminated in patients with food poisoning. Spores of *Bacillus stearothermophilus* is used to test the efficiency of sterilisation by autoclaves. Some *Bacillus* species are used to produce antibiotics, such as bacitracin, tyrothricin and polymyxin.

Bacillus cereus

B. cereus is the most important pathogen known to cause food poisoning. Episodes of *B. cereus* food poisoning occur sporadically worldwide. This occurs due to the ingestion of contaminated food in which bacteria have multiplied to high levels under conditions of improper storage after cooking. *B. cereus* is widely distributed in nature, such as soil, vegetables, milk, cereals, spices, meat and poultry. *Cereus* is a spore-forming Gram-positive bacillus. It resembles *B. anthracis* but differs by certain features, as mentioned in Table 29.3.

B. cereus produces many toxins and enzymes responsible for many syndromes. The heat-labile enterotoxin of *B. cereus* is similar to that produced by *Vibrio cholerae* and *Escherichia coli*. The toxin stimulates the adenyl cyclase-cAMP system causing profuse watery diarrhoea. Like other enterotoxins, it causes accumulation of fluid in ligated rabbit ileal loop. The toxin causes a diarrhoeal form of *B. cereus* food poisoning. The heat-stable enterotoxin of *B. cereus* causes an emetic form of *B. cereus* food poisoning, but the exact mechanism of action of the toxin is not known.

B. cereus also produces at least three other toxins (1) cerolysin, a potent haemolysin; (2) necrotic toxin, a heat-labile toxin and (3) phospholipase C, a potent lecithinase. They have been implicated in causing ocular infections such as panophthalmitis.

B. cereus causes two types of food poisoning: Diarrheal and Emetic forms. **Diarrheal form of food poisoning** is caused by *B. cereus* serotypes 2, 6, 8, 9, 10 or 12. It manifests as acute abdominal pain, diarrhoea and nausea; vomiting is rare. The disease generally lasts for 24 h or more. The faecal samples contain small numbers of *B. cereus*.

Emetic form of *B. cereus* food poisoning, caused by heat-stable enterotoxin, results from the consumption of contaminated rice. Most bacilli are killed during the initial cooking of rice, but heat-resistant spores survive. If the food is kept at room temperature, the spores germinate to vegetative forms, multiplying and producing enterotoxins. Heat-resistant enterotoxins are not killed on reheating the rice. On ingestion of the rice, the toxin causes nausea, vomiting and abdominal cramps after a short incubation period

Table 29.3 Differentiating features between *Bacillus anthracis* and *Bacillus cereus*

Characteristics	*Bacillus anthracis*	*Bacillus cereus*
Motility	Non-motile	Motile
Capsule	Capsulated	Non-capsulated
Medusa head colony	Present	Absent
Microscopic appearance	Grow in long chains	Grow in short chains
Haemolysis on sheep blood agar	Absent	Present
Growth on chloral hydrate agar	Absent	Present
Gelatin liquefaction	Slow	Rapid
Growth in broth	Absent	Present
Salicin fermentation	Negative	Positive
Phospholipase activity	Absent	Present
Growth at 45 °C	Absent	Present
String-of-pearls test	Positive	Negative
Lysis by gamma phage	Positive	Negative
Mouse pathogenicity	Positive	Variable
Penicillin Susceptibility	Sensitive	Resistant

of 6 h. Diarrhoea and fever are rare. *B. cereus* is present in large numbers in the cooked rice and the faecal samples of these patients. The disease usually lasts for a shorter duration of 24 h or less. The emetic form of food poisoning is caused by *B. cereus* serotypes 1, 3 or 5.

Food poisoning is mild and self-limiting, requiring no specific treatment. Other *Bacillus* infections need prompt treatment with antibiotics. It is resistant to penicillin and cephalosporins. It is sensitive to gentamicin, ciprofloxacin, clindamycin and vancomycin. Food poisoning is prevented by adequate cooking, avoiding recontamination of cooked food and proper storage (efficient refrigeration).

Diagnosis of the food poisoning is by the culture of the suspected food, faeces and vomitus are cultured on ordinary media or a special media. Spore-bearing Gram-positive bacilli may be seen on a smear from the colonies. It is generally motile, but non-motile strains may also occur. It is a facultative anaerobe. Mannitol, egg yolk, phenol red polymyxin agar (MYPA) is the selective media used to isolate *B. cereus* from faeces and other sources.

Case Study

A group of 35 people in a village joined in a community dinner. They consumed a mutton preparation, among other dishes. Two days after dinner, 20 of them fell sick with severe abdominal pain and fever, followed by nausea, vomiting and diarrhoea. Some patients had bloody diarrhoea. All were admitted to a local hospital. History revealed that the meat of a sheep, which died in the morning, was cooked for dinner. Few other sheep were also found to be dead. Veterinary doctors examined the carcasses of the sheep and diagnosed anthrax to be the cause of death of these animals. The patients suspected to be suffering from GI anthrax were treated with a course of ciprofloxacin. All the patients recovered fully.

- What was the infective stage of the bacteria that caused the disease?
- What investigations would be useful to diagnose the condition?
- What are the modes of transmission of anthrax?
- What are the vaccines available against anthrax in animals and humans?
- Do these bacteria show any resistance to antibiotics?

Genus *Mycobacterium* and *Mycobacterium tuberculosis*

30

The genus *Mycobacterium* included in the family Mycobacteriaceae consists of non-motile, non-spore-forming and aerobic bacilli.

Genus *Mycobacterium*

They are slightly curved rods, sometimes showing filamentous and branching forms. In liquid medium, they form mould-like pellicle; hence, the name *Mycobacterium* is derived from the word **"mould"**, meaning fungus-like bacterium.

The cell wall is thick, complex and rich in lipids, making the surface hydrophobic. The lipid-rich cell wall also makes the bacteria resistant to commonly used disinfectants and laboratory staining reagents. Due to their lipid-rich waxy cell wall, mycobacteria are difficult to stain. However, once stained, they resist decolourisation with acid solution and alcohol. They are, therefore, called acid-fast bacilli (AFB). Although they are Gram-positive, they are difficult to stain and they stain poorly with Gram stain. Most mycobacteria are slow growers, and the generation time is 2–24 h. These bacteria require 3–8 weeks of incubation to produce demonstrable colonies on the solid media.

Classification

The genus *Mycobacterium* contains 194 recognised species, many of which cause disease in humans. *Mycobacterium* species causing most human infections are summarised in Table 30.1. *Mycobacterium leprae* was the first species of the genus *Mycobacterium* to be identified by Hansen in 1874. Furthermore, Robert Koch in 1882 isolated the mammalian tubercle bacillus and established its role as a causative agent of tuberculosis, as it satisfied all the criteria of Koch postulates. Subsequently, *Mycobacterium tuberculosis* was demonstrated as the causative agent of tuberculosis.

The genus *Mycobacterium* can be broadly categorised into three groups based on their ability to cause human infections: obligate parasites, opportunistic pathogens and saprophytes.

Obligate Parasites

***M. tuberculosis* complex** is the term used currently, including four species that can cause tuberculosis in humans and other mammals. These are: (1) *M. tuberculosis*, the human type; (2) *Mycobacterium bovis*, bovine type; (3) *Mycobacterium africanum*, a species intermediates between human and bovine types and causing human tuberculosis mostly in tropical Africa and (4) *Mycobacterium microti*, another species which is pathogenic for goats and other small animals but does not cause any human infection.

Table 30.1 Human infections caused by most common *Mycobacterium* species

Mycobacterium species	Diseases
Mycobacterium tuberculosis	Tuberculosis
Mycobacterium leprae	Leprosy
Mycobacterium avium complex	Opportunistic mycobacterial infections (pulmonary disease lymphadaenopathy and disseminated disease)
Mycobacterium kansasii	Opportunistic mycobacterial infections (pulmonary disease)
Mycobacterium fortuitum	Opportunistic mycobacterial infections (post-trauma abscess)
Mycobacterium chelonae	Opportunistic mycobacterial infections (post-trauma abscess)
Mycobacterium abscessus	Opportunistic mycobacterial infections (skin infection)

M. leprae causes leprosy in humans. Despite its early discovery, very less is known about biological requirements, and it has never been cultured *in vitro* so far.

Opportunistic Pathogens

Non-tuberculous mycobacterium is isolated from different types of sources. This species has been isolated from cold- and warm-blooded animals, skin ulcers and environmental sources, such as soil and water.

Saprophytic Mycobacteria

These are the third group of mycobacterial species isolated from different sources. This group contains Mycobacteria, such as *Mycobacterium smegmatis* (present in smegma) and *Mycobacterium phlei* (present in the grass). *M. smegmatis* is a rapid grower and usually contaminates urine cultures.

General Properties

Genus *Mycobacterium* is acid-fast. They contain mycolic acid, consisting of 60–90 carbons that are split to produce fatty acid methyl esters. They contain a high guanine and cytosine content (61–71%) in the DNA.

The **cell wall of *Mycobacterium*** is complex. It is typical as that of Gram-positive bacteria. It contains a unique cytoplasmic membrane, surrounded by thick peptidoglycan and a layer of highly antigenic mycolic acid. It is thick and lipid-rich. The mycolic acid consists of peptidoglycolipids, glycolipids and lipids. The outer layer consists of peptic chains and is an important antigen of the bacteria which induces cellular immunity in infected patients. A **purified protein derivative (PPD)** of this layer is used as a source of antigen in the skin test to demonstrate the hypersensitive reaction of the host to *M. tuberculosis*.

Mycobacterium tuberculosis

M. tuberculosis is the causative agent of tuberculosis—the number one infectious killer disease worldwide.

Properties of the Bacteria

Morphology

M. tuberculosis is straight or slightly curved rods occurring singly, in pairs or small clumps. They measure about 3 × 0.3 μm in size; sometimes, long filamentous forms are also seen. *M. bovis* is usually shorter and stouter than *M. tuberculosis*.

M. tuberculosis is **acid-fast and alcohol fast**; it resists decolourisation by 20% sulphuric acid and absolute alcohol. This acid-fast staining is the most important property of the bacteria and is used widely to identify the bacteria. Ziehl–Neelsen (ZN)-staining procedure is used to demonstrate the acid fastness of the bacteria. *M. tuberculosis* stains bright red with this stain, while the tissue cells and other organisms are stained blue. These bacilli can also be stained by Auramine O stain and examined under a fluorescent microscope. They appear as bright fluorescent rods against a dark background. Many granules (Much granules) that are the non-acid-

Fig. 30.1 LJ media for growth of *Mycobacterium tuberculosis*: one medium showing colonies of *Mycobacterium tuberculosis*, while the other medium showing no growth

fast form of tubercle bacilli present in the cold abscess pus, which could produce the disease when injected into susceptible animals. They are Gram-positive but are difficult to stain. They are non-motile, non-sporing and non-capsulated.

Culture and Biochemical Properties

M. tuberculosis is an obligate aerobe and characteristically grows very slowly in media. The optimum temperature is 37 °C, and the bacteria do not grow below 25 °C or above 40 °C. The optimum pH is 6.4–7.0. It is a slow-growing bacillus with an average generation time of 14–15 h. The colonies usually appear in almost 2 weeks but sometimes require incubation up to 8 weeks to appear. *M. tuberculosis* can grow on a wide range of enriched solid and liquid media.

The examples of **solid media** are (1) egg-containing media [Lowenstein–Jensen (LJ) medium, Petragnani and Dorset egg medium]; (2) blood-containing media (Tarshis medium), serum-combining media (Loeffler serum slope) or (3) potato-based media (Pawlowsky medium). On these media, *M. tuberculosis* produces dry, rough, raised and irregular colonies with a wrinkled surface. **LJ medium** is the most widely used and is also recommended by the International Union against Tuberculosis. The medium consists of coagulated whole egg, asparagines, malachite green, mineral salt and glycerol or sodium pyruvate. Malachite green inhibits the growth of bacteria other than mycobacterium. The addition of 0.75% glycerol enhances the growth of *M. tuberculosis* but is inhibitory to the growth of *M. bovis*. After 6–8 weeks of incubation, *M. tuberculosis* produces yellowish or buff-coloured colonies on the LJ medium. They are tenacious and not easily emulsified (Fig. 30.1). The identification of tubercle bacilli and other related mycobacterium is described in the flow chart (Fig. 30.2).

The commonly used **liquid media** include Soloac solutions, Dubos medium and Middlebrook and Beck medium. In liquid media, *M. tuberculosis* produces growth that appears first at the bottom and then grips up to the sides and produces a surface pellicle that may extend along the sides above the medium. The bacteria in Dubos medium containing Tween 80 produce a diffuse growth. In glycerol broth, they form a whitish-wrinkled pellicle on the surface and a granular deposit at the bottom. Virulent strains usually form long serpentine cords in liquid media. In contrast, avirulent strains produce a relatively dispersed growth in the medium. The liquid media are not used for the routine culture of the bacilli. They are usually used to prepare mycobacterial antigens for vaccines and assess the antibiotic sensitivity of *M. tuberculosis*.

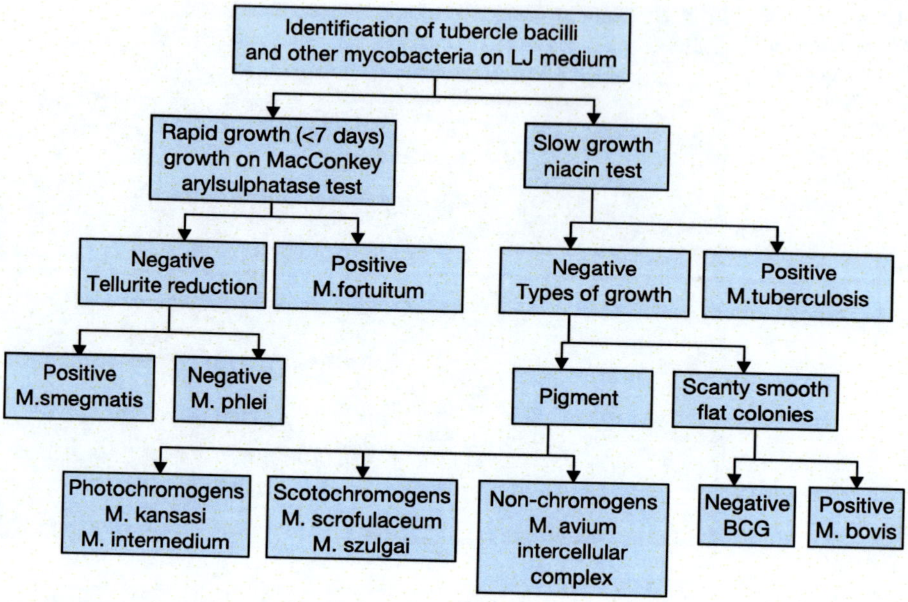

Fig. 30.2 Flow chart for the identification of tubercle bacilli and related mycobacteria

M. tuberculosis is catalase-positive. They lose their catalase activity when they become resistant to isoniazid (INH). Catalase-negative strains of *M. tuberculosis* are not virulent for guinea pigs. It is amidase-positive. It produces amidase enzymes, such as nicotinamidase and pyrazinamide, which split various amide substrates. *M. tuberculosis* is positive for the nitrate reduction test. It produces the enzyme nitrate reductase, thereby reducing nitrate to nitrite. *M. bovis* and *Mycobacterium avium* lack the enzyme nitroreductase and therefore are negative for nitrate reduction test. *Tuberculosis*, as well as *M. bovis*, is neutral red-positive. They bind neutral red in alkaline buffer solution. Niacin production, nitrate reduction and production of pyrazinamidases are useful tests to differentiate *M. tuberculosis* from *M. bovis* (Table 30.2).

Other Properties

Susceptibility to Physical and Chemical Agents

Mycobacteria are killed by heating at 60 °C for 15–20 min. The killing of the bacteria is dependent on the nature of the clinical specimen in which the bacteria are present. Mycobacteria in the sputum may survive for 20–30 h, but in dried sputum protected from the sunlight, they may survive longer, up to 6 months. In droplet nuclei, they may remain alive for 8–10 days. Bacteria are killed when exposed to direct sunlight for 2 h but remain viable at room temperature for 6–8 months.

Tubercle bacilli are sensitive to formaldehyde and glutaraldehyde. A tincture of iodine destroys them in 5 min and 80% ethanol in 2–10 min. They are generally more resistant to chemical disinfectants than other non-spore-forming bacilli. They can survive exposure to 5% phenol, 15% sulphuric acid, 3% nitric acid, 5% oxalic acid and 4% sodium hydroxide. *Tuberculosis* is susceptible to pyrazinamide, while *M. bovis* and other mycobacteria are resistant. *M. tuberculosis* is resistant to thiophen-2-carboxylic acid hydrazine (TCH), while *M. bovis* is susceptible.

Antigenic Structure

The cell wall of *M. tuberculosis* consists of four layers: (1) peptidoglycan layer, (2) arabinogalactan layer, (3) mycolic acid layer and (4) mycosides. The peptidoglycan layer is the innermost layer covalently linked with arabinogalactan (polysaccharide), and its terminal

Table 30.2 Differential features of *Mycobacterium tuberculosis* and *Mycobacterium bovis*

Characteristics	*Mycobacterium tuberculosis*	*Mycobacterium bovis*
Microscopy	Slender or slightly curved rods with barred or beaded appearance	Straight, stout, short and uniformly stained rods
Oxygen requirement	Obligate aerobe	Microaerophilic on primary isolation but becomes aerobic on sub-culture
Days for appearance of colonies at 37 °C	12–25 days (eugonic growth)	20–40 days (dysgonic growth)
Effect of glycerol (0.75%)	Enhances the growth	Inhibits the growth
Colony	White to buff coloured, rough and tough colonies	White, moist, smooth, flat and friable colonies
Nitrate reduction	Positive	Negative
Niacin production	Positive	Negative
Tween 80 hydrolysis	Variable	Negative
Susceptibility to pyrazinamide (50 μg/mL)	Susceptible	Not susceptible
Pathogenicity to guinea pigs	Non-pathogenic	Pathogenic
Pathogenicity to rabbits	Non-pathogenic	Pathogenic

ends are linked to mycolic acid. This layer is overlaid with polypeptides and a layer of mycolic acid consisting of free lipids, glycolipids and peptidoglycolipids. The peptide chains in the outer layer are important antigens that stimulate cell-mediated immunity (CMI) in infected humans. PPDs are extracted and partially purified preparation of these proteins, which are used as antigens in tuberculin skin test (TST). The mycolic acid fraction of the lipids of the cell wall is responsible for many of the characteristic properties of the bacilli. These include (1) acid fastness, (2) slow growth by delaying permeation of nutrients, (3) resistance to commonly used antibiotics, (4) resistance to detergents, (5) unusual resistance to killing by acids and alkalis and (6) clumping or cord formation.

M. tuberculosis strains are antigenically homogenous. They are antigenically similar to *M. bovis* and *M. microti* but are distinct from other species. Mycobacteria possess two types of antigens: cell wall insoluble polysaccharide antigens and cytoplasmic soluble protein antigens. Group specificity is due to polysaccharides present in the cell wall. These include arabinogalactans, lipoarabinomannan, lipids, glycolipids and peptidoglycolipids. The cytoplasmic protein antigens confer type specificity. These include antigen 5, antigen 6, antigen 14, antigen 19, antigen 33, antigen 38 and antigen 60. Mycobacteria protein (tuberculin) is responsible for developing delayed hypersensitivity in humans following infection by peptidoglycolipids. The human, bovine and murine bacilli are antigenically similar. Tuberculin also shows some antigenic relationship with that of atypical mycobacterium.

Pathogenesis

M. tuberculosis causes tuberculosis, a classic mycobacterial disease in humans.

Virulence Factors

The capability of *M. tuberculosis* for intracellular growth in alveolar macrophages is the main determinant of the virulence of the bacteria. *M. tuberculosis* does not produce any toxin.

Although cord factor and sulpholipids are toxic substances produced by mycobacterium, their role as virulence factors is doubtful. The **cord factor** was so-called because of the false belief that it is responsible for producing serpentine cords typically found on the surface of the liquid or solid media by *M. tuberculosis*. The cord factor was originally thought to be a virulent factor that no longer holds true. **Sulpholipids** are doubtful virulence factors. They are associated with the virulence of tubercle bacilli by preventing the fusion of phagosome and lysosome inside the macrophages, thereby allowing the bacteria to multiply within the macrophages. The main pathology in the infected tissue caused by mycobacterial infection is primarily due to responses of the host to *M. tuberculosis* infection rather than any virulence factor produced by it.

Pathogenesis of Tuberculosis

It depends on whether the tuberculosis is primary or post-primary, depending on the time of infection and the type of host immune response.

Primary tuberculosis represents the initial infection caused by *M. tuberculosis* in an infected host. This condition is usually seen in young children in endemic countries like India. On inhalation of the aerosolised bacteria, the bacilli reach the lower respiratory tract. The majority of the inhaled bacteria are killed by the natural defensive mechanisms of the upper respiratory tract. The bacilli that survive these defensive mechanisms reach the lungs and enter alveolar macrophages. The phagocytosed bacilli inhibit acidification of the phagosomes and prevent subsequent fusion of phagosome and lysosome. This allows the bacterial multiplication freely either in the phagosome or the cytoplasm. Multiplication inside the cells leads to the destruction of the cells and the release of mycobacteria. This is followed by further cycles of phagocytosis of bacteria by macrophages, multiplication of mycobacteria and lysis of macrophages.

Macrophages transport some bacilli to the hilar lymph nodes. They are attracted to this site by the presence of bacilli, cellular components and chemotactic factors, such as complement C5a of the serum. This leads to the formation of the focus known as Ghon focus which is formed of multinucleated giant cells known as Langerhans cells. The Ghon focus is commonly found in the lower lobe or in, the lower part of the upper lobe of the lungs. This focus is also associated with the enlargement of the hilar lymph nodes. Both Ghon focus and enlarged lymph nodes constitute the **primary complex** which usually develops in 3–8 weeks after infection by tubercle bacilli. This primary complex is associated with the development of tuberculin hypersensitivity. The lesion in the majority of cases heals spontaneously within 2–6 months.

If few bacilli are present, the bacilli are destroyed by macrophages with minimal tissue damage. However, if many bacilli are present, it leads to the development of tissue necrosis. Multiple host factors contribute to the process of tissue necrosis. These include (1) local activation of the complement, (2) exposure to macrophage-derived hydrolytic enzymes, (3) cytokine toxicity and (4) reactive oxygen intermediates. The bacilli may be present as dormant in this stage or may become re-activated in old age or the following therapy or following immunosuppressive disease. Reactivation of the site causes post-primary (secondary) tuberculosis.

Exogenous reinfection or reactivation of latent infection causes **post-primary (secondary) tuberculosis.** Immunocompromised patients like people infected with HIV patients who receive transplants are more susceptible to reactivation of the primary lesion. Secondary tuberculosis differs from primary tuberculosis by many features. (1) Granulomas of secondary tuberculosis occur more commonly in the upper lobes of the lung. The lung lesions undergo caseous necrosis and tissue destruction, forming cavities in the immunodeficient hosts. The lesions disseminate widely into lungs and other organs, such as kidneys, bones, meninges. (2) Involvement of lymph nodes is usual and (3) The cavities may rupture into blood vessels, causing dissemination of mycobacteria in the body, or these may rupture into airways, releasing mycobacterium in aerosol and sputum.

Extrapulmonary tuberculosis usually occurs due to the spread of the bacilli through blood circulation during the initial stage of multiplication at the site of primary infection, that is, the lung.

Pathology

Tubercle is the key pathology in tuberculosis. This is an avascular granulomatous condition. It comprises a peripheral zone of lymphocytes and fibroblasts and a central zone consisting of giant cells with or without caseation (Fig. 30.3). The lesions are mainly of two types, exudative and

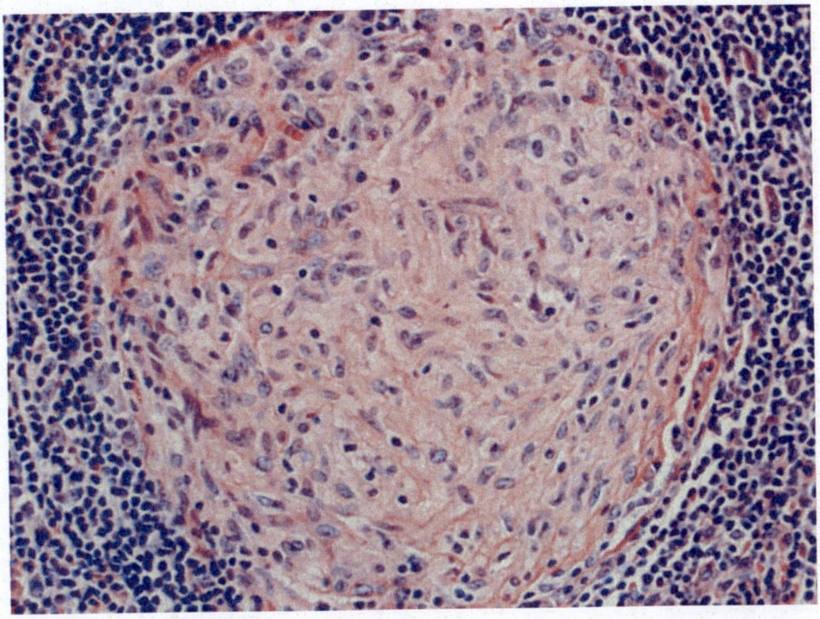

Fig. 30.3 Tubercle or granuloma with caseation in tuberculosis (x1000) (https://en.wikipedia.org/wiki/Granuloma)

productive tubercular lesions. The **exudative tuberculous lesion** is an acute inflammatory reaction. This condition is associated with an increase in the number of polymorphonuclear leucocytes and subsequently with that of lymphocytes and mononuclear cells and fluid accumulation. This lesion is seen more commonly in patients (1) infected with more virulent bacilli, (2) with an increased load of bacilli and (3) with increased delayed-type hypersensitivity (DTH) host response. **Productive tuberculosis lesion** is primarily cellular. It is associated with protective immunity rather than the DTH response.

Immunity in Tuberculosis

M. tuberculosis infection in an infected host induces **cell-mediated immunity (CMI)**. The CMI is manifested either as delayed tuberculin hypersensitivity (DTH) or as resistance to infection. The course of infection is determined by the interaction of hypersensitivity or host immunity.

Robert Koch first described tuberculin hypersensitivity reaction in experimentally infected guinea pigs. The demonstration of this tuberculin reactivity in guinea pigs is known as the Koch phenomenon. Koch phenomenon is demonstrated by subcutaneous injection of pure culture of virulent tubercle bacilli in a normal guinea pig. Initially, no immediate visible reaction is observed at the site of inoculation in the guinea pigs. However, after 10–14 days, a hard nodule appears at the site of inoculation, which soon breaks down to form an ulcer that persists till the animal dies of progressive tuberculosis. The regional lymph nodes draining the region become enlarged and caseous. In contrast, when a guinea pig already inoculated 4–6 weeks earlier by tubercle bacilli is injected with tubercle bacilli, an individual lesion develops at the site of the second inoculation within 24–48 h. In a day, the lesion undergoes necrosis to produce a shallow ulcer that heals rapidly, involving the regional lymph nodes and other tissues.

The Koch phenomenon is a combination of both hypersensitivity and immunity, which has three components: (1) local reaction (induration and necrosis), (2) a focal response (congestion and even haemorrhage around tuberculous foci) and (3) a systemic response (fever) which may even be fatal.

The DTH can be induced by live attenuated and killed bacilli, bacillary products and tubercular protein (tuberculin). Usually, live or killed bacilli or tubercular proteins are employed to demonstrate hypersensitivity reactions. This hypersensitivity reaction can be transferred passively by cells but not by serum.

Tuberculin test (Mantoux test) is performed to demonstrate delayed type IV or cell-mediated hypersensitivity reaction to tubercle bacilli. Originally, the tuberculin test was performed using a protein known as old tuberculin (OT) prepared by Robert Koch. The OT is a protein component of tubercle bacilli prepared from a 6–8-week culture filtrate of *M. tuberculosis* cultured in 5% glycerol, concentrated ten-fold by evaporation on a steam bath. The OT is a crude protein that consists of protein as a constituent but varies from batch to batch in its purity and potency. The OT has now been replaced by using PPD of tubercle bacilli.

The development of antibodies in serum characterises humoral immunity. Antibodies against polysaccharides, proteins and phosphate antigens of tubercle bacilli have been demonstrated in serum of patients with tuberculosis.

Clinical Syndromes

M. tuberculosis causes tuberculosis.

Tuberculosis

The clinical manifestations of tuberculosis depend on the site of infection. However, primary infection is usually pulmonary. *M. tuberculosis* causes pulmonary tuberculosis and extrapulmonary tuberculosis.

Pulmonary Tuberculosis

The condition manifests as fever, weight loss and productive cough. Other systematic manifestation includes night sweats, anorexia and fatigue. Bloody and purulent sputum may be seen. This is associated with cavitary lesions in the lung. Various forms in which pulmonary tuberculosis may manifest are re-activated pulmonary disease, progressive pulmonary disease and involvement of pleura. Pulmonary complications include relapse, aspergillomas, bronchiectasis, broncholithiasis and fibrothorax.

Extrapulmonary Tuberculosis

The condition varies depending on the site of infection as follows:

Genitourinary tuberculosis is one of the most common extrapulmonary manifestations of tuberculosis. The typical symptoms include dysuria, increased frequency of urination and flank pain. The condition in men may manifest as epididymitis or a growth in the scrotal area. In women, the condition may manifest as a pelvic inflammatory disease. Genitourinary tuberculosis is responsible for approximately 10% of sterility in women.

The clinical manifestation of Gastrointestinal tuberculosis depends on the site affected in the gastrointestinal tract. For example, infection of the stomach or duodenum manifests as abdominal pain mimicking peptic ulcer disease, whereas infection of the large intestine manifests as pain in the abdomen, diarrhoea etc. Tubercular meningitis is one of the most severe complications of tuberculosis. The condition may persist as a headache which is either intermittent or persistent. Tubercular lymphadenitis or scrofula most commonly involves the neck along the sternocleidomastoid muscle. The condition is usually unilateral, with little or no pain. Skeletal tuberculosis involving the spine is the most common site involved in skeletal tuberculosis resulting in Pott disease. Other conditions include miliary tuberculosis, tuberculosis of the skin and tuberculosis of the middle ear and ocular structures.

Miliary disease, disseminated tuberculosis and tubercular meningitis are the most serious complications of primary tuberculosis. Pleural effusion and pneumothorax are the noted pulmonary complications of tuberculosis. Intestinal perforation, obstruction and malabsorption are the complications of tuberculosis of the small intestine. Hydronephrosis and autonephrectomy are renal complications, whereas paraplegia is the complication of Pott disease of the spine.

HIV Patients with Tuberculosis

These are more likely to progress to disseminated disease. These patients usually do not have a cavitary pulmonary disease or upper lobe infiltrates in the lung. Patients with tuberculosis should be tested for HIV, and those with HIV need to be tested periodically for tuberculosis by TST and chest radiography. HIV re-activates latent tuberculosis infection, makes the disease more serious and renders treatment ineffective. On treatment with antiretroviral therapy, the patients with both HIV and tuberculosis develop various clinical manifestations, including fever, lymphadenopathy, and non-invasive pulmonary infiltrates. This has been suggested due to a stronger immune response to *M. tuberculosis*.

Laboratory Diagnosis

Definitive diagnosis of tuberculosis is made by detecting *M. tuberculosis* from clinical specimens by microscopy or culture.

Specimens

Sputum is the specimen of choice for pulmonary tuberculosis. Sputum, not saliva, is collected in the morning into a clean wide-mouthed container, such as a sputum cup. The collection of morning sputum specimens is ideal. If sputum is scanty, a 24-h specimen may be collected. Sputum specimen collected on three consecutive days increases the chance of detecting tubercle bacilli.

Gastric aspirate may be used in place of sputum, especially in young children who cannot produce the sputum. Propylene glycol in 10% sodium chloride (aerosol) is used for stimulating cough in elder children so that bronchial secretions are collected. Bronchoalveolar lavage is also used for the diagnosis of pulmonary tuberculosis.

Urine is the specimen of choice for the diagnosis of genitourinary tuberculosis. This is collected either as three consecutive early morning samples or as a single sample of completely voided urine in 24 h. The urine specimens are collected in large, clean, sterile containers of 500 mL or more capacity and sent to the laboratory for processing. The urine specimens are centrifuged at 300 rpm for 30 min, and the sediments are used for culture on selective media for *M. tuberculosis*.

Cerebrospinal fluid (CSF) is collected for diagnosis of tubercular meningitis. The CSF is centrifuged, and the sediment after centrifugation is stained for smears and is inoculated on the media for culture. On standing for a long time, CSF develops a spider web, the examination of which shows more bacilli.

Pleural fluid, peritoneal fluid and other exudates are collected in containers with citrate to prevent coagulation. Bone marrow and liver tissue are usually collected from patients with miliary tuberculosis, and blood is collected from patients with HIV for isolation of bacteria by culture.

It is essential to collect all these specimens before starting antitubercular therapy and be transported as soon as possible for staining and culture. In case of delay, the specimens are refrigerated at 4 °C but not more than overnight. Further delay in processing decreases the possibility of isolation of tubercle bacilli by culture. In general, few tubercle bacilli are present in extrapulmonary specimens than in sputum.

The Concentration of Specimens

Concentration methods have been described for the concentration of mycobacteria in sputum and other specimens. The concentrated sediments are used for microscopy, as well as for culture and animal inoculation. Several methods are now in use.

The Petroff method is simple and widely used. In this method, an equal volume of sputum and 4% sodium hydroxide solution are mixed and incubated at 37 °C with frequent shaking till it becomes clear, on an average of 20 min. It is then centrifuged at 3000 rpm for 30 min. The supernatant is discarded, and the sediment is neutralised by adding 8% hydrochloric acid. Phenol red is used as an indicator. Other methods of concentration of specimens use dilute acids (6% sulphuric acid, 5% oxalic acid or 3% hydrochloric acid) or mucolytic agents (*N*-acetyl-L-cysteine along with sodium hydroxide and pancreatin, desogen, zephiran and cetrimide).

Microscopy

ZN staining of sputum microscopy is the most dependable and conventional method for demonstrating AFB by the ZN-staining method. The sputum smears are made by using new slides

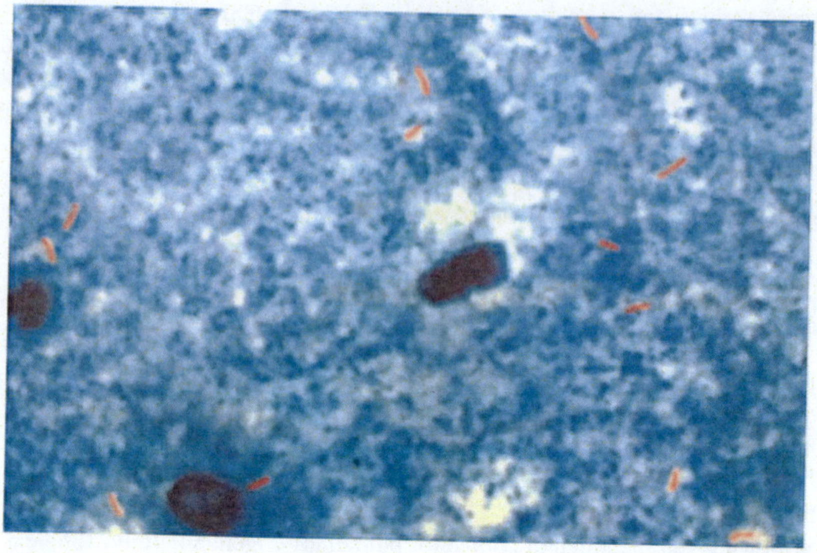

Fig. 30.4 ZN-stained smear of sputum specimen showing pink coloured acid-fast bacilli (×1000)

every time. The slides should not be re-used because AFB may adhere to the surface of the slide and may not be removed from the slide during the process of cleaning.

In this method, the smears are made from the thick purulent part of the sputum and then are air-dried fixed by heating. The carbolfuchsin is poured over the smear and heated for 5–7 min. It is then washed with water. Then the slide is washed with 20% sulphuric acid to decolourise. This is carried out till no more stain is to come off. Then it is washed with 95% ethanol for 2 min. After this washing, the slide is stained with Loeffler methylene blue, 1% picric acid or 0.2% malachite green for a minute. The stained smear is examined under an oil immersion lens (100×). Acid-fast bacteria in a stained smear appear bright red against a blue background. At least 50,000–100,000 bacilli should be present per millilitre of sputum in order to be examined, taking about 10 min before giving a negative report. The presence of at least two or more typical bacilli in the smear is reported as positive for AFB (Fig. 30.4).

The smears are graded depending on the number of tubercle bacilli present in a stained smear on examination by oil immersion lens as per **Revised National Tuberculosis Control Programme** (RNTCP) (Govt of India) guidelines. Smear grading is useful (1) in a quantitative assessment of the number of bacilli in the sputum (Table 30.3); (2) to estimate the infectiousness of the patient; (3) to monitor the effectiveness of antimycobacterial therapy and (4) to determine the discontinuation of respiratory isolation.

Microscopic demonstration of AFB in stained smears is the presumptive diagnosis of tuberculosis. This is because ZN staining of AFB smears cannot be used to differentiate *M. tuberculosis* from other acid-fast organisms, such as saprophytic mycobacteria or *Nocardia* species. Saprophytic mycobacteria may present an appearance

Table 30.3 Grading of ZN-stained smears for AFB as per RNTCP guidelines

Examination by Oil immersion field	Number of AFB	Number of fields	Interpretation
1 Field	More than 10	20	3+
1 Field	1–10	50	2+
100 Fields	10–99	100	1+
100 Fields	1–9	100	Scanty[a]
100 Fields	Negative	100	Negative

[a] Record actual number of bacilli seen in 100 fields

similar to that of *M. tuberculosis*. Nevertheless, saprophytic mycobacteria stain uniformly without any barred or beaded appearance and are usually only acid-fast but not alcohol fast. Moreover, these saprophytic mycobacteria are usually not present in the sputum and other respiratory secretions but are found in urogenital specimens, gastric aspirates and faecal specimens.

Kinyoun modification of acid-fast staining is a method in which there is no heating of the stain, and it is therefore called a cold method. Here, instead of increasing the concentration of phenol acid, increasing the duration of staining is used.

Auramine–rhodamine staining is the fluorescent stain used as a variation of the traditional ZN stain for demonstrating AFB. In this method, the smears are stained with auramine–rhodamine or auramine–phenol fluorescent dyes and are examined by fluorescent microscope under ultraviolet illumination. AFB appears as bright fluorescing rods against a dark background. Because of their contrast, the bacilli are visualised even under a high-power objective compared to the oil immersion objective in ZN stain, thus enabling rapid screening of the larger areas of the smear. The fluorescent-stained slides can be screened faster because the AFB stands out against the non-fluorescent background. This is the major advantage of this staining method; hence it is adopted in the laboratories where many smears are to be examined for AFB. However, it is always essential to confirm fluorochrome-positive smears by ZN staining.

Culture and Identification

Culture is the definitive method to detect and identify *M. tuberculosis*. The culture is also a more sensitive diagnostic method than smear microscopy. The culture may be positive with as few as 10–100 AFB per millimetre of a digested concentrated specimen. Another advantage of culture is that it helps identify specific species and determine drug susceptibility patterns of isolated strains.

LJ medium is an egg-based medium, and the Middlebrook7H10 and 7H11 media are agar-based media and these are conventionally used for culture. Since *M. tuberculosis* is a slow-growing organism, a period of 6–8 weeks is required for colonies to appear on these conventional culture media after incubation at 37 °C. Growth of most strains of *M. tuberculosis* may appear during this period. However, cultures should not be discarded as negative until they have been observed for 12 weeks.

Luxuriant eugonic growth of white to buff-coloured, rough, raised, irregular and tough colonies on LJ medium after 4–6 weeks are identified by smear, and a battery of biochemical tests, and other tests. The colony smear shows straight or slightly curved AFB by microscopy.

A **catalase test** is performed to detect the enzyme catalase produced by various *Mycobacterium* species. This test is performed by mixing equal volumes of 30% hydrogen peroxide and 10% Tween 80 in a test tube. Then a few colonies of mycobacteria are picked up by a nichrome wire loop and are inoculated in the fluid. Formation of air bubbles in the fluid is observed within 1 min. The demonstration of bubbles in the fluid indicates a positive test (Fig. 30.5). If the bubbles appear in the fluid immediately, it is considered strongly positive, but it is considered weakly positive if they appear slowly. The absence of any bubble is considered a negative test. *M. tuberculosis* and *M. bovis* are weakly catalase-positive. The bacilli lose their catalase activity when they develop resistance to INH.

Nitrate reduction test detects the presence of enzyme nitrate reductase produced by *M. tuberculosis* and other *Mycobacterium* species. It is performed by adding colonies of the mycobacteria to a buffered solution containing nitrate and incubating at 37 °C for 2 h. After incubation, sulphanilamide and *N*-naphthyl-ethylene diamine hydrochloride solution is added. Development of pink or red colour within 30–60 s is suggestive of a positive test. *M. tuberculosis* and other mycobacterial species are nitrate reductase test-positive, whereas *M. bovis* and *M. avium* are negative.

Niacin test is performed by adding 1 mL of sterile normal saline to colonies of *M. tuberculosis* in the LJ medium. The medium is kept in a slanting position so that the colonies

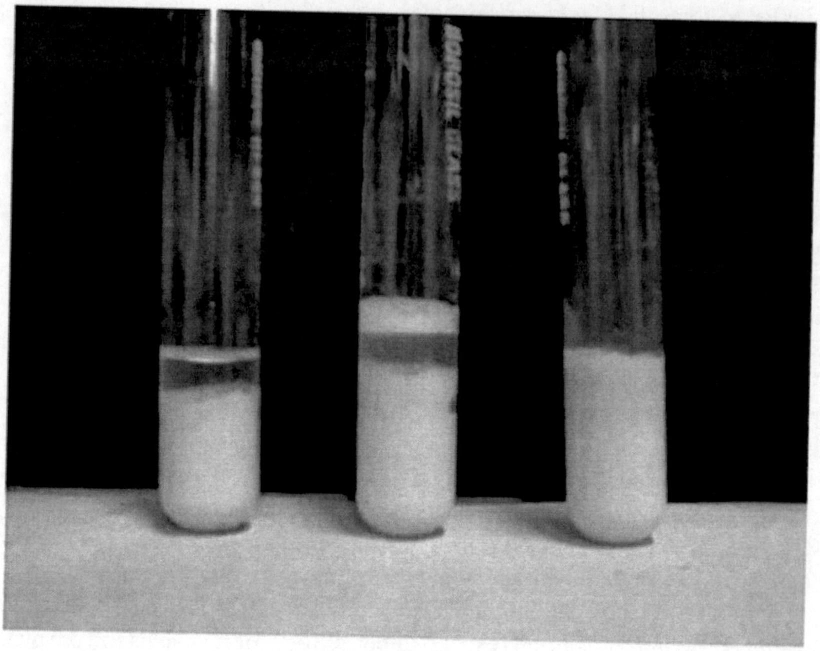

Fig. 30.5 Semi-quantitative catalase test

of *Mycobacterium* are completely covered by normal saline. After some time, 0.5 mL of the fluid is collected and is placed in a clean screw-capped test tube. 0.5 mL of alcoholic aniline and 0.5 mL of 10% cyanogen bromide are added to this fluid. A yellow colour develops in the solution within a few minutes in a positive test. It is an important test to identify niacin-positive human strains of *M. tuberculosis* and differentiate them from niacin-negative *M. bovis* strains. *M. tuberculosis* is niacin test-positive; *M. bovis*, negative.

Arylsulphatase test is performed by culturing mycobacteria in a medium containing 0.001 M tripotassium phenolphthalein disulphate. *Mycobacterium* species producing the enzyme arylsulphatase split phenolphthalein free from tripotassium phenolphthalein disulphate. The free phenolphthalein is detected by adding 2 N sodium hydroxide to the culture. The development of the pink colour indicates a positive reaction. *M. tuberculosis* is arylsulphatase-negative, while some other *Mycobacterium* species are arylsulphatase-positive.

Neutral red test is carried out to detect certain strains of mycobacteria to bind neutral red in an alkaline buffer solution. The development of pink or red-stained colonies in the amber-coloured fluid suggests a positive test. *M. tuberculosis* and *M. bovis* are neutral red test-positive.

Peroxidase test is positive for tubercle bacilli, whereas atypical mycobacteria were found to be negative peroxidase test. Similar to catalase activity, peroxidase activity of tubercle bacilli will be lost on conferring INH resistance.

Amidase test detects the ability of certain mycobacteria to split amides, such as, acetamide, nicotinamide, pyrazinamide, carbamide and benzamide. The development of blue colour indicates a positive test in the fluid. *M. tuberculosis* is amidase-positive.

Tween 80 hydrolysis is performed to demonstrate the enzyme lipase produced by certain mycobacterial species. Development of pink colour indicates a positive test. *M. tuberculosis* shows a variable result. *Mycobacterium kansasii* produces a positive test within 3–6 h, while other mycobacteria species produce a positive test within 3–10 days.

M. tuberculosis is pyrazinamidase **test-positive**. It is sensitive to 50 μg/mL pyrazinamide. *M. bovis* is pyrazinamidase test-negative. The enzyme pyrazinamidase hydrolyses pyrazinamide to produce ammonia and pyrazonic acid in this test. The pyrazonic acid can be detected upon the addition of ferric ammonium sulphate.

Mycobacteria Drug Susceptibility Testing

With the emergence of multi-drug resistant (MDR) mycobacteria, the determination of drug susceptibility testing is important for starting appropriate treatment. Phenotypic and genotyping methods can demonstrate drug susceptibility. Phenotypic **methods** include (1) resistance ratio method, (2) absolute concentration method, (3) proportion method and (4) radiometric methods. **Genotypic methods** include (1) DNA sequencing, (2) solid-phase hybridisation and iii) PCR–single-stranded combination polymorphism analysis.

Tuberculin Skin Test

The tuberculin skin test is a widely used test for the diagnosis of tuberculosis. It is used for (1) evaluation of tuberculosis cases and for diagnosis of suspected *M. tuberculosis* cases, (2) diagnosis of active tuberculosis infection in infants and young children, (3) the selection of a susceptible population for Bacille Calmette–Guerin (BCG) vaccination and (4) the prevalence measurement of tuberculosis infection in a community.

Mantoux test is the recommended method to perform a tuberculin skin test. The test is performed by intradermal injection of 0.1 mL or 5 tuberculin units (TU) PPD into the volar aspect of the forearm using a 27G needle. PPD must be injected between the layers of the skin but not subcutaneously. Development of induration of 10 mm or more at the injection site after 48–72 h is considered a positive test. Induration, less than 5 mm, is negative, while between 6 and 9 mm is considered equivocal. Erythraemia is not considered in the reading of the cutaneous. The test is negative with the PPD of 5 TU, and then the test may be repeated using a PPD of 10 or 100 TU. PPD of 1 TU is used when the recipient is considered to be extremely hypersensitive to the antigen. **Multiple puncture tests** include the Heaf test and time test. Heaf test is performed by using the hit gun and by using disposable prongs carrying dried PPD. These tests may be satisfactory for screening and surveys but are not useful as diagnostic tests because they lack sensitivity and specificity.

A positive skin test indicates hypersensitivity of the individual to the tubercle protein. This suggests infection with *M. tuberculosis* or immunisation by BCG vaccination in recent or past with or without clinical disease. The test becomes positive 4–6 weeks after infection or immunisation. The tubercle test positivity gradually disappears within 4–5 years in the absence of re-exposure to the tubercle bacilli. In endemic areas, this positivity is maintained due to repeated exposure to it. A negative tubercle test indicates that the person has never come in contact with the tubercle bacilli. Tuberculin skin test exhibits false-positive reactions in patients with infection by environmental non-tuberculous mycobacteria. False-negative reactions may occur in patients with (1) BCG vaccination, (2) immunosuppressive therapy, (3) impaired CMI, (4) lymphoreticular malignancies, (5) immunodeficiency, (6) malnutrition and (7) miliary tuberculosis. Improper administration, such as subcutaneous injection of PPD or injection of too little volume of antigen, contamination of PPD or improper storage, are the other factors that may give rise to false-negative reactions.

Interferon Gamma Release Assay (IGRA)

The test for CMI is useful to detect latent *Mycobacterium* infection. It is an *in vitro* blood test of cell-mediated immune response where it measures the T-cell release of IFN-γ following stimulation by antigens specific to the *M. tuberculosis* complex, i.e. early secreted antigenic target 6 (ESAT-6) and culture filtrate protein 10 (CFP-10). It has a specificity of >95% in settings with a low TB incidence. It has less cross-reactivity and so is not affected by BCG vaccination. Results are obtained within 24–48 h. Two commercial IGRAs kits, i.e. the QuantiFERON-TB Gold In-Tube (QFT) assay (Cellestis/Qiagen, Carnegie, Australia) and the T-SPOT.TB assay (Oxford Immunotec, Abingdon, United Kingdom) has been approved by US FDA.

Serology

M. tuberculosis infection is associated with elevated levels of antibodies in the serum. Various mycobacterial antigens have been used in these serological tests, including BCG, 5 and 6 kDa proteins of *M. tuberculosis,* 32 and 64 kDa protein of *M. bovis* etc. Detection of antibodies by serological tests, such as enzyme-linked immunosorbent assay (ELISA), is of limited value in diagnosing pulmonary tuberculosis. Latex agglutination test using latex particles coated with rabbit antibody against *M. tuberculosis* has been used to demonstrate antigen in the CSF for diagnosis of tuberculosis meningitis.

Animal Inoculation Tests

M. tuberculosis is highly pathogenic for guinea pigs, while catalase-negative INH-resistant strains and most strains isolated from south India are weakly pathogenic for guinea pigs. *M. tuberculosis* is non-pathogenic for rabbits. *M. bovis* is pathogenic for both rabbits and guinea pigs. Inoculation in guinea pigs was earlier widely used for the diagnosis of tuberculosis. However, guinea pig inoculation is now regarded as obsolete because it is cumbersome, costly and less sensitive than culture. Guinea pig inoculation is, however, superior to culture for isolation of *M. bovis* from clinical samples.

Molecular Detection

Polymerase chain reaction (PCR) and other nucleic acid amplification methods are useful for directly detecting *M. tuberculosis* in clinical specimens, such as CSF. The AMPICLOR *M. tuberculosis* direct test is an example of nucleic acid amplification tests. The AMPICLOR test targets the insertion sequence IS6110 in the DNA of the tubercle bacilli. The test's sensitivity varies from 40% to 60%, and specificity is 95% in smear-positive cases of tuberculosis. The test is also rapid. Disadvantages of the test include contamination of clinical specimen by-products of previous amplification and the presence of inhibitors that may give rise to false-negative or false-positive reactions. Nucleic acid hybridisation using DNA probes have been used to identify the *Mycobacterium* species. These commercially available probes help identify the *M. tuberculosis* complex much earlier. The DNA probes show a sensitivity and specificity of 100% when at least 100,000 tubercle bacilli are present. This test uses chemiluminescent ester-labelled single-stranded DNA probes. The chemiluminescence is determined by using a luminometer. A positive DNA probe test indicates the mycobacteria to be *M. tuberculosis*, *M. bovis* or *M. africanum.* The probe cannot differentiate between these species. CBNATT (Cartridge based Nucleic Acid Amplification is also available for simultaneous detection of *Mycobacterium tuberculosis* (MTB) and Rifampicin resistance.

As a tool for detection, **real-time PCR** is the most used, followed by conventional and digital PCRs. Many commercial systems like XpertMTB/RIF (Cepheid, United States) have been endorsed by WHO to detect the MTB complex and drug resistance screening. Likewise, TruenatMTB, TruenatMTB Plus and TruenatMTB-Rif Dx have been developed by Molbio Diagnostics, Goa, India. It is a chip-based, micro real-time PCR-based assay for TB detection that produce results in 1 h on the portable and is a more affordable alternative to Xpert.

Rapid and Automated Methods

The conventional methods are very slow, time-consuming and require 6–8 weeks to isolate *M. tuberculosis*. Hence, recently more rapid and automated methods are being increasingly used to diagnose tuberculosis. These recent methods include automated radiometric culture methods (e.g. BACTEC), SEPTICHEK, mycobacterial growth indicator tubes (MGITs).

Automated radiometric culture methods, such as **BACTEC**, employ a liquid Middlebrook7H12 medium containing radiometric palmitic acid labelled with radioactive carbon-14 (^{14}C). The medium also contains several antimicrobial agents to prevent the growth of other non-mycobacterial microbes. The test result is noted by the demonstration of radiolabelled $^{14}CO_2$ produced during the growth of mycobacteria. Growth of mycobacteria is usually detected within 9–16 days. SEPTICHEK is another rapid method for the isolation of mycobacteria. This non-radiometric method is based on a biphasic broth-based system that decreases the mean recovery time versus conventional methods. A new method employs MGITs which show microbial growth and provide a quantitative index of *M. tuberculosis* growth. Round-bottom tubes with oxygen-sensitive sensors at the bottom are used in the test.

Treatment

Chemotherapy by antitubercular drugs has revolutionised the treatment of tuberculosis. Therapy with antituberculous drugs causes the disappearance of caseous or granulomatous lesions. Antituberculous drugs are classified as first-line and second-line drugs.

First-Line Antitubercular Drugs

These include rifampicin, INH, ethambutol, streptomycin and pyrazinamide. All first-line drugs, except for ethambutol, are bactericidal. These drugs have less toxicity and show greater efficacy than second-line drugs. Both INH and rifampicin are effective against tubercle bacilli in necrotic foci and intracellular mycobacteria. In contrast, streptomycin, aminoglycosides and capreomycin show poor intracellular penetration. A combination of four drugs (INH, rifampicin, pyrazinamide and ethambutol) is given for 6–7 months for treatment of smear-positive cases of tuberculosis. These are given three times a week for the first 2 months, followed by only two drugs (INH and rifampicin) 3 times a week.

The emergence of natural drug resistance in *M. tuberculosis* is a major problem in chemotherapy of tuberculosis. This occurs by mutation with a frequency of appropriately 10^6 cell divisions. **Multiple drug resistance (MDR)** is defined as resistance to rifampicin and INH with or without resistance to one or more other drugs. This is because rifampicin and INH form the mainstay of short-term chemotherapy and *M. tuberculosis* strains resistant to both these drugs are unlikely to respond to treatment. MDR is of two types—primary and secondary. If the tuberculosis cases are treated with a single antitubercular drug, the sub-population of tubercle bacilli susceptible to that drug is killed, whereas populations not susceptible to the drug continue to multiply. Therefore, the use of multiple antitubercular agents in treating tuberculosis is useful. The emergence of multidrug-resistant tuberculosis (MDR-TB) is a very serious problem and is defined as resistance to rifampicin and INH with or without resistance to one or more other drugs. This is because rifampicin and INH form the mainstay of short-term chemotherapy, and *M. tuberculosis* strains resistant to both these drugs are unlikely to respond to treatment. Multiple drug resistance (MDR) is of two types—primary and secondary.

Primary MDR is defined as the development of resistance to antitubercular treatment in individuals with no history of antitubercular treatment. It usually occurs in a) patients residing in areas with a high prevalence of drug-resistant *M. tuberculosis*, b) those exposed to drug-resistant contagious tuberculosis and c) those with HIV infections and in the individuals using intravenous drugs. **Secondary MDR** is defined as the emergence of antitubercular resistance during infection and antitubercular treatment. It usually develops in (1) patients treated within appropriate drug regimens and (2) those not taking antituberculous drugs regularly.

Second-Line Antitubercular Drugs

These are used for cases of tuberculosis where first-line drugs become ineffective. These include many old and new drugs, such as ciprofloxacin, cycloserine, ethionamide, kanamycin, ofloxacin, levofloxacin, capreomycin.

Directly Observed Therapy (DOT)

This is a method being recently followed for the treatment of cases of tuberculosis. The DOT is extremely useful for preventing drug resistance by ensuring patient compliance. The DOT is administered under the direct supervision of healthcare workers of the patients one of the family members.

Epidemiology

Tuberculosis (TB) is the number one infectious killer disease worldwide. It is most common in Southeast Asia, Sub-Saharan Africa and Eastern Europe. According to WHO, more than 9.9 million new tuberculosis cases occur every year, with 1.5 million deaths. In addition to disease burden, it costs huge economic burden where income loss equivalent on average to more than 50% of income are documented.

Geographical Distribution

One-third of the world's population is infected with *M. tuberculosis*. Nearly two-thirds of all tuberculosis cases occur in developing countries, including India, China, Pakistan, Bangladesh, Philippines, Nigeria, Indonesia and South Africa.

Habitat

M. tuberculosis primarily inhabits the respiratory tract of the infected human host. The droplet nuclei consisting of *M. tuberculosis* have been found in the terminal air spaces of the lung.

Reservoir, Source and Transmission of Infection

The source of *M. tuberculosis* infections is only human beings. The degree of infection in infected humans is very important as it decides whether the disease would be transmissible. There are about 100–10,000 bacilli in pulmonary lesions with cavities. Cases with cavitary lesions have more potential to be highly infectious. Infection is less in cases treated with antitubercular therapy when compared to cases that are not treated. The decrease in infectiousness is primarily due to a reduction in the bacillary load in the lungs.

Humans acquire *M. tuberculosis* infection most frequently by inhalation of infectious aerosolised droplets. These infective aerosolised droplets are usually coughed or sneezed into the environment by a patient suffering from pulmonary tuberculosis. The acts of coughing, sneezing and speaking release many droplets containing as many as 3000 air-borne infectious droplet nuclei per cough. The size of the droplet nucleus is 5 μm or less. It contains about 1–10 tubercle bacilli. By theory, even a single tubercle bacillus can cause disease, but in reality, infection is caused by inhaling around 5–200 bacilli. As these droplets are small in size, they remain suspended in the air for a very long time. The infection is acquired rarely by inoculation. *M. tuberculosis* primarily causes natural infections in humans. The bacteria may also cause natural **infection in mammals** such as other primates, dogs and animals that come in close contact with infected humans. In experimental infections, *M. tuberculosis* is highly infectious for guinea pigs and hamsters but is usually non-pathogenic for bovines, goats, cats and rabbits.

M. bovis produces tuberculosis in cattle, as well as in humans and other primates. Aerosolised bacilli spread the infection amongst animals in moist cough sprays. The infected animals usually excrete the bacilli in their milk. The infection is transmitted to humans by ingestion of raw milk of the cows infected with *M. bovis*. Person-to-person transmission of *M. bovis* usually does not occur. *M. bovis* is more pathogenic for animals. Experimentally, it is highly pathogenic for guinea pigs, calves and rats and non-pathogenic for fowl.

African strains of *M. africanum* are the strains of tubercle bacilli that are isolated mostly from Africa. Asian type strains are of low virulence for guinea pigs, susceptible to hydrogen peroxide, resistant to INH and usually isolated from South India. These Asian strains have also been isolated from other Asian countries and Asians abroad. Differential features of various strains of *M. tuberculosis* are summarised in Table 30.4.

Typing Methods

Various typing methods are very useful in an epidemiological survey. Mainly, they can be categorised into molecular and phenotypic typing methods.

Molecular typing methods play an important role in a molecular epidemiological survey. Strains of the tubercle bacilli can be differentiated easily by using DNA fingerprinting technologies. IS6110 restriction fragment length polymorphism (IS6110RFLP typing) typing method is used for

Table 30.4 Differential features of various strains of *Mycobacterium tuberculosis*

Properties	Human	Bovine	Asian	African
Oxygen requirement	Aerobic	Microaerophilic	Aerobic	Microaerophilic
Growth in thiophene-2-carboxylic acid hydrazide (TCH) (5 mg/L)	Growth	No growth	No growth	No growth
Niacin test	Positive	Negative	Positive	Variable
Nitrate reduction	Positive	Negative	Positive	Variable
Phage type	A, B and C	A	I	A

distinguishing between isolates of *M. tuberculosis*. In this method, specific PCR amplicons are digested with particular restriction enzymes and run-on agarose gel to visualise distinct bands. It has been widely evaluated that the discriminatory power of RFLP typing with IS6110 is sufficiently high and can be used for clinical and epidemiological surveys. The spoligotyping (spacer oligotyping) method is based on the polymorphism present on the direct repeat locus, a highly conserved region present in all MTB complex in a unique locus. This method is very useful in the case of mycobacterial strains bearing a low copy number of the IS6110 gene.

Phenotypic typing methods were previously used for epidemiological surveys; however, they were side-lined because of their low discriminatory power in distinguishing the Mycobacterial species. Bacteriophage typing can classify tubercle bacilli into four distinct phage types, Type A, B, C and an intermediate phage-type I, which exists between phage-type A and B. These phage types have different distribution patterns across the globe. Phage type A is cosmopolitan in distribution, type B is restricted to the European continent and Type C is sparsely distributed. In India and other sub-continental countries, type I is more prevalent. Bacteriophage 33D was first isolated from an environmental mycobacterial isolate, and it was found to be lytic to all variants of *M. tuberculosis*, except for the BCG strain. Bacteriocin typing can classify *Mycobacterium* into two types. Because of this low discriminatory power, this typing method is not in use.

Prevention and Control

These include chemoprophylaxis, vaccination and general health measures.

Chemoprophylaxis

Chemoprophylaxis or preventive chemotherapy is carried out by using antitubercular drugs, such as INH. INH is usually used for the treatment of (1) persons with latent tuberculosis, (2) patients with HIV infection, (3) recent contact of patients with contagious tuberculosis in the past 3 months, even if TST result is negative, (4) unvaccinated children and (5) elderly person with radiological evidence of quiescent disease. These cases are treated by INH, given in a dose of 5 mg/kg daily for 6–12 months. Results of the study have shown that chemoprophylaxis by INH has considerably reduced the risk of acquiring the disease nearly by 90%.

Vaccination

Vaccination against tuberculosis is carried out by the administration of the **BCG vaccine**. BCG is a live attenuated vaccine prepared from an attenuated strain of *M. bovis*. This attenuated strain was originally prepared by Calmette and Guerin (1921). They sub-cultured the bacterium every 3 weeks for 239 sub-cultures on potatoes soaked in sterile bile containing 5% glycerol during a period of 13 years. The BCG vaccine is available in liquid and freeze-dried (lyophilised) form. The lyophilised form is commonly used, which is reconstituted with sterile physiological saline to make a final concentration of 0.1 mL before administration. The reconstituted vaccine has a shelf life of 3–6 h. The vaccine is given by the administration of 0.1 mL of reconstituted vaccine intradermally. In a positive test, a small nodule develops at the site of inoculation 2–3 weeks after injection. It increases slowly in size, and by about 5 weeks, it attains a diameter of 4–8 mm. It then subsides or breaks into a shallow ulcer which heals by scarring. Following BCG vaccination, a tuberculin-negative individual is converted into a positive reactor. The immunity following BCG vaccination is similar and may last for 10–15 years.

The vaccine is now being used in more than 100 countries and is given to infants, usually at birth. The WHO recommends using the vaccine in children who have asymptomatic HIV infection in areas where the risk of tuberculosis is high.

In earlier days, there were huge concerns regarding the safety of the BCG vaccine. This was mainly due to the Lubeck incidence, where BCG vaccine containing virulent, live tubercle bacilli were administered orally to children, which produced severe, fatal tuberculosis amongst immunised children. To avoid such mistakes during BCG vaccine production, a stringent safety policy has been laid down that helped in a drastic reduction of such consequences that may occur due to BCG vaccination. However, BCG vaccination is associated with complications. Local complications following BCG vaccination include abscess, keloid, indolent ulcer, tuberculides, lupoid lesions, lupus vulgaris or confluent lesions. Regional complications include enlargement and suppuration of draining lymph nodes. General complications that include fever is the most common apart from mediastinal adenitis, erythema nodosum and formation of keloids upon wounding at other parts of the body. **BCG vaccine** is contraindicated in immunosuppressed patients with (1) HIV infection, (2) primary or secondary immunodeficiency and (3) in patients receiving steroids.

Although BCG vaccination has been used since 1921, and approximately 3 billion doses of vaccines have been given, the vaccine's efficacy continues to be controversial. Several field trials have been performed to assess the efficiency of the vaccine. The results have varied widely from 80% protection to a total absence of protection. However, two meta-analyses of the vaccine trials, including those conducted in India, have shown that the BCG vaccine is effective against miliary and meningeal tuberculosis. BCG vaccination offers many added advantages. It can also induce the immune system non-specifically, conferring partial immunity against diseases like leprosy and leukaemia. Repeated BCG injections have been used as adjunctive therapy in some kinds of malignancies. Some research demonstrated that BCG could be more useful than PPD for tuberculin testing.

General Health Measures

General health measures include adequate nutrition, good housing and health education.

RNTCP is a national tuberculosis control initiative program introduced in 1993 and launched large scale in 1997 in India. Based on the WHO recommended strategy, directly observed treatment (DOTS), short course chemotherapy was implemented for tuberculosis control in India. The strategy's main aim is to offer standardised, cost-effective treatment and proper diagnostic facilities. This strategy's main aim is to reduce drug resistance by ensuring the patient's compliance. Indian DOTS strategy is the fastest-growing program globally, and treatments are given based on three categories, I, II and III. The algorithms used for diagnosis and treatment strategy have been standardised, and it also recognises standard and accredited laboratories for drug susceptibility testing.

The primary drugs used for DOTS are INH, rifampicin, pyrazinamide, ethambutol and streptomycin. Further, due to the emergence of drug-resistant strains, the DOTS plus regimen is recommended in selected areas and not for universal use. Based on the data obtained from a recent survey, RNTCP has issued a daily regimen instead of the existing alternative day regimen in selected states, and it was found to be effective in reducing relapse rates and drug resistance.

By adopting the WHO 2010 guidelines, the RNTCP has recommended three standard treatment regimens:

1. A new patient regimen containing 6 months of rifampicin: 2HRZE/4HR.
2. Re-treatment regimen with the first-line drugs 2HRZES/1HRZE/5HRE.
3. Multi-drug resistant (MDR) regimen.

The End TB Strategy of the World Health Organisation

In 2014, World Health Assembly adopted the WHO's post-2015 End TB Strategy, an ambitious blueprint to reduce TB incidence by 80%, TB deaths by 90% and eliminate catastrophic costs for TB-affected households by 2030. Being a part of newly adopted Sustainable Development Goals, the End TB strategy lays a roadmap to provide a unified response to ending TB deaths, disease and suffering by building on three strategic pillars:

1. **Pillar 1**: Integrated, patient-centred TB care and prevention. It focuses on early detection, treatment and prevention for all TB patients by ensuring equal, unhindered access to affordable services.
2. **Pillar 2**: Bold policies and supportive systems. By intensifying participation across government, communities and private stakeholders, this pillar seeks to (1) strengthen health and social sector policies and systems to prevent and end TB; (2) support the implementation of universal health coverage, social protection and strengthened regulatory frameworks; (3) address the social determinants of TB and tackle it.
3. **Pillar 3**: Intensified research and innovation by adopting new tools and technologies in operational research to design, implement and scale innovations. It also calls for an urgent boost in research investments so that new tools are developed and made rapidly available and widely accessible in the next decade. It envisages the availability of better diagnostics, including new point-of-care tests; safer, easier and shorter treatment regimens; safer and more effective treatment for latent TB infection and effective pre-and post-exposure vaccines.

Case Study

A 30-year-old male labourer working in a stone-crushing unit attended Medicine OPD with a history of fever and cough with purulent sputum for the past 3 months, associated with loss of weight and loss of appetite and cough with expectoration. Gram stain of sputum sample showed normal commensal flora. White blood cell (WBC) count and erythrocyte sedimentation rate were raised. Chest X-ray showed the presence of a cavity in the left apical lobe of the lung. ZN-stain smear of the sputum sample showed the presence of AFB. Serum was positive for HIV antibodies by ELISA.

- Diagnose the condition.
- Describe the recent methods for diagnosis of the condition.
- Discuss the antibiotic regimen for treatment of the condition.
- Describe the vaccines available against the disease.

31 *Mycobacterium leprae, Mycobacterium lepraemurium* and Non-tuberculosis Mycobacteria

Mycobacterium leprae is the causative agent of leprosy, a chronic granulomatous disease primarily affecting the skin and peripheral nervous system. *M. leprae* was the first bacterial pathogen associated with a specific human disease and the only mycobacterium known to cause infection of the nervous tissue. The Koch postulates have never been fulfilled for *M. leprae* because so far, it has not been cultivated *in vitro*.

Mycobacterium leprae

M. leprae was discovered in 1873 by Arinuer Hansen in Italy. Leprosy was not initially thought to be an infectious disease, despite the discovery of *M. leprae*. Tremendous advancement has been made on the knowledge of pathogenesis, cause, treatment and prevention of leprosy during the last three decades. Nevertheless, the lesions were responsible for the social stigma attached to the disease.

Properties of the Bacteria

Morphology

M. leprae is less acid-fast than *Mycobacterium tuberculosis*. Hence, 5% sulphuric acid, instead of 20%, is used for decolourisation after staining with carbol fuchsin. It is a straight or slightly curved rod with parallel sides and rounded ends. It measures 1–8 μm in length and 0.2–0.5 μm in diameter, showing considerable morphological variations. The bacteria exhibit cubical lateral or even branching forms. It is Gram-positive and stains more readily than *M. tuberculosis*. *M. leprae* is an obligate intracellular bacterium that multiplies preferentially in tissues at a cooler temperature. They are non-motile and non-sporing.

M. leprae bacilli are seen singly and in groups intracellularly and extracellularly lying free outside the cell. The bacilli inside the cell are usually present in parallel bundles of 50 or more. Acid-fast bacteria are bound together by a lipid-like substance known as glia. These masses of bacteria are known as globi. The parallel rows of bacilli in the globi present a cigar bundle type of appearance. These are present inside the large undifferentiated histiocytes, which have a foamy appearance. These are known as Virchow lepra cells or foamy cells.

The genomic sequence of *M. leprae* has been mapped and completed recently. The genes responsible for its major protein have been cloned and sequenced. One important finding is that *M. leprae* retains genes for the formation of the mycobacterial cell wall. Like that of other mycobacteria, the cell wall of *M. leprae* consists of four layers. Peptidoglycan is the innermost layer, which gives the bacteria rigidity and shape. Lipoarabino mannose-B (LAM-B) is a layer next to the peptidoglycan layer. Adense palisade of characteristic long-chain fatty acid known as mycolic acid is the third layer attached to the LAM-B membrane. A layer composed of mycosides is the outermost fourth layer of the cell wall. Phenolic glycolipid (PGL)-1 is the major component of the outermost layer of the cell wall.

Culture and Biochemical Properties

So far, *M. leprae* has not been cultivated *in vitro* either in bacteriological media or in tissue culture. Many attempts have been made to develop a suitable **experimental animal model** for *M. leprae*.

Footpads of mice were first employed by ***Shepard***, in 1960, to culture leprae bacilli in the footpads of mice kept at a low temperature of 20 °C. Intradermal inoculation of lepra bacilli into the footpads of mice results in granuloma development at the inoculation site in 1–6 months. The mouse footpad model has been used to test (1) the maximum required concentration of antileprosy drugs and (2) the sensitivity of the bacilli to new antileprosy drugs.

Thymectomised mouse is an experimental animal model in which cell-mediated immunity (CMI) is suppressed by thymectomy on the administration of anti-lymphocytic serum. In such mice, inoculation of *M. leprae* produces a generalised infection similar to that of lepromatous leprosy.

The nine-banded armadillo (*Dasypus novemcinctus*) is another animal highly susceptible to infection with *M. leprae*. In armadillos, *M. leprae* causes a generalised infection with extensive multiplication of the bacteria. *M. leprae* in these animals produces a lesion typical of lepromatous leprosy and survives for about 400 days in infected armadillos. This animal is now being used as the most important source of lepra bacilli for genetic studies, including the development of the vaccine. Natural infections by *Mycobacterium* organisms resembling *M. leprae* have also been observed in some wild armadillos held in captivity in Texas and Mexico.

Other animals that have been used for experimental infection by *M. leprae* include slender loris, Indian pangolin and Korean chipmunks.

M. leprae shows the longest generation time among all bacteria. It requires 12–13 days to double in experimentally infected mice compared to about 14 h in the case of *M. tuberculosis* and about 20 min in the case of coliform bacilli.

Adaptation in Cell Culture Media

The *ICRC (Indian Cancer Research Center) bacillus* is an example of an acid-fast bacillus and was first isolated from a leprosy patient employing human foetal spinal ganglion cell culture. This bacillus was reported in 1962 from the Indian Cancer Research Center, Mumbai and has been adapted for growth on Lowenstein–Jensen (LJ) medium. However, its relation to *M. leprae* is uncertain, and many studies have suggested that ICRC bacillus is not *M. leprae* but may be a variant of *Mycobacterium* belonging to the *Mycobacterium avium* intracellular group.

Other Properties

Susceptibility to Physical and Chemical Agents

M. leprae remains viable in a warm, humid environment for 9–16 days and moist soil for 46 days. They also remain viable on exposure to ultraviolet light for 30 min and direct sunlight for 2 h.

Antigenic Structure

M. leprae contains a wide variety of mycobacterial antigens. **LAM-B** is a major antigen of *M. leprae*. It is highly immunogenic and induces a high level of serum antibodies. It is distinct from the lipoarabino mannose of *M. tuberculosis*. *M. leprae* consists of a large number of **protein antigens**. The major protein antigens include 65-kDa proteins, 28-kDa protein and 18-kDa protein. The 65-kDa heat shock protein induces a humoral antibody response in the initial phase of *M. leprae* infection. The 28-kDa antigen protein is the superoxide dismutase of *M. leprae*. The protein stimulates a higher antibody response in lepromatous leprosy than tuberculoid leprosy. The 18-kDa protein is one of the most important antigens which produces both humoral and cell-mediated responses in a person.

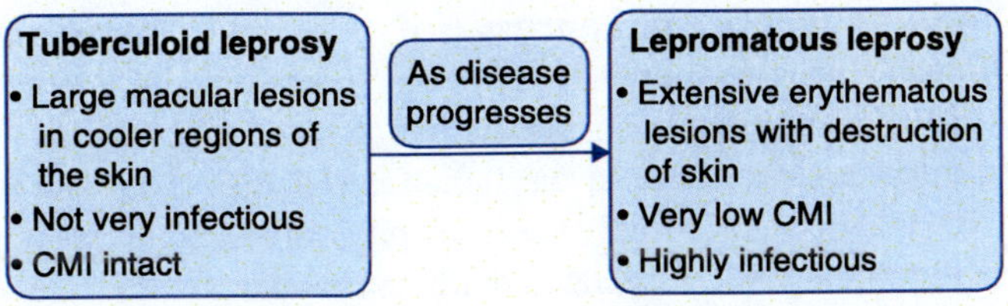

Fig. 31.1 Progression of tuberculoid and lepromatous leprosy

Host Immunity

M. leprae infection induces the development of antibody-mediated and cell-mediated immunities in the infected host. The CMI is the only protective mechanism against the disease. The progression of the disease is primarily dependent on the CMI. The CMI is intact in tubercular leprosy, whereas it is lacking or minimal in lepromatous leprosy (Fig. 31.1). The infected persons with tuberculoid leprosy show a strong delayed hypersensitivity reaction to lepromin. The macrophages phagocytose the bacilli and destroy them.

Classification of Leprosy

Ridley and Jopling (1966) introduced a system of classification of leprosy based on clinical, histopathological and immunological findings. This classification is used to differentiate types of leprosy and helps in determining the prognosis following treatment with antileprosy therapy. They classified leprosy into five groups: (1) tuberculoid (TT), (2) borderline tuberculoid (BT), (3) borderline (BB), (4) borderline lepromatous (BL) and (5) lepromatous (LL). Lepromatous leprosy and tuberculoid leprosy are the two extreme or polar forms of the disease.

In contrast, WHO has classified as paucibacillary leprosy (PB) and multibacillary leprosy (MB), focusing on the purpose of treatment. The classification is based on the number of skin lesions and the number of nerves involved. PB is characterised by 1 to 5 lesions with an asymmetrical shape and definite loss of sensation having one nerve involvement. MB is characterised by more than five symmetrical lesions and with or without loss of sensation but involving more than two nerves.

Pathogenesis of Leprosy

Leprosy is a chronic granulomatous disease exclusively of humans.

Virulence Factors

M. leprae is an obligate intracellular acid-fast bacillus with an affinity towards macrophages and Schwann cells. The virulence of *M. leprae* is mainly due to its capabilities for intracellular multiplication and growth and the host's immune system that influences the clinical form of the disease.

Phenolic glycolipid-1 is the major antigen component of the outermost layer of the cell wall. This lipid is specific to *M. leprae* and is the best-studied virulence factor of *M. leprae*. PGL-1 binds to the C3 component of the complement, which leads to phagocytosis mediated by complement receptors 1, 3 and 4 found on the cell surfaces. PGL-1 protects the lepra bacillus, once inside the phagocytic cells, from oxidative killing by macrophages by removing hydroxyl radicals and superoxide anions. The antigen also suppresses CMI and protects *M. leprae* from the action of host cell enzymes.

Toll-like receptors (TLRs) also play an important role in the pathogenesis of leprosy. TLR-1 and TLR-2 are found on the surface of Schwann cells, especially in patients with tuberculoid leprosy. *M. leprae* activates this receptor on Schwann cells, activating apoptosis genes and enhancing the onset of nerve damage seen in the mild disease. Type 1 reversal reactions occur due to a sudden increase in T-cell immunity. Type 2 reaction occurs due to activation of tumour necrosis factor-alpha and the deposition of immune complexes in tissues with neutrophilic invasion. Thus *M. leprae* causes leprosy, a chronic granulomatous disease.

Immunogenetics of Leprosy

Human leucocyte antigen (HLA) class II serological typing has shown an association of tuberculoid leprosy with DR2 in the Indian population. DRB1-1501 and -1502 are the predominant subtypes of DR2 associated with lepromatous leprosy. These two subtypes are implicated in the presentation of pathogenic peptides of mycobacteria. DQ1 presentation has been suggested to be associated with an immunosuppression gene for *M. leprae* since lepromatous leprosy and borderline leprosy has shown an increased frequency of DQ1.

In tuberculoid leprosy, *M. leprae* peptides probably bind selectively to HLA allele forms (HLA-DR2) and other positively associated DR antigens and stimulate T-cell clones that result in a harmful immune response. Complete energy to native heat shock proteins is characteristically seen in lepromatous leprosy.

Clinical Syndromes

Leprosy

Leprosy is a chronic granulomatous disease primarily affecting the skin, peripheral nervous system and nasal mucosa but can affect any organ or tissue. The clinical manifestations of leprosy depend on the immune status of the host and the spread of the bacilli. The incubation period varies from 6 months to 40 years or longer. The mean incubation period for tuberculoid leprosy is 4 years, and lepromatous leprosy is 10 years.

WHO classification of leprosy based on clinical manifestations and smear findings have classified leprosy into **multi-bacillary leprosy (MB)** (smear collected from any site positive for bacilli) or **paucibacillary leprosy (PB)** (smear is negative for bacilli). Based on Madrid classification (1953), leprosy has been classified into four groups: (1) **lepromatous**, (2) **tuberculoid**, (3) **dimorphous** and (4) **indeterminate**. Lepromatous leprosy and tuberculoid leprosy are the two extreme or polar forms of the disease.

Lepromatous Leprosy

The condition is seen in persons with low CMI. The skin lesions are usually bilaterally symmetrical and non-anaesthetic. These include macules, nodules, plaques, or papules. These are extensive and are most severe at the cooler parts of the body. This form of the disease is associated with disfigurative lesions. The skin of the face and forehead becomes thickened and corrugated, giving rise to a typical leonine face. Bacilli in large numbers are shed in oral and nasal secretions. The lateral part of the eyebrows may be lost. Other disfiguring lesions include pendulous ear lobes, hoarseness of voice, involvement of cornea (Fig. 31.2), perforation of the nasal septum and nasal collapse. Even in most advanced cases, the destructive lesions in leprosy are limited to the peripheral nerves, skin and upper respiratory passages above the larynx, anterior chamber of the eye, testis, hands and feet.

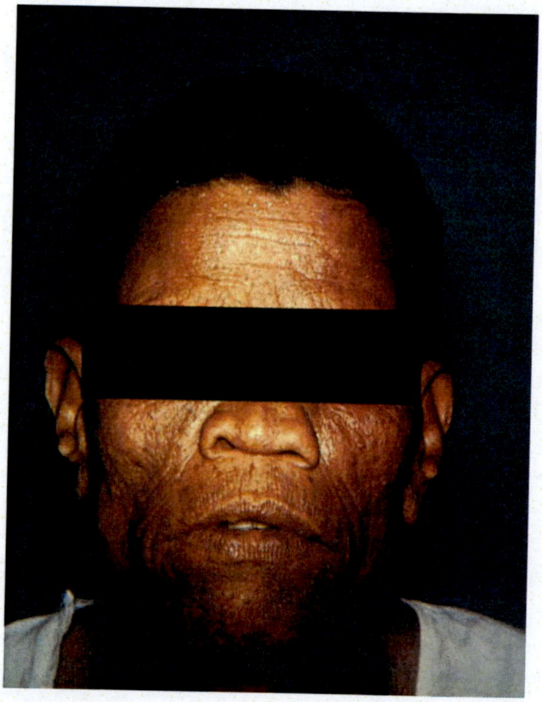

Fig. 31.2 Lepromatous leprosy with corneal involvement (left side). (Courtesy: https://phil.cdc.gov/Details.aspx?pid=15494)

Lepromatous leprosy is known as **multibacillary leprosy** because the lepra bacilli are seen in large numbers inside or outside the cells. The patients are highly infectious to others. The lepromin test is negative due to low or nearly absence CMI. In contrast, humoral immunity is stronger in lepromatous leprosy and is characterised by strong humoral antibody responses. High levels of antibodies are found in the serum. Autoantibodies are also frequently found. Tuberculin sensitivity may be negative or suppressed in untreated lepromatous leprosy, but it becomes positive after treatment.

Tuberculoid Leprosy

It is seen in patients with stronger CMI, a higher degree of resistance. Skin lesions are few and sharply demarcated. The macular anaesthetic patches have raised, and the erythematous border with a dry, scaly appearance in the centre with complete anaesthesia. They are usually circular, or serpiginous, and often hypopigmented. The skin lesions are commonly found on the face, limbs, buttocks or elsewhere. They, however, are not found in the axilla, perineum, or scalp. Bacilli are scanty or few in the lesions, hence are of minimal infectivity to others. Neural involvement is common in tuberculoid leprosy. The ulnar, peroneal and greater auricular nerves are involved, leading to tender, thickened nerves with subsequent loss of function. Damage to the nerve can result in wrist drop or foot drop.

Tuberculoid leprosy is also known as **paucibacillary disease**. It is characterised by a positive lepromin test. The humoral antibody response is weak; hence antibodies in the serum are rare. Tuberculin sensitivity is positive, and the prognosis is good.

Borderline or Dimorphous Leprosy

It is another form of leprosy, which exhibits lesions characteristics of both tuberculoid and lepromatous leprosy. Skin lesions are few, asymmetric and have nearly complete anaesthesia (Fig. 31.3). Peripheral nerves are thickened and involved asymmetrically. This form of leprosy can regress to the tuberculoid form, or it can progress to a lepromatous form, depending on chemotherapy or change in host immunity.

Indeterminate Leprosy

It is the most unstable form of leprosy. This form is not characteristic of either lepromatous or tuberculoid leprosy. In some cases, it may progress either to tuberculoid or lepromatous leprosy. In some other cases, it may undergo spontaneous healing.

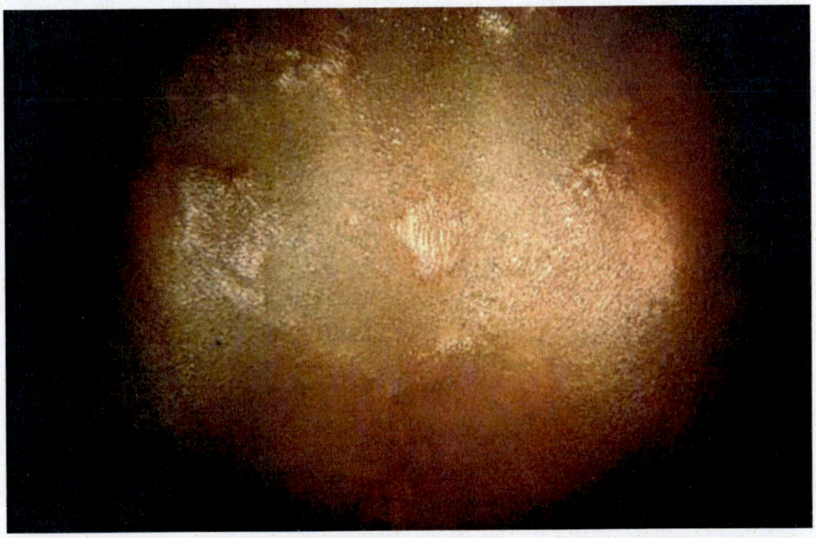

Fig. 31.3 A case of borderline tuberculous leprosy, or dimorphous leprosy. (Courtesy: https://phil.cdc.gov/Details.aspx?pid=19226)

Purely Neuritic Leprosy

It is an additional form of leprosy described recently. It is characterised by asymptomatic peripheral neuropathies without any skin lesions. These cases are negative for the bacteria.

Lepra Reactions

The course of leprosy may be interspersed with the reactional stage. These reactions, which are of allergic nature and occur nearly in one-third of the patients, are known as **lepra reactions**. These reactions are considered a medical emergency, requiring immediate treatment and management. Lepra reactions are of the following types:

Lepra type I reaction is a type IV cell-mediated allergic hypersensitivity. This reaction is seen mostly in patients with borderline leprosy, occurring spontaneously or more often during chemotherapy. These reactions are characterised by the development of skin erythema with oedema and tenderness of peripheral nerves. These reversal reactions usually suggest a shift towards tuberculoid form after the start of chemotherapy. It is precipitated by puberty, pregnancy and childbirth. This reaction usually occurs during the first 2 months of therapy to up to 12 months.

Lepra type II reactions are type III humoral hypersensitivity reactions with a systemic inflammatory response due to deposition of immune complexes. The appearance of crops of painful erythematous nodules on the skin and subcutaneous tissue is the characteristic manifestation. Other symptoms include fever, malaise, arthralgia, neuralgia, iridocyclitis, arthritis and proteinuria. This condition occurs in 20% of patients with lepromatous leprosy and 10% of patients with borderline leprosy. This reaction occurs after a few years of therapy and relapse intermittently over several years.

Lucio phenomenon is an unusual form of type II reaction. The condition manifests as cutaneous haemorrhagic infarct in patients with diffuse lepromatous leprosy. This condition is commonly documented in Mexico and Central America.

Differential features of various forms of leprosy are summarised in Table 31.1.

Laboratory Diagnosis

The diagnosis of leprosy is essentially clinical. Laboratory diagnosis of leprosy is essentially based on microscopy's demonstration of acid-fast *M. leprae* in the skin smears.

Specimens

In lepromatous leprosy, lepra bacilli are always found in large numbers in the skin and the nose. Therefore, smears collected from the nasal mucosa, skin lesions and ear lobules are the specimens of choice for microscopy. Occasionally, the specimens may be collected from lymph

Table 31.1 Differential features of various forms of leprosy

Characteristics	Tuberculoid leprosy	Borderline tuberculoid leprosy	Borderline leprosy	Borderline lepromatous leprosy	Lepromatous leprosy
Lepra bacilli in tissue	–	+/–	+	+++	++ + + +
Lepra bacilli in nasal secretions	–	–	–	+	++ + +
Antibodies to *Mycobacterium leprae*	+/–	+/–	+	+++	++ + +
Granuloma formation	++ + +	+++	+	–	–
Lepromin test	++ + +	+	+/–	–	–
Main phagocytic cell	Mature epithelioid	Immature epithelioid	Immature epithelioid	Macrophage	Macrophage
In vitro correlation of CMI	++ +	+ +	+	+/–	–

nodes and affected nerves in lepromatous leprosy. In tuberculoid leprosy, the bacilli are not found in these specimens.

Skin smears are collected from the leprous lesions, such as nodules, thick papules and areas of infiltration. In cases of patches, the samples are obtained from the edge of the lesion rather than from the centre. In patients with only diffuse infiltrations, skin smears are collected from five to six different sites, including the skin over the ear lobes, buttocks, forehead, cheeks and chin. The skin smears from these sites are collected by the slit and scrape method. In this method, the skin is pinched up and raised between the thumb and index finger of the left hand, which squeezes out blood from that part, thereby minimising bleeding when a cut is made. Then with the help of a scalpel, a cut about 5 mm is made on the pinched skin, deep enough to get into the infiltrated layers. After wiping of blood or lymph that may have exudated, the blade of the scalpel is turned at a right angle to the cut and the sides and bottom of the cut are scraped with the point of the blade several times in the same direction so that the tissue fluid and pulp but not blood are collected on one side of the blade. Obtained fluid and tissues are thickly spread on a slide and stained by Ziehl–Nielsen (ZN) technique.

Nasal smears from the nasal lesions are collected by scraping material from the mucous membrane of the internal nasal septum, particularly from inferior turbinate folds of the nasal septum. This mucous membrane is then spread on a slide, and a uniform smear is made, which is stained by the ZN method, as mentioned earlier.

Microscopy

The slit-skin smears, nasal smears, or smears from other specimens are stained using the ZN method and decolourised with 5% instead of 20% sulphuric acid. In a positive, stained smear, red-coloured acid-fast bacilli are seen on examination by oil immersion lens (Fig. 31.4). The smears are graded based on the numbers of bacilli, as mentioned in Table 31.2.

Bacillary index and morphological index are two indices based on the demonstration of *M. leprae* in smears from skin or nasal discharges. These two indices are useful for (1) demonstrating the viability of organisms, (2) assessing the amount of infection and (3) assessing patients' progress after treatment with chemotherapeutic agents.

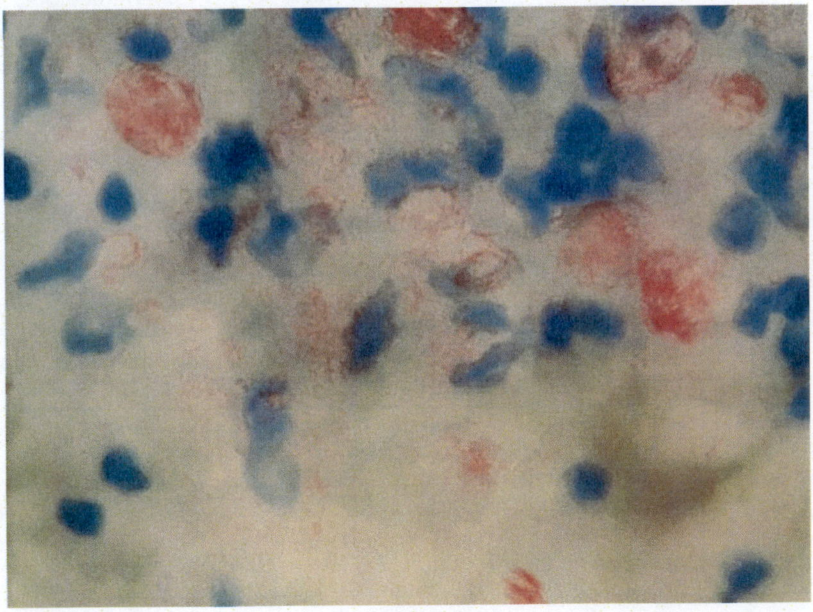

Fig. 31.4 Slit-skin smear showing acid fast *Mycobacterium leprae* (×1000)

Table 31.2 Grading of Ziehl Neelsen-stained smears for lepra bacilli

Examination by oil immersion field	Number of AFB	Interpretation
100 fields	1–10 bacilli/100 fields	1 +
10 fields	1–10 bacilli/10 fields	2 +
1 field	1–9 bacilli in every field	3 +
1 field	10–100 bacilli in every field	4 +
1 field	100–1000 bacilli in every field	5 +
1 field	>1000 bacilli in every field	6 +

AFB: acid-fast bacilli

Bacillary index (BI) is an expression of the extent of bacterial load. It is calculated by counting six to eight stained smears under the 100× oil immersion lens. This index is obtained by totalling the number of plusses (plus scored in all the smears and divided by the number of the smears). For this index, a minimum of four skin lesions, a nasal swab and both ear lobes are examined. The bacillary index can be affected by the thickness of the film, depth of the skin incision and thoroughness of the scrap.

Morphological index (MI) uses the uniformity of the bacilli as a criterion to differentiate live bacilli from dead bacilli in the stained smears. It is calculated by counting solid stained bacilli out of the total number of bacilli counted. The lepra bacilli that stain uniformly with carbol fuchsin as solid acid-fast rods are believed to be viable. The bacilli that stain irregularly are considered dead and degenerated. The percentage of uniformly stained bacilli in the tissues is known as the morphological index (MI). This index is useful for assessing the response of patients with leprosy to antileprosy agents and is more useful than the bacteriological index (BI). A decrease in MI indicates a better response of the patients to chemotherapy. In contrast, a decrease in MI followed by an increased MI indicates a poor response to chemotherapeutic agents.

Other methods that are also used to detect viable lepra bacilli include (1) fluorescent diacetate-ethidium bromide (FDA–EB) staining, (2) laser microscope mass analysis (LAMMA), bioluminescent technology and (3) macrophage-based assays.

Nasal smears are positive in lepromatous and borderline leprosy but are negative in tuberculoid leprosy, mid-borderline leprosy and borderline tuberculoid leprosy. Examination of nasal smears is useful to know whether a leprosy patient is infectious or not. The lepra bacilli usually disappear more rapidly from the nose than from the skin lesions following chemotherapy.

Biopsy

Skin biopsy is useful for the diagnosis and proper classification of leprosy. The samples are usually obtained from the edge of the lesion; a biopsy smear or scalpel may be used for the purpose. **Nerve biopsy** is a very useful method in the diagnosis of leprosy. This is useful to differentiate the lepromatous cause of thickened nerves from other non-lepromatous causes of thickened nerves in people living in the disease's endemic areas. This is also useful to rule out other diseases, such as polyradiculoneuropathy, hereditary neuropathies, or polyarteritis nodosa. Nerve biopsy is the only method to confirm the diagnosis in cases of leprosy with pure neuropathic forms.

Lepromin Test

The lepromin test is used to study host immunity to *M. leprae* and was first described by Mitsuda in 1919. It is an intradermal skin test performed using lepromin antigen, a suspension of killed *M. leprae* obtained from infected human or armadillo tissue.

Antigen

Originally, crude antigen obtained from human lepromatous tissue rich in *M. leprae* was used as antigen. Mitsuda first prepared this antigen in 1919 by using antigens extracted from skin lesions of lepromatous patients. This antigen, known as *standard Mitsuda lepromin*, contains 4.0×10^7 *M. leprae* per mL, The antigen has a shelf life of 2 years when stored at 4 °C. The standard lepromin antigen has been recently prepared from armadillo-derived lepra bacilli (Lepromin A) by replacing human-derived lepra bacilli (Lepromin H).

Dharmender antigen is an example of Bacillary lepromin type of antigen. This contains more bacillary components and fewer tissue components. Indian scientist Dharmender first prepared this antigen by separating the bacilli from finely ground lepromatous tissue and treating it with chloroform. The suspension is then evaporated to make it dry and then washed with ether to remove the lipids. The antigen, finally, is reconstituted in phenol saline for use.

Test

Lepromin skin test is performed by intradermal injection of lepromin antigen into the forearm. Two types of reactions are observed after the test. First, early or Fernandez reaction and second, late or Mitsuda reaction.

Early or Fernandez reaction is observed after 48 h of injection. A positive reaction is characterised by the appearance of a localised area of inflammation with congestion and oedema measuring 10 mm and more in diameter during 24–48 h of injection. These lesions disappear within 3–4 days. The positive Fernandez reaction, like tuberculin reaction, indicates a delayed type of hypersensitivity to antigens of *M. leprae* or mycobacteria that cross-react with *M. leprae.* The positive reaction suggests that the patient has been infected by leprae bacilli sometime in the past. **Late or Mitsuda reaction** is characterised by the development of a nodule at the inoculation site after 3–4 weeks of injection (Fig. 31.5). The nodule subsequently may undergo necrosis followed by ulceration. The ulcer takes several weeks to heal. Positive Mitsuda reaction indicates that the immune system can induce an efficient cell-mediated response. Mitsuda reaction is positive in tuberculoid leprosy.

The lepromin test is used mainly to determine the type of leprosy. A positive test suggests resistance to disease in tuberculoid leprosy. A

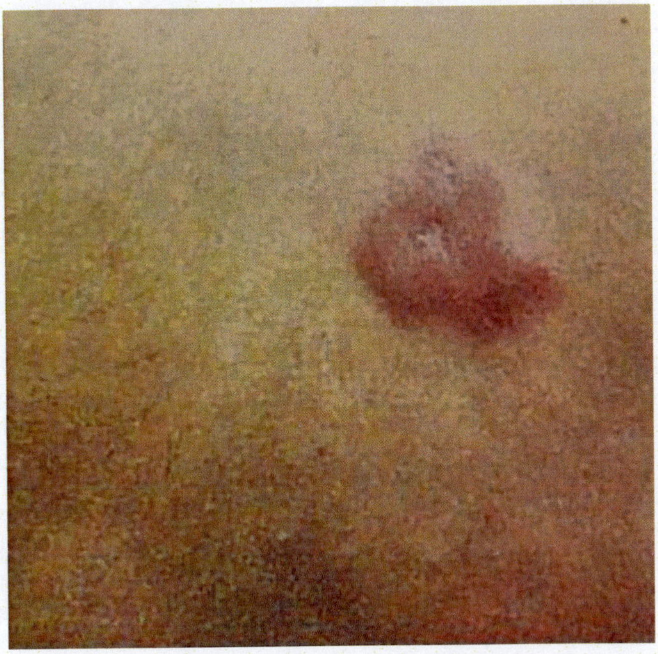

Fig. 31.5 Lepromin test (forearm) showing Mitsuda reaction in the form of a papule (https://www.researchgate.net/figure/Erythematous-lesion-in-the-dorsal-region_fig1_259368237)

negative test suggests a lack of resistance to the disease in lepromatous leprosy patients. It is also used to monitor the response of leprosy patients to treatment with chemotherapy. A positive reaction indicates a good prognosis, while a negative test suggests a very bad prognosis. It is used to assess host resistance to leprosy.

The lepromin test is not used to confirm the diagnosis of leprosy. It is also not useful to indicate prior contact with the person with leprae bacilli. Healthy people residing in areas non-endemic for leprosy with no chance of contact with the bacilli may show a positive lepromin reaction.

Serology

Serodiagnosis of leprosy is based on demonstrating antibodies to *M. leprae*, specific PGL-1 antigens. ELISA and latex agglutination tests are used to detect serum antibodies. The serology is useful primarily in patients with untreated lepromatous leprosy, as most (90%) of patients have higher levels of serum antibodies. The serology, however, is less useful for diagnosing paucibacillary disease because serum antibodies are present in only 40–60% of such patients.

Molecular Detection

Polymerase chain reaction (PCR) is used for identifying DNA that encodes 65- and 18-kDa *M. leprae* proteins and repetitive sequences of *M. leprae*. It is used to monitor treatment, diagnose relapses, or determine the need for chemotherapy. It is most useful in cases of leprosy showing atypical clinical or histopathological features but positive for acid-fast bacilli. It is not useful for diagnosing cases when acid-fast bacilli are not detected by light microscopy.

Evidence indicates that PCR is the most preferred choice in difficult-to-diagnose cases. It results from incorrect diagnosis and results in early identification. It could facilitate the identification of clinical manifestations like pure neural leprosy (PNL) and atypical skin clinical presentations. Results obtained are 100% specific, while the sensitivity is greater than 90% in patients.

Treatment

The management of leprosy depends on treatment by chemotherapeutic agents and immunotherapy followed by physical, social and psychological rehabilitation.

Chemotherapy

The goals of chemotherapy against leprosy are to (1) stop infection, (2) reduce morbidity, (3) prevent complications and (4) eradicate the disease. Dapsone was the first effective chemotherapeutic agent used against leprosy. Earlier it was used as a monotherapy, but it resulted in the development of resistance against the drug by lepra bacilli. Hence since 1981, the WHO has advocated multiple drug therapy (MDT) against leprosy, like tuberculosis.

Multiple drug therapy (MDT) against leprosy quickly decreases the contagiousness of the disease, reduces relapse, reduces reactions and reduces disabilities. Moreover, MDT prevents dapsone resistance. The duration of treatment varies from 6 months to 2 years. The recommendations of WHO for treatment of leprosy in adults are as follows:

1. Single skin lesion: A single dose of the drug, which includes rifampicin 600 mg, ofloxacin 400 mg and minocycline 100 mg.
2. Paucibacillary disease: Dapsone 100 mg daily and rifampicin 600 mg once a month was given for 6 months.
3. Multibacillary disease: Rifampin (rifampicin) 600 mg once a month, dapsone 100 mg daily, clofazimine 300 mg once a month and 50 mg daily are given for 1 year.

A minimum of 2 years of follow-up for paucibacillary and 8 years for multibacillary cases is required to detect any relapse. The treatment schedule for children consists of dapsone (2 mg/kg) daily, clofazimine (6 mg/kg) once a month under the supervision and 1 mg/kg daily self-administered and rifampin (10 mg/kg) once a month.

Reactions during treatment with MDT are a major problem. These reactions need urgent treatment, failing which they can give rise to irreversible deformities. Early diagnosis and timely initiation of anti-inflammatory measures are important. MDT is not stopped during reactions and is continued at full doses without interruption.

Immunotherapy

Many immunotherapy agents have been evaluated for the treatment of leprosy. These include immunomodulatory drugs, transfer factor, acetoacetylated *M. leprae* and delipidified cell components of *M. leprae*. These agents have been shown to enhance CMI, which results in increased killing and rapid clearing of dead lepra bacilli. Immunotherapy in combination with chemotherapy has shown to be more beneficial, better tolerated by patients and is not associated with increased lepra reactions.

Epidemiology

Leprosy is a disease prevalent worldwide.

Geographical Distribution

The worldwide prevalence of leprosy is <1 case/10,000 population. Most cases of leprosy are found in Southeast Asia, Africa and the Americas. Fifteen endemic countries in these areas have a prevalence of >1 case/10,000 populations. More than 120,000 new cases were reported globally in 2020. Among the new cases, 7198 new cases were detected with grade2 disabilities (G2D). However, the number suffered from the impact of COVID-19 that affected programme implementation resulting in a reduction in new case detection by 37% in 2020 compared with 2019.

Habitat

In an infected host, *M. leprae* is found in large numbers in infected nasal secretions of patients with lepromatous leprosy.

Reservoir, Source and Transmission of Infection

Leprosy is an exclusively human disease. Humans are the prime reservoir of *M. leprae*. Nine-banded armadillo, chimpanzee and mangabey monkeys are the three animal species known to be important animal reservoirs of leprosy. Leprosy is not a highly infectious disease. Lepromatous forms of the disease, but not the tuberculoid form, is infectious.

Leprosy is primarily transmitted by infected nasal secretions. Infections by a very large number of lepra bacilli lead to lepromatous leprosy. Nasal secretions from patients with lepromatous leprosy are an important source of infection. Very large numbers of lepra bacilli are shed in the nasal secretions. As high as 8×10^8 bacilli can be discharged during a single nose blow of the patients with untreated lepromatous leprosy.

Inhalation of infectious aerosol and skin contact with respiratory secretions or wound exudates are the most important routes of infection in leprosy. The following observations support the transmission of infection through the respiratory route: (1) a large number of morphologically intact lepra bacilli are demonstrated in the nasal discharges, (2) the organisms have not been detected on the surface of the skin, (3) lepra bacilli survive for several hours or days outside the human host and (4) experimental transmission of leprosy has been achieved by aerosols containing *M. leprae* and by topical application in immunosuppressed mice. Leprosy is usually not spread by means of direct contact through intact skin, though close contacts are more susceptible.

Prevention and Control

The preventive and control measures include: (1) early diagnosis and treatment of leprosy and surveillance of contacts, (2) health education, (3) vaccines and (4) chemoprophylaxis.

Attempts have been made to develop and evaluate **vaccines** against leprosy. The antileprosy vaccines can be used as both immunoprophylactic and therapeutic agents. The vaccines that have been evaluated currently include the *M. avium*-intracellular complex (*Mycobacterium* ICRC) vaccine, intracellular–Guerin (BCG) vaccine, *Mycobacterium* w vaccine; BCG plus heat-killed *M. leprae*, *Mycobacterium tufu* and *Mycobacterium habana* vaccine. **BCG vaccine** has shown variable results in conferring protection against certain populations. However, repeat immunisation with BCG has been shown to confer further protection. Both *Mycobacterium* w and *Mycobacterium* ICRC vaccines are used in India. *Mycobacterium* ICRC vaccine was prepared in 1979 and contained gamma-radiation-inactivated ICRC bacilli. Uses of these vaccines—both in mouse models and human studies—have shown better results.

Many studies using one or more antileprosy drugs (dapsone, acedapsone and rifampicin) as **chemoprophylaxis** have failed to show any significant protection against leprosy.

WHO Roadmap on Leprosy

In 2018, WHO had developed "WHO guidelines for the diagnosis, treatment and prevention of leprosy 2018". It recommended three drugs, namely rifampicin, dapsone and clofazimine, to both paucibacillary and multibacillary types of leprosy.

Recently in April 2021, WHO has released "*Towards Zero Leprosy – Global Leprosy (Hansen's disease) Strategy 2021–-2030*". With a goal to eliminate leprosy by achieving zeros in cases, infection, transmission, disability, stigma and discrimination, the strategy aims for a 70% reduction in the annual number of new cases, 90% reduction in rate per million population of new cases with grade-2 disability (G2D) and 90% reduction in rate per million children of new child cases with leprosy.

To operationalise the strategy, four pillars have been implemented:

1. Implement integrated, country-owned zero leprosy roadmaps in all endemic countries.
2. Scale-up leprosy prevention alongside integrated active case detection.
3. Manage leprosy and its complications and prevent new disability.
4. Combat stigma and ensure human rights are respected.

Mycobacterium lepraemurium

M. lepraemurium was first described by Stefansky as the causative agent of leprosy in rats in 1901 at Odessa. The rat leprosy was found in 4–5% of slaughtered rats investigated during an attack of human plague. The condition, subsequently, has been documented with variable incidence from different countries of the world. *M. lepraemurium* resembles *M. leprae* and is found in the mononuclear cells of the subcutaneous tissues, lymph nodes and the nodules in the lungs and liver of the infected rat. However, both *M. lepraemurium* and *M. leprae* are not related species, and they differ from each other, as shown by DNA studies. *M. lepraemurium* is cultivated with difficulty in Ogawa egg-yolk medium (pH 5.8–6.3) at 37 °C and also in rat fibroblasts cell lines. *Lepraemurium* in rats causes rat leprosy, which is characterised by two forms: glandular form and musculocutaneous form. Enlarged, hard and whitish inguinal, axillary and cervical lymph nodes are found in the glandular type and ulceration and loss of hair are found in the musculocutaneous type. The bacteria do not cause disease in humans.

Case Study

A 25-year-old male attended Dermatology OPD and complained of hypopigmented patches on limbs and a few on the buttocks. On examination, the ulnar nerve was found to be tender and thickened. Skin biopsy and nasal secretion were negative for acid-fast bacilli. The lepromin test was positive. The condition was diagnosed as leprosy.

- Diagnose the type of leprosy the patient is suffering from.
- Describe the laboratory methods for diagnosis of the condition.

- Discuss the antibiotic regimen for treatment of the condition.
- Describe the vaccines available against the disease.

Non-Tuberculous Mycobacteria (NTM)

Non-tuberculous mycobacteria (**NTM**) are the mycobacteria other than the tubercle and lepra bacilli. These are also known as mycobacteria other than tubercle bacilli (**MOTT**). These bacteria, which normally occur as saprophytes of soil and water, may occasionally cause opportunistic infections in humans that resemble tuberculosis. These bacteria, therefore, were also earlier known as atypical, environmental or opportunistic mycobacteria.

NTM or atypical mycobacteria show the following features: (1) They can grow at 25, 37 °C and even at 44 °C. (2) Some of them are rapid growers. They produce colonies within 1–2 weeks of incubation in Lowenstein–Jensen (LJ) medium. (3) Some of them may produce bright yellow or orange pigments during their growth on the LJ medium. (4) They are acid-fast as well as alcohol fast. They may differ from or may resemble the tubercle bacilli. (5) They are arylsulphatase test positive but are niacin and neutral red test negative. (6) They are usually resistant to antitubercular drugs, such as streptomycin, isoniazid (INH) and *p*-aminosalicylic acid and (7) They are non-pathogenic for guinea pigs but pathogenic for the mouse.

NTM have been classified as **rapid growers** and **slow growers**. *Mycobacterium fortuitum*, *Mycobacterium chelonae* and *Mycobacterium abscessus* are examples of more rapidly growing *Mycobacterium* species, which require incubation for 3 days or more to develop the colonies. *M. avium-intracellulare* complex and *Mycobacterium kansasii* are the slow-growing *Mycobacterium* species that require 3–8 weeks of incubation for their growth.

Most human infections are caused by *M. tuberculosis* and *M. leprae*. *M. avium* complex, *M. kansasii*, *M. abscessus*, *M. chelonae* and *M. fortuitum* are examples of MOTT associated with frequent human infections (Table 31.3).

Runyon Classification

Runyon classified saprophytic bacteria or MOTT into four groups based on their growth rate and their ability to produce pigments in the presence or absence of light. This classification system is known as **Runyon classification** (Table 31.4).

Photochromogens (Runyon Group I)

These includes mycobacteria, which produce pigment on exposure to light(photochromogens). They form pigments when the young culture is

Table 31.3 Human infections caused by atypical *Mycobacterium* species

Atypical *Mycobacterium* species	Diseases
Mycobacterium kansasii	Pulmonary disease
Mycobacterium marinum	Swimming pool granuloma
Mycobacterium simiae	Pulmonary disease (rare)
Mycobacterium scrofulaceum	Lymphadenopathy
Mycobacterium gordonae	Pulmonary disease (rare)
Mycobacterium szulgai	Pulmonary disease and bursitis (occasional)
Mycobacterium xenopi	Chronic pulmonary disease
Mycobacterium avium complex	Pulmonary disease, lymphadenopathy and disseminated disease
Mycobacterium ulcerans	Buruli ulcer
Mycobacterium fortuitum	Post-trauma chronic abscesses
Mycobacterium chelonae	Post-trauma chronic abscesses
Mycobacterium abscessus	Abscesses
Mycobacterium genavense	Disseminated diseases (AIDS-related)

Table 31.4 Runyon classification of non-tuberculous mycobacteria (NTM)

Group	Nomenclature	Growth and pigmentation on culture media	Clinically significant species	Disease
Group I	Photochromogens	Colonies are pigment less in dark but upon exposure to light colonies become pigmented	*Mycobacterium kansasii*, *Mycobacterium marinum*	Chronic pulmonary disease Cutaneous infections-Swimming pool granuloma or fish tank granuloma
Group II	Scotochromogens	Produce pigmented colonies even in dark.	*Mycobacterium scrofulaceum*, *Mycobacterium gordonae*	Scrofula (cervical lymphadenitis)
Group III	Non-Photochromogens	No pigment even after exposure of light into colonies.	*Mycobacterium avium*, *Mycobacterium intracellulare*, *Mycobacterium ulcerans*, *Mycobacterium xenopi*	Pulmonary infection in HIV infected Lymph node infection Cutaneous infections-Buruli ulcer
Group IV	Rapid growers	Colonies appear within 7 days	*Mycobacterium fortuitum*, *Mycobacterium chelonae*, *Mycobacterium smegmatis*, *Mycobacterium phlei*	Post-trauma injection abscess and catheter-related infections

exposed to light for 1 h in the presence of air and is re-incubated for 24–48 h. These bacteria produce yellow-orange pigments during such conditions. They do not produce any pigments in the colonies that are incubated in the dark. These mycobacteria are slow-growing but grow faster (after 7 days) than the tubercle bacilli. The most common species are *M. kansasii*, *Mycobacterium marinum* and *Mycobacterium simiae*. Some of these mycobacteria produce diseases similar to pulmonary tuberculosis. Most of these mycobacteria have been isolated from water and soil in the environment.

M. kansasii reduces nitrates to nitrites and shows a positive Tween 80 hydrolysis test within 3–6 h, whereas other mycobacteria species produce positive test within 3–10 days. *M. kansasii* causes a disease identical to pulmonary tuberculosis, which is associated with the formation of a cavity and scarring, usually in the upper lobe of the lungs. This species is the second most common NTM, next to the *M. avium* complex, as a causative agent of lung diseases. These strains have been frequently isolated from tap water, and infected tap water is believed to be the major source of infection. *M. kansasii* grows rapidly on LJ medium, producing visible colonies within 2 weeks. It is sensitive to rifampicin and other antitubercular drugs.

M. marinum closely resembles *M. kansasii* but is differentiated from it by its (1) poor growth at 37 °C, (2) failure to reduce nitrate to nitrite and (3) failure to produce the enzyme catalase. *M. marinum*—originally isolated from fish—is the causative agent of swimming pool or fish tank granuloma. This condition is associated with the development of superficial granulomatous lesions in the skin.

M. simiae associated with pulmonary disease in humans was originally isolated from monkeys. This species grows well at 37 °C and is niacin positive, like *M. tuberculosis*.

Scotochromogens (Runyon Group II)

These include mycobacteria, are characterised by their ability to produce yellow, orange or red pigmented colonies on the LJ medium even when incubated in the dark (scotochromogen) (Fig. 31.6). These species are widely distributed in the environment. *Mycobacterium scrofulaceum*, *Mycobacterium gordonae* and *Mycobacterium szulgai* are important species.

M. scrofulaceum causes infection localised to lymphatic tissues and is responsible for causing

Fig. 31.6 LJ medium showing the colonies of *Mycobacterium fortuitum*

scrofula or cervical adenitis in children. Morphologically, on staining, the bacillus shows short or long filaments. It is sensitive to cyclobenzaprine but resistant to INH.

M. gordonae rarely causes pulmonary disease. It is often found as a contaminant in tap water, hence called tap water scotochromogen. This is also found as a contaminant in clinical specimens. *M. gordonae* does not hydrolyse urea, nicotinamide and pyrazinamide; in this respect, it differs from *M. scrofulaceum*.

M. szulgai occasionally may cause pulmonary disease and bursitis. It is a scotochromogen when grown at 37 °C and a photochromogen when grown at 27 °C.

Non-Photochromogens (Runyon Group III)

These are the mycobacteria that do not produce pigment in the dark or exposure to light (non-photochromogens). These include *M. avium* complex, *Mycobacterium xenopi* and *Mycobacterium ulcerans*. They also include *Mycobacterium terrae*, *Mycobacterium triviale* and *Mycobacterium nonchromogenicum*, which rarely cause infections in humans.

***M. avium* complex** possesses 28 agglutination types. Types 1, 2 and 3 are considered *M. avium* and other types (Types 4–28) as *M. intracellulare*. They are found commonly in the environment, such as in soil and water (brackish, ocean and drinking water). They are strongly acid-fast mycobacteria with a cell wall rich in lipids. The ability to grow inside the cells is the main virulence determinant of the bacteria. In infected humans, the development of the disease depends primarily on the host's response to infection.

M. avium complex produces a disease similar to pulmonary tuberculosis in patients with compromised pulmonary function, such as patients with chronic bronchitis or obstructive pulmonary disease, that is, previous damage to lungs due to infection or other diseases. In patients with human immunodeficiency virus (HIV) infection, it produces a typically disseminated disease affecting virtually every organ in infected humans. The intensity of infections in these patients is very high. In some patients, tissues are overloaded with mycobacteria, and blood contains nearly 100–1000 mycobacteria/mL of blood. The disseminated infection is commonly seen in persons at the terminal stage of the immunological disorder, with their CD4 count falling below 10 cells/mm^3. The infection is acquired following inhalation of infectious aerosols. The infection is also acquired by ingesting the water or food contaminated with mycobacteria. Person-to-person transmission does not occur.

Diagnosis of infection is made by microscopy of clinical specimens and culture on LJ medium. Treatment with clarithromycin or azithromycin combined with ethambutol and rifabutin for a long period is effective. Chemoprophylaxis with clarithromycin or azithromycin in AIDS patients with low CD4 cell count has drastically reduced the incidence of infection in these patients.

M. xenopi was originally isolated from a cutaneous lesion of a South African toad (*Xenopus laevis*). The organism is a thermophilic mycobacterium, which may occasionally cause chronic pulmonary lesions in humans. Most of these infections have been documented in London. The bacteria have been isolated from mostly hot water taps and main water supplies in hospitals.

M. ulcerans is a skin pathogen that was originally isolated from human ulcerative skin lesions in Australia (1958). Subsequently, the species causing similar cutaneous lesions have been documented from Uganda (Buruli ulcer), Nigeria, Congo, Malaysia, New Guinea and Mexico. *M. ulcerans* grows slowly on LJ medium in 4–8 weeks when incubated at 31–34 °C. However, this species fails to grow when incubated at 37 °C in primary culture. *M. ulcerans* is the only known mycobacterial species to produce exotoxins. This toxin is a phospholipoprotein–lipopolysaccharide complex with a high molecular weight. The toxin produces inflammation and necrosis when inoculated intradermally into the skin of guinea pigs, and lesions similar to those seen in humans are formed. Therefore, it is believed that the toxin plays an important role in the pathogenesis of ulcers—the main pathology of the disease seen in humans.

. After an incubation period of a few weeks, the infection begins with minor injuries on the skin through which mycobacteria gain access. Indurated nodule at the inoculation site, which breaks down, forming an indolent ulcer, is the typical lesion. Ulcers of the skin are typically seen on the legs or the arms. Diagnosis is based on microscopy of the smears collected from the edge of the ulcer. Large numbers of acid-fast and alcohol-fast bacilli are seen in the stained smear by microscopy. Finally, the ulcer heals with the formation of scars. *M. marinum* and *M. ulcerans* belong to two different groups, but both are exclusively skin pathogens, and their differential features are listed in Table 31.5.

Rapid Growers (Runyon Group IV)

These are a heterogeneous group of mycobacteria that rapidly produce visible growth on LJ medium within 1 week of incubation at 37 or 25 °C. This group may also include photochromogens, scotochromogens or non-photochromogens species. Some of the rapid growers are chromogenic (*Mycobacterium smegmatis*, *Mycobacterium*

Table 31.5 Distinguishing features of *Mycobacterium marinum* and *Mycobacterium ulcerans*

Features	*Mycobacterium marinum*	*Mycobacterium ulcerans*
Runyon classification	Group I	Group III
Occurrence	Common in temperate regions	Common in Tropics
Clinical course	Self-limited ulcer	Chronic progressive ulcer
Number of bacilli in ulcer	Scanty	Abundant
Growth on culture medium	Faster, 1–2 weeks	Slower, 4–8 weeks
Growth at 25 °C	Positive	Negative
Growth at 37 °C	Positive	Negative
Cord formation in Culture film	No	Yes
Pigment upon exposure to light	Positive	Negative
Mouse footpad lesion	Marked inflammation with purulent discharge from ulcer	Oedema is common, rarely progress to ulcer.

Table 31.6 Differential properties of important atypical *Mycobacterium* species

Mycobacterium species	Growth				Pigment		Urease test	Niacin test	Nitrate reduction test
	7 days	25 °C	37 °C	45 °C	Dark	Light			
Mycobacterium kansasii	–	+	+	–	–	+	+	–	+
Mycobacterium marinum	–	+	±	–	–	+	+	–	–
Mycobacterium scrofulaceum	–	+	+	–	+	+	+	–	–
Mycobacterium avium intracellulare complex	–	±	+	±	–	–	–	–	–
Mycobacterium fortuitum	+	+	+	–	–	–	+	–	+
Mycobacterium chelonae	+	+	+	–	–	–	+	–	–

phlei) and are saprophytes. These rapid-growing mycobacteria include five species that are recognised as important opportunistic pathogens (*M. fortuitum*, *M. chelonae*, *M. abscessus*, *M. smegmatis* and *M. phlei*) of humans. All these rapid growers stain irregularly with acid-fast stains, show relatively low virulence and are relatively more susceptible to treatment with conventional antibacterial antibiotics. Differential properties of important atypical *Mycobacterium* species are summarised in Table 31.6.

Rapid growers are associated with diseases of deep subcutaneous tissues following trauma or iatrogenic infections. These species have been isolated from (1) contaminated wound dressing, (2) infections associated with intravenous catheters or (3) infections associated with prosthetic devices, such as heart-lung machines. Such infections are now being increasingly recognised from hospitalised patients due to an increase in the use of such medical devices.

M. fortuitum and *M. chelonae* cause chronic abscesses in humans following infections of vaccines and other preparations contaminated by these mycobacteria. The infection also occurs following some injury. They are non-chromogen do not produce any pigments. The bacilli are usually found in the soil. *M. fortuitum* causes a pulmonary disease similar to pulmonary tuberculosis. Both *M. chelonae* and *M. fortuitum* grow well at 35 °C and form white- to cream-coloured colonies and reveal coccoid to filamentous morphology on staining.

M. phlei is mostly non-pathogenic. It produces buff-coloured colonies, which become yellow to orange on further incubation. It differs from *M. smegmatis* by its ability to grow at 52 °C and survive heating at 60 °C for 4 h. *M. smegmatis* is normally found in smegma, a whitish secretion around the orifice of the urethra. These bacilli are slender, curved and rigid rods. These bacilli are both acid and alcohol fast, but some bacilli may only be acid-fast but not alcohol fast. *M. smegmatis* is a rare human pathogen associated with occasional cases of lung, skin and bone infections.

Treatment

Most of the environmental mycobacteria are resistant to common antituberculous drugs. Nevertheless, *M. avium* and *M. kansasii* infections may be treated with rifampicin, INH and ethambutol. Various combinations of drugs, including rifabutin, clofazimine, quinolones, macrolides and others, are also used in the treatment.

Epidemiology

NTMs as environmental bacteria are ubiquitous and are prevalent in soil, water and air. Infection by these mycobacteria may result in sensitisation to mycobacterial antigens. This results in a weakly positive Mantoux reaction due to the cross-reactivity with the tubercle bacillus protein. Protective response to BCG vaccination could be influenced by the sensitisation with environmental mycobacteria. This might be one of the reasons for a wide variation in protective responses of the BCG vaccine in field trials in different parts of the world. It has been observed that NTM diseases are more frequent in areas where tuberculosis is rare. Conversely, opportunistic mycobacterial diseases are rare in the area endemic for tuberculosis. Opportunistic mycobacterial species such as *M. avium, M. kansasii* and *M. xenopi* may colonise tap water. Use of this contaminated water during staining for mycobacteria may give false-positive results.

Case Study

A young man was admitted to the Medicine ward with a history of severe cough and fever for 15 days. Detailed history revealed that he was a lorry driver and was exposed to multiple sexual partners. His serum was positive for HIV antibodies. ZN-stained smear from early morning sputum sample was positive for acid-fast bacilli. Chest X-ray revealed diffuse haziness. Sputum culture on LJ medium showed growth within 3 days. The colonies were non-pigmented in the presence or absence of light.

- What is the most probable diagnosis?
- What are the predisposing factors in the causation of diseases due to atypical mycobacteria?
- How will you diagnose this condition in the laboratory?
- What are the antibiotics used for the treatment of the condition?

32 Actinomycetes

Actinomycetes are Gram-positive, catalase-positive, non-motile bacilli. They are considered transitional forms between bacteria and fungi. They have a cell wall with muraminic acid similar to bacteria, possess prokaryotic nuclei and are sensitive to various antibiotics. They also produce delicate filament-like hyphae, similar to fungal hyphal forms. These hyphal forms are seen in bacteria isolated by culture and are also seen in clinical specimens.

The actinomycetes include a wide range of bacteria found in soil and putrefied vegetables. They are also found in humans and animals. The actinomycetes include anaerobic or facultative anaerobic bacteria belonging to the genus *Actinomyces*. Furthermore, aerobic actinomycetes can be broadly classified into two groups depending on the presence or absence of mycolic acids in the cell wall.

1. **Actinomycetes with mycolic acids:** This group includes members of three families: Corynebacteriaceae, Mycobacteriaceae and Nocardiaceae. The family Nocardiaceae consists of four genera: *Nocardia*, *Rhodococcus*, *Tsukamurella* and *Gordonia*. Members of all these genera stain poorly with Gram stain and are partially acid-fast.
2. **Actinomycetes without mycolic acid:** This group includes many opportunistic pathogens, such as *Actinomadura*, *Nocardiopsis*, *Streptomyces*, *Dermatophilus*, *Oerskovia*, *Rothia*, *Tropheryma* and thermophilic actinomycetes—*Saccharopolyspora*, *Saccharomonospora* and *Thermoactinomyces*.

Human infections caused by actinomycetes are summarised in Table 32.1.

Actinomyces

In 1877, a mould-like organism in the lesion of "lumpy jaw" (actinomycosis) in cattle was found by Bollinger. Harz coined the name *Actinomyces*, which refers to the **ray-like appearance** of the organism in the granules that is typical of the lesions. *Actinomyces* refers to ray fungus. In 1891, an anaerobic bacillus was isolated from human lesions, which produced experimental infection in guinea pigs and rabbits by Wolff and Israel. This was known as *Actinomyces israelii*.

A. israelii is the most common *Actinomyces* causing human infection. Other species are *Actinomyces gerencsonei*, *Actinomyces turicensis*, *Actinomyces radingae*, *Actinomyces europaeus*, *Actinomyces naeslundii*, *Actinomyces odontolyticus*, *Actinomyces viscosus*, *Actinomyces meyeri* and *Propionibacterium propionicum*.

Table 32.1 Human infections caused by Actinomycetes

Bacteria	Diseases
Actinomyces species	Cervicofacial actinomycosis, thoracic actinomycosis and actinomycosis of the abdomen and pelvis
Nocardia species	Primary cutaneous nocardiosis, bronchopulmonary infection and secondary CNS infection
Rhodococcus species	Pulmonary infections and opportunistic infections (traumatic endophthalmitis, peritonitis in patients undergoing dialysis and post-traumatic skin infection)
Gordonia species	Opportunistic infections
Tsukamurella species	Opportunistic infections
Tropheryma whippeli	Whipple disease
Dermatophilus species	Exudative dermatitis
Oerskovia species	Opportunistic infections (catheter-associated bacteraemia, traumatic endophthalmitis, CNS shunt infections etc.)

Properties of the Bacteria

Morphology

Actinomyces are Gram-positive, non-motile, non-sporing and weakly acid-fast in Kinyoun acid-fast staining. They measure 0.5–1 μm in diameter. They often grow in filaments that separate into *bacillary* and *coccoid* filaments.

Culture

Actinomyces are facultative anaerobes. They grow better under anaerobic or microaerophilic conditions at an optimum temperature of 35–37 °C. The presence of 5–10% CO_2 facilitates growth. *Actinomyces* species grow slowly; they need a longer incubation period of 3–4 days. *A. israelii* may require even 7–14 days for growth.

Brain–heart infusion agar or **heart infusion agar** supplemented with 5% defibrinated rabbit, sheep or horse blood is the enriched medium used frequently for *Actinomyces*. On this medium, in anaerobic to microaerophilic conditions, the bacteria form colonies with a characteristic molar-tooth appearance.

Heart infusion blood and **thioglycolate blood** supplemented with 0.1–0.2% sterile rabbit serum are examples of **liquid media** used to culture *Actinomyces* species. In thioglycolate broth, the production of general turbidity indicates *Actinomyces bovis*, whereas the deposition of fluffy balls at the bottom of the tube indicates *A. israelii*.

Pathogenesis

Actinomyces species are present as normal flora of the oral cavity and in the lower gastrointestinal tract and female genital tract of human hosts. *Actinomyces* by themselves are not virulent. They require the presence of devitalised or dead tissue and a break in the continuity of the mucosal membranes to cause infection. *Actinomyces* always need companion bacteria to initiate infection. These accompanying bacteria help establish infection by releasing toxins or enzymes or inhibiting host immunity.

These companion bacteria are believed to act as co-pathogens that enhance the virulence of *Actinomyces*. These companion bacteria include *Bifidobacterium dentium*, *Actinobacillus actinomycetemcomitans*, *Eikenella corrodens*, *Haemophilus aphrophilus*, *Bacteroides*, *Fusobacterium*, staphylococci and anaerobic streptococci.

Once Actinomyces establish the infection, the immune system of the infected human host stimulates an intense inflammatory response in the form of a suppurative granulomatous and fibrotic reaction. Infection by *Actinomyces* typically spreads contiguously and invades surrounding tissues and organs. Finally, the infection results in the production of draining sinus tracts which contain much-damaged tissue. Bacteria from this site may disseminate through blood circulation to distant organs.

Clinical Syndromes

Actinomyces causes actinomycosis.

Actinomycosis

Actinomycosis is a sub-acute and chronic bacterial infection. It is characterised by contiguous spread and suppurative and granulomatous inflammation. The condition is associated with the formation of multiple abscesses and the development of sinus tracts discharging white to yellowish granules, known as sulphur granules (Fig. 32.1). Actinomycosis may manifest as (1) cervicofacial actinomycosis, (2) thoracic actinomycosis and (3) actinomycosis of the abdomen and pelvis.

Cervicofacial Actinomycosis

It is the most common manifestation in humans, comprising about two-thirds of reported cases. The infection occurs in the cervicofacial region, typically following oral surgery in patients with poor oral hygiene. Initially, the condition manifests as a swelling of the soft tissue of the perimandibular area. Subsequently, during the course of the infection, the disease spreads directly into the adjacent tissues and leads to the formation of **fistulas**. These fistulas or sinus tracts discharge purulent material containing yellow granules, known as sulphur granules. If left untreated, this condition may spread to the blood and eventually to the brain and the orbit.

Thoracic Actinomycosis

It is responsible for 15–20% of cases of actinomycosis. It is caused by aspiration of oropharyngeal secretions containing *Actinomyces* and occasionally during perforation of the oesophagus. The condition also occurs by direct spread from an actinomycotic lesion of the nape of the neck or the abdomen or through blood circulation from other distant sites. The condition commonly presents as a pulmonary infiltrate or mass involving the lung. This condition, without specific treatment, can rapidly spread superficially through the pleural cavity, pericardium and chest wall. Finally, the condition results in the formation of multiple sinuses, which discharge sulphur granules.

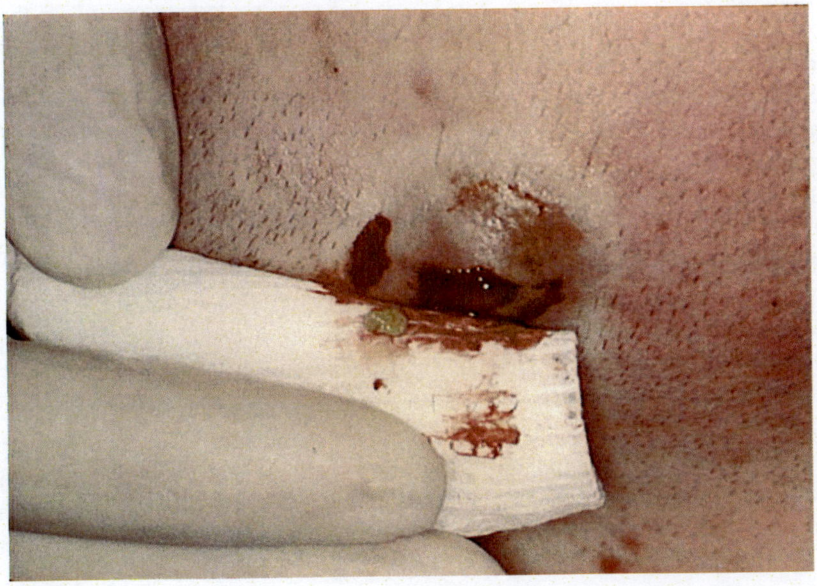

Fig. 32.1 Sulphur granules of actinomycosis. (http://phil.cdc.gov/phil_images/20030129/9/PHIL_3041_lores.jpg)

Actinomycosis of the Abdomen and Pelvis

The condition accounts for nearly 10–20% of reported cases. The ileocaecal region is the most common site involved in the condition. The condition typically presents as a slowly growing tumour. The infection subsequently spreads and involves abdominal organs, including the abdominal wall, leading to the formation of draining sinuses. Actinomycosis of the pelvis is commonly associated with prolonged (for many years) use of intrauterine contraceptive devices. The infection spreads directly from the uterus to the pelvis.

Laboratory Diagnosis

Laboratory diagnosis is made by directly detecting *Actinomyces* in specimens by microscopy and isolating the organism by culture.

Specimens

The specimens include sputum, bronchial secretions and discharges and infected tissues. All these specimens may contain a large number of sulphur granules. The sulphur granules are also present on the dressings removed from a draining sinus tract. It is essential to transport these specimens immediately to the laboratory for processing, preferably under anaerobic conditions.

Microscopy

Sulphur granules are white to yellow and vary in size from minute specs to large granules. These granules are separated from pus and other specimens, and are collected directly from draining sinuses. These are crushed between two slides and are stained by the Gram or Ziehl–Neelsen staining method, using 1% sulphuric acid for decolourisation. The stained smears show Gram-positive hyphal fragments, surrounded by a peripheral zone of swollen, radiating, club-shaped structures presenting a sun-ray appearance. These club-shaped structures are Gram-positive and acid-fast, and are believed to be antigen complexes.

Culture

Sulphur granules or pus-containing *Actinomyces* are immediately cultured under anaerobic conditions at 35–37 °C for up to 14 days. The specimens are inoculated on blood agar, BHI agar and into thioglycollate broth and incubated anaerobically at 37 °C. After 10 days of anaerobic incubation, *A. israelii* produces large (2–5 mm in diameter), white, smooth, entire or lobulated colonies resembling a molar tooth.

Actinomycetes colonies are identified by microscopy, biochemical reactions, direct fluorescent antibody test and gel immunodiffusion test. The direct fluorescent antibody test and gel immunodiffusion test are very useful for differentiating *A. israelii* from other actinomycotic species and filamentous anaerobes that produce similar granules in the tissues.

Molecular Detection

DNA probes and polymerase chain reaction (PCR) have been evaluated and used with high sensitivity and specificity to identify *Actinomyces* species in clinical specimens accurately.

Treatment

High-dose penicillin or tetracyclines given over a prolonged period are the mainstay of therapy for actinomycosis. Methicillin, oxacillin and cloxacillin are not effective against the *Actinomyces* species. Surgical therapy is included for more extensive and complicated cases of actinomycosis.

Epidemiology

Actinomycosis is distributed worldwide. The condition is more common in rural areas and farm workers. More commonly, the condition is seen in individuals with poor dental hygiene and low socioeconomic conditions. Men are affected more commonly than women (the male-to-female ratio is 4:3) except for pelvic actinomycosis. The majority of the cases are reported in young and middle-aged patients.

Prevention and Control

Good dental hygiene and oral hygiene are important for preventing the disease.

Nocardia

Nocardia are aerobic actinomycetes that are ubiquitous in the soil, decaying organic matter and water. The genus *Nocardia* has been classified into ten species based on phylogenetic analysis of 16S rRNA. These species can be differentiated based on biochemical reactions. *Nocardia asteroides*, *Nocardia brasiliensis*, *Nocardia aotitidiscaviarum* and *Nocardia farcinica* are frequently associated with human infections. The *Nocardia* species also cause infection in animals, including bovine mastitis in cattle and nocardiosis in horses.

Nocardia infection was first described by French veterinarian Edmund Nocard in a pulmonary and cutaneous disease of cattle. The genus *Nocardia* is named so in honour of the scientist.

Properties of the Bacteria

Morphology

Nocardia are Gram-positive bacilli showing a branching, beaded and filamentous form. They stain poorly with Gram stain and appear to be Gram-negative with intracellular Gram-positive granules. *Nocardia*, like mycobacteria, contain mycolic acid in their cell wall. They are usually weakly acid-fast. Acid-fastness is a property of *Nocardia* by which they can be differentiated from other similar bacteria, such as actinomycetes.

Culture and Biochemical Properties

Nocardia are strictly aerobic bacteria. They readily grow on commonly used routine media, such as nutrient agar, Sabouraud dextrose agar, BHI agar and yeast malt extract agar. The presence of 5–10% of CO_2 facilitates the growth of the bacteria. **Buffered charcoal yeast extract (BCYE) agar** is a selective medium used to isolate *Nocardia* from specimens potentially contaminated with other bacteria.

Nocardia are slow-growing bacteria and need prolonged incubation of 7 days or more for their culture and isolation in the media. *Nocardia* on these media produce colonies varying from dry to waxy and white to orange.

Nocardia are catalase positive, and they use sugars oxidatively. They produce acid from glucose, and they use urea. Biochemical reactions of different *Nocardia* species are summarised in Table 32.2.

Table 32.2 Biochemical characteristics of *Nocardia asteroides* complex and *Nocardia brasiliensis*

Tests	*Nocardia asteroides* complex	*Nocardia brasiliensis*
Decomposition of		
Urea	Positive	Positive
Gelatin	Negative	Positive
Casein	Negative	Positive
Tyrosine	Negative	Positive
Production of acid from		
Glucose	Positive	Positive
Rhamnose	Variable	Negative

Pathogenesis

Nocardia species are opportunistic pathogens. They cause suppurative necrosis with the formation of abscesses at the site of infection.

Clinical Syndromes

N. asteroides is the most important species causing approximately 90% of human infections. Clinical manifestations of disease caused by *Nocardia* depend on the site of infection.

Primary Cutaneous Nocardiosis

Primary cutaneous nocardiosis may manifest as cutaneous infection and lymphocutaneous infection.

Cutaneous infection may occur through wound contamination or trauma. Usually, feet and hands are involved. This is a painless chronic skin infection characterised by subcutaneous swelling, suppuration and multiple sinuses. This is an invasive and destructive infection in which underlying bone and connective tissues are destroyed. Multiple sinuses open to the skin's surface with exudation or pus as the infection progresses. The pus typically contains sulphur granules which consist of filamentous *Nocardia* bound together by calcium phosphate.

Lymphocutaneous infection is characterised by the presence of cutaneous nodules and ulcers associated with lymphadenopathy. The lymphadenopathy also may occasionally drain purulent pus.

Bronchopulmonary Infection

Bronchopulmonary infection is one of the important clinical manifestations in most patients with nocardiosis. The presence of inflammatory endobronchial masses, localised or generalised abscesses are the important presentations. The condition is associated with the formation of cavities, abscesses in the lungs and pleural effusion. Cough with the production of sputum and fever are the common symptoms. Clinically, this condition cannot be distinguished from infections of other microbial aetiology.

Secondary Central Nervous System (CNS) Infection

The condition manifests as chronic meningitis or a manifestation of single or multiple abscesses in the brain. CNS nocardiosis presents as a slow, progressing mass lesion and is found in 20–40% of disseminated nocardial infections.

Laboratory Diagnosis

Laboratory diagnosis of nocardiosis depends on the demonstration of *Nocardia* by microscopy and isolation by culture.

Specimens

The nature of the clinical specimens depends on the clinical presentation of nocardiosis. **Sputum is a frequently used specimen.** Other specimens include respiratory secretions, skin biopsies or pus from the abscesses.

Microscopy

Diagnosis is made by demonstration of branching actinomycotic filaments in pus or multiple sputum specimens by microscopy. Smears from the specimen are stained by the Gram or Ziehl–Neelsen method.

The granules of *N. brasiliensis*, *N. asteroides* and *Nocardia caviae* are soft and white, and measure 0.5–1 mm in size. These granules are very useful for diagnosis. The granules are collected from pus specimens with an inoculating needle and then washed with saline. The granules then are crushed between two slides, are stained with Giemsa or Ziehl–Neelsen method for the presence of typical filamentous hyphae (Fig. 32.2).

Culture and Identification

Pus, sputum and granules are cultured on nutrient agar, BHI agar and Sabouraud agar and are incubated at 36 °C for up to 3 weeks. Colonies on these media may appear as early as 7 days. The colonies are identified based on Gram-positive and weekly acid-fast positive filamentous bacilli and other biochemical properties. All the species of the genus *Nocardia* (*N. brasiliensis*, *N. otitidiscaviarum*, *N. asteroides* complex including *N. asteroides*, *N. farcinica* and *Nocardia nova*) are identified by using a battery of biochemical tests.

Treatment

Sulphonamides are the drug of choice for the treatment of nocardiosis. Sulphadiazine is useful for the treatment of CNS infections. Trimethoprim and sulphamethoxazole are considered better than sulphadiazine. *Nocardia* are also

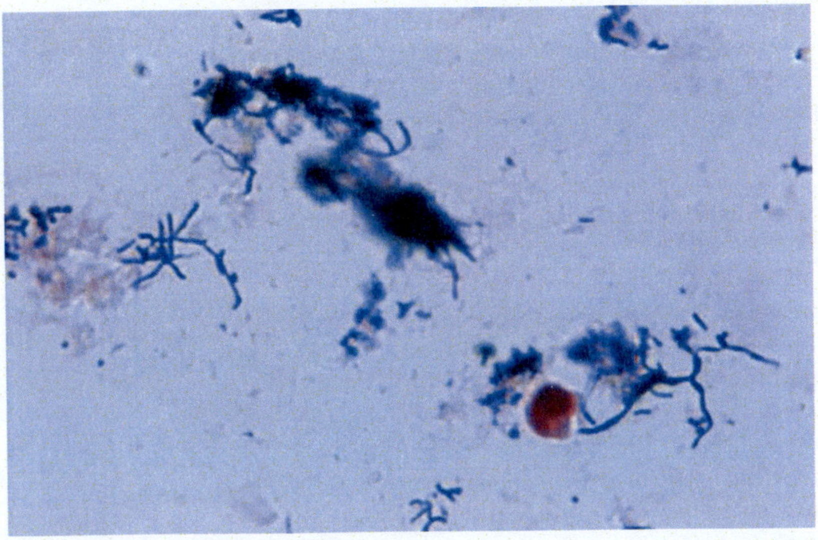

Fig. 32.2 Gram-stained pus smear sample, had been extracted from a lumpy-jaw lesion showing Grampositive, rod-shaped, darkly staining, filamentous, Actinomyces (x1000) (https://phil.cdc.gov/Details.aspx?pid21035)

susceptible to carbapenems, such as imipenem, cilastatin, cefotaxime, ceftriaxone and amikacin.

Epidemiology

Nocardial infections are reported worldwide, but no reliable estimates are available.

Habitat

Nocardia are distributed in soil rich in organic matter. They are present as transient pathogens in humans.

Reservoir, Source and Transmission of Infection

Nocardia spp., which colonise healthy humans, usually do not cause infection in the same host. They are exogenous pathogens. The infection occurs when infective aerosols are inhaled or when they penetrate through the skin. Pulmonary and disseminated nocardiosis increasingly occur in immunocompromised individuals. The infections are much more common in patients with HIV or immunocompromised patients who received an organ, bone marrow or stem-cell transplantation, and in patients with cirrhosis, SLE and lymphoreticular malignancies. The condition is also more common in patients with chronic pulmonary disease, such as emphysema, bronchitis, bronchiectasis.

Prevention and Control

Prophylactic antibiotic therapy with trimethoprim and sulphamethoxazole is of some value in preventing the disease.

Rhodococcus

Rhodococcus species are Gram-positive and strict aerobic actinomycetes. The cell wall of *Rhodococcus* organisms is similar to that of mycobacterium; it contains mycolic acid and tuberculostearic acid, which makes the bacteria acid-fast. The genus contains 20 species, out of which at least seven species are known to cause disease in humans.

Rhodococcus equi, formerly known as *Corynebacterium equi*, is the most important human pathogen. *R. equi* is a pleomorphic Gram-positive coccobacillus showing some degrees of branching. It is weakly acid-fast. *Rhodococcus* grows well on routinely used media aerobically. On blood agar, on prolonged incubation at room temperature, the bacteria produce colonies resembling those of *Klebsiella*, with the production of pink pigments. The bacteria do not ferment carbohydrates and show a variable biochemical reaction. *R. equi* is an intracellular pathogen that multiplies in macrophages. This causes a granulomatous reaction, resulting in the formation of an abscess. The bacteria cause invasive pulmonary diseases, such as pneumonia and lung abscess. Subsequently, the infection may spread to distant body sites, causing meningitis, pericarditis and so on.

In immunocompetent hosts, rhodococci cause opportunistic infections, such as traumatic endophthalmitis, peritonitis in patients undergoing dialysis and post-traumatic skin infection. The diagnosis of the condition is made by demonstration of weakly acid-fast and Gram-positive pleomorphic bacilli in clinical specimens. Definitive diagnosis by culture is often difficult. Rhodococci are difficult bacteria to treat, particularly in immunocompromised patients. They are resistant to penicillins and cephalosporins. Treatment with vancomycin or a combination of erythromycin and rifampicin is effective.

Tropheryma whippeli

T. whippeli is an actinomycete known to cause **Whipple disease**. The bacterium is yet to be cultured. It is usually identified by molecular techniques. Whipple disease manifests as abdominal pain, diarrhoea, fever, lymphadenopathy, skin pigmentation and arthralgia. The condition is diagnosed by demonstrating the periodic acid–Schiff-positive inclusion bodies found inside macrophages of the lamina propria of the small intestine. Detection of the bacterial genome by PCR in intestinal tissue confirms the diagnosis.

Dermatophilus

Dermatophilus is an actinomycete that causes exudative dermatitis affecting the skin of the hands and feet. The condition is commonly seen in individuals in close contact with infected animals or contaminated animal products. These pathogens are susceptible to commonly used antibiotics, such as penicillins and aminoglycosides. *Dermatophilus* is commonly found in the soil.

Actinomycotic mycetoma

Actinomyces (*A. israelii* and *A. bovis*), *Nocardia* (*N. asteroids*, *N. brasiliensis* and *N. caviae*), *Actiomadura* (*Actiomadura madurae* and *Actiomadura pelletierii*) and *Streptomyces someliensis* are different actinomycetes known to cause bacterial mycetomas. *Botryomycosis* or mycetoma-like lesions may be caused by *Staphylococcus aureus* and a few other pyogenic bacteria.

Actinomycetes and *Hypersensitivity Pneumonitis*

A few thermophilic actinomycetes like the *Faenia* and *Saccharomonospora* species that are found in mouldy hay produce spores that may induce allergic alveolitis when inhaled. This leads to chronic obstructive pulmonary disease (COPD) (Farmer's lung).

Case Study

A 55-year-old man attended the dental OPD with a swelling of the soft tissue of the perimandibular area with sinus tracts discharging purulent material. The material containing yellow granules was sent to the microbiology laboratory for identification. Microscopy of smears made from granules showed Gram-positive hyphal fragments surrounded by a peripheral zone of swollen, radiating, club-shaped structures presenting a sun-ray appearance.

- Identify the bacteria.
- Describe the culture methods for isolation of the bacteria.
- Describe other infections caused by the bacteria.
- Add a note on the virulence of the bacteria.

33 *Clostridium*

The genus *Clostridium* consists of Gram-positive, anaerobic bacilli capable of forming endospores. The endospores typically are wider than the bodies of the bacilli, giving the bacteria a swollen appearance resembling a spindle, hence the name *Clostridium* (*kloster* means spindle).

Genus *Clostridium*

Most *Clostridium* species are harmless saprophytes in the soil, water, sewage and decomposed (animal and plant) products. They are also present as part of the normal flora in the gastrointestinal tract of humans and animals. Only a few species cause infections in humans (Table 33.1).

Multiple factors are responsible for the anaerobic infections, including those caused by the *Clostridium* species. Anaerobic infections are more common in diabetes, malnutrition and long-term use of ineffective antibiotics such as aminoglycosides to treat anaerobic infections; these factors act as predisposing factors. A pre-existing infection with aerobes and facultative anaerobes may also create a more favourable environment for the growth of anaerobic bacteria by lowering the oxygen level at the site of infection. Normally, certain underlying conditions such as trauma, tissue necrosis, inadequate blood supply and haematoma formation result in decreased oxidation and reduction potential of the tissue, which facilitates the development of anaerobic infections. Due to the anaerobic environment, there will be impairment in the body's immune mechanisms, favouring the progression of anaerobic infection.

Tetanus, gas gangrene and botulism are three major clinical syndromes caused by the *Clostridium* species. A classification of clostridia based on morphological and biochemical classification characteristics is provided in Table 33.2.

Classification

Phenotypic classification of clostridia is based on morphology, culture and biochemical properties. Molecular **classification** is based on 16S rRNA gene sequences. A total of 19 clusters of clostridia have been identified. The type species is *Clostridium butyricum*, and the genus *Clostridium* is a mixed bag of all Gram-positive, spore-forming, anaerobic organisms. As a result, there are many phylogenetically and phenotypically diverse organisms in this genus. It consists of Gram-positive, large bacilli. The important characteristics include the following:

Endospore formation; fermentative metabolism, inability to grow in aerobic condition (a few are aero-tolerant); long survival of spores, mostly saprophytic, and they produce various extracellular enzymes active against proteins, lipids and sugars.

Table 33.1 Human infections caused by *Clostridium* species

Bacteria	Diseases
Clostridium perfringens	Soft tissue infections (cellulitis, fasciitis, suppurative myositis, clostridial myonecrosis and gas gangrene), food poisoning, necrotising enteritis and septicaemia
Clostridium novyi	Gas gangrene (rare)
Clostridium sordelli	Gas gangrene (rare)
Clostridium septicum	Gas gangrene
Clostridium tetani	Tetanus
Clostridium botulinum	Botulism
Clostridium baratii	Botulism (rare)
Clostridium butyricum	Botulism (rare)
Clostridium histolyticum	Botulism (rare)
Clostridium difficile	Antibiotics-associated diarrhoea
Clostridium tertium	Opportunistic infections

Table 33.2 Classification of clostridia based on morphological and biochemical characteristics

	Spore position		Biochemical feature	
Clostridium species	Central/sub terminal	Terminal	Proteolytic	Saccharolytic
Clostridium sordelli, Clostridium histolyticum, Clostridium botulinum A, B, F, *Clostridium sporogenes, Clostridium bifermentans*	+	–	+	–
Clostridium perfringens, Clostridium septicum, Clostridium chauvoei, Clostridium novyi	+	–	–	+
Clostridium difficile	–	+	–	+
Clostridium tertium	–	+	–	+
Clostridium tetanomorphum, Clostridium sphenoides	Spherical and terminal		–	+
Clostridium tetani	Spherical and terminal		–	+
Clostridium cochlearum	–	+	–	–

General Properties of Clostridia

Most species are anaerobic, but a few species, such as *Clostridium tertium* and *Clostridium histolyticum*, are aerotolerant and hence can grow on agar even in the presence of air.

Clostridia are rod-shaped bacilli measuring 3–8 × 0.4–1.2 μm in size. The bacilli are highly pleomorphic and show long filaments and involution forms. They are Gram-positive but old cultures may appear Gram-variable and even Gram-negative. The *Clostridium* species with a few exceptions (e.g. *Clostridium perfringens, Clostridium tetani* type VI) are motile due to the presence of peritrichous flagella. Most clostridia except *C. perfringens* and *C. butyricum* are non-capsulated.

Spores

The ability to produce endospores is the most important feature of clostridia. Sporulation takes place outside in the environment and also in animal bodies, but not in humans. The formation of spores is variable depending on the species.

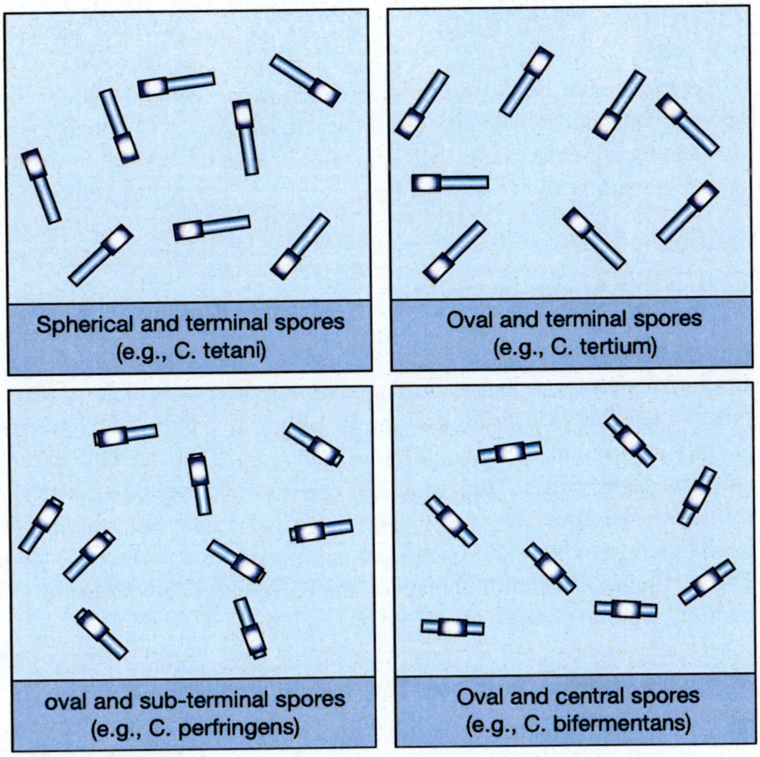

Fig. 33.1 Schematic diagram showing different types and arrangement of clostridium spores

Clostridium sporogenes and other clostridia sporulate readily, while *C. perfringens* and other species produce spores inconsistently. The shape and location of the spores (Fig. 33.1) in the bacteria vary across different clostridial species.

Central spores are present *in Clostridium bifermentans*, giving the bacillus a spindle shape. **Sub-terminal spores** are present in *C. perfringens*, giving the bacillus a club shape. **Oval terminal spores** are present in *C. tertium*, giving the bacillus a tennis racket shape, and **spherical terminal spores** in *C. tetani*, giving the bacillus a drumstick appearance.

Spores are relatively more resistant forms than the vegetative forms of the bacilli. They show a variable degree of resistance to heat, drying and disinfectants. Spores are killed by a 1% aqueous solution of iodine and 2% glutaraldehyde at pH 7.5–8.5. They are particularly resistant to phenolic disinfectants. The spores survive in 2% formaldehyde solution for up to 5 days.

Culture and Biochemical Tests

Most species are anaerobic, but a few species, such as *C. tertium* and *C. histolyticum*, are aerotolerant and hence can grow on agar even in the presence of air. Most clostridia grow at an optimum temperature of 37 °C and pH of 7–7.4. Some clostridial species produce haemolysis on blood agar. Robertson cooked meat (RCM) broth is a useful medium for the growth of clostridia. The medium contains unsaturated fatty acids, digested meat and sulfhydryl compounds. Clostridia grow well in the medium, rendering the broth turbid. Saccharolytic clostridia turn the meat pink, while proteolytic species turn the meat black and produce a foul smell. Most clostridia produce gas.

The bacteria are sensitive to metronidazole, penicillin and cephalosporins. They are less sensitive to tetracyclines and resistant to quinolones and aminoglycosides.

Clostridium perfringens

C. perfringens is the most important *Clostridium* species causing gas gangrene, a severe life-threatening disease. The bacteria also cause necrotic enteritis and food poisoning.

Properties of the Bacteria

Morphology

C. perfringens is a large, rectangular, Gram-positive bacillus measuring 4–6 μm in length. It is a straight bacillus with parallel sides, and round and truncated ends. They possess central or sub-terminal spores. The bacilli occur as single, in chains or bundles. The bacilli are capsulated. The bacilli are non-motile; however, the bacteria multiply rapidly, giving a characteristic spreading colony appearance on the media, resembling the growth of motile clostridia. The bacilli are capsulated.

Culture and Biochemical Properties

C. perfringens is an anaerobic but aerotolerant bacterium. The bacteria can grow under microaerophilic conditions and do not die on exposure to air. They grow at a temperature range of 20–44 °C (optimum temperature 37 °C) and a pH range of 5.5–8.0.

C. perfringens produces good growth in **RCM broth**. The meat present in the broth is not digested but is turned pink. It produces an acidic reaction and a sour odour in the culture. Some strains of *C. perfringens* can grow optimally at 45 °C with generation time reduced to 10 min. This property is used to isolate *C. perfringens* in the RCM medium when specimens are contaminated with other clostridial species.

Inoculating the specimens in RCM media followed by incubation at 45 °C for 4–6 min and subsequently cultured on **blood agar** produces pure and predominant colonies of *C. perfringens*. On blood agar containing human, sheep or rabbit blood, *C. perfringens* on prolonged incubation produces a dual zone of haemolysis. This is due to a narrow zone of complete haemolysis by theta-toxin and a much wider zone of incomplete haemolysis by the alpha-toxin of the bacteria.

C. perfringens is metabolically active. The bacteria ferment glucose, lactose, sucrose and maltose with acid and gas production. They produce H_2S and reduce nitrate to nitrite. They are MR positive, VP negative and indole negative.

They ferment lactose with the production of acid in litmus milk. The colour of the medium changes from blue to red. The production of acid coagulates casein of the milk, and the coagulated milk is broken down by producing a large volume of gas. Due to the gas release, the paraffin plug is pushed up with shreds of broken clots adhering to the sides of the glass tube. This reaction is known as **stormy fermentation**, characteristic of *C. perfringens.*

Susceptibility to Physical and Chemical Agents

C. perfringens spores are killed within 5 min by boiling, but spores of certain strains (food poisoning strain of type A, certain strains of type C) are killed by boiling for a longer period of 1–3 h. The spores are also killed by autoclaving at 121 °C for 15 min. *C. perfringens* spores are resistant to commonly used antiseptics and disinfectants. The bacilli are sensitive to metronidazole and penicillins.

Typing

Based on four major toxins (alpha-, beta-, epsilon- and iota-toxin) produced by *C. perfringens*, strains of *C. perfringens* are classified into five types (A, B, C, D and E). Typing of strains is carried out by *in vivo* tests in animals. This is done by neutralising the toxins with specific antitoxins via intravenous or percutaneous injections in mice.

Pathogenesis

C. perfringens produces more than 12 **toxins** (major toxins, minor toxins), enzymes and **biologically active soluble substances**. All these are

Table 33.3 Virulence factors of *Clostridium perfringens*

Virulence factors	Biological functions
α (Alpha)-toxin	Lethal, dermonecrotic, haemolytic and is a lecithinase; causes toxaemia, increases vascular permeability of blood vessels, leading to tissue destruction. Main cause of toxaemia associated with gas gangrene
β (Beta)-toxin	Lethal toxin; causes necrotic lesions in necrotising enteritis
ε (Epsilon)-toxin	Lethal protoxin; increases vascular permeability of the wall of the gastrointestinal tract
ι (Iota)-toxin	Lethal toxin; causes necrotic lesions and increases vascular permeability
δ (Delta)-toxin	Haemolytic
θ (Theta)-toxin	Oxygen-labile haemolytic and cytolytic toxin
κ (Kappa)-toxin	Collagenase
λ(Lambda)-toxin	Proteinase and gelatinase
μ (Mu)-toxin	Hyaluronidase
ν (Nu)-toxin	Deoxyribonuclease
Enterotoxin	Enterotoxic and haemolytic; alters permeability of the gastrointestinal membrane
Neuraminidase	Alters cell surface ganglioside receptors and promotes capillary permeability
Bursting factor	Causes typical muscle lesions in gas gangrene
Circulating factor	Increases adrenaline sensitivity of the capillary membrane
Hyaluronidase	Breaks down intercellular cement substance and promotes the spread of infection along tissue planes

the virulence factors that contribute to the pathogenesis of the diseases (Table 33.3).

Virulence Factors

Toxins

C. perfringens is classified into five types (A, B, C, D and E) depending on the toxins it produces.

Major Toxins

Alpha-, beta-, epsilon- and iota-toxin are the four major toxins produced by *C. perfringens*.

Alpha-toxin: It is the most important toxin produced by all strains of *C. perfringens*. The largest volumes of alpha-toxin are provided by *C. perfringens* type A strain. The alpha-toxin is a lecithinase, a phospholipase C, which breaks down lecithin into phosphorylcholine and diglyceride in the presence of calcium and magnesium ions. The toxin is responsible for toxaemia typically observed during gas gangrene. It is lethal, dermonecrotic and haemolytic. The alpha-toxin increases vascular permeability of blood vessels, thereby causing massive haemolysis and bleeding, tissue destruction and myocardial dysfunction. The toxin lyses erythrocytes, leukocytes, platelets and endothelial cells. The alpha-toxin causes haemolysis of red cells of most animal species except for goats and horses. The toxin is relatively heat stable and is partially inactivated by boiling for 5 min. The lysis of red cells is observed best on incubation at 37 °C followed by re-incubation at 4 °C (hot-cold lysis).

Beta-, epsilon- and iota-toxin: These are the other major toxins in addition to the alpha-toxin; these toxins also play a major role in the pathogenesis of the disease (Table 33.3).

Minor Toxins

These include **delta-, theta-, kappa-, lambda-, mu-** and **nu-toxins** that contribute to the disease's pathogenesis (Table 33.3).

Other Toxins

Enterotoxin: It is produced primarily by the type A strain of *C. perfringens*. This is a heat-labile protein. The toxin is produced during the sporulation stage of vegetative cells to form spores, which is stimulated by the alkaline environment of the small intestine. The toxin binds specifically to receptors in the epithelium of the small intestine. This disrupts the transport of ions in the ileum and jejunum, and alters the membrane's permeability. The toxin is antigenic, but the specific antibodies against this toxin found in the serum are not protective.

Enzymes and Biologically Active Soluble Substances

C. perfringens produces many enzymes and many biologically active soluble substances.

Neuraminidase is the most important enzyme. This alters cell surface ganglioside receptors and promotes capillary permeability. **The bursting factor** acts specifically on muscle tissue. It may be responsible for typical muscle lesions observed in gas gangrene. **The circulating factor** increases the adrenaline sensitivity of the capillary membrane. It also inhibits phagocytosis. Other enzymes include **haemagglutinin**, **fibrinolysins** and **haemolysins** (Table 33.3).

Pathogenesis of Gas Gangrene

Gas gangrene is a rapidly spreading oedematous myonecrotic life-threatening condition caused by *C. perfringens*. It occurs following road traffic accidents or other injuries (like during war) involving crushing trauma of large muscle mass contaminated with pathogenic clostridia. Rarely the condition may occur following surgical operations.

Gas gangrene is caused by the inoculation of *C. perfringens* type A into tissues during trauma or surgery, followed by replication of the bacteria and subsequent invasion of the tissues. Low oxygen tension is the most important condition, most frequently seen in war wounds with implanted bullets or shell fragments, along with soil particles and bits of clothing. This contributes to the multiplication of *C. perfringens* and other anaerobic bacteria at the wound site. The condition occurs in association with extensive muscle trauma, tearing of arteries and crushing of tissue causing anoxia of the muscles. The condition is commonly associated with anaerobic *streptococci and facultative anaerobes, such as staphylococci, Escherichia coli and Proteus* species.

Clinical Syndromes

C. perfringens produces a variety of clinical syndromes, including gas gangrene, food poisoning, necrotising enteritis and gangrenous appendicitis.

Gas Gangrene

Gas gangrene is a rapidly spreading oedematous myonecrotic life-threatening condition. It is also referred to as anaerobic myositis and clostridial myonecrosis. The gas gangrene was described as early as 1954 by Oakley in 1954, like a rapidly spreading oedematous myonecrosis. Gas gangrene is an invasive disease caused by the invasion of muscle masses by *C. perfrigens*. It occurs following road traffic accidents or other injuries (like during war) involving crushing trauma of large muscle mass contaminated with pathogenic clostridia. Rarely the condition may occur following surgical operations.

The incubation period is variable. It may be as short as 7 days or as long as 6 weeks after clostridia are introduced into the tissue by trauma or surgery. The onset of the disease is rapid and is associated with increasing pain, tenderness and oedema of the affected part and systemic signs of toxaemia. Accumulation of gas in the tissues—caused by the metabolic activity of the rapidly dividing clostridia—is characteristic of the disease, hence the name gas gangrene (Fig. 33.2). Production and accumulation of gas make the tissues crepitant. There is a thin watery discharge from the wound, and subsequently, the discharge becomes serosanguineous. In untreated cases, the condition progresses very fast, with extensive muscle necrosis, shock, renal failure and even death within 48 h of the onset of the disease. The clostridial toxins typically cause extensive haemolysis and bleeding, and finally, death occurs due to circulatory failure.

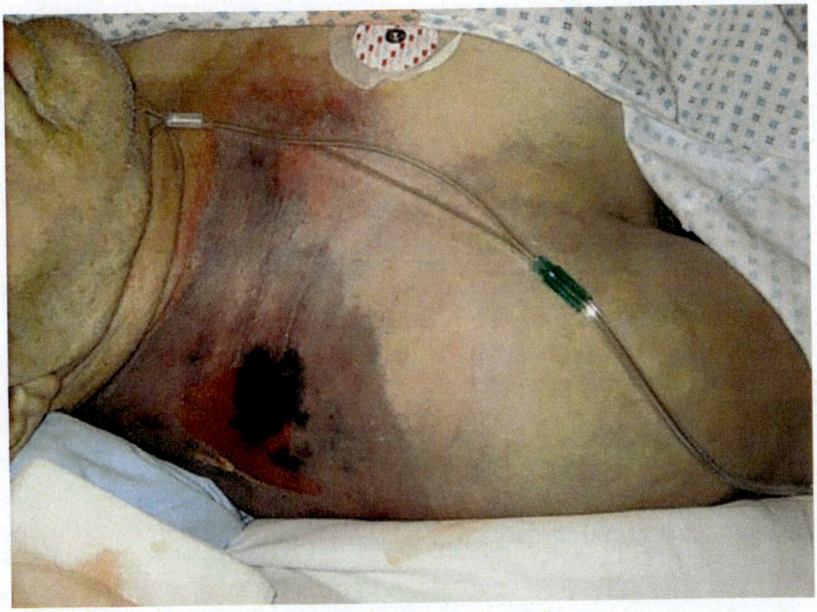

Fig. 33.2 *Clostridium perfringens* gas gangrene. (https://en.wikipedia.org/wiki/Gas_gangrene)

Food Poisoning

It is a **toxin-mediated disease** caused by a heat-labile enterotoxin of *C. perfrigens* type A. These strains produce spores, which are heat resistant. They typically produce an enterotoxin, but alpha- and theta-toxin production is very minimal.

Food poisoning is caused by the ingestion of cold and warmed up meat dishes contaminated with large numbers (10^8–10^9) of *C. perfringens* type A strains. The incubation period is short, between 8 and 24 h. Abdominal cramps and watery diarrhoea—but an absence of nausea, vomiting or fever—are characteristic presentations of this condition. The disease is self-limiting, and recovery occurs within 24–48 h.

Necrotising Enteritis

Necrotising enteritis caused by *C. perfringens* type C is an acute necrotising condition of the jejunum. The condition is characterised by abdominal pain, bloody diarrhoea, shock and peritonitis. It is severe and often fatal. Immunisation with type C toxoid has been reported to protect against this condition. This condition is known as **Pigbel** in Papua New Guinea, where it is very common, and as **Darmbrand** (meaning "fire bowels") in Germany. Sporadic cases have also been reported from East Africa, Thailand and Nepal.

Other Manifestations

These include gangrenous appendicitis, biliary tract infection, endogenous gas gangrene of intraabdominal origin, brain abscess and meningitis, panophthalmitis, thoracic infections, urogenital infections and so on. These also include septicaemia, a life-threatening situation with the isolation of bacteria in blood culture.

Laboratory Diagnosis

Laboratory diagnosis of gas gangrene and other soft tissue diseases is carried out to (1) confirm the diagnosis, (2) identify the clostridial species and (3) differentiate gas gangrene from anaerobic streptococcal myositis. Treatment has to be started immediately in suspected and clinically diagnosed cases without waiting for the laboratory results.

Specimens

The specimens for gas gangrene include (1) **pus and other exudates** from a deeper part of the wound and from the sides where infection appears to be most acute; the specimen is collected by a capillary tube and swab; (2) **smears** from the muscles at the edge of the affected site, from the pus and/or exudate from deeper parts of the wound and the necrotic tissue; and (3) **the necrotic tissue and pieces** of muscle fragments.

Microscopy

Gram staining of smears is a useful method in diagnosing *C. perfringens* infection. It provides information regarding the species of clostridia that causes the infection and also indicates a relative number of the clostridial species found in clinical specimens. Demonstration of a large number of Gram-positive bacilli without spores, and absence of leukocytes in stained smears of clinical specimens, is suggestive of *C. perfringens* infection. The presence of large Gram-positive bacilli with oval and sub-terminal spores suggests *Clostridium novyi*. Demonstration of leaf-shaped or boat-shaped pleomorphic bacilli with irregular staining suggests *Clostridium septicum*, and that of slender bacilli with oval and terminal spores suggests *C. tetani* or *Clostridium tetanomorphum*.

Culture and Identification

Specimens are inoculated simultaneously on **a set of two fresh and heated blood agar** containing 5–6% agar to prevent swarming. One inoculated medium is incubated aerobically, and the other inoculated medium is incubated anaerobically for 24 h or less. *C. perfringens* grows very rapidly; hence, colonies can be demonstrated on the media after a few hours of incubation.

A set of four tubes of RCM media are inoculated and heated at 100 °C for 5, 10, 15 and 20 min and incubated at 37 °C for 24–48 h. Subsequently, sub-cultures are made on blood agar to detect and differentiate various clostridial species. In cases of septicaemia, blood culture is often positive for *C. perfringens* and *C. septicum*.

C. perfringens grows rapidly, produces spreading colonies on blood agar and produces a double zone of haemolysis around colonies on blood agar. Gram-stained smear of the colony shows large rectangular Gram-positive bacilli. It is non-motile and capsulated bacteria. It ferments lactose with acid production in litmus milk (stormy fermentation positive) and Nagler reaction positive.

Nagler Reaction

It is a useful test for rapid detection of *C. perfringens* in clinical specimens (Fig. 33.3). This reaction demonstrates the biological property of the enzyme lecithinase—to produce opalescence in the serum and the egg yolk media. The use of specific antitoxin specifically neutralises this reaction.

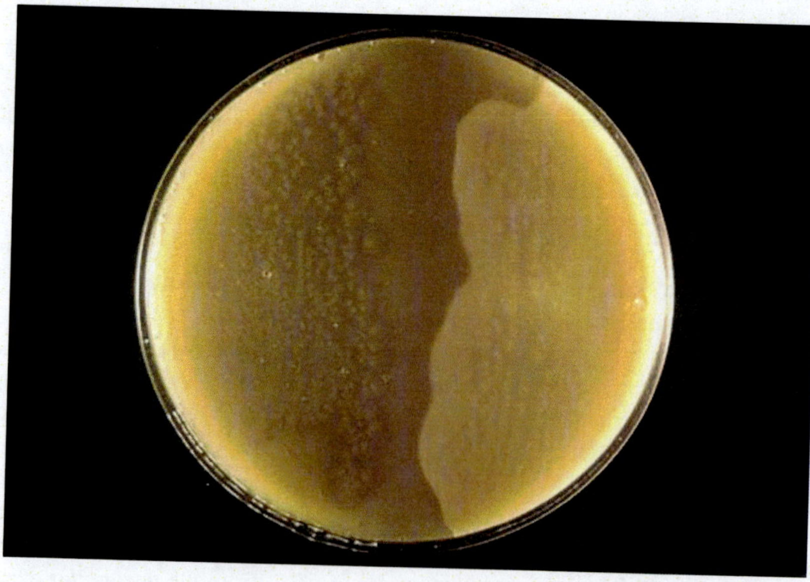

Fig. 33.3 Nagler reaction (https://phil.cdc.gov/Details.aspx?pid=3873)

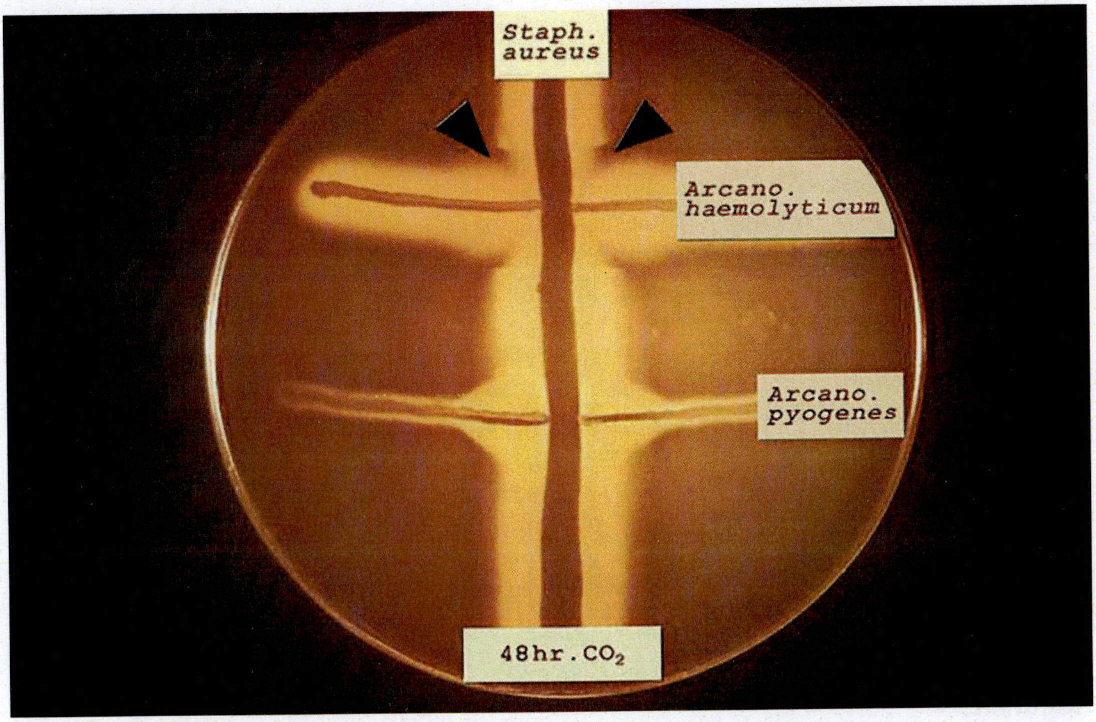

Fig. 33.4 Reverse CAMP test. (http://phil.cdc.gov/PHIL_Images/14644/14644_lores.jpg)

The test is performed by culturing *C. perfringens* on a medium containing 6% agar, 5% Fildes peptic digest of sheep blood and 20% human serum and antibiotic neomycin sulphate. One-half of the medium is inoculated with specific antitoxin raised against *C. perfringens*, while the other half of the medium does not contain any antitoxin. After 48 h of incubation, the colonies of *C. perfringens* on the part of the agar without antitoxin do not show any opacity. This is due to the neutralisation of the enzymatic activity of the enzyme lecithinase. The aerobic spore bearers and coliforms are inhibited by neomycin sulphate in the medium, making the medium more selective for *C. perfringens*.

C. novyi, *C. bifermentans*, some aerobic spore bearers and some *Vibrio* species produce the enzyme lecithinase; therefore, they also produce opalescence in the egg yolk medium. However, the reaction is not neutralised by the *C. perfringens* antitoxin but rather by the antitoxin of *C. bifermentans*, which produces a serologically related lecithinase.

Reverse CAMP Test

It is used to identify *C. perfringens*. In this test, on the blood agar plate, C. perfringens is streaked in a straight line at the centre of the plate, and near to this at the right angle, *Streptococcus agalactiae* is streaked. In the case of a positive reverse CAMP test, an arrow-shaped zone of enhanced haemolysis facing *C. perfringens* can be seen (Fig. 33.4).

C. perfringens ferments lactose with the production of acid in **litmus milk**. The colour of the medium changes from blue to red. The production of acid coagulates the casein of the milk, and the coagulated milk is broken down by producing a large volume of gas. Due to the gas release, the paraffin plug is pushed up with shreds of broken clots adhering to the sides of the glass tube. This reaction is known as **stormy fermentation**, characteristic of *C. perfringens*.

Laboratory Diagnosis of Food Poisoning

It is made by demonstrating more than 10^6 bacteria per gram of faeces collected within 48 h of the condition's onset. The specimens for investigation of food poisoning include faeces and remnants of food.

Serology

An enzyme immunoassay to detect enterotoxin in faeces has recently been developed to diagnose food poisoning. Commercial kits are available to detect enterotoxin in stool samples using the ELISA technique.

Molecular Detection

The *C. perfringens* enterotoxin is encoded by the CPE gene, which can be detected in the stool specimen by PCR. For epidemiological purposes and to pinpoint the source of infection, pulsed-field gel electrophoresis can be done to identify the presence of the same strain in the food specimen and patient sample.

Treatment

Debriding surgery is the mainstay of prophylaxis and treatment of gas gangrene. The cases are treated aggressively with prompt removal of damaged tissue and clearing the wounds to remove foreign materials, necrotic tissue and blood clots. **Hyperbaric oxygen** treatment has been suggested to be beneficial. Antiserum against alpha-toxin is no longer used.

Prophylactic use of the **antibiotic** in association with surgery is effective for gas gangrene. Metronidazole is the antibiotic of choice. The drug is administered intravenously before surgery and is given three times a day at an interval of 8 h. Antibiotic prophylaxis using broad-spectrum antibiotics, such as gentamicin, amoxicillin and metronidazole, is effective since the occurrence of mixed infections with aerobic and anaerobic bacteria is frequent.

Antibiotic therapy is not recommended for the treatment of *C. perfringens* food poisoning.

Epidemiology

Geographical Distribution

C. perfringens is distributed worldwide. *C. perfringens* type A is responsible for most human diseases, including food poisoning, soft tissue infections, gas gangrene and primary septicaemia. *C. perfringens* type C is responsible for enteritis necroticans.

Habitat

C. perfringens type A is part of the normal flora of the intestinal tract of humans and animals. These bacteria are excreted in the faeces and contaminate the perianal region's skin, buttocks and thighs. *C. perfringens* type A spores are ubiquitous. They are found in the soil, dust and air. They remain viable for a very long period, even in adverse environmental conditions. *C. perfringens* types B, C, D and F colonise the intestinal tracts of animals and occasionally humans. Their spores usually do not survive in soil.

Reservoir, Source and Transmission of Infection

Gas gangrene occurs following road traffic accidents or other injuries (like during war) involving crushing trauma of large muscle mass contaminated with pathogenic clostridia. Rarely the condition may occur following surgical operations. The condition is caused by contamination of a wound with soil, such as manure soil or cultivated soil, road dust or bits of clothing heavily contaminated with *C. perfringens* and other pathogenic clostridia (exogenous infection). *Clostridium* species on normal skin, especially on the perineum and thighs, invade through a wound and cause the infection (endogenous infection).

Clostridial food poisoning occurs following the ingestion of cold or warmed up meat dishes contaminated with spores of *C. perfringens* type A. Spores are responsible for the infection. Foods contaminated with a large number of bacteria are the sources of exogenous infection in case of necrotising enteritis. The conditions that facilitate food exposure to a large number of *C. perfringens* spores and malnutrition are the risk factors for the disease.

Prevention and Control

The use of antibiotics and wound hygiene are the key factors that help prevent *C. perfringens* infections. Wounds are cleaned and debrided, and penicillin may be given for prophylaxis. No vaccine is available against these diseases.

Clostridium histolyticum

This actively proteolytic *Clostridium* species occurs worldwide and is found commonly in soil, faeces or sewage. It is motile and forms oval, bulging, sub-terminal spores. *C. histolyticum* produces nearly five toxins, namely alpha, beta, gamma, delta and epsilon. Infrequently, in association with other bacterial species, it may cause gas gangrene.

C. histolyticum can be present in normal skin, on the perineum and on the thighs. It can enter the wounded site along with foreign particles like dust, soil, cloth bits or sharpeners. After some clean surgical procedures such as amputation, *C. histolyticum* infection may occur.

Clostridium septicum

C. septicum is ubiquitous. Thus far, six groups have been recognised based on somatic and flagellar antigens. Along with other clostridia, *C. septicum* can also be seen in gas gangrene of humans regularly in association with malignancy, pericarditis and mycotic aneurysm. It causes **braxy** (sheep) and **malignant oedema** (cattle and sheep) in animals.

C. septicum, previously known as *Vibrion septique*, is a pleomorphic Gram-positive motile bacillus. Pasteur and Joubert first described it in 1877. It measures about 3–8 × 0.6 μm in size, with oval, central or sub-terminal spores. The growth on ordinary media occurs anaerobically. On continued incubation, the colonies turn opaque, which are initially irregular and transparent. This bacterium is saccharolytic and produces abundant gas. On blood agar, beta-haemolysis is seen. It also ferments lactose, maltose and mannose. It is an etiological agent of spontaneous gas gangrene or atraumatic myonecrosis.

C. septicum produces at least four distinct toxins. The **alpha-toxin** is the most lethal and necrotising pore-forming cytolysin, similar to *Aeromonas hydrophilia*'s aerolysin, and the beta-toxin is a leucotoxic deoxyribonuclease, the **gamma-toxin** a hyaluronidase and a thiol-activated **delta-toxin** or **septicolysin**. Other enzymes, such as protease and neuraminidase, are also produced by this bacterium. *C. septicum* is found in the soil or in animal intestines. Along with other clostridia, *C. septicum* can also be seen in gas gangrene of humans regularly in association with malignancy, pericarditis and mycotic aneurysm. It causes **braxy** (sheep) and **malignant oedema** (cattle and sheep) in animals.

Clostridium novyi

C. novyi is a Gram-positive bacillus that is long, thick and pleomorphic. It is an obligate anaerobe. Spores are large, oval and sub-terminal. *C. novyi* naturally occurs in the soil. Based on toxin production, it is divided into four types (A, B, C and D), of which type A is medically important and causes gas gangrene in humans. The characteristic feature of gas gangrene is excess oedema fluid with or without observable gas formation in infected tissue. In 2000, an outbreak of *C. novyi* type A infection among heroin addicts was reported from Britain, in which *C. novyi* type A was found to be highly lethal.

Clostridium tetani

C. tetani, an obligate anaerobic Gram-positive bacillus, causes tetanus. Tetanus is an infectious disorder characterised by increased muscle tone and spasms. It is caused by the release of a neurotoxin, **tetanospasmin**, produced by *C. tetani* when it gets inoculated into humans.

Properties of the Bacteria

Morphology

C. tetani is a slender, Gram-positive bacillus measuring 4–8 μm in length. Young cultures are usually Gram-positive, but old cultures are Gram-variable and even Gram-negative.

It is a straight bacillus with parallel sides and rounded ends. The bacillus occurs in singles and occasionally in chains. The bacteria consist of round, terminal and bulging spores giving **a drumstick appearance** to the bacillus. The spores are rarely seen in clinical specimens from lesions or culture media.

All strains of *C. tetani* except type VI are motile by the presence of flagella. Type VI strain bacilli do not contain any flagella; hence they are non-motile. They are capsulated.

Culture and Biochemical Properties

C. tetani is an obligate anaerobe. The bacteria are extremely sensitive to oxygen, hence can grow only in the absence of oxygen. The bacillus grows at an optimum temperature of 37 °C and a pH of 7.4. *C. tetani* can grow on ordinary media and media enriched with serum and blood.

C. tetani grows well on **the RCM medium**. The bacteria produce turbidity with the production of some gas in the medium. Although it is not digested, the meat turns black on prolonged incubation.

On **blood agar**, *C. tetani* produce alpha-haemolysis surrounding the colonies. On prolonged incubation, the alpha-haemolysis becomes beta-haemolytic due to the production of tetanolysin, a haemolysin produced by the bacteria. Surface colonies tend to swarm over the entire surface of the agar. *C. tetani* produces an extremely fine translucent film of growth, which is difficult to visualise except at the edges of the colonies.

C. tetani liquefies gelatine anaerobically. Hence, the bacillus produces a *fir-tree type* of growth in **gelatine stab culture**.

Inoculation of the bacteria into the water of condensation at the bottom of the slope of **nutrient agar**, followed by anaerobic incubation for 24 h, yields a pure colony of *C. tetani* at the top of the slope in the tube. This method is known as the Fildes technique and is employed as a routine method for isolating pure colonies of *C. tetani*.

C. tetani does not ferment any sugars. It neither produces H_2S nor reduces nitrates. It is indole positive, but MR and VP negative. It produces a greenish fluorescence on MacConkey medium.

Susceptibility to the Physical and Chemical Agents

C. tetani spores of different strains show variable heat resistance. Most spores are killed on boiling at 100 °C for 10–15 min and autoclaving at 121 °C for 20 min. They are also killed by a 1% aqueous iodine solution and 10% hydrogen peroxide. The spores are resistant to most antiseptics. They are not killed by 5% phenol or 0.1% mercuric chloride solution. They can survive in the soil for years.

Typing

C. tetani are classified into **ten serological** types (types I to X) based on agglutination. All strains produce the same toxin.

Pathogenesis

C. tetani is a **non-invasive bacillus**. It causes disease only by producing toxins, which are the most important virulence factors (Table 33.4).

Table 33.4 Virulence factors of *Clostridium tetani*

Virulence factors	Biological functions
Tetanospasmin	Potent heat labile toxin; prevents the release of neurotransmitters (e.g. GABA, glycine), hence blocks specific synaptic inhibition in the spinal cord. Motor neurons are left under no inhibitory control and undergo sustained excitatory discharge
Tetanolysin	Heat-stable haemolysin; unknown role in pathogenesis of tetanus. Oxygen-labile haemolysin
Neurotoxin	Non-spasmogenic and peripherally active neurotoxin of unknown significance

Virulence Factors

Tetanolysin and **tetanospasmin** are two major toxins produced by *C. tetani*. These are pharmacologically and antigenically distinct. Non-spasmogenic toxin or **neurotoxin** is a recently identified toxin.

Tetanospasmin

It is the toxin responsible for the clinical manifestations of tetanus. The toxin is produced during the stationary phase of growth but is released only after the lysis of bacteria. It is a protein synthesised as a single polypeptide chain of molecular weight (MW) 151,000 Da. On release from the bacillus, the peptide is split by an endogenous protease into two chains: a light chain of 52,000 MW protein and a heavy chain of 93,000 MW protein. The two chains are joined by the non-covalent forces of a disulphide bond. The purified toxin is extremely potent. The minimum lethal dose (MLD) of toxin for animals is $50–75 \times 10^{-6}$ mg and for humans is 130 Ng. The toxin acts by preventing the release of neurotransmitters, such as gamma-aminobutyric acid (GABA) and glycine, thereby specifically blocking synaptic inhibition in the spinal cord. This leads to the unregulated spread of impulses, inhibited anywhere in the central nervous system (CNS). The binding of the toxin is irreversible.

The toxicity of tetanospasmin depends on the route of administration of the toxin. The route of administration also modifies the course of clinical manifestation of the disease. Intraneural injection and injection directly into the CNS are most lethal. Intravenous, intramuscular and subcutaneous injections are effective, but toxin given orally is ineffective because the toxin is destroyed by enzymes present in the gastrointestinal tract. Different species of animals show a wide variety in susceptibility to tetanospasmin. Birds and reptiles are highly resistant to the toxin. In descending order, horses are most susceptible to the toxin, followed by guinea pigs, goats and rabbits.

Tetanus toxoid is antigenic but non-toxic. Tetanus toxin is made into a toxoid by treating it with formaldehyde.

Tetanus in Experimental Animals

Experimental tetanus in animals such as mice depends on the administration route of toxins. Intramuscular injection of toxin in mice causes ascending tetanus. Intramuscular injection in one of the hind limbs is associated with the appearance of tonic spasm of the muscles in the injected limb first. This is due to toxins acting on the segment of the spinal cord. The toxins prevent the inhibition or remove the nerve impulse after crossing the synaptic junctions. The nerve continues to send impulses resulting in spasmodic contractions or tetani of the affected muscles. Muscle stiffness is the early symptom, with jaw muscles often developing symptoms first. This condition is called **lockjaw**; with the progression of the disease, spasm develops in other muscles. The spasm, although brief, occurs frequently and causes immense pain and exhaustion. This condition is called local tetanus. The subsequent spread of the toxin up the spinal cord causes ascending tetanus. The opposite hind limb, trunk and forelimbs are involved in this condition.

Intravenous injection of the toxin causes descending tetanus. In this condition, spasticity develops first in the head and neck muscles, and then spreads downwards, which resembles naturally occurring tetanus in humans.

Tetanolysin

It is oxygen- and heat-labile haemolysin. It is antigenically related to other clostridial haemolysins and streptolysin O. The toxin is of doubtful pathogenicity and appears not to play any role in the pathogenesis of tetanus.

Neurotoxin

It is the third toxin recently identified. It is a non-spasmogenic and peripherally active neurotoxin. The role of this toxin in the pathogenesis of tetanus is not understood.

Pathogenesis of Tetanus

Tetanus is a toxin-mediated disease. *C. tetani* spores are the infective form of the bacteria. Soil, animal faeces and, occasionally, human faeces and inanimate objects contaminated with spores are the primary sources of infection.

Tetanus is caused by the entry of the *C. tetani* spores. Under anaerobic conditions, the spores germinate to vegetative form and subsequently produce toxins, such as tetanospasmin and tetanolysin. The conditions favouring anaerobiosis in tissues include wounds with low oxidation-reduction potential, such as those with (1) dead devitalised tissue, (2) a foreign body, or (3) active infection.

On release from the bacillus, tetanospasmin is absorbed in the peripheral nervous system and at the myoneural junction. It is then transferred centripetally into the neurons of the CNS. An endogenous protease splits the toxins into two chains: a light chain of 52,000 MW protein and a heavy chain of 93,000 MW protein. The toxin's heavy chain (100 kDa) is responsible for specific binding to neural cells and protein transport. The light chain blocks the release of two major inhibitory neurotransmitters, such as GABA and glycine. This, in turn, stops the inhibition of motor reflex responses to sensory stimulation. This results in a sustained muscle contraction characteristic of a **tetanic spasm**. The peripheral nerves transmit toxin quickly to the CNS, which leads to early symptoms of facial distortion, back and neck stiffness and characteristic Trismus or **lockjaw**.

Clinical Syndromes

Tetanus caused by *C. tetani* is a disease found worldwide.

Tetanus

The incubation period of tetanus is variable, ranging from a few days to several weeks, but commonly 6–12 days. The duration of the incubation period depends on (1) distance of primary wound infection from the CNS, (2) the inoculating dosage of bacteria, (3) toxigenicity of bacteria, and (4) immune status of the host.

Trismus or **lockjaw** is the most common and early sign of the disease. Difficulty in swallowing, irritability and restlessness are the other early signs. As the condition progresses, patients have generalised muscle rigidity with intermittent reflex spasms in response to stimuli such as noise or touch. Tonic contractions cause **opisthotonus**, a condition in which the head, neck and spinal cord are arched backwards. During these periods, the patients will have severe pain. The spasms can cause acute respiratory failure, fractures and tendon ruptures. ***Risus sardonicus*** or characteristic **sardonic smile** is a typical feature due to the continued contraction of facial muscles. The condition may progress for 2 weeks even with the administration of the antitoxin because of the time needed for intraaxonal antitoxin transport.

Tetanus with a short incubation period is graver than tetanus with a long incubation period. Recovery is slow and usually takes as long as 2–4 months.

Neonatal Tetanus

Neonatal tetanus results from the infection of a neonate. The infection reported earlier occurs due to use of a contaminated blade, knife or other materials to cut or dress the umbilical cord in newborns, particularly of unimmunised mothers. The condition had a very poor prognosis. The condition has been reduced drastically due to the universal immunisation of infants and pregnant mothers.

Localised Tetanus

Localised tetanus is an unusual form of tetanus. The disease is confined to the extremity with a contaminated wound and occurs when only the nerves supplying the affected muscle are involved. The condition is characterised by rigidity of muscles, caused due to a dysfunction in the interneurons that inhibit the alpha-motor neurons of the affected muscles. In localised tetanus, there is no involvement of the CNS. The mortality rates are also very low.

Cephalic Tetanus

Cephalic tetanus is a variation of localised tetanus. The condition usually occurs following a head injury or an infection of the middle ear.

The incubation period is very short (1–2 days) and has a poor prognosis.

Laboratory Diagnosis

Laboratory diagnosis is carried out in the patients only to confirm the clinical diagnosis.

Specimens

The specimens include excised bits of tissue from the necrotic depths of wounds. Swabs from the wounds are not good specimens.

Microscopy

Gram staining of smears for *C. tetani* is frequently unsuccessful and also unreliable. Demonstration of typical drumstick bacilli in wound tissue (Fig. 33.5) is not diagnostic of tetanus. This is because *C. tetani* may be present in some wounds without causing tetanus. Furthermore, it may not be possible to distinguish by microscopy *C. tetani* from *C. tetanomorphum* and *Clostridium sphenoides*.

Culture and Identification

The specimens are inoculated on blood **agar** and incubated anaerobically for 24–48 h. *C. tetani* produces alpha-haemolysis surrounding the colonies. On prolonged incubation, the alpha-haemolysis becomes beta-haemolytic due to the production of tetanolysin, a haemolysin produced by the bacteria. Surface colonies tend to swarm over the entire surface of the agar. *C. tetani* produces an extremely fine translucent film of growth, which is difficult to visualise except at the edges of the colonies. *C. tetani* produces **swarming growth**, which spreads throughout the plate.

The specimens are also inoculated to **a set of three tubes of RCM media**. One of these is heated at 80 °C for 15 min, the second for 5 min and the third is left unheated, and all three are incubated at 37 °C for 24–48 h. The bacteria produce turbidity with the production of some gas in the medium. Although it is not digested, the meat turns black on prolonged incubation. Subsequently, the sub-culture is made daily on blood agar for up to 4 days. From the swarming edges of the colonies on blood agar, *C. tetani* is isolated in pure culture.

Culture is positive in only 30% of cases of tetanus. This is because tetanus is caused only by a few organisms, and many organisms are killed when exposed to air during the processing of specimens.

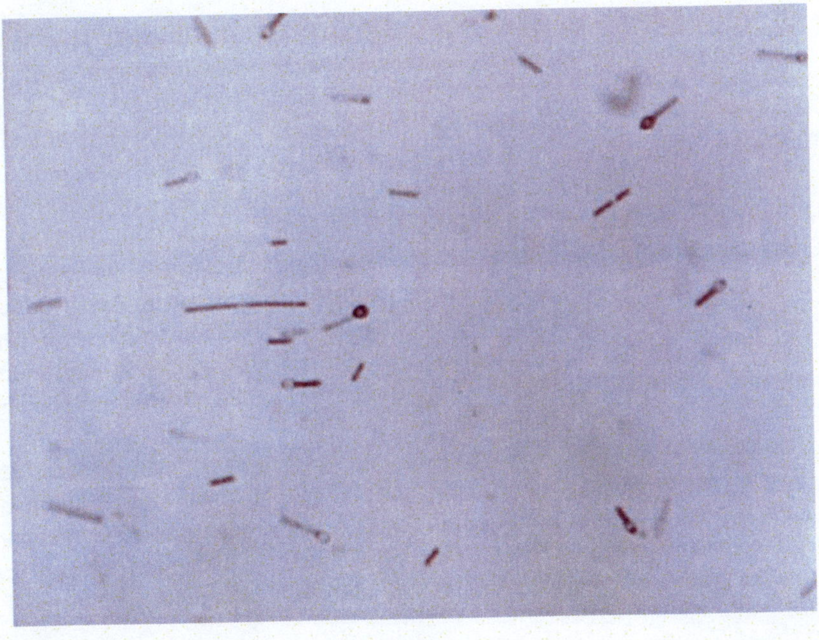

Fig. 33.5 Gram-stained smear showing drumstick appearance of *Clostridium tetani* (×1000)

Fildes technique is employed as a routine method for isolating pure colonies of *C. tetani.* In the method, a mixed bacterial culture is inoculated into condensation water at the bottom of the slope of nutrient agar. A pure colony of *C. tetani* appears at the top of the slope in the tube, followed by anaerobic incubation for 24 h.

C. tetani liquefies gelatine anaerobically. Hence, the bacillus produces a *fir-tree type* of growth in **gelatine stab culture** under anaerobic incubation.

C. tetani produces an extremely fine translucent film of growth on the blood agar, which tends to swarm over the entire surface of the agar. It produces alpha-haemolytic colonies initially on blood agar, which on prolonged incubation become beta-haemolytic due to the production of tetanolysin. Gram-stained smear of the colony reveals Gram-positive bacilli with prominent terminal spores (drumstick appearance). The motile (except type VI) and capsulated bacteria do not ferment any sugar. Toxigenicity testing in mice is a reliable method of identifying the colony as *C. tetani.*

Toxigenicity Testing

The tests are useful to demonstrate the identification of the colony as *C. tetani.*

In-vitro Neutralisation Test

It is carried out on blood agar. The test is performed on a blood agar containing 4% agar. A high percentage of agar is used to inhibit swarming by *C. tetani.* Strains of *C. tetani* are stab-inoculated on each half of the plate with and without antitoxins. The colonies of *C. tetani* show haemolysis on the part of the blood agar without any antitoxin but do not show any haemolysis on the part of the agar with tetanus antitoxins (1500 units per millilitre). This is due to the inhibition of the haemolytic activity of the toxin by the antitoxin present in the agar.

In-vitro Neutralisation Test

The test is carried out in mice. In this test, 0.2 mL of a 2–4 day old cooked meat culture of *C. tetani* is inoculated into the root of the tail of two mice, one of which is protected with 1000 units of tetanus antitoxin 1 h before the test (*control animal*). The other mouse is not protected with any antitoxin (*test animal*). Symptoms of ascending tetanus due to tetanospasmin produced by *C. tetani* develop in the test animal 12–24 h after inoculation of the bacteria. The symptoms begin in mice with stiffness in the tail, which proceed rapidly to the leg on the inoculated side, the opposite leg, trunk and forelimbs, in that order. The animal usually dies within 48 h. No symptoms appear in the inoculated control animal. It is a reliable method of identification of the colony as *C. tetani.*

Other Tests

The **spatula test** is a clinical test which is a simple bedside diagnostic test for tetanus. In this test, the oropharynx is touched with a spatula. The positive test result is that the patient develops a reflex spasm and bites the spatula. The negative result is that the patient would develop a gag reflex and try to expel the spatula; the test is 100% specific.

Treatment

Treatment of tetanus depends on the nature of the wound and the immune status of the infected host.

Antibiotics Therapy

Antibiotics are used to prevent the multiplication of *C. tetani* in the wound, thus halting the production and release of toxins. In a dosage of 5000 mg every 6 h, metronidazole is the current antimicrobial drug of choice. Penicillin given on a dosage of 100,000–2000,000 IU/Khg, injected intravenously, in 2 to 4 divided dosages is an alternative treatment. Tetracycline is an alternative drug for patients allergic to penicillin or metronidazole. Clindamycin, erythromycin and vancomycin are the other antimicrobials used to treat tetanus.

Human Immunoglobulin Therapy

Human tetanus immunoglobulin (TIG) is given to neutralise unbound tetanus toxins and prevent the circulating tetanus toxin from reaching the CNS. A single total dose of 3000–6000 IU is given intramuscularly for children and adults immediately at diagnosis. A lower dose of 500 IU for infants with tetanus neonatorum has been effective. In countries where TIG is not available, equine tetanus antitoxin (ATS) is used for the purpose. Usually, a single dose (50,000–100,000 IU) of tetanus antitoxin is administered intramuscularly after testing for sensitivity and desensitisation. A part of this dose (20,000 IU) is given intravenously.

Surgical Management

A dirty wound contaminated with soil, debris, faeces and so on could introduce the organism. Wounds containing devitalised tissue are fertile grounds for the multiplication of the tetanus bacilli. All wounds should be cleaned, and if needed, devitalised and necrotic tissues should be surgically debrided.

Supportive Management

Sedation and control of muscle spasms form the mainstay of supportive management in the case of tetanus. Benzodiazepines or opioids like morphine are used as sedatives, which are also helpful in muscle relaxation. Autonomic dysfunction is an important complication that can be tackled with fluid loading. Magnesium sulphate is a useful addition in relaxation, sedation and controlling the autonomic disturbance in tetanus.

Epidemiology

Tetanus is common in areas where the soil is cultivated, in rural areas, in warm and damp climates and during summer months.

Geographical Distribution

The condition is predominantly a disease of the underdeveloped countries. Tetanus affects all age groups, with the highest prevalence among newborns and young people. Overall, the annual incidence of tetanus is 0.5–1 million cases, mostly in underdeveloped countries. Neonatal tetanus accounts for 50% of tetanus-related deaths in developing countries.

Habitat

C. tetani organisms are found in soil, animal faeces, and, occasionally, human faeces and inanimate objects. The spores may survive for years in some environments and are resistant to disinfectants and even to boiling for 20 min.

Reservoir, Source and Transmission of Infection

C. tetani spores are the infective form of the bacteria. Soil, animal faeces and, occasionally, human faeces and inanimate objects contaminated with spores are the primary sources of infection. Risk factors for neonatal tetanus include unvaccinated mothers, homebirth and unhygienic cutting of the umbilical cord. Animal dung, clarified butter and so on applied to the umbilical stump are the other risk factors for neonates. The grossly contaminated wounds; wounds exposed to saliva or faeces; stellate, ischemic or infected wounds; deep (>1 cm) wounds; as well as avulsions, punctures or crush injuries are more susceptible to tetanus.

Maternal and Neonatal Tetanus

Maternal and neonatal tetanus are among the most common life-threatening consequences of unhygienic practices in deliveries and care of the umbilical cord. It is rare in countries with well-developed health infrastructure where tetanus vaccination is routinely given as ante-natal care. Neonatal tetanus has a very high mortality, which manifests within a week after birth with fever, vomiting and seizures. Maternal vaccination helps prevent neonatal tetanus as well. The WHO estimates that in 2018, 25,000 newborns died from tetanus, which is an 88% reduction from the 2000 figure of 200,000. This has been achieved by The Maternal and Neonatal Tetanus Elimination (MNTE) initiative of the WHO.

Prevention and Control

Active Immunisation

The prevalence of protective immunity to tetanus is greater in children and persons aged between

6 and 39 years. Protective immunity from tetanus decreases in older people. Serological immunity studies have shown a low level among elderly individuals in many countries. Clinical tetanus does not produce a state of immunity; hence, patients who survive the disease require active immunisation with tetanus toxoid to prevent a recurrence of the disease.

Tetanus is completely preventable by active immunisation. Active immunisation by vaccination with tetanus toxoid is the key to preventing tetanus. Vaccination is carried out by toxoids, available either as plain toxoids or adsorbed on aluminium hydroxide or phosphate. The toxoid is given alone or in combination with diphtheria toxoid and acellular pertussis (whooping cough) (DTaP or triple) vaccine.

Vaccine

The tetanus toxoid and the triple vaccine containing tetanus toxoid are very safe and effective. The toxoid stimulates the production of protective antibody response in virtually all immunocompetent subjects. Specific antibodies produced against the tetanus toxin are protective. Antibodies specifically combine with free toxins and prevent the action of the toxin. Many studies have shown protective serum antibody levels in up to 90% of immunised people 15 years after vaccination.

Primary prevention of tetanus is achieved by vaccination with the triple vaccine at ages 2, 4, 6 and 12–18 months and 4–6 years. A tetanus-diphtheria (Td) booster dose is administered at 11–12 years and then every 10 years. For the primary immunisation of unimmunised children 7 years or older, the tetanus toxoid vaccine is administered in two doses 4–6 weeks and the third dose at 6–12 months. Usually, booster doses are administered at 10 years once or at the time of major injury if it occurs after 5 years after a dose.

Vaccination with tetanus toxoid is recommended for (1) all adults who have not had a booster dose in the last 10 years. It has been demonstrated that approximately 50% of persons older than 50 years are non-immune because they were never vaccinated or did not receive appropriate booster doses; (2) adults who have recovered from tetanus; and (3) adults who have never received immunisation against tetanus.

Immunisation for Prevention of Neonatal Tetanus

Neonatal tetanus is prevented by increasing immunisation in women of childbearing age, especially pregnant women and by improving maternity care. For previously unimmunised pregnant women, tetanus toxoid is given twice during pregnancy, 4–6 weeks apart, preferably in the last two trimesters and again at least 4 weeks before delivery. Maternal antitetanic antibodies are passed to the foetus, and this passive immunity is effective for many months after the child's birth.

Passive Immunisation

Human tetanus immunoglobulin (TIG) is used for passive immunisation. In countries where TIG is not available, ATS is also used for the purpose. Passive immunisation is carried out by intramuscular administration of 250–500 IU of TIG or by subcutaneous or intramuscular injections of 1500 units of ATS after giving a skin test dose. Passive immunisation is recommended for (1) non-immunised individuals, and (2) those whose immunisation status is uncertain whenever a wound is contaminated or likely to have devitalised tissue.

Combined Immunisation

Secondary prevention of tetanus is achieved post-exposure through appropriate wound cleansing and debridement, and combined immunisation. Combined immunisation is carried out by simultaneously administering (1) human TIG or the ATS, and (2) tetanus toxoid. It is indicated if the patient has not been previously immunised with a series of at least three doses of toxoid.

Table 33.5 shows tetanus prophylaxis recommendations in different types of wounds.

Table 33.5 Tetanus prophylaxis in different types of wounds

Type of wound	Immunological status of an individual against tetanus toxin: Immune (past history of complete three doses of toxoid injection)	Partially immune (history of only two doses of toxoid injection)	Non-immune (status unknown or history of one or no toxoid injection)
Clean wound (fresh, uninfected, no inflammation)	A single dose of toxoid should be given only if the active immunisation has not been received during last 3 or more years.	A single dose of toxoid	Injection of three doses of toxoid
Contaminated wound (presence of soil or foreign particles or highly inflamed tissue)	A single dose of toxoid should be given only if the active immunisation has not been received during last 3 or more years.	A single dose of toxoid + tetanus immune globulin + antibiotics	Injection of three doses of toxoid + tetanus immune globulin + antibiotics
Infected wound (old traumatic wounds, devitalised tissue, infected with microbes)	A single dose of toxoid should be given only if the active immunisation has not been received during last 3 or more years + antibiotics	A single dose of toxoid + tetanus immune globulin + antibiotics	Injection of three doses of toxoid + tetanus immune globulin + antibiotics

Clostridium botulinum

C. botulinum causes botulism, a paralytic disease with the presentation of food poisoning.

Properties of the Bacteria

Morphology

C. botulinum is a heterogeneous group of spore-forming, anaerobic and Gram-positive bacteria. It is Gram-positive in young cultures less than 18 h old. The organism may be Gram-negative in cultures after 18 h of incubation. The bacillus measures 5 × 1 μm. It is motile by the presence of peritrichous flagella and possesses sub-terminal and oval bulging spores. It is non-capsulated.

Culture and Biochemical Reactions

C. botulinum is a **strict anaerobe**. The bacteria can grow on a wide range of media, including blood agar, Mueller–Hinton agar and RCM media. On blood agar, *C. botulinum* produces large, irregular and semi-transparent colonies with irregular fimbriate borders. They produce spores when grown in alkaline glucose gelatine media at 20–25 °C. Production of enzyme lipase is demonstrated by forming an iridescent film on *C. botulinum* colonies grown on egg yolk agar.

In Robertson's cooked meat broth medium, *C. botulinum* grows luxuriantly with marked turbidity and two distinct features. Serotypes A, B and F are proteolytic, which turns the meat particles black with a foul smell, while the saccharolytic C, D and E serotypes cause pink discolouration of the meat particles.

C. botulinum is identified by its ability to ferment glucose, hydrolyse gelatine, digest protein and produce the enzyme lipase. Production of enzyme lipase is demonstrated by forming an iridescent film on *C. botulinum* colonies grown on egg yolk agar.

Other Properties

Susceptibility to Physical and Chemical Agents

Both bacteria and spores are highly resistant. They are resistant to boiling at 100 °C for several hours; They are killed by pressure cooking at 120 °C for 10 min, *C. botulinum* types B, E and F spores are relatively less heat resistant. The toxins are destroyed by boiling for 10 min or cooking at 80 °C for 30 min.

Typing

C. botulinum is a heterogeneous group of spore-forming, anaerobic and Gram-positive anaerobes. They are classified into **eight antigenically distinct botulinum toxins** (A, B, C1, C2, D, E, F and G) depending on the immunological differences in the toxins. These different toxins appear identical in their biological functions but differ immunologically. The toxins are specifically neutralised only by their homologous antiserum. A single strain of bacteria produces only one toxin. In rare instances, a single strain may produce more than one toxin. All toxins except C2 are neurotoxins.

Pathogenesis

C. botulinum organisms are non-invasive. They cause disease primarily by the production of an exotoxin, the **botulinum toxin**, which is the major virulence factor of the bacteria.

Virulence Factors

Botulinum toxin, a neurotoxin, is a protein with an MW of 70,000 Da. The toxin is relatively heat stable. It is inactivated at 80 °C for 30–40 min and at 100 °C for 10 min. The toxin is similar in structure and function to the tetanus toxin, except only in the site of action.

The neurotoxin acts specifically on cholinergic nerves. It acts by preventing the release of a neurotransmitter, acetylcholine, at the synapses and the neuromuscular junctions. It causes atrophy of the muscles at the site of the infection, but the neurons regenerate in 2–4 months of recovery. Toxins A and B are the most potent toxins. The production of toxins by *C. perfringens* types C and D appears to be mediated by bacteriophages.

The botulinum toxin differs from other exotoxins in that it is produced only on death and autolysis of the bacteria, but not when the bacilli are alive. It is a **progenitor protein** synthesised intracellularly in the bacteria. It consists of two subunits, A and B. **Sub-unit A** is the light chain and is a neurotoxin. Sub-unit B is a heavy chain, which prevents the neurotoxin part (chain A) from being inactivated by the stomach's acidity. More than three B sub-units may be present in the toxin.

The botulinum toxin is the most potent naturally occurring toxin known to humankind. The lethal dose for mice is 0.000,000,033 mg and for human beings is 1–2 g. It is lethal at femtomolar level doses of 10^{-12} g/kg, making botulinum toxin 15,000–100,000 times more potent than sarin gas. The toxin can also be toxoided.

Pathogenesis of Botulism

Food-Borne Botulism

The condition occurs on ingestion of preformed toxins in food contaminated with *C. botulinum.* When food is contaminated with small amounts of toxins and consumed, it results in full-blown disease. Toxin A has been the most frequent cause of food-borne outbreaks in the past 20 years. After toxin A, toxins B and E follow in frequency. The toxins are absorbed from the stomach and small intestine, as digestive enzymes do not denature them. The toxin enters the vascular system and is transported to peripheral cholinergic nerve terminals, such as neuromuscular junctions, cholinergic parasympathetic nerve endings and peripheral ganglia. They bind to receptor sites on pre-synaptic motor nerve terminals and cause blockade of neuromuscular conduction. They then enter the nerve terminal and inhibit acetylcholine's release and transmission in cholinergic nerve fibres. The toxin irreversibly binds to the neurons.

Infant Botulism

It occurs following the ingestion of *C. botulinum* spores present in infant food such as honey. It is caused by neurotoxins produced in the intestine, *in Vivo*, by *C. botulinum* that has colonised the gastrointestinal tract of infants. The spores then germinate to toxin-producing vegetative forms that colonise the infant's gut. The toxin produced in and absorbed from the gut causes the clinical disease.

Wound Botulism

The condition occurs due to the contamination of the wound with spore-forming *C. botulinum.* The condition occurs (1) in people with a traumatic injury involving contamination with soil or water containing *C. botulinum* spores (2) in people who chronically abuse intravenous drugs (e.g. black tar heroin) and (3) in women after caesarean delivery (very rare). The condition can also occur even after antibiotics are administered to prevent wound infection.

Clinical Syndromes

Botulism is the major disease caused by *C. botulinum.*

Botulism

The condition may manifest as foodborne botulism, infant botulism or wound botulism.

Food-borne Botulism

The condition manifests from a mild to a very serious disease resulting in death within 24 h. The incubation period is short and varies from 12 to 36 h after ingestion of the contaminated food. The initial signs are vomiting, nausea, dry mouth, constipation, abdominal pain and blurred vision with fixed dilated pupils. The disease progresses to a bilateral descending weakness of the peripheral muscles, leading to flaccid paralysis. Fever typically is absent. Death is due to respiratory paralysis and occurs in 1–7 days after the onset of the disease.

Infant Botulism

Symptoms at the early stages are usually non-specific. These include constipation, lethargy, weakness, weak and altered cry, loss of head control. The condition may progress to flaccid paralysis and respiratory arrest. The infants excrete toxins in their faeces. Infant mortality due to infant botulism is relatively low (11–20%).

Wound Botulism

It occurs following heavy contamination of wounds with soil or water containing *C. botulinum* spores. The incubation period is 4–14 days with a mean of 10 days. Wound botulism is similar to food-borne botulism except that the incubation period is longer and no gastrointestinal symptoms are present.

Laboratory Diagnosis

Botulism can be diagnosed only by a high degree of clinical suspicion. The diagnosis is suspected in an afebrile patient with progressive descending paralysis, especially in the presence of gastrointestinal manifestations. Laboratory diagnosis of the condition rests on the demonstration of *C. botulinum* bacilli or toxins.

Specimens

Faeces, vomitus or gastric aspirate are collected for all forms of botulism. In addition, tissue from the wound is collected for wound botulism.

Microscopy

Gram staining of the smears made from food, faeces and other specimens may show Gram-positive spore-bearing *C. botulinum.*

Culture and Identification

Foodborne botulism is diagnosed by demonstrating *C. botulinum* in food by culture and by demonstrating toxins in food or faeces. *C. botulinum* may be isolated from suspected contaminated food and the faeces by culture. Food culture is performed first by heating the specimens at 80 °C for 10 min. This is done to destroy all the vegetative forms of the bacteria. The heated specimens are then cultured on RCM media and incubated in anaerobic conditions, followed by a sub-culture on blood agar. Anaerobic incubation facilitates germination of the spores to vegetative forms of the bacteria. Demonstration of *C. botulinum* in faeces, vomitus or gastric aspirate by culture is highly suggestive of food-borne botulism because the intestinal carriage is rare. Culture is positive in approximately 60% of cases. The toxin can also be demonstrated in food or faeces to diagnose food-borne botulism.

Demonstrations of *C. botulinum* in pus culture and detection of botulinum in wound pus and exudates are highly suggestive of wound botulism. Isolation of bacilli and detection of the botulinum toxin in infants' faeces confirm infant botulism diagnosis.

C. botulinum produces large, irregular and semi-transparent colonies with an irregular fimbriate border on blood agar. The colony smear reveals Gram-positive bacilli with sub-terminal and oval bulging spores in the Gram-stained smear of the colony. The bacteria are motile and capsulated, ferments glucose, hydrolyses gelatine, digests protein and produces enzyme lipase.

Demonstration of Toxin

C. botulinum toxin can also be demonstrated in food or faeces to diagnose food-borne botulism, the faeces for infant botulisms and the wound pus for infant botulism. Toxins in these specimens can be detected using serum toxin bioassay, enzyme-linked immunosorbent assay and polymerase chain reaction.

The toxin is demonstrated in food or faeces by **serum toxin bioassay,** a neutralisation test carried out in mice. In this method, food filtrate in sterile saline is inoculated intraperitoneally into two mice. One mouse is protected with polyvalent botulinum antitoxin (control animal), and another is not protected by any antitoxin (test animal). If the test animal dies but the control animal remains healthy, the test is considered positive and suggests the presence of toxin in the specimen. This is a useful test to demonstrate toxins during the early stages of food botulism.

In recent years, toxin detection by the ELISA method has become popular since mouse bioassay is technically demanding and not routinely available. The organism is first grown in a liquid culture medium for 5 days, and the A, B, E and F botulinum toxin-producing organisms and their toxins can be detected by amplified ELISA, which has a sensitivity of detecting about 10 MLD/mL (0.12–0.25 ng/mL) of the toxin. Digoxigenin-labelled antitoxin ELISA, a modification of the amplified ELISA, has a similar detection limit.

Molecular Detection of Toxin Genes

PCR has been developed to detect the genes encoding for the toxin types A, B, E and F, which are neurotoxins. The organism is first grown for 24 h in a culture medium before testing. Multiple botulinum toxin-producing types can also be detected within a single PCR assay. The PCR method can also be used along with the mouse bioassay to determine the specific toxin type.

Treatment

Treatment of botulism involves initial supportive therapy, neutralising unbound toxin by specific antitoxins and stopping toxin production using antibiotics. Patients with botulism should be monitored closely for neurologic, respiratory and autonomic manifestations. Intubation and mechanical ventilation may be necessary in almost half of the admitted patients. Antibiotics are not required except in the case of wound botulism.

Antitoxin Therapy

This consists of a trivalent A-B-E botulinum antitoxin serum used for the specific treatment of botulism. The sera contain antibodies against *C. botulinum* types A, B and E; these act by binding and neutralising the toxins present in the serum. Symptomatic patients with high clinical suspicion of foodborne botulism and wound botulism are administered with the trivalent botulinum antitoxin. The antitoxin can be given even after several weeks of toxin ingestion. This antitoxin can slow down disease progression.

Antibiotics Therapy

This is carried out to prevent the multiplication of *C. botulinum* in the gastrointestinal tract and the wound, thereby preventing the production and release of toxins. Metronidazole is the current antimicrobial drug of choice, with penicillin as an alternative treatment.

Epidemiology

Human botulism is found worldwide.

Geographical Distribution

C. botulinum types A, B, E and rarely F and G toxins cause human disease. Types C and D cause avian and non-human mammalian disease. Infant botulism is more common than food-borne botulism. Food-borne botulism is caused by *C. botulinum* types A, B and F, and very rarely E types. Wound botulism is very rare and is caused by *C. botulinum* type A strains.

Habitat

C. botulinum is ubiquitous. It is found in the soil and water worldwide. Spores of *C. botulinum* type A or B are distributed widely in the soil and have been found throughout the world.

Reservoir, Source and Transmission of Infection

C. botulinum spores are responsible for causing botulism. Preserved food, especially homemade, canned meat and meat products in Europe, canned vegetables in America and preserved fish in Japan, contaminated with the preformed toxin, are the main sources of infections for foodborne botulism. Honey and other food contaminated with spores of bacteria are the sources of infection for infant botulism. Soil and water heavily contaminated with spores are sources of infection for wound botulism.

Wound botulism occurs (1) in people with a traumatic injury involving contamination with soil, (2) in people who chronically abuse intravenous drugs (e.g. black tar heroin) and (3) in women after caesarean delivery (very rare). The condition can also occur even after antibiotics are administered to prevent wound infection.

Food-borne botulism occurs when food is contaminated with small amounts of toxins and consumed, resulting in full-blown disease. Toxin A has been found to be the most frequent cause of food-borne outbreaks in the past 20 years. After toxin A, toxins B and E follow in frequency. High-risk foods include home-canned or home-processed low-acid fruits and vegetables; fish and fish products; and condiments, such as relish and chilli peppers. Commercially prepared foods and improperly handled fresh foods may occasionally cause botulism outbreaks.

Infant botulism occurs following the ingestion of *C. botulinum* spores in infant food such as honey. The spores then germinate to a toxin-producing vegetative form that colonises the infant's gut. The toxin that is produced in and absorbed from the gut causes the clinical disease.

Prevention and Control

Food-borne botulism is best prevented by high-temperature pressure cooking, which kills the spores present in fruits and vegetables. Storing the food in the refrigerator at 4 °C or an acidic pH prevents germination of spores, if present, to vegetative bacterial form. Infant botulism is prevented by avoiding the administration of honey to infants. Wound botulism is prevented by immediate and thorough debridement of contaminated wounds.

Clostridium difficile

C. difficile was first described in 1935 from the faeces of healthy newborns and was initially not thought to be a pathogen. It was named *difficile* because it grows slowly and is difficult to culture. It is a long slender Gram-positive, anaerobic bacillus bearing large, oval and terminal spores. It is non-haemolytic, saccharolytic and mild proteolytic.

C. difficile is the causative agent of antibiotic-associated diarrhoea and colitis. It produces two antigenically distinct toxins: toxin A and toxin B. Toxin A is an enterotoxin, and toxin B is a cytotoxin. Both the toxins contribute to the pathogenesis of *C. difficile* colitis and diarrhoea in humans. Both are proteins that bind to specific receptors present in the intestine's mucosa.

Antibiotic therapy is the key factor that alters the normal bacterial flora of the intestine. The use of antibiotics suppresses the normal flora of the intestine. It also facilitates the colonisation and multiplication of *C. difficile* and the production of toxins that cause inflammation of the mucosa and damage. *C. difficile* causes malaise, anorexia and mild-to-moderate diarrhoea.

Diarrhoea develops in most patients during or shortly after starting antibiotics. It is associated with the formation of pseudo-membranes and, occasionally, adherent yellowish-white plaques on the intestinal mucosa.

C. difficile infection is more common in old people. This may be due to the increasing susceptibility of mucosal flora to colonisation by the bacteria and disease. The infection is uncommon in infants and young children, although they frequently harbour the bacteria and its toxins.

Any patient with diarrhoea who took antibiotic treatment during the past 2 months or diarrhoea occurring 3 days after hospitalisation would be suspected of *C. difficile* colitis. Diagnosis of *C. difficile* diarrhoea is made by the demonstration of *C. difficile* toxin in the faeces by stool cytotoxin test and by enzyme-linked immunosorbent assay:

The **stool cytotoxin test** is the test of choice. The diarrhoeal stool is filtered and then inoculated to Hep-2 and human diploid cell cultures in this test. The demonstration of a cytopathic effect that is neutralised by a specific antiserum indicates the presence of the toxin (positive test). The absence of cytopathic effect is considered a negative test.

Several enzyme immunoassays are now commercially available for rapid detection of the toxin and diagnosis. The latex agglutination test has been employed to detect the presence of glutamate dehydrogenase, produced by *C. difficile*, for diagnosing *C. difficile* diarrhoea. PCR and lateral flow assay have been used to detect *C. difficile* toxins A and B in stool samples. Stool cultures are not useful because of the presence of non-toxigenic strains of *C. difficile* in faeces.

No treatment is necessary for asymptomatic carriers. Stoppage of the use of causative antibiotics may be the only treatment necessary for those with mild antibiotic-associated diarrhoea without fever, abdominal pain or leucocytosis. Patients with more severe diarrhoea or colitis require treatment with antibiotics. Metronidazole is the drug of choice. Vancomycin and bacitracin are also effective.

Case Study

An 18-year-old college student came to a hospital with a complaint of facial spasms. He was unable to eat for 2 days due to severe pain in his jaw. Examination revealed trismus and rises sardonicus. The student gave a history of playing football for his college team, during which he had sustained a minor knee injury with abrasions 6 days earlier. He did not visit any doctor earlier or receive any toxoid for the injury. He did not remember receiving any booster vaccination with tetanus toxoid in the last many years. A clinical diagnosis of tetanus was made.

- What laboratory tests should be performed to confirm the diagnosis?
- Should treatment wait until the laboratory results are available?
- Describe the pathogenesis of tetanus.
- What are the vaccines available against tetanus in humans?

Non-Sporing Anaerobes

34

The anaerobic cocci are a heterologous group of bacteria that typically colonise the skin and the mucous membranes. The taxonomical classification of these cocci has undergone several modifications, and importantly, the classification has been simplified now. The non-sporing anaerobes are divided into anaerobic cocci and anaerobic bacilli.

Classification

There are numerous anaerobic bacteria present in nature, but few are medically important. Based on the DNA base arrangement and investigation of fatty acid end products of metabolism, medically significant anaerobic bacteria may be categorised under three broad groups (Table 34.1).

Anaerobic Cocci

Anaerobic Gram-Positive Cocci

Most of the important anaerobic Gram-positive *cocci* belong to the genus *Peptostreptococcus*. The **cocci** are small and measure 0.2–2.5 μm. Many of them are aerotolerant and grow well in air supplemented with 10% CO. These Gram-positive cocci are normal inhabitants of the oral cavity, gastrointestinal tract, genitourinary tract and skin. These bacteria, on spreading from these sites to other normally sterile sites, may cause several clinical infections: Bacteria colonising the skin can cause cellulitis and infections of the soft tissue, and bacteria in the intestine can cause intraabdominal infection. Bacteria in the genital tract can cause pelvis abscess, salpingitis and endometritis, and those invading the bloodstream can cause infections in bones and visceral organs.

Peptostreptococcus anaerobius is most commonly associated with puerperal sepsis and *Peptostreptococcus magnus* with abscess. The laboratory diagnosis of the infection caused by these cocci depends on the isolation of the bacteria by culture. The cocci are often found in large numbers in pus from suppurative lesions; hence Gram staining of pus smears is frequently helpful in diagnosis.

The cocci are generally sensitive to imipenem chloramphenicol and metronidazole. They are intermediately sensitive to broad-spectrum cephalosporins, tetracyclines and clindamycin. They are resistant to streptomycin and gentamicin. Most infections are mixed, the cocci being present along with anaerobic Gram-negative bacilli and clostridia. Hence, broad-spectrum antibiotics effective against all these pathogens are usually given for treatment.

Anaerobic Gram-Negative Cocci

The genus *Veillonella* spp. includes anaerobic Gram-negative cocci of varying sizes occurring in pairs, short chains or clusters. These cocci normally colonise the mouth, intestines and the genitourinary tract. *Veillonella parvula* is frequently reported from clinical specimens, but its role in the causation of the disease is unknown.

Anaerobic Bacilli

Anaerobic Gram-Positive Bacilli

The group includes many genera, but a few of these are associated with clinical infections (Table 34.2). *Propionibacterium* spp., *Lactobacillus* spp., *Mobiluncus* spp. and *Actinomyces* spp. are well known to cause human infections. *Eubacterium* spp. and *Bifidobacterium* spp. are isolated in clinical specimens; they rarely cause human disease.

Propionibacterium

Propionibacterium species are anaerobic or aerotolerant bacteria found as part of the normal flora of the skin, oropharynx, conjunctiva, external ear and female genital tract. They ferment carbohydrates with the production of propionic acid, hence named *Propionibacterium*. They are Gram-positive bacilli that are frequently present in short chains and clumps. They are non-motile and are catalase positive. *Propionibacterium* is not fastidious. They can grow on nutrient agar or other simple media. However, they grow slowly and take 3–5 days to produce colonies.

Propionibacterium acnes and *Propionibacterium propionicus* are two important species of clinical importance. *In teenagers and young adults, P. acnes* is the most common causative agent of acne. The lesions in acne develop within the sebaceous follicles and hence are not related to the effectiveness of the cleanliness of the skin. The condition is treated best with topical application of benzoyl peroxide and antibiotics. The bacteria are sensitive to erythromycin and clindamycin.

Lactobacillus

Members of the genus *Lactobacillus* are facultatively anaerobic bacilli. They are commonly found in the mouth, stomach, intestine and adult vagina. They are commonly isolated in blood and urine cultures. The presence of lactobacilli even in high numbers in urine culture usually represents contamination of bacteria from the urethra, in which *lactobacillus* are most commonly present. Lactobacilli usually do not cause infection of the urinary tract because they fail to grow in the urine.

Lactobacilli can invade the bloodstream during (1) endocarditis, (2) opportunistic septicaemia in the immunocompromised host, and (3) transient bacteraemia following genitourinary procedure. A combined therapy with penicillin and aminoglycoside has proved effective against infections caused by lactobacilli. Lactobacilli are resistant to vancomycin.

Mobiluncus

The *Mobiluncus* species are obligate anaerobes. They are found as part of the normal flora of the genitourinary tract. In women with bacterial vaginitis, they are found in very high numbers. The exact role of the bacteria as causative of bacterial vaginosis is not known. Typical Gram-staining features of the bacteria are diagnostic. They are sensitive to erythromycin, ampicillin, clindamycin and vancomycin but resistant to colistin.

Bifidobacterium and *Eubacterium*

Bifidobacterium spp. and *Eubacterium* spp. are commonly found in the oropharynx, large intestine and vagina. They are usually isolated as contaminants. Repeated isolation of these bacteria in the absence of other pathogens may suggest their role as possible agents of infection.

Anaerobic Gram-Negative Bacilli

Anaerobic Gram-negative bacilli of clinical importance belong to the family Bacteroidaceae. This is classified into four genera: *Bacteroides*, *Fusobacterium*, *Porphyromonas* and *Prevotella*. All these are Gram-negative. The *Prevotella* species are very small and elongated, whereas the *Fusobacterium* species are long and thin. A surface lipopolysaccharide (LPS) is the major component of the cell wall.

Properties of Anaerobic Gram-Negative Bacilli

Bacteroides species have now been reorganised. Earlier, the genus *Bacteroides* consisted of almost

Table 34.1 Classification of medically important anaerobic bacteria

Anaerobic cocci		Anaerobic bacilli		Spirochetes
Gram positive	*Peptostreptococcus*	Endospore-forming Clostridia		*Treponema*
	Peptococcus	Non-sporing anaerobes		*Borrelia*
Gram negative	*Veillonella*	Gram positive	*Eubacterium*	
			Propionibacterium	
			Mobiluncus	
			Lactobacillus	
			Bifidobacterium	
			Actinomyces	
		Gram negative	*Bacteroides*	
			Leptotrichia	
			Fusobacterium	
			Porphyromonus	
			Prevotella	

50 species. Now, a saccharolytic pigmented species have been re-classified into the genus *Porphyromonas* (Greek word, meaning "*purple*") and saccharolytic bile-sensitive species in the genus *Prevotella*.

Bacteroides

The genus *Bacteroides* now includes *Bacteroides fragilis* and closely related other species, such as *Bacteroides distasonis*, *Bacteroide svulgatus* and *Bacteroides thetaiotaomicron. B. fragilis* is the prototype of an endogenous anaerobic pathogen that colonises humans. *B. fragilis* is pleomorphic in size and shape, having a typical Gram-negative cell wall. A surface LPS is a major component of the cell wall. The LPS of *B. fragilis* lacks endotoxin activity, unlike the LPS of *Fusobacterium*. *B. fragilis* is surrounded by a polysaccharide capsule. *Bacteroides* are non-fastidious and grow rapidly in culture, whereas other anaerobic Gram-negative bacilli are fastidious and grow slowly (3 days and more) in the culture media.

The ability of the bacteria to inactivate immunoglobulins, cause tissue destruction and resist oxygen toxicity plays an important role in the pathogenesis of anaerobic infections. *B. fragilis* and other anaerobic Gram-negative bacilli produce a wide variety of virulence factors, which play a major role in the pathogenesis of the disease (Table 34.3). The presence of a polysaccharide capsule in *B. fragilis* helps the bacteria adhere to peritoneal surfaces more effectively than other anaerobes. The capsular polysaccharide is antiphagocytic; it inhibits phagocytosis by

Table 34.2 Human infections caused by anaerobic Gram-positive and Gram-negative bacilli

Diseases	Bacteria
Acne and opportunistic infections	*Propionibacterium* spp.
Bacterial vaginosis	*Mobiluncus* spp.
Actinomycosis	*Actinomyces* spp.
Endocarditis and opportunistic infections	*Lactobacillus* spp.
Opportunistic infections	*Bifidobacterium* spp., *Eubacterium* spp.
Head and neck infections	*Bacteroides ureolyticus, Fusobacterium nucleatum, Porphyromonas asaccharolytica, Porphyromonas gingivalis, Prevotella intermedia, Prevotella melaninogenica*
Intraabdominal infections and bacteraemia	*Bacteroides thetaiotaomicron, Prevotella melaninogenica, Bacteroides fragilis*
Gynaecological infections	*Prevotella bivia, Prevotella disiens, Bacteroides fragilis*

Table 34.3 Virulence factors of anaerobic Gram-negative bacilli

Virulence factors	Biological functions
Capsule (*Bacteroides fragili* and *Prevotella melaninogenica*)	Antiphagocytic and adhesins
Fimbriae (*Bacteroides fragilis* and *Porphyromonas gingivalis*), Hemagglutinin (*Porphyromonas gingivalis*), Lectin (*Fusobacterium nucleatum*)	Adhesins
Lipopolysaccharides (*Fusobacterium* spp.)	Antiphagocytic
Succinic acid (many species), Immunoglobulin (Ig) A, IgM, IgG proteases (*Porphyromonas* spp. and *Prevotella* spp.)	Antiphagocytic
Superoxide and catalase (many species)	Resist oxygen toxicity
Other enzymes, such as protease, collagenase, phospholipase, neuraminidase, heparinase, glucuronidase, haemolysins, fibrinolysins. (many species)	Cause destruction of tissues

polymorphonuclear leukocytes. Succinic acid and other short-chain fatty acids provided by these bacteria during anaerobic metabolism also exhibit phagocytosis and intracellular killing of bacteria.

B. fragilis and other species possess pili by which they adhere to the epithelial cells. *Bacteroides* produce the enzymes catalase and superoxide dismutase, which inactivate the hydrogen peroxide and superoxide-free radicals, thereby protecting these anaerobic bacteria when they are exposed to oxygen. The other factors that facilitate the spread of these anaerobic bacteria are trauma, tissue necrosis, impaired circulation and the presence of foreign bodies.

In large numbers, B. fragilis and other anaerobic Gram-negative bacteria usually colonise the skin, mouth, nasopharynx and upper respiratory tract, intestine and vagina. As resident bacterial flora, they prevent colonisation by other pathogenic organisms and help digest food. These organisms, however, cause infections when they spread from their usual sites to other parts of the body that are usually sterile.

B. fragilis causes more than 80% intraabdominal infections, even though *B. distasonis* and *B. thetaiotaomicron* are found predominantly in the gastrointestinal tract. These two species merely cause any abdominal infections. *B. fragilis* is the common anaerobic bacteria that cause an abscess in the genitourinary tract. It is the most important anaerobic bacteria causing pelvic inflammatory disease, endometritis, abscess and other infections of the female genital tract.

Other Gram-Negative Anaerobes

Other Gram-negative anaerobes, such as *Fusobacterium* spp., *Prevotella* spp., *Porphyromonas* spp. and non-*B. fragilis* spp., cause up to 50% of chronic infections of the sinuses and ears, and most periodontal infections.

Porphyromonas

This genus is separated from *Bacteroides*; it includes two major species, namely *Porphyromonas gingivalis* and *Porphyromonas endodontalis. P. gingivalis* inhabits the oral cavity and is implicated in certain periodontal diseases, whereas *P. endodontalis* is responsible for black-coloured colonies on the gingival epithelium and dental root canal infections. The genus *Porphyromonas* is non-motile and forms black-pigmented colonies on blood agar.

Prevotella

The genus *Prevotella*, which was earlier classified under *Bacteroides*, contains the moderately saccharolytic species inhibited by 20% bile. The main species include *Prevotella melaninogenica, Prevotella buccalis, Prevotella denticola. P. melaninogenica* produces black- or brown-coloured colonies because of hemin derivative but not due to the melanin pigment as suggested before. It causes multiple infections such as lung or liver abscess, mastoiditis, intestinal lesions, and mouth and gum lesions. A typical red fluorescence can be seen when *P. melaninogenica* cultures or dressings from infected patients are exposed to UV light.

Fusobacterium

Members of the genus *Fusobacterium* are long, spindle-shaped bacteria with pointed ends. *Fusobacterium necrophorum* produces a variety of exotoxins and is responsible for liver abscesses and other abdominal infections in animals more frequently than in humans. *Fusobacterium nucleatum* forms normal commensal of the mouth and is also a causative agent of oral infections and pleuropulmonary sepsis.

Leptotrichia

They are Gram-negative, non-motile, anaerobic bacilli. *Leptotrichia buccalis* (*Fusobacterium fusiforme* or Vincent fusiform bacillus) is an important species in the genus *Leptotrichia*. This genus also includes other species, including *Leptotrichia goodfellowii, Leptotrichia hofstadii, Leptotrichia shahii,* and two recently defined species, *Leptotrichia trevisanii* and *Leptotrichia ammionii. L. buccalis* is one of the first bacteria described and drawn by Antony van Leeuwenhoek. This genus is the normal flora of the oral cavity and is also found as commensal in the human intestine and urogenital tract. This genus has a significant role in the pathogenicity of a common condition called **Vincent's angina**, also called acute necrotising ulcerative gingivitis (ANUG), in which inflamed pharyngeal mucosa shows a greyish membrane that peels easily.

Laboratory Diagnosis of Anaerobic Gram-Negative Bacilli Infections

Laboratory diagnosis depends on the isolation of Gram-negative anaerobic bacteria in clinical specimens. Isolation of these bacteria needs careful interpretation since most anaerobes are commensals and constitute part of the normal flora of the skin and mucous surface. The mere presence of the bacteria does not suggest their causative role. As far as possible, the specimens are not collected from their normal resident sites. The specimens are also not allowed to dry because drying reduces the number of bacteria. Hence, specimens are always kept in a moist environment. The specimens are collected and transported to the laboratory in an oxygen-free system to minimise contact with air. The specimens are inoculated into specific media promptly and incubated in anaerobic conditions in the laboratory. Pus and other biological fluids are collected in glass vials or bottles with airtight caps. Collection of fluid and other aspirations in airtight syringes is a better collection method. After collection, the needle is plunged into a sterile rubber container and sealed. Swabs are not the specimen of choice for anaerobic infections but, if collected, are sent in Stuart transport medium.

Microscopic examination of clinical specimens by Gram staining is useful for diagnosing Gram-negative anaerobic infections. Gram-stained findings of pus smear showing pleomorphic, Gram-negative bacilli, polymicrobial flora and numerous pus cells are a useful preliminary diagnosis of anaerobic infections. A direct fluorescent examination may show the bright red fluorescence of *P. melaninogenica.*

Many selective media are used, facilitating Gram-negative anaerobic bacteria's isolation. The gas-pack system, which provides a better method of anaerobiosis, is now used widely. Most *Bacteroides* grow within 24–48 h, but some other Gram-negative anaerobes, such as *Fusobacterium*, grow slowly; hence, culture plates require a longer incubation period. The specimens are cultured in parallel because many aerobic bacteria are also involved in polymicrobial infections.

Gas-liquid chromatography of specimens is also occasionally useful for identifying metabolic end products of anaerobic bacteria to supplement laboratory diagnosis.

Treatment

This depends mainly on surgical intervention and antibiotic therapy. Metronidazole is the antibiotic of choice for treating infection caused by *Bacteroides* and other Gram-negative anaerobes. The anaerobic Gram-negative bacilli are also sensitive to penicillin, clindamycin, cephalosporins and chloramphenicol.

Most of the *B. fragilis*, *Prevotella* and *Porphyromonas* species are resistant to penicillins and many cephalosporins due to beta-lactamase production. The resistance is overcome by treating with a high concentration of carbenicillin, piperacillin, imipenem and beta-lactamase inhibitors. *Bacteroides* show plasmid-mediated resistance to clindamycin.

Case Study

A 42-year-old man was admitted to a hospital with abdominal pain, tenderness and fever. On examination, the patient was clinically suspected to be having appendicitis. He was operated on. During surgery, a ruptured appendix and a copious volume of foul-smelling pus were found. The pus was sent to the microbiology laboratory for aerobic and anaerobic culture. The culture was positive for *B. fragilis*, *Enterococcus faecalis* and *Escherichia coli*.

- Which organisms are responsible for intraabdominal infections and produce foul-smelling pus?
- How will you diagnose *B. fragilis* infection?
- What are the virulence factors of *B. fragilis?*
- Which antibiotics will be useful for treating *B. fragilis* infection?

35 Coliforms

Enterobacteriaceae organisms are ubiquitous bacteria. Members of the family show very wide biochemical and antigenic heterogeneity amongst themselves. The oxidase test is an important test by which the members of the Enterobacteriaceae can be distinguished from many other fermentative and non-fermentative Gram-negative bacilli. Enterobacteriaceae organisms cause various diseases in humans (Fig. 35.1).

1. Some members of the family, such as *Salmonella* species and *Shigella* species, always cause diseases in humans.
2. Other species, such as *Escherichia coli*, *Klebsiella pneumoniae* and *Proteus mirabilis*, found as a normal intestinal flora in humans, can also cause infections in other sites of the human body.
3. In addition to these, another group of Enterobacteriaceae bacteria are found as normal commensals in humans but become pathogenic when they acquire virulence factor genes through plasmids, bacteriophages or pathogenicity islands. *E. coli*, associated with gastroenteritis in humans, is one such example.

Classification

Earlier, the characteristics of the colonies on commonly used mediums (such as the MacConkey medium) were used to identify and classify the members of the family Enterobacteriaceae. The colonies were classified as **lactose-fermenting bacteria** (*Escherichia* spp., *Klebsiella* spp., *Enterobacter* spp., *Citrobacter* spp. etc.) or **non-lactose-fermenting bacteria** (*Salmonella* spp., *Shigella* spp., *Proteus* spp. etc.) depending on the ability of the bacteria to ferment lactose.

However, the current practice is to classify bacteria based on several morphological, biochemical, serological and DNA-based characteristics. **Bergey manual, Kauffmann and whites classification,** and **Edwards–Ewing** classifications are the three widely used methods for classifying the members of the family Enterobacteriaceae.

As per the new classification (Ewing 1986), the family Enterobacteriaceae has been classified into eight tribes, as given in Table 35.1. The genus consisting of *Yersinia pestis*, the causative agent of a major disease plague, is included under the tribe Yersinieae in the family Enterobacteriaceae. Important properties distinguishing different genera of the family Enterobacteriaceae are summarised in Table 35.2. In 2016, a new order, Enterobacterales, was proposed with seven families: *Enterobacteriaceae, Erwiniaceae, Pectobacteriaceae, Yersiniaceae, Hafniaceae, Morganellaceae* and *Budviciaceae.*

Human infections caused by common members of the family Enterobacteriaceae are summarised in Table 35.3.

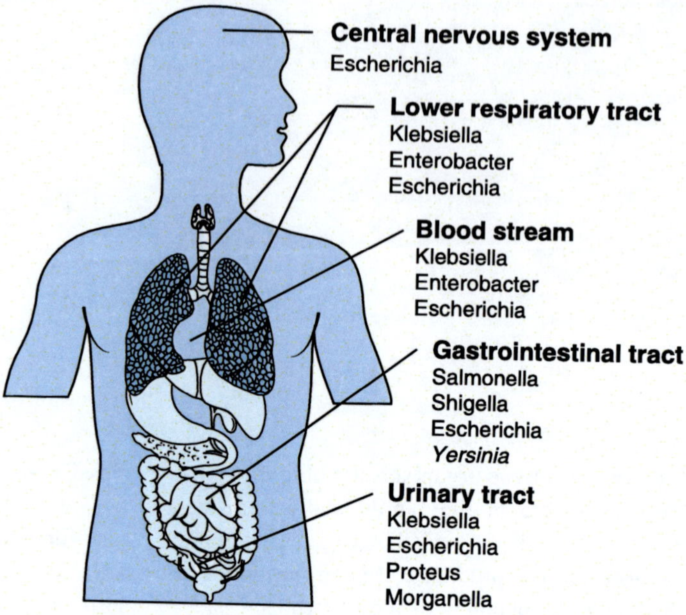

Fig. 35.1 Schematic diagram showing a variety of diseases caused by the members of the family Enterobacteriaceae in humans

Tribe Escherichieae

Escherichia is an animal and human intestinal pathogen. The genus *Escherichia* consists of five species: *E. coli*, *Escherichia fergusonii*, *Escherichia hermanii*, *Escherichia vulneris* and *Escherichia blattae*. Of these species, *E. coli* is the most common and most important species causing infection in humans (Table 35.4).

Escherichia coli

E. coli has been recognised as both a harmless commensal and versatile pathogen. *E. coli* in humans causes a broad spectrum of diseases. It is an important cause of enteric illness, urinary tract infections (UTIs), neonatal sepsis and neonatal meningitis. Haemolytic uremic syndrome (HUS) is a serious complication of enteric infection with certain *E. coli* strains.

Table 35.1 Edward and Ewing classification of Enterobacteriaceae

Tribe	Family	Genus/Genera
Tribe I	Escherichieae	*Escherichia*
		Shigella
Tribe II	Edwardsielleae	*Edwardsiella*
Tribe III	Salmonelleae	*Salmonella*
Tribe IV	Citrobactereae	*Citrobacter*
Tribe V	Klebsielleae	*Klebsiella*
		Enterobacter
		Serratia
		Hafnia
Tribe VI	Proteeae	*Proteus*
		Morganella
		Providencia
Tribe VII	Yersinieae	*Yersinia*
Tribe VIII	Erwinieae	*Erwinia*

Table 35.2 Important properties distinguishing genera of the family Enterobacteriaceae

Characteristics	*Escherichia*	*Klebsiella*	*Enterobacter*	*Salmonella*	*Shigella*	*Serratia*	*Hafnia*	*Edwardsiella*	*Citrobacter*	*Proteus*
Motility	+	–	+	+	–	+	+	+	+	+
Acid from glucose	+	+	+	+	+	+	+	+	+	+
Gas from glucose	+	+	+	+	–	V	+	+	+	V
Indole	+	–	–	–	–	–	–	+	V	V
H_2S	–	–	–	+	–	–	–	+	V	+
Urease	–	+	V			V	–	–	V	V
Citrate	–	+	V	+	–	V	+	–	+	
Phenyl–pyruvic acid (PPA)	–	–		–				–	–	+
Lysine decarboxylase	+	+	V	+	–	V	+	+	–	
Arginine dihydrolase	V	–	V	V	–					
Ornithine decarboxylase	V	–	+	+	–	V	+	+	V	V

V, Variable results in different species or strains. *Salmonella Typhi* does not produce gas from sugars. *Shigella sonnei* ferments lactose and sucrose late

Table 35.3 Human infections caused by genera of the family Enterobacteriaceae

Bacteria	Diseases
Escherichia spp.	Gastrointestinal tract infections
	Urinary tract infections
	Blood stream infections
	Lower respiratory tract infections
	Central nervous system infections
Shigella spp.	Gastrointestinal tract infections
Salmonella spp.	Gastrointestinal tract infections
	Blood stream infections
Klebsiellas spp.	Urinary tract infections
	Lower respiratory tract infections
	Blood stream infections
Proteus spp.	Urinary tract infections
Enterobacter spp.	Lower respiratory tract infections
	Blood stream infections
Morganella spp.	Urinary tract infections
Yersinia spp.	Gastrointestinal tract infections
	Blood stream infections

Table 35.4 Differentiating features of Group I and II antigens of *Escherichia coli*

Properties	Group I antigen	Group II antigen
Susceptibility to heat at 100 °C	Heat stable	Heat labile
Electrophoretic mobility	Low	High
Molecular weight	High (>100,000 Da)	Low (<50,000 Da)

Properties of the Bacteria

Morphology

E. coli is a Gram-negative bacillus measuring around 1–3 × 0.4–0.7 μm in size, arranged singly or in pairs. It is motile due to the presence of peritrichous flagella. Some strains are non-motile, and some may be fimbriated. The fimbriae are of type I (haemagglutinating and mannose-sensitive) and are present in both motile and non-motile strains. Some strains of *E. coli* isolated from extra-intestinal infections possess a polysaccharide capsule.

Culture and Biochemical Properties

E. coli is an aerobe and a facultative anaerobe. It grows in the temperature range of 10–40 °C (optimum 37 °C) and has a pH of 7.2. The bacteria grow on a wide range of media, including Mueller–Hinton agar, nutrient agar, blood agar and MacConkey agar. Primary isolation can be made on nutrient agar and blood agar.

E. coli on **nutrient agar** after 18 h of incubation at 37 °C produces large, circular, low convex, greyish-white, moist, smooth, opaque or partially translucent colonies (**smooth** or **S forms**). These smooth colonies are easily emulsifiable in saline. The **rough** or **R forms** produce rough colonies with an irregular dull surface. These colonies are often autoagglutinable in saline. The smooth to rough variation (S–R variation) is associated with the loss of surface antigens and also with the loss of virulence and occurs as a result of repeated sub-culturing.

E. coli produces bright pink, flat colonies due to lactose fermentation on **MacConkey agar**. Many strains, especially those isolated from pathological conditions, produce beta-haemolytic colonies on **blood agar**. They do not grow on selective media, such as deoxycholate citrate agar

(DCA) or *Salmonella–Shigella* (SS) agar, used for the culture of salmonellae and shigellae.

The indole, methyl red (MR)-, Voges–Proskauer (VP)- and citrate utilisation tests, generally referred to as the "IMViC" tests are four important biochemical tests widely used in the classification of enterobacteria. *E. coli* is indole- and MR-positive and VP- and citrate-negative (IMViC + + – –) (Table 35.2). Some strains of *E. coli* are late lactose- or non-lactose fermenters and even produce H_2S. The ability to produce H_2S (positive variants) and use citrate by *E. coli* is controlled by transmissible plasmids.

Other Properties

Susceptibility to Physical and Chemical Agents

E. coli is inhibited by the presence of 7% sodium chloride in salt media used for isolation of staphylococci. Growth of the bacteria is also inhibited by sodium selenite in selenite broth, sodium tetrathionate in tetrathionate broth and brilliant green in brilliant green tetrathionate broth.

Antigenic Structure

The heat-stable lipopolysaccharide (LPS) is the major cell-wall antigen of *E. coli*. LPS consists of three components: (1) the genus-specific somatic O polysaccharide, (2) a core polysaccharide common to all Enterobacteriaceae (common antigen) and (3) lipid A.

E. coli organisms possess four major antigens: H or flagellar antigen, O or somatic antigen, K or capsular antigen and F or fimbrial antigens.

H or Flagellar Antigen

These are heat- and alcohol-labile proteins present on the flagella. The H antigens are genus specific and usually are not shared by other enterobacteria. All H antigens are present as monophasic but very rarely as diphasic. A total of 75 H antigens have been recognised so far.

O or Somatic Antigen

These occur on the surface of the outer membranes and are determined by specific sugar sequences on the cell surface. O antigen is an LPS complex and is an integral part of the cell wall. It is heat stable, resistant to boiling up to 2.5 h. Till now, 173 (1–173) O antigens have been described. Somatic O polysaccharide antigen shows cross-reactions with related genera (*Shigella*, *Salmonella*, *Yersinia* and *Citrobacter*) in the family Enterobacteriaceae.

K or Capsular Antigen

It is the acidic polysaccharide, a heat-labile antigen present in the "envelope" or microcapsule (K for *Kapsel*, German for capsule) of the bacteria. K antigen encloses the O antigen and may interfere with the detection of the O antigens. This problem is overcome by boiling the bacterial suspension to remove the K antigens. A total of 103 K antigens have been recognised.

F or Fimbrial Antigens

These antigens are present on the fimbriae and are heat-labile proteins. Several filamentous protein structures resembling fimbriae have been described in *E. coli*. These are K88 and K99 antigens in *E. coli* strains causing diarrhoea in animals or colonisation factor antigens (CFA) in enterotoxigenic *E. coli* (ETEC) causing diarrhoea in humans. These fimbrial antigens also contribute to the virulence of the bacteria.

Antigenic Typing

The serotyping of *E. coli* is based on three major groups of antigens: O antigens, K antigens and H antigens. *E. coli* strains based on O antigens are initially divided into many O groups. Each O group is then further divided into sub-groups based on K antigens. Finally, each of the sub-group includes strains with different H antigens. The antigenic pattern of a strain is recorded depending on the number of the particular antigen it carries (e.g. *E. coli* O111:K58:H2).

Different serotypes of *E. coli* are found in the normal intestine of humans, and they do not have K antigens. The normal colon strains belong to the ***early*** O groups (1, 2, 3, 4 etc.). Enteropathogenic strains responsible for intestinal diseases belong to the "***later***" O groups (55, 86, 111, 112 etc.).

Table 35.5 Virulence factors of *Escherichia coli*

Virulence factors	Biological functions
Fimbriae	Adherence of bacteria to gastrointestinal tract; of importance in UTIs; and cause mannose-resistant haemagglutination
Endotoxin	Systemic manifestations of endotoxic shock; and protect the bacillus from phagocytosis and from the bactericidal effects of complement
Capsule	Antiphagocytosis protects *Escherichia coli* from phagocytosis; and protect the organism from serum killing
Sequestration of growth factors	The capability of *Escherichia coli* to compete for nutrients in host cells
Adhesins	Adhesins facilitate firm adhesion of *Escherichia coli* to the gastrointestinal or urinary tract mucosa, thereby preventing the bacteria being eliminated by the flushing action of voided urine or intestinal motility
Haemolysins (HlyA)	Important in the pathogenesis of disease caused by uropathogenic strains of *Escherichia coli*
Enterotoxins	
Heat-stable toxin (ST)	Causes increased secretion of fluids
Heat-labile toxin (LT)	Watery diarrhoea due to hypersecretion of fluid into the lumen of the gut

Pathogenesis

E. coli is an invasive bacterium.

It colonises the human intestine and, under specific conditions, directly invades the intestinal mucosa or produces toxins to cause intestinal infections. The bacteria can enter the bloodstream and cause septicaemia, meningitis and other systemic manifestations. The bacteria, under certain conditions, directly invade the urinary tract causing UTIs or cause intraabdominal infections.

Virulence Factors

E. coli produces several virulence factors (Table 35.5).

Surface Antigens

The somatic O LPS antigen protects *E. coli* from phagocytosis and the bactericidal effects of complement proteins. The capsular K antigens also protect *E. coli* from phagocytosis. They interfere with the binding of antibodies to the bacteria. Most strains of *E. coli* responsible for neonatal meningitis and septicaemia possess the KI envelope antigen, a virulence factor similar to the group B antigen of meningococci. *E. coli* and other enteric bacteria capable of producing systemic infections are frequently resistant to serum killing. **Fimbriae**: It plays an important role in the pathogenesis of UTIs caused by *E. coli*. Surface antigens, such as K88 and K99 antigens in *E. coli*, strains causing diarrhoea in animals or CFA in ETEC are examples of such fimbriae.

Exotoxins

E. coli also produces two types of exotoxins: (1) **haemolysins,** and (2) **enterotoxins.**

1. **Haemolysins** are considered important in the pathogenesis of disease caused by uropathogenic strains of *E. coli.*
2. **Enterotoxins** are important virulent factors of *E. coli*. Three distinct types of *E. coli* enterotoxins have been recognised. These include (1) Shiga toxins (Stx-1 and Stx-2), (2) STa and STb and (3) LT-I and LT-II.
   (a) **Shiga toxins (Stx)**: Stx is so named because its physical, antigenic and biological properties are similar to the *Shigelladysenteriae* type 1 toxin. The toxin is also named **verocytotoxin** or **verotoxin (VT)** because the toxin was first detected because of its cytotoxic effect on Vero cells. Stxs are of two types: **Stx-1**and **Stx-2**. Lysogenic bacteriophages encode both toxins. Although they show the same biological

activity, both the toxins are antigenically different. Shiga toxins demonstrate cytotoxicity activities in Vero or HeLa cells. The toxin also shows enterotoxicity in rabbit ileal loops and paralytic lethality in the mouse.

(b) ***E. coli* heat-stable toxin (STs)**: STs are low-molecular-weight proteins and are of two types: **STa** and **STb.** STa is associated with disease in humans. STa is a small, methanol-soluble, monomeric toxin that activates cyclic guanosine monophosphate (cGMP) in the intestine. STa binds to guanylate cyclase leading to an increase in the level of cGMP and a subsequent increase in the secretion of fluids. The toxin acts very rapidly and causes accumulation of fluid in the intestines of infant mice within 4 h of intra-gastric administration; hence infant mouse is a frequently used animal model for demonstrating STa. The toxin also causes fluid accumulation in the intestine of neonatal but not weaned piglets. STb is not associated with human diseases. STb, unlike STa, is not methanol soluble. The exact mode of action of STb is not known. The toxin causes the accumulation of fluid in ligated intestinal loops of young piglets up to 9 weeks of age but not in infant mice.

(c) ***E. coli* heat-labile toxin (LT):** LT is a heat-labile protein. The toxin was first demonstrated by De et al. in 1956 in *E. coli* isolated from cases of adult diarrhoea in adults in Kolkata. They demonstrated the toxin in the bacteria by the rabbit ileal loop method, the method used to detect the cholera enterotoxin. LT is of two types: LT-I and LT-II. LT-I, but not LT-II, is associated with human diseases. LT-I is structurally and antigenically similar to the cholera toxin. The LT-I toxin consists of one A subunit of molecular weight of 25,000 Da and five identical B subunits, each sub-unit measuring 11,500 Da.

The B subunits bind to the GM1 gangliosides, the same receptor as cholera toxin, and other surface glycoproteins on the epithelial cells in the small intestine. This binding facilitates the entry of sub-unit A into the cell by endocytosis. The A subunit has adenosine diphosphate (ADP)-ribosyltransferase activity, by which it interacts with a membrane protein (Gs) that regulates adenylate cyclase. This increases cyclic adenosine monophosphate (cAMP) levels leading to increased secretion of chloride and a decreased absorption of sodium and chloride. This ends in watery diarrhoea due to hypersecretion of fluid into the lumen of the gut. The toxin also stimulates the secretion of prostaglandin and the production of inflammatory cytokines, resulting in further fluid loss. Genes for LT-I and STa are present on the I transferable plasmid, which can also carry the genes for adhesins (CFA/I, CFA/II and CFA/III).

Adhesins

E. coli possess numerous highly specialised **adhesins**. These adhesins include (1) CFAs (CFA/I, CFA/II and CFA/III), (2) aggregative adherence fimbriae (AAF/I, AAF/II and AAF/III), (3) bundle-forming pili (Bfp), (4) intimin, (5) P pili (binds to P blood group antigens), (6) invasion plasmid antigen (Ipa) protein and (7) Dr. fimbriae (bind to Dr. blood group antigens). All these adhesins facilitate firm adhesion of *E. coli* to the gastrointestinal or urinary tract mucosa, thereby preventing the bacteria from being eliminated by the flushing action of voided urine or intestinal motility.

Siderophores

These remove iron from mammalian iron transport proteins, such as transferring a lactopherin.

Pathogenesis of *E. coli* infection: *E. coli* colonises the human intestine and, under specific conditions, directly invades the intestinal mucosa or produces toxins to cause intestinal infections. The bacteria can enter the bloodstream and cause septicaemia, meningitis and other systemic manifestations. The bacteria, under certain conditions, directly invade the urinary tract causing UTIs or cause intraabdominal infections.

Clinical Syndromes

E. coli causes (1) urinary tract infections (UTIs), (2) diarrhoea, (3) septicaemia, (4) neonatal meningitis, and (5) other infections.

Urinary Tract Infections

E. coli is the most common bacteria responsible for causing more than 80% of all community-acquired UTIs. It causes a wide range of UTIs, including uncomplicated urethritis or cystitis, symptomatic cystitis, pyelonephritis, acute prostatitis, prostatic abscess or urosepsis. Uncomplicated cystitis occurs primarily in sexually active females who are colonised by uropathogenic strains of *E. coli*. Subsequently, the periurethral region is colonised by *E. coli* due to faecal contamination, and the bacteria reach the urinary bladder during sexual intercourse.

E. coli serotypes that are normally found in the faeces are commonly responsible for UTIs. UTI is an ascending infection in which the bacteria that originate from the intestinal tract of the same host and contaminate the urethra ascend into the bladder and may spread to the kidney or prostate.

Although most strains of *E. coli* can cause UTI, the disease is more common with certain specific *E. coli* serogroups. These serogroups that cause UTI are known as *nephritogenic strains*. These include *E. coli* serotypes O1, O2, O4, O6, O7 and O18. These serotypes, particularly because of their ability to produce adhesins (primarily P pili, AAF/I, AAF/III and Dr), which bind to cells lining the bladder and the upper urinary tract, cause UTI. This prevents the elimination of the bacteria in voided urine. They also produce haemolysins, which lyses erythrocytes and also other cells, leading to the release of cytokines and stimulation of an inflammatory response.

Diarrhoea

The strains of *E. coli* that cause gastroenteritis are classified into the following six groups (Table 35.6).

Enteropathogenic *E. coli* (EPEC)

It is the major cause of **infant diarrhoea** in tropical countries. EPEC strains include O26, O55, O86, O111, O114, O119, O125, O126, O12, O128 and O142. EPEC causes infection by adhering to epithelial cells of the small intestine, followed by the destruction of the microvilli. Subsequently, the attached bacteria multiply and cause microvilli destruction, resulting in diarrhoea due to malabsorption.

Diarrhoea Caused by Enterotoxigenic *E. coli*

The condition is endemic in developing countries, amongst all population age groups. This is also responsible for causing traveller's diarrhoea in which individuals from developed countries visit

Table 35.6 Gastroenteritis caused by *Escherichia coli*

Disease	Site and mechanisms of action	Causative type
Traveller's diarrhoea	Heat stable/heat labile toxin Acts on small intestine	ETEC
Infant diarrhoea	Acts on small intestine	EPEC
Dysentery	Invasion and destruction of epithelial cells Acts on large intestine	EIEC
Haemorrhagic colitis (HC)	Shiga toxin	EHEC
HUS	Acts on large intestine	
Infant diarrhoea	Adhesion Acts on small intestine	EAEC
Infant diarrhoea	Adhesion Acts on small intestine	DAEC
Chronic diarrhoeal disease	Acts on small intestine	EAEC

endemic areas. The disease is caused by the consumption of faecally contaminated food or water. Diarrhoea is caused by certain specific ETEC serogroups (O6, O8, O15, O25, O27 and O167). These serotypes produce heat-labile enterotoxins (LT-I, LT-II) and cause diarrhoea. LT-I, which is structurally similar to cholera toxin, produces cholera-like diarrhoea in patients. The disease process is facilitated further by the presence of adhesins (primarily P pili, AAF/I, AAF/III and Dr), which bind to the intestinal mucosa.

Enteroinvasive *E. coli* (EIEC)

These strains closely resemble shigellae in many ways: (1) EIEC strains are non-motile, (2) they do not ferment lactose or ferment late with the production of acid only and (3) they do not decarboxylate lysine. These strains show cross-reactivity with the O antigen of shigellae. These **"atypical"** *E.coli* strains earlier were named *Shigellaalkalescens* under the "**Alkalescens-Dispar Group**" (resembling *Shigellaflexneri* except in fermenting dulcitol and forming alkali in litmus milk) and *Shigelladispar* (late lactose fermenter-like *Shigella sonnei* but indole-positive). These have been renamed EIEC because they can invade interstitial epithelial cells and penetrate HeLa cells in tissue culture. The EIEC strains can invade and destroy the colonic epithelium, producing a disease characterised initially by watery diarrhoea. Specific serogroups commonly associated with outbreaks of EIEC include O28 ac, O112 ac, O124, O136, O143, O114, O152 and O154.

Enterohaemorrhagic *E. coli* (EHEC)

These strains are the most common cause of gastrointestinal infections in developed countries. These strains produce diarrhoeal disease, ranging from mild, uncomplicated diarrhoea to fatal haemorrhagic conditions such as haemolytic uraemic syndrome.

EHEC disease is most common in children below 5 years and during the summer months. The condition occurs as a result of ingestion of water, unpasteurised milk or fruit juices, uncooked vegetables and fruits contaminated with human or animal faeces. The disease also occurs on the consumption of undercooked ground beef or other meat products. The condition is characterised by acute renal failure, thrombocytopenia and microangiopathic haemolytic anaemia. Serotypes O157:H7 and O26:H1 are the EHEC strains that commonly cause the disease. The infection typically lasts for 4–10 days.

Haemolytic uraemic syndrome is a serious life-threatening complication in 10% of infected children below 10 years. The ingestion of as few as 100 bacilli can cause the disease. The condition manifests as watery diarrhoea to severe HC. Watery diarrhoea is often accompanied by abdominal cramping and vomiting. In most patients, diarrhoea becomes bloody in 1–2 days but is usually not associated with faecal leukocytes.

Enteroaggregative *E. coli* (EAEC)

They are so-called because they show a typical **"stacked brick"** arrangement on Hep-2 cells or glass due to their autoagglutination. Bundle-forming fimbriae of the bacteria (such as AAF/I and AAF/II), which are carried on a plasmid, mediate this process. These strains are associated with persistent, watery diarrhoea with dehydration in infants, especially in developing countries.

Diffusely Adherent *E. coli* (DAEC)

The strains cause watery diarrhoea found primarily in children between 1 and 5 years of age. These strains are identified by their ability to adhere to cultured cells. They cause elongation of the microvilli with the bacteria trapped in the cell membrane.

Septicaemia

E. coli causes septicaemia, which is usually associated with UTI, especially in urinary obstruction of any cause. The endotoxin or LPS of the bacteria causes a systemic reaction that can lead to disseminated intravascular coagulation and even death. Mortality and morbidity associated with *E. coli* septicaemia are the same for other aerobic Gram-negative bacilli.

Neonatal Meningitis

E. coli, along with group B streptococci, are the major causes of infection of the CNS in infants who are 1 month of age. In adults, it occurs following neurosurgical procedures, CNS trauma or complicating *Strongyloides stercoralis* hyperinfection involving the CNS. Neonatal meningitis has a high mortality rate of 8%. The disease is caused by *E. coli* strains that possess the KI capsular antigen, which is commonly present in the gastrointestinal tracts of pregnant women and newborn infants.

Other Infections

These include intraabdominal infections due to *E. coli*, often resulting from a perforated appendix, diverticulum or due to intraabdominal abscess, cholecystitis and ascending cholangitis. Septic arthritis, endophthalmitis, sinusitis, osteomyelitis, endocarditis or skin and soft tissue infections are the other pyogenic infections caused by *E. coli.*

Laboratory Diagnosis of UTI

Specimens

Freshly voided urine is the specimen of choice for diagnosing UTI caused by uropathogenic *E. coli.* Clean-voided midstream samples of urine are usually employed for culture. Catheterised urine and urine collected by suprapubic aspiration are used in certain situations. Urine is a good medium for the growth of coliforms and other urinary pathogens and hence, should be sent immediately to the microbiology laboratory for processing the specimen. If a delay of more than 1–2 h is unavoidable, the specimen should be refrigerated.

Urine Culture

The most commonly used method is a semi-quantitative urine culture for *E. coli* and other Gram-negative bacteria. In this method, a predetermined quantity of urine is inoculated on MacConkey and blood agar with a standardised inoculating loop. The loop is calibrated to deliver 0.05 mL of urine; hence 200 loopful of urine will deliver 1 mL of urine. The number of colonies obtained after overnight incubation of inoculated plates multiplied by 200 will be the approximate number of bacteria/mL of urine. For example, if the number of colonies on a MacConkey agar after overnight incubation is 500, the viable bacterial count/mL of urine will be (500 × 200) 100,000 or 10^5. This forms the basis of significant bacteriuria suggested by Kass. After incubation overnight at 37 °C, pink, flat colonies of *E. coli* on the MacConkey agar (Fig. 35.2) and beta-haemolytic colonies on the blood agar are identified by various biochemical tests.

In recent years, two new strategies have been tried to increase the sensitivity of the culture method. The standard method is used in the modified urine culture, the same set of culture media (Blood agar and MacConkey's agar) and urine volume. These plates are incubated for up to 48 h in 5% CO_2 at 35 °C. In the Expanded Quantitative Urine Culture Method, a larger volume of urine (0.1 ml) is used to inoculate multiple culture medium plates are incubated in aerobic, anaerobic and microaerophilic atmospheres at 35 °C for up to 48 h. The disadvantage of this method is the use of a large number of culture plates instead of only two in standard and modified methods.

The significant bacteriuria concept suggested by Kass is based on the fact that a colony count exceeding 100,000 (10^5) bacteria/mL of urine denotes significant bacteriuria and is suggestive of an active UTI. Counts of 10,000 bacteria or less per millilitre are of no significance and are due to urine contamination during voiding. Bacterial counts between 10,000 (10^4) and 100,000 (10^5) are infrequent when the sample is collected properly and processed promptly. Such results are considered equivocal, and the culture is repeated. The concept of significant bacteriuria is applicable only for *E. coli* and other Gram-negative bacteria, and for midstream urine specimens. It is not applicable for the urine specimens collected directly

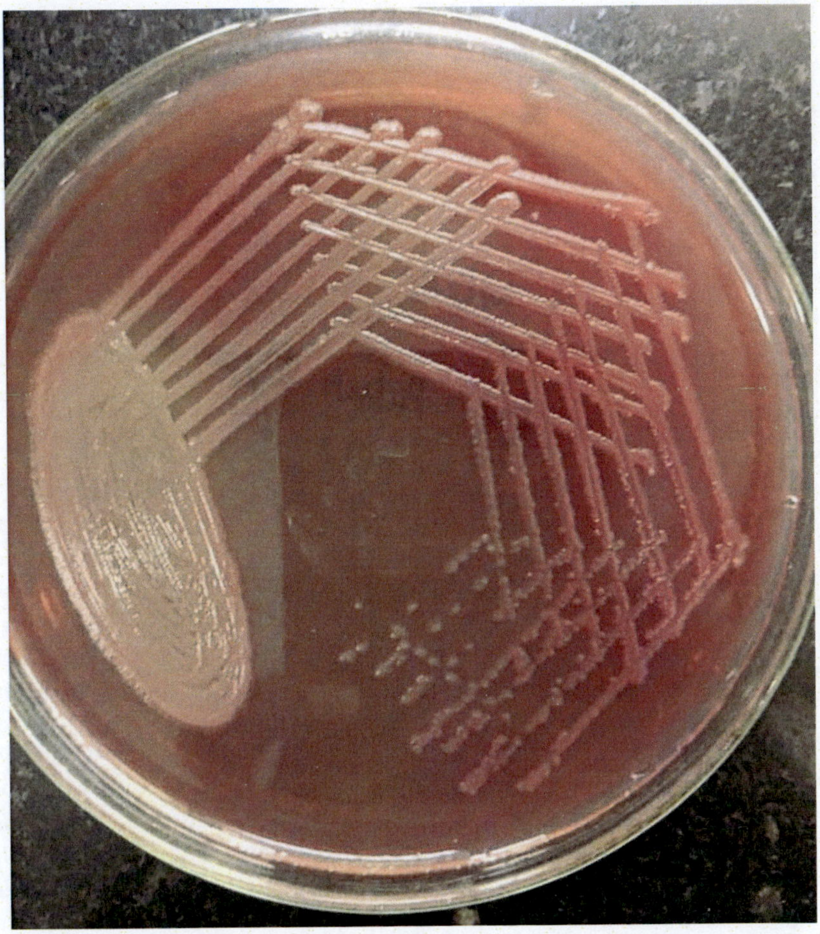

Fig. 35.2 Lactose fermenting colonies of *Escherichia coli* on Mac Conkey medium

from the urinary bladder by cystoscopy and for Gram-positive bacteria, such as *Staphylococcus aureus*, in which even low counts may be significant.

However, interpretation of bacteriuria requires caution and should always be with reference to the patient's clinical condition. Since UTI is a common problem and culture facilities are not available everywhere, several simple methods have been introduced for the presumptive diagnosis of significant bacteriuria.

Laboratory Diagnosis of Diarrhoea

Specimens

The faeces are collected from the patient in a sterile container and sent immediately to the laboratory.

Stool Culture

Stool culture is widely used to isolate diarrhoeagenic *E. coli*. The faecal samples are inoculated directly on MacConkey and blood agar media. The plates are incubated at 37 °C overnight and looked for the characteristic lactose-fermenting colonies on MacConkey and beta-haemolytic colonies on blood agar as described earlier.

Since *E. coli* is present as commensals in the intestine, hence is detected even in the normal stool. Therefore, it is essential to perform various diagnostic tests to consider a diarrhoeagenic pathogenic *E. coli* strain. These strains are identified by serotyping, animal inoculation, cytopathic effects in cell cultures or molecular methods.

Identification of EPEC

Specific serogroups of *E. coli* (O26, O55, O86, O111, O114, O119, O125, O126, O12, O128 and O142) are commonly associated with outbreaks of EPEC. Therefore, *E. coli* colonies isolated from faeces on MacConkey agar are identified by agglutination tests with specific polyvalent and monovalent antisera. In this method, a saline suspension of *E. coli* colonies is made on the slide and mixed with a drop of specific polyvalent and monovalent antisera against EPEC serogroups. At least ten colonies/plates should be tested. If isolated colonies are negative, the confluent growth is emulsified and tested. In a positive test, if *E. coli* colonies show agglutination with a specific serogroup (e.g. O111), then the isolate is identified as *E. coli* of that serogroup (O111).

Identification of ETEC

Diagnosis of ETEC diarrhoea depends on the demonstration of enterotoxin in *E. coli* isolates from stool because toxin production is not associated with specific serogroups of *E. coli.* A strain of ETEC may produce either LT or ST or both. Many tests are available to demonstrate the LT or ST produced by ETEC (Table 35.7).

LT in isolates of *E. coli* from faeces can be identified by (1) fluid accumulation in rabbit ileal loop method, (2) vascular permeability factor of the toxin in adult rabbit skin method, (3) tissue culture tests [rounding of Y1 mouse adrenal cells and elongation of Chinese hamster ovary cells (CHO) cells due to an intracellular increase in cAMP concentration], (4) enzyme-linked immunosorbent assay) and (5) DNA probes. The **Biken test** is an *in vitro* technique based on the principles of the Elek test and the Ouchterlony double gel diffusion test. Wells are cut in an agar plate, and suspected LT-producing *E. coli* is placed in one well while the other well contains anti-LT antiserum. A line of precipitation forms if the *E. coli* strain is as ETEC.

The presence of ST in isolates of *E. coli* can be demonstrated by (1) the frequently employed infant mouse test, and (2) DNA probes.

Identification of EIEC

Many EIEC strains are atypical *E. coli* strains. They are non-motile and do not ferment lactose or ferment it late with acid production but without producing any gas. They also do not decarboxylate lysine. The following tests identify EIEC.

Sereny test is carried out in guinea pigs by instilling isolated EIEC into the conjunctival sac of guinea pigs. The animal is examined after 72 h for mucopurulent conjunctivitis and severe keratitis. **Cell cultures** are used to demonstrate the toxin also. In this method, the bacterial suspension is added to a monolayer of the cells, either HeLa or HEP-2. The cells are then examined for intracellular *E. coli* because EIEC, if present, penetrates these cells and replicates inside the cells. **VMA enzyme-linked immunosorbent assays** are used to detect the plasmid, which codes for outer membrane antigens called the VMA in stool isolates of EIEC.

Table 35.7 Methods of detection of LT and ST of *Escherichia coli*

Type of toxin of *Escherichia coli*	*In-vivo* tests	*In-vitro* tests
LT	Injection into closed ligated rabbit illeal loops produces outpouring of fluid and ballooning of the loops Increases skin capillary permeability in adult rabbit skin (vascular permeability factor)	Causes rounding and steroid production from Y1 mouse adrenal cells Causes elongation of Chinese Hamster ovary cells ELISA DNA probes
ST	Induces fluid accumulation in the intestines of infant mice within 4 h of intragastric administration. Induces fluid accumulation in the intestinal loops of neonates but not weaned piglets	ST–ELISA DNA probes

Identification of EHEC

E. coli O157:H7 is the most common serotype associated with the clinical disease caused by EHEC strains. The strain typically does not ferment sorbitol; hence the sorbitol–MacConkey medium is frequently used to isolate the strain from the stool by culture. Identification of EHEC strains is by (1) demonstrating cytotoxic effects of EHEC on Vero or HeLa cells, and (2) using DNA probes for VT1, and VT2 genes in EHEC directly in faeces or in culture isolates.

Identification of Other Strains

EAEC strains are identified by agglutination tests with specific antisera. Most of them are not typed by O antisera but by specific H antisera.

In recent years, multiplex PCR has come into use which can simultaneously detect all the important diarrhoeagenic *E. coli* at the same time without the necessity of performing individual PCR for each type. The gene targets selected for each group comprise the following: *eae* or *bfpA* for EPEC isolates, CVD432 for EAEC isolates, *elt* and *stla* for ETEC isolates, *stx1* and *stx2* or *hlyA* for EHEC isolates, and *ipaH* or *ial* for EIEC isolates.

Laboratory Diagnosis of Septicaemia, Neonatal Meningitis and Other Infections

The specimens include blood for septicaemia, cerebrospinal fluid (CSF) for meningitis, sputum for pneumonia or other body fluids, such as pus from the wound, and biliary and peritoneal abscesses caused by *E. coli*. The culture of blood, CSF and other specimens is also carried out depending on the clinical diseases caused by *E. coli*.

Molecular Detection

Polymerase chain reaction (PCR) is currently being evaluated for bacterial culture isolates and directly from specimens and is commercially available. This approach is necessary because sepsis may be caused by various microorganisms, including both bacteria and fungi. Although considered the "gold standard", simple blood culture has serious drawbacks and may not support the growth of some fastidious organisms. In addition, sepsis is a rapidly developing life-threatening condition that requires prompt definitive diagnosis, and the long turnaround time of blood culture is a distinct disadvantage in such a situation.

The nucleic acid amplification tests can be divided into three types: those that target either a species or a genus (pathogen-specific); those that target pan-bacterial (16S, 5S and 23S rRNA genes) or pan-fungal (18S, 5.8S and 28S ribosomal DNAs) genes (broad specific); and finally multiplex PCR aimed at parallel detection and targeting the most important pathogens. A commercial multiplex system (BioFire™ from Biomerieux) tests for 43 targets and ten resistance genes with a turnaround time of 1 h. However, the pathogen must be grown in culture before the test can be performed. Another technique in the DNA microarray system uses microchips coated with several targets to identify a broad range of pathogens.

Treatment

E. coli isolated from community-acquired infections are usually sensitive to commonly used antibiotics except for penicillins. Hospital-acquired *E. coli* isolates, however, show multidrug resistance. Most *E. coli* infections are best treated based on antibiotics susceptibility testing results. UTI and neonatal sepsis are treated with appropriate antibiotic therapy. Treatment of bacterial gastroenteritis is primarily supportive. Antibiotic therapy is rarely indicated and is deferred until culture results are available. Presumptive therapy is not given due to the potential risks associated with antibiotic treatment for diarrhoea caused by *E. coli* O157:H7. Third-generation cephalosporins, such as ceftriaxone, are recommended for meningitis and pneumonia caused by *E. coli*.

Epidemiology

Geographical Distribution

There is a wide variation in the prevalence of diarrhoeagenic *E. coli* strains worldwide. EPEC, EAEC and DAEC have been reported to occur most often in developing countries. The WHO Global Burden of Foodborne Diseases report shows that >300 million illnesses and nearly 200,000 deaths are caused by diarrheagenic *E. coli* globally every year. ETEC is the primary cause of traveller's diarrhoea and is the major cause of infantile diarrhoea in developing countries; the majority of cases of ETEC occur in children less than 2 years of age. ETEC is a cause of traveller's diarrhoea estimated to be 30–45%. As per the WHO, it is estimated that ETEC causes about 220 million diarrhoea episodes globally, with about 75 million episodes in children under 5 years of age and up to 42,000 deaths. EHEC is an emerging cause of food-borne disease, particularly in the northern United States and Canada. Outbreaks of HUS in children that caused fatalities have also been documented in these countries.

Habitat

Most strains of *E. coli* causing diseases in humans inhabit the human intestine. Few strains, such as EHEC, inhabit the cattle intestine also.

Reservoir, Source and Transmission of Infection

Humans and also cattle (for *E. coli* O157:H7) are reservoirs of infection.

Most *E. coli* infections, except for neonatal meningitis and gastroenteritis, are endogenous. The infection is caused by *E. coli* that constitute the patient's normal microbial flora. These bacteria cause infections, such as UTIs, due to poor personal hygiene or compromised patient defences.

Diarrhoea caused by *E. coli* is an exogenous infection acquired from water, food or vegetables contaminated with faecal *E. coli*. The condition is widespread in areas with poor sanitation, and infections are exogenous. Food and water contaminated with human or cattle faeces are important sources of diarrhoeagenic *E. coli*. Ground beef, apple juice and alfalfa sprouts contaminated with cattle faeces are important sources of infection caused by EHEC (e.g. O157:H7). The usual source of *E. coli* in neonatal infections is the intestinal tract of the mother. The bacteria may also be acquired nosocomially, particularly in premature infants or those who require mechanical ventilation.

Prevention and Control

Availability of safe drinking water, proper food hygiene and sanitary disposal of excreta are the most cost-effective strategies for reducing the incidence of enteric infections caused by *E. coli*. Cooking ground beef thoroughly is the most effective way to prevent HC caused by *E. coli* O157:H7.

Disease caused by EPEC is rare in older children and adults, presumably because they have developed protective immunity. Immunity develops to ETEC surface antigens in local adult populations; hence the disease is confined to immunologically naïve travellers and weaning infants.

Tribe Edwardsiellae

The genus *Edwardsiella* differs from the genus *Escherichia* by its ability to produce H_2S. The genus *Edwardsiella* includes *Edwardsiella tarda*, the only pathogenic species for humans. *E. tarda* inhabits the intestines of snakes and other cold-blooded animals. The name *tarda* refers to the bacteria's slow or weak fermentation of sugars. *E. tarda* is a Gram-negative, non-capsulated, motile bacillus with weak fermentative powers. *E. tarda* is an occasional human pathogen isolated from wounds, blood and CSF in fatal meningitis. The bacteria have also been isolated from the stool of normal healthy people and that of patients with diarrhoea. Nevertheless, the pathogenic role of the bacteria

Table 35.8 Important tests for differentiation of *Citrobacter* species

Tests	*Citrobacter freundii*	*Citrobacter koseri*	*Citrobacter amalonaticus*
Indole test	Negative	Positive	Positive
H_2S production	Positive	Negative	Negative
Acid from salicin	Negative	Positive	Positive
Acid from malonate	Negative	Positive	Negative
Acid from adonitol	Negative	Positive	Negative

in the causation of diarrhoea is yet to be established.

Tribe Citrobacteriaceae

Citrobacter is a normal inhabitant of the intestine of humans. The genus *Citrobacter* consists of three species, *Citrobacter freundii*, *Citrobacter amalonaticus* and *Citrobacter koseri* (formerly *Citrobacter diversus*). They grow well on nutrient agar and other ordinary media, producing smooth and convex colonies. The colonies are not pigmented. On MacConkey and DCA media, they produce pale colonies. *Citrobacter* spp. are motile, H_2S positive, MR positive, citrate positive and indole variable. They do not decarboxylate lysine, but most strains decarboxylate ornithine. They ferment lactose late or do not ferment at all. Differences between the three species in their biochemical characteristics are summarised in Table 35.8. They show extensive antigenic sharing with salmonellae, hence may be mistaken for salmonellae. Certain strains possess a Vi antigen, closely related to the antigen of *Salmonella* typhi and *Salmonella* paratyphi. *Citrobacter* spp. may cause infections of the urinary tract, gall bladder, middle ear and meninges. *C. koseri* may occasionally cause meningitis in neonates.

Tribe Klebsielleae

The bacteria *Klebsiella* are named after Edwin Klebs, who demonstrated the bacteria for the first time. Based on DNA homology, the genus *Klebsiella* has been divided into seven species: *K. pneumoniae*, *Klebsiella ozaenae*, *Klebsiella rhinoscleromatis*, *Klebsiella oxytoca*, *Klebsiella planticola*, *Klebsiella terrigena* and *Klebsiella ornithinolytica.*

Klebsiella pneumoniae

K. pneumoniae, also known as *Friedlander bacillus*, was first isolated by Friedlander in 1883 from fatal cases of pneumonia. *K. pneumoniae* are Gram-negative, short and straight rods measuring about 1–2 × 0.5–0.8 μm in size. They are arranged singly or in pairs. Freshly isolated strains show a well-defined polysaccharide capsule. The capsule is often prominent and can be made out even in Gram-stained smears as haloes around the bacilli. It is produced well when grown in media enriched with carbohydrates. The capsule can also be demonstrated by India ink preparation and Quellung reaction. Accumulation of extracellular polysaccharides as a loose slime gives a mucoid appearance to *Klebsiella* colonies.

They are non-motile and non-sporing. They are lactose fermenting, urease positive and indole negative. However, some strains of *K. pneumoniae* and *K. oxytoca* are exceptions. They grow well on ordinary media, such as nutrient agar and MacConkey agar at 37 °C, forming large, dome-shaped, mucoid colonies. They produce lactose-fermenting red colonies on MacConkey agar. *Klebsiella* possesses 79 capsular polysaccharides (K antigens) and more than 12 somatic LPS (O antigens). O1, O2 and O3 and the K2 are the most prevalent types infecting humans and the serotypes K1 and K2 have been associated with bacteremia.

Klebsiella causes a variety of clinical manifestations in humans.

Community-Acquired Pneumonia

It is a serious condition with a rapid onset and often fatal outcome despite early and appropriate antimicrobial treatment. Lobar pneumonia characteristically is associated with massive mucoid inflammatory exudate of lobar or lobular distribution, involving one or more lung lobes. *K. pneumoniae* serotypes 1, 2 and 3 are usually associated with the condition. Patients present with an acute onset of high fever and chills, flu-like symptoms and productive cough with abundant, thick, tenacious and blood-tinged sputum. Blood culture is positive in about 25% of the cases.

UTI

It is caused by *K. pneumoniae* is a common problem in patients with indwelling catheters. *K. pneumoniae* are emerging as important agents of nosocomial infections in hospitals. The presence of invasive devices, contamination of respiratory support equipment, use of urinary catheters and antibiotics greatly increases the likelihood of nosocomial infections in hospitalised patients.

Bacteraemia and Sepsis

The conditions are caused by *K. pneumoniae* show clinical manifestations similar to those caused by *E. coli* and other Gram-negative enteric organisms.

In neonatal units, outbreaks of infection caused by extended-spectrum beta-lactamase (ESBL)-producing *Klebsiella* strains present a more serious problem and may be associated with high mortality. ESBL strains of *Klebsiella* are usually highly virulent, possess capsular type K55 antigen and have an extraordinary ability to spread.

Laboratory Diagnosis

Laboratory diagnosis of *K. pneumoniae* infection is made by isolating bacteria from clinical specimens obtained from possible sites (e.g. wounds, peripheral or central intravenous access sites, urinary catheters, and respiratory support equipment) by culture (Fig. 35.3). *Klebsiella* may also be isolated from urine, blood, pleural fluid and wounds. Serological tests are not useful for the detection of infection with *K. pneumoniae*.

The choice of a specific **antimicrobial agent** depends on the antibiotic-susceptibility patterns of the isolated strains. Cephalosporines are widely used as monotherapy and in combination with aminoglycosides. Cephalosporines are not used for ESBL strains of *K. pneumoniae*. The carbapenems, especially imipenem, are effective against such ESBL strains. Aztreonam and quinolones are recommended for patients allergic to penicillins.

Klebsiella rhinoscleromatis

Rhinoscleroma caused by *K. rhinoscleromatis* is a chronic inflammatory disease involving the nasopharynx. Patients present with a purulent nasal discharge with the formation of crusts and nodules that may lead to respiratory obstruction. The bacilli are seen intracellularly in lesions which can be isolated and identified by biochemical reactions. Diagnosis is by positive blood culture supplemented with histology. Rifampin has been used for the treatment of rhinoscleroma.

Klebsiella ozaenae

Ozena, caused by *K. ozaenae*, is chronic atrophic rhinitis characterised by necrosis of nasal mucosa and mucopurulent nasal discharge. It often occurs in elderly persons. Nasal congestion and a constant nasal bad smell are the common symptoms. Identification of *K. ozaenae* is difficult due to wide variations in the biochemical reactions of the isolated strains. Trimethoprim and sulphamethoxazole are used for the treatment of ozena.

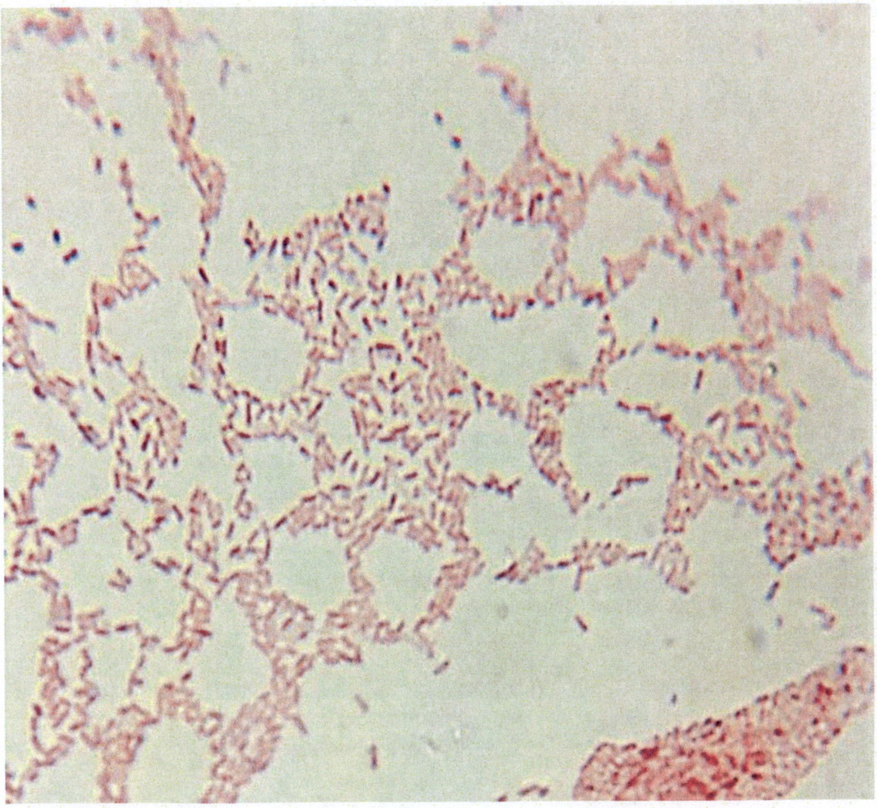

Fig. 35.3 Gram-stained smear showing capsulated *Klebsiella pneumoniae* (×1000)

Klebsiella oxytoca

K. oxytoca may be rarely isolated from clinical specimens. It is being increasingly isolated from patients with neonatal septicaemia. The bacteria are also associated with neonatal bacteraemia, especially among premature infants and in neonatal intensive care units.

The biochemical reactions of different *Klebsiella* species are summarised in Table 35.9.

Enterobacter

The genus *Enterobacter* includes 12 species, of which *Enterobacter cloacae* and *Enterobacter aerogenes* are the two most important species responsible for a variety of nosocomial infections (Table 35.10). *Enterobacter* are Gram-negative bacilli. On sheep blood agar, *Enterobacter* produces large, grey and dry or mucoid colonies; and on MacConkey agar, lactose-fermenting pink colonies. The bacteria ferment glucose with the production of acid. They differ from *Klebsiella* by being motile, urease negative and ornithine decarboxylase positive.

Endotoxin of the bacteria is known to play a major role in the pathogenesis of sepsis and its complications. *Enterobacter* species are opportunistic pathogens. Patients who stay in the hospital, especially in the ICU, for prolonged periods are at high risk to acquire *Enterobacter* infections. Patients treated earlier with antimicrobial agents and those with serious underlying conditions (e.g. diabetes, malignancies, burns and mechanical ventilation), with foreign devices (e.g. intravenous catheters) and with immunosuppression are also at increased risk of infection by the bacteria. In these patients, they cause frequent and severe nosocomial infections, such as UTIs,

Table 35.9 Important tests for differentiation of *Klebsiella* species

Tests	*Klebsiella pneumoniae*	*Klebsiella ozaenae*	*Klebsiella rhinoscleromatis*	*Klebsiella oxytoca*
Indole	–	–	–	+
Urease	+	–	–	+
Citrate	+	V	–	+
ONPG	+	+	–	+
Malonate	+	–	+	+
Lysine decarboxylase	+	V (40%)	–	+
Ornithine decarboxylase	–	–	–	–
MR	–	+	+	V (20%)
VP	+	–	–	+

V: Variable

Table 35.10 Differentiation of *Enterobacter* species

Tests	*Enterobacter aerogenes*	*Enterobacter cloacae*
Gas from glycerol	Positive	Negative
Aesculin hydrolysis	Positive	Negative
Arginine dihydrolase	Positive	Negative
Lysine decarboxylase	Positive	Negative

lower respiratory tract infections, skin and soft tissue infections, bacteraemia, endocarditis, intraabdominal infections, septic arthritis and osteomyelitis.

Carbapenems show the best activity against *E. cloacae*, *E. aerogenes* and other species. First-generation and second-generation cephalosporins are not used against infections caused by *Enterobacter*.

Hafnia

Hafnia alvei is the only species of the genus *Hafnia*. It is found in human and animal faeces, sewage, soil and water. *H. alvei* is motile. It does not ferment lactose, raffinose, sucrose, adonitol-dulcitol and inositol. It is indole- and MR negative, and VP- and citrate positive. Biochemical reactions are better read after incubation at 22 °C than at 37 °C. The bacteria have been isolated from abscesses, wounds, sputum, urine, blood and other sites, but often with other bacteria. The pathogenic role of *H. alvei* is yet to be established.

Serratia

Serratia is a Gram-negative bacterium classified in the tribe Klebsielleae. *Serratia marcescens* is the only pathogenic species causing human infection. Some strains of *S. marcescens* typically produce a non-diffusible pigment called **prodigiosin**. This pigment varies in colour from dark red to pink or magenta, depending on the age of the colonies. *S. marcescens* usually grows on starchy foodstuffs, where the production of pigmented colonies is easily mistaken for drops of blood. *S. marcescens* exists as pleomorphic with minute coccobacillary and normal bacillary forms.

Serratia usually colonises the respiratory and urinary tracts of adult patients during their period of hospital stay. The bacteria are responsible for nearly 2% of nosocomial infections of the urinary tract, lower respiratory tract, surgical wounds, blood and skin and soft tissues of the hospitalised patients. *S. marcescens* is sensitive to amikacin and quinolones but is resistant to gentamicin and tobramycin.

Table 35.11 Differentiating features of genera of the tribe Proteeae

Biochemical properties	*Proteus mirabilis*	*Proteus vulgaris*	*Morganella morganii*	*Providencia stuartii*	*Providencia rettgeri*
Indole	–	+	+	+	+
H_2S	+	+	– (20%)	–	–
Citrate	+ (65%)	V (15%)	–	+	+
Fermentation of sucrose	–	+	–	V (50%)	–
Fermentation of maltose	–	+	–	–	–
Fermentation of mannitol	–	–	–	–	+
Fermentation of trehalose	+	V (30%)	–	+	–
Ornithine decarboxylase	+	–	+	–	–

Note: V: Variable

Tribe Proteeae

The formation of the enzyme phenylalanine deaminase, which converts phenylalanine to phenyl pyruvic acid (**PPA reaction**), is the characteristic feature by which they are differentiated from other members of the family Enterobacteriaceae. The differentiating features of different genera in the tribe Proteeae are summarised in Table 35.11.

Proteus

The genus *Proteus* has four species: *P. mirabilis*, *Proteus vulgaris*, *Proteus penneri* and *Proteus myxofaciens* of which *P. mirabilis* is the most important. They are Gram-negative and non-capsulated coccobacilli measuring 1–3 × 0.6 μm in size. Many of them form long, curved and filamentous forms in young cultures. Most of them, with few exceptions, are motile due to the presence of peritrichous flagella. They are fimbriated. They are aerobic bacteria, which grow well on ordinary media, such as nutrient agar. *Proteus* colonies on the medium emit a characteristic putrefactive (“**fishy**” or “**seminal**”) odour.

P. mirabilis and *P. vulgaris* typically spread or swarm on the surface of the medium. They spread on the surface of the plate in successive waves to form a thin filmy layer in concentric circles. This is known as swarming (Fig. 35.4). The exact mechanism responsible for swarming shown by *Proteus* species is not known. Swarming, shown by *Proteus*, is a problem when mixed growth on a solid medium is obtained in which *Proteus* bacilli are present with other bacteria. Hence, several methods are available to inhibit swarming; these include the use of increased concentration of agar in the medium from 1% to 6% and the use of chloral hydrate (1:500), sodium azide (1:500), alcohol (56%), sulphonamide, surface-active agents or boric acid (1:1000). Swarming does not occur on the MacConkey medium, on which *Proteus* produces colourless non-lactose-fermenting colonies. It is because bile salts present in the MacConkey medium inhibit the swarming. The biochemical characteristics of the *Proteus* species are summarised in Table 35.11.

Motile *Proteus* strains possess somatic O- and flagellar H antigens. Thirty-two different O antigens are found in *P. mirabilis*, 22 in *P. vulgaris* and 5 in *P. penneri* and *P. myxofaciens*. Their O antigens, *P. mirabilis* and *P. vulgaris*, have been classified into 54 O groups based on their O antigens. These O groups are further sub-divided into many O types depending on their flagellar or H antigens. Non-motile strains of *P. vulgaris* (OX19 and OX2) and *P. mirabilis* (OXK) cross-react with rickettsial antigens and form the basis of the Weil-Felix test for the diagnosis of rickettsial infections.

Proteus are invasive bacteria. Fimbriae or pili are the important virulence factors that facilitate adherence of *P. mirabilis* bacteria to host tissue sites, such as the urinary tract epithelium. The ability of *Proteus* to produce urease is an

Fig. 35.4 *Proteus* spp. showing swarming on the blood agar (Courtesy of Prof S. Pramodhini)

important factor in the pathogenesis of the UTI caused by the *Proteus* species. Hydrolysis of urea to ammonia makes the urine alkaline, which provides a suitable environment for *Proteus* to survive. Subsequently, alkalinisation of urine leads to precipitation of organic and inorganic compounds, leading to renal calculi formation. These stones are composed of a combination of magnesium ammonium phosphate (struvite) and calcium carbonate–apatite.

Proteus species cause UTIs, hospital-acquired infections and other miscellaneous infections. UTIs are the most common clinical manifestation of *Proteus* infections. Hospital-acquired infections are usually transmitted from attending doctors or other healthcare workers and are caused by hospital staff's interruption of the closed sterile system. The *Proteus* species is also an important agent of wound infections. The species also cause infection of the umbilical stump in neonates, which often leads to sepsis neonatorum, bacteraemia and meningitis. Urine is the specimen of choice for the diagnosis of UTIs. Culture of blood, urine, CSF and other specimens is also carried out depending on the clinical diseases caused by *Proteus*.

Proteus species can be typed by serotyping, phage typing, bacteriocin (protein) typing and Dienes typing. Proticin typing is frequently carried out using 12 standard proticin-producing strains, and most of the *Proteus* strains are typeable by this method.

Dienes typing: It is used to determine the identity or non-identity of strains of *Proteus*. This is based on the observations that two identical strains of *Proteus* show swarming, which coalesce together without any demarcation, on a solid agar that does not inhibit swarming. However, if the strains are not identical, then the swarming strains do not coalesce together and instead are separated from each other with a clearly visible furrow between the swarming of the two. This is known as the Dienes phenomenon.

P. mirabilis is susceptible to nearly all antimicrobials except tetracycline. *P. mirabilis* is sensitive to ampicillin; broad-spectrum penicillins, such as ticarcillin and piperacillin; first-, second- and third-generation cephalosporines; imipenem; and aztreonam. *P. vulgaris* and *P. penneri* are sensitive to trimethoprim and sulphamethoxazole, quinolones, imipenem, aminoglycosides and fourth-generation cephalosporins.

Morganella

The genus *Morganella* belongs to the tribe Proteeae. The genus *Morganella* has only one species, *Morganella morganii*, with two sub-species, *morganii* and *sibonii*. *M. morganii* was classified earlier under the genus *Proteus* as *Proteus morganii*. *M. morganii* are small, Gram-negative, motile bacilli, but unlike the *Proteus* species, they do not produce swarming on solid media. *M. morganii* is commonly found in human and animal faeces and rarely causes severe invasive diseases. It is most often found as an opportunistic pathogen in hospitalised patients, particularly those on prolonged antibiotic therapy. *M. morganii* causes UTIs, which are often associated with an alkaline urine pH. Nosocomial *M. morganii* strains are usually susceptible to cefepime, imipenem, meropenem, piperacillin, aminoglycosides and fluoroquinolones. ESBL-producing strains of *M. morganii* have been reported recently.

Providencia

The genus *Providencia* consists of five species: *Providencia stuartii*, *Providencia rettgeri*, *Providencia alcalifaciens*, *Providencia rustigianii* and *Providencia heimbachae*. *Providencia* spp. is Gram-negative, motile bacilli but does not show swarming on solid media. They produce a fruity smell and are from yellow to orange colonies on DCA. All the species typically deaminate phenylalanine; only *P. rettgeri* hydrolyses urea consistently. Other biochemical properties of *Providencia* are summarised in Table 35.11. *Providencia* species have been isolated from urine, stool and blood, and the throat, perineum, axilla and wounds from humans. *P. stuartii* is the most common species causing infection in humans. *P. stuartii* is the most resistant species of all the *Providencia* species, *P. alcalifaciens* and *P. rustigianii* are usually susceptible to antibiotics.

Tribe Erwinieae

Erwinia organisms are usually found in soils, and they cause infection in plants. *Erwinia herbicola* is the only species that has occasionally been isolated from respiratory and urinary infections in chronically debilitated and hospitalised patients.

Case Study

A 19-year-old female medical student visited medicine OPD complaining of a burning sensation when passing urine and increased frequency of passing urine. The consultant advised a culture of urine before prescribing any antibiotics.

- What is the infection?
- What is the best urine specimen to be collected for culture?

Salmonella and Shigella

36

Salmonellae are ubiquitous human and animal pathogens.

Genus *Salmonella*

Salmonella spp. include Gram-negative, flagellated and facultative anaerobic bacilli characterised by the presence of O, H and Vi antigens. They colonise virtually all animals, including poultry, birds, livestock, reptiles, rodents, domesticated animals and humans. *Salmonella* infections in humans typically produce one of three clinical syndromes, such as gastroenteritis, enteric fever or focal disease. In addition, *Salmonella* infection in animals causes substantial losses of livestock.

Classification and Nomenclature

The taxonomic classification of the genus *Salmonella* is complex and has undergone many changes.

Kauffmann–White Scheme

It is a method of classification of *Salmonella* strains based on structural formulae of the O and H antigens of the strains, which are identified by agglutination with specific antisera. Salmonellae are classified into different O serogroups based on distinctive O antigen factors. Each serogroup contains many serotypes possessing a common O factor not found in other serogroups. These O factors are now designated 1, 2, 3, 4, etc.; these were originally designated by capital letters A–Z and afterwards by numbers 51–67. Some serogroups were subdivided into subserogroups (e.g. C1–C4; E1–E4). It seems logical to name each serogroup by its characteristic O antigen factor numbers rather than letters. Hence, it is now being proposed to designate group A as 2, B as 4, C1 as 7, C2 as 8, Das 9, etc.

Within each O serogroup, the different serotypes are identified by the presence of phase 1 and phase 2 of flagellar antigens. The antigenic structure of a *Salmonella* serotype is designated by an antigenic formula, which has three components demonstrating the O antigens, antigens of Phase 1 H and Phase 2 H in subsequent orders. In this, colons separate the three parts, and commas separate each part of the component antigen.

The Kauffmann–White scheme designates each serotype as a species (Table 36.1). More than 2400 serotypes have been described by Kauffmann–White scheme.

- Based on DNA homology and host range, the genus *Salmonella* is classified into *Salmonella enterica* and *Salmonella bongori*. *S. enterica* is further subdivided into six subspecies I, II, IIa, IIIb, IV and VI.
- Most of the salmonellae that are pathogenic to human beings belong to the subgroup I of *S.* Enterica subspp. Enterica. This includes

Table 36.1 Kauffmann–White classification of *Salmonella*

O Serogroup				H Antigens	
New	Old	Serotype	O Antigens (and Vi)	Phase 1	Phase 2
2	A	*Salmonella* Paratyphi A	1,2,12	a	–
4	B	*Salmonella* Paratyphi B	1,4,5,12	b	1,2
		Salmonella Typhimurium	1,4,5,12	1	1,2
		Salmonella Stanley	4,5,12	d	1,2
7	C1	*Salmonella* Paratyphi C	6,7 (vi)	c	1,5
		Salmonella Choleraesuis	6,7	c	1,5
8	C2	*Salmonella* Muenchen	6,8,20	e,h	1,2
9	D	*Salmonella* Typhi	9,12 (Vi)	d	–
		Salmonella Enteritidis	4,9,12	g,m	–
		Salmonella Gallinarum	1,9,12	–	–
		Salmonella Pullorum	1,9,12	–	–
3,10	E1	*Salmonella* Anatum	3,10	e,h	1,6

typhoid and paratyphoid bacilli and most other serotypes responsible for diseases in mammals.

- Additionally, each *Salmonella* isolates are serotyped according to the presence of particular somatic O, flagellar H and surface Vi antigens. Presently more than 2400 serotypes are described.

Salmonellae serotypes are named as, for example, *S.* Enterica subspp. Enterica serotype Enteritidis. However, for the sake of convenience, it is abbreviated as *S.* Enteritidis (this chapter follows the same system) or just Enteritidis for use in clinical situations. In addition, serotypes are not mentioned in italic but Roman.

Human infections caused by *Salmonella* spp. are summarised in Table 36.2.

Properties of the Bacteria

Morphology

Salmonella is motile with the presence of peritrichous flagella (*Salmonella* Gallinarum and *Salmonella* Pullorum are exceptions that are non-motile). They do not form spores and capsules. Some strains of salmonellae may produce fimbriae, but most strains of *Salmonella* Paratyphi A and a few strains of *Salmonella* Paratyphi B, *Salmonella* Typhi and *Salmonella* Typhimurium are non-fimbriated.

Culture and Biochemical Properties

They are aerobic and facultatively anaerobic; they grow at an optimum temperature of 37 °C at a pH of 6–8 on various selective and non-selective media.

Non-selective Media

On **nutrient agar** and **blood agar**, *Salmonella* spp. produce greyish white moist colonies with a smooth convex surface after 18–24 h of incubation. Rough strains produce opaque and granular colonies with an irregular surface. On **MacConkey agar**, they produce pale colourless colonies because they do not ferment lactose. The colonies on **deoxycholate citrate agar** are similar to those produced on MacConkey agar. Sometimes after incubation of 48 h or more, they produce colonies with a black centre.

Selective Media

Wilson and Blair bismuth sulphite agar is the medium of choice for *Salmonella* spp., especially *S.* Typhi. The growth of *Shigella* spp., *Proteus* spp. and coliforms is inhibited on this medium. On this medium, salmonellae produce jet black colonies surrounded by a metallic sheen due to hydrogen sulphide production. *S.* Paratyphi A and other species, which do not produce H_2S, form green colonies. **XLD (xylose lysine deoxycholate agar)** is another selective medium used to isolate *Salmonella* spp. On this medium,

Table 36.2 Human infections caused by *Salmonella* species

Salmonella spp.	Diseases
Salmonella Typhi	Typhoid fever, *Salmonella* Bacteraemia
Salmonella Paratyphi A, B and C	Paratyphoid fever, *Salmonella* Bacteraemia
Salmonella Choleraesuis	*Salmonella* Bacteraemia
Salmonella Typhimurium	*Salmonella* Gastroenteritis
Salmonella Enteritidis	*Salmonella* Gastroenteritis
Salmonella Hadar	*Salmonella* Gastroenteritis
Salmonella Heidelberg	*Salmonella* Gastroenteritis
Salmonella Agona	*Salmonella* Gastroenteritis
Salmonella Virchow	*Salmonella* Gastroenteritis
Salmonella Senftenberg	*Salmonella* Gastroenteritis
Salmonella Indiana	*Salmonella* Gastroenteritis
Salmonella Newport	*Salmonella* Gastroenteritis
Salmonella Anatum	*Salmonella* Gastroenteritis

Salmonella spp. produce pink colonies with black centres as a result of H_2S production. H_2S-negative *Salmonella* serotypes produce red colonies without black centres.

Enrichment Media

Selenite F broth is frequently used to enrich *Salmonella* spp. from clinical specimens. However, sometimes this medium inhibits the growth of some salmonellae, such as *S.* Paratyphi B and *Salmonella* Choleraesuis. **Tetrathionate broth**, although it is used for salmonellae at times the broth allows the growth of *Shigella* spp. and also that of *Proteus* spp. Tetrathionate broth with brilliant green, although it inhibits the growth of *Proteus* spp., sometimes is inhibitory to *Salmonella* spp.

Salmonellae are catalase positive and oxidase negative. Salmonellae ferment glucose, mannitol and maltose, forming acid and gas. *S.* Typhi is an exception that does not ferment the sugars. They do not ferment lactose. They do not grow in Simmons citrate media as they need tryptophan as the growth factor. Salmonellae decarboxylate amino acids, such as lysine, ornithine and arginine, but not glutamic acids. However, *S.* Typhi does not decarboxylate ornithine and *S.* Paratyphi A does not decarboxylate lysine. The biochemical characteristics are useful for distinguishing different *Salmonella* spp. (Table 36.3).

Other Properties

Susceptibility to Physical and Chemical Agents

The bacilli are killed at a temperature of 55 °C in 1 h or 60 °C in 15 min. They are also killed by 0.2% mercuric chloride or 5% phenol in 5 min. Boiling, chlorination of water and pasteurisation of milk kill the bacteria. They survive for weeks in polluted water and soil and months in ice. Cultures may be viable for years if prevented from drying.

Antigenic Structure

The cell wall of salmonellae-like any other Gram-negative bacilli contains a complex lipopolysaccharide (LPS) structure. LPS, an endotoxin, plays a significant role in the virulence of the bacteria. It has three regions (1) inner lipid A coat, (2) middle R core portion and (3) outer O polysaccharide coat.

1. The outer O polysaccharide chain has repeated sugar units responsible for O antigen specificity. This can determine the virulence of the bacteria. Rough strains of *Salmonella* lack a complete sequence of O sugar.
2. The endotoxin component of the cell wall is important in the pathogenesis of *Salmonella* infections. The circulatory endotoxin is also partly responsible for the development of septic shock that can occur in systemic infection.

Table 36.3 Biochemical reactions of common *Salmonella* species

Biochemical reactions	*Salmonella* Typhi	*Salmonella* Paratyphi A	*Salmonella* Paratyphi B	*Salmonella* Paratyphi C
Glucose	A	AG	AG	AG
Mannitol	A	AG	AG	AG
Lactose	–	–	–	–
Sucrose	–	–	–	–
Indole	–	–	–	–
Citrate	–	–	+	+
MR	+	+	+	+
VP	–	–	–	–
H_2S	+	–	+	+
Xylose	d	–	AG	AG
D-Tartrate	A	–	–	AG
Mucate	d	–	AG	–

A: acid; *AG*: acid and gas; *d*: delayed

3. Antibodies produced against the R core (common enterobacterial antigen) are protective against infection caused by a wide variety of Gram-negative bacteria due to sharing the common core structure. In some situations, the antibodies against the R core mediate the lethal effects of Gram-negative bacteria.

Salmonella possesses the following three major antigens **H or flagellar antigen, O or somatic antigen** and **surface antigens** based on which they are identified and classified.

H or Flagellar Antigen

This antigen is present on the flagella and is heat and alcohol labile. The H antigens of *Salmonella* are genus specific and are not shared by other enterobacteria. The antigens are destroyed by boiling or treatment with alcohols and acid, but they are preserved in 0.2–0.4% formaldehyde. The H antigen is strongly immunogenic and is associated with the formation of antibodies following infection or immunisation. H antigen may occur in either or both the forms called phase I and phase II. The organisms tend to change from one phase to another.

O or Somatic Antigen

O antigens occur on the surface of the outer membranes and are determined by specific sugar sequences on the cell surface. O antigen is an LPS complex and is an integral part of the cell wall. It is heat stable resistant to boiling for up to 2 h and 30 min. It is also alcohol stable, resistant to treatment with 96% ethanol at 37 °C for 4 h and resistant to 0.2% formaldehyde. O antigen is less immunogenic than H antigen. Generally, the O antibody titre produced after infection or immunisation is lower than H antibodies. *Salmonella* is classified based on the presence of a characteristic O antigen on the bacterial surface. Till now, 67 "O" antigens have been described. *Salmonella* has been classified into 46 "O" serogroups based on the somatic antigens.

Surface Antigens

These include Vi antigen, M and N antigens and F antigens.

1. **Vi antigen** is a surface antigen overlying the "O" antigen. Felix and Pitt, who first described this antigen, believed that it was related to virulence and gave it the name **Vi antigens**. It is analogous to the K antigens of coliforms. The antigen is present only in a few serotypes, the most important being *S.* Typhi. This antigen is also present in some strains of *S.* Paratyphi C, *Salmonella* Dublin and *Citrobacter freundii*. The presence of this antigen on the surface renders these bacteria inagglutinable by their specific O antiserum but agglutinable by Vi antisera. Vi antigen is poorly immunogenic and is associated with the production of low titre of antibodies during the infection. Antibody disappears early in convalescence. Its persistence indicates the

development of the carrier state. Demonstration of the Vi antibody is not useful for diagnosing cases of typhoid; hence, Vi antigen is not tested in the Widal test.

2. **M and N antigens** are present on the surface of bacteria and are polysaccharides in nature. These antigens prevent agglutination by O antiserum. Boiling for 2.5 h destroys these antigens. The presence of the M antigen is responsible for the mucoid nature of *Salmonella* colonies.
3. **F antigen** is present on the fimbriae. They occur in two phases: one with F antigen and the other without F antigen.

Antigenic Variations

Salmonellae antigens undergo phenotypic and genotypic variations as follows:

OH–O Variation

Some of the flagellated *Salmonella* (OH) at times gives rise to non-flagellated (O) strains. This is called OH–O variation. Sub-culture in Craigie tube is used to obtain a population of flagellated bacteria from such cultures. Craigie tube method (Fig. 36.1) uses a wide tube containing semi-solid agar (0.2%). A short, narrow tube, open at both ends, is placed into the agar of the tube in such a way that the upper end of the tube projects well above the agar. The bacterial strain is inoculated into the inner tube, and after 8–16 h of incubation, sub-cultures made from the top of the agar outside the central tube yield a population of motile cells. A U tube of soft agar instead of Craigie tube may be used for the purpose in which inoculation is made into one limb and sub-culture is taken from the other.

Rarely, salmonellae may lose their flagella by mutation. *S.* Typhi 901-O strain is an example of a stable non-motile mutant, which is used to prepare O agglutinable bacterial antigen suspensions for use in the Widal test and other serological tests.

V–W Variation

Vi antigen of *S.* Typhi completely masks the O antigen. They are called the **V form**. After sub-culturing many times, the Vi antigen is completely lost, and they are known as **W form**. Such forms are called **VW forms**.

S–R Variation

The smooth-to-rough (S–R) variation occurs due to mutation and is associated with the (1) change

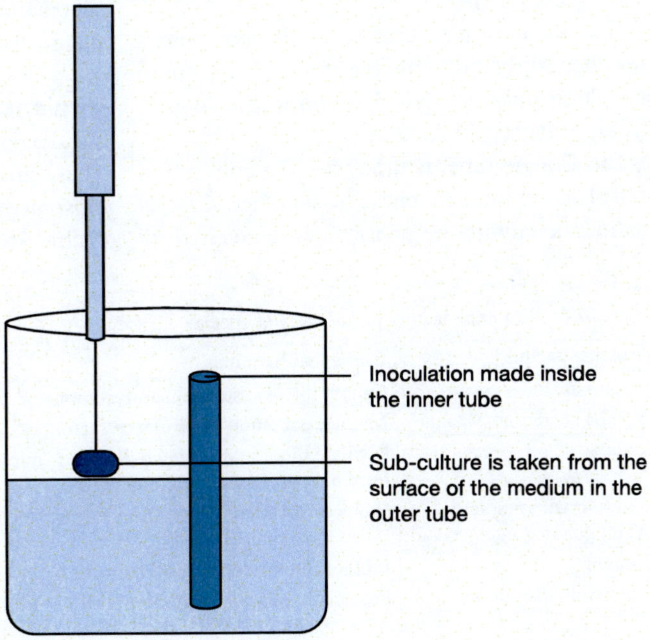

Fig. 36.1 Craigie tube method of sub-culture

in the colony morphology from smooth to rough, (2) loss of the O antigen and (3) loss of virulence. The rough colonies become large, rough and irregular and are autoagglutinable in saline. These colonies show a defective capacity to synthesise O antigen. Conversion into rough forms rarely occurs in nature but is common in laboratory strains maintained by serial sub-culture on ordinary media. The smooth S–R variation may be prevented to some extent by maintaining cultures on Dorset egg media or by lyophilisation.

Phase Variation

The flagellar or H antigens of most salmonellae occur in two alternate phases: phase 1 and phase 2. **Phase 1 antigens** are either specific for a serotype or shared by a few species only, hence known as the "specific" or species phase. A large number of flagellar antigens, more than 80, have been found in this phase. These are designated as small letters of alphabets a–z except j and subsequently z1–z68. The presumptive identification of serotype is based on identifying the antigen in phase 1. **Phase 2 antigens** are widely shared, hence known as the "non-specific" or "group" phase. Phase 2 antigens are far fewer and are designated by Arabic numerals from 1 to 12 (1, 2, 3 . . .12).

Salmonella strains that occur in both phases are called **biphasic**, while the strains that occur in one phase only are called **monophasic** (e.g. *S.* Typhi, *S.* Paratyphi A, etc.).

The Craigie tube method uses the conversion of isolates from one phase to another. In this method, a culture in phase 1 is converted to phase 2 by passing it through the Craigie tube containing specific phase 1 antiserum and the reverse conversion is achieved by using phase 2 antiserum.

O antigen Variation

When the bacterium is lysogenised with converting phages, there occurs a change in the structural formula of the O antigen, which brings alteration in bacterial serotypes. Thus, *Salmonella* Anatum *(serotype 3,10:e,h:1,6)* is converted by phage 15 into *Salmonella* Newington *(serotype 3,15:e,h:1,6)* by one phage and the latter into *Salmonella* Minneapolis *(serotype 3,15,34:e, h:1,6)* by phage 34. Serotypes Newington and Minneapolis are designated as Anatum var. 15+ and Anatum var. 15 +, 34 +, respectively.

Pathogenesis

S. Typhi is an invasive bacterium. It colonises the human intestine and, under specific conditions, directly invades the intestinal mucosa or multiplies for several days within the mononuclear phagocytic cells in the liver, spleen, lymph nodes and Peyer patches of the ileum before the invasion. The bacteria subsequently enter the bloodstream and cause the disease manifestations.

Virulence Factors

Virulence factors of salmonellae are complex. These are encoded both on the organism's chromosome and on large (34–120 kDa) plasmids (Table 36.4) and include the followings:

Table 36.4 Virulence factors of *Salmonella* species

Virulence factors	Biological functions
Endotoxin	Causes many systemic manifestations of salmonellosis
Type III secretion systems	Mediate secretion of virulence factors of *Salmonella* into host cells
Fimbriae	Mediate binding of *Salmonella* to M (microfold) cells present in Peyer patches of the terminal part of the small intestine
Acid tolerance response (ATR) gene	Protects *Salmonella* spp. from stomach acids and the acid pH of the phagosome
Catalase	Protects the bacteria from intracellular killing in macrophages
Superoxide dismutase	Protects the bacteria from intracellular killing in macrophages

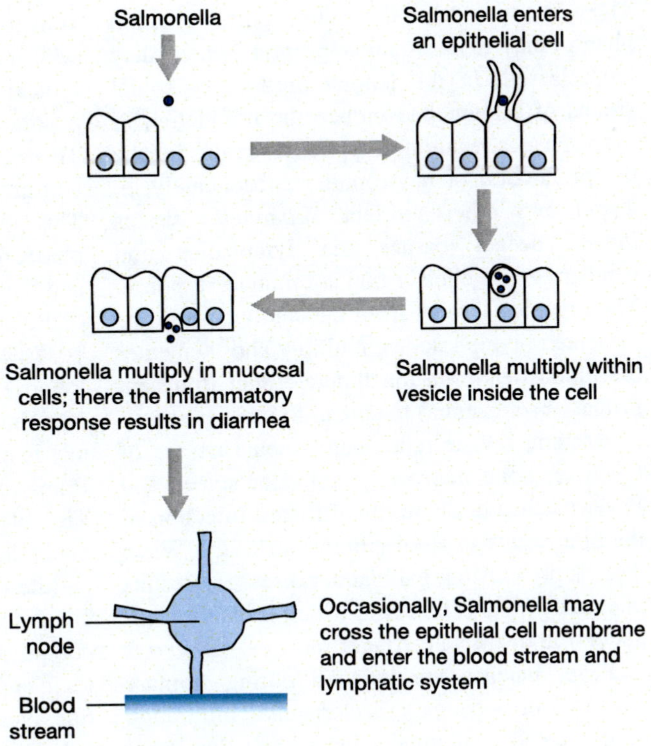

Fig. 36.2 Pathogenesis of *Salmonella* infection

Endotoxin

Endotoxin is responsible for many of the systemic manifestations of the disease caused by *Salmonella* spp.

Fimbriae

The species-specific fimbriae mediate binding of *Salmonella* to M (microfold) cells present in Peyer patches of the terminal part of the small intestine. These M cells typically transport foreign antigens, such as bacteria, to the underlying macrophages for clearance.

Acid Tolerance Response (ATR) Gene

The acid tolerance response (ATR) gene protects *Salmonella* spp. from stomach acids and the acidic pH of the phagosome, thereby facilitating the survival of bacteria in phagosomes.

Enzymes

Catalase and superoxide dismutase are the enzymes that protect the bacteria from intracellular killing in macrophages.

Pathogenesis of Enteric Fever

The infection is acquired by ingesting food or water contaminated with salmonellae (Fig. 36.2). In human volunteer experiments, the infective dose (ID50) is about 10^3–10^6 bacilli. The dose of inoculum required to produce infections varies from strain to strain. Usually, a large amount of inoculum is needed to overcome stomach acidity and normal flora competition. When the inoculum is large, the illness rate will be higher with a shorter incubation period.

The salmonellae attach themselves by fimbriae or pili to cells lining the ileal mucosa on reaching the intestine. The bacteria selectively attach to specialised epithelial cells (M cells) of the Peyer patches. Type III secretion system (TTSS-1) encoded by Salmonella pathogenicity island 1 (SPI-1) and genes mediate the initial invasion of *S.* Typhi into the intestinal mucosa. An ATR gene also protects *Salmonella* spp. from stomach acids and the acidic pH of the phagosome. Catalase and superoxide dismutase are other factors that protect the bacteria from intracellular killing.

The bacteria are then transported within phagosomes to the lamina propria, where they are released. In the lamina propria, typhoidal strains of salmonellae induce the production of macrophages, while non-typhoidal strains induce the production of neutrophils. Subsequently, *S.* Typhi and other virulent *Salmonella* strains invade deeper tissues via lymphatics and capillaries and elicit a major immune response. The organisms travel from the submucosa to the mesenteric lymph nodes, multiply and then enter the bloodstream via the thoracic duct (transient primary bacteraemia) to spread to other tissues.

Bacteria invade reticuloendothelial tissues of the liver, bone marrow, gallbladder spleen and Peyer patches in the terminal ileum. Infection of the bile results in stool culture positivity. When there is pre-existing gallbladder disease, it results in chronic biliary infection, which leads to a longer period of faecal carriage.

Peyer patches are invaded during primary intestinal or secondary bacteraemia. Infiltration of chronically inflamed cells makes the Peyer patches hyperplastic. This may lead to the necrosis of the outer layer and form an ulcer, with the possibility of haemorrhage from blood vessel erosion.

Clinical Syndromes

S. Typhi causes typhoid fever, while *S.* Paratyphi A, *Salmonella* Schottmuelleri (formerly *S.* Paratyphi B) and *Salmonella* Hirschfeldii (formerly *S.* Paratyphi C) cause a mild form of this disease referred to as a paratyphoid fever. The term enteric fever includes both typhoid and paratyphoid fever caused by these *Salmonella* spp.

Typhoid Fever

Typhoid fever is generally an acute illness manifested by fever, headache and abdominal symptoms. The incubation period is usually from 7 to 14 days but may range from 3 to 56 days. The onset of the disease is usually gradual, with headache, malaise, anorexia, a coated tongue and abdominal discomfort with either constipation or diarrhoea. Step ladder pyrexia with relative bradycardia and toxaemia is the typical feature. The condition is associated with a soft, palpable spleen and an enlarged liver. These symptoms are present for a week or more and are followed by gastrointestinal symptoms. This phase corresponds to an initial bacteraemic phase, followed by colonisation of the gallbladder and then reinfection of the intestines.

Relapse is common after initial recovery in 10–20% of patients treated with antibiotics. Relapse typically occurs approximately 1 week after the therapy is discontinued. A relapse generally is milder and shorter than the initial disease. Rarely, second or even third relapses may occur. The blood culture and serum H, O and Vi antibodies are positive in relapse cases.

Intestinal perforation, severe haemorrhage and circulatory collapse are the most important complications. Other complications include toxic encephalopathy, cerebral thrombosis, hepatitis, pancreatitis, arthritis and myocarditis. Convalescence is slow.

Paratyphoid Fever

Paratyphoid fever caused by *S.* Paratyphi A and B resembles typhoid fever but is generally milder. *S.* Paratyphi C may also cause paratyphoid fever, but it often leads to frank septicaemia with suppurative complications.

Laboratory Diagnosis

Specimens

Blood, blood clot, bone marrow and stool are common specimens used for isolation of *Salmonella* by culture. Other specimens include the cerebrospinal fluid, peritoneal fluid, mesenteric lymph nodes, resected intestine, pharynx, tonsils, abscess, bone and urine.

Blood Culture

Blood culture is useful for diagnosing enteric fever. In this method, approximately 5–10 mL of blood is collected aseptically by venepuncture and inoculated into a culture bottle containing 50–100 mL of 0.5% bile broth. This ten-fold

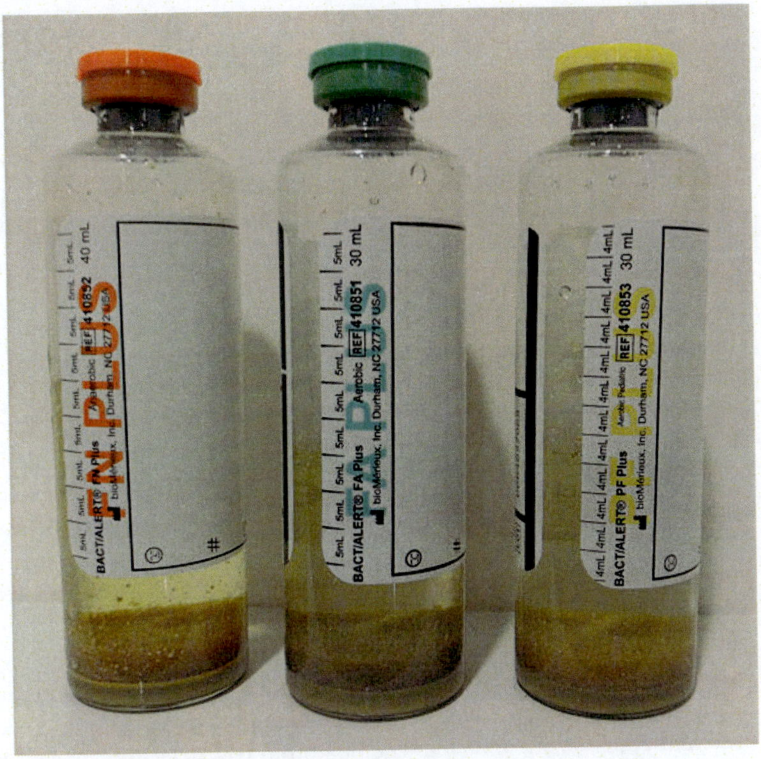

Fig. 36.3 Blood culture bottles used in automated blood culture system

dilution of blood in the bile broth is carried out to neutralise the bactericidal action of many substances present in the blood. The addition of liquid (sodium polyanethol sulphonate) further counteracts the bactericidal action of blood. The blood culture bottle is incubated at 37 °C for 7 days. After incubation overnight at 37 °C, the bile broth is sub-cultured on MacConkey agar. Salmonellae form pale non-lactose-fermenting colonies on MacConkey agar.

Blood culture is positive in approximately 90% of cases in the first week of fever, in the second week of fever with 75% of cases, in the third week of fever with 60% and with 25% of cases until pyrexia subsides. Blood cultures, however, rapidly become negative on treatment with antibiotics.

Castañeda Biphasic Method of Blood Culture is a better method of culture to reduce the risk of contamination during repeated sub-cultures. The method has the additional advantage of being more safe and economical. In this method, the culture bottle has an agar slant on one side, which is flooded with bile broth. After inoculation of blood, the bottle is incubated in the upright position, so that surface of the agar remains free without any broth covering the slant. Broth remains only in the lower part of the agar slope. For sub-culture, the bottle is simply tilted so that the broth flows over the surface of the agar slant and is re-incubated in the upright position. If salmonellae are present, colonies appear on the agar slant.

Automated Blood Culture System

At present automated blood culture systems, is available as a more reliable technique for microbial growth detection systems with specific culture bottles (Fig. 36.3). Currently, **three automated blood culture systems** are approved and function differently to detect bacteraemia and fungaemia. The BacT/Alert automated microbial detection system of bioMerieux, Durham, NC, uses an internal colorimetric sensor that changes from grey to yellow in the presence of CO2 produced by growing microorganisms. BD Bactec Plus automated microbial detection system of BD Diagnostics, Sparks, MD, uses a chemical sensor to detect increases in CO2 produced due to microbial growth. The VersaTREK automated microbial detection system of TREK Diagnostic Systems, Cleveland, OH, measures pressure changes in the sample bottle headspace.

Clot Culture

Clot culture is a more sensitive method than the blood culture because certain inhibitory substances found in the serum are absent in the clot proper.

In this method, 5 mL of blood is collected under strict aseptic conditions, from the patient, into a sterile test tube and allowed to clot. The serum is pipetted off and used for serological tests. The clot is broken up with a sterile glass rod and added to a bottle of bile broth containing streptokinase (100 units/mL). Streptokinase facilitates the lysis of the clot with the release of bacteria trapped inside the clot. The bile broth is incubated and sub-cultured on media in the same way as described for blood culture earlier.

Clot culture shows higher isolation of Salmonellae than the blood culture. Another advantage of this method is that serum eluted from the blood during clotting can be used to demonstrate *Salmonella* antigens or antibodies.

Bone Marrow Culture

Bone marrow culture is the most sensitive method. It is positive in most cases, even when blood cultures are negative. It is also positive even if patients have been taking antibiotics for several days, regardless of how long they have been suffering from enteric fever. This test is recommended for patients whose initial blood culture results are negative, possibly due to prior antibiotic therapy.

Faeces Culture

Faeces culture is also useful because salmonellae are excreted in faeces throughout the disease and even during convalescence with varying frequency. The faecal samples are inoculated directly on MacConkey, DCA and Wilson–Blair media. Relatively, a heavy inoculation of stool is made on the Wilson–Blair media because it is highly selective. For enrichment, one tube each of selenite and tetrathionate broth is inoculated and incubated for 12–18 h before sub-culture onto selective media. The plates are incubated at 37 °C overnight. *S.* Typhi produces large black colonies, with a metallic sheen on the Wilson–Blair medium. On this medium, *S.* Paratyphi A produces green colonies due to the absence of H_2S production. Salmonellae form pale non-lactose-fermenting colonies on MacConkey and DCA media. If no growth is observed after 7 days, the culture is declared negative. **Other specimens for culture:** These include urine, bile, rose spots, pus from suppurative lesions, CSF and sputum. Cultures may be obtained from the gall-bladder, liver, spleen and mesenteric lymph nodes at autopsy.

Isolation of the organism from blood, blood clots or bone marrow is definitive for the diagnosis of enteric fever. Bile culture is useful for the detection of carriers. Faecal cultures are useful, but a positive faecal culture may occur in carriers and inpatients.

Colonies are identified by carrying out motility tests, biochemical tests and slide agglutination with specific *Salmonella* antisera.

Slide Agglutination Test

The test is performed by adding a loopful of typhoid O antiserum (factor 9/group D) to one drop of bacterial emulsion on the slide. The slide is rocked gently. Development of immediate agglutination suggests that *Salmonella* strain tested belongs to *Salmonella* group D. Identity of *S.* Typhi is established by agglutination with the flagellar antiserum (anti-d serum).

Sometimes, fresh isolates of *S.* Typhi occur in the V form and do not agglutinate with the O antiserum. Such strains may be tested for agglutination against anti-Vi serum. Alternatively, the growth scraped off in a small amount of saline is tested for agglutination with the O antiserum after boiling for 20 min. If the isolate is a non-typhoidal *Salmonella*, producing gas from sugars, it is tested for agglutination with O and H antisera for groups A, B, C, etc.

Serology

Serodiagnosis of enteric fever is based on detecting specific *Salmonella* antibodies in the serum or antigen in the serum and urine by various serological tests.

Widal Test

Traditionally, the Widal test had been used for diagnosing typhoid fever by demonstration of serum antibodies. The agglutinating antibodies against the somatic antigens and the flagellar (H) antigens of *S.* Typhi are measured for typhoid and paratyphoid bacilli in the patient's sera. This is a tube agglutination test in which Dreyer narrow agglutination tube with a conical bottom is used for H agglutination and Felix short round-bottomed tube for O agglutination. The H and O antigens of *S.* Typhi and the H antigens of *S.* Paratyphi A and B are used in the test. The paratyphi O antigens are not used because they cross-react with the typhoid O antigen due to their sharing of antigenic factor 12. H antigens of *S.* Typhi and *S.* Paratyphi A and B are used individually because H antigens of these bacteria do not show any cross-reactivity with each other. *S.* Typhi 901, "O" and "H" strains are used to prepare antigen.

The O antigen is prepared by the culture of *S.* Typhi on phenol agar (1:800) and harvesting the growth in a small volume of saline.

The test is performed by taking equal volumes (0.4 mL) of serial dilutions of the serum (from 1/20 to 1/640) and H antigen of *S.* Typhi (TH), *S.* Paratyphi A (AH), *S.* Paratyphi B (BH) and O antigens of *S.* Typhi (TO) and mixing in Dreyer tubes and Felix tubes, respectively. The tubes are then incubated in a water bath at 37 °C overnight. Some recommend incubation at 37 °C for 4 h, followed by overnight incubation at 4 °C. The formation of loose woolly clumps characterises H agglutination, while O agglutination by a disc-like pattern at the bottom of the tube. The supernatant fluid remains clear in both types of agglutinations. The highest dilution of the serum showing agglutination with H or O antigens suggests the antibody titre of the patient's serum. The Widal test is positive in enteric fever in only 40–60% of patients at admission.

The results of the Widal test should be interpreted, taking into account the followings:

1. Antibodies against H and O antigens usually appear by the 7th–10th day of the illness and increase steadily till the third or the fourth week, after which it declines gradually. Hence, the blood collected before 7–10 days will be negative for antibodies.
2. Demonstration of a fourfold or more rise in titre of antibodies in a paired sample, one sample collected in the first week and second sample is collected in the third week, is more useful than a demonstration of antibodies in a single serum.
3. A titre of 1/100 or more for O antibodies and 1/200 or more for H antibodies usually suggests enteric fever. However, the results of a single test should be interpreted with caution. Moreover, it is necessary to obtain basal antibody titre levels in "normal sera" in different areas before arriving at a cut off diagnostic titre for H and O antibodies.
4. An elevated level of antibodies may be present in sera of patients suffering from enteric fever in the past and sera of individuals with inapparent infection or vaccination against the enteric fever. Therefore, the mere demonstration of antibodies in the Widal test should not be considered suggestive of enteric fever.
5. Serum from an individual vaccinated with TAB vaccine may show high titres of antibodies to *S.* Typhi and *S.* Paratyphi A and B. However, in case of infection, high titres of antibodies will be seen only against the infecting species. H antibodies persist for many months after vaccination, but O antibodies disappear earlier within 6 months. However, in case of infection, high titres of antibodies will be seen only against the infecting species.
6. Individuals who had suffered from enteric infections in the past or who have been immunised may develop an anamnestic response during an unrelated fever, such as malaria and influenza. The anamnestic response shows only a transient rise of antibodies, while the antibody rise is sustained in enteric fever.

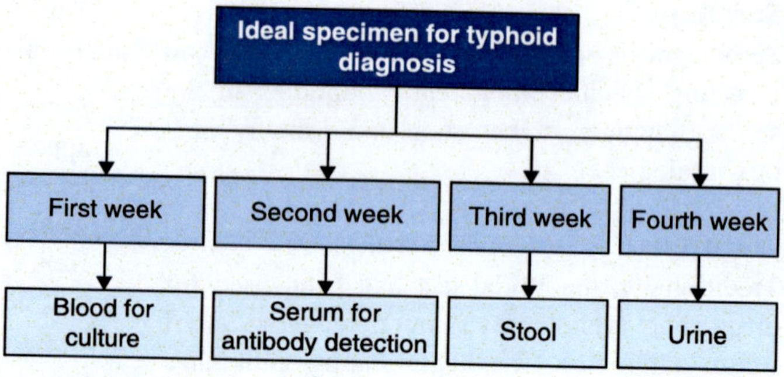

Fig. 36.4 Ideal specimen used for the diagnosis of typhoid fever

7. Patients treated early with antibiotics, such as chloramphenicol may show a poor antibody response.

The utility of the Widal test, blood culture and faeces culture for the laboratory diagnosis of typhoid fever during different periods of the disease is shown in Fig. 36.4.

Other Antibody-Based Serological Tests

These include indirect fluorescent Vi antibody and indirect enzyme-linked immunosorbent assay (ELISA) for immunoglobulin M (IgM) and IgG antibodies to *S.* Typhi polysaccharide are available for diagnosis of typhoid fever with varying sensitivity and specificity. Monoclonal antibodies against *S.* Typhi flagellin have been evaluated to increase the specificity of the ELISA for antibodies.

Antigen-Based Serological Tests

These include co-agglutination test and ELISA. These are frequently employed to detect circulating antigens in the serum and urine to diagnose typhoid fever with varying sensitivity and specificity. In typhoid fever, circulating *S.* Typhi antigen is present in the serum and urine but absent in serum of a cured case of typhoid. Hence, demonstration of serum antigen always indicates an active or recent typhoid fever.

Molecular Detection

Polymerase chain reaction (PCR) is currently being evaluated to diagnose typhoid fever. It has offered rapid detection and differentiation of Salmonella serovars directly from the sample without culturing. Primers sets targeting genes, like *viaB*, *prt*, *tyv*, *fliC-d*, *fliC-a*, *inv*, *pefA*, *sefA*, etc., are widely used to detect species. The promising aspect of such molecular detection is the multiplex PCR where multiple primers sets could simultaneously differentiate *Salmonella* spp., like S. Typhimurium and S. Enteritidis.

Although the process is simple, inexpensive and sensitive, it requires laboratory setup and technical expertise that has limited its field application, especially in resource-limited settings. In this, loop-mediated isothermal amplification (LAMP) has emerged as a promising tool that offers a rapid, reliable, sensitive, specific and easy molecular diagnostic tool for on-site pathogen detection. For example, novel technology, like triplex LAMP assay, only uses three sets of primers that only amplify the target genes without cross-reactivity with non-target genes. It has a limit of detection of 2.5 pg of *S.* Typhimurium DNA and takes 60-min sampling-to-result turnover time.

Diagnosis of Carriers

The diagnosis of carriers, especially cooks and food handlers, plays an important role in epidemiological and public health purposes; therefore, it is an important aspect of typhoid fever control.

Detection of faecal carriers can be done by isolating the bacillus from bile or faeces. It is essential to test repeated samples because the frequency and intensity of bacillary shedding vary from individual to individual. Repeated urinary cultures are required for the detection of urinary carriers. The cholagogue purgative increases the chances of isolation of *Salmonella*.

Detection of antibodies to Vi antigens plays an important role in the carrier state. Detection of Vi antibody is not useful in the diagnosis. Hence, Vi antigens are not being used in the Widal test. Therefore, the Widal test does not play any role in carrier detection in endemic countries.

Millipore membranes filtration technique is also useful in isolating salmonellae from sewage, in which the membranes are cultured on selective media, such as Wilson and Blair media.

Typing

Salmonella is typed by various methods as follows:

Bacteriophage Typing

Bacteriophage typing is the intraspecies classification of *S.* Typhi, first developed by Craigie and Yen (1937). They observed that bacteriophage acting on the Vi antigen of *S.* Typhi (Vi phage II) is highly adaptable. The parent phage is known as phage A. It could be made specifically for a particular strain of *S.* Typhi by serial propagation in the strain. Such adaptation was obtained by phenotypic or genotypic variation.

The phage typing is carried out by determining the sensitivity of the cultures to a series of variants of the Vi phase. However, most other types are sensitive to only one or a few related adaptions. As phage typing of *S.* Typhi depends on the presence of Vi antigens, a proportion of strains (Vi negative) will be untypable. At present, 97 Vi II phage types of *S.* Typhi are recognised. *S.* Typhi phage types E1, O and A is most common in India.

The phage typing is useful in (1) tracing the source of epidemics and (2) providing information on the trends and patterns of typhoid epidemiology at the local, national and international levels. Phage typing schemes have also been applied to *S.* Paratyphi A and B, *S.* Typhimurium, *S.* Enteritidis, *S.* Dublin and other serotypes.

Biotyping

Biotyping is a useful method to classify different *Salmonella* strains of a particular serotype into several biotypes by their biochemical characteristics. *S.* Typhi has been differentiated into different biotypes based on arabinose, dulcitol and xylose fermentation.

Biotyping is useful for supplementing phage typing because it subdivides a large number of untypable strains or members of common phage types. Conversely, strains of the same biotype may be reorganised into different phage types. Therefore, a combination of phage typing and biotyping provides better discrimination of *Salmonella* strains than either method used alone.

Genotyping

Genotyping methods, such as plasmid fingerprinting, multi-locus enzyme electrophoresis, IS-200 profiling and random amplified polymorphic DNA analysis, have been employed as more discriminating methods for epidemiological characterisation of *Salmonella* in advanced centres.

The central research institute of Kasauli at Himachal Pradesh has the National Salmonella Reference Centre. Indian Veterinary Research Institute of Izatnagar, Uttar Pradesh, has the Animal Origin Salmonella Reference Centre. These reference centres provide services for identifying unusual serotypes and reconfirming other serotypes of salmonellae.

Treatment

Chloramphenicol was the antibiotic of choice for the treatment of enteric fever since its introduction in 1948. It acts by binding to 50S bacterial-ribosomal subunits and inhibits bacterial growth by inhibiting protein synthesis. Because of low cost, for sensitive *S.* Typhi strains, chloramphenicol is still used to treat typhoid fever. A 14-day course of chloramphenicol, ampicillin or trimethoprim and sulphamethoxazole is indicated for *S.* Typhi infection.

Multidrug-Resistant *Salmonella* Typhi

Chloramphenicol-resistant *S.* Typhi causing an epidemic with high mortality, appeared in Mexico in 1972, spreading to North America and Europe. In India, the epidemics of multidrug-resistant *S.* Typhi first appeared in Calicut (Kerala) in 1972. It became endemic and was confined to Kerala till 1978. Subsequently, such resistant strains appeared in many other parts of India. These resistant strains were also resistant to streptomycin, sulphadiazine and tetracycline due to a transmissible plasmid carrying resistant determinants to these antibiotics. *S.* Typhi-resistant strains were sensitive to amoxicillin, cotrimoxazole, ampicillin and furazolidone. However, in the late 1980s, India reported resistant strains of *S.* Typhi to almost all the drugs mentioned above from several parts of the country.

Multidrug resistance conferred by the R factor has become a widespread phenomenon among salmonellae. The reason for the rapid development of multidrug resistance in *Salmonella* is considered to be the integron gene cassette system. There is a spread of resistance within clinical strains through horizontal transfer, which is proved by the presence of the same gene cassette in different genotypes. Some multidrug-resistant *Salmonella* is considered to be an important agent of hospital infections. Such nosocomial infections affect the neonates causing septicaemia, meningitis and suppurative lesions. Recently, multidrug-resistant *Salmonella* causing hospital outbreaks of neonatal septicaemia has been seen in India

Fluoroquinolones (e.g. ciprofloxacin, pefloxacin, norfloxacin) and the third-generation cephalosporins (e.g. ceftazidime, ceftriaxone, cefotaxime) are the antibiotics of choice for the treatment of multidrug-resistant *S.* Typhi. They are increasingly used for typhoid fever because of their efficacy and low relapse, and low carrier rates associated with their use. In addition, the emergence of domestically acquired ceftriaxone-resistant *Salmonella* infections has also been described.

Epidemiology

Geographical Distribution

Enteric fever is endemic in many countries which have poor sanitary conditions. It is more common to children of school age or younger.

Typhoid fevers are endemic in the Indian subcontinent, Southeast and Far East Asia, the Middle East, Africa, Central America and South America. Approximately 11–20 million cases of typhoid fever occur globally each year, with around 128,000 and 161,000 deaths. Typhoid fever has been virtually eliminated in developed countries due to improvements in water supply and sanitation during the last many decades. However, climate and increasing antibiotic resistance pose a greater threat to global health safety and contribute to the rise of the burden of diseases. Travellers visiting countries, particularly in Asia and sub-Saharan Africa, are susceptible to developing typhoid fever.

Paratyphoid fever continues to be a health problem in both developing and developed countries. *S.* Paratyphi is prevalent in South America, Eastern Europe and Asia. *S.* Paratyphi B is prevalent in Western Europe, Britain and America. In Eastern Europe and *Guyana*, Paratyphi C is commonly seen.

Habitat

S. Typhi and *S.* Paratyphi (A, B and C) are strictly human pathogens. They are not found in any other animal hosts. In infected human hosts, they colonise the small intestine, especially the ileac mucosa. Asymptomatic long-term colonisation occurs commonly in infected hosts.

Other salmonellae are parasitic in various domestic animals, rodents, reptiles and birds. *S.* Typhimurium have a wide host range affecting animals, birds and humans, *Salmonella* Abortusoris in sheep and *S.* Gallinarum in poultry.

Reservoir, Source and Transmission of Infection

The infected patient and, more frequently, carriers are important reservoirs of infection for enteric fever.

There are three different types of *Salmonella* carriers. **Convalescent carriers** are the patients who continue to excrete typhoid bacilli in faeces for *3 weeks to 3 months* after clinical cure. **Temporary carriers** are those who shed the bacilli for *>3 months but <1 year.* **Chronic carriers** excrete the bacilli for *>1 year*. About 2–4% of patients become chronic carriers. The bacilli persist in the gallbladder and are excreted in faeces (*faecal carrier*) or persist in the kidney and are secreted in the urine (*urinary carrier*). Urinary carriers are less frequent than the faecal carriers and usually present with urinary lesions, such as calculus or schistosomiasis.

The carrier state is more common in women than in men and older people (over 40 years) than in young people. Food handlers or cooks who become carriers are potentially dangerous to the community. Mary Mallon (**Typhoid Mary**), a New York cook, was a classical case of such a carrier who, over 15 years, caused at least seven outbreaks affecting more than 200 persons. The carrier stage also occurs with *S.* Paratyphi infections. *S.* Paratyphi A infection occurs only in humans, while *S.* Paratyphi B infection occurs in animals, such as dogs or cows.

Human faeces infested with *S.* Typhi from the major source of infection by contaminating food, water and vegetables. Improper personal hygiene and consumption of contaminated food result in infections.

There is no animal reservoir for typhoidal salmonellae. Animal-to-human zoonotic transmission is common in only non-typhoidal salmonellae. Poultry, livestock, reptiles and pets are the principal reservoirs for non-typhoidal *Salmonella* organisms. Ingestion of improperly cooked fruits, vegetables, foods of animal origin, including poultry, red meats, unpasteurised milk and eggs that infected animals have contaminated or an infected human is the mode of transmission. Contact with infected reptiles, such as iguanas, pet turtles and tortoises and ingestion of contaminated water are other modes of transmission.

Prevention and Control

Availability of safe drinking water, proper food hygiene and sanitary disposal of excreta are the most cost-effective strategies for reducing the incidence of typhoid fever in endemic countries. These measures reduce the incidence of typhoid fever and reduce other enteric infections, which are a major cause of morbidity and mortality in those areas.

Vaccines

Salmonella spp. are primarily intracellular pathogens; hence cell-mediated immunity (CMI) rather than humoral antibodies play a more important role in protection against the disease. CMI develops during the disease and appears to be common in populations in areas endemic for typhoid fever. Individuals with depressed CMI appear to be more susceptible to *S.* Typhi infections.

Immunisation with typhoid vaccines at regular intervals also considerably reduces the incidence of typhoidal *Salmonella* infections. Two types of typhoid vaccines, such as killed or inactivated vaccine and oral vaccine, are administered.

Killed or Inactivated Vaccines

1. **TAB Vaccine** TAB vaccine is a complete cell inactivated vaccine containing 0.5% phenol preserved, heat-killed *S.* Typhi 1000 million per ml and 750 million per mL of each *S.* Paratyphi A and B. Administration of the vaccine is done sub-cutaneously in two subsequent doses of 0.5 mL each and is given within an interval of 4–6 weeks, which is then continued as a booster dose for every 3 years. The overall efficiency of 70–90% has been shown through field trials in typhoid fever for the span of 3–7 years. There are two concerns about this vaccine: fever and pain at the injection site, and a larger volume of antigen needs to be injected. Due to these concerns in India, the trivalent TAB vaccine has been replaced by a divalent typhoid-paratyphoid A vaccine without *S.* Paratyphi B because infections from paratyphi are uncommon here.
2. **Vi Capsular Polysaccharide Antigen Vaccine (VPS)** The ViCPS antigen preparation involves *S.* Typhi isolated from blood cultures, and this vaccine contains purified Vi antigen and capsular polysaccharide. A single parental dose of 0.5 mL of ViCPS is administered as primary vaccination. If continued exposure to *S.* Typhi is expected, booster doses are required every 2 years to ensure protection. Two field trials showed 50–64% overall protection rates in

South Africa and 72% in Nepal, the areas endemic for the disease. Fever, headache, erythema and induration are some of the side effects. The vaccine is not recommended for children below 2 years.

Oral Vaccines

1. **Ty21a Oral Vaccine** This oral vaccine contains live attenuated strains of *S.* Typhi. It has an enteric-coated capsule, and this mutant strain lacks the enzyme UDP-galactose-4-epimerase. Upon ingestion vaccine starts infection; however, after a few cell divisions (4–5), it results in self-destruction. Hence, it cannot induce illness. The vaccine is believed to induce intestinal and serum antibodies with cell-mediated immune responses. Nevertheless, the mechanism is not completely known. Ty21a primary vaccination contains a single enteric-coated capsule. It is administered on alternate days with a cycle of four capsules. If there is continued exposure, then booster doses are required once every 5 years to protect from infection. The capsules have to be stored at four °C in a refrigerator, and all four doses must be taken to obtain complete efficiency.
   This oral vaccine may not be advised to be administered for immunocompromised individuals, such as HIV-infected individuals and infants.

Salmonella Gastroenteritis

Salmonella Gastroenteritis or food poisoning is the most common form of salmonellosis. The condition is generally a zoonotic disease caused by certain species of non-typhoidal salmonellae, which are primarily animal pathogens. *S.* Typhimurium is the most common species causing the disease in many parts of the world. Other common species include *S.* Enteritidis, *Salmonella* Hadar, *Salmonella* Heidelberg, *S.* Agona, *Salmonella* Virchow, *Salmonella* Senftenberg, *Salmonella* Indiana, *Salmonella* Newport and *S.* Anatum.

Human infection usually occurs through the consumption of contaminated foods. The most common sources of salmonellae are milk and milk products, meat, poultry and eggs. Of great concern are eggs and egg products. Salmonellae can enter the shell if eggs are left on contaminated chicken feed or faeces and grow inside. Droppings of rats, lizards or other small animals may also cause food contamination. Improperly prepared fruits, vegetables, dairy products and shellfish may cause infection if contaminated with manure or unhygienic handling.

The incubation period is 6–72 h. Nausea, vomiting and loose, watery stools are the common symptoms. Fever, abdominal cramps, myalgias and headaches are also common. Fever, which rarely exceeds 39 °C, occurs in approximately one-half of infected patients. Symptoms usually resolve spontaneously in 2–7 days.

Laboratory diagnosis is made by isolating the salmonellae from the faeces by culture. In food poisoning outbreaks, isolation of salmonellae from the food confirms the diagnosis.

Uncomplicated, non-invasive salmonellosis is treated symptomatically. Antimicrobial therapy is not indicated because it does not shorten the duration of illness. However, for serious invasive cases, antibiotic treatment is recommended. Treatment with antibiotics is recommended for *Salmonella* Gastroenteritis in infants below 3 months and infants below 12 months with high fever and unknown blood culture results. It is also recommended for patients with haemoglobinopathies, HIV infection or other causes of immunosuppression, neoplasms or chronic GI illnesses.

Ampicillin, amoxicillin, trimethoprim–sulphamethoxazole, cefotaxime and ceftriaxone effectively treat the condition.

Salmonella Septicaemia

All *Salmonella* spp. can cause septicaemia. However, *S.* Choleraesuis, *S.* Paratyphi and *S.* Typhi more commonly cause a bacteraemic disease. Paediatric and geriatric patients and patients with AIDS are increasingly susceptible to suffering from *Salmonella* Bacteraemia.

The clinical presentations of *S.* Septicaemia are like that of another Gram-negative septicaemia. However, localised suppurative infections, such as osteomyelitis, deep abscesses, endocarditis, arthritis and meningitis, can occur in as many as 10% of patients. The case fatality may be as high as 25%.

Diagnosis of the condition is made by isolating salmonellae from the blood or the pus from the suppurative lesion and rarely from faeces.

Case Study

A 35-year-old adult was admitted to the hospital with a high fever (>101 °F), headache and abdominal symptoms for the last 10 days. He gave a history of abdominal discomfort with diarrhoea. A local doctor had examined the patient 4 days back and had given a course of chloroquine to treat malaria, but the patient did not show any sign of improvement. A battery of tests was carried out to rule out typhoid fever.

- Describe the virulence factors of the bacteria responsible for typhoid fever.
- Describe the laboratory tests employed for the diagnosis of typhoid fever.
- What are the vaccines available against typhoid fever?
- Does the bacteria causing typhoid fever to show any resistance to antibiotics? Discuss that.

Genus *Shigella*

Shigella is the most common cause of bacillary dysentery, which occurs worldwide. The disease is spread through faecal-oral transmission, and humans are the only natural reservoir of the bacteria. Human infections caused by *Shigella* species are summarised in Table 36.5.

Classification

Based on a combination of biochemical and serological characteristics, Shigellae are classified into four species or subgroups, consisting of more than 45 O antigen-based serogroups and each species consisting of different serotypes.

Shigella dysenteriae **(subgroup A):** This is subdivided into 12 serotypes. Each serotype is characterised by the presence of a different type of antigen. *S. dysenteriae* serotype 1 is the bacillus originally described by Shiga, hence known as Shiga's bacillus. Serotype 1 is indole negative, and this is the only serotype to be always catalase positive in this family. *S. dysenteriae* serotypes 3–7 were described by Large and Sachs in India and hence were known as the Large–Sachs group.

Shigella flexneri **(subgroup B):** This group is named after Flexner (1900), who described the first mannitol-fermenting shigellae from the Philippines. *S. flexneri*, based on type-specific (I–VI) and group-specific (1–8) antigens, have been classified into six serotypes (1–6) and several subtypes (1a, 1b; 2a, 2b; 3a, 3b, 3c; 4a, 4b; 5a, 5b).

Shigella boydii **(subgroup C):** This group is named after Boyd, who first described these strains from India (1931). A total of 19 serotypes have been identified in this group. Members of *the S. boydii* group resemble biochemically, but not antigenically, with those of *S. flexneri*.

Shigella sonnei **(subgroup D):** This group is named after Sonne, who first described these strains from Denmark (1915). *S. sonnei* have been divided into 26 colicin types using 33 indicator strains.

Table 36.5 Human infections caused by *Shigella* species

Bacteria	Diseases
Shigella species	Shigellosis, bacillary dysentery, asymptomatic carriage and haemolytic uraemic syndrome

Properties of the Bacteria

Morphology

Shigella are short Gram-negative rods about 0.5 × 1–3 μm in size. They are non-motile, non-sporing and non-capsulated. Except for *S. flexneri* serotype 6 and some strains of other serotypes, Shigella species possess fimbriae.

Culture and Biochemical Properties

Shigella are aerobes and facultative anaerobes. They grow at an optimum temperature of 37 °C and pH 7.4.

Nutrient Agar

They grow on nutrient agar or Mueller–Hinton agar. *After overnight incubation, Shigella* colonies on nutrient agar are small, circular, convex, smooth, and translucent. Occasionally on primary isolation and frequently in sub-cultures, a proportion of the colonies may be of the rough type.

MacConkey Agar

Shigella spp. on MacConkey agar produce non-lactose-fermenting pale, colourless colonies. However, *S. sonnei* (which ferments lactose late) forms pale pink colonies on prolonged incubation.

Deoxycholate Citrate Agar (DCA), Xylose Lysine Deoxycholate (XLD) Agar, Salmonella–Shigella (SS) Agar and Hektoen Enteric (HE)

DCA, XLD agar, SS agar and HE agar are frequently used **selective media** for the isolation of *Shigella* species. DCA is a useful selective medium for isolating *Shigella* spp. from faeces. On this medium, *Shigella* spp. produce small colonies, which on prolonged incubation produce lactose-fermenting pink colonies. XLD agar is another selective medium, which is less inhibitory to *S. dysenteriae* and *S. flexneri*. *Shigella* spp. forms red colonies on this medium. *Shigella* spp. on SS agar form colourless colonies. *Shigella* spp. on HE agar forms green colonies.

Selenite F and Gram-Negative (GN) Broth

Selenite F and GN broth are commonly used enrichment media. Enrichment of faeces in GN broth for 4–6 h followed by sub-culture on XLD or HE medium is useful for isolating *Shigella* from clinical specimens.

Shigella ferments mannitol, forming acid but no gas. The mannitol fermentation test is an important biochemical test used to classify shigellae into mannitol-fermenting and non-fermenting species. *S. flexneri*, *S. boydii* and *S. sonnei* are mannitol-fermenting species, whereas *S. dysenteriae* is mannitol-non-fermenting species. However, exceptions are not that uncommon. The biochemical characteristics useful for distinguishing different *Shigella* species are listed in Table 36.6.

They are catalase positive except for *S. dysenteriae* type 1, which is catalase negative. They are MR positive, citrate negative and oxidase negative.

Susceptibility to Physical and Chemical agents

Shigellae are killed at a temperature of 55 °C in 1 h or by 1% phenol in 30 min. In faeces, they die

Table 36.6 Differentiation of *Shigella* species

Characteristics	*Shigella dysenteriae*	*Shigella flexneri*	*Shigella boydii*	*Shigella sonnei*
Number of serotypes	12	6 + 2 variants	18	2 phases; 26 colicin types
Lactose	–	–	–	–
Sucrose	–	–	–	–
Mannitol	–	+	+	+
Dulcitol	–	–	V	–
Xylose	–	–	V	V
Indole	V	V	V	–
Lysine decarboxylase	–	–	–	–
Ornithine decarboxylase	–	–	–	+

V: variable; –: negative; +: positive

Table 36.7 Biotypes of *Shigella flexneri* serotype 6

Biochemical Properties	Boyd 88	Manchester	Newcastle
Indole	–	–	–
Glucose	⊥	+	⊥ or +
Mannitol	⊥	+	–

⊥: acid only; +: acid and gas

within a few hours due to acidity produced by the growth of intestinal bacteria. They remain viable in moist environments for days but die rapidly on drying. *S. sonnei* is, in general, more resistant to unfavourable environmental conditions than the other *Shigella* species.

Antigenic structure

Other Properties

The antigenic structure of shigellae is simple, unlike the complex antigenic structure of salmonellae. Shigellae possess somatic O antigens, and certain strains possess K antigens. *Shigella* K antigens, when present, may sometimes interfere with agglutination by O antisera. Shigellae strains also possess fimbrial antigens. Common fimbrial antigens are also found, particularly in *S. flexneri*.

S. flexneri is biochemically heterogeneous and antigenically the most complex among shigellae. Two antigenic variants, called X and Y, which lack the type-specific antigens, are also recognised in addition to this. *S. flexneri* serotype 6 is always indole negative and is classified into three biotypes: Boyd 88, Manchester and Newcastle (Table 36.7).

S. sonnei is antigenically homogeneous but may occur in two forms: phase I and phase II. Phase I strains produce smooth colonies, whereas phase II colonies form large, flat and more irregular colonies. Cultures contain a mixture of both forms. Usually, phase II strains are isolated more frequently from convalescing cases and from carriers than from patients.

Pathogenesis

Virulence in *Shigella* species involves both chromosomal- and plasmid-coded genes, which express many virulence factors (Table 36.8).

Virulence Factors

Intestinal Adherence Factor

Intestinal adherence factor is a 97 kDa outer membrane protein encoded by each gene on chromosomes. This mediates the colonisation of *Shigella* spp. in infected human hosts and animal models.

Endotoxins

The cell wall of shigellae, like other Gram-negative bacilli, contains a lipopolysaccharide (LPS) structure. The LPS is liberated during lysis of the cell and, to some extent, during culture. The LPS moiety functions as endotoxin and is an important component of the virulence of the bacteria. It plays an important role in the resistance of *Shigella* to non-specific host defence encountered during tissue invasion. The toxin helps in the invasion, multiplication and resistance of *Shigella* to phagocytosis by tissue macrophages. It also increases the cytotoxic

Table 36.8 Virulence factors of *Shigella* species

Virulence factors	Biological functions
Endotoxins	Invasion, multiplication and resistance of *Shigella* to phagocytosis by tissue macrophages
Intestinal adherence factor	Colonisation of *Shigella*
Shiga toxin	Disrupts protein synthesis and produces endothelial damage

activity of Shiga toxin on human vascular endothelial cells. Chromosomal genes of the bacteria express the endotoxin.

Shiga Toxin

Shiga toxin is an exotoxin produced by *S. dysenteriae*. It is a heat-labile protein and acts as enterotoxin and neurotoxin. Shiga toxin (Stx) is a group of cytotoxins that contain two major immunologically non-cross-reactive groups called Stx1 and Stx2. Both Stx1 and Stx2 groups are encoded by a bacteriophage inserted into the bacteria's chromosome. Stx1 groups of toxins can be found in *S. sonnei* and type 4 *S. dysenteriae*. Stx1c and Stx1d are the two variants of the Stx1 toxin. Stx1c and Stx1d are rarely found in human disease and are linked with a mild disease course.

Shiga toxins have one A subunit and five B subunits. Subunit A is a 32-kDa polypeptide. The mature A subunit of Stx1 and Stx2 consists of 293 and 297 amino acids, respectively, wherein the active site of the toxin is the glutamic acid (position 167). Intrinsic enzymatic activity cleaves the A subunit asymmetrically into an A_1 subunit, and A_2 peptide held together by a disulphide bridge. It cleaves the 28SrRNA in the 60S ribosomal subunit, thereby preventing the binding of aminoacyl-transfer RNA and disrupting protein synthesis.

The B subunits of Stx1 and Stx2 consist of 69 and 71 amino acids, respectively. The main function of the B subunit is to bind toxins to the host cell glycolipid (Gb3) surface receptor, present on the brush border of the epithelial cell of the intestines. It also mediates the transfer of the A subunit into the cell.

The Shiga toxin shows three types of toxic activities:

1. **Neurotoxic activity:** Although called neurotoxin, the primary site of its action is not the neural tissue but is the blood vessels, neurological manifestations being secondary.
2. **Enterotoxic activity:** These toxins are enterotoxic for ligated rabbit intestinal segments with induction of fluid accumulation in ligated rabbit ileal loop. Two new *Shigella* enterotoxins, designated as S. ET-l and S. ET-2—the former confined to *S. flexneri* 2a and the latter more widespread—have been identified.
3. **Cytotoxic activity:** This is demonstrated by cytotoxicity of toxin for Vero, HeLa and some selected endothelial cells, such as human renal vascular endothelial cells. This appears to be the same as verotoxin 1 (or Shiga-like toxin) produced by certain strains of *Escherichia coli* (VTEe). The ability of the bacillus to penetrate cultured HeLa or HEp-2 cells or by the Congo red binding test is used to demonstrate invasiveness of the bacillus. The bacillus, which is invasive, has a large plasmid (M.W. 140 × 106) which codes for an outer membrane protein. This protein is responsible for cell penetration. These proteins are termed as "Virulence marker antigens" (VMA). VMA can be detected by ELISA, which is a virulence test for Shigellae, the same as for enteroinvasive *E. coli*.

The primary manifestation of Shiga toxin is damage to the intestinal epithelium of the infected host, causing diarrhoea and dysentery. However, Shiga toxin can mediate damage to the glomerular endothelial cells in a few patients, resulting in haemolytic urinary syndrome.

Pathogenesis of Bacillary Dysentery

Shigella species are not zoonotic pathogens. It causes natural infections in humans and non-human primates. However, in rare and unique cases, zoonotic transmission of organisms, like *S. flexneri*, *S. sonnei* and *S. dysenteriae*, from other laboratory research animals has been recorded.

Shigella spp. produces a serious disease known as bacillary dysentery. The infection is acquired by the faecal-oral route by hand-to-mouth infection through contaminated fingers. It is also caused by ingestion of food and water contaminated with human faeces containing *Shigella* spp. The infectivity dose (ID) is extremely low. As few as 10 *S. dysenteriae* bacilli can cause clinical disease, whereas 100–200 bacilli are needed for *S. sonnei* or *S. flexneri* infection.

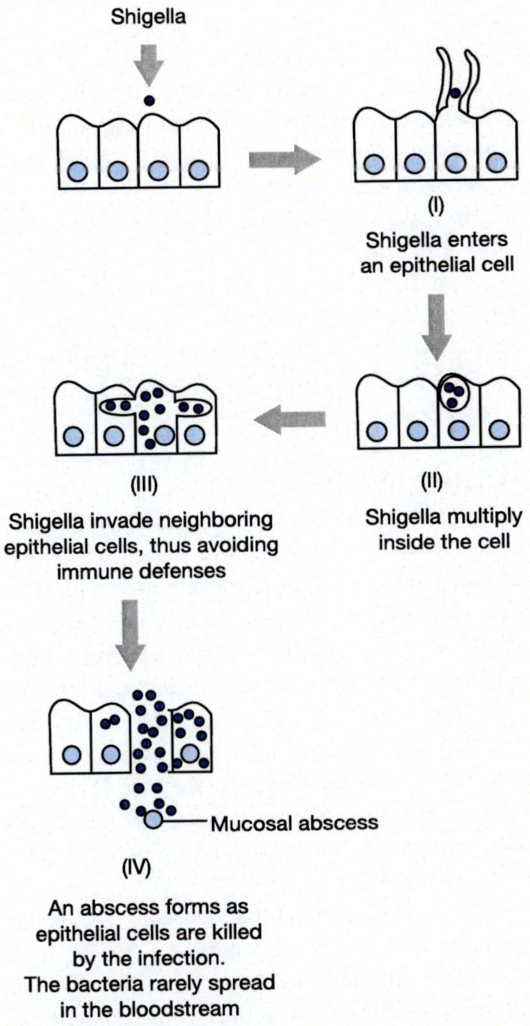

Fig. 36.5 Schematic diagram showing pathogenesis of bacillary dysentery

Shigella spp. cause bacillary dysentery by invading and replicating in cells lining the intestinal mucosa of the colon. Structural proteins, such as intestinal adhesion factor, endotoxin and exotoxin, mediate the adherence of the bacteria to the cells as well as their invasion, intracellular replication and cell-to-cell spread. The bacilli infect the epithelial cells of the villi in the large intestine and multiply inside them. Subsequently, bacteria spread laterally to involve adjacent cells and penetrate the lamina propria (Fig. 36.5). Shigellae lyse the phagocytic vacuole and multiply in the host cell cytoplasm. Shigellae survive phagocytosis by inducing programmed cell death or apoptosis. Shiga toxin produced by the bacteria also plays an important role in the progression of mucosal lesions after the invasion of the colonic cells. The toxin also induces vascular damage in the colonic mucosa. Mucosal oedema, erythema, friability, superficial ulceration and focal mucosal haemorrhage involving the rectosigmoid junction are the typical pathological features primarily observed in the condition.

Clinical Syndromes

Shigella spp. causes shigellosis, a clinical syndrome encompassing the whole spectrum of disease caused by the bacteria. *S. sonnei* may cause only mild diarrhoea. *S. flexneri* and *S. boydii* cause more severe illness than that caused by *S. sonnei*.

Bacillary dysentery is a severe clinical form of shigellosis. The incubation period is usually short. It varies from 12 h to 7 days, usually 48 h, and is inversely related to a load of ingested bacilli. Bacillary dysentery is an acute gastrointestinal illness manifested by fever, vomiting, abdominal cramps and tenesmus. The condition manifests with a sudden onset of high-grade fever along with abdominal cramp, tenesmus, urgency and passage of loose, scanty faeces containing frank blood and mucus.

Complications are most often associated with *S. dysenteriae* type-1 infection. These include arthritis, toxic neuritis, conjunctivitis and in children intussusception. Haemolytic uraemic syndrome may also occur following infection with *S. dysenteriae* because of vasculopathy mediated by Shiga toxin. In a few patients, the asymptomatic colonisation of shigellae occurs in the colon, making the patient a persistent reservoir for infection.

Laboratory Diagnosis

Specimens

The stool is the specimen of choice. Fresh faeces are inoculated without delay or transported in a suitable medium, such as Sachs buffered glycerol saline, pH 7.0–7.4. Also, rectal swabs may be taken from the ulcer site by sigmoidoscopy. However, rectal swabs that do not contain a copious stool volume or mucus are not satisfactory.

Microscopy

Routine microscopy of stool may reveal clumps of polymorphonuclear leukocytes. Faecal blood or leukocytes are detectable in the stool in approximately 70% of cases of shigellosis.

Culture and Identification

Diagnosis of shigellosis is made by isolating *Shigella* spp. from faeces by culture. For enrichment, one tube each of selenite F and GN broth are inoculated and incubated at 37 °C for 12–18 h before sub-culture onto selective media. Usually, more than one stool or rectal swab is collected and inoculated immediately on at least two different culture media, such as MacConkey, XLD, DCA or eosin-methylene blue agars. After overnight incubation, *Shigella* produces pale non-lactose-fermenting colonies on MacConkey and DCA media, red colonies on XLD medium and colourless colonies on SS agar.

Pale non-lactose-fermenting colonies on MacConkey agar are identified by carrying out motility tests, biochemical tests, such as manifold fermentation tests, and slide agglutination tests with specific *Shigella* antisera (polyvalent and monovalent sera).

Serology

Serological tests are not useful in the diagnosis of shigellosis.

Molecular Detection

For the limitations in the classical culture-based diagnosis of Shigella, molecular technologies, like PCR and real-time PCR assays, multiplex, pentaplex PCRs, have emerged as a promising tools for bacterial identification. Such PCRs generally target the invasion plasmid H (*ipaH*) gene, O antigen synthesis genes and the 16S rRNA gene for detection of *Shigella* spp. *ipaH* gene that code for invasion-associated plasmid antigen are present in multiple copies in both the chromosomes and the plasmids of all *Shigella* species and are considered hallmark genes for detecting all *Shigella* strains. Pentaplex PCR has been found promising for detecting the genus *Shigella,* in addition to separating clinically important *Shigella* spp., namely, *S. flexneri, S. sonnei* and *S. dysenteriae.*

However, considering difficulties associated with running PCR in resource-poor regions because of the requirement of specialised laboratories, trained personnel, complicated apparatus and expensive reagents, LAMP–based methodology targeting ipaH gene has shown greater promises.

Molecular Typing

Many genetic investigations on different isolates of a species of Shigella have indicated striking diversity at gene levels. It was reported within the four epidemics causing *S. flexneri* isolates when analysed using pulsed-field gel electrophoresis (PFGE) or for IS1 and IS2 patterns. Likewise, completely different DNA profiles were also revealed within *S. dysenteriae* isolates. Such clonal differentiation leads to varying degrees of disease outcome within a species and has potential usefulness in the molecular epidemiology of *Shigella* strains.

To determine such clonal diversity, molecular techniques, such as ribotyping, PFGE, and PCR-based methods, have been extensively used for Shigella. PFGE being the "gold standard" in molecular typing and MLVA (multi-locus variable number of tandem repeat analysis) are the most widely used tools for Shigella typing.

In conclusion, both PFGE and IS probing demonstrated their potential usefulness in molecular epidemiology and typing of *Shigella* strains. The degree of differentiation given by these two methods was generally comparable, although IS probes showed better discrimination of the isolates.

Treatment

Uncomplicated shigellosis is a self-limited condition, and patients usually recover spontaneously in a few days. Hence, no antibiotics are recommended for these cases. Antibiotic treatment for *Shigella* infection is recommended (1) for severe or toxic cases and (2) for the very young, debilitated and aged individuals. Antibiotic treatment is recommended to decrease the duration of illness, person-to-person spread and cases in household contacts.

Trimethoprim–sulphamethoxazole, ampicillin, tetracycline and the quinolones, such as nalidixic acid and ciprofloxacin, are frequently used antibiotics. Trimethoprim–sulphamethoxazole is very effective for shigellosis. The antibiotic sulphamethoxazole is a competitive inhibitor of para-aminobenzoic acid (PABA) and trimethoprim, reversibly inhibits dihydrofolate reductase, preventing nucleic acid biosynthesis in *Shigella* by producing a sequential blockade in the folic acid synthesis pathway. This is the drug of choice when the antibiotic susceptibility of the bacteria is not known. However, ampicillin is still the drug of choice if *Shigella* isolate is susceptible to this drug.

Antibiotics-resistant *Shigella*

Multiple-antibiotics-resistant plasmids are widely prevalent in Shigella sp. and were first documented in Japan during the early 1950s. About half of the strains reported worldwide has been found to be resistant to multiple drugs. Most of these strains were resistant to streptomycin, chloramphenicol and sulphonamides. Shigella employs various mechanisms for drug resistance, including drugs extrusion by active efflux pumps, decrease in cellular permeability, overexpression of drug-modifying and/ or -inactivating enzymes, etc. Indiscriminate use of antibiotics has worsened the problem. Hence, it is essential to treat the cases of shigellosis with the results of *in vitro* antibiotic susceptibility testing of *Shigella*.

In an alternative to control drug-resistant Shigella, the use of nanoparticles (NPs) and phage therapy is emerging as promising approaches. The use of copper oxide NPs and nanoantibiotic formulations have been used successfully to control Shigella. In contrast, phage therapy for Shigella was discovered by a French microbiologist, d'Herelle. It showed high specificity in killing Shigella with fewer side effects.

Epidemiology

Shigellosis occurs worldwide. It is a common cause of bacterial diarrhoea or dysentery. It has been found that as few as 10 to 100 organisms can cause disease to humans. An estimated 180 million cases of *Shigella* diarrhoea or dysentery occur annually worldwide, including 164,000 associated deaths. The incidence of shigellosis in developing countries is nearly 20 times more than in developed countries due to poor sanitary conditions. In developed nations, incidences of sporadic food or water-borne outbreaks have been reported.

Global epidemiological data suggest that 99% of all shigellosis is present in developing countries, with northern Africa having the highest disease burden. The US, Europe and Australia have the lowest incidences. *S. flexneri* is the most prevalent serotype worldwide and is abundant in Africa and South and East Asia. *S. sonnei* is predominant in the US and Europe.

Geographical Distribution

Shigellosis is worldwide in distribution but shows many variations epidemiologically between the nature and extent of the infection in the industrially advanced and developing and poor countries. Endemic shigellosis is found in all age groups and is caused by all species in developing countries where environmental sanitation is poor.

Habitat

Shigella species are strictly human pathogens. They are found in the large intestine of infected human hosts. They are not found in any other animal hosts.

Reservoir, Source and Transmission of Infection

Infected patients or, less often, carriers are reservoirs of infections for shigellosis. Chronic carriage is rare because the bacilli are not excreted in faeces within a few weeks, except in some malnourished children or patients with AIDS.

Shigellosis is transmitted by the faecal-oral route by hand-to-mouth infection through contaminated fingers. Because as few as 10–200 bacilli can cause disease, shigellosis spreads rapidly in areas where sanitary standards and the level of personal hygiene are low. It is also transmitted by ingestion of food and water contaminated with human faeces containing *Shigella* spp. The condition is transmitted less frequently by fomites, such as door handles, water taps, lavatory seats and sexually among young male homosexuals due to oro-anal contact. Flies may transmit the infection as mechanical vectors.

Shigellosis is primarily a disease of children. Nearly 70% of all infections occur in children younger than 15 years. *Shigella* infection in malnourished children often causes a vicious cycle of further impaired nutrition, recurrent infection and further growth retardation. Endemic disease in adults is common in household contacts of infected children and male homosexuals.

Prevention and Control

In developed countries, person-to-person transmission is the most common source of infection. Water contaminated with human excreta is the most common source of infection. Therefore, control consists essentially in improving personal and environmental sanitation. Antibiotics are not used in prophylaxis. No effective vaccine is available.

Case Study

A 7-year-old child was admitted to the causality for treatment of bloody diarrhoea and abdominal pain for 48 h. Her mother gave a history of her child vomiting a few times and mentioned that she developed these symptoms after eating at a fast-food counter. Sigmoidoscopy showed mucosal erythema petechiae with pus but no ulcer.

1. List the causative bacterial agent for the disease.
2. List virulence factors of the bacteria.
3. Describe the epidemiology of the disease caused by the bacteria.
4. Discuss the treatment of the condition.

Yersinia, Pasteurella and Francisella

37

Earlier, the plague bacillus and many other Gram-negative, short bacilli that are primary pathogens of rodents were included in the genus *Pasteurella*. This group has been divided into three genera: *Yersinia*, *Pasteurella* and *Francisella*.

Genus *Yersinia*

The genus *Yersinia* is now included in the tribe Yersinieae in the family Enterobacteriaceae and includes 11 species, 3 of which are pathogenic in humans. Human infections caused by *Yersinia* species are summarised in Table 37.1.

Yersinia pestis

The plague, caused by *Yersinia pestis*, was one of the most devastating diseases in history. In the early twentieth century, plague epidemics were responsible for about ten million deaths in India.

Properties of the Bacteria

Morphology

Y. pestis is a Gram-negative coccobacillus that measures 1.5–30.7 μm in size, with rounded ends and convex sides. The bacilli are arranged in single, in pairs or short chains and are motile at 22 °C. Pleomorphism is very common. Giemsa or methylene blue-stained smears of clinical specimens show characteristic bipolar staining (safety pin appearance) with both ends densely stained and a clear central area of the bacteria. *Y. pestis* is a capsulated bacterium, catalase positive and **coagulase positive.**

Culture and Biochemical Properties

Y. pestis is aerobic and facultatively anaerobic. It grows at a temperature range of 2–45 °C with an optimum temperature of 27 °C (unlike most bacteria), but it grows better at 37 °C in culture. The bacteria grow at a wide range of pH (5–9.6), with an optimum pH of 7.2.

Y. pestis can grow on various media, including Mueller–Hinton agar, nutrient agar, blood agar and MacConkey agar. On **nutrient agar**, *Y. pestis* produces small, delicate and transparent colonies becoming opaque on prolonged incubation. **Blood agar containing sodium azide (7 μg/mL)** is a selective medium for *Y. pestis*. On blood agar, it produces dark-brown colonies due to the absorption of the haemin pigment. The bacteria grow poorly on **MacConkey agar and deoxycholate citrate agar**, producing pinpoint reddish colonies after 24 h of incubation.

In broth, *Y. pestis* produces a flocculent growth with granular deposit at the bottom and along the sides of the tube, with little or no turbidity. On prolonged incubation, a delicate pellicle may form at the surface. *Y. pestis* produces a characteristic growth when grown in a flask of broth with oil or *ghee* (clarified butter) floated on top (*ghee* broth). The growth in the medium appears

Table 37.1 Human infections caused by *Yersinia* species

Yersinia species	Diseases
Yersinia pestis	Plague (bubonic, pneumonic and septicemic)
Yersinia enterocolitica	Enterocolitis primarily in young children Lymphadenitis with terminal ileitis (pseudo-appendicitis syndrome) Extraintestinal infections (cellulitis, conjunctivitis, meningitis, osteomyelitis, pharyngitis, pneumonia and urinary tract infection): Rare
Yersinia pseudotuberculosis	Gastroenteritis (mesenteric lymphadenitis simulating acute appendicitis) Izumi fever Far East scarlatinoid fever Other manifestations (erythema nodosum, arthralgias, reactive arthritis, ankylosing spondylitis; terminal ileitis and intussusception, especially in children)

to hang down into the broth from the surface, resembling stalactites (stalactite growth). Biochemical reactions of the bacteria are listed in Table 37.2.

Other Properties

Susceptibility to Physical and Chemical Agents

Y. pestis is sensitive to heat, sunlight, drying and chemical disinfectants. The bacilli are killed at a temperature of 55 °C in 30 min. They are also killed rapidly by disinfectants, such as 0.5% phenol in 15 min. The bacilli remain viable for long periods in sealed agar slopes, frozen tissue and cold and moist environments, such as in the soil of rodent burrows. A specific antiplague bacteriophage lyses all strains at 22 °C.

Antigenic Structure

Y. pestis is antigenically homologous; hence no serotypes are present. *Y. pestis* possess major antigens such as V and W antigens and F1 envelope antigens, which are of importance in the pathogenesis of the disease.

Pathogenesis

Y. pestis is a highly virulent bacterium that causes plague with a high mortality rate. The ability of *Yersinia* species to resist phagocytic killing is the hallmark of the pathogenesis of plague. Serum resistance and the ability of the bacilli to absorb organic iron as a result of a siderophore-independent mechanism further contribute to the pathogenesis of the disease.

Virulence Factors

Y. pestis produces many toxins and enzymes, all of which contribute to the pathogenesis of the diseases. Virulence factors of *Y. pestis* like that of *Salmonella* are complex. These are encoded both on the organism's chromosome and on large plasmids (Table 37.3).

Plague Toxin

It refers to a complex of at least two different types of toxins, endotoxin and murine toxin, found in culture filtrates or cell lysates. The **endotoxin** is a lipopolysaccharide found in the cell wall and is responsible for many of the systemic

Table 37.2 Differentiating features of *Yersinia pestis* var.

Yersinia pestis var.	Nitrate reduction	Glycerol fermentation	Geographical distribution
Yersinia pestis var. *orientalis*	Positive	Negative	Primary foci are in India, Myanmar and China. Also responsible for wild plague in the western USA, South America and South Africa
Yersinia pestis var. *antiqua*	Positive	Negative	Distributed in Mongolia, Manchuria and Transbaikalia
Yersinia pestis var. *medievalis*	Negative	Positive	Southeast Russia

Table 37.3 Virulence factors of *Yersinia pestis*

Virulence factors	Biological functions
Plague toxin	Causes systemic manifestations of plague
F1 envelope antigen	Inhibits phagocytosis
V and W antigens	Inhibit phagocytosis and intracellular killing of the plague bacillus inside macrophages
Type III secretion systems	Facilitate secretion of virulence factors of *Y. pestis* into host cells Prevent phagocytic killing of the pathogenic *Yersinia* species
Plasminogen activator (pIa) protease	Degrades C3b and C5a components of the complement Also degrades fibrin clots
Yersinia outer membrane proteins (YOPs)	Cell surface adhesion, iron acquisition, inhibition of phagocytosis and intracellular killing

manifestations of the disease caused by *Y. pestis*. **Murine toxins** exhibit some properties of both exotoxins and endotoxins. It is a heat-labile protein. The toxins may be toxoided but do not diffuse freely into the medium and are released only by the lysis of the cell. The toxin is pathogenic in rats and mice but not in guinea pigs, rabbits, humans and other primates. In experimental animal infections, the murine toxins produce local oedema and necrosis with systemic effects on the peripheral vascular system and liver. The role of murine toxins in the pathogenesis of plague in humans is unknown.

F1 Envelope Antigen (Fraction I or F–I Antigen)

It is a major virulence factor. It is a heat-labile envelope protein generally present only in virulent strains. The antigen is formed in the *in-vivo* and *in-vitro* culture of virulent strains at 37 °C. Production of the F–I antigen is mediated by a plasmid. The antigen inhibits phagocytosis and plays an important role in conferring protective immunity in humans and mice. The antibody to F–I antigen is protective in mice.

V and W Antigens

These two antigens are always present together, and a plasmid mediates their production. V antigen is a protein of molecular weight 90 kDa, while W antigen is an acidic 145 kDa lipoprotein. These proteins are produced by virulent strains of *Y. pestis* when grown at 37 °C in the presence of a low concentration of calcium. These two antigens inhibit phagocytosis and intracellular killing of the bacillus inside macrophages.

TTSS

TTSS consists of many proteins, which facilitate the secretion of virulence factors of *Y. pestis* into host cells. The TTSS mediates the ability of the pathogenic *Yersinia* species to resist phagocytic killing. The TTSS also inhibits the production of cytokines, reducing the inflammatory immune response to infection.

Other Virulence Factors

Yersinia produces the enzymes, such as coagulase, fibrinolysin and pIa protease, which contributes to the virulence of the bacterium. For example, enzyme pIa protease degrades C3b and C5a components of the complement, thereby preventing opsonisation and phagocytic migration, respectively. The enzyme also degrades fibrin clots, thereby facilitating *Y. pestis* to spread rapidly. The virulence of the bacterium is enhanced further by its ability to absorb organic iron as a result of a siderophore-independent mechanism.

Pathogenesis of Plague

Y. pestis causes plague, which occurs in three forms: bubonic plague, pneumonic plague and septicaemic plague.

Bubonic Plague

This is a zoonotic disease transmitted by rat flea *Xenopsylla cheopis* from animals to humans. *Y. pestis* is a natural parasite of rodents; infection is transmitted and maintained among them by rat fleas. The fleas acquire the infection by feeding on infected rodents. When a rat flea bites an infected and diseased rat, it sucks nearly 0.5 mL of blood per feed containing nearly 5,000–50,000 plague bacilli. In the flea's stomach, the bacilli multiply to such an extent that they block the proventriculus and prevent food entry into the stomach. The interval between the ingestion of infected blood and blocking in the proventriculus is referred to as the extrinsic incubation period, which varies from 2 to 3 weeks in *X. cheopis*.

The blocked flea, to overcome starvation, begins a blood-sucking rampage by biting other rodents. Between its attempts to swallow, the distended bacillus-packed stomach recoils, depositing the bacilli into the victim's skin, thereby transmitting the infection (blocked flea phenomenon). When the diseased rat dies, the fleas leave the carcass and, in the absence of another rat, may bite humans, causing bubonic plague.

The bubonic plague is characterised by the pathognomonic **bubo** (bubon meaning groin). The disease occurs following the bite of rat flea, depositing thousands of bacilli into the host's skin during blood meals. The bacilli migrate to the regional lymph nodes, where they are phagocytosed by the polymorphonuclear cells and mononuclear phagocytes and multiply intracellularly. The subsequent spread of bacteria occurs along the lymphatic channels towards the thoracic duct, with the eventual distribution of the bacteria in circulation leading to bacteraemia and septicaemia. In the absence of specific therapy, the bacteria potentially infect every organ, including the lungs, liver, spleen, kidneys and rarely even the meninges.

Pneumonic Plague

This is transmitted from human to human. It occurs following direct inhalation of the bacilli by droplet infection due to close contact with infected hosts or by inhalation of aerosolised bacteria, which may occur if used as a biological weapon. The bacilli spread through the lungs, producing a severe and rapidly progressive multi-lobar bronchopneumonia, subsequently leading to bacteraemia and septicaemia. It is not a zoonotic disease.

Septicaemic Plague

This is usually the terminal stage of the bubonic or pneumonic plague. It may sometimes occur primarily. Primary septicaemic plague may occur when the plague bacilli are deposited directly in the circulation, bypassing the lymphatics. Early dissemination with sepsis occurs, but without the formation of a bubo. This is usually observed in individuals bitten by the rat flea in the oral, tonsillar and pharyngeal areas due to the vascularity of the tissue and short lymphatic distance to the thoracic duct.

Clinical Syndromes

Bubonic plague is the most common clinical form of the disease. The condition is associated with a sudden onset of high fever, chills and headaches, body aches, extreme exhaustion, abdominal pain and diarrhoea. The presence of painful, swollen lymph glands (*buboes*), usually in the groin, axilla or neck, is the characteristic manifestation. The buboes are unilateral, oval and extremely painful. They measure 2–10 cm in size. Inguinal lymph nodes are most commonly involved. Enlargement of the buboes may cause rupture and discharge of foul-smelling pus. Hepatomegaly and splenomegaly may occur, causing tenderness. The mortality rate for untreated plague is 40–70%. An attack of plague provides a long-lasting immunity to infected humans.

Pneumonic plague is highly infectious and transmitted by aerosol droplets. The condition is characterised by abrupt onset of fever and chills, lymphadenopathy, chest pain, dyspnoea, purulent sputum or haemoptysis. Buboes may or may not be present in pneumonic plague. The condition, if untreated, is invariably fatal.

Septicaemic plague is associated with a high mortality rate. The condition is associated with a rapid onset of symptoms, such as nausea, vomiting, abdominal pain and diarrhoea. Diarrhoea may be the predominant symptom. Patients have a toxic appearance due to overwhelming infection with the plague bacillus and may

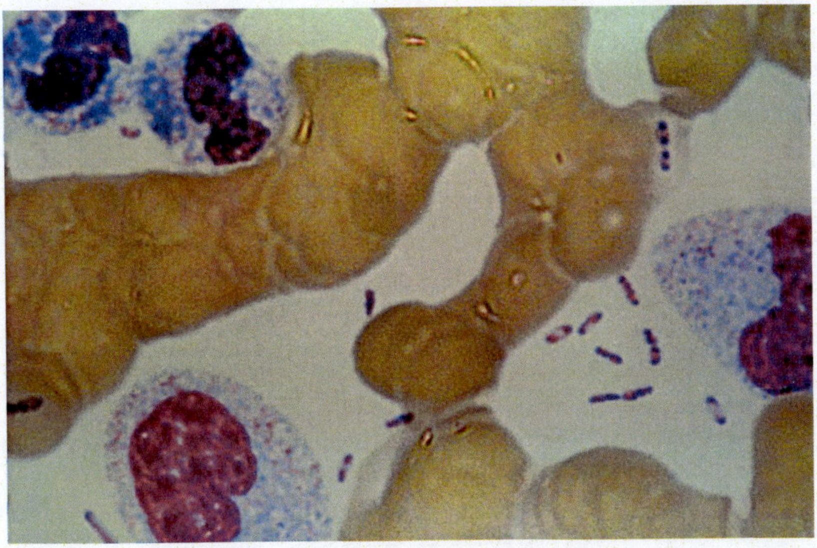

Fig. 37.1 Characteristic 'safety pin' appearance of *Yersinia pestis* in Wayson stain of bubo aspirate (×1000) (http://phil.cdc.gov/PHIL_Images/09042002/00008/PHIL_2050_lores.jpg)

present with tachycardia, tachypnoea and hypotension. Hypothermia is common. Generalised purpura leading to necrosis and gangrene of the distal extremities may be observed. Buboes are not demonstrated in the septicaemic plague. Patients usually die due to a high level of bacteraemia.

Laboratory Diagnosis

Specimens

These include bubo aspirates (in bubonic plague), sputum (in pneumonic plague) and blood and cerebrospinal fluid (CSF) (in septicaemic plague).

Microscopy

Bubo aspirates are collected by injecting 1 mL of sterile saline into the bubo with a 20G needle. The fluid is aspirated after that by withdrawing several times. The bubo aspirate smears are stained with Gram, Wright, Wayson or Giemsa stain to demonstrate the typical bipolar (safety pin) morphology of *Y. pestis*. Blood and CSF of septicaemic patients can also be stained to reveal the bacteria.

Gram stain of the aspirate of the fluid from the inguinal lymph nodes shows Gram-negative coccobacilli with a characteristic bipolar appearance that resembles a closed safety pin and also shows polymorphonuclear leucocytes. **Wayson stain** of the aspirate smear shows *Y. pestis* as light-blue bacilli with dark-blue polar bodies (Fig. 37.1). A **direct fluorescent antibody test** employing monoclonal antibodies is a rapid diagnostic method to demonstrate *Y. pestis* F1 antigen in blood or sputum samples. A positive fluorescence antibody test on smears or cultures is presumptive evidence of infection.

Culture and Identifications

Y. pestis culture to be performed only in BSL-III laboratories. Isolation of *Y. pestis* by culture confirms the diagnosis of plague. The organism can be isolated from blood, CSF, sputum and bubo aspirates, depending on the clinical presentation, whether it is bubonic, pneumonic or septicaemic. *Y. pestis* is slow-growing, but it does not require special growth media. Selective medium, such as blood agar with sodium azide, is used to isolate *Y. pestis* from sputum and bubo aspirates containing numerous other bacteria. Dark-brown colonies on blood agar and pinpoint colonies on MacConkey agar are the characteristic features of colonies of *Y. pestis*. The identifying features of *Y. pestis* colonies are summarised in the Table 37.4.

Animal Inoculation

Y. pestis can be isolated from bubo aspirate or sputum by inoculating guinea pigs or white rats.

Table 37.4 Differentiating features of *Yersinia* species

Properties	*Yersinia pestis*	*Yersinia pseudotuberculosis*	*Yersinia enterocolitica*
Growth on MacConkey agar	+	+	+
Motility at 22 °C	–	+	+
Oxidase test	–	–	+
Urease test	–	+	+
Indole test	+	–	+
Acid from sucrose	+	–	+
Acid from maltose	–	+	+
Acid from rhamnose	–	+	–
Acid from melibiose	–	+	–
Ornithine decarboxylase test	–	–	+

Y. pestis causes the death of the animal within 2–5 days. Postmortem of the dead animal shows necrosis and oedema at the inoculation site. Draining lymph nodes are enlarged, congested and show greyish white patches. Gram- and methylene blue-stained smears of lymph nodes, spleen and heart blood reveal typical bipolar-stained Gram-negative bacilli.

Sputum collected from suspected cases of pneumonic plague is inoculated by applying them over the shaven skin or to the nasal mucosa of a guinea pig. The bacteria present in the specimen penetrate the skin through minute abrasions. Bacilli are demonstrated in the sputum and in the blood from the guinea pig's heart, which dies 2–3 days after inoculation.

Serology

Serological tests usually supplement the diagnosis of plague. Enzyme-linked immunosorbent assay (ELISA) is frequently employed to detect specific antibodies in acute and convalescent sera. A fourfold or greater difference in antibody titre between acute and convalescent sera collected 10 days apart, demonstrated by ELISA, suggests plague infection. A single positive ELISA in a patient who has not received plague vaccine nor has had the previous plague is also presumptive of infection.

Molecular Detection

Polymerase chain reaction (PCR) has been widely tested for accurate and timely detection of pathogenic *Yersinia* sp. Conventional PCR techniques targeting virulence genes in *Yersinia* have been widely used. However, due to the post analytics challenges of classical PCR, RT PCR has emerged as a promising alternative. Single species targeted RT PCR and multiplex RT PCR targeting numerous species in a sample have been developed. Commercial kits like TaqMan™ *Yersinia pestis* Detection Kit are available to offer a fully integrated, single-vendor solution to identify *Y. pestis*. In addition, random amplified polymorphic DNA (RAPD-PCR) and inter-repeat PCR (IR-PCR) have also been used for PCR-based DNA fingerprinting and have enabled the identification of multiple clinical isolates of *Yersinia* sp.

Treatment

Treatment of a suspected case of plague with antibiotics should be started as early as possible, without waiting for laboratory confirmation. Immediate treatment with antibiotics has dramatically reduced mortality due to the disease. Streptomycin, doxycycline and chloramphenicol are used effectively for the treatment of plague. Treatment with antibiotics is usually given for 10 days. A two-drug regimen should be used in severe cases. Streptomycin is the drug of choice to treat plague. Doxycycline is given to patients who are allergic to streptomycin or who cannot tolerate streptomycin. Chloramphenicol is the drug of choice in septicaemic plague or patients with hypotension. In patients with hypotension, intramuscularly administered streptomycin may be poorly absorbed. Resistance of *Y. pestis* to streptomycin, gentamicin, doxycycline and chloramphenicol is very rare.

Epidemiology

Plague is globally distributed, with most human cases reported from developing countries.

Geographical Distribution

Most plague cases occur in the developing countries of Africa and Asia, including Vietnam and India. Based on the fermentation of glycerol and nitrate reduction, *Y. pestis* has been typed into three distinct types: *Y. pestis* var. *orientalis*, *Y. pestis* var. *antiqua* and *Y. pestis* var. *medievalis*. These biotypes show different geographical distributions. The primary foci of *Y. pestis* var. *orientalis* are India, Myanmar and China. This biotype is also responsible for wild plague in the western USA, South America and South Africa. *Y. pestis* var. *antiqua* is distributed in Mongolia, Manchuria and Transbaikalia. *Y. pestis* var. *medievalis* is restricted to Southeast Russia.

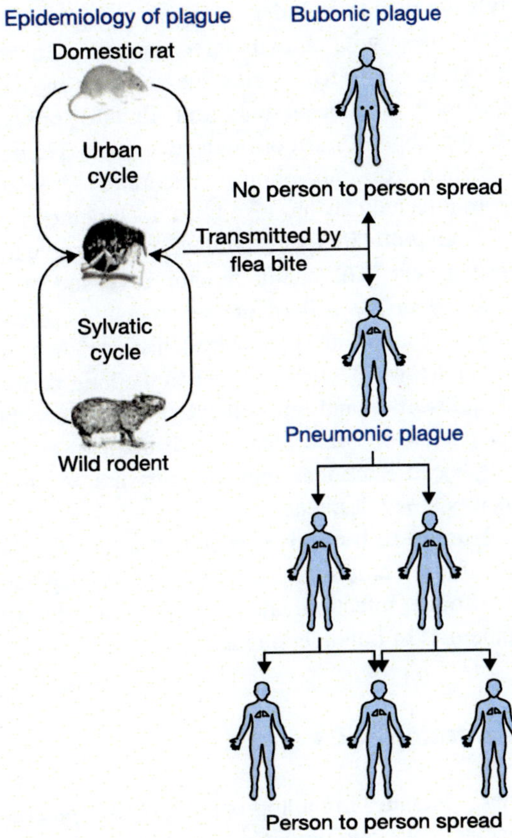

Fig. 37.2 Epidemiology of plague

Habitat

Y. pestis inhabits infected lymph nodes—most commonly inguinal lymph nodes in bubonic plague, the lower respiratory tract in pneumonic plague and the blood circulation in the septicaemic plague.

Reservoir, Source and Transmission of Infection

Bubonic plague is a zoonotic infection transmitted to humans by the bite of an infected rat flea. Infection may also be transferred by contamination of the bite wound with the faeces of infected rat fleas. There are two natural cycles of plague: the urban plague and the wild plague (Fig. 37.2).

Urban or domestic plague is maintained in rat populations and is transmitted among rats or between rats and humans by infected rat fleas. In India during earlier years, plague produced epizootics first in sewer rat, *Rattus norvegicus*. When their number decreased, the disease spread to the domestic rat, *Rattus Rattus*. Domestic rat is the important reservoir of the urban plague from which the infection spreads to human beings. The rat flea is the usual vector. Thirty different flea species have been identified as vectors, the most important being *X. cheopis*, *Xenopsylla astia* and *Ceratophyllus fasciatus*. *X. cheopis* is the predominant species in north India, while *X. astia* is in south India. Cool and humid seasons favour the multiplication of fleas, leading to a high flea index (mean number of fleas per rat). Hence, plague epidemics occur during the cold season more frequently than in the hot, dry weather, in which fleas do not thrive, and the transmission of infection is interrupted.

Wild or sylvatic plague occurs between animals and wild rodents independent of human beings. More than 200 different animal species have been identified as hosts. The gerbil (*Tatera indica*) and the bandicoot are reservoirs for the sylvatic plague in India. A sylvatic stage occurs when these wild animals infect humans. Human infection may occur during the handling of dead carcasses of infected wild animals and through the ingestion of meat of infected animals. Human infection from inhalation of respiratory droplets from infected cats has also been documented.

In **enzootic foci**, infected fleas may survive for over a year. The bacilli remain alive and even multiply in the soil of unused rodent burrows. Plague bacilli infect new rodents that live in such burrows. This contributes to the long period of quiescence and subsequent re-emergence of the plague. Sylvatic plague is difficult to eliminate because the mammalian reservoirs and flea vectors are widespread.

Human-to-human spread occurs only in pneumonic plague. Pneumonic plague is spread by droplet infection and may be seen during epidemics of bubonic plague.

Prevention and Control

These include immunoprophylaxis by plague vaccine, chemoprophylaxis by antibiotic therapy and environmental sanitation.

Immunoprophylaxis by Plague Vaccine

A killed vaccine for plague is available in India. It is a whole culture antigen prepared at the Haffkine Institute, Mumbai. The vaccine contains 2000 millions of plague bacilli per millilitre. A series of three injections of 0.5 mL of vaccine is given subcutaneously. Two doses are given at an interval of 1–3 months, followed by a third dose in the sixth month. The vaccine confers protection only for 6 months against bubonic plague. The vaccine is recommended for laboratory and field workers who come in contact with the plague bacillus or individuals residing in endemic areas who cannot avoid exposure to rodents or fleas. The vaccine is not effective against pneumonic plague.

Chemoprophylaxis by Antibiotic Therapy

Prophylactic antibiotic therapy includes prophylaxis with tetracycline or streptomycin. Trimethoprim-sulfamethoxazole has also been effective for prophylaxis. Prophylactic antibiotic therapy is recommended for contacts of pneumonic plague victims, people exposed to the bites of rodent fleas during a plague outbreak and persons who have handled an animal known to be infected with the plague bacilli. However, antibiotic prophylaxis is not recommended for contacts with bubonic or septicaemic plague.

Environmental Sanitation

A key component in preventing plague outbreaks is identifying the source of infection. Rodent control forms the mainstay of prevention of plague in an urban area. Trained professionals should apply chemicals to kill fleas and rodents. Educating the public in rural plague-endemic areas to avoid handling sick or dead rats and to avoid places where wild animals live is important.

WHO Guidelines for the Management of Plague Outbreak

WHO has programs that enable surveillance and support to at-risk countries to prepare and manage the incidences of plague? Specific guidelines have been issued for the Indian subcontinent, South America and Sub-Saharan Africa. The guidelines are region-specific and account for the risk and conditions of human transmission depending on the type of animal reservoir in a particular region. In general, these guidelines provide up-to-date guidance on the diagnosis, case management, prevention and control of the different forms of plague, both for sporadic cases and during outbreaks.

Yersinia enterocolitica

Y. enterocolitica is a well-recognised enteric bacterium that causes distinctive clinical manifestations, ranging from asymptomatic infections to life-threatening sepsis, especially in children. Schleifstein and Coleman were the first to describe *Y. enterocolitica* in 1939. *Y. enterocolitica* is a Gram-negative coccobacillus that is motile at 22 °C but not at 37 °C. This bacillus resembles *Yersinia pseudotuberculosis* in being motile at 22 °C but differs from it in fermenting sucrose and cellobiose and decarboxylating ornithine. *Y. enterocolitica* are aerobes and grow well on commonly used basic enteric media.

The antigenic structure of *Y. enterocolitica* is distinct. *Y. enterocolitica* possesses 34 different O antigen factors and 19 H factors. Based on these antigens, more than 60 O serotypes have been reported. Most human isolates belong to serotypes O:3, O:5, 27; O:8 and represent the most virulent worldwide causes of human yersiniosis. Virulence of *Y. enterocolitica* is chromosomal or plasmid-mediated.

Humans acquire infection by ingestion of food or water contaminated with the bacteria. *Y. enterocolitica* causes enterocolitis primarily in young children, pseudo-appendicitis syndrome and extraintestinal infections.

Enterocolitis is the most common clinical form of the disease, occurring primarily in young children, with a mean age of 24 months. The incubation period is short that varies from 4 to 6 days. The condition manifests as watery, mucoid diarrhoea in most patients; fever, colicky abdominal pain, bloody stools and leucocytes in the stool. Diarrhoea usually lasts from 1 day to 3 weeks. Most cases are self-limited. The bacteria also cause mesenteric lymphadenitis with terminal ileitis. The condition, known as **pseudo-appendicitis syndrome**, is more common in older children and young adults and mimics appendicitis. It is characterised by fever, abdominal pain, tenderness in the right lower quadrant and leucocytosis.

Laboratory diagnosis is made by isolating *Y. enterocolitica* from the stool by culture. On MacConkey agar, they produce pinpoint colourless colonies after 24 h of incubation. On blood agar, they form non-haemolytic smooth and translucent colonies measuring 2–3 mm in diameter after 2 days of incubation at 22 °C. Cefsulodin–irgasan–novobiocin (CIN) agar is a frequently used selective medium for *Y. enterocolitica*. Stool culture is positive within 2 weeks of the onset of the disease. ELISA and radioimmunoassay are frequently used tests to detect sera antibodies to diagnose the condition. Antibody titres rise 1 week after onset of illness and reach maximal levels at 2 weeks; raised antibody levels can be found for years after infection. Antibody titres more than 1:128 are suggestive of a previous infection with *Y. enterocolitica*.

Good nutrition and hydration are the mainstays of treatment. The role of antibiotic therapy in uncomplicated acute colitis and mesenteric adenitis is not established. Antibiotic therapy is indicated only for patients with septicaemia, focal extraintestinal manifestations and enterocolitis who are immunocompromised. Tetracycline, aminoglycosides and trimethoprim-sulfamethoxazole are the antibiotics of choice.

Yersinia pseudotuberculosis

Y. pseudotuberculosis is the least common of the three main *Yersinia* species to cause human infections. It is a small, ovoid and bipolar-stained Gram-negative bacillus. It is non-sporing, non-capsulated and slightly acid-fast. It is motile at 22 °C but not at 37 °C. It is non-lactose-fermenting coccobacillus that is differentiated from other species, such as *Y. enterocolitica*, by its fermentation of sorbitol, ornithine decarboxylase activity and other tests. The optimum growth occurs on MacConkey or eosin-methylene blue (EMB) agar medium at 20–35 °C. The growth of the bacteria is enhanced noticeably at lower temperatures (e.g. 4 °C cold enrichment in buffered saline). The bacteria grow slowly on blood and chocolate agar plates, forming small, grey and translucent colonies at 24–72 h of incubation.

Y. pseudotuberculosis has been classified into six serotypes (I–VI). Approximately 80% of human diseases are caused by type I serotypes. The bacteria show antigenic cross-reactivity with *Y. pestis* as well as salmonellae. *Y. pseudotuberculosis* in experimentally infected guinea pigs produces multiple nodules in the lungs, spleen and liver, resembling tuberculous lesions, hence named pseudotuberculosis. Humans acquire infections primarily through the ingestion of contaminated food products. An inoculum of 10^9 organisms is the infective dose.

Y. pseudotuberculosis is the causative agent of gastroenteritis, Izumi fever and Far East scarlatinoid fever. The bacteria primarily cause **gastroenteritis**. The incubation period varies from 5 to 10 days. Gastroenteritis is characterised by self-limited mesenteric lymphadenitis simulating acute appendicitis. Fever, abdominal pain (often right lower quadrant location) and rash are a major triad of *Y. pseudotuberculosis* infection. Diarrhoea is not common. Erythema nodosum and reactive arthritis are the noted late complications of the infection. **Izumi fever**, a syndrome caused by *Y. pseudotuberculosis*, is characterised by scarlatiniform rash, systemic symptoms and coronary artery aneurysms. **Far East scarlatinoid fever** is another syndrome caused by bacteria. The presence of a scarlatinoid-appearing rash on the head and neck, erythema on the upper and lower extremity, mucous membrane enanthem and strawberry tongue are characteristic features of the syndrome.

Y. pseudotuberculosis is primarily a zoonotic infection. The animal reservoirs include many mammalian and avian hosts, such as horses, cattle, dogs, deer, cats, rabbits, rodents and birds (e.g. geese, turkey, ducks, canaries, cockatoos). Butchers working in abattoirs slaughtering swine are at an increased risk of *Y. pseudotuberculosis* infection. The infection is distributed worldwide. Most cases occur in the winter months because a cold temperature favours the growth and multiplication of the bacteria. Most cases have been reported in Europe.

Laboratory diagnosis of *Y. pseudotuberculosis* infection is made by isolating the bacteria by culture from stool, blood, CSF, peritoneal fluid, synovial fluid or other organ-based biopsies (e.g. intestinal tissue, skin). Isolation of bacteria from stool is difficult, given the slow growth pattern and overgrowth of normal faecal flora. However, stool culture may be facilitated by cold enrichment, using special culture media, such as CIN agar or by alkali treatment. ELISA and agglutination tests are employed to demonstrate serum antibodies, which may appear soon after the onset of illness but decrease for 2–6 months. A rise in agglutinating antibodies in paired serum samples, collected after 2 weeks, is suggestive of the diagnosis. However, the test shows false-positive reactions in sera from patients infected with *Yersinia*, *Vibrio*, *Salmonella*, *Brucella* and *Rickettsia* species due to the presence of cross-reactive antibodies. Histological examination of specific tissue, such as mesenteric lymph nodes, may show both the pathologic and microbiologic presence of the organism.

Antibiotic therapy is indicated only for younger patients with critical illness and immunocompromised patients. Tetracycline, aminopenicillins and aminoglycosides are frequently used to treat the condition.

Genus *Pasteurella*

The genus *Pasteurella* consists of more than 17 species. *Pasteurella multocida* is the most common species causing infection in humans. Other species, which rarely cause human infections, include *Pasteurella canis*, *Pasteurella dagmatis*, *Pasteurella pneumotropica*, *Pasteurella aerogenes*, *Pasteurella haemolytica*, *Pasteurella caballi* and *Pasteurella bettyae* (Table 37.5).

Pasteurella multocida

P. multocida is a non-motile Gram-negative bacillus. The organism generally resembles *Yersinia* but differs from it in being oxidase positive and producing indole. The bacterium grows well on blood and chocolate agar but grows poorly on MacConkey agar. On blood agar, after overnight incubation, *P. multocida* produces large buttery colonies with a characteristic musty odour of the colonies due to the production of indole.

P. multocida is present in the upper respiratory tract of dogs, cats, rats, cattle and sheep. The

Table 37.5 Human infections caused by *Pasteurella* species

Pasteurella species	Diseases
Pasteurella multocida	Cellulitis and adenitis at the site of cat- or dog-bite wound Exacerbation of respiratory tract infections and systemic infection in immunocompromised patients
Pasteurella canis	Dog-bite wound infection
Pasteurella dagmatis	Dog-bite wound infection
Pasteurella bettyae	Rare opportunistic infection

bacteria sometimes are also present in the human upper respiratory tract. Human infection is rare but may occur following animal bites or trauma. After a bite or scratch by cats or dogs, localised cellulitis and adenitis occur at the site of the bite. Patients with underlying pulmonary lesions may also show exacerbation of respiratory tract infections (such as pneumonia, bronchitis and sinusitis). *P. multocida* in immunocompromised patients may cause a systemic infection. Penicillin is the antibiotic of choice. Tetracycline, cephalosporins or fluoroquinolones are other antibiotics equally effective against the bacteria.

Francisella tularensis

F. tularensis causative agent of tularaemia is one of the most infectious bacteria known. It is classified as a category A agent because of its high infectivity, ease of dissemination and ability to cause substantial illness and death. Humans become infected after introducing the bacteria by intradermal injection, inhalation or oral ingestion.

It is a minute (0.2 × 0.2–0.7 μm) and poorly stained Gram-negative intracellular coccobacillus. The bacterium is capsulated and non-motile. Like *Mycoplasma* organisms, this organism is filterable and multiplies, that is, by filament formation and budding, besides binary fission. Like *Bordetella*, the bacteria are strictly aerobic and fastidious; hence they cannot be cultured on common laboratory media. The organism requires enriched media, such as **Francis blood dextrose cystine agar** and a longer incubation period of 3–5 days. *F. tularensis* produces minute transparent colonies on these media after incubation of 3–5 days.

F. tularensis is an intracellular parasite. In infected animals, the bacteria are found in large numbers in the reticuloendothelial cells of the spleen, liver, etc. These bacteria remain viable for a longer period in macrophages of the reticuloendothelial system because the organisms inhibit phagosome–lysosome fusion. Like other Gram-negative bacilli, *F. tularensis* has endotoxin but is relatively less biologically active than that found in *Escherichia coli* and other Gram-negative bacilli. Capsulated strains of *F. tularensis* are pathogenic, while non-capsulated strains are non-pathogenic. Intradermal injection or inhalations of 10–50 bacilli are required to cause disease. Nearly 100 million bacteria are required to cause disease on oral ingestion.

F. tularensis causes **tularaemia**, also known as glandular fever, rabbit fever, tick fever and deer fly fever—an acute, febrile, granulomatous, infectious disease in humans and other mammals. The incubation period varies from 3 to 5 days. The onset of the disease is abrupt, with symptoms such as fever, chills, malaise and fatigue. Tularaemia can manifest in one to six well-recognised clinical forms. These include ulceroglandular tularaemia, glandular tularaemia, oculoglandular tularaemia, oropharyngeal tularaemia, pneumonic tularaemia and typhoidal (septicaemic) tularaemia.

Tularaemia is zoonotic inaction. Contact with infected animals or their carcasses is the primary mode of transmission. The infection is transmitted by (1) ingestion of poorly cooked meats of the infected animals, such as rabbits, (2) ingestion of contaminated water, (3) bite of hard ticks (*Amblyomma americanum*, *Dermacentor andersoni* and *Dermacentor variabilis*) or deerfly (*Chrysops discalis*), (4) direct contact with contaminated soil, water or fomites and (5) inhalation of water aerosols or dust from soil, grains or contaminated pelts. However, person-to-person transmission is rare.

F. tularensis is a highly infectious bacterium. Hence, isolation of the bacteria from clinical specimens is not attempted in routine laboratories because culture poses a danger to laboratory workers. The plates must be sealed and handled using a biosafety level (BSL)-2 facility that is essential for handling the culture plates and a BSL-3 facility for isolation and identification of *F. tularensis* isolates.

Gram staining of the aspirates from ulcer exudates or infected lymph nodes is not useful because *F. tularensis* is extremely small and stains very faintly. Detection of antigen directly in clinical specimens by direct fluorescent antibody staining, using fluorescein-labelled antibodies directed against the organism, is a more sensitive and specific method for diagnosis of the condition. The test is yet to be available for wide use.

Blood cultures in a specific medium (such as cysteine–glucose–blood agar) are usually negative unless the cultures are incubated for a week or longer. Aspirates of lymph nodes or draining sinuses are usually positive if the cultures are incubated for 3 days or longer. The identification of *F. tularensis* is confirmed by testing with specific antiserum-containing antibodies against *Francisella*.

The Tularaemia tube agglutination test is the most commonly used serological test. An agglutination titre greater than 1:160 is considered presumptively positive. Diagnosis is confirmed by a fourfold increase in titre of the serum samples collected 2 weeks apart. The test shows cross-reactivity with *Brucella* species, *Proteus* Ox-19, *Salmonella* species and *Yersinia* species. Capture ELISA using specific monoclonal antibodies raised against LPS of the virulent forms of *F. tularensis* is a sensitive method for detecting serum antigens.

PCR has been used to detect *F. tularensis* in clinical specimens even after the start of antibiotic therapy.

Streptomycin is the antibiotic of choice for the treatment of tularaemia. Gentamicin is used as an alternative to streptomycin. Fluoroquinolones are now increasingly used in the treatment of tularaemia. A live attenuated vaccine obtained from avirulent *F. tularensis* biovar *palaearctica* (type B) is available. The vaccine confers only partial protection but reduces disease severity in vaccinated people.

Case Study

A 35-year-old man was admitted to the hospital with a sudden onset of high fever, chills, headache, body aches, extreme exhaustion, abdominal pain and diarrhoea. Physical examination revealed the presence of painful, swollen lymph glands in the groin. Aspirate from the bubo on Gram staining showed pleomorphic Gram-negative coccobacilli. *Y. pestis* was isolated from the specimen by culture. The case was diagnosed to be bubonic plague.

- What is the other rapid method for demonstration of *Y. pestis* in bubonic aspirate?
- What are the serological tests available for diagnosis of the condition?
- What is the biotype of *Y. pestis* found in India?
- What are the foci of *Y. pestis* infection in India?
- What is the vaccine available for the prevention of plague? Describe the vaccines available against the disease.

Pseudomonas, Burkholderia and *Acinetobacter*

38

Pseudomonas and related bacteria belong to the family Pseudomonadaceae, which contains over 200 species. Most of them are pathogenic to plants, insects and reptiles. A few species cause disease in humans. These are *Pseudomonas aeruginosa*, *Pseudomonas fluorescence*, *Pseudomonas putida*, *Pseudomonas stutzeri*, *Burkholderia cepacia*, *Burkholderia mallei*, *Burkholderia pseudomallei*, *Stenotrophomonas maltophilia*, *Acinetobacter baumannii* and *Moraxella catarrhalis* (Table 38.1).

Genus *Pseudomonas*

Pseudomonas is ubiquitous and is found worldwide. They are found in soil, vegetation, water, plants and animals with a predilection to moist environments. They are also commonly found in the hospital environment. They are opportunistic pathogens often causing hospital or nosocomial infections. These organisms are notorious because of their potential to develop innate resistance to many antibiotics.

Pseudomonas aeruginosa

P. aeruginosa is the most important species associated with human infection. Previously it was called by the name *Pseudomonas pyocyanea* or *Bacillus pyocyaneus*. It is a most common human saprophyte, but it rarely causes disease in healthy individuals. It causes most human infections in an immunocompromised human host.

Properties of the Bacteria

Morphology

P. aeruginosa is a straight or slightly curved, Gram-negative bacillus measuring 0.5–1.0 × 1.5–5.0 μm in size arranged singly, in pairs or short chains. It is motile by the presence of a polar flagellum. Occasionally, strains may possess two or three polar flagella. *Pseudomonas* spp. is non-capsulated. Although the bacteria are non-capsulated, many strains appear mucoid by producing an abundant extracellular polysaccharide composed of alginate polymers. This slime layer forms a loose capsule or glycocalyx around the bacillus. These strains are particularly isolated from patients with cystic fibrosis. *Pseudomonas* spp. is fimbriated.

Culture and Biochemical Properties

Pseudomonas spp. are strict aerobes. *P. aeruginosa* grows over a wide range of temperatures (5–32 °C), the optimum temperature being 37 °C. It grows on commonly used routine media, including **nutrient agar, blood agar, MacConkey agar** and **cetrimide agar**, a selective medium for the culture of *P. aeruginosa*.

P. aeruginosa produces different types of pigments. Pyocyanin is specifically produced by *P. aeruginosa,* which diffuses into the

Table 38.1 Human infections caused by *Pseudomonas* and other genera of the family Pseudomonadaceae

Bacteria	Diseases
Pseudomonas aeruginosa	Respiratory tract infections, skin infections, urinary tract infections, ear infections, eye infections, bacteraemia and endocarditis
Pseudomonas fluorescence	Opportunistic infections
Pseudomonas putida	Opportunistic infections
Pseudomonas stutzeri	Opportunistic infections and urinary tract infections
Burkholderia cepacia	Opportunistic respiratory infections in patients with cystic fibrosis or chronic granulomatous disease and urinary tract infections in patients with indwelling catheters
Burkholderia mallei	Acute or chronic infection with infection localised to the skin, subcutaneous tissue or respiratory tract
Burkholderia pseudomallei	Melioidosis
Stenotrophomonas maltophilia	Nosocomial infections (pneumonia, meningitis, wound infection, etc.)
Acinetobacter baumannii	Nosocomial infections of respiratory tract, urinary tract and wound
Moraxella catarrhalis	Bronchitis and bronchopneumonia were seen in patients with chronic pulmonary disease and in elderly patients

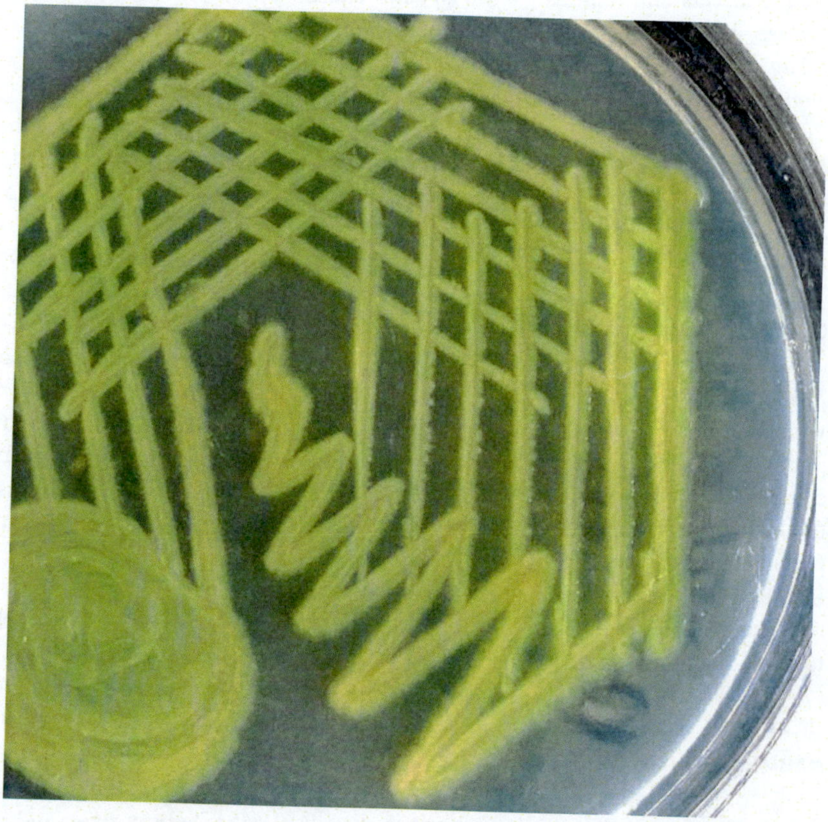

Fig. 38.1 *Pseudomonas aeruginosa* showing pigmented colonies on the nutrient agar (Courtesy of Prof S. Pramodhini)

surrounding medium (Fig. 38.1). The pigment is soluble in chloroform and water. Pyoverdine, or fluorescein, is produced by *P. aeruginosa* and many other *Pseudomonas* species. These pigments give a yellow tinge to bacteria colonies and are best demonstrated in a microscope using an ultraviolet source of light. Fluorescein is soluble in water but insoluble in chloroform. Pyorubin, a red pigment, and Pyomelanin, a brown pigment, are produced by *Pseudomonas* species.

Pseudomonas are oxidase positive. Oxidase test is an important test for the identification of *P. aeruginosa.* All the strains are oxidase positive within 30 s of performing the test. They are catalase positive. They are non-fermentative bacteria and utilise sugars by an oxidase metabolism, with oxygen as the terminal electron acceptor. They do not utilise lactose and maltose. They are indole, MR, VP and HS negative, reduce nitrates to nitrite and further broken down to gaseous nitrogen, and arginine dihydrolase positive. Special media, such as oxidation–fermentation (OF) media, are used to demonstrate the low quantity of acid produced during the oxidative breakdown of sugars. *P. aeruginosa* utilises glucose, forming acid only.

Other Properties

Susceptibility to Physical and Chemical Agents

P. aeruginosa is a heat-labile bacterium, readily killed at 55 °C in 1 h. It is also highly susceptible to acid, silver salts, 2% alkaline, glutaraldehyde and disinfectants (such as Dettol and cetrimide). However, *P. aeruginosa* is very strongly resistant to common antiseptics and disinfectants, such as chloroxylenol, hexachlorophene and quaternary ammonium compounds.

Antigenic Structure

P. aeruginosa consists of O and H antigens. **Somatic or O antigens** are the group-specific antigens. *P. aeruginosa* possesses 19 distinct group-specific O antigens based on which the organism has been classified into 19 serogroups. O antigens are heat stable and can be extracted with acid or formamide. The O antigen can be detected by precipitation reaction-in-gel or tube or slide agglutination tests using specific antisera produced against these antigens. *P. aeruginosa serogroups* O6 and O11 are isolated from the majority of the clinical specimens. Serogroup O11 has been associated with most hospital-acquired infections. **Flagella or H antigens** are found in the flagella of *P. aeruginosa*. These antigens are heat-labile. Slide agglutination tests have demonstrated two types of H antigens by using specific antibodies against these antigens.

Pathogenesis

P. aeruginosa is an invasive pathogen, and its mechanism of pathogenesis is not clearly understood, though multiple virulence factors have been identified.

Virulence Factors

P. aeruginosa invasion is mediated by many virulence factors, including toxins, enzymes and cell wall structure (Table 38.2).

Exotoxin A is one of the most important virulence factors produced by *P. aeruginosa*. The toxin is like that of the diphtheria toxin that prevents the synthesis of proteins in eukaryotic cells. It is responsible for causing tissue damage in chronic pulmonary infection, dermatonecrosis in burns wounds and the cornea's destruction in ocular infection. The toxin also causes suppression of immunity in the infected host. **Exotoxins S and T** are believed to facilitate the spread of bacteria and invasion of tissues by causing damage to the epithelial cells. The enzymes **serine protease (LasA)** and **zinc metalloprotease (LasB)** are the two types of the enzyme **elastase** that act in combination to destroy elastin present in elastin-containing tissues. These enzymes cause damage to parenchymal tissues of the lung and produce haemorrhagic lesions associated with spreading *P. aeruginosa* infections. These enzymes also facilitate the spread of infection and damage of tissues in acute infections by degrading several components of the complement and inhibiting chemotaxis activities of neutrophils. *P. aeruginosa* cell wall is surrounded by a **loose slime layer**. The loose slime layer, also known as the alginate coat or glycocalyx, protects the bacteria from phagocytosis and against the activity of many antibiotics, such as aminoglycosides. Genes mediate the production of the mucoid slime layer. These genes are activated in patients suffering from chronic

Table 38.2 Virulence factors of *Pseudomonas aeruginosa*

Virulence factors	Biological functions
Toxins	
Exotoxin A	Acts by prevention of synthesis of proteins in eukaryotic cells Causes tissue damage in chronic pulmonary infection, dermatonecrosis in burns wound and destruction of cornea in ocular infection Causes immunosuppression
Exoenzymes S and T	Exhibit adenosine diphosphate ribosyl transferase activity Inhibit protein synthesis and cause immunosuppression
Enzymes	
Elastase	Destroys elastin present in elastin-containing tissues (skin, lung tissue, blood vessels, etc.), immunoglobulins and complement factors
Alkaline protease	Causes tissue destruction, inactivation of interferon
Phospholipase C	Causes tissue destruction by breaking down lipids and lecithin
Rhamnolipid	Breaks down lecithin-containing tissues
Cell wall components	
Pili	Causes adhesion of the bacteria to the epithelial cells
Capsule	Inhibits antibiotics killing of the bacteria
Lipopolysaccharide	Shows endotoxic activity, sepsis
Pyocyanin	Causes tissue damage, inflammation

respiratory disease in those with cystic fibrosis and when colonised.

Pathogenesis of *Pseudomonas aeruginosa* Infection

P. aeruginosa are true pathogens and opportunistic pathogens. *P. aeruginosa*, as a true pathogen, cause infections of the respiratory tract in patients with underlying disease of the respiratory tract. They also cause infection of the urinary tract in patients with catheterisation. *P. aeruginosa*, as an opportunistic pathogen, causes most human infections in immunocompromised hosts.

Clinical Syndromes

P. aeruginosa causes a wide variety of clinical syndromes.

P. aeruginosa causes **infection of the lower respiratory tract**, almost exclusively in patients with malignancies and immunodeficiencies. Primary non-bacteraemic *Pseudomonas* pneumonia usually occurs in patients who have been colonised earlier with *P. aeruginosa*. This colonisation is seen in patients with cystic fibrosis, other chronic disease and neutropenia. *P. aeruginosa* can cause various **skin infections**, such as infections of burn wounds, chronic paronychia, infected toe web, pseudomonal folliculitis and pseudomonal cellulitis. Infection of burn wounds is the most common. The bacteria *P. aeruginosa* causes **urinary tract infections** in persons with indwelling urinary catheters and persons undergoing instrumentation and surgery of the urinary tract. **Other infections** include ear infections (external otitis), eye infections (pseudomonal endophthalmitis), cardiovascular infections (pseudomonal infectious endocarditis), etc.

Nosocomial infections caused by *P. aeruginosa* are opportunistic infections seen in immunocompromised hosts. Such immunocompromised hosts include a. persons with dysfunctional immune mechanisms, such as AIDS, cystic fibrosis, neonates, complement deficiency, hypogammaglobulinemia and neutropenia and b. patients with injury to the skin and mucocutaneous membrane that occurs during burn injuries and use of urinary catheters, dialysis catheters, endotracheal tubes and intravenous lines. *P. aeruginosa* causes bacteraemia—a condition clinically similar to Gram-negative bacteraemia in immunocompromised patients. It also causes pseudomonal meningitis, which occurs in immunocompromised patients who have undergone neurosurgery.

Laboratory Diagnosis

Specimens

These include pus, urine sputum, cerebrospinal fluid (CSF), blood and other specimens depending on the nature of the infections.

Microscopy

Gram staining of CSF or pus may show Gram-negative bacilli but is not useful for specific diagnosis of Pseudomonas.

Culture and Identification

Laboratory diagnosis of pseudomonal infection is based on isolation of *P. aeruginosa* from faeces or other clinical specimens containing mixed microbial flora by culture on selective medium, such as cetrimide agar. Since *P. aeruginosa* is frequently present as a contaminant in the clinical specimen, hence not a single isolation but repeated isolations are essential to confirm *P. aeruginosa* as the causative agent of the condition.

P. aeruginosa, after incubation for 24 h at 37 °C on nutrient agar, produces large (2–3 mm in diameter), opaque, translucent and irregularly round colonies. These colonies emit a characteristic musty to fruity odour due to the production of aminoacetophenone from the amino acid tryptophan. The bacteria on MacConkey agar produces colourless non-lactose-fermenting colonies. It produces haemolytic colonies on blood agar. It produces on cetrimide agar.

Non-lactose-fermenting and beta-haemolytic green pigmented colonies (Fig. 38.1), which are rapid oxidase positive, are identified by biochemical reactions, production of pyocins and susceptibility to phage typing, serotyping and molecular typing.

Typing

Recent hospital-acquired infection trends suggest that *P. aeruginosa* is one of the important pathogens; hence, it is mandatory to classify these strains for epidemiological purposes. There are various typing methods available for classifying *Pseudomonas*, such as serotyping, bacteriophage typing, bacteriocin typing and molecular typing. Out of all these typing methods, molecular methods such as pulse-field gel electrophoresis and multi-locus sequence typing are more useful since they can distinguish different clones among a particular species of *Pseudomonas*.

Treatment

Treatment of *Pseudomonas* infection is by specific antimicrobial therapy. *Pseudomonas* is susceptible to cefotaxime, ceftazidime, gentamicin, tobramycin, carbenicillin, azlocillin and ticarcillin. Ciprofloxacin is the most frequently used antibiotic because it is active against *P. aeruginosa* in most tissues. *Pseudomonas* infections are treated best with a combination of at least two antipseudomonal antibiotics. A combination of aminoglycosides or quinolone with another antipseudomonal antibiotic is effective for most *Pseudomonas* infections.

P. aeruginosa shows a considerable degree of **resistance** to many commonly used antibiotics. Most strains are intrinsically resistant to the most commonly used antimicrobial agents. They can become more resistant during therapy with antibiotics. The mutation of the genes encoding for outer membrane porin proteins is most commonly responsible for developing resistance. They also produce many beta-lactamases, which cause the inactivation of many beta-lactam antibiotics such as penicillin, cephalosporins, carbapenems, etc.

Epidemiology

P. aeruginosa is opportunistic pathogens, causing infection, distributed worldwide.

Geographical Distribution

P. aeruginosa infection is ubiquitous in most environmental places in the hospital. The bacteria are especially found in the hospital environment, such as toilets, bathtubs, washbasins, respiratory and dialysis equipment, kitchens and even antiseptic or disinfectant solutions.

Habitat

In hospitalised patients, *P. aeruginosa* transiently colonises the respiratory and gastrointestinal tract of the patients. The colonisation is common in patients, particularly those hospitalised for a very long period, treated with broad-spectrum antibiotics over a long duration and those exposed to respiratory therapy by instrumentation.

Reservoir, Source and Transmission of Infection

The hospital environment is the most common source and reservoir of infection. Pseudomonal infections are transmitted through aerosols and by contact with infectious materials. *P. aeruginosa* is responsible for nearly 5–15% of the hospital or nosocomial infections.

Prevention and Control

Prevention of *Pseudomonas* infections, particularly in hospitals, is always difficult. It requires constant monitoring and strict attention to asepsis. The measures suggested for hospital infection control practices are primarily based on the prevention of cross-infection of patients by attending medical and paramedical personnel. It also depends on preventing contamination of equipment, such as dialysis machines, respiratory instruments and restricting the use of broad-spectrum antibiotics. The latter suppresses normal microbial flora and facilitates the overgrowth of resistant *Pseudomonas* organisms. ***Pseudomonas* vaccines** have been evaluated in specialised situations, such as cases of cystic fibrosis that are highly susceptible to *Pseudomonas* infections.

Other *Pseudomonas* Species

S. maltophilia, *P. stutzeri*, *P. putida* and *P. fluorescence* are also associated with opportunistic infections in humans.

S. maltophilia was previously known as *Pseudomonas maltophilia* or *Xanthomonas maltophilia*. They are slightly curved (or) straight bacilli, Gram-negative and motile. It is an important cause of nosocomial infection, and it occurs primarily in debilitated or immunosuppressed individuals (as an opportunistic pathogen). And also, it is associated with extending spectrum of clinical syndromes such as bacteraemia, endocarditis, respiratory tract infections, urinary tract infections, skin and soft tissue and bone and joint infections and ophthalmologic infections. A majority of strains of *S. maltophilia* are resistant to many currently available broad-spectrum antibiotics, including carbapenem class.

P. stutzeri is associated with occasional urinary tract infections. It produces light-brown pigmented colonies in culture, and the pigment does not diffuse into the medium.

Genus *Burkholderia*

The genus *Burkholderia* includes four species formerly classified as *Pseudomonas* species. These include *B. cepacia*, *B. mallei*, *B. pseudomallei* and *Burkholderia gladioli*, and *B. mallei*, *B. cepacia* and *B. pseudomallei* are human pathogens, whereas *B. gladioli* and *B. mallei* are not.

Burkholderia cepacia

B. cepacia, previously known as *Pseudomonas cepacia*, is a Gram-negative, motile bacillus and stains irregularly with Gram staining (Fig. 38.2). It is an aerobe, grows well on nutrient agar at 25–30 °C. The bacteria produce non-diffusible reddish-purple pigmented colonies on prolonged incubation. The bacteria do not grow on DCA agar. It is oxidase positive. It utilises glucose, lactose, maltose and mannitol with acid production only. *B. cepacia* is motile with a polar tuft of flagella.

B. cepacia is an opportunistic pathogen found in moist environment surfaces and is associated with nosocomial or hospital-acquired infections. It causes respiratory infections in patients with cystic fibrosis or chronic granulomatous disease and urinary tract infections in patients with indwelling catheters. It has also been reported to

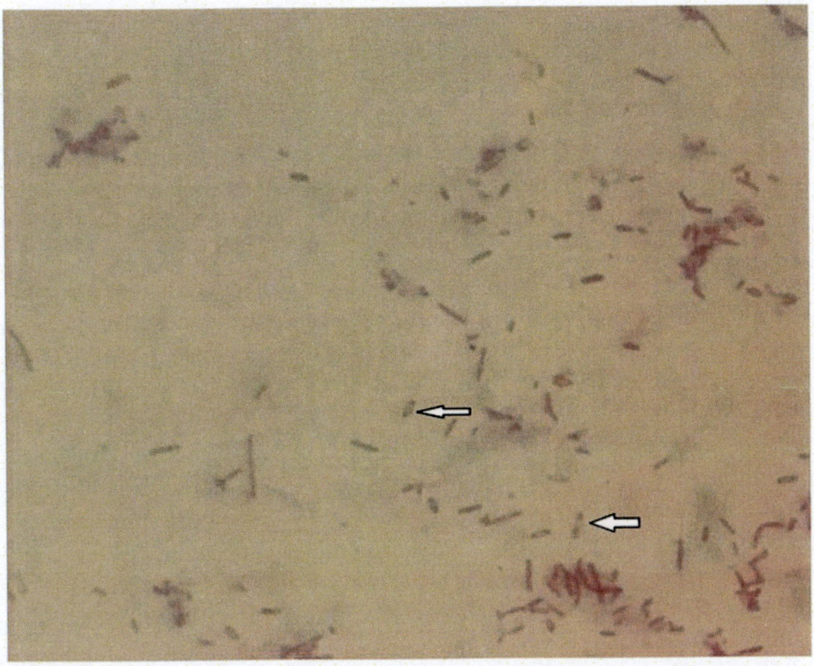

Fig. 38.2 Gram stain smear of *Burkholderia* showing safety pin appearance (x1000) (http://cid.oxfordjournals.org/content/40/7/1053.full.pdf)

cause pneumonitis, osteomyelitis, wound infections and septicaemias in patients with contaminated intravenous catheters and endocarditis in drug addicts. *Cepacia* is susceptible to trimethoprim-sulfamethoxazole, chloramphenicol and ceftazidime. But it is inherently resistant to aminoglycosides, polymyxin, penicillins and cephalosporins.

Burkholderia mallei

B. mallei, earlier known as *Pseudomonas mallei,* is a slender, Gram-negative, small bacillus measuring 2.5 μm in length. It stains irregularly, often giving a beaded appearance. It is an aerobe and facultative anaerobe, which can grow on ordinary culture media. It produces small and translucent colonies, which become yellowish on prolonged incubation. It is relatively biochemically inert, fermenting only glucose.

In animals, such as horses, mules and asses, *B. mallei* produce two types of the clinical syndrome: glander and farcy. *B. mallei* can be transmitted occasionally to humans from infected animals suffering from glanders. The infection is transmitted by inoculating injured skin or the mucous membrane with contaminated discharges. The condition is mostly occupational, found among veterinarians and persons handling horses and other equine animals. **In humans**, *B. mallei* may cause acute or chronic infection, localised to the skin, subcutaneous tissue or respiratory tract. Human cases of glanders are usually rare. Acute infection is characterised as an acute fulminant febrile disease with mucopurulent nasal discharge and severe prostration. The mortality rate is very high.

Glanders in animals may be diagnosed by a skin test known as **the Mallein test**. The skin test is similar to the tuberculin test, demonstrating a delayed hypersensitivity reaction to the protein of *B. mallei.* The condition can also be diagnosed by intraperitoneal inoculation of clinical specimens into male adult guinea pigs. The positive test is demonstrated by the appearance of swelling of the testes, inflammation of tunica vaginalis and ulceration of scrotal skin within 2–3 days of inoculation. This reaction is known as the Strauss reaction.

B. mallei infection is being documented frequently in parts of Asia, Africa, the Middle East,

Central America and South America. *B. mallei* is usually sensitive to ciprofloxacin, streptomycin, novobiocin, gentamicin, imipenem and sulphonamides. Treatment with these antibiotics is usually given for 1–2 months, along with surgical drainage of pus from the affected site. *B. mallei* is usually resistant to chloramphenicol.

Burkholderia pseudomallei

B. pseudomallei is also known as Whitmore bacillus, *Malleomyces pseudomallei* and *Loefflerella pseudomallei.* It is the causative agent of **melioidosis** (from Greek word, meaning resemblance to distemper of asses). *B. pseudomallei* resembles *B. mallei* but differs in being motile; also, it liquefies gelatin and forms acid from several sugars.

B. pseudomallei causes melioidosis in humans. **Melioidosis** manifests as acute, local and chronic infections. The incubation period is variable. It may range from as short as two days to as high as too many years. **Acute infection** is characterised by the development of a nodule at the inoculation site of the bacteria in the skin. The bacteria can subsequently spread, causing secondary lymphangitis, regional lymphangitis, fever and myalgia. Acute melioidosis may progress rapidly to acute septicaemia with a high mortality rate. Acute bloodstream infection is most commonly seen in patients with HIV, diabetes, renal failure, etc. The condition results in septic shock. **The pulmonary infection** manifests as mild bronchitis to severe pneumonia. The condition is associated with high fever, headache, chest pain, anorexia and general myalgia. Non-productive or productive cough with normal sputum is a typical manifestation of this condition. **Chronic suppurative infection** is associated with multiple caseous or suppurative foci of infection in several organs, including joints, skin, lymph nodes, spleen, lungs, liver and brain. Bacteria remain intracellular pathogens of the reticuloendothelial system, which contributes to long latency and reactivation of the infection.

Melioidosis caused by the bacteria is endemic in Southeast Asia. In India, most cases of melioidosis are reported from Kerala, Tamil Nadu, Maharashtra, Orissa, West Bengal and Tripura. The condition is not only a primary disease of rats but also occurs in guinea pigs and rabbits. Humans acquire infection either by inhalation of infected aerosols, inoculation of contaminated material or ingestion of food and water contaminated with excreta of infected animals. The infection also may be transmitted by the bite of haematophagous insects. Agricultural workers are highly susceptible to infection.

Diagnosis of the condition is made by demonstrating Gram-negative bacilli and demonstrating the typical bipolar safety pin appearance of the bacteria in the methylene blue staining. Isolation of *B. pseudomallei* by the culture of urine, sputum, pus or blood confirms the diagnosis of *B. pseudomallei* infection. Serology is frequently helpful for the diagnosis of melioidosis, especially chronic melioidosis. ELISA is used to demonstrate specific IgM and IgG antibodies in patients with melioidosis. PCR has also been evaluated to detect the *B. pseudomallei* genome in pus, sputum and other specimens.

B. pseudomallei is susceptible to ceftazidime, which is the drug of choice. Other antibiotics include tetracycline, cotrimoxazole and chloramphenicol. Prolonged therapy with antibiotics is essential to treat the patient.

Case Study

A 55-year-old man was admitted to a hospital with 65% burns following an accident in the kitchen. The infected pus from the skin showed *P. aeruginosa* by culture. The patient was treated for 7 days with broad-spectrum antibiotics. On the eighth day of hospitalisation, the patient experienced fever and chill. The blood cultures were repeatedly positive for *P. aeruginosa*. The patient succumbed to the infection and died after 15 days.

- What are the factors that increase the patient's susceptibility to *P. aeruginosa* infection?
- What are the virulence factors that make *P. aeruginosa* a serious pathogen?
- What are the mechanisms of antibiotic resistance demonstrated in *P. aeruginosa*?
- How will you prevent nosocomial infections caused by *P. aeruginosa*?

Vibrio, Aeromonas and Plesiomonas 39

Members of the family Vibrionaceae are aerobes and facultative anaerobes, non-sporing, catalase and oxidase positive. They reduce nitrates to nitrites and can grow on ordinary media. The family Vibrionaceae includes the genera *Vibrio*, *Plesiomonas* and *Aeromonas*.

Genus *Vibrio*

The genus *Vibrio* consists of at least 136 species of curved bacilli, of which 12 species have been implicated in human infections. *Vibrio cholerae*, *Vibrio parahaemolyticus* and *Vibrio vulnificus* are the most prominent species causing human infections (Table 39.1). The name 'vibrio' is derived from the characteristic vibratory motility (from *vibrare*, meaning to vibrate).

Vibrio cholerae

V. cholerae is the most important species that causes cholera, the most feared epidemic diarrhoeal disease. Dehydration and death can occur rapidly within a matter of hours of infection.

Properties of the Bacteria

Morphology

V. cholerae are Gram-negative bacilli with rounded or slightly pointed ends. They measure 1–3 μm in length and 0.5–0.8 μm in diameter. They are typically comma-shaped. S-shaped or spiral forms may be seen due to two or more cells lying end to end. They are frequently pleomorphic in old cultures. In a stained film of mucous flakes from a case of cholera, the vibrios are typically arranged in parallel rows, described by Koch as the 'fish in stream' appearance. They are actively motile in the presence of a single polar flagellum. They show typical darting type motility and appear as a 'swarm of gnats' when examined under the microscope. The bacteria are catalase positive and oxidase positive.

Culture and Biochemical Properties

V. cholerae is strongly aerobic and grows best under aerobic conditions at an optimum 37 °C. It grows better in an alkaline medium, the range of pH being 7.4–9.6 (optimum 8.2). Sodium chloride (0.5–1%) is required for optimal growth, though high concentrations (5% and above) are inhibitory for the growth of the bacteria. Unlike other halophilic bacteria, *V. cholerae* can grow in the absence of salt.

V. cholerae grows well on a wide variety of media, including non-selective media (e.g. nutrient agar, MacConkey agar, blood agar, gelatin agar and peptone water) and special media (selective media, enrichment media and transport media).

Table 39.1 Human infections caused by *Vibrio* species

Vibrio species	Diseases
Vibrio cholerae	Diarrhoea
Vibrio parahaemolyticus	Gastroenteritis and wound infections
Vibrio vulnificus	Gastroenteritis and wound infections
Vibrio hollisae	Gastroenteritis and wound infections
Vibrio mimicus	Gastroenteritis and wound infections
Vibrio metschnikovii	Bacteraemia
Vibrio alginolyticus	Wound infections and otitis externa

Non-selective Media

Vibrio grows well under aerobic conditions on routine culture media used in bacteriology laboratories.

On nutrient agar, *V. cholerae*, after overnight incubation, produces moist, translucent colonies with a bluish or greenish tinge. The colonies measure about 1–2 mm in diameter. The colonies emit a distinctive odour. The colonies **on MacConkey agar** are initially colourless after overnight incubation but become reddish on prolonged incubation due to the late fermentation of lactose. **On blood agar**, *V. cholerae* biotype Eltor produces haemolytic colonies. Biotype Classical strains, however, produce greenish discolouration around the colonies, which later becomes clear due to haemodigestion. **In gelatin stab culture**, *V. cholerae* produces infundibuliform (*funnel-shaped*) or napiform (*turnip-shaped*) liquefaction after 3 days of incubation at 22 °C. **In peptone water**, the growth of bacteria forms a fine surface pellicle in about 6–9 h of incubation.

Selective Media

The selective media characteristically have a high pH, which suppresses the growth of other intestinal bacteria but favours the growth of vibrios.

TCBS medium (pH 8.6) contains thiosulphate, citrate, bile salts, sucrose and bromothymol blue (indicator). *V. cholerae* produces large, yellow convex colonies on this medium (Fig. 39.1). This is due to the fermentation of sucrose by the bacteria, leading to the production of acid. Accumulation of acid reduces the pH of the medium, so the bromothymol blue indicator becomes yellow, thus making *V. cholerae* colonies yellow. Non-sucrose-fermenting *V. parahaemolyticus* produces blue–green colonies.

Monsur GTTA medium (pH 8.5), due to its high pH and presence of potassium tellurite, inhibits most Gram-positive bacteria and enteric bacteria, except for *Proteus* species. Hence, the GTTA medium is used to isolate *V. cholerae* and other vibrios from faeces. *V. cholerae* produces small, translucent colonies with a greyish black centre and a turbid halo after 24 h of incubation. The colonies become larger (3–4 mm in size) after prolonged incubation of 48 h.

Alkaline BSA (pH 8.2) is another selective medium used for *V. cholerae*. The colonies on BSA are similar to those on nutrient agar.

Transport Media

These media are used to transport stool samples in case of a delay in processing the specimens in the laboratory. These help maintain the viability of *V. cholerae* and prevent the bacteria from being inhibited by normal intestinal bacterial flora present in stool specimens. Venkatraman–Ramakrishnan (VR) medium and Cary–Blair medium are two frequently used transport media.

VR medium is an alkaline medium with a pH of 8.6–8.8. The medium is dispensed in screw-capped bottles in 10–15 mL volumes. About 1–3 mL stool is to be added to each bottle. **Cary–Blair medium** is a solid medium with an alkaline pH of 8. It is a suitable transport medium for *Vibrio* spp. as well as for *Salmonella* and *Shigella* spp.

Enrichment Media

The enrichment media for *V. cholerae* are liquid media with a high pH. The high pH of the media suppresses the growth of many commensal

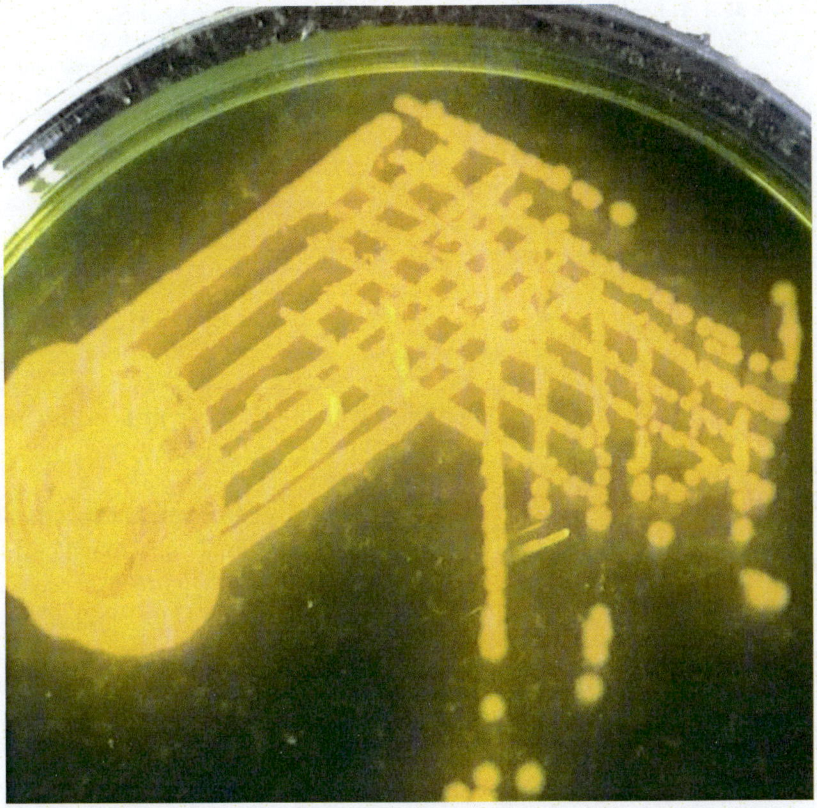

Fig. 39.1 TCBS medium showing sucrose fermenting yellow colonies of *Vibrio cholerae* (Courtesy of Prof S. Pramodhini)

intestinal bacteria but favours the growth of *V. cholerae*. Alkaline peptone water (APW) and Monsur taurocholate tellurite peptone water are two examples of transport media used for *V. cholerae*.

Alkaline peptone water (APW) (pH 8.6) is a frequently used medium to enrich stool specimens from patients convalescing from cholera, excreting a few bacteria in their stool. **Monsur taurocholate tellurite peptone water (pH 9)** is more selective for *V. cholera* due to the presence of potassium tellurite.

V. cholerae ferments sugar with the production of acid only but no gas. They ferment glucose, sucrose, maltose, mannitol and mannose. It is a late lactose fermenter and ferments lactose after several days of incubation. It forms indole and reduces nitrates to nitrites. *V. cholerae* shows positive cholera red reaction by producing a reddish-pink colour in the peptone water due to the formation of nitroso indole. The two properties of indole formation and reduction of nitrates to nitrites form the basis of the cholera red reaction. This reaction is tested by adding a few drops of concentrated sulphuric acid to a 24-h peptone water culture at 37 °C. It is methyl red positive and urease test negative. It liquefies gelatin and decarboxylates lysine and ornithine, but not arginine. The differences between Classical and Eltor biotypes of *V. cholerae* are summarised in Table 39.2.

Other Properties

Susceptibility to Physical and Chemical Agents

V. cholerae is most susceptible to heat, drying, acids and common disinfectants. *V. cholerae* are killed by heating at 56 °C for 30 min and killed in a few minutes in gastric juice of normal acidity. They survive for several days in untreated excreta of humans. They are resistant to high alkalinity. They remain viable for 1–2 weeks in fresh seawater. They survive in pure tap water for up to

Table 39.2 Differential characteristics of *Vibrio cholerae* biotypes

Properties	*Vibrio cholerae* biotype	
	Classical	Eltor
Haemolysis of sheep RBCs	Negative	Positive
Agglutination of chick erythrocytes	Negative	Positive
Voges–Proskauer test	Negative	Positive
Polymyxin B sensitivity	Positive	Negative
Susceptibility to		
Mukerjee group IV phage	Positive	Negative
Eltor phage 5	Negative	Positive
Vibriostatic (O/129) agent	Positive	Negative

30 days. However, vibrios do not survive long in grossly contaminated water in rivers, most possibly due to the presence of a large number of bacteriophages in this water.

Antigenic Structure

V. cholerae possesses two antigens; Somatic O antigen is present in the bacteria's cell wall. It is a group-specific antigen. Flagellar H antigen is a heat-labile antigen present in the flagella and is shared by all strains of *V. cholerae*.

Classification

A consensus was laid down based on different biochemical tests to differentiate vibrio and other closely related genera (Table 39.3). In the 1930s, two classification systems were employed to segregate vibrios among themselves.

Heiberg Classification

In 1934, Heiberg classified *Vibrio* into six groups based on their ability to ferment sugar such as mannose, sucrose and arabinose. Later, two more groups were added to the Heiberg grouping. *V. cholerae* fit in Group I of Heiberg classification (Table 39.4).

Serological Classification

Based on somatic carbohydrate O antigen, *V. cholerae* has been classified into Group A and B (Fig. 39.2). This serological classification was suggested by Gardner and Venkataraman (1935). Group A vibrios included Cholera vibrios sharing a common H antigen and biochemical similarities, which have been classified broadly into two groups: *V. cholerae* O1 and non-O1 *V. cholerae*:

V. cholerae O1 agglutinate with antisera to the O1 group and are called cholera vibrios or **agglutinable vibrios**. Based on their biochemical parameters, *V. cholerae* O1 is subdivided into **two biotypes**—Eltor and Classical. Each biotype has been divided into three subtypes—Ogawa, Inaba and Hikojima (Table 39.5). This classification is based on the differences in minor O antigens, such as A, B and C. Serotype Ogawa contains A and B antigens, Inaba contains A and C antigens and Hikojima contains all three antigens A, B and C. Hence, Hikojima strains are agglutinated by both Ogawa and Inaba

Table 39.3 Differentiating features of *Vibrio* and closely related genera

Genus	Oxidative and fermentative reaction (*Hugh–Liefson* test or OF test)		Amino acid utilisation test			String test
	Oxidative	Fermentative	Ornithine	Arginine	Lysine	
Vibrio	+	+ (gas production- no)	+	–	+	+
Aeromonas	+	+ (gas production- +/–)	–	+	–	Variable
Pseudomonas	+	–	Variable	Variable	Variable	–
Plesiomonas	+	+	+	+	+	–

Table 39.4 Heiberg classification of *Vibrio*

Vibrio group	Fermentation of sugar		
	Mannose	Sucrose	Arabinose
Group I	+	+	–
Group II	–	+	–
Group III	+	+	+
Group IV	–	+	+
Group V	+	–	–
Group VI	–	–	–
Group VII	+	–	+
Group VIII	–	–	+

antisera. In contrast, Ogawa and Inaba strains are agglutinated by their specific antisera only.

Non-O1 *V. cholerae*, which does not agglutinate with O1 group antisera, are designated as **non-cholera vibrios (NCV)** or **non-agglutinating vibrios (NAG)** (Fig. 39.2). Non-cholera vibrios are classified into **138 serotypes** designated from O2 to O139. Of these serotypes, only *V. cholerae* O139 causes cholera in humans. The rest of the serotypes (O2–O138) do not cause any disease in humans.

Phage Typing

V. cholerae O1 biotype Classical has been divided into five types based on four phages (I, II, III and IV) by **phage typing**. All these phages lyse biotype Classical but not Eltor strains. Based on these four phages and an additional fifth phage, *V. cholerae* biotype Eltor can also be divided into six types.

Pathogenesis

Cholera is a toxin-mediated disease.

Virulence Factors

V. cholerae has many virulence factors that include cholera toxin, toxin co-regulated pilus (TCP), adhesin factor (ACF), haemagglutination-protease (hap; mucinase), neuraminidase and siderophores that contribute to the

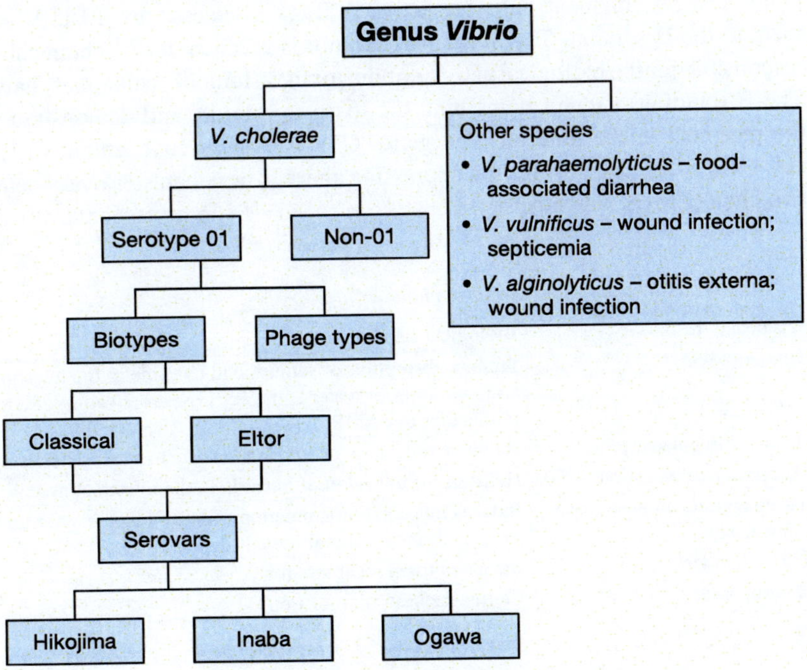

Fig. 39.2 A schematic diagram showing the classification of the genus *Vibrio*

Table 39.5 Serotypes of *Vibrio cholerae* O1

Serotype	O antigens
Ogawa	AB
Inaba	AC
Hikojima	ABC

pathogenesis of cholera (Table 39.6). Most of these virulence factors are expressed by multiple chromosomal genes present in the bacteria. *V. cholerae*, like Gram-negative bacilli, also possesses the lipopolysaccharide (LPS). The LPS has no role in the pathogenesis of cholera but is responsible for the immunity produced by the killed *V. cholerae* vaccine.

Cholera Toxin or CT (CTx)

It is an exotoxin and the most important virulence factor produced by *V. cholerae*. The toxin, known as choleragen or cholera enterotoxin, is structurally and functionally similar to the heat-labile enterotoxin of *Escherichia coli*. It is a heat-labile protein of molecular weight 85,600 kDa and is destroyed by heating at 56 °C for 30 min. The toxin consists of a **single A subunit** of molecular weight 27,215 kDa and **five B subunits** each of molecular weight 11,677 kDa. Subunit A consists of two fragments—A1 and A2, which remain covalently connected by a single disulphide bond. The A2 fragment links the active A fragment to the B subunit. The A (active) subunit is responsible for the biological activity of the toxin. The B (binding) subunit helps bind the toxin to the receptors in the mucosa. The genes CTx A and CTx B are responsible for expressing these two cholera toxin subunits.

Cholera toxin has many biological functions. This inhibits the absorption capacity and activates the excretory chloride transport in the intestinal enterocytes, eventually leading to the loss of sodium chloride in the intestinal lumen. It also inhibits the absorption of sodium and chloride in the intestine. Furthermore, it also increases skin capillary permeability. Hence, it is also called a permeability factor. A skin bluing test can detect the permeability factor. In this test, cholera toxin is injected intradermally in rabbits or guinea pigs, followed by an intravenous injection of Evans blue. In a positive test, the skin at the injection site becomes blue due to increased capillary permeability.

Many tests can demonstrate the cholera toxin. **Rabbit ileal loop test** was one of the classical methods developed by De and Chatterjee in 1952. It is a widely used method in which injection of *V. cholerae* culture or culture filtrate into the ligated ileal loop causes fluid accumulation, leading to a ballooning of the ileal loop. The toxin can also be detected by ELISA and rapid immunofluorescence tests, chemical estimation of cAMP in tumour cells that have been treated with the toxin, and demonstration of histological changes in adrenal tumour (Y1) cells, elongation of Chinese hamster ovary cells and activation of lipolysis in rat testicular tissue.

Table 39.6 Virulence factors of *Vibrio cholerae*

Virulence factors	Biological functions
Cholera toxin	Inhibits absorption of sodium and chloride in the intestine; causes hypersecretion of large volumes of water and electrolytes, activation of adenylate cyclase and overproduction of cAMP.
Toxin co-regulated pilus	Helps in adherence of *Vibrio cholerae* to mucosal cells of the intestine
Accessory colonisation	Helps in the adhesion of bacteria to the intestinal mucosa
Haemagglutination-protease (mucinase)	Induces intestinal inflammation and also helps in releasing free vibrios from the bound mucosa to the intestinal lumen
Neuraminidase	Increases toxin receptors for *Vibrio cholerae*
Siderophores	Causes sequestration of iron

The A and B subunits of CT are encoded by ctxA and ctxB genes, respectively. These genes are located on the chromosome, forming an operon where ctxA precedes ctxB. The Vibrio Pathogenicity Island (VPI) codes for a regulatory protein, ToxT, which activates the transcription of both the ctxAB. Upon activation, the CT is secreted and is transported across the outer membrane by a type II secretion system. It is known as Eps for extracellular protein secretion, which comprises 15 gene products, 12 of which are required for translocation of specific substrates, including CT and hemagglutinin, across the outer membrane. The energy for secretion is obtained from EpsE, a cytoplasmic ATPase.

Other Virulence Factors

These include **TCP** which helps in adherence of *V. cholerae* to mucosal cells of the intestines. **Accessory colonisation also** helps in the adhesion of bacteria to the intestinal mucosa. **Haemagglutination-protease (mucinase)**, formerly known as cholera lectin, is both agglutinin- and zinc-dependent protease. The enzyme splits mucus and fibronectin and a subunit of the cholera toxin. It induces intestinal inflammation and also helps in releasing free vibrios from the bound mucosa to the intestinal lumen. **Neuraminidase** destroys neuraminic acid, thereby increasing toxin receptors for the *V. cholerae*. **Siderophores** are responsible for the sequestration of iron.

Pathogenesis of Cholera

Cholera is an acute diarrhoeal disease. *V. cholerae* enters the body orally through contaminated water and food (Fig. 39.3). The infective dose of *V. cholerae* is usually high, more than 10^6 organisms/mL, because most organisms are killed by high acidity in the stomach. Vibrios are highly susceptible to the acidity of the stomach. The conditions that reduce the stomach's acidity to pH above 5 make the host more susceptible to infection by cholera vibrios.

The bacteria reach the mucous layer of the small intestine by motility, chemotaxis and production of proteolytic enzymes. A haemagglutinin-protease breaks mucin and fibronectin of the intestinal mucosa. Subsequently, bacteria adhere to the intestinal wall and colonise, facilitated by TCP, a special fimbria. The synchronous action of CTx, TCP and a few other virulence factors is regulated by Tox R gene products, known as the **master switch**. Although *V. cholerae* remains adhered to the intestinal epithelium throughout infection, it does not cause any histological changes or damage to the intestinal mucosa.

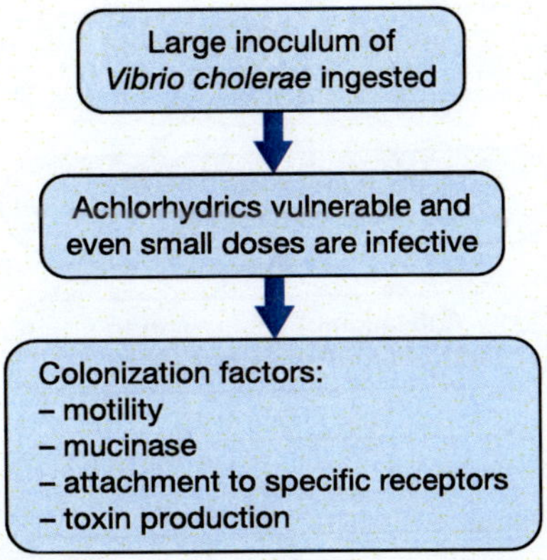

Fig. 39.3 Colonisation of *V. cholerae* in the intestine

Vibrios, once adhered to the intestinal wall, produce cholera toxin. The B **(binding) unit** of the toxin binds to the GM1 ganglioside receptors present on the surface of the jejunal epithelium cells. The **A (active) unit is** transported into enterocytes, within which it is split into two components: **A1** and **A2**. The A1 component is linked to the B subunit and activates cellular **adenylate cyclase (cAMP).** This inhibits the absorption of sodium transport and activates the excretory chloride transport in the intestinal epithelial cells. This leads to the secretion of sodium (Na), chloride (Cl) and potassium (K) and large volumes of water in the lumen of the intestine leading to watery diarrhoea. The toxin also inhibits intestinal absorption of sodium and chloride by the microvilli. This leads to the large secretion of water, which overcomes the absorptive capacity of the lumen, eventually causing diarrhoea (Fig. 39.4).

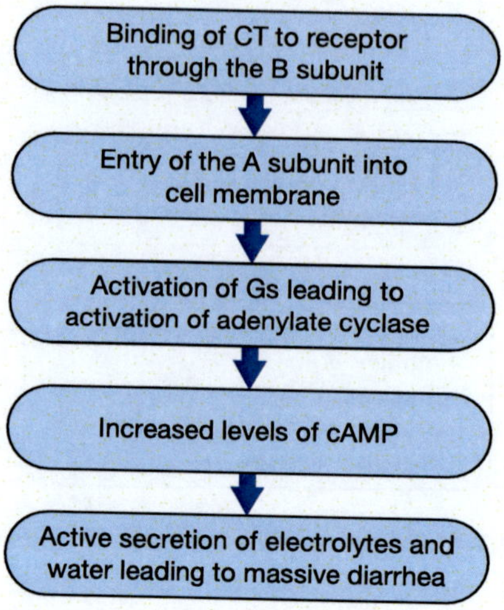

Fig. 39.4 Pathogenesis of diarrhoea caused by cholera toxin

Clinical Syndromes

V. cholerae causes cholera, an acute diarrhoeal disease, which may be rapidly fatal if not treated immediately.

Cholera

The incubation period is short and varies from 2 to 3 days after ingestion of the bacteria. The condition shows an abrupt onset of watery diarrhoea and vomiting. Profuse watery diarrhoea is the most important manifestation of cholera.
The volume of diarrhoeic stool excreted in cholera is much more than that of diarrhoea caused by any other infectious pathogen. Patients may excrete as high as 250 mL of stool per kilogram body weight in a day in a severe condition.

Cholera stool is profusely watery that is colourless and odourless, free of proteins and speckled with mucous. The cholera stool is often described as rice-water stool (in colour and consistency, stool resembles water washed off from cooked rice). It contains a few leukocytes but no erythrocytes because *V. cholerae* does not stimulate an inflammatory response in the intestinal mucosa. It is a bicarbonate-rich electrolyte fluid, which contains little protein.

Severe abdominal cramp, possibly caused by distension of the small intestine due to excretion of larger volumes of intestinal fluid, is also seen in these patients. Vomiting is another important manifestation of cholera and occurs in the early stage of the disease. This is caused by decreased gastric and intestinal motility.

Cholera, if left untreated, can lead to a severe loss of fluid and electrolytes due to diarrhoea and vomiting. This could lead to isotonic dehydration, metabolic acidosis, hypokalaemia and hypovolemic shock. Dehydration in cholera characteristically develops very fast, within hours after the onset of symptoms. In patients with severe disease, diarrhoea leads to vascular prolapse and shock, and patients may die of cardiac arrhythmia and renal failure. Other complications include electrolyte imbalance and hypoglycaemia. In untreated patients, the case fatality rate has been estimated to vary from 25% to 50%. In patients treated well with replacement of lost fluid and electrolytes, the case fatality is usually less than 1%.

V. cholerae O139 (Bengal vibrio) causes cholera, which can be as severe as that caused by *V. cholerae* O1. It may occasionally cause invasive disease-producing bacteraemia. *V. cholerae* O139 strain shows a similar pathogenic mechanism to that of *V. cholerae* O1 except that it produces a unique O139 LPS and an immunologically related O antigen capsule. These two virulence factors enhance the virulence of this organism.

Laboratory Diagnosis

Specimens

Fresh stool specimen collected before administration of antibiotics is the specimen of choice. The stool may be collected by introducing a sterile soft rubber catheter (No: 24–26), or the liquid stool may be collected directly in a screw-capped container. Collection of stools from the bedpan is not recommended due to the risk of contamination or the presence of disinfectants. **Rectal swabs** may also be used for the collection of specimens.

The collected stool or rectal swab specimen should be sent to a laboratory for immediate culture. If a delay is anticipated, the specimens may be preserved at 4 °C in a refrigerator or transport media used for *V. cholerae*. The specimen may be transported in enrichment media, such as APW or Monsur taurocholate transport media and sent to a laboratory for processing within a few hours. Stool or rectal swab is inoculated into 10 mL of APW in a screw-capped tube and into Monsur taurocholate transport medium and is transported to the laboratory. These are incubated at 37 °C for 3–6 h, and afterwards, the subculture is made on the selective media in case of a longer delay in transport; the stool samples may be preserved in Cary–Blair medium or VR fluid for a long duration. In case of non-availability of transport media, the watery stool can be soaked in a strip of thick blotting paper and sent to the laboratory—packed in a plastic envelope.

Microscopy

Dark-field microscopy is a useful method for demonstrating the characteristic darting motility of the bacilli and its inhibition by antisera. This is a rapid method of examining stool collected from cases or after enrichment for 6 h. **Direct immunofluorescence** is another rapid method used to demonstrate vibrios in the stool. However, microscopic examination of a Gram-stained stool smear is not recommended for diagnosing cholera.

Culture and Identifications

The specimen collected in holding media is inoculated first in the enrichment media and incubated for 6–8 h before being inoculated on selective (TCBS or GTTA) and non-selective media (BSA, MacConkey agar and blood agar). The specimen may also be plated directly into these media before enrichment. The specimens collected in transport media are incubated for 6–8 h, including the transit time before inoculation into the suitable media. The inoculated plates are incubated at 37 °C overnight. *V. cholerae* produces characteristic yellow colonies on TCBS and non-lactose-fermenting colonies on MacConkey agar.

String test is a presumptive test for the identification of *Vibrio* species. In this test, a loopful of the suspected colony from the culture media is mixed with 0.5% sodium deoxycholate in saline on a clean glass slide. The loop is lifted upwards slowly from this suspension to look for the string formation (Fig. 39.5). A positive string test is indicative of *Vibrio* spp., and there will not be any string formation in the case of *Aeromonas* and *Plesiomonas shigelloides*.

Colonies on media are identified by performing a series of tests. *V. cholerae*, which is catalase positive and oxidase positive, ferments sugar with a production of acid only but no gas. They ferment glucose, sucrose, maltose, mannitol

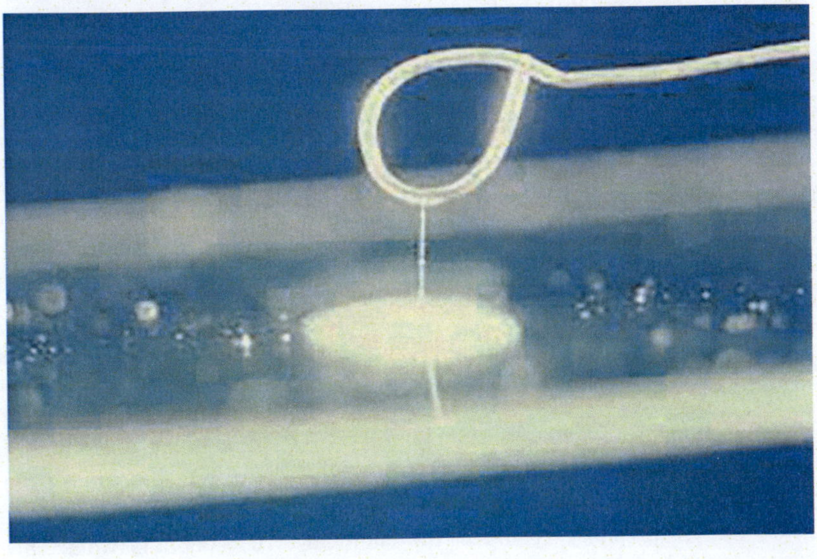

Fig. 39.5 String test for preliminary identification of *Vibrio* spp. from other closely related genera (http://link.springer.com/referenceworkentry/10.1007%2f978-03-642-30144-5_90#page-1)

and mannose. The bacteria are positive for cholera red reaction, form indole and reduce nitrates to nitrites. They utilise amino acids, such as lysine and ornithine, but not arginine. Non-O1 *V. cholerae* does not agglutinate with O1 group antisera and is designated as non-cholera-vibrios or non-agglutinating vibrios. *V. cholerae* O1 is identified as biotype Classical or Eltor depending on their biochemical reactions.

Typing

Serotyping

Suspected *V. cholerae* is tested using specific *V. cholerae* O1 antisera by slide agglutination. In this test, the colonies are picked with a straight wire and mixed with a drop of antisera on the slide. Agglutination of the bacteria shows that the test is positive for *V. cholerae* O1. If positive, agglutination of bacteria is repeated using specific Inaba and Ogawa sera for serotyping. Hikojima strains agglutinate well with both Inaba and Ogawa sera. If agglutination is negative, the test is repeated with at least five more colonies, as both O1 and non-O1 vibrios may coexist in the same specimen.

Bio Typing

If the slide agglutination is positive and the colony is identified as *V. cholerae* O1, then it is tested by various tests to determine whether isolated *V. cholerae* O1 is Classical or Eltor. These tests include haemolysis of sheep RBCs, agglutination of chick RBCs, sensitivity to polymyxin B, VP test and sensitivity to bacteriophages (Table 39.2).

Vibrio colonies that are not agglutinated with *V. cholerae* O1 antisera are usually tested with antisera to the H antigen. *Vibrio* isolates agglutinated by H antisera but not by O1 antisera are identified as non-O1 cholera vibrio. Such non-cholera-vibrios are tested for O139 using specific antisera against the O139 antigen.

Phage Typing

The isolated strains of *V. cholerae* O1 for phage typing may be sent to the International Reference Centre for *Vibrio* phage typing at the National Institute of Cholera and Enteric Diseases (NICED) in Kolkata, West Bengal.

Molecular Typing

The pulsed-field gel electrophoresis (PFGE), multilocus sequence typing (MLST) and MLST of virulence gene (V-MLST) methods are the three common methods for molecular typing of Vibrio. Housekeeping genes like genes (cat, chi, dnaE, gyrB, lap, pgm, recA, rstA and gmd) are used for the PFGE and MLST, whereas V-MLST uses virulence gene sequences like ctxAB, tcpA and toxR. Among all, PFGE had a higher resolution ability.

Serology

The complement-dependent vibriocidal antibody and antitoxin assays using live or killed vibrio suspensions are useful for demonstrating vibrio antibodies in the serum. These tests are useful for seroepidemiological studies but are of little value in diagnosing cases.

Molecular Detection

Classical, real-time, multiplex PCRs are the molecular tools of choice for detecting Vibrio sp. in samples. *V. cholerae*-specific ompW sequence and the ctxA gene of the cholera toxin are often amplified. Raid detection kits are now available that uses RT-PCR-based detection of the organisms.

Diagnosis of Carriers

Repeated stool culture is essential for isolating *V. cholerae* from carriers because vibrios are known to be excreted intermittently in their stool. Collecting a stool after using a purgative (such as mannitol or magnesium sulphate) or examining bile collected by duodenal intubation is frequently useful for diagnosing carriers.

Water Sample Testing

The water samples can be tested for the presence of vibrios, and it usually involves two methods: enrichment and filtration. In the **enrichment method**, 900 mL of the water to be tested for *Vibrio* contamination is mixed with 100 mL of 10× peptone water (pH 9.2) and incubated at

37 °C for 6–8 h. Further, a second enrichment is done by repeating the procedure mentioned above. After the second enrichment, the test sample is plated on selective media to look for the growth of *Vibrio*. **Infiltration method**, the water to be tested for *Vibrio* contamination is passed through a Millipore membrane filter. Then, the filter's membrane is directly placed on selective media and incubated at 37 °C, overnight to see any colony growth. However, sewage water needs to be diluted in saline and passed through gauze prior to testing water by either of the methods mentioned above.

Treatment

Treatment of cholera includes replacement of fluid and electrolytes, and antibiotic therapy.

Replacement of Fluid and Electrolytes

The cholera treatment is primarily based on the prompt and adequate replacement of fluid and electrolytes before massive fluid loss leads to hypovolemic shock. Replacement of fluid by oral administration of fluid containing glucose and electrolytes is the most successful and highly effective method for treating cholera.

Oral rehydration therapy (ORT) solution consisting of glucose, sodium chloride, potassium chloride and sodium citrate is widely used. The glucose facilitates the absorption of sodium in the small intestine, and salts present in the ORT restore the electrolyte balance and reverse acidosis. The ORT is highly cost-effective and is simple to use; hence it is widely used with tremendous success in developing countries endemic to cholera. **Cereal-based preparations**, such as thoroughly cooked and salted rice soup, are equally effective for treating dehydration in cholera cases. Immediate **intravenous fluid therapy** is usually recommended for more serious cases of dehydration.

Antibiotic Therapy

Antibiotic therapy is of secondary importance and is a supplement to fluid therapy. Treatment with antibiotics diminishes the duration and volume of fluid loss. It also hastens clearance of the *Vibrio* organisms from the intestine. *V. cholerae* is uniformly sensitive to tetracycline, ciprofloxacin and erythromycin. Tetracycline or doxycycline is the drug of choice for adults and trimethoprim-sulfamethoxazole for children. Pyrazolidine is the usual recommended treatment for pregnant females suffering from cholera. Of late, strains resistant to ciprofloxacin have been documented from Kolkata, India.

Epidemiology

V. cholerae is found naturally in the estuary and marine environment worldwide.

Geographical Distribution

Cholera continues to be a major health problem in many parts of the world, including the Indian subcontinent and sub-Sahara in Africa. The condition has been rare in developed and industrious nations for many decades. A majority of cases have been reported from Africa and Asia. Cholera occurs as an endemic, epidemic or pandemic disease.

Cholera Pandemics

A total of six cholera pandemics occurred in the nineteenth century, affecting Europe and the United States, causing more than 115,000 deaths in 1832 and 50,000 deaths in 1856. The seventh pandemic of cholera appeared first in the twentieth century, starting in 1961, and affected five continents by 1991. Unlike six earlier pandemics, which were caused by *V. cholerae* biotype Classical, the seventh pandemic was caused by *V. cholerae* biotype Eltor. The latter originated in Asia in the 1960s, reached Africa in the early 1970s and was responsible for epidemic cholera. In 1991, the pandemic reached Peru and subsequently spread rapidly through many countries of South and Central America and the United States and Canada. By mid-1995, more than one million cases and 10,000 deaths due to the disease were reported in the United States.

The **seventh pandemic of cholera** was different from the previous pandemics of cholera in the following ways:

1. It was the first pandemic to be originated from outside the Indian subcontinent. It originated from Sulawesi (Celebes), Indonesia. From here, it spread to Hong Kong and the Philippines and entered India in 1964. By 1966, it entered and spread throughout the Indian subcontinent and West Asia. The pandemic then spread to Africa and parts of southern Europe in the 1970s.
2. It was the first pandemic caused by *V. cholerae* biotype Eltor in contrast to the earlier six pandemics caused by *V. cholerae* biotype Classical.
3. *V. cholerae* biotype Eltor, responsible for the seventh pandemic, caused a less severe illness with a large proportion of mild and asymptomatic infections. The mortality rate was low, but characteristically carrier rate was high.
4. *V. cholerae* biotype Eltor has nearly replaced the *V. cholerae* biotype Classical strain following this pandemic spread. Hence, in countries like India, *V. cholerae* Classical strains are rarely encountered, but in Bangladesh, there has been a resurgence of cholera caused by *V. cholerae* Classical strains.

V. cholerae O139

In 1992, a new epidemic strain, *V. cholerae* O139 emerged as a cause of epidemic cholera in India. This new epidemic strain was first reported from a cholera outbreak in Chennai in 1992, followed by similar cholera outbreaks from different parts of India. The new strain entered Bangladesh and spread across the country, causing cholera epidemics by Jan. 1993. *V. cholerae* O139 is the first non-cholera *Vibrio* or non-cholera strain associated with cholera epidemics. It is also responsible for producing disease in individuals previously infected with *V. cholerae* O1 strains.

Habitat

V. cholerae is a saltwater bacterium. In association with plankton, the marine ecosystem is the primary habitat of the bacteria. The bacteria can multiply freely in the water. The number of bacteria in contaminated water increases during the warm and hot months of the year. *V. cholerae* is never found in normal humans. In infected humans, *V. cholerae* inhabits the small intestine.

Reservoir, Source and Transmission of the Infection

Humans and water are the two main reservoirs of infection for cholera. Both chronic and convalescent carriers play an important role as reservoirs in the transmission of infection. These chronic carriers continuously excrete vibrios in their stool intermittently for several years.

The infection is transmitted to humans by ingestion of contaminated food or water. Hence, the infection rate is highest in the areas where hygienic community standards are low and potable water is not available. *V. cholerae* fails to survive in an acidic environment. Therefore, any condition (e.g. achlorhydria) that reduces gastric acid production makes the person highly susceptible to infection. People of all ages are susceptible to infection by *V. cholerae*. Infants, however, are protected due to maternal antibodies transmitted during breastfeeding. Animals do not play any role in the transmission of the disease.

Prevention and Control

The preventive measures against cholera include cholera vaccination and general preventive measures.

Cholera Vaccines

Immunity in cholera may be produced against the cholera toxin or the bacteria. An infection by *V. cholerae* biotype Classical usually protects against the infection caused by *V. cholerae* biotype Classical and Eltor strains. However, infection by *V. cholerae* biotype Eltor does not confer protection against biotype Classical.

Killed Cholera Vaccines

These are the vaccines extensively used against cholera. This vaccine is a suspension of 8000 million *V. cholerae* per millilitre, consisting of equal numbers of Inaba and Ogawa serotypes. This vaccine is given either subcutaneously or intramuscularly in two doses at an interval of 4 weeks. This vaccine confers about 60% protection for a short duration of 3–6 months. The rate of protection in endemic areas increases with age. A single dose of vaccine is not effective in children younger than 5 years but is effective in adults. In order to improve the antigenic stimulus, the number of organisms in the vaccine has been increased to 12,000 million/mL.

Non-living oral B Subunit Whole-cell Vaccine (WC/rBS Vaccine)

This is a non-living oral B subunit whole-cell vaccine. It consists of heat-killed whole-cell *V. cholerae* O1 (2.5×10^{10}) combined with the purified recombinant B subunit of cholera toxoid (WC/rBS). The vaccine is administered in two doses at an interval of 1 week. This vaccine is currently available in Sweden and Latin America. Clinical trials carried out in Bangladesh, Peru and Sweden have shown that the vaccine is safe and confers 85–90% protection during 6 months in all age groups after administration of two doses.

Live Oral Cholera Vaccine (CVD 103-Hg R Vaccine)

This is an oral cholera vaccine. It consists of an attenuated live oral genetically modified *V. cholerae* O1 strain (CVD 103-Hg R). It induces local immunity by the production of IgA antibodies in the gut. This cholera vaccine developed by the Swiss Sera and Vaccine Institute is available in Europe, Latin America and Canada. The vaccine is extremely immunogenic and highly protective against moderate and severe cholera cases. The vaccine trials in adult volunteers in the United States have shown that a single dose confers high protection (95%) against *V. cholerae* Classical and 65% protection against *V. cholerae* Eltor following a booster dose given after 3 months.

General Preventive Measures

These include early identification and case management, improved water supply and sanitation, improved personal hygiene and health education.

Non-Cholera Vibrios

V. cholerae serotypes from O2 to O139 are non-cholera-vibrios, non-agglutinable vibrios or *V. cholerae* non-O1 vibrios. They are similar to *V. cholerae* O1 biochemically and genetically. As mentioned earlier, *V. cholerae* O139 is the only example of non-cholera-vibrios that causes cholera. Non-cholera-vibrios other than *V. cholerae* O139 are widely found in aquatic environments. In humans, they may cause mild to severe diarrhoeal disease resembling cholera. Occasionally, they may also cause wound infections, septicaemia and other extraintestinal infections.

Vibrio mimicus

V. mimicus is so named because it resembles *V. cholerae* in its biochemical reactions. *V. mimicus*, unlike *V. cholerae*, does not ferment sucrose, but like *V. cholerae*, it grows well at a 0.5–1% concentrated sodium chloride. Infection is acquired by eating seafood, especially oysters. The organisms cause a gastrointestinal disease, which is usually self-limited. *V. mimicus* has been associated with many sporadic cases of diarrhoeal disease on the gulf coast of the United States.

Halophilic Vibrios

Vibrios that require a higher concentration of sodium chloride are known as halophilic vibrios. They are natural inhabitants of seawater and marine life. *V. parahaemolyticus*, *Vibrio alginolyticus* and *V. vulnificus* are three important halophilic *Vibrio* species that cause human infection.

Vibrio parahaemolyticus

V. parahaemolyticus is now recognised as an important cause of sea food-associated gastroenteritis worldwide. *V. parahaemolyticus* is a curved Gram-negative bacillus that resembles

V. cholerae. However, it differs from *V. cholerae* by having a capsule showing bipolar staining and pleomorphism. Like other vibrios, it has polar flagella when grown in liquid culture but shows peritrichous flagella when grown on solid media. The organism fails to grow in the medium without sodium chloride. It can grow well in salt concentrations up to 8% and at an optimum salt concentration of 2–4%. On MacConkey agar, it produces colourless non-lactose-fermenting colonies; on blood agar, it forms beta-haemolytic colonies. The colonies on TCBS agar are non-sucrose-fermenting green colonies with an opaque and raised centre and translucent periphery. *V. parahaemolyticus* is oxidase, catalase, indole and citrate positive. It reduces nitrates to nitrites. It ferments glucose, mannose, maltose, mannitol and arabinose with acid production only. It does not ferment sucrose, lactose and salicin. It decarboxylates lysine and ornithine but not arginine. It is VP positive.

Thermostable haemolysin is the key virulence factor of *V. parahaemolyticus*. This virulence factor is found only in strains that are pathogenic to humans. Strains isolated from humans possess this toxin, hence are pathogenic, while strains isolated from environmental sources, such as water, crabs, oysters, or fish, do not possess this thermostable toxin, hence they are non-pathogenic.

The pathogenic strains of *V. parahaemolyticus*, when grown on a special high-salt mannitol medium (Wagatsuma agar), show haemolysis on blood agar. This is called **the Kanagawa phenomenon**. This haemolysis is caused by a thermostable haemotoxin released by pathogenic strains known as Kanagawa-positive strains. Non-pathogenic strains isolated from the environment that do not cause any haemolysis on the blood agar are Kanagawa-negative strains.

V. parahaemolyticus in humans causes gastroenteritis. The severity of the condition can vary from mild self-limited diarrhoea to an acute illness. After an incubation period of 5–72 h, the patients with gastroenteritis manifest as eosinophilic watery diarrhoea. The condition is associated with nausea, vomiting, abdominal pain and low-grade fever, which may be present for 3 days. Stool usually does not contain any blood or mucus but contains cellular exudates. The patient usually recovers within 3 days. *V. parahaemolyticus* also causes extraintestinal infections, such as wound infections and ear and eye infections in individuals exposed to contaminated seawater.

V. parahaemolyticus is widely distributed in the estuary and marine environments. Seafood, such as fish, crabs or oysters, is the main source and reservoir of infection. The infection is acquired through the consumption of contaminated seafood. *V. parahaemolyticus*, although widely found in sea fish in some parts of India, is relatively less common in human infections.

Vibrio alginolyticus

V. alginolyticus is another halophilic vibrio. It shows a high-salt tolerance and can grow even in the presence of 10% sodium chloride. On TCBS agar, it forms large yellow sucrose-fermenting colonies (Table 39.7). It shows swarming on the surface of the non-selective medium, such as blood agar. *V. alginolyticus* is widely distributed in seawater and seafood. *V. alginolyticus* is associated with infections of superficial wounds exposed to contaminated seawater. It also has been associated with infections of the gastrointestinal tract, ear and eye.

Vibrio vulnificus

V. vulnificus is a halophilic vibrio, formerly known as L+ vibrio or *Benekea vulnificus*. Like *V. parahaemolyticus, V. vulnificus* is a Gram-negative bacillus and facultative anaerobe that requires salt for growth. It also produces green non-sucrose-fermenting colonies on TCBS medium. It differs from *V. parahaemolyticus* and other *Vibrio* species by its ability to ferment lactose. Relatively, *V. vulnificus* is a more virulent bacterium than *V. parahaemolyticus*. The presence of capsule and the production of hydrolytic

Table 39.7 Differentiating features of *Vibrio parahaemolyticus* and *Vibrio alginolyticus*

Properties	*Vibrio parahaemolyticus*	*Vibrio alginolyticus*
Growth in 10% agar	–	+
Presence of swarming	–	+
Fermentation of sucrose	–	+
VP test	–	+
Growth in 10% NaCl	–	+

enzymes, such as cytolysins, proteases and collagenases, are responsible for the virulence of the bacteria. These organisms are also resistant to complement- and antibody-mediated lysis.

V. vulnificus is a natural inhabitant of sea water. *V. vulnificus* infections are usually food-borne diseases and are largely associated with ingestion of raw or undercooked seafood and exposure of the wound to contaminated seawater. *V. vulnificus* infection in humans produces two types of illness; wound infection and gastroenteritis rapidly progressing to septicaemia. Diagnosis of the condition is made by isolating the bacteria from wounds and blood. Since *V. vulnificus* produces life-threatening illness, immediate and prompt treatment with antibiotics is essential. Tetracyclines or aminoglycosides are the antibiotics of choice to treat the condition.

Other *Vibrio* Species

Other *Vibrio* species associated with occasional human infection include *Vibrio fluvialis*, *Vibrio furnissii*, *Vibrio damsela* and *Vibrio hollisae*. These species are associated with gastroenteritis, wound infections and bacteremia. Rare bacteremia infections by *Vibrio metschnikovii* have also been documented.

Genus *Aeromonas*

The members of the genus *Aeromonas* are Gram-negative facultative anaerobic bacilli that morphologically resemble other members of the family Vibrionaceae. At least 16 species of *Aeromonas* have been described, of which 11 species are associated with human disease. *Aeromonas hydrophila*, *Aeromonas caviae*, *Aeromonas veronii*, *Aeromonas schubertii* and *Aeromonas jandai* are most important species. *Aeromonas* are motile bacilli in the presence of polar flagella. They are catalase positive, oxidase positive and nitrate positive. They ferment sugar with the production of both acid and gas. They grow at an optimum temperature of 32 °C and a pH of 7. *Aeromonas* spp. are widely distributed in brackish water and soil. *Aeromonas* spp. in humans cause gastroenteritis and wound infections. *Aeromonas* species are known to colonise the intestinal tract of humans in approximately 3% of individuals. Therefore, isolation of the bacteria from a stool by culture does not indicate disease by *Aeromonas* species. It needs careful interpretation by considering the clinical presentation of the illness. *Aeromonas* species are susceptible to chloramphenicol, gentamicin and trimethoprim-sulfphamethoxazole. They are resistant to penicillin, mostly cephalosporins and erythromycin.

Genus *Plesiomonas*

Plesiomonas species are oxidase-positive, Gram-negative and facultative anaerobes differentiated from genera *Aeromonas* and *Vibrio* of the family Vibrionaceae by some biochemical reactions. *P. shigelloides* is the only species described in the genus *Plesiomonas*. It is serologically related to *Shigella sonnei* and shows agglutination with *S. sonnei* antiserum. However, it differs from *Shigella* by being oxidase positive and motile in the presence of polar flagella. *P. shigelloides* in humans causes gastroenteritis.

Case Study

There was an outbreak of diarrhoea during a major festival in which many people participated. More than 20 people were admitted to the local hospital with a history of severe watery diarrhoea. Most of the patients were dehydrated. They were treated with fluid and electrolytes to maintain fluid balance. All the patients recovered. The stool was positive for *V. cholerae* by culture, and the strain was found to be *V. cholerae* O1 by serotyping.

- What tests will you perform to know whether the isolated strain is Classical or Eltor?
- How does *V. cholerae* cause diarrhoea?
- What are the vaccines available against cholera?

Haemophilus and *Bordetella* 40

The members of the genus *Haemophilus* are small, sometimes pleomorphic, Gram-negative, non-motile and non-sporing organisms. They require one or both accessory growth factors, namely, X and V, present in the blood. Hence, the genus derived its name from its essential growth requirement of certain factors, such as X and V present in the blood (*Haemophilus*: *haem*, blood; *philus*, loving). In the year 1883, Koch isolated *Haemophilus aegyptius*, the first member of the genus from a case of conjunctivitis in Egypt.

Genus *Haemophilus*

Haemophilus spp. are obligate bacteria present in the mucous membranes of humans and certain species of animals. *Haemophilus influenzae* and *Haemophilus ducreyi* are two major species associated with human diseases. Other *Haemophilus* species are rarely pathogenic. They are responsible primarily for opportunistic infections (Table 40.1).

Haemophilus influenzae

H. influenzae is the species most commonly associated with human disease. It is an important cause of meningitis in children and also of respiratory tract infection in children as well as in adults.

Properties of the Bacteria

Morphology

H. influenzae is a small, pleomorphic, Gram-negative bacillus and measures 1 × 0.3 μm. In fresh cultures, the bacteria are usually coccobacilli, while long filamentous forms are seen in older cultures. Some strains of *H. influenzae* possess polysaccharides capsule. The capsule can be detected by Indian ink preparation or by capsular swelling reaction using type-specific antiserum.

Culture and Biochemical Properties

H. influenzae is a fastidious bacterium. It is a facultative anaerobe and grows better in anaerobic conditions. The optimal temperature is 35–37 °C. The presence of 5–10% CO_2 enhances the growth of the bacteria. *H. influenzae* requires two erythrocytic growth factors, X and V factors; hence they grow on **blood agar** and **chocolate agar**. The **X factor** is a heat-stable factor present in the blood. It is a protoporphyrin IX, haemin or another iron-containing porphyrin. X factor is required to synthesise iron-containing bacterial enzymes, such as catalase, cytochrome oxidase and peroxidase. **V factor** is a heat-labile factor present in red blood cells and other animal and plant cells. This factor is destroyed at 120 °C in 30 min. This factor is also synthesised by *Staphylococcus aureus* and also by some fungi. Earlier, this factor was thought to be a vitamin; hence it was named the V factor. This has now been

Table 40.1 Human infections caused by *Haemophilus* species

Haemophilus species	Disease
Haemophilus influenzae	Meningitis, epiglottitis, cellulitis, pneumonia, otitis media, bronchitis and conjunctivitis
Haemophilus ducreyi	Soft sore or chancroid
Haemophilus aphrophilus	Sub-acute endocarditis, brain abscess, pneumonia and sinusitis
Haemophilus parainfluenzae	Endocarditis, opportunistic infections
Haemophilus aegyptius	Conjunctivitis
Haemophilus haemolyticus	Occasional opportunistic infection
Haemophilus parahaemolyticus	Occasional opportunistic infection
Haemophilus segnis	Occasional opportunistic infection

identified as NAD or NADP coenzyme I. The V factor acts as a hydrogen acceptor in the oxidation-reduction process in a replicating bacterial cell.

H. influenzae produces large colonies on the blood or chocolate agar by any of the following methods:

Blood Agar

Growth of *H. influenzae* on ordinary blood agar is scanty, and the colonies are small because the V factor is not freely available to be found inside the red blood cells. *H. influenzae* does not produce any haemolysis on blood agar. Staphylococci produce V factor; hence growth on blood agar can be promoted by providing *S. aureus* as the source of V factor. This method is known as satellitism. Hence, blood agar with *S. aureus* streak is routinely used to culture and identify *H. influenzae*.

Chocolate Agar

V factor is released from erythrocytes by heating blood agar at 80–90 °C in order to prepare chocolate agar. The released V factor from the erythrocytes supplements the growth of colonies. Chocolate agar containing penicillin and bacitracin is used to isolate *H. influenzae* from clinical specimens, which consist of normal bacterial flora.

Fildes Agar or Levinthal Agar

These are clear and transparent media used for primary isolation and culture of *H. influenzae*. These are prepared by boiling and filtering a mixture of blood and nutrient broth (Levinthal agar) or adding a peptic digest of blood to nutrient agar (Fildes agar).

H. influenzae is catalase and oxidase positive. It ferments glucose and galactose, but does not ferment lactose, sucrose and mannitol. It reduces nitrates to nitrites.

Other Properties

Susceptibility to Physical and Chemical Agents

H. influenzae is a delicate bacillus. It is readily killed within 30 min by moist heat at 55 °C. It dies within a few days at 4 °C as well as in culture plates and dies in less than 2 days in dried clinical secretions. Bacteria are sensitive to commonly used disinfectants and also to desiccation. Cultures are difficult to maintain due to autolysis. Cultures may be preserved by frequent subcultures on chocolate agar. Lyophilisation is a good method for the long-term preservation of *H. influenzae* culture.

H. influenzae Antigens

Three major surface antigens are present in *H. influenzae*. They are capsular polysaccharide antigen, outer membrane protein (OMP) and lipooligosaccharide.

Capsular polysaccharide antigen is the major antigenic determinant of encapsulated *H. influenzae*. This antigen is present in the capsulated strain, which produces a capsule that is a

Table 40.2 Virulence factors of *Haemophilus influenzae*

Virulence factors	Biological functions
Capsular polysaccharide	Polyribosylribitol phosphate (PRP) of the capsule is antiphagocytic It resists phagocytosis of the bacteria by leucocytes
Lipopolysaccharide	Causes meningococcal inflammation
IgA1 protease	Causes break down of IgA Facilitates colonisation of *Haemophilus influenzae* on mucosal surface
Pili	Helps in adherence of *Haemophilus influenzae* to epithelial cells

polysaccharide in nature. This antigen is not found in unencapsulated strains of *H. influenzae*. The polysaccharide antigen confers type specificity to the organism and is the basis for the grouping of organisms into six antigenic serotypes. These six antigenic serotypes are designated as a, b, c, d, e and f. Type b capsular polysaccharide is a host-protective antigen containing polyribosylribitol phosphate (PRP), hence it is used in several PRP vaccines and PRP–protein conjugate vaccines. These proteins induce IgG, IgM and IgA antibodies which are bactericidal, opsonic and protective. *H. influenzae* serotype b was responsible for more than 95% of all invasive *Haemophilus* infections before introducing *H. influenzae* type b (Hib) vaccines. But after introducing the Hib vaccine, specifically directed against *H. influenzae* serotype b, most of the diseases caused by this serotype have been reduced. After the vaccination era, serotypes c and f and non-encapsulated, non-typeable *H. influenzae* are responsible for most *H. influenzae* diseases.

Outer membrane protein (OMP) antigens show considerable variations. The OMP antigens of Hib have been classified into 13 subtypes.

Lipo-oligosaccharide is antigenically complex.

Pathogenesis

Haemophilus are the obligate bacteria present in the mucous membranes of humans and certain species of animals.

Virulence Factors

H. influenzae produces the following virulent factors:

Capsular polysaccharide is the major virulence factor in Hib. This polysaccharide capsule, which contains ribose, ribitol and phosphate, known as PRP, is antiphagocytic. It resists phagocytosis of the bacteria by leucocytes. Loss of capsule leads to loss of virulence of the bacteria. Other virulent factors include **lipid A lipopolysaccharide, IgA1 protease** and **Pili,** functions of which are listed in Table 40.2.

Pathogenesis of *Haemophilus influenzae* Infections

The encapsulated strains, such as strain type b, are usually associated with invasive conditions, such as pneumonia, meningitis, septicaemia, cellulitis, septic arthritis, etc. (Fig. 40.1).

H. influenzae enters the human host by the respiratory route. Pili and non-pilous adhesions of the bacteria mediate colonisation of *H. influenzae* in the oropharynx or nasopharynx. Lipid A lipopolysaccharide impairs ciliary function, leading to damage to the respiratory mucosa. A large bacterial load or the presence of a concomitant viral infection can potentiate the bacterial infection that invades the mucosa and enter the bloodstream. The clearance of bacteremia is determined by the presence of antibodies, complement components and phagocytes. The absence of anti-PRP antibodies contributes to bacterial multiplication. Subsequent high-grade bacteraemia leads to the dissemination of bacteria to various sites, including meninges, subcutaneous tissues, joints and even pleura and pericardium. Bacteraemia usually precedes Hib meningitis and other invasive Hib diseases. Direct extension of infection from the sinuses or ears is rare. Meningitis occurs due to inflammation, oedema and increased cerebrospinal fluid (CSF) pressure. Brain parenchymal invasion is rare.

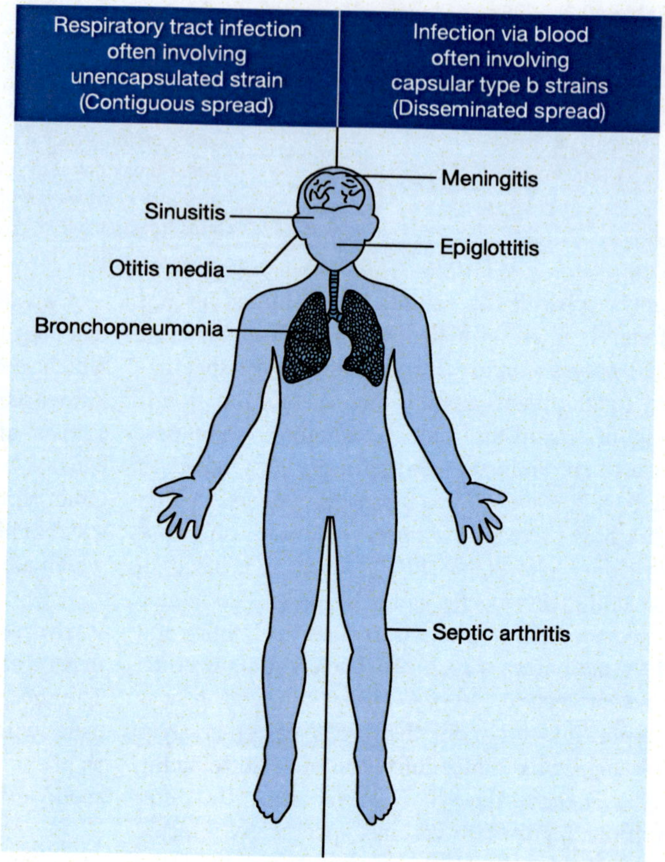

Fig. 40.1 Schematic diagram showing infections caused by *Haemophilus influenzae*

Non-capsulated or non-typeable influenza strains cause infection by direct extension from the colonised sites. The spread of bacteria by direct extension to sinuses causes sinusitis, to the Eustachian tube causes otitis media and down the respiratory tract results in bronchitis and pneumonia.

Clinical Syndromes

Infections Caused by Encapsulated *Haemophilus influenzae*

Meningitis is the most serious manifestation of Hib infection. It occurs primarily in children of 2 months to 2 years of age. Altered mental state and fever are the most common symptoms. Headache and photophobia are usually present in older children. The mortality rate is above 90% in untreated children. Hib meningitis is rare in adults. **Laryngo-epiglottitis** is the second most common infection of *H. influenzae* and is a life-threatening emergency. Since the Hib vaccine era, this condition is seen in children of 3--18 months of age. This condition is characterised by cellulitis and obstructive laryngeal oedema. **Pneumonia** typically occurs in infants and is clinically indistinguishable from other bacterial pneumonia. The condition is frequently accompanied by empyema and meningitis. **Bronchitis** is an acute exacerbation of chronic bronchitis and bronchiectasis. *H. influenzae* is considered a pathogen in association with pneumococci. Hib can cause **suppurative lesions**, such as epiglottitis, pericarditis, cellulitis and septic arthritis. Endophthalmitis, cervical adenitis, osteomyelitis and endocarditis are less common invasive infections caused by Hib.

Infections Caused by Non-encapsulated *Haemophilus influenzae*

Non-encapsulated *H. influenzae*, which colonises up to 80% of individuals along with *Streptococcus pneumoniae*, is the most common cause of **otitis media**. It is a major causative agent of **conjunctivitis** in older children, next only to *S. pneumoniae*. This is the most common cause of **community-acquired bacterial pneumonia** in adults.

Laboratory Diagnosis

Specimens

CSF and blood are the specimens of choice for the diagnosis of meningitis. One to two millilitres of CSF are usually collected from the patients for performing various tests. In meningitis, the CSF shows pleocytosis (mean 4000–5000 WBC/mL) with a predominance of neutrophils. It shows decreased CSF glucose level and elevated CSF protein. Blood is cultured to diagnose cellulitis, epiglottitis, arthritis or pneumonia caused by Hib. Throat swab, pus and aspirates from joints, middle ear and sinuses are the other specimens collected depending on the lesions caused by *H. influenzae*.

All the specimens are collected under strict aseptic conditions in sterile containers. Since *H. influenzae* is very sensitive to low temperatures, the clinical specimens are never refrigerated but are kept in an incubator at 37 °C before transporting to the laboratory. The specimens are transported immediately to the laboratory for processing and better isolation of the pathogen.

Microscopy

Gram-staining smear of the CSF shows small, Gram-negative coccobacilli in more than 80% of the CSF from cases of untreated meningitis (Fig. 40.2). Gram-stained smear is also useful as a rapid diagnostic method to demonstrate *H. influenzae* in lower respiratory tract disease and arthritis.

Culture and Identifications

Detection of *H. influenzae* in blood, CSF or other body fluids by culture confirms the diagnosis of *H. influenzae* infection. CSF culture is useful in cases of meningitis. On chocolate agar or Levinthal agar, *H. influenzae* colonies appear as 1–2 mm, smooth and opaque colonies after 24 h of incubation.

H. influenzae on blood agar can also be detected by **satellites**. In this procedure, a suspected isolate of *H. influenzae* is streaked on a blood agar plate. Then *S. aureus* is streaked across the same blood agar plate and incubated at 37 °C for 18–24 h. After incubation, the colonies of *H. influenzae* nearer to the *S. aureus*

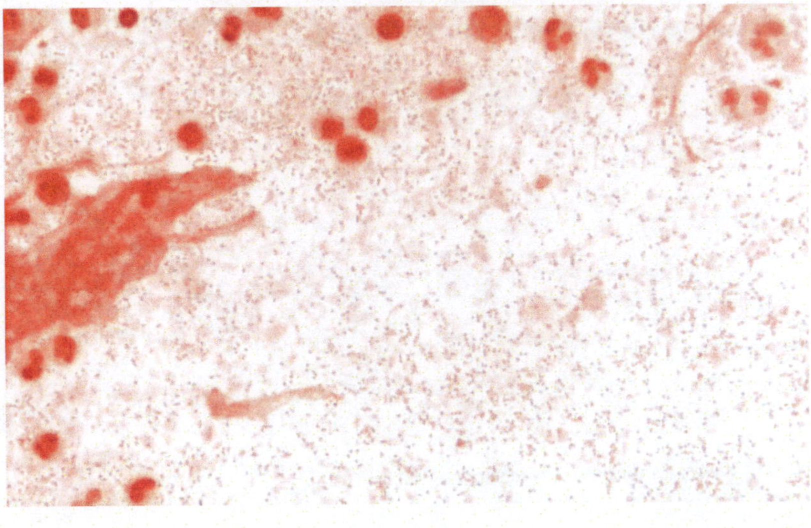

Fig. 40.2 Gram-stained smear showing Gram negative coccobacilli of *Haemophilus influenzae* (×1000) (https://commons.wikimedia.org/wiki/File:Haemophilus_influenzae_sputum_1000x_edited.jpg)

are larger than those away from it. This phenomenon is known as satellitism. This demonstrates that the V factor is available in increased concentration near the staphylococcal colony and a lower concentration away from it (Fig. 40.3).

Blood culture is useful in cases of epiglottitis and pneumonia.

Small, pleomorphic, Gram-negative coccobacilli of the bacterial isolates are catalase and oxidase positive and demonstrate satellitism on blood agar, followed by typing.

Typing

The serotyping of encapsulated *H. influenzae* is carried out by several methods, including agglutination reaction, Quellung reaction, co-agglutination and enzyme-linked immunosorbent assay (ELISA) methods.

Quellung reaction is carried out by using antisera prepared against type-specific polysaccharides. Fresh isolates from infected patients or primary isolates can be typed by this method. **DNA probe** is used currently for typing encapsulated *H. influenzae*. The probe hybridises with a fragment to chromosomal DNA, which codes for the production of capsular antigen. However, non-capsulated strains cannot be typed; hence they are called non-typable strains.

Antigen Detection

Detection of *H. influenzae* PRP capsular antigen in the CSF and urine is a rapid and sensitive method for diagnosing Hib disease. These antigens can be detected by CIEP, LAT, Co-A and ELISA. The disadvantage of the antigen detection test is that it can detect only Hib infection, but not other capsular serotypes and non-encapsulated strains of *H. influenzae* (because all the strains lack capsules, they do not have capsular polysaccharide antigen).

Molecular Detection

Quantitative real-time polymerase chain reaction (PCR) is the preferred choice for detecting the bacterium. Primers targeting omp P6, rnpB, 16S rRNA and bexA genes etc. are commonly used. A commercial kit for RT-PCR-based detection of *H. influenzae* has been developed.

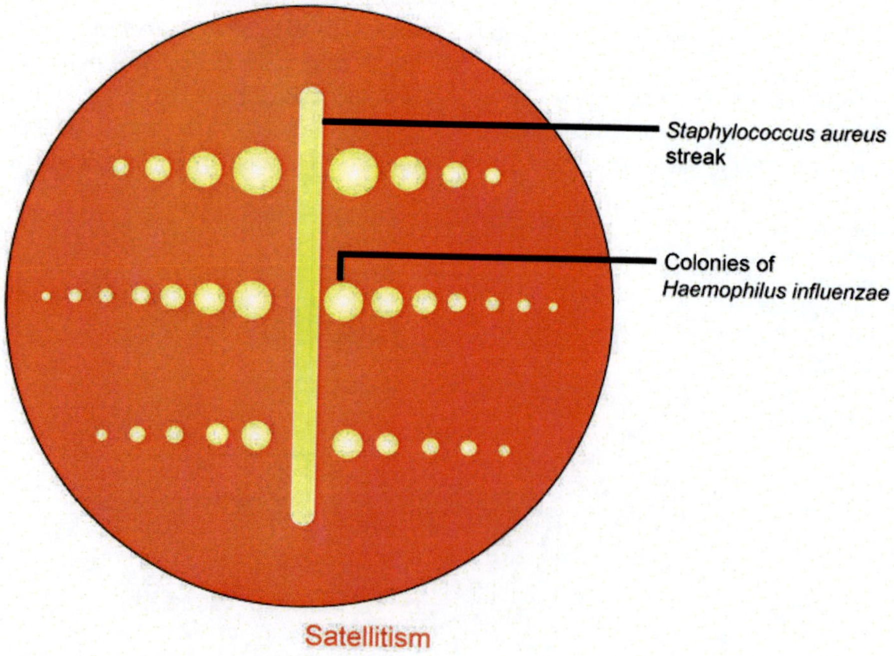

Fig. 40.3 *Haemophilus influenzae* showing satellitism

Treatment

Antibiotics are the mainstay of treatment. *H. influenzae* is susceptible to sulphonamides, chloramphenicol, ciprofloxacin, ampicillin, cefotaxime and ceftazidime. Cefotaxime and ceftriaxone are the initial drugs of choice to treat Hib meningitis. These antibiotics given parenterally to patients with uncomplicated meningitis for 7–14 days are effective.

Penicillins are useful in managing mucosal infections caused by non-encapsulated *H. influenzae*. As many as 25–50% of isolates produce beta-lactamase; therefore, they are resistant to this class of drugs. Beta-lactamase-producing oral antibiotics, with activity against *H. influenzae*, include trimethoprim-sulfamethoxazole, cefuroxime axetil, cefixime, clarithromycin, azithromycin and ciprofloxacin. These drugs are given for 10 days in otitis media cases and for at least 14 days for sinusitis.

Epidemiology

Geographical Distribution

Hib was a leading cause of infections among children worldwide before Hib vaccines became available. After the availability of the Hib vaccine, the frequency of invasive Hib disease has reduced remarkably even in developing countries. The frequency of Hib infection has come down to only 2–3 cases/100,000 children below 5 years. So, most Hib infection now occurs in non-immune children due to incomplete vaccination or poor antibody response to the vaccine and in elderly people.

In developed countries with the successful implementation of the Hib vaccination program, the disease caused by other serotypes of encapsulated bacteria and non-typeable strains of *H. influenzae* is increasingly recognised.

Habitat

H. influenzae is a strict human pathogen. Humans are the only natural hosts. Non-encapsulated strains are regular commensals of the nasopharyngeal mucosa and the oropharynx. They colonise at the site in virtually all children within the first few months of life. In contrast, *H. influenzae* serotype b is rarely found in the upper respiratory tract or, if found, is present in a few children (1–5% of cases).

Reservoir, Source and Transmission of Infection

Human cases are an important reservoir of infection. The infected nasopharyngeal mucosal secretion is the most common source of infection. Transmission is by inhalation of respiratory tract droplets or by direct contact. Transmission to neonates occurs through the maternal genital tract. Non-typable *H. influenzae* biotype 4 can colonise the genital tract and is a major cause of invasive disease in neonates.

Prevention and Control

Antibodies directed against the PRP component of the capsule play a primary role in conferring immunity. Newborns have a low risk of infection because of the passive transfer of maternal antibodies. By the age of 5 years, most children have naturally acquired antibodies. Antibodies against this capsule are protective. It promotes bacterial phagocytosis and complement-mediated bactericidal activities. These antibodies are produced during natural infection or on vaccination with purified PRP or passive transfer of maternal antibodies from mother to children. The risk of meningitis and epiglottitis is much higher in patients with no anti-PRP antibodies or patients with a low level of complement.

Vaccines

Now routinely given to infants and children, Hib conjugate vaccine is highly effective. The vaccine has reduced the prevalence of invasive Hib disease. The vaccine elicits a protective antibody response and prevents Hib disease by reducing pharyngeal colonisation with Hib. However, the Hib conjugate vaccine is not effective against disease caused by non-typable *H. influenzae*.

Currently, three types of Hib vaccines are available. These vary in a. the protein carrier used, b. the molecular size of the saccharide and c. the method of conjugating the protein to the saccharide. These vaccines are (a) HbOC (mutant diphtheria toxin as the carrier protein) vaccine, (b) PRP-T (tetanus toxoid as the carrier protein) vaccine and iii) PRP-OMP (major OMP of *Neisseria meningitidis* serogroup B as the carrier protein) vaccine.

The HbOC or PRP-T vaccine is injected in a three-dose regimen, while the PRP-OMP vaccine is given in a two-dose regimen at 2-month intervals, and the first dose is given at the age of 2 months. A booster dose is given at the age of 12–15 months. These vaccines are well-tolerated. Occasionally, they cause redness and swelling at the vaccination site in 10–15% of infants, especially after the first dose rather than after subsequent injections.

Chemoprophylaxis

Chemoprophylaxis by rifampicin is used to reduce the colonisation of Hib disease in children with high risk. Such high-risk group includes children below 2 years in a family or day care centre where a systemic disease is reported.

Haemophilus ducreyi

H. ducreyi was first described by Auguste Ducrey in 1889. It is an obligate bacterium. *H. ducreyi* grows poorly on most media, requires X factor, but not V factor, and is biochemically inert. It produces very small colonies (0.5-mm diameter) on chocolate agar after prolonged incubation of 72 h in the presence of CO_2.

In Asia and Africa, *H. ducreyi* is an important causative agent of the sexually transmitted disease called **soft sore or chancroid**. The chancroid is seen most commonly in men, and the condition remains asymptomatic in women. The incubation period varies from 5 to 7 days, after which a papule with an erythematous surface appears on the genitalia or the perianal area. Subsequently, the lesion ulcerates and becomes painful. The condition is frequently associated with inguinal lymphadenopathy. Through the process of autoinoculation, skin lesions may occur. Infection with *H. ducreyi* will not confer immunity, but hypersensitivity reactions occur upon inoculation of dead bacilli. Gram-stain smears from the ulcerative lesions show pleomorphic, Gram-negative bacilli but may appear Gram-positive with frequent bipolar staining. Usually, these bacilli arrange themselves in small groups or whorls or parallel lines (railway track or school of fish appearance). Sulphonamide and tetracycline are the drugs of choice, but cotrimoxazole, erythromycin, ciprofloxacin or ceftriaxone may be used as alternative options.

Haemophilus aphrophilus

H. aphrophilus is present as a commensal in the mouth and the throat. It forms yellowish, high convex, large colonies (1.5-mm diameter) on chocolate agar after 24 h of incubation. It requires only X factor, but not V factor, and as the name suggests, it also requires a high concentration of CO_2 for growth. *H. aphrophilus*, which is present as a commensal in the mouth, can spread to the bloodstream and infect the previously infected heart valve, causing sub-acute endocarditis. It can also cause brain abscess, pneumonia and sinusitis.

Haemophilus parainfluenzae

H. parainfluenzae is an opportunistic pathogen in the upper respiratory tract. Occasionally, the bacteria may cause endocarditis, conjunctivitis and bronchopulmonary infection. It forms relatively bigger colonies than those of *H. influenzae* on chocolate agar, which are opaque and yellowish-white. It differs from *H. influenzae* in requiring the V factor only and not the X factor, fermenting sucrose but not d-xylose.

Haemophilus aegyptius

H. aegyptius was earlier known as Koch–Weeks's bacillus. Koch observed this in 1883 in a case of acute conjunctivitis and subsequently was first

Table 40.3 Human infections caused by *Actinobacillus* species

Bacteria	Diseases
Actinobacillus actinomycetemcomitans	Sub-acute bacterial endocarditis particularly in patients with pre-existing valvular heart diseases and oral diseases, such as oral abscess, poor oral hygiene, periodontitis, etc.
Actinobacillus ureae	Opportunistic infections
Actinobacillus hominis	Opportunistic infections
Actinobacillus lignieresii	Bite wound infection

cultured by Weeks in 1887. Recent DNA-based studies revealed that *H. aegyptius* is similar to non-capsulated *H. influenzae*; hence, now it is renamed *H. influenzae* biotype *aegyptius*, and they are categorised as *H. influenza* biotype III. This organism is cosmopolitan in distribution and causes a highly contagious form of acute conjunctivitis (pink eye). It is prevalent in tropical and subtropical regions; at times, its infection results in epidemics.

H. aegyptius as the causative agent of **Brazilian purpuric fever (BPF)** has also been confirmed recently. The BPF starts with conjunctivitis, which progresses to fulminant septicaemias, especially in infants and children with a very high fatality rate. The BPF was first recorded in Brazil in 1984. Since then, it has become an endemic disease in the whole of South America.

Haemophilus haemolyticus

H. haemolyticus is present as a commensal in the upper respiratory tract. It is a non-pathogen and does not cause any human disease. On blood agar, it produces a zone of haemolysis around the colonies and may be mistaken for haemolytic streptococci. It requires both X and V factors.

Actinobacillus (Aggregatibacter)

Actinobacillus are small, non-motile and facultative anaerobic Gram-negative bacilli. The bacteria grow slowly on culture media requiring 2–3 days of incubation. *Actinobacillus* are present in the oropharynx of humans and animals and cause various infections. The genus *Actinobacillus* consists of at least six species, and *Actinobacillus actinomycetemcomitans* is the most important human pathogen. Other species cause very rare human infections (Table 40.3).

Differential features of important *Haemophilus*, *Pasteurella* and *Actinobacillus* species are summarised in Table 40.4.

Hacek Group of Bacteria

The acronym HACEK refers to a group of fastidious slow-growing Gram-negative bacilli—*Haemophilus* species (*H. parainfluenzae*, *H. aphrophilus* and *H. paraphrophilus*), *A. actinomycetemcomitans*, *Cardiobacterium hominis*, *Eikenella corrodens* and *Kingella* species. These organisms are normally present in the oropharynx, which can cause severe infections, such as endocarditis, under certain conditions. Since they are fastidious organisms to grow, they are usually incriminated as a frequent cause of culture-negative endocarditis. These organisms also cause various diseases, including otitis media, conjunctivitis, pneumonia, septic arthritis, osteomyelitis, urinary tract infections, wound infections, brain abscess, periodontal infections, etc. Infective endocarditis is a major clinical condition caused by the HACEK group of organisms.

Laboratory diagnosis of infective endocarditis depends on blood culture, followed by subculture on blood agar and chocolate agar supplemented with vitamins and other nutrients. The media are incubated in a humid atmosphere of 5% carbon dioxide or anaerobically for 24–72 h. Blood culture bottles are incubated for a minimum of 14 days, followed by a terminal subculture before declaring it negative. Antibiotics are the mainstay of treatment. Due to delays in blood culture, antibiotics are usually started empirically.

Table 40.4 Differential features of important *Haemophilus, Pasteurella* and *Actinobacillus* species

Organism	Growth factor requirement		Requirement for CO_2	Fermentation of		Catalase	Haemolysis on horse blood agar
	X	V		Glucose	Sucrose		
Haemophilus influenzae	+	+	–	+	–	+	V
Haemophilus ducreyi	+	–	–	–	–	–	–
Haemophilus aegyptius	+	+	–	+	–	–	–
Haemophilus parainfluenzae	–	+	–	+	+	V	–
Haemophilus haemolyticus	+	+	–	+	–	+	+
Haemophilus parahaemolyticus	–	+	–	+	+	+	+
Haemophilus aphrophilus	+	–	+	+	+	–	–
Haemophilus paraphrophilus	–	+	+	+	+	–	–
Pasteurella multocida	–	–	–	+	+	+	–
Actinobacillus actinomycetemcomitans	–	–	+	+	–	+	–

Note: *V* variable

Ceftriaxone is the drug of choice. Ampicillin combined with gentamicin is also equally useful. However, of late, resistance to beta-lactams has been reported among the HACEK group of organisms.

Case Study

A 2-year-old child was brought to paediatric OPD with a history of fever, headache, convulsions and vomiting. Physical examination showed the presence of neck rigidity and pus in the external auditory canal. Further enquiry revealed that the child had a bout of upper respiratory tract infection in the previous week. A provisional diagnosis of bacterial meningitis was made. CSF protein was raised, but glucose was decreased. CSF was sent for Gram stain, culture and sensitivity and biochemical analysis. Gram stain revealed the presence of pus cells and pleomorphic Gram-negative coccobacilli. *H. influenzae* was isolated by culture.

- Describe the methods followed in a laboratory for isolation of the bacteria.
- Can meningitis caused by *H. influenzae* be prevented by vaccination? Discuss the recent vaccines with their advantages and disadvantages.
- List other infections caused by this organism.
- Which is the antibiotic of choice to treat these bacteria?

Genus *Bordetella*

Genus *Bordetella* includes the bacteria that are extremely small, strictly aerobic, non-fermentative and Gram-negative coccobacilli. The genus *Bordetella* consists of 15 recognised species, of which three are responsible for human diseases (Table 40.5).

Bordetella pertussis

B. pertussis (*pertussis*, Latin for intense cough) is the causative agent of whooping cough, an infectious bacterial illness that affects the respiratory tract. Whooping cough is one of many diseases that a vaccine can prevent.

Table 40.5 Human infections caused by *Bordetella* species

Bordetella species	Diseases
Bordetella pertussis	Whooping cough or pertussis
Bordetella parapertussis	Milder form of pertussis
Bordetella bronchiseptica	Pertussis-like symptoms in humans

Properties of the Bacteria

Morphology

B. pertussis are extremely small ovoid coccobacilli measuring 0.2–0.5 × 1 μm. Characteristically, they demonstrate pleomorphism in their morphology. They are Gram-negative occurring in singles or pairs and are non-motile. Toluidine blue staining of the bacteria demonstrates characteristic bipolar metachromatic granules. The freshly isolated strain of *B. pertussis* possesses a poorly defined capsule and fimbriae. In culture smears, the bacilli are arranged in loose clumps with clear spaces in between, giving a thumbprint appearance.

Culture and Biochemical Properties

B. pertussis is a strict aerobe with an optimum temperature for growth at 35 °C. The bacteria are nutritionally fastidious. They do not grow on common laboratory media, such as blood agar and nutrient agar. Even on blood agar, the bacteria grow slowly and require 3–6 days to form pinpoint colonies. The bacteria are usually grown on a rich medium supplemented with charcoal, starch, blood, albumin and growth factors, such as nicotinamide. The latter is essential for the growth of bacteria. Blood or albumin present in the medium is used not to provide nutrition for the growth of bacteria but to neutralise toxic substances, such as fatty acids present in the agar.

Bordet–Gengou agar with 15–20% of blood is a common medium used for primary isolation of *B. pertussis*. This medium consists of glycerol, potato, agar and 15–20% blood. After 48–72 h of incubation, *B. pertussis* produces small, smooth, opaque, greyish white refractile colonies resembling bisected pearls or mercury drops (Fig. 40.4). A hazy zone of haemolysis is present around the colonies. Confluent growth of *B. pertussis* on the medium typically presents an aluminium paint appearance. **Charcoal agar with 10% blood** has also been used for primary isolation of *B. pertussis*. The bacterium does not grow on MacConkey medium, but other *Bordetella* species grow.

B. pertussis is oxidase positive and catalase positive but biochemically inert. The bacteria do not ferment carbohydrates. The bacteria also do not produce indole, do not reduce nitrate, do not split urea and do not utilise citrate.

Other Properties

Susceptibility to Physical and Chemical Agents

B. pertussis is killed by heating at 55 °C for 30 min. It is also killed by drying and also by standard disinfectants. If left on culture plates, the bacteria die within a few days, but they can survive in dry droplets for up to 5 days and 3 days on cloth.

Bordetella pertussis Antigens

B. pertussis possesses two types of antigens. These are genus-specific somatic O antigen and strain-specific capsular K antigen. The O antigen is a protein in nature, heat stable and is of a single antigenic type. It is present in most of the strains of Bordetella. The K antigen is heat-labile. These K antigens are of different types and are used for differentiating B. pertussis isolates in epidemiological studies.

A total of 14 agglutinating factors or agglutinogens have been identified. *B. pertussis* strains are classified into various types based on agglutinogens on their capsule. Factor 7 is found in all strains of *B. pertussis* (1–7, 13), *Bordetella parapertussis* (7–10, 14) and *Bordetella bronchiseptica* (7–13).

Fig. 40.4 *Bordetella pertussis* growing on Charcoal Agar supplemented with Cephalexin (https://commons.wikimedia.org/wiki/File:Bordetella_pertussis_on_Charcoal_Agar_supplemented_with_Cephalexin_-_Detail.jpg)

Pathogenesis

B. pertussis colonises the cilia of the mammalian respiratory epithelium. The organism usually does not invade the tissues but cause disease primarily by toxins.

Virulence Factors

B. pertussis produces several virulence factors (Table 40.6). These include (1) filamentous haemagglutinin (FHA), (2) pertussis toxin, (3) invasive adenylate cyclase, (4) lethal toxin and (5) tracheal cytotoxin.

FHA

This is the most important virulence factor, which mediates the attachment of *B. pertussis* to the ciliated epithelial cells of the respiratory tract. It is a large protein (220 kDa) that forms filamentous structures on the cell surface when seen under an electron microscope hence named FHA. FHA binds to galactose residues on a sulphated glycolipid called sulfatide, present on the surface of the ciliated cells. This adhesin also binds to CR3, a receptor present on the surface of polymorphonuclear leucocytes. FHA and pertussis toxin haemagglutinins of *B. pertussis* are also responsible for secondary infections. In contrast, *H. influenzae* and *S. pneumoniae* adhere to these substances, which facilitates the binding of these bacteria to the respiratory epithelium. This phenomenon, therefore, is called piracy of adhesins. The intracellular survival of the *B. pertussis* inside the leucocyte protects the bacterium from humoral antibodies against FHA. Mutations in the FHA structural gene reduce the ability of *B. pertussis* to colonise. Recently, the structural gene for FHA has been cloned, thereby raising the possibility of the production for use in a vaccine.

Pertussis Toxin

It is a major virulent factor in the initial colonisation of the bacteria in the tracheal epithelium. Pertussis toxin is a 105 kDa protein; it is

Table 40.6 Virulence factors of *Bordetella pertussis*

Virulence factors	Biological functions
Filamentous haemagglutination (FHA)	Binds to galactose residues on a sulphated glycolipid called sulfatide, present on the surface of the ciliated cells Binds to CR3, a receptor on the surface of polymorphonuclear leucocytes Mediates the attachment of *Bordetella pertussis* to ciliated epithelial cells of the respiratory tract
Pertussis toxin	Causes adhesion of *Bordetella pertussis* to tracheal epithelium S2 subunit binds to glycolipid present on ciliated epithelium S3 subunit binds to ganglioside receptor on the surface of phagocytic cells S1 subunit inhibits the eukaryotic adenyl cyclase, killing of phagocytes and migration of monocytes
Invasive adenylate cyclase (haemolysin)	It has adenylate cyclase activity and a binding component that mediates attachment to the host cell surface
Lethal toxin	Causes inflammation and lethal necrosis around the site of adherence of the bacteria
Tracheal cytotoxin	Kills ciliated respiratory cells Stimulates the release of cytokine IL-1 and inhibits ciliary movement

secreted into the extracellular fluid and bound to the cell wall. Pertussis toxin, typically composed of two major units, A and B, consists of six subunits S1, S2, S3, S5 and two S4 subunits. **Unit A** is an enzymatically active moiety and consists of one toxic subunit, S1. **Unit B** is responsible for binding the toxin to epithelial cells and consists of five binding subunits (S2, S3, S5 and two S4 subunits). The S2 subunit of the B unit binds to a glycolipid present on the ciliated epithelium. The S3 subunit of the B unit binds to the ganglioside receptor on the surface of phagocytic cells. Following the binding of the B subunit to the epithelial cells, A subunit inserts through the membrane and is released into the cytoplasm. By its enzymatic activity, the A subunit transfers the ADP ribosyl moiety of the NAD to the membrane-bound regulated protein G1, which normally inhibits the eukaryotic adenyl cyclase. The gene for pertussis toxin has been cloned and sequenced. The toxin is toxoided for use in the component vaccines.

Invasive Adenylate Cyclase or Haemolysin

This is a 45 kDa protein bound to a cell or released into the medium. This toxin is produced by all three *Bordetella* species causing human infections. The adenylate cyclase was originally identified as haemolysin because it causes lysis of blood cells. This enzyme is responsible for producing a zone of haemolysis around *B. pertussis* colonies on blood agar.

Tracheal Cytotoxin

Tracheal cytotoxin is not a classical exotoxin because it is not composed of protein but is composed of a low-molecular-weight peptidoglycan fragment that kills ciliated respiratory cells. It also stimulates the release of cytokine IL-1 and is responsible for fever. A tracheal cytotoxin is toxic for ciliated respiratory epithelium and is responsible for inhibiting ciliary movement. The toxin specifically interferes with DNA synthesis, thereby preventing the regeneration of damaged cells.

In addition to these virulent factors, B. pertussis toxins produce adhesins, enzymes and toxins, whose role in the pathogenesis of whooping cough remains unclear. For example, the bacteria possess **lipopolysaccharide (LPS)** in their cell wall, which is unusual. It is heterogeneous with two distinct types of LPSs, lipid A and lipid X. Both lipid A and lipid X can activate the alternative complement pathway and stimulate the release of cytokines. The role of this unusual LPS in the pathogenesis of whooping cough is unknown. **The lethal toxin** is a heat-labile protein. It is a 102 kDa protein composed of four subunits: two with a molecular weight of 24 kDa. The lethal toxin causes inflammation and lethal necrosis around the adherence site of *B. pertussis*.

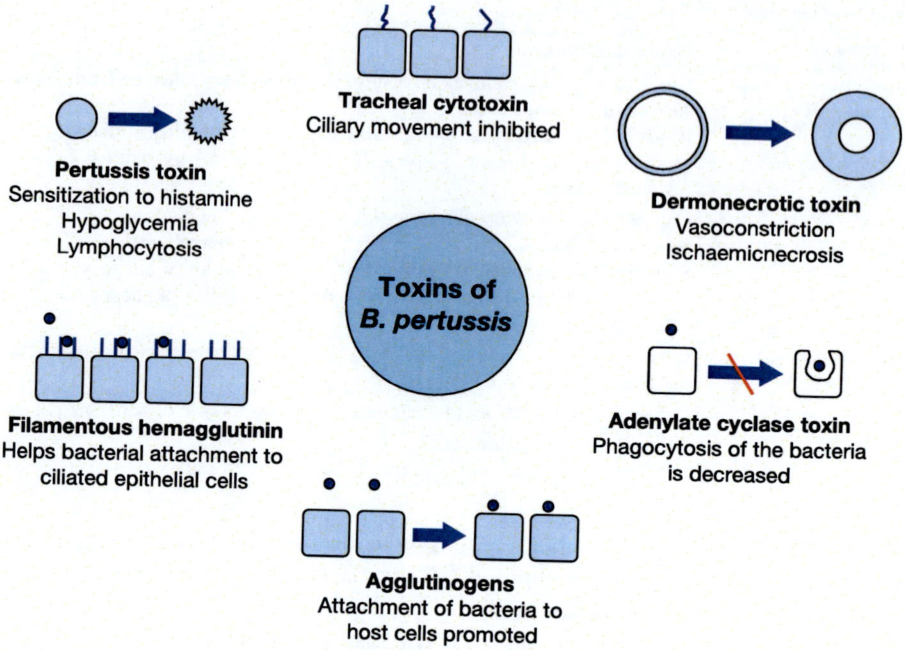

Fig. 40.5 Mechanisms of actions of various toxins produced by *Bordetella pertussis*

The toxin is lethal in high doses in experimental infection of mice. The role of the toxin in the disease is unknown. Similarly, *B. pertussis* produces at least **two other types of adhesins and pertactin**, but their role in adherence and pathogenesis remains unclear. Pertactin is present as an outer membrane protein (OMP) in all virulent strains. Antibodies against pertactin can be seen in infected or immunised children.

Pathogenesis of Whooping Cough

B. pertussis causes pertussis or whooping cough in two stages. The **first stage of colonisation** is characterised by the multiplication of bacteria and the production of localised tissue damage. Infection with *B. pertussis* is initiated by the attachment of the bacteria to ciliated epithelial cells of the respiratory tract. The attachment is mediated primarily by two bacterial adhesions, FHA and pertussis toxin. The **second stage of toxaemia** is characterised by the manifestation of systemic toxicity produced by many toxins. The bacteria multiply at the site of infection, where they produce several toxins that paralyse the tiny cilia and cause inflammation of the respiratory tract (Fig. 40.5). Subsequently, many other toxins produced by *B. pertussis* also contribute to the pathogenesis of the disease. Adenylate cyclase toxin also inhibits leucocyte chemotaxis, phagocytosis and killing of the bacteria.

Clinical Syndromes

B. pertussis causes whooping cough, the most common vaccine-preventable disease among children.

Whooping Cough

Whooping cough is also known as pertussis. The incubation period varies from 7 to 10 days. The disease typically has three stages (Fig. 40.6): catarrhal stage, paroxysmal stage and convalescent stage.

The catarrhal stage is the first stage of whooping cough. This stage resembles an upper respiratory tract infection with a running nose, nasal congestion, sneezing, malaise and occasional cough. A low-grade fever may also be seen. This phase typically lasts 1–2 weeks. This

Fig. 40.6 Clinical stages of whooping cough

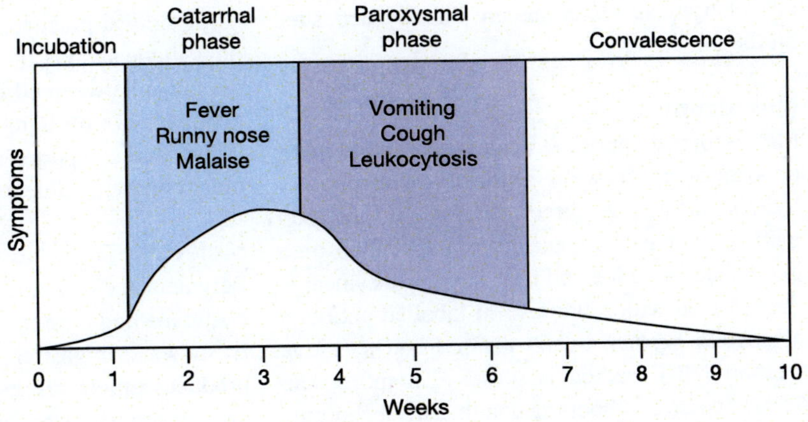

stage is highly infectious because of the production of many bacteria during this stage of the disease.

The paroxysmal stage is the second stage of the disease. This stage is characterised by classic whooping cough or paroxysms. Each paroxysm consists of a series of repetitive, intense, and drawn out bouts of cough followed by an inspiratory whoop. The attack is more frequent at night, with an average of 15 attacks in 24 h. The paroxysm may frequently end with vomiting and exhaustion. During this stage, mucus production in the respiratory tract is more frequent and partially responsible for obstructing the respiratory tract. Duration of this stage is highly variable, lasting within 1–6 weeks up to 10 weeks.

The convalescent stage is the third stage of the disease, which may last for weeks or months. It is characterised by a chronic cough that becomes less paroxysmal. **Many complications** mark this stage. These include subconjunctival haemorrhage, respiratory distress, secondary bacterial pneumonia and neurological complications, such as convulsions. These neurological complications may result in permanent sequelae, such as epilepsy, paralysis, blindness and deafness.

Laboratory Diagnosis

Specimens

Nasopharyngeal aspirates are the specimen of choice. These samples should be collected during the first stage or early in the second stage of disease because the organisms are most abundant in the respiratory secretions during these stages. Also, the specimen for culture should be collected before administering antibiotics. The specimens are usually collected using either calcium alginate or Dacron swabs but not by cotton swabs because certain fatty acids inhibit the growth of bacteria in the cotton, which are toxic to *B. pertussis*.

The specimens for culture are collected by swabs or by coughing on a culture medium. The specimen is collected by the **per-nasal or nasopharyngeal swab** method. In this method, a sterile swab on a flexible wire is passed gently along the floor of the nasal cavity, and the mucus and pus are collected by the swab. **West post-nasal or oral swab** is usually employed to collect posterior pharyngeal wall secretions through the oral cavity. Contamination with saliva should be avoided for better results. The swab containing the mucus and pus is inoculated immediately on freshly prepared medium, such as Bordet–Gengou or charcoal horse blood agar at the patient's bedside. In case of delay, they are transported in 0.2–0.5 mL casamino acid solution (pH 7.2), modified Stuart transport medium or Regan–Lowe transport medium.

In **the Cough plate method**, specimens are directly coughed out by the patient on a culture medium during a bout of spontaneous or induced coughing by an infected child. The culture plate is held 10–15 cm before the patient's mouth. During the process of coughing by children, infected nasopharyngeal secretions are directly deposited on the medium.

Microscopy

Microscopic diagnosis is made by **Gram staining** or a **direct fluorescent antibody** of respiratory secretions. In the direct fluorescent antibody method, a smear is made from a specimen on a glass slide, air-dried and heat-fixed. This smear is then stained with a fluorescent labelled antibody against *B. pertussis*. It is a rapid and sensitive method. This method is positive in more than 75% of cases of whooping cough. This technique, however, may show occasional false-positive reactions due to other bacterial infections.

Culture and Identification

The specimens are cultured on freshly prepared Bordet–Gengou medium or charcoal horse blood agar medium and incubated in a moist environment at 35 °C. Incubation even up to 7 days is required because the colonies are observed only after three or more days of incubation.

Several characters identify *B. pertussis* colonies. They show bisected pearl or mercury drop colonies on Bordet–Gengou agar. They are small, round Gram-negative coccobacilli and are catalase and oxidase positive. They are biochemically inert do not ferment carbohydrates.

Serology

Indirect haemagglutination and ELISA are frequently used to demonstrate IgG and IgA antibodies against FHA and IgG antibodies against pertussis toxin in the patient's sera. Serodiagnosis is only recommended for patients with symptoms compatible with pertussis, such as prolonged coughing with paroxysms and/or whooping or choking. ELISA for demonstrating specific secretory IgA antibodies in nasopharyngeal secretions is also useful, especially in culture-negative cases. Commercial kits or in-house validated protocols are used for the purposes. A significant rise in the titre of antibodies between acute and convalescent serum in paired sera samples by these tests suggests recent infection.

In clinical practice, diagnosis is mostly based on a single sample by using a cut-off value be in the range between 60 IU/ml and 75 IU/ml. However, it is not meaningful in neonates and young infants and is reported to have diagnostic value in older children, adolescents and adults only.

Molecular Detection

Polymerase chain reaction (PCR) is a new diagnostic method used to diagnose whooping cough. It shows a high sensitivity of 80–100% and produces timely results. Aspiration or swabbing the posterior nasopharynx is used to collect samples for PCR, wherein throat swabs and anterior nasal swabs are not used due to the lower quantity of DNA recovery. PCR tests are recommended for adolescents and adults with coughing less than 3 weeks. After the fourth week, bacterial DNA rapidly diminishes, increasing the chances of false-negative results.

Both conventional and real-time PCR are used for diagnosis. However, the latter took the commercial interests for being more affordable and less labour intensive. Several commercial RT-PCR kits are available that targets virulence and structural genes like IS481, IS1001, IS1002, ptxS1, ptx-pr, RecA, Plp1 and Plp2 etc.

Other Laboratory Tests

There is a marked leucocytosis and relative lymphocytosis in pertussis. The total leucocyte count usually ranges from 20,000 to 30,000 mm^{-3} along with increased lymphocytes. There is no increase in erythrocyte sedimentation rate except during secondary infection.

Treatment

Macrolides like erythromycin, clarithromycin and azithromycin are the drug of choice for treating pertussis in persons 1 month of age and older. Tetracycline, chloramphenicol and ampicillin are also effective. These antibiotics effectively eradicate the bacteria from the respiratory tract and reduce the duration of infectivity of the patients. However, the treatment protocol of pertussis is primarily supported by good nursing care.

Epidemiology

Whooping cough is predominantly a disease of childhood.

Geographical Distribution

Despite the vaccines and widespread global vaccination, *B. pertussis* has remained an endemic disease and sometimes leads to outbreaks and death. WHO estimated more than 151,000 cases of pertussis globally in 2018? Many recent studies have indicated a resurgence of the disease in many countries irrespective of economic status.

Habitat

B. pertussis colonises the cilia of the mammalian respiratory epithelium of infected children.

Reservoir, Source and Transmission of Infection

Whooping cough is highly contagious. Humans are the only known reservoir of *B. pertussis* infection. An infected child, particularly during the catarrhal stage of disease, is the important source of infection. The infection is transmitted primarily by inhaling airborne droplets coughed out by the infected patients. The infection is also transmitted by contact with hard surfaces recently contaminated with respiratory secretions or droplets from an infected child. Secondary attack rates are the highest (75–100%) among unimmunised household contacts. In pseudo-whooping cough, the clinical manifestations mimicking whooping cough may also be seen in respiratory infections caused by other pathogens such as *Mycoplasma pneumoniae* and adenovirus.

Initially, whooping cough was considered a disease of young children (<5 years), but now this disease is also seen in older children and adults. They contribute to 25% of cases of whooping cough. This is due to failing immunity over time in adults, even in vaccinated.

Prevention and Control

An attack of whooping cough gives lifelong immunity to a child. The secondary attack is extremely rare. Specific serum antibodies confer immunity against *B. pertussis*. Vaccination of infants and children with pertussis vaccine is very effective. The vaccination is usually given to children below 7 years of age.

Vaccines

Whole-cell inactivated vaccine has been used for the past five or more decades to protect children against whooping cough. This vaccine is a Merthiolate-killed bacterial cell suspension, administered as part of DTP (diphtheria, tetanus and pertussis) or triple vaccine. These vaccines are effective in 80–85% of the population in eliminating asymptomatic pertussis. These vaccines are given in three doses intramuscularly at an interval of 4–6 weeks before 6 months of age. A booster follows this at the end of the first year. The first dose is given at 6 weeks. Unfortunately, more than 20% of children who receive whole-cell vaccines suffer from mild side effects. Convulsions occur on 0.1% of infants soon after receiving the vaccine.

Acellular vaccines have been developed from purified components of *B. pertussis*. The multicomponent vaccine contains FHA, pertussis toxin, fibrin agglutinin and pertactin. These components are believed to confer protective immunity against whooping cough. This new acellular pertussis vaccine has been found to confer protective immunity similar to that of the whole-cell vaccine. However, the most important advantage is that these vaccines have very low side effects than whole-cell vaccines.

Bordetella parapertussis

B. parapertussis is a less frequent cause of whooping cough, with only about 5% of cases of whooping cough. They relatively cause mild disease. *B. parapertussis* organisms differ from *B. pertussis* by their ability to grow on nutrient agar and to produce larger colonies, but the former do not produce pigment (Table 40.7). They are antigenically different from those of *B. pertussis*. Pertussis vaccine does not confer any protection against *B. parapertussis*.

Table 40.7 Differential characteristics of *Bordetella* species

Characteristics	*Bordetella pertussis*	*Bordetella parapertussis*	*Bordetella bronchiseptica*
Growth on			
MacConkey agar	No growth	Variable growth	Positive growth
Sheep blood agar	No growth	Positive growth	Positive growth
Bordet–Gengou medium	3–6 days	1–2 days	1 day
Other features			
Motility	Non-motile	Non-motile	Motile
Oxidase test	Positive	Positive	Negative
Urease test	Negative	Positive	Positive
Nitrate to nitrite	Negative	Negative	Positive
Citrate utilisation	Negative	Variable	Positive

Bordetella bronchiseptica

B. bronchiseptica is a motile bacterium by virtue of the presence of peristriate flagella. The bacteria grow on nutrient agar and are antigenically related to *B. pertussis* and *Brucella abortus*. It differs from other *Bordetella* species in its biochemical and other properties (Table 40.7). *B. bronchiseptica* is responsible for causing a very small proportion (0.1%) of cases of whooping cough.

Case Study

A 3-year-old child was brought to paediatric OPD with a history of continuous dry cough for the past 15 days. On examination, episodes of continuous violent coughing and a typical 'whoop' were found. Peripheral blood smear showed leucocytosis. Chest X-ray was normal. Gram stain showed the presence of oral commensals. Ziehl–Neelsen stain from gastric aspirate was negative for acid-fast bacteria.

- What is the likely diagnosis?
- What are the likely complications of this condition?
- How will you diagnose the condition in the laboratory?
- Which is the antibiotic of choice for the treatment of this case?
- What are the vaccines available for the prevention of the condition?

Brucella

41

In 1886, British army doctor David Bruce experimentally infected the monkeys by inoculating small microorganisms isolated frosm the spleen of Malta fever case. He was able to see similar disease manifestations in the monkeys also. Hence, this organism was named *Brucella melitensis*, *Brucella* after Bruce and *melitensis* after Malta (Roman name *Melita*). Further, in 1905, Zammit elucidated the transmission of *B. melitensis* to humans by consuming contaminated goat milk. Similarly, other major species of *Brucella* were identified by Bang in 1897 (*Brucella abortus* causes contagious abortion in cattle) and Traum in 1914 (*Brucella suis* isolated from pigs in the United States). All these three species are known to cause human infections.

Genus *Brucella*

The genus consists of 25 recognised species, four of which are human pathogens. These are *B. abortus*, *B. melitensis*, *Brucella canis* and *B. suis* (Table 41.1). Human infections with *Brucella neotomae* and *Brucella ovis* have not been described. Moreover, each *Brucella* species has a characteristic predilection for infecting certain animal species where it causes a chronic disease that persists for life.

Brucellosis is one of the most important worldwide zoonoses affecting livestock and humans. Considering the potential to pose a severe threat to both human and animal health, three strains of Brucella, i.e. *B. abortus, B. melitensis* and *B. suis,* have been categorised as select agents by HHS and USDA of the USA.

Properties of the Bacteria

Morphology

Brucellae are Gram-negative but counterstain poorly and require relatively more time for staining. They are small coccobacilli measuring 0.5–0.7 × 0.6–1.5 μm in size. These are arranged singly, sometimes in pairs or short chains. They do not produce spores, flagella or capsule.

Culture and Biochemical Properties

Brucellae are strict aerobes. Most biovars of *B. abortus* require 5–10% CO_2 for their growth. No growth occurs in aerobic conditions. They grow at an optimum temperature of 37 °C and pH range of 6.6–7.4. Brucellae can grow on simple media, but their growth is slow and scanty. Growth is improved by adding serum, blood, liver extract and glucose. *Brucella* organisms grow best on trypticase soy-based or other enriched media. **Blood agar** and **trypticase soy agar** are the media of choice. **Tryptose agar, trypticase soy agar, serum potato infusion agar** and **serum dextrose agar** are the other media used for Brucellae culture. The addition of bacitracin, polymyxin and cycloheximide to these media makes them selective. On these solid media,

Table 41.1 Human infections caused by *Brucella* species

Brucella species	Diseases
Brucella abortus	Mild brucellosis with rare suppurative complications
Brucella melitensis	Severe acute brucellosis with many complications
Brucella suis	Chronic brucellosis with suppurative complications
Brucella canis	Mild brucellosis with rare suppurative complications

Brucella spp. produce small, moist, translucent and glistening colonies after three or more days of incubation. In liquid media, growth is uniform.

Brucellae are catalase and oxidase positive (except for *B. ovis* and *B. neotomae*, which are oxidase negative). They reduce nitrate to nitrite and have variable urease activity—*B. suis* becomes urease positive within 30 min, whereas *B. abortus* becomes positive within 1–2 h. Some *Brucella* species produce H_2S, whereas some species do not produce H_2S. They do not ordinarily ferment any sugar. They do not produce indole and are MR and VP test negative. Brucella does not utilise citrate.

Classification

The genus *Brucella* is considered monospecific based on gene sequencing and nucleic acid homology. However, the traditional classification based on the natural hosts and differences in pathogenicity is used in practice. *Brucella* species show a high degree of host specificity. *B. melitensis* infects goats and sheep; *B. abortus*, cattle; *B. suis*, swines and *B. canis*, dogs and foxes.

Brucella species and biovars are differentiated by the following characteristics (Table 41.2): (1) growth in the presence of aniline dyes, such as basic fuchsin dyes or thionine; (2) requirement for CO_2; (3) ability to use glutamic acid, ornithine, lysine and ribose; (4) ability to produce H_2S; (5) agglutination by antisera against specific lipopolysaccharide (LPS) and vi) susceptibility to lysis by bacteriophage.

Biotypes

B. abortus consists of seven biotypes: *B. melitensis* three biotypes and *B. suis* five biotypes. H_2S-producing *B. suis* strains are American strains, and those not producing H_2S are Danish strains.

Phage Types

Tbilisi (Tb) is the reference phage used for phage typing of *Brucella* species. This phage lyses *B. abortus* both at routine test dilution (RTD) and at 10,000 RTD. The phage lyses *B. suis* at 10,000 RTD only but does not lyse any strain of *B. melitensis*.

Other Properties

Susceptibility to Physical and Chemical Agents

The brucellae are rapidly killed at 60 °C in 10 min; hence they are killed by pasteurisation in the milk. They are also killed by disinfectants, such as 1% phenol in 15 min. They are sensitive to direct sunlight and acid. The bacteria survive for 10 days in refrigerated milk, for 1 month in ice cream and 4 months in butter. *B. melitensis* remains viable in urine for 6 days, dust for 6 weeks and water or soil for 10 weeks.

Brucella Antigen Structure

Based on these O polysaccharide linkages, somatic antigens of *Brucella* are classified into two major types: **antigen A** and **antigen M.**

B. abortus has the highest concentration of **A antigen** (about 20 times as much as M antigen), whereas *B. melitensis* has the highest concentration of M antigen (about 20 times that of **A antigen**). *B. suis* has an intermediate antigenic pattern. *Brucella* species show antigenic cross-reaction with *Salmonella* serotypes N (O: 30 antigens), *Escherichia coli* (O: 116, O: 157), *Vibrio cholerae*, *Pseudomonas multocida*, *Yersinia enterocolitica* and *Francisella*

Table 41.2 Differential features of *Brucella* species

Species	Animal reservoir	Biotypes	CO_2 requirement	H_2S requirement	Urease production	Growth in the presence of basic Fuchsin 1: 50,000	Growth in the presence of Thionine 1:25,000	Growth in the presence of Thionine 1:50,000
Brucella melitensis	Goats, sheep	1	–	–	Variable	+	–	+
	Goats, sheep	2	–	–	Variable	+	–	+
	Goats, sheep	3	+(–)	–	Variable	+	–	+
Brucella abortus	Cattle	1	+(–)	+	1–2 h	+	–	–
	Cattle	2	+(–)	+	1–2 h	–	–	–
	Cattle	3	+(–)	+	1–2 h	+	+	+
	Cattle	4	–	+	1–2 h	+	–	–
	Cattle	5	–	–	1–2 h	+	–	+
	Cattle	6	–	+(–)	1–2 h	+	–	+
	Cattle	9	–	+	1–2 h	+	–	+
Brucella suis	Swine	1	–	+	0–30 min	–	+	+
	Swine	2	–	–	0–30 min	–	–	+
	Swine	3	–	–	0–30 min	+	+	+
	Swine	4	–	–	0–30 min	+	+	+
	Swine	5	–	–	0–30 min	–	+	+
Brucella canis	Dogs		–	–	0–30 min	–	+	+
Brucella neotomae	Wood rat		–	+	Rapid	–	–	–
Brucella ovis	Sheep		–	–	–	–	+	+

tularensis. Agglutination and other features of *Brucella* species are summarised in Table 41.3.

Pathogenesis

Brucella species have a predilection for intracellular growth, hence may be demonstrated inside phagocytes. The intracellular location of the *Brucella* explains the relative resistance of the bacterium to chemotherapy.

Virulence Factors

LPS of *Brucella* is the principal virulence factor of *Brucella* spp. The LPS is unique in that the lipid A part of the LPS contains 16-carbon long fatty acids but lacks the 14-carbon myristic acid typical of lipid A of Enterobacteriaceae. The polysaccharide O portion of LPS contains an unusual sugar, 4,6-dideoxy-4-formamido-alpha-D-mannopyranoside, which is expressed either as a homopolymer alpha-1,2-linked sugar (A-type) or as 3-alpha-1,2- and 2-alpha-1,3-linked sugar (M type). **The intracellular location of the bacteria** makes them resistant to killing in serum and also by phagocytes. *B. melitensis* is the most virulent species, as it can survive in phagocytic cells and multiply in them, producing a large number of bacteria. It causes the most severe and acute cases of brucellosis.

Pathogenesis of Brucellosis

Brucellae enter through abrasion or cut in the skin, conjunctiva, respiratory tract and gastrointestinal tract. Shortly after gaining entry to the body, brucellae are rapidly ingested by

Table 41.3 Agglutination and other features of *Brucella* species

Species	Biotypes	Agglutination by monospecific serum against			Lysis by phage	
		A	M	R	RTD	10,000 RTD
Brucella melitensis	1	−	+	−	−	−
	2	+	−	−	−	−
	3	+	+	−	−	−
Brucella abortus	1	+	−	−	+	+
	2	+	−	−	+	+
	3	+	−	−	+	+
	4	−	+	−	+	+
	5	−	+	−	+	+
	6	+	−	−	+	+
	9	−	+	−	+	+
Brucella suis	1	+	−	−	−	+
	2	+	−	−	−	+
	3	+	−	−	−	+
	4	+	+	−	−	+
	5	−	+	−	−	+
Brucella canis		−	−	+	−	−
Brucella ovis		−	−	+	−	−
Brucella neotomae		+	−	−	−	+

polymorphonuclear leukocytes, which are attracted to the site of inflammation. However, inside the leukocytes, the brucellae are not killed due to the presence of enzyme superoxide dismutase, O polysaccharide of LPS and nucleotide-like substances produced by the bacteria. Brucellae that are not killed by leukocytes spread from the site of infection to the local lymph nodes, which drain the site of infection. Inside the lymph glands, the bacteria multiply and are released into the bloodstream following the rupture of the cells and are phagocytosed by the macrophages.

Bacteria present in the macrophages are then carried to the organs of the reticuloendothelial system, such as the liver, spleen, bone marrow, lymph nodes and kidney. In these organs, brucellae multiply in phagosomes of macrophages by producing adenine and guanine monophosphate. These compounds inhibit phagolysosome fusion, oxidative burst activity and tumour necrosis factor production. Apart from the organs of the reticuloendothelial system, any other organ system such as the central nervous system (CNS), heart, joints, genitourinary system, pulmonary system, and skin can also be affected by brucellosis.

Immunity in brucellosis, like other obligate intracellular pathogens, is primarily cell-mediated. Activated macrophages are primarily responsible for killing the bacteria by producing interferon-gamma and other cytokines. It is suggested that antibody, complement and macrophage-activating cytokines also play a supportive role in controlling the multiplication of extracellular bacteria or in early infection. Humoral immunity is characterised by the production of antibodies against the LPS of the bacteria. However, even in the presence of opsonic activity, these serum antibodies against LPS do not provide any protective immunity against bacteria.

Clinical Syndromes

Brucellosis is a systemic disease in humans that can involve almost any organ system. The condition is known by various names, such as Malta fever, Mediterranean fever, undulant fever, etc.

Brucellosis

The incubation period may range from 3 days to several weeks. Human infections may be of three types, as follows:

Acute Brucellosis

The condition is seen in approximately 50% of patients. Fever is an important symptom. The fever is intermittent and undulant (hence known as *undulant fever*) and can be associated with relative bradycardia. In untreated patients with the adverse disease, the patients can show respiratory tract symptoms (20%), bone and joint symptoms (20–60%), neuropsychiatric symptoms and gastrointestinal tract symptoms.

Chronic Brucellosis

The condition develops in incompletely treated patients. This condition is associated with a low-grade non-bacteraemia infection with periodic exacerbations. Symptoms often last for 3—6 months and occasionally for a year or more.

Localised and Suppurative Infection

These infections are more likely caused by *B. suis*.

Complications include infections of the heart, CNS and the skin. Brucella endocarditis is the most dangerous complication and is responsible for 80% of deaths in brucellosis. Chronic meningoencephalitis is the usual manifestation of CNS infection.

Laboratory Diagnosis

Clinical diagnosis of brucellosis is very difficult due to the protean manifestation of the disease. Hence, laboratory diagnosis plays an important role in confirming the diagnosis of brucellosis.

Specimens

Blood is the specimen of choice and is collected for culture and the serological test. Bone marrow and sometimes synovial fluid and pleural fluid are also collected for culture. Specimens such as liver and lymph nodes can also be cultured to isolate *Brucella* organisms. Rarely, the bacteria can be isolated from cerebrospinal fluid (CSF), urine, sputum, breast milk, vaginal discharge and seminal fluid.

Microscopy

Gram staining is not useful for demonstrating *Brucella* organisms in clinical specimens due to their small size and intracellular location.

Culture and Identification

Isolation of *Brucella* from blood and other clinical specimens is the definitive diagnostic procedure in brucellosis. Approximately 5–10 mL of blood is collected in the 200 mL serum dextrose broth or trypticase soy broth and incubated at 37 °C under 5–10% of CO_2. After the fourth day of incubation, subcultures are made on solid media every 3–5 days for 8 weeks before declaring the culture negative.

Castaneda's Method of Blood Culture

This is a useful method to culture blood without the need for frequent subculture on a solid medium; hence, the possibility of contamination can be avoided. This biphasic medium contains both trypticase soy broth and solid trypticase soy agar slant in the same bottle (Fig. 41.1). The blood is inoculated onto the liquid broth, and the bottle is incubated in the upright position. For subculture, one does not have to open the bottle; instead, it is tilted such that liquid broth flows over the solid medium slant. It is again incubated in the upright position. In the case of positive blood culture, colonies appear on the slant. The Castaneda method of blood culture also reduces the risk of laboratory-acquired infection to both medical and paramedical staff. The sensitivity of blood cultures ranges from 30% to 50%, depending on the *Brucella* species isolated. *B. melitensis* and *B. suis* are more likely to have detectable bacteraemia and hence are readily cultured than *B. abortus*.

Bone Marrow Cultures

These are more sensitive than blood cultures. They typically are positive in the negative blood culture and serological results. **Synovial fluid culture** is positive in 50% of patients.

Fig. 41.1 Biphasic medium

Identification of oxidase-positive *Brucella* colonies is made by microscopy of Gram-stained smears of the colonies (small, Gram-negative coccobacilli), colony morphology (growth in the presence of aniline dyes, such as basic fuchsin dyes or thionine, produce small, moist, translucent and glistening colonies after three or more days), biochemical tests (ability to use glutamic acid, ornithine, lysine and ribose and to produce H_2S), agglutinate with specific anti-brucella sera and are susceptible to lysis by bacteriophage.

Serology

Serological tests are useful for diagnosing subclinical brucellosis and for acute and chronic brucellosis cases by demonstrating specific antibodies in the patient's serum. Specific brucella antibodies, both IgG and IgM antibodies, appear in the serum 7–10 days after infection. IgM antibodies persist for up to 3 months, after which these antibodies decline. Then IgG and IgA antibodies appear after 3 weeks of infection and persist for a longer time. Hence, in the acute stage or subclinical brucellosis, both IgG and IgM can be demonstrated; in chronic brucellosis, only IgG can be demonstrated, as IgM is absent. As IgG antibodies persist for many months or years, demonstrating a significant rise in the antibody titre is the definitive serological evidence of brucellosis. A fourfold increase in the titre or a single high antibody titre of 1:160 is the presumptive evidence of *Brucella* infection.

B. abortus is used as an antigen in brucella agglutination tests because the antibodies against *B. melitensis* or *B. suis* can cross-react with *B. abortus* antigen. *B. abortus* antigen not only detects specific antibodies against its own antigens but also detects antibodies produced against *B. melitensis* and *B. suis*.

Several serological tests have been developed and tested in the serodiagnosis of brucellosis. These are as follows:

The Standard Tube Agglutination Test (SAT)

The test remains the standard method and is the most commonly used serological test for diagnosing brucellosis. It detects the presence of antibodies against the LPS component of *Brucella*. The test uses killed strains of *B. abortus* as antigens and is useful for diagnosing brucellosis caused by *B. abortus, B. melitensis* and *B. suis*. This test is not useful for the infection caused by *B. canis* because this bacterium does not have O polysaccharide on its surface; hence it does not produce antibodies against LPS.

The SAT is considered positive when antibody titres are greater than or equal to 1 in 160 or when a fourfold rise titre is demonstrated in convalescent sera. A rise in titre is especially important in patients from an endemic area in which titres of 1 in 160 are not unusual in asymptomatic persons.

Nevertheless, the SAT may show false-negative or false-positive results. **False-negative SAT** is due to the prozone phenomenon. This occurs due to the high level of brucella antibodies in the serum, caused by hyperantigenemia. The prozone phenomenon gives rise to a false-negative test. This can be avoided by routine dilution of serum to at least 1:320 because inhibition of agglutination may occur at low dilution. **False-positive SAT** is due to blocking or non-agglutinating antibodies in the serum. This can be avoided by heating the serum at 55 °C for 30 min or by using 4% saline or diluent before testing the serum. These blocking antibodies can be diluted by adding antihuman globulin to the antigen serum mixture (Coombs test). **False-positive** is also due to cross-reactivity with *V. cholerae, Y. enterocolitica* serotype O9, *F. tularensis* and *Salmonella* species can cause false-positive results due to the presence of cross-reacting LPS in these organisms.

V. cholerae-induced antibodies may be removed by 2-mercaptoethanol absorption. In **the modified tube agglutination test,** 2-mercaptoethanol is added to the patient's sera before testing it. The addition of mercaptoethanol disrupts the disulphide bond of IgM; hence only IgG is detected. This modified tube agglutination test is useful for specific detection of IgG antibodies, and titres higher than 1:80 are suggestive of active infection. A high IgG antibody titre or a titre that is higher after treatment suggests relapse or persistent infection. This test is useful for the diagnosis of brucellosis during convalescence.

Indirect Immunofluorescent Assay

The test is a sensitive method for detecting brucella antibodies in the serum, which may be positive even in agglutination-test-negative cases.

Enzyme-Linked Immunosorbent Assay (ELISA)

The test is the most sensitive test for detecting IgM, IgA and IgG brucella antibodies during acute and chronic brucellosis. In this test, the LPS-depleted cytoplasmic protein antigen of *Brucella* is used as an antigen. The use of this antigen allows accurate evaluation of relapse of brucellosis in a patient because LPS antigenaemia persists for a longer duration than protein antigen. ELISA is extremely useful for diagnosing neurobrucellosis by demonstrating brucella antibodies in the CSF.

Brucella Skin Test

It is a delayed type of hypersensitivity reaction to brucella antigen. In this test, brucellin, protein extracts of the bacteria, is used as an antigen and is administered intradermally. The presence of erythema and induration of 6 mm or more within 24 h is suggestive of a positive reaction. This test is positive only in chronic brucellosis but negative in acute brucellosis. Repeated negative skin test excludes brucellosis.

Molecular Detection

PCR has been evaluated for detection and rapid diagnosis of *Brucella* species. It exhibits excellent sensitivity for diagnosing acute disease and follow-up cases.

Conventional PCR for the detection of Brucella has been in practice since the 90s. Later it was replaced by real-time PCR for ease of use. Multiplex PCR is also available to detect multiple species in a single reaction. Commercial kits are now available that have increased the use of PCR in the diagnosis of Brucella. In most cases, primers encoding 16S rRNA, outer membrane protein (omp2a, omp2b and omp31), 31 kDa immunogenic *B. abortus* protein (BCSP 31 B4/B5), 16S-23S ribosomal DNA interspace region (ITS66/ITS279) and insertion sequence (IS711) are used.

Other Tests

Bone marrow examination shows erythrophagocytosis, and CSF cultures for neuro-brucellosis show pleocytosis, increased protein level and hypoglycorrhea. Anaemia in 75% of patients, thrombocytopenia in 40% and pancytopenia in 6% of patients are other features.

Diagnosis in Animals

The diagnosis of brucellosis in animals is based on the same method employed to diagnose human infections. In addition, the culture of milk and urine from infected animals may give positive results. Rapid latex agglutination and Rose Bengal card tests are the rapid diagnostic methods used to diagnose brucellosis in the cattle population.

The milk ring test is a frequently used serological test to demonstrate antibodies in the milk of infected cattle. This is a screening test used to detect the presence of *Brucella* infection in the infected animal. In this test, a concentrated suspension of killed *B. abortus* or *B. melitensis* stained with haematoxylin is used as an antigen. This test is performed by adding a drop of coloured brucella antigen to a whole milk sample in a test tube. Then it is mixed, and mixed suspension is incubated in a water bath at 70 °C for 40–50 min. In a positive test, if antibodies are present in the milk, the bacilli agglutinate and rise with the cream to form a blue ring at the top, leaving the milk unstained. The milk remains uniformly blue in a negative test without forming any coloured ring.

Treatment

Brucellae are sensitive to several oral antibiotics and aminoglycosides. Relapse is common in therapy with a single drug. So, a combination of antibiotics is recommended whenever possible. A combination of doxycycline with rifampicin is effective. Doxycycline given in a dosage of 200 mg/day orally with the addition of streptomycin(rifampicin) in a dosage of 1 g/day given intramuscularly for 2–3 weeks is most effective for the treatment of most forms of brucellosis in adults. Most of the patients respond to a 6-week course of therapy with a combination of rifampicin and doxycycline given orally. Dosage of doxycycline for 6 weeks with a combination of streptomycin for the first 3 weeks is also effective. Relapse in brucellosis is due to inadequate therapy and not due to the development of resistance to antibiotics.

Epidemiology

Brucellosis is primarily a zoonotic infection caused by *Brucella* organisms infecting domestic animals.

Geographical Distribution

Brucella infection is worldwide in distribution. The condition is endemic in certain Mediterranean countries. More than 500,000 cases of brucellosis are reported annually, with an estimated 2.4 billion people at risk. The prevalence is increasing due to growing international tourism, trade and migration. Different *Brucella* species cause human infection in different geographical areas. *B. melitensis* is the major species causing human infection in India.

Habitat

Brucella species are primarily obligate intracellular pathogens of the reticuloendothelial system, such as lymph nodes, liver, spleen and bone marrow.

Reservoir, Source and Transmission of Infection

Brucella species cause infection in a wide range of hosts. *B. melitensis* infects goats and sheep; *B. abortus*, cattle; *B. suis*, swines and *B. canis*, dogs and foxes. The infected animals, such as cattle, goats, sheep, buffalos and swine, are the important reservoirs of infections.

The incidence of human disease is very closely related to the prevalence of infections in cattle, goats and sheep. This is also related to activities that allow exposure of humans to potentially infected animals or their products.

Human brucellosis is always acquired from animals directly or indirectly. *Brucella* is transmitted to humans by ingestion, inhalation, direct contact or accidental inoculation (Fig. 41.2). Brucellosis results primarily from **ingestion** of

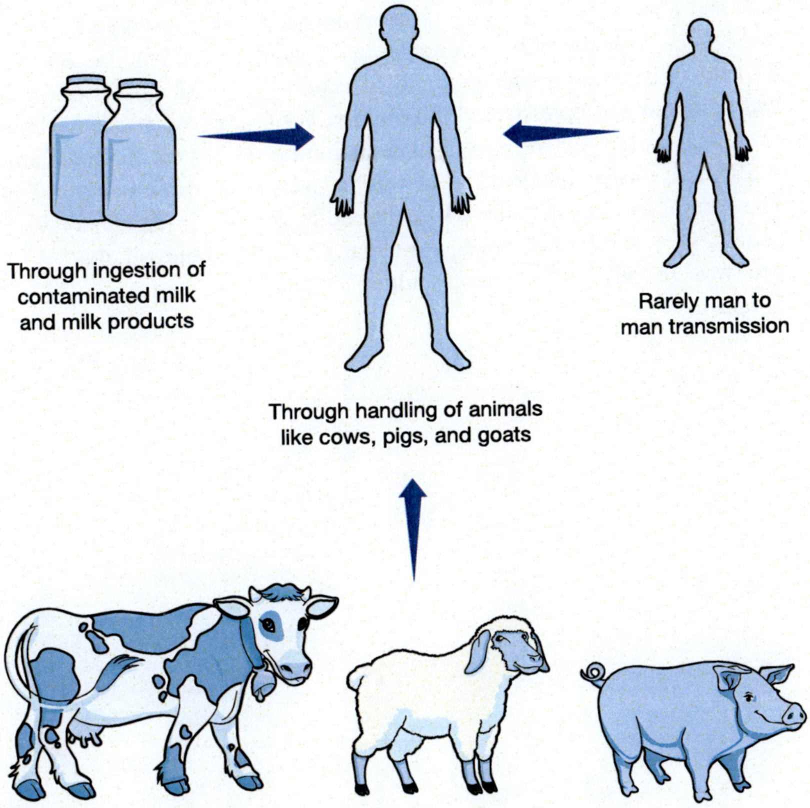

Fig. 41.2 Transmission of brucellosis

drinking contaminated unpasteurised milk or milk products. The infection can also be transmitted by drinking water or eating raw vegetables contaminated with faeces or urine of infected animals and also by eating the meat of infected animals. The infection is acquired by **direct contact** with the infected materials of septic abortion of the animals or at the time of slaughter of animals. *Brucella* species present in the infected materials (placenta, foetuses, vaginal discharge, urine or infected carcasses) enter the human host through the mucosa, conjunctiva or skin. The infection can also be transmitted by **inhaling** dust from wool or other dried material of infected animals. Infection by inhalation is important among veterinarians and laboratory workers. **Accidental inoculation** is a serious risk for laboratory workers who handle the culture of the organism. The incidence of infection in laboratory workers is reported in the Middle East countries and the United States, China, Peru, Mexico and India.

Prevention and Control

Pasteurisation or boiling milk is most important to prevent brucellosis transmission in humans, as most human infections are acquired by ingesting contaminated milk and milk products. Also, the use of protective clothing and gloves by persons handling or coming into close contact with animals prevents the transmission of infection.

Vaccination of animals is effective to control brucellosis in animals. Live attenuated *B. abortus* vaccine containing *B. abortus* 1019 strain has been used successfully to prevent infection in cattle. Vaccines are not available against either *B. suis* or *B. canis.* No vaccine is available for humans.

Case Study

A 54-year-old male patient attended Medicine OPD with a history of fever, headache, low back pain and joint pain for 1 month. History revealed that he was a shepherd by profession and was in the habit of consuming unboiled milk. The patient was admitted and investigated thoroughly. WBC count was raised. Erythrocyte sedimentation rate (ESR) was raised; the Widal test was negative. The serum was negative for HIV antibodies by ELISA. Peripheral blood smear for malaria parasite was negative. The blood culture yielded *B. abortus*.

- Diagnose the condition.
- What are the serological tests available for diagnosis of the condition?
- What are the reservoirs of the infection, and how the infection is transmitted to humans?
- How will you treat the condition?
- Is there any vaccine against the disease for human use?

Campylobacter and Helicobacter 42

The family Campylobacteraceae consists of two genera of medical importance: *Campylobacter* and *Helicobacter*.

Genus *Campylobacter*

Campylobacter infections are one of the most common bacterial infections in humans. They first obtained importance in the 1970s as a common cause of both diarrhoeal and other systemic diseases. The generic name *Campylobacter* is derived from the Greek word *kampylos*, meaning curved rod. The genus *Campylobacter* resembles the genus *Vibrio* in being motile using polar flagella and in being curved Gram-negative bacilli and oxidase positive. They, however, differ from vibrios in being microaerophilic, not fermenting carbohydrates and having a lower guanosine and cytosine content of DNA.

The genus *Campylobacter* consists of 39 species and subspecies, of which at least 15 species are known to cause infections in humans. These species cause intestinal and extraintestinal diseases (Table 42.1). *Campylobacter* is the most important veterinary pathogen. *Campylobacter jejuni* and *Campylobacter fetus* are the two most important species that cause most *Campylobacter* infections in humans.

Properties of the Bacteria

Morphology

Campylobacter is small, comma-shaped Gram-negative rods. They vary in width from 0.2 to 0.5 μm. They show rapid darting motility in cork-screw fashion utilising a single polar flagellum at one or both poles *(monotrichous/amphitrichous)*.

Culture and Biochemical Properties

Campylobacter spp. grow more slowly than other enteric bacteria. *Campylobacter* is microaerophilic, requiring an atmosphere with decreased oxygen concentration (5%) and increased carbon dioxide concentration (5–10%) for their growth. Most species, including *C. jejuni*, *Campylobacter coli*, *Campylobacter lari* and *Campylobacter hyointestinalis* grow well at 42 °C; hence they are called thermophilic bacilli.

Selective media for *Campylobacter* species are blood-based and antibiotics-containing media, such as **Skirrow, Butzler and Blaser (Campy-BAP) selective media**. *Campylobacter* on these media produces well-formed colonies after 48 h of incubation. *C. jejuni* produces moist, grey or colourless, flat and non-haemolytic colonies on this medium. Other *Campylobacter* species produce circular and convex colonies.

Campylobacter is relatively biochemically inert. They are oxidase positive, catalase positive

Table 42.1 Human infections caused by *Campylobacter* species

Campylobacter species	Diseases
Campylobacter jejuni, Campylobacter jejuni subsp. *Doyle*	Gastroenteritis, septicaemia, meningitis, proctitis and Guillain–Barre syndrome
Campylobacter fetus, Campylobacter fetus subsp. *Venerealis*	Gastroenteritis, septicaemia, meningitis and proctitis
Campylobacter coli	Gastroenteritis, septicaemia, meningitis and spontaneous abortion
Campylobacter upsaliensis	Gastroenteritis, septicaemia and abscesses
Campylobacter hyointestinalis	Gastroenteritis
Campylobacter showae	Periodontal disease
Campylobacter curvus	Periodontal disease
Campylobacter rectus	Periodontal disease
Campylobacter concisus	Periodontal disease and gastroenteritis
Campylobacter lari	Gastroenteritis and septicaemia
Campylobacter insulaenigrae	Gastroenteritis and septicaemia
Campylobacter sputorum	Enteritis, abortions in sheep
Campylobacter helveticus	Gastroenteritis and septicaemia
Campylobacter mucosalis	Mucosal infection
Campylobacter ureolyticus	Superficial ulcers, soft tissue infections, nongonococcal urethritis, perianal abscess, gangrenous lesions

(*C. jejuni* subsp. *Doyle* is an exception) and nitrate positive. *C. jejuni* is the only species that hydrolyses sodium hippurate.

Other Properties

Susceptibility to Physical and Chemical Agents

Campylobacter spp. survive at 4 °C in Cary–Blair transport medium for many weeks. *Campylobacter* spp. are sensitive to hydrochloric acid in the stomach, so conditions that decrease or neutralise gastric acid secretion reduce the amount of inoculum needed to cause disease.

Pathogenesis

Campylobacter species are invasive bacteria.

Virulence Factors

C. jejuni possesses many virulence factors, such as enterotoxins, lipopolysaccharide (LPS), adhesins and cytotoxic enzymes (Table 42.2). However, the specific role of these virulence factors in the pathogenesis of the disease is not well known.

Pathogenesis of Enteric Infections

Most human infections occur by ingesting contaminated and improperly cooked poultry meat products. Human infection follows ingestion of 1000–10,000 bacteria, but the illness is infrequent with a dose of fewer than 10,000 bacteria.

Clinical Syndromes

Campylobacter spp. infections in humans can present as either enteric infections or extraintestinal infections.

Enteric Infections

C. jejuni is the most important species associated with human enteric infections. Other *Campylobacter* species are also associated with enteric infections (Table 42.1). The clinical manifestations of enteric infections caused by all the *Campylobacter* species, including *C. jejuni*, are similar. Watery diarrhoea is the main manifestation and occurs in approximately 10% of children. The condition is also associated with abdominal pain, malaise, myalgia, headache and vomiting. The condition is self-limiting.

Table 42.2 Virulence factors of *Campylobacter* species

Virulence factors	Biological functions
Enterotoxins	Facilitate adherence to the jejunum, ileum and colon
LPS	Adhesion
PEB1	Superficial antigen that has been found to be a major adhesion protein
Adhesion proteins	Adhesion
Cytotoxic enzymes	Cytotoxicity action
S protein	Found exclusively in *Campylobacter fetus* and is the major virulence factor Inhibits C3b binding responsible for both the serum and phagocytic resistance of the bacteria

Extraintestinal Infections

Bacteraemia is the most important extraintestinal manifestation. Bacteraemia caused by *C. jejuni* is uncommon and occurs in patients with immunodeficiency, chronic disease and old individuals.

C. fetus is the most important *Campylobacter* species known to cause **perinatal infection** because of its affinity for the genital tract and its tropism for foetal tissue. Abortion, stillbirth and premature labour are the common manifestations. Premature babies show signs and symptoms of sepsis. *C. hyointestinalis* causes occasional bacteraemia in immunocompromised hosts. **Guillain–Barre syndrome** is an autoimmune disease believed to be due to molecular mimicry of terminal tetrasaccharide of LPS of *C. jejuni* and glycosphingolipid present on the surface of peripheral nerve. This is a disease of peripheral nerves and is characterised by ascending paralysis. **Periodontal disease** is caused by *Campylobacter showae, Campylobacter rectus and Campylobacter concisus.* A rare serotype of *C. jejuni* called Penner (LPS type O:19) is associated with this condition. Reactive arthritis, haemolytic uremic syndrome and Reiter's syndrome are some of the noted complications associated with *Campylobacter* infection.

Laboratory Diagnosis

Specimens

Diarrhoeic fresh stool is the specimen of choice for enteric infections. A rectal swab may also be used. Other specimens include blood, body fluids and tissues to diagnose extraintestinal infections. In case of delay, faeces or rectal swabs are transported in Cary–Blair transport medium. *Campylobacter* spp. survive for 1–2 weeks at 4 °C in this medium.

Microscopy

Presumptive diagnosis is made by examining the wet mount of stool by dark-field microscopy or phase-contrast microscopy. The bacteria are identified by their characteristic darting motility. Gram staining of the stool shows typical Gram-negative, curved rods and 'gull-wing appearance'. Faecal leucocytes and erythrocytes can also be detected in Gram-stained smear of the stool in approximately 75% of patients with *Campylobacter* enteritis.

Culture and Identification

Definitive diagnosis of *Campylobacter* enteric infection is best made on the isolation of the organism by stool culture. Fresh stool specimens collected within 24 h are inoculated directly on selective media such as Butzler medium, Skirrow medium, Preston–*Campylobacter* selective medium and Blaser medium. However, old stool specimens are first enriched in an enrichment medium, such as Preston–*Campylobacter* enrichment broth, for at least 24 h at 4 °C (cold enrichment) before inoculating on the selective media. Inoculated media are incubated in 5% O_2 and 10% CO_2 at 42 °C for 48 h. If *C. fetus* or other unusual *Campylobacter* species are suspected, stool specimens are inoculated on media without antibiotics and are incubated at 37 °C. Other extraintestinal specimens, such as blood, body fluids and tissue, can be inoculated on routine media to isolate *Campylobacter* species.

Oxidase-positive and catalase-positive *C. jejuni* produce moist, grey or colourless, flat and non-haemolytic colonies on Skirrow, Butzler and Campy-BAP selective media. It is biochemically inert and is the only species that hydrolyses sodium hippurate.

Histology

Examination by sigmoidoscopy shows widening or proctocolitis in up to 80% of patients with *Campylobacter* enteritis. The histopathological changes include faecal mucosal oedema and hyperaemia with crypt-abscess formation.

Serology

ELISA is available to demonstrate specific antibodies in the serum, and a high titre of antibodies is usually seen after the symptoms are resolved. Serology may not be useful for routine diagnosis but is useful for epidemiological studies.

Molecular Detection

Polymerase chain reaction (PCR) is used to detect enteric *Campylobacter* infection. Conventional and qPCR are used to target single species (i.e. multiplex PCR) in the sample. Common gene targets include 16S rRNA, flaA, flab, heat shock protein genes Hsp60, cytochrome oxidase ccoN, etc.

Treatment

Most *C. jejuni* infections are mild and self-limited. Therefore, they do not require antibiotic therapy. Supportive treatment is adequate. Treatment with antibiotics is recommended only for persons with fever and bloody diarrhoea and symptoms persisting for more than 7 days. Erythromycin is the antibiotic of choice. Ciprofloxacin and tetracycline are alternative antibiotics but are not recommended in young children. Recently, reports of erythromycin- and ciprofloxacin-resistant strains have increasingly been documented from many parts of the world.

Epidemiology

Campylobacter infections are extremely common worldwide.

Geographical Distribution

Campylobacter infection occurs in all age groups. *Campylobacter* enteritis is common in children younger than 5 years. In contrast, *Campylobacter* bacteraemia occurs in patients older than 65 years.

Habitat

In infected humans, *C. jejuni* inhabits the duodenum, jejunum and colon. *C. fetus* inhabits the genital tract of infected mothers.

Reservoir, Source and Transmission of Infection

Campylobacter infections are zoonotic. Infected animals are the main reservoir of infection. The animal reservoirs are the dogs, cats and other pets that carry the organism in their gastrointestinal tract. Infected animal food products are the source of infection. Most human infections occur by ingesting contaminated and improperly cooked poultry meat products. The infection also occurs by drinking contaminated raw milk or unpasteurised milk. The infection may also be transmitted from infected animals and humans by the faeco–oral route on ingestion of food and water contaminated with human or animal faeces. *C. fetus* infection may be transmitted from person to person by sexual contact.

Prevention and Control

Thorough cooking of all poultry products and strict personal hygiene (such as washing hands with soap before and after handling raw foods of animal origin) are the measures for preventing *Campylobacter* infections.

Genus *Helicobacter*

Helicobacter are *Campylobacter*-like bacteria with a spiral or helical morphology. They are motile using sheathed flagella. *Helicobacter* spp. inhabit the stomach of humans and many other mammals, such as monkeys, dogs, cats, ferrets, mice, etc. The genus *Helicobacter* consists of many species associated with human diseases

Table 42.3 Human infections caused by *Helicobacter* species

Helicobacter species	Diseases
Helicobacter pylori	Peptic ulcer, gastritis and gastric adenocarcinoma
Helicobacter cinaedi	Gastroenteritis, septicaemia and proctocolitis
Helicobacter fennelliae	Gastroenteritis, septicaemia and proctocolitis
Helicobacter canis	Gastroenteritis
Helicobacter rappini	Gastroenteritis
Helicobacter canadensis	Gastroenteritis
Helicobacter pullorum	Gastroenteritis
Helicobacter heilmannii	Gastritis

(Table 42.3). Of these, *Helicobacter pylori*, *Helicobacter cinaedi* and *Helicobacter fennelliae* are three important species that cause human infections.

Helicobacter pylori

The association of *H. pylori* with gastritis, peptic ulcer, gastric adenocarcinoma and gastric mucosa-associated lymphoid type (MALT) and B-cell lymphoma is recognised worldwide.

Properties of the Bacteria

Morphology

H. pylori is a curved, spiral or S-shaped Gram-negative bacillus measuring 3 μm in length and 0.5–0.9 μm in breadth. The organism is highly mobile and shows cork-screw motility due to the presence of a unipolar tuft of lophotrichous flagella.

Culture and Biochemical Properties

H. pylori is microaerophilic and grows in the presence of decreased oxygen concentration (5%) and increased carbon dioxide concentration (10%) in a temperature range between 30 and 37 °C. The bacteria do not grow anaerobically, grow poorly at 42 °C and do not grow at all at 25 °C. *H. pylori* grows well on freshly prepared chocolate agar and Skirrow–*Campylobacter* media. *H. pylori* produces circular, convex and translucent colonies on these media after an incubation of 3–5 days.

H. pylori produces the enzyme urease, which is almost 100 times more active than that produced by *Proteus vulgaris*. Urease production is the most important feature of *H. pylori*. *H. pylori* is catalase and oxidase positive. It is biochemically inactive and does not ferment or oxidise sugars, although it can metabolise amino acids by fermentative pathways.

Pathogenesis

H. pylori is an invasive bacterium that characteristically causes alteration of gastric acid production, gastric inflammation and tissue destruction.

Virulence Factors

H. pylori produces several virulence factors such as **urease, flagella, adhesins, Hsp-B, acid inhibitory protein** and **cytotoxin**, contributing to the disease's pathogenesis (Table 42.4). Of these, the enzyme urease is a very important factor in colonising *H. pylori* in the gastric mucosa. The enzyme produces ammonia from urea, which increases the gastric mucosa's pH near the bacterial cell. Therefore, it helps to neutralise gastric acidity and the colonisation of the organism in gastric mucosa. The enzyme also stimulates monocytes and neutrophil chemotaxis and the production of cytokines.

According to DNA-level analysis studies, *H. pylori* is more genetically diverse amongst all the bacterial species. These diversity patterns may affect the ability of the bacteria to colonise or cause disease in specific hosts. The whole-genome sequence of the bacterium revealed the presence of certain alleles of genes, such as cytotoxin-associated gene (*cag*) and vacuolating

Table 42.4 Virulence factors of *Helicobacter* species

Virulence factors	Biological functions
Urease	Helps in the colonisation of the organism in gastric mucosa; the enzyme also stimulates monocytes and neutrophil chemotaxis and the production of cytokines
Flagella	It helps the organism to penetrate the gastric mucous layer, and hence protects the bacteria from the acid environment of the stomach
Adhesins	Facilitate binding of *Helicobacter pylori* to the gastric mucosa
Enzymes	Both mucinase and phospholipase break down gastric mucus, while superoxide dismutase and catalase prevent the phagocytic killing of the bacteria
Heat shock protein (Hsp-B)	Facilitates expression of the enzyme urease
Acid inhibitory protein	Causes hypochlorhydria by blocking secretions of acid from parietal cells
Cytotoxin	Causes vacuolation in epithelial cells of the host
VacA	Causes vacuolation along with CagA

cytotoxin gene (*vac*), which plays an important role in pathogenesis.

Pathogenesis of *Helicobacter pylori* Infection

H. pylori infection begins with the colonisation of gastric mucosa, which is facilitated by many virulence factors. *H. pylori* blocks the production of acid by its acid inhibitory protein and neutralises gastric acidity via ammonia from urea by the enzyme urease. Heat shock proteins enhance the biological activity of urease further. Subsequently, *Helicobacter* passes through gastric mucosa through their active cork-screw motility and adheres to epithelial cells. The adhesion is mediated by many adhesin proteins, such as haemagglutinin, sialic acid-binding adhesin and Lewis's blood-group adhesin. *H. pylori* is protected from phagocytosis and intracellular killing by enzymes, such as elastases and superoxide dismutase. After adhesion, damage to the epithelial cells is caused by multiple factors, such as mucinases, phospholipases, vacuolating aflatoxin and the by-products of urease. Finally, *Helicobacter* infection causes atrophic and even metaplastic changes in the stomach.

Clinical Syndromes

H. pylori in humans causes peptic ulcer disease and chronic atrophic gastritis.

Peptic Ulcer Disease

H. pylori is now accepted to be the infective pathogen that causes most gastric and duodenal ulcers. Treatment of *H. pylori* infection by specific antibiotics and thereby eliminating the organism from the stomach results in healing of these ulcers and a marked reduction in the recurrence of ulcers.

Chronic Atrophic Gastritis

H. pylori infection causes chronic atrophic gastritis during the later stage of the illness. Chronic gastritis is now recognised as a risk factor for many gastric malignancies. Gastric adenocarcinoma and squamous cell oropharyngeal cancer are the most important complications of *H. pylori* infection. Gastric MALT lymphomas show a better prognosis than gastric adenocarcinoma and are found to regress after elimination of *H. pylori* by treatment.

Laboratory Diagnosis

Specimens

These include stool samples and gastric biopsy specimens.

Microscopy

H. pylori is demonstrated by microscopic examination of gastric biopsy specimen stained with Gram, Giemsa, Warthin–Starry silver and

Fig. 42.1 Positive rapid urease test shown by *Helicobacter pylori*

haematoxylin-eosin stains. A stained smear shows typical curved, spiral or S-shaped Gram-negative bacillus. All these stains show organisms of *H. pylori* adhered to the gastric mucosa. These methods are highly specific.

Culture and Identification

Diagnosis of *H. pylori* infection is made by culture of clinical specimens on freshly prepared chocolate agar and Skirrow–*Campylobacter* selective media followed by incubation at 35–37 °C in a microaerophilic environment (5% O_2, 10% CO_2 and 85% N_2) for 3–5 days. *H. pylori* produces convex, circular and large colonies on these selective media. *H. pylori* produces circular, convex and translucent colonies on freshly prepared chocolate agar and Skirrow–*Campylobacter* media.

The colonies are oxidase positive and catalase positive and show characteristic cork-screw motility. *H. pylori* produces the enzyme urease, which is almost 100 times more active than that produced by *P. vulgaris*. Urease production, detected by the urease test, is the most important feature of *H. pylori.*

Urease Test

It is the most rapid test which can be demonstrated directly in the biopsy specimen. In this test, the biopsy tissues are put in a urease indicator medium (0.5 mL urea solution with an indicator) and incubated at 37 °C. If *H. pylori* is present, the colour of the medium changes due to the change of pH of the medium within a few minutes to 2 h due to the production of ammonia. The test is highly specific (100%), but sensitivity varies from 75% to 95% (Fig. 42.1).

Faecal Antigen Test

It is an immunochromatographic test that uses monoclonal antibodies for direct detection of *H. pylori* antigen in stool samples. This test is very sensitive (94%) and specific (98%).

Serology

ELISA is used to demonstrate serum antibodies against *H. pylori* in the patient serum. However, since the antibody titres continue to remain elevated a long time after *H. pylori* eradication, the antibody-based test cannot distinguish between recent and old infections. ELISA shows high sensitivity and specificity.

Rapid Urea-Breath Test

This test is based on detecting the products of urea degraded by *H. pylori*. In this method, patients drink a beverage containing urea labelled with a carbon isotope, such as carbon 13 or 14.

After a short time, the concentration of labelled carbon is measured in the breath. The concentration is high only when urease present in *H. pylori* found in the stomach breaks down the urea. In a normal human host, the concentration of the labelled carbon in-breath is not high because the human stomach does not contain any urease. A positive urea-breath test indicates *H. pylori* infection. The breath test is very selective. The test's disadvantage is that it may show false-positive results in a) infections with coccoid forms of *H. pylori* that do not produce more urease enzymes and b) patients receiving antibiotics, such as bismuth and histamine-2 blockers, and it needs facility for isotope detection (Fig. 42.2).

Treatment

H. pylori is sensitive to several antibiotics and bismuth salts. Using a single antibiotic or an antibiotic containing bismuth is not effective. A higher success has been achieved by using a combination of bismuth, a proton-pump inhibitor (e.g. omeprazole) and one or more antibiotics (ampicillin, metronidazole, clarithromycin and tetracycline) for the treatment of gastric or peptic ulcer. Combining three types of drugs is known as triple therapy and is usually given for either 10 or 14 days. Recently, *H. pylori* strains resistant to macrolides have been reported. Also, some strains resistant to clarithromycin have been documented in children.

Epidemiology

H. pylori is ubiquitous and found worldwide.

Geographical Distribution

H. pylori infection occurs more frequently in developing countries than in developed countries.

Habitat

H. pylori colonise the gastric mucosa of healthy persons and persons suffering from peptic ulcer diseases.

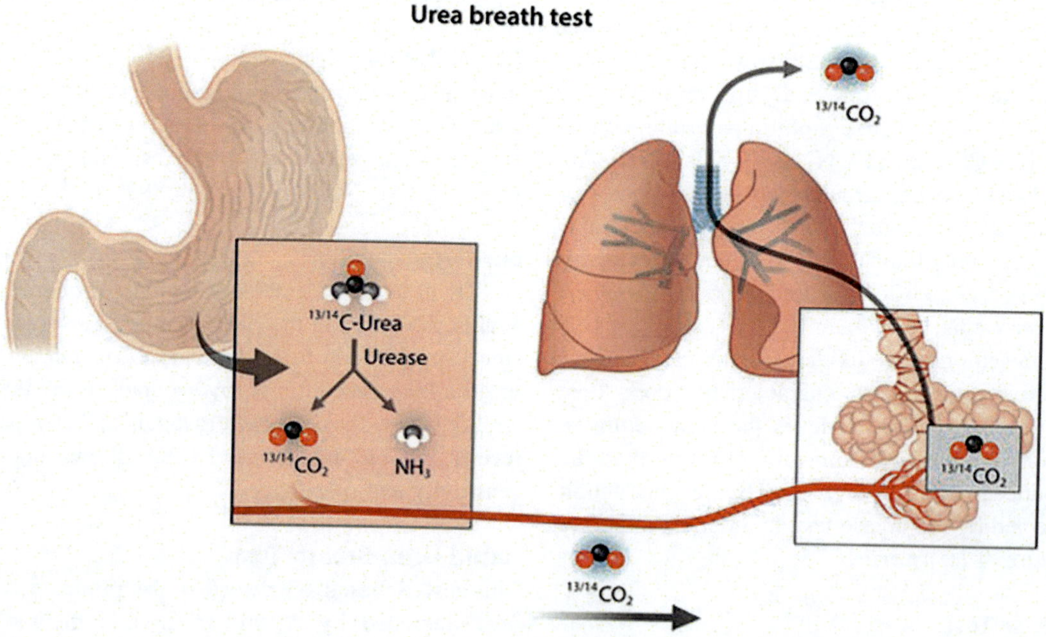

Fig. 42.2 Urea breath test (https://link.springer.com/chapter/10.1007/978-981-287-706-2_11)

Reservoir, Source and Transmission of Infection

Humans are the primary reservoirs of infection. The prevalence rate of *H. pylori* infection may differ based on geographical area and race. Poor hygiene, overcrowding and poverty facilitate transmission. The most likely route of *H. pylori* infection is either faecal-to-oral infection (from stool-to-mouth) or oral-to-oral (stomach contents transmitted from mouth-to-mouth) contact.

Prevention and Control

Improved personal hygiene is important in the prevention of the disease.

Other *Helicobacter* Species

H. fennelliae (formerly known as *Campylobacter fennelliae)* and *H. cinaedi* (formerly known as *Campylobacter cinaedi*) have been isolated from homosexual men with concurrent HIV and tuberculosis infection suffering from proctitis, proctocolitis or enteritis. *Helicobacter*-like organisms have been described in the last decade. *Helicobacter mustelae* was first described in ferrets, and subsequently, *Helicobacter hepaticus* has been detected in Syrian hamsters by using specific PCR tests. These *Helicobacter*-like organisms do not cause any infection in humans but are suggested as useful models for studying *H. pylori* infections.

Case Study

A 5-year-old child attended OPD of a hospital with his mother complaining of diarrhoea associated with abdominal cramps for the last 2 days. The child is restless with a low-grade fever and has been passing the stool mixed with blood. The child had consumed a lunch consisting of boiled rice, green salads, vegetables and tomato soup. The stool sample was sent to the microbiology laboratory for culture. *C. jejuni* was isolated from the stool through culture.

- Describe the methods for stool culture for isolation of C. *jejuni.*
- List three other *Campylobacter* species known to cause gastroenteritis.
- How will you treat this condition?
- Suggest methods for the prevention of this condition.

43 Treponema, Borrelia and Leptospira

Spirochetes have been classified in the order Spirochaetes. They vary in size from 5 to 500 μm in length. Many of them are free-living saprophytes, while a few are obligate parasites. They may be aerobic, anaerobic or facultative. Spirochetes show common morphological properties. They are thin, helical and Gram negative. They are also elongated, motile and flexible, twisted spirally along the long axis, giving these bacteria the name spirochetes (*spiral* meaning "coiled", *chait* meaning "hair"). The presence of endoflagella is the characteristic feature of spirochetes. Endoflagella is the polar flagella situated between the outer membrane and cell wall. They are responsible for the spiral shape and motility of the spirochetes. Spirochetes exhibit three types of motilities: (1) flexion and extension, (2) corkscrew-like rotatory movement and (3) translatory motion.

The order Spirochaetales is subdivided into two families: Spirochaetaceae and Leptospiraceae. The family Spirochaetaceae consists of four genera: *Spirochete*, *Cristispira*, *Treponema* and *Borrelia*. The genera *Treponema* and *Borrelia* include pathogenic species, which cause diseases in humans. Members of the genus *Cristispira* are found in molluscs, while *Spirochetes* are saprophytes found in sewage and water. Family Leptospiraceae contains only one genus *Leptospira*, which consists of species pathogenic to humans. The spirochetes causing disease in humans are summarised in Table 43.1.

Genus *Treponema*

The generic name *Treponema* is derived from the Greek word *Trepos* meaning "to turn" and *nema* meaning "thread". The treponemes are short and slender spirochetes with fine spirals and pointed ends. Some are pathogenic for humans, while others occur as commensals in the mouth, intestine and genitalia. Pathogenic members of the genus include *Treponema pallidum* (with three subspecies) and *Treponema carateum*. All these species are morphologically identical, produce the same serological responses in infected humans and are sensitive to penicillin. These species, however, differ in their (1) clinical manifestations, (2) natural history of the disease they cause and (3) epidemiological features.

As for recent nomenclature and taxonomical studies, species *T. pallidum* consists of *T. pallidum* subsp. *pallidum* causing venereal syphilis, *T. pallidum* subsp. *endemicum* causing endemic syphilis or bejel, *T. pallidum* subsp. *pertenue* causing yaws and *T. carateum* causing pinta. In this book, *T. pallidum* subsp. *pallidum* is referred to as *T. pallidum*, *T. pallidum* subsp. *pertenue* as *Treponema pertenue* and *T. pallidum* subsp. *endemicum* as *T. endemicum* for the sake of familiarity.

Table 43.1 Human infections caused by most common spirochetes

Spirochetes	Diseases
Treponema pallidum subsp. *pallidum*	Syphilis
Treponema pallidum subsp. *endemicum*	Bejel and endemic syphilis
Treponema pallidum subsp. *pertenue*	Yaws
Treponema carateum	Pinta
Borrelia recurrentis	Epidemic relapsing fever
Many *Borrelia* species	Endemic relapsing fever
Borrelia burgdorferi	Lyme disease
Leptospira interrogans	Leptospirosis

Treponema pallidum

T. pallidum is the causative agent of syphilis, the most common sexually transmitted disease.

Properties of the Bacteria

Morphology

T. pallidum is a thin, coiled spirochete. It measures 0.1 μm in breadth and 5–15 μm in length. It has a 6–10 sharp and angular coil. *T. pallidum* is too thin to be seen by microscopy in specimens stained by simple Gram or Giemsa staining. It is stained by the silver impregnation method, which makes the bacteria thickened by the deposition of silver compounds during staining. Levaditi method (Fig. 43.1) and Fontana method are examples of silver impregnation staining methods, useful for staining tissue sections and blood films, respectively.

Endoflagella is responsible for the motility of the bacteria. Three, occasionally four flagella, known as endoflagella, originate from each end of *T. pallidum*. They extend towards the opposite end of the bacteria in the space between the cell wall and outer membrane layer. This endoflagella does not extend beyond the cell wall but remains always confined within the outer membrane layer. *T. pallidum* shows three types of motilities: flexion of the whole body, backward and forward movement and rotation around the long axis. As the *Treponema* moves, it shows a series of secondary curves which appear and disappear, but its primary spiral structure remains unchanged.

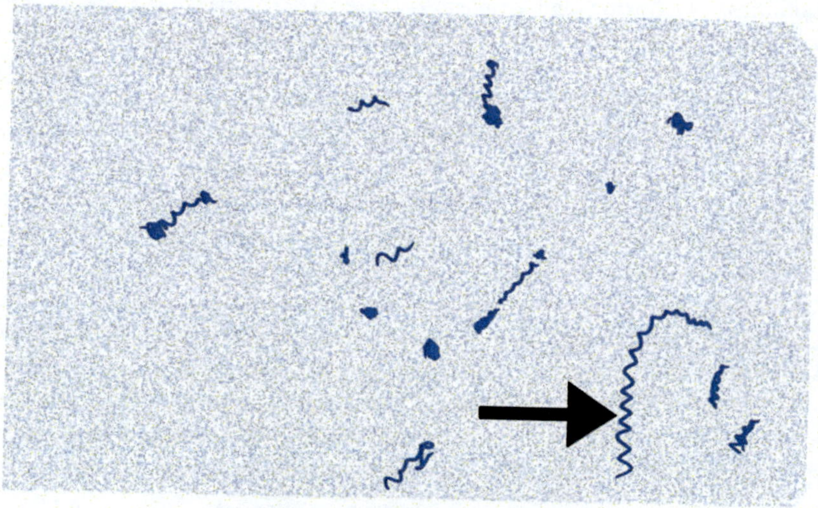

Treponema pallidum-Levaditi stain

Fig. 43.1 Levaditi silver impregnation staining method showing *Treponema pallidum* (×1000)

Dark-ground or phase-contrast microscopy is useful for demonstrating the morphology and motility of live *T. pallidum*. *T. pallidum* on electron microscopy shows a trilaminar cytoplasmic membrane surrounded by a cell wall. The latter contains peptidoglycan that confers the cell its shape and rigidity. An outer membrane layer surrounds the cell wall.

Culture

T. pallidum does not grow in artificial culture media. *T. pallidum* had been maintained for a long time by sub-culture in animals.

The Nichole strain of *T. pallidum* is a pathogenic strain that has been maintained for several decades by serial passage in rabbit testes. This strain was isolated in 1913, originally from the brain of a fatal case of a patient with general paralysis of insane. This strain is most commonly used for diagnostic and research purposes in the laboratory. Strains of *T. pallidum* have been maintained for a short period in cell lines, such as Eagle and McCoy cell lines supplemented with foetal bovine serum and reducing agent.

Treponema phagedenis (earlier known as Reiter *Treponema*) and *Treponema refringens* are examples of non-pathogenic treponemes. These treponemes can be grown in Smith–Noguchi medium or digest broth enriched with serum under strictly anaerobic conditions. Reiter strain is widely used to prepare antigen for use in group-specific treponemal tests to diagnose syphilis.

Other Properties

Sensitivity to Physical and Chemical Agents

T. pallidum is a very delicate bacterium. It is readily killed by drying or heating at 41–42 °C for 60 min, at 0–4 °C for 1–3 days. It is also readily killed on contact with distilled water, soap, arsenic compounds, mercuric compounds, bismuth compounds and common antiseptics. *T. pallidum* strains can be preserved for laboratory use by many methods. It can be stored frozen in a medium containing 5% glycerol at 70 °C or in liquid nitrogen at −130 °C for 10–15 years.

Antigenic Structure

T. pallidum is antigenically complex. Infection by the bacteria induces the production of at least three types of antibodies against (1) cardiolipin antigen, (2) group-specific antigens and (3) species-specific antigens of *T. pallidum*.

The **cardiolipin antigen** is a hapten and is chemically a phosphatidylglycerol. This hapten elicits the production of an antibody, known as reagin antibody, in the blood of a patient suffering from syphilis. Demonstration of the reaginic antibodies in the serum forms the basis of standard non-specific tests for syphilis (e.g. VDRL, Kahn and Wasserman). This cardiolipin antigen is demonstrated in *T. pallidum*. However, it is unknown whether the reagin antibody is produced by cardiolipin that is present in the *Treponema* or by the cardiolipin that is released from damaged tissues following infection by *Treponema*.

***T. pallidum* group-specific antigen** is protein in nature. It is found in *T. pallidum* and non-pathogenic cultivable treponemes, such as Reiter Treponeme. This antigen induces the production of antibodies in the serum, which is found in patients with syphilis. Reiter protein complex complement fixation test, a test used in syphilis serology, employs this antigen to detect serum antibodies in patients with syphilis.

***T. pallidum* species-specific antigen** is probably polysaccharide in nature. This antigen induces the development of antibodies, which can be detected in the serum of the patients suffering from syphilis. The specific treponemal test, such as the *T. pallidum* haemagglutination (TPHA) test, detects antibodies by using these antigens.

Immunity

Immunity in syphilis is incomplete. For example, both humoral antibodies and cell-mediated immunity prevent the formation of the chancre, a primary lesion on re-infection with *T. pallidum*, but they do not clear the organisms from the sites of infection. Their failure to kill the organism possibly (1) is due to the presence of the outer layer of the *Treponema*, which lacks immunogenic molecules or (2) may be due to downregulation of helper T cells of the T_{H1} class.

Pathogenesis

T. pallidum is a strict human pathogen.

Virulence Factors

The virulence of *T. pallidum* has recently been studied by cloning *T. pallidum* genes in *Escherichia coli* and demonstrating various proteins. Although many proteins have been isolated from pathogenic treponemes, the exact role of these proteins in the pathogenesis of the disease remains unclear.

The outer membrane protein of *T. pallidum* appears to play an important role in the virulence of the bacteria. It promotes adherence of *T. pallidum* to the surface of host cells, thereby facilitating the infection. **Enzyme hyaluronidase** produced by only pathogenic treponemes may facilitate perivascular infiltration. **Fibronectin** present in the infected host cell forms a coating on the surface of pathogenic *Treponema*, thereby preventing it from phagocytosis by macrophages.

Pathogenesis of Syphilis

T. pallidum causes disease by invasion and multiplication at the site of infection, then spreading via circulation and producing disseminated disease. The host's immune response is believed to be responsible primarily for tissue destruction and causing pathogenic lesions observed in patients with syphilis.

On sexual contact, *T. pallidum* from an infected partner is passed to another partner through an intact mucous membrane or through minor skin abrasions. The organisms invade the skin at these lesions and multiply at the site of infection. The chancre is the primary lesion, which develops at the site of infection. Subsequently, the treponemes get transmitted in the bloodstream and produce disseminated lesions (papular skin rashes, mucous patches in the oropharynx, etc.). During the course of infection, the condition may progress to the late stage of the disease, and possibly all tissues are affected by *T. pallidum*. Every stage of the disease—whether primary, secondary or late—represents localised multiplication of the treponemes and destruction of the tissues.

Clinical Syndromes

T. pallidum causes syphilis transmitted by sexual contact. Congenital syphilis and occupational syphilis are examples of non-sexual transmission. Syphilis depending on the transmission, can be venereal syphilis or non-venereal syphilis.

Venereal Syphilis

The condition is transmitted by sexual contact has an incubation period of 4 weeks, varying from 10 to 90 days. Clinical manifestation progresses through different stages (primary, secondary or tertiary) if remains untreated.

Primary Syphilis

The condition occurs within 3 weeks of sexual contact. The chancre is a primary lesion (Fig. 43.2). It is a single or solitary lesion with a surrounding areola. The edge and base of the ulcer have a button-like consistency on palpation. Hence it is called hard chancre to differentiate it from the soft chancre caused by Haemophilus *ducry*. The lesion occurs on the penis or the scrotum of 75% of men and the vulva, *cervix* or perineum of more than 50% of women suffering from syphilis. The chancre is highly infectious. The exudates of the chancre contain numerous *T. pallidum* bacteria.

Secondary Syphilis

The condition occurs 2–10 weeks after the primary chancre and is most florid 3–4 months after infection. This stage is characterised by mucocutaneous lesions, which are discrete, macular pink to red and measure 3–10 mm in diameter. The cutaneous lesions may vary from macular to papular, pustular and nodular type. The papules in the moist area of the skin of the body folds, especially about the anus and genitalia, subsequently become broad, flat and pink in colour, leading to the formation of the so-called condylomata lata. The lesions are found in the entire body, including the palms, sole and other sites. This condition is associated with mild symptoms of headache, nausea, fever and pain in the bones. Painless generalised lymphadenopathy is seen in nearly 85% of the patients.

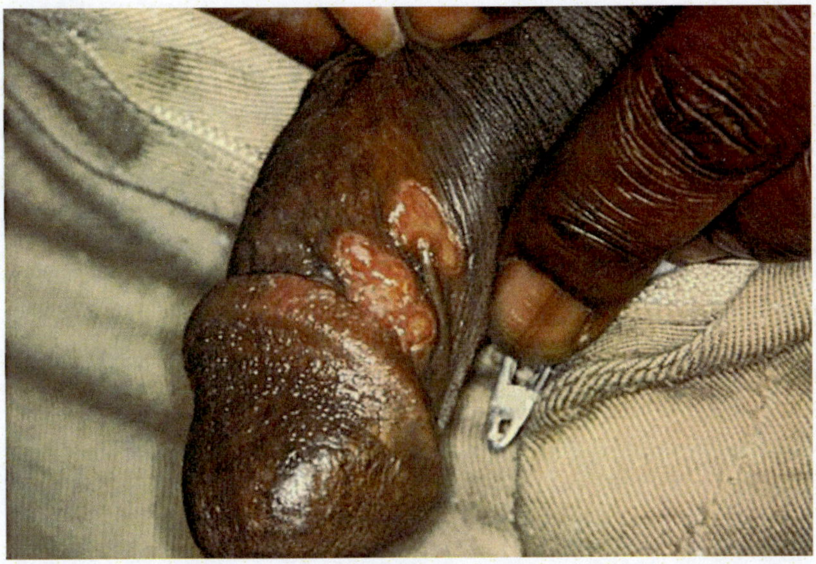

Fig. 43.2 Primary chancre. (Courtesy: https://phil.cdc.gov/Details.aspx?pid=6803)

Tertiary Syphilis

It develops within 3–10 years of infection. Gumma is a typical pathological lesion found on the skin, in the mouth and the upper respiratory tract. Gummatous lesions may be multiple or diffused but are usually single lesions, measuring from 1 cm to several centimetres in diameter. This lesion contains few spirochetes and represents manifestations of delayed hypersensitivity. Cardiovascular syphilis, chronic granulomatous and meningovascular manifestations and neurosyphilis (tabes dorsalis, general paralysis of insane) are the other manifestations of syphilis, which occur several decades after the infection.

Non-venereal Syphilis

Congenital syphilis and occupational syphilis are an example of non-venereal syphilis.

Congenital Syphilis

The condition occurs by vertical transmission from mother to foetus during pregnancy. If the mother is suffering from secondary syphilis and is not treated for the same, a higher proportion of infants are affected compared to untreated early latent syphilis. In more than 40% of untreated maternal infections, *T. pallidum* causes late abortion, stillbirth and death. Deaths in neonates may occur due to secondary bacterial infection, fulminant hepatitis or pulmonary haemorrhage. Infants suffering from congenital syphilis are born usually without any overt clinical manifestation of the disease. Poor feeding and rhinitis may be the earliest signs of congenital syphilis. Manifestations of late-onset congenital syphilis include neurosyphilis and involvement of the eighth cranial nerve, teeth and bones.

Occupational Syphilis

It is a condition that may occur in medical and paramedical workers handling a case of secondary syphilis. The lesion usually develops on the palm of infected health workers and may also occur on other exposed body parts.

Laboratory Diagnosis

Since the clinical manifestations of syphilis are protean, laboratory tests play an important role in confirming the disease's diagnosis.

Specimens

Specimens for microscopy include serous transudates from moist lesions, such as primary chancre, condyloma latum and mucous patch. The serum is used for serodiagnosis, and cerebrospinal fluid (CSF) is used to diagnose neurosyphilis. Specimens from the oral cavity cannot be used because saprophytic non-pathogenic treponemes are present as normal flora of the oropharynx.

Microscopy

Dark-field microscopy is useful for diagnosing primary, secondary or congenital syphilis by demonstrating treponemes in the clinical specimen. It is particularly helpful for diagnosis early in the disease before the appearance of serum antibodies. *T. pallidum* is identified by its slender spiral structure and slow movement. Dark-field microscopy, although useful, has many limitations. First, it is reliable only when examined by an experienced microscopist. Second, the method is of low sensitivity because as high as 10^4 bacteria per millilitre of the exudates need to be present for demonstration by dark-ground microscopy.

Direct fluorescent antibody *T. pallidum* (DFA-TP) is a sensitive and better method for direct detection of treponemal antigen in the exudates for diagnosis of syphilis. The test using fluorescent-tagged *T. pallidum* antibodies is used to detect treponemal antigen directly in the acetone-fixed smears of the exudates. The test is highly sensitive.

Culture

Since *T. pallidum* cannot be cultured on an artificial medium, culture is not used to diagnose syphilis.

Serology

Serology is the mainstay in the diagnosis of syphilis. These tests are broadly of two types: a) Standard tests of syphilis (STS) and b) Tests using *T. pallidum*.

Standard Tests of Syphilis

This group of tests uses **cardiolipin**, a non-treponemal antigen, and detects **reagin antibodies**. The Wasserman complement fixation test was the first serological test used in 1906, and the Kahn tube flocculation test is no longer used nowadays. VDRL and rapid plasma reagin (RPR) tests are examples of STS being used currently.

Reagin antibodies are produced against lipid antigens present on the surface of treponemes or produced as a result of damage of the infected host cells during the early stage of the disease. The antibodies appear 7–10 days after the appearance of the primary chancre or 3–5 weeks after active infection in primary syphilis. These antibodies also develop in syphilitic patients with secondary and latent diseases. These antibodies do not appear in early primary syphilis, latent acquired syphilis of long duration and late congenital syphilis. Therefore, Venereal Disease Research Laboratory (VDRL) and other tests may show false-negative reactions during these conditions.

Prozone phenomenon occurs in approximately 2% of patients with secondary syphilis, in patients with congenital syphilis and pregnant women. A high level of reaginic antibodies in these conditions may lead to a false-negative reaction with undiluted serum. Repeating the test with several dilutions of the patient serum excludes this phenomenon.

VDRL Test

It is a slide flocculation test used widely to diagnose syphilis. The test is so named because it was developed first in the VDRL, USPHS, New York. This is a simple and more rapid test, which uses cardiolipin antigen with added lecithin and cholesterol. In this test, the serum is inactivated at 56 °C for 30 min, and a measured volume of serum is placed on a special cavity slide. The cardiolipin antigen after preparation is added to the serum sample on a slide and is rotated on a VDRL rotator for a specified period of 4 min. The reaction is read under a low-power objective of a microscope. The cardiolipin antigen reacts with reagin antibodies in the infected serum and forms visible clumps in a positive test. Cardiolipin remains as uniform crystals in the serum in a negative test. The test result is reported as reactive, weak reactive or non-reactive depending on the extent of the formation of the clumps. The antibody titre of the serum can also be determined by testing on serial dilutions. The endpoint's reciprocal is considered the titre (e.g. reactive eight dilutions or titre 8). VDRL test can also be performed on CSF samples. Unlike serum samples, CSF samples are not heated prior to the test. The VDRL test becomes positive 4–5 weeks after exposure to *T. pallidum* and 1–2 weeks after the appearance of the chancre.

The VDRL test is highly sensitive. The sensitivity of the test depends on the stage of the disease. The test is 60–75% sensitive in primary syphilis and 100% sensitive in secondary syphilis. High antibody titres may be found in patients developing gummous lesions and neurovascular or cardiovascular complications. VDRL test is negative in the tertiary or third stage of syphilis. This test is negative in early primary syphilis, latent acquired syphilis and late congenital syphilis.

Quantitation of the sera is useful to monitor the activity of the disease and is of prognostic value. The quantitative estimation of reaginic antibodies is a good prognostic marker used for the treatment of syphilis. A fourfold decrease in titre following therapy suggests adequate therapy, whereas a fourfold increase in titre of reaginic antibodies following therapy suggests re-infection or relapse. The reaginic test becomes negative 1 year after successful therapy in primary syphilis, 2 years after successful therapy in secondary or congenital syphilis and 5 years after successful therapy in late syphilis. In some cases of neurosyphilis, this reaginic test may become negative with serum but positive with the CSF.

The main disadvantage of the VDRL test is that it shows **biological false positive (BFP) reactions.** BFP reactions are defined as positive reactions observed with a test using cardiolipin antigen but are negative with a specific test using *Treponema* antigen. These results are not due to technical faults. They represent antibody responses to non-treponemal cardiolipin antigens found in other tissues. BFP reactions are classified as acute or chronic reactions. The reactions that become negative within 6 months are called acute reactions, while those found for a longer period are called chronic reactions.

Acute BFP reactions are typically seen in acute bacterial or viral infections, injuries or inflammatory conditions and early HIV infection. **Chronic BFP reactions** last for more than 6 months. This is seen in patients with parenteral drug use, autoimmune or connective tissue diseases, such as systemic lupus erythematosus, ageing and hypergammaglobulinemia, also seen in other infectious conditions, such as tropical eosinophilia, malaria, infectious mononucleosis, hepatitis, leprosy and relapsing fever (Table 43.2). The BFP reactions can usually be ruled out by using a *Treponema*-specific test.

The VDRL enzyme-linked immunosorbent assay (ELISA) is a modification of the VDRL test, which is being used to detect both IgM and IgG antibodies separately in the patient serum. This test is used mainly for large-scale screening of patient serum.

Rapid Plasma Reagin Test

It is a popular test used to diagnose syphilis by demonstrating reaginic antibodies. The test uses VDRL antigen containing finely divided carbon particles suspended in choline chloride. The latter destroys inhibitory factors in the serum, thus avoiding the need to heat the serum before testing. The use of this antigen produces a clearer and

Table 43.2 Biological false positive (BFP) reactions of standard tests for syphilis

Acute BFP reactions
Acute bacterial or viral infections
Injuries or inflammatory conditions
Early HIV infection
Chronic BFP reactions
Patients with parenteral drug use
Autoimmune or connective tissue diseases, such as systemic lupus erythematosus
Ageing
Patients with hypergammaglobulinemia
Other infectious conditions, such as tropical eosinophilia, malaria, infectious mononucleosis, hepatitis, leprosy and relapsing fever

well-defined flocculation reaction easily observed by the naked eye. The RPR test is used for testing serum or plasma but not for the CSF. **Automated RPR test** is a modification of the RPR test, which is used for screening a large number of serum samples. **The unheated serum reagin test** is a flocculation test used to detect cardiolipin antibodies. **Toluidine red unheated serum test** is used for testing CSF samples.

Treponema pallidum-specific tests

Treponema-specific tests measure antibodies specific for *T. pallidum*. These tests use (1) live *T. pallidum* strains (*T. pallidum* immobilisation test), (2) killed *T. pallidum* (*T. pallidum* agglutination test, *T. pallidum* immune adherence test and fluorescent treponemal antibody test) or (3) *T. pallidum* extracts as antigens [TPHA test and EIA (enzyme immunoassay)].

Treponema pallidum Immobilisation Test

The test uses live *T. pallidum* strains that detects the treponemal antibodies in the patient's serum. The test, when introduced, was the most specific serological test for the diagnosis of syphilis. However, this test is no longer used because of its complexity and difficulty maintaining live treponemal strains. Similarly, ***the T. pallidum* agglutination test** that uses killed *T. pallidum* suspension inactivated by formalin is no longer used because it is non-specific and is associated with false-positive reactions. ***T. pallidum* immune adherence test** is also not used nowadays.

Fluorescent Treponemal Antibody (FTA) Test

It is the most specific and popular test used to diagnose syphilis. It is indirect immunofluorescence (IIF) test, which uses acetone-fixed smears of *T. pallidum* on the slides. The test is performed by adding a drop of test serum to the smear on the slide, followed by washing and re-incubating the smear with fluorescent labelled antihuman immunoglobulin. The slide is examined under a fluorescence microscope. The demonstration of fluorescent treponemes is suggestive of a positive FTA test.

FTA Absorption (FTA-Abs)

It is a modification of FTA-Abs, which shows high sensitivity and specificity. In this test, the patient's serum is first absorbed with a sonicated extract of non-pathogenic *T. phagedenis* (Reiter Treponeme) to remove group-specific antibodies. The test is almost as specific as the TPI test and is considered a standard reference test in syphilis serology. The FTA-Abs test is positive in 80% primary syphilis, 100% secondary syphilis and 95% tertiary syphilis. The test is highly specific (92–99%).

IgM FTA-Abs Test

It is another modification of FTA-Abs used to detect serum IgM antibodies in congenital syphilis. This test is used to differentiate seropositivity due to passively transferred IgG maternal antibodies to the foetus from IgM antibodies found in utero in congenital syphilis.

Treponema pallidum Haemagglutination Assay (TPHA)

The assay uses erythrocytes sensitised with a sonicated extract of *T. pallidum* as antigen. TPHA has now been modified to perform in microtitre plates and is called the microtitre haemagglutination *T. pallidum* (MHA-TP) test. This test is performed by incubating the serial dilution of the patient's serum with erythrocytes sensitised with *T. pallidum* antigen in a microtitre plate. If antibodies are present, haemagglutination of RBCs occurs. The serum samples before testing for TPHA are re-absorbed with a diluent containing Reiter Treponeme, rabbit testes and sheep erythrocytes.

TPHA is as specific as FTA-Abs but less sensitive than FTA-Abs in primary syphilis. It is also equally sensitive to FTA-Abs in secondary and late syphilis. After successful therapy, the haemagglutinating antibodies persist in the serum for longer; hence, this test has no prognostic value. The negative TPHA that virtually excludes the diagnosis of syphilis except during the early stage of the disease and eliminates the diagnosis of neurosyphilis's is the most widely used treponemal test available commercially. It is simple, economical and does not require expensive, sophisticated equipment.

Enzyme Immune Assays (EIA)

These assays use ultrasonicated *T. pallidum* antigen coated on tubes or ferrous metal beads as a solid-phase carrier for antigen. The serum antibodies are detected by an enzymatic reaction. The test is available commercially.

Western Blot

Western Blot is the gold standard for anti-TP antibody detection. Compared to other antibody assays, like ELISA, it has greater sensitivity and specificity to confirm samples showing inconsistent results between ELISA and TPPA.

Molecular Detection

It has emerged as a very promising tool for detecting *T. pallidum* in clinical samples mostly because of its ability to detect without culturing, wherein it is very difficult to grow the organism in culture. Five types of molecular strategy, including routine PCR, nested PCR (nPCR), real-time PCR (qPCR), reverse transcription PCR (RT-PCR) and loop-mediated isothermal amplification (LAMP), are available for diagnostic purposes. In most cases, the target genes are *polA*, *tpp47*, BMP, *16S rRNA gene, tmpC* and *tmpA*. For molecular tests, samples are often obtained from lesion sites, including genital, anal or oral ulcers, surface rashes, tissue lesions and mucosal erosion. Recently, nested PCR also uses blood, urine and saliva samples. The sensitivity and specificity of detection range from 78.4 to 89.1% and 93.1 to 100% in primary syphilis, however, much lower for secondary syphilis.

All the treponemal tests are not completely specific for syphilis. They also show false-positive reactions in patients with other spirochaetal diseases, such as leptospirosis, relapsing fever, Lyme disease, pinta, yaws and rat-bite fever. The sensitivity of common tests performed for the diagnosis of syphilis is summarised in Table 43.3. The test accuracy of various diagnostic methods in syphilis is presented in Table 43.4.

Serological Assay Diagnostic Utility

Serological screening of syphilis varies across the globe. For example, in the United Kingdom, VDRL and TPHA are being used in combination to screen syphilis. The results obtained from this serological test are more reliable since it can detect antibodies in all three stages of syphilis. However, early primary cases might show false-negative reactions, and in these cases, a repeat serological test is required after 1–3 months to rule out syphilis. In the USA, syphilis serological screening is done using VDRL and RPR. This combination of tests may fail to diagnose 1% of secondary syphilis due to excessive antibodies (prozone phenomenon), and it also gives about 30% of false-negative results in latent and late syphilis.

Table 43.3 Sensitivity of different tests for syphilis

Stage of syphilis	VDRL	RPR	FTA-Abs	TPHA	Molecular tests
Primary	±	−	+	−	+
Late primary	+	+	+	±	+
Secondary	+	+	+	+	+
Tertiary	+	+	+	+	+
Late	+	+	+	+	+
Latent	±	±	+	+	+
Treated syphilis	−	−	+	+	+
Congenital syphilis	+	+	+	+	+

VDRL: Venereal Disease Research Laboratory; *RPR*: rapid plasma reagin; *FTA-Abs*: fluorescent treponemal antibody absorption test; *TPHA*: *Treponema pallidum* haemagglutination

Table 43.4 Diagnostic tests for syphilis

Diagnostic Test	Types	Sensitivity
Microscopy	Dark-field microscopy	Less sensitive; test is reliable when examined immediately
	Direct fluorescent antibody method	Highly specific
Serology	VDRL	Sensitivity high in secondary stage and less in primary stage and late syphilis
	RPR	Sensitivity high in secondary stage and less in primary stage and late syphilis
	FTA-Abs	Highly specific and sensitive
	TPHA	Highly specific and sensitive
Molecular methods	Polymerase chain reaction (PCR)	Specific; helps distinguish between the various subspecies of *Treponema pallidum*
	DNA amplification	Specific

VDRL: Venereal Disease Research Laboratory; *RPR*: rapid plasma reagin; *FTA-Abs*: fluorescent treponemal antibody absorption test; *TPHA*: *Treponema pallidum* haemagglutination

Response to Treatment

The re-infections and treatment response can be monitored using quantitative non-specific treponemal tests, such as VDRL and RPR. Following treatment, the VDRL and RPR tests become negative. At times, early treatment administration may not raise the antibody titre at all. In primary syphilis, treatment reverses the seropositivity in 4 months. In secondary and early latent stages, it requires 12–18 months, whereas, in later stages, it may require 5 or more years for screevers. However, in a few cases, xenoreactivity with very low titre persists for a long time despite the completion of proper effective treatment. Specific treponemal tests, such as TPHA remains positive even after complete therapy; hence, they are of little use as a prognostic marker of treatment response. However, TPHA titre may drop sharply after treatment in secondary syphilis, but low syphilis antibody titre remains throughout the lifetime.

Congenital Syphilis Diagnosis

Since IgM cannot cross the placenta, in neonates' detection of IgM antibodies is the definitive diagnosis. Modifications of FTA-Abs, TPHA, EIA and VDRL tests are currently available for IgM antibody detection in syphilis. Screening of maternal and neonatal blood may be useful in congenital syphilis since neonatal serum always shows a higher titre of antibodies than maternal serum.

A histological examination of the placenta and cord are performed when congenital syphilis is suspected before or during delivery. Also, desquamation or ulcerative skin lesions or nasal discharge ("snuffles") is examined by dark-field microscopy or fluorescent antibody techniques. Infants exposed to syphilis in utero are taken for serological testing, regardless of maternal treatment. Asymptomatic infants for positive mothers, however, are monitored monthly until their non-treponemal antibody disappears.

In 2017, WHO had released the "WHO Guideline on Syphilis Screening and Treatment for Pregnant Women" that includes updated recommendations for syphilis screening and treatment for pregnant women based on the most recent evidence and available serologic tests for syphilis.

Treatment

Penicillin is the drug of choice to treat all the stages of syphilis. A single intramuscular dose of benzathine penicillin G (50,000 units per kg, not to exceed 2.4 million units) effectively treats primary, secondary and early latent syphilis. No reports of *T. pallidum* showing resistance to penicillin have been documented. Doxycycline or tetracycline may be used for non-pregnant patients allergic to penicillin. Jarisch–Herxheimer reaction is a noted complication observed following therapy with antibiotics. It is characterised by chills, rigors, an increase in temperature, a decrease in blood pressure and tachycardia. This reaction occurs due to the production of endotoxin and other toxic products released following rapid killing and lysis of the bacteria.

Epidemiology

Syphilis is a sexually transmitted disease found worldwide.

Geographical Distribution

Syphilis occurs worldwide, mostly in large cities. The disease is extremely common in areas of dry, hot climates. Also, it is common in areas of poor economic status, education and personal hygiene.

Habitat

T. pallidum inhabits the genital tract of infected males and females.

Reservoir, Source and Transmission of Infection

T. pallidum is a strict human pathogen and does not naturally occur in any animal species. Humans are the only natural hosts. Infected human hosts secreting *T. pallidum* in serous transudates from moist lesions, such as primary chancre, condyloma latum and mucous patch, are the sources of infection. Transmission of syphilis occurs i) primarily through sexual contact by inoculation of the spirochetes through mucosal membranes and abrasions on epithelial surfaces') by vertical transmission of early syphilis during pregnancy, and by transfusion of *T. pallidum*-contaminated blood. Unprotected sex, promiscuous sex and intravenous drug use are the major risk factors for syphilis. Doctors, nursing staff and other healthcare workers are at occupational risk.

Syphilis in persons with human immunodeficiency virus (HIV) infection is a problem increasingly recognised recently. Concomitant HIV and syphilis are common. The presence of HIV may result in extremely high antibody titres, which do not decrease even following adequate treatment.

Prevention and Control

Avoidance of sexual contact with infected partners avoids transmission of infection. Use of condoms during sexual intercourse or antiseptics (such as potassium permanganate) or prophylactic use of antibiotics may minimise the risk of transmission of the disease. However, penicillin prophylaxis carries the danger that it may suppress the primary lesion without eliminating the infection.

Non-Venereal Treponematosis

The condition includes three distinct entities: endemic syphilis, yaws and pinta. These conditions are found in several parts of the world associated with poor socioeconomic conditions and poor hygiene.

Endemic Syphilis

Endemic syphilis is a non-venereal treponemal infection reported from different parts of the world. The condition, also known as bejel in the Middle East, *njovera* in Zimbabwe, *dichuchwa* in Bechuanaland, *Skerjevo* in Eastern Europe and *siti* in Zambia, is caused by *T. pallidum* subsp. *endemicum*. The condition has also been documented in India.

The disease is transmitted from person to person by using contaminated utensils. The disease is commonly seen in young children. The initial lesions, such as primary chancre, are rarely seen. Secondary manifestations of syphilis are usually seen, including lesions, like oral papules and mucosal patches. Late manifestations of the disease include gummatous lesions of skin, bones and nasopharynx. The laboratory diagnosis and treatment of endemic syphilis are similar to venereal syphilis.

Yaws

Yaws is a non-venereal treponemal disease caused by *T. pertenue*. The condition, also known as pian, parangi, etc., is prevalent in the primitive tropical parts of central Africa, South America and Southeast Asia.

The causative agent, *T. pertenue,* is morphologically and antigenically similar to *T. pallidum.* The pathogen causes yaws, which is characterised by an extragenital papule, which subsequently enlarges and breaks down to form an ulcerating granuloma. Late manifestations of the disease include destructive lesions of the skin, bone and lymph nodes. The disease is transmitted by direct contact with infected skin lesions. Laboratory diagnosis and treatment are similar to that of venereal syphilis.

Pinta

Pinta is a contagious disease of the skin caused by *T. carateum.* Pinta, also known as carate, mal del pinto, is endemic in Central and South America and the neighbouring islands. An extragenital papule is the primary skin lesion observed in the condition. Small pruritic papules develop on the skin's surface after an incubation period of 1–3 weeks. These lesions do not ulcerate but enlarge and persist for months to years before resolving. If it remains untreated, the condition progresses to form disseminated recurrent and hypopigmented lesions leading to scarring and disfigurement. Pinta spreads by direct contact with infected lesions. Laboratory diagnosis and treatment are the same as for syphilis.

Non-Pathogenic Treponemes

Non-pathogenic treponemes are a heterogeneous group of treponemes found as commensals on the buccal and genital mucosa. They are also found on the surface of gastric and colonic epithelium in humans and animals. These include *Treponema denticola*, *Treponema macrodentium*, *Treponema oralis*, *Treponema socranskii* and *Treponema pectinovorum*. *T. denticola* is an oral treponeme easily cultivable on an artificial medium. *T. phagedenis* is also known as Reiter and Kazan strain, and *T. refringens* is also known as Nichol strain.

Genus *Borrelia*

The genus *Borrelia* consists of many species found as commensals on buccal and genital mucosa. Members of the genus *Borrelia* cause two important diseases in humans. Epidemic or louse-borne relapsing fever is caused by *Borrelia recurrentis*, whereas endemic relapsing fever is caused by as many as 15 species of *Borrelia.* These include *Borrelia duttoni*, *Borrelia henselae*, *Borrelia parkeri*, *Borrelia turicatae*, etc. *Borrelia burgdorferi* causes Lyme disease.

Borrelia recurrentis

B. recurrentis is the causative agent of relapsing fever, which manifests as one or more relapses of fever after the subsidence of primary paroxysm of fever.

Properties of the Bacteria

Morphology

B. recurrentis has an unequal spiral-shaped structure with one or both the ends pointed. It measures 8–20 μm in length and 0.2–0.4 μm in breadth. It possesses 15–20 endoflagella per cell and 5–10 loose, uneven spiral coils. The spiral coils are coarser and more irregular than treponemes and leptospires. It is Gram negative. It can be stained by Wright and Giemsa stain. It is actively motile in fresh blood preparation, moves forward and backward waves and exhibits corkscrew-like motility.

Culture

Borrelia is microaerophilic. The bacteria grow at an optimum temperature of 28–30 °C. They have complex nutritional requirements, hence are difficult to grow on artificial media. They can grow in the chorioallantoic membrane of the chick embryo. Mice or rats are highly susceptible to infection. Primary isolation of bacteria from clinical specimens can be made by intraperitoneal inoculation of specimens in these laboratory animals. After inoculation, borreliae appear early in the blood and also are found in the brain for a longer time.

Pathogenesis

B. recurrentis is an invasive bacterium. It enters the skin and causes disease.

Virulence Factors

The ability of *B. recurrentis* to undergo antigenic shift and escape from immune clearance of the host is the main virulence factor attributed to the bacteria. The **antigenic shift** or **antigenic variation** is a unique property exhibited by *Borrelia* in humans and is responsible for causing relapse, characteristically seen in the relapsing fever caused by the bacteria. DNA rearrangement in linear plasmid present in *Borrelia* appears to be responsible for antigenic variations shown by the bacteria.

Pathogenesis of Relapsing Fever

B. recurrentis penetrates the skin, reaches the blood circulation and invades various organs in the body and causes the disease. After the invasion of blood by bacteria, *Borrelia*-specific IgM antibodies agglutinate *Borrelia* and cause complement-mediated lysis. This leads to the rapid clearing of *Borrelia* from the bloodstream. *Borrelia* present in the internal tissue alters their serotype-specific outer proteins through gene rearrangement and exhibits a new antigen.

The IgM antibodies produced against old antigens are not effective against the *Borrelia* pathogens exhibiting new antigenic variation. The host immunity responds again by producing specific antibodies against these new antigens and clears the organism from circulation. Patients recover after several relapses due to the development of humoral immunity.

Clinical Syndromes

B. recurrentis causes relapsing fever, the epidemic louse-borne and endemic tick-borne relapsing fever.

Relapsing Fever

Clinical manifestations of epidemic louse-borne and endemic tick-borne relapsing fever are essentially similar. The incubation period is usually 7 days. Two or more episodes of high fever, headache and myalgia are the hallmarks of the disease. Splenomegaly and hepatomegaly are the associated symptoms. These symptoms correspond to the bacteraemia phase of the disease when *Borrelia* is found in large numbers in the patient's blood (Fig. 43.3).

The fever subsides after 3–7 days when the *Borrelia* is cleared from the circulation. After an afebrile period of 4–7 days, bacteraemia occurs, and fever re-appears. The *Borrelia* re-appears in the blood during relapse of the fever. The condition ultimately subsides after 3–10 relapses. A single relapse is the characteristic of the epidemic louse-borne disease and is usually associated with high morbidity of 4–50%. Repeated relapses are common in endemic tick-borne diseases and are usually associated with low mortality of less than 5%. Humoral immunity contributes to recovery in a patient after many relapses.

Laboratory Diagnosis

Specimen

Blood collected from a patient during fever, but not from an afebrile patient, is a useful specimen.

Microscopy

A wet mount preparation of the blood examined by **dark-field** or **phase-contrast microscopy** is a useful method for direct observation of *Borrelia*. The organisms are identified by their dashing movements. *Borrelia* organisms can also be demonstrated in a peripheral blood smear stained with Giemsa or Wright stains. Staining is the most sensitive method for diagnosing 70% or more patients with louse-borne relapsing fever.

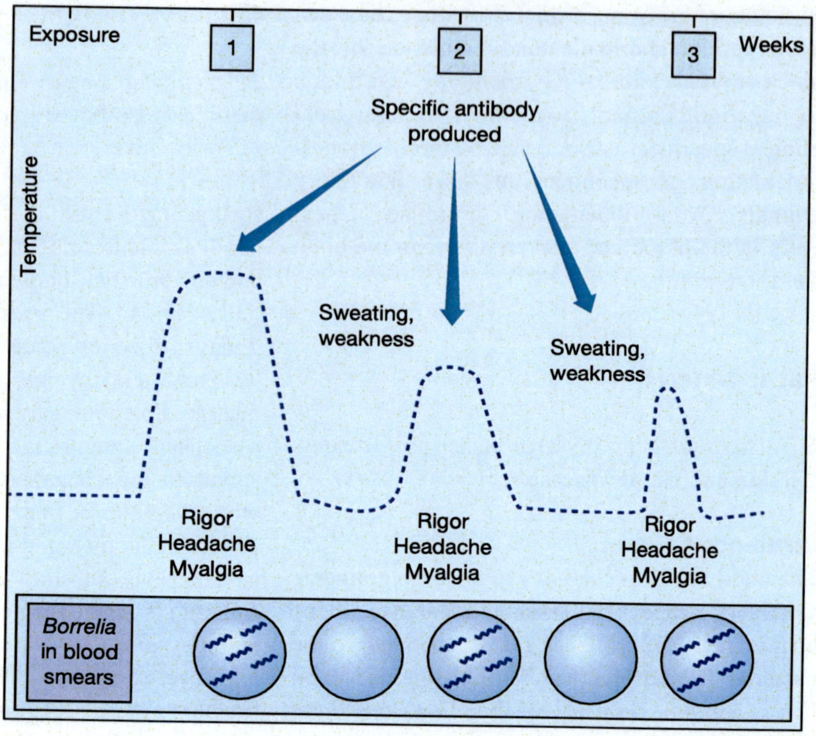

Fig. 43.3 Course of relapsing fever

Culture

Cultures are not carried out routinely for the isolation of *Borrelia* from clinical specimens. This is because *Borrelia* organisms have complex nutritional requirements for their growth, and they grow very slowly on these media.

Animal Inoculation

Animal inoculation is a more sensitive method. This test is performed by inoculating a mouse intraperitoneally with 1–2 mL of blood from the infected patient. The blood is collected from the tail vein of the mice after 1–10 days. Blood is examined for the presence of *Borrelia* daily for 3 weeks.

Serology

B. recurrentis causing relapsing fever undergoes antigenic phase variation; hence, serological tests are not useful in diagnosing the condition.

Molecular Detection

PCR-based tests have been developed for the laboratory-based diagnosis of relapsing fevers. In most cases, real-time PCR targeting the 16S rRNA gene or the *glpQ* gene is used. It improves the sensitivity and specificity, resulting in more accurate and timely detection of the organism. Commercial kits are now available for detection.

Treatment

Tetracycline or erythromycin is most effective for the treatment of relapsing fever. Doxycycline is usually recommended for children and pregnant women.

Epidemiology

Epidemic louse-borne relapsing fever now has disappeared with improvement in hygiene and the use of insecticides.

Geographical Distribution

It is an important disease only in north-eastern Africa, especially in the islands of Ethiopia and in

South America. Endemic tick-borne relapsing fever has worldwide distribution and is endemic in western states of the United States.

Habitat

B. recurrentis is found in the bloodstream and the spleen, liver, lungs, kidneys and bone marrow of the infected patients suffering from relapsing fever.

Reservoir, Source and Transmission of Infection

Humans are the only reservoir of epidemic relapsing fever. No extra-human reservoir is known. The infection is transmitted only from person to person. Persons suffering from relapsing fever are the source of infection. The human body louse (*Pediculus humanus corporis*) is the vector of the disease. No extra-human reservoir is known. *Borrelia* is found only in the haemolymph of the lice. They are not excreted in saliva or excreta. Hence, the infection is transmitted by being crushed and rubbed on the absorbed skin. The bite of the lice does not transmit them. Endemic tick-borne relapsing fever is a zoonotic disease transmitted from animals to humans, rodents or small mammals.

Soft ticks (*Ornithodoros* species) are the main vectors of the disease. Unlike the louse-borne infection, *B. recurrentis* produces a disseminated infection in ticks. However, the vector survives from the infection and becomes a reservoir of the infection by transovarian transmission. The *Borrelia* organisms are found in all parts of the tick's body; hence, they are secreted in the saliva and excreta. Therefore, the infection is transmitted to humans by the bite of the ticks. Several species of the genus *Ornithodoros* act as reservoirs in different parts of the world. *Ornithodoros tholozani*, *Ornithodoros crossi*, *Ornithodoros lahorensis* and the fowl tick, *Argas persicus,* are the vectors in India. Soft ticks are primarily nocturnal feeders. They feed while the host sleeps and remain attached for only a few minutes. Hence, the bite of the tick goes usually unnoticed.

Prevention and Control

Prevention of louse infiltration and insecticide use are the measures that prevent louse-borne relapsing fever. Prevention of tick-borne relapsing fever consists of wearing protective clothing, use of insect repellents and avoidance of ticks and their natural habitat. No vaccine is available for relapsing fever.

Borrelia vincenti

B. vincenti is a motile spirochete measuring 7–18 μm in length and 0.2–0.6 μm in breadth. It is Gram negative and is usually stained with Giemsa and Leishman stain. *B. vincenti* is an obligate anaerobe found as a normal commensal in the mouth. The bacteria under certain predisposing conditions (such as malnutrition and viral infections) produce a condition known as **Vincent angina**. Vincent angina is ulcerative gingivostomatitis or oropharyngitis often associated with fusiform bacilli (*Fusobacterium fusiforme*). It is characterised by the formation of an ulcer on the mouth and tonsillar areas.

Laboratory diagnosis of the condition is made by demonstrating *B. vincenti* and fusiform bacilli in stained smears of exudate from the lesion. This bacterium is grown with difficulty on enriched media anaerobically. Penicillin and metronidazole are effective for the treatment of the condition.

Borrelia burgdorferi

B. burgdorferi, a newly identified *Borrelia* species, is the causative agent of Lyme disease. Lyme disease was first demonstrated in children in 1975, during an outbreak of arthritis in Lyme, Connecticut, in the United States. The causative agent of the fever was isolated by Burgdorfer in 1982, after whom the species *burgdorferi* is named. *B. burgdorferi* is a fastidious bacterium, which measures 4–30 μm in length and 0.2 μm in breadth. It is helical and Gram negative. It is a microaerophilic spirochete, which can be grown on BSK (Barbour–Stoenner–Kelly) medium at 33 °C after incubation for 2 weeks or longer.

Lyme disease is a tick-borne disease transmitted to humans by ixodid ticks. The incubation period varies from 7 to 14 days. After the tick's bite, *B. burgdorferi* is inoculated through the skin and then spreads locally. The local spread of the bacteria causes erythema migrans, a rash seen in approximately two-thirds of the cases. This skin rash may be a confluent patch of erythema or may have central clearing. The lesion begins as a small macular papule and becomes larger over the next many weeks, and forms a large area of lesion of 5–50 cm in diameter.

The patient may also complain of fever, chills, myalgias and headache during the early stage of the disease, with or without rash. Subsequently, during a time ranging from days to months, the bacteria spread through blood circulation and cause disseminated disease. This disease is characterised by the presence of multiple erythema migrans, systemic complications (fever, myalgias, arthralgia, malaise and headache) and even septic meningitis; this disease usually develops 3–10 weeks after the tick bite. The pathogenesis of these late manifestations is poorly understood. It is not known whether the live organisms cause these manifestations or these manifestations occur due to an antigenic cross-reactivity to *Borrelia* antigens.

Lyme disease has been reported in the USA, Germany, Austria, Switzerland and Scandinavian countries. Lyme disease is a zoonotic disease. Rodents, bears and other mammals are the natural reservoir hosts. Hard ticks (ixodid ticks) are the vectors of the disease. The infection is transmitted by the hard tick from mice to humans and occur by regurgitation during a tick bite. Individuals exposed to hard ticks are at increased risk for Lyme disease.

Clinical diagnosis of the condition may be made by the presence of erythema migrans in the early stage of the disease. Laboratory diagnosis of the condition is primarily serological. Serodiagnosis depends on the demonstration of specific antibodies in the serum, which persist for many years even after eradicating the infection. ELISA and IIF are the most common serological tests employed to diagnose the disease. Western blot is used to confirm the specificity of serum positive by ELISA or IIF. Serology is positive in one-third of the patients with the early disease, in 90% of patients with early disseminated disease and all the patients with late disease. Microscopy is not recommended because *B. burgdorferi* is rarely seen in clinical specimens. Culture is also not used because the bacteria are difficult to culture.

Amoxicillin, tetracycline, cefuroxime or ceftriaxone are effective for treating Lyme disease. Avoidance of exposure to ticks and use of insecticides are useful methods for preventing the disease.

Genus *Leptospira*

Traditionally, the genus *Leptospira* has been grouped by serological properties and by their pathogenicity into two species: the pathogenic *Leptospira interrogans* and non-pathogenic *Leptospira biflexa*. Based on shared antigens, these species were further divided into serogroups, serovars and strains (Table 43.5). Pathogenic species *L. interrogans* currently includes more than 250 serovars. However, the classification is not consistent with recent classification based on nucleic acid analysis.

The current classification based on nucleic acid analysis divides *L. interrogans* and *L. biflexa* complex into 12 species. These species are further subdivided into serogroups, serovars and strains based on microscopic agglutination tests (MATs). *L. interrogans* complex includes pathogenic leptospires that cause leptospirosis. *L. biflexa* consists of non-pathogenic leptospires and includes more than 63 serovars. *L. biflexa* is so-called because of its double-curved structure (*Biflexa*: twice bend). These are mostly free-living saprophytes found in moist environmental areas but are not associated with disease in humans and animals.

Table 43.5 Characteristics of infections caused by important *Leptospira interrogans* serogroups

Leptospira Serogroup	Disease	Epidemiology	Reservoir	Clinical Manifestations
Icterohaemorrhagiae	Weil disease	Cosmopolitan	Rat	Fever with jaundice, haemorrhages
Canicola	Canicola fever	Cosmopolitan	Dog	Influenza-like, aseptic meningitis
Grippotyphosa	Swamp or marsh	USA, Europe, Africa, Southeast Asia	Field mice	Fever, prostration, aseptic meningitis
Pomona	Swineherd	America, Europe, Middle East, Indonesia, Australia	Pig	Fever
Hebdomadis	Seven day fever	Japan, Europe, USA	Field mice	Fever, lymphadenopathy
Fort Bragg	Pretibial fever, Fort Bragg fever	SE Asia, USA, Japan	Unknown	Fever, rash over tibia
Pyrogens	Febrile spirochetosis	SE Asia, Europe, USA	Pig	Fever
Bataviae	Indonesian Weil disease	SE Asia, Africa, Europe	Rat	Fever
Hardjo	Dairy farm fever	United Kingdom, USA, New Zealand	Cattle	Fever

Leptospira Interrogans Complex

L. interrogans complex includes pathogenic leptospires that cause leptospirosis. The disease, often referred to as swamp fever, mud fever or swineherd disease, is the most common zoonosis in the world.

Properties of the Bacteria

Morphology

L. interrogans are thin, delicate, coiled spirochetes measuring 6–20 μm length and 0.1 μm in breadth. They are actively motile by means of two periplasmic flagella, each present at the opposite end of the bacteria by continually spinning on the long axes. They possess tightly coiled spirals and hooked ends, resembling an umbrella handle. They stain poorly with aniline dyes but stain well with silver impregnation methods, such as Levaditi and Fontana (Fig. 43.4).

Culture

L. interrogans are obligate aerobes. They use fatty acids and alcohols as the source of carbon and energy. They grow at an optimum temperature of 25–30 °C and optimum pH of 7.2–7.5. Leptospires are unique spirochetes—they can be grown in an artificial medium. They grow on special media enriched with rabbit serum or bovine serum albumin. Korthof medium, Stuart medium and Fetcher medium are frequently used media for the culture of *Leptospira*. Ellinghausen, McCullough, Johnson and Harris (EMJH) medium is a semi-synthetic medium commonly used nowadays for the culture of *Leptospira*.

Leptospira can be grown on the **chorioallantoic membrane** of 10- to 20-day-old chick embryos. The bacteria are demonstrated in the blood of the allantoic sac 4–5 days after inoculation. **Animal inoculation** is a useful method for obtaining a pure culture of *Leptospira*. The bacteria are inoculated intraperitoneally in guinea pigs, followed by blood collection from the heart 10 min after inoculation. Culture of the heart blood yields pure colonies of *L. interrogans*.

Other Properties

Susceptibility to Physical and Chemical Agents

Leptospira is readily killed by heating at 50 °C for 10 min and 60 °C for 10 s. The acidity of gastric juice destroys them in 30 min and by the bile. They are also readily destroyed by chlorine and

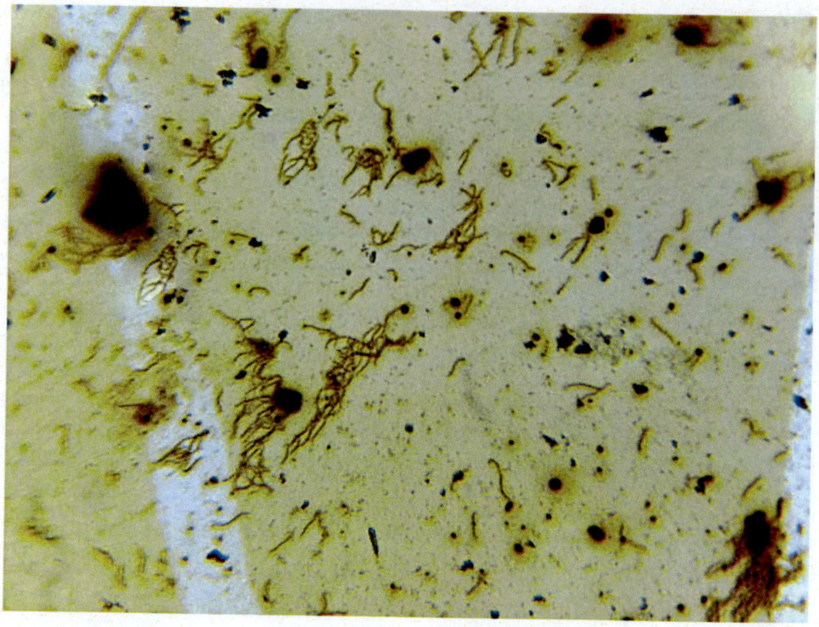

Fig. 43.4 Fontana stain showing *Leptospira* (×1000)

commonly used antiseptics and disinfectants. They die rapidly in acidic urine, salty or brackish water and non-aerated sewage.

Antigenic Structure

Leptospira organisms show a considerable degree of antigenic cross-reactivity. A genus-specific somatic antigen is present in all *Leptospira*, based on which leptospires are classified into serogroups and serotypes. This has been divided further based on DNA homology into serogroups, serovars, strains, etc.

Pathogenesis

L. interrogans is an invasive bacterium.

Virulence Factors

Direct invasion and **multiplication in the blood and tissue** is the main virulence determinant of *Leptospira*. Although the direct invasion of leptospires may cause tissue damage, it is now observed that the degree of multi-organ tissue injury does not correlate with the number of *Leptospira* in infected tissues. Therefore, it is suggested that other factors, such as the endotoxin, haemolysin and lipase, may also contribute to the pathogenicity of *Leptospira*. However, the exact virulence factors of *Leptospira* are not known. The factors responsible for causing injury to host tissues remain unclear and appear to be a complex interaction of many factors.

Pathogenesis of Leptospirosis

The severity of the leptospirosis depends on (1) host immunity, (2) virulence of infecting strain and (3) the number of infecting leptospires. After *Leptospira* enters the body through various sites, it multiplies in blood and tissues, resulting in leptospiraemia. *Leptospira* can spread to any part of the body but primarily affects the liver and kidney. In the kidney, the bacterium causes interstitial necrosis and tubular necrosis. Subsequently, the bacterium causes tubular damage, hypovolemia due to dehydration and altered capillary permeability, causing renal failure. In the liver, it causes centrilobular necrosis and hepatocellular dysfunction leading to jaundice. *Leptospira* may also invade skeletal muscle, causing oedema and focal necrosis.

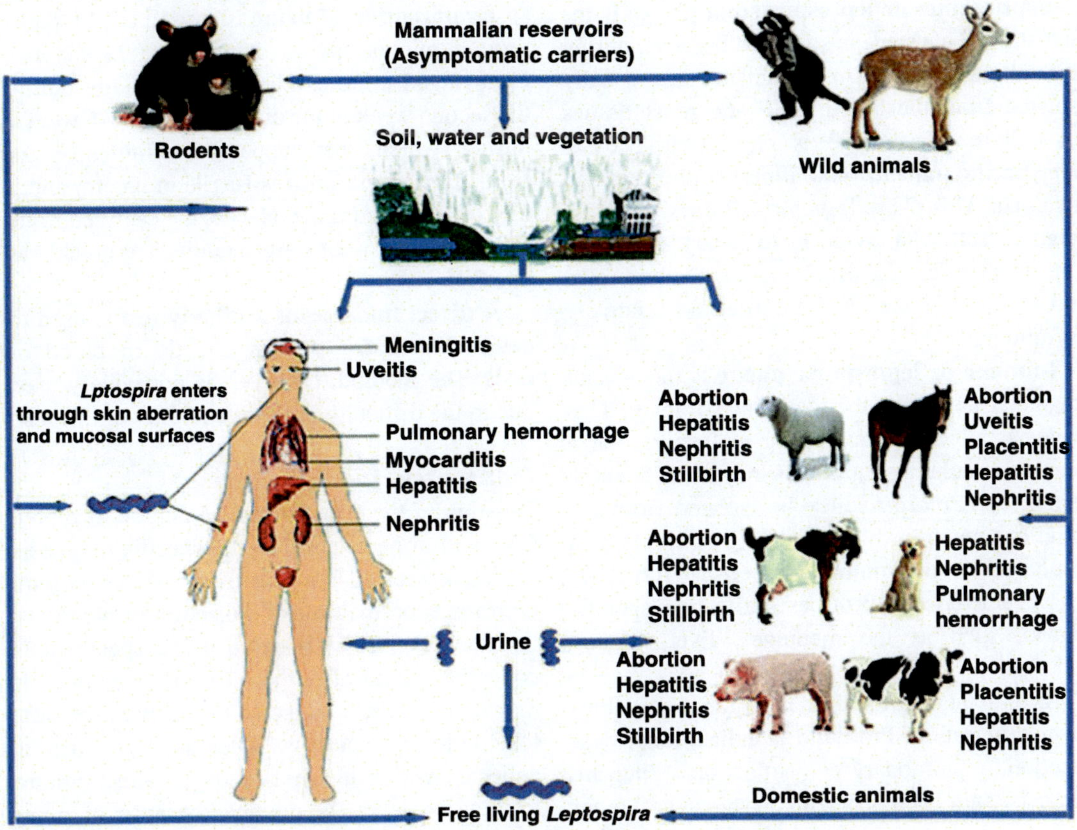

Fig. 43.5 Pathogenesis of leptospirosis (https://link.springer.com/chapter/10.1007/978-1-4614-5404-5_8#citeas)

Humoral immunity plays an important role in the clearance of leptospires from circulation. The host's immune reactions to *Leptospira* appear to be responsible for many complications, such as meningitis and other clinical manifestations seen during the later stage of the disease (Fig. 43.5).

Clinical Syndromes

L. interrogans causes leptospirosis.

Leptospirosis

Leptospirosis occurs as two recognisable clinical syndromes as anicteric leptospirosis and icteric leptospirosis. The incubation period varies from 2 to 30 days but is usually about 10–14 days.

Anicteric Leptospirosis

The condition occurs in a majority (90%) of infections with *L. interrogans*. It manifests as a mild anicteric (absence of jaundice) form of the disease. These clinically inapparent infections are diagnosed only by demonstrating specific *Leptospira* antibodies in the patient serum.

Icteric Leptospirosis

Approximately 10% of the patients develop icteric manifestations of the disease, otherwise known as **Weil disease**. The disease is characterised by the development of jaundice. The disease shows two distinct phases: septicaemic and immune (leptospuric), but these two phases of illness are continuous and indistinguishable.

Septicaemia or leptospiraemia phage is the first stage. This stage is so-called because, during this phase, *Leptospira* organisms are usually isolated from the blood, CSF and most tissues by culture. This stage is characterised by a non-specific influenza-like illness with fever and myalgias. This phase lasts for 4–7 days. The first stage is followed by a 1- to 3-day period of improvement during which the temperature falls, and the patient becomes afebrile and relatively asymptomatic.

Immune or leptospuric phage is the second stage that starts with the recurrence of fever. This stage is called the immune or leptospuric stage because circulating antibodies may be detected in serum, or *Leptospira* may be isolated from the urine but not from blood or CSF. It occurs as a result of the body's immune response to infection. It is characterised by diseases affecting specific organs, such as the meninges, liver, kidney and eye.

Aseptic meningitis is the most important clinical manifestation. Profound jaundice, renal dysfunction, pulmonary dysfunction, hepatic necrosis and haemorrhagic diastases are other severe manifestations. Mortality is very high (5–10%) with Weil syndrome.

Laboratory Diagnosis

Specimens

Blood, urine and CSF are commonly used to demonstrate leptospires by microscopy or culture.

Microscopy

Dark-field microscopy is employed to demonstrate motile leptospires in wet mount blood preparation directly. The method is useful for detecting leptospires in the blood, early in the disease, but not in the late stage. However, the dark-field microscopy is relatively insensitive and also shows false-positive results. Protein filaments from erythrocytes and other blood artefacts can be easily mistaken for *Leptospira*. Gram stain or silver impregnation staining methods are not useful to detect leptospires.

Examination of urine is carried out by dark-ground microscopy to demonstrate *Leptospira*. Centrifuged urine can be used especially during the second week of the disease and till 4–6 weeks, during which they appear intermittently for *Leptospira*. *Leptospira* is fragile in nature, and it dies off in stored urine. Hence, the captured urine should be microscopically examined as quickly as possible.

A **direct fluorescent antibody test** is used to detect *Leptospira* antigen directly in blood or CSF. The method, however, is not widely used due to the difficulty in the availability of this test.

Culture and Identification

Isolation of the organism in the culture of blood, CSF and urine establishes the specific diagnosis of leptospirosis. The positivity of the culture depends upon the time of collection of blood.

Blood culture is positive if the specimen is collected 4 days after the onset of symptoms. It may be negative if collected too early or too late. CSF culture is positive when the specimen is collected within the first 10 days. Direct culture from urine is rarely possible because of other contaminant bacteria.

The specimens are cultured on Fetcher, Stuart or EMJH media supplemented with neomycin and 5-fluorouracil. The culture bottles are incubated at 37 °C for 2 days and then in the dark, at room temperature for 2 weeks. The cultures are examined every third day for leptospires by examining in the dark-ground microscope. Primary isolation requires a longer period of incubation for many weeks to months.

The identification of *Leptospira* isolates is made by agglutination tests using type-specific sera. Identifying isolates in particular serogroups, serovars, etc., is cumbersome; hence, it is not carried out in routine laboratories but at reference laboratories.

Animal Inoculation

Leptospires can be isolated by animal inoculation. Blood is the usual specimen, which is inoculated intraperitoneally into young guinea pigs. If the blood contains virulent *L. interrogans*, the guinea pig develops fever jaundice and dies within

8–12 days with haemorrhages into the veins and serous cavities.

Serology

Serological tests are increasingly used for the diagnosis of leptospirosis.

Microscopic agglutination test is the traditional gold standard, available only at reference laboratories for serodiagnosis of leptospirosis. This test depends on the ability of the patient's serum to agglutinate live leptospires obtained by culture. The MAT uses a battery of antigens taken from common (frequently locally endemic) Leptospira serovars prepared from the Leptospira strains cultured on the media. The test is performed by mixing the *Leptospira* antigen with serial dilutions of the patient's sera and examining microscopically for agglutination of *Leptospira* antigens. A single titre exceeding 1:200 or serial dilutions exceeding 1:100 is suggestive of leptospirosis. A fourfold rise in convalescent serum is also considered positive. The MAT is highly sensitive (92%) and specific (95%).

The MAT shows a false-negative reaction with serum specimens collected in the immune phase of the disease and patients treated with antibiotics. The MAT may show false-positive reactions with serum from the cases of Lyme disease, *Legionella* infection and syphilis.

Another group of **serological tests** detects serum-specific antibodies by employing non-pathogenic *L. biflexa* Petco I strain. These tests do not employ the exact infective serovars. Examples of such tests are the microscopic slide agglutination test, indirect haemagglutination assay (IHA), the leptospirosis immunoglobulin M (IgM) dipstick test, latex agglutination test and the *Leptospira* immunoglobulin G (IgG) ELISA. IgM ELISA is of value for diagnosing new infections within 3–5 days.

Molecular Detection

PCR is common to detect leptospiral DNA. Samples obtained from animals and human has been diagnosed using convention, real-time PCR. Primers include the 16 s rRNA gene major outer membrane lipoprotein LipL32 gene. Although it is highly sensitive, yet require expensive reagents, technical expertise and equipment, such as a thermal cycler. Combining PCR with a serology-based MAT is often recommended to effectively reduce cost without affecting the outcome.

Diagnostic tests for leptospirosis are summarised in Table 43.6.

Treatment

Antimicrobial therapy is the mainstay of the treatment of leptospirosis. Oral doxycycline is the drug of choice to treat uncomplicated *Leptospira*

Table 43.6 Diagnostic tests for leptospirosis

Diagnostic Test	Method	Test Accuracy
Microscopy	Gram stain	Organisms too thin to be detected
	Dark-field examination	Insensitive, non-specific
	Silver impregnation stain	Insensitive, non-specific
	Direct fluorescent antibody	Insensitive, specific
Culture	Blood	Positive during first 10 days
	CSF	Positive during first 10 days
	Urine	Positive after first week
Serology	Indirect haemagglutination	Insensitive, non-specific
	Slide agglutination	Insensitive, non-specific
	ELISA	Insensitive, non-specific
	MAT	Sensitive, specific, serovar specific
Molecular methods	Direct hybridisation	Insensitive, specific
	PCR	Sensitive, specific

infection, not requiring hospitalisation. Intravenous penicillin or ampicillin is recommended for hospitalised patients. Streptomycin, tetracycline or erythromycin is indicated for patients allergic to penicillin.

Epidemiology

Leptospirosis is the most common zoonotic bacterial disease throughout the world.

Geographical Distribution

Leptospirosis is distributed worldwide, occurring with the greatest frequency in the tropics. This is not found in the polar regions. The condition is much more common in the Caribbean islands, the Pacific islands, Central and South America and Southeast Asia.

Habitat

In an infected host, leptospires are found to multiply in the blood and any organ of the body but most commonly in the liver, kidney and meninges.

Reservoir, Source and Transmission of Infection

Leptospirosis is a zoonotic disease. Wild mammals are the primary reservoirs of infection. Leptospires infect as many as 160 mammalian species, including rats, dogs, cats, cattle, pigs, raccoons and other animals. Rodents are the most important reservoirs of infection worldwide. Leptospirosis in animals is usually a sub-clinical infection. Leptospires are found for a long period in the kidneys of these animals; as a result, animals excrete a large number of leptospires in their urine without showing any evidence of clinical disease. The leptospires survive in the soil for as many as 24 days and in freshwater for as many as 30 days.

The urine of the infected animals containing a large number of leptospires is the most important source of infection. Direct contact with the infected urine or urine-contaminated inanimate objects results in human infections. Animal bedding, soil, mud and aborted tissues are examples of such inanimate objects contaminated with infected urine. The infection is transmitted to humans (1) through intact mucous membrane or conjunctiva; (2) through minor abrasions on the surface of the skin, waterlogged skinnier; (3) through the nasal mucosa and cribriform plate; (4) through inhalation of aerosolised body fluids and (5) by congenital infection (from infected mother to foetus through the placenta).

Differences between leptospires and treponemes are summarised in Table 43.7.

Table 43.7 Differential features of treponemes, leptospires and Borrelia

Features	Treponemes	Leptospires	Borrelia
Size	6–14 × 0.2 μm	6–20 × 0.1 μm	10–20 × 0.3 μm
Spirals	6–12 regular, close spirals at 1 μm interval; the ends of the spirals are pointed	Tightly coiled regular spirals with hooked ends; the interval and amplitude of spirals are 0.5 μm each	Loosely coiled, irregular in shape with wide coils
Number of endoflagella	3–4 at each pole	1 at each pole	7–11 at each pole
Motility	Actively motile with flexion and extension; translatory and corkscrew-like rotatory motility	Rotation around long axis forwards and backwards and bending and flexion	By periplasmic flagella
Disease	Syphilis, bejel, yaws	Leptospirosis	Relapsing fever, Lyme disease
Mode of infection	Sexual contact, intimate cutaneous contact, blood transfusion and transplacental	Water contaminated with rodent urine	Transmitted by lice and ticks
Incubation period	10–90 days	7–14 days	3–30

Prevention and Control

Preventive measures include detecting cases and detecting a common source of infection. Disinfection of water supplies, rodent control and wearing protective clothing control disease transmission. Leptospire infection in livestock is controlled by improved sanitation, immunisation of animals and proper veterinary care.

Recently, human vaccines have been available against leptospirosis. These vaccines are recommended only for high-risk workers. These vaccines are serovar specific and usually given every year.

Case Study

A 28-year-old hairdresser complained of a painless small ulcer on the penis during the last 2 weeks. He said he had repeated sexual relations with a female sex worker many months back when asked. The pus exudate from the smear did not reveal any *Treponema* on examination. The serum sample was positive for syphilis by the VDRL test. ELISA for HIV was negative.

- What are the specific tests used for the diagnosis of the condition?
- How will you demonstrate the causative bacteria in the clinical specimens?
- What are the virulence factors of the bacteria?
- What are the antibiotics used for the treatment of the condition?

Rickettsia, Orientia, Ehrlichia and *Coxiella*

44

Rickettsiae are obligate, intracellular, very small (0.3 × 1–2 μm), Gram-negative bacilli that multiply within the cytoplasm of eukaryotic cells. They have a very small genome composed of 1–1.5 million base pairs. Because of their small size, these organisms were once thought to be viruses. Nevertheless, these organisms are bacteria because they (1) have typical Gram-negative cell walls, (2) contain both DNA and RNA enzymes for the Kreb cycle and ribosomes for protein synthesis, (3) multiply by binary fission and (4) are susceptible to antibiotics.

Rickettsiae are primary pathogens of arthropods, such as lice, fleas, ticks and mites. In these hosts, they are found in their intestinal tract. They are usually transmitted to humans by arthropod vectors, such as lice, mites, ticks, etc. *Coxiella burnetii* causing Q fever is an exception, usually transmitted by airborne droplets. They also infect humans in whom they are found in the reticuloendothelial cells and vascular endothelium.

Classification

Rickettsiae are a group of bacteria that phylogenetically occupy a position between bacteria and viruses. These pathogens are included in the order Rickettsiales, tribe Rickettsiae and family Rickettsiaceae. Family Rickettsiaceae comprises three genera: *Rickettsia, Orientia* and *Ehrlichia.* Rickettsiae are strict intracellular parasites. Earlier, the genus *Coxiella* was included in the family Rickettsiaceae but now has been excluded from it because, unlike other genera of the family Rickettsiaceae, *Coxiella* organisms are not primarily obligate intracellular parasites. They can grow in cell-free media.

The cell wall of all four genera is typical of Gram-negative bacteria. They have lipopolysaccharide (LPS) and peptidoglycan layers. The LPS shows weak endotoxic activity. The bacteria are non-motile, do not have any flagella and are surrounded by a loose slime layer. Rickettsiae that cause human diseases are summarised in Table 44.1.

Genus *Rickettsia*

Rickettsia causes various diseases varying considerably in severity from self-limiting illness to fulminating, life-threatening infection.

Properties of the Bacteria

Morphology

They are small Gram-negative *Rickettsia* coccobacilli varying from 0.3–0.6 to 0.8–2 μm in size. They are non-motile and non-capsulated. They are stained poorly with Gram stain. They are stained deep red with Machiavello and Gimenez stain and bluish-purple with Giemsa and Castaneda stain.

Table 44.1 Human infections caused by *Rickettsia, Orientia, Ehrlichia* and *Coxiella* species

Bacteria	Diseases
Rickettsia prowazekii	Epidemic or louse-borne typhus Relapsing louse-borne typhus or Brill–Zinsser disease
Rickettsia typhi	Endemic or flea-borne murine typhus
Rickettsia rickettsiae	Rocky mountain spotted fever
Rickettsia akari	Rickettsial pox
Rickettsia conori	Boutonneuse fever
Orientia tsutsugamushi	Scrub typhus
Ehrlichia sennetsu	Sennetsu fever
Ehrlichia chaffeensis	Monocytic ehrlichiosis
Ehrlichia phagocytophila	Human granulocytic ehrlichiosis
Coxiella burnetii	Q fever
Rickettsia australis	Queensland tick typhus
Rickettsia sibirica	North Asian tick typhus

Culture

Rickettsiae fail to grow on cell-free media. They grow inside the cell, usually in the cytoplasm (most *Rickettsia)* or in the cell's nucleus *(Rickettsia* causing spotted fever). They grow well at an optimum temperature of 32–35 °C. They grow in various cell lines, developing chick embryos and laboratory animals.

They grow on **cell lines** such as HeLa, Hep2, Detriot-6, mouse fibroblasts and other continuous cell lines. Cultures in the cell lines are used primarily to maintain *Rickettsia* but are not useful for primary isolation *of Rickettsia* from clinical specimens.

In the developing **chick embryo** 5–6 days old, *Rickettsia* spp. grow well in the **yolk sac**. The inoculated eggs are incubated at 35 °C for most *Rickettsia* spp. and at 33 °C for the spotted group. The yolk sac is widely used as a source of *Rickettsia* for the preparation of rickettsial antigens and vaccines. *Rickettsia* shows poor growth on the chorioallantoic membrane.

Guinea pigs and mice are the commonly used **laboratory animals** to isolate *Rickettsia* organisms from animal specimens.

Other Properties

Susceptibility to Physical and Chemical Agents

The extracellular *Rickettsia* organisms are very delicate microorganisms. They are rapidly killed by heating at 56 °C and room temperature. They are destroyed by the usual strength of antiseptics, such as hypochlorite, 1% ethanol, 2% formaldehyde, 5% hydrogen peroxide and 70% ethanol. They are preserved at −70 °C or in a lyophilised state. They are preserved better in a special medium known as **SPG medium** containing sucrose, potassium phosphate and glutamate and skimmed milk.

Antigenic Structure

Rickettsia possesses three different types of antigens. The **group-specific antigen** is a soluble antigen present on the surface of the *Rickettsia.* This is extracted from rickettsial pathogens by repeated washings and centrifugations. **Species- or strain-specific antigen** is present in the bacteria's cell wall (e.g. scrub typhus). **Alkali-stable polysaccharide antigen** is a surface antigen found in some species of *Rickettsia* and some strains of *Proteus* species *(Proteus* OX19, OX2 and OK). This sharing of antigen between *Rickettsia* and *Proteus* forms the basis of the Weil–Felix test, which is employed for the diagnosis of rickettsial infections by demonstration of antibodies using *Proteus* strains.

Pathogenesis

Virulence Factors

Multiplying inside the cell is an important mechanism of the disease process caused by rickettsial. **Adhesins** are the most important virulent factor of rickettsial pathogens. These are outer membrane proteins, which facilitate the entry of rickettsiae into the host cells. Once inside the cells, they remain, multiply and accumulate in large numbers before lysing the host cell.

After inoculation from the infected sites, the rickettsiae reach the circulation, multiply and cause rickettsemia. Rickettsiae are localised in the endothelial cells of small arterial capillary and venous vessels. At these sites, they multiply and cause endothelial cellular hyperplasia resulting in multiorgan vasculitis. The process may end in thrombosis and the development of small nodules. Thrombosis of supplying blood vessels may cause gangrene of the extremities, ear lobes, nose and genitalia. The vasculitis process also may result in increased vascular permeability with consequent oedema, loss of blood volume, hypoalbuminemia, reduced osmotic pressure and hypotension.

Typhus Fever Group

Typhus refers to a group of infectious diseases caused by different rickettsial organisms. These are of three types: i) epidemic or louse-borne typhus caused by *Rickettsia prowazekii, ii)* relapsing louse-borne typhus or Brill–Zinsser disease caused by *R. prowazekii* and iii) endemic or flea-borne murine typhus caused by *Rickettsia typhi.*

Epidemic or Louse-Borne Typhus

R. prowazekii is the causative agent of epidemic or louse-borne typhus. This condition is an acute febrile illness transmitted by the human body louse *Pediculus humanus corporis. R. prowazekii* is named after the scientist Von Prowazek who died of typhus fever while studying the disease. This typhus fever is an ancient disease reported from all parts of the world. This disease was widely prevalent in Russia and Eastern Europe. This disease was also responsible for Napoleon's defeat in Russia in 1812. This was one of the three diseases responsible for misery and suffering during the Irish famine of 1845–1850.

R. prowazekii is an invasive bacterium, which characteristically multiplies in endothelial cells of the blood vessels, leading to vasculitis. The incubation period varies from as low as 2–3 days to an average of 8 days.

Epidemic typhus is characterised by high fever, severe headache and chills. The appearance of a petechial or macular rash on the fourth or fifth day—first starting on the trunk and then spreading over to the extremities but without affecting the face, palms and sole—is the characteristic feature of the condition. The patient, if left untreated, may become stuporous and delirious. The name typhus is derived from the word *typhus*, meaning cloud of smoke, which denotes a cloudy state of consciousness in the disease process. Myocarditis and central nervous system (CNS) dysfunction are the noted complications of epidemic typhus. The disease is associated with a mortality rate as high as 60% in old or debilitated persons.

Epidemic typhus is present in Central and South America, Africa (Ethiopia and Nigeria), Northern China and India. Humans are the primary reservoir of the epidemic typhus and are the sources of infection. The lice become infected with *R. prowazekii* after feeding on a person suffering from typhus fever or Brill–Zinsser disease. Body louse, *P. humanus corporis*, is the arthropod vector of epidemic typhus. Occasionally, the head louse *P. humanus capitis* may transmit the infection, but not the pubic louse. *R. prowazekii*, the causative agent of typhus, lives and multiplies in the alimentary tract of the louse. The bacteria are excreted in faeces within 3–5 days of infection. The lice are the only arthropod vector for rickettsial infection, which die after the infection (Fig. 44.1). Infection is transmitted to humans when a rickettsia-harbouring louse bites a human during a blood meal.

Typically, lice defecate while feeding. When the host scratches the bite, the louse is crushed and the contaminated louse faeces are inoculated into the minute lesions of the bite wound. The rickettsiae reach the circulation from the infected sites, multiply and cause rickettsemia. Rarely, the

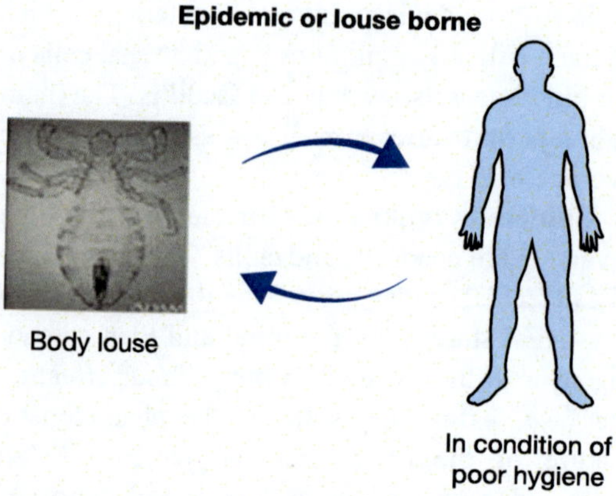

Fig. 44.1 Transmission of louse-borne typhus fever

infection may also be transmitted through the conjunctiva or inhalation of aerosols of dry louse faeces.

Relapsing Louse-Borne Typhus

Relapsing louse-borne typhus or **Brill–Zinsser disease** is an example of recrudescent typhus fever. Brill first noticed this disease in 1988, and *R. prowazekii* were isolated from the areas in 1934. Hence, the disease is also called Brill–Zinsser disease. The exact mechanism responsible for the recrudescence is not known. This condition was seen in some patients treated with antibiotics and has been apparently cured of the disease. *R. prowazekii* in such patients may persist in their body tissues, may re-emerge and cause a recurrence of typhus fever months, years or even decades after antibiotic treatment.

Endemic or Flea-Borne Murine Typhus

Endemic or flea-borne murine typhus caused by *R. typhi* is a milder disease than the epidemic typhus and has a shorter duration. The incubation period varies from 7 to 14 days. The condition has a sudden onset of symptoms with fever, headache, malaise and myalgia. A rash develops on the third to the fifth day of infection in approximately half of the infected patients. The rash is typically present on the chest and abdomen but may spread to the palms and soles. The untreated course of the disease may last up to 3 weeks.

Endemic typhus occurs mainly in sporadic forms. Rats (*Rattus rattus*), mice and cats are the natural reservoirs of infection. Humans are the accidental hosts. Rat flea (*Xenopsylla cheopis*) or cat flea (*Ctenocephalides felis*) are the main vectors responsible for disease transmission. Endemic or flea-borne murine typhus is transmitted from rats to rats by a rat flea and accidentally to humans by the faeces of infected fleas. Fleas become infected by feeding on mice, cats or other natural hosts. These infected fleas may subsequently transmit the disease to humans during the act of biting. During the bite, they transmit the disease by direct inoculation or indirect inoculation of the infected faeces into the site of the bite. The cat flea *C. felis* may also transmit the disease. The infection can also be transmitted by inoculation or inhalation of aerosolised infectious specimens. The infection may also be transmitted by ingestion and food contaminated with infected rat urine or flea faeces.

Spotted Fever Group

Spotted fever group of rickettsial diseases includes (1) rocky mountain spotted fever caused by *Rickettsia rickettsiae*, (2) rickettsial pox caused by *Rickettsia akari* and (3) boutonneuse fever (i.e. Kenya tick-bite fever, African tick typhus, Mediterranean spotted fever, Indian tick typhus and Marseilles fever) caused by *Rickettsia conori*

Rickettsiae of this group possess a common soluble group antigen. They also multiply in the nucleus and the cytoplasm of the infected cells. All these species except *R. akari* are transmitted by ticks. A total of 12 species of rickettsiae have been associated with humans causing spotted fever, and seven species have been isolated from arthropod vectors. *R. rickettsiae* is the most common species belonging to the spotted fever group and is responsible for Rocky Mountain spotted fever.

Rocky Mountain Spotted Fever

Rocky Mountain spotted fever is the most serious type of spotted fever. It was the first among the spotted fever group to be described. This disease was earlier called Mediterranean disease and later boutonneuse fever by Connor, who described this condition for the first time in 1910. Megaw, in

1917, first described this disease in the foothills of the Himalayas in India.

The incubation period is 7 days. The condition is characterised by fever, severe headache, chills and myalgia. A rash may develop after three or more days and typically appears initially on the wrist, ankles and palms and soles and then spreads to the trunk. The rash is maculopapular early in the disease but may later become petechial and haemorrhagic. This is a serious disease associated with many complications, such as respiratory failure, encephalitis and renal failure. The patient may die within 5 days of the onset of symptoms.

Rocky Mountain spotted fever is distributed in southern Canada, Central America, Mexico and South America. Ticks are the natural hosts, reservoirs and vectors of *R. rickettsiae*. Different tick hosts are associated with infection in different parts of the world. These vectors include the wood tick (*Dermacentor andersoni*) in the western United States and western Canada, the American dog tick (*Dermacentor variabilis*) in eastern Canada and the eastern United States and Lone star tick (*Amblyomma americanum*) in the southwestern United States. Rickettsia multiply in these ticks and are transmitted transovarially to the next generation (Fig. 44.2).

R. rickettsiae is transmitted to humans through saliva during the bite of a tick. It usually takes 6 h of attachment and feeding before rickettsiae are transmitted to the host. Infection can also be transmitted occasionally by scratching and rubbing infectious tick faeces into the abraded skin.

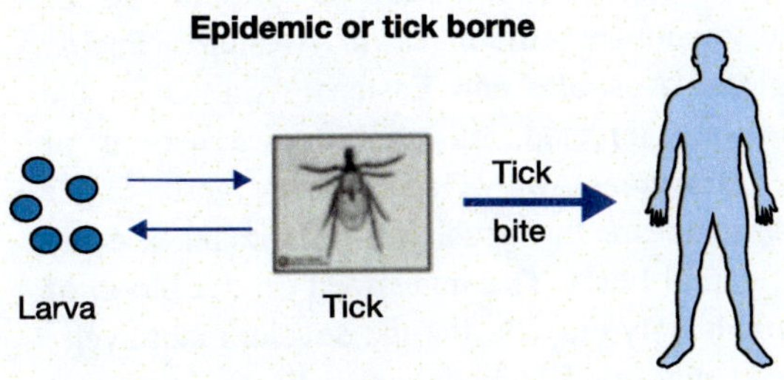

Fig. 44.2 Transmission of tick-borne typhus fever

Rickettsial Pox

Rickettsial pox caused by *R. akari* is prevalent worldwide. It has been reported in Russia, South Africa and Korea. The common house mouse (*Mus musculus*) is the natural reservoir. *R. akari* is transmitted from mouse to mouse by the bite of the mouse mite (*Liponyssoides sanguineus*). The infection in the mite is transmitted to the progeny transovarially (Fig. 44.3).

The incubation period is 7 days. The condition manifests by the development of papule at the bite site by the mite. Subsequently, the papule progresses to an ulcer, leading to the formation of eschar. This is followed by fever, headache, malaise and myalgia in 3–10 days. A generalised papular vesicular rash usually appears 3–4 days after the emergence of fever. The illness lasts for a short duration of 10–14 days, after which recovery occurs. Complete healing of rash occurs within 2–3 weeks without treatment.

Clinically rickettsial pox is a mild form of infection. Like other rickettsial infections, diagnosis of the condition is clinically supported by serology. This disease is differentiated from other rickettsial infections by (1) the presence of an *eschar* at the site of bite of mouse mite, (2) the presence of a *vesicular pustular eruption* and (3) a negative Weil–Felix reaction. Rodents are the natural host of the mite transmitting rickettsial pox in different parts of the world. Treatment with doxycycline or chloramphenicol is highly effective for rickettsial pox infections.

Other Rickettsial Species in the Spotted Fever Group

Six other rickettsial species in the spotted fever group have been associated with human disease. These include tick-borne diseases, such as boutonneuse fever caused by *R. conori*, Australian tick typhus caused by *Rickettsia australis* and Siberian tick typhus caused by *Rickettsia sibirica.* All these rickettsial pathogens are maintained in wild animals and ixodid ticks. Humans are the accidental hosts. The clinical diseases produced by these rickettsiae are similar to the Rocky Mountain spotted fever but are

Fig. 44.3 Comparison of Rickettsial pox and Rocky Mountain spotted fever

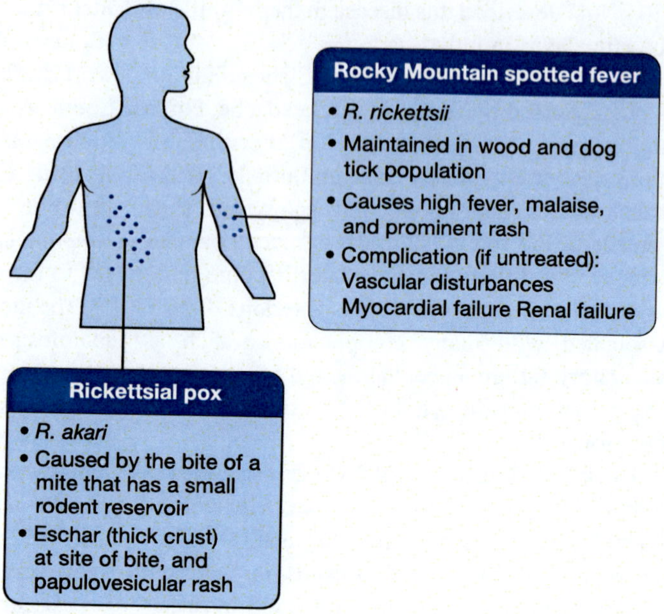

Table 44.2 Epidemiology features of *Rickettsia* species causing human diseases

Species	Vector	Reservoir	Disease
Rickettsia prowazekii	Louse	Humans	Epidemic typhus and Brill–Zinsser disease
Rickettsia typhi	Rat flea	Rat	Endemic typhus
Rickettsia rickettsiae	Tick	Ticks, wild rodents	Rocky Mountain spotted fever
Rickettsia akari	Gamasid mite	Mouse	Rickettsial pox
Rickettsia conori	Tick	Rodents	Boutonneuse fever
Rickettsia australis	Tick	Unknown	Queensland tick typhus

relatively milder. Indian tick typhus transmitted by *R. conori* is a milder disease (Table 44.2).

Genus *Orientia*

Genus *Orientia* contains *Orientia tsutsugamushi*, formerly known as *Rickettsia tsutsugamushi*, the causative agent of **scrub typhus**.

The scrub typhus is transmitted to humans by the mite *Leptotrombidium akamushi* and possibly by *Leptotrombidium deliense*. Four stages of development usually occur in the mite's life cycle. Of these stages, only the larval stage (*chigger*) is infectious to humans and other mammals because these stages require a blood meal for further development. *R. tsutsugamushi* shows a remarkable antigenic heterogeneity. Five major serotypes are recognised. These are Karp, Gilliam Kato Boryon and Kawazaki types. The organism is transmitted to humans by chiggers that live and reproduce in the soil and in the scrub vegetation (this is called a zoonotic thread). The clearing of forests has led to the spread of the disease to urban areas of the country.

Scrub typhus is restricted in its geographical distribution. The condition is present in the Tsutsugamushi triangle of western Pacific regions, Australia and Eastern Asia. Once the mice are infected, they act as a reservoir of *O. tsutsugamushi*. Rats, mice and larger mammals are the usual hosts. Humans are the accidental hosts. The mites feed on the blood of animals only once during the development cycle; hence they are not believed to be an important reservoir for human diseases. The infection is transmitted from rats to the progeny by transovarial transmission.

Scrub typhus is a mild and self-limiting disease but, if untreated, is fatal. The incubation period, on average, varies from 10 to 12 days. Usually, a characteristic eschar will develop at the site of a mite bite, accompanied by regional lymphadenopathy and maculopapular rash. A thorough physical examination is required to identify the eschars. Further, the condition manifests with severe headache, fever, myalgia and rash. The maculopapular rash develops initially on the trunk and later on the limbs and is present in nearly half of the patients. In untreated patients, fever disappears after 2–3 weeks. Splenomegaly, CNS complications, generalised lymphadenopathy and heart failure are the condition's complications.

Tetracyclines, doxycycline or chloramphenicol are highly effective in treating scrub typhus. An effective vaccine is not available. Using insect repellents and wearing protective clothing to avoid exposure to chiggers prevent the disease.

Genus *Ehrlichia*

Ehrlichia are small, Gram-negative, obligate, intracellular bacilli that parasitise mononuclear and granulocytic phagocytes but not erythrocytes. The bacteria multiply in the cytoplasm of infected phagocytic cells as mulberry-like clusters called the *morula*. The genus *Ehrlichia* consists of three species that cause infections in humans. These are (1) *Ehrlichia sennetsu* causing sennetsu fever, (2) *Ehrlichia chaffeensis* causing monocytic ehrlichiosis and (3) *Ehrlichia phagocytophila* causing human granulocytic ehrlichiosis.

The genus *Ehrlichia* is named after the scientist Paul Ehrlich. The first case of *Ehrlichia* infection in humans was reported in Japan in 1954. The cases resembled glandular fever with the serum positive for antibodies against the agents of canine ehrlichiosis. The causative agent isolated from the cases was named *E. sennetsu*. *Sennetsu* is a Japanese word meaning glandular fever. *Ehrlichia* species cause three clinical syndromes, namely sennetsu fever, human monocytic ehrlichiosis and human granulocytic ehrlichiosis.

Sennetsu fever is caused by *E. sennetsu*. The condition presents typically as an acute febrile illness and resembles glandular fever. Cervical lymphadenopathy, atypical lymphocytosis and an increase in peripheral mononuclear cells are the typical manifestations of the disease.

Human monocytic ehrlichiosis is caused by *E. chaffeensis,* which infects mostly monocytes. Human diseases are associated with headache, malaise and myalgia after 1–3 weeks of a tick-bite. Leukopenia, thrombocytopenia and elevated liver enzymes are other manifestations. A rash is found in only 20% of the patients. Mortality due to condition is less and is seen primarily in patients with immunocompromised status and in elderly patients.

Human granulocytic ehrlichiosis is caused by *E. phagocytophila* and *Ehrlichia ewingii,* which infect mostly granulocytes. This condition is characterised by a febrile illness with similar manifestations to human monocytic ehrlichiosis. The condition is associated with leukopenia and thrombocytopenia, similar to that in human monocytic ehrlichiosis.

Sennetsu fever is restricted to Japan. Human monocytic ehrlichiosis is distributed mainly in the south-eastern, mid-Atlantic and south-central parts of the United States. There are areas in which large numbers of lone star ticks (*A. americanum*) are found. Human granulocytic ehrlichiosis is distributed in the Northeast and Central Atlantic states and midwestern states of the United States and Europe. Reservoirs are not known for sennetsu fever. It is an exception, which is transmitted by ingestion of raw fish infected with flukes but not by ticks.

Ehrlichiosis is a tick-borne disease. Ticks are the vectors for all *Ehrlichia* species causing diseases in humans and animals. White tick deer and domestic dogs are reservoirs for *E. chaffeensis*, the lone star (*A. americanum*) tick that transmits the bacteria. Dogs are the vectors for *E. ewingii*. The lone star tick also transmits the infection. Small mammals, such as white-footed mice, chipmunks and whorls, are the reservoirs for *E. phagocytophila*. *Ixodes* species, including *Ixodes capillaris*, *Ixodes ricinus* and *Ixodes pacificus* are the vectors for the pathogen.

Table 44.3 Epidemiology features of *Ehrlichia* species causing human diseases

Ehrlichia species	Vector	Reservoir	Disease
Ehrlichia sennetsu	No tick vector; ingestion of raw fish infected with flukes	Unknown	Sennetsu fever
Ehrlichia chaffeensis	Lone star tick	White tick deer and domestic dogs	Monocytic ehrlichiosis
Ehrlichia phagocytophila	*Ixodes* ticks	Small mammals (white-footed mouse, chipmunks and whorls)	Human granulocytic ehrlichiosis
Ehrlichia ewingii	*Amblyomma* ticks	Unknown	Granulocytic ehrlichiosis

The epidemiology of *Ehrlichia* species causing human diseases is summarised in Table 44.3.

Laboratory Diagnosis of Rickettsia, Orientia and Ehrlichia Infections

Laboratory diagnosis is made either by isolation of rickettsia or by serological tests.

Specimens

These include blood and tissue samples for culture and serum for serological tests.

Microscopy

Cytoplasmic inclusion bodies in rickettsial infections can be demonstrated by microscopy. They are stained well with the following stains: deep red with Machiavello and Gimenez stain and bluish-purple with Giemsa and Castaneda stain. Morula forms are demonstrated in infected phagocytic cells stained by Giemsa stain in ehrlichiosis.

Culture

Rickettsiae are highly infectious pathogens; therefore, isolation of these pathogens from clinical specimens is carried out only in the BSL III safety laboratory equipped with high safety provisions.

Animal Inoculation

Rickettsiae isolation is carried out in male guinea pigs or mice by inoculating the clinical specimen intraperitoneally. After inoculation, the animals are observed for 3–4 weeks. The response of animals to different rickettsial infections may vary. *R. prowazekii* infection produces fever without any testicular inflammation. Infection with *R. typhi*, *R. conori* and *R. akari* develops fever and tunica reaction. Rickettsia causing Rocky Mountain spotted fever causes fever and scrotal necrosis in guinea pigs. The guinea pigs may even die of the disease. Smears from the infected animals' peritoneum, tunica and spleen may be stained by Giemsa and Gimenez methods to detect these pathogens.

R. typhi is differentiated from *R. prowazekii* by **Neil Mooser reaction** and partial DNA homology. Neil Mooser or tunica reaction is performed in an adult male guinea pig. The specimens (e.g. blood from the case of endemic typhus or *R. typhi* culture) are inoculated intraperitoneally. An inoculated animal is observed for the development of fever and a characteristic scrotal inflammation. In a positive test, the scrotum becomes enlarged and the testis cannot be pushed back into the layer of tunica vaginalis. This reaction is known as **Neil Mooser or tunica vaginalis reaction**. *R. typhi* shows a positive Neil Mooser reaction. *R. prowazekii* shows a negative Neil Mooser reaction.

Rickettsiae may be isolated by growing in the cell culture. They grow better on Vero cells in three and half days. Immunofluorescence identifies the rickettsiae in these infected cell lines using group-specific and strain-specific monoclonal antibodies.

Serology

Weil–Felix test is a heterologous agglutination test used for a long to diagnose rickettsial

Table 44.4 Weil–Felix test in Rickettsial diseases

Rickettsial Diseases	Agglutination pattern with OX19	OX2	OXK
Epidemic typhus	Positive (+++)	Positive(+)	Negative (−)
Brill–Zinsser disease			Negative (−)
Endemic typhus	Positive (+++)	Variable (+/−)	Negative (−)
Spotted fever	Positive (++)	Positive (+ +)	Negative (−)
Scrub typhus	Negative (−)	Negative (−)	Positive (++ +)

infections. The test detects anti-rickettsial antibodies that cross-react with O antigens of certain non-motile strains of *Proteus*. In this test, non-motile strains of *Proteus vulgaris* OX19 and OX2 and *Proteus mirabilis* OXK are used as antigens. The Weil–Felix test becomes positive 10–20 days after infection; sera from epidemic typhus strongly react with OX19 antigen and weakly agglutinate with OX2 antigen. They do not agglutinate with OXK antigen. Weil–Felix test is negative or weakly reactive in Brill–Zinsser disease. Reactions of the Weil–Felix test in other rickettsial infections are summarised in Table 44.4.

Complement fixation test, indirect haemagglutination, indirect immunofluorescence, latex agglutination and enzyme immunoassay are the **other serological tests** used to diagnose rickettsial infections, including *R. prowazekii*. These tests employ rickettsial antigens for the demonstration of rickettsial antibodies. IFA test using *R. typhi*-specific antigen is employed as a specific test for serodiagnosis of endemic typhus. A single titre of 1:128 or a fourfold rise in antibody titre in paired sera is diagnostic of the disease. It is considered the gold standard.

Molecular Detection

Polymerase chain reaction (PCR) is used to detect rickettsia in blood or tissue for early diagnosis of the condition.

Treatment

Tetracycline and chloramphenicol are the drugs of choice to treat epidemic typhus. Antibiotic therapy in combination with the treatment of louse infestation of the human host is effective. Tetracyclines, chloramphenicol and fluoroquinolones, such as ciprofloxacin, are effective against *R. rickettsiae*. Doxycycline is the drug of choice for the treatment of human ehrlichiosis. Chloramphenicol and fluoroquinolones are not that effective.

Prevention and Control

Control of the human lice population and sanitation is the effective measures to control epidemic typhus. Control of endemic typhus is difficult because the fleas causing the disease are distributed widely. The control measures are essentially based on the control of the rodent and flea populations in the area endemic to the disease. The use of protective clothing and insect repellents and avoidance of tick-infested areas are the preventive measures against *R. rickettsiae* infection. No vaccine is available against the Rocky Mountain spotted fever. Use of insect repellents, wearing protective clothing, etc. are preventive measures against bites by ticks. A live vaccine containing the attenuated 'E' strain of *R. prowazekii* is available and is recommended for use in a high-risk population.

Genus *Coxiella*

The genus *Coxiella* was originally classified with rickettsia because the bacteria showed features of a rickettsial pathogen. Like rickettsiae, they (1) stain poorly with Gram staining, (2) multiply intracellularly in eukaryotic cells and (3) are transmitted by arthropods. *Coxiella* differs from rickettsiae by being not transmitted by arthropod vectors but are transmitted by inhalation or ingestion. They show relatively more resistance to the actions of dry heat.

The genus *Coxiella* now has been separated from rickettsiae and is placed in the group Protobacteria along with other genera, such as *Legionella* and *Francisella*. This is because *Coxiella* was more closely related to these two genera. The genus *Coxiella* includes the species *C. burnetii,* which causes Q fever (Table 44.4).

Coxiella burnetii

C. burnetii is the causative agent of Q fever, a zoonotic disease transmitted from animals to humans. Q fever was first studied in an experimental infection of guinea pigs by inoculation of blood from patients suffering from typhus-like fever to guinea pigs by Derrick in 1935. As the etiological agent of the disease was not known, the condition was referred to as query or Q fever. Subsequently, Burnett identified the causative agent as a rickettsial, after which the pathogen was named *Rickettsia burnetii*. Cox in the United States demonstrated the agent in ticks and named it *Rickettsia diaphorica*. Later, both the rickettsial strains were shown to be identical. The organism is now named *C. burnetii* and reclassified in the group Proteobacteria.

Properties of the Bacteria

Morphology

C. burnetii is an obligate intracellular bacterium. It is pleomorphic and measures 1 × 0.3 μm in size. The bacterium is Gram-negative but stains poorly with Gram staining. These bacteria, like other rickettsiae, stain best with Giemsa or Gimenez stains.

Culture

C. burnetii primarily infects the monocytic macrophage cell. It multiplies in phagolysosomes of the infected cell. It grows well in various **cell lines** and the **yolk sac**.

Other Properties

Susceptibility to Physical and Chemical Agents

Coxiella is the most resistant pathogen to heat and dryness. They are not completely killed at 60 °C or by 1% phenol in 60 min. Hence, pasteurisation by holding method is not effective for killing *C. burnetii,* but the flash method of pasteurisation is effective. The bacteria survive for a month in meat and a year or more in dried faeces at 4 °C.

Antigenic Structure

Phase variation: Characteristically, *Coxiella* exists antigenically in two forms: phase 1 antigen and phase 2 antigen. **Phase 1 antigen** is a cell wall LPS antigen with a complex carbohydrate structure. This antigen usually prevents the interaction of antibodies with surface protein. *Coxiella* in phase 1 antigen is highly infectious. Phase 1 is highly immunogenic and elicits a strong antibody response to both 1 and 2 antigens. **Phase 2 antigen** occurs by following repeated passage in the yolk sac in which the LPS is modified by exposing the surface proteins to antibodies. In this phase, the surface proteins act with host antibodies. *Coxiella* occurring in phase 2 is a less infective form of the bacterium.

C. burnetii either in phase 1 or phase 2 does not show any antigenic sharing with rickettsia or *Proteus* bacillus antigen. Therefore, the Weil–Felix test is not useful for detecting *C. burnetii* infection. The antibodies in serum of Q fever do not react with *Proteus* antigens used in the Weil–Felix properties stage of Q fever is characterised by the presence of antibody titres against phase 2 antigen. In contrast, chronic disease is characterised by the presence of antibodies against both phase 1 and phase 2 antigens. These high titres of antibodies appear to be responsible for pathological changes in various organs during the disease. Cell-mediated immunity is associated with recovery from clinical illness.

Pathogenesis

C. burnetii causes infection by penetrating abraded skin, mucosa, lung or intestinal tract. They typically grow and multiply in the acidic environment of fused phagosome and lysosome. Phase 1 form of *C. burnetii* is protected from the action of antibodies.

Coxiella affects lungs and heart valves and has also been found in macrophages in the lungs and vegetations of heart valves. The bacteria also cause granulomatous changes in reticuloendothelial organs, such as the liver, spleen, etc. Host-mediated pathogenic mechanisms are believed to play an important role in the pathogenesis of Q fever.

Clinical Syndromes

C. burnetii causes Q fever, which can occur in acute or chronic forms.

Q Fever

The incubation period is 20 days. The acute stage of Q fever manifests as sudden onset of severe headache, high fever, chills and myalgia. The bacteria produce a respiratory infection mimicking atypical pneumonia. They also cause hepatosplenomegaly, which is found in approximately half of the patients.

The chronic form of Q fever has a long incubation period varying from months to years. The onset of the disease is usually insidious. Sub-acute endocarditis on a previously damaged heart valve or a prosthetic heart valve is the most common manifestation of the condition. Hepatitis and meningoencephalitis are the other manifestations of the disease. Uncomplicated acute disease is a self-limiting disease that lasts for 1–2 years. Complications of chronic diseases may increase the mortality rate.

Laboratory Diagnosis

C. burnetii can be diagnosed by isolating the bacteria from clinical specimens in guinea pigs, mice and developing chick embryos. But the isolation of the bacteria by these methods is not attempted because these procedures are biohazardous and are also not required.

Serology

Serology is the mainstay of diagnosis of Q fever. Serological tests include immunofluorescence test, complement fixation test and ELISA. These tests detect IgM or IgG antibodies by using phase 1 and 2 antigens.

Serodiagnosis of acute Q fever is made by demonstration of (1) IgG titre of 1:200 or more, or IgM titre of 1:50 or more in a single serum specimen or (2) a fourfold rise of antibody titre between acute and convalescent sera. The IgG antibodies are present in serum for more than 1 year in 90% of patients, whereas IgM antibodies are present only for 2 weeks and become negative after 2 weeks. Serodiagnosis of chronic Q fever is made by demonstrating high antibody titre against phase 1 and 2 antigens. Antibodies against phase 1 antigen are always higher. Weil–Felix test is not used for the diagnosis of Q fever. Molecular Diagnosis.

Molecular Detection

PCR is currently evaluated to detect *C. burnetii* in clinical specimens; however, it still remains to be a research tool and is not widely used in the diagnosis.

Treatment

Treatment with tetracyclines, such as doxycycline, is effective for acute Q fever. A combination of tetracycline and cotrimoxazole or rifampicin for a prolonged period is essential for treating sub-acute endocarditis and other chronic diseases caused by *C. burnetii*.

Epidemiology

Like other human rickettsial infections, Q fever is a zoonotic disease transmitted from infected animals to humans.

Geographical Distribution

C. burnetii infection is common in domestic livestock in many parts of the world. *C. burnetii* infection in birds and animals has been documented by serological studies. Sporadic cases of human infections have been reported.

Habitat

C. burnetii are intracellular pathogens in infected livestock and also in humans.

Reservoir, Source and Transmission of Infection

Q fever is a zoonosis distributed worldwide. Farm animals, such as sheep, cattle and goats, and cats, dogs and rabbits are the primary reservoirs of infection. Ixodid ticks are responsible for transmitting the disease to rodents and domestic animals and for maintaining infection in these animals.

C. burnetii infection in ticks is transmitted transovarially to progeny. *Coxiella* are excreted in the faeces and survive in dried faeces for a long time. In infected animals, *Coxiella* are excreted in the milk and particularly in their products of conception during parturition of these animals. These bacteria contaminate the animals' surroundings, where they remain as a potential source of infection to humans for months. Ticks do not play any role in transmitting Q fever from infected animals to humans. Humans acquire diseases by inhalation of infected aerosols. They also possibly acquire the disease from milk consumption infected with *C. burnetii*. Person-to-person transmission does not occur.

Prevention and Control

Preventive measures for Q fever include (1) wearing protective clothing (mask, gloves) while handling carcasses and animal hides, (2) pasteurisation of milk by the flash method instead of the holder method and (3) isolation of facilities for the parturition of animals.

Case Study

A 30-year-old farmer from Himachal Pradesh attended the Medicine OPD with symptoms of fever, headache and appearance of rashes over the body's surface. He gave the history that the rash appeared on the fifth day of the fever, which first appeared on the trunk and then spread to the extremities. On examination, the rashes were found to be maculopapular. These rashes were not found on the face, palm and sole. Serum was sent for febrile agglutination tests. Widal and standard agglutination tests for brucellosis were negative. Paul–Bunnell's test was also negative. ELISA for HIV was also negative. Weil–Felix test showed agglutination with OX19 and OX2 but not with OXK.

- What is the most probable diagnosis?
- What other tests can be carried out to diagnose the condition?
- How will you treat the condition?
- Is this condition found in India?

Mycoplasma and *Ureaplasma* 45

Mycoplasma belongs to class Mollicutes (*Mollis*, soft; *cutis*, skin) and order Mycoplasmatales. The family Mycoplasmataceae includes parasite mycoplasmas, which require cholesterol or other sterols as growth factors. It consists of two genera—*Mycoplasma* and *Ureaplasma*. At least 13 species of genera, *Mycoplasma* and *Ureaplasma*, are known to colonise or cause diseases in humans (Table 45.1).

Genus *Mycoplasma*

Mycoplasmas, previously known as 'Eaton agent', Pleuro-Pneumonia-like Organisms (PPLO), are prokaryotes. However, they differ from prokaryotes by having a unique cell membrane that contains sterols. Also, mycoplasmas do not have a cell wall. *Mycoplasma* pathogens, when first discovered, were thought to be viruses or L-forms of bacteria. *Mycoplasma* species were also considered an L-form of bacteria because they lack a cell wall. However, they differ from bacteria, including L-forms, in many properties. They (1) have sterols in the cell membrane, (2) do not show any reversion to structure with cell walls, (3) do not share any DNA homology with other bacteriomes, (4) have a genome with low molecular weight and (5) have low guanine and cytosine contents. The absence of a cell wall makes *Mycoplasma* resistant to penicillin, cephalosporins, vancomycin and other antibiotics that interfere with the synthesis of the cell wall.

The genus *Mycoplasma* contains pathogens that use glucose or arginine but do not utilise urea. The genus contains 200 validly published species. These species occur as commensals, parasites and pathogens of various plant insects and mammalian hosts. *Mycoplasma pneumoniae* is the most important species known to cause upper respiratory tract disease. It is also known for causing walking pneumonia or primary atypical pneumonia. *Mycoplasma hominis*, *Mycoplasma fermentans*, *Mycoplasma pirum*, *Mycoplasma salivarium*, *Mycoplasma orale* and *Mycoplasma genitalium* are the other pathogens commonly associated with human infections.

Properties of the Bacteria

Morphology

Mycoplasmas are very small bacteria measuring 150–250 nm in dimension. They do not have a cell wall, and typically, their cell membranes contain sterols. Many mycoplasmas can pass through a 0.45-μm filter and were once believed to be viruses. However, they differ from viruses because they (1) contain both ribonucleic acid (RNA) and deoxyribonucleic acid (DNA), (2) can grow on cell-free media *in vitro* and (3) show both intracellular and extracellular parasitism *in vivo*.

Mycoplasma species typically show pleomorphism and occur as granular filaments of various sizes. The filaments are slender of varying lengths

Table 45.1 Human infections caused by *Mycoplasma* and *Ureaplasma* species

Mycoplasma and *Ureaplasma* species	Diseases
Mycoplasma pneumonia	Primary atypical pneumonia
Mycoplasma hominis	Pelvic inflammatory disease and post-partum fever
Mycoplasma fermentans	Opportunistic infections in patients with HIV
Mycoplasma pirum	Septicemia in patients with HIV
Mycoplasma salivarium	Infection unknown
Mycoplasma orale	Infection unknown
Mycoplasma genitalium	Infection unknown
Ureaplasma urealyticum	Chorioamnionitis and pelvic inflammatory disease

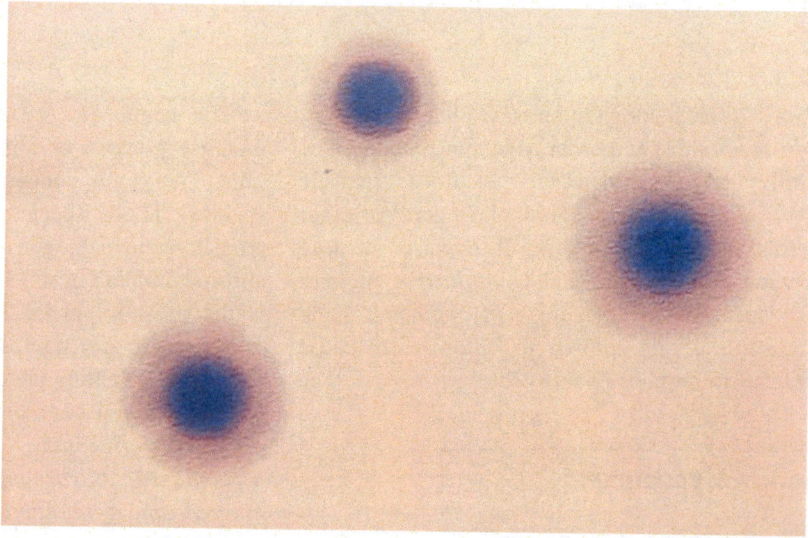

Fig. 45.1 Fried-egg appearance of *Mycoplasma* colonies. (https://www.pheculturecollections.org.uk/services/mycotestingeradication/mycocultureisolation.aspx)

and show true branching. They multiply by binary fission. However, genomic replication and cell division are often asynchronous, resulting in the production of multi-nucleate fragments and other body forms and chains of beads. They do not possess flagella or pili, but some *Mycoplasma* species, including *M. pneumoniae*, show gliding motility on liquid-covered surfaces. *Mycoplasma* stain poorly by Gram stain and are Gram negative. They are better stained by Giemsa and Diene stain.

Culture and Biochemical Properties

Mycoplasmas are aerobic and facultative anaerobes. *M. pneumoniae* is an exception, which is a strict aerobe. They grow at 37 °C and a pH range of 7.3–7.8.

Pleuropneumonia-like organism (PPLO) broth is a medium widely used to isolate mycoplasma. This medium is supplemented with 20% horse serum, 10% yeast extract and glucose. Phenol red is used as a pH indicator. The high concentration of animal serum (horse serum) is used as a source of exogenous sterols (cholesterol and other lipids). The medium is supplemented with penicillin, ampicillin and polymyxin B to inhibit the growth of contaminating bacteria. Amphotericin B is used to inhibit contamination with fungi.

PPLO agar is made by the addition of agar in the PPLO broth. The mycoplasmas are typically slow growers with a 1–6 h of generation time. Mycoplasmas on PPLO agar produce small colonies, typically described as a fried-egg appearance. This consists of a central, opaque, granular growth area surrounded by a flat, translucent, peripheral area. Initially, the mycoplasmas multiply within the agar to produce opaque, ball-shaped colonies that grow up to the surface of the agar and then spread around it, forming a translucent peripheral zone (Fig. 45.1). *M. pneumoniae* does not produce fried-egg appearance colonies but instead produces colonies, known as mulberry-shaped colonies. These colonies do not show any thin hallow, unlike that of fried-egg appearance colonies.

M. pneumoniae are very slow-growing bacteria. They require 1–4 weeks to produce colonies on agar. Most *Mycoplasma* colonies produce a zone of haemolysis on blood agar. Mycoplasmas do not have the capability to synthesise cholesterol and related sterols; hence these are supplied from outside for the growth of mycoplasmas. They also lack the ability to synthesise purines and pyrimidines.

M. pneumoniae and other species (*M. fermentans*, *M. genitalia* and *M. agalactiae*) utilise glucose and other carbohydrates as the major energy source. *M. salivarium* and other species (*M. orale*, *M. hominis* and *M. fermentans*) utilise arginine as a major energy source. Mycoplasma are chemo-organotrophs, the metabolism being mainly fermentative.

Other Properties

Susceptibility to Physical and Chemical Agents

Mycoplasmas are readily killed by heating at 56 °C for 30 min. The bacteria are sensitive to antiseptic solutions, such as cycloheximide and cetrimide, which inhibit their growth. They are resistant to UV light and the photodynamic action of methylene blue. Hence, *M. pneumoniae* can grow in the presence of 0.002% of methylene blue in agar, while many other *Mycoplasma* species are inhibited at this concentration.

Immunity

Membrane glycolipids and proteins are the major antigenic determinants of mycoplasmas. Membrane glycolipid antigens show cross-reaction with human tissues and other bacteria. Glycolipids with similar antigenic structures have been demonstrated in neurons in the human brain. The antibodies against *M. pneumoniae* glycolipid may cross-react with brain cells, causing damage to neuronal cells. This cell damage possibly is responsible for neurological manifestations of *M. pneumoniae* infection.

M. pneumoniae possesses multiple surface proteins like P1, P30 and P116, which are responsible for the attachment of bacteria to cell structures. The P1 protein is the major adhesin and immunogenic protein present on the attachment organelle of *M. pneumoniae*. The P1 protein induces the production of antibodies, which not only react with P1 protein but also react with antigenic determinants of RBCs, leading to lysis of erythrocytes in the autoimmune disease process.

Pathogenesis

Mycoplasmas are primarily extracellular pathogens.

Virulence Factors

The adhesion protein called **P1 antigen** is the key virulence factor of mycoplasma. It is a membrane-associated protein that helps in the adhesion of mycoplasmas to epithelial cells. This protein or adhesin combines specifically with sialated glycoprotein receptors present at the base of the cilia on the epithelial surface. This same receptor is also present on the surface of the erythrocytes. The antibodies against the P1 antigen can also act as an autoantibody against RBCs, causing agglutination.

Pathogenesis of Respiratory Infections

Mycoplasmas adhere to the surface of ciliated and non-ciliated epithelial cells. Following attachment, *Mycoplasma* causes direct damage to the epithelial cells in which first cilia and then ciliated epithelial cells are destroyed. Loss of the cells interferes with the normal functioning of the upper respiratory tract. This results in the lower respiratory tract becoming infected with microbes and mechanically irritated. The mechanical irritation causes persistent cough, typically seen in patients with respiratory infection caused by *M. pneumoniae*.

M. pneumoniae behaves as a superantigen. This causes migration of inflammatory cells to the site of infection and produces cytokines, such as tumour necrosis factor-alpha, interleukin-1 and interleukin-6. These cytokines help in clearing the bacteria and the disease. The bacteria usually do not cause invasion of the blood to produce a systemic manifestation of the disease. *Mycoplasma* rarely penetrates the submucosa except in rare cases of immunosuppression or following instrumentation. Under these conditions, they may invade the bloodstream and cause infection in different body organs.

Clinical Syndromes

M. pneumoniae primarily causes respiratory infections in humans.

Respiratory Infections

A majority of respiratory infections are mild and self-limiting.

M. pneumoniae typically causes mild **upper respiratory tract infections**. The condition is characterised by low-grade fever, malaise and headache. A non-productive cough is a typical manifestation, which appears 2–3 weeks after exposure. The cough is initially non-productive but may later produce small to moderated quantities of sputum, which may become mucopurulent and even blood-tinged in more severe cases.

Lower respiratory tract infections include tracheobronchitis and bronchopneumonia. The condition is characterised by primary infection of bronchi with infiltration of bronchial epithelial cells by lymphocytes and plasma cells.

Primary atypical pneumonia is also known as walking pneumonia. The incubation period varies from 2 to 3 weeks. Patients suffering from atypical pneumonia usually do not appear ill. Hence, the illness is often referred to as walking pneumonia. The pharynx affected becomes oedematous but without any cervical adenopathy. The condition is associated with the presence of patchy bronchopneumonia seen on the chest X-ray. The disparity between physical findings and radiological evidence of the chest is the hallmark of infection. The infection is usually self-limiting. In 5–20% of patients, pleural effusion may occur.

Extrapulmonary Infections

Extrapulmonary manifestations are not rare. Cardiac abnormalities, such as myocarditis and pericarditis, are the most frequently reported extrapulmonary manifestations. Other manifestations include neurological abnormalities, otitis media and erythema multiforme (Stevens-Johnson syndrome) (Fig. 45.2).

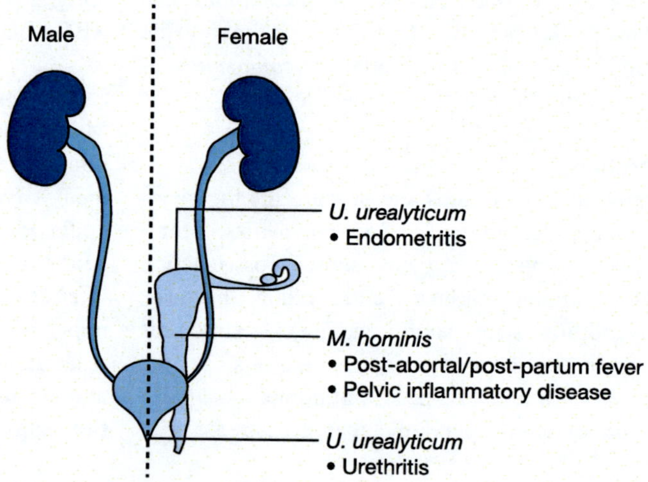

Fig. 45.2 Infections caused by *Mycoplasma* and *Ureaplasma* species

Laboratory Diagnosis

Initial treatment of *M. pneumoniae* infection is based on the clinical diagnosis of the condition. The definite diagnosis of the condition usually takes 3–4 weeks. Hence, treatment is started without waiting for the result of laboratory tests.

Specimens

Respiratory specimens include bronchial washings, throat washings and expectorated sputum. Tracheal washings are more useful than sputum specimens. This is because most patients with respiratory tract infections do not produce any sputum as they have a dry and non-productive cough.

The specimens are collected and transported to the laboratory immediately. If a delay is anticipated, they are usually inoculated in suitable transport media, such as SP4 transport medium, to prevent desiccation. The specimens can be stored at −70 °C if these cannot be sent immediately to the laboratory after collection.

Microscopy

Microscopy is of no value in the diagnosis of *M. pneumoniae* infections. Mycoplasmas are stained poorly because they lack cell walls.

Direct Antigen Detection

Antigen capture immunoassay has been used to directly detect *M. pneumoniae* in sputum specimens with high sensitivity and specificity. Direct immunofluorescence and immunoblotting with monoclonal antibodies are also used to detect antigens in clinical specimens.

Culture and Identification

Bacterial culture is of little practical value because of its fastidious growth requirements and the need for 3–4 weeks for culture. Isolation of *M. pneumoniae* from clinical specimens by culture confirms the diagnosis of *M. pneumoniae* respiratory illness. The specimens are inoculated into a mycoplasma medium, such as PPLO agar. *M. pneumoniae* grows very slowly on this medium, and colonies are demonstrated at 37 °C. The growth is facilitated by incubation in the presence of 95% of N_2 and 5% of CO_2. The bacteria produce small colonies with a homogenous appearance, typically described as *mulberry-shaped colonies.*

On agar, mycoplasmas other than *M. pneumoniae* produce fried-egg appearance colonies with an opaque central zone of growth within the agar and a translucent peripheral zone on the surface.

Suspected *M. pneumoniae* colonies are identified by colony characters and a battery of tests. Microscopy reveals very small bacteria, poorly stained with Gram stain but stain well with Diene stain.

The colonies are identified by noting a **colour change** from red to yellow of phenol red due to fermentation of glucose resulting in the production of acid, which makes the pH of the media acidic.

Mycoplasma colonies are identified by **the Diene test**. In this method, the Diene stain (diluted 1:10 with distilled water) is added directly to the plate containing suspected colonies *of Mycoplasma.* The plate is then immediately rinsed with distilled water to remove the stain. This is followed by decolourising the medium by adding 1 mL of 95% ethanol and keeping it for 1 min. The plate is re-washed with distilled water and is allowed to dry. After drying, the colonies are observed under the low power of the microscope.

By Diene staining, the fried-egg appearance colonies of *Mycoplasma* appear highly granular. The centre of the colonies is stained dark blue, and the periphery of the colonies staining light blue. The agar in the medium appears clear or slightly violet. *Mycoplasma* other than *M. pneumoniae* becomes colourless after some time because it reduces the methylene blue. The identified colonies are confirmed by inhibiting their growth with specific *M. pneumoniae* antisera.

M. pneumoniae colonies are identified by a positive **hemadsorption test.** The test is performed by flooding *M. pneumoniae* colonies grown on the surface of agar with 2 mL of 0.2–0.4% suspension of guinea pig erythrocytes in *Mycoplasma* growth medium. The plate is incubated at 35 °C for 35 min followed by washing with 3 mL *of Mycoplasma* growth medium. The washing fluid is gently removed by aspiration with a pipette. *M. pneumoniae* colonies adsorb guinea pig erythrocytes, which is facilitated at 37 °C. *M. pneumoniae* colonies adsorbing erythrocytes on their surface are demonstrated under the 40× magnification of the microscope.

Colonies are also positive for the **tetrazolium reduction test**. The test is based on the principle that *M. pneumoniae* can reduce triphenyl tetrazolium, a colourless compound, to formazan, a red-coloured compound. This test is performed by flooding the colonies of *M. pneumoniae* on agar with a solution of 2-*p*-iodophenyl-3-nitrophenyl-5-phenyltetrazolium chloride and incubating it at 35 °C for an hour. In a positive test, *M. pneumoniae* colonies appear reddish after 1 h and may appear purple to black after 3–4 h of incubation.

Serology

Serodiagnosis is based on demonstrating specific antibodies in serum using *Mycoplasma* antigens. *M. pneumoniae* glycolipid antigen extracted by chloroform and methanol is also widely used for serodiagnosis of atypical pneumonia. Complement fixation test and ELISA are the tests frequently used. IgM ELISA is the most frequently used test to detect specific IgM antibodies in a single-serum specimen.

Non-specific Serological Tests

These tests are called non-specific because they do not use specific *Mycoplasma* antigens; instead, they use cross-reacting and non-specific antigens. These tests include the *Streptococcus* MG agglutination test and the cold agglutination test.

Streptococcus MG agglutination test uses a heat-killed suspension of *Streptococcus* MG as an antigen. The antigen is mixed with serial dilution of the patient's unheated serum. The agglutination is observed after overnight incubation at 37 °C. An antibody titre of 1:20 or more is suggestive of *M. pneumoniae* infection.

Cold agglutination test is based on the principle that autoantibodies that agglutinate human O group cells at low temperatures appear in most cases of atypical pneumonia. In this test, human O group erythrocytes are used as antigens. The patient's blood, which is collected for testing, should never be refrigerated before the separation of serum because homologous RBCs readily absorb agglutinins at low temperatures. The patient's serum is mixed with an equal volume of 0.2% washed suspension of human O erythrocytes. The suspension is incubated at 4 °C overnight and observed for clumping. The clumping is disassociated at 37 °C. The titre of 1:32 or more is suggestive of *M. pneumoniae* infection. A fourfold rise in cold agglutinin titre of the paired serum and convalescent sera or a single titre of 1:32 is suggestive of *M. pneumoniae* infection.

Cold agglutinins usually appear by the second week of infection and peak at 4–5 weeks. After that, the antibody titre declines rapidly, and the test becomes negative in about 5 months. Since the test is non-specific, it shows false-positive reactions in sera of other diseases, such as rubella, infectious mononucleosis, adenovirus infections, psittacosis, tropical eosinophilia trypanosomiasis, cirrhosis of the liver and haemolytic anaemia.

Molecular Detection

A semi-nested polymerase chain reaction (PCR) assay using 60S ribosomal DNA has been developed to demonstrate P1 adhesin in protein for diagnosing *M. pneumoniae* infection. These tests are of high sensitivity and specificity. They have added advantage of detecting *M. pneumoniae* infections at an early stage.

With time, convention PCR has been replaced by numerous real-time PCR assays for research and diagnostic purposes. The predominant real-time PCR assay is based on the TaqMan® probe, and common target genes are 16S rRNA, the P1 gene or the ATPase operon. More recently, the gene encoding the community-acquired respiratory distress syndrome (CARDS) toxin has been identified as a potential target for detection.

Treatment

M. pneumoniae remains susceptible to tetracyclines and erythromycin because these antibiotics act on the mycoplasmas by inhibiting protein synthesis. Tetracycline has the additional advantage of also being active against most other mycoplasmas and chlamydiae, the common causes of non-gonococcal urethritis. *Mycoplasma* is resistant to penicillin and cephalosporins because these antibiotics act on the cell wall, which is lacking in mycoplasmas (L-forms).

Epidemiology

Geographical Distribution

M. pneumoniae infection is distributed worldwide. The organism is responsible for up to 20% of community-acquired pneumonia, which requires hospitalisation. *M. pneumoniae* infection occurs both in epidemic and in endemic forms.

Habitat

M. pneumoniae is a strict human pathogen. The bacteria inhabit the upper respiratory tract in an infected host. They are usually present in the mucosa, residing extracellularly on the upper respiratory tract.

Reservoir, Host and Transmission of Infection

Humans are the usual host of *M. pneumoniae* and thus a significant reservoir of infection. Patients with active infection are more likely to transmit *M. pneumoniae* infection. *M. pneumoniae* is most commonly transmitted by close contact through nasal secretions. The infection is transmitted by inhalation of aerosolised droplets. *M. pneumoniae* is usually associated with pneumonia, and the highest rate of infection is found in children between 9 and 10 years and young adults. Infection is common among school-going children.

The infection is also being increasingly documented in people older than 65 years in recent years. In the old population, *M. pneumoniae* is responsible for causing nearly 15% of community-acquired pneumonia and is the second pathogen next to *Streptococcus pneumoniae* as a cause of pneumonia.

Prevention and Control

Isolation of the patients infected with *M. pneumoniae* is the best way to prevent the spread of the disease. Antibiotic prophylaxis with tetracyclines or erythromycin is also useful. No vaccine is available against *Mycoplasma* infections.

Genital *Mycoplasma* Infections

M. hominis and *M. genitalium* are the genital *Mycoplasma* species that inhabit the mucosa of the urogenital tract.

M. hominis is associated with infection of the genitourinary tract and reproductive disease. It causes genital infection, resulting in diverse manifestations, such as salpingitis, pelvic abscess, puerperal infection and septic abortion. *M. hominis* typically produces large fried-egg appearance colonies on mycoplasma medium. Inhibition of the growth of the bacteria with specific antisera to *M. hominis* is used to differentiate it from another genital mycoplasma. *M. hominis* organisms, unlike other mycoplasmas, are resistant to erythromycin and occasionally to tetracyclines. Clindamycin is useful to treat infections caused by such resistant strains of *M. hominis*. *M. genitalis* is another species that has also been implicated as a cause of non-gonococcal urethritis and pelvic inflammatory disease (PID).

The colonisation of infants by *Mycoplasma* species usually occurs during the baby's passage through the birth canal. Colonisation with these genital mycoplasmas occurs only for up to 2 years. Since genital mycoplasmas have been transmitted sexually, avoidance of sexual activity or the use of proper safety procedures prevents the disease caused by these genital mycoplasmas.

Mycoplasmas and L-Forms in Bacteria

In 1935, the first pleuropneumonia-like forms were recovered from *Streptobacillus moniliformis* cultures by Klieneberger, and it was designated as L-forms after Lister Institute, London, where the discovery was made. Later, it was shown that many of the bacteria lose their cell wall and develop into L-forms either spontaneously or induced by certain substances such as penicillin. Two types of L-forms are unstable- L-forms that are capable of dividing but can revert to the original morphology, and stable L-forms that are unable to revert to the original bacteria, but they continue in the cell wall-deficient state permanently. These cell wall-deficient forms (L-forms, protoplasts, spheroplasts) cannot initiate disease. They may help in bacterial persistence and subsequent recurrence of infection during antibiotic therapy. Mycoplasmas may represent stable L-forms of bacteria. However, the antigenic, genetic and biochemical evidence are against this possibility.

Mycoplasmas as Cell Culture Contaminants

Different species of mycoplasma are often found in research laboratories as contaminants in continuous cell culture. The mycoplasmal cell culture contamination occurs from individual workers or animal sera or contaminated cell culture medium ingredients such as trypsin. Generally, the contamination does not produce destruction of cell lines but might affect the growth of the desired viruses, and it may lead to wrong interpretation of results. Most of the time, mycoplasmas in cell cultures have often been mistaken for viruses. It is very difficult to eradicate the mycoplasma contamination of cell line cultures. Desirably a change of cell line is the best solution.

Genus *Ureaplasma*

Genus *Ureaplasma* consists of five species, of which *Ureaplasma urealyticum* is the most important species isolated from infections of the genital and respiratory tract in humans. It is one of the most important causes of non-gonococcal urethritis.

Mycoplasma medium is supplemented with urea is useful for the growth of *U. urealyticum*. *U. urealyticum* typically hydrolyses urea as an essential growth factor for the bacterium. The bacteria like those of mycoplasmas produce very minute colonies measuring 15–50 μm in size. Hence, they were earlier called T strain or T strain *Mycoplasma.* This human T strain *Mycoplasma* is now re-designated as *U. urealyticum.*

U. urealyticum is a sexually transmitted pathogen and causes chorioamnionitis, pre-maturity, vaginitis, cervicitis, acute salpingitis and pelvic inflammatory disease in women. *Ureaplasma* organisms are most frequently isolated from infants, particularly girls. In men, the bacteria also cause urethritis, proctitis, balanoposthitis and Reiter's syndrome.

Atypical Pneumonia

Atypical pneumonia is a term used to distinguish from typical pneumonia caused by *S. pneumoniae*. Earlier, any patient with sudden onset of fever, chills, pleuritic pain and expectoration of blood-tinged sputum was believed to present with typical pneumonia caused by *S. pneumoniae*. The patients not showing these characteristic manifestations were referred to as having atypical pneumonia. *M. pneumoniae*, *Chlamydia pneumoniae* and *Legionella pneumophila* are three important agents responsible for atypical pneumonia. Under this condition, these pathogens are usually not isolated from sputum by routine culture methods.

Beta-lactam antibiotics are usually ineffective against *Mycoplasma* because of the absence of cell walls because these antibiotics usually act by inhibiting the synthesis of the cell wall. Penicillin and cephalosporins are ineffective against all these pathogens because they do not show intracellular penetration in these pathogens.

Case Study

A 3-year-old boy attended the Medicine OPD with symptoms of sudden onset of fever, chills, pleuritic pain and expectoration of blood-tinged sputum. The routine bacteriological culture was negative. Serum for cold agglutination test was positive.

- Diagnose the disease.
- What are the other laboratory tests used to diagnose the condition?
- List other pathogens causing the disease.
- What are other genital mycoplasmas known to cause human infections?

Chlamydia and *Chlamydophila* 46

The taxonomy of *Chlamydia* has undergone extensive revision recently based on genomic studies of this microorganism. They are included in the family Chlamydiaceae, which includes two genera: *Chlamydia* and *Chlamydophila*. The species *Chlamydia trachomatis* is included in the genus *Chlamydia*, whereas *Chlamydophila psittaci* and *Chlamydophila pneumoniae* are included in the new genus *Chlamydophila*. A fourth species, *Chlamydophila pecorum*, is normally present in the intestine, and vagina of ruminants is responsible for various pathological conditions in ruminants, swine and koalas.

The three species of importance to humans resemble each other by their similar morphology, a common group antigen, and identical developmental cycle in the host cell. They have been differentiated based on antigenic composition, intracellular inclusions, sulphonamide susceptibility and disease production. Glycogen containing inclusion bodies that can be stained by iodine are found only in *C. trachomatis*, and this is also the only species among the three which is susceptible to sulphonamides because of the presence of the folate biosynthesis pathway.

Genus *Chlamydia*

Chlamydiae are obligate intracellular parasites of humans and animals with a marked affinity for the squamous epithelial cells of the gastrointestinal and respiratory tracts. The chlamydiae were once considered viruses due to their filterability through 0.45-μm filter and their failure to grow in cell-free media. Earlier, they were named *Psittacosis lymphogranuloma trachoma* (PLT) viruses or PLT agents. However, the *Chlamydia* differs from viruses. They, like bacteria, (1) contain both DNA and RNA, (2) possess cell wall like that of Gram-negative bacteria, (3) contain prokaryotic ribosomes, (4) multiply by binary fission, (5) produce and synthesise their own nucleic acid, lipids and proteins and (6) are susceptible to a wide range of antibiotics, such as tetracyclines, erythromycin, macrolides and rifampicin. Species causing human infections are summarised in Table 46.1.

General Properties

Morphology

Chlamydiae occur in two morphologically distinct forms: elementary body (EB) and reticulate body (RB).

Elementary body (EB) is a small, extracellular, infective form. It is a round particle measuring 300–400 nm in size. The cell wall possesses a rigid trilaminar structure, as seen in the cell wall of Gram-negative bacteria. These bacteria lack the peptidoglycan layer found in other bacteria. However, their outer membrane proteins, due to the presence of an extensive cross-linked protein of the outer membrane, confer rigidity to

Table 46.1 Human infections caused by *Chlamydia* species

Chlamydia species	Serotypes	Diseases
Chlamydia trachomatis	A, B, Ba, C D, E, F, G, H, I, J, K L1, L2, L3	Lymphogranuloma venereum (LGV), ocular LGV, trachoma, adult inclusion conjunctivitis, neonatal conjunctivitis, infant pneumonia and urogenital infections
Chlamydophila pneumoniae	Single serotype	Pharyngitis, sinusitis, bronchitis and pneumonia
Chlamydophila psittaci	Multiple serotypes	Psittacosis

the cell wall. The chlamydiae do not multiply in the EB form but are infectious. They cause infections by binding to receptors on the epithelial cells and stimulate uptake of the bacteria by infiltration.

Reticulate body (RB) is a large, non-infectious form of *Chlamydia*. It measures 500–1000 nm in size. It is a metabolically active and replicating form of *Chlamydia*. The extensively cross-linked proteins that confer rigidity are absent in the RBs. Hence, this form of *Chlamydia* is osmotically fragile and friable. This form, however, is protected by its intracellular location.

A genus-specific lipopolysaccharide (LPS) is present in the cell wall of *Chlamydia*. This LPS can be detected by a complement fixation test (CFT) and species-specific and strain-specific outer membrane protein.

Growth and Multiplication

Chlamydia multiplies by a characteristic growth cycle within susceptible host cells (Fig. 46.1). Elementary bodies are the infective form that initiates the cycle. The infection is initiated by the attachment of the EB to the microvilli of susceptible epithelial cells, followed by penetration into the host cell. Inside the host cells, EB

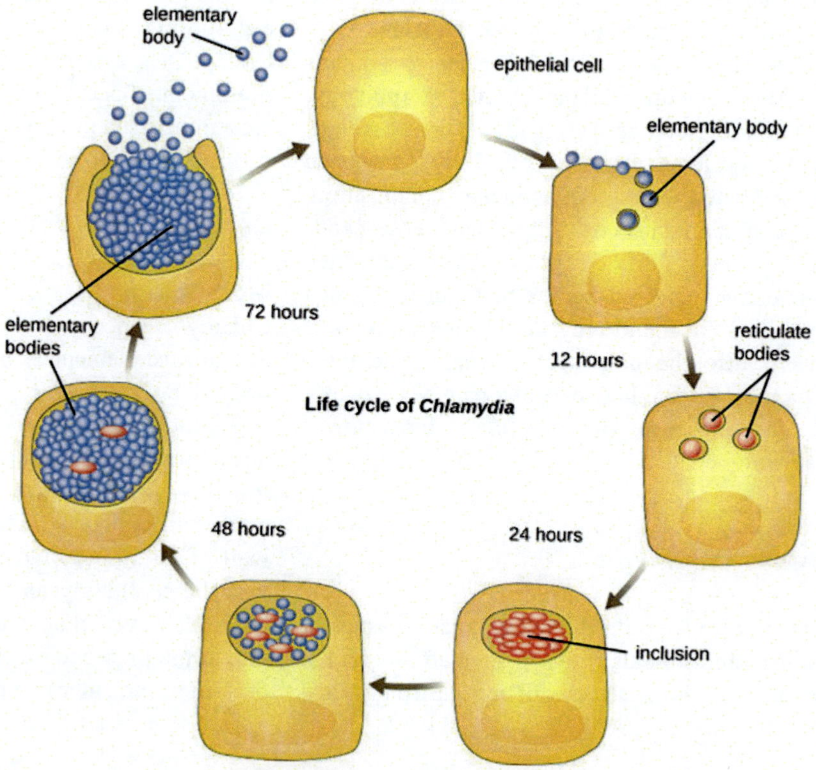

Fig. 46.1 Life cycle of *Chlamydia*. EB: elementary body; RB: reticulate body (https://link.springer.com/chapter/10.1007/978-3-030-56978-5_27)

remains within the cytoplasmic phagosomes, in which EBs begin to multiply. The fusion of EB containing phagosome with cell wall lysosome is prevented; thus, the intracellular killing of EB is inhibited. This phagolysosomal fusion is usually prevented in the host cells with the intact outer membrane.

Within 6–8 h after entering the cell, the EBs within the phagosome are transformed to large, metabolically active RBs. These RBs synthesise their own proteins and nucleic acids but cannot produce their own high-energy phosphate compound. Chlamydiae are called energy parasites because of this deficiency. Some strains of *Chlamydia* also depend on the host for the requirement of their amino acid.

The RBs multiply by binary fission, which continues for the next 18–24 h. The developing phagosome with accumulated reticulated bodies within the host cell is called the inclusion body. The mature inclusion body contains nearly 100–500 EBs, which various staining procedures can readily demonstrate. Ultimately, the host cell ruptures, releasing the Eosin *C. trachomatis* infection, the release of EBs occurs within 70–96 h. The release of the host cell is marked by the presence of a scar. The release of EBs in *C. psittaci* infection occurs within 48 h by lysis of the host cell leading to severe damage of the infected host cell.

Chlamydiae during the active growth express the *Chlamydia*-specific LPSs on the cell surface of the infected host cell. These LPSs expressed on the cell's outer surface are highly antigenic and induce immunological and inflammatory responses. The properties that differentiate the three *Chlamydia* species are summarised in Table 46.2.

Chlamydia trachomatis

C. trachomatis is a strict human pathogen.

Properties of the Bacteria

Morphology

C. trachomatis are Gram-negative bacteria. However, they can be stained better by Giemsa, Castaneda, Macchiavello or Gimenez stains. *C. trachomatis*, like other chlamydiae, occurs in two morphologically distinct forms: EB and RB.

EB is an extracellular infectious particle. It is small and spherical and measures 800–1200 nm in diameter. These inclusion bodies in the infected cells, such as conjunctiva, urethra and corneal smears, can be demonstrated after staining with Giemsa, Castaneda or Macchiavello methods. These inclusion bodies are large-sized particles, easily demonstrated under a light microscope. These bodies consist of a glycogen matrix; hence they are demonstrated on staining with Lugol's iodine.

RB is metabolically active and replicating form of *Chlamydia*.

Culture

C. trachomatis grows better in various tissue cultures, using non-replicating stationary-phase cells. The bacteria can grow in a few **cell lines**, such as HeLa-229, McCoy, BHK-21 and buffalo green monkey kidney cells. McCoy and HeLa cells are frequently used for the isolation of bacteria.

C. trachomatis can be grown by inoculation into embryonated eggs and by experimentally producing infection in animals, such as mice.

Table 46.2 Differential features of *Chlamydia* species causing human infections

Properties	*Chlamydia trachomatis*	*Chlamydia pneumoniae*	*Chlamydia psittaci*
Elementary body	Round; narrow periplasmic space	Pear-shaped; large periplasmic space	Round; narrow periplasmic space
Inclusion body	Single inclusion per cell	Multiple, uniform inclusions per cell	Multiple, variable inclusions per cell
Plasmid DNA	Yes	No	Yes
Iodine staining	Stains inclusion body	Does not stain	Does not stain

Chlamydia spp. grow in **the yolk sac** of 6- to 8-day-old chick embryos. The growth of *Chlamydia* is demonstrated by the presence of elementary and inclusion bodies as well as group-specific complement-fixing antigens in the yolk sac. *C. trachomatis* strains differ in their infectivity (L1, L2 and L3) and cause infection in **mice** when injected intracellularly.

Other Properties

Susceptibility to Physical and Chemical Agents

Chlamydiae are heat-labile bacteria and are readily killed within minutes by heating at 56 °C. They are susceptible to ethanol, ether, phenol, formalin, iodine, potassium permanganate, sodium hypochlorite, silver nitrite and chlorite. They remain fully viable for several days at 4 °C. Moreover, they can be preserved at −70 °C or in liquid nitrogen for a long period.

Antigenic Structure

Chlamydiae possess three types of major antigens.

The genus-specific antigen is a heat-stable, complement-fixing and genus-specific antigen. It is an LPS-protein complex resembling the LPS of Gram-negative bacilli. It is present in EBs and RBs. The antigen can be extracted by ether, chloroform or methanol. The antigen is identified by CFT. **The species-specific antigen** is present at the envelope surface and is species-specific. This antigen is present in all the strains of *Chlamydia*. **Serotype-specific antigens** is present only in a few species of Chlamydiae. They are located in the major outer membrane proteins (MOMPs) and are useful for intraspecies typing of *Chlamydia* species.

Typing

Chlamydia species are classified into various **serovars** and **serologic variants** based on these antigens. *C. trachomatis* has been subdivided into three biological variants known as biovars: i) trachoma biovar causing trachoma and inclusion conjunctivitis (TRIC), ii) LGV biovar causing LGV and iii) serovars causing mouse pneumonitis. These biovars based on antigenic differences in the MOMPs have been further classified into 20 serotypes. The trachoma biovar consists of 13 serotypes (A, B, Ba, C, D, Da, E, F, G, H, I, Ia, J, Ja and K). The LGV biovar consists of only five serotypes (L1, L2, L2a, L2b and L3).

Pathogenesis

C. trachomatis is an obligate intracellular bacterium that causes diseases of many systems in humans.

Virulence Factors

The ability to multiply intracellularly in the infected cell is the key mechanism of virulence of *C. trachomatis*. The bacteria prevent phagolysosome fusion with cellular liposomes, thereby preventing the intracellular killing of the bacteria by the host cell. Repeated infections caused by *C. trachomatis* contribute to the pathology seen in the infected eye in trachoma.

Pathogenesis of *Chlamydia trachomatis* Infections

C. trachomatis causes disease mainly by (1) direct destruction of infected host cells during multiplication and (2) inducing inflammatory responses in the host. *C. trachomatis* enters the host through minute abrasions or injuries in the skin. The bacteria react specifically with the receptors found on the non-ciliated columnar, cuboidal or transitional epithelial cells. These epithelial cells are typically found on the mucous membranes of the conjunctiva and genitourinary systems, such as the urethra, endocervix, endometrium, fallopian tube and respiratory tract.

The LGV biovar multiplies in mononuclear phagocytes found in the lymphatic system. The pathological lesions are typically found in the lymph nodes draining the site of primary infection. Granuloma is a characteristic pathological lesion. Subsequently, the inflammatory process spreads to other surrounding tissues, and finally, rupture of the lymph nodes leads to the formation of abscesses or sinus tracts. Infections with trachoma serovars are associated with severe inflammatory reactions consisting of neutrophils, lymphocytes and plasma cells, as seen in the case of trachoma.

Clinical Syndromes

C. trachomatis causes a variety of diseases. It is an important cause of genital and ocular infections worldwide.

Chlamydia trachomatis LGV biovar Infections

C. trachomatis **LGV biovar** Causeslymphogranuloma Venereum (LGV) and Ocular LGV.

Lymphogranuloma Venereum (LGV)

It is caused by *C. trachomatis* LGV biovar (serotypes L1, L2, L2a, L2b and L3). Serotype L2 is most commonly responsible for the condition. LGV is a sexually transmitted disease that affects the cervix, urethra, salpinges and epididymis. The incubation period varies from 1 to 4 weeks. A small painless papule or ulcerative lesion is the primary lesion, which appears on the external genitalia in patients with LGV. Penis in males (Fig. 46.2), fourchette in females and rectum in homosexuals are the common sites of infection. Occasionally, primary lesions may be found on the extragenital sites (e.g. fingers and palms). The absence of pain in ulcers is the typical manifestation, which differentiates the ulcers from the painful ulcers found in herpes simplex virus infection and syphilis. Proctitis is a common manifestation in women with LGV. This occurs due to the lymphatic spread of the bacteria from the infected cervix or the vagina. Proctitis also occurs in homosexual men.

Ocular LGV

It is caused by *C. trachomatis* LGV biovar. This causes Parinaud oculogenital conjunctivitis. It is a condition characterised by inflammation of the conjunctiva associated with periauricular, sub-mandibular and cervical lymphadenopathy.

Chlamydia trachomatis trachoma biovar Infections

C. trachomatis **trachoma biovar** causes trachoma, adult inclusion conjunctivitis, neonatal conjunctivitis, infant pneumonia and urogenital infections.

Trachoma

It is a communicable disease of the eye caused by *C. trachomatis* serotypes A, B, Ba and C. This is a condition characterised by follicular hypertrophy, papillary hyperplasia, pannus formation and in late stages cicatrisation (Fig. 46.3). The condition

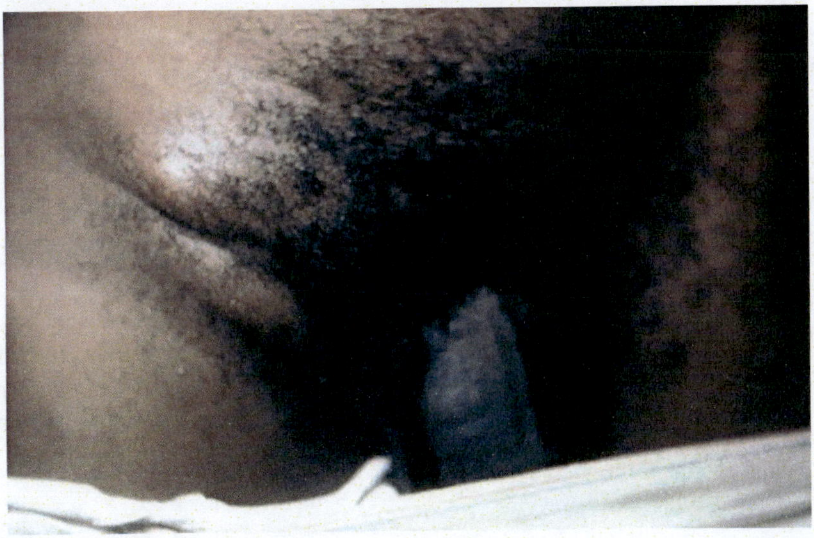

Fig. 46.2 Sign of groove and ruptured buboes in LGV. (Courtesy: https://phil.cdc.gov/Details.aspx?pid18038)

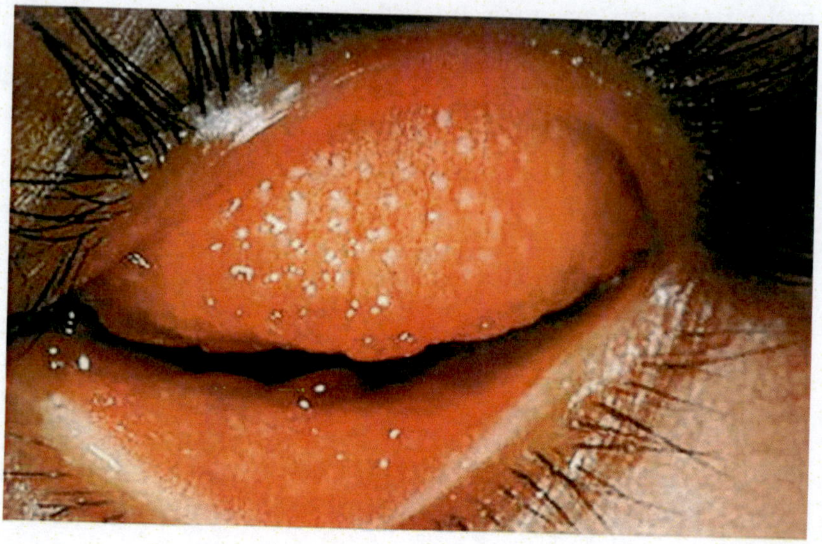

Fig. 46.3 Trachoma (https://link.springer.com/chapter/10.1007/978-3-7091-1613-5_4)

begins in patients with follicular conjunctivitis with diffuse inflammation that affects the entire conjunctiva. Subsequently, the condition progresses with the formation of pannus, which indicates invasion of blood vessels of the cornea and finally loss of vision. The loss of vision is the most important and serious complication of trachoma.

Adult Inclusion Conjunctivitis

The condition, also known as swimming pool conjunctivitis, results from the infection with C. trachomatis strains associated with genital infection (A, B, Ba and D–K). This infection is more frequently seen in sexually active adults.

Neonatal Conjunctivitis

It is the neonatal form of inclusion conjunctivitis. The condition develops in infants acquiring the infection from an infected birth canal. The infection is usually seen in infants born to pregnant mothers who have chlamydial infections of the cervix.

Infant Pneumonia

It is caused by *C. trachomatis* is seen in infants between 4 and 16 weeks of age. It is one of the most common causes of pneumonia in newborns. This infection is seen in 60% of neonates born to infected mothers.

Urogenital Infection

It is the most common infection caused by *C. trachomatis*. Approximately 80% of infected females and 50% of infected males are asymptomatic. Genital chlamydiasis is the most common sexually transmitted disease worldwide. Women with chlamydial infection with *C. trachomatis*, especially serotype C, are at a greater risk of developing cervical cancer, nearly 6.5 times more than women without infection. Reiter's syndrome is a triad of recurrent conjunctivitis, polyarthritis and urethritis or cervicitis and is most commonly associated with genital infection caused by *C. trachomatis*. This condition is seen in both men and women. Most patients with Reiter's syndrome have genital infections with *C. trachomatis* at the onset of arthritis. EBs have been demonstrated in synovial fluid of patients suffering from arthritis.

Laboratory Diagnosis

However, the sensitivity of various laboratory methods depends on (i) the nature of the disease, (ii) the site of infection from where the specimen is collected and (iii) the population of the patient examined.

Specimens

Specimens from the urethra, cervix, rectum oropharynx and conjunctiva are the frequently collected specimens. In addition, other specimens such as blood, respiratory secretions, sputum, lung and other tissues are collected and examined. Pus from bubo is also useful for the diagnosis of LGV.

Microscopy

Laboratory diagnosis of chlamydial infection is made by microscopic examination of various clinical specimens to demonstrate chlamydial inclusion bodies. These inclusion bodies (Halberstadter–Prowazek or HP bodies) can be demonstrated in specimens stained by Giemsa, Castaneda, Macchiavello or Gimenez stains. *C. trachomatis* infections of conjunctiva, urethra and cervix are diagnosed by demonstrating typical reniform inclusion bodies surrounding the nucleus in the stained smear of conjunctiva, urethra and cervical smears.

Iodine staining of conjunctival scrapings is a simple and rapid method for diagnosing Trachoma and Inclusion Conjunctivitis (TRIC). Iodine stains the glycogen matrix of inclusion bodies of *C. trachomatis*. However, iodine staining is a method of poor sensitivity and is positive only in certain stages of development of inclusion bodies in infected cells. This is more useful for demonstrating chlamydial inclusion bodies in the cell cultures inoculated with clinical specimens.

Culture

Isolation of *C. trachomatis* in cell cultures is the more specific method for diagnosing *C. trachomatis* infection. Clinical specimens are inoculated in different cell lines for isolation of *Chlamydia*. The sensitivity of the cell cultures for isolation of *C. trachomatis* is increased by (1) pretreatment with cycloheximide (i.e. a metabolic inhibitor, which inhibits the metabolism of host cells) and (2) use of irradiated cell lines (treated McCoy cells are most commonly used cell lines for isolation of *C. trachomatis*).

C. trachomatis infection in cell culture is demonstrated by the presence of intracellular inclusion bodies. These are detected by the use of iodine stains or fluorescence-conjugated antibodies. The culture methods are difficult and expensive. These are specifically preferred for the isolation of *C. trachomatis* from rectal specimens because non-cultural methods are usually negative. The culture shows high sensitivity and specificity.

Chlamydial Antigen Detection

The chlamydial antigen can be detected in clinical specimens by direct fluorescent antibody (DFA) staining and enzyme immunoassay (EIA). Antibodies prepared against either the *Chlamydia* MOMP or the cell wall LPS are used to detect antigens in clinical specimens by these two methods.

Serology

Antibody-based tests are useful for the diagnosis of LGV. CFTs, microimmunofluorescence and ELISA are employed to detect specific antibodies in the sera. **CFT** uses a genus-specific LPS antigen for the detection of antibodies. A single serum showing a high antibody titre of 1:256 or more or a fourfold increase in antibody titre of a paired sera sample is highly suggestive of LGV. **ELISA** is a genus-specific test that uses LPS antigens like CFT. **The micro-immunofluorescence test** is a specific test that utilises the species- and serovar-specific antigens, such as *Chlamydia* MOMPs. Confirmation of LGV is carried out by the MIF test.

The antibodies-based serological tests are of limited value in diagnosing urogenital infections caused by *C. trachomatis*. This is because antibody titres cannot differentiate between recent and past infection since antibodies remain in circulation for a long time. However, detection of IgM antibodies by ELISA is very useful for the diagnosis of *Chlamydia* pneumonia in infants.

Frei's Skin Test

Frei's skin test is an intradermal skin test used to diagnose LGV. A heat-inactivated *C. trachomatis* LGV serovar grown in the yolk sac of an embryonated egg is used as an antigen. 0.1 mL of antigen is injected intradermally in the forearm.

A control antigen prepared from an uninfected yolk sac is injected into the other forearm.

Frei skin test is a delayed hypersensitivity reaction. A positive reaction is shown by developing an inflammatory macule, measuring >7 mm in diameter, on the test arm after 2 days. The nodule reaches maximum size within 4–5 days. The skin test becomes positive 2–6 weeks after infection and remains positive for several years. However, nowadays, the skin test is rarely used for diagnosis.

Molecular Detection

Molecular diagnosis of *C. trachomatis* infection is made by DNA probes and polymerase chain reaction (PCR). DNA probes are currently available, which detect the presence of a species-specific sequence of 16S rRNA. This method is not highly sensitive. PCR is useful specifically for testing urine specimens but is less sensitive for detecting the bacteria in genital swabs collected from infected women.

Treatment

Azithromycin is the drug of choice for the treatment of genital chlamydial infections. The antibiotic has the advantage of being given in a single-dose regimen and has high tolerability and few contraindications. Tetracyclines are usually recommended for treating patients with LGV for at least 3 weeks. Children below 9 years, pregnant women and patients unable to tolerate tetracyclines are treated with a macrolide, such as erythromycin or azithromycin combined with sulfisoxazole.

Doxycycline for 7 days or a fluoroquinolone (e.g. ofloxacin) for 7 days is also effective for treating genital and ocular infections. Erythromycin given for 10–14 days is very useful for treating conjunctivitis in infants and infant pneumonia. Erythromycin may be administered orally and topically for the treatment of ophthalmia neonatorum.

Local application and oral administration of erythromycin and tetracycline gave for several weeks are effective for the treatment of trachoma.

Epidemiology

C. trachomatis is distributed worldwide.

Geographical Distribution

LGV caused by *C. trachomatis* is highly prevalent in Asia, Africa and South America. Trachoma is distributed worldwide, and nearly 500 million people are infected worldwide. The infection is endemic in the Middle East, Africa, Far East and India. The condition is responsible for blindness in 7–9 million patients. Trachoma is particularly prevalent in these countries because of overcrowding, poor sanitation and poor personal hygiene. All these factors facilitate the transmission of infection. Moreover, *C. trachomatis* is the common cause of infant pneumonia worldwide.

Habitat

C. trachomatis is a strict human pathogen. It is found in the conjunctiva and genitourinary tract in an infected host. *C. trachomatis* also inhabits the respiratory and gastrointestinal tracts of humans.

Reservoir, Source and Transmission of Infection

Humans are the only natural host of *C. trachomatis* and thus are the only significant reservoir of infections. Ocular discharges from infected cases are the common source of eye infection for trachoma. Occasionally, respiratory discharges and human faeces can be a source of infection.

Trachoma is transmitted by eye-to-eye contact through droplets, contaminated hands and contaminated clothing. All these methods facilitate the transmission of ocular discharges from the eyes of infected children to those of normal children. Genital discharges are the source of infection for adult inclusion conjunctivitis. Adult inclusion conjunctivitis is usually transmitted by orogenital contact and also by autoinoculation. Although rare, eye-hand transmission has been reported for adult inclusion conjunctivitis.

Inclusion conjunctivitis in newborns is acquired by the infants born vaginally from mothers infected with *C. trachomatis*. *C. trachomatis* eye infection occurs in approximately one-fourth of infants whose mothers' genital tract is infected with C. trachomatis. Genital discharge is also the source of infection for pneumonia of the newborn. Pneumonia develops in some infants during their birth through an infected birth canal. Chlamydial genital infections are usually caused by vaginal, anal and oral sexual contact.

Prevention and Control

It is difficult to prevent *C. trachomatis* infection. However, morbidity (such as blindness associated with trachoma) can be prevented by immediate treatment at an early stage of the disease. It can also be prevented by taking measures to prevent re-exposure to the bacteria. Chlamydial genital infections are prevented by using safe sexual practices and prompt treatment of symptomatic patients and their partners.

Genus *Chlamydophila*

C. pneumoniae was first isolated from the conjunctiva of a child in Taiwan in 1965 and subsequently from the throat of a patient with pharyngitis in 1983. These organisms were initially believed to be *C. psittaci* strains **TWAR** (from Taiwan acute respiratory) because they produced inclusion bodies in cell cultures similar to those produced by *C. psittaci*. Taiwan isolate was designated as TW-183, and the pharyngeal isolate was designated as AR-9. These two strains were initially believed to be related to *psittaci* strains, but later on, were found to be different. These two organisms were then classified as *Chlamydia pneumoniae* but now have been re-classified and finally placed in a new genus *Chlamydophila*.

Chlamydophila pneumoniae

C. pneumoniae is the third common cause of pneumonia following *Streptococcus pneumoniae* and *Haemophilus influenzae*. It is an important cause of respiratory disease in older children and adults worldwide. It causes mostly asymptomatic infections. Pharyngitis, sinusitis, bronchitis and pneumonia are the common manifestations of symptomatic cases in severe respiratory tract infections.

C. pneumoniae is also a strict human pathogen and does not have animal reservoirs. Infection is transmitted by respiratory secretions from person to person without any avian or animal host. Approximately 200,000–300,000 cases of *C. pneumoniae* occur every year, and most of these cases occur in adults. The infection mostly remains asymptomatic, where an individual can be reinfected.

Diagnosis is made by demonstrating *C. pneumoniae* antigen in specimens by EIA or DFA test. Diagnosis by culture is not made as *C. pneumoniae* grows poorly in cell cultures. Serodiagnosis is carried out by using CFT, ELISA or MIF. Antibodies have been demonstrated in sera of more than 50% of people. PCR is also used to detect bacterial genomes in the specimen. Tetracycline, doxycycline, erythromycin and azithromycin are given for 10–14 days to treat *C. pneumoniae* infection. Control of *C. pneumoniae* infection is difficult.

Chlamydophila psittaci

C. psittaci is the causative agent of psittacosis, a disease of parrots and psittacine birds which can be transmitted to humans. The disease was first observed in parrots, hence named psittacosis (Greek word *psittakos* means 'parrot'). A similar disease of non-psittacine birds was called ornithosis (the Greek word *ornithos* means 'bird'). Now both conditions are called psittacosis.

Infection in birds may occur as asymptomatic or symptomatic. Symptomatic infection may manifest as respiratory infection, diarrhoea and emaciation. In the infected birds, the bacteria are present in the blood, tissue and feathers and are also excreted in the faeces, nasal discharges, aerosols, etc. *psittaci* can be transmitted from birds to humans as well as sheep, cows and goats. Humans acquire the infection mostly from psittacine birds, such as parrots, parakeets, cockatiels and macaws. They get infected by inhaling dried bird droppings, urine or respiratory secretions of these infected birds.

Human infections are mostly occupational, and veterinarians, poultry workers, pet-shop workers, pigeon farmers, etc. are at increased risk for this infection. Person-to-person transmission is rare. Ingestion of poultry birds also does not cause any infection. Laboratory infection is well documented. *C. psittaci* in humans causes a clinical syndrome, which varies from a mild influenza-like illness to fatal pneumonia. The untreated cases may be severe and progress to encephalitis, endocarditis, pericarditis and even death.

Laboratory diagnosis is based mainly on serological tests. Demonstration of a fourfold rise in antibody titre in the paired acute and convalescent sera by the CFT is suggestive of *C. psittaci* infection. MIF test is a specific test used to confirm the diagnosis.

Leventhal–Cole–Lille (LCL) inclusion bodies are demonstrated in the mouse brain, yolk sac and cell cultures. They are also demonstrated in infected cells, including alveolar macrophages from patients suffering from *C. psittaci* infection. These inclusion bodies are stained with iodine and are more diffuse and irregular. They are not inhibited by sulphadiazine or cycloserine. *C. psittaci* can be grown in cell cultures, such as L cells, after 5–10 days of incubation but is not used routinely.

Tetracycline, doxycycline, erythromycin and azithromycin are useful for treating *C. psittaci* infection. Infection is prevented by controlling it in birds.

Case Study

A 6-year-old boy attended the ophthalmology OPD with symptoms of conjunctivitis of the right eye. Examination showed follicular hypertrophy with diffuse inflammation that had affected the entire conjunctiva along with pannus formation. Iodine staining of conjunctival scrapings demonstrated inclusion bodies of *C. trachomatis*. The condition was diagnosed as trachoma.

- List *C. trachomatis* serovars that cause the disease.
- Describe laboratory diagnosis of the condition.
- Describe other infections caused by *C. trachomatis*.
- What is Frei's skin test? Describe the test. For which *C. trachomatis* infection, this test is used?

Miscellaneous Bacteria

47

This chapter includes a wide range of various groups of bacteria that have been associated with human infections. Important bacteria are *Listeria monocytogenes, Erysipelothrix rhusiopathiae, Alcaligenes faecalis, Chromobacterium violaceum, Flavobacterium meningosepticum, Calymmatobacterium granulomatis, Streptobacillus moniliformis, Spirillum minus, Legionella pneumophila, Bartonella* species, *Capnocytophaga* species and *Gardnerella vaginalis*.

Listeria monocytogenes

Morphology and Other Properties

L. monocytogenes is a small (1–3 × 0.5 μm), Gram-positive bacillus with a tendency to occur in chains. Typically, the bacteria show a slow tumbling type of motility by means of peritrichous flagella at 22–25 °C but are non-motile at 37 °C. The bacterium is aerobic or microaerophilic. It grows between 35 and 37 °C and in the presence of reduced oxygen tension and 5–10% CO_2. It is non-sporing, non-capsulated and non-acid fast. *L. monocytogenes* grows on ordinary medium but grows better on blood agar and tryptose phosphate agar. The bacteria on blood agar after 24 h of incubation produce smooth, round, translucent, easily emulsifiable and non-pigmented colonies. On blood agar, they produce a narrow zone of weak beta haemolysis. Due to this haemolysis, the colonies of *L. monocytogenes* may be confused with those of group B streptococci.

L. monocytogenes differs from group B streptococci by various tests. It is motile catalase positive and shows positive CAMP (Christie, Atkins and Munch–Peterson) reaction and negative hippurate hydrolysis Hippurate CAMP reaction. *L. monocytogenes* produces a block type of haemolysis unlike that of arrow-head haemolysis produced by group B streptococci. *L. monocytogenes* ferments sugars such as glucose, saline and aesculin with acid production. It is MR (methyl red) and VP (Voges–Proskauer) test positive but oxidase urease indole and H_2S negative.

L. monocytogenes produces marked monocytosis on experimental inoculation into the rabbits, hence the name **monocytogenes**. This feature can also be observed in human infections. **Anton test** is used to identify this organism, wherein a culture of *L. monocytogenes* is inoculated into a rabbit or guinea pigs' eyes to develop keratoconjunctivitis.

Pathogenesis

Haemolysin, a protein, is an important virulence factor of *L. monocytogenes*. This is also known as histolysin. Histolysin is an extracellular oxygen-labile and cholesterol-sensitive protein of molecular weight 60,000 Da. It shows antigenic cross-reactivity with pneumolysin and streptolysin O.

Clinical Syndromes

L. monocytogenes produces a variety of infections in pregnant women, neonates, adults and elderly people. The bacteria in non-pregnant ladies may cause asymptomatic or mild febrile illness. However, infection in pregnant ladies may cause abortion or stillbirth of the child. In non-pregnant females, asymptomatic infections of the genital tract may lead to **infertility**.

L. monocytogenes in adults may cause meningitis, meningoencephalitis, septicaemia, endocarditis and brain abscess. Patients receiving immunosuppressive therapy, and suffering from immunosuppressive disease, are more susceptible to the infection. *L. monocytogenes* may cause occupation-related infections in poultry workers, butchers and veterinarians. Due to direct contact with infected animals, birds or other products, these people may develop cutaneous infections.

Laboratory Diagnosis

This depends on the isolation of *L. monocytogenes* from clinical specimens by culture.

Specimens

Cervical and vaginal secretions, meconium, blood, cerebrospinal fluid (CSF), amniotic fluid, placenta and biopsy are frequently used specimens for diagnosis.

Gram staining of the CSF smear is useful for the detection of bacteria. This is positive in nearly 50% of the cases. Specimens are cultured on blood agar, chocolate agar and tryptose phosphate agar and incubated for 24–72 h at 35–37 °C. Isolation of bacteria is much more increased by cold enrichment. In this method, the specimens are stored in thioglycolate broth or tryptose phosphate at 4 °C, and subcultures are carried out at weekly intervals for 1–6 months.

Treatment

L. monocytogenes is sensitive to ampicillin, gentamicin and cotrimoxazole. Cephalosporins are not useful for infections caused by *Listeria*.

Epidemiology

L. monocytogenes is distributed worldwide. Many outbreaks of listeriosis caused by food contaminated with *Listeria* have been documented in Europe and North America. Nosocomial outbreaks of *Listeria* have also been documented in hospitals.

L. monocytogenes occurs as a saprophyte in soil, water and sewage. It is also widely prevalent in different mammals or birds, fish, ticks and crustaceans. The bacteria have been isolated from milk, cheese and other milk products. The infection is transmitted by (1) ingestion of milk, milk products and other food preparations contaminated with the bacteria; (2) direct contact with an infected animal or animal products and (3) inhalation of contaminated dust.

Prevention and Control

Infections can be prevented by thoroughly washing and cooking food material before consumption. In pregnant women, early diagnosis is necessary to avoid meningitis in newborn.

Erysipelothrix rhusiopathiae

E. rhusiopathiae is a slender, straight or slightly curved, Gram-positive bacillus with a tendency to form long filaments. It is non-motile, non-sporing and non-capsulated. It is catalase negative. It ferments glucose and lactose, producing acid but no gas. It is H_2S negative. It is negative for MR, VP, indole, urease and nitrate reduction test.

It is aerobic and facultative anaerobic. The growth is improved in the presence of 5–10% CO_2. It can grow on nutrient agar and also on blood agar. After 24–48 h of incubation, it produces convex and translucent colonies surrounded by a variable zone of alpha-haemolysis. On tellurite agar, it produces black colonies.

E. rhusiopathiae in humans causes septicaemia, endocarditis and erysipeloid. The latter is a localised infection of the skin that resembles streptococcal erysipelas. These cutaneous lesions are painful, oedematous and erythematous. They usually occur on the hands or fingers of persons handling animals, fish or animal products. The cutaneous lesions are usually associated with arthritis, lymphopharyngitis or lymphadenitis.

Diagnosis is made by isolating *E. rhusiopathiae* from pus and other clinical specimens by culture. *E. rhusiopathiae* occurs as a saprophyte in soil, food and water. It occurs as a natural parasite of swine, mice, rabbits, turkeys and many others. The bacillus is sensitive to penicillin, erythromycin, ampicillin, methicillin, ciprofloxacin and clindamycin.

Alcaligenes faecalis

A. faecalis are Gram-negative, small, non-sporing bacilli. They are motile due to the presence of peritrichous flagella. They are strict aerobes and do not ferment sugars. They are usually oxidase positive, citrate positive and urease negative. *A. faecalis* occurs as a saprophyte in water and soil contaminated with putrefied decayed organic matter. The bacteria are also found as commensals in the gastrointestinal tracts of humans and animals. The bacterium causes urinary tract infection, infantile gastroenteritis and typhoid-like fever in humans.

Chromobacterium violaceum

C. violaceum is a Gram-negative bacillus. It is catalase and oxidase positive but indole and urease negative. It is a motile bacterium due to the presence of polar and lateral flagella. It is aerobic and facultative anaerobic. It grows on nutrient agar producing a non-diffusible violet pigment, which is soluble in ethanol and insoluble in chloroform and water. *C. violaceum* is found as saprophytes in water, soil and tropical and subtropical countries. Intestinal and genitourinary infections and septicaemic illnesses with pneumonia are rare human infections associated with the bacteria. The bacillus is sensitive to erythromycin, tetracycline, carbenicillin, cefoxitin, etc.

Flavobacterium meningosepticum

F. meningosepticum is a Gram-negative, non-sporing and non-motile bacterium. It is catalase and oxidase positive and weakly fermentative. It can grow on ordinary mediums, such as nutrient agar. It produces smooth, circular and glistening colonies after 24 h of incubation at 37 °C. On incubation at room temperature for 48 h, it produces yellow, non-diffusible pigment on the nutrient agar. *F. meningosepticum* is a ubiquitous saprophyte present in the soil, water, foodstuff and moist environment. It can cause opportunistic infections, neonatal meningitis in premature infants and pneumonia in immunocompromised hosts. In immunocompetent adults, it causes a mild illness. Novobiocin, rifampicin, clindamycin and cefoxitin effectively treat *F. meningosepticum* infections.

Klebsiella granulomatis

Klebsiella granulomatis, formerly known as *Donovania granulomatis*, is the causative agent of **granuloma inguinale**, a granulomatous disease, which involves the genital and inguinal part of the infected host. *K. granulomatis* was first discovered by Donovan in 1905. Donovanosis, a sexually transmitted disease caused by *K. granulomatis*, was first described in India by Mac Leod. He demonstrated the characteristic intracellular bodies in stained smears of pus from an ulcer of a patient suffering from donovanosis. *K. granulomatis* is a small Gram-negative coccobacillus, which measures 1–2 × 0.5–1.5 μm in size. The bacterium often shows bipolar staining, giving rise to a safety-pin appearance. It is a capsulated bacterium but is non-motile, non-sporing and non-acid fast. It can be grown with difficulty on egg yolk medium or modified Levinthal agar. The bacterium grows readily in the yolk sac of an embryonated egg.

K. granulomatis causes granuloma inguinale, which has an incubation period of a few weeks to months. The condition manifests as subcutaneous nodules on genital organs or the inguinal area. Subsequently, these nodules suppurate and break down, leading to the formation of painless granulomatous lesions. The specimen is collected by scraping the border of the ulcerative lesion and making a smear on a glass slide. The smears are stained with Giemsa or Wright stains. In the stained smear, the body of the bacillus is stained blue, and the capsule is stained pink. Laboratory diagnosis is made by demonstrating the characteristic organisms in the cytoplasm of histiocytes polymorphonuclear leukocytes or remain free outside the cell. These are called **Donovan bodies**. About 1–25 bacteria can be found within a mononuclear phagocyte.

Granuloma inguinale is seen in Caribbean Islands and New Guinea. The condition is transmitted through sexual contact. Tetracyclines, erythromycin and trimethoprim-sulfamethoxazole are effective for the treatment of the condition.

Genus *Acinetobacter*

The genus *Acinetobacter* consists of Gram-negative bacterium that is strictly aerobic and non-motile. They are oxidase and nitrate negative and non-fermentative. They appear as Coccobacillary rods, often occurring in pairs, ranging from 1–1.5 × 1.5–2.5 μm in size. They mimic the appearance of Neisseria and hence were named *Mimeae*. The first member of the family was *Micrococcus calcoaceticus*, a soil bacterium identified by Beijernick in 1911.

As classification based on routine identification in the clinical laboratory is erroneous, classifying *Acinetobacter* has undergone many changes. Recently, with the help of molecular studies, such as DNA hybridisation, and different DNA homology studies, the genus *Acinetobacter* has been reclassified. It consists of 23 named and 11 unnamed species. The following are the three major complexes studied within the genus *Acinetobacter:* (1) *Acinetobacter calcoaceticus-baumannii complex*: glucose-oxidising, non-haemolytic strains. (2) *Acinetobacter lwoffii*: glucose-negative, non-haemolytic strain and iii) *Acinetobacter haemolyticus*: haemolytic strain.

A. baumannii grows on MacConkey medium, producing pinkish colonies. It produces acid in the presence of glucose, arabinose, xylose and occasionally rhamnose. The acid formation seen in 10% lactose but not in 1% lactose is a characteristic feature of this organism. However, molecular methods are employed for final identification. ***A. lwoffii*** forms yellow-coloured colonies on MacConkey medium. It is not known to acidify sugars. Few strains may be oxidase positive.

Acinetobacter species are widely spread in nature and commonly occur in soil. They are opportunistic pathogens that can survive on moist and dry surfaces and are frequently present on normal skin. They are an important cause of nosocomial infections such as ventilator-associated pneumonia, skin and wound infections, meningitis and bacteraemia. The strains isolated from clinical specimens are often multi-drug resistant, posing a great challenge for treatment and patient care. Following standard precautions or preventive measures can curtail healthcare-associated infections, thus reducing the use of antibiotics.

Streptobacillus and *Spirillum*

S. moniliformis and *S. minus* are the natural pathogens of rodents. Both these agents cause the same condition called **rat-bite fever**.

Streptobacillus moniliformis

S. moniliformis is a pleomorphic, Gram-negative bacillus (0.1–0.5 × 1–5 μm) that stains poorly with Gram staining. The bacteria may appear as a string of beads containing granules. The bacterium may lose its cell wall and readily develop into L forms. *S. moniliformis* is an aerobe and facultative anaerobe, and a highly fastidious bacterium. It is catalase, oxidase, indole, urease and nitrate negative. It ferments glucose with the production of acid but no gas. The bacterium is a slow grower. It grows on enriched media supplemented with 15% blood, 20% horse or calf serum or 5% ascitic fluid. It produces discrete, granular or smooth, grey colonies after 3 days of incubation at 35–36 °C in a humid aerobic condition. *S. moniliformis* causes rat-bite fever, a condition characterised by relapsing fever, rash and arthralgia. The incubation period varies from 2 to 10 days. The condition manifests with sudden onset of fever, headache and myalgias followed by a petechial rash and arthritis. In untreated cases, relapses are common.

Rat-bite fever is usually caused by the bite of a rat or occasionally by the bite of a mouse, cat, squirrel or dog. The disease also occurs without rat bite and is believed to occur due to ingestion of raw meat or water contaminated by the excreta of the rat. This condition is known as *Haverhill fever* or *erythema arthriticum epidemicum.*

Laboratory diagnosis of the condition is made by culture of the bacillus from blood or other body fluids. Serological tests, such as indirect fluorescent antibody (IFA) test, complement fixation test and agglutination test, are frequently useful for diagnosis. All these tests use *S. moniliformis* as an antigen. A serum titre of 1: 80 or more or a fourfold rise between acute and convalescent sera is considered diagnostic of the disease.

Penicillin is the antibiotic of choice for treating cases of rat-bite fever. Cephalosporins, erythromycin, clindamycin and tetracycline are also used to treat the condition.

Spirillum minus

S. minus is a short, spiral, Gram-negative bacillus and measures 0.2–0.5 × 3–5 μm in size. It is a motile bacterium with the presence of 1–5 amphitrichous flagella. *S. minus* was first observed in a rat by Carter in 1888 from India. Later, researchers from Japan identified it as the causative organism of a type of rat-bite fever (**Sodoku**), acquired by the infected rat bite or scratches. The bacteria stain poorly with Gram stain but stain better with Giemsa or Fontana stains. *S. minus* also causes rat-bite fever similar to that caused by *S. moniliformis,* but with few differences in the clinical manifestations. Rat-bite fever caused by *S. minus* has a long incubation period of 2–4 weeks. The condition may manifest as suppuration of the rat-bite wound, and the onset of fever is associated with regional lymphadenopathy. Mortality is relatively less (6%) in untreated cases.

Other clinical manifestations of rat bite are similar to those caused by *S. moniliformis. S. minus*, unlike other organisms such as the *S. moniliformis*, is found in the nasopharynx of rats and other small rodents. They are transmitted to humans by the bite of these animals but not by contaminated water or milk, as seen with *S. moniliformis.*

Darkfield microscopic examination of blood, pus or lymph node aspirate is useful to demonstrate *S. minus.* Microscopy of Giemsa- or Wright-stained blood smear is also useful. Animal inoculation is a frequently used method for diagnosis. In this method, clinical specimens are inoculated intraperitoneally into rodents, following which these pathogens can be demonstrated in the blood and the peritoneal fluid of these rodents, 1–3 weeks after inoculation.

Serology is not useful. Rat-bite fever caused by *S. minus* can be treated with both penicillin and tetracycline. Table 47.1 shows a comparison between rat-bite fever caused by *Streptobacillus* and *Spirillum* species.

Genus *Legionella*

Legionella is ubiquitous aquatic saprophytes present worldwide. Many of these species are found in the environment and are also associated with human infections. *L. pneumophila* is an important species that causes both community-acquired and nosocomial pneumonia.

Legionella as the causative agent of pneumonia was first identified in 1976 during the annual convention of the American Legion held at a hotel in Philadelphia. The infection was believed to be

Table 47.1 Distinguishing features of rat-bite fever caused by *Streptobacillus* and *Spirillum* species

Features	*Streptobacillus moniliformis*	*Spirillum minus*
Incubation period	Short (10 days)	Long (15 days)
Symptoms	High fever, headache, chills, myalgia, rash, arthritis, arthralgia and recurrent fever if untreated	Fever, headache, chills, rash, lymphangitis, lymphadenitis; and recurrent fever if untreated
Diagnosis	Culture and serology	Dark ground microscopy, microscopy of Giemsa-stained blood smear, and animal inoculation
Antibiotic therapy	Penicillin (L forms not sensitive to penicillin). Both forms sensitive to streptomycin and tetracycline	Penicillin In the case of endocarditis, the addition of an aminoglycoside
Mortality	Higher if untreated	Lower if untreated

transmitted from the contaminated water in the hotel's air conditioning system. A total of 182 cases were detected, of which 29 patients died because of the disease. The disease was characterised by fever, cough and chest pain, leading to pneumonia and often the patient's death. This condition was referred to as **Legionnaire disease**, and the causative agent isolated from the fatal cases was designated as *L. pneumophila*.

The genus *Legionella* contains 39 genetically defined species and more than 60 serogroups. Many of these species cause infections in humans. **Legionellosis** is the term used for infections caused by different *Legionella* species. **Legionnaire disease** is the term used to describe pneumonia caused by *L. pneumophila*.

Legionella pneumophila

L. pneumophila is the causative agent of Legionnaire disease and Pontiac fever.

Properties of the Bacteria

Morphology

L. pneumophila is a small, slender, pleomorphic, Gram-negative bacillus. *L. pneumophila* organisms in clinical specimens stain poorly with Gram staining. They can be stained better by the silver impregnation method using Dieterle silver stain. It measures 0.3–0.9 × 2–5 μm with pointed ends and a tendency to form filaments in culture on solid media. It is motile by means of one or more polar or sub-polar flagella. It is non-acid fast.

Culture and Biochemical Properties

L. pneumophila is a strict aerobe, that grows at an optimum temperature of 35 °C and an optimum pH of 6.9. It is nutritionally fastidious and grows on the medium supplemented with iron salts and l-cysteine.

Buffered charcoal yeast extract (BCYE) agar containing buffered charcoal, yeast extract and cysteine is a useful medium for the growth of *L. pneumophila*. It is a slow-growing bacterium and may take 3–5 days to produce visible colonies. The colonies on BCYE agar are circular, grey or grey-blue and low convex with a slightly irregular margin. Under the microscope, the colonies typically have a ground-glass appearance.

L. pneumophila is mostly non-fermenter and derives its energy from the metabolism of amino acids. It does not ferment any sugars and is urease and nitrate test negative. It is catalase-positive, oxidase variable and liquefies gelatin. It hydrolyses starch and hippurate.

Other Properties

Susceptibility to Physical and Chemical Agents

L. pneumophila is readily killed by 70% ethyl alcohol, 2% glutaraldehyde and 1% formaldehyde. It is also killed by residual chlorine (5 ppm of chlorine) in chlorinated water in

1 min, by 1.25 ppm of chlorine in 15 min and by 0.65 ppm of chlorine in 1 h.

Pathogenesis

L. pneumophila is an obligate intracellular pathogen. Based on direct immunofluorescence and slide agglutination tests, a total of **15 serogroups** (1–15) of *L. pneumophila* have been identified. Of these, three serogroups (1, 4 and 6) are primarily associated with human disease. Nearly 80% of the cases of legionellosis disease caused by *L. pneumophila* are believed to be due to serogroup 1.

Virulence Factors

The capability of the bacteria to multiply in macrophages and their ability to prevent phagolysosome fusion are the main determinants of the virulence of the bacteria. *L. pneumophila* organisms possess many virulence factors, such as protease, haemolysin and cytotoxin. Protease is an important virulence factor, which plays an important role in causing infection.

Pathogenesis of Legionnaire's Disease

The bacteria infect human monocytes and alveolar macrophages. The bacteria multiply in these cells by binding complement to an outer membrane porin protein and subsequent deposition of C3b, a component of the complement on the surface of the bacteria. This facilitates them to combine with CR3 complement receptors on phagocytes and enter the cell through endocytosis. Once inside the cell, the bacteria are not killed because they prevent phagolysosome fusion. Subsequently, the bacteria multiply in large numbers in the intracellular vacuoles and produce many enzymes, such as phosphatase, lipase, nuclease and proteolytic enzymes. Finally, they lyse and kill the infected host cell.

Clinical Syndromes

L. pneumophila causes two distinct clinical entities—Legionnaire's disease and Pontiac fever.

Legionnaire's Disease

The incubation period varies from 2 to 10 days. Pneumonia is the primary manifestation of multilobular consolidation, inflammation and abscesses in the lung. The common symptoms are fever, chills, dry or non-productive cough and pleuritic or non-pleuritic chest pain. Haemoptysis is rare. The condition becomes more severe and is associated with morbidity and high mortality if not treated immediately. Progressive respiratory failure is the most common cause of death in patients with Legionnaire's disease. The mortality may be as high as 80% in untreated patients.

Pontiac Fever

Pontiac fever is a mild, self-limiting condition. The condition presents as fever and myalgia that resolve without treatment. Table 47.2 compares the Legionnaire disease and Pontiac fever caused by *L. pneumophila*.

Laboratory Diagnosis

Specimens

Sputum, lung fluid, pleural fluid, transtracheal aspirations and bronchoalveolar lavage fluid are the respiratory specimens frequently used for the diagnosis. Blood is also used for culture.

Microscopy

Gram staining of the sputum or other respiratory specimens may show many leukocytes, but the organisms are rarely demonstrated. Bacteria, if found, are very poorly stained. Dieterle silver or Gimenez stain can be used to demonstrate bacteria.

Table 47.2 Distinguishing features of Legionnaire disease and Pontiac fever caused by *Legionella pneumophila*

Characteristics	Legionnaire disease	Pontiac fever
Incubation period	2–10 days	1–2 days
Disease	More severe, much morbidity, and causes death if therapy not started promptly	Mild febrile illness, minimal morbidity and causes no death if resolved spontaneously
Mortality	Higher if delay in treatment (15–20%)	Negligible (less than 1%)
Antibiotic therapy	Needed	Not needed, self-limited
Distribution	Epidemic and sporadic	Epidemic
Source	Contaminated water source	Contaminated water source
Pathogenesis	Legionellae engulfed by monocytes and survive as intracellular parasites. Subsequently, lyse and kill infected host cell	Pathogenesis not well understood

Direct Fluorescent Antibody (DFA) Test

It is a sensitive method to detect *Legionella* spp. in sputum and other specimens. This test uses fluorescent-labelled monoclonal or polyclonal antibodies produced against *Legionella* species. This is a rapid test; the result can be obtained within 2–4 h. The sensitivity of DFA depends on the presence of a large number of bacteria in the specimen. Therefore, sensitivity increases if samples from the lower respiratory tract are used. A positive DFA test becomes negative in 4–6 days of treatment.

Culture

Isolation of *Legionella* spp. from sputum or other respiratory secretions (bronchoalveolar lavage, transtracheal aspiration and bronchoscopic specimens) by culture is the definitive method for diagnosing *Legionella* infection. Blood culture shows very low sensitivity.

BCYE agar is the medium most frequently used for the culture of the bacteria. The bacteria produce small colonies after 3–5 days of incubation on BCYE agar at 35 °C in the presence of 3–5% CO_2. Bronchoalveolar lavage, transtracheal aspiration and bronchoscopic specimens increase the sensitivity of the culture. However, blood culture shows very low sensitivity.

Serology

IFA test and ELISA are the most commonly used antibody-based serological tests to diagnose *Legionella* infection. IFA titre of 1:256 or more on a single serum is diagnostic. A fourfold or greater increase in antibody titre between acute and convalescent serum to 1:128 or greater in titre is also considered diagnostic of the disease.

ELISA, radioimmunoassay (RIA) and the latex agglutination test are used to detect *Legionella* lipopolysaccharide antigen in the urine. Urinary antigen detection is relatively rapid and simple and can even detect antigen excreted in patients on antibiotic therapy. The disadvantage of the test is that it can detect only *L. pneumophila* (serogroup 1 antigen). The antigens remain in urine even for months after the infection has resolved.

Molecular Detection

Polymerase chain reaction (PCR) is a specific method for detecting *Legionella* spp. in urine, serum and bronchoalveolar lavage fluid. The main advantage of this method is that it can detect infections caused by *Legionella* species other than *L. pneumophila* serogroup 1.

Treatment

Erythromycin or tetracycline is useful for the treatment of community-acquired infections. Fluoroquinolones, such as doxycycline, telithromycin and azithromycin, are recommended to treat severe cases. Beta-lactam antibiotics and aminoglycosides are not useful.

Epidemiology

Geographical Distribution

Legionnaire's disease is prevalent worldwide. This is responsible for 2–15% of all community-acquired pneumonia that requires hospitalisation.

Habitat

L. pneumophila is an obligate intracellular pathogen found in alveolar macrophages in the infected human host.

Reservoir, Source and Transmission of Infection

L. pneumophila is a saprophyte widely found in aquatic environments. The bacteria have been isolated from various man-made water environments, such as hot water systems in hotels and hospitals, cooling water systems, air conditioning cooling towers, showerheads, industrial coolants and respiratory ventilators. The factors that contribute to the presence of these bacteria in these environments include (1) stagnation of water, (2) temperature of water varying from 25 to 40 °C and (3) presence of free-living amoebae, such as *Acanthamoeba* and *Bartonella* species, which support the growth of *Legionella*. Therefore, contaminated water is the primary source of infection.

Transmission to humans occurs by inhaling aerosolised mist from a contaminated water source, contaminated with either the bacteria or free-living amoeba infected with the bacteria. Nosocomial infection is transmitted through aspiration, contaminated water or respiratory equipment. Infection is also transmitted by using nebulisers, humidifiers and other instruments that have been washed with contaminated tap water. Person-to-person transmission does not occur. Animals also do not play any role in the transmission of the infection. The elderly people, people with chronic heart or lung diseases and the immunocompromised patients or are receiving immunosuppressive therapy are at the greatest risk for the disease. This is because of their depressed cell-mediated immunity and altered pulmonary functions.

Prevention and Control

Hyperchlorination of water and superheating of water to 70–80 °C have proved to be of value in preventing the disease. Complete elimination of bacteria from water, however, is difficult.

Eikenella corrodens

Eikenella corrodens is a Gram-negative, facultative anaerobic, capnophilic bacillus. This bacillus is positive for nitrate and oxidase test, whereas negative for catalase and indole test. The pitting or corroding of blood agar by the colonies is the characteristic feature of this bacterium and hence the name **'corrodens'**. It is a commensal of the human mouth, upper respiratory and gastrointestinal tract. Infection occurs due to salivary or faecal contamination mainly through the skin's subcutaneous tissues and is typically indolent. It rarely causes osteomyelitis, pneumonia, endocarditis and meningitis. It is sensitive to penicillin, tetracycline and cephalosporins and resistant to macrolides and clindamycin.

Cardiobacterium hominis

Cardiobacterium hominis is a pleomorphic, oxidase and indole positive, catalase negative, Gram-negative bacillus. It is present commonly in the human mouth and upper respiratory tract, mainly the nose and throat. It rarely causes infection such as endocarditis, particularly for those with previous cardiovascular disease. It grows on a blood agar medium under high humidity and 3–5% CO_2. The recommended drugs for treatment are penicillin and streptomycin.

Genus *Bartonella*

Members of the genus *Bartonella* are very small, Gram-negative bacilli transmitted by arthropods. The bacteria are Gram-negative aerobic bacilli and are fastidious in growth. They cause

Table 47.3 Human infections caused by *Bartonella* species

Bartonella species	Diseases
Bartonella bacilliformis	Oroya fever
Bartonella quintana	Trench fever
Bartonella clarridgeiae	Endocarditis
Bartonella elizabethae	Endocarditis
Bartonella henselae	Bacillary peliosis Bacillary angiomatosis Cat-scratch disease

infections in a mammalian host by invading endothelial cells and red blood cells. *Bartonella* species are found in various animal hosts and are typically present without evidence of the disease. The genus *Bartonella* consists of 11 species, of which five are associated with human diseases (Table 47.3).

Bartonella bacilliformis

Bartonella bacilliformis is the causative agent of **Oroya fever**, an acute febrile illness consisting of severe anaemia. This condition was first identified in the mountainous parts of Peru in 1870 during the laying of railway lines from Lima to Oroya in Peru. The outbreak of Oroya fever killed 1000 workers associated with this railway project.

B. bacilliformis are short, Gram-negative coccobacilli measuring 0.3–0.5 × 1.0–1.7 μm. The bacteria occur singly, in pairs, in chains or clumps. They are motile by the presence of as many as 10 flagella at one pole of the bacteria. They are aerobic and require an optimum pH of 7.8 and an optimum temperature of 25–28 °C for their growth.

Oroya fever is restricted in its geographical distribution and is mainly confined to Peru, Columbia and Ecuador in South America. The laboratory diagnosis is made by demonstrating intracellular bacteria in a blood smear stained by Giemsa. The bacteria are found in the cytoplasm. Bacteria can be isolated from blood in enriched blood agar.

Penicillin, streptomycin, tetracycline and chloramphenicol effectively treat *B. bacilliformis* infection. The use of insecticides, such as DDT, to kill the sand fly prevents transmission of the disease.

Bartonella quintana

Bartonella quintana, earlier known as *Rochalimaea quintana*, is a causative agent of **trench fever** or **5-day fever**. This condition was first recognised in the soldiers fighting in trenches in Europe during the First World War. The causative agent was earlier identified as rickettsia and named *Rickettsia quintana* because it caused a 5-day fever (from *quintana,* means fifth), a synonym for trench fever. The organism subsequently was found not to be a rickettsia and differed from the latter by its ability to grow in cell-free media, such as blood agar; hence it was separated into a new genus *Rochalimaea.* Currently, it has undergone further taxonomical classification and has been reclassified as *B. quintana.*

B. quintana is a small, Gram-negative bacillus (0.3–0.5 × 1.0–1.7 μm) and is non-motile. It grows slowly on sheep or rabbit blood agar at 37 °C in the presence of 5% CO_2. It requires prolonged incubation of 1–2 weeks to produce demonstrable colonies.

B. quintana is transmitted by the body louse. The lice become infectious 5–9 days after feeding on a trench fever patient, after which the lice remain infectious throughout their life and excrete organisms in their faeces. When they bite a new host, the infected lice defecate on the skin's surface. When these faeces come in contact with minor scratches or abrasions on the skin's surface, the bacteria present in the faeces enter the skin and initiate the infection.

Trench fever has an incubation period of 14–30 days. The condition can vary from asymptomatic to symptomatic infection. Severe headache, fever (giving the name of the disease as 5-day fever), chills, weakness and severe pain in the back and legs are the common manifestations of symptomatic cases. The condition is also associated with rashes on the chest, abdomen or back. The infection does not cause the patient's death but can produce a very severe debilitating illness. Trench fever is an exclusively human disease. No animal reservoir for this disease has been identified. The disease is transmitted from humans to humans by the human body louse vector.

Laboratory diagnosis is made by isolating the bacteria from the patient's blood on blood agar after 2 weeks of incubation. The organism can also be demonstrated in lice after allowing the healthy lice to feed upon the patient *(xenodiagnosis)*. Weil–Felix test used for diagnosis of rickettsial infection is negative in trench fever. The condition can be treated with gentamicin alone or with erythromycin.

Bartonella henselae

Bartonella henselae is the causative agent of cat-scratch disease. The disease follows scratches and the bite of the cat or with the bite of cat fleas. *B. henselae* is a small, Gram-negative bacillus measuring 2.0–2.5 × 0.5–0.6 μm. Like other *Bartonella* species, it can grow on chocolate agar or Columbia agar supplemented with 5% sheep or rabbit blood. *B. henselae* produces dry white cauliflower-like colonies after 5–15 days of incubation at 35–37 °C in the presence of 5% CO_2.

B. henselae typically causes **cat-scratch disease**. It is a benign condition characterised by regional lymphadenopathy and fever. The cervical and axillary lymph nodes are most commonly affected. Usually, a single lymph node is affected in 50% of the cases. It also causes **bacillary angiomatosis**, a condition resembling Kaposi sarcoma in patients with AIDS. The skin, lymph nodes, liver and spleen are primarily involved in this condition. The bacteria also cause sub-acute bacterial endocarditis.

Laboratory diagnosis is confirmed by demonstration of clusters of *B. henselae* in lymph node biopsy smears stained with Warthin–Starry impregnation stain. The bacteria can also be isolated by culture on chocolate agar or Columbia agar enriched with sheep or rabbit blood on incubation for 3 weeks or more. Gentamicin and erythromycin are effective for the treatment of the condition.

Genus *Capnocytophaga*

Members of the genus *Capnocytophaga* are filamentous Gram-negative bacilli. They are slow-growing aerobic and anaerobic bacteria and require the presence of CO_2 for their growth. The genus is classified into two groups: (1) dysgonic fermenter 1 (DF-1) and (2) dysgonic fermenter 2 (DF-2). **DF-1 group** consists of three species: *Capnocytophaga ochracea*, *Capnocytophaga gingivalis* and *Capnocytophaga sputigena*, which are found as members of the normal flora of the oropharynx in humans. These species have been associated with periodontitis and bacteremia. In an immunocompromised host, they occasionally cause severe systemic disease, such as endocarditis. **DF-2 group** bacteria are found only in dogs and not in humans.

Most *Capnocytophaga* strains are resistant to aminoglycosides but are sensitive to broad-spectrum cephalosporins, fluoroquinolones and penicillins.

Gardnerella vaginalis

G. vaginalis was earlier known as *Corynebacterium vaginalis* or *Haemophilus vaginalis*. Since the bacterium does not require X and V growth factors, it has been shifted from the genus *Haemophilus* and placed in the genus *Gardnerella*. The bacterium is present as a commensal in the male urethra and vagina. *G. vaginalis* are small, pleomorphic, Gram-negative rods, which are sometimes Gram variable. They are non-sporing, non-motile and non-capsulated. *G. vaginalis* causes non-specific vaginitis and cervicitis, frequently associated with anaerobic vaginosis. The condition is characterised by foul-smelling vaginal discharge with a fishy odour and acidic pH.

A high vaginal swab or endocervical swab with exudate material is the specimen for diagnosis. The specimens are transported in the Stuart medium. Gram staining of the smear shows Gram-variable small bacilli and the presence of clue cells. These clue cells are the vaginal epithelial cells covered with many small Gram-variable rods. The bacteria are cultured on blood agar and incubated anaerobically at 37 °C for 48 h for colonies. The colonies of *G. vaginalis* are identified by Gram staining, negative catalase test, and positive hippurate hydrolysis test.

G. vaginalis is sensitive to penicillin, ampicillin and trimethoprim. It is resistant to gentamicin, nalidixic acid and colistin. Metronidazole is the drug of choice.

Moraxella lacunata

These are thick to plump rods. They are Gram-negative occurring in pairs and chains. They do not have flagella but are reported to be sluggishly motile. They grow on ordinary media and are strict aerobes. They are positive for catalase and oxidase tests. They are negative for indole production. Morax and Axenfeld were the first to report *Moraxella lacunata* as the cause of **angular conjunctivitis,** which is also known as Morax–Axenfeld bacillus.

Kingella

They were previously grouped under the genus *Moraxella*. They are non-motile, catalase negative, oxidase positive and urease negative. Glucose and other carbohydrates are fermented with the production of acid. Cells are Gram-negative rods that occur as diplococcal and coccobacillary forms. They are present in the normal flora. They are associated with infections of the bones, joints, tendons and endocarditis. *C. hominis*, *E. corrodens* and *Kingella* belong to the **HACEK group**, slow-growing bacteria causing endocarditis.

Case Study

A 29-year-old software engineer working in Peru for the past year came back to India with complaints of myalgia, arthralgia, headache, severe anaemia and hepatosplenomegaly. Routine laboratory studies were negative. A blood smear stained by Giemsa stain showed intracellular bacteria. Blood culture showed colonies, which were identified as *B. bacilliformis*.

- Diagnose the disease.
- What are the other laboratory tests used to diagnose the disease?
- How will you treat the condition?
- Is this condition present in India?

Part IV

Virology

48 Introduction to Viruses

The viruses are too small to be seen with a light microscope. Their small size allows them to pass through filters used to retain bacteria in contaminated fluids. Hence, they were first described as filterable agents. Like other microorganisms (e.g. bacteria, fungi and parasites), viruses are the infectious agents associated with disease in humans. Unlike other infectious agents (Fig. 48.1), viruses are obligate intracellular parasites; that is, they require living host cells to multiply. In addition, viruses replicate by assembling the individual components rather than binary fission.

The viruses show the following features:

1. They are filterable agents.
2. They are obligate intracellular parasites.
3. They contain a single type of nucleic acid, either DNA or RNA, but not both.
4. The virion of the virus particle consists of a nucleic acid genome packaged into a protein coat (capsid), which itself is sometimes enclosed by an envelope of lipid, proteins and carbohydrates known as an envelope.
5. They multiply inside the living cells by using the synthesising machinery of the host cell.
6. They replicate by the assembly of the individual components and do not replicate by division, such as binary fission.
7. They have few or no enzymes for their own metabolism. They always use host cell machinery to produce their components, such as viral messenger RNA (mRNA), protein and identical genome copies.

The differences between bacteria and viruses are summarised in Table 48.1. The viruses affect a wide range of hosts. Some viruses infect invertebrates, vertebrates, plants, protists, fungi and bacteria. We are concerned mainly with viruses that infect either humans or bacteria in medical microbiology. The viruses that infect bacteria are known as **bacteriophages** or **phages**.

Morphology of Viruses

The extracellular, infectious viral particle is called the **virion.** It is a complete, fully developed infectious viral particle composed of nucleic acid surrounded by a protein coat. The latter protects it from the environment and is a vehicle of transmission from one host to another. The viruses are classified based on differences in the structure of these coats. The virion may have a lipid envelope. The virion may also contain essential or accessory enzymes or other proteins.

Size

The clinically important viruses vary widely in their size (Fig. 48.1). They range from as small as 20-nm viruses (picornaviruses) to as large as 300-nm viruses (poxvirus). Electron microscopy

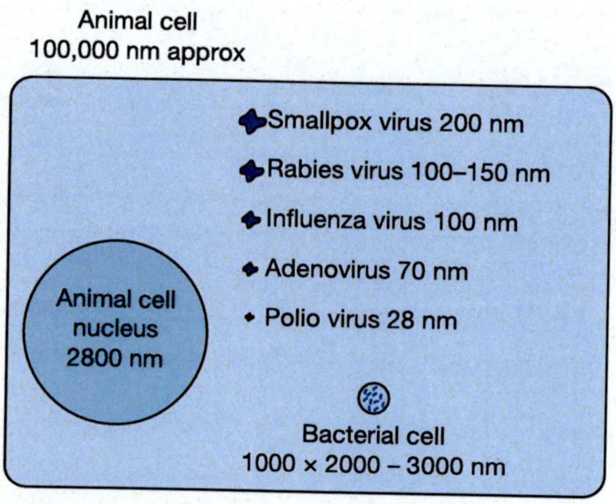

Fig. 48.1 Size of the nucleic acid of viruses

is the most current method for determining the size and shape of the virus.

Structure and Symmetry of Virus

Viral Structure

The virion consists of a nucleic acid core, the genome, surrounded by a protein coat and the capsid (Fig. 48.2). Together with the enclosed nucleic acid, the capsid is known as the nucleocapsid.

Capsid

The nucleic acid of a virus is surrounded by a protein coat called the capsid. Each capsid is composed of a large number of protein subunits (polypeptides) called **capsomeres**, which form its morphological units. The polypeptide molecules composing the capsomeres are of a single type in some viruses, while several types may be present in other viruses. The arrangement of capsomeres is characteristic of a particular type of virus.

Functions of Capsid

The capsid facilitates entry of the viral genome into the host cells by adsorbing readily to cell surfaces. The capsid of the virus protects its nucleic acid from the activity of nuclease enzymes in biological fluids and thereby facilitates the attachment of the virus to target cells in the host. Symmetrically arranged polypeptide molecules of capsid form an impenetrable shell around the nucleic acid core. The viral capsid structure is best demonstrated by X-ray crystallography or electron microscopy.

Envelope

In some viruses, the capsid is covered by an envelope; such viruses are called enveloped viruses. All of the negative-stranded RNA viruses are enveloped. The viruses that lack an envelope are called non-enveloped or naked viruses.

The virion envelope usually consists of lipoproteins and glycoproteins. It has a membrane structure similar to the cellular membrane of the host cell. The viral envelope does not contain any cellular proteins, even though viruses are released from the host cell by an extrusion process that coats the virus with a layer of host cell plasma membrane that becomes the viral envelope. The envelope contains proteins determined and

Table 48.1 Distinguishing features of bacteria, Rickettsiae, Chlamydiae and viruses

Characteristics	Typical bacteria	Rickettsiae	Chlamydiae	Viruses
Intracellular parasite	No	Yes	Yes	Yes
Cellular organisation	Yes	Yes	Yes	No
Plasma membrane	Yes	Yes	Yes	No
Replication by binary fission	Yes	Yes	Yes	No
Growth in inanimate media	Yes	No	No	No
Pass through bacteriological filters	No	No	Yes	Yes
Possess both DNA and RNA	Yes	Yes	Yes	No
ATP-generating metabolism	Yes	Yes	No	No
Ribosome	Yes	Yes	Yes	No
Sensitive to antibiotics	Yes	Yes	Yes	No
Sensitive to interferon	No	No	Yes	Yes

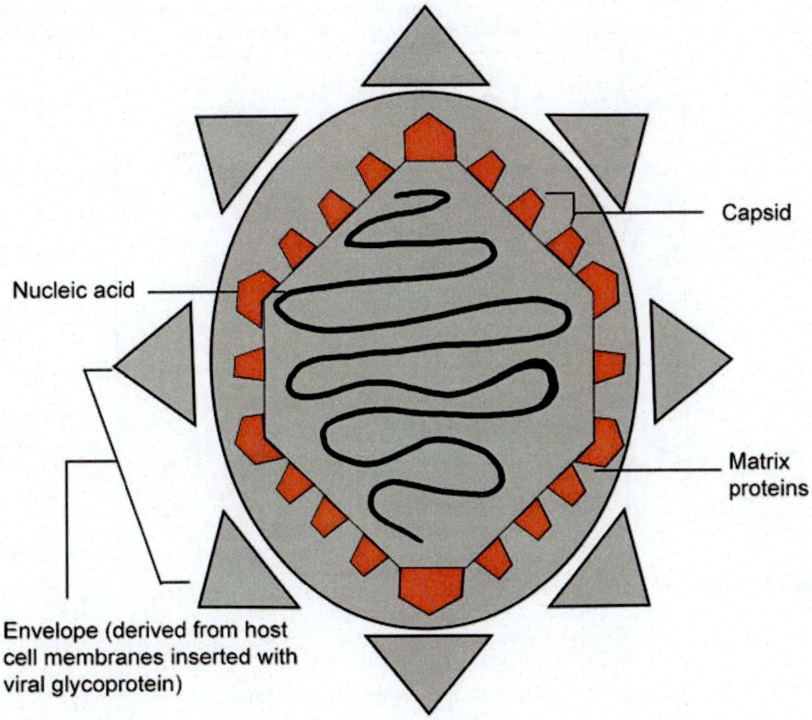

Fig. 48.2 Schematic diagram of the structure of a virus

encoded by viral nucleic acid in most cases. The lipid component of the envelope is usually of host cell origin.

Depending on the virus, the envelopes of the viruses may or may not be covered by spikes. The spikes are glycoprotein-like projections on the outer surface of the envelope. Most spikes act as viral attachment proteins (VAPs). The VAP that binds to red blood cells is called haemagglutinin. The ability of certain viruses, such as the influenza virus, to agglutinate red blood cells is due to the presence of these haemagglutinins. The process is called haemagglutination (Fig. 48.3), and it forms the basis of the haemagglutination inhibition test used in viral serology (Table 48.2). The VAPs in some viruses perform different functions, such as neuraminidase activity of influenza virus, fusion glycoprotein of paramyxovirus and C3b receptor associated with Herpes simplex virus.

The structural components of the envelope remain biologically active only in aqueous solutions and are readily destroyed by drying or on treatment with acids, detergents and solvents, such as ether, leading to the inactivation of the virus. They are rapidly killed in the stomach due to the sensitivity of enveloped components to gastric acidity. Therefore, most enveloped viruses are usually transmitted through body fluids, such as blood, respiratory droplets, tissue exudates, etc.

Viral Symmetry

Three types of symmetry are observed depending on the arrangement of the capsid around the nucleic acid core (genome). These are (1) icosahedral (cubical), (2) helical and (3) complex symmetry.

Icosahedral Symmetry

Two types of capsomeres constitute the icosahedral capsule. They are the pentagonal capsomeres or the vertices (*pentons*) and hexagonal capsomeres making up the facets (*hexons*) (Fig. 48.4). There are always 12 pentons, but the number of hexagons varies with the virus group.

Each penton has fivefold symmetry (pentamer or pentagon) in the shape of an equilateral

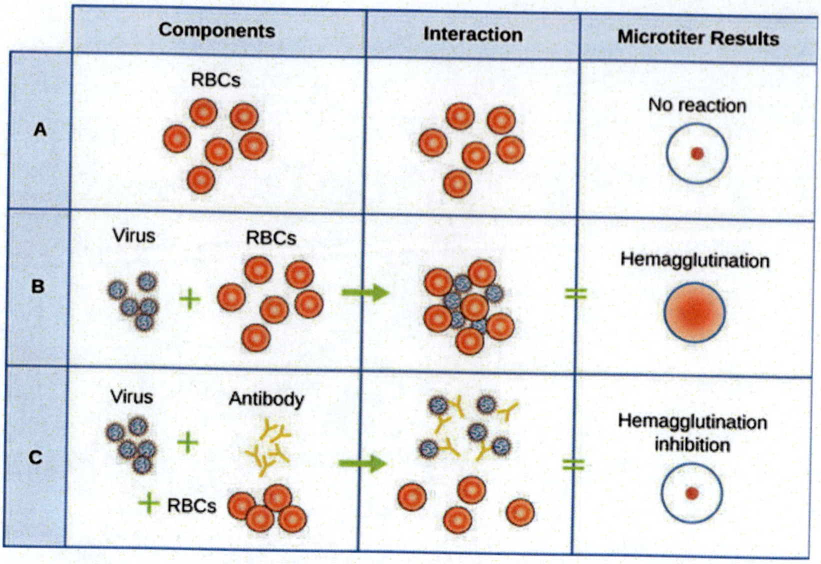

Fig. 48.3 Haemagglutination reactions of virus. (http://phil.cdc.gov/PHIL_Images/17270/17270_lores.jpg)

Table 48.2 Haemagglutination of erythrocytes by various viruses

Virus	Haemagglutination of erythrocytes of different species and temperature
Enteroviruses, coxsackie, echoviruses Rhinovirus and some serotypes	Human erythrocytes 4 °C and 37 °C Sheep erythrocytes at 4 °C
Reoviridae	Human erythrocytes 37 °C
Rabies	Goose at 4 °C
Influenza A Influenza B Influenza C	Only Guinea pig cells at room temperature and 4 °C Both fowl and Guinea pig red cells Only fowl erythrocytes
Parainfluenza and mumps Measles	Guinea pig erythrocytes Monkey erythrocytes at 37 °C
Rubella virus	Human, chick, goose and pigeon erythrocytes at 4 °C

Fig. 48.4 Icosahedral symmetry of a virion

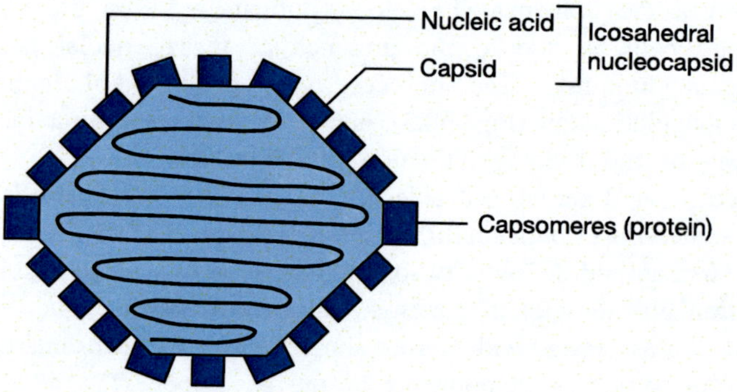

triangle. This pentamer symmetry is found in simple viruses, such as the picornaviruses and parvoviruses. In picornaviruses, each pentamer is made of five protomers, each of which is composed of three subunits of four different proteins.

The hexamer symmetry is usually found in large capsid virions, such as herpes viruses and adenoviruses. The presence of hexon extends the icosahedral and is called an icosadeltahedral. The adenovirus nucleocapsid has 12 pentons and 240 exons, whereas the herpes virus nucleocapsid has 12 pentons and 150 hexons surrounded by an envelope.

Helical Symmetry

The nucleic acid and the capsomeres are wound together to form a spherical or spiral tube. The viruses with helical structures usually appear as rods and their capsomeres self-assemble on the RNA genome into rods extending to the length of the genome. These capsomeres cover and protect the RNA. The tubular nucleocapsid structure may be rigid as in tobacco mosaic virus, but maybe pliable and may be coiled on itself in case of some other animal viruses. Helical nucleocapsids are usually demonstrated within the envelope of most negative-stranded RNA viruses.

Complex Symmetry

Some viruses may not exhibit either icosahedral or helical symmetry but instead may exhibit a complex symmetry. For example, poxvirus shows complex symmetry.

Shape

Most of the enveloped viruses are round or pleomorphic, except for poxvirus and rhabdovirus. Rhabdovirus is a bullet-shaped virus, whereas poxvirus is brick-shaped.

Viral Components (Nucleic Acid, Proteins and Lipids)

Viral components include nucleic acid, proteins and lipids

Viral Nucleic Acid

The virus's genome consists of either DNA or RNA but never both. The DNA can be single-stranded or double-stranded. Depending on the virus, the DNA can be linear or circular. The RNA can be either positive sense (+) like mRNA or negative-sense (−), double-stranded (+/−) or ambiguous (containing + and − regions of RNA attached to it). In some RNA viruses, such as the influenza virus, the RNA genome is in several separate segments, each segment encoding an individual gene. The total amount of nucleic acid may vary from a few thousand nucleotides to as many as 250,000 nucleotides.

Viral Proteins and Lipids

Viruses contain proteins, which constitute capsids. The viral protein protects the nucleic acid and determines the antigenic specificity of the virus. In addition, the enveloped viruses contain lipids, which are derived from the host cell membrane.

Susceptibility to Physical and Chemical Agents

Disinfectants

The viruses are usually more resistant than bacteria to chemical disinfectants. Most viruses are relatively resistant to phenol. The most active antiviral disinfectants are oxidising agents, such as hydrogen peroxide, potassium permanganate, hypochlorite and organic iodine compounds. Formaldehyde and β-propiolactone are active virucidal agents commonly used to prepare killed viral vaccines. The chlorination of drinking water is useful for killing most common viruses, except for hepatitis A and polioviruses. These two viruses are relatively resistant to chlorination.

Temperature

Most of the viruses, with few exceptions, are highly heat-labile. They are inactivated within seconds at 56 °C, within minutes at 37 °C and within days at 4 °C. Viruses such as influenza, measles and mumps are very labile and may survive outside the host only for a few hours. Other viruses, such as polio and hepatitis A, are relatively stable and may survive for many days, weeks or even months in the environment. Viruses, such as hepatitis B, show resistance to heating at 60 °C for 60 min; slow viruses, such as scrapie virus, are resistant to autoclaving at 121 °C for 15 min.

The viruses are stable at low temperatures. They can be stored by freezing at −35 or −70 °C. Lyophilisation or freeze drying is useful for the long-term storage of viruses. The poliovirus is an exception, as it does not withstand freeze drying.

pH

The viruses usually remain viable in a pH range of 5–9 but are sensitive to extremes of acidity and alkalinity. Rhinoviruses are very susceptible to acidic pH, whereas enteroviruses are highly resistant.

Lipid Solvents

Ether, chloroform and detergents are active against enveloped viruses but not against non-enveloped, naked viruses.

Radiations

The viruses are readily inactivated by sunlight, ultraviolet (UV) radiations and ionising radiations.

Antibiotics

Antibiotics are inefficient against viruses. This property of antibiotics has been exploited in viral culture, in which, before virus isolation, clinical specimens are treated with antibiotics to remove all the bacterial contaminants.

Replication of Viruses

The replication of viruses in the host cell depends upon the synthesis mechanism of the host cell for the manufacture of different viral components. The genetic information for viral multiplication is present in the viral nucleic acid. Multiplication of viruses follows the basic pattern of bacteriophage multiplication but has several important differences.

The multiplication of viruses, both DNA and RNA containing viruses, is divided into six phases as follows: (1) attachment, (2) penetration, (3) uncoating, (4) biosynthesis, (5) maturation and (6) release (Fig. 48.5).

Attachment

Attachment or adsorption is the first event in the infection of the cell by a virus. The viruses have attachment sites that attach to the complementary receptor sites on the host cell surface. These receptor proteins in the virus are distributed on the virus's surface. These receptor proteins vary from one virus to another (Table 48.3). For example, in the influenza virus, these receptor proteins are the spikes present on the surface of the envelope, whereas, in adenovirus, these receptor proteins are small fibres present at the corner of the icosahedron.

The attachment sites of the virus bind specifically to the complementary receptors on the surface of the host (Table 48.4). These receptor sites on the cell vary depending on the nature of the virus. Rabies virus binds specifically to the acetylcholine receptors found on neural cells. HIV-1 binds specifically to the CD4, a60-kDa glycoprotein, on the surface of mature T lymphocytes. Influenza virus binds specifically to the sialic acid residue of glycoprotein receptor sites on the surface of the respiratory epithelium. Therefore, the host's susceptibility to virus infection depends on the presence or absence of receptors on the cell surface.

Penetration

Whether enveloped or non-enveloped, the viruses penetrate cells by different mechanisms depending on their nature. A non-enveloped virus enters the cell by a process known as endocytosis. Endocytosis is an active process that brings nutrients and other molecules into a cell. This process is also known as viropexis. The enveloped viruses enter the cell by an alternate method called fusion, in which the viral envelope fuses with the plasma membrane and releases the capsid into the cell cytoplasm. Human

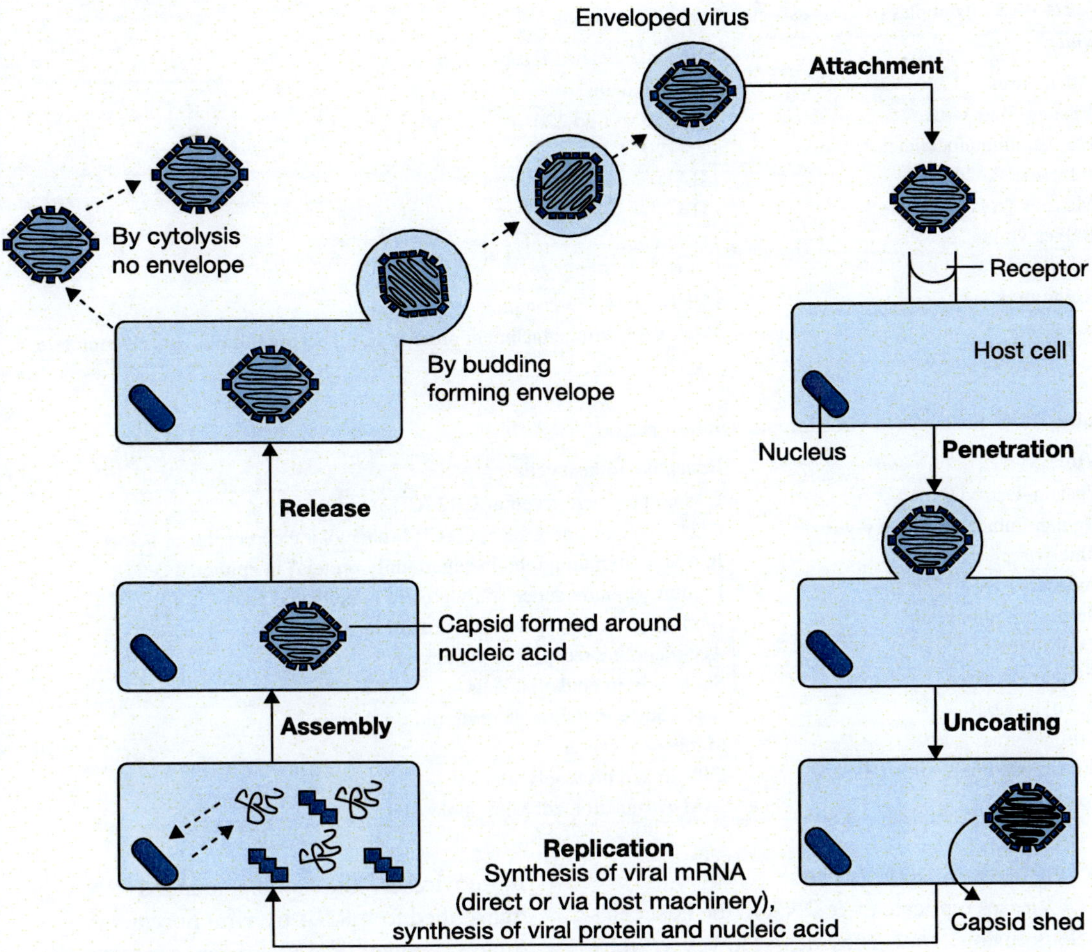

Fig. 48.5 Replication cycle of viruses

immunodeficiency virus (HIV) is a classical example of an enveloped virus, which enters the host cell by this method.

Uncoating

Uncoating is the process of the separation of viral nucleic acid from its protein core. This process apparently varies depending on the nature of the virus causing infection. With most viruses, the uncoating is accomplished by the action of lysosomal enzymes present inside the phagocytic vacuoles and Golgi vesicles of the host cell. These lysosomal enzymes degrade the protein of the viral capsid. In some other viruses, uncoating is caused exclusively by enzymes present in the host cell cytoplasm. In the case of poxvirus, the uncoating is affected by the action of a specific enzyme encoded by the viral DNA, which is synthesised soon after infection.

Biosynthesis

The sites of viral synthesis depend on the type of virus, whether RNA virus or DNA virus. Biosynthesis begins with the transcription of viral nucleic acid to produce mRNA. Translation of mRNA into early or non-structural proteins is the components of a virus. These viral proteins can even shut down the translation and replication machinery of the host cell, followed by viral nucleic acid replication. Subsequently, structural protein synthesis acts as daughter virion capsids.

Table 48.3 Examples of virus-associated receptor proteins

Virus	Virus-associated receptor proteins
Adenovirus	Fibre protein
Epstein–Barr virus	gp 350 and gp 220
Human immunodeficiency virus	gp 120
Influenza A virus	HA
Measles virus	HA
Rabies virus	G protein
Reovirus	—1
Rhinovirus	VP1–VP2–VP3 complex
Rotavirus	VP7VAP, viral attachment protein; HA, haemagglutinin; gp, glycoprotein.

Table 48.4 Examples of virus receptors in host cells

Virus	Receptors in host cells
Epstein–Barr virus	C3d complement receptor CR2 (CD21) in B cell
Human immunodeficiency virus	CD4 molecule and CCR5, CXCR4 (chemokine coreceptor in helper T-cell)
Rhinovirus	ICAM-1 (immunoglobulin superfamily protein) in epithelial cells
Poliovirus	Immunoglobulin superfamily protein in epithelial cells
Herpes simplex virus	Immunoglobulin superfamily protein in many cells
Rabies virus	Acetylcholine receptor in the neuron
Influenza A virus	Sialic acid in epithelial cells
Parvovirus B19	Erythrocyte P antigen in erythroid precursors
Dengue virus	Mannose receptor
Human papilloma virus (HPV 16)	GPI- linked proteoglycans
Reovirus	JAM-A (junction adhesion molecule)

Biosynthesis of DNA Viruses

DNA viruses replicate their DNA in the host cell's nucleus using viral enzymes. However, they synthesise their capsid and other proteins in the cytoplasm by using host cell enzymes. Subsequently, the proteins enter into the nucleus and assemble with the newly synthesised DNA to form virions, and then these virions are transported along with the endoplasmic reticulum to the membrane of the host cell for release.

Most DNA viruses (herpes virus, papovavirus, adenovirus and hepadnavirus) synthesise their nucleic acid in the host cell nucleus. Poxvirus is an exception because its components are synthesised in the cytoplasm.

Biosynthesis of RNA Viruses

The RNA viruses multiply in the host cell's cytoplasm. The major differences in the multiplication process of various RNA viruses lie in how mRNA and viral RNA are produced, as follows.

1. In reoviruses, the double-stranded RNA is transcribed to mRNA by viral polymerase.
2. Single-stranded RNA viruses are classified into two categories, depending on the method of mRNA transcription. In a positive-strand (+ strand, positive sense) RNA virus, the viral RNA itself acts as the mRNA. The viral RNA is translated directly into viral proteins in the host cell cytoplasm. In the negative-strand (—sense) RNA viruses, the RNA is antisense, with polarity opposite to mRNA. They possess their own RNA polymerase for the transcription of mRNA.
3. The single-stranded RNA genome of Retroviridae shows a unique replication. First, the RNA genome is converted into an RNA–DNA hybrid by the viral enzyme reverse transcriptase (RNA-directed DNA polymerase). The double-stranded DNA is synthesised from the RNA–DNA hybrid, after which it is integrated into the host cell

chromosome. Integration of viral DNA into the host cell may lead to the transformation of the cell and the development of malignancies.

Mechanism of Virus Replication

Baltimore classified viruses into six classes depending on the method of mRNA transcription.

Class 1: Fully double-stranded DNA viruses like adeno, herpes and papovaviruses, where the DNA is transcribed with the help of host cell machinery after it enters the nucleus. The DNA thus obtained is found to be infectious. However, hepadnavirus has partially double-stranded DNA, which converts into fully double-stranded DNA with the help of viral DNA polymerases in the host cell's cytoplasm. Further, they are transcribed in the host cell's nucleus with the help of host transcription factors. Hence, hepadnavirus DNA extract is not infectious. Poxviruses undergo replication in the cytoplasm by using mRNA with the help of polymerases produced by virions. Because of this, Poxvirus DNA is also not infectious.

Class 2: In the case of single-stranded DNA viruses like parvoviruses, DNA transcription occurs only after its conversion into duplex form inside the host cell nucleus.

Class 3: In reoviruses, mRNA is formed only after transcription of double-stranded RNA by viral polymerases.

Class 4: Based on the method of mRNA transcription, double-stranded RNA is divided into two classes, positive strand and negative strand. In the positive strand, the viral RNA by itself behaves as an mRNA. The viral RNA by itself is infectious. Translation of viral RNA into viral proteins directly takes place inside the host cell cytoplasm.

Class 5: In the case of negative-sense RNA like orthomyxoviruses, paramyxovirus and rhabdoviridae, the RNA is antisense, having polarity just opposite mRNA. They have their own RNA polymerases for mRNA transcription. The DNA extracted from these viruses is found to be non-infectious.

Class 6: In the case of retrovirus, a distinct replicative phenomenon occurs in which single-stranded RNA is transformed into RNA–DNA hybrid with the help of viral reverse transcriptase. From this RNA–DNA hybrid, double-stranded DNA is produced, which is known as provirus. This provirus gets integrated into the host cell chromosome, which leads to cell transformation and the development of neoplasia.

Maturation

The assembly of the protein capsid is the first step in viral maturation. The capsids of many enveloped viruses, such as orthomyxoviruses and paramyxoviruses, are enclosed by envelopes consisting of proteins, lipids and carbohydrates. During maturation, the envelope protein is encoded by the viral genes and is incorporated into the plasma membrane of the infected host cell. In contrast, the envelope lipids and carbohydrates are encoded by host cell genes but not by viral genes and are present in the plasma membrane. The envelope develops around the capsid by a process called *budding*. The assembly of various viral components into virions may occur in the nucleus (e.g. herpes virus and adenoviruses) or cytoplasm (e.g. picornaviruses and poxviruses).

The non-enveloped viruses are present intracellularly as fully developed viruses, but in the case of enveloped viruses, only the nucleocapsid is complete. Subsequently, the nucleocapsid is surrounded by an envelope, which is derived from the host cell membrane during the process of budding.

Release

The release of the completed viruses is the last step in the replication of viruses.

In the case of animal viruses, the release of progeny virions usually occurs without cell lysis. The host cell is unaffected and goes on dividing into the daughter cells, which continue to release virions. Progeny virions are released into the surrounding environment and may affect new host cells. Budding does not immediately kill the host cell, and in some cases, the host cell survives.

Non-enveloped viruses, however, release through the rupture in the host cell plasma membrane. This type of release usually results in the death of the host cell.

A single infected cell may release a large number of daughter virions. The time taken for a single replication cycle is about 15–30 h. In a replication cycle, the virus cannot be demonstrated inside the host cell from the stage of penetration until the appearance of mature daughter virions. This period is known as the **eclipse stage**.

Abnormal Replicative Cycles

During replicative cycles of the virus, many types of abnormal replicative cycles do occur. These are as follows:

Incomplete Viruses

Due to defects during the assembly of viral components, some of the daughter virions that are produced may not be infective. An example of such an incomplete virus is the influenza virus which shows a high haemagglutination titre but low infectivity. This is known as the **von Magnus phenomenon**.

Pseudovirions

Pseudovirions are viruses that occasionally enclose host cell nucleic acid instead of viral nucleic acid. These, therefore, are non-infective and cannot replicate.

Abortive Infections

In this type of infection, the virus components may be synthesised, but the maturation is defective. The progeny virions are either not released or are non-infectious. This type of infection occurs due to infection of the wrong host cells by the virus.

Defective Viruses

Some genetically defective viruses do not produce fully formed virions. They produce fully mature virions only in the presence of helper viruses which supplement the genetic deficiency in the defective viruses. Hepatitis D virus and adeno-associated satellite viruses are examples of defective viruses, which replicate only in the presence of helper viruses, such as hepatitis B virus and adenovirus, respectively.

Viruses of Medical Importance

They can be classified broadly as DNA viruses and RNA viruses.

DNA Viruses

These belong to the following families (Table 48.5):

Adenoviridae

The members of the family Adenoviridae are medium-sized viruses measuring 20–90 nm in size. These viruses are non-enveloped, icosahedral viruses with 252 capsomeres. They are so named because they were first isolated from adenoids. These viruses are mostly associated with acute respiratory diseases.

Poxviridae

These are large-sized, brick-shaped viruses measuring 300 × 240 × 100 nm in size. They have a complex structure with a core containing a single linear molecule of double-stranded DNA genome. The pox (pox: pus-filled lesions) viruses are associated with skin lesions. The viral components are synthesised and assembled in the cytoplasm of infected host cells.

Herpesviridae

These are medium-sized icosahedral nucleocapsid viruses (100 nm). They have enveloped viruses containing linear, double-stranded DNA. They are named after the spreading (herpetic) appearance of cold sores. The viruses multiply in the nucleus of the host cells.

Table 48.5 Properties of human DNA viruses

Family	Genera (Members)	Shape	Size (nm)		Nature
Poxviridae	*Orthopoxvirus* (smallpox virus and vaccinia virus), *Molluscipoxvirus* (Molluscum contagiosum)	Brick-shaped, enveloped	300 × 240 × 100	85–140	Double-stranded, linear
Herpesviridae	*Herpes simplex virus* (herpes simplex virus types 1 and 2), *Varicellovirus* (varicella zoster virus), *Lymphocryptovirus* (Epstein–Barr virus), *cytomegalovirus* (human cytomegalovirus) *Roseolovirus* (HHV-6 and – 7), *Rhadinovirus* (Kaposi sarcoma-associated herpes virus)	Icosahedral, enveloped	Capsid: 100–100 envelope: 120–200	100–150	Double-stranded, linear Presence of terminal and internal reiterated sequences
Adenoviridae	*Mastadenovirus* (adenovirus)	Icosahedral cubic symmetry with 252 capsomeres Fibres protrude from the vertex capsomeres	70–90	20–25	Double-stranded, linear
Hepadnaviridae	*Hepadnavirus* (hepatitis B virus)	Spherical, enveloped	42	1.8	Double-stranded, circular, virion carries a DNA polymerase able to make fully double-stranded molecules
Papovaviridae	*Papillomavirus* (*human papillomavirus*) and *polyomavirus* (JC virus, BK virus and SV40)	Icosahedral	45–55	3–5	Double-stranded, circular
Parvoviridae	*Erythrovirus* (parvovirus B19) and *Dependovirus* (adeno-associated virus)	Icosahedral cubic symmetry and 32 capsomeres	18–26	1.5–2	Single-stranded, linear and double-stranded

Papovaviridae

These are small (40–55 nm) viruses containing double-stranded DNA. They are non-enveloped viruses. They replicate in the host cell nucleus and the host cell chromosome. This may cause host cells to proliferate, resulting in a tumour. Papovaviruses are acronyms for papillomas (warts), polyomas (tumours) and vacuolation (cytoplasmic vacuolation produced by some of these viruses).

Hepadnaviridae

Hepadnaviridae (*hepa:* liver; *DNA:* DNA core) are so named because they cause hepatitis and contain DNA as a genome. These viruses differ from other DNA viruses by synthesising their DNA by copying RNA using reverse transcriptase. Human hepatitis B virus, an important virus associated with human disease, is included in this family.

RNA Viruses

These belong to the following families (Table 48.6):

Togaviridae

These viruses include arboviruses and alphaviruses. Most of these viruses multiply in arthropods as well as invertebrates. Togaviridae (toga: envelope) are enveloped viruses containing a single-stranded RNA genome. These viruses are small spherical viruses measuring 40–70 nm in size.

Rhabdoviridae

Rhabdoviruses (rhabdo: rod) are bullet-shaped viruses. They are enveloped, measure 130–300 × 20 nm in size and contain a single-stranded RNA.

Reoviridae

They are icosahedral, non-enveloped viruses measuring 60–80 nm in size. They contain a double-layered capsid enclosing 10–12 segments of double-stranded RNA. Their name is derived from the first letters of respiratory, enteric and orphan. When first discovered, the viruses were not associated with any diseases; hence, they were called orphan viruses. These viruses are now known to cause respiratory and intestinal infections.

Retroviridae

Retroviruses (*re*: reverse, *tr*: transcriptase) viruses are so named because, characteristically, they possess the enzyme reverse transcriptase RNA-dependent DNA polymerase. They are icosahedral, enveloped viruses measuring 100 nm in size. Many of these viruses are associated with tumours in infected hosts. One genus, lentivirus includes the subspecies HIV-1 and HIV-2, the causative agents of acquired immunodeficiency syndrome (AIDS).

Picornaviridae

Picornaviruses (pico: small) are the smallest viruses, measuring 20–30 nm in size. They are non-enveloped, icosahedral viruses with a single-stranded RNA genome. These include three genera (*Enterovirus*, *Rhinovirus* and *Hepatovirus*) of medical importance.

Orthomyxoviridae

These are medium-sized (80–120 nm) viruses. They are spherical and elongated, enveloped viruses consisting of a single-stranded but segmented (eight segments) RNA genome. The influenza virus is the only virus of medical importance belonging to this group.

Paramyxoviridae

These are pleomorphic, enveloped viruses measuring 150 nm in size. They contain non-segmented, single-stranded, linear RNA. Three genera have been described: *Paramyxovirus*, *Morbillivirus* and *Pneumovirus*.

Bunyaviridae

These are enveloped, spherical viruses measuring 90–100 nm in size. The genera of medical importance include *Bunyavirus*, *Hantavirus*, *Uukuvirus*, *Phlebovirus* and *Nairovirus*.

Arenaviridae

They are spherical, pleomorphic viruses with variable sizes (50–300 nm). They contain electron-dense, chromosome-like particles giving a sandy

Table 48.6 Properties of human RNA viruses

Family	Genera (Members)	Shape	Size (nm)		Nature
Paramyxoviridae	*Respirovirus* (Sendai virus, human parainfluenza virus types 1 and 3), *Rubulavirus* (mumps virus), *morbillivirus* (measles virus) and *Pneumovirus* (human respiratory syncytial virus)	Spherical	150–300	5–7	Single-stranded, linear, negative sense and non-segmented
Orthomyxoviridae	*Influenzavirus A* (human influenza A virus), *Influenzavirus B* (human influenza B virus) and *Influenzavirus C* (human influenza C virus)	Spherical	80–120	5	Single-stranded, segmented, linear negative sense
Coronaviridae	*Corona virus* (corona virus)	Spherical	80–130	6	Single-stranded, unsegmented positive-sense RNA
Arenaviridae	*Arenavirus* (Lassa fever virus, Junin and Machupo viruses, lymphocytic choriomeningitis virus)	Spherical	50–300	3–5	Single-stranded, segmented, circular and negative-sense
Rhabdoviridae	*Vesiculovirus* (vesicular stomatitis virus) and *lyssavirus* (rabies virus)	Bullet-shaped	180 × 75	4	Single-stranded, non-segmented negative-sense RNA
Filoviridae	*Filovirus* (Ebola virus and Marburg virus)	Filamentous	800 × 80	4	Single-stranded
Bunyaviridae	*Orthobunyavirus* (California encephalitis virus and La Crosse virus), *Phlebovirus* (sandfly fever virus) and *Hantavirus* (Hantaan virus)	Spherical	90–100	4–7	Single-stranded
Retroviridae	Deltaretroviruses (human T-cell leukaemia virus types I and II), lentiviruses (human immunodeficiency virus), alpharetroviruses, betaretroviruses, gammaretroviruses and epsilon retroviruses	Spherical	80–110	2 × (2–3)	Single-stranded
Reoviridae	*Rotavirus* (rotavirus), *Orbi virus* (Colorado tick fever virus) and *Orthoreo virus*	Icosahedral	60–80	11–15	Double-stranded, segmented
Picornaviridae	*Enterovirus* (poliovirus, echovirus, Coxsackie virus), *rhino virus* (human rhinovirus) and hepatovirus (human hepatitis A virus)	Icosahedral	25–30	2.5	Single-stranded and positive sense
Togaviridae	*Rubivirus* (rubella virus) and *alphavirus* (Western, eastern and Venezuelan equine encephalitis virus, Ross River virus, Sindbis virus and Semliki forest virus)	Icosahedral	60–70	4	Single-stranded
Flaviviridae	*Flavivirus* (yellow fever virus, dengue virus and St. Louis encephalitis virus), *Pestivirus* and *hepatitis C virus* (human hepatitis C virus)	Spherical	40–50	4	Single-stranded
Caliciviridae	Norovirus (Norwalk virus), sapovirus (Sapporo virus), lagovirus and vesivirus	Icosahedral	35–40	2.6	Single-stranded

appearance; hence, they are named arenaviruses (*arena*: sand).

Caliciviridae

These are naked non-enveloped viruses. They are small and spherical and measure 35–39 nm in size.

Filoviridae

They are long filamentous, enveloped viruses with variable sizes. They contain a single-stranded RNA genome. The Marburg and Ebola viruses are viruses of medical importance.

Prions

Prions are infectious particles that can transmit disease. These prions are composed chiefly of a protein without any detectable nucleic acid. This is a new name widely accepted for the old name slow viruses. Prusiner coined the term prion for proteinaceous infectious particles. Unlike conventional viruses, the prions apparently have no virion structure of genomes and evoke no immune response in the infected host. These are extremely resistant to inactivation by heat, disinfectants and radiation.

The prions are causative agents of slow viral infections, such as sub-acute spongiform encephalopathy. After a long incubation period of years, they produce a progressive disease that causes damage to the central nervous system, leading to sub-acute spongiform encephalopathy. A detailed description of prions and slow virus diseases is provided in Chap. 62.

Viroids

Viroids are protein-free fragments of single-stranded, circular RNA that cause disease in plants. The term viroids were coined by Triener (1971) to describe a new subclass of subviral agents having a genome very much similar to those of known viruses; they consist of very short pieces of naked RNA: 300–400-nucleotides long with no protein coat. The RNA does not code for any proteins. The low-molecular-weight RNA is resistant to heat and organic solvents but sensitive to nucleases. So far, viroids have been identified as pathogens of only plants; these were first identified in the potato spindle tuber disease. The viroids are yet to be linked to any disease in humans.

Pathogenesis of Viral Infections

Viruses cause infections in the host first by breaking the natural protective mechanisms of the body, then evading the host's immune system and finally by killing off the host cells and triggering immune and inflammatory responses. The outcome of viral infection depends on the host's response to the viral infection and the nature of the host–virus interaction. Viruses replicate only inside the living cells; hence, the primary pathogenic manifestations are seen at the cellular level.

Stages of Viral Infections

Viruses initiate infection and may cause disease through many defined stages. These include (1) entry into the body, (2) initiation of infection at a primary site (infection of the target tissue), (3) replication of the virus (Fig. 48.6) and spread to the secondary site and (4) manifestations of the disease.

Entry into the Body

Entry of the viruses into the body is the first step of viral infections. The skin is the best barrier to infections. In addition to the skin, mucus, ciliated epithelium, gastric acid, bile, tears, etc., confer basic natural protection against many viruses. The viruses enter the body through the respiratory tract, skin, conjunctiva, alimentary tract and genital tract and initiate the infection by breaking these natural barriers to infection.

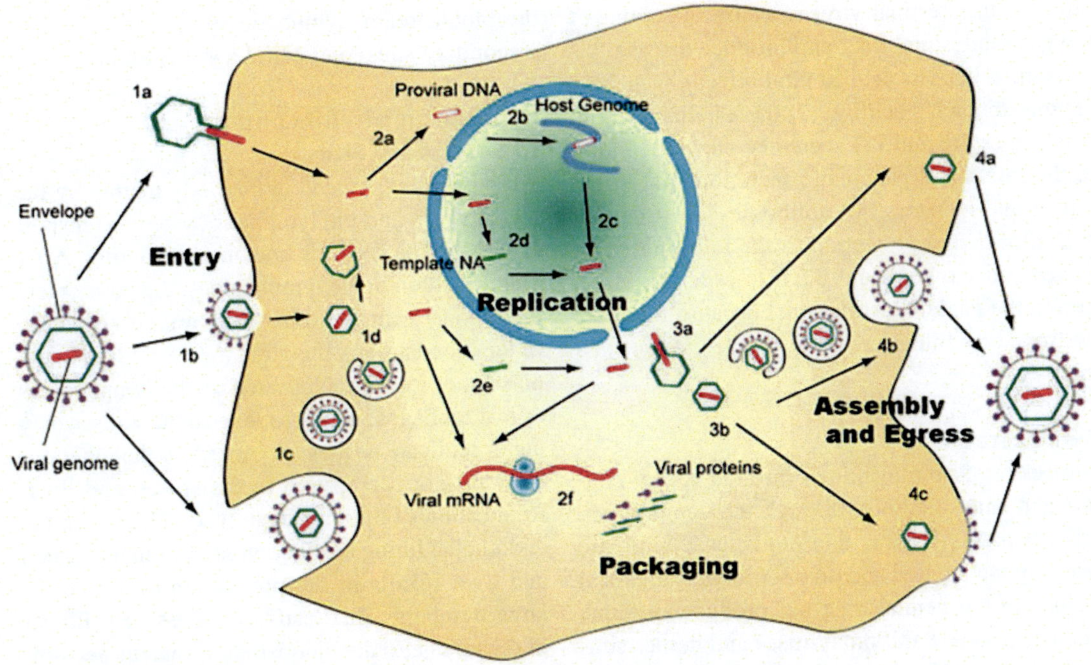

Fig. 48.6 Replication of virus (https://link.springer.com/article/10.1007/s40042-020-00030-w)

Respiratory Tract

Many viral infections are caused by the entry of the virus through the respiratory tract. The viruses enter the respiratory tract by droplets containing the viruses expelled from the nose and mouth of infected individuals during the act of coughing, sneezing or talking.

After entry into the body, some viruses are confined to the respiratory tract, where they multiply and produce local diseases. These are known as respiratory viruses. Examples of these viruses are influenza virus, respiratory syncytial virus, rhinovirus, coronavirus, adenovirus and Coxsackie virus A. Other viruses, after entry, initially multiply at the site of the respiratory tract followed by haematogenous or lymphatic spread to other sites of the body. The viruses replicate in large numbers at these sites and cause systemic manifestations of the disease. Examples of such viruses include measles, mumps, rubella, varicella-zoster, cytomegalovirus and Epstein–Barr virus.

Skin

Many viruses enter the skin through abrasions or breaks in the skin. Molluscum contagiosum, cowpox, vaccinia and variola viruses enter the skin through minor lesions and produce cutaneous lesions at the entry site. Other viruses, such as papillomavirus, enter the skin through minor injuries on the skin's surface, arboviruses by the bite of insects, rabies virus by the bite of dogs and other animals and hepatitis B virus and HIV by injection.

Conjunctiva

Some viruses may enter through the conjunctiva and may cause the disease. For example, adenovirus causes local manifestations and measles virus causes the systemic manifestation of the disease by entering through the conjunctiva.

Alimentary Tract

The viruses also cause infection by entering through the alimentary tract, which is another important route of infection by viruses. The viruses, such as rotaviruses, enteroviruses, adenoviruses, reoviruses, hepatitis viruses and other gastrointestinal viruses, cause infections of the gastrointestinal tract and produce disease.

Natural barriers against viral infections include (1) the stomach's acidity, (2) the alkalinity of the small intestine and (3) secretory enzymes found in the saliva and pancreatic secretions. Intestinal mucus and secretory IgA antibodies are important and offer partial protection to the intestinal tract. Enveloped viruses usually fail to establish infection in the gastrointestinal tract because these are destroyed by bile secreted in the gastrointestinal tract.

Genital Tract

Viruses are also transmitted through sexual contact and enter the body through the genital tract. HIV, hepatitis B virus and hepatitis C virus are sexually transmitted and do not produce any local lesions in the genital tract but produce systemic manifestations. Papillomaviruses and herpes simplex viruses (HSVs) are transmitted sexually and produce lesions locally in the genital tract and perineum.

Congenital Infections

Few viruses may cause congenital infection in the foetus in utero being transmitted from the infected mother. Rubella and cytomegaloviruses are viruses that are more commonly associated with congenital infections. Depending on the age of the foetus, these viruses may cause malformations or even foetal death and abortion.

Initiation of Infection at Primary Site (Infection of the Target Tissue)

On entry into the human host, the viruses may remain at the primary site of infection, replicate and cause infection of the target tissue. The specificity of the virus-attachment proteins and tissue-specific expression of receptors during replication are two important properties of viruses to cause infection of target tissues.

Many viruses cause infection in the oral mucosa or upper respiratory tract. Replication of viruses at the primary sites may be accompanied by clinical symptoms. On the other hand, some viruses may disseminate to other tissues through the bloodstream, lymphatic system and neurons (secondary sites) and cause a systemic infection.

Replication of Virus and Spread to Secondary Sites

The viruses are spread in the body mainly by the bloodstream and the lymphatic system. Transport of virus in the blood is known as **viraemia**. After multiplication in the lymph nodes, the virus enters the bloodstream, resulting in **primary viraemia**. In the bloodstream, the virus may exist free in the plasma, or it may be ingested by the lymphocytes or macrophages. In macrophages, the viruses may die or replicate or may be carried by the mononuclear phagocytic system to the spleen and liver. Replication of the viruses in macrophages, in the endothelial lining of blood vessels or in the spleen and liver results in the production of viruses in large numbers. This leads to a massive spillover of the virus into the bloodstream, causing **secondary viraemia**. This heralds the onset of clinical symptoms of viral infections, including the prodromal phase in eruptive fever. Subsequently, it is carried by the bloodstream to reach target organs (skin, brain, liver, etc.) in which the viruses replicate, producing characteristic distinctive lesions.

Viruses enter the brain or central nervous system (1) through the bloodstream, (2) through the infected cerebrospinal fluid or meninges and/or (3) through the infection of the peripheral and sensory (olfactory) neurons.

Manifestations of the Viral Diseases

The clinical manifestations of viral diseases depend on the complex interaction of virus and host factors. The outcome of the infection, that is, the disease manifestation, depends on the (1) age, general health and immune status of the person; (2) dose of the infective virus and (3) genetics of the host and the virus.

After the virus infects the host, the host's immune status plays an important role and determines the outcome of the viral infection, whether it will be an asymptomatic infection, a benign disease or a life-threatening disease.

Viral Pathogenesis at the Cellular Level

The host cells show various responses to viral infection depending on the virus causing the infection and the type of the cell infected. Moreover, the outcome of infection is determined by the properties of the virus and the cell. The host cells may be (1) permissive, (2) semi-permissive or (3) non-permissive.

A **permissive cell** is a cell that allows replication of a particular type of strain of virus by providing biosynthesis compounds, such as transcriptional factors and post-translational enzymes. A **non-permissive cell** does not provide any biosynthesis compound, hence does not support replication of the viruses. A **semi-permissive cell** may support some but not all the stages of viral infection.

Replication of viruses in cells may cause a broad spectrum of effects, ranging from non-apparent cellular damage to rapid cell destruction. Some viruses may cause cell death or even cell lysis. For example, polioviruses cause the death of cells (*cytocidal effect*) and even lysis of the cells (*cytolysis*), molluscum contagiosum causes proliferation of cells and oncogenic viruses cause malignant *transformation of cells*. Some other viruses may cause alterations in cells' morphology, functional properties, malignant transformation or antigenicity (Table 48.7). In some conditions, viruses infect cells and replicate independently within the cells without causing any cellular injury to the infected cells. This is known as *steady-state infection*. The virus produces demonstrable cellular changes in the infected cells in cell cultures, known as cytopathic effects *(CPEs)*.

The cellular changes produced by the virus may be due to many factors. These cellular changes may be caused (1) by the viral takeover of macromolecular synthesis of proteins and enzymes by viruses instead of the host cell, (2) by the accumulation of virus proteins or particles or (3) by alteration or disruption of cellular structure.

Production of early or non-structured viral proteins often inhibits the production of host protein and the synthesis of host DNA. The accumulation of large amounts of viral proteins or particles in the infected cells may modify or disrupt the cellular architecture.

At the cellular level, depending on the nature of the virus and the cell infected, the virus infection in a cell can produce any of the three infections (Fig. 48.7): failed infection (abortive infection), cell death (lytic Infection) or infection without cell death (persistent viral infection).

Abortive Infection

Some viruses cause failed or abortive infections in which those viruses do not multiply, and therefore disappear from the cell.

Lytic Infection

Some viruses cause cell death or lytic infection. This occurs due to the replication of the virus in an infected cell which kills the target cell. The cellular injury caused by viruses may be due to a

Table 48.7 Cytopathic effects (CPEs) of viruses

Viruses	Cytopathic effects (CPEs)
Enterovirus	Rapid crenation and degeneration of entire cell sheet—Cell death
Measles virus	Syncytium formation—Cell fusion
HSV	Diffuse roundening and ballooning of cell line
SV40	Cytoplasmic vacuolations
Polyomavirus	Transformation
HIV	Destruction of T-cells
Adenovirus	Large granular clumps forming grape like clusters—Inclusion bodies in nucleus
Papovavirus	Inclusion bodies in nucleus
Rabies virus	Inclusion bodies in cytoplasm (Negri bodies)
Cytomegalovirus	Inclusion bodies in cytoplasm and nucleus

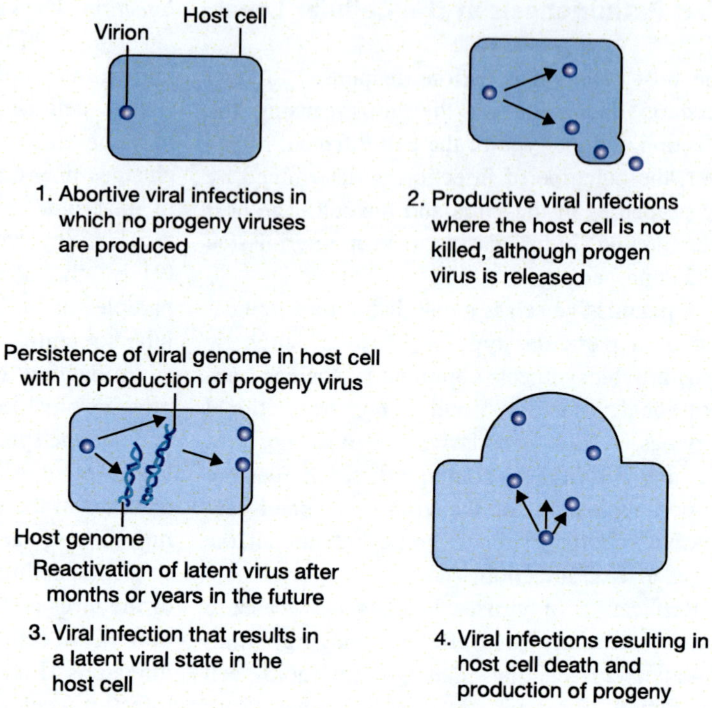

Fig. 48.7 Viral infection at the cellular level

combination of several factors. Viruses adopt different mechanisms (Table 48.8) for preventing cellular growth and causing cell death as follows.

Inhibition of Cellular Protein Synthesis

HSV, poliovirus, togaviruses and poxviruses cause cell death by inhibiting cellular protein synthesis. These viruses, during replication, produce proteins that inhibit the synthesis of cellular DNA and/or RNA. These viruses also produce other proteins, which break down host cellular DNA to components that serve as substrates for replication of the viral genome. The accumulation of large amounts of viral macromolecules and viral progeny in infected cells during virus replication may disrupt the cellular structure or functions of the cell or disrupt lysosomes, causing autolysis.

Cell Fusion

Infections of cells with certain viruses cause the cells to fuse, resulting in the production of giant multi-nucleated cells. Infection by paramyxoviruses, herpes viruses and retroviruses

Table 48.8 Mechanisms of cytopathogenesis of viral infections

Viruses	Mechanisms of cytopathogenesis
Herpes simplex virus, varicella zoster virus, paramyxovirus and Human immunodeficiency virus	Alteration of the cell membrane by syncytia formation
Non-enveloped viruses	Alteration of the cell membrane by disruption of cytoskeleton
Togaviruses and herpes virus	Alteration of cell membrane by permeability
Adenovirus	Toxicity of virion components
Poliovirus, Herpes simplex virus, togavirus and poxvirus	Inhibition of cellular protein synthesis
Herpes viruses	Inhibition of degradation of cellular DNA
Rabies and cytomegalovirus	Accumulation of viral proteins and structures (inclusion bodies) in the cytoplasm or nucleus

results in the expression of glycoproteins on the cell surface. This triggers the fusion of neighbouring cells, resulting in the formation of large multi-nucleated giant cells called syncytia. The formation of syncytia facilitates the spread of the virus from cell to cell and allows the virus to escape from the lytic effect of antibodies.

The cell-to-cell fusion of infected cells is of two types:

Fusion from without In this type, cell fusion occurs in the absence of new protein synthesis (e.g. Sendai viruses and other Paramyxoviruses).

Fusion from within: In this type, cell fusion requires new protein synthesis (e.g. HSV).

Infection by viruses causes a variety of changes in infected cells as follows.

- Certain viruses cause alterations in the cytoplasm of infected cells or lead to the expression of virus-coded antigens on the surface of infected cells. For example, the respiratory syncytial virus causes fusion of adjacent cell membranes, resulting in fusion or syncytia formation.
- The infected cell may confer new properties on the cells. For example, infection by the influenza virus results in the appearance of viral haemagglutinin on infected cell surfaces, leading to hemadsorption or adsorption of erythrocytes to the cell surface. Some viruses cause *apoptosis* in the infected cells. Apoptosis leads to cellular suicide that facilitates the release of viruses from the cell.
- Certain viral infections produce characteristic changes in the properties and appearance of target cells. For example, measles viruses, mumps viruses, adenoviruses, cytomegaloviruses and varicella viruses cause chromosomal aberrations and degradations. Infection by adenovirus types 12 and 31 and HSV causes marginated chromatin rings in the nuclear membrane.

Transformation

Transformation is an irreversible genetic process caused by the integration of viral DNA with the DNA of the host cell. Infection of cells with certain viruses causes the transformation of normal cells into malignant cells. The viruses that stimulate uncontrolled cell growth causing the transformation of immortalisation of the cell are known as oncogenic viruses.

Different oncogenic viruses cause transformation or immortalisation of the cells by different mechanisms as follows: (1) they promote or provide growth-stimulating genes, (2) they prevent apoptosis and (3) they bring in mechanisms that limit DNA synthesis and cell growth.

The DNA oncogenic viruses infect semi-permissive cells, which express only selected viral genes but do not produce complete viruses. Most oncogenic DNA viruses integrate with the host cell chromosome during transformation. For example, Epstein–Barr virus causes the transformation of the cells by stimulating cell growth and inducing the expression of an oncogene (Bcl-2 oncogene) that prevents apoptosis or programmed cell death. Other viruses, such as adenovirus, papillomavirus and SV40, induce expression of the genes. These genes, such as P53 and neuroblastoma gene products, encode proteins that bind and inactivate all growth regulatory proteins. Loss of the p53 gene increases the cell's susceptibility to mutation, thus releasing the mechanisms that inhibit cell growth.

The Human T-cell lymphotropic virus (HTLV-1) is the only oncogenic retrovirus known to cause oncogenesis in humans. It encodes a protein (*tax*) that transactivates gene expression, including genes for interleukin-2 and other growth-stimulating lymphokines. The virus continues to produce viruses in the transformed cells. Leukaemia associated with HTLV-1 develops very slowly, 20–30 years after infection.

Cytopathic Effect

CPE is a term that denotes any visible change in the appearance of target cells infected by viruses. Some viruses can be tentatively identified by the nature of CPE in cell culture and the types of cells in which these viruses cause CPE.

Inclusion bodies are the characteristic histological feature in virus-infected cells reflecting the change in the appearance of the target cells. They may result from virus-induced changes in the membrane or chromosomal structure. They also represent the sites of viral replication or accumulation of viral capsids. Inclusion bodies are the structures with distinct size, shape, location and staining properties found in virus-infected cells. The inclusion bodies may be found in the nucleus (e.g. herpes virus), cytoplasm (e.g. poxvirus) or both (e.g. measles virus).

Intranuclear inclusion bodies are of two types: Cowdry type A and Cowdry type B. Cowdry type A inclusion bodies are variable in size and are produced by infection with the herpes virus and yellow fever virus. Cowdry type B inclusion bodies are often multiple and more circumscribed and are produced by adenovirus and poliovirus infections. *Owl eye* appearance is typical intranuclear inclusion bodies demonstrated in cytomegalovirus infection. **Negri bodies** are the intracytoplasmic eosinophilic inclusion bodies in the neural cells and are pathognomonic of rabies.

These inclusion bodies may be acidophilic, which appear as pink structures when stained with Giemsa or eosin–methylene blue stains. They may be basophilic as produced by some viruses, such as adenovirus. Demonstration of inclusion bodies is characteristic of a particular viral infection; the presence of such bodies facilitates laboratory diagnosis. **Guarneri bodies** are small, multiple inclusions found in vaccinia-infected cells. **Bollinger bodies** are large inclusions seen in fowl pox. **Molluscum bodies** are very large inclusion bodies seen in the molluscum contagiosum infected cell (Table 48.9).

Persistent Viral Infection

The persistent viral infection is characterised by viral infection without cell death. The persistent viral infection is caused by the slow and gradual release of viruses from the infected cell through exocytosis or budding from the plasma membrane. The persistent viral infections may be (1) non-lytic and productive (chronic), (2) latent, limited production of viral macromolecules in the absence of synthesis of the virus, (3) recurrent or (4) transforming.

HSV typically causes a latent infection in neurons lacking the machinery to transcribe the early viral genes essential for virus replication. However, stress and other stimuli activate the successful replication of the virus, resulting in the infection.

Table 48.9 Inclusion bodies produced in viral infections

Inclusion body	Virus
Intracytoplasmic inclusion bodies	
Negri bodies	Rabies virus
Molluscum bodies	Molluscum contagiosum virus
Guarnieri bodies	Vaccinia virus
Bollinger bodies	Fowl poxvirus
Perinuclear cytoplasmic acidophilic bodies	Reovirus
Intranuclear inclusion bodies	
Acidophilic	
Cowdry type A inclusion bodies	HSV and measles virus Yellow fever, varicella zoster
Acidophilic inclusion bodies	Papovavirus
Basophilic	
Owl eye inclusion bodies	Cytomegalovirus
Cowdry type B inclusion bodies	Adenovirus, poliovirus
Intracytoplasmic and intranuclear inclusion bodies	
Warthin–Finkeldey	Measles

Certain DNA viruses may cause infection of the cell, resulting in a latent or immortalising infection. This occurs due to the failure of the biosynthesis machinery of viruses for transcribing all the viral genes. The specific transcription factor needed of such a virus may be expressed in growing cells or specific tissues or after treatment with hormones or cytokines.

Host Response to Viral Infections

The main goal of the host response in viral infection is to eliminate the virus and eliminate the infected cells harbouring or replicating the virus. The host responds by immunological and non-specific means. The immune response is the best and most important means of controlling virus infections.

Immunity in Viral Infections

Protection against viral infections is induced by both humoral and cell-mediated immunities (CMI). Infection by the virus and subsequent replication induces immunity in the host against a wide range of virus antigens, including the viral internal and surface antigens, non-structural antigens and early proteins (Table 48.10).

Humoral Immunity

The humoral immunity mediated by antibodies plays an important role against infections caused by extracellular viruses and viraemia's caused by those viruses. Humoral immunity also inhibits the multiplication of cytolytic viruses. The circulating IgG and IgM antibodies are effective against viruses in blood and tissue spaces, while IgA plays an important role against viruses replicating on the mucosal surfaces.

The antibodies act against viruses in many ways. They prevent the adsorption of the virus to the cell receptor at the primary site of infection. They also cause rapid degradation of viruses and prevent the release of viral progeny from infected cells. The antibodies in combination with complement cause damage to the enveloped viruses resulting in cytolysis of virus-infected cells.

The antibodies against surface antigens are usually more effective than against internal antigens. The antibodies produced against surface antigens also vary in their ability to reduce the infectivity of viruses. For example, the ant haemagglutinin antibodies produced against the haemagglutinin antigen of the influenza virus neutralise infectivity of the influenza virus, while ant neuraminidase antibodies against neuraminidase antigen are not effective in neutralising the infectivity of the virus. However, the anti-neuraminidase antibody helps prevent the release of the virion progeny from infected cells.

Although protective, humoral antibodies may cause injuries to host cells and contribute to the pathogenesis of the disease. For example, antibodies in combination with complements initiate an immune complex disease in infections caused by hepatitis B virus and rubella viruses. Interferons and lymphokines stimulate flu-like symptoms in respiratory virus infections and viraemia caused by arboviruses.

Cell-Mediated Immunity (CMI)

CMI plays a very important role in viral infections. CMI is essential for the lysis of target cells in the infections caused by enveloped

Table 48.10 Immunopathogenesis of viral infections

Viruses	Pathology	Immune mediators
Enveloped viruses	Delayed type hypersensitivity and inflammation	T-cells, macrophages and polymorphonuclear leucocytes
Respiratory viruses and arboviruses	Flu-like symptoms	Interferon and lymphokines
Dengue virus	Haemorrhagic disease	T-cell, antibody and complement
Hepatitis B virus and rubella	Immune complex disease	Antibody and complement
Human Immunodeficiency Virus, cytomegalovirus, measles virus and influenza virus	Immunosuppression	T-cells and antibody

viruses and non-cytolytic infections caused by the hepatitis A virus. Individuals with deficient cellular immunity are more susceptible to infection by viruses, such as herpes virus, measles virus and poxvirus. CMI also plays a major role in recovery from viral infections. A deficient CMI results in failure to resolve the infection, which may lead to persistent viral infection, chronic disease and even death of the patient.

CMI also contributes to the pathogenesis of diseases by inducing T-cell-induced inflammatory and hypersensitivity reactions. For example, CMI-induced inflammatory response is responsible for the typical clinical manifestation of measles and mumps rather than CPEs of viruses. In dengue haemorrhagic fever, antigen-specific T-cells and specific antibodies induce significant inflammatory and hypersensitivity damage to infected endothelial cells, leading to haemorrhage. Antigen-specific T-cells are also responsible for inducing post-infection cytolysis in the infection caused by enveloped viruses, such as the measles virus. T-cells, macrophages and polymorphonuclear (PMN) leucocytes induce a delayed type hypersensitivity and inflammation in many viral infections caused by enveloped viruses.

Infection by some viruses is usually associated with suppressed host immunity; for example, infection by measles virus causes a temporary depression of delayed hypersensitivity to tuberculin antigen, and infection by HIV causes a depressed CMI following depletion of CD4+ helper T-cells.

Once infected by a virus, it usually confers lifelong immunity to re-infection. In some viral infections, such as influenza or the common cold, the immunity is not lifelong because the infection is caused by antigenically different strains of the same virus.

Non-immunological Responses

These include many factors as follows:

Phagocytosis

During viral infections, the role of polymorphonuclears (PMN) is insignificant. In fact, during most viral infections, decrease in PMN count is obvious. However, macrophages help clear off viruses from the bloodstream by phagocytosis.

Age

Age is the most important factor in determining the host's susceptibility to viral infections. Most viral infections are common and more serious in persons at extremes of age, such as infants, children and the elderly. Infants and children are more susceptible to various respiratory and exanthematous viral diseases. They are prone to more serious diseases caused by a paramyxovirus, respiratory infections, etc. Similarly, elderly people are susceptible to new viral infections (the new strains of influenza A and B virus) and re-activation of the latent viruses (e.g. varicella-zoster virus).

Poor Nutrition

Poor nutrition or malnutrition affects the infected person's immune system because tissue-regenerating capacity is usually decreased in a person with malnutrition. In such persons, some viral infection (such as measles) causes immunosuppression, thereby facilitating the replication of viruses in large numbers. This results in a serious disease associated with more serious and fatal complications.

Genetic Makeup

A person's genetic makeup also plays an important role in determining the outcome of viral infection. Genetic differences in immune response genes and genetic defects in other genetic loci affect the susceptibility of the host to viral infection, as well as the severity of infection.

Body Temperature

Fever is a common manifestation of many viral infections. Fever acts as a natural defence mechanism against viral infections as a higher temperature above 39 °C inhibits many viruses.

Interferons

Interferons were initially observed more than centuries ago by Isaacs and Lindenmann in 1957. They noticed the production of a diffusible substance by the chicken chorioallantoic membrane treated with live or inactivated influenza virus, which had antiviral activity. Subsequently, these virus-infected cells were found to be resistant to viral infection. These interferons are a heterogeneous and complex group of glycoproteins produced by cells on induction of infection by viral or non-viral microbial agents. These interferons show antiviral activity by inhibiting protein synthesis. Later studies on interferon demonstrated that interferon production is a natural defence mechanism by vertebrate hosts against viral infection.

Interferons' antiviral action occurs by complex interactions with host cell functions, which may vary based on the type of virus being inhibited. Host cells produce a protein called **translation inhibiting protein (TIP)** when exposed to interferon, which specifically inhibits translation of viral mRNA without disturbing the cellular mRNA. TIP is a mixture of three enzymes: protein kinase, an RNase and oligonucleotide synthetase, which block the mRNA translation.

Interferons exert antiviral effects by several pathways. First, they cause increased expression of class I and class II major histocompatibility complex (MHC) glycoproteins, thereby facilitating the recognition of viral antigens by the immune system. Second, they activate the cells, such as natural killer (NK) cells and macrophages, that can destroy virus-infected target cells. Finally, they directly inhibit the replication of viruses.

Interferons are also produced by leucocytes and many other cells in response to infection by virus, double-stranded RNA (dsRNA), endotoxin and mutagenic and antigenic stimuli. The dsRNA is a potent stimulator. The viruses that replicate slowly and those that do not inhibit the synthesis of host proteins are usually good inducers of the interferons. IFN-7 differs from other interferons in being released as lymphokines from activated T-cells, natural killer cells and occasionally macrophages. Interferon production is inhibited by steroids and increased oxygen tension and temperature (40 °C). Synthesis begins in less than an hour and peaks in 8–12 h.

Types of Interferons

Interferons have been classified based on the type of receptor through which they signal antigenic character, cell of origin and other properties. There are three types of interferons: **alpha**, **beta** and **gamma**. The abbreviation of IFN designates interferon, and species of origin is indicated as a prefix. For example, human interferon-alpha is usually abbreviated as HuIFN-α.

IFN-α is a non-glycosylated protein produced by leucocytes following induction with suitable viruses, and it was previously called leucocyte interferon. It has at least 16 antigenic subtypes. **IFN-β** is a glycoprotein molecule produced by fibroblasts and epithelial cells after the stimulation by viruses or polypeptides; it is commonly known as fibroblast interferon. T lymphocytes and NK cells produce IFN-γ after stimulating sensitised lymphocytes with antigen or non-sensitised lymphocytes with mitogens. It has lymphokine and immunoregulatory functions rather than antiviral activity. It enhances the MHC antigen and activates cytotoxic T lymphocytes, macrophages and NK cells. It is commonly known as immune interferon, and it is a glycoprotein molecule. It has a separate cell receptor, but **IFN-α** and **IFN-β** do not possess separate cell receptors.

The proteolytic enzymes help in the inactivation of interferons, but other enzymes, such as nucleases or lipases, do not have this property. IFN-α and -β are potent antiviral agents; when a cell becomes infected with a virus, they are synthesised. They induce the production of three enzymes, namely synthetase, RNase L and protein kinase. This leads to inhibition of viral protein synthesis but does not affect the host protein synthesis. The main properties of the interferons include resistance to heating at high temperatures (56–60 °C for 30–60 min) and stability over a wide range of pH (2–10), but the exception is IFN-γ which is labile at pH 2. The molecular weight of interferons is about 17,000 Da, and they are non-dialysable and nonsedimentable (100,000 *g*). There is no routine serological test available for their detection and estimation because they have very poor antigenicity. Interferons are estimated based on biological activity. For example, in plaque assay, the ability of a sensitive virus to inhibit plaque formation can be calculated. International units (IU)/mL are used to evaluate the potency of IFN.

Clinical Applications of Interferons

Interferons are increasingly being used to treat chronic hepatitis B and C virus carriers at risk for progression to cirrhosis and hepatocellular carcinoma. When given with ribavirin, the interferon has proved to be more effective than interferon alone to treat hepatitis C virus infection. Currently, synthetic IFN-α is actively used against hepatitis A, B and C viruses, papillomavirus; HSV and rhinovirus. It is also used for the treatment of condylomata acuminatum. IFN-β has been used in the treatment of multiple sclerosis. IFN-γ has been tried to treat the chronic granulomatous disease characterised by severe impairment of the ability of phagocytes to kill the ingested organism.

Clinical Manifestations of Viral Diseases

The clinical manifestations of viral diseases depend on a complex interaction of virus and host factors. These include (1) the age, general health and immune status of the person; (2) the dose of the infective virus and (3) the genetics of the host and the virus.

After a virus infects a host, the host's immune status plays an important role and determines the outcome of the viral infection—whether it will be an asymptomatic infection, a benign disease or a life-threatening disease.

Incubation Period

It is the time taken by the virus to spread from the site of entry to the organs of virus replication and then to the target organs for the production of lesions. Therefore, the duration of the incubation period depends on the (1) time taken for the site of entry, (2) multiplication and (3) lesions. The incubation period is the important initial period before the manifestation of characteristic symptoms of the disease. During this period, virus replicates in the host's body but has not reached the target tissue or caused sufficient damage in the host's tissues to cause the disease.

Significance of pncubation period: The incubation period for many viral diseases is variable. It is **short (1–3 days)** in the infections caused by respiratory viruses or gastroenteritis caused by an enteric virus, in which both the site of entry and lesions are the same. On the other hand, the incubation period is **longer (usually 10–20 days)** in systemic diseases in which viruses enter through the respiratory or alimentary tract and produce lesions in different target tissues away from the site of entry. However, it may be relatively shorter in some systemic viral diseases, such as yellow fever or dengue fever (5–6 days). The incubation period may **still be longer (2–6 months)** in hepatitis B virus infection or even years in slow virus infection.

Incubation periods of common DNA and RNA viruses' viral infections are summarised in Tables 48.11 and 48.12, respectively. In some viral infections, the incubation period may proceed without symptoms, causing asymptomatic infections or may cause non-specific early symptoms known as the prodrome.

Severity of the Viral Disease

It depends on the site of viral infection and the immunopathological responses induced following the viral infection. It may be (1) an inapparent infection, (2) an acute infection or (3) a chronic or persistent infection.

Inapparent Infection is usually caused if (1) the infection is controlled before the virus reaches the target tissues, (2) the infected tissue is not damaged, (3) the tissue is damaged but it is rapidly repaired or (4) the severity of damage is below a functional threshold for the particular disorder. The inapparent infections are detected usually by serological tests detecting specific viral antibodies in the infected host.

Table 48.11 Incubation periods of common DNA virus infections

DNA virus infections	Incubation period
Adenovirus	5–7 days
Herpes simplex	5–8 days
Smallpox	12–14 days
Chickenpox	13–17 days
Infectious mononucleosis	30–50 days
Hepatitis B	50–150 days
Papilloma (warts)	50–150 days

Table 48.12 Incubation periods of common RNA virus infections

RNA virus infections	Incubation period
Common cold	1–3 days
Influenza	1–2 days
Dengue	5–8 days
Enteroviruses	6–12 days
Poliomyelitis	5–20 days
Measles	9–12 days
Mumps	16–20 days
Rubella	17–20 days
Hepatitis A	15–40 days
Rabies	30–100 days
HIV	1–10 years

Epidemiologically, the patients with inapparent or asymptomatic infection serve as the major source of infection.

The onset of the acute or chronic viral disease depends on the ability of the immune system of the infected host to control and resolve the viral infection. If the immune system acts rapidly, the infection may result in acute disease. If the immune system fails to act effectively or slowly, it may result in chronic viral disease. Slow virus infections have typically long incubation periods of many years before progressing to the manifestation of the disease.

Epidemiology of Viral Infections

Geographical distribution of the viral infections depends on (1) the presence of vectors for the transmission of disease, (2) co-factors facilitating the transmission of disease and (3) the presence of susceptible populations in the particular area. For example, the presence of suitable arthropod vectors for a particular virus in large numbers increases the risk of a high prevalence of viral disease in that community. Viral diseases also show seasonal differences; for example, enteric viruses are more prevalent in the summer due to poor hygiene and contaminated water because of a shortage of water supply. Similarly, respiratory viruses are more prevalent during winter due to overcrowding which facilitates the spread of the viruses.

The viral infection in a community may occur as (1) outbreak, (2) epidemic or (3) as a pandemic. The viral disease **outbreak** usually occurs from a common source, for example, infected food and is seen only in clusters of people. **Epidemics** of viral infection, in contrast, occur in a large geographical area. An epidemic usually occurs due to introducing a new strain of virus into an immunologically susceptible population (e.g. influenza epidemic). **Pandemics** are usually worldwide epidemics resulting from introducing a new virus, for example, HIV. Pandemics of influenza have occurred due to the introduction of new strains of influenza viruses.

Infected humans are usual reservoirs of viruses. They serve as a source of infection for other susceptible hosts (HIV, poliomyelitis, etc.). Asymptomatic patients are an important source of infection. In HIV and varicella-zoster virus infections, the viruses are usually released before the manifestations of the symptoms. Viruses changing the antigenic structure of their genome, such as influenza and HIV, infect many immunologically naive populations. Rhinovirus and other viruses with many different types similarly cause infection in a large number of susceptible persons.

For many other viral diseases, animals also act as reservoirs. Viral infections transmitted from these animal reservoirs to humans are called zoonotic viral diseases. For example, rabies is a zoonotic viral disease, and the infection is transmitted from the infected dogs, bats, foxes, cats, etc., to humans. Mosquitoes, ticks and sand flies are the arthropods that act as vectors for transmission of togaviruses, flaviviruses, bunyaviruses and reoviruses. These viruses, hence, are called a*rbo*v*iruses* (from *ar*thropod-*bo*rne viruses).

The mode of transmission of the virus depends on the (1) site of virus replication and secretions and (2) the presence or absence of an envelope in the viruses.

1. **The site of virus replication and secretions:** The viruses that replicate in the intestinal tract are excreted in the faeces and are transmitted by the faecal–oral route (e.g. picornavirus and reovirus). The viruses that replicate in the respiratory tract, such as the influenza virus, are secreted in aerosol droplets and transmitted by inhalation.
2. **Presence or absence of envelope in the viruses:** The enveloped viruses are relatively fragile viruses that require the presence of intact envelopes for their infectivity. These viruses are usually transmitted by respiratory droplets, saliva, mucus, blood or semen, organ transplantation and ingestion. As they are sensitive to the presence of acid and detergent, many of the enveloped viruses are destroyed rapidly in the gastrointestinal tract of the infected humans and hence are not transmitted by the faecal–oral route. Hepatitis B virus and coronavirus are exceptions transmitted by the faecal–oral route.

Non-enveloped viruses are relatively sturdy viruses. These can resist drying, extremes of pH and temperature and the effects of detergents. Therefore, these non-enveloped viruses withstand the acidity of the stomach and the lytic effect of bile in the intestine and mild disinfectants and insufficient sewage techniques. Therefore, these viruses are generally transmitted by the respiratory and faecal–oral routes and often by contaminated objects (fomites). Hepatitis A virus is a common example of an enveloped virus transmitted by the faecal–oral route. Rhinoviruses and many other non-enveloped viruses are examples of viruses that are transmitted by contact with fomites, such as handkerchiefs, towels, bed linens, etc.

Prevention and Control of Viral Diseases

The viral diseases can be prevented, and the spread of viruses can be controlled by (1) adoption of good hygiene, (2) control of the vectors and (3) immunisation of the population. Immunisation of the population by vaccines is the best means to control many diseases. Vaccines are available against many viral diseases (poliomyelitis, rabies, etc.). These protect the population against infection by viruses.

Immunoprophylaxis of Viral Diseases

Immunoprophylaxis indicates developing immunologic protection against the viral disease either by vaccination (active prophylaxis) or antibodies cocktail (passive prophylaxis).

Active prophylaxis: Different types of antiviral vaccines have been developed that specifically induce the body's cell-mediated immune response and provide steady protection against invading viruses. Most of the first-generation vaccines were attenuated or killed types. However, novel advancements in this field have led to the development of recombinant protein DNA vaccines. Latest to the vaccine depository, mRNA vaccines against SARS-CoV-2 have emerged strong and showcased excellent protection against the viruses. Many antiviral vaccines are now approved for use globally and have been recommended as essential for inclusion in the childhood vaccination schedule.

Passive prophylaxis: Passive anaphylaxis by **antibodies cocktail**, on the other hand, provides humoral protection against a viral disease at the time when an individual has been infected or expected to be infected soon (for, e.g. due to travel to virus prone region). Cocktails of antiviral antibodies are administered that provide short-term protection till the antibodies circulate in the body and do not contribute to the generation of memory response of the immune system.

Antiviral Agents

49

The viruses, unlike most bacteria, are obligate intracellular pathogens that use biosynthesis mechanisms and enzymes of the host cells for replication. Earlier, it was feared that it might not be possible to inhibit virus replication without also being toxic to host cells. Current antiviral drugs are used successfully to treat a few viral diseases without causing much toxicity or serious side effects.

The first antiviral drugs to be used were like selective poisons that targeted cells with intensive DNA and RNA syntheses. However, recently used antiviral drugs act specifically against virus-coded enzymes or structures of the virus that are important for replication of the viruses.

Methisazone (Marboran) was the first antiviral drug used clinically to treat poxvirus infection in 1960s effectively. The compound was used successfully against eczema vaccinatum and smallpox. Subsequently, in 1962, an antineoplastic agent, idoxuridine, was found to be effective for treating herpetic eye infection, and amantadine (a molecule with an unusual structure) was found effective for the treatment of influenza A virus. In the 1970s, acyclovir (ACV) was used most successfully to treat herpes virus infection by administering the drug parenterally. Subsequently, many antiviral agents have been distributed for use against many viral infections, including the human immunodeficiency virus (HIV).

Mechanism of Action of Antiviral Drugs

Recently, many antiviral drugs have been developed against viruses associated with high morbidity and mortality in humans. These viruses provide potential targets during the replication cycle for action by antiviral drugs. They may inhibit the viral infection at the level of (1) attachment, (2) penetration and uncoating, (3) transcription of viral nucleic acid, (4) translation of viral mRNA, (5) protein synthesis, (6) replication of viral DNA, (7) nucleoside biosynthesis and nucleoside scavenging and (8) assembly and release of viral progeny. Mechanisms of action of antiviral agents against the possible targets are summarised in Table 49.1.

Classification of Antiviral Drugs

The antiviral compounds vary greatly in complexity and include nucleoside analogues, synthetic oligonucleotides, oligosaccharides and also natural products of plants and some inorganic and organic compounds. The antiviral agents available against viruses can be classified as (1) nucleoside analogues, (2) non-nucleoside polymerase inhibitors, (3) protease inhibitors, (4) neuraminidase inhibitors, (5) M2 channel blockers and (6) interferons.

Table 49.1 Mechanisms of action of antiviral drugs

Virus	Antiviral drug	Mechanism of action
Herpes simplex virus	Acyclovir Vidarabine	Viral polymerase inhibitors (nucleoside analogue)
Varicella-zoster virus and herpes simplex virus	Valacyclovir	Viral polymerase inhibitors (nucleoside analogue)
Cytomegalovirus	Ganciclovir Foscarnet	Viral polymerase inhibitors
Human immunodeficiency virus	Zidovudine Didanosine Zalcitabine Stavudine (d4T)	Nucleoside reverse transcriptase inhibitors
	Lamivudine (3TC) Nevirapine Delavirdine	Non-nucleoside reverse transcriptase inhibitors
	Saquinavir Ritonavir Indinavir Nelfinavir	Protease inhibitors
	Raltegravir	Integrase inhibitor
	Enfuvirtide	Fusion inhibitor
Influenza A virus	Amantadine Rimantadine	Blocks viral uncoating
	Oseltamivir	Neuraminidase inhibitor
Hepatitis B virus	Lamivudine Adefovir	Viral polymerase inhibitors
Hepatitis C virus	Ribavirin Interferon-alpha	Inhibits protein synthesis
Respiratory syncytial virus	Ribavirin	Blocks capping of viral mRNA

Nucleoside Analogues

Numerous analogues of naturally occurring nucleosides have been synthesised in the laboratory to use against viruses possibly. These nucleoside analogues that act by inhibiting the enzyme viral polymerase are generally activated by phosphorylation by cellular or viral kinases.

Nucleoside analogues cause selective inhibition of virus replication (1) by binding better to viral DNA polymerase rather than to the cellular DNA polymerase and (2) by being utilised more extensively in virus-infected cells due to the more rapid synthesis of DNA in infected cells. The commonly used nucleoside analogues are acyclovir, valacyclovir, penciclovir, famciclovir, ganciclovir, azidothymidine (AZT), ribavirin and dideoxynucleosides (dideoxyinosine, dideoxycytidine, stavudine and lamivudine).

Acyclovir (ACV) is a synthetic guanine nucleoside analogue. It differs from the nucleoside guanosine by having an acyclic (hydroxyethoxymethyl) chain instead of a ribose or deoxyribose sugar. The ACV has selective action against herpes viruses, such as herpes simplex virus (HSV) and varicella-zoster virus. It acts through the viral enzyme thymidine kinase, encoded by these herpes viruses.

The ACV is used to treat HSV infections, such as HSV encephalitis, disseminated herpes and other serious life-threatening manifestations of herpes infections. The compound inhibits HSV replication in infected host cells but is ineffective to resolve the latent HSV infection. The ACV at higher doses is also effective against varicella-zoster virus (VZV) infections. The compound is less effective against this virus because ACV is phosphorylated less efficiently by the VZV thymidine kinase.

Valacyclovir is the valyl ester derivative of ACV that is well absorbed. Its bioavailability is

2–5 times more than ACV and is usually recommended to treat and suppress genital herpes infection.

Penciclovir is a guanosine analogue. It has a higher affinity for HSV thymidine kinase than ACV but has a lower affinity for HSV DNA polymerase than ACV. It acts by inhibiting viral DNA polymerase and also as a chain terminator. It is used for the treatment of genital herpes infection. Penciclovir is effective for the treatment of both HSV and VZV. After administration, the drug is available in more quantity and persists longer in infected cells than ACV. The penciclovir also has some antiviral activity against cytomegalovirus (CMV) and Epstein–Barr virus.

Ganciclovir is a guanine nucleoside chemically related to ACV. It differs from ACV in having a single hydroxymethyl group in the acyclic side chain. It acts as a chain terminator in the subsequent termination of viral DNA replication. It is highly effective against all herpes viruses, including cytomegalovirus. It is more active against CMV than ACV. It is more useful in treating severe CMV infections in immunocompromised hosts, such as acquired immunodeficiency syndrome (AIDS). Ganciclovir resistance is a noted problem of therapy.

Azidothymidine (AZT) is the synthetic analogue of thymidine and was the first useful antiviral agent to be reported to treat HIV infection. It acts by blocking the synthesis of pro-viral DNA by inhibiting viral reverse transcriptase. The latter is 100 times more susceptible to inhibition by AZT than host cellular DNA polymerase.

AZT is widely used for the treatment of HIV infection. It is currently used to manage HIV with reduced CD4 T-cell counts (400–500 or less) to prevent the progression of the disease. It is also used for the treatment of pregnant HIV-infected women. It has been shown to reduce or prevent the transmission of HIV from the mother to the baby. The drug, however, is toxic and costly. The emergence of resistance to AZT is also a worrisome problem.

Ribavirin is a synthetic analogue of the nucleoside guanosine. Ribavirin triphosphate is the active form of the drug. It differs from guanosine by having a base ring that is incomplete and open. Ribavirin is effective against many DNA and RNA viruses. It acts mainly by preventing replication of the viruses by inhibiting nucleotide biosynthesis, mRNA capping and other processes essential for virus replication.

When administered as an aerosol, ribavirin has been shown to effectively treat a severe respiratory syncytial viral infection in children and treat severe influenza and measles in adults. Intravenous ribavirin is also effective for treating infections caused by influenza B virus, Rift Valley virus, Lassa, Crimean–Congo and other haemorrhagic fevers. In combination with interferon-alpha (IFN-α), Ribavirin is shown to be effective against the infection caused by the hepatitis C virus.

Dideoxynucleosides, the analogues of nucleosides, have been synthesised for use against HIV. These agents (e.g. dideoxyinosine, dideoxycytidine, stavudine and lamivudine) inhibit HIV replication by blocking the enzyme reverse transcriptase. These compounds inhibit the enzyme reverse transcriptase by preventing DNA chain elongation. These compounds are usually recommended for the treatment of AIDS in patients not responding to therapy with AZT. These are also used in combination with AZT for treatment of the AIDS cases.

Other nucleoside analogues consist of several compounds, including idoxuridine, trifluorothymidine, fluorouracil and adenine arabinoside. These are analogues of thymidine. They inhibit synthesis and replication of viruses (1) by inhibiting the synthesis of thymidine, a nucleic acid essential for the synthesis of viral DNA or (2) by replacing thymidine with itself in the viral DNA. These are effective against viruses such as herpes simplex virus (HSV), replication of which is associated with the synthesis of a large volume of viral DNA. Recently, many other nucleoside analogues have been evaluated as antiviral agents for the treatment of infections caused by HIV, hepatitis B virus and herpes viruses.

Non-Nucleoside Polymerase Inhibitors

Non-nucleoside polymerase inhibitors include **foscarnet** and related phosphonoacetic acid.

These inhibitors inhibit the replication of viruses by binding to the pyrophosphate-binding site of the DNA polymerase to block the binding of nucleotides. Foscarnet specifically inhibits DNA polymerase of all herpes viruses and reverse transcriptase of HIV. The compound has also shown antiviral activity against the hepatitis B virus.

Nevirapine, delavirdine and **efavirenz** are the other non-nucleoside polymerase inhibitors with different mechanisms of action. They bind to sites on the enzyme different from the substrate. These compounds are usually given in combination with other nucleoside analogues to delay or prevent the emergence of drug resistance in HIV.

Protease Inhibitors

Saquinavir, indinavir, ritonavir, nelfinavir and **amprenavir** are some of the examples of protease inhibitors. These agents act specifically on the unique structure of HIV protease, which is essential for the production of a functional HIV. Human immunodeficiency virus strains showing resistance to these drugs occur through mutation of the HIV protease. Hence, a combination of protease inhibitor with AZT and a nucleoside analogue is usually recommended to reduce replication of viruses to minimum undetectable levels.

Other Antiviral Drugs

Amantadine (Adamantanamine hydrochloride, symmetrel) and **rimantadine** are anti-influenza drugs useful for treating influenza virus infections. These are not effective for the treatment of influenza B or C viruses. These act specifically against the influenza A virus by their ability to bind and block protein channels by the matrix protein (M2) of the influenza A virus. Resistance to these drugs occurs due to mutations, resulting in changed M2 matrix protein or haemagglutinin protein.

Amantadine and rimantadine are useful in reducing the severity of influenza A infection if taken within 48 h of exposure. They are also useful as prophylactic agents in treating influenza A infection. Amantadine is also used for the treatment of Parkinson's disease. The drug, however, is toxic to the central nervous system.

Zanamivir (Relenza) and **oseltamivir** (Tamiflu) are antiviral compounds with clinical efficacy against influenza A and B viruses. They are potent inhibitors of influenza neuraminidase. Without the production of the enzyme neuraminidase, the haemagglutinin of the virus binds to sialic acid on other viral particles, forming clumps and thereby preventing the release of virus particles. Both the agents have demonstrated efficacy in several clinical trials with minimal side effects. If taken within 48 h of infection, these drugs reduce the duration of illness.

Interferon

There are three classes of interferon: (1) interferon-alpha (IFN-α), (2) interferon beta (IFN-β) and (3) interferon-gamma (IFN-γ). IFN-α occurs as at least 15 subtypes, the genes for which show 85% homology. Interferons are produced by leukocytes and many other cells in response to infection by virus, double-stranded RNA (dsRNA), endotoxin and mutagenic and antigenic stimuli. The dsRNA is a potent stimulator. The viruses that replicate slowly and viruses that do not inhibit the synthesis of host proteins are usually good inducers of the interferons. IFN-γ differs from other interferons in being released as lymphokines from activated T cells, natural killer cells and occasionally from macrophages.

Interferons exert antiviral effects by several pathways. They cause increased expression of class I and class II MHC (major histocompatibility complex) glycoproteins, thereby facilitating the recognition of viral antigens by the immune system. They also activate the cells, such as natural killer cells and macrophages, the cells that can destroy virus-infected targets. In addition to these, they directly inhibit the replication of viruses.

Interferons are increasingly being used to treat chronic hepatitis B and C virus carriers at risk of progressing to cirrhosis and hepatocellular carcinoma. When given with ribavirin, the interferon has proved to be more effective than interferon alone to treat hepatitis C virus infection. Currently, synthetic IFN-α is actively used against hepatitis A, B and C viruses, papilloma virus;

HSV and rhinovirus. It is also used for the treatment of condylomata acuminatum.

Resistance to Antiviral Agents

Reduced susceptibility of a particular antiviral during laboratory tests indicates the development of resistance phenotype in the virus. Resistance can be defined as a reduction in susceptibility to the drug in a laboratory culture system. It is expressed as an altered IC50 or IC90. It causes great public health concerns for infection management, reducing efficiencies of medical countermeasures. Soon after the emergence of antiviral resistance viruses, it replaces susceptible phenotype from the environment and become predominant.

The development of the resistance is multifaced and depends on the virus and the change in antiviral targets. Some of the potential mechanisms of antiviral resistances are mentioned.

1. **Resistance to the acyclovir, valacyclovir and famciclovir drugs** (choice for Herpes simplex virus treatment) targets viral DNA due to mutations in the viral TK or DNA polymerase. A frameshift mutation in the genome due to the addition or deletion of nucleotides results in TK function loss. It makes the target modified where the antiviral could not work, thereby resisting the virus.
2. **Resistance to the ganciclovir** (used to treat CMV infection) is due to the mutations in the viral UL97 kinase gene. The mutation decreases the phosphorylation of ganciclovir and gives protection to the virus from its action; and
3. **Conformational changes** also confer drug resistance. For example, reverse transcriptase domains of Hepatitis B virus polymerase gene when changes conformation effectively prevents actions of nucleotide and nucleoside analogue antivirals like lamivudine, entecavir, adefovir and tenofovir.

Laboratory Diagnosis of Viral Diseases 50

The clinical manifestations may be protean and nonspecific for many of the viral infections. However, they provide a clue in diagnosing viral infections by excluding common bacterial, parasitic, fungal diseases, etc. Laboratory diagnosis, therefore, plays an important role in confirming the viral aetiology of suspected viral diseases.

Laboratory diagnosis for viral diseases is performed to (1) confirm the viral aetiology of diseases for which antiviral chemotherapeutic agents are available, (2) screen blood donors for blood-borne viral pathogens, such as hepatitis B and C, human immunodeficiency virus (HIV), etc., to prevent transmission of infection by transfusion of blood and blood-borne products contaminated by viruses, (3) define the viral disease process, (4) perform seroepidemiological studies of viral infections and (5) monitor and detect epidemics of serious virus infections, such as influenza, encephalitis and poliomyelitis, etc., earlier to initiate appropriate control measures to prevent further spread of these diseases.

Methods of Laboratory Diagnosis

Laboratory diagnosis of viral infections can be carried out by many methods. These methods include: (1) direct detection of viruses/viral antigens, (2) demonstration of virus-induced cytopathic effects (CPEs) in the cells, (3) virus isolation, (4) viral assays, (5) detection of viral proteins and other enzymes, (6) detection of the viral genome and (7) viral serology (Fig. 50.1).

Specimens

The collection of appropriate clinical specimens depends on the type of the viral disease. The tests of clinical specimens to be collected for diagnosis of other types of viral diseases are summarised in Table 50.1.

The timing of the collection of specimens is important. The specimens collected early in the acute stage of infection—before the shedding of the virus is stopped—are most important. For example, enteroviruses are present in the CSF for only 2–3 days after the onset of CNS manifestations. Herpes simplex virus and varicella-zoster virus are found in lesions only within the first 5 days of onset of symptoms, and respiratory viruses are present in respiratory secretions during only the first 3–7 days of onset of symptoms.

Immediate transport of the specimen to the laboratory for processing also facilitates better isolation of virus from clinical specimens. The viruses are usually heat-labile, and the clinical specimens may be infected secondarily by contamination with bacteria and fungi. Hence, clinical specimens for viruses are usually transported and stored in ice. They are transported in special transport media containing antibiotics to inhibit bacterial and fungal contaminants and contain

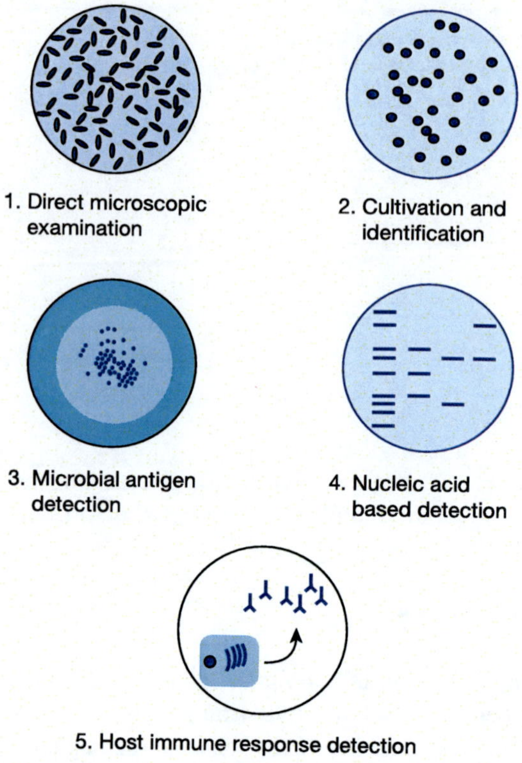

Fig. 50.1 Approaches to laboratory diagnosis of viral infections

proteins, such as serum albumin or gelatin. Some of the viruses, such as influenza virus, HSV and VZV, lose their infectious titre when clinical specimens are stored at room temperature or kept frozen at −20 °C.

Direct Detection of Viruses/Viral Antigens

Viruses/viral antigens can be detected directly in clinical specimens by electron microscopy (EM), fluorescence microscopy and light microscopy.

Electron Microscopy

Electron microscopy (EM) is the most commonly used method for detecting viruses in clinical specimens and biopsies to diagnose many viral diseases. Electron microscopy is used for detection of (1) rotaviruses and herpes virus in stool samples, (2) herpes virus in brain biopsy specimens, (3) poxvirus in vesicle fluid, (4) virus in cell cultures and (5) Nipah viruses in infected tissues by immune electron microscopy.

Fluorescence Microscopy

Using specific monoclonal or polyclonal antibodies, direct fluorescence microscopy is employed to detect viral antigens on the cell surface or within the cells infected by viruses. The viral antigens can be detected in the acetone-fixed cell smears, the frozen tissue sections of the cells from virus-infected cells and the vesicular fluid. The test has been employed for (1) antemortem diagnosis of rabies by detection of rabies virus antigen in smears from the face or nape of the neck; (2) demonstration of viral antigens in brain biopsy specimens for diagnosis of herpes simplex encephalitis and subacute sclerosing encephalitis

Table 50.1 Laboratory specimens for diagnosis of viral infections

System	Direct examination	Isolation	Serology
Respiratory system	Nasopharyngeal aspirate (IF, EM)	Throat swab, throat washings	Paired sera
CNS	Brain biopsy (IF, EM), CSF (EM, IF)	Faeces, blood (for arbovirus), CSF and brain biopsy	Paired sera
Skin	Vesicular/pustular fluid (EM, ID), ulcer scrapings (EM) and crusts (EM, ID)	Macular/papular scrapings, vesicular/pustular fluid, ulcer scrapings, crust, urine	Paired sera
Eye	Conjunctival scrapings and smears (LM, IF)	Conjunctival scrapings or swabs	Paired sera
Liver	Serum and faeces	Blood (for yellow fever)	Serum
Congenital infections	Nil	Throat swab, products of conception	Single sera (mother and baby)
Gastro-intestinal tract	Stool (antigen detection, EM for rotavirus)	Not cultured	Paired sera (ELISA)

IF, immunofluorescence; EM, electron microscope; ID, immunodiffusion; LM, light microscopy and (3) rapid diagnosis of infections caused by adenoviruses, paramyxoviruses and orthomyxoviruses.

Light Microscopy

Guarnieri bodies of smallpox virus can be seen in the vesicular fluids by Giemsa stain. Viral antigens in infected cell cultures are demonstrated by immunoperoxidase staining. The Negri bodies can also be demonstrated in neural tissues of the brain by Sellers' stain.

Demonstration of Virus-induced CPEs in the Cells

Many viruses produce characteristic morphological changes in the cells they infect. These changes are known as CPEs, and the viruses causing the CPEs are known as cytopathogenic viruses. The characteristic CPEs include (1) change in cell morphology, such as rounding of cells or rounding and aggregation of cells, (2) syncytia formation and (3) inclusion bodies formation.

Replication of virus in infected cells may cause rounding, refractility, degeneration and nuclear pyknosis. This may finally lead to complete or partial cell lysis and vacuolation, as seen in picornaviruses. Viruses, such as adenoviruses, may cause rounding of cells and aggregate to form grape-like clusters.

Viruses, such as paramyxovirus, varicella-zoster, respiratory syncytial virus and herpes simplex virus (HSV), cause syncytia formation, containing several (up to 100) nuclei in infected cells. These syncytia are multinucleated giant cells formed by the fusion of virus-infected cells with neighbouring or uninfected cells.

Inclusion bodies are intranuclear or cytoplasmic bodies produced by viruses in infected cells. They are produced due to histological changes in infected cells caused by viral components. These are also produced as a result of virus-induced changes in cell structures. These inclusion bodies can be demonstrated on staining by light microscopy. These inclusion bodies may be acidophilic or basophilic, small or large, round or irregular, and single or multiple. These bodies may be present either in the nucleus or cytoplasm, or both. Many viruses produce different types of inclusion bodies. For example, the rabies virus produces intracytoplasmic inclusion bodies called Negri bodies in the neural tissue of the brain. Cytomegalovirus (CMV) produces intranuclear owl eye inclusion bodies (Fig. 50.2) in infected cells or cell sediments excreted in the urine of the patient infected with CMV. Human herpes viruses produce Cowdry type A inclusion bodies. Measles virus produces inclusion bodies that are seen both in the cytoplasm and nucleus of infected cells (Table 50.2).

Virus Isolation

Demonstration of virus in appropriate clinical specimens by culture establishes the diagnosis of viral diseases. Most viruses can be cultivated in (1) laboratory animals, (2) embryonated eggs or (3) tissue culture.

Laboratory Animals

The mouse is most frequently used to isolate viruses by animal inoculation. In addition, rabbits, hamsters, newborn or suckling rodents are also used. Experimental animals are rarely used to cultivate viruses but play an essential

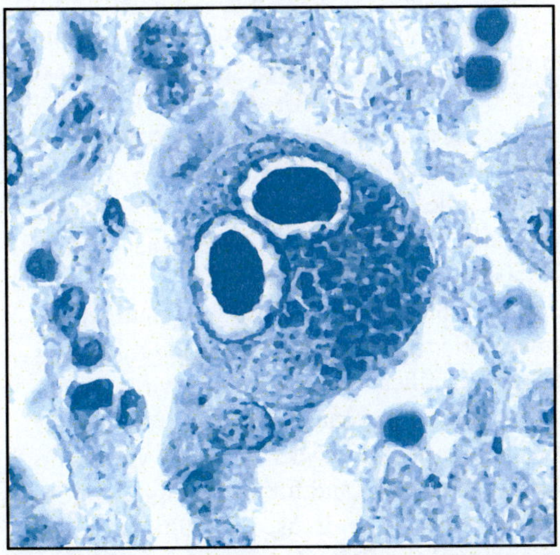

Fig. 50.2 Cytomegalovirus inclusion body (×1000)

Table 50.2 Laboratory procedures for diagnosis of viral infections

Approaches	Methods
Demonstration of virus-induced cytopathic effects (CPEs) in the cells	Characteristic CPEs include (1) change in cell morphology, such as rounding of cells, or rounding and aggregation of cells, (2) syncytia formation and (3) inclusion bodies formation
Direct detection of viruses	Electron microscopy, fluorescence microscopy and light microscopy
Virus isolation	Animal inoculation, embryonated egg inoculation and cell culture
Detection of viral proteins and other enzymes	The enzyme-linked immunosorbent assay (ELISA), direct immunofluorescence assay, radioimmuno assay (RIA), etc.
Detection of the viral genome	DNA probes, dot blot or southern blot analysis, northern blot or RNA: DNA probe hybridisation, polymerase chain reaction (PCR) and reverse transcriptase PCR (RT PCR)
Viral serology	Haemagglutination inhibition (HI) test, neutralisation test (NT), indirect fluorescent antibody (IFA) test, ELISA, RIA, latex agglutination test (LAT) and Western blot

role in studying the pathogenesis of viral infections and that of viral oncogenesis.

Intracerebral, subcutaneous, intraperitoneal or intranasal routes are various routes of inoculation. After inoculation, the animals are observed for signs of disease or death. The infected animals are then sacrificed, and infected tissues are examined for the presence of viruses by various tests and inclusion bodies in infected tissues. Furthermore, infant (suckling) mice are used to isolate the coxsackie virus and rabies virus.

Embryonated Eggs

Embryonated eggs were used initially for the growth of viruses. Embryonated chick egg was used first to cultivate viruses by Goodpasture in 1931. The method further developed by Burnet was used to cultivate viruses in different sites of the embryonated egg. Usually, 8–11 day-old chick eggs are used to culture viruses. The viruses are isolated in different sites of the egg, such as the yolk sac, amniotic cavity and allantoic cavity and chorioallantoic membrane (CAM) (Fig. 50.3).

Many of these viruses cause well-defined and characteristic foci, providing a method for identification, quantification or assessing virus pathogenicity. The embryonated egg is also used for growing higher titre stocks of some viruses in research laboratories and for vaccine production. **Yolk sac inoculation** is used to cultivate Japanese encephalitis, Saint Louis encephalitis and West Nile virus. It is also used for the growth of chlamydia and rickettsia. **Inoculation in the amniotic cavity** is used mainly for primary isolation of the influenza virus. **Inoculation in the allantoic cavity** is used for serial passages and for obtaining large quantities of viruses, such as influenza virus, yellow fever (17D strain) and rabies (Flury strain) viruses to prepare vaccines. For the production of the rabies virus, duck eggs were used due to their bigger size than that of hen's egg. This helped produce large quantities of the rabies virus, which are used to prepare the inactivated non-neural rabies vaccine.

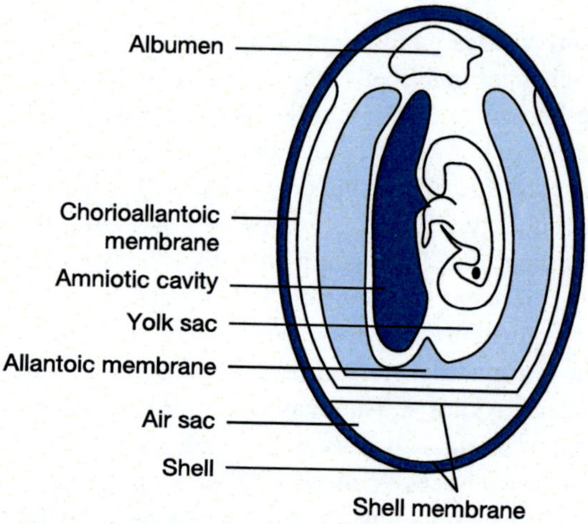

Fig. 50.3 Embryonated egg inoculation

Inoculation of some viruses on the Chorioallantoic membrane (CAM) produced visible lesions known as *pocks*. Each infectious virus particle produces one pock. The poxviruses, such as variola or vaccinia, are identified by demonstrating typical pocks on the CAM inoculated with the pox virus. Nowadays, in a virology laboratory, chick embryo inoculation has been replaced by cell cultures for routine isolation of viruses.

Tissue Culture

Tissue culture is most widely used in diagnostic virology for the cultivation and assays of viruses. Different types of tissue cultures are used to grow viruses. Tissue culture can be of three different types as follows:

Flurry Culture

This was used earlier to isolate some viruses, which show an affinity for certain tissue organs. For example, coronavirus, a respiratory pathogen, was isolated in the tracheal ring organ culture. In this method, small bits of the organs are maintained *in vitro* for days and weeks, preserving their original morphology and function. Nowadays, organ culture is not used.

Explant Culture

In this method, components of minced tissue are grown as explants embedded in plasma clots. Earlier, adenoid tissue explant cultures were used for the isolation of adenoviruses. This method is now seldom used in virology.

Cell Culture

Cell culture is now routinely used for growing viruses. In this method, tissues are dissociated into component cells by treatment with proteolytic enzymes (trypsin or collagenase) followed by mechanical shaking. The cells are then washed, counted and suspended in a growth medium containing essential amino acids and vitamins, salts, glucose and a buffering system. This medium is supplemented by up to 5% of foetal calf serum and antibiotics. The cell suspension is dispensed in glass or plastic bottles, tables or Petri dishes. On incubation, the cells adhere to the glass surfaces and divide to form a confluent monolayer sheet of cells covering the surface within a week (Fig. 50.4). The cell culture may

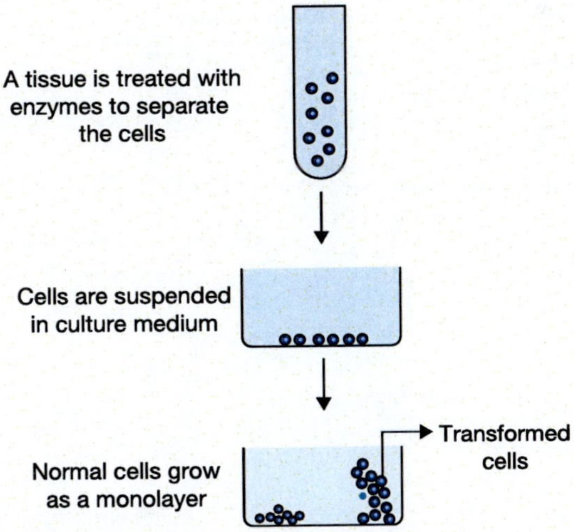

Fig. 50.4 Schematic diagram showing the steps in preparation of cell lines for culture

be incubated either as a stationery culture or as a roller drum culture. The latter is useful for the growth of some fastidious viruses due to better aeration by rolling the culture bottle in special roller drums. The cell cultures are classified into three different types based on their origin, chromosomal characters and number of generations for which they can be maintained as follows (Table 50.3):

Primary cell culture: This is a culture of normal cells obtained freshly from the original tissues that have been cultivated *in vitro* for the first time and that have not been subcultured. These cell cultures can be established from whole animal embryos or selected tissues from adults, newborns or embryos. These cells have the normal diploid chromosomal number and only limited growth (5–10 divisions) in culture. They cannot be maintained in serial culture but can be subcultured to obtain a large number of cells. Monkey kidney cell culture, human embryonic kidney cell culture and chick embryo cell culture are the common examples of primary cell culture. Primary monkey kidney cell cultures are highly useful for the primary isolation of myxovirus, paramyxovirus, many enteroviruses and some adenoviruses.

Table 50.3 Cell lines in use in virology

Cell lines	Examples of cell lines
Primary cell culture	Rhesus monkey kidney cell culture Human amnion cell culture Chick embryo fibroblast cell culture
Diploid cell strains	Human embryonic lung cell strain (WI-38) Fibroblast from lung tissue (MRC-5) Rhesus embryo cell strain (HL-8)
Continuous cell lines	Human carcinoma of cervix cell line (HeLa) Human epithelioma of larynx cell line (Hep-2) Human carcinoma of nasopharynx cell line (KB) Human synovial carcinoma cell line (McCoy) Sternal marrow cell line (Detroit-6) Human conjunctiva (C), intestine (I), liver (L), Kidney (K) cell line (Chang C/I/L/K) Vervet monkey kidney cell line (Vero) Baby hamster kidney cell line (BHK-21)

Diploid cell strains: Diploid cell strains are of a single cell type that retains their original diploid chromosome number and karyotype. However, they have specific characteristics and compositions and are usually composed of one basic cell type. They are usually fibroblasts and can be cultured for a maximum of 50 serial passages before they undergo senescence (die-off) or a significant change in their characteristics. Diploid cells derived from human fibroblasts are useful for isolating some fastidious viruses. They are also used to produce vaccines; for example, WI-38 human embryonic lung stem cell is used to cultivate fixed rabies virus and human foetal diploid cells for isolation of adenovirus picornaviruses, HSV, CMV and VZV.

Continuous cell lines: Continuous or immortal cell lines are cells of a single type derived from cancerous tissue and capable of continuous serial cultivation indefinitely without senescing. The cells are usually derived from diploid cell lines or malignant tissues and have an altered and irregular number of chromosomes. Immortalisation may occur spontaneously or induced by chemical mutagens, tumourigenic viruses or oncogenes. Hep-2, HeLa and KB derived from human carcinoma cervix, human epithelioma of the larynx and human carcinoma of the nasopharynx and other cell lines are excellent for the recovery of a large number of viruses. These cell lines have been used extensively to grow a number of viruses. These cell lines are usually stored at −70 °C for use when necessary or are maintained by serial subculture. The type of cell line used for virus culture depends on the sensitivity of the cells to a particular virus; for example, the Hep-2 cell line is excellent for the recovery of respiratory syncytial viruses, adenoviruses and HSV. Most viruses can be isolated by using one of these cell lines.

Detection of Virus Growth in Cell Cultures

The following methods can detect the growth of viruses in cell cultures.

Cytopathic Effect (CPE)

Many viruses can be detected and initially identified by observing the morphological changes in the cultured cells they replicate. The CPE produced by different types of viruses is characteristic and helps in the initial identification of virus isolates. Nuclear shrinking, vacuoles in the cytoplasm, syncytia formation, rounding up and detachment are the examples of alteration of morphology of the cells. Most CPEs can be demonstrated in an unfixed and unstained monolayer of cells under the low power of the microscope. For example, adenoviruses produce large granular changes resembling bunches of grapes,

SV-14 produces well-defined cytoplasmic vacuolation, measles virus produces syncytium formation, herpes virus produces discrete focal degeneration and enteroviruses cause crenation of cells and degeneration of the entire cell sheet.

Haemadsorption

Haemadsorption is the process of adsorption of erythrocytes to the surfaces of infected cells, which serves as an indirect measurement of viral protein synthesis. This property is used to detect infection with non-cytocidal viruses and the early stage of cytocidal viruses. Viruses, such as influenza virus, parainfluenza virus, mumps virus and togavirus, when infecting cell lines, code for the expression of red cell agglutinins, which are expressed on the infected cell membrane during infections. These haemagglutinins bind some erythrocytes to the infected cell surface. Sometimes, viruses can be detected by agglutination of erythrocytes in the culture medium.

Heterologous Interference

This property is used to detect viruses that do not produce classic CPEs in the cell lines. In this method, the growth of non-CPE-producing virus in cell culture can be tested by a subsequent challenge with a virus known to produce CPEs. The growth of the first virus will inhibit infection by the cytopathic challenge virus by interference. For example, the rubella virus usually does not produce any CPE but prevents the replication of picornaviruses, which is inoculated as a cytopathic challenge virus.

Transformation

Oncogenic viruses associated with tumour formation induce cell transformation and loss of contact inhibition in the infected cell lines. This leads to surface growth that appears in a piled-up fashion producing micro-tumours. Examples of oncogenic viruses that produce transformation in cell lines are herpes viruses, adenoviruses, hepadnaviruses, papovavirus and retroviruses.

Detection of Viral Antigen/ Virus in Cell Lines

Light Microscopy

Viral antigens in infected cell cultures are demonstrated by staining virus-infected cells of tissue sections with specific viral antibodies conjugated with horseradish peroxidase. This is followed by the addition of hydrogen peroxide along with a benzidine derivative substance. A red insoluble precipitate is deposited on the cell line in a positive reaction, which is demonstrated by examination under an ordinary light microscope.

Immunofluorescence

Direct immunofluorescence using specific antibodies is frequently used to detect viral antigens in inoculated cell lines to identify viruses.

Electron Microscopy

EM can also demonstrate the viruses in infected cell lines.

Viral Assays

The virus content of a specimen can be assessed either by estimating the total virus particle count or by the assay of infectivity of viruses.

Total Virus Particles Count

Electron microscopy and haemagglutination are used to estimate total virus particles.

Electron microscopy (EM) is useful to count virus particles directly in a negatively stained viral suspension. In this method, the virus suspension is mixed with a known concentration of latex particles and the number of virus particles in the suspension is estimated by a ratio between the virus and latex particles demonstrated by EM.

Quantitation of haemagglutinating viruses is carried out by the determination of **haemagglutination** titres. Although it is not a sensitive method, it is used as a convenient method of virus assay. For example, approximately 10^7 influenza virions are essential to produce microscopic agglutination in cultured cells.

Assay of Infectivity of Viruses

Quantitative and quantal assays are the two types of essays that are carried out to determine the assay of infectivity of viruses.

Quantitative Assay of Infectivity

A quantitative assay is used to estimate the presence of an actual number of viable infectious viral particles in the inoculum. Two methods are available for the purpose, including plaque assay in monolayer cell culture and pock assay on chick embryo CAM.

1. **Plaque assay**: Dulbecco introduced it in 1952 as a modification of bacteriophage plaque assay. This is based on the principle that each infectious viral particle gives rise to a localised focus of infected cells that the naked eye can visualise. Such foci are called ***plaques***, and each plaque indicates an infectious virus. The test is performed by adding a viral inoculum to a monolayer of culture cells in a bottle or Petri dish. After some time, this allows the adsorption of viruses. The liquid medium is removed and replaced with a solid agar gel to ensure that the spread of progeny virus is confined to the immediate vicinity of infected cells.
2. **Pock assay**: Viruses that form pocks on CAM can be assayed by counting the number of pocks formed on the inoculated CAM. Each pock on CAM arises from a single virus particle. This is known as pock assay. Vaccinia and variola viruses can be assayed by pock assay.

Quantal Assays of Infectivity

Quantal assays of infectivity can be carried out to quantitate a virus by quantitating the virus particles in animals, embryonated eggs or tissue culture. This assay method of infectivity only indicates the presence or absence of infectious viruses, but it does not indicate an actual number of viruses. The endpoints used for infectivity titrations are estimated by the (1) development of CPE in cell cultures, (2) production of haemagglutination in allantoic fluid of embryonated egg, or (3) death of experimentally infected animals. The titre of the virus is usually expressed as the '50% infectious dose (ID_{50})' per mL, which indicates the highest dilution of the inoculum that initiates detectable symptoms, antibodies or other responses in 50% of inoculated test animals, eggs or cell cultures.

Detection of Viral Proteins and Other Enzymes

During the replication of viruses in host cells, viral proteins, antigens and other enzymes are produced. These viral products and components can be detected by many methods, including biochemical, immunological and molecular methods in clinical specimens.

Enzyme-linked immunosorbent assay (ELISA), direct immunofluorescence assay and radioimmunoassay (RIA) are widely used methods. These methods are used for the detection and identification of viruses and viral antigens in clinical specimens, as well as in cell cultures. These tests use specific monoclonal or monospecific antibodies raised against specific viral antigens.

Identifying and quantifying specific viruses can also be carried out by detecting and assaying characteristic viral enzymes. For example, detection of reverse transcriptase in serum or cell culture suggests the presence of a retrovirus. Similarly, demonstration of haemagglutinins or haemadsorption in the cell culture indicates the presence of the influenza virus.

Detection of Viral Genome

The unique genomic structure and genetic sequences are the most important characteristics of the type and family of viruses. Therefore, molecular methods are increasingly used to diagnose viral diseases. The restriction endonuclease fragment lengths from the genome of DNA viruses, such as HSV-1 and HSV-2 or the electrophoretic pattern of RNA viruses, such as influenza and reoviruses, is considered as genetic fingerprints for these viruses.

The methods for detection of viral genome include: (1) DNA probes, (2) dot blot or Southern blot analysis, (3) Northern blot or RNA-DNA probe hybridisation, (4) polymerase chain reaction (PCR) and (5) reverse transcriptase PCR (RT PCR).

DNA Probe

The DNA probes are highly sensitive and specific tools used to detect viruses in clinical specimens, even in the absence of replication of the virus. In this method, two strands of the target DNA in clinical specimens are first separated and then allowed to hybridise with a labelled single-stranded DNA probe. Enzyme-labelled or fluorescent-labelled DNA probes with sequences complementary to unique nucleic acid sequences of a viral genome can be used for detecting a virus. The hybridised labelled probe, after hybridisation, can be detected by enzymatic or fluorescence method depending on the label used.

The commercially available DNA probes are available to detect CMV, papillomavirus and Epstein–Barr virus in clinical specimens. *In situ* hybridisation is also used to detect integrated or non-integrated copies of viral genomes in fixed permeability biopsy specimens in persistent infections or viral malignant conditions.

DNA probe is especially useful for (1) studying the expression of specific viral genomes on the different cell types; (2) detecting, in clinical specimens, viral genomes that do not provide any CPE or for which no serological test is available and (3) detecting slowly replicating or non-productive viruses, such as human papillomavirus or CMV.

Southern Blot Analysis

Southern blot analysis or dot blot is used to detect viral genomes in clinical specimens. In this method, the viral genome or restriction endonuclease cleavage fragments of the genome are blotted onto nitrocellulose papers and then detected on the paper by their hybridisation to DNA probes.

Northern Blot Analysis

Northern blot analysis or RNA: DNA probe hybridisation is a method also followed to detect viral genomes. In this method, electrophoretically separated viral RNA is transformed onto a nitrocellulose paper and then detected by their hybridisation to DNA probes.

Polymerase Chain Reaction (PCR)

It is a DNA amplification system that allows a millionfold amplification of a target sequence of nucleic acids, at least 100,000-folds in a few hours. This method is useful for diagnosing viral infections caused by HIV-1, HIV-2, CMV, human papilloma virus, HSV, hepatitis B, C and D viruses, morbillivirus, echovirus, rhinovirus, measles virus, rotavirus, etc. The PCR is useful specifically for the detection of latent and integrated sequences of viruses, such as retrovirus, herpes virus, papilloma viruses and other viruses and viruses present in low concentration in clinical specimens.

Viral Serology

Viral serology is based on detecting specific viral antibodies in the serum of the infected human host. Many serological tests are used to demonstrate specific viral antibodies in patients' sera. These include haemagglutination inhibition (HI) test, neutralisation test (NT), Western blot, ELISA, indirect fluorescent antibody (IFA) test, RIA and latex agglutination test (LAT).

The detection of virus-specific immunoglobulin (IgM) antibody in serum indicates a recent primary infection. This is because IgM antibodies appear in the serum during the first 2 or 3 weeks of primary infection. A fourfold increase in the antibody titre between the serum collected during the acute and convalescent phase, 2–3 weeks after the acute phase, suggests seroconversion. An anamnestic or secondary antibody response occurs during reinfection or recurrence of viral infection later in life. The serum antibody titre usually remains high in individuals who suffer a frequent disease recurrence, such as the herpes virus.

Haemagglutination Inhibition (HI) Test

The test is used to detect viruses that agglutinate red blood cells of chickens, guinea pigs, humans or other mammals. Antibodies in serum prevent viruses from binding and agglutinating the new blood cells. The HI test is used for the diagnosis of infections caused by orthomyxoviruses.

Neutralisation Test (NT)

The test is based on the inhibition of infection by the antibody and that of CPEs of the viruses in tissue culture cells. The neutralisation antibody response is usually virus and strain-specific. The highest dilution of serum preventing infectivity is 50% of the virus. The serum mixture tested is considered the titre of the test.

Western Blot

It is most commonly used to confirm infection caused by HIV, initially diagnosed by ELISA. **ELISA** is most commonly used for screening blood samples for hepatitis B and C viruses, Dengue and HIV. **IFA, RIA** and **LAT** are also used to diagnose many common viral infections.

Viral serology has many uses. It is used (1) to identify the virus and its strain or serotype, (2) to determine whether the viral infection is acute or chronic, or primary infection or reinfection, (3) for diagnosis of viral infections that cause diseases of long duration and (4) or diagnosis of infections caused by viruses that are difficult to culture.

The limitations of serological tests in viral diseases are the following:

1. The presence of antiviral antibody in serum only indicates infection but cannot determine whether it is recent or old. Demonstration of IgM antibodies or demonstration of a fourfold increase in the antibody titre between acute and convalescent sera indicates only recent infection.
2. The serological tests may be associated with false-positive or false-negative reactions. The serological cross-reaction may occur between different viruses, leading to false-positive reactions. The formation of immune complexes in serum may give rise to false-negative reactions observed in viral infection caused by the hepatitis B virus.

Molecular Diagnosis

The increase of instances of emerging and re-emerging viruses and their deleterious effects on human health, including the potential for outbreak and pandemic, has necessitated an early warning system for rapid and highly specific detection of the viruses. As such, a greater need has been felt to develop novel, powerful diagnostic assays where molecular tools have found perfect application. The advantages of molecular tools compared to traditional systems are rapid and accurate detection, higher sensitivity, specificity to differentiate closely related species, detection of very low concentration of samples (suitable for early warning), quantitative and qualitative detection, etc. In contrast, the current disadvantages that are mostly procedural include the requirement of highly trained staff, higher processing cost, the requirement of sophisticated laboratories, the absence of a uniform globally acceptable protocol and reagents for many pathogenic viruses.

The breakthrough in molecular diagnosis was initially achieved using conventional gel-based PCR that qualitatively detects the specific viral genes by using designed primers and visualisation of amplification in an agarose gel. However, the process was complex and time-consuming. It was later replaced with other types of PCR, including real-time PCR, nested PCR, multiplex PCR, hybridisation and in situ PCR. Most of these such techniques could quantitatively detect viruses, even more than once, in a single reaction and without the need for gel-based visualisation. Even today, real-time PCR is considered the gold standard for virus detection in a quantitative approach. Several commercial kits targeting different viruses have been developed to eliminate the process complexities and make way for the technology to be applied widely in clinical laboratories.

However, one of the major complexities of these processes was the requirement of trained staff, equipped laboratories and the inability to process many samples at a time. Scientific advancements in the recent time have introduced PCR robotics, portable PCR machines, amplification strategies without thermocycling, etc. These have effectively facilitated bench to field transition of molecular diagnostics and have opened the avenue for these tools for wider use.

In addition to the DNA-based molecular detection methods, antibody-detecting molecular methods are also under development and finding ways to reach clinical laboratories. Some of such new tools include improved ELISA systems, PCR for protein detection (DNA tags, proximity ligation), pen-side tests (e.g. dip-stick assays).

Bacteriophages

51

Bacteriophages are viruses that infect bacteria. They are obligate intracellular parasites that multiply inside bacteria by using some or all of the host biosynthesis machinery. They are also called **phages**. These extrachromosomal genetic elements usually survive outside a host cell due to the presence of a nucleic acid genome surrounded by a protein coat. Phages occur widely in nature in close association with bacteria and are distributed widely in the soil, faeces and other substances in the environment. They are associated with the transmission of genetic material from one bacterium to another. Twort first described the activities of bacteriophage in 1915, who described it as an infectious agent that distorted the appearance of the colonies of staphylococci. Subsequently, d'Herelle in 1979 demonstrated the lytic activities of the culture filtrate on bacterial colonies. He suggested that the lytic agent was a virus and gave it the name bacteriophage (*phage* meaning to 'eat').

Morphology

Most phages consist of single, linear and double-stranded DNA genomes (Fig. 51.1). This genome is surrounded by a protein coat known as phage capsid. Large phages usually consist of a head and a tail. The head encloses the genome, and the tail is used as an organ of attachment and the conduit through which phage DNA passes into the host cell. The best-studied phages are the T-even phages (T2, T4, T6, etc.) that infect the bacteria *Escherichia coli*.

Phages are tadpole-shaped with a hexagonal head and a cylindrical tail.

The hexagonal head of all phages varies in size and shape. It is icosahedral (20 sides) or filamentous in shape. The size of the head measures between 28 and 100 nm in length. The capsid of the bacteriophage head encloses a compartment that contains the nucleic acid, which in most of the phages is a double-stranded DNA molecule. This DNA is protected against degradation by environmental nucleases by virtue of its location inside the bacteriophage head. However, a group of phages that specifically infect male strains of *E. coli* contains only RNA. The capsid or protein coat of a virus is composed of individual proteins known as capsomeres.

Many but not all phages have tails attached to the phage head. The **tail** consists of a hollow tube through which the nucleic acid passes during infection. The tail of a complex phage like T4 is surrounded by a contractile sheath, which contracts during infection of the bacterium and a terminal tail plate. The spikes and tail fibres both originate from the tail plate and specifically bind to receptors on the outer membrane of the bacterial cell wall.

Phages vary considerably in their morphology. Phages that are spherical or filamentous and possess single-stranded DNA or RNA have also been described. Phages show high host specificity. They usually inhabit the intestinal microbial

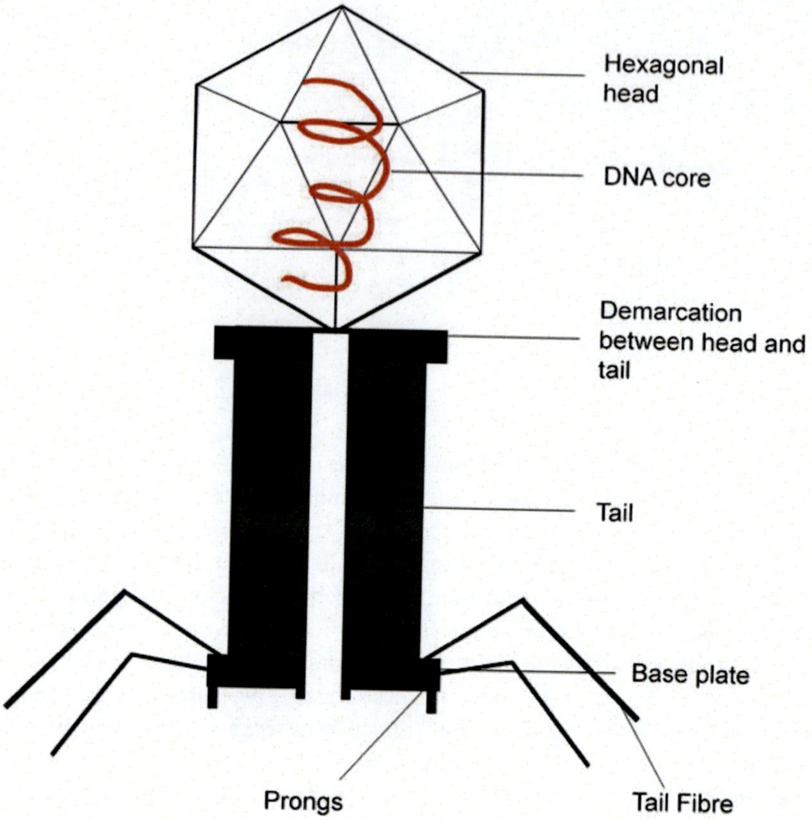

Fig. 51.1 Schematic diagram of a bacteriophage

flora of humans and animals. They pass through filters, which hold back the bacteria. These are inactivated by boiling.

Life Cycle

Phages exhibit two different types of life cycles: **The lytic cycle**, also known as a virulent cycle, and the **lysogenic cycle, also** known as the temperate cycle.

Lytic Cycle (Virulent Cycle)

In this cycle, intracellular multiplication of the phage results in the lysis of host bacteria, resulting in the release of progeny virions. The lytic cycle of a virulent bacteriophage shows the stages of adsorption, penetration, synthesis of the phage components, maturation and release of progeny phages.

1. **Adsorption** is the first stage and is a specific process. It is a very rapid process and, under optimal conditions, is completed within minutes. The specificity of the phage is determined by the presence of chemical receptors on the surface of the bacteria. The sites of the receptors for bacteriophage in bacteria are different and may be situated in different cell wall layers or on the surface structure. For example, bacteriophage receptors are found on the outer membrane of the cell wall of *E. coli*, in the Vi antigen of *Salmonella*, in glycosylated teichoic acid of *Bacillus subtilis*, and portions of the outer membrane of the cell wall of Gram-negative bacteria.

Adsorption is a specific biochemical reaction; hence bacteriophages can infect only certain host bacterial cells. Infection usually does not occur in the absence of adsorption. However, infection of the bacterium directly by naked phage nucleic acid can occur, and this process is known as **transfection**.

2. Adsorption is followed by **penetration** of phage nucleic acid into the bacterial cell. Phages with contractile tails behave as hypodermic syringes, injecting the phage DNA into the bacterial body through the hollow core (Fig. 51.2). The phage DNA passes through the central tail core into the host cytoplasm, while the empty protein capsid remains attached to the cell surface. More than one phage can adsorb to a host cell and inject its DNA. If more bacteriophages are adsorbed, they make so many holes in the cell wall that it causes lysis of the cell, a phenomenon called lysis from without.

   The empty head and tail of the phage, after penetration, remain outside the cell surface as the empty shell or ghost. After entry, the phage DNA is re-arranged to a circular form, and this process is known as the circulation of phage DNA.

3. **Synthesis of phage components** begins immediately after the phage nucleic acid penetration. **Early proteins** are the first products to be synthesised. These early proteins include the enzymes, such as nucleases, synthetic enzymes and RNA polymerase. These early proteins are essential for building complex molecules peculiar to the phage. Subsequently, **late proteins** are synthesised, which include (1) protein subunits of phage, (2) structural proteins necessary for the virion self-assembly, (3) enzymes involved in maturation and (4) proteins used in the release of bacteriophages from the cell. During this period, the synthesis of bacterial proteins, DNA and RNA stop. It is noteworthy that phage DNA, head proteins and tail proteins are synthesised separately in the bacterial cell.

4. **Maturation begins with** the protein subunits of the phage head and tail aggregating spontaneously to form the compact capsid. The DNA is condensed into a compact polyhedral shape and packaged into the head. The assembly of the phage component into the mature infective phage particles is known as maturation.

5. **Release of progeny phages** is a process in which new mature progeny phages are released by lysis of the bacterial cells. During the replication of phages, the bacterial cell wall is weakened by lysozyme. Finally, the cell wall bursts as a result of osmotic pressure. This is known as **lysis from within**.

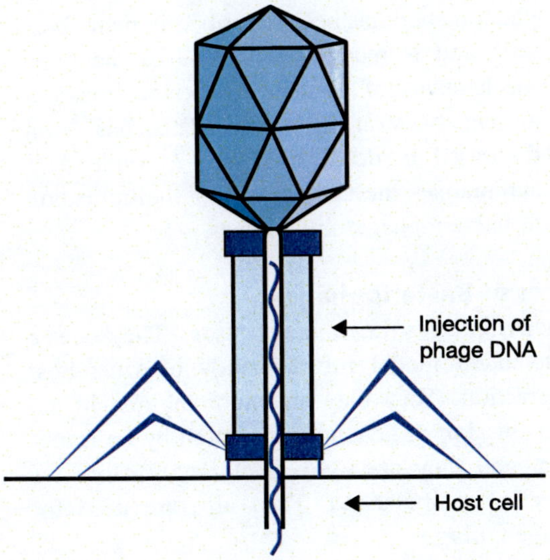

Fig. 51.2 Process of injection of phage DNA into the host cell

The interval between the infection of a bacterial cell and the first release of mature daughter phages is known as the latent period. The interval between the entry of phage nucleic acid into the bacterial cell and the appearance of the first infectious intracellular phage particle is known as the eclipse phase. The average yield of daughter phages from the infected bacterial cell is known as the burst size. The period during which the number of infectious phages released rises is known as the rise period. A one-step growth curve is obtained when the bacteriophage growth titre is plotted against time. A plaque assay enumerates the lytic cycle. A plaque is a clear area that results from the lysis of bacteria. Each plaque arises from a single infectious phage.

Lysogenic Cycle (Temperate Cycle)

In this cycle, phage DNA becomes integrated with the bacterial genome and replicates with the bacteria synchronously without causing any harm to the host cell.

Infection with every phage does not result in the lysis of the host cells. Unlike virulent phages, which cause lysis of the host cell, some phages (such as temperate phages) integrate into the genome of the bacterial chromosome without causing any lysis of the bacteria. The integrated phage nucleic acid is known as the **prophage**. The prophage behaves like a segment of the host chromosome and multiplies synchronously with it. This phenomenon is known as **lysogeny**. The bacterium that carries a prophage within its genome is called **lysogenic bacterium**, and such phages are called **lysogenic** or **temperate phages**.

Lysogeny: After infection by a temperate phage, the DNA from the temperate phage enters the host cell. This phage inside the cell forms a circle and, by a single recombinant event, becomes integrated into the DNA of the bacterium. The host bacterial cell, now called the lysogenic cell, replicates the phage DNA or prophage every time it divides. Sometimes, the phage may become excised from the bacterial chromosome during the multiplication of lysogenic bacteria. The excised prophage can initiate a new lytic cycle resulting in the release of daughter phage particles. These daughter phage particles also infect other bacterial cells and make them lysogenic. This is known as spontaneous induction of prophage.

All lysogenic bacteria on exposure to ultraviolet rays, hydrogen peroxide and nitrogen mustard can also be induced to shift to the lytic cycle. A lysogenic bacterium is resistant to re-infection by the same or related phages. This is called superinfection immunity. During lysogeny, the prophage confers certain new properties on the lysogenic bacteria. This is known as **lysogenic** or **phage conversion**. The examples of phage conversions are as follows:

1. **Phage-mediated conversion of somatic antigens of *Salmonella:*** This occurs when a large number of temperate phages of *Salmonella* modify antigenic properties of somatic O antigen. For example, *Salmonella anatum* (antigenic formula: 3,10: e, h:1,6), when lysogenised by a temperate phage, shows a new antigenic formula (3,15: e, h:1,6), which is an antigenic formula of *Salmonella newington*.
2. **Toxigenicity of *Corynebacterium diphtheriae:*** The presence of prophage beta (β) is responsible for the production of toxins in *C. diphtheriae*. Removal of the prophage beta makes the bacteria non-toxigenic. Similarly, non-toxigenic *C. diphtheriae* can be made toxigenic by lysogenisation with (β) phage.
3. **Toxicity of *Clostridium botulinum:*** Toxin production in *C. botulinum* types C and D is determined by the presence of phages CE(β) and DE(β), respectively. Elimination of the phages abolishes the toxigenicity of the bacillus.
4. **Transduction:** Bacteriophages may act as carriers of genes from one bacterium to another. This is known as transduction. This phenomenon has been described in many bacteria and is recognised as one of the main mechanisms of transferring genetic material among bacteria in nature. This has been discussed in detail in Chap. 7. Table 51.1 summarises the comparison of multiplication of bacteriophages and viruses.

Uses of Bacteriophages

Bacteriophages have many uses. They are a convenient model for the study of virus–host interaction. They play an important role in the transmission of genetic information by transduction, and they are used as cloning vectors for genetic manipulations. They also are used for phage typing.

Table 51.1 Distinguishing features of multiplication of bacteriophages and viruses

Stage of multiplication	Bacteriophages	Animal viruses
Attachment	Tail fibres attach to cell wall proteins	Attachment sites are plasma membrane proteins and glycoproteins
Penetration	Viral DNA is injected into the host cell	Capsid enters by endocytosis or fusion
Uncoating	Not required	Enzymatic removal of capsid proteins
Biosynthesis	In the cytoplasm	In the nucleus (DNA viruses) or cytoplasm (RNA viruses)
Genome integration	Lysogeny	Latency; slow viral infections; cancer
Release	Host cell lysed	Enveloped viruses bud out; non-enveloped viruses rupture plasma membrane

Bacteriophage Typing

Different strains of a serologically or otherwise identical species of bacteria are susceptible to one or more different bacteriophages. When a suspension of phages is deposited on the lawn culture of a susceptible bacterium, an area of clearing occurs after incubation due to the lysis of the susceptible bacteria by the phages. These zones of lysis are called *plaques*. The shape, size and nature of plaques are characteristic of different phages. Since a single phage particle can produce one plaque, a plaque assay can be used for titrating the number of viable phages in preparation. Many bacterial species can be divided into various phage types based on this phenomenon. Phage typing has been used in the epidemiological study of infections or outbreaks caused by *Staphylococcus aureus*, *Salmonella* spp., *Vibrio cholerae* and many other bacteria.

Different phages are available, which show differences in their specificity for genus, species or strains. Examples are (1) genus-specific bacteriophages for *Salmonella*, (2) specific bacteriophages for all members or strains of *Bacillus anthracis* and (3) for all members of *V. cholerae* biotype Classical (e.g. Mukherjee phage IV). Mukherjee phage IV lyses all strains of *V. cholerae* biotype Classical, but not *V. cholerae* biotype Eltor.

Phage typing of *S. aureus* is a pattern method in which a set of standard phages is employed for intraspecies typing of staphylococci. *Several phages* may lyse a strain of Staphylococcus. Hence, the phage type of a strain is designated by the number of the different phages that lyse it. Phage typing of *Salmonella typhi* is carried out by using prophage, which is active against only fresh isolates of *S. typhi* possessing the Vi antigen.

Phage Therapy

Phage therapy or the practice of using bacteria-specific viruses to treat bacterial infections has been in vogue for many years. With the emerging burden of antibiotic-resistant microorganisms and lack of therapeutic options, phage therapy has recently gained importance as an alternative treatment option. These bacteria eating viruses are quite specific for the species and strain of the bacteria. The phages are dependent on the host bacteria for nutrition and survival and, in the process, often cause the death of the host cell. The phage lysins can destroy bacterial cells. Depending on their development cycles, phages are either virulent (lytic life cycle) or temperate (lysogenic life cycle). In therapeutic use, virulent phages are more promising agents.

Therapeutic uses of phages have diverse differences. While few European countries have continuously used them to treat infections in humans with multidrug-resistant bacteria and with reports of therapeutic failures with antibiotics, other countries among the developed regions have limited the therapeutic use to the veterinary and agriculture sectors. Due to the lack of robust data on several issues like safety, specificity, biofilm penetration ability, antibacterial spectrum etc., phage therapy has been limited to certain centres around the globe with focal reports of their effectiveness. Nevertheless, with increased use of this mode of therapy and better-bioengineered phage products, it might be possible to use phage therapy on a large scale in the near future.

Poxviruses

52

Poxviruses are the largest and most complex viruses that occur in humans, birds, animals and insects. These viruses, which belong to the family Poxviridae, include a large group of DNA viruses that are morphologically similar and share a common nucleocapsid protein.

Poxviruses

Poxviruses (family *Poxviridae*) are a group of morphologically complex, ovoid or brick-shaped viruses containing a large (130–260 kb) linear double-stranded DNA genome. The genomic constituency places this family in Group I of Baltimore Classifications of Viruses. To date, this family is represented by 83 species that have global footprints and can infect a wide range of hosts, including humans, animals, birds, reptiles and insects.

This group is best known for the variola virus that caused smallpox, one of the deadliest diseases of human history. However, it was completely eradicated in 1980, marking landmark success in international public health. Another member of this family, the Vaccinia virus, is the source of the smallpox vaccine and is widely used as a DNA vector in genetic engineering research and development.

Classification and Host Range

International Committee on Taxonomy of Viruses (ICTV), authorised for virus taxonomic classification and nomenclature, has placed the viruses in 22 genera distributed into two subfamilies. The family Poxviridae, based on whether they infect insect or vertebrate hosts, has been classified into two subfamilies: Chordopoxvirinae and Entomopoxvirinae. The subfamily Chordopoxvirinae contains viruses that infect vertebrate hosts and include eight genera. Of these, at least four genera cause diseases in humans resulting in the development of characteristics lesions, skin nodules or disseminated rash.

The genus *Orthopoxvirus* includes the poxviruses of mammals, such as smallpox (variola), vaccinia, monkeypox, cowpox, buffalopox, rabbitpox, mousepox and Camelpox viruses. *Parapoxviruses* include viruses of ungulates that may cause occasional infections in humans. These viruses are Orf viruses, pseudocowpox virus, deerpox viruses and bovine papular stomatitis virus. *Yatapoxviruses* include tanapox viruses and yabapox viruses that are found mainly in Africa. *Molluscipoxviruses* include the Molluscum contagiosum virus. The poxviruses causing infections in humans are listed in Table 52.1.

Table 52.1 Human infections caused by poxviruses

Poxviruses	Human infections
Variola	Smallpox
Vaccinia	Used for smallpox vaccination In patients with eczema causes *Eczema vaccinatum* and in patients with immunodeficiency causes progressive vaccinia
Monkeypox	A disease similar to smallpox
Cowpox	Lesions on hand
Pseudocowpox	A localised disease, such as milker's node
Orf	Localised lesion, such as a single chronic granulomatous lesion with a central ulcer
Tanapox	Vesicular skin lesions
Yabapox	Rare
Molluscum contagiosum	A self-limiting condition with localised lesions of the skin causes chronic and extensive skin lesions in patients with HIV
Tanapox	Localised lesions, such as a single, pock-like vesicular lesion on the skin
Yabapox	Localised lesions

Except for smallpox and Molluscum contagiosum viruses that have a human reservoir, animals are the natural reservoir for other poxviruses. As such, poxvirus infection in humans is mostly zoonotic due to the transmission of the virus from animals to humans. Animal-to-human transmission occurs through bite or scratch, direct contact with body fluids or lesion material, or indirect contact with lesion material. However, persons acquiring one zoonotic poxvirus may not transmit to others. For example, person-to-person transmission of cowpox has not been observed, whereas the monkeypox virus can transmit across humans.

Smallpox (Variola) Virus

Smallpox caused by the variola virus was eradicated from the world in 1980 following an intensive worldwide campaign by the World Health Organisation (WHO). However, the possible use of the smallpox virus as an agent of bioterrorism is the main concern now.

Morphology

The variola virus is large and brick-shaped. It measures 300 × 200 × 200 nm, hence is visible by light microscopy. The virus is enveloped. The extracellular virion possesses two envelopes. The outer envelope encloses the extracellular virion consisting of host cell lipids and several virus-specific proteins, including haemagglutinins. The intracellular virion has only one envelope. The viral genome consists of a large, double-stranded DNA that measures 130–375 kbp in size.

Physical and Chemical Properties

Variola virus is most stable at low temperatures and low humidity. It remains viable for months at room temperature if protected from sunlight. It is also stable in the cold or when freeze-dried for years. It is resistant to the action of 50% glycerol and 1% phenol. Even though enveloped, the virus is not susceptible to ether; hence it is not inactivated by ether.

The virus could be inactivated by moist heating at 60 °C for 10 min; however, the viral antigenic proteins can withstand 100 °C. It is susceptible to ultraviolet light and other irradiations and readily inactivated by formalin and oxidising disinfectants.

Antigenic Properties

Smallpox virus has a single serotype responsible for successful vaccination against the disease.

All poxviruses, including vaccinia viruses, have a common nuclear protein antigen (NP). These poxviruses contain at least 20 different antigens, such as LS antigen, agglutinogen and haemagglutinin.

Laboratory Isolation

The 'Regulations and Guidelines for Recombinant DNA Research and Biocontainment' issued by the Department of Biotechnology in 2018 that mandates research and handling of infectious microorganisms in India have classified the Poxvirus under Risk Group 2. As such, isolation and handling of the viruses for research and clinical purposes requires Biosafety Level 2 laboratory facility and related practices to protect workers and the environment.

Chick Embryo

Inoculating vesicular fluid into the chorioallantoic membrane (CAM) of the chick embryo is reliable. Both variola and vaccinia viruses grow in the chick embryos. They produce pocks on CAM of 11- to 13-day-old chick embryos within 48–72 h. Variola virus produces small, shiny, white, convex, non-necrotic and non-haemorrhagic pocks. Vaccinia viruses produce larger, irregular, greyish, necrotic, pock lesions, some of which may be haemorrhagic. The monkeypox and cowpox produce well-marked haemorrhagic lesions, whereas Tanapoxvirus, Molluscum contagiosum, and the parapoxvirus do not show any growth on the CAM.

Tissue Culture

Human and non-human primate cells, such as monkey kidneys and HeLa cells, are used to isolate viruses from clinical specimens. Variola virus can be grown in chick embryo cells, monkey kidneys and HeLa cells. Unlike the vaccinia virus, which rapidly produces cytopathic effects (CPEs) within 24–48 h, the variola virus produces CPEs very slowly. The CPEs consist of Guarnieri bodies, which are eosinophilic inclusion bodies that consist of an aggregation of virus particles in a matrix. These can be demonstrated in stained preparations.

Tanapoxvirus and parapoxviruses grow poorly, whereas Molluscum contagiosum does not grow in cell cultures.

Laboratory Animals

Variola virus causes experimental infection only in monkeys. Intranasal infection of monkeys by variola virus causes smallpox in monkeys with generalised skin lesions. Scarification of the cornea of rabbits with variola virus causes keratitis with the demonstration of characteristic Guarnieri bodies in the stained smears of the cornea.

Pathogenesis of Smallpox

Transmission

Smallpox was a highly infectious disease. Respiratory secretions and exudates of the skin lesions were the most common sources of infections. The disease was most commonly transmitted by inhaling respiratory droplets from smallpox patients.

Compared to smallpox that has spread through principally respiratory routes, whereas vaccinia, monkeypox can also be transmitted through contacts, i.e. direct physical contact with infectious skin lesions, fluids and indirect contact with lesion material, such as contaminated clothing or linens.

Virus Replication

Poxvirus replication is complex. The virus replication starts within 2 h after transmitting and entering the host through the respiratory route. Among the DNA viruses, they are unique in that the complete replication cycle of the virus occurs in the cytoplasm of the infected host cell. The virus encodes the enzymes required for mRNA and DNA synthesis essential for genetic replication. The virus also encodes proteins for other functions, such as DNA synthesis and immune escape mechanisms. Finally, the virions are assembled and acquire their envelopes by budding from the cell membrane as they are released from the cell.

The virus replicates at the inoculation site and spreads to lymph nodes draining the mucosal entry site. The virus replicates in the lymphoid tissues causes transient viremia and infection of the reticuloendothelial cells. This is followed by a secondary phase of multiplication in these cells, leading to a secondary viremia. This results in clinical manifestations of the disease, which manifest as fever and other toxic manifestations. The viruses then enter the skin, localise in the skin's blood vessels, and produce the characteristic rash (*pocks*) of smallpox. The pocks are due to the replication of the virus in the skin followed by the damage caused by cytotoxic T cells, damaging the virus-infected cells.

Clinical Syndromes

Variola virus causes smallpox, a disease that is now extinct.

Smallpox

The incubation period varied from 10 to 14 days. The prodromal phase, which correlated with the viremia phase, was the first to appear. The onset of the disease was sudden. The condition manifested as sudden onset of fever, severe headache, nausea, pharyngitis, body malaise and backache. An exanthematous rash would appear on the palate, tongue and pharynx during the latter part of the prodromal stage.

The smallpox rash was characterised by skin lesions that are in the same stage of evolution, unlike those lesions seen in chickenpox. The lesions in chickenpox appear in successive waves and various stages, such as vesicles, pustules and scabs. The skin lesions first appear on the face and extremities and then spread centrifugally to the trunk. These lesions begin as macules and then develop into papules, vesicles, pustules and finally crusts during a period of approximately 17 days. The lesions heal with the formation of characteristic scarring.

Overwhelming toxaemia was the usual cause of death in patients with smallpox. There were two variants of smallpox: variola major and variola minor. The variola major was associated with a 25–30% fatality rate, whereas the variola minor was associated with a low fatality rate of less than 1%. In addition, flat smallpox and haemorrhagic variola were the unusual manifestations of smallpox in some patients and were usually fatal.

Humoral antibodies are not protective. Cell-mediated immunity plays an important role in controlling and resolving the disease. Virus-specific T cells control the spread of viruses by causing lysis of infected cells in the reticuloendothelial cells in the skin. An attack of smallpox gives complete protection against re-infection. Vaccination confers immunity, which lasts about 10 years.

Laboratory Diagnosis

Suspected cases of smallpox were usually diagnosed clinically. The main criteria for clinical diagnosis of smallpox included (1) febrile prodromal phase occurring 1–4 days before the onset of rashes; (2) characteristic smallpox lesions of the skin (deep, firm, round), which could be umbilicated or confluent and (3) skin lesions in some stage of development on any part of the body.

Specimens

Skin lesions, such as vesicular fluid, were the specimens of choice.

Microscopy

The Guarnieri bodies, the characteristic eosinophilic inclusion bodies, can be demonstrated in the stained preparation of the clinical specimens by light microscopy. The electron microscope, if available, was used for direct demonstration of typical virus particles in clinical specimens. It was useful to differentiate a smallpox virus from that of chickenpox. Negative-contrast electron microscopy result of the virus shows a surface membrane that encloses a biconcave or cylindrical core. One or two lateral bodies appear to be present between the core wall and a membrane.

Virus Isolation

Isolation of the virus either in CAM of embryonated egg, or tissue culture, as described earlier, is necessary for research, and rapid and accurate identification of poxvirus infections.

Serology

Serological tests were used to confirm the diagnosis of poxvirus infection. Indirect immunofluorescent antibody tests and HI, CF and NT tests are available to demonstrate serum antibodies that appear after the first week of infection.

Molecular Detection

PCR-based detection of the poxvirus is the confirmatory way to understand the species involved in infection. A PCR test can identify the infection among a range of suspected lesions and even before the onset of symptoms. Primers targeting poxvirus housekeeping genes like 18S rRNA, globin and DNA polymerase have been widely used. Real-time polymerase chain reaction (PCR) is more sensitive and can detect <10 genome equivalents per PCR.

Treatment

No specific antiviral agents are available against the variola virus. Methisazone is of some value against some poxviruses. It is recommended only for chemoprophylaxis but not for treatment. Vaccinia immune globulin is recommended to treat all complications except post-vaccinated encephalitis. Although smallpox has been eradicated worldwide, concern still exists about the possible occurrence of smallpox through bioterrorism.

Epidemiology

Smallpox was a highly infectious disease caused by the variola virus. It was one of the deadliest diseases that claimed more than 300 million lives before it was completely eradicated. The disease had about a 30% mortality rate. India, which accounted for 60% of global smallpox infection, witnessed a viral epidemic in 1974. For survivors, permanent skin scarring was common. Other deformities like loss of lip, nose and ear tissue and corneal scarring leading to blindness were also reported. Currently, the world is free of smallpox.

Geographical Distribution

In the 1960s, the WHO formulated a concept programme for the global eradication of smallpox. The disease was then present in 44 countries with a global incidence of around 10 million cases annually. Using a highly potent and stable vaccine, rapid identification of outbreaks and carrying out ring vaccination in all contacts of an infected person, the disease was finally eradicated in 1977 when the last case of endemic smallpox was detected in Somalia.

The last case of naturally occurring smallpox was detected in Somalia in 1977. The last recorded case in humans, which was due to an accidental laboratory infection, was reported in England in 1978. The WHO in 1980 declared that smallpox was eradicated from the world. The only remaining known virus isolates are stocked in the Center for Disease Control and Prevention (CDC) laboratories in the United States and the Vector Institute in Russia.

Reservoir, Source and Transmission of Infection

Smallpox was a highly infectious disease. Respiratory secretions and exudates of the skin lesions were the most common sources of infection. The disease was most commonly transmitted by inhaling respiratory droplets from smallpox patients. The patients were highly infectious during the first week of rash, once fever had started.

The features of smallpox that contributed to its total eradication include the followings:

1. Smallpox was the first disease to be eradicated by a vaccination programme due to the availability of an effective vaccine.
2. A safe, stable, economical and easy to administer smallpox vaccine was available.
3. Presence of a scar indicated the successful vaccination against smallpox.
4. There was only a single serotype of the variola virus; hence immunisation by vaccines protected against all cases of smallpox.

5. Human and other animal pox viruses share common antigenic determinants. Safe vaccines prepared from animal poxviruses could also be used for vaccination against humans.

Prevention and Control

Almost after two centuries of Jenner's discovery of vaccination, the 33rd World Health Assembly declared complete eradication of smallpox on eighth May 1980. In international public health, it is considered one of the biggest achievements of human history. However, its origin is still unknown. The first documented evidence dates back to the fourth and seventh centuries in China and India, respectively. The last case of naturally occurring smallpox was detected in Somalia in 1977. The last recorded case in humans, which was due to an accidental laboratory infection, was reported in England in 1978.

Vaccines

Immunisation against the poxvirus is the only way to protect individuals from getting poxvirus infection. Vaccination confers immunity, which lasts about 10 years. However, before the vaccination was introduced, empirical preventive measures against smallpox through variolation had been used in India and China in the fifteenth century and later became very popular in Europe and UK. In variolation, healthy individuals were deliberately exposed to material from smallpox sores (pustules) by scratching the material into their arm or inhaling it through the nose. Compared to naturally getting the virus, the violation process resulted in fewer deaths and the same symptoms. The discovery of vaccines started from the pioneering work of Edward Jenner on the control of smallpox infections by the use of the cowpox virus as the vaccine. Later in the 1800s, the virus used to make the smallpox vaccine changed from cowpox to vaccinia virus.

The smallpox vaccine is a live preparation of the vaccinia virus propagated on the skin of cows. The vaccine is applied by scarification, which causes a local pustule that subsequently heals with the formation of scabs. Subsequently, the virus is grown on the skin of calves to reduce the risk of transmitting syphilis and other diseases during vaccination. Vaccinia virus used as a vaccine replicating at the inoculation site produces local erythematous maculopapular. These lesions subsequently vesiculate and finally heal by producing a scab over 10–14 days. Finally, resolution of the lesions occurs by the formation of pustules followed by scaling and healing. The resolution of the disease is associated with the development of immunity to variola infection that persists for up to 10 years.

Even after smallpox has been completely eradicated, the vaccinia vaccine still finds its use. The vaccine used during the smallpox eradication program also offers protection against monkeypox. A new third-generation vaccinia vaccine is now available to prevent smallpox and monkeypox.

Future Risks

Even though the world is now free from the smallpox virus, it is considered a serious threat to biological terrorism and so is designated as a Category A agent by the Center for Disease Control and Prevention (CDC), Atlanta, USA. In the past, during the eighteenth century, the virus was used as a tool of bioterrorism in a few instances. Recent research suggests that recreating the smallpox virus from its genomic information available in the database enhances risks for bioterrorism.

Other Poxviruses

Vaccinia Virus

Vaccinia virus is unique because it is an artificial virus and does not occur in nature. Vaccinia virus was used as the smallpox vaccine. Vaccinia virus is the source of the modern smallpox vaccine used to eradicate smallpox. However, the virus's origin has remained quite mysterious in the absence of records. However, certain researchers have made it clear that the cowpox virus that Jenner initially used for the smallpox vaccine is related but not similar to the vaccinia virus. Recent findings suggest a common ancestral origin of smallpox, cowpox and vaccinia, whereas the ancient vaccinia share high genetic similarities (>99%) with the horsebox virus.

Irrespective of the debate on the origin of this virus, the virus has found tremendous applications in the field of vaccination and genetic engineering and so mostly maintained across laboratories for research and development purposes. Due to its relatively large genome size natural gene delivery methods, the vaccina virus has been widely used as a DNA vector in the genetic engineering field. The environmental presence of the virus in wild-type form, although not common, yet was also reported in very limited cases of infection. For example, the USA reported cases of vaccinia infection in 2007 and 2010.

This vaccinia virus could be readily differentiated from fresh isolates of cowpox and smallpox viruses. Vaccinia virus is now being extensively studied in greater detail than variola, as it is safer to work with (Fig. 52.1). The virus is also being used as a vector for developing recombinant vaccines.

Both vaccinia and variola viruses show similarities in their morphology. They, however, can be differentiated by their growth properties and host range. Vaccinia virus can be cultured *in vitro* on CAM. The virus is identified by its characteristic pock lesion on the CAM. The vaccinia virus produces relatively larger, irregular, flat and necrotic lesions, some of which may be haemorrhagic (Table 52.2). Vaccinia virus has recently been shown to be closely related to the New World orthopoxviruses, Cantagalo and Aracatuba viruses.

Vaccinia virus, which is used for vaccination, can also cause infection in humans. Vaccinia causes infection by accidental inoculation of the

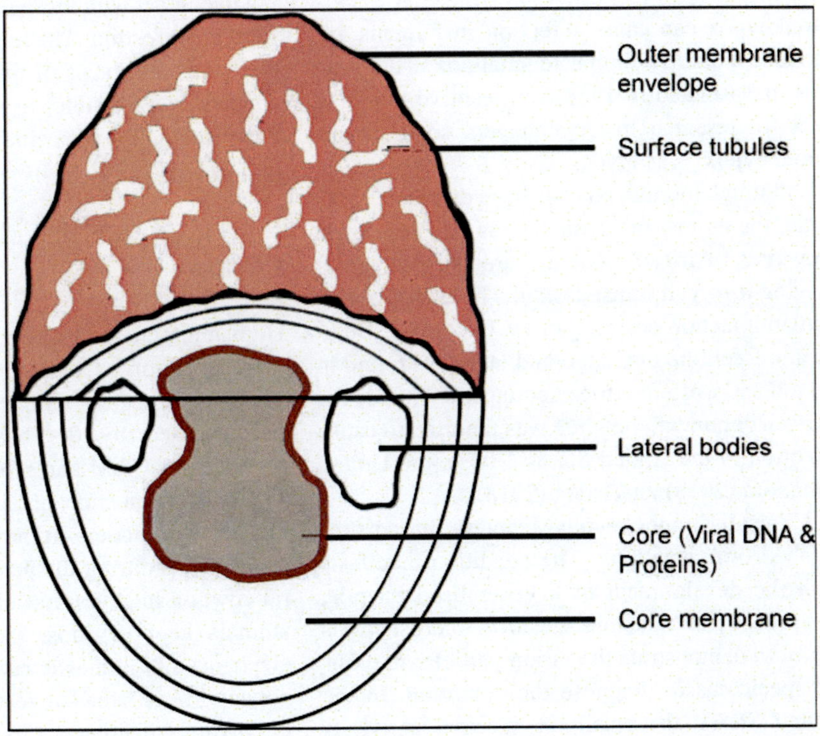

Fig. 52.1 Schematic diagram of structure of a vaccinia virus

Table 52.2 Differentiating characteristics between variola virus and vaccinia virus

Characteristics	Variola virus	Vaccinia virus
Isolation	Humans	Origin unknown
Pocks on CAM	Small, white	Large, greyish, and haemorrhagic
Ceiling temperature on CAM	37.5–38.5 °C	41 °C
Growth on rabbit skin	–	+ or + +
Thymidine kinase sensitivity	Positive	Negative
Pathogenicity for baby mice	Low	High
Polypeptide pattern	Character of variola	Character of vaccinia
Host range	Narrow host range (only humans and monkeys)	Broad host range (includes rabbits and mice)

skin. At the inoculation site, the lesion begins as a maculopapular rash, progressing to vesicle pustules and finally leading to the formation of a scab. The lesion heals with the formation of marked scarring. Vaccinia virus in patients with eczema causes *e. vaccinata*, and in patients with immunodeficiency causes progressive vaccinia.

Monkeypox

Monkeypox can cause infection in humans and produce a disease similar to smallpox. This virus was first isolated in 1958 from captive monkeys kept for research in Copenhagen, so the name 'monkeypox' was given.

Although monkeypox has been defined as zoonotic, the animal reservoir host is still unknown. However, African rodents are suspected of playing a part in transmission. The first reported human infection occurred in 1970 from the Democratic Republic of Congo and later was found in central and western African countries. Outside the African continent, the virus was reported to cause infection in the United States (2003), the United Kingdom (2018) and Israel (2018).

Clinically, monkeypox cannot be distinguished from smallpox. The condition manifests with the development of a generalised pustule, rash, fever and toxaemia. Electron microscopy is useful to demonstrate the virion particles in clinical specimens to diagnose the condition. In the United States, the smallpox vaccine, antivirals and vaccinia immune globulin (VIG) are suggested to control monkeypox infection.

Buffalopox

Buffalopox, believed to be caused by the vaccinia virus, occurs among buffaloes in India. In the infected buffaloes, the viruses produce pustular lesions on the teats and udders of lactating buffaloes. The individuals, such as milkmen coming in contact with these infected buffaloes, suffer from the infection. The lesions usually develop on the hands and faces of the milkmen. Buffalopox appears to be distinct from variola and vaccinia viruses. Smallpox vaccine does not confer protection against infection by buffalopox.

Cowpox

Human infection with cowpox is usually an occupational hazard. Cowpox resembles variola and vaccinia viruses antigenically but can be differentiated by the haemorrhagic lesions it produces on CAM and rabbit skin.

Cowpox infection in cows produces ulcers on the teats and udders. It may spread to other cows and humans during the process of milking. Natural cowpox infection has also been noted in wild animals kept in zoos, including cheetahs and elephants and domestic cats. Rodents or cats but not cows have been suggested as the primary host of cowpox infection.

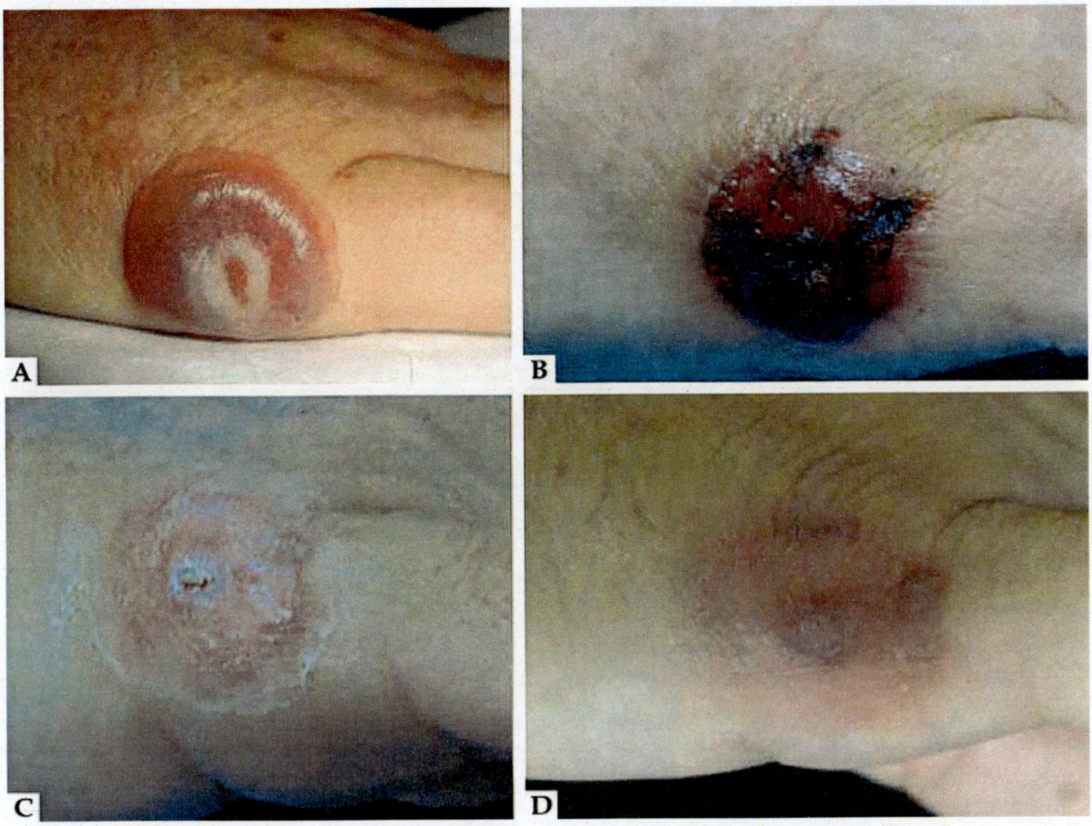

Fig. 52.2 Umbilicated nodular lesions over the finger and wrist—Milker nodules. (https://commons.wikimedia.org/wiki/File:0365-0596-abd-90-03-0407-gf02.jpg)

Milker node or paravaccinia is an occupational disease acquired by humans during milking infected cows. The common lesions include small ulcerating nodules (Fig. 52.2). The condition presents as painful and haemorrhagic lesions on the hands and face associated with pyrexia and other constitutional disturbances. The virus is distinct from cowpox and grows in bovine kidney cell culture but not in embryonated eggs.

Orf

Orf is an infection of humans caused by a virus of contagious pustular dermatitis of sheep and goats. The disease occurs as a single chronic granulomatous lesion with a central ulcer in humans. The lesions are usually present on the hands, forearm or occasionally on the face. The lesions heal without forming any scars. The virus resembles the paravaccinia virus morphologically.

Molluscum Contagiosum

Molluscum contagiosum is a poxvirus unique to humans. After smallpox was eradicated, this is the only poxvirus having a human reservoir and not zoonotic. The virus causes a disease of the skin, often spread through sexual contact among young adults. It causes small, pink, papular pearl-like benign tumours of the skin or mucous membranes (Fig. 52.3). These lesions are present on the epidermis and occur in most body parts except on soles and palms. Lesions are more commonly found in the trunk and anogenital areas. This is a self-limiting condition, and the condition generally resolves over some time.

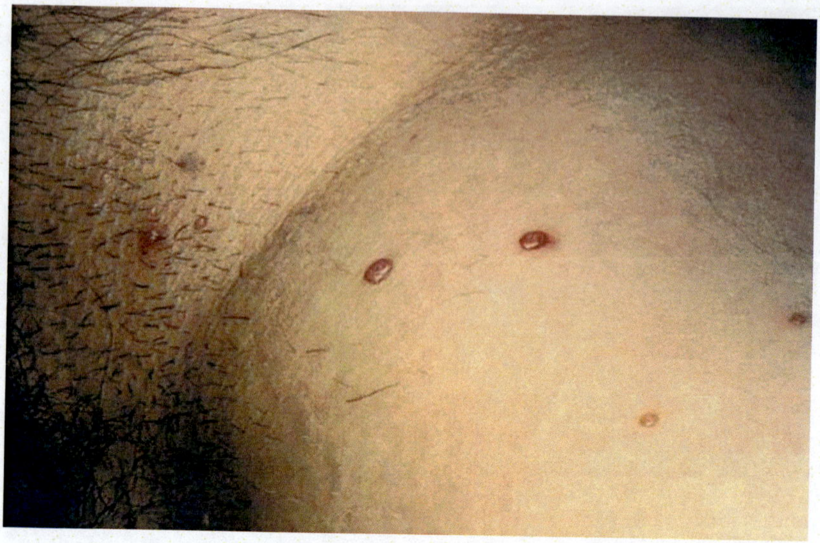

Fig. 52.3 Umbilicated papules over the face of a female child—molluscum contagiosum. (https://phil.cdc.gov/Details.aspx?pid=16687)

Molluscum contagiosum in patients with HIV may cause chronic and extensive skin lesions. Sections of the nodular lesions show hyaline acidophilic inclusion bodies called **molluscum bodies**. These inclusion bodies can be seen in the cells of the stratum corneum and the stratum granulosum. Their bodies are large and measure 20–30 μm in size and are composed of large numbers of virion particles embedded in a protein matrix (Fig. 52.4).

Tanapox

Tanapox virus was first isolated from the cases of febrile illness that occurred along the Tana River in Kenya during the 1950s. The virus produces a single, pock-like vesicular lesion on the skin, which usually does not progress to form pustules. This lesion is most commonly seen on the upper part of the body. The virus is acquired

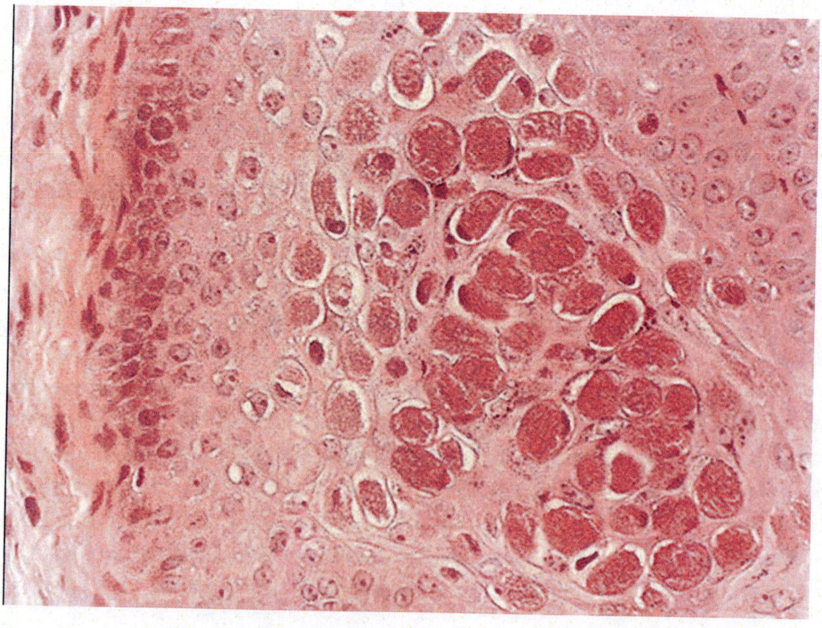

Fig. 52.4 Molluscum contagiosum inclusion bodies (×1000). (https://commons.wikimedia.org/wiki/File:Molluscum_contagiosum_1.jpg)

by insect bites from monkeys and other wild animal reservoirs. It is antigenically not related to other poxviruses. The virus grows in human and monkey tissue cultures but not in embryonated eggs.

Yabapox

Yabapox virus produces large benign tumours in monkeys. It causes benign histiocytomas 5–20 days after muscular or subcutaneous inoculation in monkeys. Such lesions have been reported in persons handling affected monkeys.

At present, the treatment of poxvirus infection in humans is supportive in the absence of any licensed drugs. Cidofovir has been approved for use in the emergency treatment of orthopoxvirus infections in humans and complications post-vaccination. Poxvirus infections may result in secondary infections where the use of antibiotic ointments is suggested. Curettage and cryotherapy of large lesions are often considered for infection due to Molluscum contagiosum.

Case Study

All doctors and nurses want to be vaccinated with the smallpox vaccine in a hospital because of a sudden threat of bioterrorism this year. Since there is a limited supply of smallpox vaccine, the hospital authorities formulate a guideline to vaccinate only those medical and paramedical staff who were not vaccinated earlier.

- Which year was the vaccination by smallpox vaccine stopped in India?
- Which age population in India will be the most susceptible group to smallpox in case of a potential threat of bioterrorism?
- After vaccination, at what point the staff receiving the smallpox vaccine will be considered to be fully protected from smallpox?
- How will you diagnose smallpox that occurs in susceptible persons in the laboratory?

Herpesviruses

53

The herpesviruses are large, enveloped DNA viruses, measuring 120–200 nm in diameter. They exhibit many common features, such as similar morphology of virions, the basic mode of replication in the host cells and the capability to establish latent and recurrent infections. All herpesviruses are structurally similar.

Herpesviruses

Herpesviruses are a group of viruses that are placed in the family in Group I of Baltimore Classifications of Viruses. They exhibit many common features, such as similar morphology of virions, the basic mode of replication in the host cells and the capability to establish latent and recurrent infections.

Some members of the family can infect humans and are prevalent worldwide. Common infections include oral and genital herpes and chickenpox. Historically the infection was traced back to ancient Greek times where cutaneous spreading of lesions was defined with the Greek word "herpes" to mean "to creep or crawl".

All herpesviruses are structurally similar. The virus has a characteristic morphology and consists of four distinct structural elements: DNA core, capsid, tegument and envelope. **DNA core** consists of a linear double-stranded DNA (dsDNA) molecule surrounded by a protein coat. DNA genome size ranges from 125 to 240 kbp with 32–75% G + C content depending on the virus species. **Capsid** surrounds the DNA core. It consists of an icosahedral capsid containing 162 units known as capsomeres. **The tegument** is an amorphous material present between the capsid and envelope. It contains viral proteins and enzymes that initiate the replication of the virus. **The envelope** surrounds the nucleocapsid. The glycoprotein-containing envelope is derived from the modified host cell nuclear membrane. It carries surface spikes.

Herpesviruses replicate in the host cell nucleus, and both replication and assembly occur in the nucleus. The virus buds from the nuclear membrane are released by exocytosis and cell lysis. This family of viruses are known for the epitome of host specificity, co-evolution with the host and the immune evasion mechanism that permit them to remain latent for lifelong and recur under stress.

As such, detailed and extensive research has been done on these viruses in basic and applied fields. However, handling these viruses during research poses varying degrees of health risks to the worker, so members were placed in different risk groups that enable the adoption of laboratory containment features for safe handling. The "Regulations and Guidelines for Recombinant DNA Research and Biocontainment" issued by the Department of Biotechnology, Govt of India in 2018 have classified the human herpesviruses 1, 3, 6 and 7 in Risk Group 2, whereas human herpesvirus 2, Herpesvirus simiae (Herpes B;

Table 53.1 Classification of human herpesviruses

Subfamily	Scientific name	Common name	Primary target cell	Site of latency
Alphaherpesvirinae	Human herpesvirus 1	Herpes simplex virus type 1	Mucoepithelial	Neurons
	Human herpesvirus 2	Herpes simplex virus type 2	Mucoepithelial	Neurons
	Human herpesvirus 3	Varicella-zoster virus	Mucoepithelial	Neurons
Gammaherpesvirinae	Human herpesvirus 4	Epstein–Barr virus	B cell and epithelial cells	Lymphoid tissues
	Human herpesvirus 8	Kaposi sarcoma-related virus		–
Betaherpesvirinae	Human herpesvirus 5	Cytomegalovirus	Monocytes, lymphocytes and epithelial cells	Monocytes and lymphocytes in secretory glands
	Human herpesvirus 6	Human B-cell lymphotrophic virus	T cells	Lymphoid tissue
	Human herpesvirus 7	RK virus	T cells	Lymphoid tissue

Cercopithecine herpesvirus 1) is placed under Risk Group 3.

Classification and Host Range

International Committee on Taxonomy of Viruses (ICTV) has recently updated the classification of the family Herpesviridae by introducing new genera and species in the family. At present, the family is classified into three subfamilies, 17 genera and 115 species.

Herpesviruses are highly host specific and evolve synchronously with their hosts. Species of the family are known to infect many hosts, including humans, animals, birds, fish, reptiles, amphibians and molluscs. However, out of the 115 species known to date, only eight species have human hosts, i.e. capable of infecting humans and are epidemiologically important. They are officially designated as human herpesvirus types 1–8, but the common names are still used (Table 53.1). The diseases caused are mentioned in Table 53.2.

Despite strong host specificity, species of the family share many properties that potentially make them capable of crossing the species barriers, although not natural. In very rare cases, species zoonotic transmission (animal to human) and vice versa of the viruses have been noted. *Cercopithecine herpesvirus* 1 (CeHV-1, Herpes B Virus, B virus) is considered the most significant zoonotic herpesvirus that can jump from macaques to humans. In contrast, Human herpesvirus 1 is capable of causing severe disease in primates.

Herpes Simplex Virus

Herpes simplex viruses (HSVs) are extremely host-adapted viruses. They can cause a wide variety of illnesses in infected human hosts. There are two types of HSVs: herpes simplex virus type 1 (HSV-1) and herpes simplex virus type 2 (HSV-2). Both types are closely related in their DNA homology, antigenic determinants, tissue tropism and disease symptoms but differ in epidemiology. HSV-1 is transmitted primarily by contact with infected saliva, whereas HSV-2 is transmitted by sexual contact or genital tract infection to the newborn from an infected mother.

Table 53.2 Human infections caused by human herpesviruses

Human herpesviruses	Diseases
Herpes simplex virus (HSV) HSV-1	Acute herpetic gingivostomatitis, acute herpetic pharyngotonsillitis, herpes labialis, herpes encephalitis, eczema herpeticum and herpetic whitlow
HSV-2	Genital herpes, neonatal infection and aseptic meningitis
Varicella-zoster virus (VZV)	Chickenpox and herpes zoster
Epstein–Barr virus (EBV)	Infectious mononucleosis, EBV-induced tumours, such as Burkitt lymphoma, other B-cell lymphoma and nasopharyngeal carcinoma. Duncan syndrome, lymphoproliferative syndrome, oral hairy leukoplakia
Cytomegalovirus (CMV)	Congenital CMV infection, acquired CMV infection, CMV infection in immunocompromised patients and CMV infection in immunocompetent adult hosts
HSV-6	Roseola infantum or exanthem subitem (sixth disease) may also cause mononucleosis syndrome and lymphadenopathy
HSV-7	Roseola infantum and febrile seizures in children
HSV-8	Associated with Kaposi sarcoma, body cavity lymphoma and Castleman disease

Morphology

HSVs are large, enveloped, icosadeltahedral viruses containing a dsDNA genome (Fig. 53.1). The genomes of both HSV-1 and HSV-2 are similar in organisation and show a higher degree of sequence homology. They, however, can be differentiated by restriction enzyme analysis of viral DNA. The unique feature of the DNA genome is that it encodes for as few as 80 polypeptides. Half of the proteins are required for the replication of viruses, whereas other proteins help in the interaction of the viruses with different host cells and immune responses.

Physical and Chemical Properties

Like other enveloped viruses, HSVs are sensitive to treatment with acid, fat solvents, detergents and drying. They are readily inactivated in the conditions prevalent in the gastrointestinal tract.

Laboratory Isolation

Chick Embryo

The virus grows on the chorioallantoic membrane of the embryonated egg and produces small, white, shining, non-necrotic pocks measuring less than 0.5 mm in diameter. Both HSV-1 and HSV-2 show cross-reaction.

Tissue Culture

HSV grows in a variety of primary and continuous cell lines. They grow readily on HeLa cells, Hep-2 cells, human embryonic fibroblasts and rabbit kidney cells. HSV typically produces **Cowdry type A** acidophilic inclusion bodies. These are well-defined foci with heaped-up cells in the infected cell lines. Many strains of HSV also produce syncytial and giant cells in the infected cells. The virus kills the cells rapidly in tissue culture.

A cell line that expresses beta-galactosidase after infection with HSV is a new approach for isolating and identifying the virus in the cell lines. The addition of a chromotropic substrate allows the detection of enzymes in the infected cell.

Pathogenesis of Herpes Simplex Virus Infection

HSV shows three unique biological properties: (1) neurovirulence, (2) latency and (3) reactivation and replication.

The viruses show **neurovirulence,** which can invade and multiply in the nervous system. The establishment and maintenance of latent infection

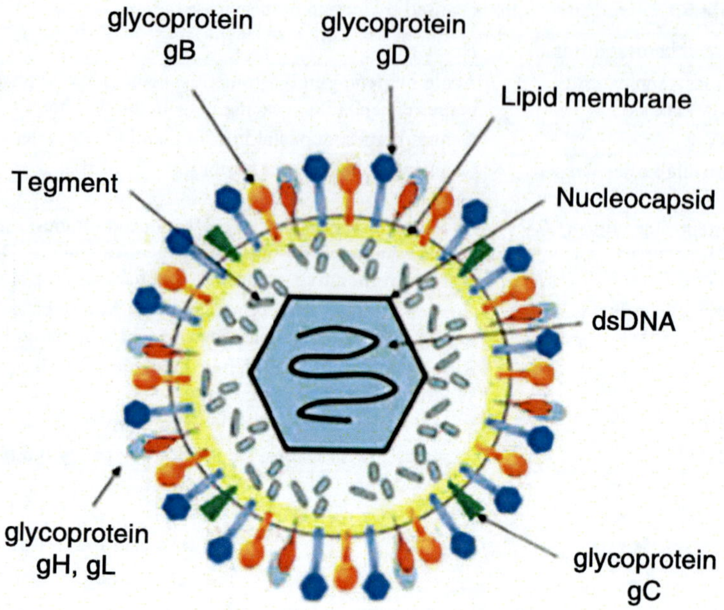

Fig. 53.1 Diagrammatic representation of herpes simplex virus. (https://en.wikipedia.org/wiki/Herpes_simplex_virus)

in the nerve cell ganglion (**latency**) is the unique property of the virus. HSV-1 affects most commonly the trigeminal ganglion, whereas HSV-2 affects the sacral nerve root ganglion (S2–S5). **Reactivation** and **replication** of latent HSV can lead to overt or covert recurrent infections and excretion of HSV. Reactivation can be caused by various stimuli, including fever, trauma, emotional stress and sunlight. In immunocompetent patients, HSV-1 reactivates more frequently in the oral cavity than on the genital parts, but HSV-2 reactivates eight to ten times more frequently in the genital parts compared to orolabial infection. In immunocompromised patients, reactivation by either HSV-1 or HSV-2 is more frequent and more severe. The viruses cause disseminated infection in individuals with HIV-related disease and people with impaired T-cell immunity, such as organ transplant recipients.

HSV Transmission

HSV-1 infection is mainly transmitted orally through saliva and leads to oral herpes. It is usually transmitted by oral contacts, such as kissing or sharing toothbrushes or other saliva-contaminated items. Virus transmission can occur from asymptomatic individuals. However, the risk increases when there are active sores. HSV-1 can also be transmitted to the genital area through oral–genital contact and leads to genital herpes. In rare circumstances, a mother with genital HSV-1 infection can transmit the virus to the infant during delivery leading to neonatal herpes. Patients having oral herpes infection are unlikely to be subsequently infected with HSV-1 in the genital area. Autoinoculation may also cause infection of the eye.

HSV-2 infection is a sexually transmitted disease. The virus is transmitted during sexual activity during contact with genital surfaces, skin, sores or fluids of an infected individual. Virus transmission can occur from asymptomatic individuals. However, the risk increases when there are active sores. The infection can also be transmitted from an infected mother to a child during delivery and by autoinoculation in rare cases. A positive correlation has been found between HSV-2 and HIV infection in terms of virus transmissibility. HSV-2 infection increases the risk of acquiring a new HIV infection by approximately threefold. As such, HSV-2 is the most common infection in HIV-positive patients.

Replication of Virus

HSV infection is initiated by direct contact and depends on the infected tissues, whether oral, genital or brain. The infection occurs by inoculating the virus into susceptible mucosal surfaces, such as the oropharynx, conjunctiva or cervix or through small abrasions on the skin.

The viruses infect cells of ectodermal origin, the cells coming in contact with infected material, such as saliva or genital secretion. Viruses replicate at the site of entry in the skin or mucous membrane. The neuroinvasiveness, neurotoxicity and latency properties of HSV influence the course of infection in an infected host.

The virus multiplies locally with cell-to-cell spread. The virus replicates in the infected cells at the base of the lesion and infects the innervating neuron. HSV grows very rapidly in infected cells, requiring only 8–16 h for completion and usually causes lytic infection of the fibroblasts and epithelial cells. After entry into the cell, the virion is uncoated, the genome is released and the genome DNA enters into the nucleus. The mRNA is transcribed by host cell RNA polymerase and then translated into early non-structural proteins. Subsequently, the viral DNA polymerase replicates the genome DNA, during which the synthesis of early proteins is stopped, but the synthesis of late structural proteins begins. These late proteins are transported to the nucleus, where virion assembly occurs. The virion acquires its envelope by budding through the nuclear membrane.

Subsequently, the viruses travel by retrograde transport to the ganglion, such as the trigeminal ganglion for HSV-1 and the sacral ganglion for HSV-2. The virus then returns to the initial site of infection and may cause inapparent infection or produce vesicular lesions. Thin-walled umbilicated vesicles—the roof of which breaks down, leaving tiny superficial ulcers—are the typical lesions caused by HSV. These vesicles heal without forming any scars. The vesicle fluid contains infectious virions.

After this retrograde axonal flow from neurons, the viral genomes become latent in the ganglia, particularly those of the trigeminal (HSV-1) and sacral (HSV-2) nerves. No viral particles are produced during this phase of latency. Also, latent infection in neurons does not cause any demonstrable damage in neurons. This latency phase may be reactivated periodically in some individuals, causing recurrent oral and genital lesions (Fig. 53.2).

Various stimuli, such as physical or emotional stress, trauma, fever and sunlight, can induce a recurrence in which the virus travels back down the nerve, leaving lesions to develop on the skin at the same spot each time. These stimuli trigger reactivation by (1) facilitating virus replication in the nerve; (2) inhibiting cell-mediated immunity (CMI) transiently or (3) inducing both the mechanisms.

Host Immunity

Cell-mediated immunity plays an important role in conferring immunity to HSV infection and facilitates recovery from the infection. Hence, the virus tends to cause more frequent and severe infections in patients with altered CMI, such as HIV and other CMI-deficient diseases. Tissue damage occurring in HSV infection is due to a combination of viral pathology and immunopathology. These immunopathologic effects of CMI and inflammatory responses are the major causes of symptoms seen during the disease.

Thus, recurrences or reactivation of infection can occur even in the presence of specific antibodies. However, the recurrent infections are more localised, less severe and of short duration than the primary infection due to the presence of past memory immune responses.

Clinical Syndromes

HSV causes a wide variety of clinical manifestations. The clinical manifestations depend on (1) age and immune status of the host, (2) antigenic type of the virus and (3) anatomical site of the infection. Generally, HSV-1 produces lesions above the waist, and

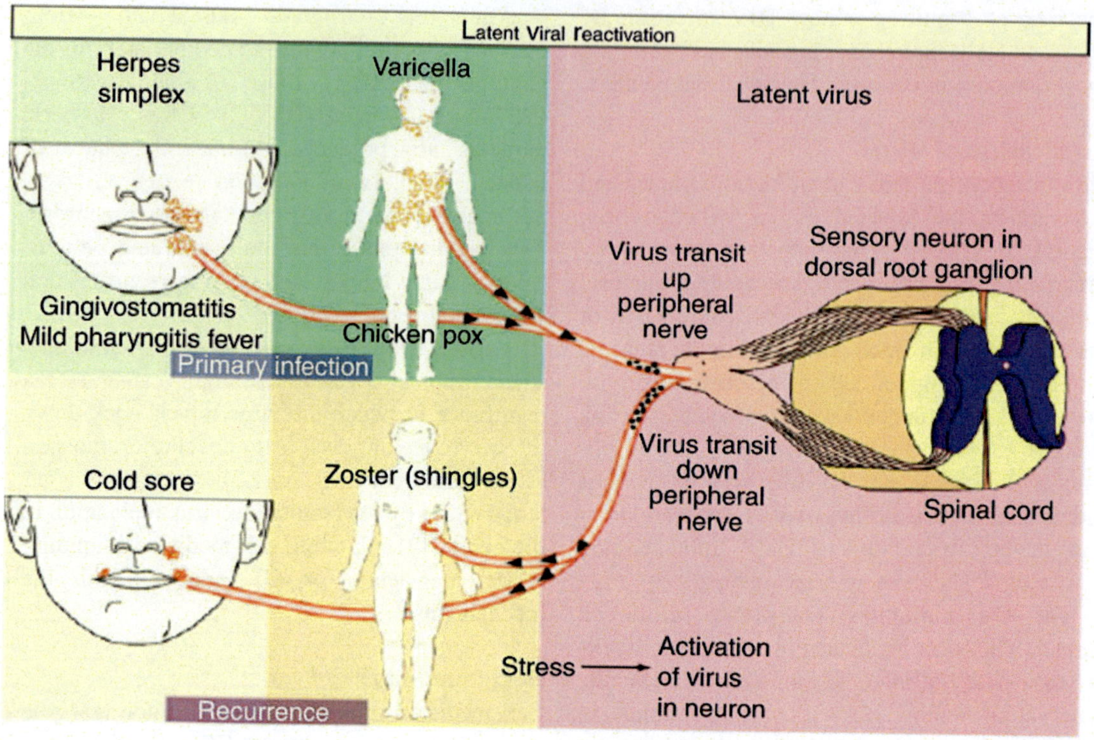

Fig. 53.2 Herpes virus infection (https://link.springer.com/chapter/10.1007/978-3-030-16996-1_19)

HSV-2 produces lesions below the waist, whereas HSV-2 is associated with genital infections. Primary infection with either virus is typically associated with systemic signs, prolonged duration, increased severity of illness and more complications.

HSV-1 Infection

HSV-1 can cause a wide variety of clinical entities.

Acute Herpetic Gingivostomatitis

This is the manifestation of primary HSV-1 infection occurring in children between 6 months and 5 years. Saliva from an infected child or an adult is the source of infection. Gingivitis is the most important manifestation with marked lip swelling and erythematous and friable gums. Vesicular lesions appear on the oral mucosa, tongue and lips. These lesions subsequently rupture and coalesce together, leaving behind ulcerated plaques. Viral shedding from the saliva may continue for 3 weeks or more.

Acute Herpetic Pharyngotonsillitis

The condition manifests as pharyngitis and tonsillitis in adults more frequently than gingivostomatitis. The vesicular lesions on the tonsils and the posterior pharynx usually rupture to form ulcers.

Herpes labialis

It is the recurrence of oral HSV-1 infection (Fig. 53.3). Pain, burning and tingling sensation at the site of infection are presenting features. An intraepidermal vesicle that later becomes pustular and ulcerate is the typical lesion.

Other HSV-1 Infections

These include **Herpetic whitlow, an** infection of the finger of health workers, including doctors and nurses. The entry of the virus causes the infection through cuts or abrasions on the skin. Other infections include **Eczema herpeticum** seen in children with active eczema. **Herpes febrilis or fever blisters** are infections of the skin that appear on the forehead and around the

Fig. 53.3 Grouped vesicles and erosions over the angle of the mouth—recurrent herpes labialis. (https://en.wikipedia.org/wiki/Herpes_labialis)

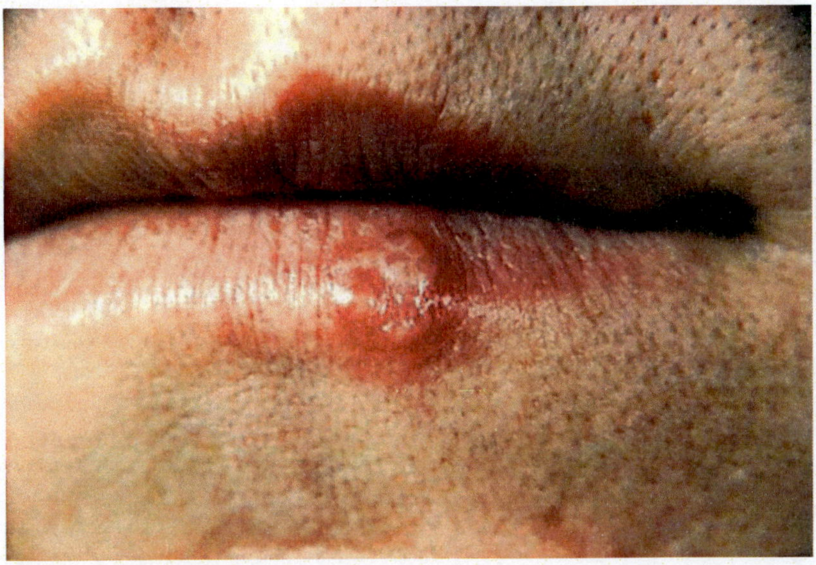

mouth in the face. These lesions subsequently rupture and coalesce together, leaving behind ulcerated plaques. **Herpes encephalitis** is an acute febrile disease that is usually caused by HSV-1. This may be a manifestation of primary or recurrent infection with the virus. **Ocular herpes** can be a primary or recurrent HSV-1 infection. Primary ocular herpes may manifest keratoconjunctivitis, dendritic ulcers in the cornea and follicular conjunctivitis with the formation of vesicles on the eyelids. The recurrent ocular herpes manifests as conjunctivitis, keratitis, iridocyclitis, cataract, etc.

HSV-2 Infection

HSV-2 causes following clinical conditions.

Genital Herpes

Genital herpes is mostly caused by HSV-2 but can also be caused by HSV-1. Most primary genital infections are asymptomatic. The clinical manifestations of primary genital herpes caused by HSV-1 and HSV-2 are similar. In symptomatic men, the herpetic vesicles appear in the glans penis, the prepuce, shaft of the penis and sometimes on the scrotum, thighs and buttocks. In women, the vesicles appear on the external genitalia, labia majora, labia minora, vaginal vestibule and introitus. Recurrent genital herpes is more common with HSV-2 as compared to HSV-1. The condition is shorter in duration and less severe than primary genital herpes.

Neonatal Infection

Neonatal infection is a most serious and usually fatal disease caused mostly by HSV-2. It usually occurs due to the shedding of HSV-2 from the cervix during vaginal delivery. It can also occur from an ascending *in-utero* infection during the primary infection of the mother. The infection can also be acquired postnatally from family members or hospital staff. Neonatal herpes is confined to the mouth, eyes or skin. Since CMI is poorly developed in neonates, the virus causes a disseminated disease with involvement of the liver, lung and organs of the CNS. The condition has a high mortality of 80%. Progression of the infection to the CNS results in mental retardation or neurological disabilities even with treatment, or may finally lead to death.

Mother acquiring HSV infection for the first time in late pregnancy poses a higher risk of neonatal herpes, whereas women with genital herpes before pregnancy pose a very low risk of transmitting HSV to their infants.

Aseptic Meningitis

Aseptic meningitis may occur as a complication of genital HSV-2 infection. It is a self-limiting

Table 53.3 Distinguishing features of HSV-1 and HSV-2

Characteristics	HSV-1	HSV-2
Site of latency	Trigeminal ganglia	Sacral ganglia
Age of primary infection	Young children	Young adults
Vesicular lesions in the skin	Above the waist	Below the waist, especially the genitalia
Gingivostomatitis	Common	Rare
Pharyngotonsillitis	Common	Rare
Keratoconjunctivitis	Common	Rare
Infection of the CNS	Encephalitis	Meningitis
Infection in neonates	Rare	Skin lesions and disseminated infections
Infection in immunocompromised hosts	Dissemination to visceral organs	No dissemination
Transmission of infection	Contact (often saliva)	Sexual
Growth on CAM	Small pocks	Large pocks
Growth in chick embryo fibroblast	Replicates poorly	Replicates well
Laboratory animals	Less neurovirulent	More neurovirulent
Recurrence	Less chance of recurrence	More recurrence irrespective of site
Resistance to antiviral agents in culture	Less resistant	More resistant

Note: *HSV-1*: herpes simplex virus type 1; *HSV-2*: herpes simplex virus type 2; *CAM*: chorioallantoic membrane

disease, with CSF showing lymphocytic pleocytosis and viruses by culture.

Differences between HSV-1 and HSV-2 infections are summarised in Table 53.3.

Laboratory Diagnosis

Specimens

These include saliva, vesicle fluid, conjunctival fluid, corneal scraping, skin swab, skin scrapings and cerebrospinal fluid (CSF).

Microscopy

Light microscopy of the stained infected cell may show ballooning of cells, ground glass nuclei and eosinophilic intranuclear inclusions and multinucleated giant cells. **Giemsa-stained smears** show type A inclusion bodies in the nuclei of the infected cell. **Electron microscopy** is used for direct demonstration of virions in the negatively stained smears of the vesicle fluid.

Tzanck Smear

It is a rapid and sensitive method. In this procedure, the vesicular lesion is aseptically ruptured, and the base of the lesion is scraped with a scalpel. The scraping from the base of the lesion is smeared on a glass slide, air dried, fixed and usually stained for 15 seconds with1% aqueous solution of toluidine blue. Demonstration of ground glass chromatin (Tzanck cells) with eosinophilic intranuclear inclusions and multinucleated giant cells is characteristic of blue. HSV 1, HSV 2 or varicella-zoster infection (Fig. 53.4).

Direct Antigen Detection

Direct enzyme immunoassay and direct fluorescent antibody tests are useful to demonstrate HSV antigens directly in vesicular fluid, tissue smear or biopsy.

Tissue Culture

A definitive diagnosis of HSV infection is made by isolating the viruses in tissue cultures. For culture, scrapings of skin vesicles and mucosal lesions are inoculated in HeLa cells, Hep-2 cells or human embryonic fibroblasts. After inoculation, HSV produces cytopathic effects (CPEs) within 1–3 days on HeLa cells, Hep-2 cells and human embryonic fibroblasts. Cowdry type A acidophilic inclusions are produced by HSV. These cells become enlarged and appear ballooned. HSV-2 causes the fusion of infected cells, leading to syncytium formation. Immunofluorescent staining of infected tissue culture is

Fig. 53.4 Tzanck smear showing intranuclear type A inclusion bodies. (https://en.wikipedia.org/wiki/Human_cytomegalovirus)

useful to confirm the diagnosis within 24 h. Polymerase chain reaction (PCR) is a useful tool to distinguish HSV-1 from HSV-2.

Virus Typing
Isolated HSV strains can be typed by biochemical, immunological and molecular methods.

Serology
Serodiagnosis is of little value for the diagnosis of primary HSV infection. It is mainly used for epidemiological studies and to determine post-exposure to HSV. It is not useful for diagnosing recurrent infections because the rise in antibody titre does not usually correlate with recurrent disease. Also, the serological tests cannot distinguish between HSV-1 and HSV-2 antibodies due to cross-reactivity.

Molecular Detection
Real-time PCR is useful for the demonstration of viral genome directly in vesicular fluid, skin scraping, CSF, etc., for diagnosis of the disease. PCR is also used to detect asymptomatic shedding of HSV in skin lesions and genotyping the virus. Amplification of herpes genomes by PCR is also being increasingly used for diagnosing HSV infection of the brain by examining the brain biopsy specimen. HSV-1 and HSV-2 types can be distinguished by nucleic acid hybridisation, restriction endonuclease cleavage and PCR.

Laboratory diagnosis is summarised in Table 53.4.

Treatment

HSV infections are among the few non-HIV viral infections that can be treated with antiviral therapy. Treatment with specific antiviral chemotherapy is used to (1) prevent disease and recurrence, (2) treat the infection and (3) reduce the clinical course of infection.

Antiviral drugs, like acyclovir, famciclovir and valacyclovir, are the choice for treating HSV infection. Oral antivirals are used to treat orolabial and genital infections and topical for herpetic keratoconjunctivitis. Intravenous treatment is used for life-threatening and serious HSV infections, such as encephalitis, and occasional severe orolabial or genital cases.

Treatment is most effective when started within a few hours after symptoms appear. Penciclovir cream effectively reduces the healing time and duration of symptoms of a cold sore by about a day. However, such treatment can only

Table 53.4 Laboratory tests for diagnosis of herpes simplex virus infections

Diagnostic methods	Diagnostic features
Microscopy	
Tzanck smear	Presence of multinucleated giant cells with faceted nuclei and homogenously stained ground glass chromatin (Tzanck cells)
Light microscopy	Eosinophilic intranuclear inclusions
Electron microscopy	Virus particle
Direct antigen detection	
Direct fluorescent antibody and direct enzyme immunoassay	HSV antigens directly in vesicular fluid, tissue smear or biopsy
Cell culture	
Culture on HeLa cells, Hep-2 cells and human embryonic fibroblasts	CPEs within 1–3 days
Serology	
ELISA, CFT, and neutralisation test	Not useful for primary infection; only used for seroepidemiological studies
Molecular diagnosis	
DNA probe and PCR	Detects viral genome directly in vesicular fluid, skin scraping, CSF and other specimens. Standard assay for specific diagnosis of HSV infections of the central nervous system

reduce the severity and frequency of symptoms but cannot cure the infection. As such, both HSV-1 and HSV-2 infections are considered lifelong.

Epidemiology

HSV infections are exclusively human diseases. They are distributed worldwide. HSV-1 infection is more common than HSV-2 infection. An infected person is a lifelong source and reservoir of the virus due to the latency of the infection. Vesicle fluid, saliva and vaginal secretions are the important sources of infection for both types of HSV.

Geographical Distribution

In a study published in 2016 on estimating the global prevalence of HSV, it was noted that infections are common across the globe and mostly endemic. An estimated half a billion people had genital infections with HSV type 2 or type 1, and several billion had oral HSV type 1 infection. The infections varied across age, sex and geographical regions. Most genital HSV-1 infections were common in the Americas, Europe and Western Pacific, whereas HSV-2 was highest among women in African Region. HSV-1 infection is more common than HSV-2 infection.

Reservoir, Source and Transmission of Infection

HSV infections are exclusively human diseases. Humans are the only natural reservoirs. No vectors are involved in the transmission of the disease. An infected person is a lifelong source and reservoir of the virus. Vesicle fluid, saliva and vaginal secretions are the important sources of infection for both types of HSV.

On coming in contact with oral and genital secretions, medical and paramedical staff are at increased risk for acquiring herpetic whitlow. Immunocompromised individuals and neonates are at high risk for developing the life-threatening disseminated disease.

Prevention and Control

Prevention of HSV infection is difficult because most transmission occurs during sub-clinical viral shedding. At present, no vaccine is available for use against HSV. Current preventive strategies are dependent on good practices. Recurrences of infection can be prevented by avoiding the trigger activities. Contracting the infection can be prevented by avoiding skin to skin oral contacts and non-sharing of glass, saliva-contaminated utensils and unprotected sexual activities.

Despite the HSV infection affected almost two-thirds of the global population below 50, there are no clinical preventive strategies, including vaccines. This is mostly because (1) the transmission may occur even from asymptomatic individuals, makes the early preventive measure difficult, and (2) owing to the complex genetic functions, HSVs can remain hibernated within our body and then periodically reoccurs, all without alarming the host immune system by taking refuge in the nervous system. This particularly makes vaccine development challenging. However, in recognition that better prevention interventions are needed, WHO and other global partners are working to accelerate research to develop new strategies for preventing and controlling HSV infections.

Herpesvirus Simea: B Virus

Unlike other humans infecting herpesviruses, Herpesvirus simiae, recently designated as *Cercopithecine herpesvirus* 1, has a zoonotic origin and is by far the most important zoonotic threat to human health caused by a member of the herpesvirus family. It was first isolated in 1934 by Sabin and Wright. The virus was demonstrated in the brain of laboratory staff, who died from ascending myelitis following a bite by an apparently healthy monkey. This virus was named as B virus from the initial name of this patient. Since then, many cases have been reported in the literature.

The virus naturally causes a virtually asymptomatic to a very mild infection in natural hosts, i.e. Asian macaques (Macaca mulatta and Macaca fascicularis). It is transmitted to humans when bitten or scratched by an infected macaque monkey or individuals coming in contact with the monkey's eyes, nose or mouth, but needle-stick injuries and even airborne infections were also reported. Person-to-person transmission is extremely rare, and only one case has been reported to date.

In humans, the disease is usually fatal. Neurological sequelae are seen in patients who rarely survive the condition. For that reason, B virus-infected animals are culled.

Varicella-Zoster Virus

Human herpesvirus 3 (commonly known as varicella-zoster virus (VZV)) belong to the genus Varicellovirus of subfamily Alphaherpesvirinae.

It causes two distinct clinical entities in humans: (1) chickenpox (varicella) and (2) herpes zoster or shingles. Chickenpox is acquired by VZV transmission from an infected host to a susceptible host, whereas herpes zoster occurs due to reactivation of the latent virus and subsequent replication and viral spreads along the nerve fibres to the skin. Contact with a case of zoster or chickenpox may transmit chickenpox but not herpes zoster.

Immunity following varicella is lifelong; the individual does not suffer from varicella again but can suffer from herpes zoster despite having immunity to varicella. Immunity is also lifelong in herpes zoster.

Morphology

Varicella-zoster virus is an enveloped, dsDNA virus showing many similarities with the HSV. It has the smallest genome of all the human herpesviruses and shows fascicular is a replicative cycle similar to HSV.

Laboratory Isolation

The virus grows readily in cultures of human fibroblasts, human amnion or HeLa cells. These cell cultures show an apoptotic effect similar to that seen in HSV-infected cells, but the lesions are less marked than those produced by HSV. Also, the CPE appears after a longer incubation period.

The viruses remain associated with the cells in culture and are not found free in the media. The virus does not grow in chick embryos or experimental animals.

Pathogenesis of VZV Infection

Chickenpox is the primary VZV infection, whereas herpes zoster represents reactivation of a previous infection.

Transmission

Only chickenpox is transmitted. It is transmitted through direct contact with an infected person's blisters, saliva or mucus or indirectly by touching contaminated items freshly soiled, such as clothing, from an infected person. Blisters that are dry and crusted could not spread chickenpox. The virus can also be transmitted through the air by coughing and sneezing.

Primary VZV infection occurs in humans when the virus comes into contact with the mucosa of the respiratory tract or conjunctiva. From these sites, the virus enters the bloodstream and lymphatic system to the cells of the reticulo-endothelial system. After 11–13 days, a secondary viraemia occurs, and the virus spreads throughout the body and to the skin. VZV spreads from cell to cell via direct contact to produce its effects in tissues. After primary infection, the virus migrates along the sensory nerve fibres to the satellite cells of the dorsal root ganglia or cranial nerve ganglia, where it becomes latent. This latency may be permanent, or the virus may become reactivated in old adults or patients with impaired cellular immunity.

Herpes zoster occurs due to reactivation of the VZV, which has remained latent in one or more sensory ganglia following primary varicella many years earlier. The viruses travel down along the sensory nerve to produce painful vesicles in the skin (*dermatome*) areas innervated by the nerves from the affected ganglia. The virus replicates and spreads along the nerve fibres to the skin on reactivation, known as *herpes zoster* or *shingles*.

Immunity

Cell-mediated immunity is important in controlling the infection. It limits the progression of the disease and results in early resolution of the disease. The virus causes a disseminated life-threatening and more serious disease in immunocompromised patients with a deficient CMI. Depression of the CMI during life in old age may cause VZV and herpes zoster recurrence. Humoral antibodies also play an important role by limiting the spread of the virus in circulation. CMI also contributes to the immunopathology and symptomatology in the infected host. In some infected adults, it is responsible for more severe manifestations and more extensive cell damage than seen in children.

Clinical Syndromes

Varicella-zoster virus causes two distinct clinical syndromes: varicella (chickenpox) and (herpes zoster (shingles).

Varicella (Chickenpox)

Varicella (Chickenpox) is one of the five childhood exanthemata along with measles, rubella, rubeola and fifth disease. It is mainly a disease of children, characterised by an exanthematous varicella rash that occurs following infection with VZV.

After an incubation period of around 2 weeks, a secondary viraemia occurs, and the virus spreads throughout the body and to the skin. The condition manifests as fever and maculopapular rash that progresses to thin-walled vesicle on an erythematous base within a few hours. This **vesicle**, the hallmark of chickenpox, is characteristically surrounded by a red ring. The vesicle, which measures approximately 2–4 mm in diameter, becomes a pustule within 12 h and begins to ulcerate, after which scabs appear. All the stages, including macule, papule, vesicle, pustule and scab, evolve very rapidly, especially in children. As a result, at any given time, all stages of the lesions are seen on the skin.

The skin lesions are characteristically **centripetal** in distribution. They are more commonly present on the trunk than on the extremities. Typically, they are also present on the scalp. Lesions typically occur in the mucous membrane in the mouth, conjunctiva and vagina. All these features help distinguish chickenpox from many other exanthematous diseases. Secondary bacterial infection followed by scarring is the noted complication in the children.

Chicken Pox in Pregnancy
Varicella-zoster virus tends to cause more serious disease in pregnancy if the mother has not been infected during childhood. The virus crosses the placenta and causes congenital infection in the foetus. Congenital malformation with hypoplasia of limbs, chorioretinitis and skin scarring is seen in the foetus infected during pregnancy.

The complications in the baby may vary according to the period of gestation when the woman contracts chickenpox. If the mother develops chickenpox during the first half of the pregnancy, the foetus will usually be asymptomatic. However, in some infants, it progresses into foetal **varicella syndrome**. The condition is characterised by cicatrising skin lesions, limb hypoplasia, chorioretinitis and CNS defects.

If the mother contracts varicella close to the delivery period, the babies might develop **neonatal varicella** within 2 weeks after birth. If the mother develops chickenpox a week or more before the delivery, she passes VZV and IgG antibodies to her baby through the placenta. In such cases, the baby will be infected with the virus, but clinical symptoms may not appear. This is because VZV-specific antibodies will inhibit VZV multiplication. If the mother develops chickenpox less than a week before delivery, she passes only VZV to her baby but not the antibodies. Hence, the babies develop neonatal varicella. This is a serious disseminating disease, mostly causing pneumonia and encephalitis. Such babies are treated with VZV antiserum and chemotherapy soon after birth.

Herpes Zoster (Shingles)
Herpes zoster (Shingles) is typically a disease of older adults aged 50 years and more, although it can be seen at any age. The condition occurs following a recurrence of chickenpox acquired many years earlier. This occurs due to the reactivation of the virus, which has remained latent in one or more sensory ganglia due to the residual immunity. This occurs due to the waning of immunity, followed by some precipitating factors.

Severe pain in the area of mucosa or skin preceding the appearance of the rash is the hallmark of the disease. The rash, unlike in varicella infections, has limited distributions. They are restricted to a dermatome covering either head, neck or trunk. For example, if thoracic nerve supplies are affected, rashes commonly appear on the chest walls. The rashes are distributed on the scalp and forehead when the ophthalmic nerve of trigeminal ganglion is affected. The accompanying pain and herpetic neuralgia are very much severe at times.

Herpes zoster is associated with many complications. **Zoster of the seventh nerve ganglion** may cause Bell palsy and Ramsay Hunt syndrome. This is caused by the reactivation of the virus involving the facial and auditory nerves. **Post-herpetic neuralgia** is a chronic pain syndrome caused by the virion. **Herpes zoster ophthalmicus** is the most serious and devastating form of acute herpes zoster (Fig. 53.5). The condition occurs due to reactivation of the fifth cranial (trigeminal) nerve.

Laboratory Diagnosis

In almost all cases, the diagnosis of chickenpox and herpes zoster is always clinical.

Specimens
These include skin lesion, respiratory secretions, organ biopsy and serum.

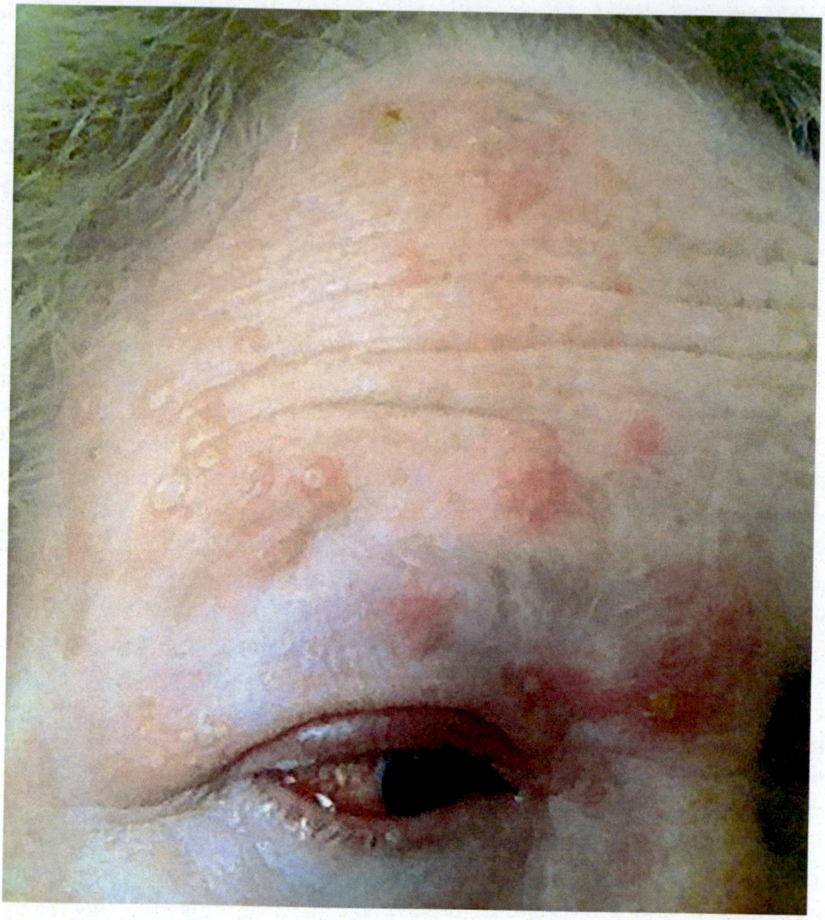

Fig. 53.5 Right side of the forehead affected by grouped vesicles and necrosis—herpes zoster ophthalmicus. https://commons.wikimedia.org/wiki/File:Day06_shingles_or_Herpes_Zoster_Virus_attacking_forehead_and_eye.jpg

Direct Antigen Detection

Direct fluorescent antibody test employing antibodies against viral membrane antigen (fluorescent antibody to membrane antigen, FAMA) is used to detect membrane antigen of the virus directly in the skin scrapings and biopsy specimens. The counter current immunoelectrophoresis (CIEP) using zoster immune serum I was also used earlier to detect antigen directly in the vesicular fluid.

Virus Isolation

Specimens from buccal or skin lesions collected during the early stage of the disease are cultured on human amnion, human fibroblast and HeLa or Vero cells. VZV produces a CPE similar to HSV but is less conspicuous and takes a long time to develop. VZV does not grow in the chicken embryo or laboratory animals.

Serology

Both indirect fluorescent antibody and ELISA are employed to detect serum antibodies. The antibody-based tests are useful mainly for epidemiological studies rather than for the diagnosis of individual cases.

Molecular Detection

PCR is a highly sensitive method for detecting viral genomes in skin scrapings or biopsy specimens.

Treatment

Acyclovir, famciclovir and valacyclovir are antiviral agents available for treatment. Treatment is recommended for adults and immunocompromised patients with varicella infection and herpes zoster infections. It is usually not indicated for the treatment of varicella infection in children.

Epidemiology

Chickenpox is mainly a disease of children. In herpes zoster infection, people of all ages are affected. However, the incidence of the disease increases with age due to the decrease in immune functions that occurs during ageing.

Geographical Distribution

Varicella-zoster virus is a highly infective pathogen distributed worldwide.

Reservoir, Source and Transmission of Infection

Chickenpox is exclusively a human disease. No animal reservoirs are present. Chickenpox or herpes zoster patient is the source of infection. VZV is shed mainly in the respiratory secretions and the vesicular lesions of the skin. The infectivity is maximum during the early stage of the disease when the virus is shed in large numbers in respiratory secretions. The buccal lesions and vesicular fluid, which are rich in virus content, are highly infectious and are sources of infection. Scabs do not contain any viruses, hence are non-infectious. Chickenpox is transmitted primarily by inhalation of respiratory secretions. The infection may also occur by direct contact with vesicular lesions on the skin. Herpes zoster is a reactivation of latent VZV infection; hence it is not acquired from an infected patient.

Prevention and Control

Prevention of transmission of varicella infection in children is difficult.

Varicella-Zoster Immunoglobulin

Varicella-zoster immunoglobulin (VZIg) is prepared by pooling sera from patients convalescing from herpes zoster. High-risk patients, such as immunosuppressed children, are protected by administering VZIg. The VZIg is also used to treat medical and paramedical personnel exposed to the virus and for the treatment of newborns of infected mothers showing symptoms within 5 days of delivery. It is, however, not useful as a therapy for the treatment of patients suffering from active varicella or herpes zoster disease.

Vaccines

A live attenuated **varicella vaccine** (*Oka strain*) is being used for protection against VZV. This vaccine was first developed by Takahashi in Japan in 1974 by attenuating strains of the varicella virus by serial passage in human and guinea pig fibroblast cell lines. The vaccine is given in a single dose for children aged 1–12 years and in two doses at 6–10 weeks' intervals for older children.

Currently, single-antigen varicella vaccine and combination measles, mumps, rubella and varicella (MMRV) vaccine are available commercially. At present, two doses of vaccination schedules are approved. For children of age group of 1–12 years, the first dose at age 12 through 15 months is followed by the second dose at age 4 through 6 years. However, for kids above 13 years, two doses of the single-antigen varicella vaccine are recommended at 4–8 weeks apart. The MMRV vaccine is not approved for people in this age group.

The vaccine is safe, protective and provides long-lasting immunity. It stimulates the production of protective antibodies and CMIs. The vaccine even confers protection in immunodeficient children. It is also used as a prophylactic agent for the treatment of people even after exposure to VZV.

Epstein–Barr Virus

Epstein–Barr virus (EBV) or human herpesvirus 4 (HSV-4) is a gamma herpesvirus belonging to the subfamily Gammaherpesvirinae. It shows very limited host specificity. It is the causative agent of infectious mononucleosis. It is also the first virus associated with human malignancies, such as Burkitt lymphoma, another B-cell lymphoma and nasopharyngeal carcinoma.

Morphology

Epstein–Barr virus is an enveloped DNA virus. It consists of a genome, a capsid and an envelope. The genome consists of a 172 kbp linear dsDNA. It is surrounded by an icosahedral capsid composed of capsomeres. Both the capsid and the genome, which form the nucleocapsid, are enclosed by an envelope containing glycoproteins. The envelope is derived from the outer nuclear membrane of the host cell and contains many proteins, which are expressed for different types of infections.

Laboratory Isolation

The virus can grow in human embryonic lung fibroblast cultures. The CPE produced by the virus is focal and is associated with refractile ballooning of the cells. The CPE, unlike other viruses, takes 2 or more weeks to develop.

Antigenic Properties

Epstein–Barr viruses contains many viral antigens of diagnostic importance. These can be classified as follows:

1. Late antigens, such as viral capsid antigen (VCA) and viral envelope glycoprotein. The VCA is the most important. The demonstration of the antigen indicates the presence of the virus and active replication of the virus.
2. Early antigens (EAs) are always expressed early in the lytic cycle. The presence of EAs indicates virus replication.
3. Latent phase antigens include Epstein–Barr nuclear antigen (EBNA) and LMPs. The presence of these antigens indicates the latency of the virus in the infected cells.

EBV consists of two types of EBV-1 and EBV-2, known as type A and type B strains, respectively, based on differences in the latency nuclear antigen genes. Both types cause the same degree of clinical illness. They are prevalent throughout the world and can simultaneously infect the same person.

Other Properties

Epstein–Barr viruses are sensitive to the action of ether and bile salts. They are relatively fragile and do not survive for long outside the human body fluids.

Pathogenesis of Epstein–Barr Virus Infection

Transmission

The infection is transmitted by (1) saliva through close contact during the act of kissing; (2) sharing of items, such as drinking glasses, cups and toothbrushes; and (3) blood transfusion and by organ and bone marrow transplantation. The EBV enters B lymphocytes at the receptor site for the C3 component of the complement. The virus replicates in the cytoplasm of the infected epithelial cells, and the replicative cycle is similar to that of HSV.

Epstein–Barr virus in humans can cause either (1) primary infections by multiplication in B cells and epithelial celosia; (2) latent infection of B cells or (3) transformation of B cells to immortal plasmacytoid cells leading to malignancies.

Primary EBV infection first occurs in the oropharynx and then spreads to the blood, causing infection of B lymphocytes. The virus is, therefore, present in large numbers in oropharyngeal secretions. Hence, the infection is most commonly transmitted through infected saliva, often due to kissing. Therefore, infectious mononucleosis is also commonly called the *kissing disease*.

The virus infection induces a strong immune response, comprising circulating antibodies against many virus-specific proteins, cell-mediated immune responses and production of lymphokines. Humoral immunity is characterised by the appearance of IgM antibodies first against the viral membrane antigen, followed by IgG antibodies, which persist for life. These antibodies against viral membrane antigen confer lifelong immunity against the second attack of infectious mononucleosis. The CMI plays an important role in controlling chronic infection and limiting primary infection.

Clinical Presentations

EBV is associated with infectious mononucleosis, EBV-induced tumours and EBV infection in an immunocompromised host.

Infectious Mononucleosis

This is a classical clinical syndrome that represents the immunopathogenic response of the host to infection with EBV. Cytomegalovirus (CMV) is most commonly associated with EBV-negative cases of infectious mononucleosis. The condition is most often asymptomatic in children younger than 4 years. In symptomatic cases in adolescents and adults, the condition is characterised by high fever, sore throat, headache, myalgia, nausea and abdominal pain. The condition is associated with pharyngitis, generalised lymphadenopathy, hepatosplenomegaly and fatigue. The symptoms are usually present for 2–3 weeks. But fatigue, which is the major complaint, continues to persist.

Epstein–Barr virus infection rarely causes death in immunocompromised patients but may cause neurological complications (meningoencephalitis, Guillain–Barre syndrome, etc.), upper airways obstruction or splenic rupture.

Epstein–Barr Virus-Induced Tumours

Endemic Burkitt lymphoma is associated with EBV. It is a poorly differentiated monoclonal B-cell lymphoma of the jaw. It is the most common tumour of childhood in Africa associated with both EBV and falciparum malaria. *Plasmodium falciparum* malaria causes stimulation and proliferation of polyclonal B cells in the presence of EBV infection. The malaria parasite also alters the response of T lymphocytes to EBV, apparently contributing to the pathogenesis of the tumour.

Epstein–Barr virus infection is associated with **nasopharyngeal carcinoma** endemic in Asia and is common in males of Chinese origin. EBV DNA EBNA 1 and LMP 1 antigens can be demonstrated in the nasopharyngeal carcinoma cells. In addition, this virus has been linked to a wide range of malignancies, such as post-transplant lymphoproliferative diseases (TLDs), nasopharyngeal carcinoma (NPC), Hodgkin's lymphoma and gastric carcinoma.

Epstein–Barr Virus Infection in an Immunocompromised Host

Lymphoproliferative diseases are observed in patients with immunodeficiency diseases. The EBV causes several syndromes and proliferation disorders in these patients, including Duncan syndrome, ataxia-telangiectasia and Wiskott–Aldrich syndrome.

Diagnosis

The laboratory diagnosis of EBV-induced infectious mononucleosis is based on the three classic criteria: (1) presence of lymphocytosis; (2) presence of at least 10% atypical lymphocytes in peripheral blood smear and (3) presence of heterophilic antibodies and antibodies to viral antigens. These criteria are supplemented by direct detection of viral antigens or EBV genomes in clinical specimens.

Specimens

These include lymphoid tissues, nasopharyngeal carcinoma tissue and saliva (for direct detection of viral antigens or EBV genomes) and blood (for serological tests) and peripheral blood (for blood smear).

Direct Antigen Detection

A direct immunofluorescent antibody test is used to detect viral antigens in lymphoid nasopharyngeal carcinoma tissues.

Virus Isolation

EBV can be isolated from saliva, peripheral blood or lymphoid tissue by transforming normal human lymphocytes obtained from umbilical blood. However, the procedure is tedious and time consuming, hence seldom used. An attempt has also been made to culture **spontaneously transformed** B lymphocytes from EBV DNA or virus-infected patients.

Serology

Epstein–Barr virus infection stimulates the production of a wide range of antibodies, including heterophilic antibodies. These heterophilic antibodies are IgM antibodies that recognise Paul–Bunnell antigen on sheep, horses and bovine RBCs. Therefore, serum from patients with acute mononucleosis agglutinates sheep, horse or bovine erythrocytes.

Heterophilic Antibody-Based Tests

Paul–Bunnell test was first described by Paul and Bunnell to demonstrate heterophile antibodies in patients with infectious mononucleosis in 1932. This heterophile antibody, which is an IgM antibody, is not directed against EBV or EBV-infected cells but is produced due to polyclonal activation of B cells by EBV. Such heterophilic antibodies are also found in serum sickness during drug reaction and naturally occurring antibodies to the Forssman antigen. These antibodies are absent or present in a very low titre in the serum of healthy individuals.

The human blood to be tested is first absorbed by the guinea pig kidney, and then it is tested for agglutination activities directed against horse, sheep or cow erythrocytes. A serum titre of 100 or greater is considered a positive test and is suggestive of infectious mononucleosis. Differential absorption of agglutinin with guinea pig and cow red cells is necessary to confirm the diagnosis. Cow RBCs absorb infectious mononucleosis heterophile antibodies from serum but not Forssman antibodies. Guinea pig kidney cells absorb Forssman antibodies, but not the infectious mononucleosis heterophile antibodies. Both guinea pig kidney cells and bovine RBCs absorb the antibodies produced in serum sickness. A serum titre of 1:100 after adsorption with guinea pig cells is considered positive and is suggestive of acute infectious mononucleosis. These heterophile antibodies are usually present for 3 months but may persist for even as long as a year.

Monospot test is a rapid slide agglutination test, which uses either horse RBCs or bovine RBCs. Bovine RBCs are more specific for acute infectious mononucleosis heterophile antibodies; they do not require differential absorption. Monospot test shows a specificity of 63–80%. It may show a rare false-positive reaction in patients with lymphoma or hepatitis.

EBV-Specific Antibody Tests

Epstein–Barr virus infection is characterised by the development of specific antibodies against different EBV antigens (VCAs); EA and EBNA. ELISA and Western blot are frequently used methods to detect antibodies against these antigens.

1. **Antibodies against VCAs:** IgM VCA antibodies appear early and circulate in the serum for 30–45 days. Hence, demonstration of IgM VCA antibodies in the serum indicates current EBV infection. IgG VCA antibodies appear during the acute phase. These antibodies reach a peak within 2–4 weeks but decline and persist in the circulation throughout life. Demonstration of IgG VCA antibodies, therefore, indicates past infection.
2. **Antibodies against early antigens:** Demonstration of anti-EA Ig in the serum indicates current infection with B.Ths. is because anti-EA antibodies appear early and disappear from the serum after 3–6 months. Nevertheless, a high titre of antibodies to EA has been demonstrated in EBV-associated lymphomas.
3. **Antibodies against EBV Nuclear antigen:** Demonstration of antibodies to EBNA antigens suggests past infection.

The serological profile of EBV infections is summarised in Table 53.5.

Other Tests

The presence of atypical lymphocytes is the characteristic feature of infection with EBV. These atypical lymphocytes of Downy type III are larger and have a less dense mucus than normal

Table 53.5 Serologic profile of EBV-infected patients

Serological parameters	Patient's clinical status: Acute primary infection	Chronic primary infection	Past infection	Reactivation	Burkitt lymphoma	Nasopharyngeal carcinoma
Heterophile antibodies	Positive	Negative	Negative	Negative	Negative	Negative
VCA-IgM	Positive	Negative	Negative	Negative	Negative	Negative
VCA-IgG	Positive	Positive	Positive	Positive	Positive	Positive
EA	Variable	Positive	Negative	Positive	Positive	Positive
EBNA	Negative	Negative	Positive	Positive	Positive	Positive

VCA-IgM: viral capsid IgM antibodies; *VCA-IgG*: viral capsid IgG antibodies; *EA*: early antigen; *EBNA*: Epstein–Barr nuclear antigen

lymphocytes. Lymphocytosis (more than 50% lymphocytes) is seen in approximately 80–90% of patients. Lymphocytosis is well marked during the second or third week of illness and persists for 2–6 weeks.

Molecular Detection

DNA probes for Epstein–Barr EAs have been used to detect and identify the EBV mRNA in the nuclei of EBV-infected lymphoid cells. PCR has been developed to detect EBV DNA in plasma during acute infectious mononucleosis. These EBV DNA levels decline during convalescence. They are not demonstrated in latently infected individuals.

Treatment

No specific treatment is available against acute infectious mononucleosis. Acyclovir has little activity against EBV. Treatment of the condition is mainly symptomatic.

Epidemiology

Epstein–Barr virus infection occurs worldwide. EBV was first discovered in cells isolated from African Burkitt's lymphoma. Reinfection is of relatively low transmissibility and does not occur as epidemics. Epidemiological studies suggest its prevalence in more than 90% of the world's population. EBV infection is exclusively human disease. Humans are the only known reservoirs of the virus. It is present in oropharyngeal secretion, saliva, peripheral blood or lymphoid tissue of the infected human host. Saliva is the main source of infection.

Malaria has been suggested as an important co-factor in the progression of chronic or latent EBV infection to acute Burkitt lymphoma in Africa. The genetic predisposition of the people in China or the presence of co-factors in the food or environment is believed to be important in facilitating the incidence of nasopharyngeal carcinoma in people living in certain parts of China. Immunocompromised people, such as patients with AIDS, transplant recipients and genetically immunodeficient people, are at high risk for lymphoproliferation disorders caused by EBV.

Prevention and Control

Prevention and control of infection are difficult due to the widespread prevalence of the virus and intermittent shedding by the infected individuals throughout their lives. No vaccine is available against EBV infection.

Cytomegalovirus

CMV or human herpesvirus 5 (HSV-5) is the causative agent of mononucleosis syndrome in symptomatic infection in immunocompetent hosts. The name is originated from the word *cytomegalo* (large cell virus) and is derived from the swollen cells containing large multinuclear inclusions that characterise these infections. CMV is a member of the Betaherpesvirinae in the subfamily Herpesviridae. The virus shares similar common features in the structure, genome and potency to cause latent and persistent infections, like those of other herpesviruses.

Morphology

CMV is the largest herpesvirus measuring 150–200 nm in size. It has the largest genome of all herpesviruses, ranging from 230 to 240 kbp. It is a double-stranded linear DNA virus. It contains hexagonal protein capsomeres surrounded by a lipid membrane. It is the only betaherpesvirus with the only class E genome, similar to that present in HSV.

Antigenic Properties

Human CMV has a single serotype. It does not show any antigenic cross-reaction with CMV or other mammalian species. It shows only minor antigenic cross-reaction with simian CMV.

Laboratory Isolation

The virus grows only in diploid human fibroblast cell culture but not in epithelial cell culture. The virus grows very slowly in these cell lines, taking 4–6 weeks to produce characteristic CPEs. The CPE includes foci of swollen refractile cells with cytoplasmic granules. These cells show multinucleated giant cells containing acidophilic inclusions in the nuclei and cytoplasm in a stained smear.

Pathogenesis of CMV Infection

CMV infection is acquired by contacting blood, tissue and most body secretions containing viruses. On entering the human, the virus primarily affects epithelial cells and causes infection of those cells. Subsequently, the infection progresses to establish persistent and latent infection in T cells, macrophages and in different organs. Activation and multiplication of the virus in salivary glands, other secretory glands, kidneys and heart lead to their secretion in saliva, urine, milk, etc.

The replicative cycle of CMV in the infected cells is similar to that of other herpesviruses, except that some of the immediate early proteins of the CMV are translated from mRNAs incorporated into the infected cells by the parental virion but are not translated from the mRNAs synthesised in the newly infected cells. In the infected cells, CMV infection produces characteristically enlarged cells with viral inclusion bodies. This change, commonly referred to as *owl's eye,* is diagnostic of the infections. However, these histological findings may be minimal or absent in infected organs.

Cell-mediated immunity is more important and essential for the resolution of infection and controlling the progression of CMV infection. The production of cytotoxic T cells against CMV is crucial to control the infection-specific CD4, and CD8 lymphocytes play an important role in protecting after primary infection or reactivation of latent disease. Studies have shown that patients who do not develop CMV-specific CD4 or CD8 cells are at higher risk for developing CMV pneumonitis. Suppression of CMI, which is caused by therapy with corticosteroid or infection with HIV, induces reactivation of the latent viral infection. Recent studies have also shown that many viral gene products appear to modulate host inflammatory response and are responsible for symptoms of the disease. The antibodies have some role in controlling the severe disease but do not prevent transplacental infection, which can occur even in pregnant mothers who are CMV seropositive.

Clinical Syndromes

CMV causes infections of different types as follows:

Congenital Infection

CMV is the most common viral cause of congenital infection.

Asymptomatic Congenital Infection

It occurs in infants born to women who have preexisting immunity to CMV. These infants appear normal at birth but are at risk of developing growth retardation and other neurodevelopmental abnormalities during the later stage of life.

Cytomegalic Inclusion Disease of Newborns

The cytomegalic inclusion disease occurs in infants born to women who acquire primary CMV infection during pregnancy. The disease is characterised by microcephaly, intracerebral calcification, hepatosplenomegaly and rash and may be fatal. Mental retardation and unilateral or bilateral hearing loss are seen in adults surviving from the disease.

Acquired CMV Infection

Acquired CMV infection occurs postnatally:

Perinatal Infection

Infection of newborns occurs following exposure to infected genital secretions in the birth canal or milk during breastfeeding. Most perinatal infections are asymptomatic. Some infants may manifest lymphadenopathy, hepatitis and pneumonitis. However, these patients do not show any neurological or neurodevelopmental complications.

CMV Mononucleosis

It is a disease of young adults acquired by person-to-person transmission. This condition is transmitted by blood transfusion or organ transplantation. Fever and severe malaise are typical symptoms. Clinically, it is difficult to distinguish CMV mononucleosis from EBV-induced mononucleosis.

Transfusion-Acquired CMV Infection

This condition occurs following transfusion of infected blood after an incubation period of 20–60 days. This condition is mostly asymptomatic. In symptomatic cases, clinical manifestations of transfusion-acquired CMV are similar to CMV mononucleosis.

CMV Infection in Immunocompromised Hosts

CMV is the most important opportunistic agent in immunocompromised hosts, in whom it causes a variety of clinical syndromes. The virus causes latent infection; hence, reactivation may result in disease in immunocompromised patients. These patients include those with HIV and those receiving organ and bone marrow transplantation. It causes pneumonia and life-threatening intestinal pneumonitis. It causes graft atherosclerosis following heart transplantation, bronchitis in lung transplants, virus-related rejections of kidneys, etc.

Infection of Adults

CMV mononucleosis is a disease in immunocompetent hosts. The infection is transmitted by blood transfusion or organ transplantation. Fever and severe malaise are typical symptoms. Clinically, it is difficult to distinguish CMV mononucleosis from EBV-induced mononucleosis. Transfusion-acquired CMV infection occurs following transfusion of infected blood after an incubation period of 20–60 days.

Laboratory Diagnosis

Specimens

Saliva, urine, throat washing, blood, CSF, cervical secretions, bronchoalveolar lavage fluid and tissue bits are the specimens that can be used for the culture.

Microscopy

Demonstration of typical *owl's eye*-like basophilic intranuclear inclusion body is characteristic of CMV. These are demonstrated in the centrifuged deposit of saliva or urine and in infected tissues on staining with haematoxylin–eosin or Papanicolaou staining.

Direct Antigen Detection

A direct immunofluorescence test using specific fluorescein-labelled monoclonal antibodies has been used. This test is used to detect CMV pp65 and pp67 proteins of the virus directly in the clinical specimens, such as blood leukocytes. These proteins are expressed only during the replication of the virus in the tissues.

Direct immunofluorescence test is very useful for detecting sub-clinical disease in immunocompromised patients at high risk. These are also useful to start specific antibiotic therapy in transplantation.

Cell Culture

It is the definitive method for establishing the diagnosis of CMV infection. However, cell culture is highly sensitive but takes a longer time of 4–6 weeks to identify the virus.

To avoid that, a **shell vial assay** is carried out to detect viruses rapidly. This is an adaptation of tissue culture, which provides results more rapidly. In this method, the clinical specimen is added to a vial containing a monolayer of the cell line. The shell vial is centrifuged at a low speed and incubated for 24–48 h. The tissue culture medium is removed, and the cells are stained using fluorescent-labelled anti-CMV antibodies and examined by a fluorescent microscope. The positive shell vial assay is the presumptive evidence of active CMV infection and has been found to be as sensitive as traditional tissue culture.

Serology

ELISA is a frequently used test to detect anti-CMV IgM antibodies. Detection of serum anti-CMV IgM antibodies indicates current infection and congenital infection. Detection of anti-CMV IgG antibodies indicates past infection. The serology is useful for primary CMV infection but not reliable for diagnosing congenital CMV infection.

Molecular Detection

PCR is currently used as a routine diagnostic test to detect CMV DNA in tissue, spinal fluid and amniotic fluid. The test is highly sensitive and becomes positive even before the viraemia in those receiving organ transplants.

Treatment

Ganciclovir is a nucleoside analogue that inhibits DNA synthesis. It is the drug of choice for the treatment of CMV. It is a ganciclovir treatment useful for gastrointestinal disease in patients who have received organ transplants and those who are HIV positive. Ganciclovir has also been used as a prophylactic treatment of CMV disease in patients who have received transplants. In addition, scarlet is another drug that has been used to treat the infection that is resistant to ganciclovir.

Epidemiology

CMV infection is worldwide.

Reservoir, Source and Transmission of Infection

Humans are the natural hosts. An infected human is the only reservoir of CMV infection. CMVs are found in the urine, blood, saliva, tears, throat swab, stool, semen, milk, amniotic fluid, cervical and vaginal secretions and tissue obtained for transplantation.

Saliva, tears, urine and breast milk are the common sources of infection for a baby or child. Cervical secretions are the source of transmission of infection to neonates. Blood, organ graft and semen are the other sources of infection in the adult population.

CMV causes latent infection; hence, reactivation may result in disease in immunocompromised patients. These patients include those with HIV and those receiving organ and bone marrow transplantation. Immunodeficiency caused by antineoplastic compounds and ionising radiation may also cause reactivation of CMV infection.

Prevention and Control

No vaccine is available for scarlet transmission of CMV can be prevented by following safe sexual practices. Transmission of the virus is also reduced by regular screening of blood and organ donors for CMV seronegative. Prophylaxis with ganciclovir prevents reactivation of latent CMV infection in immunocompromised patients.

Human Herpesvirus 6

Human herpesvirus 6 (HSV-6) was first isolated from the peripheral blood of patients with lymphoproliferative disease in 1986. HSV-6 is a lymphotropic virus and is ubiquitous. The virus replicates well in B cells, CD$ T lymphocytes, megakaryocytes and fibroblasts. The virus is spread by saliva, the main source of infection. HSV-6 has two variants: A and B. Variant 6B causes roseola infantum, also known as the **sixth disease**. This is a febrile infection associated with a rash that affects young children.

Like other herpesviruses, HSV-6 causes a latent infection in T cells and monocytes. The replication of the virus in CD4+ T lymphocytes is controlled by CMI. The reactivation of the virus occurs in immunocompromised patients (patients with AIDS or other immunosuppression disorders). HSV-6 causes roseola infantum, a febrile infection that affects young children. The virus may also cause mononucleosis syndrome and lymphadenopathy.

Human Herpesvirus 7

Human herpesvirus 7 (HSV-7) was first isolated from peripheral CD4 cells of healthy persons in 1990. It is a lymphotropic virus. HSV-7, like HSV-6, has also been isolated from the saliva of healthy adults. It has been implicated as one of the causative agents of rubeola infantum and febrile seizures in children. The virus has also been associated with acute hemiplegia of childhood, hepatitis and respiratory tract infections.

Human Herpesvirus 8

DNA sequences of human herpesvirus 8 (HSV-8) were first identified in tissues from Kaposi sarcoma in patients with AIDS in 1994. This virus, also known as KS-associated herpesvirus, is a lymphotropic virus. It is believed to be important in causing and/or maintaining lesions in Kaposi sarcoma. The lesion in Kaposi sarcoma begins as a red-purple or dusky area that enlarges into a plaque and progresses subsequently into a tumorous growth. HSV-8 also causes lymphoma, Castleman disease and vascular tumours.

HSV-8 is secreted in saliva. It is sexually transmitted, most commonly among homosexuals. The virus is also transmitted through organ transplants. PCR is used to detect viral DNA in saliva and other specimens. Ganciclovir and foscarnet show some activity against EBV.

Case Study

A 21-year-old college student attended medicine OPD with a history of sore throat, low-grade fever for several days, followed by swollen cervical lymph nodes and increasing fatigue. Blood film examination revealed lymphocytosis with atypical cells. The serum for heterophile antibodies was positive by Paul–Bunnell test.

- Name the most likely diagnosis of the condition.
- What other laboratory tests can be performed to diagnose the condition?
- Describe the methods of transmission of the condition.
- Describe the preventive measures for the condition.

Adenoviruses

54

Adenoviruses belong to the family Adenoviridae, consisting of a group of medium-sized, non-enveloped, double-stranded DNA (dsDNA) viruses that share a common complement-fixing antigen and multiply in the nucleus of the infected cell. The family consists of two distinct genera: *Mastadenovirus* and *Aviadenovirus*, consisting of mammalian and avian adenoviruses, respectively.

Adenoviruses

The genus *Mastadenovirus* comprises at least 49 serotypes that infect humans. These human serotypes have been divided into six groups (A–F) based on their DNA homology and other physical, chemical and biological properties (Table 54.1).

Morphology

Adenoviruses are double stranded DNA (dsDNA), non-enveloped viruses measuring 80–110 nm in diameter (Fig. 54.1). Adenoviruses have a characteristic morphology. They are the only viruses having a fibre protruding from each of the 12 vertices of the capsid. The capsid is composed of 252 capsomeres which consist of hexons and pentons. An apical fibre, 9–31 nm in length and 62 kDa in molecular mass projects from each penton. This apical fibre helps to bind specifically adenoviruses to receptor sites on the host cells. The fibre contains the viral attachment proteins and can act as a haemagglutinin. The penton base and fibre are toxic to cells. The type-specific antigens are present in the pentons and also in the fibres.

Antigenic Properties

The viral group-specific antigen is present on the inner surface of the hexon capsomere. These group antigens can be detected by immunofluorescence or enzyme-linked immunosorbent assay (ELISA). The type-specific antigen is located in the other region of the hexon and the fibre. The fibre protein is the main type-specific antigen. Fibre determinant (gamma) is a type-specific haemagglutinin. Serotypes are identified by a neutralisation test.

Human adenoviruses are classified into six groups (A–F), also called subgroups or subgenera, based on the properties, such as DNA fragment analysis, fibre length haemagglutination and oncogenic potential. These subgroups are further classified into 49 different serotypes based on the differences in their penton bases and fibre proteins which determine the nature of tissue tropism and disease.

Table 54.1 Classification of human adenoviruses

Groups	Serotypes	Main site of tropism	Haemagglutination with red blood cells	Oncogenicity in hamsters	Transformation of cells
A	12, 18 and 31	Gastrointestinal tract	Rat (partial)	High	+
B	3, 7, 11, 14, 16, 21, 34 and 35	Respiratory tract, urogenital system, eye	Monkey (complete)	Weak	+
C	1, 2, 5 and 6	Respiratory tract	Rat (partial)	None or low	+
D	8–10, 13, 15, 17, 19, 20, 22–30, 32, 33, 36–39 and 42–47	Eye	Rat (complete)	None but can transform cells	+
E	4	Respiratory tract	Rat (partial)	None or low	+
F	40 and 41	Gastrointestinal tract	Rat (partial)	None	+

Other Properties

Adenoviruses are relatively stable and remain viable for about a week at 37 °C but are readily inactivated at 50 °C. They are resistant to ether and bile salts. They resist drying, detergents, gastrointestinal tract secretions including acid, protease and bile and even mild chlorine treatment.

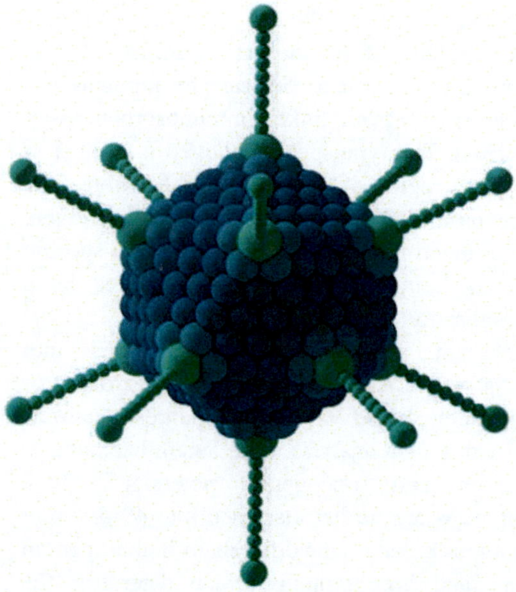

Fig. 54.1 Morphology of adenovirus. (http://link.springer.com/article/10.1007/s004280100534)

Laboratory Isolation

Human adenoviruses are highly host-specific; hence laboratory animals are not susceptible to infection by the virus.

Human adenoviruses can be grown only in tissue cultures of human origin. They grow well on a monolayer of HeLa, Hep-2, KB and human embryonic kidney cells. Cytopathic changes are observed after 1–4 weeks of inoculation and consist of cell rounding and aggregation into grape-like clusters. Infected cells swell, become ballooned and show intranuclear inclusions in stained preparations.

Primary human embryonic kidney cells support the growth of many fastidious adenoviruses. The viral antigens in the cell culture can be detected by a direct immunofluorescent test using polyclonal or monoclonal antibodies.

Pathogenesis of Adenovirus Infection

Adenoviruses are transmitted mainly by respiratory or faeco–oral contact between humans. The infection is acquired by (i) aerosol droplets, (ii) faeco–oral route, (iii) direct inoculation of conjunctiva by contaminated fingers, (iv) contaminated fomites including towels and medical instruments and (v) swimming in an improperly disinfected swimming pool.

They infect the conjunctiva or the nasal mucosa. They may multiply in the conjunctiva, pharynx or small intestine and then spread to pre-auricular, cervical and mesenteric lymph nodes, where epithelial cells are infected.

Adenoviruses attach to the surface of the cells by their fibres, then penetrate the cell and, once inside the cell, uncoat the viral DNA. The viral DNA is then transported into the cell's nucleus and initiates the replication cycle. Host cell DNA-dependent RNA polymerase transcribes the early genes leading to the formation of functional mRNA. Then in the cytoplasm, the early mRNA is translated into non-structural proteins. After viral DNA replication, late mRNA is transcribed and then translated into structural virion proteins in the nucleus. This is followed by assembly of virions in the nucleus and release of virions by lysis of the cells, but not by budding.

Adenoviruses may cause three different types of interaction with the infected cells. These are lytic infection, latent infection and transforming infection.

Lytic Infection

Adenoviruses infect mucoepithelial cells in the respiratory tract, gastrointestinal tract and conjunctiva or cornea, causing damage to these cells directly. After local replication of the virus, viraemia follows with subsequent spread to visceral organs. Dissemination occurs more commonly in immunocompromised patients than in immunocompetent individuals.

Latent Infection

The adenovirus has a unique ability to become latent in lymphoid and other tissues, such as adenoids, tonsils and Peyer patches. The exact mechanism of latency of adenoviruses in these tissues is not known. These latent infections can be re-activated in patients infected with other agents or in immunocompromised patients.

Oncogenic Transformation

Some adenoviruses belonging to groups A and B have the property for oncogenic transformation in rodent cells. However, the oncogenesis of human cells has not been demonstrated.

Adenovirus infection is characterised by developing effective and long-lasting immunity against re-infection. Both cell-mediated immunity (CMI) and humoral immunity are important. CMI plays an important role in limiting the proliferation and outgrowth of adenoviruses. Therefore, a different disease is seen in persons whose CMI is immunocompromised. Serum antibodies play an important role in resolving adenovirus infection. Resistance to clinical disease is believed to be directly related to the presence of circulating neutralising antibodies. The antibodies protect the person from re-infection with the same serotype but no other serotypes. Moreover, adenovirus infection confers lasting immunity to re-infection with the same serotype.

Clinical Syndromes

Tissue tropism of the specific groups or serotypes of adenovirus strain determines the type of disease caused by adenoviruses in humans. Certain serotypes of adenovirus are associated with specific syndromes. Respiratory infections are caused by low-numbered serotypes (1, 2, 3, 5 and 7) and gastrointestinal infections by high-numbered serotypes (40, 41, 42). Adenoviruses primarily infect children. Adults are also infected. The incubation period varies from 5 to 8 days.

Respiratory Infections

Acute Respiratory Disease

It is caused by adenovirus serotypes 4 and 7. Fever, rhinorrhoea, cough and sore throat are the typical symptoms that last for 3–5 days. This syndrome most often affects military recruits living in crowded conditions.

Pharyngoconjunctival Fever

The condition occurs primarily in school-going children. Fever, sore throat, coryza and red eye are the classic presentations of the condition. These symptoms may precede ocular findings, or they may not be present. Acute conjunctivitis may occur as a separate entity with or without pharyngitis. The condition is usually self-limiting. This condition is caused predominantly by serotypes 3, 4 and 7.

Other Respiratory Tract Diseases

These include bronchiolitis, **croup, laryngitis** and cold-like symptoms are caused by adenoviruses.

Eye Infections

Acute Follicular Conjunctivitis

The condition manifests as a non-suppurative eruption of the conjunctiva with enlargement of the sub-mucous lymphoid follicles, and swollen pre-auricular lymph nodes are features. Commonly associated serotypes are 3, 4 and 11. Chlamydial conjunctivitis is clinically similar to adenoviral conjunctivitis.

Epidemic Keratoconjunctivitis

It is a highly contagious condition and has an insidious onset of unilateral red eye. Subsequently, both eyes are involved. Patients complain of photophobia, tearing and pain. Fever and lymphadenopathy may be present in some children. Inflammation of the conjunctiva may persist for a week, accompanied by residual scarring and visual impairment. The condition may occur as an epidemic, usually caused by serotype 8 and less often by serotypes 19 and 37.

Gastrointestinal Infections

Gastroenteritis and diarrhoea are most commonly associated with serotypes 40 and 41, but other serotypes may also be involved. The enteric adenovirus infection is a common cause of infantile diarrhoea in daycare centres. The condition manifests as fever and watery diarrhoea, which resolves within 1–2 weeks.

Other Manifestations

Adenoviruses have also been associated with acute haemorrhagic cystitis caused by serotypes 11 and 21, intussusception in young children, pertussis-like illness, musculoskeletal disorders and genital and skin infections.

Adenovirus Infections in an Immunocompromised Host

Multiple serotypes of adenoviruses cause these infections. These serotypes cause disease during the post-transplant period in the patients who have received a haematopoietic stem-cell transplant. Patients receiving stem-cell transplantation and T-cell deletion, and lymphopenia are at risk of serious adenovirus infections. Infections in these patients may cause variable manifestations but include haemorrhagic cystitis, pneumonia, nephritis and gastroenteritis. Infection occurs either by re-activating old infection or by exogenous transmission from other infected hosts.

Adenovirus serotypes associated with various human illnesses are summarised in Table 54.2.

Table 54.2 Human infections associated with various serotypes of adenoviruses

Human infections	Adenoviruses serotypes
Acute respiratory disease	4, 7, 14 and 21
Pharyngoconjuctival fever	3, 4 and 7
Endemic pharyngitis	1, 2, 5 and 6
Epidemic pharyngitis	3, 4 and 7
Pneumonia	1–3 and 7
Pharyngoconjunctival fever	3, 4 and 7
Acute follicular conjunctivitis	3, 4 and 11
Epidemic keratoconjunctivitis	19 and 37
Gastroenteritis and diarrhoea	40 and 41
Intussusception	1, 2 and 5
Acute haemorrhagic cystitis	11 and 21
Disseminated infection	5, 34, 35 and 43–47

Laboratory Diagnosis

Specimens

Specimens used to depend on the nature of the clinical illness caused by adenoviruses. These include throat swab, nasopharyngeal aspirate, bronchial lavage, conjunctival swab, corneal scraping, urine and faeces.

Microscopy

Certain serotypes of enteric adenoviruses may be seen directly in stool specimens by electron microscope. These serotypes (40, 41 and 42) are difficult to grow in cell cultures, and hence are referred to as non-cultivable adenoviruses.

Direct Antigen Detection

ELISA or **direct fluorescent antibody test** using specific antibodies raised against adenoviruses is used to detect viral antigens in faeces and nasopharyngeal secretions.

Virus Isolation

Isolation of the virus by cell culture establishes the specific diagnosis of the condition. Viruses from clinical specimens may be isolated on primary human embryonic kidney cells, HeLa, Hep-2 and KB cells. Within 2–20 days, the viruses produce a lytic infection with characteristic inclusion bodies. Infected cells show rounding and grape-like clusters with intranuclear inclusions.

The viral isolates are usually classified into subgroups by haemagglutination with rat and monkey erythrocytes. Further serotyping is carried out by a neutralisation test.

Serology

Serology is rarely used for the diagnosis of cases. It is usually used for epidemiological studies. They are also used to confirm the causative role of a faecal or respiratory tract isolate by identifying its serotype.

Molecular Detection

DNA probe and polymerase chain reaction (PCR) are used to identify enteric adenovirus serotypes 40, 41 and 42 directly in stool specimens that do not grow in cell cultures. These methods also detect, type and group adenovirus isolates in tissue culture and stool specimens.

Treatment

At present, no specific antiviral agents are available to treat adenovirus infections. Ribavirin has been used with variable success in treating adenovirus infections in immunosuppressed hosts. Adenovirus has shown the capacity to carry DNA inserts of up to 8 kb but is yet to be proven as a potential vector in gene therapy.

Epidemiology

Adenovirus infections are found worldwide.

Geographical Distribution

More than one serotype of the virus may produce the same clinical syndrome, and one serotype of the virus may cause clinically different diseases. Adenoviruses 1–7 are the common serotypes worldwide and are responsible for most cases of adenovirus-associated infections.

Reservoir, Source and Transmission of Infection

Adenovirus infections are exclusively human infections. No animal reservoirs are present. Infected symptomatic, as well as asymptomatic humans who excrete adenoviruses intermittently in their respiratory secretions and also in their stool, are the sources of infection. Overcrowding, poor hygiene and close contact facilitate the transmission of infection. Adenovirus typically affects children, from infants to school-going ones, though children of any age may be affected.

Prevention and Control

Three live, non-attenuated vaccines against serotypes 4, 7 and 21 are available as oral vaccines. All three vaccines are monovalent, and each vaccine contains only one serotype. These vaccines are given orally in a gelatin-coated capsule, protecting the live virus from inactivation by stomach acidity. The virus is released in the intestine, where it replicates and induces neutralising antibodies. The vaccine strain does not spread from person-to-person contacts. The vaccine is yet to be available widely.

Adeno-Associated Viruses (AAVS)

AAVs are small, icosahedral viral particles measuring 20–25 nm in diameter. They are so called because they can multiply only in cells simultaneously infected with adenovirus but cannot multiply independently because they lack enough DNA. These viruses have been classified in the genus *Dependovirus*. Four serotypes (1–4) have been recognised, of which types 1, 2 and 3 are of human origin, and type 4 is of simian origin. The pathogenic role of the human serotypes of AAVs is not clear.

Case Study

A 3-year-old child in a nursery school suffered from a mild respiratory tract infection. Subsequently, many other children studying in the same nursery suffered from the same illness. A viral infection is suspected.

- Name the adenovirus types frequently associated with this illness.
- List other types of adenoviruses and the diseases caused by them.
- Describe the laboratory methods for diagnosis of the condition.
- Describe the vaccines available against the condition.

Picornaviruses 55

The family Picornaviridae is one of the largest families of viruses and includes a large number of very small (*pico:* measuring small; *rna: RNA* virus) RNA viruses. They are non-enveloped viruses measuring 27–30 nm in size. The capsid is a naked icosahedral made up of 60 protein subunits. The genome consists of a single linear molecule of single-stranded RNA. The genome RNA is unusual because it has a protein at the 5′ end that serves as a primer for transcription by RNA polymerase.

Classification

The family Picornaviridae consists of more than 230 members and is divided into five genera: *Enterovirus*, *Rhinovirus*, *Hepatovirus*, *Arthrovirus* and *Cardiovirus*. The first three genera (*Enterovirus*, *Rhinovirus* and *Hepatovirus*) are the important viruses that cause disease in humans. Two other genera (*Arthrovirus* and *Cardiovirus*) are of veterinary importance and cause foot-and-mouth disease in cattle and meningoencephalomyelitis in mice. The picornaviruses causing human infections are summarised in Table 55.1. The important features of picornaviruses are summarised in Table 55.2.

Enteroviruses

Human enteroviruses consist of at least 72 serotypes, which include poliovirus types 1–3, coxsackie viruses A types 1–24, coxsackie viruses B types 1–6 echoviruses types 1–34 and enteroviruses 68–71. Originally hepatitis A virus (enterovirus 72) was included in this group but has been reclassified as a hepatovirus in the genus *Hepatovirus* (Chap. 66). Most enteroviruses are host-specific; they infect only one or a few related species.

Poliovirus

Poliomyelitis is an enteric infection caused by polioviruses transmitted by the faecal–oral route.

Morphology

Poliovirus was the first animal virus to be purified and obtained a crystalline form. The viruses are spherical particles about 27 nm in diameter (Fig. 55.1). The virion comprises 60 subunits, each consisting of four viral proteins (VP1–VP4). The viral protein VP1 contains the major antigenic site for combination with type-specific neutralising antibodies. The viral genome is a single-stranded positive-sense RNA, which can be directly translated by host ribosomes to form

Table 55.1 Human infections caused by picornaviruses

Picornaviruses	Serotype	Clinical syndrome
Poliovirus	1–3	Aseptic meningitis, paralysis and undifferentiated febrile illness
Coxsackie virus group A	1–24	Aseptic meningitis and exanthemas
	7, 9	Paralysis
	2, 5–7, 9	Encephalitis
	2–6, 8, 10	Herpangina
	5, 10, 16	Hand-foot-and-mouth disease
	4, 9	Hepatitis
	9, 16	Pneumonitis of infants
	18, 20–22, 24	Diarrhoea
	21, 24	Cold
	24	Acute haemorrhagic conjunctivitis
Coxsackie virus group B	1–6	Aseptic meningitis, febrile illness
	2–5	Paralysis
	1–5	Encephalitis, pleurodynia, myocarditis, generalised disease of infants, pericarditis
	5	Exanthemas and hepatitis
	1, 3–5	Cold
	4, 5	Pneumonia
	3, 4	Diabetes mellitus
Echovirus	1–34	Aseptic meningitis and diarrhoea
	2, 4, 6, 9, 11, 30	Paralysis
	2, 6, 9, 19	Encephalitis
	2, 4, 6, 9, 11, 16, 18	Exanthemas
	1, 6, 9	Pleurodynia
	1, 6, 9, 19	Myocarditis and pericarditis
	4, 9, 11, 20, 25	Cold
	4, 9	Hepatitis
	11	Generalised disease of infants
Enterovirus	68	Pneumonia
	70	Acute haemorrhagic conjunctivitis
	71	Pulmonary oedema, herpangina, aseptic meningitis and hand-foot-and-mouth disease
	70, 71	Paralysis and encephalitis
Rhinoviruses	>100	Common cold

Table 55.2 Classification and important properties of picornaviruses of medical importance

Virus	Antigenic types	Serotypes	Stability at acid pH	Optimal temperature for growth (°C)	Sites of isolation		
					Nose	Throat	Intestine
Poliovirus	1–3	3	Stable	37	–	+	+
Coxsackie A virus	1–22, 24	23	Stable	37	–	+	+
Coxsackie B virus	1–6	6	Stable	37	–	+	+
Echovirus	1–9, 11–27, 29–34	30	Stable	37	–	+	+
Enterovirus	68–71	4	Stable	37	–	+	+
Rhinovirus	1–113	113	Labile	37	+	+	–

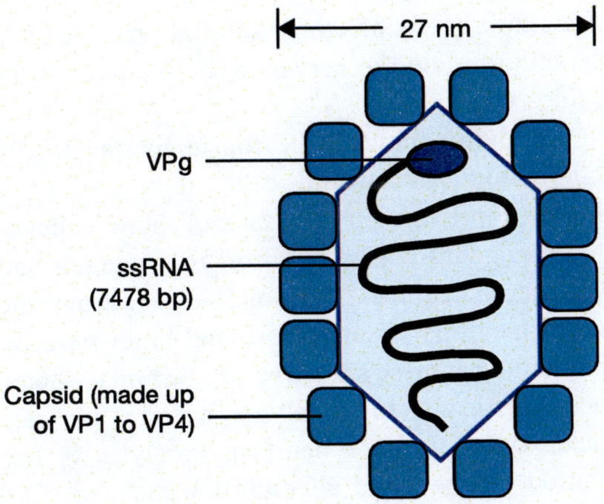

Fig. 55.1 Schematic diagram of structure of a poliovirus

a polyprotein, which is divided into 11 different proteins.

Virus Replication

Polioviruses are cytolytic; they replicate in the cytoplasm of the host cell. The virion causes infection first by binding to the specific receptor on the cell membrane and entering the cell. Once inside the cell, the virion uncoats by removing the capsid and releasing the RNA genome. The RNA serves as the mRNA and is translated into a very large polypeptide known as a non-capsid viral protein. Subsequently, the viral protein is utilised by the viral enzyme protease to form capsid proteins of the progeny virions and several non-capsid proteins, including the RNA polymerase.

RNA polymerase initiates the synthesis of progeny RNA genomes. This is followed by the assembly of the progeny virions by coating the genome RNA with capsid proteins in the cell cytoplasm. The release of progeny virions occurs by the lysis and death of the cells but not by budding from the cell membrane.

Antigenic Properties

Based on the neutralisation test, the polioviruses have been classified into three serotypes (types 1, 2 and 3). The prototype strains for type 1 poliovirus are *Brunhilde and Mahoney strains*, for type 2 poliovirus are *Lansing and MEFI strains* and for type 3 poliovirus are *Leon and Saukett strains*.

Type 1 is the most common poliovirus and causes most poliomyelitis epidemics; type 2 strain is usually associated with endemic infections, whereas type 3 strains occasionally cause epidemics. Two antigens, C (coreless or capsid) and D (dense), are recognised by enzyme-linked immunosorbent assay or precipitation test. The C antigen, also known as heated or H antigen, is associated with the empty, non-infectious virus. It is less specific and reacts with heterotypic sera. The D antigen, also known as the nature or N antigen, is associated with the whole virion. It is more specific and is type-specific. Antibodies against D antigen are protective; hence the potency of injectable polio vaccine is measured in terms of the D antigen unit.

Other Properties

Poliovirus is sensitive to heat and is readily inactivated at 55 °C for 30 min. It is also sensitive to formaldehyde and oxidising disinfectants. The viruses are readily destroyed by chlorination in water, but the presence of organic matter delays inactivation.

Polioviruses are resistant to lipid-soluble agents (ether, chloroform and bile), the intestine's proteolytic enzymes and detergents. They survive in faeces for months at 4 °C and years at −20 °C. They survive for days to several weeks in faeces at room temperature depending on the environmental conditions, such as temperature, moisture, acidity, etc. Polioviruses are killed by lyophilisation.

Laboratory Isolation

Chick Embryos

Laboratory-maintained strains may grow in chick embryos, but fresh poliovirus isolates cannot grow.

Tissue Culture

Polioviruses grow readily in cell lines of primate origin. They can be cultured on monkey kidney cell lines, human amnion, HeLa, Hep-2, MRC-5 and other cell lines. The cytopathic effects that consist of cell retraction, increased refractivity, cytoplasmic granularity and nuclear pyknosis and are observed within 48 h of inoculation. Eosinophilic intranuclear inclusion bodies may be demonstrated in stained preparation of the infected cell lines. Well-formed plaques develop in infected monolayers with agar overlay.

Laboratory Animals

Polioviruses show high host specificity restricted to primates, including humans and non-human primates, such as monkeys and apes. This is due to the presence of a specific receptor that is present only on the primate cell membrane. Cynomolgus and rhesus monkeys are commonly used animals. They can be experimentally infected by intracerebral or intraspinal inoculation. Cynomolgus monkeys and chimpanzees can also be infected orally.

Pathogenesis of Poliomyelitis

Poliovirus is transmitted primarily by the faecal–oral route by ingestion of food and water contaminated with human faeces. It is also transmitted by inhalation or through fomites contaminated with respiratory secretions. The viral particles initially multiply in the nasopharynx and the gastrointestinal mucosa. The virions are resistant to the stomach's acidity and lytic activities of the protease and other enzymes of the intestinal tract and bile.

On entering the body, the virus infects and multiplies in the tonsils and Peyer patch of the ileum. It then spreads to regional lymph nodes and enters the blood, causing a primary viremia. On continued infection and multiplication of virus in the reticuloendothelial cells, it invades the bloodstream again and causes secondary viremia. During this period of viremia, the poliovirus crosses the blood–brain barrier or gains access to the brain by infecting skeletal muscle and travelling upon the nerves to the brain as in the rabies virus.

Poliovirus shows tissue tropism by specifically combining with neural cells. It recognises a receptor present on the anterior horn cells of (1) the spinal cord, (2) dorsal root ganglia and (3) motor neurons. The poliovirus destroys the motor neurons, anterior horn and brain stem by combining at these sites. The destruction of motor neurons leads to paralysis of the muscles innervated by those neurons. Paralysis is not caused by the virus infecting the muscles. The poliovirus also infects the brain stem, causing bulbar poliomyelitis associated with respiratory paralysis.

Poliovirus causing the pathological changes in the central nervous system (CNS) is usually responsible for causing symptoms of poliomyelitis. Immune mechanisms do not play any role in the pathogenesis of the disease.

Host immunity in poliomyelitis is mostly dependent on humoral antibodies. Both the serum and secretory antibodies play an important role in conferring protection against poliomyelitis.

Serum IgM antibodies appear within a week of infection and persist for nearly 6 months. IgG antibodies, which develop subsequently, persist lifelong. The neutralising antibodies in the blood protect against disease by the same serotype of poliovirus. These antibodies, however, do not prevent infection of the intestinal neuron and excretion of virus in the faeces.

Secretory IgA antibodies provide mucosal immunity against the virus (1) by preventing intestinal infections and (2) by preventing the shedding of the virus. The IgA antibody in breast milk protects infants from infection. Therefore, poliomyelitis tends to be more severe, and the shedding of the virus is more prolonged in the infected human host with an altered humoral immune response.

Host immunity is type-specific. However, there is a significant amount of cross-protection between types 1 and 2 and types 2 and 3 polioviruses. There is little or no cross-protection between types 1 and 3 strains.

Cell-mediated immunity plays a little or very insignificant role in the immunity against poliovirus.

Clinical Syndromes

Poliovirus causes a wide spectrum of illnesses in unvaccinated people. The incubation period is usually 10 days but may vary from 4 days to 4 weeks.

Asymptomatic Illness

Most patients (90–95%) infected with poliovirus develop inapparent infection and are asymptomatic. This asymptomatic illness is caused by a viral infection confined to the oropharynx and the intestine.

Abortive Poliomyelitis

Abortive poliomyelitis is a minor illness occurring in a smaller number of people. This is a febrile illness characterised by fever, headache, sore throat, loss of appetite, vomiting and abdominal pain. Neurological symptoms are typically absent. The duration of this illness is usually less than 5 days.

Non-paralytic Poliomyelitis

Non-paralytic poliomyelitis, also known as aseptic meningitis, is caused by the invasion of the virus into the CNS. The symptoms of non-paralytic poliomyelitis are similar to those of abortive poliomyelitis but are more intense. Stiffness of the posterior muscles of the neck, trunk and limbs is present in addition to the symptoms of the minor illness. This condition occurs in 1–2% of infected patients.

Paralytic Polio

This is the major illness, which occurs in 0.1–2% of individuals infected with poliovirus. This is the most severe manifestation of poliomyelitis and appears 3–4 days after abortive poliomyelitis has subsided. This condition is caused due to invasion of the virus from blood to the anterior horn cells of the spinal cord and the motor cortex of the brain. Depending on the various tissues or organs affected and the intensity of neuronal infection, Paralytic polio may be of two types, as follows:

Paralytic poliomyelitis or **spinal paralysis** is characterised by spinal paralysis involving one or more limbs. Asymmetrical flaccid paralysis with no sensory loss is the typical manifestation. This condition is caused mostly by poliovirus type 1. This condition is seen in some of the vaccinated individuals following vaccination. This occurs due to the reversion of the attenuated vaccine virus types 2 and 3 to virulence types. In this condition, there may be paralysis of only one limb, such as one leg, or there may be complete flaccid paralysis of both the legs and hands. The condition may progress to death or may recover completely with residual paralysis.

Bulbar poliomyelitis is caused due to involvement of the cranial nerves, most commonly 9th, 10th and 12th. This condition tends to be more severe with involvement of the pharynx muscles, vocal cords and respiration. The condition may cause death in the majority of the patients.

Post-poliomyelitis Syndrome

This condition is sequelae of poliomyelitis, which may develop 20–40 years after infection with the poliovirus. This condition is seen in 20–80% of patients who have recovered from poliomyelitis. Recurrence of weakness or fatigue is observed in this condition, and it usually involves the muscles that were initially affected by the poliovirus.

Laboratory Diagnosis

Specimens

Stool, throat swab and cerebrospinal fluid (CSF) are the specimens used to isolate the viruses. Viruses can be isolated from faeces for more than 30 days during the illness and from the throat swab during the first few days of the illness. The virus is isolated rarely from the CSF specimens.

Microscopy

Microscopy of the CSF shows a predominantly lymphocytic pleocytosis with the presence of 25–500 cells/mm^3. The virus is rarely demonstrated in the CSF.

Virus Isolation

Isolation of viruses from the clinical specimen by tissue culture is the most specific method in diagnosing poliomyelitis.

Virus isolation from faeces and throat swabs is carried out by cultivating on monkey kidneys, human amnion, HeLa, Hep-2, Buffalo green monkey (BGM), MRC-5 and other cell cultures. The cytopathic effects produced by the virus are observed within 48 h. These consist of cell retraction, increased refractivity, cytoplasmic granularity and nuclear pyknosis. Eosinophilic intranuclear inclusion is seen.

Isolation of poliovirus from faeces does not indicate a diagnosis of poliomyelitis, as large numbers of asymptomatic illnesses are seen. Hence, the isolation of poliomyelitis virus from faeces needs to be interpreted carefully along with the clinical presentation of the disease.

Identification of serotype is carried out by performing a neutralisation test. Differentiating wild virulent virus strains from an attenuated vaccine strain can be done by virulence test in the monkeys and by performing polymerase chain reaction (PCR).

Serology

Serology is based on demonstrating a fourfold increase in the antibody titre of the serum collected at the time of acute illness and the period of convalescence. Neutralisation or complement fixation test is carried out to demonstrate antibodies. Neutralising antibodies appear early and are present throughout life. However, serodiagnosis is less frequently used for the diagnosis of poliomyelitis.

Molecular Detection

PCR is a highly sensitive and specific test for the diagnosis of poliomyelitis. **Sequencing** could be used to identify the three different strains of poliovirus that usually circulates in the population that is the wild virus, OPV strain and vaccine-derived polioviruses. Identification of strains present in a population would help understand the molecular epidemiology of poliovirus in a given population, which is useful from a public health perspective.

Treatment

No antivirals are available for the treatment of poliomyelitis.

Epidemiology

Poliovirus mainly affects children. However, individuals of any age may also develop the disease. Most epidemics of paralytic poliomyelitis are caused by poliovirus type 1, less commonly by type 3. Poliovirus type 2 is a common cause of paralytic poliomyelitis in India and is the causative agent of inapparent illness in Western countries.

Geographical Distribution

The worldwide prevalence of poliomyelitis has decreased by more than 99% due to improved socio-economic conditions and the availability of vaccines. Significant progress has been made in reducing the prevalence of poliomyelitis in India. India was declared to be free in the year 2014.

Reservoir, Source and Transmission of Infection

Natural infection occurs only in humans. Infected humans excreting poliovirus in their stool are the major reservoir of infection. Infected stool containing poliovirus is the major source. The viruses secreted in throat secretions during the early stage of the illness may also be another source of infection.

Poor sanitation, low socio-economic status and crowded living conditions facilitate the transmission of infection. Immunocompromised patients, such as those with HIV infection, B cell dysfunction and IgA deficiency, are particularly at high risk of developing poliomyelitis when exposed to poliovirus.

Prevention and Control

Poliovirus vaccines are the key component in preventing polio and have played an important role in the effort to eradicate polio worldwide.

Vaccines

Two types of vaccines are used currently against poliomyelitis.

Inactivated Polio Vaccine

Jonas Salk developed inactivated polio vaccine or Salk vaccine in 1956 for immunisation against poliomyelitis. The vaccine contains formalin-inactivated strains of the three serotypes of the poliovirus grown in monkey kidney cell culture. The vaccine is given to children by deep subcutaneous or intramuscular injections in three doses. The vaccine is given at 2 months, 4 months and at school entry, usually 4 years. It does not cause the development of vaccine-associated poliomyelitis (VAP).

The injection of the vaccine elicits the production of higher IgG antibody titres in the serum, but it does not stimulate the production of detectable levels of secretory IgA in the gut, hence does not prevent alimentary tract infection. The main advantage of this vaccine is that since it uses killed viruses, it can be safely used for immunisation in immunocompromised hosts. The disadvantages of this vaccine are that (1) it is not as immunogenic as the oral polio vaccine, (2) it does not induce local mucosal immunity in the gut and (3) it needs to be administered parenterally, which shows poor compliance.

In 1955, there was a major setback as more than 100 cases of paralytic poliomyelitis occurred with this vaccination. Cutter was the manufacturer of this vaccine; hence the incident came to be known as the Cutter incident. The company's vaccines had inactivated virus, which led to this incident. After this incident, many modifications were made to the vaccines to make them safer.

Live Attenuated Oral Polio Vaccine

Live attenuated oral polio vaccine or Sabin vaccine was developed by Albert Sabin in the year 1962. It contains live attenuated strains of the three serotypes of poliovirus cultured in monkey kidney cells or human diploid cells. The vaccine is given orally, which causes natural infection. It elicits the production of both local secretory IgA antibodies in the pharynx and alimentary tract and humoral IgG antibodies in the serum. The virus cannot multiply in the CNS; hence it is non-neurovirulent. The vaccine virus is excreted for several weeks in the faeces, during which the vaccine virus spread may occur to close contacts, inducing or boosting immunity in them. The first dose of the vaccine is given to infants at the age of 1½ months along with DPT. Second, third and fourth vaccine doses are given when children are raised to 2½, 3½ and 16–24 months, respectively (Fig. 55.2).

The oral polio vaccine has been used since the 1960s. This vaccine has been mainly responsible for a significant decrease in disease prevalence throughout the world. The advantages of this vaccine are many: first, it induces local mucosal immunity; second, it provides appropriate herd immunity and finally, it is cost-effective, especially in developing countries.

OPV, used in India, contains 10 lakh type 1 virus, 2 lakh type 2 virus and 3 lakh type 3 virus TC ID50 per dose (0.5 mL). $MgCl_2$, which maintains pH below seven, is thermostabilised along with the liquid vaccine. The vaccine is kept in airtight containers, so the pH is maintained. The shelf life of vaccines is 4 months when stored in 4–8 °C and 2 years when kept in −20 °C. Not storing the vaccines in proper conditions will lead to the failure of OPV in controlling poliomyelitis in developing countries.

The major disadvantage of this vaccine is the loss of efficacy of vaccine due to improper cold chain. Association of the vaccine with VAP is another major disadvantage. Although the virus content in this vaccine is attenuated, it may become neurotropic and may produce disease similar to wild type virus. The risk of VAP has been estimated at between 0.5 and 3.4 cases per million.

The advantages and disadvantages of Sabin and Salk vaccines are summarised in Table 55.3. These include the following:

1. **Ease of administration:** OPV is preferred to killed vaccines that are administrated through injections. However, an advantage of the killed vaccine is that it can be administered with the DPT vaccine as a quadruple vaccine.
2. **Cost-effectiveness:** Live vaccine is very affordable, which is an important factor in

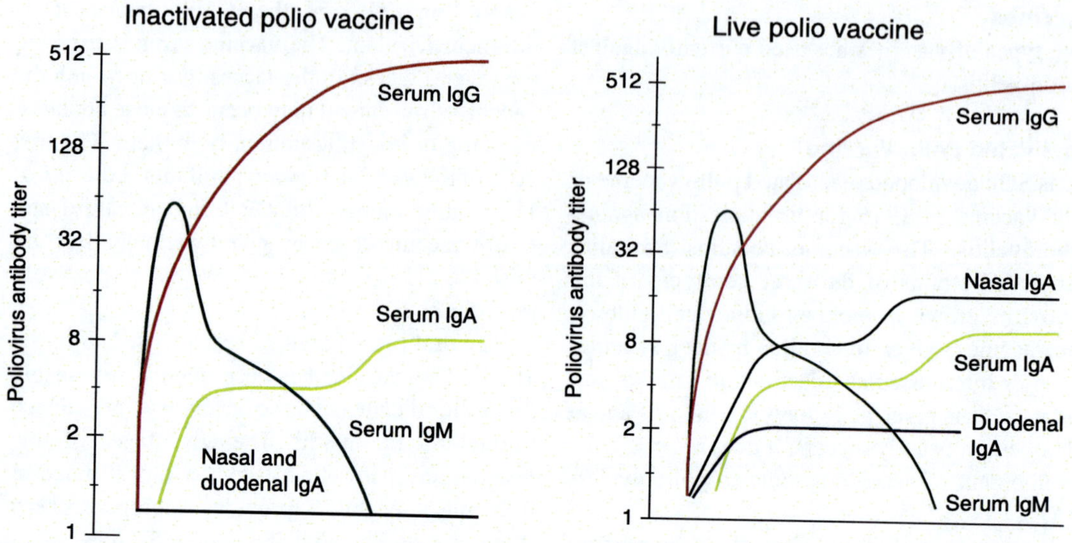

Fig. 55.2 Serum and secretory antibody response to IPV and three doses of OPV (WHO, 1993 immunological basis of immunisation series module 6: poliomyelitis global programme for vaccines and immunisation EPI, WHO)

providing vaccination to large populations in developing countries

3. **Nature of immunity:** Nature of immunity is one of the essential differences between the two vaccines. A killed vaccine stimulates only the systemic antibody response without intestinal immunity. Even though the vaccination is administered, infection with a wild strain can result in the intestinal multiplication of the virus. Only the individual is protected

Table 55.3 Comparative evaluation of polio vaccines

Polio vaccine	Advantages	Disadvantages
Live (oral) vaccine (**Sabin vaccine**)	Effective	Risk of vaccine-associated poliomyelitis in vaccine recipients
	Lifelong immunity	Not safe for administration to immunodeficient patients
	Induction of local immunity	Unstable—To be stored in cold conditions
	Induction of herd immunity	
	Ease of administration	
	No need for repeated booster vaccine	
	Cheap	
	Useful in epidemics	
Killed (parenteral) vaccine (**Salk vaccine**)	Effective	Lack of induction of secretory antibody
	Stable	Booster dose needed for lifelong immunity
	Safe in immunodeficient patients	Costly
	No risk of vaccine-related disease	No community protection
		Not useful in epidemics

by the antibodies which are being circulated. When it comes to live vaccines, it stimulates local immunity within the gut, which results in the wild viruses being unable to multiply in the intestine. This helps in protecting the individual and the community.

4. **Persistence of immunity:** Administering booster doses periodically may be required in maintaining the immunity caused by a killed vaccine. However, immunity from a live vaccine remains as a natural active immunity and lasts longer.
5. **Use in epidemics:** Administering OPV at the community level, OPV, which is a monovalent vaccine of the same type which causes the epidemic at the onset of paralytic poliomyelitis, can effectively stop the epidemic. In many parts of the world, this is being followed successfully.

Recently, a new monovalent oral poliovirus type I vaccine (MOPVI) has been introduced. This is used to eliminate some of the last poliovirus reservoirs in the endemic countries. The vaccine is contraindicated in children who are immunocompromised and also in those whose caretakers are immunocompromised.

Criteria for attenuation of poliovirus strains include: (1) the strain used should not be neurovirulent when tested by inoculating intraspinally in monkeys, (2) this should cause intestinal infection after feeding and induce an immune response, (3) must not obtain neurovirulence after serial enteric passage and (4) should be genetically stable so that it can be differentiated from the wild strains.

Differentiating markers for wild and attenuated polio strain include (1) **d marker**: Avirulent strains will not grow in a low level of bicarbonate while wild strains will grow, (2) **rct 40**: At 40 °C avirulent strains grow poorly while wild strains grow well, (3) **MS**: When stable cell line of monkey kidney is used, wild strains grow well, but avirulent cells grow poorly and (4) **McBride intratypic antigenic marker** uses the rate at which specific antiserum is inactivated.

Safety of Polio Vaccine

Killed and live vaccines are both safe. Few suggest that following vaccination, the attenuated strains may acquire neurovirulence on serial enteric passage. Since there are only a few reports on vaccine-induced poliomyelitis threat is negligible. The United States reports 5–10 paralysed poliomyelitis cases each year due to vaccine strain. Oral Polio Vaccination has been widely administered in various parts of the world. Nevertheless, OPV is not considered harmless in immunosuppressed patients, but there is no significant harm from the killed vaccine.

The Efficiency of Polio Vaccine

A complete course of killed vaccine stimulates an acceptable immune response. In developed countries, one or two doses of OPV have produced 90–100% seroconversion in children. However, in developing countries, especially in tropical regions, there are dissatisfactory responses, particularly in connection with Polio type I. More than half of the people administered with vaccines failed to respond serologically even after multiple doses, and the reasons are uncertain.

Issues Related to Vaccine Failure in Tropical Regions

In tropical countries, the vaccines are sometimes interfered with other enteroviruses. Experiments have shown that coxsackie B has interfered with the poliovirus. Frequently, vaccine virus colonisation is prevented by frequent diarrhoeal diseases. Another reason is breastfeeding soon after or before the administration of vaccines. The antibodies present in the breast milk may neutralise the vaccine virus. It has been proposed in India and other tropical regions that primary immunisation with killed vaccines along with triple antigen followed by live oral vaccines can be given because the rate of seroconversion after oral immunisation is poor.

Spread of Vaccine-Derived Polioviruses

In many developed countries, the propensity of the vaccine virus to spread naturally among children is a major disadvantage, but it can help expand the vaccine coverage in countries with wild virus endemicity. Nevertheless, it is advisable to vaccinate the entire community at one go which helps in preventing its natural dissemination.

Poliomyelitis Eradication

Eradication of poliomyelitis was considered possible through global immunisation with OPV. World Health Organisation assembly proposed to target this objective in the year 1988. WHO planned to eradicate poliomyelitis by the year 2000 globally; however, due to reduced progress of immunisation in many countries, this objective could not be accomplished.

Pulse Polio Immunisation Program (PPI)

PPI is being used in India to eradicate poliomyelitis. This program includes mass administration of OPV on a particular day to all the children aged between 0 and 5 years within the community. The program follows the strategy of administering two rounds of doses at an interval of 4–6 weeks from November to February, which is considered a low polio transmission period in winter. Whatever the child's immunisation status, the dosage can be administered. This does not replace the national universal immunisation program. By doing so, increasing the immunity of the majority of the children of a particular area can be achieved.

Coxsackie Viruses

Coxsackie viruses are so named because the viruses were first isolated in Coxsackie village in New York, United States, by Dandruff and Sickel in the year 1985. Based on the pathological changes produced in suckling mice, these viruses are classified into two groups: Coxsackie viruses A and Coxsackie viruses B. **Coxsackie A virus** produces generalised myositis, flaccid paralysis followed by death within a week in suckling mice. **Coxsackie B virus** produces patchy focal myositis, spastic paralysis, localised lesions in the liver, myocardium, pancreas and brain and brown fat pads.

Properties of the Virus

The structure and morphology of the coxsackie viruses and the RNA genome are similar to poliovirus. However, unlike poliovirus, they can infect mammals other than primates. The replication cycle is similar to that of poliovirus. Coxsackie A viruses consist of 24 serotypes and are associated with diseases involving vesicular lesions, such as herpangina. Coxsackie B viruses consist of six serotypes and are most frequently associated with myocarditis and pleurodynia.

Coxsackie A viruses show tropism for skin and mucous membranes, whereas coxsackie B viruses show a predilection for visceral organs, such as the liver, heart, pancreas and pleura. Both the viruses infect anterior horn cells of the motor neurons and meninges and cause paralysis. They replicate first in the oropharynx and gastrointestinal tract, from where they spread by the blood circulation. Host immunity is IgG-antibody mediated and is type-specific.

Coxsackie virus infection occurs worldwide. Humans are the only natural hosts. The viruses are transmitted primarily by the faecal–oral route.

Clinical Syndromes

Coxsackie A Virus

Coxsackie A viruses produce the following conditions.

Herpangina

This condition is caused by coxsackie A virus serotypes 2, 4, 5, 6, 8 and 10. The infection is most commonly seen in children between 1 and 7 years. The symptoms include sudden onset of fever, sore throat and difficulty in swallowing. The classical finding is a painful vesicular eruption of the oral mucosa around the soft palate and uvula. Less commonly, these lesions may be found on the hard palate. The condition is self-limiting. The viruses are isolated from the lesions or the faeces.

Hand-Foot-and-Mouth Disease

This is a vesicular exanthema usually caused by coxsackie virus serotypes 5 and 16. This is mainly a disease of children, seen most commonly in patients younger than 10 years. After an incubation period of 3–6 days, the infection begins with pro-dromal symptoms, such as fever, cough, sore throat, malaise and anorexia. After 12–36 h, vesicular eruptions appear on the hands, feet and oral cavity. This is a self-limiting condition, and the illness subsides in a few days.

Coxsackie B Virus

Coxsackie B viruses produce the following conditions.

Pleurodynia

This condition, also known as epidemic myalgia or Bornholm disease, is caused by coxsackie B virus serotypes 3 and 5. This condition is an acute illness, which manifests with a sudden onset of fever accompanied by muscular pain and pain in the chest and abdomen. The pain is spasmodic. The condition usually lasts for 4 days but may relapse again after the condition has been asymptomatic for several days.

Myocarditis

This is a serious condition caused by the coxsackie B virus, mostly in newborn infants. Shortness of breath, dull or sharp chest pain and fever are the common manifestations of the condition. This is a life-threatening condition, mostly in newborns. In adults, the condition presents with myocardial infection with fever. The condition is associated with high morbidity.

Juvenile Diabetes

Juvenile diabetes has been alleged to associate with coxsackie B4 infection. However, this role in causing the disease has not been proved.

Orchitis

This disease has also been reported to be associated with the coxsackie virus.

Transplacental and Neonatal Transmission

This transmission has been exhibited along with coxsackie B viruses. It results in serious disseminated diseases, including hepatitis, meningoencephalitis and adrenocortical involvement.

Type B viruses have been found to be linked with a condition called **post-viral fatigue syndrome**. However, the condition and the association have not been clearly defined.

Coxsackie A and B Viruses

Aseptic meningitis is the condition produced by both coxsackie A and B viruses. The condition is caused by coxsackie A serotypes 2, 4, 7 and 9 and coxsackie B viruses. The laboratory diagnosis of coxsackie virus infection is made by isolating the virus from the faeces or lesions by intracerebral or intraperitoneal inoculation into a 1-day-old mouse. Identification of the virus is carried out by histopathology study of the infected mouse after sacrifice. Some of the coxsackie viruses can also be isolated by culture in human diploid embryonic lung fibroblasts, BGM and HeLa cell lines. Cytopathic effects resemble those of polioviruses but develop more slowly, beginning as foci of rounded refractile cells, which die and fall off the glass. Serology by neutralising antibodies is strain specific. Other autoantibodies are also detected by immunofluorescences. TR PCR is also used to detect nucleocapsid of the virus.

Differences between Coxsackie viruses A and B are summarised in Table 55.4.

Echoviruses

Echoviruses was originally isolated from the faeces of an individual who had no clinical illness and caused a cytopathic effect in the cell culture.

Table 55.4 Distinguishing features of Coxsackie A and Coxsackie B viruses

Features	Coxsackie A virus	Coxsackie B virus
Serotypes	23 (1–24 except 23)	6 (1–6)
Pathological changes produced in suckling mice	Generalised myositis, flaccid paralysis and death within a week	Patchy focal myositis, spastic paralysis, necrosis of brown fat, pancreatitis, hepatitis and myocarditis

The prefix ECHO is an acronym for *e*nteric *c*yto-pathogenic *h*uman *o*rphan viruses (ECHO viruses). They were earlier called 'orphans' because these viruses were not associated earlier with any disease, but now, they are known to cause various human illnesses.

The echoviruses resemble other enteroviruses in their properties. They are classified into 32 serotypes (1–34 except 10 and 28) based on their type-specific neutralisation capsid antigen. Some serotypes of echoviruses (types 3, 6, 7, 11, 12, 13, 19, 20, 21, 24, 29, 30 and 33) cause haemagglutination of human red cells. Agglutination of red cells is followed by elution, making the red cells inagglutinable by echoviruses or Coxsackie viruses but not by myxoviruses. All echoviruses grow in human and monkey kidney cell lines, producing cytopathic effects.

Most of the echoviruses cause asymptomatic infections in humans, but some of them have been associated with many clinical syndromes. Non-specific febrile illness associated with rash, headache and common-cold-like symptoms is caused by many serotypes of echoviruses. Echoviruses also cause aseptic meningitis. Some strains of echovirus have been associated with gastroenteritis (serotype 8) and respiratory diseases in children (serotypes 1, 11, 19, 20 and 22).

Laboratory diagnosis of echovirus infection is made by isolating viruses from faeces, throat swabs or CSF. They are cultured using human diploid embryonic lung fibroblast and human rhabdomyosarcoma cell lines. The growth of the virus is detected by cytopathic changes, which is similar to that of other coxsackie viruses.

Humans are the natural hosts for these viruses. Echoviruses are found in the intestinal tract of infected humans. The infection, like another enterovirus, is transmitted by the faecal–oral route. The summer season is common for its epidemics. So far, vaccination for this virus has not been tried.

Other Enteroviruses

Four new enteroviruses have been described that include enteroviruses 68, 69, 70 and 71. Of these, enterovirus 69 does not cause many human diseases, but the rest of the three enteroviruses cause diseases in humans. Type 68 causes pneumonia and bronchitis, type 70 causes acute haemorrhagic conjunctivitis (AHC) and type 71 causes meningoencephalitis and paralysis.

Enterovirus 70

Enterovirus 70 is the causative agent of acute haemorrhagic conjunctivitis (AHC).

Acute haemorrhagic conjunctivitis is a highly contagious ocular infection, which can cause large-scale epidemics. The incubation period is about 24 h, and the onset is abrupt. The most common symptoms include pain in the eyes, burning sensation, swelling of the eyelids and foreign body sensation in the eye. Photophobia and watery discharge may also be found in some cases. The other eye becomes infected a few hours after infection of the first eye. Fever, malaise and headache are the non-specific symptoms. AHC is a self-limiting condition. The symptoms typically improve by the second or third day of infection, and the recovery is complete within 7–10 days.

Isolation of the virus is made by culturing in human diploid embryonic kidney and HeLa cell lines. Transmission is facilitated by overcrowding and non-sanitary conditions. The infection is transmitted directly from the finger or fomite to the eye.

Acute haemorrhagic conjunctivitis was first recognised in 1969 in Ghana and Indonesia. Subsequently, the condition spread widely, involving several parts of Japan, England, Europe, Africa, the Middle East and Southeast Asia, including India. Coxsackie virus A 24 has also been known to cause similar diseases. The condition is most prevalent in adults between 20 and 50 years.

Rhinoviruses

Rhinoviruses are the most important causative agents of **common cold** and upper respiratory tract infections. Earlier, many viruses isolated from the cases of common cold were known as common cold viruses, Salisbury viruses or Muri viruses. All these viruses have been given the name rhinoviruses (*rhino* referring to the organ specifically affected).

Properties of Virus

Rhinoviruses are small RNA viruses morphologically resembling other picornaviruses. Rhinoviruses differ from other picornaviruses in being more acid-labile and more heat stable. They are inactivated at a low pH below 5 and become completely inactive at pH 3. For several days, they are relatively stable at 20–37 °C and can survive on environmental surfaces, such as doorknobs. Some serotypes may survive heating at 50 °C for 1 h. They are resistant to 20% ether and 5% chloroform but are sensitive to hypochlorite and aldehyde. Differences between enteroviruses and rhinoviruses are summarised in Table 55.5.

The rhinoviruses have been classified into more than 102 serotypes based on a type-specific antigen in their capsids.

Laboratory Isolation

Rhinoviruses grow at 37 °C in cell lines of human or simian origin in the presence of good oxygenation, low pH (7 or less) and low temperature. They can grow in human embryonic tracheal epithelial, human diploid cell lines and HeLa cells, but without producing conspicuous cytopathic effects. Some strains grow well in both monkey and human cell lines and are designated as M strains, and those that grow only on human cells are known as H strains.

Pathogenesis of Rhinovirus Infections

Infection is transmitted to other susceptible human hosts by nasal secretions expelled from a patient's nose during sneezing and coughing. Infection can be initiated by as little as one infectious viral particle. Rhinoviruses grow better *in vivo* at a temperature of 30 °C than at 37 °C, a property that contributes partially to the predilection of these viruses for the cooler environment of the nose and conjunctiva. The infection occurs mostly by multiplication of the virus in the nose. The onset and severity of the symptoms usually correlate with the quantity of viruses excreted in the nasal secretion and the time of viral shedding.

During the acute phase of illness, nasal secretions contain approximately 500–1000 infectious virions/mL of secretions. Also, these secretions may contaminate hands, fingers, handkerchiefs or paper tissues and the normal susceptible hosts who come in contact with these contaminated surfaces acquire the virus. The virus enters through the nose, mouth or eyes and causes upper respiratory tract infections, including the throat.

Host immunity in rhinoviruses is characterised by both humoral and cell-mediated immunities.

Table 55.5 Differentiating features of Enteroviruses and Rhinoviruses

Characteristics	Enteroviruses	Rhinoviruses
Size	22–30 nm	30 nm
Capsid		
Symmetry	Icosahedral	Icosahedral
Genome	RNA, single-stranded, positive sense	RNA, single-stranded, positive sense
Polypeptides	VP1–VP4	VP1–VP4
Optimum temperature for growth	37 °C	33 °C
Effect of acid	Stable (pH 3.9)	Labile (pH 3.5)
Density in caesium chloride (g/mL)	1.34	1.39–1.42

Humoral immunity is characterised by the production of both nasal secretory IgA and serum IgG antibodies within a week of infection. The cell-mediated immunity does not play an important role in controlling rhinovirus infection. Immunity is transient and is usually serotype-specific. Therefore, one attack of the common cold by rhinovirus does not confer immunity against subsequent attacks due to the presence of numerous serotypes of the virus.

Clinical Syndromes

The incubation period of common cold caused by Rhinovirus is 2–4 days. The condition begins as an upper respiratory infection with sneezing, followed by a running nose (*rhinorrhoea*). Rhinorrhoea then increases and is associated with symptoms of nasal obstruction. Mild sore throat, headache, malaise, fever and rigor sometimes accompany rhinovirus infection. Usually, symptoms subside in about a week, but in some cases, the cough and nasal symptoms may persist for 7–10 days or longer.

Laboratory Diagnosis

The clinical manifestations of the common cold are characteristic; hence laboratory diagnosis is only supportive.

Laboratory diagnosis is made by isolating the virus from nasal washing by culture on human diploid fibroblast cells (e.g. WI-38 at 33 °C). The rhinoviruses are identified by the typical cytopathic effects and demonstration of acid lability. Serotyping of the strains is performed using pools of specific neutralising antisera.

Epidemiology

Rhinoviruses are the major cause of the common cold and are responsible for more than 50% of all colds worldwide. Other viruses which may also cause the common cold include coronaviruses, coxsackie viruses A21 and A24, echoviruses 11 and 20 and parainfluenza viruses.

The common cold occurs most commonly in infants and children and occurs most frequently in the early autumn and the late spring in people living in temperate climates. The non-enveloped rhinoviruses are extremely stable and survive on external objects, such as fomites, for many hours, thereby facilitating the transmission of infection.

Infected humans suffering from acute illness are the only reservoirs and sources of rhinovirus that cause the common cold. Asymptomatic people also play an important role in transmitting the infection even though they shed a few viruses in their respiratory secretions. Rhinovirus is transmitted mainly by two routes: first, by direct inhalation of infectious droplets and second, by direct contact with hands or fomites contaminated with respiratory droplets. Contaminated hands appear to be the major transmission route, and direct person-to-person contact is the major mode of spread of infection.

Treatment

Treatment of the condition is symptomatic.

Prevention and Control

No vaccine is available against rhinovirus. Handwashing and disinfection of contaminated objects are the best methods of preventing transmission of the virus.

Hepatitis A Virus

Hepatitis A virus, formerly known as enterovirus, has now been accorded the status of a separate genus *Hepatovirus* of the family Picornaviridae. The virus is discussed in detail in Chap. 63.

Case Study

A 7-year-old girl came to a hospital with flaccid paralysis of the right leg. Her parents provided a history of paralysis for the past 2 weeks. On examination, no sensory loss was elicited in the affected right leg. The girl was suspected to be suffering from paralytic poliomyelitis. On enquiry, the parents gave the history of the girl receiving oral polio vaccine during her childhood.

- What is the cause of vaccination failure in this girl?
- What are the countries in the world in which polio continues to be endemic?
- What are the mechanisms of pathogenesis of poliomyelitis?
- How would the diagnosis be confirmed in the laboratory?

56 Orthomyxoviruses

Myxovirus was coined for a group of enveloped RNA viruses that can adsorb onto mucoprotein receptors on erythrocytes, causing haemagglutination.

Influenza, mumps, parainfluenza and Newcastle disease viruses were initially grouped in the class myxovirus *(myxa* meaning mucus) due to their affinity to mucins. However, two separate families were created subsequent to recognising clear differences between these viruses. These two families are Orthomyxoviridae, consisting of influenza viruses, and Paramyxoviridae, consisting of parainfluenza, mumps, measles, respiratory syncytial and Newcastle disease viruses. Differences between Orthomyxoviruses and Paramyxoviruses are summarised in Table 56.1.

Influenza Viruses

The genus *Orthomyxovirus* includes influenza viruses, the classic respiratory viruses, and the causative agents of worldwide influenza epidemics.

Morphology

Influenza viruses are spherical or filamentous, enveloped particles 80–120 nm in diameter. It comprises a characteristic segmented single-stranded RNA genome, a nucleocapsid and an envelope (Fig. 56.1).

The viral genome is a single-stranded anti-sense RNA. The genome consists of an RNA-dependent RNA polymerase, which transcribes the negative-polarity genome into mRNA. The genome, therefore, is not infectious. Characteristically, it is segmented and consists of seven or eight segments. These segments code for different proteins, which are NS1, NS2, NP, M1, M2, M3, HA and NA.

The genome is present in a helically symmetric nucleocapsid surrounded by a lipid envelope. The envelope has an inner membrane protein layer and an outer lipid layer. The membrane proteins are known as matrix or M proteins and are composed of two components, M1 and M2. Two types of spikes XE or peplomers project from the envelope: i) the triangular *haemagglutinin* (HA) peplomers and ii) the mushroom-shaped *neuraminidase* (NA) peplomers.

Virus Replication

Influenza virus, hepatitis delta virus and retroviruses are the only RNA viruses with an important stage of their replication cycle in the nucleus. Infection of the host cell begins by adsorption of the cell by the influenza virus, which is mediated through the HA. HA is first cleaved by an extracellular protease to a modified HA that mediates the attachment of the virus to

Table 56.1 Differentiating features of Orthomyxoviruses and Paramyxoviruses

Properties	Orthomyxoviruses	Paramyxoviruses
Size	80–120 nm	100–300 nm
Shape	Spherical, filaments in fresh isolates	Pleomorphic
Genome	Segmented—Eight pieces of RNA	Single, linear RNA
Gene re-assortment	Common	Not reported
Antigenic stability	Variable	Stable
Haemolysis	Absent	Present
Site of synthesis of ribonucleoprotein (RNP)	Nucleus	Cytoplasm
DNA-dependent RNA synthesis	Required for multiplication of the virus	Not required for the multiplication of the virus
Effect of actinomycin D	Inhibits multiplication	Does not inhibit

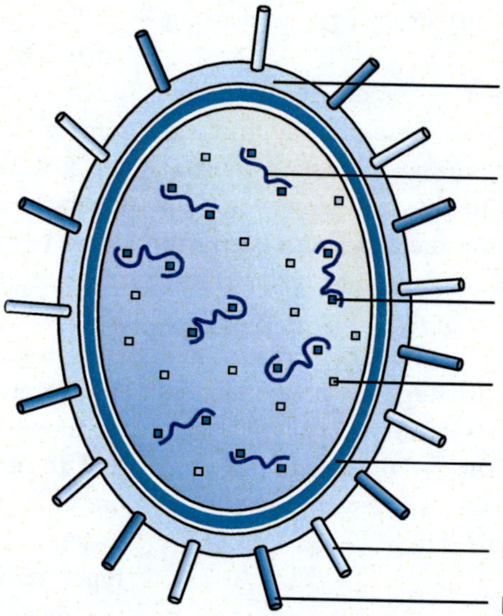

Fig. 56.1 Schematic diagram of structure of an influenza virus

the cell surface. Once inside the cell, the virus uncoats within the endosome.

In the host cell nucleus, the virion RNA polymerase transcribes the eight genome segments into eight viral mRNAs. However, most of the viral mRNAs move out of the nucleus into the cytoplasm, where they are translated into viral proteins. Some of the viral mRNAs remain in the nucleus and serve as the templates for synthesising the negative-strand RNA genomes for the progeny virions. In the nucleus, two proteins, namely, nucleoprotein (NP) and matrix protein, are synthesised, which then bind with the RNA genome of the viral progeny and form a complex, which is subsequently transported to the cytoplasm.

Matrix protein mediates the interaction of the nucleocapsid with the envelope, and finally, the virion is released from the cell by budding from the cell membrane at the site where the HA and NAs are present.

Antigenic Properties

Influenza viruses have two types of antigens: Group-specific antigen and type-specific antigen.

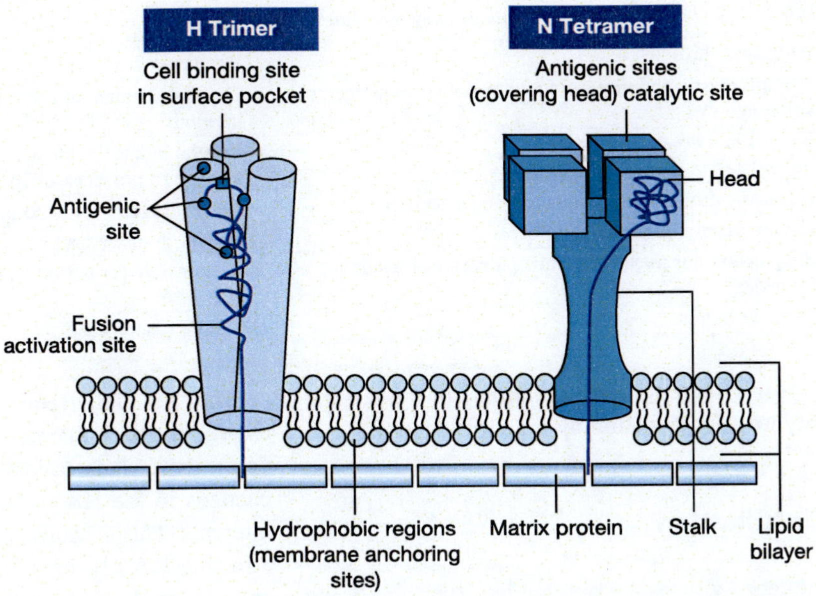

Fig. 56.2 Schematic diagram of H trimer and N tetramer of influenza virus

Group-Specific Antigen

The RNP antigen, the soluble antigen, or the internal antigen is the group-specific antigen.

Surface Antigen

Surface antigen or viral antigen, or V antigen is composed of two virus-encoded proteins, Haemagglutinin (HA) HA and NeuraminidaseNA, which are the type-specific antigens (Fig. 56.2).

Haemagglutinin (HA)

It is a trimer composed of two polypeptides, HA1 and HA2, responsible for haemadsorption and haemagglutination. The HA consists of 500 spikes, each measuring 12 nm in length. The triangular-shaped HA is inserted into the virus membrane by its tail end. The distal end, which contains five antigenic sites (designated as HA1–HA5), is responsible for binding the virion to host cells.

Influenza viruses adsorb many avian and mammalian erythrocytes. HA binds with the neuraminic acid (sialic acid) cell receptor, the cell surface glycoprotein receptor and initiates the infection in the host cell. Adsorption of erythrocytes occurs at 4 °C, but at 37 °C there is detachment of the red cells due to the destruction of the glycoprotein receptors by the viral enzyme NA. The HA agglutinates certain red blood cells, which is inhibited by the neutralising antibodies. This forms the basis of the haemagglutination inhibition test used in the serodiagnosis of influenza.

HA has the potency to undergo antigenic variations. The nucleotide and amino acid sequences of the polypeptides, HA1 and HA2, undergo radical changes in an antigenic shift. In antigenic drift, only minor changes occur in HA antigenic sites' compositions.

Neuraminidase (NA)

It is a glycoprotein and tetramer. It consists of 100 mushroom-shaped spikes. The mushroom-shaped NA is inserted by its hydrophobic tail end into the virus membrane. The distal end contains antigenic as well as enzymatically active sites. The NA causes hydrolysis of *N*-acetyl neuraminic acid or sialic acid residues present on the glycoprotein receptors on red cells, hence causing elution or detachment of the cells adsorbed to virion particles. The function of the NA is to cleave the neuraminic acid and release progeny virions from the infected host cells. The NA also degrades the mucus layer, thereby

Table 56.2 Differentiating features of antigenic shift and antigenic drift

Antigenic shift	Antigenic drift
Abrupt, drastic and discontinuous variation in antigenic structure	Gradual, sequential and regular variation at periodic intervals
Results in a new novel strain Not related to predecessor strain	Results in a different strain having new antigens Related to predecessor strain
Antigenic drift is due to mutation and selection	Antigenic shift is due to gene re-assortment
Occurs mainly in influenza virus A	Occurs in influenza virus A, B and C
Responsible for major periodic epidemics or pandemic of influenza	Responsible for periodic epidemics/pandemics of influenza
Difficult to treat	Can be treated easily

exposing the epithelial membrane of the respiratory tract for infection by the virus.

Classification

Influenza viruses are divided into types A, B and C based on variation in this RNP antigen.

All these strains of influenza A, B and C share the common internal proteins, such as NPs and membrane proteins. They, however, differ in their surface proteins, such as HAs and NAs. Influenza virus type A is subdivided into 15 HA subtypes (H1-H15) based on HA and to 9 NA subtypes based on the NA. The strains that cause infection in humans belong to subtypes H1-H3, H5 and N1 and, N2.

Antigenic Variations

Antigenic variation is a unique feature of the influenza virus. The surface antigens HA and NA show variations and are primarily responsible for antigenic variations exhibited by influenza viruses. Antigenic variations are of two types: antigenic shift and antigenic drift.

The abrupt, drastic, discontinuous change in antigenic structure is called the **antigenic shift**. This occurs due to major antigenic changes in HA or NA antigens and is caused by replacing the gene for HA by one coding for a completely different amino acid sequence. The antigenic shift is characterised by altering virtually all the antigenic sites of the HA. This occurs independently in the HA and NA.

The gradual, sequential, regular antigenic change in the influenza virus is known as **antigenic drift.** This occurs due to minor antigenic changes in the HA or NA occurring at frequent intervals. This is caused even by a single mutation affecting HA glycoprotein. The antigenic drift is characterised by changes in certain epitopes in the HA, while others are conserved. Differences between antigenic shift and antigenic drift are summarised in Table 56.2.

Influenza A virus shows maximum antigenic 3. B virus does not undergo antigenic shift because influenza B virus is the only human virus for which there is no animal source of new RNA segments. However, the influenza B virus undergoes antigenic drift. Antigenic variation never occurs in type C influenza virus.

Gene Re-assortment

As the influenza virus genome is segregated, genetic re-assortment shall occur when a host cell is affected concurrently with viruses belonging to two-parent strains, which are different. This process of genetic re-assortment accounts for the periodic appearance of the novel types of influenza A strains that cause influenza pandemics.

Influenza viruses of animals, such as aquatic birds, chickens, swine and horses, show high host specificity. These animal viruses are the source of the RNA segments that encode the antigenic shift variants that cause epidemics amongst humans. For example, if a person is infected simultaneously by an avian and human influenza strain, then it is possible that a genetic re-assortment

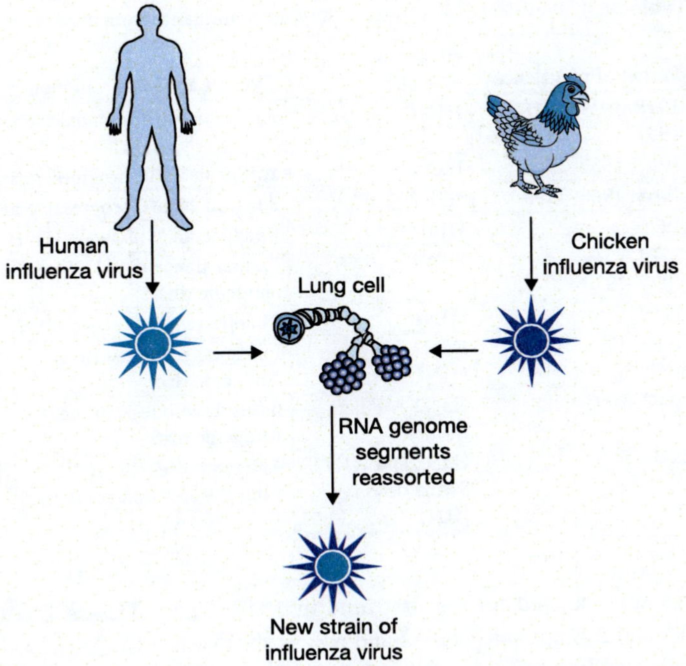

Fig. 56.3 Genetic re-assortment between a human and animal virus

could occur in infected cells in humans. The re-assortment could lead to the emergence of a new human influenza A virus, the progeny of which will contain a mixture of genome segments from the two strains (e.g. a new variant of human influenza A virus bearing the avian virus HA) (Fig. 56.3).

Many studies have conclusively demonstrated that aquatic birds (such as waterfowl) are a common source of these new genes. The pigs act as mixing vessels, where these virulent genes of waterfowl mix with the genome of the influenza virus, giving rise to a new variant of the influenza virus.

Nomenclature of Influenza Viruses

Influenza virus type A can be classified into subtypes based on the variations in their surface antigens (Table 56.3). The WHO proposed a new classification system in 1971 and was later modified, which considers the nature of both the surface antigens. According to this, the complete designation of a strain will include the (1) type, (2) place of origin, (3) serial number and (4) year of isolation, followed by (5) antigenic subtypes of the HA and NA in parentheses. For example, influenza A/Singapore/1/57 (H2N2) indicates that influenza was first originated from Singapore

Table 56.3 Types and subtypes of influenza virus

Types (Genus)	Subtypes (Species)
Influenza virus A	15 HA subtypes: H1 –H15 9NA subtypes: N1-N9 H1N1 (A1 human, $H_{SW}N1$), H2N2 A2, H3N2 (A3, A Hong Kong) H5N1
Influenza virus B	B (human)
Influenza virus C	C (human)

Table 56.4 Prevalence of antigenic subtypes of influenza virus type A

Period of prevalence	Antigenic structure	Type of antigenic variation
1918–1919; 1918–1957	H1N1	Antigenic shift, progressive antigenic drift
1957; 1957–1968	H2N2	Antigenic shift, progressive antigenic drift
1968; 1968–1990	H3N2	Antigenic shift, progressive antigenic drift
1976	H1N1	Reappearance of swine H1N1 virus
1977 1977–1978	H1N1	Reappearance of H1N1 virus (resembling 1950 strain) progressive antigenic drift
1989	H1N1	Reappearance of swine H1N1 virus
1997–1999	H3N2 A Sydney H1N1	Intrasubtypic reassortment Antigenic drift
2003–2004	H3N2 A Fujian H1N1	Intrasubtypic reassortment Antigenic drift
2009	H3N2 and H1N1 Swine origin H1N1	Antigenic drift Viral introduction(pandemic)

and was isolated for the first time in 1957. The HA and NA antigens are H2 and N2, as shown in the parentheses (Table 56.4).

Other Properties

Influenza viruses are inactivated by ether, formaldehyde, phenol, salts of heavy metals and common chemical disinfectants. Heating at 50 °C for 30 min inactivates most strains of the virus. Influenza viruses are resistant to slow drying. They remain viable in dust for up to 2 weeks, at −70 °C for a longer period, and by freeze-drying indefinitely.

Laboratory Isolation

Influenza viruses are isolated from respiratory secretions by growing in chick embryos or tissue cultures.

Chick Embryos

The influenza viruses grow in the allantoic and amniotic cavity of the chick embryos. After an incubation period of 3 days, the fluid is tested for haemagglutination activities of the viruses.

Tissue Culture

Cell lines are widely used for the culture of influenza viruses. They can grow in several primary and continuous cell lines. Rhesus monkey kidney cell lines (LLC-MK2) and Madin–Darby canine kidney (MDCK) are the continuous cell lines frequently used to isolate influenza viruses.

Laboratory Animals

The human influenza virus causes experimental infections in a variety of animals. Intracerebral inoculation of mice by neurotrophic strains produces fatal encephalitis. It causes an acute respiratory disease on intranasal inoculation in ferrets.

Pathogenesis of Influenza

Influenza virus is transmitted from person to person primarily in droplets released by sneezing and coughing. Inhaled influenza viruses reach the lower respiratory tract, tracheobronchial tree, the primary site of the disease. They attach to sialic acid receptors on epithelial cells by HA present on the viral envelope. Relatively few viruses are needed to infect the lower respiratory tract than the upper respiratory tract.

The viral envelope contains NA, which may act on the *N*-acetyl neuraminic acid residues present in mucus to cause liquefaction. With the help of mucociliary transport, the mucus which was liquefied spreads the virus throughout the respiratory tract. Mucosal cells which are infected result in destruction at a cellular level and desquamation of the mucosa superficially. Oedema caused leads to mononuclear cell infiltration of the associated areas. They are combined with the signs, which include a cough, sore throat and nasal discharge.

The antibody is the primary defence in immunity to re-infection. IgA antibody, which predominates in upper respiratory secretions, is less persistent than secretory IgG but contributes to confer immunity. Secretory IgG antibody, which predominates in lower respiratory secretions, appears most important.

Antibodies provide long-lasting immunity against the infecting influenza strain. Only antibodies directed against HA can prevent infection neutralising the virus's infectivity, thereby preventing the disease. Antibodies against NA do not neutralise infectivity but reduce the severity of the disease. The antibodies against the internal RNP group-specific antigen do not confer any immunity.

Immunity to the influenza virus is strain-specific and lasts for many years. Recurrent cases of influenza are caused primarily by antigenically different strains. The role of cell-mediated immunity in conferring protection against influenza is not clear.

Clinical Syndromes

The virus causes classic influenza (Table 56.5).

Influenza

The incubation period is short (1–3 days). Even though non-productive cough seems to be the strong predictor, the most important symptoms of influenza can be classified into (1) systemic—fever, muscle aches and (2) general—prostration. The virus remains localised to the respiratory tract; hence, viremia does not occur. Influenza presents as febrile illness of rapid onset, associated with headache, chills, fever, malaise, myalgias, anorexia and sore throat. The body temperature increases rapidly to 101–104 °F (38.3–40.0 °C), and respiratory symptoms arise.

The common symptoms are sneezing, rhinorrhoea and nasal obstruction. Patients may also complain of nausea, vomiting, diarrhoea, hoarseness, photophobia and abdominal pain. They appear severely ill and are continuously coughing. Mild to moderate obstruction of nasal passages, nasal discharge and pharyngeal inflammation may also be encountered. The examination of the lung is essentially performed. The viruses mostly cause subclinical infections. They do not display the classic syndrome described previously.

Complications

Interstitial Pneumonia

Occasionally, in the patients suffering from heart or lung disease, the infection abundantly affects the alveoli, causing interstitial pneumonia, which may be seen with considerable aggregation of oedema and haemorrhage of the lung. Viral pneumonia of mentioned type is a serious illness with increased mortality. The titres of virus in secretions are more, and viral shedding is extended. In many cases, pneumonia associated with influenza is produced by bacteria,

Table 56.5 Infections associated with influenza virus

Infections	Clinical manifestations
Influenza in adults	Fever, malaise, myalgia, sore throat and non-productive cough
Influenza in children	Acute disease similar to that in adults but with high fever, abdominal pain, vomiting, otitis media, myositis and croup
Complications	Primary viral pneumonia; secondary bacterial pneumonia; neurological complications, such as Guillain–Barre syndrome, encephalopathies, encephalitis and Reye syndrome

predominantly staphylococci, pneumococci and *Haemophilus influenzae*. These bacteria cause disease by invading as the advancing viral infection attacks the normal defence mechanisms of the lung.

Pneumonia may be fatal, particularly (1) in elderly persons above 60 years with underlying chronic disease, (2) in people with impaired resistance (chronic cardiorespiratory disease, renal disease, etc.), and (3) in pregnant women.

Guillain–Barre Syndrome

It is a rare complication of influenza and is characterised by encephalomyelitis and polyneuritis. This condition was documented in the United States in 1976, following extensive vaccination with inactivated H3N2 influenza virus.

Reye Syndrome

It is a noted complication of influenza B infection. The condition is seen most commonly in young children and is associated with degenerative changes in the brain, liver and kidney.

Laboratory Diagnosis

During an influenza epidemic, a clinical diagnosis can be made, but a definitive diagnosis depends on the laboratory methods.

Specimens

Specimens include nasal or throat washings or sputum for viral antigen and viral RNA, throat gargles for isolation of viruses and serum for viral antibodies. Throat gargles is collected in saline broth or a buffered salt solution and is sent immediately to the laboratory, or if delayed is, stored at −4 °C. The virus is isolated from the specimen by inoculation into embryonated eggs or certain cell cultures.

Direct Antigen Detection

Immunofluorescence (IF) or enzyme-linked immunosorbent assay (ELISA) using specific monoclonal antibodies are used commercially to detect viral antigens directly on cells obtained from the nasopharynx. The results of the rapid tests are useful to start treatment with the NA inhibitors within 48 h of the onset of symptoms.

Chick Embryo Inoculation

The specimen is inoculated into the amniotic cavity of the chick embryo. Except for type C, the virus also grows in the allantoic cavity after a few passages. After incubation at 35 °C for 3 days, the amniotic fluid and allantoic fluid are harvested and tested for the presence of viral HA. The highest dilution of the virus suspension that causes red cells' agglutination is called Haemagglutination (HA) titre. This is carried out using fowl and guinea pig red cells parallel and incubating at room temperature and 4 °C. Usually, influenza A viruses agglutinate only guinea pig cells, influenza B both fowl and guinea pig red cells, and influenza C agglutinate only fowl cells at 4 °C. If the test is positive, the isolate is then typed by a serological test (e.g. haemagglutination inhibition test) using specific antisera to types A, B and C.

Cell Culture

Influenza virus is usually isolated from respiratory secretions by growing in shell vial cultures of monkey kidney or baboon kidney cell lines. The cell cultures are incubated at 33 °C in the roller drums in the presence of trypsin but without serum. Virus growth is detected by direct demonstration of viral antigen in infected cell cultures by IF, RT-PCRor by testing for haemadsorption with human O, fowl and guinea pig red cells. In a positive haemadsorption test, red cells adhere to the virus budding from infected cells. If the culture tests positive, serologic tests with specific antisera may be used to identify the virus.

Serology

Demonstration of a rise in serum antibody titre between acute-phase and convalescent-phase sera by a serological test is diagnostic of infection. The acute-phase sera are collected within a few days of illness and the convalescent sera 7–10 days after the illness.

Haemagglutination Inhibition Test

It is a sensitive serological diagnostic test of influenza. This is a suitable method used to deduce and quantify the antibody to the virus. At times, certain substances in the sera cause non-specific inhibition of haemagglutination. This is a drawback of this technique. Alpha (Francis), Beta (Chu) and Gamma (Shimojo) are the names given to different kinds of non-specific inhibitors found in the sera. Many techniques are used for inactivating them without affecting the antibody content of sera. Treatments with potassium periodate kaolin, trypsin, RDE and CO_2 are used to inactivate the inhibitors. There is no single method that has been found effective in completely inactivating the inhibitors to all types of viruses from all kinds of sera.

Enzyme Neutralisation Test and ELISA

These are the other tests also used to demonstrate antibodies. However, none of these techniques is useful to identify all infections. Various approaches followed for laboratory diagnosis of influenza are summarised in Table 56.6.

Molecular Detection

The reverse transcriptase-polymerase chain reaction is used to detect viral nucleic acid in the nasopharyngeal cells for the diagnosis of influenza. The use of different target genes in the RT-PCR assays is most appropriate for the correct identification of this virus. The following gene targets are often used, namely type A influenza matrix gene, haemagglutinin gene-specific for influenza A subtypes like A (H1N1) pdm09 virus, A(H3N2), A(H1N1), A(H5N1) viruses, avian influenza A (H7N9) virus and other zoonotic subtypes. Amongst other targets, type B influenza targeting matrix or NP or NS genes and haemagglutinin gene-specific for influenza B virus are used. The WHO gives several RT-PCR protocols to detect influenza viruses. Human specimens with laboratory results indicative of influenza A that are unsubtypeable or detected/diagnosed as non-seasonal subtypes should be sent to a WHO referral centre immediately for confirmation.

Treatment

Amantadine and rimantadine are the specific antiviral agents available for the treatment of influenza. These drugs are effective against the influenza A virus but not against the influenza B virus. When given within 1–2 days of the onset of illness, these drugs reduce the severity of the disease and hasten the disappearance of fever and other symptoms. The drugs are recommended to prevent influenza in the elderly unimmunised population in whom influenza can cause a life-threatening infection. The emergence of viral resistance can occur during treatment.

Zanamivir (*Relenza*) and oseltamivir (*Tamiflu*) are newer drugs for the treatment of influenza and are effective against both influenza A and B

Table 56.6 Laboratory diagnosis of influenza virus infection

Laboratory diagnosis	Test	Detection
Direct antigen detection	IF, ELISA	Influenza virus antigen in respiratory secretions
Cell culture	Primary monkey kidney cell, Madin–Darby canine kidney cell and haemadsorption to infected cells	Limited cytopathic effects, presence of HA protein on cell surface
Chick embryo	Allantoic and amniotic cavity	Fluid is tested for haemagglutination of virus
Serology	Haemagglutination inhibition, Haemadsorption inhibition, ELISA, CFT and IF	Demonstration of a rise in serum antibody titre
Molecular diagnosis	RT-PCR	Viral nucleic acid in the nasopharyngeal cells

Note: *ELISA*: enzyme-linked immunosorbent assay; *CFT*: complement fixation test; *IF*: immunofluorescence; *RT-PCR*: reverse transcriptase–polymerase chain reaction viruses. These are the NA inhibitors, which inhibit the release of viruses from infected cells. These drugs also prevent the spread of the virus from one cell to another. Relenza is used in the form of nasal spray, whereas Tamiflu is given orally.

Epidemiology

Influenza occurs worldwide. The worst, severe **pandemic (Spanish flu)** occurred in 1918–1919, which caused about 20 million deaths worldwide. Influenza **epidemics** have been recorded throughout history. Influenza A is generally more pathogenic than influenza B. Influenza A is a zoonotic infection, and more than 100 types of influenza A viruses have been described which affects birds, horses, pigs, dogs and seals. The pandemic of influenza that occurred in 1918 and resulted in millions of human deaths has been believed to have commenced from a virulent strain, that is, H1N1 from pigs or birds (Table 56.7).

Geographical Distribution

Influenza viruses cause epidemic, endemic and pandemic influenza. Influenza A virus causes epidemics and occasionally pandemics; influenza B virus only causes epidemics, and influenza C viruses only cause minor respiratory illnesses and do not cause epidemics.

Reservoir, Source and Transmission of Infection

Infected humans are the main reservoir of infections for the influenza A virus. The respiratory secretions of infected persons are the important source of infection. The virus is excreted in respiratory secretions immediately before the onset of illness and for 3–4 days after that. Wild aquatic birds are known reservoirs of influenza A. They secrete the viruses in their faeces, which contaminates ponds and lakes. The virus is spread from person to person primarily by airborne respiratory droplets released during sneezing and coughing.

Influenza Epidemics and Pandemics

Influenza epidemics encountered are two types: influenza epidemics every year caused by influenza A and B viruses. The rare, severe influenza pandemics are always caused by the type A virus.

Antigenic Shift

Antigenic shift is the process by which a major change occurs either in one or both of the surface antigens. This change forms an antigen that shows no relationship serologically with the antigen of the strains present at that time. An antigenic shift has been demonstrated in type A influenza virus only.

Influenza A pandemics are caused by the virus strains undergoing an antigenic shift. Antigenic shift variants appear less frequently, about every 10 or 11 years. It is demonstrated that pandemic strains are the recombinant strains that originated from some animal or bird reservoir, either spreading to humans directly by host range mutation or as a result of a recombination between human and non-human strains. The pandemic strains also show the capability to spread rapidly amongst the population.

The entirely new antigens that emerge during the antigenic shift are accomplished by **genetic re-assortment**. An animal influenza virus apparently becomes the donor of the novel antigens formed. Type A viruses are explored in animals

Table 56.7 Influenza virus A subtypes—Classification and nomenclature

Older classification	1971 classification	Current classification	Reference strains
A swine	HSWN1	H1N1	A/swine/Wisconsin/15/30
AO	HON1	H1N1	A/PR/8/34
A1	H1N1		A/FM/1/4
A2	H2N2	H2N2	A/Singapore/1/57
A2 (Hong Kong)	H3N2	H3N2	A/Hong Kong/1/68

such as pigs, horses and birds, and influenza viruses affecting animals possess antigens closely similar to human viruses. Fourteen different HA and nine NA antigens are recognised. Due to advancing surveillance of animal influenza viruses in late years, the discoveries of new antigens have not been successful. This represents the entire variety of major influenza virus surface antigens, which can be grouped into many subtypes.

Antigenic Drift

On the other hand, repeated minor antigenic changes generate strains that retain a degree of serologic relationship with the currently prevailing strain. This is called antigenic drift. The epidemics are caused by the influenza A virus undergoing antigenic variations due to antigenic drift resulting from mutations and selections. Antigenic drift variants occur very frequently, virtually every year. When individuals are again infected with drift viruses, the antibodies present in serum respond to the surface antigens common to earlier strains, of which the person who had been exposed before are generally stronger and are of greater avidity than those reactions to the new antigens. This process, called **'original antigenic sin'**, is now consistently used in serological diagnosis.

Avian Influenza (H5N1 Bird Flu)

The H5N1 flu, caused by an avian subtype influenza virus, which has been associated with bird flu in domesticated birds, can be transmitted from birds to humans. The H5N1 was first described in Hong Kong in 1997. Human infections caused by this virus were established in only 18 individuals, of which six died.

An epidemic of bird flu occurred in domesticated birds in Southeast Asia, primarily Vietnam, in January 2004. More than 240 human cases have been documented, and more than 140 persons have died due to the poultry outbreaks and bird-to-human transmission. Most deaths have been reported in Vietnam and Indonesia. Sporadic outbreaks have continued to occur since the 2004 outbreak, even outside Southeast Asia, including Turkey. Thousands of birds died due to H5N1 in 14 states and Union territories of India during 2020–2021. Till now, no conclusive evidence is available to show the human-to-human transmission of H5N1.

H1N1 (Swine Flu)

A novel H1N1 swine-origin influenza virus (SOIV) H1N1 emerged in 2009 to produce the first human influenza pandemic of the twenty-first century. This virus spread to 214 countries within 1 year and caused more than 18,000 confirmed deaths worldwide. First described in April 2009, the novel SOIV emerged due to a 'triple re-assortment' of influenza viruses of avian, swine and human origin. This led to the popular term 'swine flu' being used for this virus.

The pandemic H1N1 in 2009 virus contained i) PA and PB2 segments originated from avian viruses of North American origin; ii) the PB1 segment from human H3N2 viruses; iii) haemagglutinin (HA; of the H1 subtype), the neuraminidase (NA; of the N1 subtype), NS segments and NP evolved from classical swine H1N1 viruses and iv) M segments from Eurasian 'avian-like' swine viruses. Unlike previous strains of the influenza virus, the novel SOIV did not have a predilection for populations more than 60 years of age. Rather, a greater number of younger people less than 40 years of age were found to be affected. A small percentage of those affected were found to have developed pneumonia or acute respiratory distress syndrome. The number of cases decreased greatly by May 2010, and the pandemic was officially declared to be at an end in August 2010. For diagnosis of H1N1 influenza, testing of samples like nasopharyngeal swabs or oropharyngeal swabs is essential. Real-time RT-PCR for viral nucleic acid is the recommended method for its diagnosis, and other rapid tests like antigen detection were not found to be useful.

The treatment for severe cases of H1N1 influenza is the NA inhibitor, oseltamivir. Two types of influenza vaccines have been developed against H1N1 influenza—a trivalent inactivated vaccine and a live attenuated influenza vaccine, which can be administered intranasally.

Prevention and Control

This is based on immunoprophylaxis by vaccines and chemoprophylaxis.

Immunoprophylaxis by Influenza Vaccines

Influenza A subtypes H1N1 and H3N2 are the most common prevailing human influenza viruses. The trivalent vaccine used worldwide contains influenza A strains from H1N1 and H3N2, along with an influenza B strain. Influenza virus vaccines have been used for about 40 years to prevent influenza, primarily influenza A and B.

Inactivated Influenza Vaccines

These vaccines were used Initially. The viruses grown for these vaccines are cultivated in chick embryos which are inactivated by formalin purified to some extent and altered to a dosage noted to evoke a response by an antibody in the majority of individuals. The vaccines contain the strains of types A and B viruses that are believed most likely to produce epidemics during the following winter. Local and systemic reactions to the vaccine are minor and occur in the first day or two after vaccination.

Live Attenuated Influenza Vaccines

Live vaccines are now being developed as alternatives to inactivated vaccines. These vaccines induce the production of specific secretory antibodies in the respiratory mucosa.

These vaccines are developed by considering the naturally occurring mutations in the influenza virus genome that result in less-virulent strains. Live vaccines can also be developed by manual passage of a parental influenza virus multiple times in virus-susceptible cells so that influenza vaccine strains with low virulence and pathogenicity are generated. Live influenza vaccines mainly consist of those for influenza A(H1N1), A(H3N2) subtype, and Victoria strains of influenza B virus. Intranasal inoculation of live vaccines could induce mucosal immunity, humoral immunity, and cellular immunity. Few live vaccines like Quadrivalent one have proven effective for children 2–17 years old. Recently, *temperature-sensitive (ts) mutant strains* have been used in the live attenuated vaccine preparations. These avirulent mutants can grow at 32–34 °C in the nasopharyngeal secretions but not at 37 °C in the lungs. These vaccines are useful to protect children from clinical disease.

However, concerns regarding host replication and possible mutations remain with these live vaccines.

Indications for the live influenza vaccine Internationally, as per recommendations, two types, namely injectable and intranasal influenza vaccines, are available. **Injectable vaccines** protect by producing antibodies only, while the **intranasal vaccine** offers protection at the local level of mucosal immunity (the nasal tract) and by producing antibodies. It has been proven that the clinical efficacy of an intranasal vaccine is superior to that of an injectable vaccine. Based on epidemiological evidence from the WHO and Indian Council of Medical Research, the Government of India recommends vaccination of 'High-Risk Groups' with Seasonal Influenza Vaccine. The high-risk groups comprise the healthcare personnel, persons with comorbidities, the elderly, etc. ICMR recommends egg-based quadrivalent vaccines for use in the 2020–2021 northern hemisphere influenza season. The composition of this vaccine is as follows:

(a) an A/Guangdong-Maonan/SWL1536/2019 (H1N1)pdm09-like virus;
(b) an A/Hong Kong/2671/2019 (H3N2)-like virus;
(c) a B/Washington/02/2019 (B/Victoria lineage)-like a virus; and
(d) a B/Phuket/3073/2013 (B/Yamagata lineage)-like a virus.

Recombinant Vaccines

These are now being evaluated in influenza. Recombinant strains used in the vaccine are produced by hybridisation between the *ts* mutants and new antigenic variants of the strains. These strains carry the surface antigens of the new variants and growth properties of the old established strains. The recombinant vaccines are indicated most importantly for

Table 56.8 Vaccines against influenza virus infection

Type of vaccine	Advantages	Disadvantages
Whole virus vaccines	Confers protection in 60–90% of vaccinated individuals and the protection lasts for 1–5 years	The consequent virus which is infecting may show sluggish antigenic drift and the antibody formed by vaccine shall be less potent in negotiating protection against the new strains formed
Split virus vaccines	Induces immunity Fewer vaccination reactions	Just as immunogenic as whole virus vaccine
Subunit virus vaccines	Induces a strong immunity fewer vaccination reactions	No significant disadvantage
Live attenuated vaccines	Induces a strong immunity than do inactivated vaccines when given intranasally. Fewer vaccination reactions	No significant disadvantage

immunoprophylaxis of a new strain of influenza virus, threatening to cause pandemics of influenza.

Subunit Vaccines

Subunit vaccines against influenza viruses are mostly under development. Some, including H5N1 and H7N9, are under development. M2e, HA and NP are attractive vaccine targets, amongst other viral structural proteins. Proteins such as M1 and NA also have the potential for development as influenza subunit vaccines.

Table 56.8 summarises vaccines available against the influenza virus.

Chemoprophylaxis

Chemoprophylaxis by amantadine and rimantadine hydrochloride is more successful. These two drugs effectively prevent infection and illness caused by type A, but not by type B viruses (because they lack M2 components). The persons with high-risk can be protected by administering a dosage of 100 mg/day. When administered for 10 days to household contacts of a person with influenza, these drugs protect up to 80% of the person from illness.

Case Study

There was a scare of bird flu infections in poultry birds in Maharashtra (India) during 2006. Newspapers reported that thousands of poultry birds were culled, and the poultry industry suffered a major economic loss due to that. People stopped taking chicken and chicken products.

- What was the reason for such a scare amongst the poultry industry?
- Which influenza virus causes infection in poultry birds, and can it be transmitted to humans?
- Describe the possibility of genetic re-assortment amongst bird flu and human influenza viruses.
- What are the prophylactic measures against the influenza virus in humans?

57 Paramyxoviruses

Paramyxoviruses are roughly spherical-shaped viruses and usually vary in size from 100 to 300 nm. Sometimes, long filaments and giant forms of the virus measuring up to 800 nm are also found. These viruses consist of a negative-sense single-stranded RNA genome enclosed in a helical nucleocapsid surrounded by a pleomorphic envelope. They resemble orthomyxoviruses in morphology but are larger, surface spikes are different, and their genomes are not segmented. Hence unlike the Orthomyxoviruses, there is no re-assortment, no antigenic shift or antigenic drift is seen in Paramyxoviruses.

Classification

The family Paramyxoviridae consists of four genera (*Rubulavirus, Morbillivirus*, *Parainfluenza virus* and *Pneumovirus*). All these genera contain important human pathogens responsible for causing most acute respiratory infections and contagious diseases of children and infants. The genus *Rubulavirus* includes mumps virus, and the genus *Morbillivirus* includes the measles virus, the genus *Parainfluenza virus* includes parainfluenza types 1–4 genus *Pneumovirus* includes respiratory syncytial virus (RSV), which is responsible for the majority of acute respiratory infections in infants and children.

Antigenic Structure

The nucleocapsid is associated with the nucleoprotein (NP), polymerase phosphoprotein (P) and larger (L) protein. The NP maintains the genomic structure, the P facilitates RNA synthesis, and the L protein is an RNA polymerase. The envelope surrounds the nucleocapsid. The virion envelope contains two glycoproteins, (1) a fusion (F) protein and (2) a viral attachment protein called haemagglutinin neuraminidase (HN), haemagglutinin (H) or G protein.

F protein facilitates fusion of the viral and host cell membranes, and HN promotes adsorption of the virus to the host cell surface. The binding of the HN, H or G protein on the virion envelope to these cell surface glycoproteins containing sialic acid is the first step in the replication of paramyxoviruses. This is followed by the fusion of the viral envelope with the cell's plasma membrane, promoted by the F protein of the virion envelope. In addition, paramyxovirus also induces cell-to-cell fusion, resulting in the production of multinucleated giant cells. Paramyxoviruses associated with human diseases are summarised in Table 57.1.

Measles Virus (Rubeola)

Measles or Rubeola is a highly communicable acute viral disease characterised by fever, conjunctivitis and pathognomonic Koplik spots. It is

Table 57.1 Diseases associated with paramyxoviruses

Paramyxoviruses	Diseases
Measles virus	Measles, atypical measles, giant cell pneumonia, broncho pneumonia and subacute sclerosing panencephalitis
Human parainfluenza virus (HPIV)	Croup - laryngotracheobronchitis
	Pneumonia
	Bronchiolitis and tracheobronchitis; other infections: Otitis media, pharyngitis, conjunctivitis and coryza
Mumps virus	Mumps
Respiratory syncytial virus	Upper respiratory tract infection, common cold; lower respiratory tract illness, resulting in bronchiolitis or pneumonia in infants
Nipah virus	Encephalitis
Hendra virus	Zoonotic paramyxovirus Severe respiratory disease and respiratory tract disease in children and adults
Human metapneumovirus	Respiratory tract infections in children, especially in the infants

one of the five classic exanthematous diseases of childhood, others being chickenpox, rubella, roseola and fifth disease.

Morphology

Measles virus is spherical but is often pleomorphic, measuring 120–250 nm in diameter. It contains a negative-sense RNA genome. The helical nucleocapsid is surrounded by an envelope carrying H and F proteins on its surface. The virus causes haemagglutination of monkey erythrocytes, but it is not followed by elution as the virus does not has any neuraminidase activity.

Virus Replication

Measles virus replicates in the cell cytoplasm. The virus first adsorbs the cell surface by its haemagglutinin, then enters the cell and uncoats inside the cell's cytoplasm. The viral RNA polymerase transcribes the negative-strand genome into mRNA. Multiple copies of mRNAs are produced, each of which is translated into specific viral proteins. This is followed by nucleocapsid assembly, and the virus is released by budding from the cell membrane.

Antigenic and Other Properties

The measles virus has only one serotype and infects only humans, not other mammals. The virus is antigenically uniform; it shares antigens with the canine distemper virus. The measles virus is heat-labile. It is readily inactivated by ether, formaldehyde, high temperature and ultraviolet light. The virus is stabilised by molar $MgSO_4$, following which the virus resists heating at 50 °C for 1 h.

Laboratory Isolation

The measles virus is very difficult to grow, although it can be grown in primary human or monkey kidney cell cultures. The isolates can be adapted for growth on HeLa or Vero cell lines.

Pathogenesis of Measles

Measles is highly contagious and is spread from person-to-person by aerosols. The infection is transmitted from person-to-person by inhalation of large droplet aerosols produced during coughing and sneezing.

Measles virus enters the susceptible host by the respiratory route. The virus initiates infection and replicates locally in the respiratory tract's trachea and bronchial epithelial cells. After 2–4 days, the virus spreads systemically in lymphocytes, perhaps carried by pulmonary macrophages, and causes viremia. Wide dissemination of the virus causes infection of the conjunctiva, respiratory tract, urinary tract, lymphatic system, blood vessels and the central nervous system (CNS). The characteristic rash in measles is caused primarily by cytotoxic T cells attacking the measles virus-infected epithelial cells in the skin.

Encephalitis is one of the most important sequelae of the infection caused by measles. The virus causes (1) encephalitis by direct infection of the neurons; (2) immune-mediated conditions, such as post-infectious measles encephalopathy, believed to be immune-mediated, and (3) subacute sclerosing panencephalitis (SSPE) caused by an infective mutant of measles, produced during the acute stage of the disease.

Measles causes immunosuppression, characterised by a decrease in eosinophils and lymphocytes (both B and T cells) and depression of their response to activation by mitogens. Immunosuppression caused by measles in the infected individuals may predispose individuals to bronchopneumonia, severe bacterial infection and a major cause of measles-related mortality among young children.

Immunity

Cell-mediated immunity (CMI) plays an important role in controlling measles infection. Therefore, the measles virus in individuals with a deficiency in cellular immunity causes progressive and often fatal giant cell pneumonia without a rash. The antibodies do not have any role in conferring protection against the measles virus because the viruses spread from cell-to-cell. However, maternal antibodies in infants protect against measles during the first 6 months of life. Nevertheless, one attack of measles confers lifelong immunity.

Clinical Syndromes

Measles virus is associated with measles, atypical measles and subacute sclerosing panencephalitis.

Measles

The incubation period varies from 8 to 12 days. Measles is a highly contagious febrile illness. The prodromal phase is characterised by high fever, malaise, anorexia, conjunctivitis, cough and coryza. Koplik spot is the typical pathogenic lesion found in the mucous membrane.

Koplik spots are bluish-grey specks or grain substances on a red base, usually appearing on the buccal mucosa opposite the second molar. They appear at the end of pro-drome, just before the appearance of the rash. These Koplik spots may also appear on the mucous membrane of the conjunctiva and vagina. Koplik spots usually appear after 2 days of illness. These spots, which last for 24–48 h, are pathognomonic of measles.

An erythematous maculopapular rash appears within 12–24 h of the appearance of the Koplik spots. The rash usually begins on the face, then spreads extensively and appears on the trunk, extremities, palms and soles and lasts for about 5 days. Desquamation of the rashes except those of palms and soles may occur after 1 week. Patients appear highly sick during the first or second day of the appearance of the rash. The rash is typically absent in patients with defective CMI.

Complications of measles include otitis media, bronchopneumonia, laryngotracheobronchitis (*croup*) and diarrhoea. Bronchopneumonia is a most serious condition and is responsible for 60% of deaths caused by the measles virus. Encephalitis is one of the most dangerous complications.

Atypical Measles

The condition is characterised by prolonged high fever, pneumonitis and rash. The rash characteristically begins peripherally and may be urticarial, maculopapular, haemorrhagic or vesicular. The condition appears to be immune-mediated and to occur due to hypersensitivity to the measles virus in a partially immune host.

Subacute Sclerosing Panencephalitis (SSPE)

The SSPE is a degenerating disease of the CNS caused by persistent measles infection. It is the most serious and late neurological sequelae of measles that affects the CNS. The disease is characterised by behavioural and intellectual deterioration and seizures after many years (mean incubation period is 10.8 years) of infection by measles. The condition occurs in about seven in every one million patients. The condition occurs most commonly in children initially affected when they were younger than 2 years. The condition is associated with the presence of an extremely high measles antibody titre in the blood and cerebrospinal fluid (CSF).

Laboratory Diagnosis

The clinical manifestations of typical measles cases are so characteristic that the diagnosis is self-evident. The laboratory diagnosis is frequently helpful in diagnosing atypical measles and differentiating from rubella.

Specimens

Respiratory specimens, conjunctival specimens, urine, blood and brain tissues are the frequently used specimens. The respiratory specimens and blood collected during the prodromal stage and the period following until 2 days after the appearance of the rash are the specimens of choice for isolation of viruses by culture.

Microscopy

Demonstration of multinucleated giant cells, measuring up to 100 nm in diameter, in Giemsa-stained smears is diagnostic of measles (Fig. 57.1) in biopsies of Koplik spots, nasal secretion smears and dermal rashes.

Direct Antigen Detection

A direct immunofluorescence antibody test can detect measles antigen in nasal secretions, pharyngeal secretions or urinary sediments.

Cell Culture

The measles virus is difficult to isolate, but it can be grown in primary human or monkey kidney cell cultures. The virus produces cytopathic effects (CPEs), the characteristic multinucleated giant cells with cytoplasmic and nuclear inclusion bodies very slowly, usually 1 week after

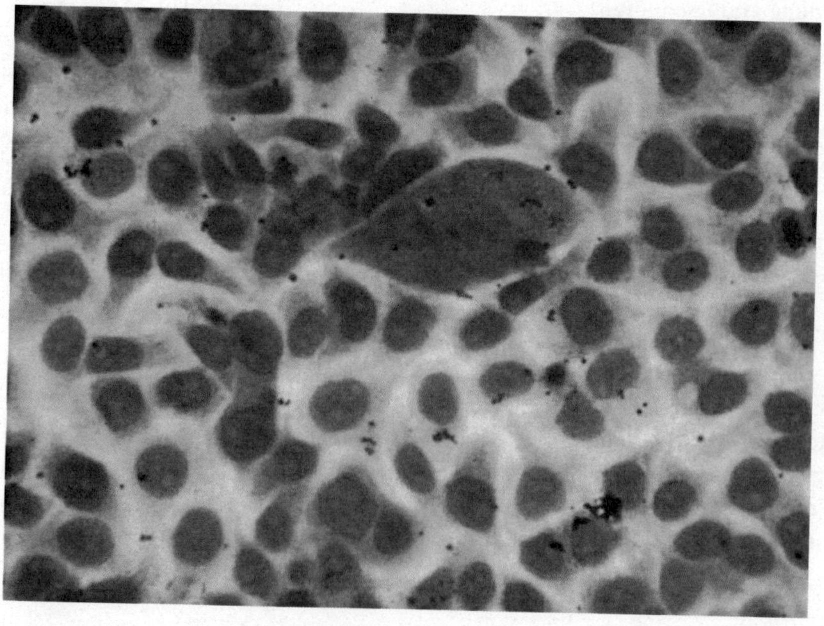

Fig. 57.1 Multi-nucleated giant cells of measles in Giemsa-stained smears (×1000)

inoculation. **Warthin–Finkeldey cells** are produced by the measles virus in the lymphoid tissue of the patients.

However, the viral antigen appearing in the infected cell cultures can be detected much earlier using a direct immunofluorescence antibody test.

Serology

Sandwich capture IgM enzyme-linked immunosorbent assay (IgM ELISA) is the quickest method to diagnose acute measles by detecting IgM antibodies in the serum. The IgM antibodies begin to appear in the serum on the third day after the appearance of the rash and remain positive 30–60 days after an illness in most individuals suffering from measles. The test is therefore positive during these days. IgM ELISA is highly sensitive (nearly 100%) during this period of illness.

Demonstration of more than a fourfold rise in IgG antibody titre between acute and convalescent sera confirms the diagnosis of measles. IgG antibodies usually appear 4 days after the onset of rash and persist for a longer period even after the patient has been cured of the illness. Demonstration of unusually high titre of measles antibody in the serum and CSF is diagnostic of SSPE.

Molecular Detection

Reverse transcriptase PCR is used as a diagnostic method because of its specificity and sensitivity.

Treatment

Ribavirin given either intravenous or in aerosol form is being now evaluated to treat severely affected adults and immunocompromised individuals with acute measles or SSPE. The measles virus has been shown to be susceptible to ribavirin *in vitro*. However, the drug is yet to be used for regular treatment of measles cases since no controlled trials have been conducted to test the efficacy of the drug in patients with measles.

Epidemiology

Measles is a disease reported throughout the world. A considerable number of people succumb to the infection globally each year. However, widespread vaccination has resulted in a decrease in mortality due to measles. The decrease has been as high as 73% in a span of nearly 18 years. The WHO has remarkably endeavoured to increase the vaccination coverage of the measles vaccine to a targeted 90% population in every nation. In 2018, about 86% of the world's children had received one dose of measles vaccine by the age of 1 year through routine health services, which has been a great achievement towards containment of the disease. Despite, breakthrough infections and outbreaks occur due to unregulated movement of the unvaccinated people, thus causing failure in efforts towards control.

Geographical Distribution

Epidemics of measles occur every 2–3 years. Approximately 30 million measles cases are reported annually, most cases being from Africa. The condition has also been well documented in America, Europe, the Eastern Mediterranean region, the Western Pacific region and Southeast Asia.

Reservoir, Source and Transmission of Infection

Measles is exclusively human disease. Infected respiratory droplets are the primary source of infection. Patients are infectious from 3 days before the onset of illness until the rash desquamates. Infectivity is maximum at the pro-drome and diminishes rapidly with the onset of the rash. Children with immunodeficiency due to leukaemia, corticosteroid therapy, or HIV are at increased risk for infection regardless of their immunisation status. Unvaccinated people are also at-risk.

Prevention and Control

Measles vaccine, along with mumps and rubella (MMR) vaccine, is currently used for universal immunisation of children.

Measles Vaccine

It is a live attenuated vaccine, which now uses Schwartz and Moraten attenuated strain of the original Edmonston B strain. The first dose of vaccine is given to children older than 12 months in the United States but are used for children aged 9 months in developing countries like India with high endemicity. The second vaccine dose is usually given to school-going children aged 4–6 years. Immunisation with two vaccine doses has shown a seroconversion of 99% after 1 year.

In India, under the **Universal Immunisation Programmes**, the first dose of measles vaccination is given as **Measles-Rubella (MR) vaccine** at 9–12 months of age and second dose at 16—24 months of age. If given alone, the measles vaccine is the first dose at 9 months of age followed by a second dose as MMR (measles mumps rubella) vaccine at 12–15 months.

The CDC recommends two doses of **MMR vaccine** to protect against measles, mumps, and rubella. The first dose is at 12 to 15 months of age, and the second dose is at 4 to 6 years of age. Adults should also be up to date on their MMR vaccination. The two doses of the MMR vaccine should be separated by at least 28 days. It is usually said that one dose of MMR vaccine is 93% effective against measles, 78% effective against mumps, and 97% effective against rubella, while both the doses of MMR vaccine are 97% effective against measles and 88% effective against mumps. MMR is a live attenuated vaccine causing harmless infection in the vaccinated person.

Post-Exposure Prophylaxis

It is carried out by serum human gamma globulin if administered within 6 days of exposure to measles, which may prevent or attenuate the disease. This is of immense value in children with immunodeficiency, pregnant women and others at increased risk of measles.

To eliminate measles, the WHO has laid down several regional elimination goals based on the occurrence of the disease. Amongst the major goals, achievement and sustenance of very high vaccination coverage (≥95%) with two doses of measles, supplementing additional coverage for the at-risk population and strengthening the surveillance system are the prominent ones.

India has performed extremely well in controlling measles cases and reducing the burden through effective vaccination. In collaboration with ten other WHO South-East Asia Region member countries, India has resolved to eliminate measles. Recently, the Rubella vaccine has been introduced in the universal immunisation programme (UIP) as the Measles-Rubella vaccine. To meet the aim of eradication, several strategies such as strengthening of routine immunisation services for all vaccines under UIP, including the first dose of measles vaccine at appropriate ages, providing a second opportunity of receiving a measles vaccine to children who may have missed the first dose have added to the population immunisation coverage.

Parainfluenza Virus

Parainfluenza are the pathogens that primarily affect young children, in whom the viruses cause upper and lower respiratory tract infections. The nomenclature and taxonomic relationship of HPIVs have changed considerably over the last decade. There are four types of parainfluenza viruses, Type 1,2,3 and 4.

Morphology

Parainfluenza viruses are pleomorphic viruses measuring 150–200 nm in diameter. The virus contains a single-stranded, non-segmented, negative-sense RNA genome with nucleoproteins P and L. It is surrounded by a helical nucleocapsid, which contains glycoprotein spikes. The surface spikes consist of H, N (neuraminidase) and F proteins. Both H and N proteins are present on the same spike, whereas the F protein is present on a separate spike. The F protein mediates the formation of the multinucleated giant cell.

Laboratory Isolation

HPI's grow in primary monkey kidney (PMK) cells (Rhesus and African green monkeys). The virus also grows in continuous monkey kidney cell lines (LLC-MK2), which is excellent for continuous passage and almost as good as PMK cells for primary isolation.

Pathogenesis of Human Parainfluenza Virus Infection

Human parainfluenza virus infection is acquired by inhaling respiratory droplets or by direct person-to-person contact with infected secretions. The virus is site-specific, and infection is confined to the respiratory tract. Respiratory epithelium appears to be the important site of virus binding and subsequent infections.

The virus adsorbs to the respiratory epithelial cells by specifically combining with neuraminic acid receptors in the cell through its haemagglutinin. Subsequently, the virus enters the cells following fusion with the cell membrane, mediated by F1 and F2 receptors. The virus replicates more rapidly than mumps and measles viruses in the cell cytoplasm and causes the formation of multinucleated giant cells. These giant cells, each containing two to seven nuclei, usually develop late in the infection. The virus also causes the formation of single and multi-locular cytoplasmic vacuoles and basophilic or eosinophilic inclusions. The virus causes inflammation of the respiratory tract, leading to secretions of high levels of inflammatory cytokines, usually 7–10 days after initial exposure.

Immunity

Humoral immunity plays a major role in defence against HPIVs. The antibodies are produced against both surface glycoproteins (HN and F) of the virus. Neutralising antibodies to all types of viruses are seen in most children at the time of birth, but the titre falls rapidly after 6 months. Most children and adults develop a detectable level of neutralising antibodies in the serum following natural infection with the virus. Secretory IgA antibodies also appear to play an important role. The CMI appears to play some role in containing the disease, as HPIV infection tends to be more severe in persons with defective CMI.

Immunity to HPIV is long-lasting, but re-infection may occur many times throughout life, most probably due to the presence of multiple serotypes of HPIV.

Clinical Syndromes

The human parainfluenza virus causes croup, pneumonia, bronchiolitis, and other infections.

Croup or **laryngotracheobronchitis** is a heterogeneous group of illnesses that affects the larynx, trachea and bronchi. It is a type of common cold disease that manifests as fever, cough, laryngeal obstruction and expiratory stridor. Human parainfluenza **pneumonia** is characterised by fever, rales and evidence of pulmonary consolidation. All types of HPIVs cause bronchiolitis. Most cases of bronchiolitis occur in infants. The condition manifests as fever, expiratory wheezing, tachypnoea and rales. **Other infections** include otitis media, pharyngitis, conjunctivitis, and coryza are the other infections caused by the virus.

Laboratory Diagnosis

Specimens

Respiratory specimens include nasopharyngeal aspirations, nasal washings and nasal aspirations.

Direct Antigen Detection
ELISA, immunofluorescence assay and fluor immunoassays are used to detect HPIV antigen directly in urine specimens. These methods are sensitive and specific. Shell viral assay is a sensitive method for rapid demonstration of the virus.

Cell Culture
The virus can be isolated from clinical specimens by culture in PMK and LLC-MK2 cell lines. The CPEs are rarely demonstrated during primary isolation of the virus in tissue culture. The virus causes the formation of syncytium in the infected cells. Hemadsorptions inhibition using guinea pig erythrocytes is also used to detect virus antigen in cell lines within 3–10 days of incubation.

Serology
Haemagglutination inhibition, neutralisation, ELISA and Western blot are frequently used antibody-based serological tests for diagnosis. A fourfold rise in antibody titre of acute and convalescent sera is diagnostic of acute infection.

Molecular Detection
A multiplex reverse transcriptase-polymerase chain reaction (RT-PCR) has been developed for simultaneous detection of the viral genome in clinical specimens.

Treatment

Ribavirin has been shown to be effective against HPIV infection *in vitro*. Uses of ribavirin aerosols or systemic therapy to treat HPIV infection in children and adults who are severely immunocompromised have shown mixed results with uncertain clinical benefits.

Epidemiology

The human parainfluenza virus is ubiquitous and has worldwide distribution. Parainfluenza viruses cause disease exclusively in humans. No animal reservoirs are present. Respiratory secretions from infected humans are the source of infection.

Prevention and Control

Live attenuated vaccine is available. Field trials of formalin-killed whole HPIV have proved to be ineffective in children. These vaccines failed to protect against natural infection by the parainfluenza virus.

Newcastle Disease Virus

Newcastle Disease Virus (NDV) is a poultry pathogen that causes severe outbreaks of pneumoencephaltis or influenza, resulting in high death rates. It is known Ranikhet virus in India. Suitable vaccination and slaughtering of the infected birds are the control measures. Poultry workers and others in contact with infected birds become subjugated to self-limited conjunctivitis.

Mumps Virus

Mumps is an acute infectious disease of children, characterised by acute, non-suppurative, painful swelling of the salivary glands.

Morphology

Mumps virus is a typical paramyxovirus containing a single-stranded, negative-sense RNA surrounded by an envelope. It has two major surface glycoproteins: (1) one with both haemagglutinin and neuraminidase and (2) the other with cell fusion protein (F). The haemagglutinin agglutinates the RBCs of fowl, guinea pigs, humans and many other species. The haemagglutination is followed by haemolysis and elution at 37 °C. The F protein is responsible for the fusion of the lipid membrane of the virus to the host cell.

Antigenic and Other Properties

Only one antigenic type of mumps virus is known. Neutralising antibodies are produced against haemagglutinin. It has an internal nucleocapsid soluble (S) antigen, which is detected by the complement fixation test.

The mumps virus is heat-labile. It is sensitive to heat and rapidly inactivated at room temperature. Treatment with formaldehyde, ether or ultraviolet light also inactivates the virus. The virus can be stored for a longer period by lyophilisation at −70 °C.

Laboratory Isolation

Chick Embryo

The virus grows well in the amniotic cavity of 6- to the 8-day-old embryonated egg.

Tissue Culture

Mumps virus grows well in PMK cells, Hep-2 cells and H292 cells. The CPE consists of multinucleated giant cells and acidophilic cytoplasmic inclusions. The growth of the virus in the cells can be detected by direct immunofluorescence and hemadsorptions.

Pathogenesis of Mumps

The infection is acquired from an infected person by direct person-to-person contact and inhalation of respiratory droplets. Infection by mumps virus begins after the entry of the virus into the respiratory tract. The virus then replicates locally and disseminates by the blood circulation to target tissues, such as the CNS and salivary glands, particularly the parotid glands. The virus causes infection of the parotid gland either through the Stensen duct or by viremia. Salivary glands, such as parotid glands, show desquamation of necrotic epithelial cells lining the ducts.

The virus replicates in these target tissues and then causes a secondary viremia phase. The virus is spread by viremia throughout the body to kidneys, testes, ovary, pancreas and other organs. Infection of the CNS, especially meninges, causes meningitis, or meningoencephalitis (Fig. 57.2).

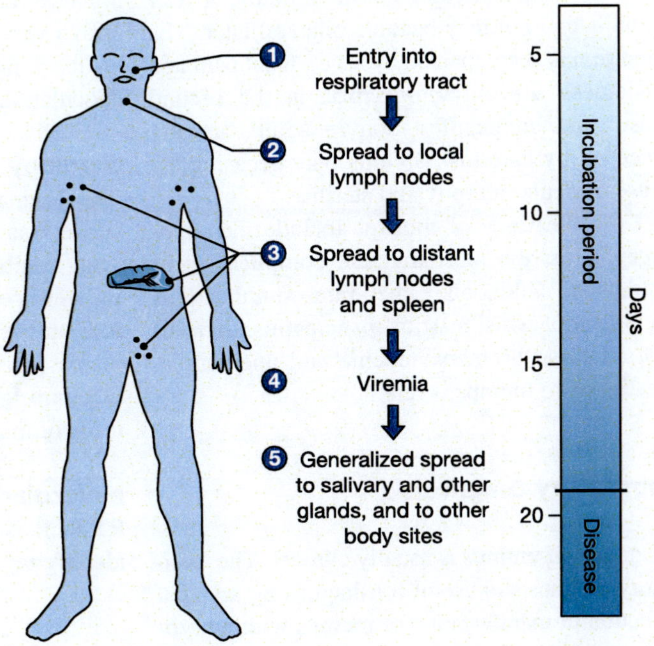

Fig. 57.2 Pathogenesis of mumps

Immunity

Humoral immunity is characterised by the appearance of antibodies against the soluble S antigen and haemagglutinating antibodies. The antibody against S antigen is the first to appear within 3–7 days after the onset of symptoms. The haemagglutinating antibodies directed against haemagglutinin confer lifelong immunity against the mumps virus. CMI is essential for the control of infection. This also contributes to the pathogenesis of the disease and is responsible partially for the symptoms observed during the course of clinical illness. Immunity in mumps is lifelong.

Clinical Syndromes

Mumps virus causes mumps.

Mumps

The incubation period is long and varies from 12 to 25 days. Most of the infections are asymptomatic. The onset of mumps is sudden. Fever, headache and earache are the initial symptoms. These symptoms are followed by **painful swelling of the parotid gland**. Initially, it may be unilateral but may become bilateral later. The condition is accompanied by fever, local pain and tenderness. **Epididymo-orchitis** is the second most common manifestation in adults, usually preceded by parotitis. Orchitis may occasionally cause testicular atrophy and sterility.

Complications of mumps include meningo-encephalitis, the most frequent complication of mumps in childhood. Other rare complications include oophoritis, mastitis, pancreatitis, thyroiditis, arteritis, thrombocytopenia and pneumonia. Death due to mumps is rare.

Laboratory Diagnosis

Diagnosis of mumps is usually clinical. The laboratory diagnosis is useful for diagnosing atypical infection or manifestation of mumps without typical symptoms.

Specimens

The specimens include the saliva, urine, secretions from the Stensen duct and the CSF. The virus is present in the saliva for 4–5 days, in the CSF for 8–9 days and the urine for 15 days after the onset of the symptoms.

Chick Embryo

Chick embryo inoculation is a less sensitive method than the cell culture. The viruses can be detected in the amniotic fluid after inoculating 6- to 8-day-old embryonated eggs. The virus is identified in amniotic fluid 5–6 days of incubation at 35 °C for haemagglutinins by haemagglutinin inhibition assay using specific antisera.

Cell Culture

Mumps virus can be isolated from clinical specimens by inoculation into monkey kidney cells, human amnion cells or HeLa cells. The growth of the virus in monkey kidney cells is detected by the presence of multinucleated giant cells. The CPE may take as long as 1–2 weeks to appear. Virus growth in the cells can be detected much earlier by haemadsorption of guinea pig erythrocytes adsorbing the surface of virus-infected cells. The immunofluorescence test is also a very rapid test used to detect viral antigen in the infected cells as early as 2–3 days after inoculation.

Serology

The haemagglutinin inhibition, immunofluorescence assay and ELISA are used to demonstrate viral antibodies in the serum. Detection of mumps-specific IgM antibodies by IgM ELISA indicates recent and active infection. A fourfold increase between acute and convalescent phases in serum IgM antibody levels confirms the diagnosis of mumps.

Molecular Detection

RT-PCR is a very sensitive and specific test for identifying mumps virus in clinical specimens.

Treatment

No specific antiviral agents are available against mumps.

Epidemiology

Mumps is a highly communicable disease occurring worldwide.

Geographical Distribution

Mumps continues to remain endemic in many countries throughout the world, as the mumps vaccine is used in only 57% of the countries. In the absence of a vaccination program, it often occurs as epidemics in children 5–15 years of age.

Reservoir, Source and Transmission of Infection

Humans are the only natural hosts of the mumps virus. The infected patients are the source of infection. A patient remains infectious usually from 9 days prior to the onset of parotid swelling as long as 7 days after onset of the swelling. Unvaccinated and immunocompromised people are at more risk of infection by the mumps virus.

Prevention and Control

Achieving and maintaining high immunisation levels, primarily in infants and young children, is the principal strategy to prevent mumps.

Vaccines

An effective **live attenuated vaccine** is available against mumps. The vaccine consists of the **Jeryl–Lynn strain of mumps virus** attenuated by serial passage in eggs and chick fibroblasts. The vaccine confers 95% protection, which lasts for 12 years. The vaccine is recommended for children of both sexes at 12 months of age. The vaccine can be given to all asymptomatic HIV-infected children at the same age. It is given as a part of the MMR vaccine. The vaccine is contraindicated in pregnancy and patients with severe febrile illness. A quadrivalent MMRV vaccine effective at preventing measles, mumps, rubella, and varicella has also been recommended. The first dose of this vaccine is advocated at 12–15 months, followed by the second dose at 4–6 years. The second dose can be given 3 months after the first dose.

Respiratory Syncytial Virus (RSV)

Respiratory syncytial virus (RSV) is the leading cause of respiratory tract infection in infants and young children.

Morphology

RSV is pleomorphic and measures from 150 to 300 nm in size. It has a small nucleocapsid, measuring 13 nm in diameter, unlike large nucleocapsid (18 nm) of other paramyxoviruses. The viral envelope contains a surface glycoprotein G, by which it is attached to the cell surface of the host cells. It contains F protein but lacks both H and N proteins. The F protein induces the fusion of the infected cells with adjoining cells, resulting in the formation of large multinucleated syncytia, from which the virus derives its name.

Antigenic and Other Properties

The virus is antigenically stable. It has only one antigenic serotype with the subtypes A and B. Subtype A appears to be more virulent than subtype B. The virus is highly labile; it is easily inactivated by dryness at room temperature and by acid. It is preserved by lyophilisation.

Laboratory Isolation

RSV is difficult to grow in tissue culture in PMK or human cell (HeLa cells, Hep-2) lines. The formation of large multinucleated syncytia is the characteristic CPE produced by the virus, but slowly in 2–10 days.

Pathogenesis of Respiratory Syncytial Virus Infection

RSV infection is acquired usually by inhalation of large droplet aerosols, coughed out by the patient during the act of coughing or sneezing. It is also acquired by direct contact with contaminated hands and fomites. The infection is restricted to the respiratory tract.

The virus initiates infection in the epithelial cells of the upper respiratory tract. Spread of the virus down the respiratory tract occurs by the cell-to-cell transfer of the virus along the syncytia from the upper respiratory tract to the lower respiratory tract, resulting in pneumonia. The virus usually does not cause any viremia or systemic spread. The virus causes necrosis of the small airway epithelium, plugging of the lumens with exudates and oedema, leading to obstruction of the normal airways of the young infants. The host's immune response plays a major role in the pathogenesis of bronchiolitis observed during illness.

Humoral antibody plays a minimal role in the host immunity against RSV. Maternal antibodies do not protect the infant from infection. CMI appears to play an important role in recovery from infection. Natural infection by RSV does not prevent re-infection by the virus.

Clinical Syndromes

RSV primarily causes infection of the respiratory tract, ranging from the common cold to pneumonia.

Common Cold

The illness begins with the upper respiratory tract infection, which manifests as a common cold with marked **rhinorrhoea** (running nose). This condition is most common in older children and adults. The incubation period varies from 4 to 5 days.

Bronchiolitis

RSV causes a more severe lower respiratory tract illness in infants, resulting in bronchiolitis or pneumonia. Clinically, the condition presents as cough, coryza, wheezing, rales and low-grade fever (<101 °F). Bronchiolitis is usually self-limiting but may cause serious disease in premature infants, immunocompromised hosts, and individuals with underlying lung diseases. Re-infection with the virus occurs throughout life. However, with advancing age and recurrent infection, the RSV infection is confined more to the upper respiratory tract than the lower respiratory tract.

Laboratory Diagnosis

Specimens

These include respiratory secretions obtained by washing, suctioning, or swabbing the nasopharynx.

Direct Antigen Detection

The viral antigens in the nasal washings or nasopharyngeal aspirates can be detected using ELISA or direct immunofluorescence antibody test using specific monoclonal antibodies.

Cell Culture

The virus can be isolated from clinical specimens by inoculating HeLa cells, Hep-2 cells or monkey kidney cells. After 2–10 days of incubation, RSV is identified by the characteristic syncytium formation in the cells. Direct immunofluorescence can also be used to detect viral antigens in infected cells.

Serology

The ELISA is used for the demonstration of antibodies in the serum. Demonstration of a four-fold or more increase in the antibody titre of acute and convalescent sera confirms the diagnosis of RSV infection. A commercial Dot-ELISA is available for RSV detection in the aspirates.

Treatment

Ribavirin, a broad-spectrum antiviral agent, has been recommended for the aerosolised treatment of children with severe RSV disease. This has shown beneficial effects by decreasing the duration of illness and decreasing the shedding of viruses in respiratory secretions.

Epidemiology

Geographical Distribution

RSV infection is prevalent worldwide.

Reservoir, Source and Transmission of Infection

RSV is highly contagious. Humans are the only hosts. Infected patients are the source of infection. They continue to excrete viruses for several days or weeks in their respiratory secretions. The virus is highly prevalent in young infants and children, with a peak incidence in 2- to 8-month-old infants. RSV infects virtually all children by the age of 4 years. The viral infection is usually seasonal and mostly occurs during winter. Nosocomial infections caused by the virus are frequent in nurseries and paediatric wards.

Prevention and Control

Passive immunisation with anti-RSV immunoglobulin has proved beneficial for prophylaxis in high-risk infants such as premature babies or babies with chronic lung disease. The anti-RSV immunoglobulin is a pooled polyclonal human immunoglobulin product with high titres of RSV antibodies.

Attempts to use a vaccine against RSV have been proved unsuccessful to date. Previously, a formalin-inactivated RSV vaccine was used in children, but the children who received this vaccine developed more severe diseases than those exposed to natural RSV infection; hence, this vaccine is no longer used. Recently, a live attenuated RSV vaccine that can be used as a nasal spray is under evaluation.

Nipah Virus

Nipah virus was reported to be the causative agent of encephalitis in Malaysia and Singapore in 1998 and 1999. People rearing pig populations were increasingly susceptible to encephalitis to this previously unrecognised virus. No specific antiviral agent or vaccine is available against the Nipah virus. Outbreaks of Nipah virus infection occur mostly in parts of Asia, primarily Bangladesh and India. Infection occurs by exposure to sick pigs and bats in areas where the virus is present and drinking raw date palm sap which can be contaminated by an infected bat. Outbreaks occur almost annually in these regions. Standard infection control practices should be adopted during an outbreak to prevent person-to-person spread in hospital settings.

Five outbreaks have been reported in India. The first Nipah virus disease outbreak was reported in Siliguri town in 2001, followed by a second outbreak in Nadia district in 2007 in West Bengal state. In 2018, an outbreak was reported in Kozhikode district, and in 2019, another outbreak in Kochi district, both in Kerala state. The recent outbreak was also reported from Kerala in 2021. Nipah has a relatively high case fatality rate, and immediate epidemiological investigations, including contacts identification, including health care workers and close family members, is required and should be placed under quarantine.

Hendra Virus

Hendra virus is a recently described, new zoonotic paramyxovirus. Hendra virus was first isolated from the cases of severe respiratory disease in Hendra, Australia, in 1994; hence the name Hendra virus. Infection is transmitted from infected horses to humans. Fruit bats are the natural reservoir. No specific antiviral agent or vaccine is available against the Hendra virus. Hendra virus infection remains a rare entity, and only 7 cases have been reported in humans though the infection has been reported in as many as 70 horses. Early recognition of the disease in the

Table 57.2 Characteristics of important viruses of the family paramyxoviridae

Virus	Serotypes	Membrane Fusion	Haemolysin	Haemagglutination	Haemadsorption	Neuraminidase	Inclusions
Measles	1	+	+	+4	+	–	Nuclear, cytoplasmic
Mumps	1	+	+	+3		+ 3	Cytoplasmic
Parainfluenza	4	+	+	+3	+	+ 3	Cytoplasmic
Respiratory syncytial virus	2	+	–	–	–	–	Cytoplasmic

horses is probably the most important step in limiting human cases. The infection can be prevented by avoiding sick horses and using appropriate personal protective equipment in veterinary procedures. A commercial vaccine has been recently licensed in Australia for horses only.

Human Metapneumovirus

Human metapneumovirus is a newly described virus in the family Paramyxoviridae. The virus was first reported as a cause of respiratory illness in children in 2001. The clinical manifestations of HMPV are closely similar to that of RSV infection in children. Like other members of the family Paramyxoviridae, human metapneumovirus is a single-stranded RNA virus. The virus is a respiratory pathogen and is associated as the cause of respiratory tract disease in children and adults worldwide.

The virus appears to be prevalent worldwide. The first report of the virus in 2001 demonstrated that all Dutch children by the age of 10 years were seropositive for the virus. Similar studies conducted in Australia, Canada, Japan and Israel have shown the high seroprevalence of HMPV antibodies in the population. Infants, elderly persons and immunocompromised individuals appear to be more susceptible to infection by the virus.

The virus is an important cause of respiratory tract infection in children, particularly infants. It causes disease, clinical manifestations of which are similar to those caused by human RSV. It causes clinical disease, ranging from mild respiratory symptoms to severe cough, bronchiolitis and pneumonia. The condition is also associated with high-grade fever, vomiting and myalgia. The virus also causes RSV-like diseases in adults, especially in those with chronic obstructive lung disease.

The virus is very difficult to grow in cell culture, and currently, serodiagnosis tests are not widely available to diagnose the condition. PCR appears to be the most specific test for diagnosing the condition by examining respiratory secretions. These include nasopharyngeal swabs, nasopharyngeal aspirates or bronchoalveolar lavage specimens.

No specific antiviral treatment is available. Hence, the treatment is majorly supportive. No vaccine is currently available. Characteristics of various genera in the family Paramyxoviridae are summarised in Table 57.2.

Case Study

A 3-year-old girl attended the Paediatrics OPD with complaints of high fever, malaise, anorexia, conjunctivitis, cough and coryza. Parents of the child gave the history of the appearance of red-coloured rash initially on the face and then spreading to the trunk, extremities, palms and soles. On examination, bluish-grey specks on a red base were found on the buccal mucosa opposite the second molar.

- What is the clinical diagnosis of the condition?
- Are there any laboratory tests to confirm the diagnosis of the condition?
- Is the child contagious, and how is the virus transmitted?
- What are the vaccines available to prevent the condition?

Arboviruses (Arthropod-borne Viruses) and Rodent-borne Viruses 58

Arthropod-Borne Viruses

Arthropod-borne viruses or arboviruses are RNA viruses that are transmitted by arthropods. The word *arbovirus* is an acronym for arthropod-borne viruses transmitted by arthropods from one vertebrate host to another. Arbovirus is a collective name for a large group of viruses, of which about 100 are known human pathogens. Most of the arboviruses causing infections in humans are zoonotic, with humans acting as accidental hosts. Yellow fever and dengue viruses are two exceptions that are not zoonotic.

The arboviruses have a special affinity to **arthropod vectors**, which can multiply and spread to humans. Some arboviruses are maintained in nature by transovarial transmission in arthropods. The **mosquitoes** are the best known and important vector of arboviral infection, following which many **ticks,** such as *Phlebotomus, Culicoides* and *Cimicidae,* also act as a vector for some of the arboviral diseases.

Important Properties

Arboviruses were named initially, according to the disease they cause, such as yellow fever virus, or depending upon the place from where they were first isolated, for example, West Nile fever and St. Louis encephalitis. Taxonomically, most arboviruses and arboviruses are recently classified into five families, namely, Togaviridae, Flaviviridae, Bunyaviridae, Reoviridae and Rhabdoviridae. Most viruses of medical importance are included in the family Togaviridae, and some belong to the family Flaviviridae and some to families Reoviridae and Rhabdoviridae.

Togaviruses, flaviviruses, bunyaviruses, reoviruses and rhabdoviruses associated with human infections are summarised in Table 58.1.

The antigenic structure of arboviruses basically comprises major proteins, like haemagglutinins, complement-fixing and neutralising antigens. Structural proteins, like E protein in the viral envelope, capsid proteins (C), membrane proteins and several non-structural (NS) proteins, contribute to the pathogenesis of infections by these viruses. The C proteins help information of infectious viral particles, E proteins help in viral assembly, attachment and internalisation of the virus inside the host cell, and M proteins help in membrane formation and maturation of the virus particles. The NS proteins play different roles in virus replication, immune suppression and evasion.

As RNA viruses, arboviruses are labile and easily inactivated at room temperature. Disinfectants and solvents, like alcohol and ether, inactivate these viruses by dissolving the lipid coat of the viruses.

The life cycle of the arboviruses is characterised by their ability to multiply in both the vertebrate hosts and blood-sucking

Table 58.1 Reservoir hosts, geographical distributions and vectors of arboviruses and rodent-borne viruses

Virus	Genus	Reservoir/Host	Geographic distribution	Vector
Viruses causing encephalitis				
Eastern equine encephalitis virus	Alpha virus	Birds	America	Mosquito
Western equine encephalitis virus	Alpha virus	Reptiles	America	Mosquito
Venezuelan equine encephalitis virus	Alpha virus	Rodents	America	Mosquito
Japanese B encephalitis virus	Flavivirus	Birds	East and Southeast Asia	Mosquito
St. Louis encephalitis virus	Flavivirus	Birds	America	Mosquito
West Nile encephalitis virus	Flavivirus	Birds	Africa, Europe, USA, West Asia	Mosquito
Murray Valley Alpha virus encephalitis virus	Flavivirus	Birds	Australia	Mosquito
Russian spring-summer encephalitis virus	Flavivirus	Rodents, birds, ticks	East Europe	Tick
Powassan virus	Flavivirus	Rodent	North America	Tick
Viruses causing haemorrhagic fever				
Chikungunya virus	Alpha virus	Not known	Africa, Asia	Mosquito
Dengue virus (1–4)	Flavivirus	Not known	Widespread, especially in Asia	Mosquito
Yellow fever virus	Flavivirus	Monkey, human	Africa, South America	Mosquito
Kyasanur Forest disease virus	Flavivirus	Rodents, shrew, monkey and ticks	India	Tick
Omsk haemorrhagic fever	Flavivirus	Small mammals	The(former) USSR	Tick
Crimean Congo	Nairovirus	Small mammals	The(former) USSR, Central Asia	Tick
Viruses causing febrile illness with or without rash and arthralgia				
Ebola virus	Filoviridae	Chimpanzees, gorilla, fruit bat, monkeys	Africa	Mosquito
O'Nyong-nyong v virus	Alpha virus	Not known	Africa	Mosquito
Mayaro virus	Alpha virus	Monkeys. marsupials	South America	Mosquito
Ross River virus	Alpha virus	Small mammals	Australia	Mosquito
Sindbis virus	Alpha virus	Birds, mammals	Asia, Africa	Mosquito
West Nile	Flavivirus	Birds	Asia, Africa, USA	Mosquito
Sandfly fever virus	Bunyavirus	Not known	Mediterranean countries, Asia, Tropical parts of America	Sandfly
Rift Valley fever virus	Bunyavirus	Sheep, cattle	Africa	Mosquito
Colorado tick fever	Orbivirus	Rodents	USA	Tick
Rodent-borne haemorrhagic fever				
Hantavirus pulmonary syndrome	Hantavirus	Deer mouse	South America and United states	Deer mouse
Haemorrhagic fever with renal syndrome	Hantavirus	Rodents	China, Russia and Korea	Rodents
Lassa fever	Lassa virus	Rodents	West Africa	Rodents

arthropods. Nevertheless, for effective transmission to vectors, the virus must be present in sufficiently higher numbers in the bloodstream of the vertebrate hosts, which can be taken up in a small volume of blood ingested during the bite of the arthropod. After ingestion, the virus replicates in the gut of arthropods and then spreads to other organs, including the salivary gland. Only the females of the species serve as the vector of the virus because female species require a blood meal to produce the progeny. Most of these viruses in their life cycle show an *extrinsic incubation period* in the vectors, during which viruses are replicated sufficiently in the vector and are present in sufficient numbers in the saliva to transmit the infection to the vertebrate host. The extrinsic period for most of the viruses ranges from 7 to 14 days. Some arboviruses are transmitted by vertical transmission through the transovarial passage from a mother tick to baby ticks.

Humans in the transmission cycle of arboviruses are usually the dead-end hosts. Usually, human-to-human transmission of infection by the arthropods does not occur due to the very low titre of viruses in human blood. In some diseases, such as dengue and yellow fever, there is a very high titre of viruses in the blood; hence, they act as a reservoir of infection to other humans.

Disease caused by occurs primarily in the dead-end host, but neither in the arthropod vector nor in the vertebrate animal that acts as the natural host. For example, after infection yellow fever virus causes disease in humans, who are a dead-end host, but causes a harmless infection among the jungle monkeys in South Africa, which serve as the natural host for the yellow fever virus.

Clinical presentations caused by arboviruses in humans may manifest as (1) fever with or without maculopapular rash, (2) encephalitis, often with high mortality or (3) haemorrhagic fever, also frequently severe and fatal. However, some arboviruses can cause more than one syndrome in the infected human host, for example, the dengue virus. Recovery from the disease usually confers lifelong immunity (Table 58.2).

Epidemiology of arbovirus infection shows distinct geographical distribution and vector patterns. The viruses occur in the tropics as well as in temperate countries. They show a tendency to cause a sudden outbreak of disease that usually occurs in the adjoining areas between human dwellings in jungle or forest areas.

Diagnosis of the condition is made by isolating the virus and viral serology. Specimens include blood collected during the acute phase of the disease. CSF can also be collected for viral isolation. The serum is collected for serodiagnosis. Some viruses may be isolated by growth in tissue culture or less commonly in embryonated eggs. Isolation from insect vectors and animal or avian species also supplements the identification of arboviruses prevalent in the study area.

Table 58.2 Togaviruses, flaviviruses, bunyaviruses, reoviruses and rhabdoviruses associated with human infections

Virus group	Human pathogens
Togaviruses	
Alphaviruses	Chikungunya, O'nyong-nyong, Mayaro, Semliki Forest, Sindbis, Ross River, eastern, western and Venezuelan equine encephalitis viruses
Rubiviruses	Rubella virus
Flaviviruses	Hepatitis C virus, Japanese encephalitis, Yellow fever, Murray valley encephalitis, West Nile, Ilheus, Spring-summer encephalitis complex, Louping ill, Powassan, KFD, Omsk haemorrhagic fever, Dengue virus, St. Louis encephalitis viruses
Bunyaviruses	Arboviruses, California encephalitis, Oropouche and Turlock viruses
Reoviruses	
Orbiviruses	Colorado tick fever, African horse sickness and Bluetongue viruses
Rhabdoviruses	
Vesiculovirus	Vesicular stomatitis virus, Chandipura virus

An indirect fluorescent antibody, enzyme-linked immunosorbent assay and other tests are employed to detect serum antibodies in arbovirus infections. These tests are useful mainly for epidemiological studies rather than for diagnosing individual cases. Antigenic cross-reaction between the related viruses is a noted problem.

Control methods essentially include vector controls and vaccination. The vaccination has been effective only in yellow fever and a few other diseases but not most arboviral diseases. Hence, vector control continues to be a mainstay in the prevention of arboviral diseases.

Roboviruses

A new group of viruses called *roboviruses* has been described recently. This is an acronym for *a* *ro*dent-*bo*rne viral disease transmitted from rodents to rodents and directly from rodents to humans without the participation of arthropod vectors. Important properties of roboviruses are summarised in the Table 58.3.

Roboviruses are spherical, 90 nm in diameter and contain single-stranded RNA, linear, non-segmented, negative sense, 8.9 kb, MW 3 million Da. The viruses consisting of six structural proteins replicate in the nucleus. They are enveloped viruses, neurotropic and are associated with neurobehavioural abnormalities. They have a broad host range. Transmission occurs by direct contact with body fluids or excreta of rodents. Hantavirus infection, Lassa fever and South American haemorrhagic fever are the major rodent-borne viral diseases worldwide.

Togaviruses

Togaviruses are the largest family of arboviruses and contain two genera: *Alphavirus* and *Rubivirus*. The genus *Alphavirus* includes 32 different viruses, of which at least 13 viruses are associated with human diseases. They are transmitted mainly by mosquitoes. The genus *Rubivirus* contains rubella virus that causes rubella, which is not arthropod borne. Rubella is not transmitted by an arthropod vector. Rubella virus is described in detail in Chap. 70.

Alphaviruses

Alphaviruses are mosquito-borne arboviruses, which can be classified as febrile illness group and encephalitis group. The febrile group includes Chikungunya consistently virus, O'nyong-nyong virus, Semliki Forest virus, Sindbis virus and Ross River virus. The encephalitis group includes the Western equine encephalitis virus, Eastern equine encephalitis virus and Venezuelan equine encephalitis virus.

Encephalitis Group

Encephalitis is an acute inflammation of the brain parenchyma, often with secondary meningeal involvement.

Western Equine Encephalitis Virus

This virus causes disease more frequently than the eastern equine encephalitis virus, but the infection is less severe. It is a neuropathic *Alphavirus,*

Table 58.3 Important properties of roboviruses

1. Spherical, 90 nm in diameter
2. Single-stranded RNA, linear, non-segmented, negative sense, 8.9 kb, MW three million Da.
3. Six structural proteins
4. Enveloped
5. Replicate in nucleus
6. Broad host range
7. Neurotropic
8. Cause neurobehavioural abnormalities

which causes encephalitis and viral symptoms with an associated rash. The virus is spread primarily by the vector mosquito *Culex tarsalis*. The condition is most commonly seen in the United States and is sub-clinical. The isolation of the virus from the cerebrospinal fluid (CSF) or demonstration of a rise in specific antibody titre in the serum confirms the diagnosis. No antiviral therapy is available. A killed vaccine is available to protect horses but not humans.

Eastern Equine Encephalitis Virus

The virus causes the most severe disease and is associated with the highest fatality. The virus is transmitted primarily by *Culiseta* mosquito. Wild boars are the reservoir host and horses, and humans are the dead-end hosts. The diagnosis is made by isolating the virus or demonstrating a rise in antibody titre in the serum. There is no antiviral therapy. There is a killed virus vaccine for horses, but not for humans.

Venezuelan Equine Encephalitis

This virus is transmitted primarily by *Culex* and *Culiseta* mosquitoes. The condition that manifests as a mild systemic infection and may lead to encephalitis is prevalent in North and South America. Birds are the reservoir hosts.

Febrile Illness with arthritis Group

Chikungunya Virus

Chikungunya virus **(CHIKV)** was first described in humans and *Aedes aegypti* mosquito from Tanzania in 1954. The name *Chikungunya* means "*doubled up*", which describes the clinical manifestation in which a patient lies doubled up due to severe joint pains. The disease is transmitted primarily by *A. aegypti* mosquitoes. Humans are dead-end hosts, and no animal reservoirs are seen.

Chikungunya virus has **three main genotypes**. These are West African, Asian and East/Central/South African (ECSA), which has at least two sublineages, Indian Ocean and Indian. Epidemiological differences exist between the genotypes based on their origin. The ECSA genotype has travelled from East Africa to the Indian Ocean and India, followed by Europe, Asia and North America. Presently, in a global scenario, multiple genotypes can be simultaneously present during CHIKV outbreaks.

The virus causes a severe disease with a sudden onset of fever, frequent joint pains, lymphadenopathy and conjunctivitis. Fever is typically biphasic with a revision period after 1–6 days. The condition is often associated with maculopapular rash. Haemorrhagic lesions were also observed in the patients in Kolkata during 1963. The debilitating joint pain can last for a few days, weeks, months or even years. CHIKV can cause acute, sub-acute or chronic disease. Polyarthralgia is a characteristic feature. The case definition of Chikungunya fever as proposed by the WHO is (1) a suspected case as a patient presenting with acute onset of fever for 3–5 days, usually with chills and rigors, with pain in multiple joints/swelling of extremities that may continue for weeks to months; (2) a probable case as one with features of a suspected case along with either history of travel/residence in areas reporting outbreaks or with the exclusion of malaria, dengue and any other known cause of fever with joint pains. A confirmed case is supported by laboratory investigations for CHIKV irrespective of symptoms.

Chikungunya virus was first demonstrated in India in 1963 during an extensive epidemic caused by the virus in Kolkata, Chennai and other areas. Outbreaks of Chikungunya have also been documented at regular intervals along the east coast of India and in Maharashtra till the 1960s. In 2006, an outbreak occurred in India, with 2500 cases reported. About 70% of cases were from Andhra Pradesh and Maharashtra. This re-emergence of the virus occurred almost after a gap of 32 years, causing an extensive outbreak in 13 states, including Andhra Pradesh, Karnataka, Maharashtra, Madhya Pradesh, Tamil Nadu, Gujarat and Kerala. Since then, there has been an increase in cases reported, particularly during 2016–2017.

Among other factors, increased case detection due to the availability of CHIKV-specific IgM detection kits developed by the National Institute of Virology and made available through NVBDCP contributed significantly. Few other factors could have been differences in climatic conditions, mosquito and human population density. Change in genotype and mutations in the virus's genome and increase in efficiency of mosquitoes to transmit the virus are other major reasons for the resurgence of CHIK in India.

However, often, there are no outbreaks. Serodiagnosis is the most common diagnostic approach. Usually, paired serum samples will be tested with IgM or IgG ELISA. RT-PCR can also be used for the detection of viral RNA. No antiviral therapy or vaccine is available.

O'nyong-nyong Virus

This virus is restricted to Africa and was first reported from Uganda. The virus, closely related to the Chikungunya virus antigenically, is transmitted by *the Anopheles* mosquito.

Semliki forest Virus

This virus is yet to be associated with any human disease, although serum antibodies against this virus have been demonstrated in an African population.

Sindbis Virus

Sindbis virus was first reported from *the Culex* mosquito in Egypt in 1952. Since then, the virus has been reported from other parts of Africa, the Philippines and Australia. Serological evidence of the Sindbis virus by demonstration of antibodies in the serum has been demonstrated in India.

Ross River Virus

The virus has been associated with epidemic polyarthritis in Australia.

Flaviviruses

Flaviviruses were originally named group B arboviruses. They are similar to togaviruses in that they are also single-stranded RNA viruses and have an icosahedral nucleocapsid surrounded by an envelope. They are slightly smaller (40–50 nm in diameter) than those of an *Alphavirus*, and the RNA does not have a polyadenylate sequence. The virus also lacks a conspicuous capsid structure in the virion. All flaviviruses are serologically related, and antibodies to one virus cross-react with another virus.

The genus Flavivirus contains more than 70 viruses, of which nearly 13 viruses can cause infection in humans. Most of these infections are transmitted either by mosquitoes or by ticks. However, the hepatitis C virus is transmitted neither by mosquitoes nor by ticks. The flaviviruses causing infections in humans are classified broadly as (1) mosquito-borne flaviviruses and (2) tick-borne flaviviruses.

Mosquito-Borne Flaviviruses

Mosquito-Borne Flaviviruses include Encephalitis Viruses and yellow fever.

Encephalitis Viruses

St. Louis Encephalitis Virus

St. Louis encephalitis virus is the most important mosquito-borne viral disease in the United States. It is transmitted by *Culex* mosquitoes. Wild birds, such as English sparrows, are the reservoir hosts. Humans are the dead-end host. These viruses are mostly seen in urban areas because these mosquitoes breed in stagnant wastewater. The virus causes encephalitis, which is associated with a mortality rate of 10%. Isolation of the virus is difficult; hence, diagnosis is mostly based on serology. No antiviral therapy or vaccine is available.

Ilheus Virus

The virus occurs mostly in South and Central America. The infection is maintained in forests by bats and mosquitoes. The virus usually causes asymptomatic infection and rarely causes encephalitis.

West Nile Virus

West Nile virus causes encephalitis. The virus is endemic in the Middle East, Africa and Asia. Wild birds are the main reservoirs of the virus. The infection is transmitted from wild birds to humans by *Culex*, *Aedes* and *Anopheles* mosquitoes. Humans are the dead-end hosts. The virus first causes asymptomatic or symptomatic illness in the wild migratory birds. These infected birds contain a high titre of the viruses and remain viraemic for 1–2 weeks. These serve as important reservoirs of infection for transmission to humans, horses, dogs and other small animals after being bitten by mosquitoes.

The virus-like other arthropod-borne viruses cause viraemia, cross the blood–brain barrier and infect the brain parenchyma causing viral encephalitis. Patients with this virus may present with features of encephalitis, aseptic meningitis or both. This condition has also been reported from India, and the virus has been isolated from *Culex* mosquitoes. The virus has been isolated from the brain of these fatal cases of encephalitis in Karnataka in 1980–1981. Mental confusion and disorientation with decreased consciousness are some of the manifestations of patients with encephalitis. The infection usually occurs in the summer months when humans, mosquitoes and wild migratory birds remain in close proximity outdoors.

Laboratory diagnosis of the condition is made by isolating the virus from the blood within the first 2 weeks of infection; the virus cannot be isolated from CSF. Enzyme immunoassay and plaque reduction utilisation tests are available for the specific diagnosis of the condition by demonstration of specific antibodies in serum. A fourfold or greater rise of antibody titres between acute and convalescent titre is diagnostic of the infection. Polymerase chain reaction (PCR) is also available. No antiviral therapy or vaccine for the virus is available.

This is an infection that can be transmitted by transfusion of contaminated blood. Hence, PCR and nucleic acid probes specific for the virus are used to screen the blood for the virus.

Murray Valley Encephalitis Virus

This virus is endemic in New Guinea and Australia. The virus occurs in the enzootic cycle involving mosquitoes and wild birds. *Culex annulirostris* is the main vector. The virus was first reported during an encephalitis epidemic in Murray River Valley in 1951 and hence was called Murray Valley encephalitis virus.

Japanese Encephalitis Virus

JE virus is a flavivirus that causes JE, the most common cause of epidemic encephalitis. JE virus is a single-stranded RNA virus related to St. Louis encephalitis and West Nile virus. The virus was first isolated in Japan during an encephalitis epidemic in 1935. The virus initially was called Japanese B encephalitis to distinguish it from the Japanese A encephalitis virus (Wang Ecunoms type A encephalitis), which had different epidemiological properties. In India, the virus was first reported in 1955 during an outbreak of encephalitis in Tamil Nadu.

Pathogenesis: The bite of infected mosquitoes transmits the JE virus to humans. The virus initially replicates at the site of the bite and in the regional lymph nodes. Subsequently, the virus spreads through circulation, causing viraemia and may produce inflammatory changes in the heart, lungs, reticuloendothelial system and liver. Usually, these infections are removed before the virus affects the central nervous system (CNS), causing underlying disease. Even though neurological protrusion can probably be caused by the virus's growth within endothelial cells vascularly, thereby affecting the large parts of the brain, including the thalamus, basal ganglia, brain stem, cerebellum, hippocampus and cerebral cortex. Circulating antibodies play an important role in conferring heterologous immunity against JE.

Clinical presentations: Most of the infections are asymptomatic. Only 1 in 250 infections results in symptomatic disease. Most symptomatic infections occur in young children between 2 and 10 years and elderly people in endemic areas. In non-endemic areas, the infection occurs in all age groups.

The incubation period of JE varies from 4 to 15 days. The condition has an abrupt onset with the appearance of symptoms, which include fever, headache, nausea, diarrhoea, vomiting and myalgia, which may last for several days. This is followed by an altered state of consciousness, tremors, incoordination and convulsions. Seizures develop in two-thirds of people—most often in children, while headaches and meningismus are more common in adults.
The mortality rate is high; in less-developed areas with poor medical care, the mortality rate may exceed 35%. More than 10,000 reported deaths occur per year due to JE worldwide. The neurological sequelae are severe and can occur in many patients who suffer from the disease. The neurological sequelae include seizure disorder, movement disorder or motor or cranial nerve paralysis.

Epidemiology: This is a seasonal disease with more cases seen in temperate areas from the duration of June to September. Transmission starts as early as March and ends until October in subtropical areas. In a few subtropical areas, such as Indonesia, transmission can start at the earliest. Around 35,000–50,000 symptomatic cases are being reported worldwide per year, even though there is lightly a miscalculation of the actual prevalence of the disease. Approximately, 3 billion people currently live in areas endemic for JE, which extend from India to Maritime, Siberia and Japan. The disease occurs throughout Asia but is most prevalent in Southeast Asia.

Birds and pigs are the two main **reservoir hosts**. JE is transmitted to humans by certain species of *Culex* mosquitoes widely prevalent in rice fields in Asia. The disease is spread throughout mostly rural areas of Asia by culicine mosquitoes, most often *Culex tritaeniorhynchus*. Natural infection occurs in ardeid birds, such as herons and egrets, and the virus is transmitted from bird to bird by *C. tritaeniorhynchus*.

Transmission of the JE virus occurs primarily in rural agricultural areas, often related to rice production and flooded fields with heavy irrigation. In some areas of Asia, even the surroundings of urban areas have similar geographical conditions. Transmission of JE is seasonal in temperate areas of Asia, with peaks in JE cases in the summer and fall. However, infections occur in tropical regions round the year, with a clustering of cases in the rainy season. Humans are incidental or dead-end hosts. Otherwise, the transmission cycle is maintained between mosquitoes and vertebrate hosts, like pigs.

Humans, cattle, buffaloes and pigs are the **vertebrate hosts**. Birds act as reservoir hosts. Pigs are usually the amplifier host. Pigs have persistent, high-grade viraemia and help transmit the virus in humans living close to the pig-rearing areas (Fig. 58.1). *C. tritaeniorhynchus* is a major

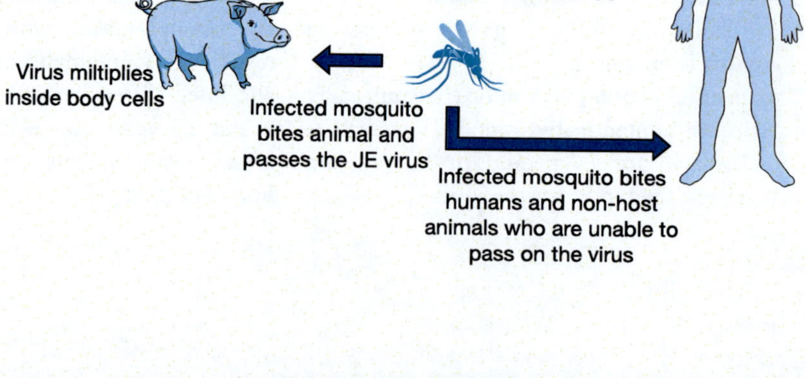

Fig. 58.1 JE virus transmission pattern

vector that has a predilection for cattle and bites them in preference to humans or pigs, but since the cattle do not develop viraemia, this does not contribute to the spread of the disease. The high cattle to pig ratio in India is a major factor suggested to control infection in India.

Indian Scenario: JE in India was first reported from Vellore, Tamil Nadu and Pondicherry in the 1950s. Since then, the condition has been reported in all the states and union territories of India. An estimated 378 million people live at the risk of JE in 12 states, union territories that are frequently affected. Several outbreaks of JE have occurred in Bankura and Burdwan in West Bengal, Dibrugarh in Assam, South Arcot in Tamil Nadu, Kolar in Karnataka, various areas in Andhra Pradesh and Pondicherry and Gorakhpur in Uttar Pradesh. Rapid agricultural development and intensive rice cultivation supported by irrigation systems facilitating breeding of culicine mosquitoes are primarily responsible for spreading the condition to newer areas in India. An epidemic of the JE virus occurred in Uttar Pradesh and Bihar in India in 2005 and throughout Nepal. More than 5000 cases were reported during this epidemic, and approximately 1000 deaths were documented.

C. tritaeniorhynchus is the most important mosquito species that transmits JE. This mosquito has a peculiar habit of biting outdoors and is extremely active in the evening and night. These mosquitoes breed in the collection of water, typically rice fields, thereby increasing the risk of infections in rural areas. Low-grade, short-term viraemia occurs in humans and other mammals; hence, these are dead-end hosts for the virus.

Laboratory diagnosis is made by isolating the virus from the blood during the first week of illness. The CSF rarely yields viruses except in severe or fatal cases. Serodiagnosis is made by using ELISA to demonstrate IgM antibodies in serum or CSF. This test has a high sensitivity, nearing 100%. The test may show a false-positive reaction with serum from other febrile illnesses, such as dengue and West Nile fever or with the serum collected after vaccination with yellow fever and JE. The IgM antibodies are present in serum or CSF in approximately three-fourths of patients within the first 4 days after onset of illness and in nearly all the patients after 7 days of illness. The direct fluorescent antibody test is a rapid method for detecting viral antigens directly in the brain tissue. Reverse transcriptase–polymerase chain reaction (RT-PCR) is used to demonstrate the viral genome in the CSF and blood for diagnosis of JE.

There is no specific antiviral therapy available for the treatment of JE.

Prevention consists of the use of the vaccine. The vaccine using the Nakayama strain is approximately 100% immunogenic after three doses. Two doses of the vaccine are administered at 2 weeks intervals followed by a booster dose after 6–12 months. The vaccine is recommended for persons living in epidemic and endemic areas and for travellers visiting these areas. The vaccine is also useful in the pig population to prevent epidemics.

A live attenuated vaccine is now widely used in India, Korea, Sri Lanka and Thailand. It is a JE strain SA 14-14-2-based vaccine. This virus strain is usually grown in the Vero cell line and further inactivated and alum adjuvanted. The primary two doses of this vaccine are administered 4 weeks apart. Following this, a booster dose is recommended approximately within 1–2 years of primary doses. The efficacy of this vaccine was found to be high and helps in preventing clinical diseases.

There are **four main types of JE vaccines** currently in use worldwide. These are inactivated mouse brain-derived vaccines, inactivated Vero cell-derived vaccines, live attenuated vaccines and live recombinant vaccines. Of these, the live attenuated cell culture-derived SA14-14-2 vaccine is the most widely used vaccine in endemic countries. In India, besides the above vaccine, two other vaccines are Inactivated SA-14-14-2 vaccine (IC51) (IXIARO® by Intercel & JEEV® by Biological Evans India Ltd.) and Inactivated Vero cell culture-derived Kolar strain, 821564XY. JE vaccine (JENVAC® by Bharat Biotech) has been in use.

Mosquito eradication and measures to prevent mosquito bites, such as the use of mosquito nets, mosquito repellents and insecticides, are useful to prevent transmission of infection.

Yellow Fever Virus

It is a virus that is type-specific belonging to the family Flaviviridae. The virus is a single-stranded RNA virus that is enveloped. The envelope contains a lipid bilayer surrounding a glycoprotein layer and a matrix protein. The single-stranded RNA is compounded with a capsid protein.

Pathogenesis of Yellow Fever

Yellow fever, caused by the virus, is transmitted by the bite of the *Aedes* mosquito. The virus replicates locally and in regional lymph nodes draining the wound. Subsequently, the virus spreads by blood to the bone marrow, liver, myocardium and spleen, where further virus replication occurs. The condition is associated with haemorrhagic manifestations occurring due to disseminated intravascular coagulation. This occurs due to bleeding from the gastrointestinal mucosa and abdominal and pleural serous layers. This is associated with a reduction in coagulation factor synthesis and altered platelet function. The condition progresses to shock and death due to multiple organ failures involving the liver, kidney, brain and heart.

Host immunity is characterised by the presence of viral neutralising antibodies by the end of the first week, during which the virus is rapidly cleared from circulation. An attack of yellow fever confers lifelong immunity.

Clinical Syndromes

The incubation period is short and varies from 3 to 6 days. No prodromal symptoms occur. As the name suggests, yellow fever is characterised by jaundice and fever. This is an illness characterised by an acute onset of fever followed by jaundice within 2 weeks of the onset of symptoms. Typically, symptoms begin suddenly with fever, chill, malaise, prostration, headache, giddiness, myalgia, anorexia, nausea and vomiting. This may be associated with bleeding from the nose, gums, gastrointestinal tract or skin within a few weeks of illness.

Haemorrhagic diathesis progressing to disseminated intravascular coagulation is the most important complication of the disease. Multiple organ failure involving the liver, kidney and heart is responsible for the patient's death. Yellow fever is a severe and life-threatening disease.

Epidemiology

Yellow fever is a mosquito-borne acute febrile illness that occurs in Africa and South America. The disease is not reported in India. Yellow fever is endemic in South America and the Caribbean enzootic countries of Africa. Although yellow fever can be transmitted in Asian countries, no documented transmission of yellow fever has been reported from Asian countries, including India. It has been suggested that previous infection with another flavivirus, such as the dengue virus, may confer protection from the yellow fever virus in Asian countries.

Yellow fever is transmitted from infected humans to mosquitoes, which are diurnal feeders. Humans suffering from the disease during the initial 3–4 weeks of illness are the source of infection. The extrinsic incubation period is 12–21 days. Vertical transmission of yellow fever takes place from female mosquitoes to the female progeny in 1% of cases. This is the important reason for the survival of the virus in the dry season. Approximately 3–10 variants are necessary to infect a mosquito. Seasonal transmission of yellow fever occurs during the mid-rainy season and early dry season in Africa. In South America, this occurs from January to March. An estimated 200,000 cases of yellow fever occur in Africa and South America, causing 20,000–30,000 deaths annually.

Yellow fever shows different transmission cycles.

The Jungle (sylvatic) Cycle

It is present in rain forests. In this cycle, the mosquito transmits the virus to wild, non-human primates; subsequently, it is transmitted to another mosquito from these hosts. Humans are the incidental hosts in the cycle, for example, humans clearing the trees in the forest when they are bitten by the mosquito. *Aedes* species are the main vectors in Africa. Wild *Aedes* species are the vectors in South America.

The Urban Cycle

It is confined to urban areas in which the mosquitoes transmit the virus to a human host, and then it is transmitted to another mosquito. *A. aegypti*, a domestic mosquito, is the primary vector.

The Intermediate Cycle

It is confined to the Savanna forest area of Africa and is the most common cycle responsible for outbreaks of yellow fever in Africa. In this cycle, the mosquito transmits the virus to wild non-human primates and human hosts and then it is transmitted to another mosquito.

Laboratory Diagnosis

Laboratory diagnosis of yellow fever can be made by (1) isolation of yellow fever virus from the clinical specimens, (2) demonstration of IgM-specific antibodies in the serum or demonstration of fourfold or more rise in serum IgG, (3) detection of yellow fever antigen in tissues by immunohistochemistry) detection of viral genomes by PCR and (4) elevated transaminase and bilirubin levels, demonstrated during the toxic stage of illness.

Prevention and Control

Prevention is by vaccines. Vaccination is the most widely used preventive measure against yellow fever.

The 17 D strain of yellow fever virus is a widely used attenuated live virus vaccine. This avirulent 17 D strain vaccine was recovered during the serial passage of a pantropic strain of yellow fever virus through tissue culture. The strain that has lost its capacity to induce viscerotropic or neurotropic disease has been used as a vaccine for more than 40 years. The vaccine, which is prepared in eggs and is available as a dried powder, is maintained in a cold chain for effective immunisation. The vaccine is highly effective. The vaccine provides immunity mainly for 30–35 years and probably for the whole life after a single dose. The vaccine is highly safe associated with very few side effects.

The vaccine is indicated for (1) all the individuals residing in countries endemic for yellow fever and (2) travellers visiting the countries South America and Africa. The vaccine is contraindicated for (1) infants below 9 months, (2) pregnant mothers and (3) individuals with immune deficiency diseases, such as HIV or persons receiving organ transplantation, etc.

Other preventive measures include mosquito eradication programs and personal protective measures.

Dengue Virus

Dengue is the most common mosquito-borne arboviral illness caused by the dengue virus. The name dengue is derived from the Swahili word *ki-dinga-pepo*, meaning a sudden seizure by a Demon. The earliest known documentation of symptoms of dengue-like illness was described in the Chinese Encyclopaedia during 265 AD. In 1780, Rauss coined the term "break-bone fever" based on the description of symptoms reported by patients during Philadelphia epidemics of probably dengue fever. The possible outbreak of dengue fever epidemics was documented sporadically every 10–30 years until World War II. Subsequently, after World War II, the dengue fever spread and became worldwide. The first epidemic of dengue haemorrhagic fever (DHF) was described in 1963 in Manila. In 1979–80, the first reported outbreak of dengue fever occurred simultaneously in Asia, Africa and North America.

Morphology

Dengue virus is a small, spherical and enveloped virus. It is a flavivirus having a cubic symmetry and measures 40–50 nm in diameter. It is a single-stranded RNA virus of 11 kb size. It has an icosahedral nucleocapsid and is covered by a lipid envelope. The virus is inactivated by diethyl ether and bile salts, such as sodium deoxycholate.

Antigenic Properties

Dengue virus has four distinct closely related serotypes: dengue 1 (DEN-1), dengue 2 (DEN-2), dengue 3 (DEN-3) and dengue 4 (DEN-4). The speciation was done by Albert Sabin in 1944. Each serotype is known to have several different genotypes.

Pathogenesis of Dengue

Humans acquire infection and become infected with the virus by the bite of *the Aedes* mosquito vector. The leakage of plasma caused by increased capillary permeability is the major pathophysiological abnormality that occurs in DHF and dengue shock syndrome (DSS). Bleeding, which is a most important manifestation in patients with DHF, is caused due to capillary fragility and thrombocytopenia, and it manifests in various ways ranging from petechial skin haemorrhages to life-threatening gastrointestinal bleeding.

It is postulated that prior infection with one or more dengue serotypes causes the development of low levels of neutralising antibodies and a high level of non-neutralising antibodies. In the same patient on reinfection with another serotype of dengue virus, the virus antibody complexes are formed within a few days of second dengue infection. This increases viral entry and replication of more mononuclear cells, followed by the release of cytokines, vasoactive mediators and a few coagulants. This phenomenon is called antibody-dependent enhancement. It is responsible primarily for the disseminated intravascular coagulation seen in patients with DHF. Significantly, certain dengue strains, especially dengue two strains, are considered to be most virulent.

Immunity

Infection with the dengue virus confers lifelong immunity. The immunity is serotype specific. Infection by one serotype does not confer protection against other serotypes. The infection with the dengue virus of different serotypes may cause more severe diseases, such as DHF. Although dengue and yellow fever viruses are antigenically related, infection with the dengue virus does not result in significant cross-immunity against the yellow fever virus.

Clinical Syndromes

The dengue virus causes classic dengue, DHF and DSS.

Classic Dengue Fever

The incubation period varies from 2 to 7 days. The onset of the disease is sudden, which begins as an influenza-like illness manifesting as fever, malaise, cough and headache. The fever, which may be as high as 41 °C, typically begins on the third day and lasts for 5–7 days. The fever is typically biphasic, coinciding with the absence of a virus in the blood. A maculopapular rash usually appears on the third or fourth day of the illness. The rash lasts for 1–5 days, fading with desquamation. Severe muscle pain, deep bone pain and joint pain (*break-bone fever*) are characteristic. Enlarged lymph nodes and leucopenia are also seen. Classic dengue fever is a self-limiting disease. It is rarely fatal and has few sequelae and complications. The convalescent phase may last for 2 weeks.

Dengue Haemorrhagic Fever

Dengue haemorrhagic fever is the most severe disease manifestation. The initial classic phase of DHF is similar to dengue fever and other febrile viral illnesses. However, subsequently, the patient's condition suddenly worsens with shock and haemorrhage, especially into the gastrointestinal tract and skin. The haemorrhagic manifestations include bleeding from the nose or gums, melena and haematemesis. This condition shows a high fatality rate as high as 10%. It occurs in children with passively acquired maternal antibodies. DHF or DSS is seen in most patients who have had prior infections with one or more dengue serotypes.

Dengue Shock Syndrome

It is the most severe form of the disease caused by the dengue virus. This is most commonly seen in untreated cases of DHF. The common symptoms include abdominal pain, vomiting and restlessness, and finally, the patient may die of circulatory failure and shock.

Laboratory Diagnosis

Specimens

Blood collected during the first 3–5 days of illness is useful for isolating viruses, and serum is useful for serological tests.

Cell Culture

Diagnosis of dengue is confirmed by isolation of virus from blood during the first 3–5 days of illness. The virus can be isolated in various cell cultures. Several mosquito cell lines have been in use. The C6/36 clone of *Aedes albopictus* cells for dengue viruses has been well in use. These cells provide a simple and rapid method for dengue virus isolation. They also provide serotype-specific monoclonal antibodies for identification and help in dengue virus surveillance. Other mosquito cell lines, like TRA-284-SF cells AP-61, are also used for viral culture. These are suitable for routine virological surveillance because of the ease of processing large numbers of specimens for virus isolation in a short period.

Serology

The serodiagnosis of dengue fever is based on the demonstration that IgM immunoglobulin is confirmatory for dengue. There is a rise in NS-1 antigen, which will be the first to appear from day 1 to day 9 and also a rise in IgG antibodies. IgM antibodies against the dengue virus are detectable after 4–5 days after onset of symptoms. The IgM capture ELISA (MAC-ELISA) is the most widely used test for demonstrating IgM antibodies in the serum. Neutralisation test, haemagglutination inhibition and IgG ELISA are the other serological tests used to diagnose the condition. The Plaque Reduction Neutralization Test (PRNT), which measures the titre of the neutralising antibodies in the infected person's serum, can detect the dengue virus and other flaviviruses. However, this test is only performed in the CDC referral laboratories presently.

The dengue MAC-ELISA, which is based on capturing human IgM antibodies on a microtitre plate using an anti-human-IgM antibody followed by the addition of dengue virus antigens, has been used for the qualitative detection of dengue virus IgM antibodies. The antigens used are E proteins of four dengue virus serotypes (DENV-1-4). They are detectable for approximately 12 weeks. Combined testing approaches using nucleic acid amplification test (NAAT) and MAC-ELISA usually gives a diagnosis in the first 1–7 days of illness.

Molecular Detection

RT-PCR is being used for genomic detection of dengue virus in the clinical specimen, but the test is now restricted only for use in research laboratories. However, CDC has devised a multiplex assay named CDC DENV-1-4 rRT-PCR multiplex assay to detect and type dengue virus. This assay detects DENV serotypes 1, 2, 3 or 4 from human serum or plasma collected from human patients with signs and symptoms consistent with dengue infection. This assay is devised to be used as a diagnostic test in patients. It is not approved for blood bank screening. Besides, it also provides epidemiologic surveillance data on circulating dengue viruses.

Treatment

No specific antiviral therapy is available to treat dengue infection. For cases with haemorrhagic manifestations and shock, supportive care is advised. Hydration in patients should be specially emphasised. Patients should be asked to avoid aspirin (acetylsalicylic acid), aspirin-containing drugs and other non-steroidal anti-inflammatory drugs (such as ibuprofen) owing to their anticoagulant properties. Close observation and frequent monitoring in an intensive care unit may be required in those with severe dengue. There is no role in prophylactic platelet transfusions in dengue patients, which might lead to fluid overload. Platelet transfusions are only indicated when the count reduces to 10,000 or below, or there is active bleeding in the patient. Other than autoimmune-related complications, corticosteroids should not be used.

Epidemiology

Dengue virus is distributed worldwide. DHF is primarily a disease of children and a leading cause of death in Southeast Asia.

Geographical Distribution

The estimated 3 billion people living in approximately 110 countries worldwide are at risk of dengue infection. Every year, approximately 100 million people are infected with dengue worldwide. The estimated 500,000 people with severe dengue require hospitalisation each year, and about 2.5% of those affected die. The actual number of dengue cases is underreported. The incidence of dengue has increased globally in recent decades. The disease is now endemic in more than 100 countries in the WHO regions of Africa, the Americas, the Eastern Mediterranean, South-East Asia and the Western Pacific, with the America, South-East Asia and Western Pacific regions being the most seriously affected. Not only is the number of cases increasing, but also explosive outbreaks are occurring. Among travellers returning from low- and middle-income countries, dengue is the second most diagnosed cause of fever after malaria. The years 2015–2016 have been marked by large dengue outbreaks worldwide, including India, Brazil, the Philippines and Malaysia. However, for the increase in cases, a potential reason is the initiation of activities to record all dengue cases also.

Reservoir, Source and Transmission of Infection

Humans are reservoirs of infection. The human host serves as a source of viral amplification. Humans are infectious to mosquitoes during viraemia for 3–5 days. The infection is transmitted by the bite of *A. aegypti* mosquitoes. After feeding, the virus shows an extrinsic incubation period of 10–14 days in the mosquito before the mosquito becomes infectious. The mosquitoes are vectors as well as sources of viral amplification. *A. aegypti* are small and highly domesticated mosquitoes, which breed on artificial water sources, and they prefer to bite humans, typically at the back of the neck and the ankles.

The initial dengue epidemics were documented in urban areas, but with the recent increase in dengue epidemics, the condition has also spread to suburban and rural areas. Dengue has been spreading from Southeast Asia to the subtropical and tropical Asian countries, the Indian subcontinent, including India, Sri Lanka, Malaysia, Philippines, New Guinea, Australia and several pacific islands. Unplanned urban development with population explosion and inadequate public health facilities, poor vector control and increased travel to the endemic areas have contributed to the spread of dengue worldwide.

The first epidemic of DHF was described in Manila, the Philippines, in 1953. Subsequently, DHF has spread rapidly, with epidemics occurring yearly and major outbreaks occurring every 3 years. Currently, DHF is one of the leading causes of death and hospitalisation in children in many Southeast Asian countries.

As per WHO, dengue is among the top ten diseases prioritised by the World Health Organization from 2019 to 2024. This is because the incidence of this viral disease has increased over

30-fold in the last five decades, through activities to record all dengue cases have also contributed to this number. However, in this global burden, one-third is contributed by India. Of the 96 million cases reported each year, 33 million of them are in India. A community-based survey in India showed that almost half of the population (48.7%) had been exposed to dengue infection in India at some point in their lives. The highest numbers are in the southern states (76·9%), followed by the western (62·3%) and northern (60·3%) states. Therefore, it could be said that urbanisation is one of the main drivers for rising dengue incidence in India, as no urban part of India is now not affected by the disease.

Prevention and Control

No vaccine is available for the prevention of dengue infection. The preventive measures are based mostly on mosquito control activities. These include the use of insecticides and clearing the stagnant water and artificial collections of water that serves as the breeding ground for the mosquitoes. Personal control measures include wearing good protective clothing and use of mosquito nets, mosquito repellents, etc.

Zika Virus

The Zika virus is transmitted through the bite of an infected Aedes mosquito (*Ae. aegypti* and *Ae. albopictus*), which can be found worldwide. These mosquitoes bite during the day and night. Infection is seen mostly in tropical and subtropical areas, including the Pacific Islands, several countries in Central, South and North America, and islands near West Africa. Zika can be passed from a pregnant woman to her fetus. Infection during pregnancy can cause certain birth defects, like microcephaly. Brain damage, joint problems and body movement disorders can also occur as complications. The virus can also spread through sexual contact. People contract the virus through blood transfusion or organ donation in rare cases.

Nearly 4 out of 5 people infected with the Zika virus have no signs or symptoms. If symptoms occur, they usually show between 2 and 14 days after an infected mosquito bites a person. Signs and symptoms of the Zika virus most commonly include mild fever, rash, joint pain and conjunctivitis. Other less common symptoms might include muscle pain, headache, abdominal pain and fatigue. There is no vaccine or treatment for Zika. Restricting travel to endemic areas and using protection during sex are some methods of preventing Zika virus infection.

Tick-Borne Flaviviruses

These viruses are broadly of two types: tick-borne encephalitis viruses and tick-borne haemorrhagic fever.

Tick-Borne Encephalitis Viruses

Russian Spring-summer Encephalitis (RSSE)

The condition caused by the Russian spring-summer encephalitis virus is transmitted by the bite of ixodid ticks. This condition, prevalent in Central Europe, Eastern Europe and Russia, is transmitted mainly by the bite of ticks. Transovarial transmission of the virus occurs in ticks and makes it an important reservoir host. Wild rodents and birds are the other reservoirs. Ingestion of infected meat obtained from infected goats may also transmit the infection to humans. Avoidance of tick bites is an important control measure. A formalin-treated Russian spring-summer encephalitis vaccine has also been found to be useful.

Powassan Virus

Powassan virus is another tick-borne virus causing Powassan encephalitis. This has been reported from Canada and the northern United States.

Tick-Borne Haemorrhagic Fever

Kyasanur Forest Disease (KFD)

It is an example of tick-borne haemorrhagic fever. This disease is called so because this was first reported in the Kyasanur Forest of Karnataka in 1957. The virus was associated with a fatal epizootic, affecting monkeys along with severe illness in some of the villages in the area. The causative virus was isolated from the patients and dead monkeys.

After an incubation period of 3–7 days, the KFD manifests with a sudden onset of fever associated with headache, vomiting, conjunctivitis, myalgia and severe prostration. In some cases, bleeding through the gastrointestinal tract and chest cavity may occur. The condition is associated with a case fatality of 5%.

The KFD is mostly localised in distribution and confined to the Shimoga district of Karnataka, India (Fig. 58.2). The disease remained restricted to north Karnataka between 1972 and 1975. Subsequently, the disease outbreaks were reported in Veltha belthangady in South Karnataka between December 1982 and May 1983. During this outbreak, 1142 human cases were reported, with 104 deaths. The infection is mostly transmitted by the bite of the tick *Haemaphysalis spinigera*. The tick also acts as the reservoir host because the virus is

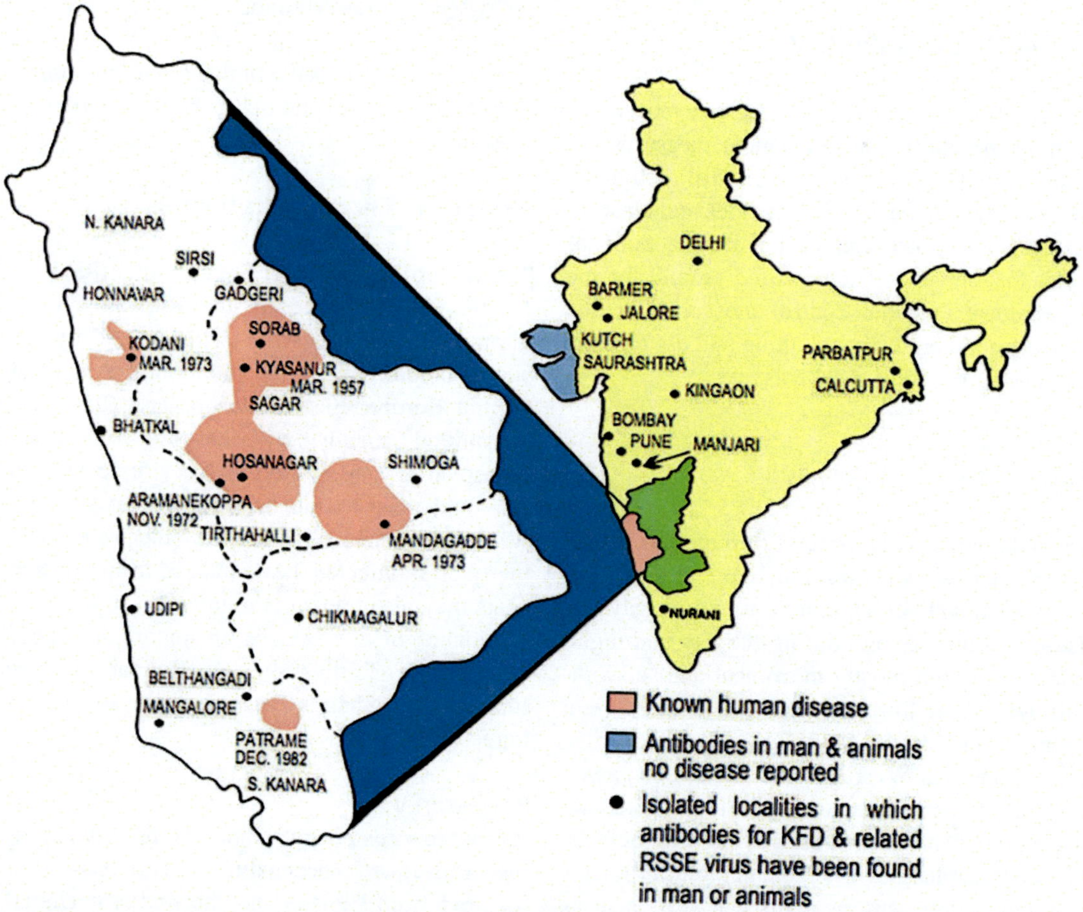

Fig. 58.2 Localised distribution of KFD. (Sources: K.G. Ajithkumar, Reghu Ravindran, and S. Ghosh. Dermacentor auratus Supino, 1897 (Acarina, Ixodidae) reported from Wayanad, India. Indian J Med Res 2012; 135: 435–436; http://www.cdc.gov/vhf/kyasanur/outbreaks/distribution-map.html)

transmitted transovarially. The monkeys appear to act as amplifier hosts but not reservoir hosts because monkeys die of the infection.

No specific antiviral treatment is available. Prevention of the disease is based primarily on the control of ticks and personal protection by using adequate clothing and insect repellents. A killed KFD vaccine is also available for use against the KFD.

Omsk Haemorrhagic Fever

This fever is caused by a virus related to KFD. Bites of Dermacentor ticks transmit the condition, and it occurs in Russia and Romania. Clinical manifestations of the disease are similar to that of KFD.

Filoviridae

Ebola Virus

This virus is a Flavivirus that consists of a group of viruses. These are Ebola virus (species Zaire ebolavirus), Sudan virus (species Sudan ebolavirus), Taï Forest virus (species Taï Forest ebolavirus, formerly Côte d'Ivoire ebolavirus), Bundibugyo virus (species Bundibugyo ebolavirus), Reston virus (species Reston ebolavirus) and Bombali virus (species Bombali ebolavirus). Of these, only four (Ebola, Sudan, Taï Forest and Bundibugyo viruses) have been associated with diseases in humans. Reston virus can cause disease in non-human primates and pigs, but there has been no case reports in humans. Bombali virus was reported in bats in 2018, and not much information is available regarding its disease-causing potential in animals and humans.

Infection with Ebola virus, known as Ebola virus disease (EVD), is a fatal disease in humans and non-human primates that presents with occasional outbreaks, mostly in the African continent. EVD most commonly affects people and non-human primates (such as monkeys, gorillas and chimpanzees). Infections due to the Ebola virus is rare. The viruses are located mainly in sub-Saharan Africa. The transmission mode is through direct contact with an infected animal (bat or non-human primate) or a sick or dead person infected with the Ebola virus.

The **infection cycle** begins with an infected animal, like a fruit bat or non-human primate. This is called a "spillover event". After that, the virus spreads from person to person and quickly affects many people. The route of transmission is through direct contacts, such as through broken skin or mucous membranes in the eyes, nose or mouth or with blood or body fluids, like urine, saliva, sweat, faeces, vomitus, breast milk, amniotic fluid and semen of a person who is sick with or has died from Ebola virus infection. Personal items, like clothes, bedding, needles and medical equipment contaminated with body fluids from an infected person, can also serve as the vehicle for transmission. Ebola virus is not transmitted through food, though in certain parts of the world, the virus may spread through the handling and consumption of infected wild animal meat. Patients with the disease can easily infect healthcare workers without proper infection control practices.

Clinical manifestations usually appear from 8 to 10 days though they may vary from 2 to 21 days after contact with the virus. There can be two distinct groups of symptoms designated as wet and dry symptoms. Diarrhoea and vomiting are common among the wet symptoms, while dry symptoms initially present as fever, aches and pains, and fatigue and then progresses to wet symptoms. Primary signs and symptoms of Ebola often present as fever, body ache, fatigue, bleeding manifestations, gastrointestinal symptoms, etc.

Laboratory diagnosis of EVD sooner after infection can be difficult. To justify whether EVD is a possible diagnosis, there must be a combination of symptoms and possible exposure to EVD within 21 days before the onset of symptoms. If a person shows signs of EVD and has had a possible exposure, he or she should be isolated, and public health authorities should be notified. Blood samples from the patient should be collected. PCR is one of the most commonly used diagnostic methods to detect a few virus particles in small amounts of blood. Other methods, based on the detection of antibodies an EVD case produces to an infection, can then be used to confirm a patient's exposure and infection by the Ebola virus.

There are currently two drugs approved by the U.S. Food and Drug Administration (FDA) to treat EVD caused by the Ebola virus, species Zaire ebolavirus, in adults and children. The first drug Inmazeb™ is a combination of three monoclonal antibodies. The second drug Ebanga™ is a single monoclonal antibody. Monoclonal antibodies (mAbs) are proteins that act like natural antibodies to stop them from replicating in an infected person. These particular mAbs binds to a portion of the Ebola virus's surface called the glycoprotein and prevent it from entering a person's cells. However, neither of the drugs have been evaluated for efficacy against species other than Zaire ebolavirus. Even if treatments are not available, supportive care can significantly improve chances of survival when provided early. These provide fluids and electrolytes, constant monitoring of blood pressure, and fever and pain management.

The general **preventive measures against EVD** are to avoid contact with blood and body fluids of infected person contact with their personal items or hospital items in common use. It is advisable to avoid funeral or burial practices that involve touching the body of the infected. Contact with suspected animals should also be avoided. Those dealing with Ebola management in outbreaks should monitor their health for 21 days after returning from outbreak areas and seek medical care immediately if they develop symptoms of EVD.

Ebola Vaccine: In 2019, The U.S. Food and Drug Administration (FDA) has approved the Ebola vaccine rVSV-ZEBOV, which is the first FDA-approved vaccine for Ebola. This vaccine is given as a single dose and is safe and protective against Zaire ebolavirus, which has caused the largest and most deadly Ebola outbreaks to date. On February 26, 2020, the Advisory Committee on Immunization Practices (ACIP) recommended pre-exposure prophylaxis vaccination with rVSV-ZEBOV for adults ≥ 18 years of age in the U.S. population at potential occupational risk of exposure to Zaire ebolavirus. This recommendation includes adults who are at risk of infection, like laboratory personnel, healthcare personnel or those involved or likely to be involved in Ebola outbreak management.

Bunyaviridae Viruses

Bunyaviridae viruses are large, spherical (80–120 nm in diameter) and have a triple-segmented and a single-stranded RNA. They have a helical nucleocapsid surrounded by an envelope. They replicate in the cytoplasm, and the assembly of the virion occurs by budding on the smooth membrane of the Golgi system. Bunyaviridae viruses consist of more than 200 viruses. Based on structural and biochemical properties, they are classified into four genera: *Bunyavirus*, *Hantavirus*, *Nairovirus* and *Phlebovirus*. All these viruses, except for hantaviruses, are arboviruses that are transmitted by mosquitoes, ticks and flies. The hantaviruses, however, are transmitted by rodents.

1. **Genus *Bunyavirus*** includes more than 150 members, including the La Crosse virus, California encephalitis virus and Chittoor virus. These three viruses are related antigenically and produce similar clinical diseases. Mosquitoes transmit these viral infections.
2. **Genus *Nairovirus*** includes six members consisting of Nairobi sheep disease virus, Ganjam virus and Crimean-Congo haemorrhagic fever (CCHF). Haemaphysalis ticks transmit all these viruses. Nairoviruses are primarily pathogenic for sheep. They cause infection in sheep and are responsible for mortality and morbidity, causing a severe economic loss for the farmers. CCHF virus is the most important *Nairovirus,* which causes CCHF in humans. Ganjam virus is a related virus that causes infection of sheep and goats in Odisha, India. The virus has also been isolated from humans closely related to the Nairobi sheep disease virus.
3. **Genus *Phlebovirus*** includes more than 36 members, including sandfly (*Phlebotomus*)

Table 58.4 Diseases associated with Bunyaviridae viruses

Virus	Diseases
Bunyavirus	Encephalitis, aseptic meningitis and fever
Phlebovirus	Sandfly fever, haemorrhagic fever, encephalitis and conjunctivitis myositis, rift valley fever
Nairovirus	Crimean-Congo haemorrhagic fever and Nairobi sheep disease
Hantavirus	Haemorrhagic fever with renal syndrome, adult respiratory distress syndrome and hantavirus pulmonary syndrome (Sin Nombre virus), shock, pulmonary oedema

fever virus and Rift Valley fever virus. Different arthropods transmit these viruses, such as ticks, sandflies and mosquitoes.

As the name suggests, the sandfly fever virus or pappataci fever virus is transmitted by female species of sandfly *Phlebotomus papatasi.* The sandfly fever virus has 20 serotypes, of which only five cause human disease. These are Sicilian, Naples, PantaToro, Chagrss and Candine.

Rift Valley fever virus is a primary pathogen of sheep and other domestic animals. This fever was originally described in Rift Valley, Kenya, and hence was designated as Rift Valley fever. Humans are secondarily infected on contact with infected sheep and other domestic animals. Humans acquire the infection from these reservoir animals through mosquito bites. Diseases caused by Bunyaviridae viruses are listed in the Table 58.4.

Morphology

Bunyaviridae virus is surrounded by an envelope that contains two glycoproteins, G1 and G2. The capsule encloses three unique negative RNAs (large, L; medium, M; and small, S) that are associated with proteins to form nucleocapsid. Unlike other negative-stranded RNA viruses, the Bunyaviridae viruses do not have any matrix protein. They replicate in the cytoplasm of the infected cell similar to that of other enveloped and negative-stranded viruses.

Pathogenesis of Bunyaviridae Virus Infections

Bunyaviridae viruses can infect humans and arthropods. Bite by an arthropod vector, such as mosquito, transmits the infection to humans. The virus from the site of infection spreads to the blood causing initial viraemia, which may manifest in flu-like symptoms. Further dissemination of the viruses during the phase of secondary viraemia may allow the viruses to reach target sites, such as the liver, kidney, vascular endothelium and CNS. In the CNS, many viruses cause neural and glial damage and cerebral oedema, leading to encephalitis. Other Bunyaviridae viruses, such as Rift Valley fever virus, may cause hepatic infection, whereas hantana haemorrhagic disease and CCHF viruses cause haemorrhagic tissue destruction accompanied by haemorrhagic necrosis of the kidney.

Immunity

Infection by Bunyaviridae viruses leads to the development of both antibodies and cell-mediated immunities. The antibody-mediated immunity may control viraemia, whereas the cell-mediated immunity and interferon may prevent the further spread of the infection.

Clinical Syndromes

Bunyaviruses usually cause a non-specific febrile flu-like illness indistinguishable from illness caused by other viruses. The incubation period is usually short (48 h), and the fever lasts approximately 3 days.

California encephalitis virus, La Crosse virus and Rift Valley fever cause encephalitis. The condition begins with a sudden onset after an incubation period of nearly 1 week. Fever, lethargy, headache and vomiting are the common symptoms. Seizers usually occur in nearly half of the patients with encephalitis. The condition may be associated with meningitis. The illness lasts for approximately 7 days. The condition is fatal in less than 1% of patients.

The sandfly fever has an incubation period of 3–5 days, after which the patients develop headache, malaise, fever and nausea, which last for 3–4 days. CCHF and Rift Valley fever viruses cause haemorrhagic fever. The condition is characterised by petechial haemorrhage, ecchymosis, haematemesis and melena. Death occurs in 75% of patients with haemorrhagic manifestations.

Laboratory Diagnosis

Serology

It is most important in the laboratory diagnosis of Bunyaviridae virus infection. ELISA for demonstration of serum IgG or IgM antibodies is useful for diagnosing bunyavirus infections. Demonstration of specific IgM antibodies in a single serum or demonstration of a fourfold increase in the IgG antibody titre indicates recent infection. IgM ELISA is used for the serodiagnosis of infection caused by hantaviruses. Demonstration of rising titres of haemagglutinating antibodies in paired sera is diagnostic of the disease. The ELISA for demonstration of antigen in the blood is also useful for diagnosing patients with intense viraemia, observed in CCHF, hantana haemorrhagic disease with renal syndrome and Rift Valley fever.

Molecular Detection

The molecular diagnosis is the recent method. RT-PCR has been used to detect viral genomes in the tissues during the infection caused by the Sin Nombre and Convict Creek hantaviruses. RT-PCR can be used to detect hantaviral RNA in fresh frozen lung tissue, blood clots or nucleated blood cells. However, RT-PCR is very prone to cross-contamination, limiting its use for diagnosis. Immunohistochemistry (IHC) testing of formalin-fixed tissues with specific monoclonal and polyclonal antibodies can be used to detect hantavirus antigens and has proven to be a sensitive method for laboratory confirmation of hantaviral infections.

Treatment

No specific antiviral therapy is available for bunyavirus infection.

Epidemiology

Most bunyaviruses are distributed in Africa, South America, South-eastern Europe and southern Asia. The sandfly fever is distributed in Mediterranean countries and Central Asia, including India. These viruses are transmitted by arthropod vectors, such as mosquitoes, ticks or *Phlebotomus* flies to rodents, birds and larger animals, which are the usual reservoirs of infection.

Humans acquire infection from the reservoir animals by arthropod vectors. The arthropods transmit the infection by biting on the skin. *Aedes* mosquitoes are responsible for transmitting the California encephalitis virus and La Crosse virus. Flies, such as sandfly, transmit the Rift Valley fever virus and the sandfly fever virus. Ticks are the vectors for CCHF.

The incidence of the disease caused by bunyaviruses correlates well with the distribution of vectors. The disease is more common during the summer months.

Prevention and Control

Avoidance of close contact of humans with arthropod vectors, such as mosquitoes, ticks or *Phlebotomus* flies, prevents transmission of infection caused by bunyaviruses. Arthropod vector control measures include the use of insecticides, wearing protective gloves and clothing, use of net and screens and control of tick infestation of animals. Control of the rodent population minimises the transmission of infection caused by hantaviruses.

Rodent-borne Bunyaviridae Viruses

The genus Hantavirus consists of 4 species, namely (1) Hantaan virus, which causes severe haemorrhagic fever with renal failure and is prevalent in North Asia and Soviet Russia; (2) a milder form, Seoul virus prevalent worldwide; (3) Puumala virus prevalent in Europe and (4) Prospect Hill virus that has not been associated with human infections. This virus is mainly associated with haemorrhagic fever with renal symptoms.

The natural reservoirs of the virus are rodents, mainly being the field mice and rats. The virus is shed in animals' urine, faeces and oral secretions.

Among the risk factors, it is seen that disturbing or inhabiting closed, actively rodent-infested areas may be an important risk factor for infections. Travel to areas where hantavirus infection has been reported is not a risk factor for infection.

ELISA targeting IgM antibodies is used to diagnose Hantavirus infections. An IgG test is used in conjunction with the IgM capture test.

Hantaviruses are spherical viruses measuring 90–120 nm in diameter. They include the **Hantaan virus (Korean haemorrhagic fever), Seoul virus** and **Sin Nombre virus.** All these viruses cause haemorrhagic fever with nephropathy. These viruses are transmitted by close contact with rodents, such as mice and rats and not by arthropod vectors. Transmission of infection occurs by inhalation of aerosolised excreta of the rodents.

The hantavirus typically causes hantavirus pulmonary syndrome, a more serious disease consisting of prodromal fever and muscle aches. The condition progresses rapidly to interstitial pulmonary oedema, respiratory failure and death within a few days. The hantaviruses cause haemorrhagic fever with renal syndrome. The cases resemble those of leptospirosis and scrub typhus.

Any arthropod vectors do not transmit hantaviruses. Humans are infected by close contact with rodents or inhaling aerosolised rodent urine.

A newly identified hantavirus, the **Sin Nombre virus**, is the causative agent of the hantavirus pulmonary syndrome, which was first identified in the United States in 1993. The infection is acquired by inhalation of aerosols of rodent excreta. No arthropods are responsible for the transmission of the disease.

Reoviridae

The family Reoviridae consists of four genera (*Orthovirus*, *Coltivirus*, *Orbivirus* and *Rotavirus*), of which arthropods transmit only orbiviruses. These viruses have a double-stranded RNA genome by which they differ from other arboviruses. Colorado tick-borne virus is the only orbivirus that causes infection in humans. The virus causes Colorado tick fever, transmitted by the tick *Dermacentor andersoni*. The virus causes natural infection in rodents. The disease is restricted in its distribution to the western United States. The virus causes a self-limiting disease in humans. *Kasba*, *Vellore* and *Palyam* are the arboviruses that have been isolated from mosquitoes in India. Their association with the disease in humans is yet to be ascertained.

Rhabdoviridae

The genus *Vesiculovirus* of the family Rhabdoviridae consists of the Chandipura virus, which has been associated with human infection in India. This is an arthropod-borne virus transmitted by both sandfly and *Aedes* mosquito. The virus was first reported from Nagpur in 1967 from patients' blood during the epidemic of dengue and Chikungunya fever. Their pathogenic role in human disease is not known.

Ungrouped Arboviruses

Wanowrie Virus

This virus has been isolated from the brain of a girl who died of a fever after 2 days in Sri Lanka.

The same virus has also been isolated from Hyalomma ticks in India. Wanowrie virus causes infections in both animals and humans with fever, photophobia and meningitis, like symptoms in humans. Other than symptomatic treatment, no other virus-specific treatment is indicated.

Bhanja Virus

Bhanja virus has been isolated from goats in Ganjam (Odisha) from India. The same virus has been shown to be associated with human infection reported from Yugoslavia. It is a tick-borne virus first discovered in a tick (*Haemaphysalis intermedia*). Usually, the Bhanja virus is not known to be transmitted between humans. Human infections cause febrile illness with symptoms, including photophobia, vomiting, meningoencephalitis and slight or partial paralysis. Symptomatic treatment is advocated for the infection.

Case Study

There was an outbreak of JE in and around Tamil Nadu a few years back. As a preventive measure to stop further dissemination of the disease, rural people engaged in pig farming were advised to be away from pig herds and measures were initiated for control of mosquito population.

- What is the basis of advising people to be away from the pig population?
- How is JE transmitted in a community?
- What are the laboratory tests used for establishing the diagnosis of JE?
- Is there any vaccine against JE?

Rhabdoviruses

59

Rabies virus is the most important member of the Rhabdoviridae family, which causes disease in humans. The family Rhabdoviridae is classified into two genera: *Lyssavirus* and *Vesiculovirus*. The genus *Lyssavirus* consists of more than 80 viruses and includes a rabies serogroup consisting of 10 viruses, including the classic rabies virus. The rabies virus is the prototypical human *Lyssavirus* pathogen. Other viruses included in this group are Mokola virus, Duvenhage virus, Obodhiang virus, Kotonkan virus, Rochambeau virus, European bat *Lyssavirus* types 1 and 2 and Australian bat *Lyssavirus*. These viruses rarely cause human disease. The genus *Vesiculovirus* includes vesicular stomatitis virus and other viruses infecting vertebrates and invertebrates.

Rabies Virus

Rabies virus causes rabies, a viral infection of the central and peripheral nervous systems that causes encephalitis with or without paralysis. It is the most fatal infection in humans.

Morphology

Rabies virus is a bullet-shaped virus with one end rounded or conical and the other end planar or concave (Fig. 59.1). It is a negative-sense, non-segmented, single-stranded RNA virus measuring approximately 60 × 180 nm. It is composed of an internal protein core or nucleocapsid containing the nucleic acid. It also consists of an outer envelope, a lipid-containing bilayer covered with trans membrane glycoprotein spikes. The nucleocapsid shows helical symmetry containing a linear negative-sense RNA with an RNA-dependent RNA transcriptase.

The rabies virus genome is unsegmented. The genome encodes for the production of five proteins the ribonucleoprotein (RNP) complex or the viral envelope. The RNA complex constitutes viral RNA along with N (nucleoprotein), L (transcriptase) and NS (transcriptase-associated) proteins. These proteins are aggregated together in the cytoplasm of virus-infected neurons and constitute **Negri bodies**, the typical histopathological finding of rabies virus infection. The lipid envelope contains M (matrix) and G (glycoprotein). The G proteins are the projections present on the outer surface of the virion envelope. It is the only rabies viral protein that can induce neutralising antibodies against this virus.

Antigenic Properties

Rabies viruses of humans and animals are of a single antigenic type. Rabies virus contains the following antigens:

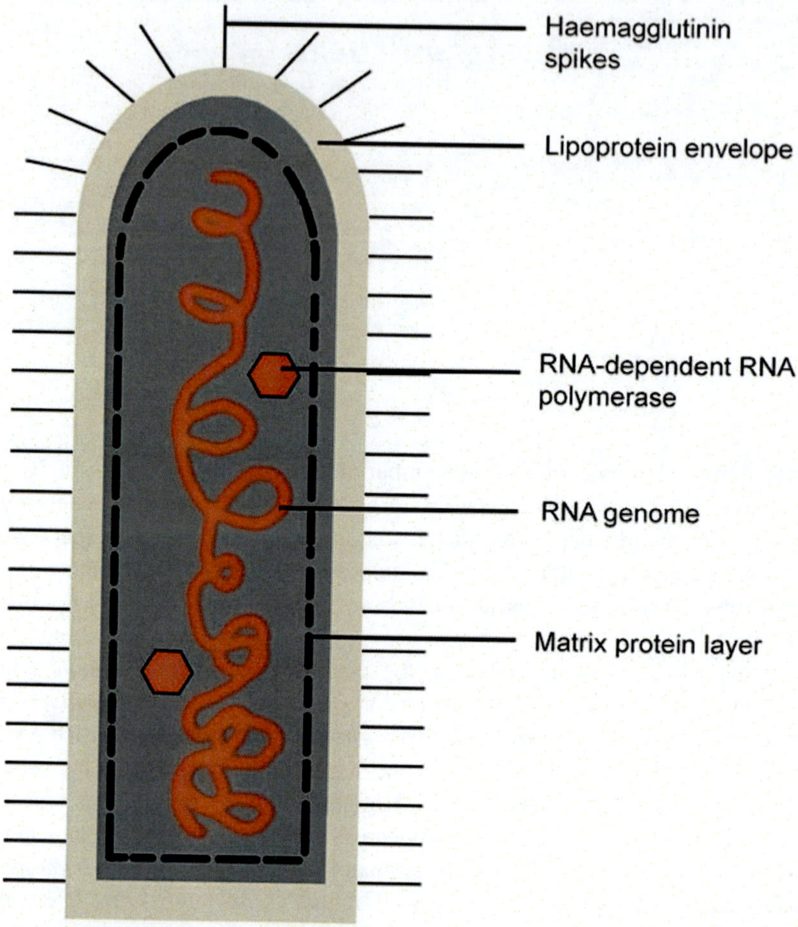

Fig. 59.1 A schematic diagram of rabies virus

The glycoprotein or **G protein** is present on the surface spikes present on the outer lipoprotein envelope of the virion. It mediates the attachment of the virus to the acetylcholine receptors of neural tissues. The G protein is important in the pathogenesis and virulence of the virus. It is strongly antigenic and elicits the production of neutralising antibodies, which are protective. It also induces haemagglutination-inhibiting antibodies and stimulates cell-mediated immunity. G protein is a serotype-specific antigen.

Nucleoprotein or **N protein** is a group-specific antigen. It shows cross-reaction with some rabies-related viruses. It is antigenic and produces antibodies that are not protective but of diagnostic value. These antibodies can be demonstrated by serological tests, such as the immunofluorescence (IF) test for the purpose of diagnosis.

Other antigens include membrane proteins glycolipid and RNA-dependent RNA polymerase.

Other Properties

Rabies virus possesses haemagglutinating property. The virus agglutinates goose erythrocytes at 0–4 °C and pH 6.2.

Virus Susceptibility

Rabies virus is sensitive to ethanol, iodine, soap, quaternary ammonium compounds, detergents, ether, chloroform, acetone, etc. It is inactivated by formalin, phenol, beta-propiolactone, ultraviolet radiation and sunlight. It can be inactivated at 50 °C for 1 h or at 60 °C for 5 min. It can also be inactivated on exposure to CO_2. The virus is resistant to drying and cold. It survives at 4 °C for weeks; hence viruses can be preserved by storing at −70 °C or by lyophilisation.

Typing of Rabies Viruses

Rabies viruses may be typed as fixed or street viruses. Fixed viruses are so-called because they show a fixed incubation period of 6–7 days. This occurs due to several passages in these isolated viruses in laboratory animals or cell cultures. The street viruses are those viruses that are isolated from natural human and animal infections and show a variable incubation period of 1–12 weeks. Differences between fixed and street viruses are summarised in Table 59.1.

Typing of street viruses helps identify viral variants originating in major host reservoirs worldwide. This is also useful to suggest the likely sources of human exposure when a history of the definitive animal bite was otherwise missing from a patient's case history. The monoclonal antibodies and genetic sequencing are now used to differentiate and type street rabies viruses.

Laboratory Isolation

Chick Embryo

The rabies virus grows in the yolk sac of the chick embryo. The rabies vaccine strains, such as Flury and Kelev strains, are grown in the yolk sac.

Tissue Culture

The rabies virus can grow in several primary and continuous cell cultures. These are WI-38, BHK (baby hamster kidney)-21 and chick embryo-related (CER) chick embryo fibroblast and porcine, mouse, or hamster kidney cell lines. The fixed virus strains are adapted to grow in vero and human diploid cell lines. The cytopathic effect produced by the virus in the cell lines is minimal.

Laboratory Animals

The mouse is the animal of choice. The intracerebral route is the frequently used route of infection. Intracerebral inoculation causes encephalitis, and the inoculated mouse die within 5–30 days.

Pathogenesis of Rabies

Rabies virus has a broad host range. The virus can infect all mammals, although certain mammals (such as dogs, foxes, wolves and bats) are important for transmitting infection.

The virus is excreted in the saliva of infected dogs, foxes, wolves, jackals, vampire bats, raccoons and skunks. Bite of a rabid dog or other infected animals is the main route of transmission of infection. Contacting saliva with broken skin or mucous membranes, exposure to aerosolised secretions from an infected animal and corneal transplant is a rare and uncommon transmission. The virus may enter the peripheral nervous system directly at the site of the bite. In some cases, after virus entry, replication occurs in muscle tissue at the same site or adjacent to the site of entry during the incubation period. The virus infects the sensory neurons and moves rapidly by axonal transport centripetally to the central nervous system (CNS) for replication. Rabies virus replicates in the cytoplasm of the infected cell.

Table 59.1 Distinguishing features of street and fixed rabies viruses

Street virus	Fixed virus
The virus is isolated from natural human or animal infection	The virus is isolated after several serial intracerebral passages in rabbits
Can cause fatal encephalitis in laboratory animals after a long incubation period of 1–12 weeks	Neurotropic but much less infective. Can cause fatal encephalitis in laboratory animals after a short and fixed incubation period of 6–7 days
Negri bodies can be demonstrated in the brain of infected animals	Negri bodies cannot be demonstrated in the brain of infected animals
Cannot be used for vaccine production	Can be used for vaccine production

It is protected from the host immune system during its transport within the neurons. To gain its entry into the spinal ganglion, the virus spreads along the axons at a rate of 12–24 mm/day and multiplies there, which is marked by the onset of pain or paraesthesia at the site of the entry earliest and characteristic symptoms. From here, the virus travels faster and reaches the CNS at a speed of 200–400 mm/day, and the spread is indicated by rapidly progressive encephalitis. Finally, the virus reaches the periphery and salivary glands (Fig. 59.2).

Clinical Syndromes

Rabies virus causes rabies, the most fatal infection in humans. No specific ant rabies agents are useful once clinical signs or symptoms develop.

Rabies

In general, four stages of rabies are recognised in humans. These are incubation, pro-dromal period, acute neurologic period and coma, leading to death.

The incubation period ranges from 20 to 90 days. It rarely can be as long as 19 years. In more than 90% of cases, incubation is less than 1 year. During incubation, the virus travels from peripheral areas to the CNS. The patients remain asymptomatic during the period. When the patient presents with bites on the head or neck or if there are deposition of heavy inoculum through multiple bites, large or deep wounds, the incubation period is less than 50 days. The symptoms of rabies develop later in a person with a scratch in hand than in a person who is bitten on the head. During this period, there is no antibody response as the virus is masked from the immune system.

The virus enters the CNS during the pro-dromal period. The duration of this period is 2–10 days. Non-specific symptoms and signs characterise the period. Pain or paraesthesia develops at the inoculation site and is characteristic of rabies. During this phase, paraesthesia may be seen in 50% of cases, which may be the patient's only presenting sign. Other symptoms may include fever, chills, headache, anorexia, malaise, pharyngitis, nausea, vomiting, diarrhoea, anxiety, agitation, lack of sleep and depression.

An acute neurologic period is associated with objective signs of developing CNS disease. The period lasts for 2–7 days and may manifest as furious rabies or paralytic rabies.

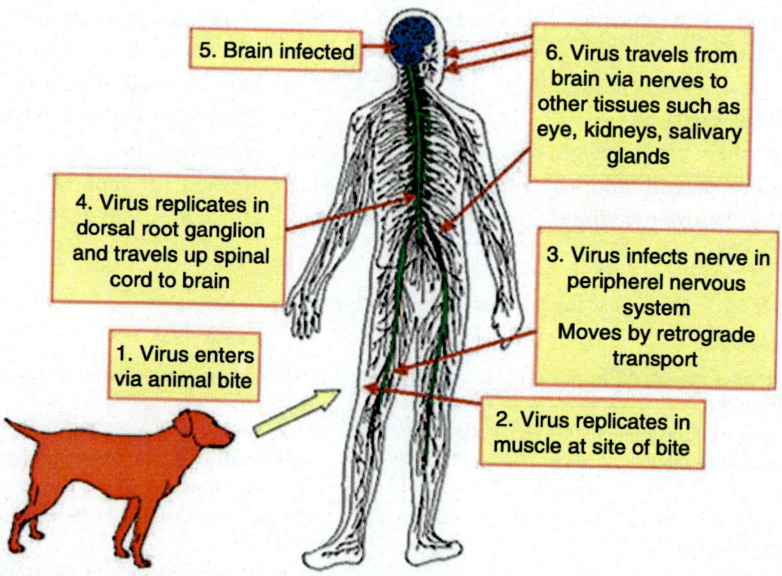

Fig. 59.2 Pathogenesis of rabies in humans (https://link.springer.com/chapter/10.1007/978-81-322-1605-6_2)

Patients develop hyperactivity, agitation, restlessness, biting, thrashing, confusion, or hallucinations in the stage of **furious rabies**. After several hours to days, these symptoms are episodic and are alternative with calm, cooperative, lucid periods. Furious episodes remain not more than 5 min. The triggers of these episodes are visual, auditory, or tactile stimuli. They may be sometimes spontaneous. Seizures may also be present. The pathognomonic signs are hydrophobia and aerophobia. If there is an attempt to drink or the air is blown in the face, severe spasms of the larynx or diaphragm and choking sensations are produced. This may be due to a violent response of the airway irritant mechanisms. Hydrophobic spasm is induced even when the patient is suggested to drink.

Autonomic instability is observed, including fever, tachycardia, hyperventilation, hypertension, drooling of saliva, anisocoria, lacrimation, salivation, perspiration, mydriasis and postural hypotension. Other neurologic signs include cranial nerve involvement with diplopia, facial palsy and optic neuritis. This phase may lead to a cardiorespiratory arrest or may advance to paralysis.

Paralytic rabies is otherwise known as dumb rabies or apathetic rabies because the patient is relatively quiet compared to a person with the furious form. Paralysis develops from the outset. Paralysis is symmetric and may be either generalised or ascending and mistaken for Guillain–Barré syndrome. Calmness and clarity gradually deteriorate to stupor, delirium and then coma.

The patient may go into a coma within 10 days of onset, but the duration is variable. Coma leads to respiratory failure within a week of neurologic symptoms. Hypoventilation and metabolic acidosis predominate. Acute respiratory distress syndrome is common. Without intensive supportive care, respiratory depression, arrest and death occur shortly after coma. Despite intensive supportive care, most cases result in death within 14 days because of complications. Recovery is very unlikely. A few reports indicate that those patients who survived had pre-exposure or post-exposure prophylaxis supported by the most advanced life-support system.

Laboratory Diagnosis

Specimens

Saliva, serum, cerebrospinal fluid (CSF), blood, urine and skin and brain biopsy are the frequently used specimens for diagnosis of rabies.

Microscopy

Demonstration of Negri bodies by microscopy is the characteristic histopathological feature of rabies (Fig. 59.3). Impression smears of the human brain tissue collected at post-mortem are stained by Seller technique for demonstration of

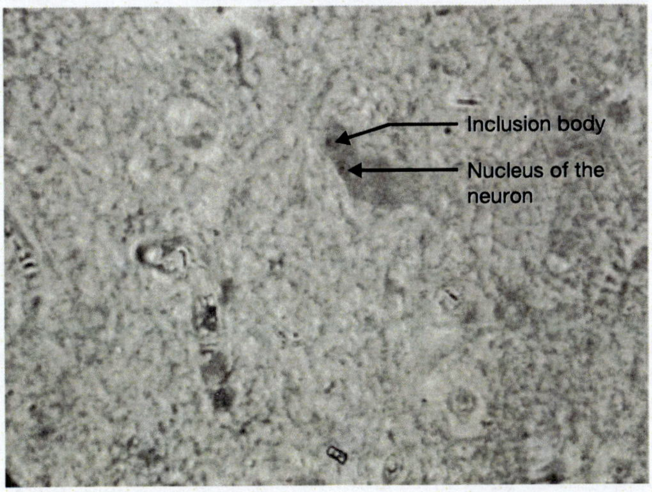

Fig. 59.3 Negri bodies in neural tissue (×1000)

Negri bodies. The stain contains methylene blue alcohol as a fixative and basic fuchsin as a staining reagent. This is still a useful method in the laboratories lacking facilities for cell culture and antigen detection methods.

Negri bodies are made up of a finely fibrillar matrix and rabies virus particles. They appear as intracytoplasmic, round or oval, purplish-pink structures with characteristic basophilic inner granules on staining by Seller technique. They measure 3–27 μm in size. They are found in the neural tissues and are found more in the cerebellum and hippocampus (Purkinje cells of the cerebellum and pyramidal cells of the Ammons horn, of the cerebellum). They are not found in non-neural tissues, such as corneal specimens, saliva, or skin. Post-mortem diagnosis is made by demonstrating Negri bodies in most human cases of rabies. Nevertheless, failure to demonstrate Negri bodies in neural tissue does not rule out the diagnosis of rabies.

Viral Antigen Detection

Viral antigens can be demonstrated in the corneal smear, skin biopsy collected from the face or neck and saliva (*antemortem*) or in the brain tissue (*post-mortem*) for diagnosis of rabies. Direct fluorescent antibody (DFA) test using specific monoclonal antibodies is a specific method to demonstrate rabies antigen in clinical specimens.

Cell Culture

Isolation of virus in cell lines (WI-38, BHK-21 and CER) is a sensitive method. The cytopathic effect produced by the virus is minimal; hence, the presence of the virus in the inoculated cell cultures is made by demonstration of viral antigen within 2–4 days. The DFA test using monoclonal rabies antibodies tagged with fluorescein isothiocyanate is a frequently used specific method to demonstrate the antigen.

Serology

Rabies antibodies are found in the serum and the CSF of human cases. A high titre of antibodies is found in the CSF. Antibodies in the serum are demonstrated by a rapid fluorescent focus inhibition test titre which gives positivity in 50% of cases. In the cerebrospinal fluid, positive antibody titres (2–25% of serum titre) are found at the end of the first week of infection.

Molecular Detection

A polymerase chain reaction is being increasingly evaluated for the diagnosis of rabies. The nucleic acid sequence-based amplification (NASBA) on saliva and CSF can be used for rapid diagnosis as early as 2 days after symptom onset. The NASBA technique on urine samples may be used in the future.

Ante-mortem Diagnosis of Human Rabies

Rabies infections are always fatal. But with intensive support care and better management of patients, few cases have survived. Corneal smears, skin biopsy from the nape of the neck or face and saliva are the usual specimens. Diagnosis is made by demonstrating rabies virus antigen by IF or the viral DNA by molecular methods. The NASBA method on saliva and CSF can be used for rapid diagnosis as early as 2 days after symptom onset.

Treatment

No specific antirabies agent is available. Although rabies was considered to be invariably fatal until recently, it has been demonstrated that complete recovery can occur from established rabies with intensive supportive care and management of complications.

Epidemiology

Rabies is a recognised zoonotic disease worldwide.

Geographical Distribution

Rabies is a zoonotic disease. Dogs are the important reservoir of infection. Other animal reservoirs include silver-haired bats, eastern pipistrella, raccoons, skunks, foxes, cats, ferrets, cattle, opossums, fowl, etc. Foxes are more infectious than dogs and other animals because larger amounts of virus (up to 10^6 infectious doses/mL) are present in their saliva.

Rabies has been recognised for over 4000 years. Today, it is found in most countries, except many Australian islands, Great Britain, Japan, Hawaii and most Caribbean islands. The risk of rabies is highest in countries with hyperendemic canine rabies, including most of Asia, Africa and Latin America.

Rabies is endemic in India. It has been estimated that more than 30,000 people die of rabies in India every year. More than 70,000 people in India receive ant rabies vaccination per year. With a dog population of over 16 million, the problem of rabies is huge. The virus is usually transmitted from a rabid animal to humans, mostly by bites or other forms of traumatic contact.

Reservoir, Source and Transmission of Infection

Dogs are the important reservoir of infection. Other animal reservoirs include silver-haired bats, eastern pipistrella, raccoons, skunks, foxes, cats, ferrets, cattle, opossums, fowl, etc. Foxes are more infectious than dogs and other animals because larger amounts of virus (up to 10^6 infectious doses/mL) are present in their saliva.

The virus is excreted in the saliva of infected dogs, foxes, wolves, jackals, vampire bats, raccoons and skunks. The virus is found in the salivary gland of these infected animals. Infected saliva or infected CNS tissue, including corneal transplants in humans, are the sources of infection. The virus is primarily transmitted to humans by the bite of a rabid dog or other infected animals. Contact of saliva with broken skin or mucous membranes, exposure to aerosolised secretions from an infected animal and contact with unpasteurised milk from dairies are very uncommon modes of transmission.

Corneal Transplants

Eight recipients of transplanted corneas are the only documented cases of rabies in case of human-to-human transmission. Presently, donor corneas are rejected if they die of encephalitis that may be consistent with rabies. In India, few such cases have also been documented. Except for corneal transplants, man-to-man infection is rare.

Rabies in Dogs

Rabies in dogs can present as a sudden change in behaviour with irritability and restlessness gradually progressing to aggressiveness. Other signs like excessive drooling, difficulty in swallowing, seizures might be seen with the progression of the disease. The usual incubation period is 21 days to 42 days but can extend up to a year. Rabid dogs can shed the virus in their saliva in 60% of the cases. However, once symptoms set in, rabid dogs hardly survive long and die within a week. The manifestations of the disease can take either of the two courses, namely furious rabies and dumb rabies. In furious rabies, the dog runs uncontrollably in any direction, biting anyone without provocation, excessive drooling with jaw drooping is seen followed by convulsions and death. This is the more common form. In dumb rabies, paralysis sets in quickly, and the animal limits itself to some secluded place. Though unprovoked biting is not seen, feeding the animal forcibly can lead to biting.

Laboratory Diagnosis of Rabies in Dogs and other Animals

It is useful to assess the risk of infection and monitor post-exposure prophylaxis in humans bitten by the animals.

Demonstration of Negri Bodies

The dead animal's brain is the specimen of choice. The brain is collected carefully from the dead animal. Part of the specimen is collected in 50% glycerol saline (preservative) to isolate the virus. The other part of the brain, including the hippocampus and cerebellum (abundant in Negri bodies), is collected in Zenkers fixative to demonstrate Negri bodies. Demonstration of Negri bodies by microscopy is the characteristic histopathological feature of rabies (Fig. 59.3). But, failure to demonstrate Negri bodies in neural tissue does not rule out the diagnosis of rabies.

Impression smears of the human brain tissue collected at post-mortem are stained by Seller technique for demonstration of Negri bodies.

The stain contains methylene blue alcohol as a fixative and basic fuchsin as a staining reagent.

Direct Antigen Detection

Demonstration of viral antigens in the infected brain tissue and the saliva of the animal by DFA is a useful method for diagnosing rabies in animals. Detection of antigen in the saliva shows whether the animal was excreting the virus in the saliva or not.

Isolation of the Virus

In a reference laboratory, viruses can be isolated from animal brain tissue and saliva by culture in cell lines or animals, as described earlier for diagnosis of human rabies.

Prevention and Control

Depending upon whether given before or after exposure, specific prophylaxis in rabies can be discussed as pre-exposure prophylaxis and post-exposure prophylaxis.

Pre-exposure Prophylaxis

Pre-exposure prophylaxis is given to persons at high risk, such as dog handlers, other animal handlers and veterinarians. This is achieved using cell culture vaccines, which are safer. Pre-exposure prophylaxis requires three doses of cell culture vaccines, given on days 0, 7 and 21 or 28, both by intradermal or intramuscular route. A booster dose is recommended after 1 year and then one dose every year.

Post-exposure Prophylaxis

Post-exposure prophylaxis is started immediately after exposure to infection. After exposure to a possibly infected dog or other rabid animals, immediate preventive actions are taken up, which consist of local treatment, passive immunisation with rabies immunoglobulin and active immunisation with the antirabies vaccine.

Local Treatment

Animal bites deposit the virus in the wounds. The wound should be immediately scrubbed with soap and water followed by treatment with quaternary ammonium compounds, such as cetavlon, tincture, or an aqueous solution of iodine or alcohol.

Passive Immunisation with Rabies Immunoglobulin

It is done by administering purified equine rabies immune globulin (ERIG) and human rabies immune globulin (HRIG). Administration of HRIG is promptly made to ensure passive immunisation against rabies. The recommended dose of HRIG is 20 IU/kg body weight. Fifty percent of the dose is given into the wound and 50% intramuscularly. It is usually given before or simultaneously with the first dose of the rabies vaccine. It is not given after injection of rabies vaccine. The recommendations for using RIG are for all type III exposures and for bites due to any wild animal. The classifications for the category of exposures required for post-exposure prophylaxis are presented in Table 59.2.

Table 59.2 Immunoprophylaxis in rabies

Category of rabies	Type of contact	Types of exposure	Recommended post-exposure prophylaxis (PEP)
I	Touching, feeding of animals or licks on intact skin No exposure	None	None, no PEP required
II	Minor scratches or abrasions without bleeding or licks on broken skin. Nibbling of uncovered skin	Minor	Administer vaccine immediately
III	Single or multiple transdermal bites, scratches or contamination of mucous membrane with saliva (i.e. licks)	Severe	Administer rabies immunoglobulin and vaccine immediately

Antirabies Vaccine

Rabies is the only disease where post-exposure vaccination is employed extensively and successfully. This is due to the long incubation period of the disease. The chances of preventing rabies are more when vaccination is given to humans as early as possible after exposure. In the past neural vaccines were used. Currently, the neural vaccine has been abandoned because of many complications and the availability of cheaper vaccines.

In 1971, Sikes, Koprowsky, Kaplan et al. developed the first **cell culture vaccine (human diploid cell)** for rabies. They demonstrated effective protection of monkeys against death from the street virus by post-exposure administration of tissue culture vaccine. Tissue culture vaccine is prepared by growing Pitman-Moore fixed rabies strain on human diploid cells (WI 38 or MRC). Further, it is purified, concentrated and inactivated with beta-propiolactonen or *n*-butyl phosphate. This vaccine is highly immunogenic and devoid of serious adverse effects, but this vaccine is costlier than other available vaccines.

To overcome the cost burden of tissue culture vaccines, equally efficient and economical vaccines are now available in India. These include commercially available purified chick embryo cell vaccines (PCECV), purified duck embryo vaccine (PDEV), Human diploid cell vaccine (HDC), purified vero cell rabies vaccine (PCECV), and chromatographically purified rabies vaccine (PVRV). All these vaccines are equally effective and safe. The dosage schedule is the same for children and adults. These are given intramuscularly or subcutaneously in the deltoid region in adults and on the anterolateral side of the thigh in children. Post-exposure prophylaxis requires a course of five to six doses, starting as soon as possible.

Rabies virus surface possesses glycoprotein subunit vaccine is available which is protective in nature. This gene was cloned, and proteins were used for vaccine preparation. This vaccine is still under trial. The list of such vaccines with their advantages and disadvantages are summarised in Table 59.3.

WHO Guidelines on Post-Exposure Prophylaxis

The WHO recommends three intramuscular regimens and one intradermal regimen as postexposure prophylaxis for rabies for category II and II types of exposures. The first one, 'The 5-dose intramuscular regime: (1-1-1-1-1)' recommends one dose of the vaccine to be administered on days 0, 3, 7, 14 and 28 in the deltoid region in adults and anterolateral thigh in children. This is currently the only intramuscular regimen recommended in India. The second one, 'The 2-1-1 regimen: (2-0-1-0-1)', recommends two doses to be given on day 0 in the deltoid muscle, right and left arm and an additional one dose to be administered in the deltoid muscle on day seven and day 21. The third regimen, 'The 4-dose regimen with RIG in both categories II and II'. The intradermal regimen with the added advantage of requiring less volume of vaccine, thus affecting cost, is suitable for rural areas where vaccine or money is in short supply. In this, 'The 2-site intradermal method: (2–2–2-0-2)' is recommended where one dose of the vaccine of

Table 59.3 New rabies vaccines

Characteristics	Tissue culture vaccines	Primary cell culture vaccines	Subunit vaccines
Type	HDC vaccine from WI 38, MRC-5	Purified chick embryo cell (PCEC) vaccine, purified Vero cell (PVC),chromatographically purified rabies vaccine (CPRV),purified duck embryo vaccine (PDEV)	Surface glycoprotein cloned and recombinant vaccine
Antigenicity	High	High	High
Immunogenicity	Low	High	High
Side effects	Nil	Negligible	Nil
Cost	High	Low	Experimental stages

0.1 ml is given intradermally at two different lymphatic drainage sites like the deltoid muscle on the left and right upper arm and suprascapular area. This is given on days 0, 3, 7 and 28.

Red Cross Schedule

The modified red cross schedule incorporating the elimination of the requirement of 90 days booster dose has been revised in the updated Thai Red Cross schedule. In this regimen, two injections of 0.1 mL of any WHO recommended tissue culture rabies vaccine is to be given intradermally at two sites on days 0, 3, 7 and 28. This regimen is also strongly advocated by the NRCP.

Re-Exposure Prophylaxis

Vaccination with complete dosages gives virtually complete protection for at least 5 years. During this period of 5 years, if any further exposure occurs, only one or two booster doses (on days 0, 3) may be given. After 5 years, a full course of five injections is given if the patient is again exposed to infection.

Animal Vaccines

Pre-exposure prophylaxis is recommended in animals, and post-exposure prophylaxis is not of use. As the neural vaccines are less immunogenic, the animal requires multiple vaccination doses (6 months once). Hence, purified, concentrated cell culture vaccines with the inactivated virus are recommended for animal vaccination. Cell culture vaccine provides better protection following a single dose of intramuscular injection. The first dose of this injection is usually administered at 12 weeks of age, and a booster should be given at every 1–3 years intervals.

Vaccine Failures

In vaccine failure cases, persons develop rabies despite administering the full course of vaccination. Neural vaccine failures are quite high compared to non-neural vaccine failures. In extremely rare cases, even after administering an immediate local treatment with rabies immunoglobulins and a full dose of cell culture vaccine, dog bite patients may develop rabies. It is always recommended to have a complete course of antirabies vaccination, even for individuals with minimal risk of contracting rabies.

Of late, neural tissue vaccines are gradually being replaced by cell culture vaccines that are grown in human diploid cells and then inactivated with propiolactone.

National Rabies Control Programme (NRCP)

It was initiated in India in 2013 to tackle the endemicity of rabies in India. The programme has both animal and human components, and it lays down training and capacity building of healthcare workers on appropriate management of animal site including post-exposure prophylaxis and its implementation, strengthening surveillance systems and regional laboratories for accurate diagnosis and creating awareness in the society through effective information, education and communication as its objectives. Based on recommendations from expert groups from WHO, the prophylaxis regimen has been revised and updated on a timely basis.

WHO has included rabies in its new 2021–2030 road map, owing to the huge burden of deaths annually due to rabies worldwide. In collaboration with FAO (Food and Agriculture Organization) and OIE (World Organisation for Animal Health), it has prioritised rabies under a One Health approach and together has launched 'United Against Rabies Forum' (UAR), a multistakeholder platform for global containment. The drive has been undertaken towards "Zero human deaths from dog-mediated rabies by 2030" through various long-term measures at the national, regional and global levels.

Rabies-Related Viruses

The genus *Lyssavirus* consists of more than 80 viruses, including the rabies virus, which is the prototypical human *Lyssavirus* pathogen. Other viruses included in this group rarely cause human disease. *Lyssavirus* of human importance has been classified into seven serotypes (Table 59.4).

Table 59.4 *Lyssavirus* serotypes, source and distribution

Lyssavirus	Serotypes	Source and distribution
Rabies	1	Warm blooded animals. Cosmopolitan in distribution with very few exceptions
Lagos bat/Natal bat	2	Bat and cat. Restricted to Nigeria, central and South Africa
Mokola	3	Shrew/dog/cat/human. Restricted to Nigeria and other African countries
Duvenhage	4	Bat/human. Distributed in South Africa
European bat *lyssavirus*: Type I	5	Bat/human. Distributed in Europe
European bat *lyssavirus*: Type II	6	Bat/human. Distributed in Europe
Australian bat *lyssavirus*	7	Bat/human. Distributed in Australian

Duvenhage Virus

The virus is classified as *Lyssavirus* serotype 4. The virus has been reported in bats from Southern Africa and many European countries. The first human case was reported in 1971 from South Africa, who died of clinical rabies after being bitten by a bat.

Lagos Bat Virus

The virus is classified as *Lyssavirus* serotype 4. The virus has been reported in bats from Southern Africa and many European countries. The first human case was reported in 1971 from South Africa, who died of clinical rabies after being bitten by a bat.

Mokola Virus

The virus is classified as *Lyssavirus* serotype 3. The virus has been isolated from many domestic and wild animals (shrew, cat and dog) in Africa. It was first reported from shrews in Africa in the year 1968. Later, the virus was isolated from animals in many European countries. The first report of human infection was from two children with CNS infection.

The virus is classified as *Lyssavirus* serotype 2. The viruses have been reported from bats and cats in Nigeria and Central Africa. The virus was initially isolated in 1956 from the pooled brains of frugivorous bats from Lagos Island, Nigeria. The virus causes a rabies-like illness following intracerebral inoculation in infected monkeys.

Other Rabies-Related Viruses

Other viruses, such as Rochambeau virus and Australian bat *Lyssavirus*, have been reported to cause occasional rabies-like fatal diseases in humans. Obodhiang and Kotonkan viruses have only been reported from mosquitoes in Sudan and Ibadan. No human infection by these viruses has been documented.

Case Study

A 35-year-old hunter from rural India came with a sore throat and had trouble swallowing. The man was treated as usual with antibiotics by a local physician. That night, the man began to pace and spit frequently, his facial muscles twitched uncontrollably, and he became anxious and fearful. He was hospitalised. Physicians thought he might be suffering from a drug overdose or viral encephalitis. Although he remained alert for several days, he became increasingly agitated, he vomited, his body temperature rose to 106 °F, and he died. His friend on enquiry revealed that the victim had a tussle with a dog while he was hunting 2 months ago.

- What is the possible cause for his death?
- What are the reservoirs for the organism?
- What are the presenting symptoms in case of this suspected infection?
- How can you confirm this condition in the laboratory?

Hepatitis Viruses

60

Hepatitis is a clinical syndrome caused by many pathogens, including viruses. Viral hepatitis denotes a primary infection caused by a heterologous group of hepatitis viruses. There are six medically important viruses called hepatitis viruses because their main site of infection is the liver and cause clinically similar acute illnesses. These viruses are hepatitis A virus (HAV), hepatitis B virus (HBV), hepatitis C virus (HCV), hepatitis D virus (HDV), hepatitis E virus (HEV) and the newly described G virus (HGV).

Although these viruses infect the liver as a common target organ, they differ greatly in their morphology, replication pattern and severity of infection. Nomenclature and definitions of different hepatitis viruses, antigens and antibodies are listed in Table 60.1. Human infections associated with hepatitis viruses are summarised in Table 60.2.

Types of Hepatitis Viruses

Hepatitis A virus (HAV): HAV causes infectious hepatitis. The condition occurs mostly in children and young adults. The infection is transmitted primarily by the faecal-oral route through faecally contaminated water.

Hepatitis B virus (HBV): Serum hepatitis is caused by HBV. The parenteral route mainly transmits the infection. This condition has been found in people receiving blood or serum transfusions.

The term **non-A non-B (NANB) hepatitis** was used for the infectious and post-transfusion hepatitis that was not caused either by type A or type B. The term is not used nowadays as it is possible to identify specific hepatitis infections.

Hepatitis C virus (HCV): The virus was identified as the major causative agent of non-A non-B transfusion-associated hepatitis.

Hepatitis D virus (HDV): This is a defective virus that relies on the helper functions of the Type B virus. This virus is also called as delta virus. It causes acute hepatitis.

Hepatitis E virus (HEV): The virus causes hepatitis transmitted through the faecal-oral route.

Hepatitis A Virus (HAV)

Hepatitis A Virus (HAV) causes infectious hepatitis. This is a subclinical infection affecting mostly children and young adults. The infection is unique because it does not cause chronic disease or fatal disease. Chronic carrier state does not occur with HAV infection.

Morphology

It is a small, non-enveloped RNA virus measuring 27–32 nm in size (Fig. 60.1). It is spherical with icosahedral symmetry. It has a single-stranded, positive-sense RNA genome. It was originally

Table 60.1 Nomenclature and definitions of Hepatitis viruses, antigens and antibodies

Hepatitis viruses	Abbreviations	Definition
Hepatitis A virus	HAV IgM HAVAb	Hepatitis A virus IgM antibody to HAV
Hepatitis B virus	HBV HBsAg HbcAg HbeAg HBsAb HBcAb HBeAb	Hepatitis B virus Hepatitis B surface antigen Hepatitis B core antigen Hepatitis B e antigen Antibody to HBsAg Antibody to HBcAg Antibody to HBeAg
Hepatitis C virus	HCV Anti-HCV	Hepatitis C virus Antibody to HCV
Hepatitis D virus	HDV HDAg	Hepatitis D virus Delta antigen
Hepatitis E virus	HEV	Hepatitis E virus

Table 60.2 Diseases associated with Hepatitis viruses

Hepatitis viruses	Diseases
HAV	Acute hepatitis
HBV	Acute and chronic hepatitis
HCV	Acute HCV infection, chronic HCV infection and cirrhosis and other complications
HDV	Acute self-limited infection to acute fulminant liver failure
HEV	Serious infection in pregnant women
HGV	Co-infection in chronic HBV and HCV infection

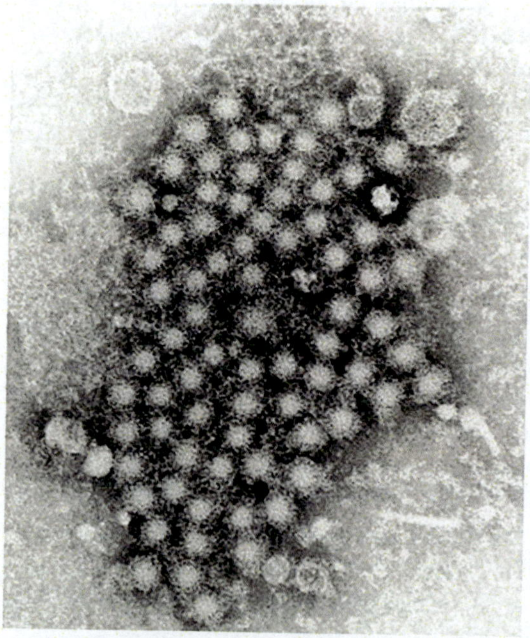

Fig. 60.1 Electron micrograph of Hepatitis A virus (https://www.cdc.gov/vaccines/vpd/hepa/public/photos.html)

designated as enterovirus, but it is recognised as a prototype of *Hepatovirus*. The naked capsid is more stable than other picornaviruses to acid and other treatments.

Virus Antigens

HAV has only one serotype. The virus does not show any antigenic relationship with hepatitis B or other hepatitis viruses.

Virus Susceptibility

The virus is highly resistant. It is resistant to heat at 60 °C for 1 h, acidic pH (at pH 1.0) for 2 h, lipid solvents, such as ether and chloroform, and drying. The virus is inactivated by formalin at 37 °C for 72 h, peracetic acid (2%) for 4 h and beta-propiolactone (0.25%) for 1 h. It is also inactivated by exposure to ultraviolet radiation (2 μW/cm^2/min). The virus is inactivated by routine chlorine treatment (1 ppm for 30 min) of drinking water.

Other Properties

The virus replicates in the cytoplasm of the infected cell. It has a replicative cycle similar to that of other picornaviruses. However, unlike other picorna viruses, it is not cytolytic and is released by exocytosis.

Laboratory Isolation

Hepatitis A is the only hepatitis virus that can be grown in human and simian cell cultures.

Pathogenesis of HAV Infection

HAV infection is acquired primarily by the faecal-oral route and is associated with hepatic injury. Most commonly, the virus spreads from person to person through contaminated water and food, including shellfish collected from sewage-contaminated water.

The virus appears to replicate first in the gastrointestinal tract and then spreads to the liver (Fig. 60.2). The viruses in the liver infect hepatocytes and cause damage to hepatocytes resulting in hepatitis. However, the mechanism by which HAV causes the cytopathic effect is unknown. Cytotoxic T cells appear to cause damage to hepatocytes; hence once the infection is cleared, the cell damage is repaired, and no chronic infection occurs. The classic findings in the hepatocytes include mononuclear infiltrate, ballooning, degeneration and acidophilic (Councilman-like) bodies. The liver pathology caused by HAV cannot be distinguished histologically from that caused by other hepatitis viruses.

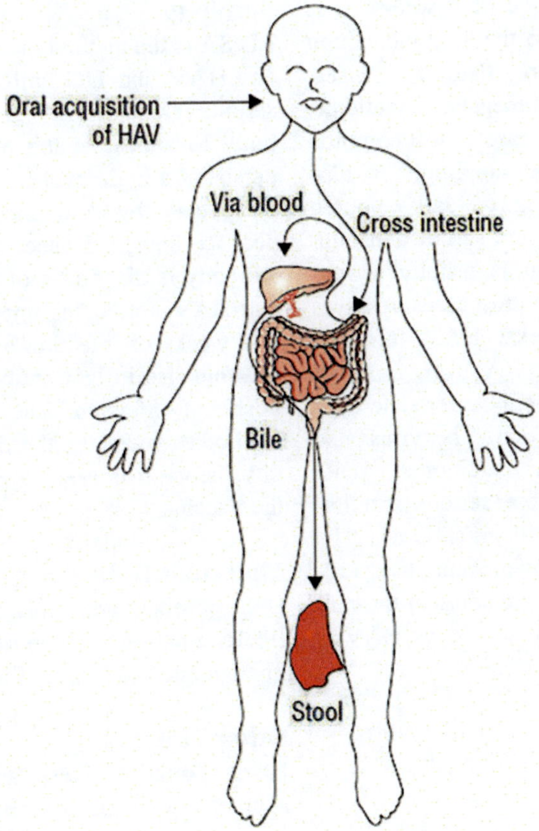

Fig. 60.2 Pathogenesis of HAV infection

Immunity

Host immunity is mediated primarily by circulating antibodies. Acute infection is characterised by the appearance of IgM anti-HAV, which is detectable at the time of appearance of jaundice in the initial stage of infection. However, the IgM antibody disappears several months after jaundice. The IgG antibody appears 1–3 weeks after the appearance of IgM antibodies. IgG antibody appears to provide lifelong immunity against recurrent HAV infection.

Clinical Syndromes

HAV causes infectious hepatitis.

Infectious Hepatitis

A majority (90%) of the HAV infections in children are subclinical. The incubation period is 15–45 days, with an average of 4 weeks. It is relatively short compared to the long incubation period of HBV infection. Fatigue, nausea, vomiting, fever, hepatomegaly, jaundice, anorexia and rash are the disease's most common signs and symptoms. The condition is also associated with the passing of dark-coloured urine, pale faeces and elevated serum transaminase levels. HAV infection is usually a self-limiting mild disease and, in most cases, resolves spontaneously in 2–4 weeks. HAV infection confers lifelong immunity to HAV.

Chronic hepatitis or chronic carrier state does not occur with HAV infection. The virus also never causes hepatocellular carcinoma. Acute hepatitis A is relatively more serious and has higher mortality in adults than children. The exact cause for this is unknown. Acute liver failure and cholestatic hepatitis are some of the rare complications. The mortality caused by HAV is very low, approximately 0.01%.

Laboratory Diagnosis

Specimens

These include (1) serum for serology anti-HAV IgM antibody detection and liver function tests, and (2) stool and blood for HAV antigen and genome.

Microscopy

The virus is present in stools for 2 weeks prior to the onset of jaundice and up to 2 weeks after the onset of jaundice. The virus can be demonstrated in the stool during this period using immunoelectron microscopy. However, it is not routinely used.

Cell Culture

Although the virus has been grown in human and simian cell culture, its isolation by the culture of faeces in the cell line is not routinely done.

Serology

ELISA is the method of choice for detecting IgM anti-HAV and IgG anti-HAV antibodies in the serum. IgM anti-HAV antibody is the first antibody to appear at the onset of symptoms and persist at a high level for 1–2 months. It usually disappears by 4–6 months but occasionally persists longer. Hence, demonstration of IgM antibody is diagnostic of a recent infection. IgG anti-HAV antibodies appear in the serum after 3–4 weeks of infection and persist for years. A fourfold rise in IgG antibody titres is also diagnostic of infection. The demonstration of IgG antibodies indicates the development of immunity; hence it is used to assess the efficiency of the vaccine.

Molecular Detection

DNA probes and polymerase chain reaction (PCR) are used to demonstrate the HAV genome in stool and the serum of infected patients.

Other Tests

Liver function tests are highly useful for supplementing the diagnosis of HAV infection.

HAV infection is associated with a consistent increase in serum alanine aminotransferase (ALT) and aspartate aminotransferase (AST). An increase in ALT and AST levels is nearly 4–100 times more than the normal levels. An increase in ALT and AST serum levels are usually seen 1 week before. The ALT and AST remain at peak level within 3–10 days after the onset of clinical illness. Serum bilirubin level is also increased, and it increases with the appearance of jaundice. A decrease in serum albumin level and prolongation of prothrombin time indicates a severe hepatocellular infection.

Treatment

No antiviral therapy is available against HAV infection. Treatment of the condition is always supportive.

Epidemiology

Hepatitis viruses are the most common cause of hepatitis in the world. HAV causes nearly two-thirds of acute cases of hepatitis.

Geographical Distribution

Infection with HAV occurs throughout the world. It is common in the developing countries of Asia, Africa and Central and South America. A high prevalence of HAV infection has been documented in the Middle East. The infection is more common in areas of low socioeconomic status and poor sanitation. Several outbreaks of HAV have been reported in India. Though a majority of the outbreaks are sporadically associated with a contaminated common source of water, there has been a recent change in the epidemiology of HAV in the Indian scenario. An increased number of infections is now being noted in the adult and adolescent population, not children. The epidemiological link between improved standard of living and decrease in HAV cases has also been noted in India.

Reservoir, Source and Transmission of Infection

Infected humans are the reservoirs and are the important source of infection. The virus is excreted in the stool during the first 2 weeks of infection, prior to the onset of symptoms; hence the quarantine of patients is not useful. The infected children and adults appear to be non-infectious after the appearance of jaundice. Humans infected with HAV are the important reservoir of infection. Infection is transmitted primarily by the faecal-oral route. Most commonly, the virus spreads from person to person through contaminated water and food, including shellfish collected from sewage-contaminated water (Fig. 60.3). This results in sporadic outbreaks of infection in the community. The virus is not transmitted by blood or blood products because the level of viremia in HAV is low and chronic infection does not occur.

International travellers to areas endemic for HAV, drug addicts using injectable drugs and homosexual men are at high-risk for HAV infection. Close contacts of infected individuals are also at high-risk. Infection is not transmitted from infected mothers to neonates. Secondary infection occurs at a high rate in household contacts of acute HAV (20%). It is the most common cause of acute hepatitis in children in India in developing countries. HAV is responsible for 10–30% of acute hepatitis and 5–15% of acute liver failure cases in India. In developed countries, infection occurs most commonly amongst adults.

Prevention and Control

Prevention of HAV infection depends on (1) vaccines, (2) specific passive prophylaxis, and (3) general prevention measures.

Vaccines

Active immunisation with vaccines containing formalin-inactivated and alum conjugated HAV is available. The vaccine is prepared from HAV cultured in human diploid cell culture. The vaccine is highly efficacious and safe. It is given in

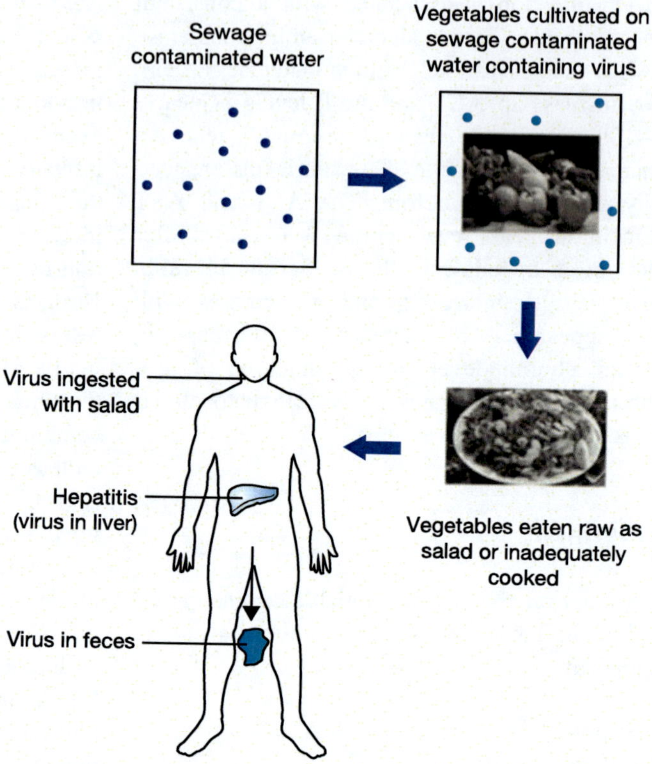

Fig. 60.3 Sewage-contaminated vegetables transmitting HAV infection

two intramuscular doses; an initial dose followed by a second dose after 6–12 months. No booster dose is given. The vaccine is recommended for children above 2 years, travellers to an endemic area, and food handlers. Twinrix is another vaccine containing inactivated HAV and non-infectious surface antigen of HBV. The recombinant vaccine administered through the intramuscular route is used to immunise against both HAV and HBV.

Specific Passive Prophylaxis

It consists of administration of IgG-containing anti-hepatitis A immunoglobulin to contacts within 2 weeks of exposure or before exposure. The immunoglobulin is given in a dose of 0.02 mL administered intramuscularly in a single dose. Specific passive prophylaxis is recommended for household contacts of infected patients and contacts in the child-care centre during outbreaks.

General Prevention Measures

These consist of handwashing, drinking of safe drinking water and good personal hygiene and sanitation. All these contribute to the prevention of the faecal-oral spread of infection of HAV.

Hepatitis B Virus (HBV)

Hepatitis B virus (HBV), a DNA virus, is a major cause of infectious hepatitis worldwide. The virus usually causes chronic disease and is associated with hepatocellular carcinoma.

The term *serum hepatitis* was used after an outbreak of hepatitis amongst American soldiers in 1942. The cause of the outbreak was linked to the yellow fever vaccine that was given to the soldiers, which was contaminated by human serum. Blumberg and co-workers in 1965 described the Australia antigen, recovered from aborigins, for which he was awarded the Nobel prize. The name Australia antigen was later called hepatitis B surface antigen (HBsAg).

Morphology

Hepatitis B virus virion is a spherical and enveloped DNA virus showing icosahedral symmetry. It is a double-walled, spherical structure and measures 42 nm in diameter. By electron microscopy, three types of particles can be seen in the serum from patients with hepatitis B. The first type is the spherical particles measuring 22 nm in diameter. The second type is the filamentous or tubular particles with a diameter of 22 nm and varying length. These two particles are antigenically identical and are known as *hepatitis B surface antigen*, or Beathard type is the double-walled, spherical structures measuring 42 nm in diameter. This article is a complete hepatitis viral particle. It was first demonstrated in human serum by D.S. Dane in 1970 and so is known as Dane particle (Fig. 60.4).

The HBV consists of nucleocapsid, which surrounds the virion DNA and DNA polymerase with reverse transcriptase and ribonuclease activity. The nucleocapsid also encloses a protein attached to the genome, surrounded by hepatitis B core antigen (HBcAg). The envelope which encloses the virus consists of HBsAg and also HBeAg.

HBsAg

This antigen consists of three glycoproteins, namely S, L and M. These glycoproteins are encoded by the same gene. The S glycoprotein is the major component of HBsAg, which self-associates into 22-nm spherical particles released from the cells. The filamentous particles of HBsAg found in the serum contain a large amount of S and only a small amount of L and M glycoprotein and other proteins and lipids. The glycoprotein L is essential for virion assembly and the formation of filamentous particles, and retention of the structures in the cell.

Virus Genome

HBV is a small (3.2 kb) enveloped DNA virus. The genome is a small, circular, partially double-stranded DNA. It is partially double-stranded because its positive-strand is incomplete. The complete negative-strand possesses four genes: **genes S, C, P** and **X** (Fig. 60.5). The gene S codes for HBsAg and also for HBeAg (hepatitis B e antigen).

The S gene has three regions: Pre S (Pre S1 and Pre S2) and S region. Surface antigen (S) or major protein is coded by the S region. S proteins are present in the circulating HBsAg. **C gene** consists of two regions in the C gene, namely Pre-C and C regions. Core antigen (HBcAg) is formed when only the C region is translated. If translation starts from the Pre-C region, a soluble antigen HBeAg is formed. This circulates in the

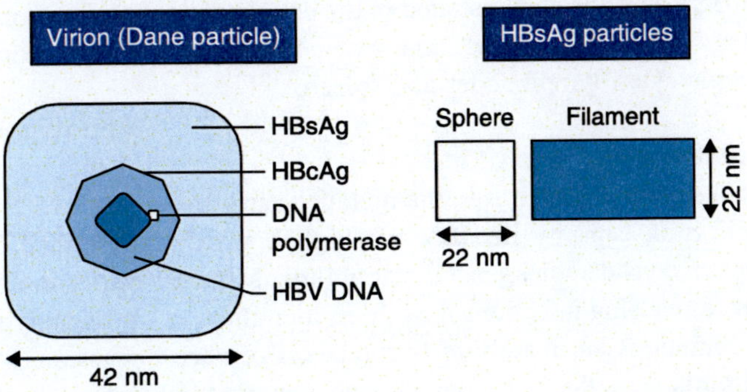

Fig. 60.4 A schematic diagram of Dane particle and HBsAg

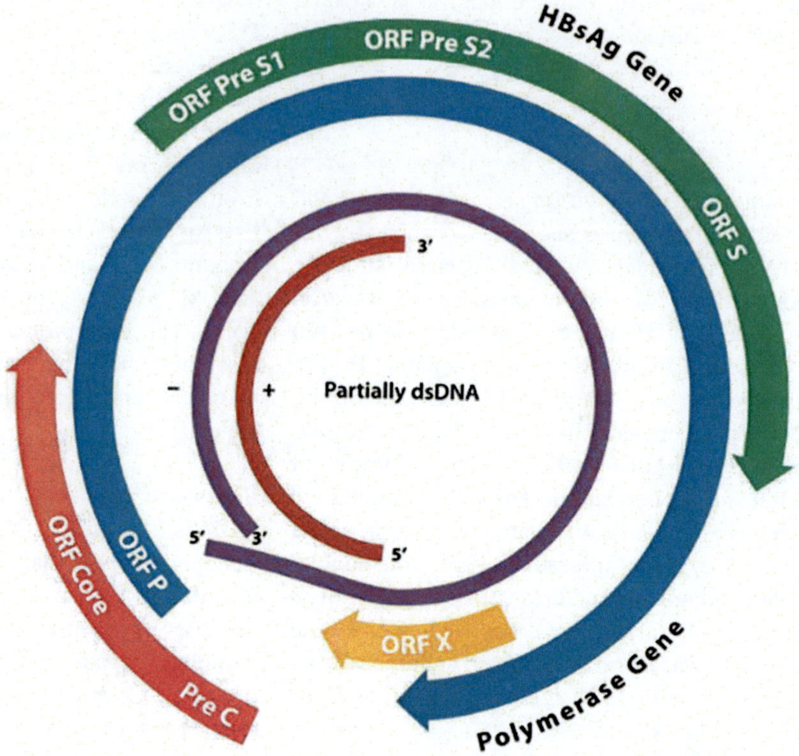

Fig. 60.5 HBV genome organisation. https://commons.wikimedia.org/wiki/File:HBV_Genome.svg

blood and is the detectable marker of HBV. **P gene** codes for DNA polymerase that has reverse transcriptase activity, although HBV is a DNA virus, the gene. **X gene** codes a tiny non-particulate protein (HBxAg). This has a transactivating effect on viruses and a few cellular genes. The antibody of HBxAg seems to be present in patients affected by chronic hepatitis and hepatocellular carcinoma.

HBV DNA is demonstrated in the extrahepatic sites such as bone marrow, circulating lymphocytes, lymph nodes and spleen.

Mutations

Unlike other DNA viruses, the mutation rates and replication capacity in HBV are higher, giving rise to considerable genetic variability. Mutant viruses causing infections have been identified.

Mutations are mainly of two types. **Pre-core mutants** were the first type that was identified in Mediterranean countries. This mutation occurs due to the inability of the pre-core mutants to synthesise HBeAg. Pre-core mutants are present in patients with chronic hepatitis. These patients may be anti-HBe and HBc positive. **Escape mutants** are the second type of mutant. They are found in liver transplant recipients immunised with anti-HBV immunoglobulin and infants born to HBeAg-positive mothers. Viral DNA is found in the hepatocyte nucleus either in a free extra-chromosomal state or together with a cell chromosome.

Genotypes

A total of eight genotypic variants (genotypes A, B, C, D, E, F, G and H) of HBV have also been described. The prevalence of different genotypes varies in different countries. The infection caused by genotype C is associated with the rapid progression of the disease and poor response to antiviral treatment than that caused by genotype D.

Virus Replication

HBV replication shows many unique features. First, the virus shows a well-defined tropism for replication in the liver. Second, a DNA virus encodes reverse transcriptase and replicates through RNA intermediates.

The virus replicates in the nucleus of the cell. The virus infects hepatocytes, the attachment of which is mediated by HBsAg glycoprotein. The infection is initiated by the binding of HBsAg to serum albumin and other serum proteins, which subsequently causes the virus to infect the liver. Inside hepatocytes, the partial DNA strand of the genome is converted to covalently closed circular double-stranded DNA. Later on, this genome is transported to the nucleus of the cell. The DNA acts as a template for all viral transcription, including 3.5 kb pre-genomic RNA. Subsequently, the genome is enclosed by a virus capsular antigen. Within the core, both negative- and positive-strand DNA viruses are synthesised. Finally, the core buds from the membrane acquire an HBsAg-containing envelope and are released from the infected cell.

The HBV cannot be conventionally cultured. However, its proteins have been cloned in yeast and bacteria.

Virus Antigens

HBsAg is antigenically complex. The HBsAg glycoprotein contains a group-specific antigen termed *a*, and type-specific antigens termed *d* or *y* and *w* or *r*. The combination of these antigens results in four major subtypes of HBV (*adw*, *adr*, *ayw* and *ayr*). These subtypes show different geographic distributions. **Subtype *ayw*** is common in Asia, the Middle East and western and northern India. **Subtype *adr*** is common in South and East India and far East countries. **Subtype *adw*** is prevalent in Europe, Australia and America, and Sub**type *ayr*** is very rare. HBcAg is found on the surface of the nucleocapsid core of the virus. Unlike HBsAg, it is not secreted and not seen in general circulation. It is found in hepatocytes. HBeAg is a translation product of the pre-core and core regions and is found in the serum of the infected individual in the free form in the initial phase of infection.

Susceptibility of Virus

HBV is an extremely resistant virus capable of withstanding extreme temperature and humidity. HBV is stable even for 15 years when stored at −20 °C, 24 months at −80 °C, 7 days at 44 °C, 60 min at 37 °C and for 6 months at room temperature. HBV, however, is sensitive to higher temperatures and is killed rapidly after heating at 100 °C for 1 min and 60 °C for 10 h.

HBsAg is a stable antigen. It is stable at pH 2.4 for up to 6 h but is associated with loss of infectivity of the virus. HBsAg is also destroyed by treatment with 0.5% sodium hypochlorite within 3 min. However, the HBsAg is not destroyed by ultraviolet irradiation of plasma or other blood products, thereby retaining the infectivity of the virus.

Pathogenesis of HBV Infection

Perinatal, sexual or parenteral methods transmit the HBV. After entering the blood, the virus infects the hepatocytes in the liver with viral antigen expression on the surface of infected cells. Cytotoxic T cells, such as activated CD4 and CD8 lymphocytes, recognise various HBV-derived proteins present on the hepatocytes' surface, resulting in an immunological reaction.

The virus by itself does not cause any cytopathic effect in the infected liver cells. The injury or cytopathic effects most probably occur due to cell-mediated injury. The formation of antigen-antibody complexes is responsible for some symptoms, such as arthralgia arthritis seen during an early stage of the disease. The immune complex is also responsible for some of the complications associated with chronic hepatitis, such as immune complex glomerulonephritis, cryoglobulinemia and vasculitis. A restricted T-cell-mediated lymphocyte response occurs against HBV-infected hepatitis.

In an infected host, whether the person will become a chronic carrier state or free of infection depends on the cytotoxic T-cell response. If the cytotoxic T-cell response is strong, the infection is cleared in the person, but if the response is inadequate, the person becomes a chronic carrier. The majority (10–70%) of infections in neonates, children and adults recover from acute infection.

A chronic infection is an important event in the pathogenesis of HBV infection. This is caused by a persistent infection of the hepatocytes that leads to the presence of HBV and HBsAg in the blood. HBsAg persists in the blood for at least 6 months. Chronic infection occurs more frequently in a newborn than in adults. It has been observed that approximately 2–10% of the infected neonates become chronic carriers. During the chronic stage, the HBV DNA is present in the cytoplasm of persistently infected cells, and in some cells, the viral DNA is integrated with cellular DNA.

Approximately 20% of HBsAg carriers, nearly 1% of all adult patients infected with HBV, and a high percentage of neonates infected with the virus progress to develop **hepatocellular carcinoma or cirrhosis**. The hepatocellular carcinoma appears to result from persistent cellular regeneration that tends to replace the dead hepatocytes. However, the HBV genome has no oncogene, which can cause hepatocellular carcinoma.

Immunity

HBV natural infection induces a lifelong immunity. Humoral antibodies primarily mediate the immunity against HBsAg. Antibodies to HBsAg are protective. These antibodies bind to surface antigens or with the virus and prevent it from interacting with receptors on the hepatocytes. These antibodies appear to neutralise the infectivity of HBV. However, the antibodies against core antigen HBcAg are not protective because the antibodies cannot act with HBcAg present inside the cells.

Clinical Syndromes

HBV is one of the most important causes of acute and chronic hepatitis. The clinical manifestations vary from subclinical hepatitis to symptomatic and icteric hepatitis. The incubation period varies from 6 weeks to 6 months. The clinical manifestations depend on the age of infection, immune status of the host and the level of HBV.

Acute Hepatitis

The prodromal or pre-icteric phase is characterised by the gradual onset of anorexia, malaise and fatigue. During the icteric phase, the liver becomes tender with the development of jaundice. Nausea, vomiting and pruritus with the passing of dark-coloured urine are the common symptoms. Clinical manifestations of acute hepatitis B are similar to that of hepatitis A. However, the symptoms tend to be more severe and life-threatening with HBV infection.

Chronic Hepatitis

Chronic infection is one of the major complications of HBV infection. The risk of chronic infection is also higher in those infected at birth (90%) and in immunocompromised patients. Only 5–10% of older children or adults progress to develop chronic infection.

Complications of HBV Infection

Patients with chronic HBV infection have a very high-risk of developing cirrhosis and hepatocellular carcinoma. The cancer is caused due to repeated episodes of chronic inflammation and cellular regeneration. Cancer appears to be due to repeated episodes of chronic inflammation and cellular regeneration and develops an average of 25–30 years after initial infection. Fulminant hepatic failure is another major complication of HBV infection.

Fulminant hepatic failure is another major complication of HBV infection. The case fatality rate of these patients is very high, nearing 80%. Glomerulonephritis, polyarteritis nodosa, varieties of skin manifestations, cardiopulmonary manifestations and joint and neurologic manifestations are other important complications of HBV infection.

Laboratory Diagnosis

Laboratory diagnosis plays an important role to confirm the HBV aetiology of hepatitis.

Specimens

The serum is an important specimen.

Serology

The definitive diagnosis of HBV depends on detecting serum markers of HBV. These include viral antigens and antibodies.

Viral Antigens

Diagnosis of acute infection is made by demonstrating HBsAg and HBeAg in the serum indicate virus replication (Fig. 60.6). Diagnosis of chronic infection is characterised by the presence of circulating HBsAg, HBeAg usually with HBcAg.

HBsAg appears in the blood even during the incubation period. It is detectable in most patients during pro-drome and acute phases of the disease. The antigen typically disappears within 2 months of the appearance of clinical disease. However, the persistent presence of HBsAg in blood for at least 6 months indicates the carrier state. It also indicates the risk of chronic hepatitis and hepatic carcinoma. It is not detectable in the serum during the convalescent stage.

HBeAg is present in the blood during the incubation period, the pro-drome and early acute disease. This is also present in certain chronic carriers. The presence of HBeAg indicates a high likelihood of infectivity and transmissibility.

HBcAg is not demonstrated in the serum. This is because it is present inside the HBsAg coat. A direct immunofluorescence test can demonstrate the antigen in the liver cells.

Viral Antibodies

Anti-HBs, the antibody against HBsAg, appear after HBS ag clearance in the serum. It is usually not detectable during the acute disease and in the chronic carrier stage. Anti-HBs is protective and remains in the serum remains for a very long period.

HBcAb, The antibody against HBcAg, can be demonstrated in the serum to diagnose HBV infection. This antibody appears earlier than the anti-HBs or anti-HBe antibodies. The HBcAb are present in individuals with acute infection, chronic infections and also in those who have recovered from acute infection. Total HBcAb, including IgM and IgG antibodies, indicates exposure to the virus and virus replication. Detection of IgM HBcAb is diagnostic of acute HBV infection during the window phase. The presence of HBcAb IgG does not differentiate between acute and chronic infection.

Interpretation of common serological markers in HBV infection is summarised in Table 60.3.

Serum marker	HBeAg	HBsAg	Anti-HBc	Anti-HBs
Resolved	–	–	+	+
Chronic[1]	+	+	+	–
Vaccinated	–	–	–	+

The absence of anti-HBs is an indication that the infection has become chronic

The currently used vaccine, containing recombinant hepatitis surface antigen, elicits only anti-HBsAg antibody, which is the neutralizing antibody

[1]One year after initial infection

Fig. 60.6 Serum markers in hepatitis B infection

Table 60.3 Interpretation of serological parameters in Hepatitis B virus infection

Virus antigen/antibody markers						
		HbcAb				
HBsAg	HbeAg	IgM	IgG	Anti-HBs	Anti-HBe	Interpretation
+	+	−	−	−	−	Late incubation period or early hepatitis
+	+	+	−	−	−	Acute HBV infection; highly infectious
+	±	−	+	−	−	Late or chronic HBV infection
−	−	−	+	+	+	Past infection
+	−	−	+	−	−	Simple carrier
+	+	−	+	−	−	Super carrier
−	−	−	−	+	−	Immunity following vaccination

Molecular Detection

HBV PCR, DNA-DNA hybridisation, etc., are used to detect the viral DNA and DNA polymerase activity during the incubation period and in the early disease. Detection of viral DNA in the serum indicates acute infection. DNA levels are higher in patients with chronic hepatitis and are associated with increased infectivity. HBV DNA levels are typically low or absent in inactive carriers. HBV PCR is highly valuable to monitor the treatment of chronic HBV infection with antiviral therapy. This is also useful to identify HBV as the cause of liver infection in HBsAg-negative patients.

Other Tests

These tests include elevation of ALT and AST. High levels are found in acute hepatitis (1000–2000 IU/mL). Estimation of serum bilirubin indicates the intensity of jaundice.

Treatment

No specific antiviral treatment is available for patients with acute HBV infection. Supportive and symptomatic care continues to be the mainstay of therapy for most patients with acute infection. Therapy by interferons and nucleoside analogues is recommended for patients with chronic hepatitis B infection. No effective treatment is available for the carrier stage.

Interferons

Interferon-alpha (IFN-α) has been the mainstay in treating chronic hepatitis B since its introduction in the mid-1980s. Interferon acts by immunomodulation and prevents the progression of acute hepatitis to the chronic stage. It also promotes more rapid resolution of viremia and normalisation of serum aminotransferase levels.

Nucleoside Analogues

These block the replication of viruses by directly blocking the replication of HBV. These nucleoside analogues are highly effective against HBV and are bioavailable and extremely well-targeted. Tenofovir and entecavir are highly effective for the treatment of HBV chronic hepatitis. In HBV/HIV-coinfected adults, for adolescents and children aged 3 years or older, tenofovir in combination with lamivudine (or emtricitabine) along with efavirenz as a fixed-dose combination is recommended by the WHO. However, in persons with confirmed or suspected antiviral resistance to lamivudine, entecavir, adefovir or telbivudine, tenofovir is recommended as the second line of treatment.

However, neither interferon nor nucleoside analogues cure HBV infection. The goal of antiviral therapy is only to reduce morbidity due to HBV and prevent complications. These antiviral drugs achieve viral suppression as demonstrated by loss of HBeAg in serum and suppression of HBV DNA.

Epidemiology

Hepatitis B is a major cause of infectious hepatitis worldwide.

Geographical Distribution

HBV is the leading cause of chronic hepatitis, cirrhosis and hepatocellular carcinoma worldwide. Nearly one-third of the world population is believed to be infected with HBV. More than 10% of people living in sub-Saharan Africa and East Asia are infected with HBV. Approximately 500,000 to 1 million persons die annually from HBV-related liver disease. Estimates suggest that 400–500 million people worldwide are HBV carriers. The HBV carrier rate varies from 1% to 20% worldwide.

Reservoir, Source and Transmission of Infection

Individuals with chronic HBV infection are the major reservoir of HBV infections. These people with HBeAg in their serum tend to have high viral titres and thus greater infectivity. HBV is present at a high level in serum. The virions are also present at very low levels in semen, vaginal mucosa, saliva and tears, and all are infectious. The virus is not detected in urine, stool or sweat; hence these specimens are not infectious.

Transmission of Infection

The HBV can be transmitted in the following ways:

1. **Perinatal transmission** is the major route of transmission of the virus. This occurs from infected mother to child due to contact with mother's infected blood during delivery instead of transplacental passage of the virus. Although HBV is found in breast milk, the role of breastfeeding in transmission is unclear.
2. **Parenteral transmission** occurs due to transfusion of HBV-infected blood and blood products. This was one of the important modes of transmission before the 1970s, but with the starting of screening of blood donors for HBsAg, the rate of blood-transfusion-associated HBV infection has reduced considerably in India and other parts of the world. Patients with haemophilia, renal dialysis and those receiving organ transplantation and intravenous drug users remain at increased risk of infection. Ear piercing, tattooing, and use of improperly sterilised scalpels or needles are associated with the risk of transition of HBV infection. The risk of acquiring HBV amongst health workers after needle stick injury from infected individuals is estimated to be as high as 5%.
3. **Sexual transmission of HBV** occurs more easily than HCV or HDV. The infection is associated with vaginal intercourse, genital–rectal intercourse and non-genital intercourse. However, the virus is not transmitted by hugging and kissing or by sharing towels, eating utensils or food. Health workers with exposure to infected blood or body fluids, heterosexual persons with multiple partners, household contacts or sexual partners of HBV carriers are other at-risk groups.

High-Risk Groups

These include Medical and paramedical people, individuals employed in blood bank and dialysis units, and barbers. These also include newborn infants to mothers with HBV, and patients receiving organ transplantations, multiple transfusions and haemodialysis. Parenteral drug abusers and sex workers have a high-risk of infection. In countries like Britain, which are non-endemic, HBV carriers are not allowed in medical practices and not allowed to be medical students.

Hepatitis B Virus Carriers

When a person is detected with HBsAg in blood for more than 6 months, he becomes a carrier. Ninety percent of the neonates, 30% of children and 5–10% of adults become carriers. Males become more susceptible as carriers. There are two categories of carriers.

Super Carriers

These are highly infectious. They have high HBsAg titre along with DNA polymerase,

HBeAg and HBV in circulation. Few carriers have high levels of antigenemia and viremia. India has a quarter of the carriers who are HBeAg-positive.

Simple Carriers

These carriers have relatively low infectivity and less titre of HBsAg in blood. They are negative for DNA polymerase, HBeAg and HBV. Sometimes many supercarriers become simple carriers.

Prevention and Control

Hepatitis B infection can be prevented using either vaccine or hyperimmunoglobulin, or both.

Vaccines

This is considered very effective. In 1982 the first vaccine was prepared from the plasma of healthy human carriers having increased antigenemia levels. Even though it was immunogenic, it was not accepted due to its limited availability, and its source was human plasma which was not free from unknown pathogenic risks. S gene HBV in baker's yeast was cloned through genetic engineering, and the vaccine was extracted. It is made of non-glycosylated HBsAg particles alone.

Plasma-derived vaccines and recombinant DNA HBV vaccines are the two types of vaccines that use HBsAg to stimulate the production of anti-HBs in non-infected individuals. Earlier Plasma-derived vaccines were used. They were prepared by purifying HAIg from healthy HBsAg –positive carriers and treating with formalin, urea, heat, etc., to inactivate the virus. Nowadays, these have been replaced by recombinant DNA-derived vaccines.

The vaccines are highly effective with more than 95% of seroconversion. The vaccine for adults is recommended at 0, 1 and 6 months and for infants at the time of birth, at 1–2 months and 6–18 months. The vaccine is indicated for all infants and people at high-risk of infection. The high-risk group includes the people frequently exposed to blood and blood products, patients receiving multiple transfusions or dialysis, patients suffering from sexually transmitted diseases and intravenous drug users.

In India, vaccination against HBV has been included in the Universal Immunisation Programme. The vaccination schedule followed is the first dose given at birth or as early as possible within 24 h of birth. Subsequently, three doses are given at 6, 10 and 14 weeks in combination with diphtheria, pertussis and tetanus (DPT), and *Haemophilus influenzae* type b (Hib) in the form of pentavalent vaccine. The route of vaccination is intramuscular injection is given at the anterolateral side of mid-thigh. No boosters are recommended.

For pre-exposure prophylaxis, in normal-risk recipients of complete vaccination, no follow-up testing is recommended. However, in persons at high-risk of exposure to HBV, anti-HBs titre should be assessed after 1 month of the third dose. Anti-HBs titre of 10 IU/L (or 10 mIU/mL) is protective. If the adequate titre is present, no further booster doses are needed. For post-exposure prophylaxis, for the unvaccinated, a combination of passive (single dose of HBIG at 0.06 mL/kg) and active immunoprophylaxis (complete three-dose vaccination) should be given. For vaccinated people, the exposed person is to be tested for anti-HBs titre. If the titer is adequate, nothing is needed. If the titre is less than 10, one dose of HBIG and a booster dose of hepatitis B vaccine is given.

Hepatitis B Immunoglobulin (HBIg)

HBIg is used for passive immunisation of patients after or just before exposure. This immunoglobulin is derived from human plasma and contains a high titre of HBsAb. It is prepared from plasma from patients who have recovered from a hepatitis B infection. Passive immunisation with HBIg is recommended for (1) people who have a history of recent exposure to the patient(s) infected with HBV, (2) a household contact with an acutely infected patient, (3) sexual contact with an acutely infected patient, and (4) an infant born to HBsAg-positive mother.

Screening blood and blood products for HBsAg is important to prevent transfusion-related HBsAg.

Combined Immunisation

Combined immunisation by HBIG, followed by the vaccine, is suggested for those non-immune persons exposed to HBV. A single injection of 0.5 mL of HBIG should be administered intramuscularly soon after the babies are born to carrier mothers. This can be followed by a full course of vaccines at the different anatomical sites.

Hepatitis C Virus (HCV)

Hepatitis C Virus (HCV) is the most important cause of parenteral NANBH worldwide. It is a flavivirus with an RNA genome. Prior to identifying the virus, it was known as NANBH to differentiate it from viral causes of non-alcoholic hepatitis. Most patients infected with HCV have chronic liver disease, which progresses to cirrhosis and hepatocellular carcinoma.

HCV is the only member of the genus *Hepacivirus* in the family Flaviviridae of RNA-containing virus. HCV appears to be closely related to hepatitis D, dengue, and yellow fever virus. It resembles flavivirus in structure and organisation and hence has been classified as a new genus *Hepacivirus* in the family Flaviviridae.

Morphology

HCV is an RNA virus. It is a spherical, enveloped, and single-stranded 55 nm RNA virus.

Virus Genome

The virus shows wide genomic variations. There are six major genotypes (genotypes 1–6). Molecular differences between these genotypes are relatively large, and more than 80 subtypes are described. The genetic variability is the main stumbling block against the effort to develop an anti-HCV vaccine. Genotype 1 is the vaccine. Genotype associated with more severe liver diseases and high-risk of hepatocellular carcinoma. It is also the main HCV genotype prevalent worldwide. Genotypes 2 and 3 are also found globally but to a lesser extent.

Other Properties

The viruses are ether sensitive and acid-sensitive.

Pathogenesis of HCV Infection

HCV is transmitted by parenteral, perinatal or sexual routes. In the infected hots, Hepatocytes are the natural targets of HCV. These viruses, like other flaviviruses, replicate in the endoplasmic reticulum of hepatocytes. In most infected people, viremia persists and is associated with a variable degree of hepatic inflammation and fibrosis. The ability of HCV to remain cell-associated and prevent host cell death is the main determinant of viral pathogenicity. These cause persistent infection in the liver.

Chronic hepatitis is characterised by lymphocyte infiltration either within the portal tract or in the liver lobule and portal and periportal fibrosis. Portal inflammation, interphase hepatitis and lobular necrosis are the main histopathological features of chronic hepatitis caused by HCV. Cell-mediated immunity, mainly cytotoxic T lymphocytes, contributes primarily to liver inflammation and ultimately to tissue necrosis.

Immunity to HCV may not be lifelong, and serum antibodies to HCV are usually protective.

Clinical Syndromes

HCV can cause acute HCV infection, chronic HCV infection, cirrhosis, and other complications induced by hepatitis. The incubation period varies from 15 to 60 days, with an average period of approximately 8 weeks.

Most patients with **acute HCV infections** are symptomatic (80%) and do not develop any jaundice. In symptomatic cases, jaundice occurs in less than 25% of acutely infected patients, whereas hepatomegaly is seen in one-third of cases. **Chronic HCV infection** is a major cause of chronic hepatitis worldwide. Most patients with chronic hepatitis are asymptomatic and may have non-specific symptoms, such as fatigue or malaise. HCV is now a leading cause of hepatitis and cirrhosis worldwide. This process may take an average of 20 years after the initial infection. **Hepatocellular carcinoma** is one of the most important complications. This occurs in around 1–5% of patients with underlying cirrhosis.

Laboratory Diagnosis

Serology

HCV antibodies are usually demonstrated in acute infections 6–8 weeks after initial infection. Then antibodies that are produced persist throughout life in chronic infection. Antibodies are directed against core envelope and NS3 and NS4 proteins and tend to be relatively low in titre.

ELISAs, including second-and third-generation ELISAs, are useful for screening serum for anti-HCV antibodies. The ELISA employs antibodies against core proteins and non-structural proteins (NS3, NS4 and NS 5) and is the most recent third-generation enzyme immunoassay (EIA). The test can detect antibodies in the serum 8 weeks after the onset of infection. Recombinant immunoblot assay using recombinant HCV antigen is a highly specific test to detect HCV infection. This test, which is more specific than the ELISA, has been used to confirm positive ELISA results. These assays are highly specific but cannot differentiate acute infection from chronic infection.

Molecular Detection

RT-PCR and branched DNA assays are used to detect HCV RNA in the serum. HCV RNA testing is the most specific. It is useful in diagnosing acute HCV infections even before antibodies are developed. This is also helpful to (1) assess the HCV genotype, (2) to confirm false-positive cases, such as autoimmune hepatitis and (3) to predict the response to interferon therapy.

HCV genotyping is a recent method, which is frequently helpful for predicting the likelihood of response and duration of treatment. The genotyping is performed by direct sequence analysis of restriction fragment length polymorphism.

Treatment

Combination therapy of pegylated interferon and antiviral agent ribavirin is the current treatment option for patients with chronic HCV infections. Currently, sofosbuvir, daclatasvir and sofosbuvir/ledipasvir are frequently used antiviral agents for most patients with chronic HCV infections. Other therapeutic options include the use of protease inhibitors, ribozymes and viral vaccines.

Epidemiology

HCV is prevalent worldwide and is exclusively a human disease. More than 3% of the world's population is infected with HCV. Worldwide 170 million people are estimated to be infected with HCV.

Geographical Distribution

It is the most important cause of parenteral NANBH worldwide. The prevalence rates in healthy blood donors are also equally high. HCV is highly prevalent in central Europe, the Middle East, Spain, Italy and Japan.

Reservoir, Source and Transmission of Infection

Hepatitis C is exclusively human disease. Patients who are infected with the virus are the important reservoir of infection. Blood or blood products and also organs of infected patients are the major sources of infection.

Transmission

The following methods can transmit hepatitis C virus:

Blood transfusion is the most important route of transmission of HCV. The current risk of transfusion-derived HCV is estimated to be one case in every 100,000 units transfused. **Parenteral transmission** is equally important. HCV is transmitted parenterally, (1) through transfusion of infected blood or blood products, (2). transplantation of organs from infected donors and (3) also, by sharing contaminated needles amongst intravenous drug users. The use of intravenous drugs is the most important risk factor responsible for around 50% of both acute and chronic infections. **Sexual transmission** is believed to be responsible for approximately 20% of cases of hepatitis C. The presence of co-existing sexually transmitted diseases, such as HIV, appears to increase the risk of transmission. **Perinatal transmission** is possible and is observed in fewer than 5% of children born to HCV-infected mothers. The risk of perinatal transmission of HCV is higher in children born to mothers co-infected with HCV and HIV.

Haemodialysis, tattooing, body piercing and acupuncture with unsterile equipment are **other modes of transmission** of HCV, though less frequent. Needle stick injury amongst healthcare workers exposed to infected blood accounts for nearly 4% of new infections.

Prevention and Control

Immunity to HCV may not be lifelong, and serum antibodies to HCV are usually protective. No vaccine against HCV is available. Immunoglobulin is not useful in preventing transmission. Transmission of HCV can be prevented by screening and preventing donation of blood, organs or semen from HCV-positive donors.

Differentiating features of HAV, HBV and HCV are summarised in Table 60.4.

Table 60.4 Distinguishing features of Hepatitis A, Hepatitis B and Hepatitis C Viruses

Features	Hepatitis A virus	Hepatitis B virus	Hepatitis C virus
Virus			
Diameter	27 nm	42 nm	50–60 nm
Genome	RNA	DNA	RNA
Envelope	Non-enveloped	Enveloped	Enveloped
Mode of infection	Faeco–oral	Parenteral	Parenteral
Immunity			
Homologous	Yes	Yes	No
Heterologous	No	No	No
Duration of immunity	Probably lifetime	Probably lifetime	–
Clinical features			
Incubation period (days)	10–50 (short)	50–180 (long)	15–160 (long)
Onset	Acute	Insidious	Insidious
Fever (>38 °C)	Common	Less common	Less common
Chronicity	Rare	5–10%	70–90%
Mortality rate (icteric cases)	<0.5%	<1–2%	0.5–1%
Age incidence	Children, young adults	15–29 years, babies	Adults
Seasonal incidence	Peak in autumn	Throughout the year	Throughout the year
Laboratory tests			
HBsAg	Absent	Present	Absent
IgM levels	Elevated	Normal to slightly elevated	Normal to slightly elevated
Duration of amino transferase levels	1–3 weeks	1–6+ months	1–6+ months
Virus in faeces	Present early	Absent	Absent
Sequelae			
Carrier state in blood	Up to 8 months	Up to 5 years	–
Association with hepatocellular carcinoma	No	Low	High

Hepatitis D(Delta) Virus (HDV)

Hepatitis D Virus (HDV) is the smallest of known human pathogens that cause infections in humans. It is an RNA virus, which is structurally unrelated to hepatitis A, B or C virus. HDV is unique in being an incomplete virus and requires the presence of HBV to replicate and infect other hepatocytes. Hence, HDV infection occurs only in those patients who suffer from HBV infection.

Rizzetho and co-workers first reported HDV in Italy in 1977. They demonstrated the viral antigen in nuclei of hepatocytes of patients infected with HBV and suggested it to be a new hepatotropic virus delta or HDV.

Morphology

It is a spherical, enveloped virus measuring 85 nm in diameter. It contains a single-stranded, negative-sense 1.7 kb RNA. In blood, HDV or delta agent contains delta Ag (HDAg) surrounded by HBsAg envelope. HBsAg is required for HDV replication, but it may be suppressed to undetectable levels with active HDV replication. The single-stranded RNA is circular and surrounded by a delta-antigen core, which is surrounded by an envelope containing HBsAg. Delta antigen may occur in two sizes—small (24 kDa) or large (26 kDa). Delta antigen is the only protein coded for HDV RNA, and it is distinct from antigenic determinants of HBV.

Pathogenesis of HDV Infection

The patient may acquire HDV infection in two ways—by co-infection or superinfection. Since the same route transmits both HDV and HBV, a person can be co-infected with HBV and delta agents simultaneously. A patient with chronic HBV infection may also be superinfected with a delta agent.

HDV causes a more rapid and severe disease with rapid progression in HBV carriers superinfected with delta and HBV. Delta agent replicates in the liver, causing liver damage and cytotoxicity. Chronic HBV carriers superinfected with HDV usually also develop chronic HDV infection. Chronic co-infection often leads to a rapidly progressive sub-acute or chronic hepatitis, resulting in more rapid progression to cirrhosis.

The virus causes an acute and chronic inflammatory disease of the liver. Although HDV can replicate independently within the hepatocytes, it requires HBsAg for its propagation. Death of hepatocytes in the liver may occur due to the direct cytotoxic effect of HDV or through a host-mediated immune response.

Clinical Syndromes

Clinically, HDV infection is indistinguishable from other forms of viral hepatitis.

The incubation period varies from 21 to 45 days but may be shorter in cases of superinfection. The clinical course of disease caused by HDV varies and ranges from acute self-limited infection to acute fulminant liver failure. Clinically, HDV infection is indistinguishable from other forms of viral hepatitis. Patients co-infected with HBV and HDV (Co-infection) show a more severe course of the disease than those infected with HBV alone. Complete clinical recovery and clearance of HBV and HDV co-infection is the most common outcome. Nearly 1% of the patients with co-infections progress to develop fulminant hepatitis resulting in more rapid progression to cirrhosis.

Laboratory Diagnosis

Diagnosis of HDV infection is usually carried out by serological and molecular tests.

Antigen Detection

Direct immunofluorescence has been evaluated to demonstrate the delta-antigen expressed in the nuclei of the liver cell.

Serology

ELISA for antigen is usually positive in 20% of patients. IgM ELISA for demonstration of anti-HDV IgM is positive in the early stage of infection, whereas IgG ELISA for anti-HDV IgG is positive during the later course of infection. Serum antibodies against HDAg are almost exclusively associated with chronic HDV infection. In superinfections, a high level of both IgM and IgG antibodies, as well as a high level of Haig and HDV RNA, is demonstrated.

Molecular Detection

Reverse transcriptase-PCR (RT-PCR) is the most sensitive method (90%) for detecting HDV RNA in blood in the stage of co-infection.

Treatment

No specific therapy is available to treat HDV infection of the liver. Lamivudine and ribavirin appear to be ineffective against HBV and HDV co-infection. Antiviral therapy with interferon is also ineffective in patients with chronic infections.

Epidemiology

HDV is distributed worldwide. The highest prevalence has been reported in **the endemic areas** of Italy, North Africa, the Middle East, West Africa and Central Asia, including China, Japan, Taiwan and Myanmar. It is believed to infect approximately 15 million of the world's population. HDV, like HBV, is a blood pathogen and is transmitted mostly by blood and vaginal secretions. It is most commonly transmitted by non-percutaneous routes, especially by close intimate contact in Mediterranean countries, endemic for the disease. Infection appears to be more commonly transmitted through contaminated blood and blood products in **non-endemic areas** of northern Europe and North America. The sexual and perinatal transmission of HDV is also described. Sharing of contaminated needles in intravenous drug users is believed to be the most common method of transmitting a million infections is most common amongst adults and children.

Prevention and Control

No vaccine is available against HVD. However, vaccination with the HBV vaccine protects against subsequent HDV infection. Avoiding the use of HDV-contaminated blood or blood products prevents the disease.

Hepatitis E Virus (HEV)

Hepatitis E virus (HEV) HEV is the primary cause of enterically transmitted viral infection. Hence the virus is named enterically transmitted non-A and non-B virus (NANB or E-AANB). The virus has many similarities with HAV. The virus was first observed during the electron microscopy of faeces contaminated with enteric non-A and non-B virus hepatitis (NANBH). HEV is currently classified in the family Caliciviridae. It resembles caliciviruses, such as the Norwalk virus. However, the HEV genome is different from the genome of other caliciviruses, and analysis of the genome sequence suggests that it is more similar to the rubella virus. Therefore, HEV remains to be classified.

Morphology and Other Properties

HEV is a non-enveloped, spherical, icosahedral virus measuring 32–34 nm in diameter. The surface of the virion shows indentations and spikes. It is icosahedral. The virus contains a positive-sense, single-stranded RNA, approximately 7.6 kb in size. The viral genome contains three open reading frames (ORFs). ORF1 is the largest, which codes for a non-structural protein responsible for virus replication. ORF2 encodes for the capsule. The function of ORF3 is not known. The viral genome is of 4 genotypes, namely 1, 2, 3 and

4. Of these, genotypes 1 and 2 are found in humans. The virus is heat stable and yet to be cultured.

Pathogenesis of HEV Infection

The virus is transmitted primarily by the faecal-oral route, Infiltration of the portal tract by lymphocytes, polymorphonuclear leucocytes, balloon hepatocytes, formation of acidophilic bodies, and intralobular necrosis of hepatocytes is the classic pathological finding of HEV infection. The IgM antibody titre appears first, which falls rapidly after infection, becoming virtually undetectable within 6 months. Anti-HEV IgG, however, persists longer for more than 6 months. The IgG antibody appears to offer protection against re-infection by HEV.

Clinical Syndromes

HEV usually causes an acute, self-limiting disease similar to HAV. Earlier it was mistaken for HAV due to clinical and epidemiological similarity. HEV infection now, called **type E hepatitis**, has been recognised as a distinct clinical entity, different from the infection caused by HAV.

Type E hepatitis is self-limiting and resolves within 15–45 days. The incubation period of HEV infection varies from 2 to 9 weeks, with an average of 35 days. The virus causes a serious infection such as fulminant hepatitis in pregnant women, especially in the last trimester of pregnancy. It has a high fatality rate of 15–20%.

Laboratory Diagnosis

Immune–electron microscopy is also used to demonstrate the virus in the stool and bile of infected persons. Western blot and ELISAs using recombinant and synthetic antigens are used for the serodiagnosis of HEV infection. These tests detect anti-HEV IgM and IgG antibodies that differentiate acute infections from recent infections. PCR is also used to detect HEV RNA in serum and stool of infected patients.

Treatment

Treatment of HEV infection is mainly supportive. No specific treatment is available.

Epidemiology

HEV is distributed worldwide. It is most commonly found in developing countries. The epidemics of HEV have been recorded in India, Pakistan, Nepal, China, Burma, North Africa and Mexico. Tropical climate, poor sanitation and poor personal hygiene all contribute to the epidemic of the disease. During 1986–88, one such large outbreak was reported in northeast China affecting nearly 10,000 people. In India, the largest epidemic of HEV occurred in Delhi during the winter of 1955–56, affecting more than 30,000 persons within 6 weeks. The virus is transmitted primarily by the faecal-oral route due to faecal contamination of water in endemic areas. The reservoir of HEV is unknown, but it may be transmitted by animals.

Prevention and Control

No vaccine is available now for the prevention of HEV. The administration of immunoglobulin does not prevent the development of clinical disease. HEV infections respond poorly to treatment with serum IgG.

Hepatitis G Virus

Hepatitis G Virus (HGV) is similar to viruses of the Flaviviridae family, which includes HCV. The flavivirus-like isolates were first demonstrated in Tamarin monkeys inoculated with blood from a surgeon with acute hepatitis in 1995. A similar virus was isolated from another patient during the same year. These viral isolates

Table 60.5 Epidemiological features of hepatitis viruses

Hepatitis virus	Mode of transmission	Vaccine	Immunoglobulins
HAV	Faecal–oral transmission	Available	Useful
HBV	Blood, sexual, vertical transmission	Available	Useful
HCV	Blood, sexual, vertical transmission	Not available	Not useful
HDV	Blood, sexual transmission	Not available	Not useful
HEV	Faecal–oral transmission	Not available	Not useful

were designated GB viruses A, B and C, respectively. A similar virus resembling GBVC (GB virus C) was isolated from a patient with chronic hepatitis in 1996. The virus has been designated as the virus as it is not cultured in cell lines, but its RNA genome has been cloned.

HGV is an RNA virus, and its genome codes for 2900 amino acids. It is a blood-borne virus, has 20% homology with HCV, is transmitted by transfusion of contaminated blood or blood products. HGV co-infection is observed in 6% of chronic HBV infections and 10% of chronic HCV infections. HGV RNA has been demonstrated in serum of patients with acute, chronic and fulminant hepatitis, patients with multiple transfusions and haemodialysis, blood donors and intravenous drug addicts. However, its role in the pathogenesis of hepatitis is yet to be elucidated.

Laboratory diagnosis of HGV is very difficult as no serological test is available. Consequently, detecting HGV RNA by RT-PCR seems to be the only method of demonstrating circulating virus inappropriate samples like blood. However, relying heavily on a single marker for diagnosis is not sufficient. An ELISA has been developed using anti-E2 envelope protein, though its use is not widely present. The only test widely performed is HGV RNA PCR.

Epidemiological features of hepatitis viruses are summarised in Table 60.5.

Case Study

A 25-year-old resident in surgery was admitted to the hospital with jaundice, nausea and vomiting. His liver was enlarged. The serum test showed elevated aminotransferase. He gave a history of vaccination with hepatitis B vaccine 6 years back. His serum tested for HAV IgM was negative, HAV IgG positive, HBsAg-negative, HBsAb positive, HBcAb negative and HCV-antibody positive.

- What is the possible hepatitis virus responsible for the condition?
- What is the most likely method of acquiring this infection?
- What are the complications of this infection?
- How can you prevent this infection?

Retrovirus

61

Human Immunodeficiency Virus (HIV)

Human immunodeficiency virus (HIV) is a retrovirus that causes acquired immunodeficiency syndrome (AIDS). AIDS is one of the most devastating epidemics ever recorded in the world. AIDS was first recognised in Los Angeles in 1981 when five cases of *Pneumocystis carinii* (now called *Pneumocystis jirovecii*) pneumonia in homosexual men and drug addicts were reported. At that time, AIDS was considered to be associated with the "4 H group-Haemophiliacs, Heroin addicts, Homosexuals and Hispanics".

Luc Montagnier and co-workers first reported the causative agent of AIDS from the Pasteur Institute, Paris, in 1983. They isolated a retrovirus from a West Asian patient with persistent generalised lymphadenopathy and named it *Lymphadenopathy-associated virus* (LAV). In 1984 Robert Gallo and co-workers from the National Institute of Health, USA, reported the isolation of a retrovirus from a patient with AIDS and called it *human T-cell lymphotropic virus-III* (HTLV-III). The International Committee on Virus Nomenclature in 1986 gave the name *human immunodeficiency virus*, or HIV, for the same virus. HIV-1 is the first isolated virus from the cases of AIDS, and HIV-2 has been isolated from some cases of AIDS from West Africa.

Classification of Lentiviruses

HIV is a *Lentivirus*, a subfamily of Lentiviridae in the family retrovirus. This family includes the viruses known for (1) poor host immune responses, (2) latency, (3) persistent viraemia and (4) infection of the central nervous system. HIV, like other retroviruses, are enveloped RNA viruses, characteristically possessing an RNA-dependent DNA polymerase called reverse transcriptase. Animal lentiviruses are bovine immunodeficiency virus (BIV), caprine arthritis encephalitis virus (CAEV), equine infectious anaemia virus (EIAV), feline immunodeficiency virus (FIV), puma lentivirus (PLV), simian immunodeficiency virus (SIV) and visna/maedi virus (VMV). Among the human species, HIV 1 and HIV 2 are common. The SIV has been associated with AIDS in primates.

Members of the lentivirus group that cause immunodeficiency in primates and non-primates.

Morphology

HIV is a spherical, enveloped virus, which measures up to 120 nm in diameter (Fig. 61.1). It has a unique three-layered structure, the innermost genome layer, middle cone-shaped

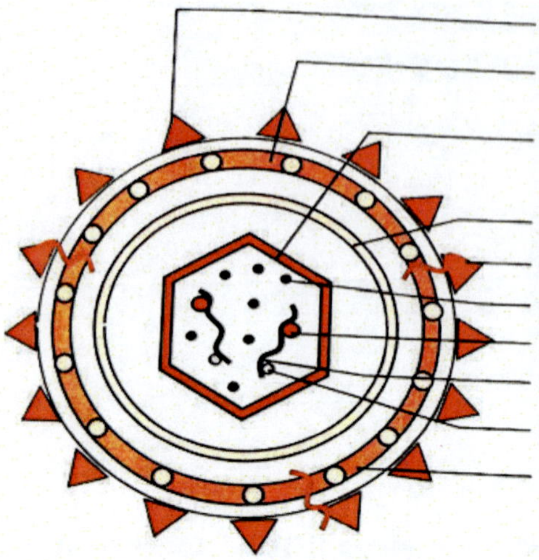

Fig. 61.1 Schematic diagram of HIV

nucleocapsid and an outer glycoprotein membrane surrounded by a lipoprotein envelope.

HIV genome is the most complex of human retroviruses. The genome is diploid and consists of two identical copies of the single-stranded positive-sense RNA genome.

The genome is surrounded by a nucleocapsid consisting of proteins. Reverse transcriptase, integrase and protease are the three enzymes located in the nucleocapsid. Reverse transcriptase performs two important functions. First, it transcribes the RNA genome into the proviral genome. Second, it also has a ribonuclease H activity, which degrades the RNA when it is in the form of an RNA–DNA hybrid molecule. Degradation of the viral RNA genome is an essential step in synthesising double-stranded proviral DNA. Integrase is another important enzyme that facilitates the integration of proviral DNA into the host cell DNA. Protease is another enzyme that splits precursor polyproteins into functional viral polypeptides.

A lipoprotein envelope surrounds the virus. The lipid component is derived from the host cell membrane and glycoproteins, which are virus coded. The major virus-coded envelope glycoproteins are the projecting spikes on the surface and the anchoring transmembrane pedicles. The projecting spikes combine with the CD4 receptors on susceptible host cells, and transmembrane pedicles cause cell fusion.

Virus Genome

HIV genome is diploid and consists of two identical copies of the single-stranded positive-sense RNA genome.

It contains three major genes, *gag*, *pol* and *env*, characteristic of all retroviruses. All these genes encode for the structural proteins.

The ***gag*** gene encodes for internal capsid and matrix "core" proteins (p15, p18 and p24). Of these three proteins, p24 is the major antigen, demonstrated in serum of HIV patients during the early stage of infection, and persists until the appearance of serum antibodies. Detection of p24 antigen in serum, therefore, is of diagnostic value.

The ***pol*** gene encodes for several proteins, including the enzymes reverse transcriptase, integrase and protease. The enzyme reverse transcriptase synthesises DNA by using the genome RNA as a template. The enzyme integrase integrates the viral DNA into the cellular DNA, and the enzyme protease cleaves various viral precursor proteins. The *pol* gene expresses precursor protein p160, which is cleaved into three proteins: p31, p51 and p64.

The ***env*** gene codes gp160, a precursor glycoprotein that is split to form two envelope glycoproteins, gp120 and gp41, which form the surface spikes and transmembrane tissue proteins, respectively.

Apart from these genes, it also consists of six other non-structural and regulatory genes (*tat*, *rev*, *nef*, *vif* and *vpr*), including an additional gene *vpu* in HIV-1 and *vpx* in HIV-2.

These genes encode several proteins, which are essential for transcription and invasion of the virion into host cells. The ***Tat*** gene is the most important one, which encodes a protein called Tat protein that facilitates viral gene transcription. Tat protein, along with other HIV-encoded regulatory proteins called Nef, suppresses the synthesis of class I MHC (major histocompatibility complex) proteins, thereby reducing the ability of cytotoxic T cells to kill the HIV-infected cells. The ***rev*** gene encodes another regulatory protein, which controls the transport of mRNA from the nucleus into the cytoplasm. The functions of the genes are summarised in Table 61.1.

Virus Replication

HIV replication is similar to that of other retroviruses. The virus binds to the CD4 proteins on the cell surface with the help of its gp120 envelope protein (Fig. 61.2). This protein also interacts with chemokine receptors on the cell surface. Then gp41 of the virus mediates fusion of the viral envelope with the cell membrane followed by entry of the virus into the cell.

CXCR4 and CCR5 are the chemokine receptors, which are very much essential for the entry of HIV into CD4+ cells. The T-cell-tropic strains of HIV bind to CXCR4, and the macrophage-tropic strains bind to CCR5 receptors. Once inside the cell, after uncoating, the virion RNA-dependent DNA polymerase transcribes the genome RNA into double-stranded DNA, which subsequently integrates with the host cell DNA. Integration is mediated by the viral enzyme integrase. The host cell RNA polymerase transcribes the viral mRNA from the proviral DNA.

The viral mRNA encodes several proteins, which are cleaved by many enzymes. For example, the Gag protein is cleaved by the virus-encoded protease to form the main core protein (p24), the matrix protein (p17) and several smaller proteins. The Pol protein is also cleaved by a protease to produce reverse transcriptase, integrase and protease. The Env protein is cleaved by a cellular protease.

The immature virions containing the precursor polyproteins are assembled in the cytoplasm, and cleavage by the viral protease occurs as the immature virion buds from the cell membrane, resulting in the production of mature infectious HIV.

Virus Antigens

HIV shows two distinct antigens: group-specific antigen and type-specific envelope glycoproteins.

Group-specific Antigen

The protein p24 is the group-specific antigen present in the virus's core. p24 is stable and does not show variations. Serum antibodies against the p24 antigen are not protective and do not neutralise the infectivity of HIV. However,

Table 61.1 HIV genes and their products

HIV genes	Gene product
Gag p24	Nucleocapsid core protein
p15	Nucleocapsid core protein
p55	Precursor of core protein
p18	Polyprotein from *gag* gene
*env*gp120	Outer envelope glycoprotein
gp41	Transmembrane envelope glycoprotein
gp160	Precursor of envelope glycoproteins
pol p31	Reverse transcriptase
p51	Reverse transcriptase
p64	Reverse transcriptase

Fig. 61.2 Invasion by HIV

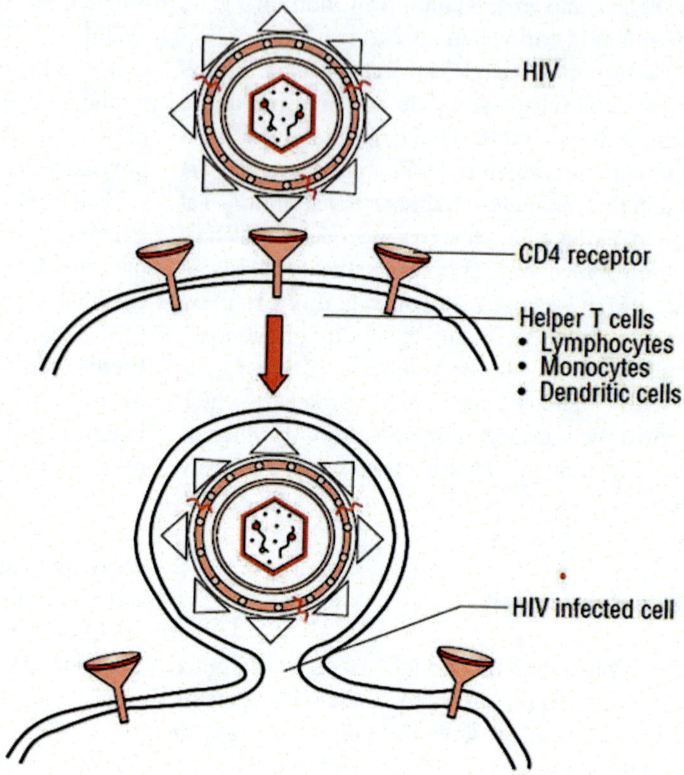

p24 is an important serological marker for diagnosis of HIV, even during the "Window period".

Type-specific Envelope Glycoproteins

gp120 and gp41 are the type-specific envelope glycoproteins present on the virus's surface. gp120, which protrudes from the surface, is the most important that combines with CD4 receptors and the chemokine receptors on the surface of CD4 cells. Antibodies against gp120 neutralise the infectivity, but the rapid emergence of gp120 variants does not confer protection against new strains; hence, it is difficult to produce an effective vaccine. gp41, which is embedded in the envelope, mediates the fusion of the viral envelope with the cell membrane at the time of infection (Fig. 61.3).

HIV-1 and HIV-2, two distinct antigenic types, based on molecular and antigenic differences have been recognised. The envelope antigens of both types are different. Their core polypeptide shows some degree of cross-reactivity.

HIV-1 and HIV-2 strains based on sequence analysis of either *gag* or *env* genes have been classified into three groups: **M (major/main), N (non-M, non-O/new)** and **O (other)**. The M group is the most prevalent group, which causes most HIV-1 infections. Group M consists of nine subtypes: A–D, F–H, J and K; all of which have originated from Central Africa and are prevalent worldwide. Groups N and O include only some HIV-1 isolates from Central Africa, which do not belong to group M.

HIV-1 represents the original isolates from the Americas, Europe and other Western countries, whereas HIV-2 isolates have originated from Western Africa. The strains of HIV-1 can be classified into four groups. These are groups M, N, O and P. Of these, M is the "major" group and is responsible for the majority of the global HIV epidemic. Within group M, there are nine genetically distinct subtypes of HIV-1. These are subtypes A, B, C, D, F, G, H, J and K. Additionally, different subtypes can combine genetic material to form a hybrid virus, known as

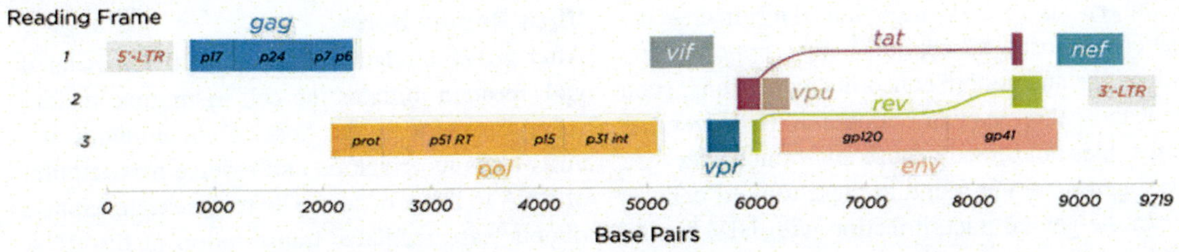

Fig. 61.3 Diagrammatic representation of HIV genome. (https://commons.wikimedia.org/wiki/File:HIV-genome.png)

a "circulating recombinant form" (CRF). Around 89 of these are known to exist. The dominant HIV subtype in the Americas, Western Europe and Australasia is subtype B. Subtype C is very common in the high-prevalence countries of Southern Africa and the horn of Africa and India. The other three groups—N, O and P—are uncommon. Group O constitutes up to 5% of infections in many west and central African countries, and Group N and P have been rarely identified in Cameroon. Recently, it has been seen that a greater risk of transmission or faster disease progression is not associated with various subtypes.

HIV-2 is more closely related to the simian immunodeficiency virus than to HIV-1. There are eight known HIV-2 groups (A to H). Of these, groups A and B are pandemics. Group A is found mainly in West Africa but has also spread globally to Angola, Mozambique, Brazil, India, Europe and the US. Group B is mainly confined to West Africa. Groups C and D have been found rarely in people from Liberia, groups E and F in Sierra Leone and groups G and H in two people from the Ivory Coast.

Virus Susceptibility

HIV is a thermolabile virus. It is readily inactivated at 60 °C in 10 min and 100 °C in seconds. The virus in dried blood, at room temperature (20–25 °C), may survive for up to 7 days. The virus has been isolated from various tissues even up to 16 days at the autopsy of the patient infected with HIV.

HIV is inactivated by treatment with 50% ethanol, 35% isopropanol, 0.5% Lysol, 0.5% formaldehyde, 0.3% hydrogen peroxide and 10% bleaching powder in 10 min. Bleaching powder is an effective disinfectant for use as a surface decontaminant at a concentration of 0.5%, with free chlorine 5 g/L (5000 ppm). A 2% glutaraldehyde solution is effective for the disinfection of medical instruments.

In liquid plasma or lyophilised blood products, HIV can be inactivated by heating at 56 °C for 30 min. It is also inactivated at a very low pH (1) and high pH (13).

Laboratory Isolation

Tissue Culture

The virus can be isolated mostly from lymphocytes in the peripheral blood and occasionally from bone marrow, plasma and other body fluids. In patients with AIDS, a high titre of the virus is found in the plasma and lymphocytes than in the peripheral blood. Viruses are found mostly within CD4 cells. **Co-cultivation** is the most sensitive method for the isolation of the virus.

Pathogenesis of HIV Infection

HIV is primarily a sexually transmitted pathogen transmitted by high-risk behaviours, such as unprotected intercourse, male homosexual intercourse and also by intravenous (IV) drug abuse.

Binding of Virus to Cell Receptors

HIV shows tropism for all the cells expressing CD4 antigens on their cell surfaces. The CD4 antigens act as receptors for HIV. The virus infects helper T cells and kills them, resulting in HIV-induced immunosuppression, leading to full-blown AIDS—a key feature of the pathogenesis of HIV infection. This makes the patient most susceptible to opportunistic infections and certain cancerous conditions, such as Kaposi sarcoma and lymphoma. However, the virus does not directly cause any tumour because HIV genes are not found in these tumour cells.

In the genital tract, infection with HIV begins in Langerhans cells, the dendritic cells that line the mucosa. This is followed by infection of the local CD4+ helper T cells in the genital tract and by the appearance of the virus in the blood 4–11 days after infection. The CD4 receptors are present on CD4 T lymphocytes and also on the cells of the macrophage lineage, such as monocytes, macrophages and alveolar macrophages of lungs, dendritic cells of the skin and microglial cells of the brain (Fig. 61.4). The gp120 protein is the principal determinant of the pathogenicity of HIV. The Vy region of the gp120 determines the cellular tropism of the virus. After HIV enters the host, gp120 binds selectively to the CD4 cell surface receptors and CCR5 or CXCR4 chemokine receptors expressed on macrophage lineage cells.

Virus Replication

After gp120 binds to the receptor, the associated gp41 protein initiates the cell membrane fusion. After fusion with the host cell membrane, the virus loses its envelope and reverse transcription of RNA to DNA occurs. The reverse transcriptase ribonuclease mediates transcription of the RNA into double-stranded DNA provirus. The provirus is integrated into the infected cell's genome, causing latent infection. The long and variable incubation period of HIV is because of this latency. During the latency period, there is a high level of virus replication. It is estimated that 10 billion HIV particles are produced and destroyed each day. The virus replicates continuously in the lymph nodes, releasing the virions and infected T cells into the blood.

Lysis of Infected Cells

Fusion of the cells, followed by lysis of fused cells, results in the replenishment of many uninfected cells from the circulation. The virus causes a drastic reduction in the number of CD4 T cells, which may occur due to HIV-induced cytolysis,

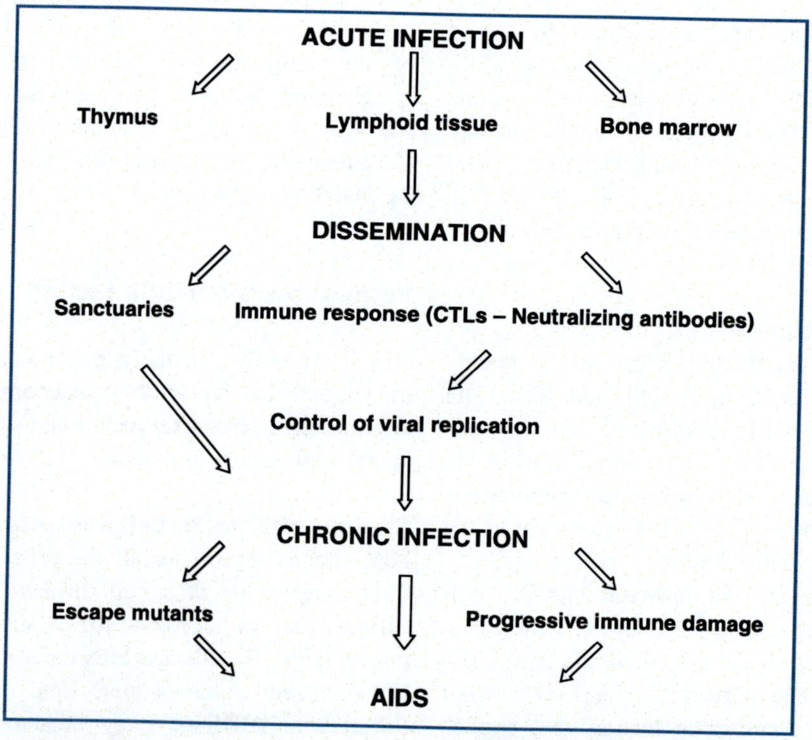

Fig. 61.4 Immunopathogenesis of AIDS. (https://link.springer.com/chapter/10.1007/978-88-470-2091-7_4)

cytotoxic T-cell immune cytolysis or natural terminal differentiation of T cells. The CD4 cells are decreased in number, and the *CD4:CD8 cell ratio is reversed.* Viral infection also suppresses the function of the infected cell without causing structural damage.

The virus also infects brain monocytes producing multinucleated giant cells and significant central nervous system manifestations. The fusion of the HIV-infected cells in the brain and other sites is the key pathological finding.

Immunity

HIV is characterised by developing both cell-mediated and humoral immunities against HIV-related proteins.

Cell-mediated Immunity

It is characterised by the development of cellular responses produced against HIV proteins. Suppression of cell-mediated T-cell immunity is the most profound consequence of HIV infection. CD4 helper T cells, monocytes and macrophages are important components of CMI against HIV infection. The CD4 helper T cell plays a very important role in the outcome of immune responses. The CD4 helper cells are essential (1) for activation of macrophages and (2) induction of functions of cytotoxic T cells, natural killer cells and B cells and a variety of soluble factors that stimulate growth and differentiation of lymphoid cells. HIV binds directly to the CD4 receptors of the T helper cells, resulting in gradual depletion of the T-cell population. The reduction of CD4 T cells causes CMI to fail gradually (1) to mount a cytotoxic T-cell response to virally infected cells, (2) to form delayed-type hypersensitivity reaction and (3) to process new foreign substances presented to the immune system.

Monocytes and macrophages also play an important role in dissemination and pathogenesis of HIV infection. CCR5 chemokine receptors are the major HIV co-receptors present on monocytes and macrophages, which appear to be the major cell types infected with HIV in the brain. These cells, therefore, contribute immensely to the development of neurological manifestations associated with HIV infection. The macrophage-tropic strains of HIV are seen in large numbers early in the infection, and these strains are responsible even for transmission of the infection. The virus may enter the brain through infected monocytes and release toxins to neurons and chemotactic factors that lead to infiltration of the brain with inflammatory cells.

Humoral Immunity

It is characterised by the development of neutralising antibodies produced against p24, gp120, gp41 and various proteins in most individuals infected with HIV. However, the level of neutralising activities is low. In adults, antibody to gp120 develops several months after the initial viraemia. The development of neutralising antibodies is associated with the slow progression of disease in adults, children and infants. HIV escapes from the host's immune system in many ways (Table 61.2).

HIV escapes from the immune system through its ability (1) to integrate viral DNA with the host cell DNA, resulting in persistent infection, (2) to undergo a high rate of mutation of the *env* gene, (3) to produce Tat and Nef proteins that downregulate class I MHC proteins required for cytotoxic T cells to recognise and kill HIV-infected cells and iv) to infect and kill CD4+ helper T cells. Continuous virus reproduction in macrophages and CD4 T cells also maintain the virus in an immune-privileged state.

Clinical Syndromes

The course of untreated HIV infection is usually 10 years or longer. The disease progresses through the stages of (1) primary infection, (2) dissemination of virus to lymphoid organs, (3) clinical latency and (4) a late stage of profound immunosuppression known as full-blown AIDS (Table 61.3). HIV is associated with the following clinical syndromes:

Table 61.2 Mechanisms of HIV escape from the immune system of the host

Mechanism	Method
Inactivation of key element of immune defence	Infection of lymphocytes and macrophages
Loss of activator of the immune system and delayed-type hypersensitivity	Inactivation of CD4 helper cells
Evasion of antibody detection	Antigenic drift of gp120
Evasion of antibody detection	Heavy glycosylation of gp120

Table 61.3 Classification of HIV infection summarised as per CDC, Atlanta, USA

Group	Subgroup	Clinical manifestations
Group I		Acute HIV syndrome
Group II		Asymptomatic infection
Group III		Persistent generalised lymphadenopathy
Group IV		Other diseases
	A	Constitutional disease—AIDS-related complex
	B	Neurological diseases
	C	Secondary infectious diseases
	Category C1	The specified infectious diseases mentioned in CDC surveillance definition for AIDS (*Pneumocystis carinii* pneumonia, toxoplasmosis, cryptosporidiosis, generalised strongyloidiasis, cryptococcosis CMV or herpes diseases)
	Category C2	The other specified secondary diseases (oral hairy leukoplakia, *salmonella* bacteremia, tuberculosis, nocardiosis, thrush)
	D	Secondary cancer (Kaposi sarcoma, lymphomas, etc.)

Acute HIV Infection

Acute HIV infection is characterised by the rapid rise in plasma viraemia with a concomitant drop in CD4 count after an incubation period of 3–6 weeks. The symptoms of HIV are non-specific and include low-grade fever, fatigue, malaise, rash, headache and lymphadenopathy; spontaneous resolution may occur within weeks. HIV antibodies are usually absent in the serum at the onset of illness but begin to appear after 3–4 weeks of the infection. This condition is referred to as **seroconversion illness**. Serum antibodies are not demonstrated, but the p24 antigen can be demonstrated during the beginning of the infection.

Asymptomatic or Latent Infection

This period is followed by an asymptomatic or clinically latent stage during which the patient continues to remain asymptomatic for several months to years. This stage is characterised by a low level of virus replication and a gradual fall in CD4 count. The serum is positive for HIV antibodies in these patients.

In this context, long-term progressors term was used for those who have been infected for over 10 years and remained asymptomatic with normal and stable immunologic profiles and who had not received any anti-HIV treatments. In long-term non-progressors, the viral load is low, but virus replication persists. Lymph node architecture and immune function appear to remain intact. The CD8 cell anti-HIV factor (CAF) appears to be a strong response in non-progressors.

Another characteristic of the latency stage is persistent generalised lymphadenopathy, which may last for several years or a period of asymptomatic infection. During this stage, the virus continues to replicate in the lymph node. This is a benign condition but may progress to AIDS-related complex (ARC) or AIDS.

Table 61.4 Characteristic immunological abnormalities during HIV infection

Characteristic immunological features of AIDS
Lymphopenia
CD4 count reduction, inversion of T4:T8 ratio
Decrease in delayed hypersensitivity (low skin test response)
Polyclonal activation of B cells results in increased spontaneous secretion of antibodies
Hyperimmunoglobulinaemia (increased IgG and IgA levels, IgM level will be high in children)
Other common features
In vitro proliferation of lymphocyte will be decreased in response to mitogens and antigens
Cytotoxic T cells and NK cells show decreased activity
Immune system show decreased antibody response to antigens
Monocyte and macrophages show decreased function
Serum will contain elevated levels of immune complexes

AIDS-Related Complex (ARC)

ARC is characterised by lymphadenopathy and fever. This has an insidious onset and may be associated with malaise and weight loss. Diarrhoea, night sweats, fatigue and opportunistic infections are the presenting symptoms. The patients with ARC may progress to AIDS in a few months.

AIDS

AIDS is the end-stage disease of HIV infection. It denotes the irreversible breakdown of the host's immune system, making the infected host highly susceptible to a wide range of progressive opportunistic infections or unusual malignancies, such as Kaposi sarcoma (Table 61.4). AIDS is characterised by deterioration of immune response as evidenced by CD4 T cell decrease response. The onset of clinical manifestations correlates with (1) a reduction in the number of CD4 T cells to less than 450 cells/μL, (2) an increased level of virus in the blood and (3) presence of p24 antigen in the blood.

When CD4 count falls less than 200 cells/μL, the patient develops full-blown AIDS. This stage is characterised by the development of HIV wasting syndrome with weight loss and diarrhoea for 1 month. This is also associated with many opportunistic infections. Patients with AIDS show clinical manifestations in different ways as follows:

Opportunistic Infections

These are usually associated with HIV-infected patients when their CD4 cell count falls to less than 200 cells/μL. The indicator diseases caused by various pathogens, such as protozoa, bacteria, viruses and fungi, are summarised in Table 61.5. These infections manifest as diseases of various systems.

1. **Respiratory** system **infections:** Lower respiratory tract infections (LRTIs), mostly in the form of pneumonia, are the most common respiratory diseases and are frequently the first clinical manifestations of HIV infections. The spectrum of HIV-associated opportunistic pneumonias includes bacterial, mycobacterial, fungal, viral and parasitic pneumonias. Bacterial pneumonia is the most frequent opportunistic pneumonia in the Western countries of the USA and Europe, while tuberculosis (TB) is the dominant pathogen in sub-Saharan Africa and Asia. The incidence of *Pneumocystis pneumonia* has declined with the use of antiretroviral therapy though it continues to occur in persons who do not have access to medical care and diagnostics. Other infectious causes, like *Cryptococcus neoformans, Histoplasma capsulatum, Coccidioides immitis*, cytomegalovirus (CMV) and *Toxoplasma gondii*, are less frequent though they are often seen in disseminated infections and are associated with significant mortality.
2. **Gastrointestinal** system **infection**: The spectrum of gastrointestinal infections ranges from oral lesions of Candidiasis, various lesions of viral infections, hepatobiliary lesions, pancreatitis and anorectal lesions. The disease manifestations depend on the level of

Table 61.5 Indicator diseases of AIDS

Pathogen/Condition	Disease
Bacteria	*Mycobacterium avium-intracellulare* complex Extrapulmonary tuberculosis Atypical mycobacterial disease Salmonellosis Pyogenic bacterial infections *Campylobacter* infections *Nocardia* infection and actinomycosis Legionellosis
Virus	CMV disease Herpes simplex virus infection Hairy leukoplakia by Epstein–Barr virus Progressive multifocal leukoencephalopathy
Parasite	Toxoplasmosis Cryptosporidiosis Isosporiasis Generalised strongyloidiasis
Fungus	*Pneumocystis jirovecii* pneumonia Candidiasis Cryptococcosis (extrapulmonary) Aspergillosis Histoplasmosis (disseminated) Coccidioidomycosis (disseminated)
Malignancies	Kaposi sarcoma Lymphomas: Hodgkin and non-Hodgkin types
Others	HIV wasting syndrome HIV encephalopathy Lymphoid interstitial pneumonia HIV dementia

immunosuppression, as determined by the CD4 counts. Oral Candidiasis (thrush) is seen in most patients with advanced HIV. Esophagitis is common in advanced diseases and frequently occurs in at least one-third of the patients. Candidal esophagitis is the most frequent, occurring either alone or in association with other opportunistic pathogens, like CMV, HSV and *Mycobacterium avium-intracellulare* (MAI) complex.

3. **Central Nervous system infections:** Cerebral toxoplasmosis, cryptococcal meningitis, tuberculous meningitis, progressive multifocal leukoencephalopathy (PML) and CMV encephalitis are the most common opportunistic infections of the central nervous system (CNS) in HIV-infected patients. Concurrent neurological HIV-CNS disease (such as the AIDS dementia complex) is also frequently seen.

Malignancies

Patients with AIDS show a marked susceptibility to the development of malignancies. Human herpesvirus-8-associated Kaposi sarcoma is the most noted malignancy associated with AIDS. Kaposi sarcoma is much more common in untreated AIDS patients than in the general population. It is a vascular tumour suggested to be of endothelial origin, which is found in the skin, mucous membrane, lymph node and visceral organs.

Other AIDS-associated malignant conditions include non-Hodgkin lymphoma, Hodgkin lymphoma, cervical cancer and anogenital cancer. Burkitt lymphoma has been shown to be even much more common in AIDS patients than in the general population.

1. **Neurological diseases**: AIDS patients are associated with several distinct neurological syndromes. These include AIDS dementia complex, sub-acute encephalitis, vacuole

myelopathy, aseptic meningitis and peripheral neuropathy. AIDS dementia complex is the most common neurological manifestation of HIV and occurs due to HIV infection of the microglial cells and neurons of the brain. This condition is characterised by poor memory, inability to concentrate, apathy, automotor retardation and behavioural changes.

2. **Cutaneous diseases:** Cutaneous manifestations of HIV infection may result from HIV infection itself or from opportunistic infections. Kaposi sarcoma can occur prior to the onset of immunosuppression. With the onset of immunosuppression, non-specific skin changes occur, such as recurrent varicella-zoster, numerous hyperkeratotic warts, treatment-resistant seborrheic dermatitis and oral hairy leukoplakia. Verrucous herpes infection, leprosy, condyloma-like molluscum contagiosum and AIDS-associated pigmented or non-pigmented erythroderma may be seen in early HIV disease. Chronic HSV, molluscum contagiosum (MC), and CMV infections appear later. Cutaneous forms of Mycobacterial infections and mucocutaneous candidiasis also occur frequently. The most common cutaneous infections in children with HIV are impetigo and cellulitis caused by *Staphylococcus aureus*. Other cutaneous manifestations of HIV infection in the pediatric age group are recurrent herpetic gingivostomatitis, scabies, severe atopic dermatitis, drug eruptions and leukocytoclastic vasculitis.

Toxoplasma encephalitis, cryptococcal meningitis and John–Cunningham (JC) virus-induced PML are some of the common infectious diseases of the brain associated with patients of AIDS.

Paediatric AIDS

Paediatric AIDS is an important condition acquired from infected mothers. AIDS in paediatric populations usually occurs by vertical transmission of HIV from infected mothers or by the perinatal transmission of HIV through breastfeeding.

Children develop clinical manifestations by 2 years of age and subsequently die of AIDS in the following 2 years. The condition is more severe in neonates because the immune system is very poor during the time of birth. Clinical manifestations of AIDS in children include pneumonia, severe oral candidiasis, interstitial pneumonitis, encephalopathy, wasting, generalised lymphadenopathy, hepatosplenomegaly, diarrhoea, growth retardation and bacterial sepsis.

Children with perinatally acquired HIV infection, if they remain untreated, show a very bad prognosis. The progression of the infection is very fast in the first year of life and is believed to be associated with a higher viral load of HIV-1 in the blood. The replication of the virus in children shows a different pattern than that seen in adults. Viral RNA load is usually low at birth, but the level of the virus then increases rapidly within the first 2 months of life, followed by a slow decline at the age of 2 years.

Progression of infection from acute HIV infection to AIDS usually occurs at a median of 11 years after the infection. Earlier, many of the patients used to die within 1–2 years following diagnosis of AIDS. However, now with the introduction of highly active antiretroviral therapy (HAART) and better management of opportunistic infections, the death rates from AIDS have begun to decline significantly. In India, the mean duration of survival of a patient after diagnosis of HIV is 7 years 8 months.

Laboratory Diagnosis

Laboratory diagnosis is useful (1) to confirm the diagnosis of AIDS, (2) to detect the individuals with HIV infection, (3) to identify carriers who may transmit infections to others, (4) to screen blood donors and high-risk groups, (5) in mental screening of mothers and (6) to perform seroepidemiological studies in the community.

It is based on cell culture for virus isolation, serology for detection of virus antigen and antibodies, and detection of virus DNA.

Specimens

These include serum and plasma for HIV serology and lymphocytes for isolation of HIV.

Cell Culture

Virus isolation is carried out only in containment laboratories by **Co-cultivation**. Co-cultivation is the most sensitive method for the isolation of the virus. It is performed by co-cultivating potentially infected and uninfected mononuclear cells to facilitate the replication of HIV. The viral growth in culture fluid is demonstrated by the presence of HIV p24 antigen and HIV reverse transcriptase. The test becomes positive after 7–14 days of culture or may require a longer period of 28 days. The virus titre in asymptomatic infection is low, hence may not be positive for virus by culture. Viral culture is useful for detecting high virus titre, which is found early in infection before the presence of HIV antibodies.

Serology

Serodiagnosis includes the demonstration of viral antigens and viral antibodies.

Detection of Viral Antigen

The antibody sandwich ELISA using specific monoclonal antibodies to HIV p24 is used to detect the viral capsid core antigen (p24 antigen) in blood. The p24 antigen appears much earlier than HIV antibodies during acute HIV infection, hence can be detected during the window period, during which antibodies are not demonstrated in the serum. The p24 antigen usually appears 16 days after infection. The p24 antigen often becomes undetectable after HIV antibodies develop in the blood. This is due to the formation of immune complexes by p24 with the antibodies.

Seroconversion is the period during which the body starts producing detectable levels of HIV antibodies. This usually occurs several weeks after initially contracting the virus. The window period is different from seroconversion and refers to the time after infection and before seroconversion, during which markers of infection (p24 antigen and antibodies) are still absent or too scarce to be detectable. Therefore, IgM antibodies are detectable during seroconversion.

Detection of Specific Antibodies to HIV

Detection of specific antibodies to HIV in serum is the most commonly used method for serodiagnosis of patients with HIV and AIDS. A detectable level of antibodies is demonstrated in most individuals within 6–12 weeks after infection and in all the individuals within 6 months of infection. The time interval before an antibody appears in the serum is known as the *window period*, and it may vary from 3 to 4 weeks. The patient's serum tested during this period is negative for serum antibodies but positive for viral antigens.

The antibody-based serological testing in HIV is of two types: (1) screening tests and (2) supplementary or confirmatory tests.

1. **Screening tests**: These are otherwise known as ERS, an acronym for enzyme-linked immunosorbent assay (ELISA), rapid test and simple test. These tests are usually highly sensitive and are used for initial screening of the serum samples for the presence of HIV antibodies. They are simple to perform, highly sensitive and can be automated. Nevertheless, all the tests positive by ERS need to be confirmed before being declared positive.

ELISA is the most frequently used test for detecting both HIV-1- and HIV-2-specific antibodies in the serum. This test is highly sensitive and specific, and commercial ELISA kits are available, which detect both HIV-1 and HIV-2 antibodies in the serum. ELISA can also be used for the demonstration of antibodies in the saliva. This is very useful for testing injectable drug users from whom it may be difficult to collect blood due to collapsed blood vessels. ELISA in HIV serology is of four types: first generation, second generation, third generation and fourth generation, depending on the nature of the antigen used and detecting both antibodies and antigen in the serum (Table 61.6).

Rapid Tests

Rapid tests include dot blot assay, latex agglutination, gelatin agglutination, casette ELISA, immunochromatographic tests (ICTs), dipstick tests and HIV spot tests. These tests are simple

Table 61.6 ELISA in HIV

Types of ELISA	Source of antigen	Antigen/Antibody detection
First-generation ELISA	Cultured virus lysate	Detection of antibodies
Second-generation ELISA	Recombinant antigen	Detection of antibodies
Third-generation ELISA	Synthetic peptides	Detection of antibodies
Fourth-generation ELISA	A mixture of synthetic peptides and recombinant glycopeptides	Detection of both antigens and antibodies

tests that can be performed in any laboratory without requiring any expensive instrument or skilled workforce. Moreover, test results can be read rapidly within 30 min of receiving the specimens, such as saliva, urine, blood from a finger prick and dried blood on filter paper.

Supplementary or Confirmatory Tests

These tests are used as confirmatory tests to detect HIV antibodies. These tests are designed for a higher specificity than the screening tests, hence are used as the test of choice to verify the results of screening tests.

Western blot is the most common confirmatory test used in HIV serology. In this test, HIV viral antigens are separated as gp160, gp120, p66, p55, p51, gp41, p31, p24, p17 and p15 depending on their electrophoretic mobility by polyacrylamide gel electrophoresis. These antigens are then blotted onto strips of nitrocellulose paper. These strips are treated with test serum. Antibodies to these HIV proteins, if present in test serum, combine with different fragments of HIV and then react with enzyme-conjugated anti-human globulin. These strips are washed, followed by the addition of a suitable substrate, which produces coloured bands. The position of the coloured band on the strip indicates the antigen with which the antibody has reacted. The demonstration of multiple bands indicates a positive test.

Western blot is considered positive if it shows bands against at least two of the three viral proteins, namely, p24, gp41 and gp120 or gp160. The test is also considered positive if multiple bands are seen with multiple proteins, which are encoded by three genes (*gag*, *pol* and *env*). This represents p24 of *gag* gene core protein, p31 of *pol* gene reverse transcriptase, gp41, gp120 or gp160 of *env* gene surface antigens. The development of bands at one site, as with p24 or gp120, is considered equivocal and may happen in early infection and also may be non-specific. The absence of any band indicates a negative test.

It was an earlier practice to confirm the positive result of anyone screening test by Western blot, but since the test is cumbersome, expensive and not readily available, other strategies are available to confirm the cases. So the practice followed now is to perform either two different types of ELISA or an ELISA with any of the rapid tests. If serum is positive for HIV antibodies by both these tests, the serum is then considered positive for HIV. In case of doubt, serum samples are re-tested after a period of 1 or 2 months.

In principle, other confirmatory tests, such as line immunoassay, are similar to Western blot; however, it differs from Western blot by having artificial HIV antigens used on the strips than using viral lysis protein separated by polyacrylamide gel electrophoresis. Indirect fluorescent antibody (IFA) is another test used as a confirmatory test.

Applications of Serological Tests

Serological tests are helpful in the following circumstances:

Screening

Screening is the systematic checking (voluntary or mandatory) for the presence or absence of HIV infection among the entire population or defined target population. Screening of a target population is particularly useful as iatrogenic transfer of

HIV is an important mode of spread of the infection to unsuspecting recipients; thus, it should be mandatory that blood, blood products, semen, cells, tissues and organs be screened. Antenatal screening is useful as infection could be transmitted from mother to baby before, during or after birth. Some countries have procedures requiring the screening of incoming foreign visitors.

Seuroepidemiology

Serological tests involving antibody surveys are most useful in identifying the source and geographical spread of HIV infection among the population in a defined area.

Diagnosis

Serology is the most reliable and useful method of diagnosis as it is most of the time positive in persons with clinical features of AIDS. It may, however, show negative during acute illness and sometimes in the later stage of the disease where the immune system is non-reactive. Serological tests are also helpful in checking the occurrence of infection followed by exposures, such as sexual contact, blood transfusion or needle prick injury. Two-time serology tests, one after 2 months and the other after 6 months, would be sufficient for checking the infection. After exposure, if serology is negative, then the occurrence of infection could be ruled out.

Molecular Detection

The molecular methods include reverse transcription-polymerase chain reaction (RT-PCR), nucleic acid-based amplification (NASBA), transcription-mediated amplification (TMA) and branched-chain DNA (bDNA). These methods are useful for quantitatively estimating viral load in the plasma. In a positive test, a five- or threefold change in the viral load indicates reliably a significant change in children younger than 2 years or those older than 2 years, respectively. These tests should not be used until the diagnosis of HIV has been confirmed with a non-quantitative serodiagnostic method because these tests may falsely show low viral loads in individuals who are HIV negative.

HIV DNA PCR is a sensitive method for detecting HIV provirus present inside mononuclear cells by employing oligonucleotide directed at highly conserved regions of the virus genome. The advantage of this test is that it can be used to diagnose HIV within 24 h of infection, and it has a sensitivity and specificity of 95 and 97%, respectively. The RT-PCR and NASBA for plasma are sensitive methods for detecting HIV-1 subtype B viruses, whereas the bDNA method is sensitive for the detection of other HIV subtypes.

Few other test formats, like the FDA-approved rapid self-test (OraQuick) used for home testing of HIV, are available. For this test, one must swab the gums to collect an oral fluid sample and use the materials in the kit to test the oral fluid sample. The results are obtained within 20 min. Similarly, mail-in self-testing services are available through online merchants, where kits provide the tools needed to safely prick one's finger and collect a very small blood sample on a card. The sample is then mailed to a laboratory for testing. When the testing is completed, a health care provider is contacted with the results. However, these facilities are CDC approved and in use mostly in the USA.

Specific tests useful for laboratory diagnosis of HIV infection and AIDS are summarised in Table 61.7.

Table 61.7 Specific tests for laboratory diagnosis of HIV infection and AIDS

State of infection	Antibody		Antigen p24	Viral genome	Virus isolation
	ELISA	Western blot			
Window period	–	–	+	+	++
Acute infection	+	+ (Partial p24 and/or gp120)	+	+	±
Asymptomatic infection	+	+ (Full pattern)	–	+	+
ARC and AIDS	+	+ (Absence of p24 antibody)	+	+	+

The reasons for HIV screening are not limited to epidemiological purposes only to find the prevalence of the disease in particular geographical locations. It extends to safety concerns involving transfusion and transplantation procedures to diagnose infection in asymptomatic high-risk individuals and to prevent disease transmission, especially from mother to child. HIV screening is also done to initiate PEP prophylaxis at the earliest to limit transmission.

Laboratory Monitoring Status of HIV Infection

The laboratory monitoring of the status of HIV infection can be carried out by analysis of (1) CD4+ T-cell count, (2) measurement of HIV RNA and (3) measurement of B2 microglobulin and neopterin.

CD4+ T-cell Count

This is an important indicator for monitoring HIV infection. This count reflects the immunological competence of the patients with HIV or AIDS. A rapid decrease in the CD4 count in adults and infants is a poor prognostic sign and requires the initiation or alteration of antiviral therapy. The absolute number of CD4 lymphocytes and the ratio of helper T-inducer lymphocytes (CD4:CD8 ratio) are very low in HIV-infected people. The CD4:CD8 T-cell ratio is reversed to 0.5:1 from a normal level of 2:1. If the CD4 count is below 500 cells/μL, it indicates progression of the disease and hence requires specific therapy against HIV. When the count falls below 200 cells/μL, it indicates a very poor prognosis and shows the increased risk for serious infection of the patient, particularly opportunistic infections.

Measurement of HIV RNA

HIV RNA level in serum is an important predictive marker of disease progression and is used as a prognostic marker to monitor the effectiveness of anti-HIV therapies. The test is also useful for the early diagnosis of HIV infection in infants born to infected mothers.

Measurement of B2 Microglobulin and Neopterin

B2 microglobulin and neopterin can be demonstrated in the serum or urine of HIV-infected people. The titre is low in asymptomatic HIV infection but is elevated with the progression of HIV disease.

Other Tests

Tuberculin test or other tests for cell-mediated immunity is negative, indicating a depressed cell-mediated immunity. Both IgE and IgA serum levels are elevated in the blood. Haematological parameters are altered in patients with HIV. There is a leukopenia with a lymphocyte count of fewer than 400 cells/μL and thrombocytopenia in patients with AIDS. Hyperimmunoglobulinaemia is associated with the progression of the disease. Hypoimmunoglobulinaemia is observed in end-stage disease and is associated with poor prognosis. Thrombocytopenia is a common finding in patients with HIV. Anaemia occurs in as many as 25% of cases at diagnosis and occurs in 80% of cases after some time. Neutropenia is observed in 10% of early asymptomatic HIV infections and 50% of patients with AIDS.

Strategies for HIV Testing in India

The following are the three different strategies followed for HIV testing in India as per the National AIDS Control Organisation (NACO), Ministry of Health and Family Welfare, Govt of India guidelines:

Strategy 1

In this strategy, test serum is subjected once to ERS test, and if reactive, the sample is considered as HIV infected, and if non-reactive, the serum is considered to be free of HIV. For this purpose, highly sensitive and reliable diagnostic kits are usually used. This strategy is used to screen blood, organ and tissue before transfusion and transplantation. The unit of blood that tests reactive (positive) is discarded.

Strategy 2

In this strategy, a serum sample is considered negative if the first ERS test report reports it so. However, if reactive, it is re-tested with a second ERS test based on different antigen preparations and different test principles. If also found reactive with the second ERS test, the serum specimen is reported as positive, and otherwise, it is considered negative. This strategy is used for HIV surveillance.

In this strategy, when the purpose of testing is for diagnosis in symptomatic patients, the algorithm slightly differs. If the result is also reactive with the second test kit, the specimen is considered to be positive for HIV in asymptomatic AIDS cases. If a specimen is reactive with the first test kit and non-reactive with the second test kit, the specimen is subjected to a third tiebreaker test. If the third test is reactive, the specimen is reported as indeterminate and follow-up testing is undertaken after 2–4 weeks. If the third tiebreaker test is non-reactive, the specimen is reported negative.

Strategy 3

It is similar to strategy two but with an added confirmation by a third ERS test. The third test should be based on different antigen preparations or test principles. A serum showing a positive result on all three ERS tests is reported positive. The test is considered equivocal if the serum is negative in the third ERS. In such cases, serum specimens are re-tested on collection after 3 weeks. If this specimen also shows an equivocal result, the person is considered to be negative for HIV antibodies. This strategy is used for the diagnosis of HIV infection in asymptomatic persons. The tests used in the second and third strategies are of higher specificity to rule out any false positivity.

Interpretation of various laboratory tests used to diagnose HIV infections is summarised in Table 61.8.

Treatment

Antiretroviral treatment is the mainstay in HIV treatment. The goals of antiretroviral therapy are to inhibit replication of HIV and to reduce morbidity and death.

Anti-HIV Drugs

Antiretroviral agents against HIV (Table 61.9) can be broadly classified as follows:

Nucleoside Analogue Reverse Transcriptase Inhibitors (NRTIs)

Azidothymidine (AZT), didanosine (DDI) and other nucleoside analogues inhibit the enzyme reverse transcriptase and alter their incorporation into DNA to cause chain termination. These agents prevent the spread of the virus to uninfected cells. AZT is recommended to treat

Table 61.8 Interpretation of laboratory tests in diagnosis of HIV

Test	Purpose
Serology	
Antibody demonstration	
ELISA	Screening test
Rapid tests: Dot blot assay, latex agglutination, HIV spot and comb test	Screening test
Western blot	Confirmation test
Indirect immunofluorescence test	Confirmation test
Antigen detection	
p24 antigen	Early marker of infection
Molecular diagnosis	
Branched-chain DNA	Detection of virus in blood
Viral RNA RT-PCR	Detection of virus in blood
Isolation of virus	Test not readily available
CD4: CD8 T-cell ratio	Correlates of HIV disease

Table 61.9 Antiretroviral drugs against HIV

Nucleoside reverse transcriptase inhibitors
Zidovudine (AZT)
Didanosine (ddI)
Zalcitabine (ddc)
Lamivudine (3TC)
Stavudine (d4T)
Abacavir (1592)
Non-nucleoside reverse transcriptase inhibitors
Nevirapine (NVP)
Delaviridine
Efavirenz
Protease inhibitors
Ritonavir
Indinavir
Saquinavir
Nelfinavir
Amprenavir
Nucleotide reverse transcriptase inhibitor
Tenofovir
Fusion transcriptase inhibitor
Enfuvirtide

asymptomatic or mildly symptomatic people with a CD4 count of fewer than 500 cells/mm^3. This is also recommended for the treatment of pregnant women to reduce the possibility of transmission of the virus to the foetus. The toxicity associated with high doses of AZT and the emergence of resistance to AZT is the main disadvantage of monotherapy with AZT. Zidovudine is also used effectively to reduce the significant transmission of HIV from mother to infant. The treatment decreases vertical transmission at all levels of maternal viral load.

Non-nucleoside Reverse Transcriptase Inhibitors (NNRTIs)

NNRTIs, such as nevirapine, delavirdine and efavirenz, inhibit the enzyme by blocking the morphogenesis of the virion by inhibiting the cleavage of the Gag and Gag core polyproteins. This, in turn, prevents activation of the virion.

Protease Inhibitors

Protease inhibitors, such as ritonavir, indinavir, saquinavir, nelfinavir and amprenavir, prevent the maturation of virus particles during the late stage of virus replication. Monotherapy with antiretroviral therapy failed to produce significant clinical benefits, including patient survival. The failure is partly due to the development of drug-resistant variants of HIV because resistance develops rapidly during monotherapy, and cross-resistance between related drugs is also being increasingly reported.

Therapy with a combination of antiretroviral drugs referred to as HAART effectively inhibits HIV replication. HAART, a strategy similar to the treatment of TB, has improved the efficacy of the therapy, minimised toxicity following therapy and delayed emergence of drug resistance. It is usually recommended to initiate double and triple-drug therapy with two NRTIs or two NRTIs + an NNRTI or a protease delavirdine drug therapy reduces morbidity due to the disease, and death in many patients due to AIDS is delayed.

In HAART, each class targets a different step in the viral life cycle as the virus infects a CD4 T lymphocyte or other target cells. Besides the classes mentioned above of drugs, Integrase inhibitors (INSTIs), Fusion inhibitors (FIs), Chemokine receptor antagonists (CCR5 antagonists) and Entry inhibitors (CD4-directed post-attachment inhibitors) have also been included in

HAART. Adherence to the HAART regimen is very vital in this treatment because any irregularity in following the prescribed regimen can lead to resistance to HIV drugs. NACO in India has made the drugs available at the treatment (ART) centres to ensure this. The major goals for HAART are to reduce HIV infection-related morbidity, improve quality of life, restore and preserve immunologic function, suppress viral load (plasma HIV RNA) and prevent HIV transmission. Suppression of viraemia, which is ensured through HAART, reduces cardiovascular, renal and hepatic events, which are in turn due to ongoing inflammation and immune activation from uncontrolled viraemia. Improved immunity may also reduce the risk for both AIDS-related and non-AIDS-associated malignancy. The guidelines recommend that therapy be initiated in all persons living with HIV irrespective of CD4 count and viral load. The course of untreated infection is depicted in the Fig. 61.5.

Epidemiology

HIV infection is an epidemic throughout the world.

Geographical Distribution

HIV-1 is the most common cause of HIV infection in the Americas, Europe, Africa and Asia. HIV-1 subtypes show differences in their geographical distribution. Subtype A predominates in West Africa; subtype B in the United States, Europe and Australia and subtype C in India, China and Southern Africa. HIV-2 is the most common cause of HIV epidemics in West Africa. This virus is also found in European countries. Since the first recognition of AIDS in the United States in 1981, AIDS has become a worldwide disease, affecting a large number of populations.

India

The first incidence of HIV infection was documented among sex workers in Chennai, Madurai and Vellore of Tamil Nadu in 1986–87. The first case of AIDS was detected in the year 1986 in Mumbai. Since then, HIV infection has been documented in each part of India. The National AIDS Control Organisation (NACO), India, has estimated that by the year 2008, nearly 2.4 million people were living with HIV, next only to South Africa. The national HIV prevalence is varied from 0.4% to 1.3%.

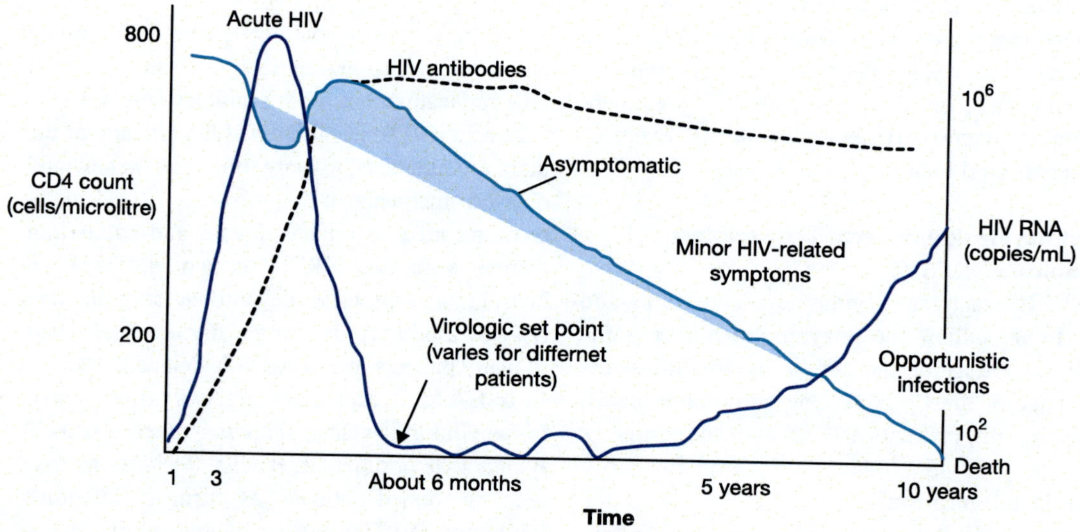

Fig. 61.5 Course of untreated HIV infection

The distribution of the HIV epidemic in India is varied and is based on the prevalence of HIV infection in the high- and low-risk groups. Different states in India have been classified as high-, medium- and low-prevalence areas. Tamil Nadu, Andhra Pradesh, Karnataka, Maharashtra and Goa are classified as high-prevalence states with HIV prevalence over 1% even in a low-risk population.

Both HIV-1 and HIV-2 have been shown to occur in India. Various studies have shown a high prevalence of subtype C in India. Nearly 78.4% of strains in North India and 95% of strains in Kolkata and South India are HIV 1C subtype.

Reservoir, Source and Transmission of Infection

HIV is primarily a human infection. Humans infected with HIV and AIDS are the reservoir of infection. The high titre of HIV is found in the infected people's blood, semen and vaginal secretions; hence, these are important sources of infection. The virus is also present in the breast milk of an infected mother.

Transmission

HIV infection occurs either by transferring HIV-infected cells or free HIV not associated with cells. HIV transmission occurs in the following ways (Fig. 61.6):

1. **Sexual transmission**: HIV is transmitted primarily through sexual contact and constitutes more than 70% of the HIV transmission. Sexual transmission is more common in heterosexual women and men than in homosexual men worldwide. Varied sexual behaviours, such as (1) more sexual partners, (2) sex with commercial sex workers, homosexuals and (3) receptive anal sex, have been reported to

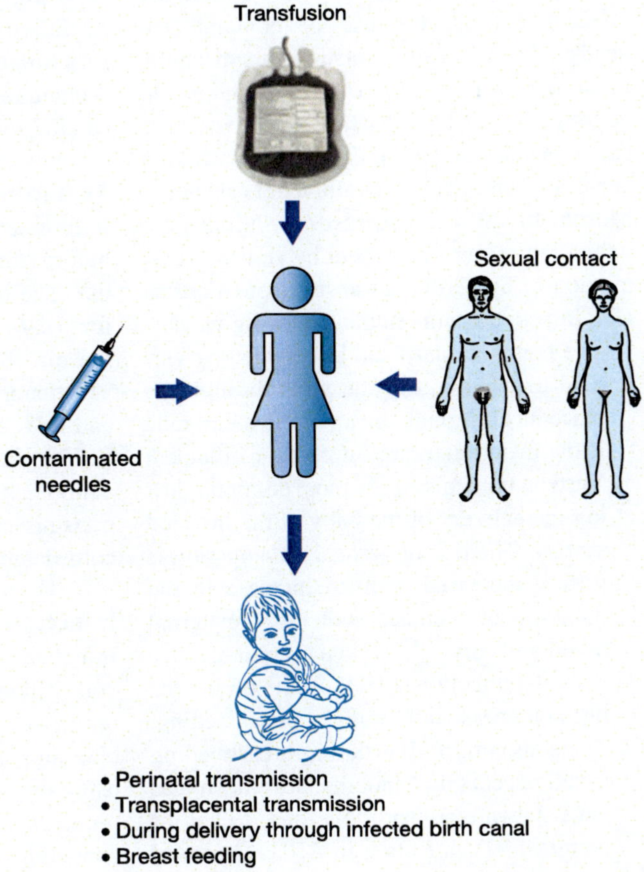

Fig. 61.6 Transmission of HIV infection

be increasingly associated with HIV infection. The presence of other sexually transmitted disease, such as gonorrhoea, syphilis or herpes simplex virus type 2 infection increases the risk of sexual HIV transmission by more than 100 times. This is due to pre-existing inflammation and ulcers caused by these diseases, which facilitate the transfer of HIV through the mucosa.

2. **Transmission by blood transfusion**: HIV is also transmitted by transfusion of infectious blood or blood products, such as serum, plasma and cells from HIV-positive individuals. It can also be transferred by the organs donated from HIV-positive individuals. HIV among health workers has also been documented following a needle stick injury with contaminated blood.
3. **Parenteral transmission**: Parenteral transmission occurs largely among IV drug users. Injection users of illicit drugs are commonly infected through the use of contaminated needles. IV drug users constitute a substantial proportion of new cases of HIV in the north-eastern states of India Manipur. This is due to the transmission of HIV through sharing needles when injecting equipment is shared.
4. **Mother-to-child transmission**: Mother-to-infant transmission can occur by vertical transmission or by perinatal transmission. Vertical HIV infection occurs in the following ways:
   1. The foetus *in utero* can be infected by vertical transmission of the virus through the placenta or through the amniotic membrane if the membranes are infected or inflamed. Vertical transmission is most common during the delivery of the baby of the infected mother. The risk of vertical transmission is greatly increased with an increase in the duration of contact with the maternal blood and cervical, vaginal secretions.
   2. Perinatal infection in HIV occurs during the birth process or through breastfeeding. Transmission of HIV during breastfeeding usually occurs by 6 months. Several studies have shown that nearly one-third to half of perinatal HIV infection in Africa is caused by breastfeeding.

HIV is not transmitted by touching, hugging, kissing, casual contact, coughing and sneezing. It is also not transmitted by the bite of insects or through water, food and utensils or by swimming in swimming pools.

Prevention and Control

These include health education, screening of blood and blood products, infection control and post-exposure prophylaxis.

Health Education

Health education plays a key and important role in preventing AIDS in the absence of a suitable vaccine. Health education is aimed at behavioural changes and maintenance of a lifestyle that minimises or eliminates the risk of transmission. Health education includes (1) safe sexual practice by using a condom, which prevents transmission of the virusoid, (2) not sharing unsterile needles or syringes and (3) information to HIV-positive women regarding the risk of vertical transmission of HIV to infants.

Screening of Blood and Blood Products

It is essential to screen potential blood donors before donating blood or blood products before use. The infected persons who are tested positive for HIV should refrain from donating blood, plasma, body organs, other tissues or sperm. The donors' blood should be screened for HIV-1 and HIV-2 antibodies by screening tests, such as ELISA. Regular screening of blood for HIV antibodies before transfusion has reduced the transmission of HIV by contaminated blood considerably.

Infection Control

Infection control methods include the use of universal blood and body fluids precautions. These universal precautions include wearing protective clothing, such as gloves, masks and gowns, and using other barriers to prevent exposure to blood products. These also include disinfection of contaminated surfaces with 10% household bleach, 70% ethanol or isopropanol, 2%

glutaraldehyde, 4% formaldehyde or 6% hydrogen peroxide. Washing clothes in hot water with adequate detergents are effective to kill HIV.

Post-Exposure Prophylaxis (PEP)

"Exposure" that places healthcare personnel (HCP) at risk of acquiring blood-borne infections could be defined and categorised as (1) percutaneous injury (e.g. needle stick injury), (2) contact with non-intact skin, (3) contact with intact skin (the duration of the contact is prolonged with blood or other potentially infectious body fluids) and (4) contact with the mucous membrane of eye or mouth. Despite the risk of acquiring HIV infection by these routes being less than 1%, PEP is recommended within 72 h of exposure, based on the exposure category of HCP and the HIV status code of the patient as per NACO guidelines (Fig. 61.7).

Vaccine Development

A safe and effective vaccine is yet to be available against HIV. An ideal HIV vaccine would: (1) prevent the acquisition of the virus by an adult during sexual intercourse, (2) prevent transmission of the virus to infants of HIV-positive mothers and (3) also block the progression of the disease.

There are many scientific obstacles to the development of the AIDS vaccine. This is mainly due to various problems unique to HIV, such as (1) antigenic diversity and hypervariability of the virus, as the antigenicity of the virus changes readily through mutation, (2) transmission of the disease by mucosal route—initial protection should require the production of secretory antibodies to prevent sexual transmission and acquisition of the virus, (3) transmission of the virus by infected cells, (4) latency of the virus—the virus can spread through syncytium and remain latent, thereby remaining protected from antibodies, (5) integration of the virus genome into the host cell chromosome and (6) rapid emergence of viral escape mutants in the host—the high error rate of the viral reverse transcriptase leads to continuous mutations in the HIV genome.

Most of the **HIV vaccines** have been evaluated by using gp120 or its precursor gp160 as an immunogen. The gene for this protein has been cloned, expressed in different eukaryotic systems and developed as a subunit vaccine. In India, the National AIDS Research Institute (NARI), Pune, is evaluating DNA+ pox virus (canary pox, fowl pox) and modified vaccinia Ankara (MVA), a highly attenuated host-range-restricted strain of vaccinia virus. The vaccine underwent a phase I clinical trial with adeno-associated virus (AAV) AIDS vaccine in 2005. The second AIDS vaccine clinical evaluation is being set up at Tuberculosis Research Centre (TRC), Chennai. Topical antiviral vaginal gels, such as Tenofovir, are being also evaluated.

Case Study

A 25-year-old college student, resident of an area endemic for HIV in Manipur, was admitted with several complaints. He had serious bouts of diarrhoea with a history of marked weight loss during the past year. He was found to be a drug addict. His stool samples were positive for oocysts of *Cryptosporidium parvum*. The serum was tested positive for HIV antibodies by ELISA. The person was diagnosed to be suffering from AIDS.

- What other laboratory tests will you perform to confirm the diagnosis?
- What are the other opportunistic infections associated with AIDS?
- How did the boy acquire the infection?
- What do you mean by the term ART. Describe the recent developments in the field?

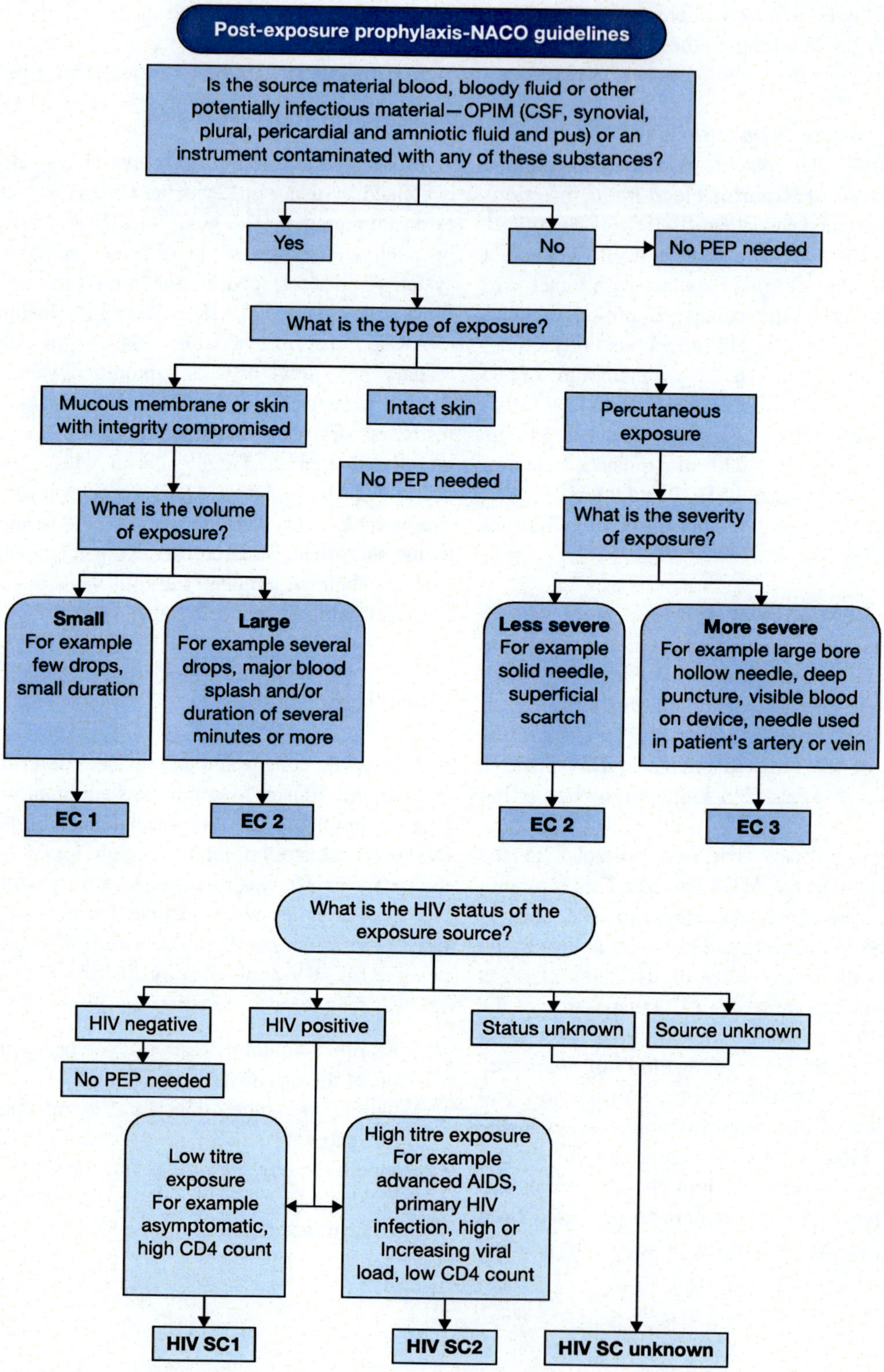

Fig. 61.7 HIVPEP leaflet from NACO. (http://www.naco.gov.in/NACO/National_AIDS_Control_Program/PEP_full/)

Slow Virus and Prion Diseases

62

Slow virus diseases are a large group of related neurodegenerative conditions which affect both humans and animals. The diseases are caused by a heterogeneous group of agents consisting of conventional viruses and unconventional agents called *prions*. The term 'slow' refers to the disease process and not to the replication of the viruses that cause the slow disease.

The slow virus diseases are characterised by (1) chronic infection, (2) a very long incubation period in years, (3) a slow and progressing course of infection and (4) being usually fatal. The concept of slow infection was originally suggested for the scrapie, visna and maedi, the slowly progressing infection of the sheep, by a veterinary pathologist, Sigurdsson, in the year 1854.

The slow viruses' diseases, depending on agents, can be classified broadly into two groups: (1) prion diseases and (2) Slow virus diseases caused by conventional viruses. A list of slow infectious diseases caused by prions and conventional viruses is presented in Table 62.1.

Slow Virus Diseases Caused by Prions

The agents causing slow diseases were earlier called slow viruses, infectious protein, infectious amyloids or crystal protein. These agents are now being named prions, and this term has gained wide acceptance, replacing all the previously used terms. The prions are unconventional agents that are not viruses. They are small protein-containing infectious particles with no detectable nucleic acid. The protein is yet to be characterised fully and appears to be a modified host protein.

The diseases caused by prions are a large group of related neurodegenerative conditions which affect both humans and animals. These diseases belong to a family of diseases known as transmissible spongiform encephalopathies (TSEs).

Human TSEs caused by prions include Kuru, Creutzfeldt–Jakob disease (CJD), variant CJD (cud), Gerstmann–Straussler–Scheinker (GSS) syndrome, fatal familial insomnia (FFI) and sporadic fatal insomnia. Animal TSEs caused by prions include (1) scrapie and visna (diseases of sheep), (2) bovine spongiform encephalopathy (BSE; mad cow disease), (3) transmissible mink encephalopathy and (4) chronic wasting disease of deer, mule and elk.

Kuru was the first human disease known to be caused by a slow virus or prion (1959). This condition was found among 'Fore' people, a tribe living in the remote highlands of New Guinea. This virus was linked to ritualistic cannibalism prevalent among these tribal people. Hallow (1959) suggested that kuru could be a possible human form of scrapie, a slow viral disease seen in sheep. Gajdusek and his colleagues (1966) first demonstrated that kuru was transferable to chimpanzees after a long incubation period, for which they were awarded the Nobel Prize. Since then, many TSEs linked to

Table 62.1 Slow diseases caused by prions and conventional viruses

Agent	Slow diseases
Humans	
Prion	Kuru
	Creutzfeldt–Jakob disease (CJD)
	Variant CJD
	Gerstmann–Straussler–Scheinker syndrome (GSS)
	Fatal familial insomnia and sporadic fatal insomnia
Virus	Sub-acute sclerosing panencephalitis (SSPE)
	Progressive multifocal leukoencephalopathy (PML)
	Acquired immunodeficiency syndrome (AIDS)
Animals	
Prion	Scrapie
	Bovine spongiform encephalopathy
	Feline spongiform encephalopathy
	Transmissible mink encephalopathy
	Chronic wasting disease of deer, mule and elk
Virus	Visna

slow viruses have been described both in humans and animals.

Prions

Prions are small protein-containing infectious particles with no detectable nucleic acid. They were suspected to be viruses but otherwise do not conform to the standard definition of viruses. They differ from the viruses in their many properties (Table 62.2). The prions like viruses are filterable. They apparently lack any virion structure or genome. Unlike viruses, they are unusually resistant to inactivation by heat, disinfectants and radiation. They do not elicit any specific immune response in the infected host.

Morphology

The prion does not contain any nucleic acid. It contains an aggregate of hydrophobic glycoprotein resistant to protease. The protein present in humans and other animals is called cellular prion protein. It is similar to glycoprotein in scrapie, known as scrapie prion protein, which has a molecular weight ranging from 27,000 to 30,000 Da. However, it differs from scrapie prion protein by being sensitive to protease (scrapie protein is protease resistant) and being present

Table 62.2 Distinguishing features of prions and viruses

Feature	Prion	Virus
Presence of nucleic acid	No	Yes
Defined morphology	No	Yes
Disinfection by		
Formaldehyde	No	Yes
Heat (80 °C)	No	Most are affected
Ionising and UV radiation	No	Yes
Disease		
Incubation period	Long	Depends on virus
Immune response	No	Yes
Cytopathologic effect	No	Yes
Inflammatory response	No	Yes

in the plasma membrane (scrapie prion protein is present in cytoplasmic vesicles).

Prion Replication

In experimental animal studies carried out in chimpanzees, Prusiner demonstrated conclusively that the prion, an aberrant protein, could cause disease, for which he was awarded Nobel Prize in 1997. Prusiner hypothesis suggests that prions do not contain any nucleic acid. The normal prion protein known as *prion protein cellular* or PrP^C has a significant amount of helical configuration. In its alpha-helical configuration, PrP^C is usually sensitive to degradation by the activity of enzyme protease. The disease occurs when the PrP^C is reconfigured into the beta-sheet configuration known as *prion protein scrapie* or PrP^{SC}, which is resistant to degradation by protease.

These abnormal proteins aggregate into filaments that disrupt neuron functions and cause the death of cells. The normal alpha-helical and abnormal beta-pleated sheet forms have the same amino acid sequences but differ in their configuration.

Other Properties

Prions are highly resistant to heat at 80 °C, ultraviolet radiation and disinfectants, such as formaldehyde. They are, however, inactivated by phenol, ether, sodium hydroxide and hypochlorite.

Prion Isolation

Prions cannot be cultivated in any tissue culture or isolated in experimental animal models.

Pathogenesis of Prion Diseases

Prion diseases are transmitted either orally or transcutaneously. A functional immune system is believed to be essential for replicating prions and their transport to the reticuloendothelial cells. Large amounts of prions are present in the follicular dendritic cells and the sympathetic nerve endings in these cells. The prions reach the CNS from these sites, but how they reach the CNS is still an important issue. It is suggested that the prions may reach the CNS via splanchnic nerves at the level of the thoracic spinal cord and via parasympathetic fibres connecting with the brain. The other possible route is blood circulation.

In experimental animal studies carried out in chimpanzees, Prusiner demonstrated that scrapie protein (PrP^{SC}) binds to the normal cellular prion protein (PrP^C) of the host. PrP^C is found in most tissues of the body but is expressed in higher quantities in the CNS, particularly the neurons, the main site of action of prions. PrP^{SC} is ingested by neurons and phagocytic cells but remains intact without being degraded. This may contribute to vacuolation of the neuron, a key pathological change observed in encephalitis caused by prions. In addition, the accumulation of prions in high concentrations causes damage to the brain tissue. These include the formation of amyloid-containing plaques and fibrils, proliferation and hypertrophy of astrocytes and fusion of neurons and adjacent glial cells.

Pathogenesis and transmission of prion diseases in humans are described in Table 62.3.

Clinical Syndromes

Diseases Caused by Prions in Humans

Prion-mediated diseases in humans are called transmissible spongiform encephalopathies (TSEs). Currently, the most widely accepted hypothesis is the '*protein-only hypothesis*', first suggested by Griffith (1965) and re-established subsequently by Prusiner to indicate that scrapie is related to a proteinaceous infectious particle (PrP).

The conditions are characterised by: (1) long incubation period of several years, (2) course of illness lasting for months to years, (3) a progressive, debilitating neurological syndrome that is invariably fatal, (4) associated with pathological changes typically restricted to the central nervous system (CNS), (5) absence of specific immunological response in the host and (6) the agents are resistant to conventionally used inactivation methods.

Table 62.3 Pathogenesis and transmission of prion-related slow diseases in human

Slow disease	Pathogenesis	Transmission
Kuru	Cannibalism	Ingesting or handling brain tissue
Creutzfeldt–Jakob disease		
Infectious	Contact with or handling prion-containing materials	Cuts in the skin Transplantation of contaminated tissue, such as cornea Use of contaminated medical devices, such as brain electrodes Possibly through ingestion of infected tissue
Hereditary	Mutations in the germ cells	Hereditary predilection
Sporadic	No relationship to any cause	Spontaneous PrP^C conversion to PrP^{SC}
Variant Creutzfeldt–Jakob disease	Infection from BSE	Probably by eating meat or nervous tissue from animals with mad cow disease Blood transfusion
Gerstmann–Straussler–Scheinker (GSS) syndrome	Mutations in the PrP gene	Hereditary/genetic
Fatal familial insomnia	D 178 N mutation in the PrP gene, with M129 polymorphism	Hereditary/genetic
Sporadic fatal insomnia	No relationship to any cause	Spontaneous PrP^C conversion to PrP^{SC}
Progressive multifocal leukoencephalopathy	Immunosuppression/malignancy	Central nervous system demyelination

Kuru

It was the first human disease known to be caused by a slow virus or prion (1959). This condition was found among 'Fore' people, a tribe living in the remote highlands of New Guinea. This virus was linked to ritualistic cannibalism prevalent among these tribal people. Hallow (1959) suggested that kuru could be a possible human form of scrapie, a slow viral disease seen in sheep. Gajdusek and his colleagues (1966) first demonstrated that kuru was transferable to chimpanzees after a long incubation period, for which they were awarded the Nobel Prize. Since then, many TSEs linked to slow viruses have been described both in humans and animals.

Kuru is a fatal disease characterised by progressive cerebellar ataxia and tremors. The condition manifests initially as difficulty in walking, followed by cerebellar tremor, hence the name *kuru*, which means 'trembling in fear'. Eventually, the tremor worsens, followed by progressive cerebellar ataxia and eventual death within a year of onset of symptoms. The clinical course lasts from 3 months to 2 years. Kuru has largely disappeared today because cannibalism has been abolished now among the Fore tribal people.

Creutzfeldt–Jakob Disease

Creutzfeldt–Jakob disease (CJD) is the most common prion disease responsible for nearly 85% of all human prion diseases. The condition was described by Creutzfeldt (1920) and Jakob (1921), after whom the disease is named. It is transmitted by i) cuts in the skin, ii) transplantation of contaminated tissue (e.g. cornea), iii) use of contaminated medical devices (e.g. brain electrodes), and iv) possibly through ingestion of infected tissue.

CJD is a sub-acute progressive encephalopathy characterised by rapidly progressive dementia associated with myoclonic jerks. Memory loss, behavioural changes and confusion are the common clinical manifestations. The condition progresses, and the patient becomes mute and comatose in the terminal stage of the disease. The condition is invariably fatal, and death occurs in about 8 months. An autosomal dominant pattern of inheritance linked to a mutation in the PrP gene is found in nearly 10% of cases of familial CJD.

CJD is rare but is found worldwide, with an incidence of about one case per million populations. Till 2003, a total of 167 definite and probable cases of vCJD have been reported worldwide. Most of the cases of vCJD have been reported in the United Kingdom. The condition is yet to be documented in Asia.

Variant CJD: The vCJD is a new disease affecting mostly adults aged 45 years. The condition was first documented in Britain in 1985. It is BSE zoonoses, probably originating from a double species switch from sheep scrapie to BSE and then from BSE to human variant. The vCJD is believed to be caused by ingestion of BSE-infected beef products contaminated by neural tissues. This is because neural tissues have a much higher concentration of PrP^{SC} compared with any other non-neural tissue.

Gerstmann–Sträussler–Scheinker Syndrome and Fatal Familial Insomnia

These are other rare conditions caused by prions in humans.

Diseases Caused by Prions in Animals

Scrapie

Scrapie is a prototype prion disease of animals. It is a slow virus disease of the sheep, known for centuries. The infection is transmitted vertically from ewe to lamb and less frequently by direct contact. The incubation period is nearly 2 years. The condition manifests as intense irritation and to relieve that the infected animals scratch themselves, against trees and rocks, hence the name *scrapie*. The condition progresses to emaciation and paralysis, leading to the animal's death.

Bovine Spongiform Encephalopathy (BSE)

Also known as mad cow disease. It is a disease of cattle. The condition has an incubation period of 2–8 years. The disease is relentlessly progressive until the animal dies. The cattle feed containing prion-contaminated meat and bone meal, which was used as a protein source, was the source of infection for transmission of BSE to cattle. The causative agent is believed to be originated from either scrapie-affected sheep or cattle with unidentified prion-associated slow virus disease. The mad cow disease, documented in 1986, has been described in cattle in the European countries and the United States. Until 2004, nearly 190,000 confirmed clinical cases of BSE in cattle have been reported worldwide, most of which were from the United Kingdom alone.

Transmissible Mink Encephalopathy

It is a scrapie-like disease of mink. The causative agent is believed to be a strain of scrapie virus, which is transmitted to mink by feeding the animals on scrapie-infected sheep meat.

Chronic Wasting Disease of Deer, Mule and Elk

Chronic wasting disease is a prion disease of deer, mule and elk. It is a progressive wasting disease of these animals in the United States.

Pathogenesis and transmission of prion diseases in animals have been shown in Table 62.4.

Laboratory Diagnosis

The diagnosis of prion diseases is always clinical. It is usually confirmed by the histopathology of the brain tissue showing characteristic histological changes. No serological tests are available, as prions are inert agents and do not elicit any immune response in infected hosts. The prions cannot be detected in the infected tissue by any

Table 62.4 Hosts and pathogenesis of prion-related slow diseases in animals

Slow disease	Host	Pathogenesis
Scrapie	Sheep	Infection in susceptible sheep
Bovine spongiform encephalopathy	Cattle	Infection from contaminated food
Feline spongiform encephalopathy	Cats	Infection from contaminated food
Transmissible mink encephalopathy	Mink	Infection from sheep or cattle in food
Chronic wasting disease	Deer, mule and elk	Not known

method, including electron microscopy, antigen detection and genomic methods, such as DNA probe or polymerase chain reaction (PCR).

Detection of protein 14–3–3 in the CSF by Western blot is a sensitive and specific method in sporadic CJD and vCJD cases. A specific reduction in uric acid level in the CSF has been shown in vCJD but not in sporadic CJD, thus facilitating the differential diagnosis of vCJD.

Treatment

No specific treatment is available for any prion disease.

Epidemiology

Prion-related diseases, although rare, are present worldwide.

Geographical Distribution

Creutzfeldt–Jakob disease is rare but is found worldwide, with an incidence of about one case per million populations. An autosomal dominant pattern of inheritance linked to a mutation in the PrP gene is found in nearly 10% of cases of familial CJD. Till 2003, a total of 167 definite and probable cases of vCJD have been reported worldwide. Most of the cases of vCJD have been reported in the United Kingdom. The condition is yet to be documented in Asia. Creutzfeldt–Jakob disease syndrome and fatal familial insomnia (FFI) in humans are rare but occur in different parts of the world.

Reservoir, Source and Transmission of Infection

The spongiform encephalopathies caused by prions may be infectious, hereditary and sporadic. The CJD, GSS and FFI are inheritable. The CNS tissue in CJD has the highest concentration of prion agents, hence the most important infection source. Unlike the prion disease, in vCJD, lymphoid tissues show the highest concentration of prions. Contaminated blood is another source of infection.

Creutzfeldt–Jakob disease is transmitted by (1) cuts in the skin and transplantation of contaminated tissue (e.g. cornea), (2) use of contaminated medical devices (e.g. brain electrodes), (3) possibly through ingestion of infected tissue and (4) blood transfusion. Brain surgery patients, transplant patients and surgeons also are at risk for CJD and GSS disease. Sporadic CJD is a type of infection that refers to the occurrence of disease in the absence of an infectious or hereditary form of the disease.

Prevention and Control

Prions are resistant not only to protease degradation but also to radiation, heat and other agents that destroy proteins. They are, however, inactivated by phenol, ether, sodium hydroxide and hypochlorite. Hence, the materials from patients with CJD or vCJD must be handled with special care. To prevent these diseases, special disinfection protocols have been developed by the World Health Organisation (WHO). These include autoclaving at 15 lbs for 1 h (instead of 20 min) or treatment with 0.1 M sodium hydroxide and 5% hypochlorite solution.

Slow Virus Diseases Caused by Conventional Viruses

Humans

Sub-acute sclerosing panencephalitis (SSPE) and progressive multifocal leukoencephalopathy (PML) are the slow virus diseases caused by conventional viruses in humans.

Sub-acute Sclerosing Panencephalitis (SSPE)

The sub-acute sclerosing panencephalitis (SSPE) is a degenerating disease of the CNS caused by persistent measles infection. It is persistent infection by a variant measles virus that fails to complete its replication. The disease is characterised by the development of behavioural and intellectual deterioration and seizures after many years (mean incubation period is 10.8 years) of infection by measles. This is a serious and late neurological sequel of measles that affects the CNS.

The condition occurs in about seven in every one million patients. The condition is associated with the presence of an extremely high measles antibody titre in the blood and CSF. Inclusion bodies, which react with antibodies to the measles virus, are present in the affected neurons.

Progressive Multifocal Leukoencephalopathy (PML)

Progressive multifocal leukoencephalopathy (PML) is a fatal demyelinating disease of the white matter of the brain. The condition is usually fatal. It occurs primarily in persons with compromised cell-mediated immunity, such as patients with AIDS and those receiving immunosuppressive drugs and anticancer therapy.

The virus following the infection kills oligodendroglia and causes syncytia in astrocytes. The condition progresses rapidly to blindness, dementia and coma. Laboratory diagnosis of the condition is made by PCR for viral genome in the brain tissue or CSF.

Animals

Visna

Visna is a disease of sheep caused by the visna virus, a lentivirus. It is a single-stranded RNA virus with an RNA-dependent DNA polymerase in the virion. Integration of the DNA provirus into the host cell DNA appears to be responsible for the persistence of the virus within the host. Clinically, the condition manifests after a long incubation period with a prolonged progressive course. It has an insidious onset with paresis, progressing to total paralysis and death. The viruses are present in the infected animal's saliva, blood, and CSF. High levels of serum antibodies are present in the serum, but these are not protective.

Maedi

This is also a disease of sheep caused by the same lentivirus. The difference in the site of infection results in two different clinical entities, Unlike visna, which affects the animal's brain, maedi is manifested as interstitial pneumonia as the virus affects the lungs. The major mode of transmission of the disease in animals is via infected body fluids and colostrum. It affects animals older than 2 years, and once the symptoms set in, life expectancy is reduced to a year. Besides pulmonary disease, it can also affect the mammary glands and joints. Presently, there are no appropriate treatment modalities or vaccines against the disease, thus causing a huge economic loss in the animal population.

Case Study

A 45-year-old adult, resident of the United Kingdom, was admitted with presentations of memory loss, behavioural changes and confusion. Ataxia, aphasia, visual loss and hemiparesis were the other manifestations. Subsequently, the patient became comatose and died within a week of admission. The patient was strongly suspected to be suffering from CJD.

- How will you confirm the diagnosis of the disease in the laboratory?
- How this disease is transmitted to humans?
- What is a variant form of CJD?
- Describe the pathogenesis of prion-mediated diseases.

Oncogenic Viruses 63

Oncogenic viruses can produce tumours in their natural host or experimental animals or malignant transformation of the cell culture. In 1911, Rous first time showed the ability of a virus to cause malignancy in fowl; this virus is called as fowl sarcoma virus. For this discovery, he was awarded the Nobel Prize in 1966. In 1932 and 1933, Shope isolated rabbit fibroma virus and papilloma virus, respectively. He was the first person to demonstrate the implication of viruses in animal malignancies. In 1957, Stewart and Eddy demonstrated the oncogenic capability of the polyoma virus. They injected polyoma virus into newborn mice to produce various neoplasms, and the results of these experiments arose the interest of many scientists to further conduct studies on oncogenic viruses. The Epstein Barr virus causes the Burkitt lymphoma in African children.

At present, it is regarded that oncogenic viruses are responsible for 10–20% of human malignancies. Thus far, approximately 150 animal viruses are known to possess oncogenic potential, which contains both DNA and RNA viruses. Important oncogenic viruses are listed in Table 63.1.

Oncogenic DNA Viruses

Some oncogenic DNA viruses are associated with malignancies in humans and animals. DNA of oncogenic viruses or a part of this DNA is found to be integrated with the host cell genome. Since the viral DNA is either incomplete or defective, there is no possibility of producing an infectious virus; nevertheless, the host cell will undergo malignant changes under the influence of the virus. The cancer cell formed as the result of the virus is analogous to the lysogenised bacterium by a defective phage. Yet the cell is not destroyed in both cases.

Papovaviruses

Human papillomavirus (HPV) 16 and 18 types have a significant association with cervical and uterine cancer. HeLa, a continuous cell line used by cell culture laboratories, was derived decades back from a cervical cancer patient. This cell line was found to have HPV-18 DNA in it.

Polyoma virus that includes BK and JC viruses, usually causes asymptomatic human infections. However, these viruses can induce tumour formation in immunosuppressive individuals.

Simian virus 40 transforms cultured cells, including cell lines derived from human cells.

Poxvirus

Among poxviruses, rabbit fibroma, molluscum contagiosum and Yaba virus are known to induce benign tumours. Experimentally tumours can be produced in animals as well as humans. However,

Table 63.1 Oncogenic viruses of animals and humans

Oncogenic DNA viruses	
Papovavirus	BK and JC viruses
	Polyomavirus
	Papilloma viruses of rabbit, other animals and of human beings
	Simian virus 40
Adenovirus	Many human and non-human types
Poxvirus	Yaba virus
	Shope fibroma
	Molluscum contagiosum
Herpesvirus	Herpes simplex virus types 1 and 2
	Herpes virus pan, papio, ateles and saimiri
	Lucke's frog tumour virus
	Marek's disease virus
	Cytomegalovirus
	Epstein–Barr virus
Hepatitis viruses	Hepatitis B and C viruses
Oncogenic RNA viruses	
Retroviruses	Leukosis-sarcoma virus of various animals
	Human T cell leukaemia viruses
	Avian leukosis viruses
	Human T cell leukaemia viruses
	Murine leukosis viruses

the tumour will regress in a week without any treatment.

Adenovirus

Adenovirus types 12, 19 and 21 may induce sarcomas in suckling mice after experimental inoculation, but they do not have any role in human malignancy.

Herpes Virus

It has been proven that many herpesvirus associations in various types of human and animal cancers.

Marek disease is an infectious disease of chickens caused by a herpes virus. It is contagious neurolymphomatosis and causes paralysis or initiates widespread tumour formation. The disease can be prevented by vaccination (a live avirulent vaccine). Interestingly, it is the only available viral vaccine that can prevent malignancy.

Lucke tumour of frogs is a renal adenocarcinoma of frog regarded to be caused by herpesvirus.

Herpesvirus saimiri virus was first isolated from the squirrel monkey kidney cell line. When experimentally injected into monkeys or rabbits, it causes fatal lymphoma or reticulum cell sarcoma.

Epstein–Barr virus (EBV) or human herpesvirus 4 (HSV-4) is the causative agent of infectious mononucleosis. Epstein–Barr virus is also the first virus known to be associated with human malignancies, such as Burkitt lymphoma, other B-cell lymphoma and nasopharyngeal carcinoma.

Human herpesvirus 8 (HSV-8) is associated with Kaposi sarcoma and also with body cavity lymphoma and Castleman disease. DNA sequences of HSV-8 have been identified in patients with Kaposi sarcoma. This virus, also known as KS-associated herpesvirus, is believed to be important in causing and/or maintaining lesions in Kaposi sarcoma. The lesion in Kaposi sarcoma begins as a red-purple or dusky area that enlarges into a plaque and progresses subsequently into a tumorous growth. This condition is observed most commonly not only in immunosuppressed patients but also occurs in children, although rarely.

Cytomegalovirus infection was found to have an association with Kaposi sarcoma and prostate cancer.

Hepatitis B Virus

The virus usually causes chronic disease and is associated with **hepatocellular carcinoma**. Approximately 20% of HBsAg carriers, nearly 1% of all adult patients infected with HBV and a high percentage of neonates infected with the virus develop hepatocellular carcinoma or cirrhosis. The hepatocellular carcinoma appears to result from persistent cellular regeneration that tends to replace the dead hepatocytes. Also, it is suggested that the integration of HBV DNA with hepatocytes DNA could activate a cellular oncogene, resulting in loss of control of the growth of hepatocytes. However, the HBV genome has no oncogene, which can be responsible directly for causing hepatocellular carcinoma.

Oncogenic RNA Viruses

Retroviruses

Retroviruses are enveloped, positive-stranded, spherical RNA viruses showing a characteristic morphology and unusual replication mode. The presence of an unusual enzyme—the RNA-dependent DNA polymerase or *reverse transcriptase*, giving the virus its name (*retro* meaning reverse)—is the unique feature of the viruses belonging to the family Retroviridae.

The enzyme reverse transcriptase prepares a DNA copy of the retroviral RNA genome—initially RNA–DNA hybrid and subsequently forms double-stranded DNA. The DNA copy of the viral genome is known as **provirus**. The provirus is then integrated into the host cell DNA to become a cellular gene for the rest of the cell's life. This is in contrast to the classical transcription of the genetic information from DNA to RNA and then to proteins. Baltimore and Tenin were the first to demonstrate that the retroviruses encode an RNA-dependent DNA polymerase and replicate through a DNA intermediate.

Properties of various retroviruses are summarised in Table 63.2.

Classification

All oncogenic RNA viruses are classified in the family Retroviridae, but all retroviruses are not oncogenic. The family Retroviridae are classified into three subfamilies, as follows:

Oncovirinae

The oncovirinae or oncoviruses include only the retroviruses that can transform target cells. The viruses depending on their core and capsid, as demonstrated in electron microscopy, are classified further into types A, B, C or D. These include human T-lymphotropic viruses (HTLV-1, HTLV-2, HTLV-3 and HTLV-4) (Table 63.3).

Table 63.2 Characteristics of retroviruses

Retroviruses	Characteristics
Mouse mammary tumour virus	Have eccentric nucleocapsid core in mature virion
Human T-lymphotropic virus (HTLV-1, HTLV-2, HTLV-5), Rous sarcoma virus (chickens)	Have centrally located nucleocapsid core in mature virion
Human immunodeficiency virus (HIV-1, HIV-2), Visna virus (sheep), caprine arthritis/encephalitis virus (goats)	Slow onset of diseases; cause neurological disorders and immunosuppression; have D-type, cylindrical nucleocapsid core
Human foamy virus	Shows vacuolated foamy cytopathology; no clinical disease
Human placental virus	Have retrovirus sequences that are integrated into human genome

Table 63.3 Retroviruses causing infections in humans

Retroviruses	Disease
Human T-lymphotropic virus (HTLV-1, HTLV-2)	Adult T-cell leukaemia/lymphoma
Human immunodeficiency virus (HIV-1, HIV-2)	Acquired immunodeficiency syndrome

Lentivirinae

The lentivirinae or lentiviruses are slow (*lent*: slow) viruses associated with neurological and immunosuppressive diseases in humans and animals. These include human immunodeficiency viruses (HIV-1, HIV-2), sheep's Visna virus and caprine arthritis/encephalitis virus of goat.

Spumavirinae

These consist of spumaviruses that contain non-oncogenic 'foamy viruses' (*spuma*: foam). These viruses are associated with asymptomatic infection in animals but are not associated with human diseases. The idea that a virus could cause cancer was first hypothesised by Peyton Rous (1911) when he transmitted solid tumours (sarcomas) of chicken by transplanting tissue. The belief that retroviruses could be associated with human leukaemia was not in favour till the discovery of the bovine leukaemia virus (BLV) and gibbon ape leukaemia virus (GALV) in the 1970s.

In the 1960s, Tenin was the first to predict that retroviruses could replicate their RNA genome by transcribing it into DNA by the enzyme reverse transcriptase. The results of a study supported this observation reported separately by David Baltimore. The discovery of reverse transcriptase was revolutionary because it contradicted the central dogma of molecular biology that genetic information is passed in one direction from DNA to RNA and then to protein. Baltimore and Tenin were awarded the Nobel Prize for their revolutionary work, which subsequently contributed immensely to the progress and developments in molecular biology.

Minna and Gazdar were the first to detect the first human leukemic retrovirus in 1979. However, it was Poiesz and Gallo in 1980 who conclusively demonstrated the infectivity, antibody response and evidence of provirus from the first human T-cell lymphotropic virus, or HTLV-1, isolated from a patient with a cutaneous T-cell malignancy. Subsequently, many human cancers caused by HTLV-1 were reported from different parts of the world.

Morphology

The retroviruses are enveloped, mostly spherical viruses that measure 80–120 nm in size. The virus's envelope is acquired by budding from the plasma membrane and contains viral glycoprotein.

Virus Genome

The virus consists of two identical copies of the linear, single, positive-stranded RNA genome. The virus also contains the cellular transfer RNAs (tRNAs), and the enzymes reverse transcriptase and integrase. The tRNAs are used as primers for the enzyme reverse transcriptase.

The presence of the enzyme reverse transcriptase—an unusual RNA-dependent DNA polymerase enzyme—is the characteristic feature of retroviruses. The enzyme prepares a copy from RNA to DNA, unlike the classical transcription of genetic information from DNA to RNA. Reverse transcriptase prepares a DNA copy of the retroviral RNA genome which is initially an RNA–DNA hybrid but later forms a double-stranded DNA provirus. Infection with oncogenic retroviruses leads to the integration of the provirus randomly with the host chromosome, and it becomes a cellular gene for the rest of the life of the cell.

The genome of the simple retroviruses consists of three major genes that encode glycoproteins of the virus. These are *gag*, *pol* and *env* genes. The *gag* gene encodes for group-specific antigens,

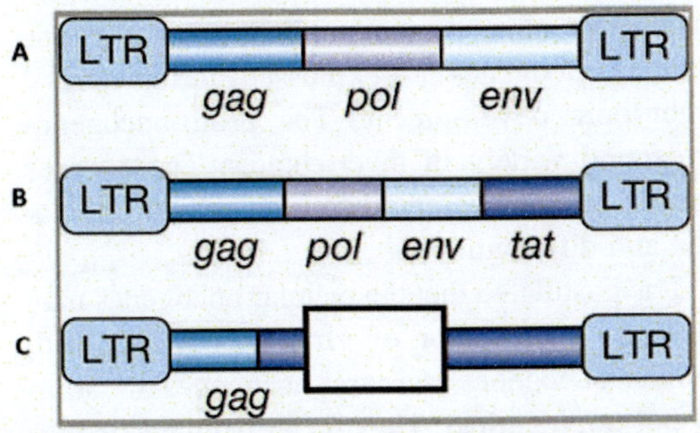

Fig. 63.1 Viral transformation: Different types of retroviruses and their provirus genomic structure. **(a)** Basic retroviral genome (avian leukaemia viruses), **(b)** Trans-regulating retroviruses (HTLV, HIV) **(c)** Acute transforming retrovirus oncogene replacing the basic genome

capsid proteins; *pol* gene for enzymes polymerase, protease and integrase and *env* gene for envelope glycoproteins. Proteolytic cleavage of the polyprotein encoded by the *env* gene leads to the production of the viral glycoproteins (Fig. 63.1).

The complex retroviruses, HTLV and the lentiviruses, including HIV, consist of several accessory genes, such as *tat*, *rev*, *nef*, *vif* and *vpu* for HIV. These genes also encode several regulatory proteins that require more complex transcriptional processing than the simple retroviruses. Various retrovirus genes and their functions are summarised in Table 63.4.

Host Specificity

Retroviruses are mostly found to infect only one host species. Retroviruses can be classified based on their ability to grow in different species' cells. **Amphotropic** retrovirus multiplies in the cells of native and foreign species. **Ecotropic** retroviruses multiply in cells of native host species only, whereas **xenotropic** retroviruses can multiply only in foreign species but not native species.

Virus Antigens

Two types of antigens are present in retroviruses. These are group-specific nucleoprotein antigen in the virion core and type-specific glycoprotein antigen in the envelope.

Other Properties

Retroviruses are heat labile, readily destroyed on exposure to heat at 56 °C for 30 min. The viruses are also readily inactivated on treatment with mild acids, ether and formalin. They are stable on storage at 30 °C.

Pathogenesis of Retrovirus Infections

Provirus and oncogenes are two key components that play a very important role in the pathogenesis of tumours induced by viruses. In the **provirus model**, the gene enters the cell at the time of

Table 63.4 Retrovirus genes and their functions

Retrovirus	Gene	Function
Retroviruses	*gag*	Group-specific antigen: Core and capsid protein
Retroviruses	*pol*	Polymerase: Reverse transcriptase, protease, integrase
Retroviruses	*env*	Envelope: Glycoprotein
HTLV	*tax*	Transactivation of viral and cellular genes
HIV-1	*tat*	Transactivation of viral and cellular genes
HTLV	*rex*	Regulation of RNA splicing and promotion of export to cytoplasm
HIV-1	*rev*	Regulation of RNA splicing and promotion of export to cytoplasm
HIV-1	*nef*	Alteration of cell activation signals and progression to AIDS
HIV-1	*vif*	Viral infectivity and promotion of assembly
HIV-1	*vpu*	Facilitation of release of virus and decrease of cell surface CD4
HIV-1	*vpr*(*vpx*)	Transport of complementary DNA to nucleus and arrest of cell growth
Retroviruses	LTR	Promoter, enhancer elements

Note: *HTLV*: human T-lymphotropic virus; *HIV*: human immunodeficiency virus; *AIDS*: acquired immunodeficiency syndrome; *LTR*: long terminal repeat (sequence)

infection by the tumour virus. In the **oncogene model**, the genes for malignancy are already present in all body cells by virtue of being present in the sperm and egg.

The oncogenic viruses can transform the cells in cultures and induce tumours in animals. The retroviruses may induce tumours by either i) by inducing or altering the expression of a pre-existing cellular gene or ii) by introducing a new transforming gene (*oncogene*) into the cellular genome.

Oncogenes

Oncogenes may be of two types: viral oncogenes and cellular oncogenes.

Viral Oncogenes

These are the genes that encode proteins that induce the transformation of normal cells into malignant cells. The viral oncogenes are usually of host cell origin. These oncogenes encode proteins that promote cell growth and several functions. Viral oncogenes initiate inappropriate cell growth and malignant transformation but are not required to replicate viruses. Viral oncogenes of DNA and RNA viruses are summarised in Table 63.5.

Cellular Oncogenes

These genes resembling viral oncogenes may be of two types: cellular oncogenes and proto-oncogenes. The oncogenes isolated from cancer cells are known as **cellular oncogenes (*c-onc*)**, while similar genes found in normal cells are called **proto-oncogenes**. More than 20 cellular oncogenes have been identified to date. Some cellular oncogenes are expressed during normal embryonic development. The proto-oncogenes are found widely in invertebrates. These genes encode proteins essential for cell growth regulation and differentiation.

It is postulated that the cellular oncogenes may serve as a precursor of viral oncogenes. Both cellular oncogenes and viral oncogenes are similar but not identical. They differ from each other by their base sequences. Moreover, cellular oncogenes consist of introns characteristic of eukaryotic genes, but viral oncogenes do not contain any of these eukaryotic genes.

Transfection

It is the method of study of oncogenes that can be demonstrated in mouse fibroblast cell lines, such as NIH 3T3. Foreign DNA is taken up by these cells, incorporated into their genome and transfected into the host cells, resulting in the formation of transforming genes. These genes are identical to that of cellular oncogenes. Retroviruses (such as HTLV or HIV) carry a fourth gene, namely, *tax* or *tat*, next to the *env* gene. The *tax* or *tat* is a transactivating gene that regulates the function of the viral genes. The oncogenic retroviruses may be slow transforming viruses or acute transforming viruses.

Slow transforming viruses, such as chronic leukaemia viruses, have low oncogenic potency and induce malignant changes, usually only in blood cells and after a long latent period. They replicate normally and do not transform cultured cells.

Table 63.5 Viral oncogenes of DNA and RNA viruses

Characteristics	DNA virus	RNA virus
Prototype virus	SV40 virus	Rous sarcoma virus
Name of gene	Early-region A gene	*src* gene
Name of protein	T antigen	src protein
Location of protein	Primarily nuclear, but some in plasma membrane	Plasma membrane
Function of protein	Protein kinase, ATPase activity, binding to DNA and stimulation of DNA synthesis	Protein kinase that phosphorylates tyrosine
Required for virus replication	Yes	No
Required for cell transformation	Yes	Yes

Acute transforming viruses, in contrast, are highly oncogenic. They induce malignancy in infected cells after a short latent period of weeks or months. They transform cells in culture and can cause different types of malignancies, such as sarcoma, carcinoma and leukaemia. Unlike slow transforming viruses, the acute transforming viruses are unable to replicate normally because they carry an additional gene, the viral oncogene (*v-onc gene*), on their genome, which replaces some of the genes necessary for the replication of viruses. The *v-onc gene* can replicate only if co-infected with a standard helper retrovirus. Therefore, most of these acute transforming viruses are replication-defective viruses (Table 63.6). Various viruses associated with human cancers are listed in Table 63.7.

Clinical Syndromes

Humans

HTLVs, which are described later, are the viruses that have been associated with **T-cell lymphoma** and **leukaemia**.

Animals

Sarcoma and Acute Leukaemia Viruses

The sarcoma and acute leukaemia viruses cause infection primarily in animals. They are not associated with any human infections. The avian leukosis complex virus includes a group of antigenically related viruses which induce avian leukosis or sarcoma in fowls.

Leukaemia Viruses

These are slow oncogenic viruses that induce malignancies after a long latency period of even 30 years. These viruses promote cancerous growth indirectly compared to the oncogene-encoded acute leukaemia or sarcoma viruses. These viruses specifically have a *tax* gene, which is a transcriptional regulator and is capable of activating LTR sequences that enhance the expression of cellular growth genes. This leads to uncontrolled cell growth, transforming the cells neoplastically or promoting other genetic changes over a long period.

Table 63.6 Oncogenes and their chromosomal location in humans

Oncogene	Description	Function	Locus	Cancer
Alk	Receptor tyrosine kinase	Tyrosine kinase	2p23	Lymphoma
Axl	Receptor tyrosine kinase	Tyrosine kinase	19q 13.1-q13.2	Leukaemia
Bc12, 13	B cell leukaemia-associated oncogenes	Apoptosis regulation	18q21; 19q 13.1-q13.2	B cell leukaemia
Ras	Harvey sarcoma viral oncogene	G protein	Several	Myeloid neoplasm
Src	Rous sarcoma virus	Tyrosine kinase	20q11.2-q12	Lymphoma
Jun	Avian sarcoma virus	Transcription factor	1p32-p31	Ovarian, breast, colon, lungs, leukaemia
Fos	Murine osteosarcoma virus	Transcription factor	14q24.3	Several human neoplasms

Table 63.7 Viruses associated with human cancer

Virus	Cancer	Latency (in years)
Human papillomavirus (5,8,14 and 17)	Skin	3–12
Human papillomavirus (16, 18, 31, 35)	Cervical, vulvar, penile, anal, perianal	20–50
Hepatitis C virus	Hepatocellular carcinoma	30–50
Epstein–Barr virus	Nasopharyngeal, Burkitt's lymphoma	3–50
Human T-lymphotropic virus type 1	Adult T cell leukaemia	30–50
Kaposi's sarcoma herpesvirus	Kaposi's sarcoma	3–10
Human polyomavirus 2 (JCV)	Giloma	7–13

Other Viruses

Murine leukosis viruses, which consist of several strains of murine leukaemia and sarcoma viruses, are associated with infections in mice. Mammary tumour virus of mice is associated with mammary cancer only in certain susceptible strains of mice having a high natural incidence of breast cancer. Leukosis sarcoma viruses of other animals consist of a large number of viruses that have been associated with leukosis and sarcoma in cats, hamsters, rats, guinea pigs and monkeys.

Laboratory Diagnosis

Most endogenous retroviral infections can be detected by molecular techniques, such as nucleic acid hybridisation or by activation after exposure to radiation or chemicals.

Treatment

No specific treatment is available for the management of retroviral infections.

Epidemiology

Retroviruses are documented from different host species worldwide. Retroviruses show host specificity, the specificity being determined mainly by the presence of viral receptors on the host cell surfaces.

Retroviruses can be transmitted in two ways: exogenous and endogenous. Most oncogenic retroviruses are exogenous. They are transmitted horizontally. Endogenous retroviruses are transmitted vertically from parents to offspring, and the process is mediated by the provirus integrated with the germline cell genome. This provirus behaves like a cellular gene and is under the regulatory control of the host cell.

Human T-Lymphotropic Viruses (HTLV)

Human T-lymphocytic virus (HTLV) and HIV are two important members of the retrovirus family. Both the viruses are enveloped viruses. Both have enzyme reverse transcriptase in the virion and two copies of a single-stranded, positive-polarity RNA genome. HIV has been described in Chap. 65.

HTLV has six distinct retroviruses (HTLV-1 HTLV-2, HTLV-3, HTLV-4, HIV-1 and HIV-2), but only HTLV-1 has been associated with human disease.

Morphology

HTLV is an enveloped, spherical virus measuring 100 nm in diameter. The virus consists of a single-stranded RNA with a diploid genome, which has the property to replicate through DNA intermediary and can integrate into the host T cell genome as a *provirus*. It is morphologically similar to HIV, but it has a centrally located nucleocapsid core in a mature virion.

Virus Genome

HTLV has a unique genome. The genome contains LTRs, which encode many polyproteins processed by virally encoded protease and cellular protease into functional peptides. The gene px encodes for Tax and Rex proteins, which regulate gene expression in the virus. The genome has the same *gag-pol-env* motif with LTR sequences as HIV, but it differs from HIV by having a fourth sequence *(xp)*, which participates in transcription.

The oncoviruses include HTLV-1, HTLV-2 and HTLV-3, but only HTLV-1 has been associated with human disease.

Virus Replication

The HTLV has two important genes, *tax* and *rex*, regulating viral replication.

The main function of ***the tax* gene** is the transactivation of viral and cellular genes. It is a transcriptional activator, which enhances the transcription of the viral genome from the promoter gene sequences in the LTR. It also activates other genes, including those for interleukin (IL)-2, IL-3, GM-CSF (granulocyte-macrophage colony-stimulating factor) and the receptor for

IL-2. Activation of this gene promotes the growth of infected T cells. The main function of *the tax* gene is the transactivation of viral and cellular genes.

***Rex* gene** regulates RNA splicing and promotion of export to the cytoplasm. It encodes two proteins that bind to the structure of the viral mRNA, preventing splicing and promoting mRNA transport to the cytoplasm. Later during the course of infection, the *rex* gene selectively enhances the expression of the singly sliced structural genes, which are required in large quantities.

HTLV primarily infects CD4+ cells. The virus enters the cell, and within the cytoplasm, reverse transcriptase of the virion synthesises a DNA copy of the genome. The genome is then transported into the nucleus and is integrated with the cell DNA by the virally encoded integrase, resulting in an infection. The host cell RNA polymerase makes viral RNA, and transcription is upregulated by Tax protein. The latter facilitates transcription of the viral genome, and Rex protein modulates the processing of viral pre-mRNA.

Pathogenesis of HTLV Infection

HTLV-1 is a lymphotropic virus. The HTLVs are distinguished from HIV as they cause lymphoproliferative disorder, whereas HIV causes lymphocytosis. It enters the host by sexual intercourse, breastfeeding or blood transfusion.

After entry in humans, the virus spreads through circulation and infects the CD4 helper and DTH (delayed-type hypersensitivity) T cells. These T cells are most commonly present in the skin and the neurons. Recent evidences show that glucose transporter (GLUT-1) is a supportive receptor for HTLV-1 cell entry. The ubiquitous distribution of GLUT-1 may help explain the viral tropism of HTLV-1. HTLV predominantly infects and integrates lymphocytes.

The virus predominantly infects and integrates lymphocytes. Once the infection is transmitted, the *tax* gene (which encodes Tax protein) transactivates the cellular genes for T-cell growth factor and IL-2 and their receptors, all of which activate growth in the infected cells.

There is a long latency period of approximately 30 years before the onset of leukaemia. Although the exact mechanism of viral pathogenesis of T-cell leukaemia that occurs in 3–5% of individuals is unknown, *in vitro* studies have shown that the *tax* gene plays an important role. It has been demonstrated that HTLV can spread by cell-to-cell contact without replication, but it is not clear how HTLV crosses the blood–brain barrier to cause neurological sequelae of HTLV infection.

Clinical Presentations

HTLV-1 is associated with the following clinical conditions.

Adult T-cell leukaemia is seen in approximately 1 in 20 persons infected by HTLV-1 over 30–50 years. Lymphadenopathy (both in the periphery and within the body cavities) represents the classic form of ATL. Hepatosplenomegaly, hypercalcemia and lytic bone lesions are the other manifestations.

Myelopathy/tropical spastic paraparesis is a slowly progressing degenerative disease that primarily affects the corticospinal tract of the thoracic spinal cord. Patients with HAM/TSP may present with weakness and stiffness in lower limbs, urinary incontinence and lower back pain.

Uveitis is a less frequent manifestation. The condition is associated with the presence of HTLV-infected lymphocytes in the eye's vitreous fluid. **Infectious dermatitis** is a chronic and severe form of dermatitis seen during childhood.

Laboratory Diagnosis

A biopsy is useful for the definitive diagnosis and classification of HTLV. The presence of convoluted nuclei, also known as *cloverleaf* or *flower lymphocytes*, in the peripheral blood is typical of ATL. Provirus can also be demonstrated in these malignant cells. Western blot, immunofluorescence antibody test and HTLV-ELISA are serological tests used to diagnose HTLV-1 infection. HTLV-ELISA is a frequently used test to demonstrate antibodies in the serum to diagnose HTLV-1 infection. High false positivity is a noted problem with ELISA, particularly in low prevalence areas. Hence, all HTLV-ELISA-positive cases must be confirmed with Western blot, immunofluorescence antibody test or polymerase chain reaction (PCR).

PCR is useful to distinguish between HTLV-1 and HTLV-2 infections. It is also useful in infants who are suspected of having false-positive results due to the presence of circulating maternal anti-HTLV antibodies.

Treatment

No specific treatment is available for the management of HTLV-1 infection. But a, limited success has been obtained in chemotherapy of some patients with a combination of AZT and interferon-alpha. However, no known regimen increases the survival time of the patients beyond 2 years. Relapse is common.

Epidemiology

HTLV-1 infection is endemic in Japan, South America, Northern Oceania and Tropical Africa and the Caribbean ocean. Few foci of infection have also been documented in Asia in the Philippines. HTLV-1 is transmitted from mother to child through breast milk during breastfeeding or childbirth. In adults, transmission is primarily through sexual contact. Apparently, a prolonged period of sexual contact is necessary to transmit the disease. Homosexual contact, however, appears to play a less important role in the transmission of the disease. Blood transfusion and intravenous drug abuse are also other important modes of transmission of HTLV-1.

Prevention and Control

Preventive measures include health education, which should focus on i) avoidance of breastfeeding by known infective mothers, ii) use of safe sexual practices by couples and iii) screening of blood donors for HTLV antibodies.

Endogenous Retroviruses

Endogenous retroviruses are those retroviruses that are integrated into and become part of chromosomes in humans and animals. These proviruses are similar to those of HTLV, mouse mammary tumour virus and other retroviruses that have been detected in humans. These viruses cannot usually replicate because they are poorly transcribed or because of the deletion of the insertion of termination codons.

Case Study

A 35-year-old man, a resident of an area endemic for HTLV infection in Japan, was admitted with the presentation of weakness and stiffness in lower limbs, urinary incontinence and lower back pain for 3 months. The person gave a history of receiving blood transfusion nearly 2 years back. Serum was positive by HTLV-ELISA. The person was suspected to be suffering from HTLV-1-associated myelopathy.

- What are the other tests you will perform to confirm the diagnosis?
- What are the other clinical syndromes associated with HTLV-1 infection in humans?
- What are the countries in which HTLV-1 infection is endemic?
- Describe the pathogenesis of HTLV-1 infection in humans.

Miscellaneous Viruses

64

Miscellaneous viruses described in this chapter include a diverse group of viruses that have been increasingly associated with infections in humans. These viruses include papovaviruses, parvoviruses, rubella virus, Norwalk virus, viral haemorrhagic fever (VHF) viruses, arena viruses, members of Reoviridae, coronaviruses, including the most recently described severe acute respiratory syndrome (SARS)-associated coronavirus (SARS-CoV), and Zika virus (ZIKV).

Papovaviruses

The family Papovaviridae consists of papovaviruses, which are a group of viruses containing small, enveloped, icosahedral capsid with a double-stranded, circular DNA. The term *papova* is derived from the first two letters of the names of the viruses (*pa*, papillomaviruses; *po*, polyomaviruses; and *va*, vacuolating agents). The family is divided into two genera *Papillomavirus* and *Polyomavirus*. Currently, the genus *Papillomavirus* is separated into a new family, Papovaviridae, and the genus *Polyomavirus* into a new family Polyomaviridae.

The papovaviruses encode proteins that promote cell growth. They induce both lytic infections and tumours (benign or malignant). These viruses are capable of causing lytic, chronic, latent or transforming infections depending on the host cells infected (Table 64.1).

Human Papillomaviruses (HPVs)

Human papovaviruses (HPVs) are the causative agents of warts and are also associated with other cancerous conditions.

Morphology and Other Properties

It is a double-stranded DNA virus without any envelope (50–55 nm).

The viral genome is a supercoiled, double-stranded DNA composed of approximately 3000 bp. The DNA encodes seven to eight early genes (E1–E8) and two late (L1 and L2) genomes. The viruses replicate and assemble in the nucleus. The early genes of the virus are responsible for the growth of the cells and facilitate replication of the viral genome during cell division. They show widespread diversity; more than 100 distinct HPV have been recognised.

The viruses do not grow on any cell cultures.

Pathogenesis of HPV Infection

HPV show a high degree of host specificity. They show a high predilection for the skin and also for the mucous membrane. HPV infection is acquired by close contact, and it infects the skin or mucous membranes. Tissue tropism and the manifestation of the disease depend on the HPV type that causes the disease. The virus replicates in the squamous

Table 64.1 Human infections caused by papovaviruses

Papovavirus	Manifestations
Papillomavirus	Skin warts (Verruca vulgaris, plantar warts)
	Benign head and neck tumours
	Anogenital warts (*Condyloma acuminatum*)
	Cancerous conditions (oral pre-malignant lesions and oral squamous cell carcinoma, intraepithelial cervical neoplasia and cancer) Recurrent respiratory papillomatosis
Polyomavirus	
BK virus	Renal diseases (haemorrhagic cystitis and urethral stenosis)
JC virus	Progressive multifocal leukoencephalopathy (PML)

epithelium of the skin to cause warts and in the mucous membrane to induce cancerous epithelial proliferation. HPV-16 and HPV-18, HPV-33 and HPV-35 have been linked with verruciform proliferation in the oral cavity, oral pre-malignant lesions and oral squamous cell carcinoma. The presence of the koilocytes is the hallmark of HPV infections of the skin. The cell-mediated immunity (CMI) plays an important role in the resolution of the disease.

Clinical Syndromes

HPV infection causes cutaneous warts, benign head and neck tumours, anogenital warts, cancerous shedding, etc. (Table 64.2).

Laboratory Diagnosis

The diagnosis of the wart is made by histology. Demonstration of hyperplasia of the prickle cells and excessive production of keratin (hyperkeratosis) is the diagnosis of the condition. DNA probe and PCR are useful for detecting HPV in cervical swabs of tissue specimens. Cell cultures are not useful because HPVs do not grow in cell lines. Serology is rarely used.

Treatment

Warts can be removed by surgical cryotherapy, electrocautery or chemical reagents.

Table 64.2 Clinical syndromes associated with human papillomavirus

Clinical syndromes	Human papillomavirus types
Cutaneous syndromes	
Skin warts	
Plantar wart	1, 2, 4
Common wart	1, 2, 4, 7, 26, 29
Flat wart	3, 10, 27, 28, 41
Epidermodysplasia verruciformis	5, 8, 9, 14, 15, 17, 19, 20, 21–25, 36, 38, 46
Warts in organ transplant patients	75, 77
Hand warts of meat and animal handlers	7
Mucosal syndromes	
Benign head and neck Tumours	
Laryngeal papilloma	6, 11
Oral papilloma	2, 6, 11, 16
Conjunctival papilloma	11
Keratoacanthoma	37
Anogenital warts	
Condyloma acuminatum	1, 2, 6, 10, 11, 16, 30, 44, 45
Cervical intraepithelial neoplasia, cancer	11, 16, 18, 31, 33, 35, 42–44

Epidemiology

HPV infections are found worldwide. HPVs are found in genital secretions, skin shedding, etc. Infected humans are the main source of infection. HPV infection is transmitted by direct contact through abrasions in the skin and mucosa, passage through the infected birth canal and sexual contact.

Prevention and Control

There are no specific preventive measures available against HPV infection. Avoidance of direct contact with infected warts may prevent transmission. Safe sexual practice is useful to prevent sexual transmission of HPV.

Vaccines are available against HPV. Three HPV vaccines, namely 9-valent HPV vaccine (Gardasil 9, 9vHPV), quadrivalent HPV vaccine (Gardasil, 4vHPV) and bivalent HPV vaccine (Cervarix, 2vHPV), have been licensed and are available. All three HPV vaccines protect against HPV types 16 and 18 that cause most HPV cancers. CDC recommends that 11 to 12 years olds receive two doses of HPV vaccine 6 to 12 months apart. Young adults within the age of 26 need three doses of the HPV vaccine. Vaccination is not recommended for those older than age 26 years.

Human Polyomaviruses

SV40 is the prototype polyomaviruses. BK and JC viruses are the other human polyomaviruses. Various polyomaviruses show different host specificities. Pathogenesis of the human polyomavirus infection in humans depends on the host's immune status. In the immunocompetent host, the replication of viruses is inhibited.

Suppression of immunity in patients receiving an organ transplant or suffering from AIDS results in reactivation of latent JC and BK viruses. Reactivation of BK viruses causes severe urinary tract infection, and that of JC viruses causes infection of the CNS. The BK, JC and SV40 viruses are known to cause tumours in animals but not in humans. The BK virus is associated with haemorrhagic cystitis in the bone marrow recipients and with urethral stenosis in kidney transplant recipients. JC virus is the causative agent of PML.

Electron microscopy (EM) is useful to detect the JC virus in brain tissue from the cases of PML and in the urine of a kidney transplant recipient. Immunoperoxidase and in situ immunofluorescence are also used to detect viral antigens in brain tissue. The BK polyomavirus is isolated from urine by culture in human diploid fibroblasts. JC virus is isolated from urine and brain tissue by culture in human foetal glial cell culture. Both DNA probe and PCR are newer, sensitive methods for detecting BK virus and JC virus in the urine, cerebrospinal fluid (CSF) or biopsy tissue.

No specific antiviral treatment is available for polyomavirus infections in humans. Prevention is difficult due to the ubiquitous presence of these viruses. The differences between the human papillomaviruses and polyomaviruses are listed in Table 64.3.

Parvoviruses

Parvoviruses are the smallest of the DNA viruses belonging to the family Parvoviridae. They are icosahedral, non-enveloped viruses containing single-stranded DNA. Human parvovirus B19 (B19) is the only known parvovirus that is pathogenic for humans, and it shows tropism for erythroid progenitor cells.

Parvovirus B19

The B19 viruses are extremely small viruses (18–26 nm) and possess a non-enveloped, icosahedral capsid. The viral genome contains single-stranded DNA. Replication of the virus occurs in the nucleus. Thus far, only one serotype of the B19 virus is known to occur.

B19 virus shows tropism for bone marrow cells, erythroid cells from foetal liver and

Table 64.3 Distinguishing features of human papillomaviruses and human polyomaviruses

Properties	Papillomavirus	Polyomavirus
Virion size (diameter)	55 nm	45 nm
Genome size	8×10^3 bp	5×10^3 bp
Coding information	On one strand	On both strands
Oncogenic potential	Yes	Uncommon
Result of natural infection	Benign tumour	Inapparent
Target tissue	Surface epithelia	Internal organs
Transform cells *in-vitro*	Rarely	Yes
Genome in transformed cells	Episomal in warts; integrated in carcinoma	Integrated
Viruses infecting humans	Human papillomavirus	BK and JC viruses; SV40
Clinical syndromes	Skin warts, genital warts, laryngeal papillomas, cervical carcinoma	PML and nephropathy in transplant recipients

erythroid leukaemia cells. The virus enters cells through the P blood antigen receptors on the erythrocyte precursors. Host immunity to B19 virus infection is primarily antibody mediated. The role of CMI in the B19 virus is unknown.

The virus causes flu-like illness, erythema infectiosum or fifth disease, infection in pregnant women and chronic B19 infection. The human infections caused by the B19 virus are listed in Table 64.4.

B19 virus is distributed worldwide, and it is strictly a human disease. The infection is transmitted by the vertical transmission during birth, the respiratory route through respiratory secretions and percutaneous exposure to blood and transfusion of blood and blood products.

Enzyme-linked immunosorbent assay (ELISA), radioimmunoassay (RIA) and immunofluorescence assay (IFA) are used to demonstrate IgG and IgM antibodies in diagnosing erythema infectiosum. PCR is available to demonstrate the B19 virus genome in the blood.

No specific antiviral therapy is available. No specific measures are available for the prevention of the infection.

Bocavirus

Bocavirus belongs to the family Parvoviridae, which is a newly identified member of this family. They are small and non-enveloped, containing single-stranded DNA. This virus is commonly isolated from infants and children presenting with pneumonia or diarrhoeal symptoms. It has also been isolated in patients suffering from other viral infections.

Although it has been associated with many clinical symptoms, its definitive role in human infections or disease is not well established. So far, no diagnostic tests are available for bocavirus detection, and also there are no established guidelines to treat bocavirus-associated diseases. In the research setup, this virus can be identified by PCR.

Table 64.4 Human infections caused by parvovirus B19

Parvovirus B19 infections	Manifestations
Flu-like illness	Malaise, headache, myalgia and rhinorrhoea
Erythema infectiosum	Cutaneous rash, arthralgia; aplastic crisis 'fifth disease' in children
Infection in pregnant women	Re-infection in a pregnant mother: No adverse effects; non-immune seronegative pregnant mothers: Hydrops fetalis and foetal death
Chronic B19 infection	Chronic anaemia, leukopenia and thrombocytopenia
Transient aplastic crisis	Severe acute anaemia

Rubella Virus

Rubella is an RNA virus classified as a rubivirus in the family Togaviridae. The name rubella, given by Henry Veale in 1866, is derived from the red-coloured rash accompanying the illness. Owing to similarity in presentation, rubella was distinguished from a more serious infectious disease, measles, in the early nineteenth century. In the nineteenth century, it came to be called German measles, when German physicians closely studied the disease. In 1942, Norman Gregg noticed that first-trimester maternal rubella caused serious birth defects. The full spectrum of rubella embryopathy remained unclear until rubella virus was isolated in tissue culture in 1962 by two independent groups comprising Parkman, Buescher and Artenstein in one of the groups and Neva and Weller in the other.

Morphology

Rubella is a pleomorphic, roughly spherical virus measuring 50–70 nm in diameter (Fig. 64.1). It consists of a single-stranded RNA genome, an icosahedral nucleocapsid and a lipoprotein envelope. Unlike measles and mumps viruses, this virus has a positive-strand RNA; therefore, it does not contain any virion polymerase. The envelope contains haemagglutinin peplomers. It agglutinates human, pigeon, goose and 1-day-old chick RBCs; both replication and assembly of the virion occur in the cytoplasm of the infected cell.

Other Properties

Rubella has only one serotype. The virus agglutinates erythrocytes of humans, chick, goose and pigeon at 4 °C. The virus is heat sensitive. It is readily inactivated by heating at 56 °C but survives for several years at −60 °C. The virus is inactivated by chemicals, such as beta-propiolactone, formaldehyde, chloroform and ether.

Laboratory Isolation

The virus is grown with difficulty in the cell lines. The virus is not cytolytic but produces limited

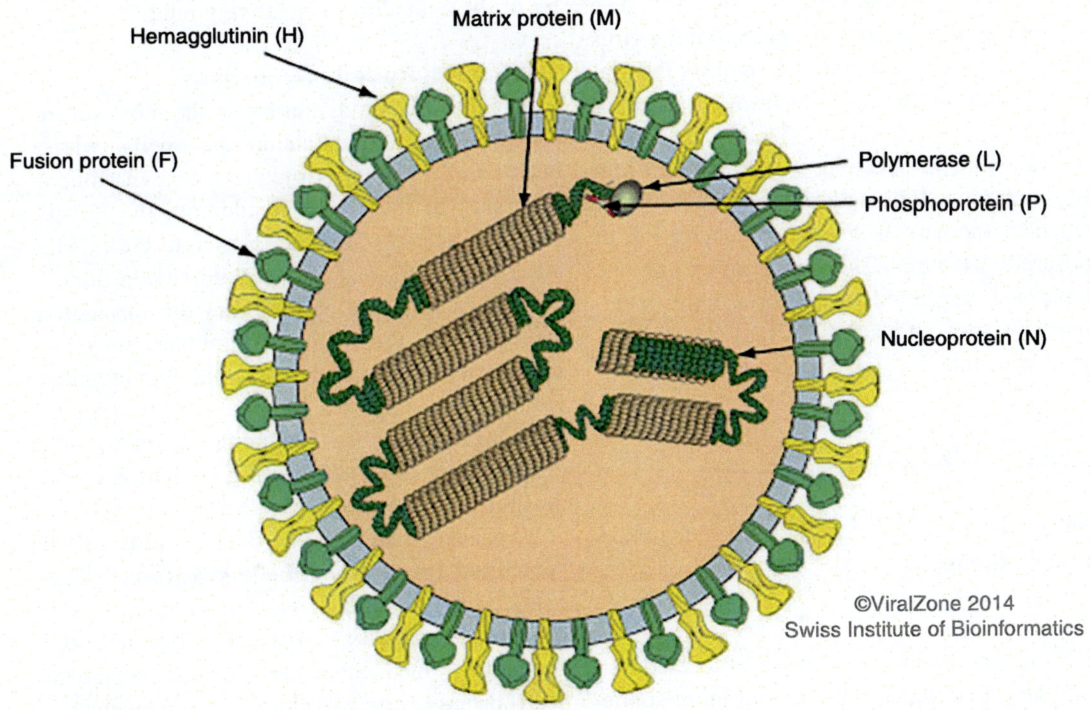

Fig. 64.1 Rubella virus. (http://phil.cdc.gov/phil/details.asp)

cytopathologic effects in certain cell lines, such as Vero and RK-13 cell lines. In the cell line, the replication of rubella prevents the replication of super infectious picornavirus. This phenomenon is known as **heterologous interference**, which is applied in detecting the rubella virus.

Pathogenesis of Rubella

The virus infection is acquired by inhalation of respiratory droplets from infected persons. Initially, the rubella virus infects the nasopharynx of the upper respiratory tract and then spreads to local lymph nodes. From there, the virus spreads by the blood throughout the body to the internal organs and skin. The occurrence of mild rash is characteristic. During the prodromal period and for nearly 2 weeks after the onset of rash, the infected persons continue to secrete virus in the respiratory droplets.

Immunity

Rubella infection is characterised by the development of circulating antibodies produced after the phase of viraemia, and their development correlates with the appearance of the rash. The circulating antibodies limit the spread of the virus in the blood. These antibodies also cross the placenta and protect the newborn from an attack of rubella.

Natural rubella infection usually confers lifelong immunity. Reinfection may occasionally occur after the natural disease or vaccination or exposure to the virus. The formation of immune complexes is suggested to contribute to the development of rash and arthralgia associated with rubella infection.

Clinical Syndromes

Rubella virus causes rubella and congenital rubella syndrome.

Rubella

The incubation period varies from 14 to 21 days. In children, Rubella is a milder and more subtle disease than measles. A 3-day maculopapular or macular rash, which starts on the face and progresses downwards to involve the extremities, is the characteristic presentation in symptomatic cases. The rash typically lasts 3 days. Tender lymphadenopathy that affects all the nodes but most commonly affects sub-occipital, post-auricular, anterior and posterior cervical nodes are the hallmark of rubella.

In adults, rubella produces a more severe disease with manifestations of arthralgia and polyarthritis and rarely thrombocytopenia or post-infectious encephalopathy.

Congenital Rubella

The severity of the disease depends on how early the infection occurs. Maternal immunity to the virus due to prior exposure or vaccination prevents the spread of the virus to the foetus. The foetus is at major risk until the fifth month of pregnancy. Rubella infection in very early pregnancy causes abortion, In the first trimester, 80% of the infants would be affected, and cataracts, mental retardation and deafness are the most common manifestations of congenital rubella infection. The infection in the last trimester of the animalities is relatively mild.

Congenital Rubella Syndrome

Congenital rubella syndrome is the most severe and important complication of rubella, which occurs in the foetus of pregnant women without immunity to the virus. In addition, the infants infected in utero continue to excrete the rubella virus for up to 1 year. These children constitute a public health hazard because they are considered an exposure threat to non-immune pregnant women. The virus can be transmitted to pregnant women from children. Common manifestations are congenital defects, including cataracts, congenital heart diseases, hearing impairment and developmental delay.

Expanded rubella syndrome is also called "**extended congenital rubella syndrome**". This complex of symptoms may affect infants with congenital rubella syndrome. Besides the above-mentioned congenital defects, congenital heart disease, corneal clouding, microcephaly, mental retardation, deafness, hepatosplenomegaly, thrombocytopenic purpura, intrauterine growth retardation, interstitial pneumonia, myocarditis and metaphyseal bone lesions also occur.

Laboratory Diagnosis

Specimens

These also include throat swabs, urine, bone marrow or CSF to isolate the virus by cell culture. These include serum for serology.

Cell Culture

Viruses can be isolated in rabbit kidneys (RK-13), rabbit cornea (SIRC) and Vero cells. The virus produces little cytopathic effects. In Vero and rabbit cornea cell lines, the virus is usually identified by its ability to interfere with the echovirus cytopathic effect. In this method, if the rubella virus is present in the patient's specimen and has grown in cell culture, no cytopathic effect will appear when the culture is superinfected with an echovirus, such as echo 11. Rubella virus grown in cell lines can also be detected by detecting antigen in cell lines using a direct fluorescent antibody method. Virus isolation is rarely used because isolation of the virus is difficult.

The virus can be isolated from a throat swab, urine, bone marrow, leucocytes or CSF by cell culture to diagnose congenital infection. Demonstration of rubella virus in amniotic fluid collected by amniocentesis indicates a definite congenital rubella infection in the foetus.

Serology

The diagnosis of rubella in **pregnant women** by serology is usually confirmed by demonstrating rubella antibodies in the serum. Demonstration of specific IgM rubella antibodies in a single acute-phase serum sample by IgM ELISA is diagnostic of rubella and indicates recent infection. Demonstration of 1:8 or greater titre of IgG antibody in serum without IgM rubella antibodies indicates past infection, immunity and consequent protection of the foetus and vaccination. Diagnosis can also be made by demonstrating a fourfold or greater rise in IgG antibody titre between acute-phase and convalescent-phase sera by using a haemagglutination inhibition test or an ELISA.

The diagnosis of **congenital rubella in an infant** by serology is made by the presence of IgM antibodies in the serum by IgM ELISA. This indicates a recent infection because IgM does not cross the placenta from the mother, as does IgG. Confirmation after 1 year with serology alone is very difficult.

Treatment

There is no antiviral therapy. Pregnant women who acquired virus are treated with acetaminophen to relieve their symptoms. In serious cases, blood transfusion or steroids are administered.

Epidemiology

Humans are the natural host. The disease is prevalent worldwide. In countries where vaccination is not routinely used, rubella epidemics occur every 6–9 years. Respiratory droplets are the source of infection. The infection is usually acquired during childhood. Patients are most contagious during the time of appearance of the rash. The viruses are excreted in the pharynx and the respiratory droplets from 7 days before until 7 days after the rash. The infection is usually acquired during childhood. The infection can also be transmitted transplacentally from mother to foetus, resulting in congenital infection.

Prevention and Control

Preventive measures include vaccination and administration of immunoglobulins.

Rubella Vaccines

Prevention of rubella is best carried out by immunisation with a live attenuated vaccine. The main objective of the rubella vaccination programme is to prevent congenital infection by decreasing the number of susceptible people, especially children. Vaccination causes a significant reduction in the likelihood of exposure of pregnant women to the virus.

Live rubella vaccine contains live, cold-adapted RA-27/3 vaccine strain grown in human diploid cells. The virus is usually administered as measles, mumps and rubella vaccine (MMR vaccine). The vaccine is very effective and induces long-lasting humoral and cellular immunity. It causes few side effects, such as transient arthralgia in some women. The vaccine also induces the production of respiratory IgA, thereby interrupting the spread of the virus by nasal droplets. The use of the vaccine has resulted in a significantly reduced incidence of rubella and congenital rubella syndrome worldwide.

The vaccine is usually given subcutaneously to children 15 months of age. It is also recommended for unimmunised young adult women if they are not pregnant and for those women who will not conceive for the next 3 months. It is contraindicated for use in pregnant women and immunocompromised patients.

Immunoglobulins

Immunoglobulins can be given to pregnant women in the first trimester of pregnancy who have been exposed to a known case of rubella, and for whom termination of pregnancy is not an option. Nevertheless, cases of congenital rubella syndrome have occurred in infants born to mothers who received immunoglobulins shortly after exposure. No adequate treatment is available for pregnant women exposed to rubella.

Norwalk Virus

Norwalk virus is one of the most common causes of viral gastroenteritis in adults.

Norwalk virus is a calicivirus in the family Caliciviridae. The caliciviruses are small, non-enveloped, single-stranded RNA viruses. Although the caliciviruses show many features similar to those of picornaviruses, they differ from those with a large genome and distinct spikes on the surface. Two important caliciviruses cause infection in humans: the Norwalk virus and the hepatitis E virus. The latter virus is described in Chap. 60.

Morphology

Norwalk viruses are round viruses with a ragged outline. The genome is a non-segmented, single-stranded, positive-polarity RNA genome. The RNA genome is present in a 27-nm naked capsid consisting of one 60,000-Da capsid protein. The genome does not contain any virion polymerase. Ten prominent spikes and 32 cup-shaped depressions can be seen on the virion by microscopy.

Other Properties

The virus **replicates in the cytoplasm** with the release of viral particles on cell destruction. The virus is presumed to replicate like that of picornaviruses. They are stable and can survive with freezing or at temperatures of up to 140 °F. They are also resistant to inactivation by chlorination of water and drying in the environment. The number of serotypes is not known.

Pathogenesis of Norwalk Virus Infection

Norwalk viruses are highly contagious. Infection is acquired by ingestion of food or water contaminated with the virus. As low as 100 viral particles can establish infection; this is typically limited to the mucosal cells of the intestinal tract. Infection is characterised by damage to microvilli in the small intestine, causing malabsorption. The virus-mediated changes in gastric mucosa and delayed gastric emptying cause vomiting. It is a non-invasive virus and does not cause invasion of the colon; therefore, leukocytes or erythrocytes in the faeces are typically absent. No histopathological lesions are seen in the gastric mucosa.

Norwalk virus infection confers a brief and short immunity. Recurrent infection occurs throughout life because of the absence of long-term immunity, lack of cross-strain immunity and because of the diversity of Norwalk virus strains.

Clinical Syndromes

It causes gastroenteritis in adults. The illness in symptomatic cases typically begins after an incubation period of 24–48 h. It is characterised by sudden onset of nausea, vomiting and watery diarrhoea. It is accompanied by low-grade fever, abdominal cramps and myalgia. Faecal leukocytes are absent. Norwalk virus gastroenteritis is short lived and typically lasts for 24–48 h.

Laboratory Diagnosis

Immunoelectron microscopy (IEM) is used to detect the virus in the stool. RIA or ELISA are also used to detect the virus and viral antigen in the stool. Both ELISA and RIA are the serodiagnostic tests frequently used to detect specific antibodies to the Norwalk virus in the serum. Polymerase chain reaction (PCR) is also being evaluated to demonstrate the viral genome in the infected stool specimens.

Treatment

No specific treatment is available.

Epidemiology

Norwalk virus gastroenteritis is found worldwide in adults. It is a strict human infection, and humans are the major source of infection. The virus is excreted in the vomitus and faeces for several weeks after recovery; hence, vomitus and faeces are important sources of infection. The infection typically occurs in group settings, such as schools, hospitals and nursing homes.

Prevention and Control

No vaccine is available against the virus. Strict personal hygiene, such as handwashing, is useful to prevent the transmission of the disease.

Viral Haemorrhagic Fever

Most viral haemorrhagic fever is caused by enveloped RNA viruses that belong to two families: Arenaviridae and Filoviridae. With the recognition of the outbreak of infection caused by the Ebola virus near the city of Kikwit, Zaire (Africa), the condition has now received worldwide attention.

The viruses causing viral haemorrhagic fever exhibit many characteristics. They are usually arthropod borne. Mosquitoes are primarily responsible for transmitting the disease. Person-to-person transmission may occur in many VHFs by direct contact with infected patients, their blood or their secretions and excretions. Rats and mice are the usual animal reservoirs for many of the VHFs. However, domestic livestock, monkeys and other primates may also serve as intermediate hosts. The manifestations of the disease vary depending upon the agent causing the disease. However, circulatory dysfunction, increased vascular permeability and diffused haemorrhage are the serious and terminal manifestations of the disease. Haemorrhage is typically present in many organs, and effusions are common in serous cavities. Widespread necrosis may be present in any organ system and varies from modest and focal to massive in extent. Liver and lymphoid systems are usually involved.

Arenavirus

Arenaviruses are enveloped viruses with a helical nucleocapsid. They are spherical to pleomorphic in shape, varying in size from 50 to 300 nm. The lipid envelope contains two major glycoprotein components, gp1 and gp2, which appear as spike-like or club-like projections. The name *arena* means sand, which refers to granules at the surface of the virion that is non-functional ribosomes. The arenaviruses include lymphocytic choriomeningitis (LCM) and haemorrhagic fever viruses, such as the Lassa, Junin and Machupo viruses.

Morphology

Arena viruses are pleomorphic, enveloped viruses (50–300 nm) with a sandy appearance. The genome is a negative-sense RNA that contains two subgenomic segments, namely, S (small segment, measuring 1.3 million bases) and L (large segment, measuring 2.4 million bases). The virion consists of beaded nucleocapsid with two single-stranded RNA circles. The strand is a negative-sense RNA that encodes for a Z protein and for the viral RNA-dependent RNA polymerase. The segment encodes for the glycoprotein precursors and the N protein that binds to the positive-sense RNA segments. The presence of internal granular structures 20–25 nm in size is the distinctive property of arenavirus. These granular structures on EM appear sand like.

Laboratory Isolation

The arenavirus, such as the Lassa virus, can be isolated in cell culture using the E6 clone of Vero cells or in suckling mice.

Pathogenesis of Arenaviruses Infections

Arenaviruses primarily infect macrophages and cause vascular damage. Pathogenesis of arenaviruses infection is largely attributed to T-cell immunopathogenesis. T-cell-induced immunopathogenic effect contributes to the exacerbation of tissue destruction.

Clinical Syndromes

Arenaviruses cause LCM, Lassa fever and South American haemorrhagic fever.

Lymphocytic Choriomeningitis

This benign infection usually begins with fever, myalgia and headache. The illness can be biphasic. Fever and more severe headaches may recur 2–4 days after recovery from the first phase.

Lassa Fever

Most infections caused by the Lassa fever virus are mild or sub-clinical. The incubation period varies from 7 to 18 days. The infection begins insidiously with fever, malaise, joint pain, cough and severe headache. Prostration, dehydration, abdominal pain, facial or neck oedema may be seen in severe cases.

South American Haemorrhagic Fever

Lassa, Junin and Machupo viruses cause South American haemorrhagic fever, which is similar in severity. This has an insidious onset with fever, malaise, myalgia and lumbar pain. A petechial and/or vesicular palatal enanthem and skin petechiae are found in many patients. The condition may progress within 3–4 days with prostration and remitting fever and mucosal bleeding. Gingival haemorrhage is the characteristic feature. Most patients improve after 1–2 weeks, but one-third of patients may show many complications. These complications include profound mucocutaneous and mucosal haemorrhages, delirium and convulsions.

Laboratory Diagnosis

There is a great risk for laboratory workers handling body fluids for diagnosis. Hence, it is mandatory to process the specimen only in facilities with biosafety level 4 for Lassa fever and other arenaviruses and biosafety level 3 for LCM virus.

Specimens

The specimens include blood and CSF.

Cell Culture

The virus can be isolated easily from the blood and CSF of the individuals suffering from LCM virus infection and the CSF in case of meningitis.

The virus is also isolated from blood or tissue in the cases of South American haemorrhagic fever. Viruses can be isolated using tissue culture using the E6 clone of Vero cells or suckling mice.

Serology

Serodiagnosis is the most important method for diagnosing arenavirus infection in humans. Serodiagnosis of South American haemorrhagic fever is carried out by both ELISA and indirect fluorescent antibody tests to demonstrate IgM antibodies in serum and also by antigen capture ELISA is used to detect antigens in blood and tissues. IgM ELISA is frequently used to detect antibodies in the serum and CSF for the diagnosis of LCM.

Treatment

Ribavirin is the antiviral drug used for the treatment of Lassa fever and also for South American haemorrhagic fever. The treatment is mostly supportive of maintaining fluid and electrolyte balance, which can be lifesaving.

Epidemiology

Geographical Distribution

Most arenaviruses are found in Africa and South America, except for LCM viruses.

Reservoir, Source and Transmission of Infection

Most of the arenaviruses are zoonotic, and disease is transmitted primarily from rodents to humans. These viruses cause infection in specific rodent species and are endemic in the areas inhabited by rodent species. The viruses that cause chronic infection in these animals are excreted in saliva, urine and faeces over a very long time. Humans usually acquire infection from these animals through (1) inhalation of infectious aerosols of dried excreta, especially urine of the rodents that have been passed in the soil; (2) ingestion of food contaminated with rodent faeces or urine or (3) contact of the abraded skin with rodent blood or serum. The bite of rodents does not usually transmit infections.

The LCM viruses are distributed in Europe, Australia, Japan and the Americas. This infection in this area is closely related to the presence of infection in common mice, such as *Mus musculus* and *Mus domesticus*. The mice are the most important reservoir of infection for humans. Hamsters are important reservoirs for pet owners and laboratory workers.

Lassa fever is the disease of West Africa. The outbreaks of Lassa fever have been reported from Nigeria, Sierra Leone, Liberia and Guinea. The rodents of the genus *Mastomys* are the reservoirs of infection. The infection is transmitted from these infected mice to humans. The infection can also be transmitted from humans to humans.

South American haemorrhagic fever is endemic in Argentina, Bolivia and Venezuela, respectively. The rodent *Calomys musculinus* is the main reservoir for Argentine haemorrhagic fever. These rodents, commonly found in the cornfield, transmit the virus to farmers harvesting the corn. The infection is transmitted by inhalation of infectious aerosols. The infection can also be transmitted by ingestion of contaminated food and also by direct contact of blood or tissue from rodents on the abraded finger.

The rodent *Calomys callosus* is the major reservoir for Bolivian haemorrhagic fever, which transmits the infection to humans. Transmission of infection occurs through these rodents by inhaling aerosols from infected organisms and ingesting food contaminated by rodent urine. The cane mouse is the main reservoir for Venezuelan haemorrhagic fever.

Prevention and Control

The prevention of rodent-borne arenavirus infection depends on the rodent control measures and avoidance of high-density rodent-infested areas. No vaccines are available against arenaviruses. Clinical trials of the attenuated Junin virus are in progress.

Filoviruses

Ebola and **Marburg** are the two most important causative agents of VHFs, with a mortality of 25–100%. Both viruses are found in Africa and possibly in the Philippines. Zaire subtype of Ebola virus has been associated with a high infection rate, especially in Zaire, Africa.

Clinical presentation of Ebola virus is non-specific. The incubation period varies from 2 to 14 days. Some of the common symptoms are insidious or sudden onset of fever, chills, malaise, generalised myalgias and arthralgias, headache, anorexia and cough. The condition may also be associated with a sore throat, epigastric pain, vomiting and diarrhoea. Ebola infection during pregnancy has been consistently fatal.

Laboratory diagnosis is essential to establish the viral aetiology of the infection. EEM (Fig. 64.2) and enzyme immunoassay are the sensitive and specific assays for directly detecting specific viral antigens in blood and tissues. ELISA for virus-specific immunoglobulin M (IgM) and immunoglobulin G (IgG) is the test of choice for diagnosis. The test, however, is negative during the acute stages of Marburg and Ebola virus infections. PCR on serum is very useful for detecting infection during the acute stages of infection during which antibodies are not demonstrated in the serum.

The vector responsible for the Ebola virus transmission is unknown, but infected primates appear to be responsible. Later close contact among humans or primates appears to spread the infection. Aerosol transmission is suggested to occur in monkeys.

No specific antiviral agents are available for the treatment of Ebola Avoidance of insect bites from the vectors and from exposure to rodent sources of infection is the most important measure for preventing the condition. Barrier nursing and needle sterilisation in African hospitals are important to eliminate epidemics of Ebola and Marburg diseases.

Reoviridae

The family Reoviridae is divided into nine genera, of which only four genera cause human diseases. These genera are *Rotavirus, Orbivirus, Coltivirus* and *Orthoreovirus*. Other genera infect only plants, insects and fish. The Reoviridae viruses are non-enveloped viruses with double-layered protein capsids. The inner capsid layer shows well-defined subunits, whereas the outer layer of some viruses (e.g. rotaviruses and orbiviruses) lacks these well-defined structures.

The genome consists of a 10- to 12-segmented double-stranded RNA with a total genome size of 16–27 kbp. The virus core contains many enzymes essential for the transcription and capping of viral RNA. These viruses are unusually resistant to heat, have a wide pH (3.0–9.0), and to lipid solvents but are sensitive to 95% ethanol, phenol and chlorine. Reoviridae viruses

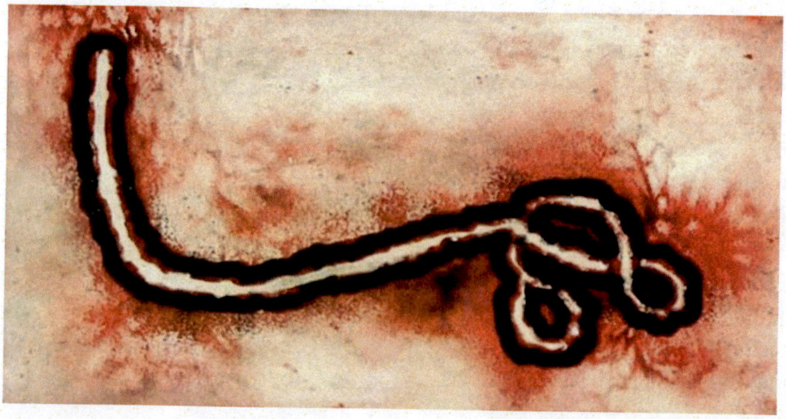

Fig. 64.2 Ebola virus under electron microscope. (http://phil.cdc.gov/phil/details.asp?pid=10815)

Table 64.5 Diseases associated with reoviruses

Reoviruses	Diseases
Orbiviruses	
Kemerovo virus	Oklahoma tick fever in Oklahoma and Texas in the United States Neurological infections in Russia and Central Europe
Orungo virus	Acute illness with headache and myalgia in Africa
Lebombo virus	Clinical illness in humans in Africa
Changuinola virus	An acute self-limited febrile illness in Panama (a single case)
Coltiviruses	
Colorado tick fever virus	Colorado tick fever
Orthoreoviruses	Mild upper respiratory tract illness, gastrointestinal tract illness and biliary atresia
Rotavirus	Diarrhoea in infants and young children

associated with human diseases are listed in Table 64.5.

Rotavirus

Rotavirus is the most common agent of gastroenteritis in children aged 6 months to 2 years. The virus was identified by Bishop and his associates in the diarrhoeic stool of the infants.

Morphology

Rotavirus has a distinctive morphological appearance by the negative stain in EM. The virus measures 70 nm in diameter and possesses a three-layer icosahedral capsid without an envelope (Fig. 64.3). The sharply defined circular outline of the outer capsid gives the appearance of the **rim of a wheel** placed on short pokes radiating from a wide arm. This appearance gives the virus the name “rota” (Latin word which means “wheel”).

Virus Genome

The genome consists of a segmented double-stranded RNA genome. The outer shell comprises a major glycoprotein with a molecular weight of 34,000 Da. This protein is known as a viral protein (VP7). It also consists of a minor trypsin-sensitive protein with a molecular weight of 84,000 Da. This protein is now designated as VP4, which was earlier called VP3. The minor shell or core protein of the virus core consists of four proteins: VP1, VP2, VP3 and VP6. The virus also consists of six non-structural proteins (NS53, NS34, NS35, NS28, NS26 and NS12).

Virus Replication

Rotaviruses replicate in the cytoplasm of the host cell. Infection of the cell begins by attachment to the cell surface at the site of the beta-adrenergic receptor. After entry into the cell, the virion uncoats and in the cytoplasm, the RNA-dependent RNA polymerase synthesises mRNA from each of the 10–11 segments.

Virus Antigens

Human rotaviruses have been classified into a total of 11 G serotypes primarily based on the VP7 (glycoprotein G) and VP4 (protease-sensitive protein P) proteins. Serotypes G1, G2, G3 or G4 have been identified with most infections and are designated as serotypes 1, 2, 3 or 4, respectively. There are six different HRP-P types. P type 1 is usually associated with G type 1, 3 or 4, whereas P type 1b is usually associated with a type 2.

Rotaviruses are classified into seven groups (from A to G) based on the antigenicity of VP6 and the electrophoretic mobility of the genomic segments. Human disease is caused mostly by group A and occasionally by group B and C rotaviruses.

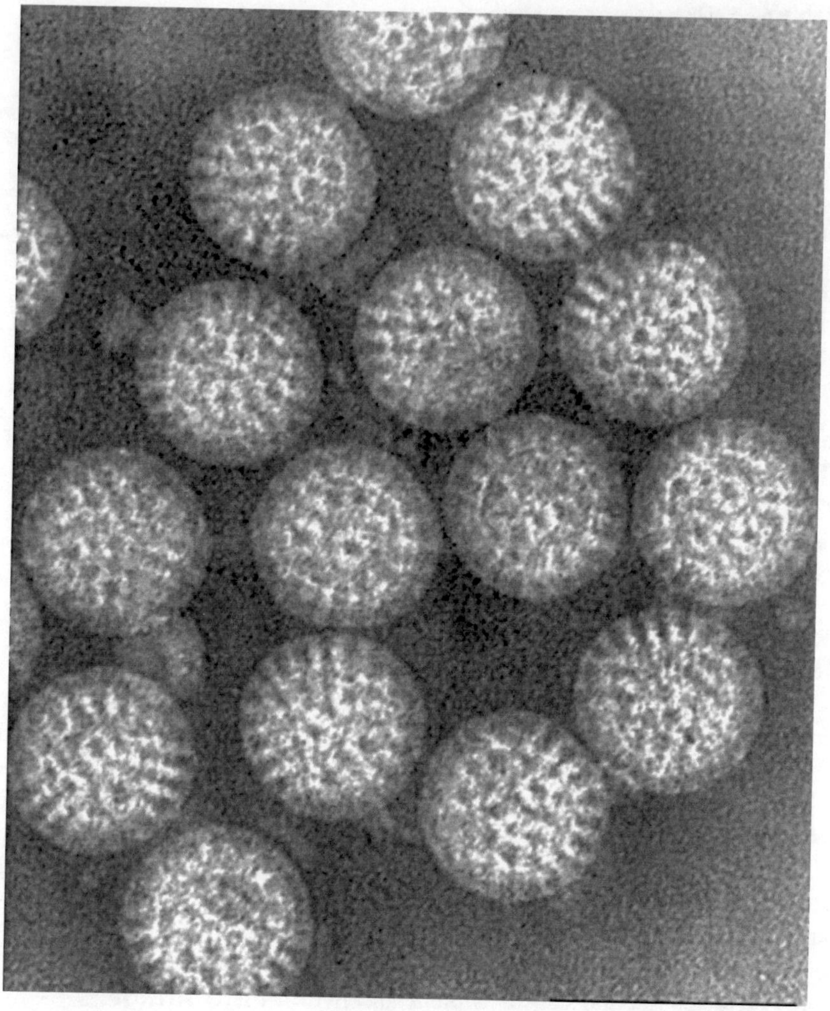

Fig. 64.3 Structure of rotavirus

Other Properties

Rotaviruses are relatively stable at low temperatures. They are also stable at pH extremes (3.5–10) and repeated freezing and thawing. Treatment with proteolytic enzymes, such as trypsin enhances the virus's infectivity.

Laboratory Isolation

Tissue Culture

Human rotaviruses are difficult to grow. They do not grow readily in cell cultures. Rotavirus growth is facilitated if the viruses are pre-treated with proteolytic enzyme trypsin and if low levels of trypsins are included in the tissue culture medium. This causes cleavage of the outer capsid protein of the virus and thereby facilitates uncoating.

Pathogenesis of Rotavirus Diarrhoea

Rotavirus is acquired by faeco–oral transmission in the infected host. It survives the acidic environment in the stomach and initiates infection in the mucosal cells of the small intestine. It does not cause infection in the mucosa of the stomach and large intestine. After absorption, the viruses replicate in the cytoplasm of the enterocytes and damage their transport mechanism.

Non-structural protein 4 (NSP4) of the rotavirus may act as a viral enterotoxin, which causes the secretion of fluids by stimulating a signal transduction pathway. The toxin induces an influx of calcium ions into enterocytes and release neuronal activators, and alters sodium and glucose absorption. The resulting diarrhoea is due to impaired sodium and glucose absorption, as damaged cells on the villi are replaced by non-absorbing immature crypt cells.

After replicating in the cell, the rotavirus causes damage to the cell. The damaged cells are released into the lumen of the intestine, releasing large quantities of viruses in the diarrhoeic stool. It takes around 3–8 weeks for restoring the normal function of the cell. Hence, the rotavirus produces watery diarrhoea similar to that seen in cholera.

Immunity

Rotavirus infection is characterised by the presence of a high quantity of immunoglobulin A (IgA) in the intestinal secretions. The IgA plays an important role in conferring gut immunity against rotavirus. It protects newborns up to the age of 6 months. The serum antibodies usually reduce the severity of the disease but do not necessarily prevent reinfection. Even a small quantity of a virus may cause infection and diarrhoea in the absence of specific antibodies in the serum.

Clinical Syndromes

Rotavirus is a major cause of diarrhoea in infants and young children.

Diarrhoea

The incubation period is short, being less than 48 h. The condition manifests commonly as fever, vomiting, diarrhoea and occasionally dehydration. Vomiting is usually of short duration and can occur before or after the onset of diarrhoea. The diarrhoeic stool may be watery, green or yellow but does not contain mucus. Rotavirus diarrhoea is a self-limiting disease, and patients recover completely within 5–10 days without any complications on sequelae. Also, rotavirus has been reported as an agent of traveller's diarrhoea in adults.

The virus has also been reported to cause gastroenteritis in adults. There have been increasing reports of outbreaks of rotaviral diarrhoea in adults and older children from many parts of China, and this rotavirus is named **adult diarrhoea rotavirus (ADRV)**.

Laboratory Diagnosis

Specimens

Diarrhoeic stool is the specimen of choice to demonstrate rotavirus and viral antigens.

Microscopy

Rotavirus can be demonstrated in stool by direct EM and IEM. **Direct EM** is a rapid method for detecting rotavirus in the faecal specimen. EM detects typical 70-nm cart-wheel-shaped viruses in the stool. The sensitivity of this method, however, is low. **On-grid IEM** is a three times more sensitive and reliable method than the conventional EM for demonstrating rotaviruses in the stool.

Antigen Detection

Enzyme immunoassays, such as Rotazyme and latex agglutination test, are useful tools to detect rotavirus antigen directly in the stool to diagnose diarrhoeal illness. This is a method used for routine diagnosis of rotavirus diarrhoea.

Cell Culture

Rotaviruses are difficult to grow in cell culture, hence are not routinely used to diagnose rotavirus diarrhoea.

Treatment

No specific antiviral therapy is available for rotavirus infection. The treatment of the condition is mostly supportive. It consists of restoring the fluid loss in dehydrated patients. Oral rehydration fluid using glucose and electrolyte solution is currently preferred over intravenous fluids to restore hydration. Intravenous fluids are used only in severely dehydrated children.

Epidemiology

Rotavirus is found worldwide. The viruses are excreted in the diarrhoeic stool of the children 2–5 days after the start of diarrhoea. Infected children are the common source and reservoir of infection. The virus is transmitted from person to person by the faecal–oral route. The virus is resistant to environmental conditions hence survives well on fomites, as well as on hands. The rotavirus diarrhoeal disease shows a seasonal variation, the disease being more common in autumn, spring and winter. A rotavirus called ADRV has been isolated from the outbreaks of diarrhoea in children and adults in China.

Prevention and Control

Few vaccines have been evaluated to protect children from rotavirus diarrhoea. Improved personal hygiene, including handwashing and isolation of known cases of rotavirus, are the best modes of control of the rotavirus diarrhoea.

Vaccines

A rhesus–human reassortant rotavirus tetravalent vaccine (RRV-TV; Rota shield) is a recent vaccine being used since 1998.

The vaccine consists of a polyvalent preparation consisting of the VP7 protein of each of the four clinically important serotypes with the attenuation phenotype of the rhesus rotavirus. This has successfully prevented severe rotavirus diarrhoea in two different studies conducted in Finland with protection rates of 80–91%. In developing countries, such as Peru and Brazil, the efficacy of this vaccine has been less impressive, varying between 20% and 30% protection rates in children. Several rotavirus vaccine trials have been conducted across the country. The ROTAVAC rotavirus vaccine study, a multi-centre Phase 3 efficacy and safety clinical trial, was India's first and largest efficacy clinical trial on vaccines. It was successfully completed in 2013. Presently Rotavac is licenced for use in India. Another trial initiated in 2014 in India as a Phase 3 trial was the bovine-human reassortant pentavalent rotavirus vaccine (BRV-PV) targeting the globally most common strains, namely Rotasiil. Both the vaccines were safe, well tolerated and provided significant efficacy against severe rotavirus gastroenteritis.

Orbiviruses

Orbiviruses are primarily animal pathogens that cause disease mainly in animals. They are named for their ring-shaped (Latin word *orbit* means "ring") structure. These viruses are differentiated from the orthoreovirus by their protein structure and their transmission by arthropod vectors. The genus *Orbivirus* consists of 19 species and at least 130 subspecies.

The orbiviruses are non-enveloped viruses with two-layered capsids. The genome consists of 10 segments of double-stranded RNA. The virion measures 70–80 nm in diameter. The virus multiplies inside arthropod vectors and vertebrates; hence, they come under arboviruses.

Only a few orbiviruses are linked to disease in humans. Kemerovo virus has been implicated in neurological infections in Russia and Central Europe. Lebombo virus, isolated in Africa, has been implicated with clinical disease in humans. Orbiviruses cause severe flu-like disease, encephalitis and polyradiculitis in humans. Fever, headache and myalgia are the common manifestations.

Coltiviruses

Coltiviruses resemble the orbiviruses in their morphology and in having two capsids. The genome consists of a 12-segmented double-stranded RNA. The coltiviruses associated with the human disease include Colorado tick fever, Salmon River virus, Banna virus, Beijing virus, Gansu virus and Eyach virus.

Orthoreoviruses

Orthoreoviruses are non-enveloped viruses, measuring 80 nm in diameter. It is composed of an inner protein shell, that is, core and an outer protein shell known as an outer capsid. The inner core comprises three major lipids, 71, 72 and cr2 and many minor proteins. The core consists of a 10-segmented double-stranded RNA. The reoviruses have three serotypes, namely, type 1, 2 and 3, based on neutralisation and haemagglutination inhibition test. All these serotypes share a common complement fixation antigen.

Coronaviruses

Coronaviruses were so named for the crown-like appearance of their virions on EM. These viruses are the second most important cause of the common cold, rhinoviruses being the first cause. The coronaviruses have also been reported to cause gastroenteritis in children and adults. SARS-CoV was a newly described coronavirus in 2002. The virus causes atypical pneumonia called SARS, a serious and potentially life-threatening viral infection of humans.

SARS-CoV causing atypical pneumonia called severe acute respiratory syndrome or SARS was first reported in 2002 and 2003 outbreak. This outbreak caused by the virus is believed to have originated in Guangdong province, South China. It predominantly affected mainly China, Hong Kong, Singapore and Taiwan. Subsequently, this outbreak spread to neighbouring countries in Asia, Canada and the United States.

Morphology

Coronaviruses are enveloped viruses measuring 80–160 nm in size on EM. The glycoproteins appear as club-shaped projections (20 nm long and 5–11 nm wide) on the surface of the envelope. The genome in association with N protein forms a helical nucleocapsid. The virus contains the glycoproteins E1 and E2 and a core nucleoprotein N. Some strains also contain a glycoprotein E3, which is haemagglutinin neuraminidase. E1 glycoprotein is a transmembrane matrix protein; E2 glycoprotein mediates viral attachment and membrane fusion (Fig. 64.4).

Virus Genome

SARS-CoV is a single-stranded, non-segmented, plus-sense RNA virus. It measures approximately 30 kb in length. The genomic sequence of SARS-CoV is different from other coronavirus strains. These strains of SARS-CoV are quite stable, unlike other coronaviruses in which mutations in the RNA sequence during virus replication are common. Such mutations contribute to the appearance of a new virus to be either more or less virulent. In contrast, the genome sequence of different isolates of SARS-CoV is very similar.

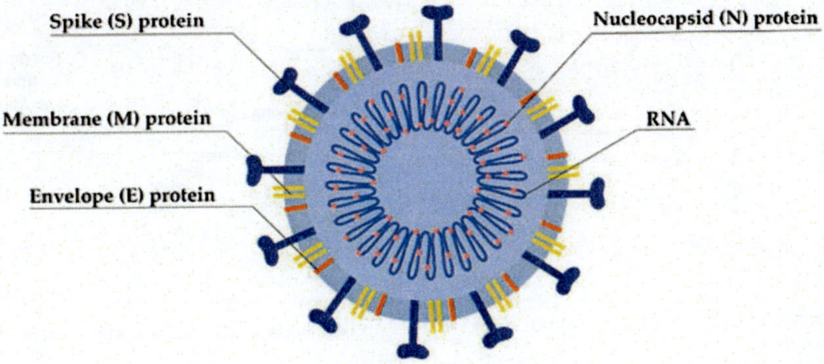

Fig. 64.4 Schematic diagram of SARS (https://commons.wikimedia.org/wiki/File:Schematic_structure_of_SARS-CoV-2.webp)

Virus Replication

The attachment of the virus to the cell is mediated by haemagglutinin. The virus enters the cell and uncoats. The replication of the virus occurs in the cytoplasm. Synthesis of protein occurs in two stages similar to that of togaviruses. The assembly of viruses occurs and derives its envelope from the endoplasmic reticulum but not from the plasma membrane (Fig. 64.5).

Other Properties

Coronaviruses have two serotypes, namely, 229E and OC43. SARS virus is relatively stable. They are sensitive to acid, ether and bile. They are difficult to grow in routine cell cultures.

Pathogenesis of Coronavirus Infections

The coronavirus infection is acquired by inhaling infectious droplets during coughing or sneezing from a SARS patient. These viruses typically cause infections in the upper respiratory tract because the optimal temperature for replication of viruses is 37–38 °C. SARS-CoV causes infection in the respiratory tract by binding to angiotensin-converting enzyme 2 receptors on the surface of the respiratory epithelium. This causes alteration in fluid balance and leads to the development of oedema in the alveolar space. Diffused oedema resulting in hypoxia is characteristic of pneumonia caused by SARS-CoV. Infections caused by coronaviruses produce a short and brief immunity, but reinfection can occur.

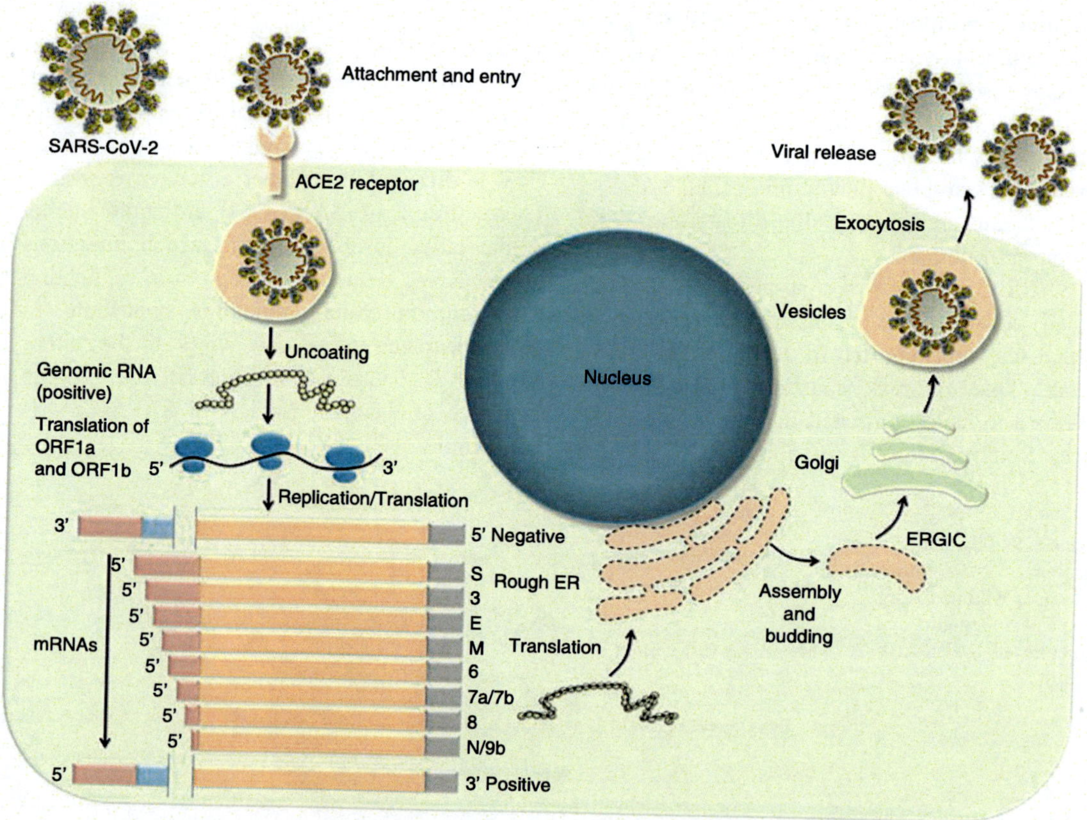

Fig. 64.5 Entry and replication of SARS-CoV-2 in host cells. (https://link.springer.com/chapter/10.1007/978-981-15-4814-7_3)

Clinical Syndromes

Coronaviruses cause SARS, common cold and gastroenteritis.

SARS

SARS is a potentially life-threatening infection associated with the onset of flu-like syndrome, which may progress to pneumonia, respiratory failure and in some cases death. The incubation period varies from 2 to 7 days, although it may be as long as 2 weeks.

Flu-like prodrome is the first stage of the disease, which is characterised by fever (>30 °C), fatigue, chills, malaise, anorexia and myalgia. The second stage represents the lower respiratory tract illness that begins three or more days after incubation. Hypoxia, cough, dyspnoea and breathing difficulties are common findings. The cough typically tends to be dry and non-productive and may range from mild to severe. Fever is typically higher than 30 °C. Chest X-ray shows interstitial ground-glass infiltrates that do not show cavitation. The condition is associated with lymphadenopathy and thrombocytopenia.

Morbidity and mortality due to SARS is more in the elderly population and also seen in more individuals with co-existing chronic illness and immunosuppression. The mortality rate of SARS is higher than influenza and other respiratory tract illnesses. The overall mortality is more than 10% and is more than 50% in elderly individuals above 65 years.

Common Cold

The coronaviruses (229E and OC43) cause more commonly upper respiratory tract and less commonly lower respiratory tract illnesses in humans. The common cold caused by coronaviruses has an incubation period of 3 days. The condition is characterised by rhinorrhea, sore throat and low-grade fever. The condition typically lasts for several days.

Gastroenteritis

The coronaviruses have also been reported to cause gastroenteritis in children and adults. The symptoms are mild, and the condition is self-limiting.

Laboratory Diagnosis

Laboratory diagnostic tests are most essential to confirm the diagnosis of SARS.

Specimens

Specimens include respiratory secretions for isolation of the virus and serum for testing of antibodies.

Cell Culture

SARS-CoV can be isolated in viral cultures. Isolation of virus is attempted only in class III laboratories.

Serology

Serodiagnosis of SARS depends on detecting specific antibodies to SARS-CoV in serum obtained during acute illness or 28 days or more after the onset of disease. IFA and ELISAs are the most frequently used tests for detecting specific antibodies in patients with SARS. Demonstration of a fourfold rise of antibody titre between acute and convalescent (>28 days after symptom onset) serum samples is suggestive of SARS.

Molecular Detection

Reverse transcriptase-polymerase chain reaction (RT-PCR) is employed to detect SARS-CoV RNA in serum, stool and nasal secretions. The test is positive in some patients within the first 10 days of fever.

Treatment

No specific antiviral treatment is available against SARS. Treatment is mostly symptomatic as given for serious community-acquired pneumonia. Persons suspected of SARS should be hospitalised immediately and kept in isolation under close observation. Supportive treatment may include assisted breathing using a ventilator, antibiotics to treat secondary infections due to bacteria and steroids to reduce lung inflammation. None of these is full proved modalities. The antiviral medicine ribavirin is known to be ineffective at treating SARS.

Epidemiology

SARS is restricted in its geographical distribution.

Geographical Distribution

The SARS outbreak in 2002–03 predominantly affected mainland China, Hong Kong, Singapore, Taiwan and Canada. The SARS strain is believed to have originated in Guangdong province in southern China. It was first reported in February 2003. Over the next few months, the illness spread to more than two dozen countries in North America, South America, Europe and Asia before the SARS global outbreak of 2003 was contained. The disease was epidemiologically linked to the National Institute of Virology in Beijing, where the outbreak is thought to have originated. A total of 8098 cases, 774 deaths and 7324 recoveries from SARS were documented between November 2002 and 21 April 2004.

Reservoir, Source and Transmission of Infection

SARS is believed to be primarily transmitted by close person-to-person contact. Most cases of SARS were seen in individuals who lived with or cared for a patient with SARS or who had exposure to contaminated secretions from a SARS patient. The infection would possibly have been acquired by inhaling infectious droplets during coughing or sneezing from a SARS patient. Moreover, another possible mode of transmission is direct contact of the eyes, nose or mouth of susceptible individuals with infectious secretions of SARS patients.

It is suggested that SARS-CoV may have originated in chickens, ducks or small mammals. In these animals, the virus would have mutated, subsequently being transmitted, causing human infection. The transmission to humans is facilitated possibly by the proximity in which humans and livestock live in rural southern China. The avian flu epidemic in Hong Kong in 1977, which originated in poultry and spread to humans (resulting in the slaughter of millions of chickens), is a classic example of this type of zoonotic transmission.

It is also postulated that the SARS-CoV initially originated in civet cats, which were sold in a Guangdong marketplace in rural southern China as a food delicacy. Close contact with their saliva or faeces or with the animals themselves possibly has transmitted a mutated SARS-CoV to humans.

Prevention and Control

Isolation of patients and strict barrier nursing is crucial to prevent transmission of SARS to others. Moreover, airport screening for potentially sick and/or febrile passengers is being carried out in SARS-affected regions in Asia by using infrared scanners.

Middle East Respiratory Syndrome (MERS-CoV)

Middle East respiratory syndrome (MERS) is a viral respiratory disease caused by a novel coronavirus (Middle East respiratory syndrome coronavirus, or MERSCoV). It was first identified in Saudi Arabia in 2012. MERS-CoV is a zoonotic virus and is transmitted between animals and people. The origins of the virus are not fully understood but are believed to have originated in bats and transmitted to camels sometime in the distant past. Humans are infected through direct or indirect contact with infected dromedary camels. MERS-CoV has been identified in several countries in the Middle East, Africa and South Asia. Since 2012, 27 countries have reported cases of MERS. Dromedary camels are the major reservoir host for MERS-CoV. The virus does not pass easily from person to person unless there is close contact, such as providing unprotected care to an infected patient.

The clinical spectrum of MERS-CoV infection varies from asymptomatic or mild respiratory symptoms to severe acute respiratory disease and death. A typical presentation of MERS-CoV disease is fever, cough and shortness of breath with or without pneumonia or gastrointestinal symptoms. Approximately 35% of patients with MERS have died. No vaccine or specific treatment is currently available. Treatment is mainly supportive. The consumption of raw or undercooked animal products, including milk and meat of camel, carries a high risk of infection. Preventive measures include general hygiene after visiting animal product selling markets, regular hand washing before and after touching animals and avoiding contact with sick animals.

COVID-19 and SARS-CoV-2

Since December 2019, the world has faced a major pandemic with another coronavirus. Severe acute respiratory syndrome coronavirus 2 (SARS-CoV-2) is the virus responsible for the outbreak of coronavirus disease (COVID-19), which was first identified in Wuhan, China, following reports of serious pneumonia. The new SARS-CoV-2 is most closely related to a group of SARS-CoVs found in humans, bats, pangolins and civets. It is very contagious and has quickly spread around the world. An estimated total number of deaths worldwide due to COVID-19 have exceeded 5 million.

Most people infected with the virus experience mild to moderate respiratory illness and recover without requiring special treatment. However, serious illnesses requiring medical interventions, like assisted breathing oxygen support, have also been noted. Aged people and those with underlying comorbidities, like cardiovascular disease, diabetes, chronic respiratory disease or cancer are more likely to develop serious illnesses. Some people, including those with minor or no symptoms, may suffer from post-COVID conditions or "long COVID".

Transmission

The virus is thought to be of natural animal origin. Evidence suggests that the virus is a descendant of a coronavirus that infects wild bats and spreads to humans through an intermediary wildlife host. The virus can spread from an infected person's mouth or nostrils via small liquid particles when they cough, sneeze, speak, sing or breathe. These particles range from larger respiratory droplets to smaller aerosols. Besides, surfaces and fomites have a minimal role in the transmission.

Laboratory Diagnosis

Several testing methods have been developed to diagnose COVID-19. The standard diagnostic method is by detection of the virus nucleic acid by real-time reverse transcription-polymerase chain reaction (rRT-PCR), transcription-mediated amplification (TMA) or by reverse transcription loop-mediated isothermal amplification (RT-LAMP) from a nasopharyngeal swab. Besides, serological testing identifying one of the outer proteins of the viral shell or envelope, called antigen testing, has also been used. Antibody detection has also been used to test for the immune response against COVID-19 vaccines and rarely for diagnosis. Chest CT scans may be helpful to diagnose COVID-19 in individuals with a high clinical suspicion of infection but are not recommended for routine screening.

Prevention and Control

Several COVID-19 vaccines have been approved and used in various countries in the form of mass vaccination campaigns. Among various vaccines, the two major vaccines used in India are COVISHIELD and COVAXIN. COVISHIELD is a recombinant, replication-deficient chimpanzee adenovirus vector encoding the SARS-CoV-2 Spike (S) glycoprotein. Following administration, the genetic material of the part of coronavirus is expressed, which stimulates an immune response. It comprises 2 doses. The Indian government recommended an interval of 12–16 weeks between the first and second dose. COVAXIN is India's indigenous COVID-19 vaccine developed by Bharat Biotech in collaboration with the Indian Council of Medical Research (ICMR)-National Institute of Virology (NIV). The vaccine is developed using Whole-Virion Inactivated Vero Cell-derived platform technology. Inactivated vaccines do not replicate and are therefore unlikely to revert and cause pathological effects. They contain dead viruses, incapable of infecting people but still able to instruct the immune system to mount a defensive reaction against an infection. It is a 2-dose vaccination regimen given 28 days apart.

General preventive measures include physical or social distancing, quarantining, ventilation of indoor spaces, respiratory and cough etiquettes, hand washing, etc. The use of face masks or coverings has been recommended in public settings to minimise the risk of transmission. Management involves the treatment of symptoms, supportive care, isolation, etc.

Variants

It is normal for viruses to change and evolve as they spread between people over time. When these changes become significantly different from the original virus, they are known as "variants", which are identified by mapping the genetic material of viruses and comparing them for differences. Since the SARS-CoV-2 virus, the virus that causes COVID-19, has been spreading globally, variants have emerged. A variant is considered a variant of interest if it has mutations that are suspected or known to cause significant changes, and the mutant is circulating widely. On the contrary, a variant of interest becomes a variant of concern if it is known to spread more easily, cause more severe disease, escape the body's immune response, change the clinical presentation or decrease the effectiveness of known surveillance, testing and management tools. One such variant is the Omicron.

Omicron Variant

The Omicron variant, variant B.1.1.529, was first reported to WHO on 24 November 2021 and was classified as a variant of concern by WHO on 26 November 2021. The classification was based on information from South Africa that the variant has a large number of mutations and has caused a significant change in COVID-19 epidemiology. This variant has many mutations, with preliminary evidence suggesting an increased risk of reinfection with this variant. This variant has been detected faster than previous surges in infection, suggesting that this variant may have a growth advantage. The pandemic continues presently, with this variant being reported in more than 77 countries worldwide.

ZIKA Virus

ZIKA virus belongs to the Flaviviridae family, and it is transmitted through *Aedes* mosquitoes, rarely from mother to baby and through infected blood or by sexual means. Out of five ZIKV-infected people, one may suffer from fever, rash, joint pain or conjunctivitis. Additionally, the patient may suffer from muscle pain and headache. The incubation time of this virus is unknown. Symptoms may persist for several days to a week, and deaths are rarely reported. Recent findings suggest that the ZIKA virus is associated with the development of microcephaly in newborns due to transmission of the virus from mother to baby during pregnancy. However, the actual causal role of this virus is yet to be proven, but CDC recommends special precautions for pregnant women who are planning a visit to endemic areas.

ZIKA virus outbreaks have been documented in Africa, Southeast Asia and the Pacific islands. In 2015 the Pan American Health Organization confirmed the first ZIKV infection in Brazil, and at present many other countries also have reported ZIKV outbreaks (Fig. 64.6). Currently, no vaccines are available to prevent ZIKV disease. Prevention can be done by avoiding day-biting mosquitoes.

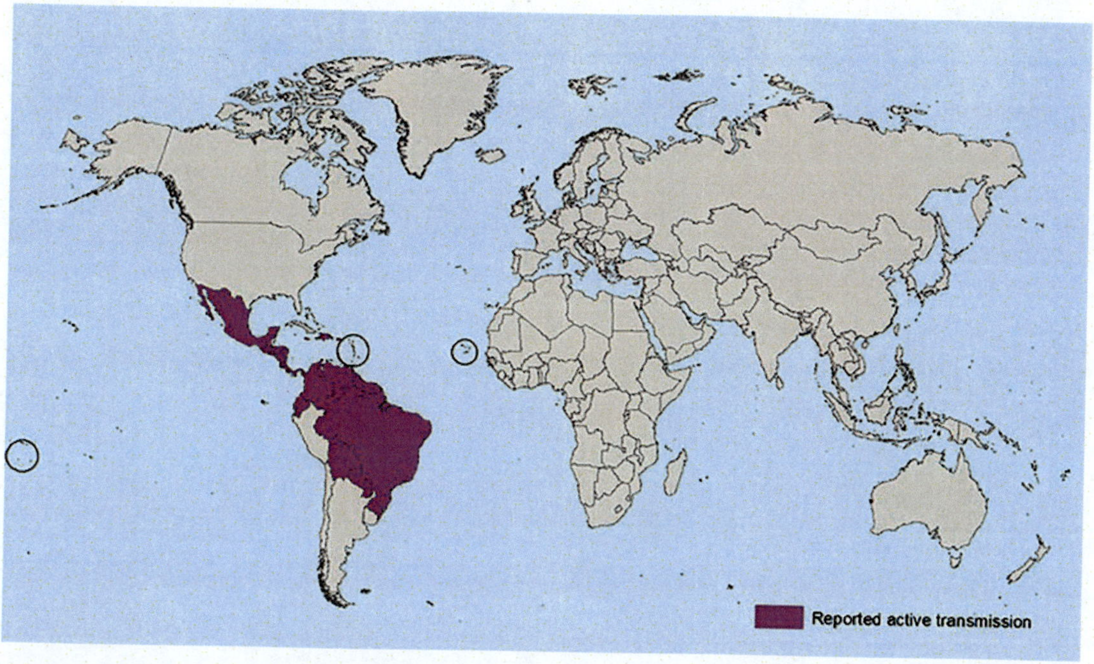

Fig. 64.6 ZIKV active transmission areas (https://commons.wikimedia.org/wiki/File:CDC_Countries_and_territories_with_active_Zika_virus_transmission_February_2016.jpg)

Case Study

A group of medical college students went for dining in a hotel on the last day of their final university examination. After 2 days of dining in the hotel, they all complained of mild fever, nausea, vomiting and watery diarrhoea. The symptoms lasted only for 24 h. The symptoms were not suggestive of bacterial gastroenteritis or food poisoning.

- What is the most likely viral aetiology of the condition?
- How is this viral agent transmitted?
- What are the laboratory tests used to diagnose the condition?
- How can you prevent the condition?

Part V

Mycology

Introduction to Mycology

65

The term mycology is used for the study of fungi. The term fungi originated from the Greek word 'mykos' which pertains to mushroom. Fungi belong to Eukaryotes and based on nutrition requirement, is classified as chemoheterotorphs. They exist both in unicellular and/or multicellular forms. The salient differentiating features of fungi and bacteria are provided in Table 65.1. The characteristics features of fungi are the presence of chitin and ergosterol in their cell wall.

Cell wall: Chitin is one of the major constituents of the fungal cell wall. It is a polysaccharide with long chains of *N*-acetylglucosamine. It provides rigidity to the fungal cell wall. The amount of chitin and another major constitute glucan varies amongst different fungal species. The fungal cell wall also contains mannoproteins, chitin synthase and glucan synthase enzymes. Of these, beta-glucan is the most important, because it is the target of antifungal drugs, echinocandins and is used as a biomarker for the diagnosis of invasive fungal diseases.

Cell membrane: The cell membrane of fungi is rich in lipids such as glycerophospholipids, sphingolipids and sterols. Fungal sterols are typically present in the form of ergosterol, which is in contrast with the human cell membrane which is made up of cholesterol. The antifungals such as polyenes (amphotericin B), triazoles (itraconazole, voriconazole and fluconazole) and allylamines (terbinafine) all target the cell membrane either killing or inhibiting the fungi. These antifungals selectively target the ergosterol present in the fungi and not the cholesterol that is present in humans.

Classification of Fungi

The fungi can be classified as follows:

Taxonomical Classification

Fungi belong to monophyletic clade classified as the kingdom 'Fungi'. Fungi causing human and animal infections belong to four phyla such as Ascomycota, Basidiomycota, Glomeromycota and Microsporidia and subphylum Mucormycotina and Entomophthoromycotina. Earlier some fungal species were arbitrarily classified as Deuteromycetes or Fungi Imperfecti.

Mucormycotina

Lower fungi have non-septate hyphae and form endogenous asexual spores known as sporangiospores which are inside a swollen-sac-like structure called sporangia. Sexual spores termed as zygospores are also produced. In contrast, the higher fungi have well-developed septum in the hyphae and produce conidia exogenously. Human infection (mucormycosis) is caused by fungi in the order Mucorales such as *Rhizopus, Mucor, Apophysomyces, Absida, Cunninghamella,* etc.

Table 65.1 Comparison of fungi and bacteria

Feature	Fungi	Bacteria
Classification	Eukaryote	Prokaryote
Mode of nutrition	Heterotrophic	Both autotropic and heterotrophic
Morphological forms and size in diameter	Yeast 3–6 μm Moulds – 3–12 μm	Cocci, bacilli, spiral and branching filaments ~1–5 μm
Cell wall content	Chitin, 1,3-beta D glucan, mannoproteins	Peptidoglycan
Cell membrane	Mainly sterols like ergosterol	No sterols with the exception of *mycoplasma*
Cytoplasm	Mitochondria and endoplasmic reticulum present	No mitochondria or endoplasmic reticulum Contains actin like proteins
Nucleus	Eukaryotic	Prokaryotic
Reproduction	Both sexual and asexual Asexual by budding, binary fission and free cell formation	Only asexual by binary fission

Note: *PAS*: periodic acid–Schiff; *GMS*: Gomori's methenamine silver

Ascomycota

Ascospores or sexual spores are formed within the ascus or sac. Both filamentous fungi and yeasts are known to produce ascospores.

Basidiomycetes

Basidiospores or sexual spores are formed on a base or basidium.

Fungi Imperfecti

It is also known as deuteromycetes or hyphomycetes, and sexual phases of this group of fungi have not been identified. Hyphomycetes are further classified as phaeohyphomycetes in which the hyphae appear black due to pigment called melanin and hyalohyphomycetes in which the hyphae are colourless or hyaline. Many medically important fungus group belongs to this family.

Morphological Classification

Based on the microscopic and macroscopic morphology, fungi are classified under three groups as described below.

Yeast

Yeasts are unicellular cells that reproduce by budding. The shape and size vary among the different species. For example, *Cryptococcus neoformans* appears round, whereas *Candida albicans* are oval. Some yeasts produce elongated cells called pseudohyphae (earlier referred as yeast-like fungi (e.g. *C. albicans*, *Candida tropicalis*, etc.) and other do not produce pseudo hypha (earlier referred as true yeasts) e.g. *Cryptococcus* and *Saccharomyces cerevisiae*. On culture medium, such as Sabouraud dextrose agar (SDA), they produce creamy mucoid colonies. Example: *C. neoformans.*

Moulds

Moulds grow as long filaments called *hyphae.* They usually measure 2–10 μm in width. Some hyphae form transverse walls and hence they are called septate hyphae, whereas others do not produce walls, hence are called non-septate hyphae. Non-septate hyphae are multinucleated. The hyphae on their continuous growth form a mat known as *mycelium.* The part of the mycelium that projects above the surface in the culture medium is called aerial mycelium. Examples include *Aspergillus, Penicillium, Rhizopus,* etc.

Dimorphic Fungi

Many of medically important fungi are dimorphic. They exist as hyphal/mycelial forms at lower temperatures, for example 22–25 °C in the environmental niche such as air and soil. They occur as yeasts or other structures in human tissue and in the culture at 37 °C

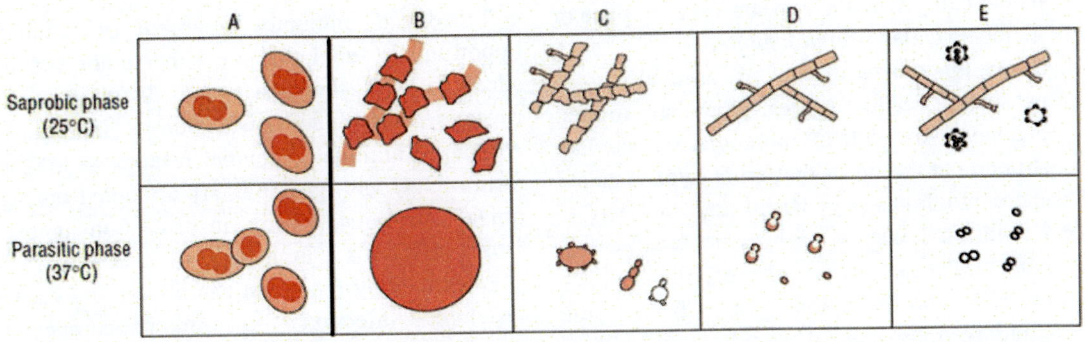

Fig. 65.1 Dimorphic fungi. (**a**) Cryptococcus; (**b**) Coccidioides; (**c**) Paracoccidioides; (**d**) Blastomyces; (**e**) Histoplasma

(Fig. 65.1). Examples include *Coccidioides immitis*, *Paracoccidioides brasiliensis*, *Histoplasma capsulatum*, *Blastomyces dermatitidis*, *Talaromyces marneffei* and *Sporothrix schenckii* species complex.

Pathogenic Classification

Fungi can be classified into the following types based on their pathogenicity.

Primarily Pathogenic

Fungi like thermally dimorphic fungi can cause infections in apparently healthy individuals.

Opportunistic Pathogens

This fungus group can cause infections in patients who already have defective immune mechanism.

Individuals infected with AIDS, taking immunosuppressive and suffering from diabetes mellitus are some of the examples.

Reproduction of Fungi

Fungi have the ability to reproduce both sexually and asexually. The end product of sexual forms is called spores and asexually products are called conidia. Zygospores, ascospores and basidiospores are produced by Mucoromycotina, Ascomycota and Basidiomycota, respectively. Ascospores are formed within the sac-like structures called ascus which may be present naked or many times in closed structures such as cleistothecia, apothecia, etc. The basidiospores are produced on the surface of *basidium* through the four sterigmata present on its surface. Zygospores are generally single large spores and are ornamented with thick walls. Sexual reproduction occurs mainly by the process of meiosis and can later be followed by mitosis. Fungi that do not produce sexual spores are earlier classified as *Fungi imperfecti*. The end product of the **Asexual reproduction is called** conidia which are produced by the process of mitosis. Conidiogenesis, the process of conidia production, conidial shape, colour and arrangement are used for morphological identification of the fungi. Some of the asexual spores or conidia are described below.

1. **Arthrospores/arthroconidia** are formed by formation of the septa within the existing hyphae resulting in fragmentation into multiple cells. They are rectangular in shape and have a thick cell wall. The arthrocondia is the infective form of *C. immitis*.
2. **Chlamydospores/chlamydocondia** are the thick-walled conidia, round to oval that are produced asexually *in between (intercalary) or at the tip of the hyphae or pseudohyphae (terminal)*. The terminal chlamydospores help in the differentiating *C. albicans* and *C. dubliniensis* from other *Candida* species.
3. **Blastospores/blastocondida** are formed by the process called budding or blasting out from the yeast cells. After budding, the cells may get elongated and remain attached to the

parent cell giving the appearance of hyphae or filaments called *pseudohyphae*.

4. **Sporangiospores** are the only asexually product that is called spores. They are formed endogenously with in sporangia, the sac-like structure at the tip of the fertile hyphae called sporangiophores (e.g. *Mucor* and *Rhizopus*).
5. **Conidia** are asexual product produced externally on the sides or tips of specialised hyphae called conidiophores or from the conidiogenous cell of conidiophores. They are further classified as microconidia (single cell and small) and macroconidia (large and multicellular due to internal septation).

Pathogenesis of Fungal Infection

Most fungi are strict aerobes or facultative anaerobes, and they do not grow in strict an aerobic state. Majority of the fungi are associated with environmental substrates which are rich in organic matter. Being heterotrophic it is essential for them to acquire nutrition from external source. They have ubiquitous distribution and are found more in the air, soil and decaying matter, etc. Fungi can be seen as colonisers as well as can cause serious diseases. Fungi also can cause allergic disease and toxins released from them may lead to mycotoxicosis.

Colonisation and Disease

Fungi are seen as a coloniser in the gastrointestinal tract, upper respiratory tract, skin and other mucosal surfaces such as the vagina. Infection by colonised fungi is restricted due to the presence of innate immunity of the epithelial and other cells. Any break in the mucosal or epithelial barrier along with reduced innate immune response will allow the fungi to invade and cause disease. In addition to the innate immune response high fatty acid content, pH, frequent epithelial shredding and bacterial flora help in providing resistance to fungal infections. For example, mucosal surfaces trap the fungal spores that are inhaled restricting them to reach the lower respiratory tract.

Cell-mediated immunity plays a key role in protection against fungal infections. When the cell-mediated immunity is absent or reduced opportunistic fungi that are either colonised or silent invades the tissue and disseminate to cause serious opportunistic fungal infections. Although humoral immune responses in the form of IgG and IgM antibody are noted during fungal infections, its role in protection against fungal infections is limited.

Infections by fungi that can invade and cause systemic infection in immunocompetent individuals are generally referred to as systemic fungal infections. Common systemic infections include those caused by endemic fungi such as *Histoplasma, Blastomyecs, Coccidioides* and *Paracoccidioides*. Whereas, opportunistic fungal infections are referred as infections caused by *Candida, Cryptococcus, Aspergillus, Mucorales, Fusarium* etc. especially, in an immunocompromised individuals.

Allergic Fungal Diseases

As the fungi are ubiquitously abundant in the environment including air, everyday humans inhale plenty of fungal conidia, spores or fungal fragments present in the air. *Aspergillus* is the most common fungi causing allergic fungal disease. *Penicillium, Curvularia, Alternaria* and *Bipolaris* are other species involved. Allergic fungal disease affects either the upper airway in the form of allergic fungal rhinosinusitis (AFRS) or the lower airways in the form of allergic bronchopulmonary mycosis (ABPA) and bronchial asthma. Patients with AFRS and ABPA are atopic with type I IgE hypersensitivity to the fungal antigen along with increased levels of total IgE and *Aspergillus*-specific IgE.

AFRS patients usually present with nasal blockage. Histologically, the surgical specimen obtained from the paranasal sinus of those patients shows a characteristic feature called ‘allergic mucin’ which is composed of amorphous mucous and necrotic eosinophils. *Aspergillus* spp. especially *Aspergillus flavus* is the most commonest cause of AFRS.

Allergic bronchopulmonary aspergillosis is also a hypersensitive disease and patients presents with wheeze and productive cough associated with thick brown sputum. Generally, *Aspergillus*

Table 65.2 Transmission and habitat of some important fungi

Fungi	Habitat	Form transmitted	Portal of entry
Cryptococcus	Soil (pigeon faeces), Eucalyptus trees	Yeast	Inhalation into lungs
Aspergillus	Soil, vegetation	Conidia	Inhalation into lungs
Candida	Human body	Yeast	Normal flora of the skin, mucosa, gastrointestinal tract and vagina
Mucorales	Soil, vegetation	Sporangiospores	Inhalation
Talaromyces marneffei	Soil	Conidia	Inhalation of conidia

conidia that are inhaled will be cleared by the mucociliary action, but in patients with bronchial asthma and cystic fibrosis due to lack of ciliary clearance, the conidia will germinate and cause epithelial damage and initiate T-helper cell type-2 inflammatory responses. *Aspergillus fumigatus* is the most common causative agent of ABPA and recent reports suggest that *A. flavus* is also an important agent. Diagnosis can be made a clinical feature such as worsening of symptoms in cystic fibrosis and bronchial asthma patients, serum eosinophilia increase in total and *Aspergillus* specific and infiltrates on the chest radiography. High-attenuation mucus on CT scan is the pathognomic feature of ABPA. Allergic mucin noted in AFRS is also noted with ABPA.

Mycotoxicosis

Mycotoxicosis is the disease caused by ingestion or inhalation of the excessive toxins produced especially by the plant pathogen. Generally, the agricultural harvest gets contaminated with the fungi and releases different mycotoxins which when ingested causes acute or chronic mycotoxicosis. The symptoms of mycotoxicosis depend on the type of the mycotoxin, duration of exposure to mycotoxin and the organ involved. It affects the kidney, liver, lungs, skin, etc., again depending on the toxin. Aflatoxin is one of the most common mycotoxins affecting humans and animals. Aflatoxin, produced by *A. flavus,* causes hepatic carcinoma by imparting mutation in the tumour suppressor gene-P53 followed by the unchecked multiplication of hepatocytes. The association of this mutation has been noted in liver cancer patients from China and Africa. It has been hypothesised that kwashiorkor and Rey's syndrome are caused by aflatoxins. Ochratoxin is another major toxin produced by *A. flavus* and is known as nephrotoxin, hepatotoxin, mutagenic and carcinomatous. Citrinin, another mycotoxin produced by various species, is a nephrotoxin species. The yellow rice toxicosis reported is the classical example of mycotoxicosis.

The transmission and habitat of some important fungi in India are described in Table 65.2.

Laboratory Diagnosis

Laboratory diagnosis of fungal infections can be made either by conventional phenotypic methods like direct microscopic examination of specimens and isolation of fungi by culture or by using non-culture methods such as serology, detection of biomarker/s or DNA-based molecular methods.

Direct Microscopy

Direct microscopic examination is an important method that can be used for rapid diagnosis. Demonstration of the fungal elements such as hyphae and yeast forms in the clinical samples suggests either colonisation or infection. While dealing with opportunistic fungal infections caused by saprophytic fungi, it becomes more important especially to correlate with culture. Direct microscopy sometimes give a definitive diagnosis by demonstrating the size, shape and other morphological features in the specimens such as in coccidioidomycosis, histplasmosis, paracoccidioidomycosis, blastomycosis, mucormycosis, eumycetoma, sporotrichosis, etc.

Potassium hydroxide (KOH) wet mount examination is commonly used in the concentration of 10–40% depending on the specimens subjected to examination. For bone and nail samples, 40% KOH is desirable compared to 10% used for scrapings and biopsies. KOH dissolves all the protein material in the specimen but chitin present in the cell wall is resistant allowing identification of the fungal elements.

For more rapid detection of fungi in the tissue fluorescent dyes such as Calcofluor or blankophore which specifically binds to chitin present in the fungal are used and examined under the fluorescent microscope. Some special stains such as Gomori's methenamine silver stain, Periodic acid Schiff's stain can be used to demonstrate fungi from the histopathological tissues. Negative stains such as India ink or Nigrosin preparation are used to demonstrate the capsule of *Cryptococcus* spp. in the cerebrospinal fluid (CSF) or other body fluids. Gram stain can also be a useful technique especially to demonstrate *Candida* species in the exudates, scrapings or swabs.

The major disadvantage of direct microscopy is its low sensitivity and requires experience to interpret the results accurately.

Culture

Fungal culture, direct microscopic examination and histopathological examination of sterile specimens are the gold standards for the diagnosis of fungal infections. For primary isolation of the fungi from the tissues, the most commonly used media is Sabouraud's dextrose agar (SDA) which supports the growth of all medically important fungi. Instead of using the standard SDA, Emmons modification of SDA with less glucose and neutral pH should be preferred. Other primary isolation media that should be routinely inoculated include brain heart infusion agar (for blood culture and isolation of dimorphic fungi), bird seed agar (for selective isolation of *Cryptococcus*), CHROM agar (as differential media to differentiate *Candida* species), blood agar, etc. The media should be supplemented with antibiotics such as chloramphenicol or gentamycin to prevent bacterial growth, especially for non-sterile specimen. The addition of cycloheximide to the media will inhibit the saprophytic fungi and bacteria allowing the pathogen to grow. Two sets of SDA should be inoculated, one set with cycloheximide and chloramphenicol/gentamicin should be incubated at 25° and other set without these antibiotics at 37°.

Growth Characteristics

The growth characteristics generally helpful in the identification of fungi include: (1) morphology of the colonies/presence of hyphae or pseudohyphae (diameter, septa), conidia (spores) on the obverse of the culture plate, (2) colour of colonies on the obverse, (3) pigmentation on the reverse and (4) time taken for colony growth (slow/rapid). These characteristics can be observed in tease mount, slide culture or adhesive tape preparations using lactophenol cotton blue (LPCB).

Teased Mount Preparation

Several methods are available for microscopic preparation of fungi, of which tease mount is traditionally followed. It is an easy and rapid method to perform and often sufficient enough to identify fungi encountered commonly in clinical setups. Tease mounts are prepared for the purpose of demonstrating conidia, other reproductive structures or morphological forms may aid in the identification of the fungi. The characteristic shape, arrangement of conidia and size of hyphae are add-on information for identification of moulds. The major setback of this method is that the arrangement of conidia may be dislocated by the pressure applied on the cover slip during mount preparation.

Slide Culture

It has been an important diagnostic aid for many years and is being used worldwide. The major disadvantage of tease mount preparation can be resolved by this method. The strength of slide culture is that the fungi are grown and maintained in their characteristic shapes. The principle behind this technique is that growing, maintaining and periodically observing the colonies may help in accurate identification. Technical expertise and the time taken for colony growth and incubation duration are the demerits of this preparation.

An SDA plug is transferred to the centre of a sterile slide using a sterile scalpel. The slide is then set atop a piece of filter paper placed in the bottom of a sterile Petri dish. The fungal strain to be identified is inoculated to the edges of the agar plug in 3 or 4 areas. A sterile coverslip is placed on top of the agar plug. Moistened filter paper in the Petri dish prevent drying. Culture is incubated at 25 °C. After growth (approximately 48 h), the coverslip is then placed on a fresh slide with a drop of LPCB and microscopically observed. Sporulation-inducing media like potato dextrose agar, oatmeal agar and corn meal agar are alternatives to SDA.

Cellophane Tape Preparation

It is said to be a technically easier and more economical method for microscopic identification of fungi. A small length of thetape is touched to the surface of the colony. It is then adhered to a microscopic slide with a drop of LPCB, followed by microscopic observation. The advantage of this technique is that the conidia are generally intact, thus identification can be made with ease.

Though culture based method is gold standard for diagnosis of fungal infections, it has some limitation in terms of turnover time because of slow growing nature of some fungi and low yield sensitivity. The sensitivity of the technique can be improved by concentrating the specimens before inoculation of the specimens such as blood (lysis centrifugation) and body fluids by taking as much large volume as possible and subjecting it to centrifugation. However, culture can help in management of the case as antifungal susceptibility testing can guide accurate institution of the antifungals.

Serology

Serological techniques, especially antibody detection, have limitation in the diagnosis of fungal infections as raised antibodies can be noted in colonisation or prior exposure to opportunistic fungi. Antibody-based tests are, in general, useful for the diagnosis of endemic fungal infections such as histoplasmosis, coccidioidomycosis and blastomycosis. However, rise in antibody titre should be demonstrated in the paired sera to confirm the diagnosis. The complement fixation test is still used for the diagnosis of endemic fungal infections. Highly sensitive tests such as ELISA (enzyme-linked immunosorbent assay), Western blot and radioimmunoassay are developed for the diagnosis of these infections, interpretation becomes difficult as it may indicate prior exposure or previous infection. Antibody-based tests are mainly applicable for the diagnosis of allergic fungal diseases such as ABPA and allergic fungal sinusitis.

Antigen Detection

It is an useful approach for the diagnosis of fungal infections. Antigen detection can be done from serum, CSF or urine specimen. Detection of polysaccharide capsular antigen of *C. neoformans* in CSF is one of the best antigen detection tests and has the sensitivity and specificity of more than 95%. Latex agglutination test has been commonly used to detect cryptococcal antigen. However, recently, lateral flow assay is increasingly being used as it can be used as point-of-care test at the bedside as it does not demand technical expertise. Urine histoplasma antigen is one of the highly sensitive and specific techniques for the diagnosis of disseminated histoplasmosis. Detection of mannan, the component of Candida cell wall is another approach for diagnosing invasive candidiasis. However, this test has low sensitivity (~50%) and may not detect the infection by all the species of Candida. Hence, a combination of mannan antigen and mannan antibody detection tests are used which have high sensitivity and specificity (70–80%).

Galactomannan (GM) is a heat-stable heteropolysaccharide antigen found in the cell walls of all *Aspergillus* species and released into the blood only when the hyphae are invading the tissues. Hence only a low amount of GM is present in the circulation. The development of highly sensitive technique such as sandwich ELISA that uses rat monoclonal antibody raised against galactomannan antigen has allowed the lower detection of GM. The sensitivity of this test is one nanogram. This test is currently used for the diagnosis of invasive aspergillosis. Galactomannan (GM) detection can also allow monitoring the aspergillosis as the cut-off level will decrease if the patient responds to the antifungal therapy.

Non-culture Methods

Detection of Fungal Cell Wall Markers

Detection of the cell wall component, 3-beta-D-glucan (BDG) of the pathogenic fungi is another approach for the diagnosis of fungal infections, especially invasive candidiasis. BDG is released into the circulation during the invasive fungal infections in low quantity hence its detection can also be used as an indicator of invasive fungal infections. BDG has the ability to coagulate the amoebocytes of the horseshoe crab and this coagulation can be detected using a kinetic ELISA reader. As the amount of BDG present in the serum sample suggests invasive disease, the cut-off has been described to interpret the results. The cut-off of >80 picogram is considered significant. The sensitivity of diagnosing invasive candidiasis is >90%. As *C. neoformans* and *Mucorales* have very less BDG in their cell wall, this test may not be useful to diagnose infection by these agents.

Detection of Fungal Metabolites

However, detection of fungal metabolites such as D arabinitol can be used for the diagnosis of invasive candidiasis. However, this approach is not been employed in the routine diagnostic laboratory as detecting this metabolite requires sophisticated instruments such as gas liquid chromatography which is cumbersome and difficult to run the instrument.

Molecular Diagnosis

Detection of fungal DNA in the clinical sample is another interesting approach for the diagnosis of fungal infections.

The next generation of molecular diagnosis which is gaining importance is polymerase chain reaction where amplification of target gene and its detection are done by conventional gel electrophoresis or by real-time automated format. One approach has been to detect pan-fungal markers in diverse specimen types to establish the fungal aetiology. The internal transcribed spacers (ITS) region ITS1 and ITS 2 and D1/D2 regions of the 28*S* rRNA gene are commonly targeted. This method is also useful to identify novel fungi. The pan fungal PCR can be followed by a second PCR directed against a genus or species-specific target gene.

Presently, most molecular techniques available are mainly used for the identification of fungi rather than direct detection from clinical specimens. Sequencing of ribosomal DNA using pan fungal primers is the gold standard for the identification of fungi. Presently, Matrix-associated laser desorption ioniation–Time of Flight (**MALDI_TOF**) is increasingly being used for the identification of the yeast and also to some extent for moulds.

DNA sequencing can also be used for precise species identification.

Antifungal Drugs

Unlike the large plethora of antibiotics available for treating bacterial infection, the number of antifungal class available are few. The class of antifungals available for treating human fungal infections are poleyenes (amphotericin B), azoles (fluconazole, voriconazole, itraconazole, posaconazole and isavuconazole), echinocandins (caspofungin, anidulafungin and micafungin) and allylamines (terbinafine) and pyrimidine analogue of flucytosine (5-flurocytosine). Ketoconazole, miconazole and griesofulvin which were earlier used to treat systemic infection are not being used presently due to their toxicity. However, ketoconazole, miconazole and many other antifungals are available for topical treatment of fungal infections. Amphotericin B is a broad spectrum antifungal and it is available in different

Table 65.3 Antifungal agents and primary sites of activity

Group of compounds	Antifungal agent	Mechanism of action	Application
Polyenes	Amphotericin B nystatin	Bind to ergosterol	Systemic disease
Azole derivatives	Miconazole ketoconazole fluconazole Sertaconazole Luliconazole Itraconazole Voriconazole Posaconazole	Inhibit cytochrome P-450-dependent enzymes	
Nucleoside analogues	5-fluoro-cytosine	Inhibits DNA and RNA syntheses	Used along with amphotericin B
Grisans	Griseofulvin	Inhibits microtubular function	Dermatophytoses
Allylamines	Naftifine Terbinafine	Squalene epoxidase inhibitors	Topical disease
Thiocarbamates	Tolnaftate Tolciclate	Squalene epoxidase inhibitors	Topical disease
Morpholines	Amorolfine	Inhibits ergosterol biosynthesis	Fungal nail infections
Echinocandins	Caspofungin, Anidulafungin Micafungin	B-1,3 glucan synthetase inhibitors	Systemic disease

formulations such as liposomal amphotericin, lipid complex, etc. The first line of drugs for the treatment of mucormycosis, aspergillosis and candidiasis is liposomal amphotericin B, voriconazole and echinocandins, respectively. Table 65.3 summarises the common antifungal agents and their primary sites of activity.

Superficial, Cutaneous and Subcutaneous Mycoses

66

Fungal infections, depending on the tissue part involved, can be classified into three major groups as follows:

Superficial mycoses: These are surface infections of the skin, affecting the outermost layers of skin, hair and mucosa.

Cutaneous mycoses: These are skin infections involving the epidermis and its integuments, the hair and nails.

Subcutaneous mycoses: These are dermis infections, subcutaneous tissue, muscle and fascia.

Superficial Mycoses

Superficial mycosis caused by fungi is restricted to the outermost layers of the skin and hair. The condition usually causes cosmetic problems, quickly diagnosed and treated. It includes four important conditions: (1) pityriasis versicolor, (2) tinea nigra, (3) black piedra and (4) white piedra.

Pityriasis Versicolor

Pityriasis versicolor or tinea versicolor is a superficial skin infection caused by *Malassezia* species. *Malassezia furfur, Malassezia globosa* and *Malassezia restricta* are the common species implicated with the condition.

All *Malassezia* species, except *Malassezia pachydermatis*, require fatty acids for growth; hence it is cultured on the Sabouraud dextrose agar (SDA) overlaid with a layer of olive oil or special media containing fatty acids. On incubation at 37 °C, the fungus produces creamy colonies within 5–7 days. The lactophenol cotton blue (LPCB) wet mount shows budding yeast cells and some bottle-shaped cells. The fungus is found on parts of the body rich in sebaceous glands.

Lesions of pityriasis versicolor are found most commonly on the upper tissue, arms and abdomen. They appear as hypopigmented (often) or hyperpigmented macular lesions often associated with slight scaling or itching. The condition is mostly asymptomatic. It occurs most frequently in hot and humid weather.

Laboratory diagnosis of the condition is usually made by demonstrating both budding yeast cells and hyphae in the KOH preparation of skin scrapings. Characteristic **'spaghetti and meatballs'** appearance of fungus is demonstrated in the microscopy of KOH preparation of the skin. Culture is not carried out routinely for diagnosis. Topical miconazole is the treatment of choice.

Tinea Nigra

Tinea nigra is an infection of the keratinised layer of skin caused by *Exophiala werneckii* or

Cladosporium werneckii. *C. werneckii* is a dimorphic fungus that produces melanin. The fungus on the SDA grows as yeast with many cells in various stages of cell division, producing a typical two-celled oval structure on primary isolation from a clinical specimen. On prolonged incubation, elongated hyphae develop, and mycelia and conidia are predominantly found in older cultures.

A **well-demarcated brown–black macular lesion**, which appears as a brownish spot on the skin, is a typical manifestation of the condition. These brownish to black lesions are most commonly seen on palms and soles.

Laboratory diagnosis of tinea nigra is made by microscopy of the KOH preparation of skin scrapings collected from the affected part. Typical darkly pigmented yeast-like cells and hyper fragmented hyphae are demonstrated. The culture of the skin scraping on the SDA confirms the diagnosis.

Black Piedra

Black piedra is a superficial infection of the hair caused by *Piedraia hortae*, a **dematiaceous fungus**. The fungus occurs in the perfect state when it colonises the shaft of hairs. The culture of specimens on SDA shows slow-growing brown to reddish black mycelium, which is considered an asexual or **anamorphic stage** of the fungus. The **teleomorphic state**, which is the perfect state of the fungus, is occasionally found in old cultures. At this state, **asci**, which contain spindle-shaped ascospores, develop within specialised structures.

Infection of shaft of hairs of beard and scalp is the major clinical feature of black piedra. **Laboratory diagnosis** of the condition is made by demonstrating nodules containing asci with spindle-shaped ascospores in 10% KOH mount of the hair.

White Piedra

White piedra is an infection of the hair caused by the yeast-like organism *Trichosporon* species. *Trichosporon asahii, Trichosporon inkin, Trichosporon cutaneum, Trichosporon mucoides* and *Trichosporon ovoides* are the species implicated with white piedra. The fungus can be grown on SDA and other media containing cycloheximide. On SDA, it forms green-coloured colonies, which subsequently become yellowish-grey and wrinkled. Microscopic examination of the colony shows septate hyphae that break rapidly to form arthroconidia. The latter subsequently become round and develop to blastoconidia.

White piedra is commonly found in South America, Central and Eastern Europe and Japan.

The **hair of the scalp, moustache and beard** are commonly affected in white piedra. The development of a soft, pasty, cream-coloured growth along the infected hair shaft characterises the condition. The initial growth of *Trichosporon* occurs beneath the epidermis of hair. The infected hair shaft consists of mycelium that rapidly fragments to arthroconidia.

Laboratory diagnosis of the condition is made by demonstrating fragmented hyphae that develop into arthroconidia or produce blastoconidia in 10% KOH mount of hair. The culture of the fungus from the clinical specimen confirms the diagnosis. Features of the organisms causing superficial mycoses are summarised in Table 66.1.

Cutaneous Mycoses

Dermatophytosis or cutaneous mycoses are diseases of the skin, hair and nail. These infections are caused by a homogeneous group of closely related fungi known as dermatophytes. These dermatophytes infect only superficial keratinised structures, such as skin, hair and nail, but not deeper tissues.

Dermatophytes

The most important dermatophytes that cause infection in humans are classified into the following three genera: *Trichophyton* (causes infection

Table 66.1 Features of the fungi causing superficial mycoses

Superficial mycoses	Fungus	Site of infection	Diagnosis
Pityriasis versicolor	*Malassezia furfur*	Skin	Microscopy of skin scraping with 10% KOH shows 'spaghetti and meatballs' appearance
Tinea nigra	*Exophiala werneckii*	Skin	Culture on SDA yields black yeast and hyphae
Black piedra	*Piedraia hortae*	Hair	Direct microscopy; culture yields asexual phase
White piedra	*Trichosporon beigelii*	Hair	Direct microscopy; culture on SDA; hyphae, arthroconidia and blastoconidia

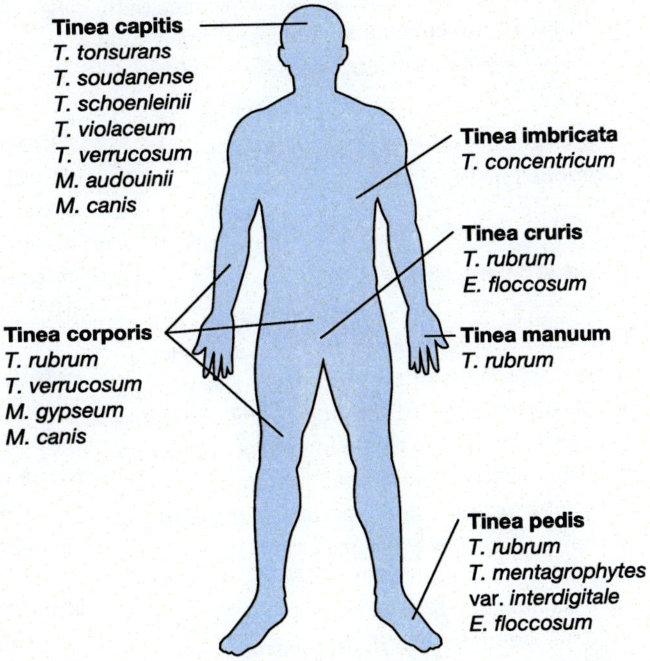

Fig. 66.1 Infections caused by different dermatophytes

of hair, skin and nail), *Microsporum* (infection of hair and skin) and *Epidermophyton* (causes infection of skin and nails, but not hair) (Fig. 66.1).

The dermatophytes, based on their natural habitat and host preferences, can be classified into the following groups:

1. **Anthropophilic species**: These dermatophytes are typically adapted to live on a human host. They are transmitted from human to human through fallen hairs, desquamated epithelium, combs, hairbrushes, towels, etc. Examples are *Trichophyton rubrum*, *Microsporum audouinii* and *Epidermophyton floccosum*.
2. **Zoophilic species**: These dermatophytes live on animals and often cause infection in their animal host. These zoophilic species are transmitted from infected animals to humans by direct and indirect contact with domestic animals (e.g. cat and dog) and occasionally wild animals. Examples are *Trichophyton violaceum* and *Microsporum canis*.
3. **Geophilic species**: These are saprophytic fungi found in soil or dead organic substances. They occasionally cause infection in humans and animals. Examples are *the Microsporum gypseum* and *Trichophyton ajelloi*. Dermatophytes usually grow only on keratinised skin and appendages and do not penetrate the living tissues. In some infected persons, hypersensitivity to fungus antigen may cause secondary eruptions, such as

Table 66.2 Asexual state of some dermatophytes

Microsporum	*Trichophyton*	*Epidermophyton*
Microsporum audouinii	*Trichophyton mentagrophytes*	*Epidermophyton floccosum*
Microsporum canis	*Trichophyton rubrum*	*Epidermophyton stockdaleae*
Microsporum gypseum	*Trichophyton tonsurans*	
Microsporum gallinae	*Trichophyton verrucosum, Trichophyton schoenleinii*	
Microsporum nanum	*Trichophyton violaceum*	

vesicles on the finger. This reaction is known as **the dermatophytid (*id*) reaction**. This reaction occurs due to a hypersensitivity response to circulating fungal antigen, and these lesions do not contain any fungal hyphae.

The asexual state of selected dermatophytes is mentioned in Table 66.2.

Clinical Syndromes

The skin infections caused by dermatophytes are chronic infections of the skin often found in the warm, humid areas of the body, such as athlete foot and jock itch. Typical **ringworm lesions** are circular, which have an inflamed border containing papules and vesicles surrounding a clear area of relatively normal skin. These lesions are associated with variable degrees of scaling and inflammation. Broken hair and thickened broken nails are often seen in this lesson.

Clinically, ringworm can be classified depending on the site affected. These are (1) *Tinea capitis* involving the scalp, (2) *Tinea corporis* involving non-hairy skin of the body, (3) *Tinea cruris* affecting groin, (4) *Tinea pedis* affecting the foot and (5) *Tinea barbae* affecting beard areas of the face and neck.

Favus is a chronic ringworm infection affecting hair follicles. It leads to alopecia and scarring.

Laboratory Diagnosis

Laboratory diagnosis is based on demonstrating a fungal element in a clinical specimen by microscopy and confirmation by culture.

The **specimens** include skin scrapings and nail clippings or hair taken from the areas suspected to be infected by dermatophytes. These entire specimens are treated with an alkali solution to clear epithelial cells and other debris.

Direct Microscopy

Direct microscopy is useful only for diagnosis. Examination of 10% direct KOH mount may show fungal hyphae. Three types of hair infections can be demonstrated in microscopy of 10% KOH wet mount as follows (Fig. 66.2).

Ectothrix

Ectothrix infection is characterised by the presence of a layer of arthrospores on the surface of the hair shaft (Fig. 66.3). It is caused by *M. audouinii*, *M. canis* and *Trichophyton mentagrophytes*.

Endothrix

The clusters of arthrospores are found entirely within the hair shaft in endothrix infection. It is caused by *Trichophyton tonsurans*, *T. violaceum* and *Trichophyton schoenleinii*.

Favus: In favus, there is sparse hyphal growth and formation of air spaces within the hair shaft. It is caused by *T. violaceum*, *T. schoenleinii* and *M. gypseum*.

Culture

Culture is always carried out to identify the specific causative fungal agent. The clinical specimens are cultured by inoculation on SDA-containing antibiotics like cycloheximide. The media after inoculation are incubated at 25–30 °C for 3 weeks. In the media containing cycloheximide and incubated at 25 °C, most pathogenic fungi, including dermatophytes, grow

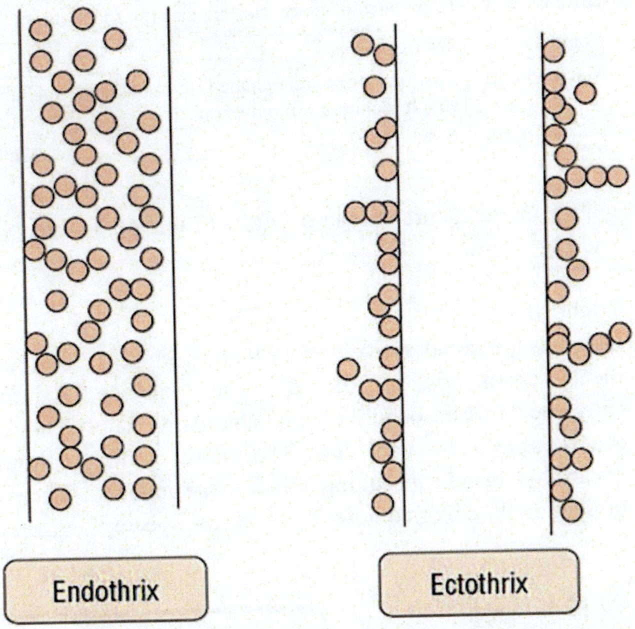

Fig. 66.2 Ectothrix and endothrix infections

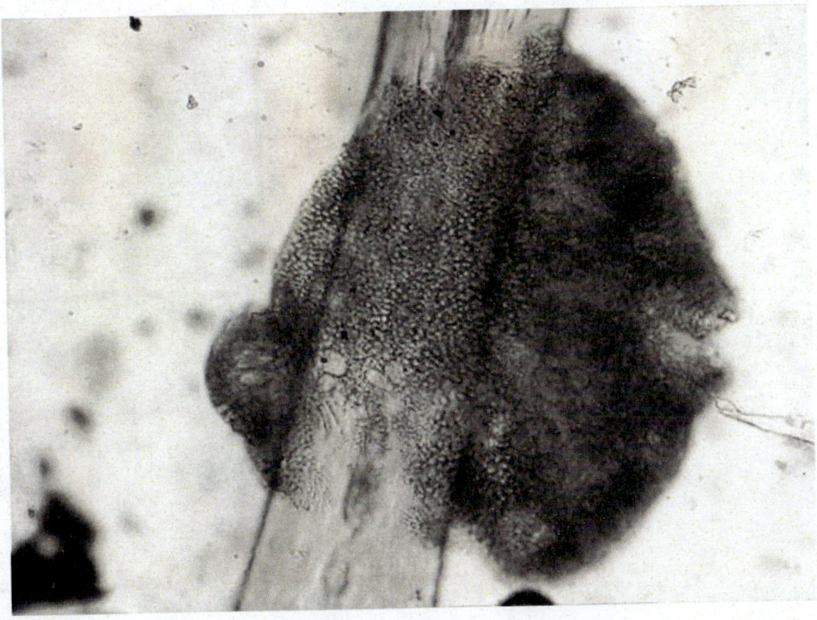

Fig. 66.3 Hair shaft showing ectothrix infection (x 100)

well, while saprophytic fungi and bacteria are inhibited.

The cultures are examined at regular intervals, and dermatophytes are identified based on (1) colony morphology, (2) pigment production and (3) the presence of microconidia and macroconidia. The LPCB preparation of the colonies shows microconidia, macroconidia or both. Only macroconidia are present in the *Epidermophyton* infection. Few macroconidia and more microconidia are present in *Trichophyton* infection. Macroconidia are predominantly present in *Microsporum* infection. The differentiation of the three genera is based

Table 66.3 Characteristics of macroconidia and microconidia of dermatophytes

Conidia	*Microsporum*	*Epidermophyton*	*Trichophyton*
Macroconidia	Fusiform or spindle shaped, thick-walled, rough (echinulate), numerous	Club shaped, smooth-walled, numerous	Clavate or cigar-shaped, thin-walled, smooth, rare
Microconidia	Rare	Absent	Abundant

mainly on the nature of macroconidia (Table 66.3 and Fig. 66.4).

Trichophyton

Colonies are powdery, velvety or waxy with pigment characteristics of different species (Fig. 66.5). Microconidia are abundant, but **macroconidia are very few**. The microconidia are arranged in clusters along with hyphae and are present on the tip of conidiophores. Some species may have a special type of hyphae, such as spiral hyphae or racket hyphae. *Trichophyton* species infect hairs, skin and nails.

Microsporum

It infects both skin and hair, but not the nails. They produce cottony, velvety or powdery colonies with white-brown pigment. Microconidia are relatively scanty, but macroconidia are abundant (Fig. 66.6).

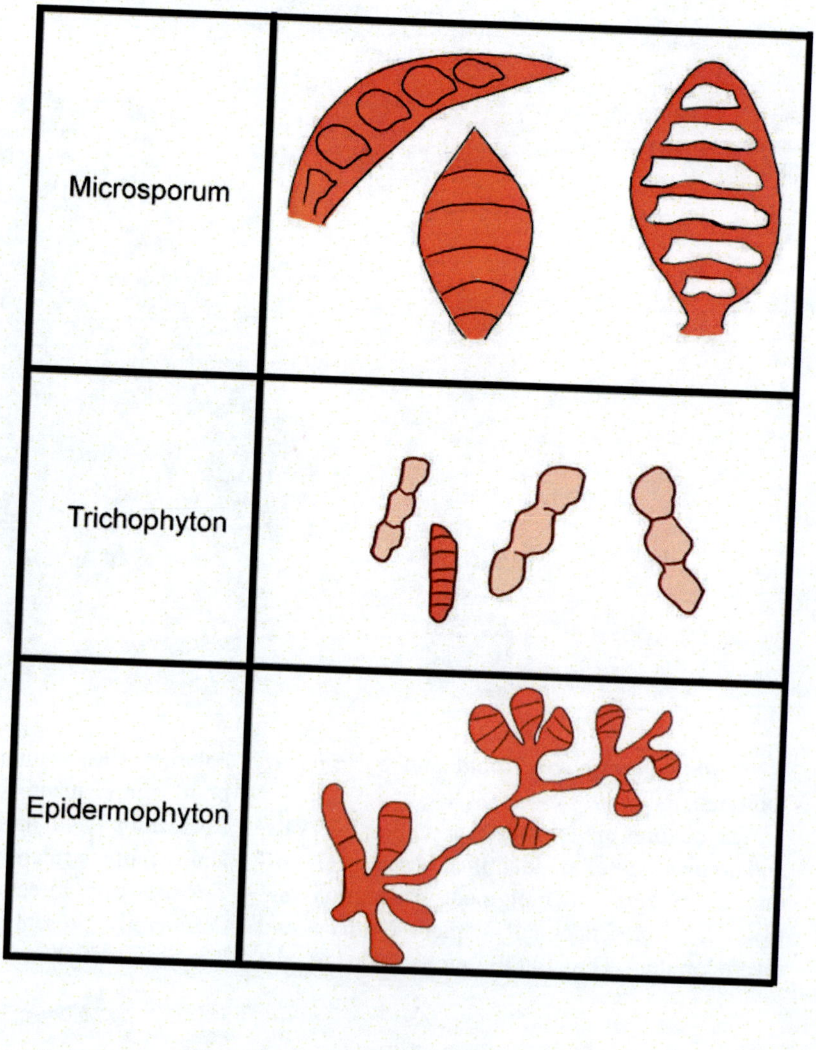

Fig. 66.4 Macroconidia in genera *Trichophyton*, *Microsporum* and *Epidermophyton*

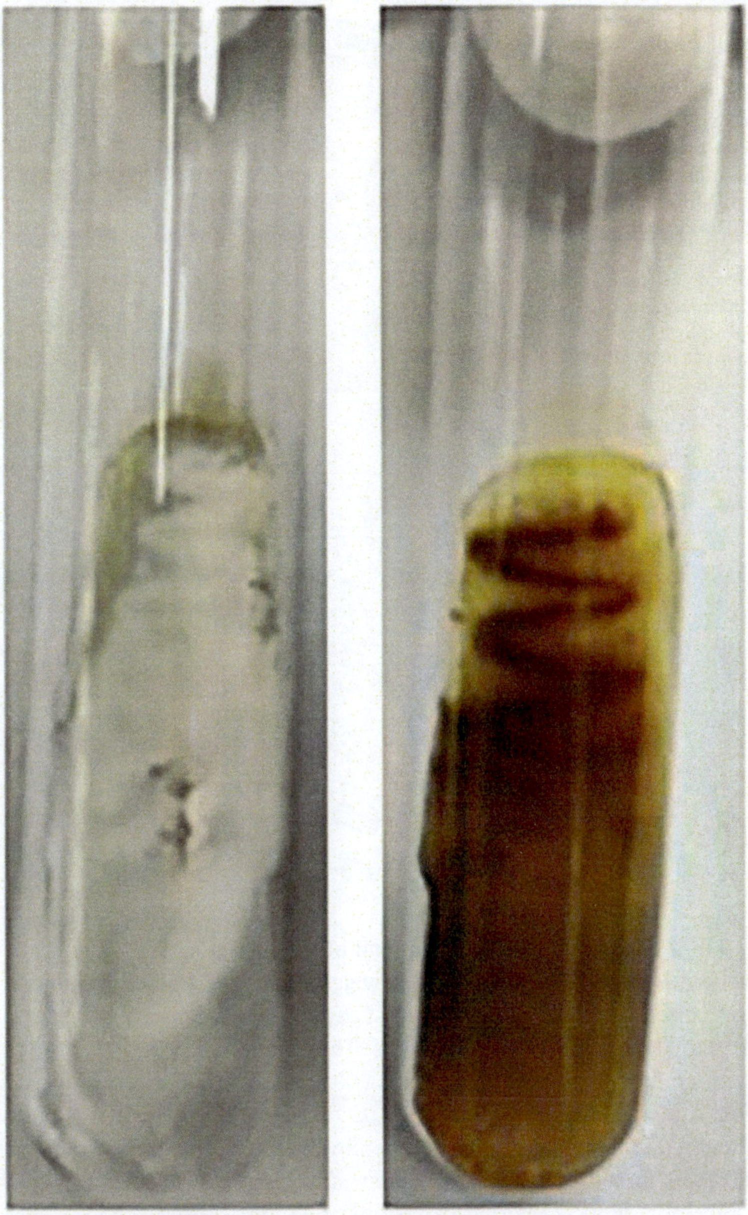

Fig. 66.5 SDA showing white, powdery with concentric and radial fold colony of *Trichophyton mentagrophyte* with reverse side coloured yellow (Courtesy of Prof. MR Shivaprakash)

Epidermophyton

It infects skin and nails but not hair. It produces powdery and greenish-yellow colonies. **Macroconidia are numerous**. The macroconidia are club-shaped, multi-septate, and arranged in groups of two to three. Chlamydospores are numerous, but microconidia are absent.

Other Tests

Hair Perforation Test

This test is performed to differentiate *T. rubrum* from *T. mentagrophytes*. The test is also used to differentiate *M. canis* from *Microsporum equinus*. This test is performed by collecting a

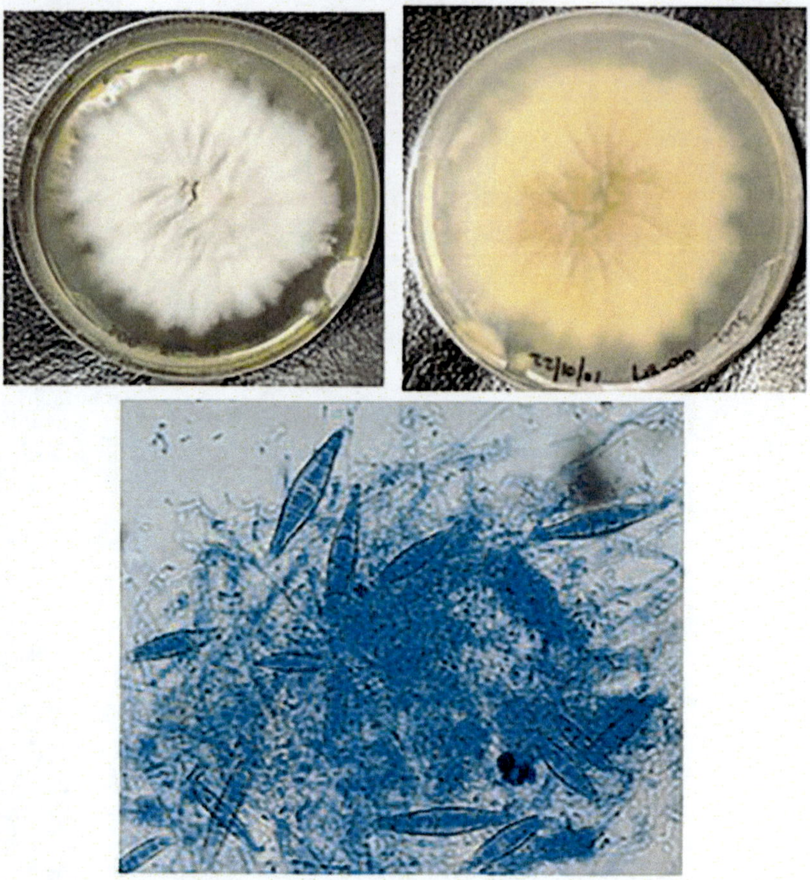

Fig. 66.6 SDA showing (**a**) white surface, hairy to silky, closely spaced radial grooves of *Microsporum canis* on Sabourauds dextrose agar and reverse appearing yellow. (**b**) LPCB mount showing spindle shape, rough, thick walled, taper to beak like end with >6 compartments of macroconidia of *Microsporum canis* (×400) (Courtesy of Prof. MR Shivaprakash)

5–10 mm short shaft of human hair and placing it in a Petri dish with 20 mL of distilled water. Then, two to three drops of 10% sterile yeast extract are added to the Petri dish with hair shafts. These hair shafts are inoculated with test fungus cultured on SDA. The culture is incubated at 25 °C for up to 1 month, during which the hairs are removed, and the LPCB mount of the hair is examined microscopically at weekly intervals to demonstrate the perforation of hair.

T. mentagrophytes show a positive hair perforation test characterised by a wedge-shaped hair perforation. This test is negative for *T. rubrum*, in which only surface eruption of the hair shaft is demonstrated.

Urease Test

The urease test is carried out to differentiate *T. mentagrophytes* from *T. rubrum*. This test is performed by inoculating a tube of Christensen medium with the fungus and incubating at 25 °C for 5 days at room temperature. Most *T. mentagrophytes* are urease positive within 5 days.

Growth on Rice Grains

This test is useful to differentiate *M. canis* from *M. audouinii*. In this test, sterile, non-fortified rice is inoculated with the hyphae of the fungi to be tested. The medium is observed for growth after 10 days of incubation at room temperature. *M. canis* usually grows well and forms many conidia on rice grain, whereas *M. audouinii* fails to grow.

Treatment

Terbinafine is the first drug of choice, followed by itraconazole or fluconazole for treating dermatophytosis. These drugs can be given as topical application or orally, depending on the extent of the lesion. Griseofulvin is another systemic antifungal used for treating dermatophytosis. Nail infection should be treated for a longer time. Local application of antifungal lacquer is effective. Other topical agents such as miconazole, clotrimazole, econazole, amorolfine, luliconazole and sertaconazole are effective.

Subcutaneous Mycosis

Subcutaneous mycosis is defined as a fungal infection associated with the development of a characteristic lesion in the subcutaneous tissue and overlying skin with or without extension to bone and muscle. This is caused by a heterogeneous group of fungi of low pathogenic potential introduced in the body percutaneously from a trivial trauma. Table 66.4 shows the classification of subcutaneous mycoses.

Mycetoma

The condition was described as early as in *Atharva Veda*: *padmavalmikam* means 'anthill foot'. John Gill first described the condition in Madurai (India) in 1842. Chalmer and Christopherson coined the term *madura mycosis* in 1916.

Table 66.4 Aetiological agents of subcutaneous mycoses

Subcutaneous mycoses	Agents
Mycetoma	Bacteria *Actinomadura madurae* *Actinomadura pelletieri* *Nocardia* spp. *Streptomyces somaliensis* Fungi *Madurella mycetomatis* *Madurella grisea* *Exophilia jeanselmei* *Scedosporium apiospermum* *Acremonium falciforme*
Chromoblastomycosis	*Fonsecaea pedrosoi* *Fonsecaea compactum* *Phialophora verrucosa* *Cladophialophora carrionii* *Rhinocladiella aquaspersa*
Phaeohyphomycosis	*Exophiala jeanselmei* *Bipolaris spicifera* *Wangiella dermatitidis*
Sporotrichosis	*Sporothrix schenckii*
Rhinosporidiosis	*Rhinosporidium seeberi*

Mycetoma is a slowly progressive, chronic granulomatous infection of the skin and subcutaneous tissues with occasional underlying fascia and bone involvement, usually affecting extremities. The condition is characterised by a triad of tumefaction, draining sinuses and grains or granules. It is caused by several actinomycetes (**actinomycotic mycetoma**) and filamentous fungi (**eumycotic mycetoma**) that enter through penetrating injuries resulting from thorn pricks, splinters, etc. Lower extremities are most commonly involved. Microabscesses are formed in subcutaneous tissues surrounded by a polymorphonuclear inflammatory reaction. The centre of the lesion consists of tangled filaments of these organisms (Fig. 66.7).

During the course of infection, **microabscesses** burst open with the formation of chronic multiple sinuses discharging copious, seropurulent fluid-containing granules. The colour and consistency of these granules vary depending on the fungi or actinomycetes that cause the disease (Table 66.5). The condition is characterised by the formation of painless, localised, swollen lesions on the affected limbs. **Multiple discharging sinuses** are present. Discharge is purulent, mucopurulent or serosanguinous and usually contains granules of varying size, colour and consistency. Fascia and bone may be affected, but underlying tendons and nerves are never affected.

The condition during the period spreads slowly but over the years by contiguity. Sometimes the infection becomes extensive by involving the buttock and trunk. Actinomycetes spread faster than eumycetes. Haematogenous spread is rare.

Mycetoma is documented worldwide but is common in tropical and subtropical countries. Actinomycetoma accounts nearly for 65% of all mycetomas in India. *Actinomadura madurae* is the major causative agent. *Streptomyces somaliensis* infection is seen more frequently in eastern India. The condition is more common in

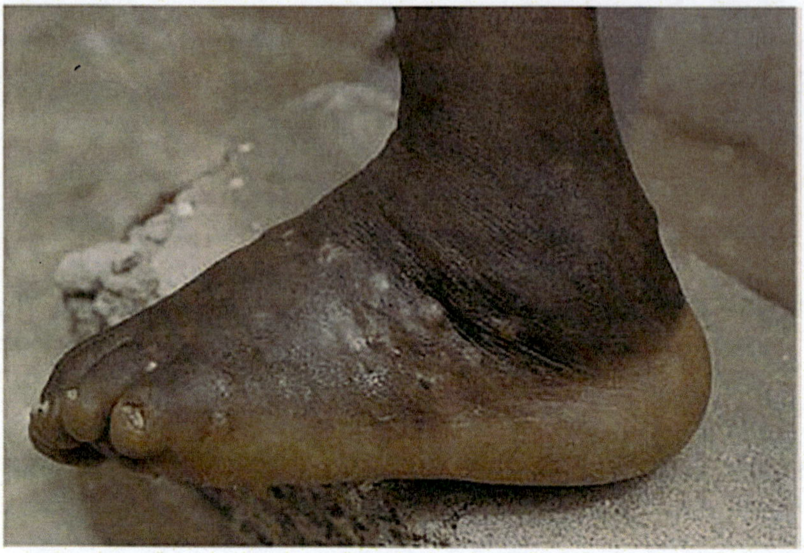

Fig. 66.7 Mycetoma foot. (https://link.springer.com/chapter/10.1007/978-3-642-76200-0_7)

Table 66.5 Important causative agents of grains in mycetoma

Grain	Causative agents
Black grain	*Madurella mycetomatis* *Madurella grisea* *Exophiala jeanselmei* *Curvularia geniculata*
White grain	*Pseudoallescheria boydii* *Acremonium falciforme* *Actinomadura madurae* *Nocardia brasiliensis*
Red grain	*Actinomadura pelletieri*

rural areas and males (3.5:1) than females. It affects mostly highly active persons aged 20–40 years. It is an occupational hazard affecting farmers, herdsmen, carpenters, builders, land workers, field workers, etc. **Euycotic mycetomas** require surgical treatment by amputation, while actinomycotic mycetomas are treated well by sulphonamides and antibiotics.

A high index of clinical suspicion and proper history is essential for the clinical diagnosis of mycetoma. Laboratory diagnosis depends on the demonstration of the fungi and fungal filaments in granules, pus and biopsy tissue by microscopy. The granules on naked eye examination may be hard or soft, round or lobulated and vary in size from 0.5 to 3 mm. These may be black, white, cream or red. **Microscopy of the granules** may reveal very thin filaments, measuring less than 1 μm in diameters in cases of actinomycotic mycetomas. However, the fungal filaments may be broader and often show septate hyphae and chlamydospores in eumycotic mycetoma (Fig. 66.8).

Chromoblastomycosis

Chromoblastomycosis is a slowly progressing granulomatous infection caused by several soil fungi such as *Fonsecaea pedrosoi*, *Fonsecaea compactum*, *Cladophialophora carrionii* and *Phialophora verrucosa*. These fungi are called dematiaceous fungi because they have brown to black melanin pigment in their cell wall, and their conidia or hyphae are dark coloured, either grey or black. All the fungi causing chromoblastomycosis appear morphologically similar in tissues stained by haematoxylin and eosin (H and E) or other stains.

The infection occurs following the introduction of any of the dematiaceous fungi into the skin through trauma. The development of warty nodules that appear at the inoculation site characterises chromoblastomycosis. During the course of infection, these lesions vegetate and develop into a cauliflower-like lesions. The

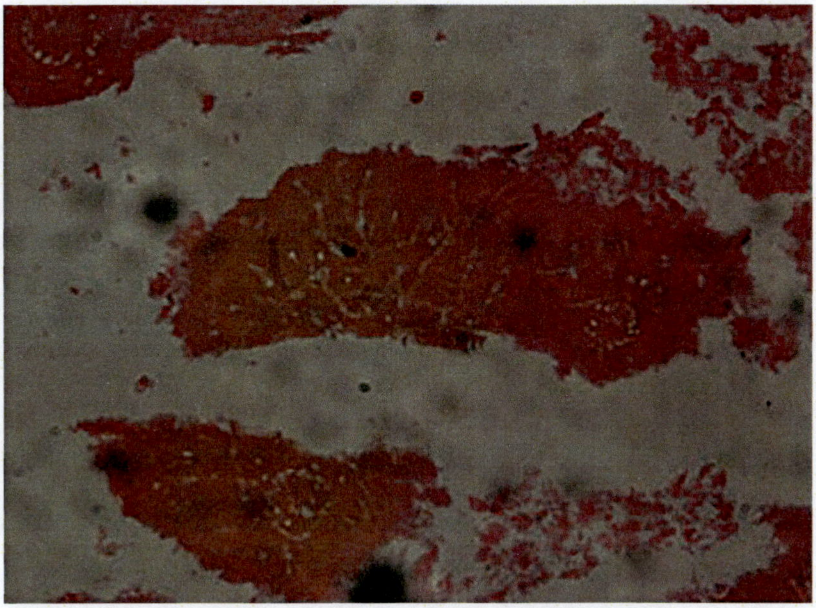

Fig. 66.8 Sections of granules from a case of eumycetoma. Haematoxylin and Eosin stain showing thin septate hyphae may be noted inside the granules. (x1000) (Courtesy of Prof. MR Shivaprakash)

disease is more common in tropical and subtropical countries (Fig. 66.9).

Laboratory diagnosis of chromoblastomycosis is made by demonstrating small clusters of round and dark brown, thick-walled cells (8–12 μm) in tissues. The fungi are found in the dermis but are occasionally seen in the subcutaneous tissues. Direct KOH wet mount of skin is a rapid method for diagnosing the condition by demonstrating these fungi in scales removed from the surface of these lesions. In addition to these, copper-coloured spherical cells in various stages of cell division are seen. These are the tissue forms of the fungus and called sclerotic or medlar bodies (Fig. 66.10) of the fungus.

Phaeohyphomycosis

A phaeohyphomycosis is a heterogeneous group of cutaneous diseases caused by various dematiaceous fungi. The phaeohyphomycotic cyst is the most common form of presentation. A wide number of dematiaceous fungi has been associated with various types of phaeohyphomycosis. Some of the common causes of phaeohyphomycosis are *Exophiala jeanselmei*, *Bipolaris spicifera*, *Wangiella dermatitidis*, *Curvularia lunata*, *Alternaria alternata*, etc.

Diagnosis of phaeohyphomycosis is made by demonstrating darkly pigmented, septate hyphae measuring 5–10 μm in diameter in tissues. The drug of choice for the treatment of phaeohyphomycosis is itraconazole along with 5-fluorocytosine. Cautery and surgical removal of the early lesion are also useful.

Sporotrichosis

Sporotrichosis is a chronic pyogenic granulomatous lesion of the skin and subcutaneous tissue caused by the *Sporothrix schenckii* complex. *S. schenckii*, *Sporothrix globosa*, *Sporothrix braziliensis* and *S. mexicana* are common species causing human infections. *Sporothrix* is a dimorphic fungus found worldwide, but some species (*S. braziliensis* and *S. mexicana*) have geographically restricted distribution. The fungus is found in soil, decaying woods, thorns and infected animals, including rats, cats, dogs and horses.

Spore is the infective stage of the fungus. It causes infection primarily on the extremities through direct contact of the skin by spores. Typically, infection is introduced in the skin through

Fig. 66.9 Chromomycosis: Verrucous lesion on the right lower extremity (https://link.springer.com/chapter/10.1007/978-3-642-76200-0_7)

the penetration of thorns. At the site of thorn injury, it causes a local pustule or ulcer with the nodules along the draining lymphatics. Frequently, the regional lymph nodes draining the ulcer enlarge, suppurate and ulcerate (Fig. 66.11). The primary lesion may remain localised or in the immunocompromised individuals and may disseminate to involve the bones, joints, lungs and the central nervous system.

Sporothrix spp. occurs in two phases. The yeast phase occurs in tissue and the culture at 37 °C, while the mycelium phase occurs in the culture at 22–25 °C. The yeast appears as round, oval or cigar-shaped cells with irregular borders in infected tissue. Periodic acid–Schiff (PAS) or Gomori methenamine silver (GMS) stain is useful to demonstrate these structures in the stained smears. The fungus on SDA at 25 °C produces black and shiny colonies, which become wrinkled and foggy over time. The mould contains hyphae-bearing flower-like structures of small conidia on delicate sterigmata.

Fig. 66.10 Sclerotic bodies

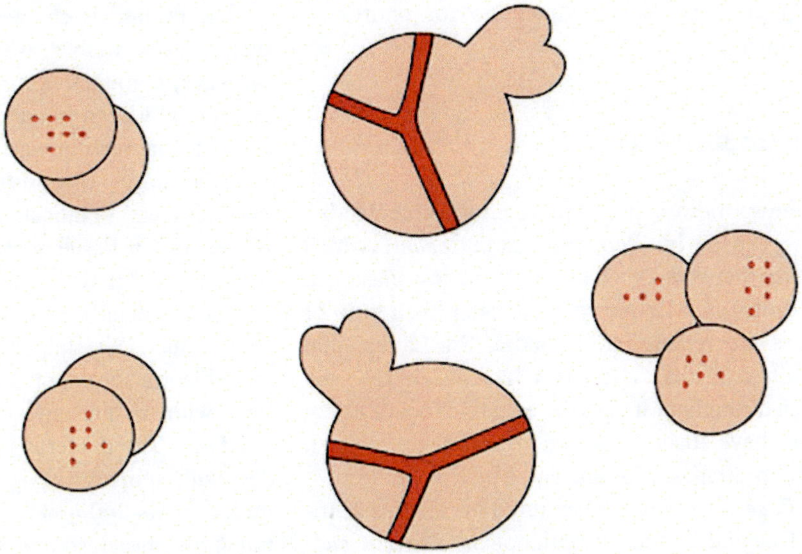

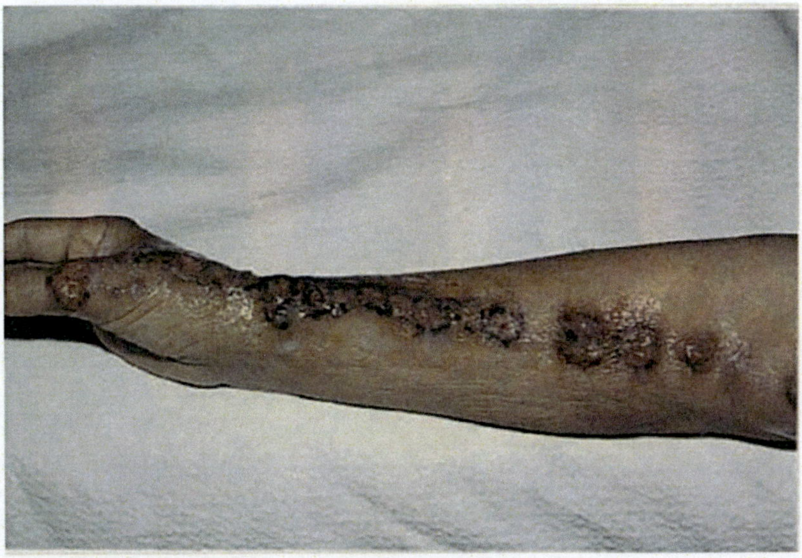

Fig. 66.11 Sporotrichosis: Lymphangitic form (https://link.springer.com/chapter/10.1007/978-3-642-76200-0_7)

Laboratory diagnosis of sporotrichosis is made by demonstrating asteroid bodies in the pus of the abscesses. Asteroid bodies consist of a central basophilic budding yeast cell with eosinophilic material radiating from the centre.

Isolation of the fungi by the culture of ulcer exudates or pus aspirated from a subcutaneous nodule or biopsy material confirms the diagnosis of sporotrichosis. LPCB wet mount of the colony shows thin, delicate hyphae bearing conidia occurring in a rosette pattern at the end of delicate conidiophores. Conidia are also demonstrated along the sides of the hyphae. Repeated subculture of mycelium of fungi on blood agar tubes at 37 °C induces the formation of yeast colonies. The LPCB wet mount of these colonies shows cigar-shaped yeast-like cells.

Itraconazole is the drug of choice for the treatment of the condition. Potassium iodide drops help and add as an adjunctive for antifungal treatment.

Rhinosporidiosis

Rhinosporidiosis is a chronic granulomatous disease caused by *Rhinosporidium seeberi* which was earlier considered as fungi and now classified as Protozoa, Mesomycetozoa. More than 90% of cases are reported from India, Sri Lanka, and South America. The cases have been reported from throughout India. The endemic foci of infection have been reported from Tamil Nadu, Andhra Pradesh, Kerala, and Madhya Pradesh.

R. seeberi cannot be cultured in cell-free artificial media. Animal inoculation is also not successful. The mode of infection is not known. However, it is suggested that it is transmitted in dust and water. Fish is believed to be the natural host. Infection is seen most commonly in persons taking bath in stagnant pools and in individuals who dive in streams to collect sand from river beds.

The disease is characterized by the development of large friable polyps or wart-like lesion in the nose, conjunctiva, or eye. The lesion is also occasionally seen in ears, larynx, bronchus, urethra, vagina, rectum, and skin.

Laboratory diagnosis of rhinosporidiosis depends on demonstration of sporangia of *R. seeberi* in tissue sections. *R. seeberi* can be demonstrated in tissue sections stained with H&E or other special stains, such as GMS stain and PAS stain. The stained smear on microscopic examination shows a hyperplastic surface epithelium with the presence of chronic inflammatory exudates. The epidermis and stroma are embedded with sporangia (Fig. 66.12). The sporangia measure 10–200 μm in diameter and contain hundreds of endospores, each measuring 6–7 μm in diameter. The spores when released develop into new sporangia.

Treatment of the condition is carried out by surgery or cauterization. Chemotherapy with dapsone is also useful.

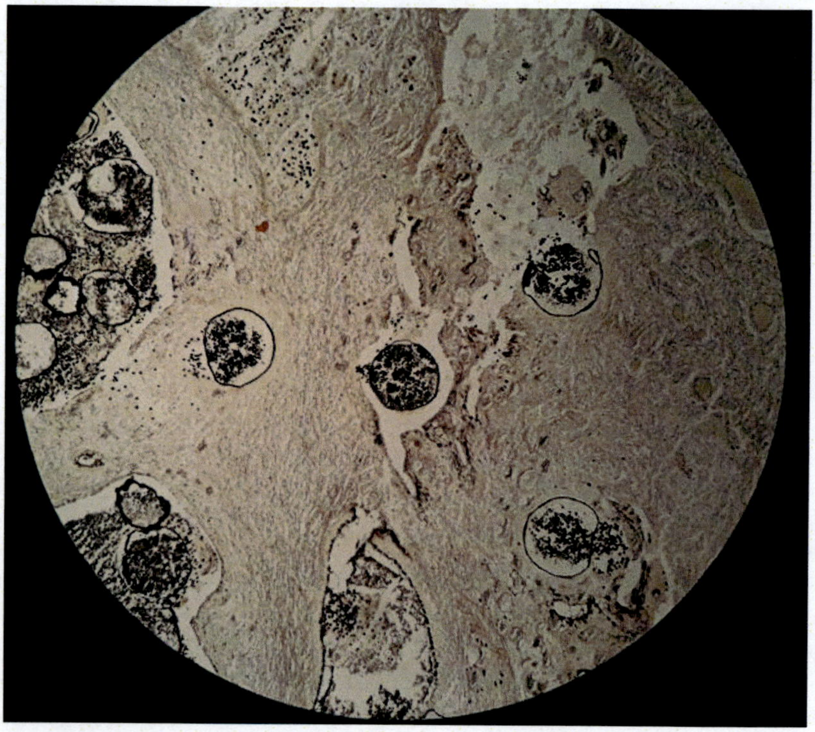

Fig. 66.12 Histology of *Rhinosporidium seeberi* granules (×1000) (Courtesy of Prof. MR Shivaprakash)

Entomophthoromycosis

The entomophthorales, *Conidiobolus coronatus* and *Basidiobolus ranarum*/*Basidiobolus haptosporus* are the pathogenic fungi that cause entomophthromycosis. It is generally found in decaying fruit and vegetable matter and commensal in the intestines of reptiles and amphibians. It has a global distribution. It was firstly reported from Indonesia, followed by reports from tropical Africa, India, Southeast Asia and northern parts of Australia.

Entomophthoromycosis is a chronic, inflammatory or granulomatous, slowly progressive disease often restricted to subcutaneous tissues. It first develops as a localised subcutaneous nodule/lesion and spreads slowly to form a hard, painless mass involving the cutaneous and subcutaneous tissues.

Diagnosis includes deep skin biopsy from the subcutaneous mass that is required for microscopy, culture and histopathological examination. Colonies grow rapidly and are flat, cream-coloured, glabrous becoming radially folded and covered by a fine-powdery, white mycelium and conidiophores. It shows the presence of primitive aseptate or pauciseptate hyphae. When incubated at 30 °C, culture shows white to grey waxy colonies for *Conidiobolus* spp. and cream or yellow waxy colonies for *Basidiobolus* spp.

Diminutions of the lesions sometimes occur spontaneously. The use of potassium iodide has been shown to be satisfactory in many cases. Oral itraconazole and terbinafine hydrochloride may also be effective. Surgery may be required in a few cases.

Case Study

A mother comes to dermatology with her 10-year-old child with complaints of scaly and patchy alopecia. On examination by a dermatologist, the condition was suspected to be tinea capitis.

- How will you confirm the diagnosis of tinea capitis?
- What are the clinical features of tinea capitis, tinea pedis and other taeniae due to anthropophilic, zoophilic and geophilic fungi?
- How will you treat the condition?

Systemic Mycoses and Opportunistic Infections

67

Systemic mycoses are generally geographically restricted diseases caused by fungi collectively referred to as endemic fungi. These fungi can cause infection in apparently immunocompetent individuals.

Systemic Mycosis

The systemic mycoses include coccidioidomycosis, histoplasmosis, blastomycosis and paracoccidioidomycosis. In most cases, the infection develops initially in the lungs; later, the skin and other organs may be involved. All these fungi exhibit **thermal dimorphism**, growing as mould or mycelial form in their natural habitat and seen as yeast form in tissues.

Coccidioidomycosis

Coccidioidomycosis caused by *Coccidioides immitis* was first recognised as a distinct disease entity in 1892. The disease is distributed mainly in southwestern ca. Its two species are *C. immitis* and *Coccidioides posadasii*. Most of the cases caused by *C. immitis* have been reported from San Joaquin Valley, California, whereas *C. posadasii* has been implicated from patients in other states of the United States and Mexico. *Coccidioides* spp. is a dimorphic fungus, which occurs as a mould in soil and the culture at 25 °C and as a spherule in tissue and the culture at 37 °C. The spherule is oval with a thick, double-refractile wall filled with endospores. Each endospore, measuring 2–5 µm in diameter, gives rise to a new spherule.

The Fungus

C. immitis grows in the mycelial form in the soil of endemic areas. Subsequently, the hyphal cells either develop into barrel-shaped structures producing the characteristic arthroconidia. In pulmonary acinus (where alveoli are located), the arthrospore uncovers its external layer, expands its size and develops into a round structure called '**spherule**'.

Pathogenesis of Coccidiomycosis

The arthroconidia are the infective stage of the fungus. Arthroconidia turn airborne when the soil is disrupted. Inhalation of such arthroconidia by the susceptible host or immunocompromised host initiates the infection. The transport of arthroconidia, either in soil on fomites or due to unusually severe dust storms, has produced infections in persons without endemic exposure. The spherule replicates by a process called endosporulation. Disruption of the spherule induces the release of endospores, which further develops into spherules again, and this cycle repeats.

The **spherule** is the characteristic tissue form of the organism. Resistance of the spherule to eradication by host defences is the main factor in the pathogenesis of the disease. Spherules cause progressive suppuration and tissue necrosis.

Clinical Syndromes

More than half of the cases are asymptomatic. In symptomatic cases, *C. immitis* causes a primary pulmonary disease and disseminated disease. Pulmonary infection is the most frequent presentation in symptomatic patients, and the first symptoms of the primary infection usually appear 7–21 days after exposure. Common symptoms include cough, chest pain, shortness of breath, fever, fatigue, weight loss and even headache. Skin manifestations may develop as part of the primary illness as a nonpruritic fine papular rash, which is usually transient and easily missed. Other complaints include erythema nodosum, erythema multiforme and migratory arthralgia (especially in knees and ankles), which is called *desert rheumatism.*

Extrapulmonary dissemination is less common but can occur in immunodeficiency conditions. The most common site of dissemination is the skin presenting as superficial maculopapular lesions, verrucose ulcers and subcutaneous abscesses. Joints and bones also are common sites of dissemination and are associated with prominent synovitis and effusion. Coccidioidal meningitis is the most serious form of disseminated infection, which is nearly always fatal within 2 years of diagnosis.

Laboratory Diagnosis

Microscopy

Laboratory diagnosis of coccidioidomycosis is made by microscopy. This is based on the demonstration of spherules containing endospores in (i) sputum or smears from the lesion stained by calcofluor white and (ii) biopsy material stained by haematoxylin and eosin, silver or periodic acid-Schiff stains.

Culture

It is the most definitive method for diagnosis. The fungus grows well on Sabouraud dextrose agar (SDA) and other media producing white and cottony colonies within 5 days. Identification of colonial morphology is not adequate because other fungi show similar mycelial forms. Therefore, demonstrating typical arthroconidia is useful for identifying the organism. However, arthroconidia are infectious, hence posing a significant risk to laboratory personnel.

Serology

Serodiagnosis of coccidioidomycosis is the most frequent means of diagnosing primary coccidioidal infections because the patients may not be able to produce sputum, and fungal cultures often are not practical. Serodiagnosis is based on the demonstration of antibodies to coccidioidal antigens in patient's serum. A negative serologic test result never excludes the presence of a coccidioidal infection. However, performing one or more repeated serologic tests over 2 months increases the sensitivity of serologic diagnosis, especially for recently acquired infections. The two major antigens used to detect antibodies are the tube-precipitating antigen and the complement-fixation antigen. Enzyme immunoassay (EIA) is the most frequently used test to detect serum tube-precipitating antibodies that are IgM antibodies to mycelial phase antigens. These IgM antibodies appear in more than 85% of patients with primary infection and are found within the first week after the onset of symptoms. In most patients, these antibodies disappear within 6 months. IgG antibodies detected by complement fixation appear later, resulting to be positive in 85–90% of patients.

Skin Test

A skin test using coccidioidal antigens is used for the diagnosis of coccidioidomycosis. It is a delayed-type hypersensitivity reaction that appears 2–21 days after the onset of symptoms and precedes the appearance of serologic markers. The skin test has epidemiologic and prognostic implications. The absence of delayed-type hypersensitivity indicates a poor prognosis. However, the skin test shows very less sensitivity and specificity in endemic areas. Because of this, it has a very minimal role to play in diagnostics. A negative skin test can be seen in infected individuals due to the absence of immune response, and a positive skin test can be seen in healthy individuals due to the earlier infection.

Molecular Detection

DNA probe is a recent method used for accurate identification of the fungus.

Treatment

Amphotericin B is the drug of choice for the treatment of the condition. Fluconazole can be used to treat mild-to-moderate disease and, occasionally, for the treatment of life-threatening disease in patients in whom amphotericin B is contraindicated. It is used as the drug of choice for long-term therapy of meningeal infection.

Epidemiology and Prevention

C. immitis has a distribution restricted primarily to areas of the Western Hemisphere. The southern-central part of California (San Joaquin Valley), Arizona, southern New Mexico, western Texas and northern Mexico are the areas of highest endemicity.

As the conidia of the *C. immitis* are mainly present in the soil and dispersed in high concentration during a dust storm, susceptible individuals in endemic regions and even non-susceptible individuals from the non-endemic area should strictly avoid exposing and inhaling the conidia during storms. All measures to decrease the inhalation of the spores, such as using an effective mask filtering the air inside the residence during the storm, help reduce the acquisition of the infection by this agent. Avoiding activities such as gardening and agriculture also reduces the chance of infection.

Paracoccidioidomycosis

Paracoccidioidomycosis, also known as South American blastomycosis and Lutz-Splendore-Almeida disease, is a chronic progressive systemic mycosis caused by *Paracoccidioides brasiliensis. P. brasiliensis* is a thermally dimorphic fungus found in the soil. It occurs as a mould in the soil and as yeast in tissue.

The Fungus

These yeasts are unique in that one large mother cell produces multiple blastoconidia (daughter cells) that arise from multiple sites, resembling a *'Mickey Mouse head'* or a *'pilot wheel'*. This characteristic appearance helps differentiate this yeast in the tissues of secretion from *Blastomyces dermatitidis* and all other yeasts.

Pathogenesis of Paracoccidiomycosis

Infection occurs by inhalation of conidia or mycelial fragments. The lungs are the primary site of infection. The fungus then disseminates to other organs through the venous and lymphatic systems from this site. Cell-mediated immunity (CMI) is the most important defence mechanism in an immunocompetent host.

Clinical Syndromes

Most initial infections are subclinical. The overt disease has two main clinical presentations, a chronic adult form where the course of the disease is long term, and the outcome is better with appropriate therapy, and the juvenile form, which is an acute or subacute more severe disease and carries a worse prognosis.

Hallmarks of the **chronic adult type** of the disease are significant lung involvement and extrapulmonary lesions, **while the juvenile type** is characterised by marked involvement of the reticuloendothelial system with hypertrophy of various lymph node chains and liver and spleen enlargement; respiratory complaints are minimal. Cutaneous and mucosal involvement with infiltrated, ulcerated, and painful skin lesions usually occurs in the head and neck region.

Laboratory Diagnosis

Laboratory diagnosis depends on the demonstration of large, multiple budding yeasts (blastoconidia) in 10–30% KOH wet mount preparation of sputum. Gomori methenamine silver (GMS) staining of biopsy specimens shows yeast cells measuring 2–30 μm in diameter. Tissue reactions are diffuse in the juvenile form, while granuloma formation is typical in the adult form. The organism is isolated by culture on SDA at 37 °C after 20–30 days. Direct fungal DNA detection in exudates and tissue samples by polymerase chain reaction (PCR) assay has been reported to be helpful.

Serodiagnosis depends on the demonstration of *P. brasiliensis* antibodies or antigens in the serum. Enzyme-linked immunosorbent assay (ELISA) is used to detect antibodies to gp43, the main antigenic determinant of *P. brasiliensis*, with a reported sensitivity of 95% and a specificity of 93%. The test shows cross-reactivity with sera from patients of histoplasmosis or lobomycosis. Western blot detects circulating antibodies to gp43 in the serum in 100% of patients and to gp70, another antigen of the fungus, in 96% of patients. Newly developed test can detects aAg7, antigen after 1 week of growth. This diagnostic method has sensitivity and specificity of 84% and 99%, respectively. It is mainly useful in monitoring the patient's response to drugs and to find out recurrence. However, skin tests are rarely helpful.

Treatment

Paracoccidioidomycosis is the only endemic mycosis amenable to treatment with sulpha drugs. Itraconazole is the drug of choice for treating paracoccidioidomycosis and is effective in 95% of patients. Imidazole is also an effective agent with an 85–90% cure rate. Amphotericin B is used as a reserve for treating severe cases that are refractory to treatment with these drugs.

Epidemiology and Prevention

Paracoccidioidomycosis has a restricted geographical distribution. It is found in South and Central America. The highest incidence is in the southeast province of Brazil, followed by Colombia, Venezuela, Ecuador and Argentina. *P. brasiliensis* causes natural infection only in armadillos. Bats and saguis may serve as reservoirs.

The infection is transmitted by inhalation of conidia or mycelial fragments that are found in the soil. Direct inoculation of conidia or mycelial fragments onto the skin tissue or oral mucous membrane is uncommon. Nevertheless, it has been observed in rural Brazil because of the use of twigs to clean teeth, which is a very common practice in those regions.

Prevention and Control

P. brasiliensis is known to be present in the soil, especially near armadillo's burrows. Preventing inhalation of these fungal forms is difficult. However, visitors from non-endemic areas should take precautions, such as using an appropriate mask to avoid inhalation during a visit to the endemic areas.

Histoplasmosis

Histoplasmosis is primarily a disease of the reticuloendothelial system caused by an intracellular fungus *Histoplasma capsulatum. H. capsulatum* is a dimorphic fungus, which occurs in two stages: as a mould in soil and as a yeast at body temperature in mammals. On SDA medium at 37 °C, this fungus produces cottony mycelial growth.

The Fungus

The colony is characterised by thin, branching septate hyphae that produce tuberculate macroconidia and microconidia. The macroconidia are thick-walled, spherical spores measuring 8–20 μm in diameter and have finger-like projections. These are the diagnostic forms of fungus. The microconidia are smaller and thin-walled spores and are the infectious form of the fungus.

The parasitic or tissue phase of the fungus is a yeast that measures 2–4 μm in diameter. It is exclusively found within macrophages in the infected host.

Pathogenesis of Histoplasmosis

H. capsulatum infection is acquired by inhalation of conidia and mycelial fragments from contaminated soil. Once inhaled, it is deposited in the lung's alveoli, and transformation from the mycelial to the pathogenic yeast form occurs intracellularly inside the macrophages. The yeast inside the macrophages multiplies in approximately 15–18 h. Multiplication continues within the phagosomes despite fusion with lysosomes. This is possibly due to the production of certain proteins by yeasts that inhibit the activity of lysosomal proteases.

Cell-mediated immunity (CMI) is a key host defence component against the fungus. T lymphocytes are crucial in limiting the extent of infection. In persons with impaired CMI, *Histoplasma* species that remained latent in healed granulomas may be reactivated, resulting in histoplasmosis.

Clinical Syndromes

Most infected individuals are asymptomatic. Nevertheless, *H. capsulatum* causes acute pulmonary histoplasmosis, chronic pulmonary histoplasmosis and progressive disseminated histoplasmosis.

The majority of patients with **acute pulmonary histoplasmosis** are asymptomatic. The incubation period varies from 3 to 14 days. Fever, headache, malaise, myalgia, abdominal pain and chills are common symptoms. Cough, haemoptysis, dyspnoea and/or chest pain may be present. **Chronic pulmonary histoplasmosis** is seen in patients with underlying pulmonary disease. Cough, weight loss, fevers and malaise are symptoms. **Progressive disseminated histoplasmosis** is seen in immunocompromised patients, such as patients with AIDS. Cases of isolated adrenal histoplasmosis and mucocutaneous form (especially oral) are reported mainly from India. Adrenal histoplasmosis many of the times mimics carcinomas. Hence, it is essential to specifically investigate *Histoplasma* when isolated adrenal mass is detected (unilateral or bilateral) in a patient.

Laboratory Diagnosis

Laboratory diagnosis depends on the demonstration of oval yeast cells within macrophages in bone marrow aspirates and **tissue biopsy** specimens (Fig. 67.1). Biopsy of oropharyngeal ulcers is usually diagnostic. The sputum and blood culture on SDA at 25 °C show hyphae with tuberculoid macroconidia and at 37 °C show yeasts (Fig. 67.2). It generally grows as a mould at 25 °C, and for confirmation of the identification, this mould needs to be converted to yeast forms using special media incubated at 37 °C. A tissue biopsy smear stained by haematoxylin and eosin may show the presence of yeast forms in tissue. Yeast may be detected in areas of caseation necrosis and calcified lymph

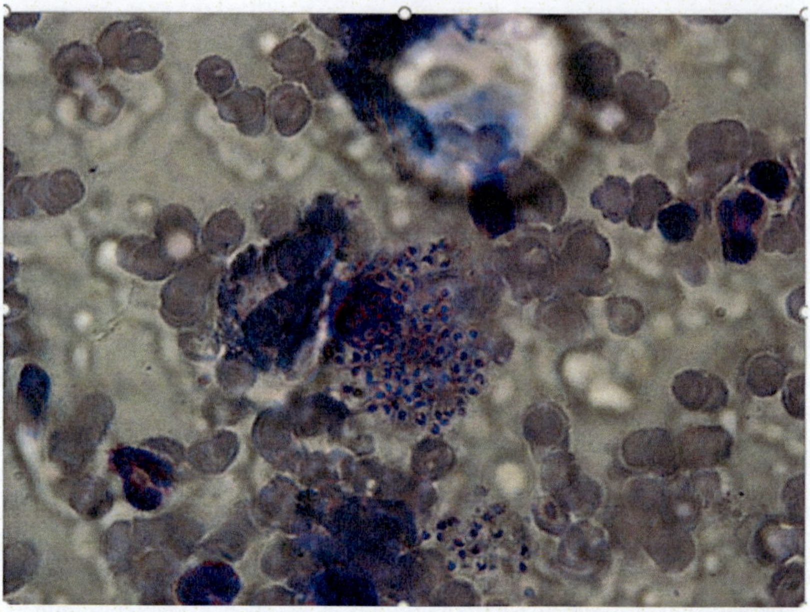

Fig. 67.1 Yeast form of *Histoplasma capsulatum* inside macrophages (×1000) (Courtesy of Prof. MR Shivaprakash)

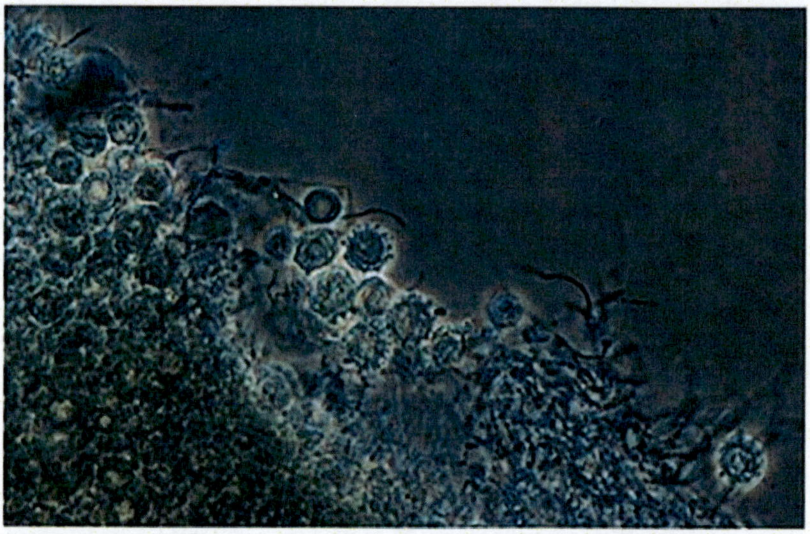

Fig. 67.2 Characteristic tuberculate macroconidia of *Histoplasma capsulatum* in the mould grown at 25 °C (×1000) (Courtesy of Prof. MR Shivaprakash)

nodes by using Grocott–Gomori methenamine silver stain. Yeast forms are rarely seen in stained peripheral blood smears.

Sputum culture is positive in approximately 10–15% of patients with acute pulmonary histoplasmosis and 60% of patients with chronic pulmonary histoplasmosis.

Complement fixation test (CFT) and immunodiffusion (ID) are useful **serological tests** for demonstrating specific antibodies in serum. The CFT titre greater than 1:8 is considered positive. A titre of 1:32 or more suggests active histoplasmosis. Cross-reactivity with antigens from *B. dermatitidis* and *C. immitis* is a noted problem that may lead to a false-positive reaction. The ID test detects antibodies to two glycoproteins, H and M, of *H. capsulatum*. Anti-H antibody is more specific for active histoplasmosis, is positive in 50–80% of patients and remains elevated for years. Detection of specific *H. capsulatum* antigen in serum and urine is useful in immunocompromised hosts when antibody production may be impaired. They are usually positive in case of acute progressive disseminated histoplasmosis.

Histoplasmin skin test detects a delayed-type hypersensitivity to histoplasmal antigens that occur 3–6 weeks after exposure. A positive skin test indicates a past or present infection but does not differentiate between active and past infections. It is positive in approximately 85–90% of immunocompetent individuals. DNA probes and polymerase chain reaction (PCR) are also used to diagnose the condition.

Treatment

Liposomal amphotericin B is the drug of choice to treat disseminated disease. Fluconazole is often recommended for meningitis due to its better penetration of cerebrospinal fluid (CSF). Itraconazole is used to treat mild-to-moderate disease and cutaneous and mucocutaneous forms. Asymptomatic or mild conditions need no treatment.

Epidemiology and Prevention

H. capsulatum is predominantly found in river valleys in North and Central America. It is endemic in the Ohio, Missouri and Mississippi River valleys of the United States.

In endemic areas, the soil provides an acidic, damp environment with high organic content, favouring fungal mycelia growth. Bats play an important role in the transmission of the disease. They can become infected, and they transmit the fungus through their droppings. The soil contaminated with birds' excretions and droppings of bats remains highly infectious for years. Moreover, human infection occurs following inhalation of spores from contaminated soil.

Prevention and Control

Preventing the inhalation of the microconidia of the *H. capsulatum* is difficult. Susceptible individuals residing or visiting endemic areas of this disease should avoid the activities such as disturbing the soil or the wood where the bat droppings are present to prevent this infection. Explorative activities such as visiting the caves pose a major risk and hence should be avoided. In addition, simple measures such as cleaning chicken coops and cleaning or demolishing the old buildings in which bats thrive may also help reduce acquiring this infection.

Blastomycosis

Blastomycosis is a granulomatous fungal infection caused by *B. dermatitidis.*

The Fungus

B. dermatitidis is a dimorphic fungus that occurs in two stages: mould in soil and yeast in tissue. On culture at 37 °C and tissue, the yeast is a round structure with a double-refractile wall and a single broad-based bud (Fig. 67.3). This appearance helps differentiate it from the *Cryptococcus neoformans* or *H. capsulatum*, which has a narrow-based bud. On culture at 25 °C, the fungus produces a mycelial growth showing typical pyriform microconidia, measuring 2–4 μm in diameter.

Pathogenesis of Blastomycosis

B. dermatitidis infection occurs by inhaling microconidia found in soil, its natural habitat. The conidia are transformed to the yeast phase at body temperature inside the lungs. It is resistant to phagocytosis due to its thick cell wall, and it also stimulates Blastomyces adhesin 1 (BAD1) expression, which is an immune-modulatory virulence factor lying on the cell wall. Furthermore,

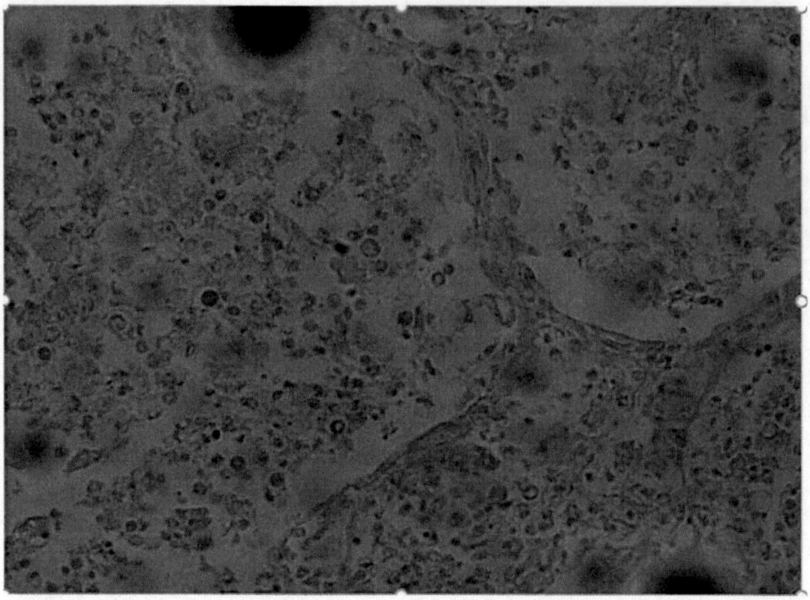

Fig. 67.3 Yeast form of *Blastomyces dermatitidis* (×1000) (Courtesy of Prof. MR Shivaprakash)

the yeast replicates and may spread through the blood and lymphatics to other organs.

Clinical Syndromes

Pulmonary Infection

It is the major clinical manifestation. A flulike illness with fever, chills, myalgia, headache and a non-productive cough may occur, which resolves within days. In some cases, it may develop into a progressive pulmonary disease or result in extrapulmonary dissemination. Patients with chronic pulmonary illness and immunocompromised individuals are more prone to extrapulmonary dissemination involving other organs.

Extrapulmonary Infection

It is most common during the chronic form of illness. Skin involvement with prominent pseudoepitheliomatous hyperplasia is present with micro-abscess formation that may mimic squamous cell carcinoma. Skeletal blastomycosis manifesting as osteomyelitis and septic arthritis can also occur. In men, genitourinary tract involvement, primarily in the prostate and epididymis, is also reported. Rarely, CNS disease may occur.

Laboratory Diagnosis

Sputum or Broncho-Alveolar Lavage (BAL) Microscopy

It is a very useful test for demonstrating characteristic broad-based budding yeasts of *B. dermatitidis*. KOH wet mount of pus aspirated from skin micro-abscesses, fistulae or subcutaneous abscesses is useful to demonstrate the yeast. Sputum or BAL culture on SDA medium confirms the diagnosis of the condition. However, it may take as early as 5 days or as late as 30 days for colony formation.

In the histological examination, the presence of pyogranulomas should alert one to the possibility of blastomycosis. Yeast forms can be difficult to visualise with routine haematoxylin and eosin stains, and special stains like Gomori methenamine silver stain should be used for screening tissue.

Serology

It is useful. ELISA is a useful test to detect antibodies against an antigen of *B. dermatitidis* in the patient's serum. A titre of greater than 1:32 is suggestive of the condition. Radioimmunoassay (RIA) and Western blot techniques are the newer tests that use more specific cell wall antigens of *B. dermatitidis* to detect antibodies in the serum.

Molecular Detection

Molecular detection of *B. dermatitidis* in culture has been accomplished but is used primarily to supplement traditional methods. Skin tests are not reliable for diagnosis.

Treatment

Amphotericin B and itraconazole are the drugs of choice to treat blastomycosis. However, itraconazole and fluconazole are now considered effective alternatives for immunocompetent patients with mild-to-moderate disease.

Epidemiology and Prevention

Blastomycosis is documented in the United States and from Canada and the African continent. In endemic areas, blastomycosis is a common infection among dogs. It is also reported in horses, cows, cats, bats and lions. Microconidia are the infective stages that are present in the soil. Infection is transmitted to humans through inhalation of aerosolised conidia. Middle-aged men with outdoor occupations that exposed them to soil are at the greatest risk for blastomycosis. Recreational activities in wooded areas along waterways have been identified as the major risk factor in outbreak settings.

Preventing inhalation of *B. dermatitidis* conidia is difficult. However, individuals who are immunosuppressed and residing in the endemic areas should avoid soil-disturbing and recreational activities in wooded areas.

Case Study

A 32-year-old lady with AIDS was admitted to a hospital with fever and other non-specific symptoms. Examination of her tongue revealed a white, thick patch. HIV serology was positive. Microscopy of the scrapings collected from the patch showed Gram-positive yeast. Her CD4 count was 200 cells/μL of blood.

- How will you confirm the species identification of *Candida*?
- What are the non-culture detection methods of *Candida*?
- Describe the pathogenesis of infection caused by *Candida* species.
- Describe the treatment of the infection caused by *Candida*.

Opportunistic Fungal Infections

The opportunistic fungi usually cause infections in persons with impaired host defence but do not cause disease in most immunocompetent hosts. Since these fungi become pathogens in individuals with impaired immunity by taking advantage of the host's debilitated conditions, they are called opportunistic fungi.

In recent times, there has been an increasing list of exotic and rare fungi, which have been associated with causing opportunistic infections. However, most opportunistic infections are caused by *Candida* spp., *Aspergillus* spp., *Mucorales*, *Cryptococcus* and *Talaromyces marneffei* (previously known as *Penicillium marneffei*) (Table 67.1).

Candidiasis

Candida species are the most common fungal pathogens that affect humans. The genus *Candida* includes more than 100 species, of which few cause disease in humans. *Candida albicans* was the most frequent cause of infection, which has been taken over by non-albicans Candida worldwide.

C. albicans, Candida tropicalis, Candida glabrata, Candida parapsilosis and *Candida krusei* are the five most common species, which cause opportunistic infections among immunocompromised hosts. Multidrug-resistant species *Candida auris*, which was first isolated in 2009 from Japan, has now been reported from more than 35 countries, causing serious concern.

Table 67.1 Common opportunistic infections

Opportunistic infections	Causative fungus
Candidiasis	*Candida albicans*, *Candida tropicalis* and other species
Aspergillosis	*Aspergillus fumigatus*, *Aspergillus flavus*, *Aspergillus niger* and other species
Zygomycosis	*Mucor*, *Rhizopus* and *Absidia* species
Pneumocystis carinii pneumonia (PCP)	*Pneumocystis jiroveci*
Penicilliosis	*Penicillium marneffei*
Pseudoallescheria boydii infection	*Pseudoallescheria boydii*
Fusariosis	*Fusarium solani*
Cryptococcosis	*Cryptococcus neoformans*

The Fungus

Candida species is a general ovoid or spherical budding yeast with or without pseudohyphae. These form part of the normal flora of the mucous membrane of the gastrointestinal, genitourinary and respiratory tract. *Candida* generally produces **pseudohyphae** in the cultures and tissues. Pseudohyphae are elongated yeast that may resemble hyphae morphologically but are not true hyphae. It grows readily on Sabouraud dextrose agar (SDA) and bacteriological culture media.

C. albicans produces smooth, creamy white colonies with a yeasty odour. It can be differentiated from other *Candida* species by carbohydrate fermentation reaction and characteristic growth on cornmeal agar (CMA). Only *C. albicans* and *Candida dubliniensis* produce chlamydospores on cornmeal agar culture at 25 °C. Germ tube test is another screening test used to differentiate *C. albicans* from non-albicans yeast.

Pathogenesis of *Candida* Infection

In a susceptible host, *Candida* gains access to systemic circulation from the oropharynx of the gastrointestinal tract. The colonisation of the mucocutaneous surface is the first stage in the pathogenesis of *Candida* infection. The fungus causes invasion in human tissue through different routes. Disruption of the skin or mucosa allows the organism access to the bloodstream. Massive colonisation with large numbers of *Candida* also permits the organism to pass directly into the bloodstream, causing the infection.

Candida may disseminate to many organs in immunocompromised hosts, such as the lung, spleen, liver, heart, kidney and brain. *Candida* may induce inflammation of the eye, causing endophthalmitis, and may involve skin in 10–30% of patients with disseminated infection. Deficiency in host defence mechanisms plays a significant role in developing *Candida* infection.

Clinical Syndromes

Candida causes a wide spectrum of clinical illnesses in humans.

Cutaneous Candidiasis

Candida species in an immunocompetent host can cause infection of any warm and moist part of the body exposed to the environment. It causes infection of the nail (paronychia—infection of the nail fold and white superficial onychomycosis) and other skin folds (*Candida* intertrigo—affects groin or axilla often due to excessive sweating and diaper rash).

Mucocutaneous Candidiasis

Oral thrush, vulvovaginitis, perianal disease, etc. are the commonest manifestations of mucocutaneous candidiasis but do not cause any mortality. In immunocompromised individuals, mostly in HIV infection, *Candida* species can cause extreme oropharyngeal and oesophageal candidiasis (AIDS-defining illness).

Chronic Mucocutaneous Candidiasis

These are the different clusters of clinical syndromes mainly indicated by superficial *Candida* infection of the skin, nails and oropharynx with the chronic presentation, and generally, they are refractory to the routine antifungal therapy. Nevertheless, it does not show any evidence of disseminated candidiasis.

Systemic Candidiasis

This involves candidiasis of the respiratory tract, gastrointestinal tract, genitourinary tract, liver, spleen and heart (endocarditis). It also includes **candidemia** and **disseminated candidiasis**. Among HIV patients, oral thrush and candidial oesophagitis are the commonest presentations compared to candidemia or disseminated candidiasis. Central nervous system candidiasis may present with scattered brain microabscesses or larger lesions, vasculitis or thrombosis.

Increased incidence of disseminated candidiasis is commonly seen in patients on long-term immunosuppressive drugs for serious haematological malignancies. Marked neutropenia in these patients is the main reason *Candida* spreads through the blood to most of the organs leading to disseminated candidiasis. Patients usually present with the clinical features of sepsis. Hepatosplenic candidiasis is another form of disseminated candidiasis that occurs in severe neutropenia patients. Most candidemia cases are being reported in critically ill patients admitted to the intensive care units. Long stay in the hospital, central line, long course of multiple antibiotics, burns and abdominal surgery are the important risks of acquiring candidemia in critically ill patients.

Laboratory Diagnosis

Specimens

These include exudates or tissues for microscopy obtained from skin or nails examined by microscope for demonstration of pseudohyphae or budding yeast cells of *Candida*. Blood collected in the blood culture bottle is one of the key samples for detecting candidemia.

Microscopy

Gram-stained smear of the exudates or tissue shows Gram-positive, oval, budding yeast and pseudohyphae (Fig. 67.4). It is preferred to perform the KOH-calcofluor wet mount preparation and examine under a fluorescent microscope to increase the chance of detecting yeasts and

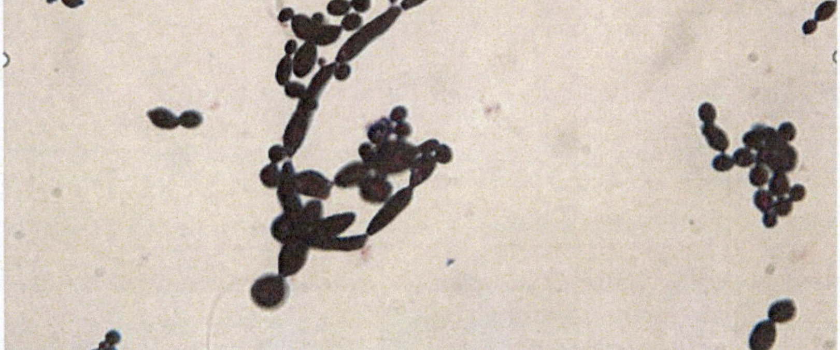

Fig. 67.4 Gram-stained smear showing Gram positive, oval, budding yeast and pseudohyphae (×1000) (Courtesy of Prof. MR Shivaprakash)

moulds. Since *Candida* is found as a part of normal flora on normal skin or mucosa, interpretation should be made in conjunction with clinical presentation. Tissue invasion is of more diagnostic value.

Culture

Aspirates, exudates and tissues are inoculated on the SDA slants and incubated at 37 and 25 °C for up to 4 weeks with regular inspection for colonies. For diagnosis of candidemia, at least 8–10 ml (in adults) of peripheral blood should be collected, taking strict aseptic precaution. The bottles will be incubated for up to 4 weeks with frequent subculture onto SDA plates to recover *Candida* species. Any number of colonies isolated from blood culture is considered significant and warrants antifungal therapy. Culture on SDA produces typical creamy white, smooth colonies (Fig. 67.5). Different *Candida* species are identified by their growth characteristics, sugar fermentation and assimilation tests. **Germ tube test** is a rapid method for identifying *C. albicans* and *C. dubliniensis*. This test depends on the ability of *C. albicans* to produce germ tube within 2 h when incubated in human serum at 37 °C. This phenomenon is called **Reynolds–Braude phenomenon** (Fig. 67.6). Terminal chlamydospores are typically produced by *C. albicans* (mostly single) and *C. dubliniensis* (in bunches) on cornmeal agar at 25 °C, but not by other *Candida* species.

Moreover, CHROM agar allows for the presumptive identification of several *Candida* species by using colour reaction in specialised media, thereby showing different colours of the colonies depending on the *Candida* species. Different *Candida* species can also be identified with more accuracy by automated biochemical assays, such as AP120C, Vitek and AP131C. These assays evaluate the assimilation of various sugars to identify different fungal species.

Non-culture *Candida* Detection

These include (a) 1,3-beta-D-glucan assay, (b) *Candida* mannan antigen and/or antibody assay, (c) *C. albicans* germ tube antibody (CAGTA) assay and (d) d-arabinitol assay.

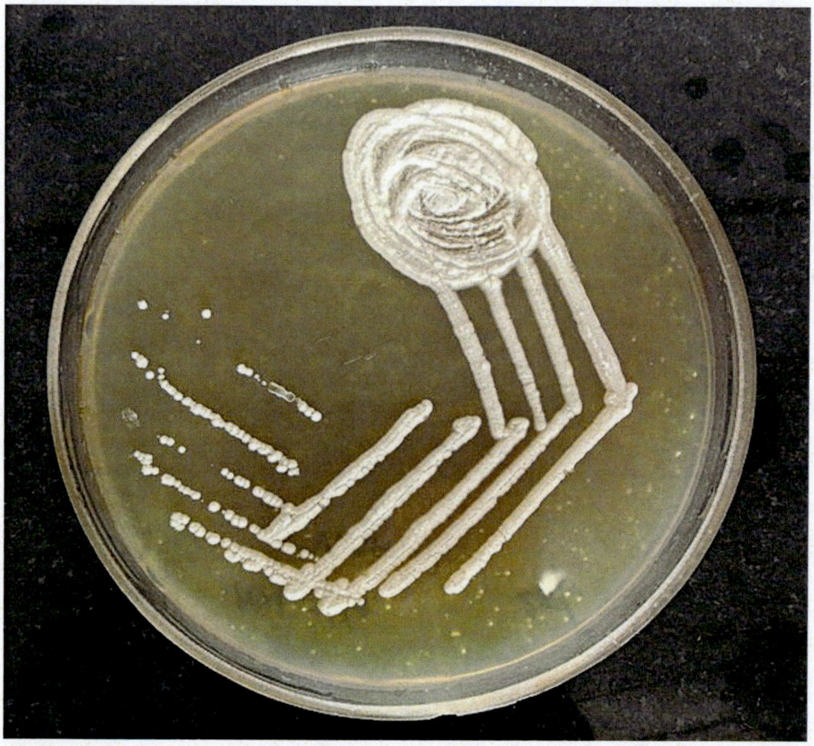

Fig. 67.5 SDA medium showing creamy white colonies of *Candida albicans* (Courtesy of Prof. S. Pramodhini)

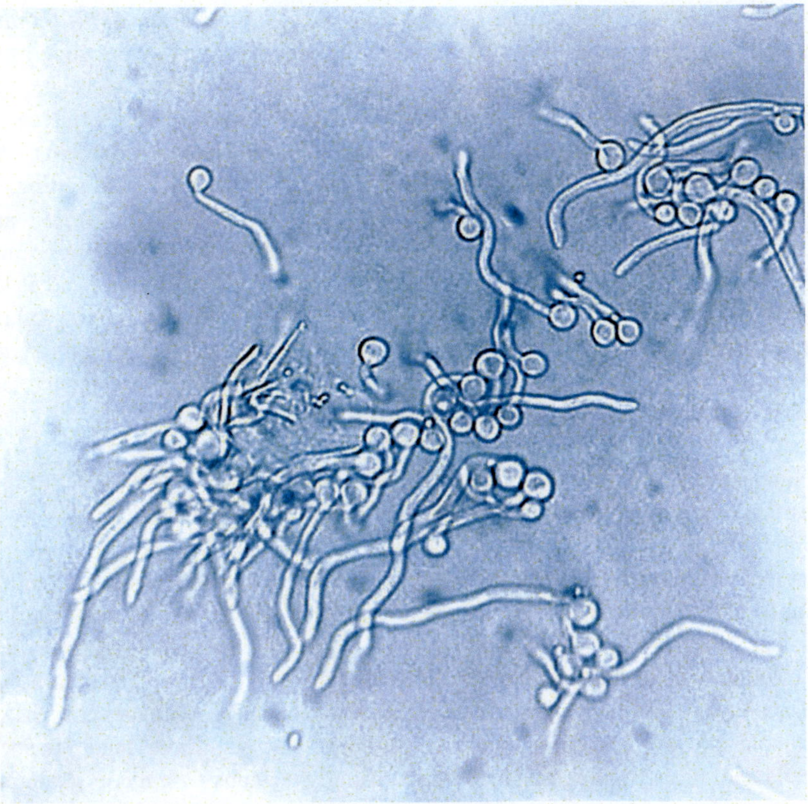

Fig. 67.6 *Candida albicans* showing formation of the germ tube (×400) (Courtesy of Prof. MR Shivaprakash)

Beta-D-glucan assay is a broad-spectrum test that detects *Candida* and *Aspergillus*, *Fusarium*, *Acremonium*, *Pneumocystis* and *Saccharomyces* species. This test depends on the principle that beta-d-glucan is a component of the cell wall of these fungi, which can be detected by its ability to activate factor G during the coagulation cascade of amoebocyte lysate of the horseshoe crab. This test is highly specific and sensitive in diagnosing invasive candidiasis.

With the advent of MADLI-TOF, yeast identification can be made within 2–3 h. Worldwide, this has become the main tool in identification. In addition, many advanced laboratories use DNA sequencing, which is now considered the gold standard in identifying yeasts and bacteria.

Serology

Antibody-based tests are not useful for diagnosing invasive candidiasis as *Candida* is a commensal yeast, and antibodies are noted in the sera of the assay of healthy individuals. As mentioned above, only antibodies against the mannan antigen can help diagnose at least 50% of invasive candidiasis.

Skin Test

A skin test with *Candida* antigen is a delayed hypersensitivity skin test, which is used as an indicator of functions of the CMI. The skin test is uniformly positive in immunocompetent adults and indicates that the person has intact CMI. The skin test is negative in individuals who have deficient CMI. Such a person is anergic and is negative to other skin tests, such as purified protein derivative (PPD) skin test for tuberculosis.

Molecular Detection

DNA probe and polymerase chain reaction (PCR) are still under evaluation but appear promising. T2 Candida is the automated diagnostic platform used to diagnose candidemia using whole-blood specimens. In this platform, *Candida* cells are lysed by mechanical bead beating, then its DNA is amplified by PCR and the amplified *Candida* DNA is detected using amplicon-induced agglomeration of supermagnetic particles and T2 magnetic resonance (T2MR). This helps in the rapid diagnosis of candidemia within 2–3 h after collecting the blood samples.

Treatment

Antifungal therapy forms the mainstay of treatment of the infections caused by *Candida*. Echinocandins are the first drug of choice for treating invasive candidiasis. Caspofungin, anidulafungin and micafungin are echinocandins that have fungicidal action against yeast. Liposomal amphotericin B is the next drug of choice.

Azoles such as fluconazole, voriconazole and posaconazole are other alternatives. Fluconazole ketoconazole and nystatin can be used for superficial candidiasis. Azole resistance is increasingly reported with *C. glabrata*, and *C. krusei* is intrinsically resistant to fluconazole. *Candida lusitaniae* is intrinsically resistant to amphotericin B but susceptible to azoles and echinocandins. *C. auris* is multidrug resistant as it exhibits variable resistance to all the major classes of antifungals such as azoles (fluconazole >90%, amphotericin ~40%, voriconazole ~20%, echinocandin ~15%). Hence, treatment of *C. auris* infection should be done based on the antifungal susceptibility profile.

Epidemiology and Prevention

Candida species are distributed worldwide. *C. albicans*, *C. glabrata*, *C. tropicalis*, *C. krusei* and *C. parapsilosis* are responsible for causing the majority (~80%) of invasive candidiasis. *C. tropicalis* is an important cause of candidemia in patients with leukaemias and those who have undergone bone marrow transplantation. *C. parapsilosis* is commonly associated with invasive devices such as vascular catheters.

Multidrug-resistant *C. auris* is an emerging *Candida* species that can spread from patients or hospital environments. Once *C. auris* is isolated from the hospital setting, it is difficult to eradicate it. *Candida utilis* and *Wickerhamomyces anomalus* are opportunistic yeasts especially reported among paediatric patients.

Antifungal prophylaxis is indicated for patients at high risk of developing invasive candidiasis. There is no vaccine available against candidiasis.

Cryptococcosis

Cryptococcosis, also called **European blastomycosis**, is an acute-to-chronic disease caused by an encapsulated yeast, *C. neoformans*. Cryptococcosis is the most common life-threatening fungal disease in patients with AIDS.

Of more than 30 species that comprise the genus *Cryptococcus*, human disease is associated with *C. neoformans var. grubii* (serotype A), *C. neoformans* (serotype D) and *Cryptococcus gattii* (serotype B/C strains). *C. neoformans* was first described by Busse, a pathologist, in 1984. Kwon-Chung (1976) has reported the exact (i.e. **sexual, teleomorphic**) structure of *C. neoformans*, which the author named *Filobasidiella neoformans*. Prior to the documentation of *F. neoformans* (mycelial form), *C. neoformans* was regarded as a monomorphic fungus.

The Fungus

C. neoformans is considered a **true yeast** as it does not produce pseudohyphae. It is an oval and budding yeast and measures 3–6 μm in diameter. The yeast may be single or may have a single budding daughter cell. Within the host and in certain culture media, the yeast is surrounded by a wide polysaccharide capsule.

The polysaccharide capsule is composed of mannose, xylose and glucuronic acid. This fungus forms a narrow-based bud in contrast to *B. dermatitidis*, which forms a broad-based bud. Rarely, long filamentation may be noted.

On culture, *C. neoformans* on SDA medium forms smooth, convex, cream-coloured mucoid colonies at 20–37 °C. Lactophenol cotton blue (LPCB) wet mount of the colony shows budding yeast cells.

Based on the antigenic specificity of the capsular polysaccharide, the species was classified into four serotypes. At present, based on molecular phylogeny, these serotypes are classified into three species *Cryptococcus grubii* (serotype A), *C. neoformans* (serotype D) and *C. gattii* (serotype B/C strains).

Pathogenesis of Cryptococcosis

The host's immune status is the crucial factor in the pathogenesis of cryptococcosis. *C. neoformans* and *C. grubii* usually cause the most serious infections in patients with impaired CMI. However in addition, *C. gattii* can cause infection even in immunocompetent individuals.

Cryptococcus is primarily transmitted by inhalation (Fig. 67.7). Following inhalation, the yeasts are deposited into the pulmonary alveoli, in which they survive before alveolar macrophages phagocytose them. Glucosylceramide synthase has been identified as an essential factor in the survival of *Cryptococcus* in pulmonary alveoli. Cryptococcal polysaccharide capsule has antiphagocytic properties. Hence, unencapsulated yeast is readily

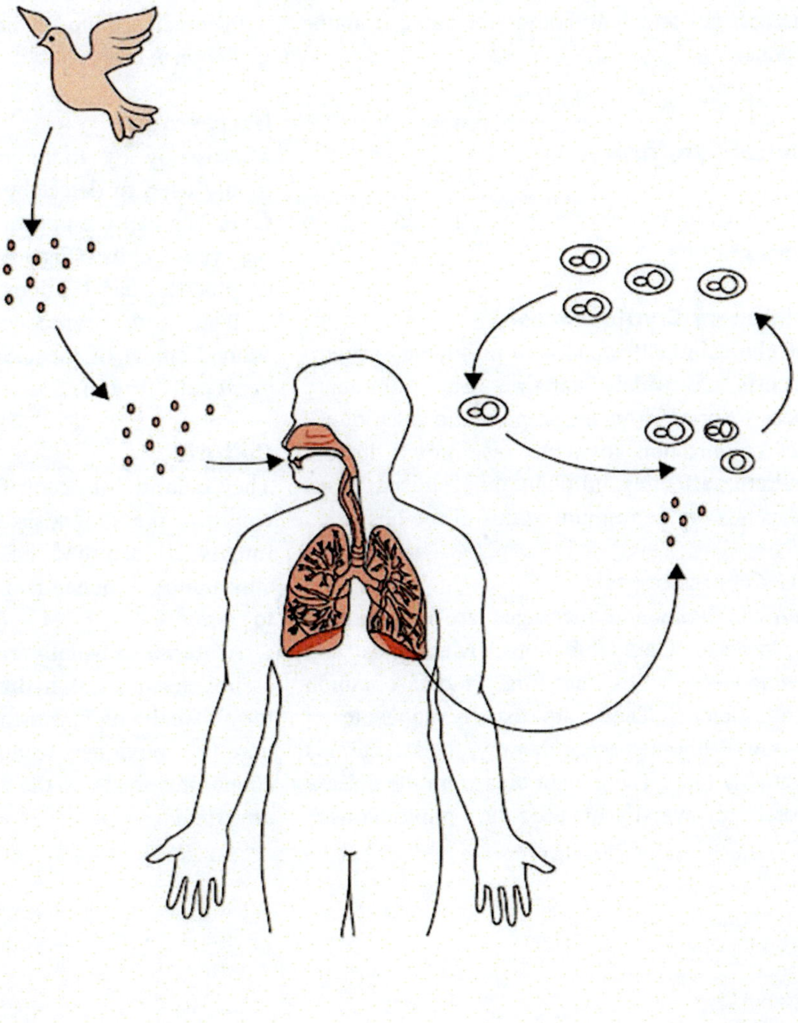

Fig. 67.7 Transmission of *Cryptococcus neoformans*

phagocytosed and destroyed, and then the encapsulated organisms are more resistant to phagocytosis. The capsule is antiphagocytic. Hence, the phagocyte cannot recognise the encapsulated yeast, thereby obstructing the migration of the leukocyte into the fungal replication zone.

Both the cellular and humoral responses mediate the host immunity in cryptococcal infection. CMI is mediated by natural killer cells, and T lymphocytes can inhibit or kill cryptococci. An increase in helper T-cell activity, skin test conversion and a reduction in the number of viable organisms in the tissues indicate a successful host response against the fungus. Anticryptococcal antibodies and soluble anticryptococcal factors mediate humoral immunity. Both anticryptococcal antibodies and the complement play a crucial role in facilitating the organism's macrophage- and lymphocyte-mediated immune response.

Clinical Syndromes

Cryptococcus causes a variety of infections in humans.

Pulmonary Cryptococcosis

The clinical manifestations of pulmonary cryptococcosis are widely variable. The pulmonary disease varies from asymptomatic respiratory tract colonisation to acute respiratory distress syndrome affecting immunocompromised hosts. It depends on the immune status of the host.

CNS Cryptococcosis

Both the brain and meninges are involved in cryptococcal CNS infections. Meningitis and meningoencephalitis are the most common manifestations. These are usually subacute or chronic. Without specific therapy, the infection is invariably fatal. The patient dies from the disease 2 weeks to several years after the symptom onset.

Disseminated Non-pulmonary and Non-CNS Cryptococcosis

Disseminated cryptococcosis includes the skin, prostate and medullary cavity of the bones, next only to the lungs and CNS, and it occurs most commonly in patients with AIDS and other immunosuppressed conditions.

Laboratory Diagnosis

Specimens

Laboratory diagnosis of cryptococcal infection is made by demonstrating the yeast in CSF, sputum, pus and brain biopsy tissue by smear and culture. Gomori methenamine silver or periodic acid-Schiff stains are used to stain the tissue specimens to demonstrate the capsule of *C. neoformans*. Fixed tissue may also be stained with mucicarmine, which preferentially stains *C. neoformans.*

Microscopy

Microscopy by India ink preparation is commonly used to detect budding yeast cells in the CSF. The capsule appears as a clear halo around the yeast cells (Fig. 67.8). By this method, cryptococci can be demonstrated in 25–50% of patients with cryptococcal meningitis. Gram-stained smear of the CSF shows Gram-positive yeast cells (Fig. 67.9).

Culture

The culture of centrifuged CSF specimens confirms the diagnosis of the condition. This fungus is identified based on its microscopic appearance, biochemical test results and ability to grow at 37 °C (98.6 °F). Pathogenic *C. neoformans* usually grows at 37 °C (98.6 °F), assimilates inositol, hydrolyses urea and does not form mycelia on cornmeal agar. Melanin production is obvious when incubated on agar containing seeds of the common weed *Guizotia abyssinica*.

Fig. 67.8 Capsulated yeasts of *Cryptococcus neoformans* in the India ink preparation of CSF (×400) (Courtesy of Prof. MR Shivaprakash)

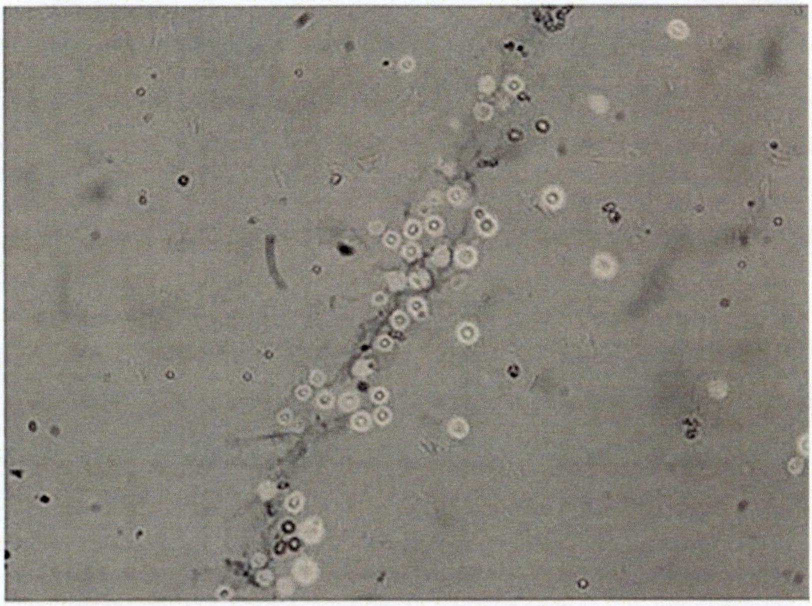

Fig. 67.9 Gram staining showing *Cryptococcus neoformans* (×1000) (Courtesy of Prof. MR Shivaprakash)

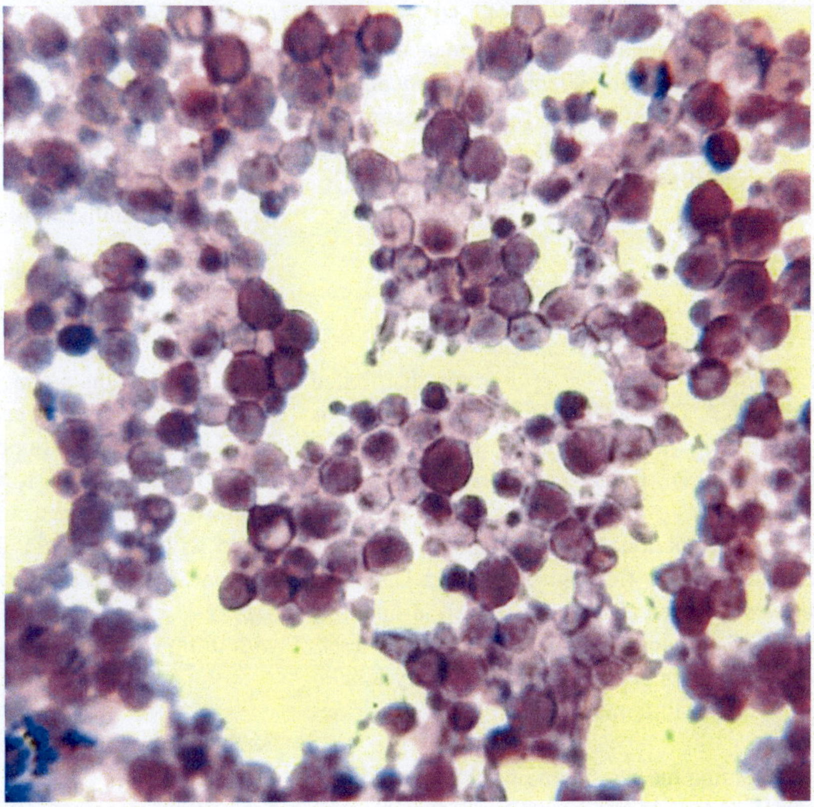

Serology

Latex agglutination test (LAT) is a frequently used serological test to detect cryptococcal polysaccharide antigen in the serum or CSF to diagnose meningitis. LAT is a highly sensitive test in the pathogenic diagnosis of *Cryptococcus* infection. Recently, antigen detection using lateral flow has become popular due to ease of testing.

Treatment

The standard treatment constitutes liposomal amphotericin B at 3–4 mg/kg/day IV plus flucytosine (100 mg/kg/day orally in four divided doses) for 2 weeks followed by maintenance therapy with fluconazole (400 mg daily) for 6–12 months. *C. gattii* causes severe infections. Hence, the duration of treatment is for an extended time period.

The therapeutic goal for patients with cryptococcal disease not complicated by HIV infection is to cure the fungal infection permanently. The therapeutic goal for patients with concomitant HIV infection without a CD4 count of greater than 100 cells/μL is to control the acute infection, followed by lifelong suppression of *C. neoformans.*

Epidemiology and Prevention

C. neoformans and *C. grubii* are distributed worldwide. The incidence of cryptococcosis is increasing, and now it represents a major life-threatening fungal infection in patients with AIDS. *C. gattii* causes disease in immunocompetent individuals and is known to cause outbreaks. While *C. neoformans* and *C. grubii* are closely associated with pigeon droppings, *C. gattii* is associated with the eucalyptus trees. *Cryptococcus* is primarily transmitted by inhalation. Human-to-human transmission does not occur.

Preventing the inhalation of cryptococci is difficult. Limited studies have shown that *Cryptococcus* is present in chicken, pigeon and other birds' dried bird droppings. Hence, susceptible individuals, especially those with HIV/AIDS, should avoid such exposure.

Aspergillosis

More than 100 species of *Aspergillus* have been identified to date. Among these, *Aspergillus fumigatus* and *Aspergillus flavus* are the causative agents of most human diseases, followed by *Aspergillus niger* and *Aspergillus terreus.*

The Fungus

Aspergillus species are moulds, and they have septate hyphae that form V-shaped dichotomous branches (Fig. 67.10). The *Aspergillus* species are identified by (i) their morphological features, (ii) the pattern of conidiophore development and (iii) the colour of the conidia.

The presence of septate hyphae that branch at 45-degree angles is the typical feature of *Aspergillus* species hyphae on histopathology. The hyphae in tissues are best demonstrated with silver stains. The walls of the hyphae are more or less parallel, unlike those of *Mucorales*, which tends to fold.

Lactophenol cotton blue (LPCB) preparation of the colonies shows septate hyphae and branching conidiophores. Asexual conidia arranged in chains are present on elongated cells known as sterigmata. The latter is present on the vesicle of the conidiophores. The conidia of *Aspergillus* typically form radiating chains in contrast to *Rhizopus* and *Mucor*, which are found within sporangia. The fungus grows rapidly on SDA and other culture media at 25 °C. *Aspergillus* produces colonies within 1–2 days with different colours based on the species.

Pathogenesis of Aspergillosis

Aspergillus species, unlike *Candida* species, do not form part of the normal flora of humans. They are ubiquitous in the environment;

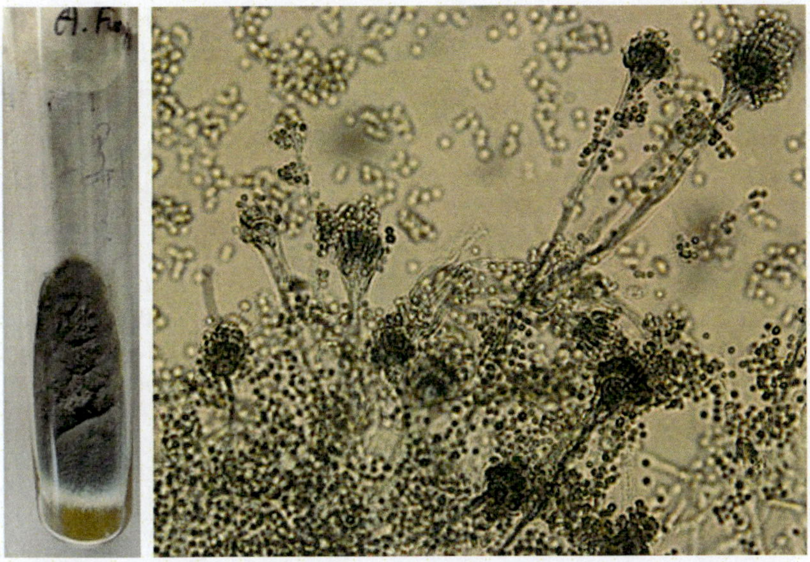

Fig. 67.10 SDA medium showing bluish green mycelial colonies of *Aspergillus fumigatus* and on microscopy uniseriate phialides that cover only upper two-thirds of vesicle and conidia in chains (×400) (Courtesy of Prof. MR Shivaprakash)

hence, transmission of infection is mostly exogenous.

Aspergillus species rarely cause infections in immunocompetent individuals. They cause invasive infections, mostly in the patients who are immunocompromised. In these hosts, *Aspergillus* species cause invasion of the blood, thereby causing infarction, haemorrhage and necrosis of lung tissues. *Aspergillus* spp. also produces toxic metabolites like elastolytic proteases, gliotoxin and hyaluronic acid that inhibit macrophage and neutrophil phagocytosis, facilitating the dissemination of the infection.

The size of the conidia is an important factor that affects virulence in some aetiologic agents. The small conidia of *A. fumigatus* allow accessibility to the lower respiratory tract, making it the main agent in invasive pulmonary aspergillosis. However, *A. flavus*, with a large conidial size, is an important etiologic agent in sinusitis, cutaneous and wound aspergillosis.

Clinical Syndromes

In immunocompetent hosts, *Aspergillus* species may primarily affect the lungs, causing four main syndromes discussed as follows.

Allergic Bronchopulmonary Aspergillosis

It is a hypersensitivity reaction to *A. fumigatus* organisms, colonised in the tracheobronchial tree. This condition often occurs in association with asthma and cystic fibrosis.

Chronic Necrotising Pulmonary Aspergillosis

It is a subacute infection seen in patients with some degree of immunosuppression. The condition occurs in conjunction with alcoholism, underlying lung disease or chronic corticosteroid therapy.

Aspergilloma

It is a condition that occurs in a pre-existing cavity in the lung parenchyma. This cavity may have been caused earlier by tuberculosis, sarcoidosis, cystic fibrosis and emphysematous bullae. The condition is characterised by the presence of a ball of fungus within the cavity. The fungus, however, does not invade the cavity. It may cause haemoptysis.

Invasive Aspergillosis

It is a rapidly progressive infection in patients who are severely immunocompromised. The condition is mostly fatal. In an immunocompromised host, *Aspergillus* organisms cause a disseminated disease, leading to endophthalmitis, endocarditis and abscesses in the viscera, such as liver, spleen, kidney, soft tissues and bone.

Laboratory Diagnosis

Laboratory diagnosis of invasive aspergillosis or chronic, necrotising aspergillus pneumonia depends on the demonstration of septate hyphae in tissue by direct microscopy and isolation of *Aspergillus* in culture.

Specimens

Respiratory specimens such as BAL, sputum and lung biopsy are the main samples for diagnosing invasive pulmonary aspergillosis and other pulmonary manifestations of aspergillosis. Tissue exudates or aspirates, or scrapings from other affected organs, will help diagnose different forms of aspergillosis.

Direct KOH Preparation

Direct KOH preparation of the specimen shows non-pigmented, septate hyphae with characteristic dichotomous branching at an angle of 45°. Biopsy specimens show septate, branching hyphae invading the tissues.

Culture

It is carried out by inoculation of specimens on SDA without cycloheximide and is incubated at 25 °C for 1–2 days. Different *Aspergillus* species produce colonies of various colours. *A. fumigatus* (Fig. 67.10) produces grey colonies, whereas *A. niger* (Fig. 67.11) and *A. flavus* (Fig. 67.12) produce black and yellowish-green colonies, respectively. The isolation of *Aspergillus* by culture needs to be interpreted with care because they are also common laboratory contaminants. Repeated isolation of the fungi may help suggest the pathogenic role of the *Aspergillus*.

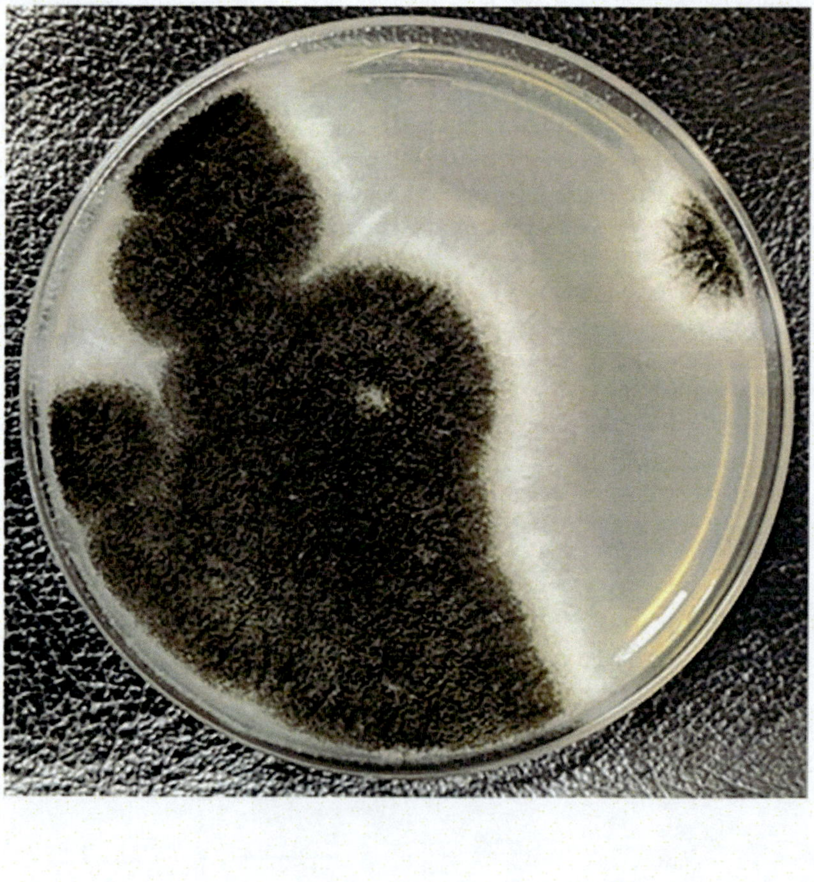

Fig. 67.11 SDA medium showing black colonies of *Aspergillus niger* (Courtesy of Prof. MR Shivaprakash)

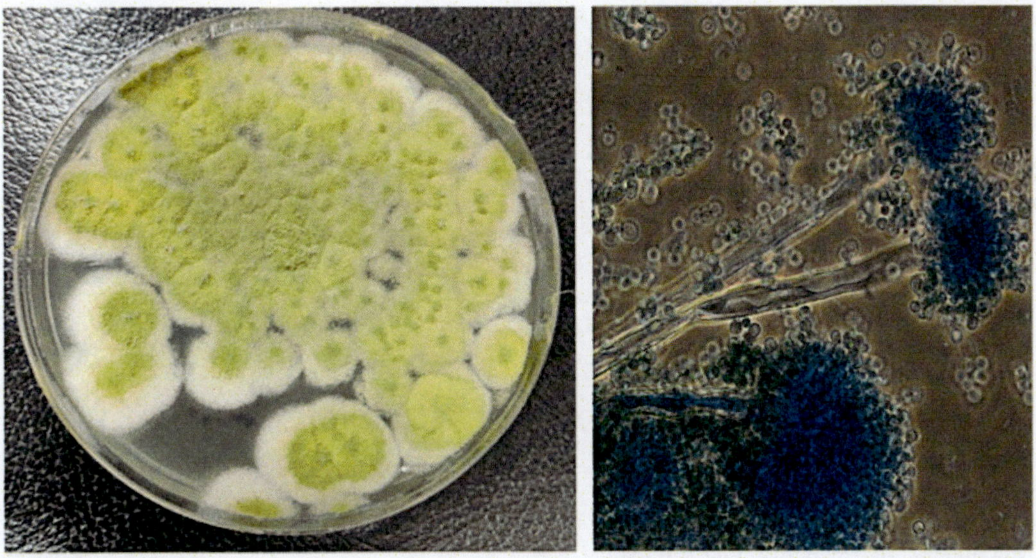

Fig. 67.12 SDA medium showing yellow colonies of *Aspergillus flavus* and on microscopy uniseriate/biseriate phialides that cover entire vesicle and conidia in chains (×400) (Courtesy of Prof. MR Shivaprakash)

Serology

Serology is by estimation of serum galactomannan is useful for diagnosing invasive aspergillosis. Demonstration of elevated galactomannan in bronchoalveolar lavage and blood is also useful for diagnosing invasive pulmonary aspergillosis. Highly increased levels of serum IgE (>1000 IU/cm^3) and positive serology for *Aspergillus* precipitation or *Aspergillus*-specific IgG or IgE antibodies are suggestive of the allergic form of aspergillosis. Serum IgG levels are usually positive in aspergilloma.

The diagnosis of allergic bronchopulmonary aspergillosis can also be established by a positive **skin test** for *A. fumigatus*.

Treatment

Voriconazole is the recommended drug for the treatment of invasive aspergillosis, whereas itraconazole is given for ABPA/CPA patients. Liposomal amphotericin B is an alternative drug given to patients. Aspergilloma is treated best by removing the fungus ball from the cavity by surgery.

Epidemiology and Prevention

Various clinical manifestations of *Aspergillus* infection have been documented worldwide. As this agent has ubiquitous distribution, there are no specific preventive measures available against aspergillosis.

Mucormycosis (Zygomycosis)

Mucormycosis, also known as **zygomycosis**, is an infection caused by saprophytic moulds belonging to *Mucorales* such as *Mucor* spp. (Fig. 67.13), *Rhizopus* spp. (Fig. 67.14) and *Absidia* spp. These fungi are ubiquitous in the environment and generally saprophytic. They are the third most common fungal infection in immunocompromised individuals.

The Fungus

Rhizopus species are the most common causative agents of mucormycosis in humans. Of many *Rhizopus* species, *Rhizopus arrhizus* is the most

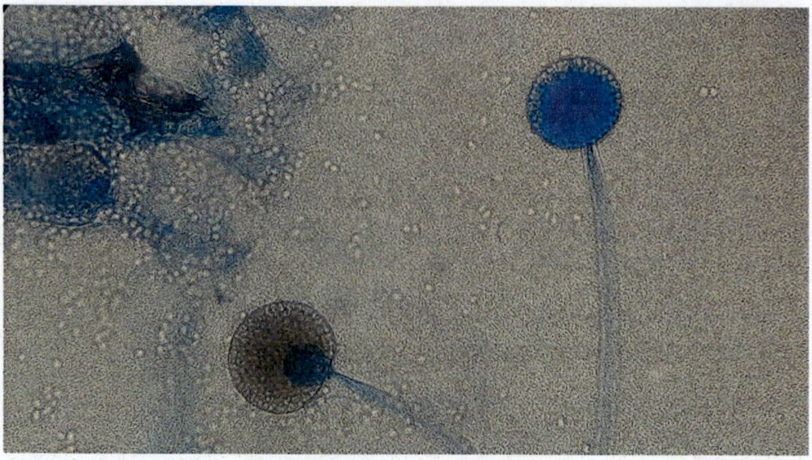

Fig. 67.13 LPCB mount of *Mucor* showing broad hyaline aseptate hyphae, branched sporongiospores , with terminal round spore filled sporangia (×400) (Courtesy of Prof. MR Shivaprakash)

common agent of mucormycosis. The hyphae of the Mucorales are generally characteristically aseptate (absence of septa), but occasionally few septa may be noted (pauciseptate). The growth of the Mucorales appears typically cottony. For example, *R. arrhizus* typically have a salt-and-pepper appearance. The genus and species can be differentiated based on the size, shape of sporangia and sporangiospores, branching or non-branching of sporangiophores, presence or absence of rhizoids and their origin (nodal or internodal), etc.

Pathogenesis of Mucormycosis

Mucorales have a high degree of predilection to invade major blood vessels, leading to ischaemia, necrosis and infarction of adjacent tissues. Uncontrolled diabetes is one of the commonest risk factors for acquiring mucormycosis. They also affect patients with metabolic acidosis secondary to renal insufficiency, diarrhoea and aspirin intake. Most of the infections caused by Mucorales are acute and are usually fatal despite

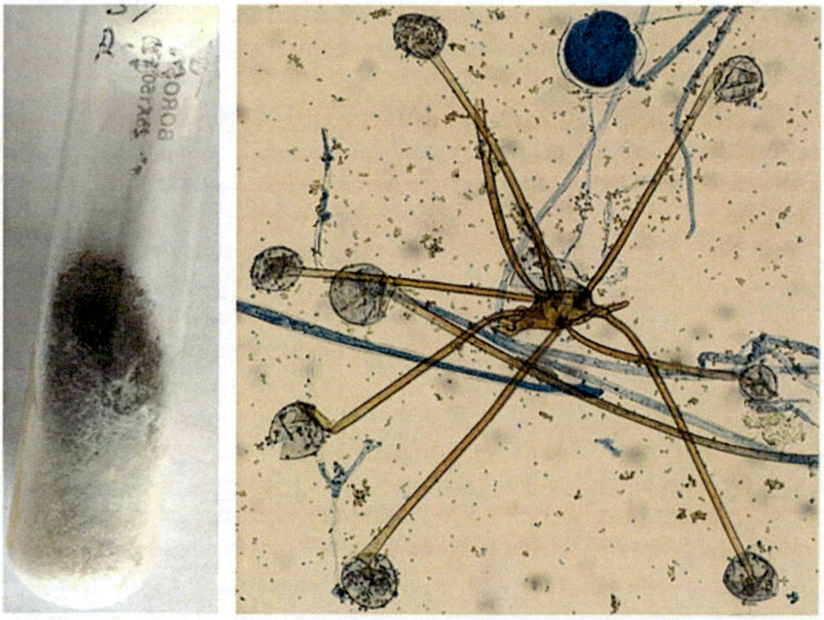

Fig. 67.14 SDA medium showing salt and pepper appearance, dark brown colonies of *Rhizopus* and on microscopy broad hyaline aseptate hyphae, unbranched sporongiosphores , stolon and rhizoids (×400) (Courtesy of Prof. MR Shivaprakash)

early diagnosis and treatment. The agents of mucormycosis are transmitted by air through their asexual spores. Recently, a large outbreak of COVID-19-associated mucormycosis led to declaring this disease as a notifiable disease in India.

Clinical Syndromes

Mucorales cause a spectrum of diseases in humans depending on the host's immune status and the portal of entry. These cause rhino-orbitocerebral mucormycosis, pulmonary mucormycosis and gastrointestinal mucormycosis. Rhino-orbitocerebral mucormycosis is the most common manifestation of disease seen in patients with diabetic acidosis. Cutaneous or subcutaneous mucormycosis occurs after trauma and spreads very rapidly if not treated.

Laboratory Diagnosis

Clinical diagnosis of mucormycosis is frequently difficult. It requires a high degree of suspicion and a host with appropriate risk factors. Laboratory diagnosis is by microscopy, culture and histopathology.

Microscopy
It depends on demonstrating the characteristic appearance of broad, non-septate hyphae with right-angled branches in the specimen. KOH mount of tissue material exhibits broad, irregularly set hyphae with right-angled branches.

Culture
Culture on SDA medium without antibiotics is dense and cottony. The LPCB preparation of the colony shows the coenocytic nature of the hyphae and the characteristic sporangia that contain sporangiophores. Fungal cultures are negative in about 30% of cases. The fungi are difficult to culture because they are fragile due to thin cell walls, and individual cells are long (due to the absence of septa or having rare septa) and can get damaged easily, losing their viability. Blood cultures are of no value.

Histopathology
Histopathology of affected tissue is very useful to confirm the diagnosis. A fungal stain of biopsy material obtained from the affected tissue showing non-septate hyphae with broad, irregular branches that form more or less at right angles suggests the diagnosis.

Serology
No serological tests are available.

Treatment

Liposomal amphotericin B is the drug of choice for mucormycosis. Posaconazole is also given to patients with acute renal failure and who cannot tolerate amphotericin B. Surgical removal of necrotic, infected tissues should be done if possible.

Epidemiology and Prevention

The agents of mucormycosis are transmitted by air through their asexual spores.

Mucormycosis is reported globally, although more reports are seen from tropical and subtropical regions. The main burden of mucormycosis is seen among individuals with uncontrolled diabetes and diabetic ketoacidosis. However, patients with haematological malignancies, those who have undergone haematopoietic or solid-organ transplants, those on voriconazole therapy and iron chelation therapy, those with malnutrition and trauma (like a road traffic accident or burns), or those on immunosuppressants are at risk of developing mucormycosis. Recently, a large outbreak of COVID-19-associated mucormycosis led to declaring this as a notifiable disease in India, particularly among those receiving unsupervised or inappropriate corticosteroid therapy. The clinical presentations of mucormycosis are classified based on the anatomic localisation, such as rhino-orbital-cerebral

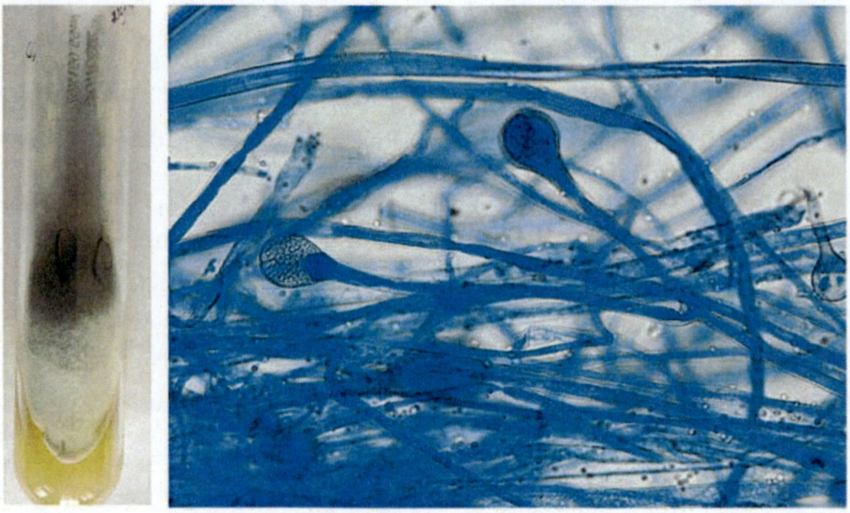

Fig. 67.15 SDA medium showing the growth of *Lichtheimia corymbifera* that resembles grey cotton candy, and on microscopy aseptate hyphae branched sporangiophores arise at points on stolon. (Courtesy: Prof. MR Shivaprakash)

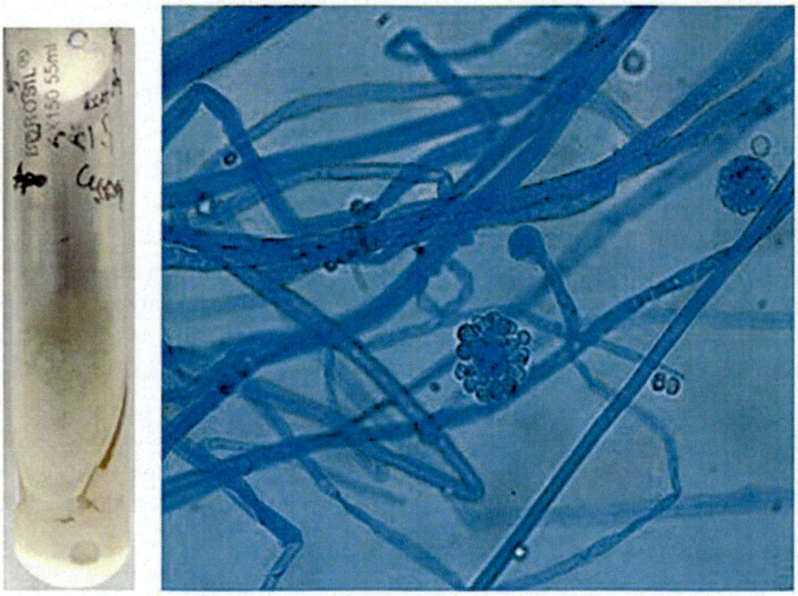

Fig. 67.16 SDA medium showing colonies of *Cunninghamella bertholletiae* appearing grey and cottony, and on microscopy hyphae are aseptate with branched sporangiophore ending up with vesicle with denticles supporting oval sporangium. (Courtesy: Prof. MR Shivaprakash)

(ROCM), pulmonary, gastrointestinal, cutaneous, renal and disseminated mucormycosis. Globally, *R. arrhizus* is the commonest cause of mucormycosis, followed by *Lichtheimia* (Fig. 67.15) and *Cunninghamella* (Fig. 67.16) species. In India, *Apophysomyces* spp. is the second most common agent, infection due to *Rhizopus microsporus*, and *Rhizopus homothallicus* are also rising.

Prevention and Control

Though no specific preventive measures are described for mucormycosis, some small steps may reduce the chance of acquiring mucormycosis:

- Those suffering from diabetes mellitus should regularly monitor the sugar level and maintain the sugar level.

- Avoid areas such as construction or excavation sites to reduce inhalation of Mucorales spores. If not possible, at least wear a mask while working in those areas.
- Wash the wound with soap and clean water after injury, especially when the wound is contaminated with soil or dirt.

Pneumocystosis

Pneumocystis jiroveci, previously known as *Pneumocystis carinii*, is the causative agent of **pneumocystis pneumonia (PCP)**. PCP is the most common opportunistic infection in HIV-infected patients.

Pneumocystis is a unicellular fungus found in the respiratory tracts of many mammals and humans. The genus *Pneumocystis* was initially mistaken for trypanosome, and later as a protozoan. Biochemical analysis of the nucleic acid composition of *Pneumocystis* rRNA and mitochondrial DNA in the 1980s established it as a fungus. The cyst wall closely resembles that of fungi. However, it does not have ergosterol in its membrane as do the fungi but instead has cholesterol.

The Fungus

The organism presents in three morphological forms. Trophozoite or trophic stage presents commonly as clusters. The sporozoite comprises precystic forms. The cyst comprises many intracystic bodies called spores.

Pathogenesis of Pneumocystosis

Transmission of infection occurs by inhalation. Once inhaled, the tropic form of the organism attaches to the alveoli. Defects in both cellular immunity and humoral immunity allow for uncontrolled replication of the organism and development of the disease. Activated alveolar macrophages without CD4+ cells fail to eradicate the organism.

Clinical Syndromes

P. jiroveci causes PCP in HIV patients with their CD4+ cell count below 200 cells/μL. It also causes PCP in other patients with primary immune deficiencies, including hypogammaglobulinemia and severe combined immunodeficiency, an organ (e.g. heart, lung, liver, kidney)-transplant recipient's on long-term immunosuppressive regimens and in patients with haematologic and non-haematologic malignancies. *P. jirovecii* is also seen as a coloniser in the respiratory tract.

In symptomatic cases, sudden onset of fever, non-productive cough, dyspnoea and tachypnoea are typical manifestations. Bilateral rales and rhonchi are present. On different radiological investigations, it presents classically as interstitial pneumonia. Extra-pulmonary manifestations are rare. It occurs in AIDS patients during their advanced stage.

Laboratory Diagnosis

Chest radiographs in most patients show diffuse bilateral infiltrates extending from the perihilar region. Patchy asymmetric infiltrates and pneumatoceles are less common findings.

Unlike other fungi, *P. jiroveci* cannot be cultured in any media used for fungal culture. Sputum induced by inhalation of a hypertonic saline solution is a useful specimen for demonstration of the organism by **microscopy**. *Pneumocystis* is less frequently demonstrated in expectorated sputum. Bronchoalveolar lavage is the most common invasive procedure used to diagnose PCP. This is recommended when PCP is strongly suspected, but the induced sputum sample is repeatedly negative. Bronchoalveolar lavage has a high sensitivity of more than 90%.

Wright, Cresyl violet and Giemsa stains are taken up by both trophozoite and cyst forms but do not stain the cyst wall. Toluidine blue and Gomori methenamine silver stain only the wall of *Pneumocystis* cysts. Papanicolaou stain is valuable in detecting the foamy eosinophilic material present around the *P. jirovecii*. **Direct immunofluorescence** using monoclonal antibodies is a more sensitive method than histologic techniques for detecting *Pneumocystis* in histopathologic specimens.

Open lung biopsy has 100% sensitivity and specificity because it provides the greatest amount of tissue for diagnosis. This is an invasive procedure and hence reserved for rare cases when bronchoscopy is non-diagnostic. Presently, diagnosis of PCP is made by **real-time PCR** as it can differentiate colonisation from infection.

Treatment

P. jiroveci, although considered a fungus, does not respond to treatment with antifungal agents. A combination of trimethoprim and sulfamethoxazole is the drug of choice for the treatment of PCP. Pentamidine and atovaquone are alternative drugs.

Epidemiology and Prevention

Pneumocystosis was first noticed when it produced interstitial pneumonia among severely malnourished and premature infants in parts of Central and Eastern Europe during the World War II. Prior to 1981, there were only a few reports (less than 100) from the United States. However, in 1981, the CDC in the United States reported the occurrence of PCP in five healthy homosexual males residing in Los Angeles. Now, *P. jiroveci* is recognised as one of the common causes of life-threatening opportunistic infections in patients with HIV infection worldwide.

In developing countries, many of the cases are not being reported due to the non-availability of modern healthcare. *Pneumocystis* is responsible for up to 80% of HIV-infected infants with pneumonia in Africa.

Chemoprophylaxis with trimethoprim and sulfamethoxazole or aerosolised pentamidine is useful for preventing infection in patients with CD4 counts below 200 cells/μL.

Talaromycosis

Talaromycosis (**formerly penicilliosis**) is a life-threatening disease caused by a thermally dimorphic fungus *Talaromyces (Penicillium) marneffei*, which was first isolated from a bamboo rat in Vietnam in 1956.

Capponi and co-workers (1956) first demonstrated *P. marneffei* in the liver of a bamboo rat. Gabriel Segretian documented a laboratory-acquired case due to accidental inoculation of the skin. Gabriel Segretian named the organism in honour of Hubert Marneffei. The first case of natural human infection was reported by DiSalvo and co-workers (1973) in an American minister with Hodgkin's disease.

The Fungus

T. marneffei is the only dimorphic fungus in the genus *Talaromyces*. The fungus grows rapidly on SDA. At 25 °C, it grows as a mould and produces greenish-yellow colonies with a dark green edge and pink or red centre. It exhibits brick red pigment on the reverse. At 37 °C, it grows as a small yeast, producing cream-coloured mucoid colonies resembling *H. capsulatum*.

Pathogenesis of Talaromycosis

Inhalation of its conidia is the main route of infection, followed by invasion through the skin. It is recognised by the host immune cells as a pathogen leading to phagocytosis by macrophages. However, instead of being killed by macrophage, *T. marneffei* uses it as a host for cell multiplication and transformation into yeast form.

The most frequent organs involved in this disease are the liver and lungs. However, infection in these two organs also affects lymph nodes, bone marrow, skin and intestines. Talaromycosis typically develops as an opportunistic infection in the immunocompromised hosts, especially in patients with HIV when CD4+ T cells are <100 cells/mm^3.

Clinical Syndromes

In talaromycosis-associated HIV infection patients, the severity of disease likely depends on CD4 number with low CD4 count associated with more invasive and necrotising lesions. Talaromycosis can present as a local or disseminated infection. Mucocutaneous lesions are frequently present in the disseminated form.

Laboratory Diagnosis

Microscopy

Direct examination of the tissue or secretions by microscopy after staining with Wright stain and another histological staining like periodic acid-Schiff (PAS), haematoxylin and eosin (H&E), Giemsa or Gomori methenamine silver (GMS) can show yeast form with pleomorphic, sausage-like shape and cross-walls.

Culture

Culture from bone marrow or blood samples can be used to cultivate the fungus after 2–7-day incubation at 37 °C.

Serology

Serology, both antibody and antigen tests, has been developed. Mp1p mannoprotein antigen detection by 4D1 monoclonal antibody with ELISA or strip tests in plasma and urine can be used. Galactomannan assay for *Aspergillus* is another positive diagnostic tool in talaromycosis patients.

Molecular Detection

Molecular based assay or polymerase chain reaction (PCR) for *T. marneffei* identification is now available but limited in research use only.

Treatment

The therapeutic strategy for talaromycosis comprises three phases: induction phase with amphotericin B, consolidation phase with itraconazole and maintenance phase with continuation with itraconazole until the CD4 count rises above 100 cells/mm^3 for at least 6 months.

Epidemiology and Prevention

T. marneffei causes endemic infection in several countries, mostly in South and Southeast Asia, including Thailand, Vietnam, Hong Kong, Southern China, Taiwan, Northeast India and Laos. It is an AIDS-defining illness. Regarding the natural reservoir of *T. marneffei*, only human and bamboo rats have been reported so far.

Prevention and Control

As talaromycosis is mainly seen in immunocompromised individuals such as with HIV/AIDS, itraconazole prophylaxis is advised to prevent this infection in the endemic zones.

Pseudallescheria boydii and *Scedosporium apiospermum* Infection

Pseudallescheria boydii (**teleomorph** or **sexual stage**) and *Scedosporium apiospermum* (Fig. 67.17) (**asexual stage**) are moulds that cause opportunistic infection in immunocompromised persons such as patients with HIV. These cause a disease that clinically resembles aspergillosis.

The diagnosis is made by demonstration of septate hyphae in tissue that resemble those of *Aspergillus*. It can be differentiated from *Aspergillus* by culture on SDA medium. In culture, *S. apiospermum* produces pear-shaped conidia, whereas *P. boydii* produces fruiting bodies and ascospores. The fungus causes cutaneous and disseminated infections. Blood culture is frequently useful for the diagnosis of disseminated infection.

Voriconazole and itraconazole are the drugs of choice to treat *P. boydii* infection. Resistance to

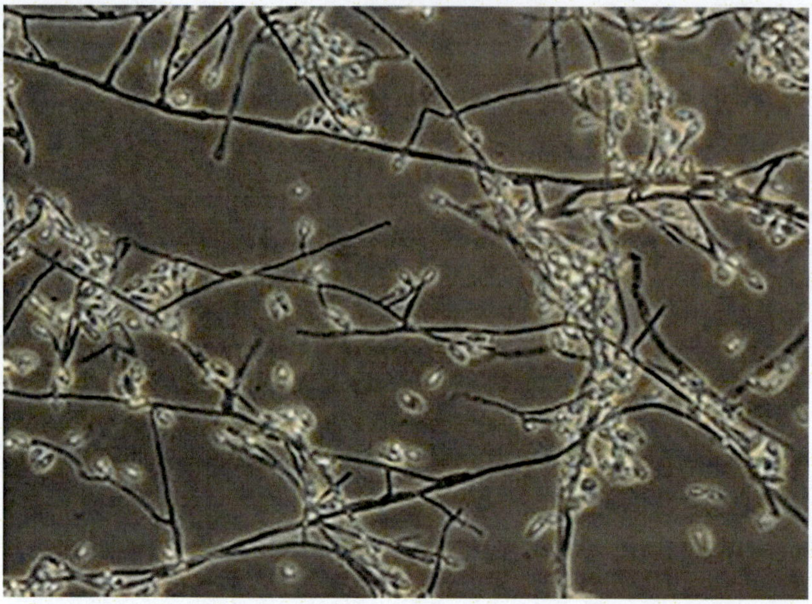

Fig. 67.17 Microscopy showing septate hyphae conidiophores bearing oval-shaped unicellular conidia of *Scedosporium apiospermum*. (Courtesy: Prof. MR Shivaprakash)

amphotericin B has been reported. Surgical removal of infected necrotic tissue is also useful.

Fusariosis

Fusarium solani (Fig. 67.18) is the most common agent that causes fusariosis, primarily in neutropenic patients. It resembles *Aspergillus* in that it is a mould with septate hyphae, which tends to invade blood vessels. Fever associated with lesions in the skin is the common clinical presentation. The fungus differs from *Aspergillus* by culture in which it produces banana-shaped conidia. *F. solani* causes a disseminated disease primarily in neutropenic patients. Blood culture is frequently useful for the diagnosis of the condition. Amphotericin B is the drug of choice.

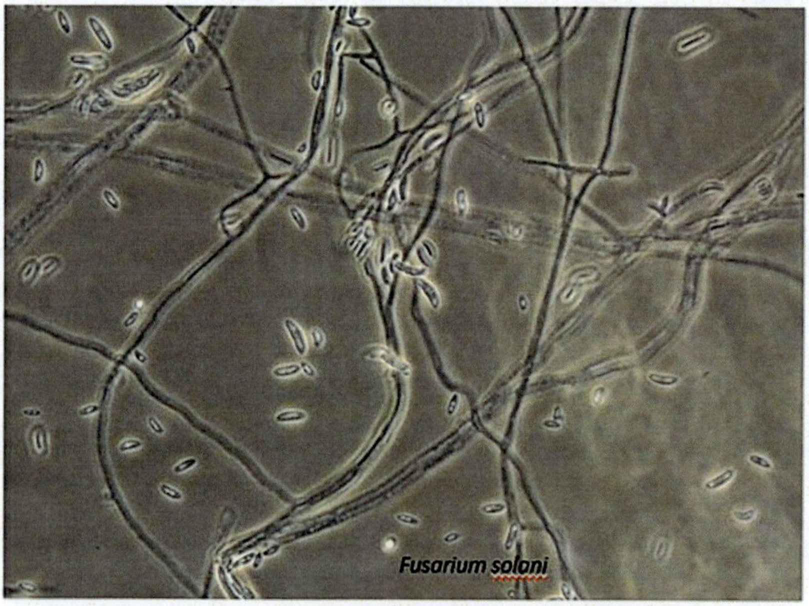

Fig. 67.18 Microscopy showing falcate or fusiform macroconidia with 3–5 septa and cylindrical, one- or two-celled microconidia of *F. solani*. (Courtesy: Prof. MR Shivaprakash)

Other Fungal Infections

Otomycosis

Otomycosis is an infection caused by a fungus belonging to species of *Aspergillus* and *Penicillium* on the external ear. It is a very common disease characterised by itching, inability to hear and pain. It can be diagnosed by demonstrating the fungi in scrapings and cultures.

Oculomycosis (Keratomycosis, Fungal Keratitis, Mycotic Keratitis)

Several saprophytic fungi cause ocular infections. Some of them are *Fusarium*, *Aspergillus* spp., *Curvularia*, *Alternaria*, *Acremonium* and *C. albicans*. Mycotic keratitis is a corneal infection caused by fungi, mostly leading to trauma in the cornea. The condition is characterised mainly by corneal injury. Extended usage of corticosteroids in ophthalmology has become a major cause of keratomycosis. The injured tissues are colonised with the fungal spores resulting in inflammatory reactions leading to endophthalmitis and hypopyon ulcer. Deep corneal scraping specimens collected under a slit-lamp microscope can be cultured for microscopy. Application of nystatin, natamycin and amphotericin B can be used locally to treat the condition.

Mycotic Poisoning

Mycetism and mycotoxicosis are examples of mycotic poisoning caused by toxic substances produced by the fungi.

Mycetism

Many varieties of poisonous mushrooms have been found inedible for many years. Mycetism may cause dermatitis and gastrointestinal disease and sometimes may lead to death. In recent years, hallucinogenic substances produced by *Psilocybe* species have become attention-seeking. Some of them are *Coprine* species causing coprine poisoning, *Claviceps* species causing ergot poisoning and *Inocybe* species causing muscarine poisoning.

Mycotoxicosis

It is caused by ingesting food substances contaminated with fungal toxins called mycotoxins. These are naturally produced by fungi and are seen in certain food particles.

Aflatoxin is the best-known type of mycotoxin. These are commonly seen in mouldy foods, specifically ground peas and corn. *A. flavus* secretes aflatoxin B1. *Aspergillus parasiticus* secretes aflatoxins B1, B2, G1 and G2. These toxins are very toxic to human beings, animals and birds. They are also of great concern because they cause carcinogenic effects in human beings and hepatoma in ducklings and rats. Several cases of aflatoxicosis in both humans and animals have been reported from India. Some examples include **fumonisin**, a toxin produced by *Fusarium* species present in maize, **ochratoxin**, released by *Aspergillus* species in cereals; and toxin produced by *Penicillium* species in bread.

Ergo toxicosis: It is caused by ergot alkaloids produced by the fungus *Claviceps purpurea*, which grows on the fruiting heads of rye. Some species of *Fusarium* produce toxins called **trichothecenes**. *Fusarium graminearum* produces a toxin called **zearalenone**, which, when consumed by animals, may develop symptoms of mimicking an osteogenic disorder.

Case Study

A 32-year-old lady with AIDS was admitted to a hospital with fever and other non-specific symptoms. Examination of her tongue revealed a white, thick patch. HIV serology was positive. Microscopy of the scrapings collected from the patch showed Gram-positive yeast. Her CD4 count was 200 cells/μL of blood.

- How will you confirm the species identification of *Candida*?
- What are the non-culture detection methods of *Candida*?
- Describe the pathogenesis of infection caused by *Candida* species.
- Describe the treatment of the infection caused by *Candida*.

Part VI

Applied and Clinical Microbiology

68 Bacteriology of Water, Milk, Air and Food

Ideally, drinking water or potable water should be not only safe but also pleasant to drink. It should be clear, colourless and devoid of disagreeable taste or smell. It should be free from pathogenic microorganisms and chemical substances.

On the other hand, milk remains sterile at secretion from the udder but immediately gets contaminated even before it is secreted out of the udder. Apart from mastitis cases, these milk-contaminant bacteria are found to be harmless and occur in very few numbers. In addition, infection of the milk by microorganisms may occur during milking, handling, storage and all the downstream pre-processing events.

Innumerable microorganisms are present all over the air and in the environment. Apart from bacteria, moulds and viruses are also present and can be transmitted from person to person in the form of aerosols. The proportion of the dust particles or aerosols reaching the lung depends on their size. All particles over 5 μm are retained in the nose, and those of 1 μm reach the lung and are retained in the alveoli.

Bacterial analysis of water, milk and air is therefore carried out to indicate the quality of these items used in day-to-day life.

Bacteriology of Water

Water is said to be contaminated or polluted when contaminated with sewage or other excreted matter from humans and animals that contain infective and parasitic agents, poisonous chemical substances and industrial or other wastes.

Bacterial Flora in Water

Bacterial flora in water (Table 68.1) can be classified into three groups as follows:

Natural water bacteria: These are the bacteria that are commonly found in water free from gross pollution.

Soil bacteria: These are the bacteria that are not normal inhabitants of water but are found after being washed into the water during heavy rains.

Sewage bacteria: These are the bacteria that are not normal inhabitants of water but are found in the water after being contaminated with sewage. These bacteria include those which are the normal inhabitants of the intestine of humans and animals. These also include the bacteria that live mainly on the decomposed organic matter of either plant or animal origin.

Factors Determining the Number of Bacteria in Water

The following factors determine the number of bacteria in water:

Salinity: The number of bacteria present in water depends on the salinity of water; more is the salinity, lesser is the number of bacteria.

Table 68.1 Bacterial flora in water

Source of water	Bacterial flora
Natural water bacteria	*Micrococcus, Pseudomonas, Acinetobacter, Serratia, Alcaligenes* and *Flavobacterium*
Soil bacteria washed into the water	*Bacillus subtilis, Bacillus megaterium, Enterobacter cloacae* and *Enterobacter aerogenes*
Sewage bacteria	
Proper sewage bacteria	*Clostridium sporogenes, Proteus* species
Intestinal flora through sewage	*Escherichia coli, Enterococcus faecalis, Proteus* species, *Klebsiella* species, *Clostridium perfringens*

However, some halophilic bacteria survive better in saline water.

Acidity: Acidity of water has a deleterious effect on most bacteria.

Temperature: Low temperature usually favours the survival of the bacteria. However, in the presence of organic materials, bacteria tend to multiply even at high temperatures.

Light: Sunlight with the wavelength of 300–400 nm is highly bactericidal, provided water is clear and static. The bactericidal effect is reduced due to the presence of organic matter and due to movement in the water.

Storage: Water storage decreases bacterial count in stored water due to sedimentation and revitalisation.

Organic matter: When organic matter is plenty, the microorganisms tend to multiply and are present in large numbers, whereas when it is less, the organisms are few.

Type of water: Surface water is more likely to be contaminated than deep water. The latter is usually pure.

Water-Borne Microorganisms

The supply of drinking water is contaminated with sewage or other excreted matter from humans and animals; it can transmit many pathogens, including bacteria, viruses and parasites (Table 68.2).

Table 68.2 Water-borne pathogens in the community and hospital

Pathogen	Disease	Transmission	Clinical features
Campylobacter spp.	Campylobacteriosis	Drinking and recreational water	Fever, headache, diarrhoea, muscle pain, vomiting
Escherichia coli O157:H7	Haemorrhagic colitis	Faecal oral route through water	Watery/ bloody diarrhoea, severe abdominal cramp, vomiting
Shigella spp.	Shigellosis	Faecal oral route through drinking and recreational water	Diarrhoea with blood loss
Vibrio cholerae	Cholera	Faecal oral route through water	Severe watery diarrhoea
Salmonella Typhi	Typhoid	Faecal oral route through water	Fever, abdominal pain, malaise
Giardia duodenale	Giardiasis	Faecal oral route by ingestion	Diarrhoea, malabsorption
Entamoeba histolytica	Diarrhoea	Faecal oral by ingestion	Diarrhoea, dysentery
Cryptosporidium parvum	Cryptosporidiosis	Faecal oral route by ingestion	Watery diarrhoea, cramps, nausea, vomiting, dehydration, low-grade fever
Toxoplasma gondii	Toxoplasmosis	Faecal oral route by ingestion	Lymphadenopathy, fever, congenital infection
Rotavirus	Diarrhoea	Faecal oral contamination	Diarrhoea
Hepatitis A virus, hepatitis E virus	Diarrhoea	Faecal oral contamination	Infective hepatitis
Norwalk virus	Diarrhoea	Faecal oral contamination	Diarrhoea

Bacteriological Examination of Water

A wide number of pathogens can be transmitted by contaminated water. Hence, the main purpose of water testing is to know whether a water-borne pathogen contaminates the water or not. Therefore, in the interests of public health, water supplies should be tested regularly to confirm their freedom from such contamination.

It is impracticable to attempt to detect all types of water-borne pathogens, any of which may be present only intermittently. Instead, an attempt is made on testing the water for microorganisms that indicate faecal contamination of water. These are known as indicator organisms. These are usually common intestinal commensal bacteria, which are universally present in and excreted in large numbers in human and animal faeces. They are rarely found in other sources.

Indicator Organisms

The presence of indicator organisms specifies that (1) faecal matter has contaminated the water supply, (2) the faecal bacteria have not been removed by purification processes and (3) hence, the water supply may be a source of contamination with dangerous intestinal pathogens. The indicator organism should have the properties of (1) being present in human faeces in large numbers, (2) not multiplying in water to any extent and (3) being more resistant than pathogens to the stresses of the aquatic environment and disinfection processes, such as the process of chlorination.

Some of the organisms that are used as indicator organisms are as follows:

Coliforms

These are the bacteria that occur in large numbers in sewage and faeces. They are also found in the environment in the absence of faecal contamination. Traditionally, these are members of the Enterobacteriaceae, which grow in the presence of bile salts and produce acid and gas from lactose within 48 h at 37 °C. The total coliform count is widely regarded as the most reliable indicator of potable water quality.

Thermotolerant Coliform Bacteria

These are the coliform organisms that can ferment lactose at 44–45 °C; the group includes the genus *Escherichia* and some species of *Klebsiella*, *Enterobacter* and *Citrobacter*.

Faecal Streptococci

These bacteria are found in the intestinal tract of humans and animals. They belong to the genus *Enterococcus* and are catalase-negative and Gram-positive cocci. They can survive at 60 °C for 30 minutes and grow at a pH of 9.6 and in the presence of 6.5% sodium chloride.

Sulphite-reducing Clostridia

These are members of the genus *Clostridium*, which can reduce sulphite to sulphide. The most important of the group is *Clostridium perfringens*. Although it is less numerous in human faeces than other indicator organisms, its spores can survive in the environment and treatment processes better than most indicator bacteria.

Collection of Water Samples for Analysis

Water samples should be collected in heat-sterilised bottles of 250 mL capacity with ground glass stoppers. A 0.23 mL of fresh 1.8% solution of sodium thiosulphate crystal is added into the bottle before sterilisation to neutralise the bactericidal effect of any chlorine or chloramine in the water.

Sampling from a Tap or Pump Outlet

When collecting from taps, allow the water to run to waste for 2–3 min before running it into the bottle. Then the stopper is opened, the bottle is filled with water and the stopper is replaced.

Sampling from a Reservoir, such as a Stream, River, Lakes and Tanks

When sampling from streams or lakes, the bottle is opened at a depth of about 30 cm with its mouth facing the current, and it is ensured that the water entering the bottle has not been in contact with hands.

Sampling from a Well

The sampling bottle is tied with a stone and a clean cord of suitable length and is lowered to the required depth in the well. The bottle is completely immersed in the water. When the bottle is filled, it is pulled out and stoppered. It is ensured that the bottle does not touch the side of the well at any time. After the collection of water, the bottle is wrapped in Kraft paper and is labelled with the following details: the source, time and date of collection of water samples. The bottle is transported immediately or preferably within 6 h of collection in a cool container, being protected from the sunlight.

Sample Storage

Collected samples should be analysed within 6–24 h. Collected samples should be transported immediately with ice packs. If ice packs are not available, transportation time must not exceed 2 h.

Methods of Water Analysis

Following tests are generally done for the routine bacteriological analysis of water.

Presumptive Coliform Count

The test is called presumptive coliform count because the reaction demonstrated may occasionally be due to the presence of some other microorganisms. This is a **multiple tube method** in which measured volumes of water and water dilutions are added to a series of tubes or bottles containing a liquid indicator medium. The most common medium used is double-strength and single-strength MacConkey broth containing bromocresol purple and an inverted Durham tube to detect gas production (Fig. 68.1).

The measured volume of water to be tested is added by sterile graduated pipettes to tubes or McCartney bottles as follows: (1) 50 mL of water is added to one bottle of 50 mL double-strength medium, (2) 10 mL of water each to five bottles of 10 mL double-strength medium, (3) 1 mL of water each to five tubes of 5 mL single-strength medium and (4) 0.1 mL of water each to five tubes of 5 mL single-strength medium.

The inoculated bottles are incubated at 37 °C for 48 h. The presumptive coliform count per 100 mL of water is determined by demonstrating acid and gas in MacConkey broth using the McCrady probability. The water sample showing a presumptive coliform count of 0 is considered excellent, whereas water samples showing coliform counts of 1–3, 4–10 and more than 10/ml are considered satisfactory, suspicious and unsatisfactory, respectively.

Differential Coliform Count

Differential coliform count is carried out by the **Eijkman test**. Some spore-bearing bacteria give false-positive reactions in the presumptive coliform test. The Eijkman test determines whether the coliform bacilli detected in the presumptive test are *Escherichia coli* or not. After the usual presumptive test, the test is carried out and is performed by inoculating sub-cultures from the positive presumptive tests at 44 and 37 °C into fresh tubes of single-strength MacConkey broth with the Durham tube. Those tubes showing gas in the Durham tube after incubation at 44 °C after 24 h indicate the presence of *E. coli*. Confirmation of *E. coli* is made by sub-culturing on the solid media and testing the colour for indole production and citrate utilisation.

Detection of Faecal Streptococci

Detection of faecal streptococci in water samples also indicates faecal contamination of water. After the usual presumptive test, the test is carried out after the usual presumptive test is performed by incubating sub-cultures from the positive presumptive tests at 45 °C for 18 h. The presence of acid in the medium indicates the presence of *Enterobacter faecalis*. Further confirmation of *E. faecalis* is made by sub-culturing on the MacConkey medium.

Detection of *C. Perfringens*

It is detected in water samples by inoculating it in a litmus milk medium and incubating it

Fig. 68.1 Multiple tube method

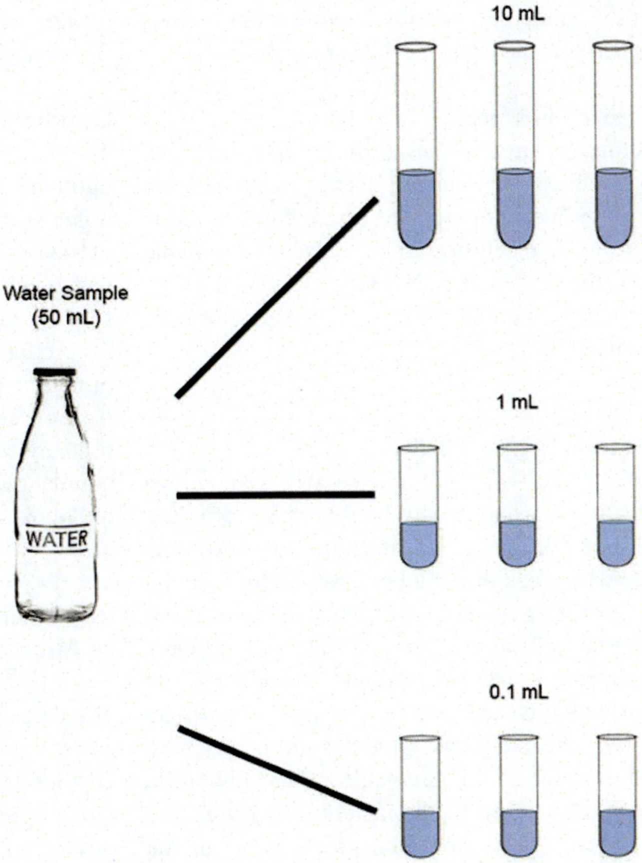

anaerobically at 37 °C for 5 days. A typical stormy clot reaction along with the formation of the acid indicates the presence of *C. perfringens*. Further confirmation of the bacteria is made by motility and nitrate reduction test.

Membrane Filtration Method

In this method, a measured volume of the water sample is filtered through a membrane with a pore size small enough to retain the indicator bacteria to be counted. The membrane is then placed and incubated on a selective indicator medium so that the indicator bacteria grow to produce colonies. These colonies are then identified and counted to measure the contamination of the water.

Detection of Specific Pathogens

V. cholerae, *Salmonella* Typhi, etc., are the specific pathogens which can be detected in a water sample to know the contamination. Specific selective media for these bacteria are employed to detect these in contaminated water. For example, double-strength selenite F broth is used for *S.* Typhi, whereas alkaline peptone water is used for *V. cholerae*. Subsequently, these bacteria are identified by their cultural and biochemical characteristics.

Viruses in Water

Over 100 enteric viruses and other cytopathogenic viruses (e.g. Hepatitis A virus, Hepatitis E virus, poliomyelitis virus, rotavirus and Norwalk virus) are said to be present in both sewage and drinking water. Viruses can be removed from water by either coagulation or disinfection procedures. By chlorination, with a concentration of 0.5 mg/L free residual chlorine, pH <8, turbidity of 1 or less nephelometric turbidity units and minimum contact period of 30 min, viruses are assumed to be destroyed. Although chlorination is used as the best possible treatment for removing viruses, the exact mechanism is unknown.

Protozoa in Water

Protozoans are also said to be the most common contaminants in drinking water. There are no sensitive indicators for protozoan contamination of water. And coliform counts are also not reliable indicators as these protozoans are found to be resistant to normal levels of chlorine used for disinfection.

Bacteriology of Milk

Milk is an opaque white liquid, which provides the primary source of nutrition for newborns before they can digest other types of food. It is a water-based fluid that is primarily an emulsion of butterfat globules. Every single fat globule encompasses a membrane made up of phospholipids and proteins. These emulsifiers prevent the individual globules from attaching with each other into observable grains of butterfat and also protect the globules from the fat-digesting activity of enzymes found in the fluid portion of the milk.

Bacterial Flora of Milk

Lactic Acid Bacteria

This group of bacteria can produce lactic acid from lactose. They are generally present in the milk, and they have been used as starter cultures in the manufacturing of cultured dairy products like yoghurt. Some of the lactobacilli present in milk are mentioned as follows:

1. Lactococci, such as *Lactobacillus delbrueckii* subsp. *lactis* (*Streptococcus lactis*) and *Lactococcus lactis* subsp. *cremoris* (*Streptococcus cremoris*),
2. Lactobacilli, such as *Lactobacillus casei*, *L. delbrueckii* subsp. *lactis* (*L. lactis*), *L. delbrueckii* subsp. *bulgaricus* (*Lactobacillus bulgaricus*) and *Leuconostoc*.

Coliforms

These are facultative anaerobes that grow at an optimum temperature of 37 °C. They are indicator organisms that are closely correlated with the existence of pathogens but are not compulsorily pathogenic themselves. They are capable of spoiling milk rapidly as they can ferment lactose by producing acid and gas and can disintegrate proteins in milk. They can be killed by pasteurisation; hence, their presence after pasteurisation is usually a clear indication of contamination. *E. coli* is an example that can act as a milk-contaminating microbe.

Importance of Presence of Microorganisms in Milk

The data about milk's microbial content can help assess its sanitary quality and the sterile conditions maintained during its production. Bacteria present in the milk can cause spoilage of the product if allowed to multiply. Milk is easily susceptible to being contaminated with pathogenic microorganisms. Therefore, precautions must be taken to prevent this possibility and eradicate pathogens that may gain access. Certain microorganisms produce changes chemically that is beneficial in producing dairy products, such as yoghurt and cheese.

Microorganisms Causing Spoilage of Milk

The quality of microbes in raw milk is vital for producing dairy foods of the utmost quality. Spoilage describes the degradation of a food's texture, odour, flavour or colour to the extent where it is unappetising or unacceptable for human consumption. The spoilage of food by

Table 68.3 Milk-borne diseases

Diseases	Pathogen	Source	Symptom
Q fever	*Coxiella burnetii*	Infected cattle, goats and sheep	Fever, headache, endocarditis, chills
Gastroenteritis haemolytic uremic syndrome	*Escherichia coli* O157:H7	Faecal contamination	Bloody diarrhoea, abdominal pain, kidney failure
Tuberculosis	*Mycobacterium tuberculosis, Mycobacterium bovis*	Infected animal	Lung diseases
Listeriosis	*Listeria monocytogenes*	Water, soil, environment	Flu-like symptom, spontaneous abortion, miscarriage, stillbirth
Gastroenteritis typhoid	*Salmonella* sp.	Faeces and environment	Disease, nausea, fever
Brucellosis	*Brucella* sp.	Unpasteurised dairy products	Joint and muscle pain, fever, weight loss and fatigue

microbes often involves the disintegration of protein, carbohydrates and fats by either microorganisms themselves or by their enzymes.

Psychrotrophic organisms are microbes that are involved in the spoilage of milk. Most of the psychrotrophic organisms are killed by pasteurisation temperatures; some microorganisms like *Pseudomonas fluorescens* and *Pseudomonas fragican* produce lipolytic and proteolytic extracellular enzymes, which are stable in heat and capable of producing spoilage. Some species such as *Arthrobacter*, *Bacillus*, *Clostridium*, *Corynebacterium*, *Microbacterium*, *Micrococcus*, *Lactobacillus* and *Streptococcus* species can withstand pasteurisation and can grow at refrigeration temperatures and thus can cause spoilage problems.

Pathogenic Microorganisms in Milk

Hygienic milk production involves the proper handling and storage of milk, and compulsory pasteurisation shall decrease the hazard of milk-borne diseases, such as brucellosis, tuberculosis and typhoid fever. There have been reports of several food-borne illnesses that resulted from various issues such as ingestion of raw milk or poorly handled dairy products or not properly pasteurised, which cause post-processing contamination of milk.

Milk-borne diseases are of three types: (1) infections primarily of humans but transmitted through milk, (2) infections primarily of animals that can be transmitted to humans and (3) infections transmitted by milk contaminated with excreta of ticks and rats (Table 68.3).

Moulds of the species such as *Aspergillus*, *Fusarium* and *Penicillium* can also grow in milk and dairy products. If the conditions are permissible, these moulds produce mycotoxins which cause a serious health hazard.

Pasteurisation

Pasteurisation is the process of heating the milk by either of the below-mentioned methods: (1) a high temperature for a short time (at least 72 °C for 15 s. This method is known as the **'Flash' method**. (2) a low temperature for a long time (at least 63 °C for 30 min). This method is known as the **'Holder' method**. Certain bacteria such as *Bacillus* spp. and *Coxiella burnetii* resist pasteurisation by the 'Holder' method.

Ultra-High-Temperature Treatment

Ultra-high-temperature processing (UHT) of milk involves a continuous flow of heat at a high temperature (135 °C) for a short time (1–8 s).

Boiling

Boiling milk boiling at 100 °C for 15 min destroys all vegetative cells leaving behind spores of spore formers. Hence this method is not as efficacious as the methods mentioned above.

Bacteriological Examination of Milk

Following groups of tests can carry out a bacteriological examination of milk.

Colony Counts

This test is carried out by plate dilution methods. Raw milk may contain 500 to several million bacteria per millilitre of milk.

Coliform Counts

Coliforms in milk indicate improper pasteurisation of milk or post-pasteurisation contamination of milk. This is because all coliforms are destroyed during the pasteurisation of milk. This test is carried out by inoculating varying milk dilutions into MacConkey medium and noting the production of acid and gas after 48 h of incubation at 37 °C. The presence of acid and gas indicates the presence of coliforms in milk.

Methylene Blue Test

This has been a test used for a long to demonstrate bacterial contamination of milk. It indicates the number of viable bacteria present in the milk. It is a rapid and inexpensive way of indicating poor-quality milk that had been unrefrigerated. The basis of the test is that the presence of viable bacteria in milk reduces methylene blue and decolourises the milk when kept in a dark place. The test is performed by using a 1:300,000 solution of methylene blue. One millilitre of the methylene blue solution is added to 10 mL of the milk sample in a test tube. Both the milk and methylene blue solution are mixed by shaking and then placing the mixture in a water bath at 37 °C for 30 min in the dark. Untreated milk can be considered satisfactory if it fails to decolourise the dye within 30 min.

Phosphatase Test

Alkaline phosphatase is normally present in milk and is inactivated if pasteurisation has been carried out effectively. Successful pasteurisation, which kills non-sporing pathogens, also inactivates alkaline phosphatase. The test depends on the ability of the enzyme to liberate *p*-nitrophenyl phosphate after breaking down disodium *p*-nitrophenyl phosphate. The test determines the amount of alkaline phosphatase present after pasteurisation by measuring the amount of *p*-nitrophenyl phosphate it liberates, which is known by the development of a yellow colour quantitated by a colourimeter.

Turbidity Test

This is the definitive test for checking the sterilisation of milk, thereby distinguishing it from the untreated milk and milk that has been merely pasteurised. The degree of heating necessary for sterilisation causes all the heat-coagulable proteins in milk to become precipitable by ammonium sulphate. If the amount of heat applied to milk is insufficient for sterilisation, some of its protein will not be precipitated by ammonium sulphate and will be detected by its coagulation, resulting in turbidity when a filtrate of ammonium sulphate-treated milk is boiled. The absence of turbidity indicates that the milk has been boiled or heated to at least 100 °C for at least 5 min.

Detection of Specific Pathogens

Mycobacterium tuberculosis and *Brucella* species are the specific pathogens that can be detected in milk to know the transmission of these bacteria through milk. Specific selective media for these bacteria are employed to detect these in infected milk. For example, *M. tuberculosis* can be isolated from the centrifuged milk deposit by inoculation in Lowenstein–Jensen (LJ) medium or by inoculation in guinea pigs. Similarly, *Brucella* spp. maybe isolated from milk samples by inoculating in serum dextrose agar or intramuscular inoculation in guinea pigs. Subsequently, these bacteria are identified by their cultural and biochemical properties. *Brucella* spp. in infected animals can also be demonstrated by milk ring test, whey agglutination test and demonstration of *Brucella* antibodies in serum.

Bacteriology of Air

The load of microorganisms present in the air depends on whether the air is indoors or outdoors. The number of bacteria at any time is dependent on many factors, the most important of which are the number of persons present, the amount of their body movements and the amount of disturbance of their clothing.

Bacteriological Examination of Air

Air-borne infection denotes the infection caused by the transmission of respiratory droplets or aerosols less than 5 μm in size. Droplet infection denotes the infection caused by the transmission of respiratory droplets or aerosols larger than 5 μm size.

Observations of the number of bacteria carrying particles in the air may be required in premises where safe working depends on the air's content of bacteria being kept at a very low level, for example, operation theatres. Bacteriological examination of air is also necessary for monitoring air quality in hospital wards, storehouses of food and pharmacy, etc. Ideally, the bacterial count should not exceed 1 per cubic foot of air in operation theatre for neurosurgery, 10 per cubic foot in operation theatre for other surgery and 50 per cubic foot in homes, offices and factories.

Microbial Quality of Outdoor Air

The microbial quality of outdoor air depends on various factors such as (1) the density/mass of both human and animal populations, (2) the nature of vegetation/plantations, (3) the content of the soil and (4) atmospheric conditions including humidity, temperature, wind, rainfall and sunlight.

The four main areas are known to contribute to the presence of pathogens in the outdoor air. The occurrence of these microbes that have been associated with human health is natural environments, engineering environments, agricultural and waste treatment plants. Fragments of moulds and spores naturally occur high in number than bacteria in the atmosphere. Most of the bacteria that may be present in the atmosphere are non-pathogenic, and even if some pathogens are present, they may just contaminate the air and are incapable of causing any disease due to the adverse conditions of the outdoor air. It is also known that pathogenic microbe does not multiply in air. Aerobic spore-bearing bacilli and some *Achromobacter*, *Sarcina* and *Micrococcus* found in the upper atmosphere are mostly isolated from soil and surface dust, and they can be carried over for miles. Some infective microbes are carried only for short distances, and during the course, their infectious ability is diminished, except in rare cases such as the foot and mouth disease virus.

Microbial Content of Indoor Air

This may contain particles of biological origin suspended in the air in an indoor environment. This depends on the following sources:

Dust Particles

The dust particles of varying sizes originating from animals (pets), plants or vegetables, ventilation or air conditioning system.

Humans

They are an important source of common pathogens. Nasal secretions dispersed by fingers/hand to any other part of the body or the skin, bedding or clothing form dust particles when detached. Pathogenic microbes form dust particles when detached directly from different parts of the body or the skin, such as septic wounds, perineum, etc. Intestinal pathogens can be disseminated through dried faecal particles from napkins used for infants. Heavy particles fall to the ground and form settled house dust, while smaller particles of 1 mm or less in diameter remain suspended in the air. The suspended biological particles are commonly associated with respiratory illness in humans. The layer of air surrounding the human body serves as a source of pathogenic microbes. Physical activity liberates the desquamated epithelial cells into the environment, contributing to the microbes in the dust. Hospital wards, where patients are treated, are also sources of potential infectious agents like haemolytic streptococci, staphylococci, tubercle bacilli, diphtheria bacilli, etc. Bed clothes are also a rich source of pathogen-loaded dust. Tuberculosis, different forms of bacterial pneumonia, coccidioidomycosis, influenza, measles and gastrointestinal illness are caused by indoor aerosols.

Droplets and Droplet Nuclei

Human generates a varying number of droplets during talking, sneezing and coughing. These droplets are expelled from the body that ranges in size from less than 1–15 um (Table 68.4). These particles may fall to the ground or stay suspended in the air based on the different sizes. In the process, it might get evaporated; the smaller the particle, the faster the evaporation rate and vice versa. Droplet nuclei are formed when droplets are converted to minute particles during evaporation. The viability of the microbes in the droplet nuclei depends on numerous factors. Particles of 5 um are retained in the nose, while those of 1 mm reach the lung and are retained in the alveoli. Droplets containing infectious agents may also be released from various laboratory practices, dental manipulations, washing or flushing water closets.

Measuring Bacterial Content of the Air

Various methods have been devised to measure the bacterial content of the air. A primary distinction must be drawn between the methods that measure the rate at which bacteria-carrying particles are settling by gravity on exposed surfaces and those that count the number of bacteria-carrying particles in a given air volume. The settle plate method (Passive sampling) and the slit sampler method (Active sampling) are used for bacteriological examination of air.

Settle Plate Method (Passive Sampling)

The settle plate method is used for testing the bacteriological quality of air in surgical operation theatres and hospital wards. In this method, Petri dishes containing nutrient agar and blood agar (for detecting pathogenic staphylococci and streptococci) must be left open to the air for 1 h, 1 m above the floor, 1 m from the wall. During this exposure period, large bacteria-carrying dust

Table 68.4 Air-borne pathogens and their transmission

Properties	Air-borne transmission	Droplet transmission
Droplet size	Less than 5 μm	More than 5 μm
Droplet origin	Produced during invasive procedures and also during coughing, sneezing, talking	Produced during invasive procedures like bronchoscopy, suction aspiration and sneezing, talking, coughing
Droplet characteristics	Transmission occurs through close contact Can be spread by tiny particles which remains suspended in the air produced by talking Present in the air for brief time and travel just for a short distance	Immunocompromised individuals can be infected even if less distance from infected person Can be spread by droplets of saliva and mucus by coughing and sneezing
Bactewrial pathogens	*Bordetella pertussis, Corynebacterium diphtheriae, Yersinia pestis* *Haemophilus influenza* type B *Mycoplasma pneumoniae, streptococcus* species *Neisseria* species	*Mycobacterium tuberculosis*
Viral pathogens	Adenovirus Rubella virus Influenza virus Parvovirus B19 Mumps	Measles virus Varicella-zoster virus Influenza virus

particles settle onto the media. The plates are then incubated at 37 °C for 24 h, following which the colonies formed on the media are counted.

Slit Sampler Method (Active Sampling)

The slit sampler method is the most efficient and convenient method used to estimate the number of bacteria present in a measured volume of air. In this method, 1 $ft.^3$ volume of air is directed onto a plate containing a culture medium through a slit 0.25-mm wide. The plate is then rotated to allow the microorganisms present in the air to spread out evenly on the medium. The culture medium is incubated, and the number of colonies formed on the medium indicates the number of bacteria present in the air.

Bacteriology of Food

One of the major causes of morbidity, particularly due to the public health problems, results from food-borne infections. Morbidity is not the only problem, but sometimes it might result in mortality. Several species of microorganisms can contaminate food, and it becomes a good medium for the growth of these microorganisms.

There are different, and variety of aetiological patterns concerning food-borne infections and these patterns may vary around the world. Factors that influence these aetiological patterns include food preferences, ability of the laboratories, public and medical practitioner's awareness. For example, in the United States, more than 50% of the food-borne diseases are caused by *Staphylococcus aureus* and *Salmonella*, but in the United Kingdom, more than 90% of food-borne diseases are caused by *C. perfringens*. In Japan, more than 50% of the diseases are caused by *Vibrio parahaemolyticus*.

Food-borne outbreak infections refer to a condition when two or more people get the same illness from the same contaminated food or drink. Usually, the illness is restricted to gastrointestinal manifestations, and food is incriminated as the source of illness. Food-borne illness, most commonly referred to as food poisoning, occurs due to the consumption of contaminated, spoiled or toxic food. The commonest symptoms of food poisoning are nausea, vomiting and diarrhoea. The causative agents are usually gastrointestinal microorganisms which may also produce harmful toxins.

Source of Food Contamination

Food ingested must be free of microorganisms. When microbes contaminate the food, it indicates a lack of sanitation and precautions while handling and preparing cooked food. The food can be contaminated in various ways.

Faeces

The contaminations of faeces in food are mainly due to the poor hygiene of food handlers.

Insects

The housefly has been identified as an important vector in propagating various infectious diseases such as shigellosis, cholera and salmonellosis. They act as a cross-contamination vector for other foodborne pathogens as well.

Air

The food uncovered may attract a lot of microorganisms in the air and invite dust to settle on it.

Domestic animals

They contaminate through their excreta when handled by people who do not practise proper sanitary precautions.

Laboratory Diagnosis

Laboratory diagnosis of suspected food-borne infection or food poisoning depends on the following:

Clinical history

A careful and detailed history is usually obtained to rule out the possible causes altogether.

Specimens

Samples include the consumed food and vomitus and faeces from affected persons.

Detection of Specific Pathogen

These can be detected by (1) direct microscopic examination of samples and (2) culture of samples. This is also supplemented by detecting bacterial toxins and microbial and chemical analysis of food samples for the possible presence of gastrointestinal pathogens.

Prevention of Food-Borne Infections

The four steps to food safety and prevention of food-borne infections are Clean, Separate, Cook and Chill. The first step is cleaning and washing hands, utensils, cutting surfaces and the raw food itself. The use of clean and safe water is essential, or else the water ends up contaminating the food material. Cross-contamination from food materials that are potential sources of organisms like meat, poultry, fish and unwashed fruits and vegetables should be separated and not allowed to contact other foodstuffs. Cooking should be adequate at the correct temperature to destroy disease-causing organisms while preserving the nutrition and palatability of the food. Finally, to prevent bacterial multiplication and growth of other organisms, cooked or raw food should be properly refrigerated, and if necessary, it should be frozen.

In recent years, consuming exotic dishes made of raw or semi-cooked fish or meat products poses a special risk for the development of food-borne diseases. Experimenting with such food should be avoided. It is also to be kept in mind that there are vulnerable populations like the immunocompromised, infants and elderly population, and pregnant women for whom the microbiological safety of food is essential to minimise food-borne infections.

Biomedical Waste Management 69

Biomedical Waste

A vast amount of waste is generated in the process of healthcare, research, testing or related procedures on human beings or animals conducted in hospitals, clinics, labs or similar establishments. This waste is called **biomedical waste**.

Potential Hazards of Biomedical Waste

The biomedical wastes contain infected materials and other items like blood, urine, tissue etc., which are excellent substrates for the growth of many microorganisms. These are potential sources of infection to others who come into contact with it. In addition, the chemicals and toxic drugs and substances can cause pollution of water, air or the soil or toxicity to the individuals and the biological forms which may come in contact with it. Radiation wastes and sharp instruments or glass wastes can cause injury.

Due to a higher concentration of infectious and other such hazardous substances, this form of waste requires special attention for its safe disposal. If the waste finds its way back to the community, it would be disastrous. It can act as a source of infection to residents or visitors of hospitals, and the chemicals contained in the waste may seep into the surrounding soil and contaminate wells and tanks, thereby polluting them.

Most governments in the world have laws and guidelines ensuring that institutions, where biomedical waste is generated take appropriate measures for the safe disposal of such waste. An ordinance to this effect was promulgated in India in 1998, under which the institutes where such waste is generated have been held legally responsible for the safe disposal of biomedical waste. The new Biomedical Waste Management Rules 2016 (amended in 2018) replaced the earlier Rules (1998) and amendments. It has simplified the process of categorisation of waste and authorisation. In 2018, the rules were further amended with the specific purpose of environmental protection. It stipulates that all healthcare facilities should phase out and finally discontinue the use of chlorinated plastic bags and gloves (except blood bags).

Types of Biomedical Waste

Handling of biomedical waste is increasingly associated with the risk of laboratory-acquired infections. Most of these infections are caused by hepatitis B virus, *Bacillus anthracis, Mycobacterium tuberculosis, Francisella tularensis, Shigella* species and *Brucella* species. Biomedical wastes are of the following types:

Human Anatomical Waste

This category includes human organs, tissues and other body parts. Hair, teeth and nails are not included in this category.

Animal Waste

All the animal tissues, body parts, organs, bedding, blood products and body fluid fall under this category. Teeth, hair, hooves and feathers are not included in this category.

Microbiology Laboratory Waste

All laboratory wastes like stock or specimen microorganisms, cultures either live or dead and vaccines. Cell lines used for research come under this category.

Human Blood and Body Fluid Waste

Human blood and body fluid waste include blood, human body fluids and fluids removed during surgery or autopsy. Urine and faeces are excluded.

Waste Sharps

Waste sharps include all things capable of causing punctures and cuts like syringes, needles, blades and laboratory glass.

Cytotoxic Waste

This includes pharmaceutical wastes generated from drugs used for treating cancer, for example, chemotherapy agents.

The categories of biomedical waste as defined by the Pollution Control Board are listed in Table 69.1.

Waste Treatment and Disposal

Waste treatment and disposal form a very important component of the proper waste management program of the hospital. Segregation of infectious waste from non-infectious waste is the most important stage required before treating and disposing of biomedical wastes. This is required to

Table 69.1 Categories of biomedical wastes (Biomedical rule 1988, As per the National Pollution Control Board, India)

Categories	Type	Description
Category 1	Human anatomical waste	Human tissues, organs and body parts
Category 2	Animal waste	Animal tissues, organs, body parts carcasses, bleeding parts, fluid, blood and experimental animals used in research, waste generated by veterinary hospitals, colleges, discharge from hospitals and animal houses
Category 3	Microbiology and biotechnology wastes	Wastes from laboratory cultures, stocks or specimens of microorganisms live or attenuated vaccines, human and animal cell culture used in research and infectious agents from research and industrial laboratories, wastes from the production of biologicals, toxins, dishes and devices used for transfer of cultures
Category 4	Waste sharp	Needles, syringes, scalpels, blades, glass, etc., may cause punctures and cuts. This includes both used and unused sharps
Category 5	Discarded medicines and cytotoxic drugs	Wastes comprising of outdated, contaminated and discarded medicines
Category 6	Soiled waste	Items contaminated with blood and body fluids, including cotton, dressings, soiled plaster casts, lines beddings and other material contaminated with blood
Category 7	Solid waste	Wastes are generated from disposable items other than the waste (sharps) such as tubings, catheters and intravenous sets
Category 8	Liquid waste	Waste generated from laboratory and washing, cleaning, housekeeping and disinfecting activities
Category 9	Incineration ash	Ash from incineration of any biomedical waste
Category 10	Chemical waste	Chemicals used in the production of biomedicals, chemicals used in disinfection as insecticides, etc.

protect the health of medical personnel coming in contact with infectious biomedical wastes (Table 69.2).

Infectious wastes include microbiology and biotechnology waste, waste sharps, animal wastes, solid wastes and liquid wastes, as mentioned above. These are segregated and collected in different coloured plastic bags/containers (Fig. 69.1):

Red coloured non-chlorinated plastic bags/ containers are used for infected plastic waste

- Yellow coloured non-chlorinated plastic bags/ containers are used for infectious non-plastic waste.
- White: puncture-proof, leak- proof containers are used for disposal of sharps.
- Blue: puncture-proof, leak- proof containers are used for disposal of glassware and implants.

General Principles of Waste Management

The main objectives of bio-waste management area are as follows: (1) to minimise the biomedical waste generated, (2) to ensure that bio-waste is handled and disposed efficiently, (3) to reduce the incidence of disease or injury among healthcare workers and (4) to retrieve usable materials.

Steps Involved in Waste Management

The following are the steps in waste management:

Reduction

It focuses on the use of replacement of medical and surgical equipment with reusable ones. Utilisation of reusable or recyclable products or products containing recycled materials should be preferred wherever possible and with patients' safety in mind.

Segregation

It refers to the separation of waste at the point of generation and placed in appropriate containers based on waste category (Table 69.2). Segregation is required because if biomedical waste is mixed with general waste, then the total waste requires special treatment and handling.

Storage

It refers to waste being safely contained during handling and to the point of its disposal. After the biomedical waste has been collected and moved from its point of generation, it may be held in storage areas to await disposal. Important factors to be considered are the type of waste being contained, appropriate colour coding and labelling, special transportation requirements; the method of disposal and the disposal facility. These storage areas must be covered and separated from food preparation areas.

Transportation

This refers to transporting waste from the point of generation to the treatment or disposal area. The handling, sending for transport and transport of biomedical waste must be organised accurately. Packaging containers should be checked to meet requirements for all modes of transport.

Treatment

There are different procedures to reduce the bulk of waste to less or non-infectious materials. Appropriate treatment options should be followed for various types of waste. For example, steam autoclaving can be carried out for microbial waste, whereas it may not be useful for human or anatomical waste. The other treatment option available is chemical decontamination.

Disposal methods available for different types of waste have to be followed accurately. The final disposal can be made either on the premises or away from crowded areas.

Table 69.2 Biomedical waste segregation and disposal as per Biomedical Waste Management Rules 2016

Category	Type of bag/container used	Type of waste	Treatment/ disposal options
Yellow	(a) Non-chlorinated plastic bags: Separate collection system leading to an effluent treatment system	Yellow coloured non-chlorinated plastic bags	Incineration or plasma pyrolysis or deep burial
	(b) Animal anatomical waste: Experimental animal carcasses, body parts, organs, tissues, including the waste generated from animals used in experiments or testing in veterinary facilities		
	(c) Soiled waste: Items contaminated with blood, body fluids like dressings, cotton swabs and bags containing residual/ discarded blood and its components		Incineration or plasma pyrolysis or deep burial* in the absence of the above facilities, autoclaving or microwaving/ hydroclaving followed by shredding or mutilation or combination of sterilisation and shredding
	(d) Expired or discarded medicines: Pharmaceutical waste like antibiotics, cytotoxic drugs, all items contaminated with cytotoxic drugs along with glass or plastic ampoules, vials, etc.	Yellow coloured non-chlorinated plastic bags or containers	Incineration at temperature > 1200 0C or encapsulation or plasma pyrolysis at >12000C
	(e) Chemical waste: Chemicals used to produce biological and used or discarded disinfectants	Yellow coloured containers or non-chlorinated plastic bags	Disposed of by incineration or plasma pyrolysis or encapsulation in hazardous waste treatment, storage and disposal facility
	(f) Chemical liquid waste: Liquid waste generated due to the use of chemicals in the production of biological and used or discarded disinfectants, silver X-ray film developing liquid, discarded formalin, infected secretions, aspirated body fluids, liquid from laboratories and floor washings, cleaning, house-keeping and disinfecting activities, etc.	Separate collection system leading to an effluent treatment system	Chemical liquid waste shall be pretreated before mixing with other wastewater
	(g) Discarded linen, mattresses, beddings contaminated with blood or body fluid	Non-chlorinated yellow plastic bags or suitable packing material	Pretreat with non- chlorinated chemical disinfection followed by incineration or plasma pyrolysis or for energy recovery. In the absence of the above facilities, shredding or mutilation or combination of sterilisation and shredding
	(h) Microbiology, biotechnology and other clinical laboratory waste: Blood bags, laboratory cultures, stocks or specimens of microorganisms, live or attenuated vaccines, human and animal cell cultures used in research, industrial laboratories, production of biological, residual toxins, dishes and devices used for cultures	Autoclave safe plastic bags or containers	Pretreat to sterilise with non-chlorinated chemicals on-site followed by incineration
Red	Contaminated waste (recyclable): (a) wastes generated from disposable items such as tubing, bottles, intravenous tubes and sets, catheters, urine bags, syringes (without needles and fixed needle syringes) and vaccutainers with their needles cut and gloves	Red coloured non-chlorinated plastic bags or containers	Autoclaving or micro-waving/ hydroclaving followed by shredding or mutilation or a combination of sterilisation and shredding

(continued)

Table 69.2 (continued)

Category	Type of bag/container used	Type of waste	Treatment/ disposal options
White	Waste sharps including metals: Needles, syringes with fixed needles, needles from needle tip cutter or burner, scalpels, blades or any other contaminated sharp object that may cause puncture and cuts. This includes both used, discarded and contaminated metal sharps	Puncture-proof, leak-proof, tamper-proof containers	Autoclaving or dry heat sterilisation followed by shredding or mutilation or encapsulation in a metal container or cement concrete; a combination of shredding cum autoclaving and sent for final disposal to iron foundries or sanitary landfill or designated concrete waste sharp pit
Blue	(a) Glassware: Broken or discarded and contaminated glass, including medicine vials and ampoules, except those contaminated with cytotoxic wastes	Cardboard boxes with blue coloured marking	Disinfection by soaking the washed glass waste after cleaning with detergent and sodium hypochlorite treatment or through autoclaving, microwaving or hydroclaving
	(b) Metallic body implants	Cardboard boxes with blue coloured marking	

Reference: BMW Rules, 2016_1 (1).pdf [Internet]. [cited 2022 Jan 4]. Available from: https://www.aiims.edu/images/pdf/Departments_Centers/BiomedicalWaste/BMW%20Rules,%202016_1%20(1).pdf

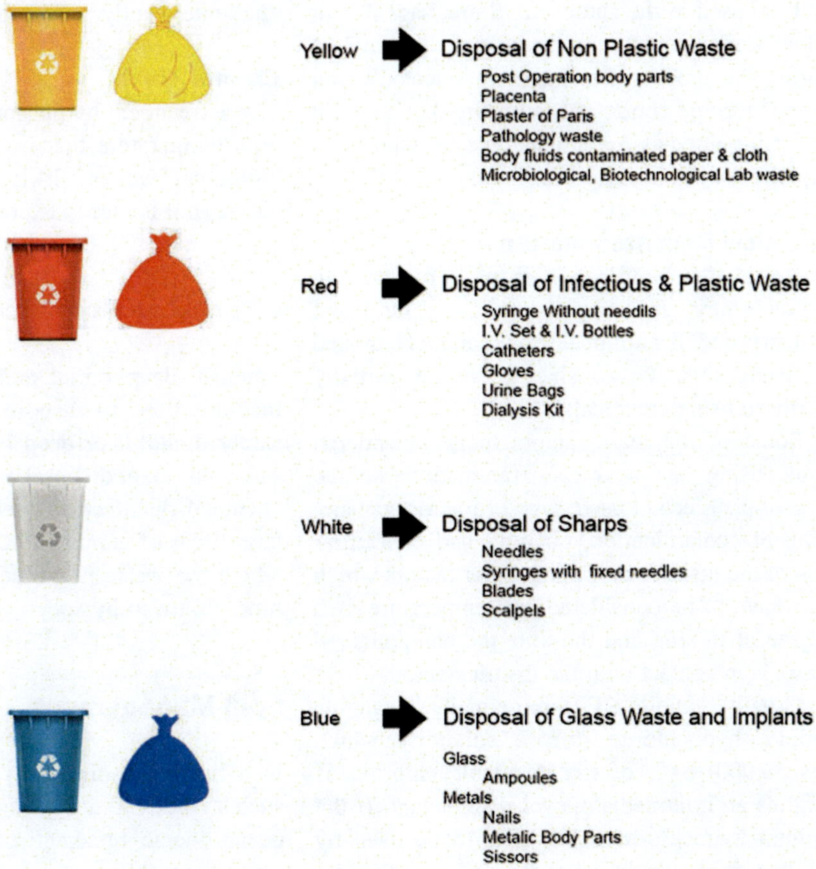

Fig. 69.1 Different colour containers used for the disposal of biomedical wastes

Methods of Waste Treatment

There are various methods of management of biomedical wastes.

Autoclaving

Steam autoclave is one of the best methods to treat microbiological laboratory waste. This is used to treat non-anatomical waste like waste sharps, etc.

The person handling the equipment should be well trained. The period and temperature are very important as they decide how effectively the substances are decontaminated. Steam should penetrate the waste for effective autoclaving results. Containers that have good steam permeability are used to achieve good results. Special attention should be given to packing.

Special care should be given for selecting plastic bags used in the autoclave. Some bags do not allow steam penetration, and some get melted during the process. So, bags are selected using actual working conditions to ensure they are best suited for autoclaving. Organic wastes which can explode are not autoclaved.

Chemical Decontamination

Chemical decontamination is used for treating microbiology laboratory waste, body fluid and human blood. Anatomical waste cannot be treated by using this. Waste sharps can be partially sterilised using chemicals.

Some of the main factors to be considered while using chemical decontamination are the type of disinfectant used, type of microorganism, level of contamination, quantity and concentration of the disinfectant. Some other factors which also have to be considered are temperature, pH, degree of mixing and the time the contaminated waste is in contact with the disinfectant.

Usually, undiluted commercially available sodium hypochlorite (5.25% solution), which has 50,000 mg/L of free available chlorine, is used as an intermediate-level disinfectant. If the sodium hypochlorite solution is to be used by diluting, it should be prepared daily, so there is no loss of germicidal action.

Incineration

In this process, the waste is combusted to ash. This process achieves a reduction of 75% by weight and 90% by volume. This method is widely used in medical hospitals and other healthcare institutes. This can be used to treat anatomical and non-anatomical biomedical wastes. Incineration is the most successful technology used to treat all kinds of biomedical waste till now.

Microwave Treatment

Radiations produced by microwaves are increasingly used nowadays to treat infectious hospital wastes. The microwaves heat the biological wastes to a temperature of 97–100 °C in a 40–45 min cycle. The advantage of the method is that it is not associated with any hazardous emissions, but it cannot be used to treat anatomical body parts and tissues.

Plasma Torch

Waste treatment by plasma torch produces a very high temperature that is very useful for treating biological wastes. It is a very safe method, although it is highly expensive.

Disposal of Biomedical Wastes

Landfill, deep burial and sewage are different methods used to dispose of biomedical wastes safely. Landfill or deep burial—although it is a safe method for disposal of sharps and products of chemical disinfection—requires the presence of large areas of uninhabited land. Trenches are dug where the waste is buried and are covered with lime, followed by soil.

Spill Management

In a hospital, spillage of hazardous substances such as body fluids, drugs, chemicals etc., can occur due to breakage of containers, defective equipment or human error. Any such spill poses a risk to the staff, visitors and patients susceptible to infection. Thus, every healthcare facility should have a clear-cut policies, proper standard operative procedures (SOP) and trained staff to manage such an event immediately and effectively. There should be separate SOPs for blood and body fluids spill, chemical spill, flammable and non-flammable liquid spill, solid spill and mercury spill.

Management of Blood and Body Fluids, Including Infectious Materials Spill

For **small spills**, the following procedure is adopted:

1. The persons managing the spill should apply workman's gloves and other personal protective equipment, including masks and face shields.
2. Any sharp object like glass pieces should first be picked up with forceps and discarded in an appropriate container.
3. The spill is covered with blotting paper or newspaper or paper towel.
4. The spill should be wiped with paper soaked in 1% hypochlorite solution containing 500 ppm chlorine. The paper is then discarded in a biohazard bag.
5. The area is again wiped with 1% hypochlorite solution and allowed to dry.

For **larger spills of more than 10 ml**, the procedure is more or less the same as above except for the following points:

1. The area should be confined, and all entry of unnecessary persons is restricted.
2. Instead of 1% hypochlorite, a 10% solution should be used.

70 Healthcare-Associated Infections

The term hospital infection, hospital-acquired infection or nosocomial infection has now been known as healthcare-associated infection (HAI) to include infections acquired from clinics, ambulatory healthcare services, health camps, etc. Around 5% of the patients who are hospitalised experience nosocomial infection. Nosocomial infection is a public health threat in developing and developed countries. These infections are the important cause of morbidity, mortality and economic problems in hospitalised patients.

Healthcare-Associated Infections

The impact of hospital-acquired infections is considerable—the patient may need longer hospital treatment, readmission or even further surgery, increasing the time of absence from work and use of hospital and community resources. Minimised stay in hospital these days shows that infections acquired from the hospital have increased, and such cases present themselves to the community practitioners or private health sectors. As the administration of intravenous medication at households has increased, infections that are cannula-associated, once which were confined to patients in the hospitals, are seen at the community level.

Definition of Healthcare-Associated Infection

Healthcare-associated infections (HAIs) are infections that first appear 48 h or more after hospitalisation. HAI applies to infections occurring in hospitalised patients who were neither infected nor were in incubation at the time of their admission to the hospital.

Iatrogenic Infections

Iatrogenic infection is defined as an infection after medical or surgical management, whether or not the patient was hospitalised. Blood-borne virus infections such as HIV and hepatitis virus due to unsafe injection drug use and unsterile medical procedures are examples of iatrogenic infections.

Laboratory-Acquired Infections

Laboratory-acquired infections (LAIs) are defined as all infections acquired through laboratory or laboratory-related activities resulting in occupational exposure to infectious agents. Brucellosis, hepatitis and typhoid fever are some examples of laboratory-acquired infections.

Table 70.1 Biosafety level laboratories and risk groups of organisms

Risk groups	Biosafety level	Equipment
1	Biosafety level I	Open benchwork
2	Biosafety level II	Open bench plus biosafety cabinet
3	Biosafety level III	Biosafety cabinet
4	Biosafety level IV	Class III BSC or positive pressure suits in conjunction with class II BSCs, double-ended autoclave (through the wall), filtered air

They can be acquired through (1) inhalation such as aerosols, (2) percutaneous inoculation (needle and syringe, cuts or abrasions from contaminated items and animal bites), (3) contact between mucous membranes and contaminated material (hands or surfaces) and (4) ingestion (aspiration through a pipette, smoking or eating).

Biosafety Level

Biosafety is the application of safety precautions that reduce a laboratorian's risk of exposure to a potentially infectious microbe and limit contamination of the work environment. There are four biosafety levels. Biosafety level designations are based on a composite of the design features, construction, containment facilities, equipment, practices and operational procedures required for working with agents from the various risk groups.

Biosafety Level I

Laboratory facilities are designated as BSL-I handles the microorganisms which are not known to consistently cause disease in healthy adults and present minimal potential hazard to laboratorians and the environment. An example of such microorganisms is the non-pathogenic strain of *Escherichia coli*.

Biosafety Level II

Laboratory facilities are designated as BSL-II deals with microorganisms such as *Staphylococcus aureus*, which pose moderate hazards to laboratorians and the environment. These microorganisms are typically indigenous and associated with diseases of varying severity.

Biosafety Level III

Laboratory facilities are designated as BSL-III deals with microorganisms such as *Mycobacterium tuberculosis*, which can be either indigenous or exotic, and they can cause serious or potentially lethal disease through respiratory transmission.

Biosafety Level IV

Laboratory facilities are designated as BSL-IV deals with microorganisms posing a high risk of aerosol-transmitted infections. Infections caused by these microbes are frequently fatal and without treatment or vaccines, for example, Ebola, Marburg viruses.

Biosafety level laboratories and risk groups of organisms are listed in Table 70.1.

Factors Affecting Healthcare-Associated Infection

HAIs as a group differ from other community-acquired infections, both in their patient profile and the severity and treatment of the disease caused by them. Various factors influence HAIs.

Hospitalised Patients

Susceptible hospitalised patients are one of the many factors responsible for healthcare-associated infections. This is because most patients admitted in wards have impaired immunity either as a part of their pre-existing disease processes or, in some instances, due to the treatment they have received in the hospital. They are, therefore, highly susceptible to infection.

Hospital Environment

The hospital environment harbours a higher load of microorganisms due to the multitudes of infected patients visiting the hospital. These organisms manage to infect the susceptible hosts through the medium of fomites in certain cases and human carriers in others. Contamination of food, water and in a few cases, the air has also been implicated in outbreaks of hospital infections.

Invasive Devices

Modern healthcare employs many types of invasive devices and procedures to treat patients and help them recover. Infections can be associated with the devices used in medical procedures, implants and prostheses. These include central line-associated bloodstream infections (CLABSIs), catheter-associated UTIs and ventilator-associated pneumonia (VAP).

Antibiotics Resistance

The infectious agents present in the hospital environment also possess the dubious distinction of being more difficult to treat as they are usually resistant to a range of commonly used drugs. The initial resistant strains of bacteria are present in large numbers due to the constant selection pressure exerted because of the necessary antibiotic usage inside the hospitals, and these then spread, replacing the other strains in the hospital.

Diagnostic or Therapeutic Procedures

Diagnostic or therapeutic interventions, such as the insertion of intravenous or urethral catheters, may introduce the infection to susceptible patients and cause iatrogenic infections. There is a good chance that the patient's flora causes infections as many invasive procedures are carried out within the hospitals, which exposes them to this risk.

Transfusion

Transfusion of blood, blood products and intravenous fluids, if not properly screened, may transmit blood-borne pathogens to the recipients who are hospitalised.

Epidemiology of Healthcare-Associated Infection

Sources of Infections

The sources of healthcare-acquired infection may be exogenous or endogenous.

Exogenous Infection

It is most important and occurs mostly from (1) another patient, (2) a member of the medical and paramedical staff harbouring the pathogens or (3) from the environment. The latter includes inanimate objects, such as medical equipment (e.g. catheters, endoscopes, cystoscopes, etc.), bedpans and surfaces contaminated by the patient's secretions, excretions, blood and other body fluids. It includes hospital food, water and environmental air.

Endogenous Infection

It is due to microbes present in the patient's flora present in the body. Microorganisms from these flora may cause infections in different patient tissues during certain surgical operations, manipulation by instruments or nursing procedures.

Microorganisms Causing Infections

Initially, much of the hospital outbreaks were caused by Gram-positive organisms like *Streptococcus pyogenes* and methicillin-resistant *S. aureus* (MRSA). Currently, Gram-negative enteric pathogens, including *E. coli*, *Klebsiella*, *Enterobacter* species, etc., have emerged as important causes of these infections.

Pseudomonas and *Acinetobacter* species are also common nosocomial pathogens because they can survive as saprophytes in the environment and are extremely hardy, developing resistance to most of the commonly used antibiotics and in some cases managing to grow even inside bottles of disinfectant solutions. *Candida auris* has been causing severe illness in hospitalised patients. *Clostridioides difficile* causes diarrhoea and fever due to the overuse of antibiotics. Viruses like hepatitis, HIV and noroviruses are also implicated in healthcare-associated infections.

The new and emerging pathogens have several features in common. a) They survive for a long period in the hospital environment; b) they are resistant to commonly used antibiotics and disinfectants and c) they cause both endemic and epidemic hospital outbreaks, resulting in significant patient morbidity and mortality.

The microorganisms causing healthcare-associated infections are listed in Table 70.2.

Table 70.2 Common healthcare-associated infections

Hospital-acquired infection	Causative organisms	Predisposing factors	Preventive measures
Catheter-associated urinary tract infections	*Escherichia coli*, *Klebsiella* spp., *Enterobacter* spp., *Serratia* spp., *Proteus* spp., *Pseudomonas* spp., *Enterococcus faecalis*, *Candida* spp., *Staphylococcus epidermidis* and other coagulase-negative *staphylococci*	Bacteriuria develops in at least 10–15% of hospitalised patients with indwelling urethral catheters Issues related to the high risk of catheter-associated urinary tract infections are feminine sex, long-standing catheterisation, serious underlying illness, disconnection of catheter and drainage tube	Restrict the use of an indwelling catheter in difficult situations and use it for less time to avoid manipulation
Surgical site infections	*Staphylococcus aureus* (MRSA), *Escherichia coli*, *Enterococcus faecalis*	Surgical wound infections account for nearly 30% of all hospital infections Contamination of surgical wounds with the patient's flora or with people in the operating room or the room itself at the time of surgery results in most surgical wound infections. It is uncommon for haematogenous seeding of the wound site	Strict aseptic measures during surgery Use chemoprophylaxis if required
Respiratory infection (ventilator-associated aspiration in comatose patients, etc.)/pneumonia/ lower respiratory tract infections	*Klebsiella* spp., *Enterobacter* spp., *Staphylococcus aureus*, *Escherichia coli*, *Pseudomonas* spp., *Acinetobacter* spp., *Proteus* spp. and *Legionella pneumophila*	Nosocomial pneumonia accounts for 15–20% of nosocomial infections Increased risk of colonisation: Prior antimicrobial therapy, contaminated ventilator circuits Processes that increase the possibility of aspiration of oropharyngeal contents to LRT: Intubation, presence of a nasogastric tube	Respiratory assistance apparatus should be strictly maintained under sterile conditions The patient should be maintained in the gatch position to avoid aspiration pneumonia
Bacteremia/septicaemia	*Staphylococcus aureus* *Staphylococcus epidermidis* *Enterococcus faecalis* *Candida* species	Indwelling intravenous catheter	Plastic catheters should be avoided and replaced by stainless steel needles Intravenous fluid therapy should be used only when it is required

Transmission of Microorganisms

Several organisms enter the body surface either by evasion or breaching the body's self-defence. These include epithelial integrity breaching such as wounds from surgeries, drain tubes and intravascular cannulas, failure to wash the body fluids (e.g. infections from the urinary catheter) and intervention with respiratory defences like anaesthesia effects. Infections can be transmitted in various ways.

Airborne Transmission

Infections may be transmitted by airborne droplets, dust particles and aerosols. Inhalation of airborne droplets or droplets' nuclei transmits *M. tuberculosis* and many other respiratory pathogens. Hospital dust generated from beddings, hospital floors, from the skin by natural shedding of skin scales are rich sources of many pathogens, such as *S. aureus*, *Pseudomonas aeruginosa*, etc. Therefore, inhalation of these infective dust particles transmits infections. Aerosols produced by nebulisers, humidifiers and air conditioning apparatus transmit certain pathogens, such as *Legionella pneumophila*.

Transmission by Direct Contact

Direct contact is the principal transmission route of healthcare-associated infections. Direct contact with the hands and clothing of medical personnel harbouring microorganisms may transmit infection from one person to another. *S. aureus* and *S. pyogenes* are two important pathogens transmitted by hand contact. The certain pathogen, such as *P. aeruginosa*, can also be transmitted by contact with certain hospital instruments (e.g. endoscopes, bronchoscopes, cystoscopies, etc.) if not properly disinfected.

Transmission by Oral Route

Hospital foods served to hospitalised patients may transmit certain gastrointestinal pathogens. The food may be the source of antibiotic-resistant *P. aeruginosa*, *E. coli*, *Klebsiella* spp. and other pathogens. These pathogens may colonise the intestine and later cause infection in susceptible patients.

Transmission by Parenteral Route

Blood-borne pathogens, such as hepatitis B and C and HIV (human immunodeficiency virus), may be transmitted by transfusion of blood or blood products through contaminated needles or sharp instruments that have not been properly disinfected.

People with long-term care in hospital facilities are prone to several infections. Some factors which play an important role in the initiation of infection are age, prolonged medical conditions, staying with catheters, living in a closed community and moving between hospitals and care facilities. Respiratory tract infections, urinary tract infections (UTIs), gastroenteritis and scabies outbreaks have been reported. Evidence from the MRSA suggests increased concern regarding the organisms that spread from hospitals to the community at large, specifically involving prolonged care facilities, such as nursing homes, thereby causing a more serious health problem in the community.

Common Types of Healthcare-Associated Infections

Common types of hospital infections can be grouped as follows:

Catheter-Associated Urinary Tract Infections (CAUTI)

Catheter-associated urinary tract infections (CAUTI) account for 40–45% of nosocomial infections. These infections are the major source of the spread of antibiotic-resistant bacterial strains in hospitals, though it contributes only to 10–15% of an extended stay in hospital. Most of the nosocomial UTIs are due to instrumentation, bladder catheters. There is a 3–10% of infection risk each day. The most common pathogens causing UTI are *E. coli*, nosocomial Gram-negative bacilli, enterococci and *Candida*.

CAUTI is usually caused by pathogens that spread the periurethral space from the patient's perineum or gastrointestinal tract, which is seen most commonly in women. Contamination of urinary catheters intraluminally causes this infection. This is mainly due to cross-infection by healthcare workers who irrigate catheters or empty the bags. Sometimes the pathogens are transmitted from improper disinfected urological equipment. There is also a chance this comes from a contaminated supply (e.g. dilute aqueous benzalkonium chloride).

Patients with a catheter in place may develop infection and complain of fever, urinary tract discomfort and peri catheter discharge. Urinary tract infection is the common symptom after the removal of the catheter. Antimicrobial therapy is not necessarily required to treat catheter-associated asymptomatic bacteriuria. However, immunocompromised individuals and patients under urological surgery should be exempted. The gastrointestinal and vaginal carriage may be responsible for asymptomatic catheter-associated candiduria.

If catheter-related symptoms are reported, the catheter must be removed immediately, urine culture should be done if possible, and if the infection is diagnosed, then an antibiotic course must be administered. There are chances for the organism to persist even after the antibiotic course with just relief in symptoms if the catheter is still kept in. When the catheter is still kept in, and the antibiotic is administered, there are chances for the organism to emerge as an antibiotic-resistant organism.

Health-Care Associated Bacteremia (Bloodstream Infection)

Central Line-Associated Bloodstream Infections (CLABSIs)

Central line-associated bloodstream infections (CLABSIs) are commonly caused by infected intravenous cannula, which may be a consequence of infections at any site. The longer the cannula are in contact, the greater is the risk of infection. Central line-associated bloodstream infection (CLABSI) is defined as a laboratory-confirmed bloodstream infection not related to an infection at another site that develops within 48 h of central line placement. After central venous catheter placement, bacteria on the skin surface migrate along the external surface of the catheter from the skin exit site towards the intravascular space. Contamination of the hub (intraluminal) hematogenous seeding of bacteria from a contaminated infusate or another source is some of the mechanisms of CLABSI. The femoral central venous catheters are associated with the highest risk of CLABSI, followed by the internal jugular and subclavian catheters.

Catheter-Related Bloodstream Infections (CRBSI)

Catheter-related bloodstream infection (CRBSI) is defined as the presence of bacteraemia originating from an intravenous catheter. CRBSI is a clinical definition used when diagnosing and treating patients that require specific laboratory testing that more thoroughly identifies the catheter as the source of the BSI. It is not typically used for surveillance purposes. Following microbiological methods are used to detect (1) quantitative paired blood culture, (2) differential time to positivity and (3) semiquantitative culture of the catheter tip.

Ventilator-Associated Pneumonia (VAP)

Ventilator-associated pneumonia (VAP) accounts for 15–20% of healthcare-associated infections. Hospital-acquired oropharyngeal flora or endogenous aspiration is the major cause of most bacterial nosocomial pneumonia infections. A majority of the deaths are reported due to nosocomial pneumonia more than any other body site infections. Deaths resulting from nosocomial pneumonia are influenced by antibiotic treatment inadequacy, comorbidities and specific pathogen involvement, specifically *Acinetobacter* and *P. aeruginosa*. Examination and laboratory diagnosis of pneumonia become a major concern in hospitals because of a large number of patients. Diagnosis is even more difficult for those in intensive care units due to deteriorating body conditions such as fever, abnormal chest roentgenographs and leucocytosis.

Risk factors are many for ventilator-associated pneumonia. Occasions that increase the danger of colonisation by potential pathogens decrease gastric acidity, ventilator circuits or equipment contaminations and previous antimicrobial therapy. Aspiration of oropharyngeal substances into the lower respiratory tract may be increased by following processes such as usage of a nasogastric tube, intubation and decreased levels of consciousness. The host defence mechanism of the lung is reduced, which allows the overgrowth of the aspirated pathogens caused by diseases such as chronic obstructive pulmonary disease, the result of upper abdominal surgery and factors of old age.

Hospital-Acquired Pneumonia (HAP)

Community-acquired pathogens such as *Streptococcus pneumoniae* and *Haemophilus* species cause Hospital-acquired pneumonia (HAP) manifesting within the first 4 days of hospital care. Pathogens like *Enterobacter* species, *S. aureus, Klebsiella pneumoniae*, *P. aeruginosa* and *Acinetobacter* species cause late-onset of pneumonia. This results in increasing concern for many ICUs. Infection is polymicrobial in as many as 20–40% of cases. The role of anaerobic bacteria in ventilator-associated pneumonia is not well defined. The appropriate duration of therapy for ventilator-associated pneumonia, although it usually ranges from 10 to 21 days, is not well studied.

Surgical Site Infections (SSI)

Thirty per cent of all healthcare-associated infections are due to infections from surgical wounds. These infections are common in 10% of patients who have had clean surgery. The frequency of the infections varies with surgery complexity, surgical skills and inherent patient risk. Contamination of the surgical wound is the primary cause of most surgical site infections. The contamination may be due to the patient's flora or from the environment during the surgical procedures or the personnel in the operating room. Haematogenous seeding of the wound site is not common in post-operative cases.

Infections of the surgical wound are characterised by a few common clinical features such as redness, pain in the site and discharge. If there is a fever within 4 weeks after surgery with no other localising features, it may be caused by surgical site infection. Infection of the external wound may be healed with or without administering antibiotics. Surgical drainage, imaging and antimicrobial therapy with the help of culture are required for diagnosing organ space and deep infection.

Organisms that are responsible for causing surgical wound infections depend on the location and type of surgery. MRSA is the most common cause in some hospitals. Gram-negative bacilli (e.g. *E. coli*) are more common infection causes following gastrointestinal or genitourinary tract surgery.

Healthcare-Associated Infections Due to Hepatitis Viruses B and C

Hepatitis viruses B and C infections are of serious risk for surgeons and healthcare workers from cardiac surgery and dialysis units and unvaccinated healthcare personnel. It is also a major risk for patients receiving blood transfusions or undergoing renal dialysis. At the time of any surgical or dental procedures, healthcare workers who are carriers of HBV pose a risk of transmission to the patients. The transmission of HCV from patients to healthcare personnel has been widely documented. With widespread awareness of screening for hepatitis surface antigen (HbsAg) and anti-hepatitis C virus antibody, there is a reduced risk of acquiring such infections, although there are other predisposing factors.

Adherence to strict aseptic conditions may further reduce the risk for healthcare workers.

Healthcare-Associated Episodes of Tetanus

The use of contaminated/infected or improperly sterilised sutures and dressing materials may cause this infection. It can also be caused by improper disinfection of the site of intramuscular injections (in the gluteal region) and if adequate measures are not taken while cutting the newborn's umbilical cord.

Healthcare-Associated Episodes of Acute Gastroenteritis

Outbreaks of *Salmonella* spp., causing food poisoning, contamination of cooked food by enterotoxins produced by *Staphylococci* spp. and few enteroviruses, *Shigella flexneri*, *Shigella dysenteriae*, *Shigella sonnei*, *E. coli* (in children and neonates) and *Clostridium difficile* are common healthcare-associated episodes of acute gastroenteritis.

Antibiotics Associated Diarrhoea

Antibiotic-associated diarrhoea (AAD) arises when the antibiotic disrupts the ecology of the intestinal microbiota by altering the diversity and numbers of bacteria in the gut. These changes can affect the capacity of the resident microbiota to resist the invasion of pathogenic microorganisms or the overgrowth of opportunistic pathogens species that are endogenously present in the microbiota. *C. difficile* is held to be the causative agent of approximately 20% of AAD. AAD is most frequently associated with cephalosporins, clindamycin and broad-spectrum penicillin, like amoxycillin and piperacillin. Treatment of antibiotic-associated diarrhoea involves discontinuation of the offending antibiotics and oral metronidazole. Oral vancomycin is used for severe cases of AAD.

Needlestick and Sharps Injury

Healthcare professionals at the highest risk for needlestick injuries are surgeons, emergency room workers, laboratory room professionals and nurses. In most cases, needlestick injuries occur chiefly because of unsafe practices and gross negligence on the part of the healthcare workers. The most important organisms that can be acquired after a needlestick injury include HIV, hepatitis B and hepatitis C. Tetanus should always be considered when a needlestick injury has occurred, and the patient's vaccination history must be obtained.

Prevention of Needlestick and Sharps Injury

These include various approaches such as (1) eliminating unnecessary needle use, (2) using devices with safety features, (3) developing safe work practices for handling needles and other sharp devices and (4) safely disposing of sharps and blood-contaminated materials.

Needlestick Injury Management

This comprises immediate management, postexposure treatment and laboratory follow-up of cases.

Immediate Management

It includes washing with soap and running water and avoiding the use of antiseptics or skin washes. All needlestick/sharp injuries must be reported to the hospital infection control officer.

Postexposure Treatment (PEP)

The decision to start PEP is made based on the degree of exposure to HIV and the HIV status of the source from where the exposure/infection has occurred. More so, it should begin as soon as possible, preferably within 2 h, and is not recommended after 72 h. Postexposure prophylaxis for HIV involves the use of three antivirals for 4 weeks.

If the patient is HBsAg positive, the recipient's serology must be assessed. If the post-vaccination anti-HBs level is high (greater than 10 mIU/mL), this is known to be protective. However, if the post-vaccination anti-HBs titre is low, the healthcare worker should be administered hepatitis B immunoglobulin, and if the healthcare worker is not vaccinated against hepatitis B, then a course of active immunisation starting 14 days later should be given. If the patient is HBsAg negative, the healthcare workers need to be observed, and his or her anti-HBs levels should be monitored.

Laboratory Follow-up of Cases

Exposed persons should have post-PEP HIV tests. HIV test at 3 months and again at 6 months is recommended. If the test at 6 months is negative, no further testing is recommended.

Diagnosis of Healthcare-Associated Infections

Diagnosis of healthcare-associated infections is made by routine bacteriological methods, such as direct demonstration of microorganisms in specimens by microscopy, isolation by culture and testing bacterial isolates for their antibiotics' sensitivity pattern. The specimens are usually sampled from possible sources of infections, such as environment, inanimate objects, water, air, food, hospital personnel, etc. The source of infection may be traced by performing phage typing, bacteriocin typing, biotyping or molecular typing.

Prevention and Control of Health Care-Associated Infections

One-third of healthcare-associated infections can be prevented. Effective prevention means achieving the two of the objectives. First is, the development of antibiotic-resistant organisms should be arrested, and second is that resistant organisms spread between patients should be prevented.

There are well-known principles that help in preventing hospital-acquired organisms spread. Every hospital has a customised approach to prevent acquired infections; it might be based on the available resources or the organism's epidemiology. For example, the most important measures for preventing catheter-associated UTIs are: i) the placement of catheters only when necessary and not solely for the convenience of caregivers, ii) the use of the aseptic technique for catheter insertion and urinary tract instrumentation, iii) the manipulation and opening of drainage systems as infrequently as possible and iv) the removal of catheters as soon as feasible.

Various approaches have been made to minimise the risk of surgical wound infections. These include a smaller number of preoperative stays in hospitals, antibiotic prophylaxis, optimising patient's risk factors and monitoring surgical site infections with timely feedback given to the physician's teams. Components of a program to prevent health-care-associated infection are many. These are listed in Table 70.3.

Intubation and mechanical ventilation increase the risk of hospital-acquired pneumonia and therefore need to be avoided, or their duration of use needs to be reduced whenever possible. Limiting the use of sedative and paralytic agents that depress cough and other host-protective mechanisms and maintaining endotracheal cuff pressure at greater than 20-cm H2O helps to reduce aspiration pneumonia. Continuous aspiration of subglottic secretions, through the use of a specially designed endotracheal tube, has significantly reduced the incidence of early-onset VAP.

For reducing the incidence of CLABSI, precautions include: (1) following hand hygiene and following aseptic techniques, (2) handling and maintaining central lines appropriately and (3) promptly removing unnecessary central lines.

Standard Precautions

It is a known fact that almost all living beings are likely to harbour infectious agents. Standard precautions are the basic infection processes that

Table 70.3 Suggested components of a program to prevent healthcare-associated infection in a hospital

Components
1. Compliance with handwashing protocols
2. Utilising aseptic methods while inserting intravenous and urinary catheters
3. Guidelines on antimicrobial usage compliance
4. Appropriate care for patients
5. Short stays in hospitals
6. Removing invasive devices at an earlier stage
7. Patients with infections should be isolated
8. Proper hospital infrastructure and policies
9. Timely vaccination of the staff (e.g. tuberculosis, hepatitis B and influenza)
10. Sharps policy
11. Complete sterilisation of all surgical instruments
12. Program to control infections
13. Hospital-acquired surveillance system

apply to every patient despite perceived or confirmed infectious status in any setting where healthcare is delivered. These precautions achieve a basic level of infection prevention and control in any healthcare setting. Adaptation of these precautions may help prevent or reduce the possibility of transmission of infectious agents from one person or place to another.

Infection Control Committee

In the twenty-first century, the speciality of infection control requires an extent of expertise that no sole practitioner can possess. Therefore, every hospital should have an infection control committee. The major responsibilities of the committee are the control of hospital-acquired infection and the monitoring of hygienic practices in the hospital. The committee is a multi-disciplinary team consisting of clinicians, infection control practitioners, a hospital epidemiologist, a biostatistician, a medical records officer, an infectious diseases physician, a blood bank officer and a microbiologist. The medical superintendent usually chairs the committee, with the medical microbiologist being the member secretary.

The committee (1) regularly reviews the infection control activities of the hospital, (2) monitors the emergence of drug resistance, (3) formulates the antibiotics policy of the hospital, (4) recommends suitable sterilisation and disinfection procedures in a hospital and (5) maintains data on the incidence and types of infections and antibiotics susceptibility patterns of the common prevalent pathogens.

Active Surveillance for Healthcare-Associated Infections

Strict surveillance of the hospital is the key to successful hospital infection control. To achieve this, measures such as high-quality data collection, data analysis and timely feedback to healthcare practitioners are essential. The surveillance system of the hospital should be potential, targeted, adjusted to risk, should validate methods and definitions and also should validate hospital to hospital comparison.

Usage of molecular methods has been contributing immensely to hospital infection surveillance. These molecular methods are used to identify the type of organisms and to clear the polyclonal or clonal character of evident outbreaks. Few strains of MRSA have been found to cause endemic infections and are confirmed through molecular testing.

Hand Hygiene

Hands become an important source of infectious agents by contact with a patient or their surrounding environment or from other healthcare workers. To reduce the risk of cross-contamination, practising a few hand hygiene precautions are advised before every episode of patient contact (including caring between patients and between various procedures for the same patient) and after any activity or contact that may result in hands becoming contaminated (such as removal of gloves).

Good hand hygiene practice includes alcohol-based hand rubs, handwashing with soap and clean water and single-use clean towels. The **5 Moments for Hand Hygiene** approach was designed by the World Health Organization to minimise the risk of transmission of microorganisms among a healthcare worker, the patient and the environment. It states that hand hygiene is very much essential in situations such as: (1) before touching a patient, (2) before a procedure (e.g. placing an intravenous line or preparing an injection), (3) after a procedure or body substance, body fluid or excretions or wound dressings exposure risk, (4) after touching a patient and (5) after touching a patient's immediate surroundings (Fig. 70.1).

Hand hygiene is also followed when caring between various procedures on different body sites of the same patient and after removing gloves.

Personal Protective Equipment (PPE)

Personal protective equipment (PPE) refers to the use of wearable equipment to protect the mucous membranes, skin and clothing of health workers from exposure to or contact with infectious agents. These include a variety of wearable PPEs, such as aprons, gowns, gloves, surgical masks, protective eyewear/goggles and face shields. The selection of a proper PPE depends on: (1) nature of the patient to be examined, (2) possibility of exposure to body fluids, (3) known or possible infectious agents and (4) and/or the likely modes of transmission of infection.

While using the PPE, (1) apron/gown has to be worn to protect skin and clothing when close contact is expected with the patient, materials or equipment, and where there is a risk of contamination with blood, body substances, secretions or excretions, (2) the same gown should not be worn while caring between different patients, (3) during procedures that can cause splashes of blood or other body fluids, the mouth, nose and eye protection must be worn and (4) mask should be in place when placing a catheter or injecting materials into the spinal canal or subdural space; masks should be changed when they become soiled or wet.

The sequence for putting on **(Donning)** personal protective equipment (PPE) includes gown, mask or respirator, goggles or face shield and gloves. The sequence for removing on **(Doffing)** PPE include gloves, goggles or face shield, gown, mask or respirator and washing hands or using an alcohol-based hand sanitiser immediately after removing all PPE.

Hand hygiene is followed between steps if hands become contaminated and immediately after removing all PPE.

Injection Safety

These are simple measures intended to prevent the transmission of infectious diseases between patients or between a patient and a healthcare worker during the preparation and administration of medication. Injection safety is a must to be followed in a hospital setup. It is recommended that before injecting a device into a medication vial, the diaphragms should be cleaned with 70% alcohol, and the same syringe should not be used between multiple patients, even if the needle is changed. The syringe should never be reused in a medication vial. Single-use or single-dose vials, bags or bottles of intravenous solution should not be administered for more than a patient; where possible, multi-dose vials should be used for a single patient and fluid infusions or intravenous tubing should not be used for more than a single patient.

Environmental Cleaning

This refers to the removal of soil or other organic contaminants from a device or environmental surfaces (beds, bedrails and other frequently touched surfaces, etc.) using the physical actions of scrubbing with a surfactant or detergent and water with appropriate chemicals. These cleaning procedures remove a large number of microbes from the surface and are always essential to be performed before disinfection. Compared to sterilisation, disinfection is considered a less toxic process capable of eliminating almost all

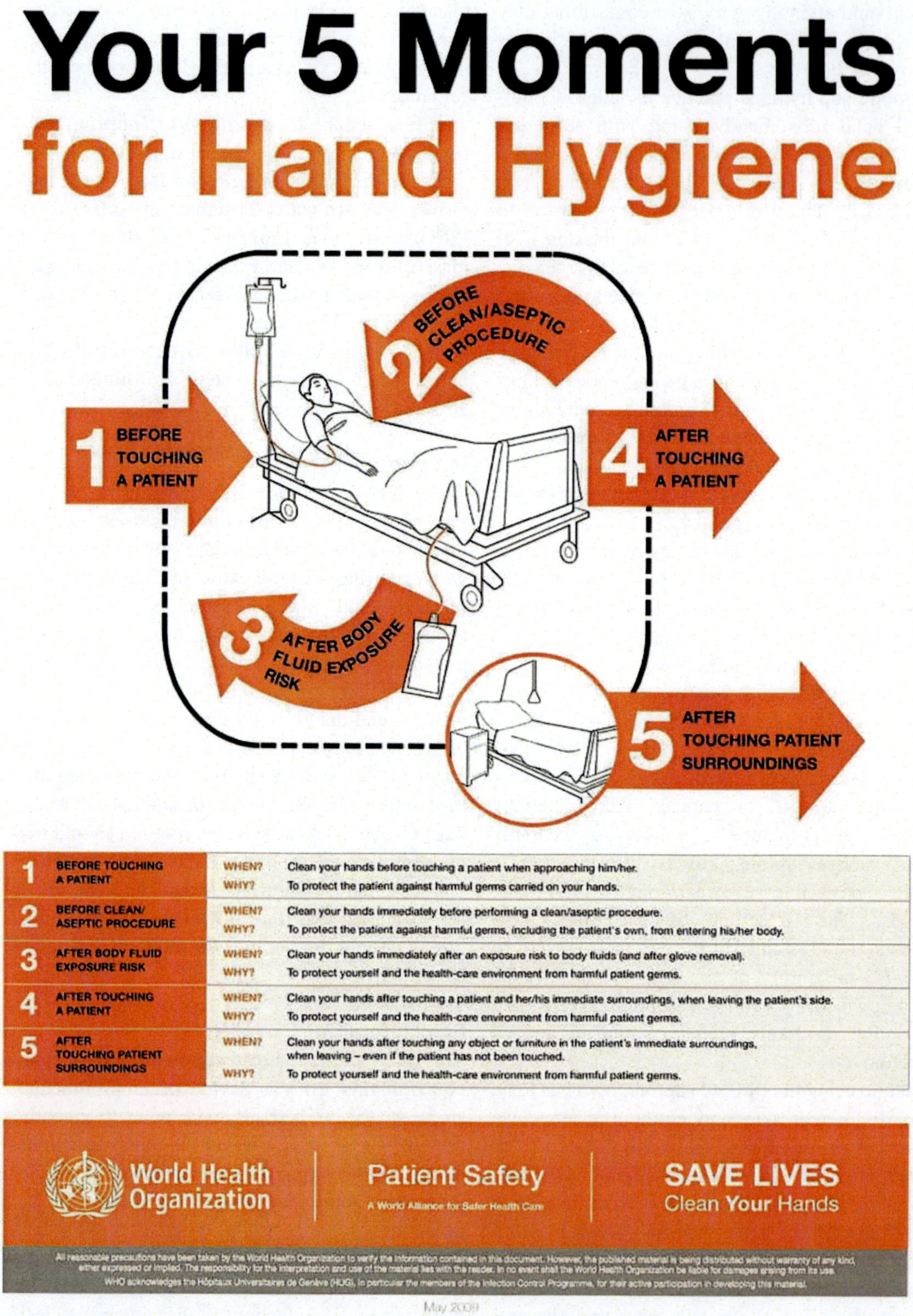

1	BEFORE TOUCHING A PATIENT	WHEN?	Clean your hands before touching a patient when approaching him/her.
		WHY?	To protect the patient against harmful germs carried on your hands.
2	BEFORE CLEAN/ ASEPTIC PROCEDURE	WHEN?	Clean your hands immediately before performing a clean/aseptic procedure.
		WHY?	To protect the patient against harmful germs, including the patient's own, from entering his/her body.
3	AFTER BODY FLUID EXPOSURE RISK	WHEN?	Clean your hands immediately after an exposure risk to body fluids (and after glove removal).
		WHY?	To protect yourself and the health-care environment from harmful patient germs.
4	AFTER TOUCHING A PATIENT	WHEN?	Clean your hands after touching a patient and her/his immediate surroundings, when leaving the patient's side.
		WHY?	To protect yourself and the health-care environment from harmful patient germs.
5	AFTER TOUCHING PATIENT SURROUNDINGS	WHEN?	Clean your hands after touching any object or furniture in the patient's immediate surroundings, when leaving – even if the patient has not been touched.
		WHY?	To protect yourself and the health-care environment from harmful patient germs.

Fig. 70.1 Indications of hand hygiene

known pathogens, but it may not be helpful in the inactivation of spores.

Medical Equipment

Adequate care is ensured while using any medical equipment, reusable, recyclable or single-use. Those instruments marked as single-use by the manufacturer need to be used only once; re-processing of these instruments should be strictly restricted. Reusable or recyclable equipment (e.g. blood glucose meter, surgical instruments, etc.) need to be cleaned and re-processed appropriately while using between multiple patients.

Respiratory Hygiene/Cough Etiquette

Control measures must be implemented to prevent infection at the first point of encounter with the facility and or ambulatory settings. Individuals visiting healthcare facilities with any signs of cough, sneeze, congestion, rhinorrhoea and increased respiratory secretions should be identified and monitored. When coughing/sneezing, the mouth/nose should be covered with tissue or a mask when coughing/sneezing. Hand hygiene has to be performed if contacted with respiratory secretions (Fig. 70.2).

Isolation Precautions

Isolation precautions are needed to prevent the transmission of pathogenic microorganisms within the healthcare setting. Patients infected with COVID-19, severe influenza cases, Sub-acute respiratory syndrome (SARS), open case of tuberculosis, anthrax, diphtheria, pertussis, pneumonic plague, chickenpox and patients infected with multidrug-resistant bacterial pathogens require isolation. Appropriate patient placement is decided based on route(s) of transmission of the infectious agent. Use of appropriate personal protective equipment should be made.

Precautions in Operating Theatre

The conventional or plenum type ventilation maintains approximately 20 air changes per hour in operating theatres. However, there may be as many as 300 air changes per hour in laminar flow-based operating theatres. The bacteria-carrying particles per cubic metre or bioload should not exceed 35–180 while empty or during operation, respectively. Airborne organisms or colony-forming units (cfu) typically occur between

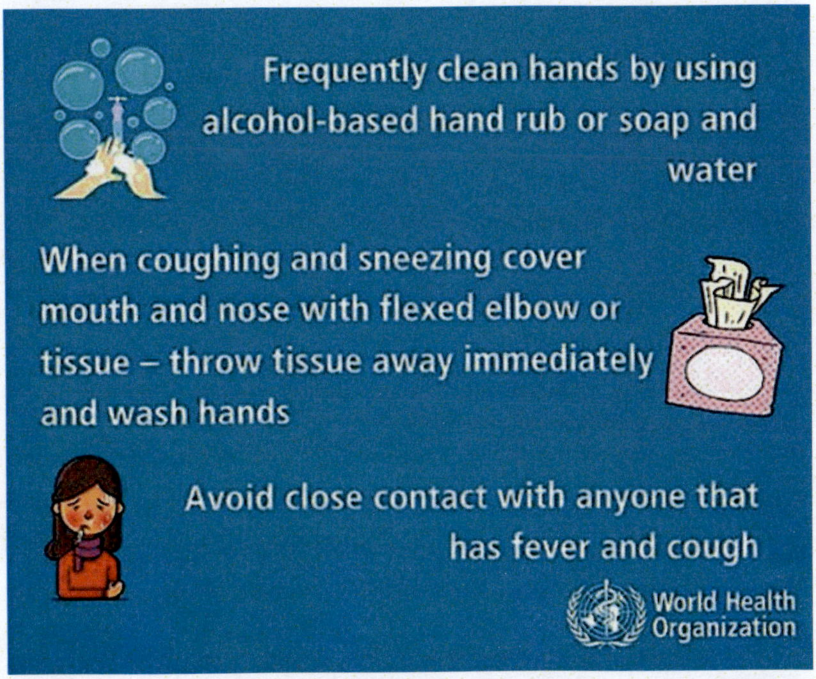

Fig. 70.2 Etiquettes of sneezing and coughing

150 and 300 cfu/m^3 in conventional operating theatres and 10 cfu/m^3 for the laminar flow ventilations.

Investigation and Follow-up Outbreaks of Disease

Even if strict preventive measures are followed, outbreaks associated with healthcare infections may occur. When such outbreaks occur in any area of the hospital, immediate investigations are required to resolve (1) if there has been a single common source of the pathogen, for example, food served at a specific meal; (2) if carriers providing a continuous reservoir of an infectious agent are responsible for outbreak and (3) if, in affected areas, there is any deviation from the daily protocol or general preventive measures.

71 Infective Syndromes: A System and Organ-Based Approach

'Infective syndrome' is synonymous with the term 'Syndromic infection'. An infection manifesting as a single syndrome could be caused by a single pathogen or many different pathogens. Conversely, many clinical syndromes could be caused by a single pathogen.

Pyrexia of Unknown Origin (PUO)

An increase in body temperature which is above the normal daily variation accompanied by an increase in hypothalamic set point is defined as fever. On the other hand, when there is an uncontrolled elevation of temperature over and above the normal ability of the human body to lose heat but with unchanged hypothalamic set point, it is termed hyperthermia (Fig. 71.1).

Definition

Peters Dorf and Beeson defined pyrexia of unknown origin (PUO) in 1961 as '(1) body temperature of more than 38.3 °C (>101 °F) on several occasions, (2) duration of fever of more than 3 weeks and (3) failure to reach a diagnosis despite one week of inpatient investigation'. Durack and Street classified PUO into four types: (1) classic PUO, (2) nosocomial PUO, (3) neutropenic PUO and (4) PUO associated with HIV infection (Table 71.1).

Aetiologies of PUO

PUOs are most often due to infections (30–40%). Other conditions responsible for the condition are neoplasms (20–30%), collagen vascular diseases (10–20%) and miscellaneous diseases (15–20%).

Infections

Infections are the most common causes of PUO, especially in more than 50% of children suffering from PUO. In adults, infection accounts for 30–45% of cases of PUO. The infectious causes of PUO are summarised in Table 71.2.

Neoplasms

These include malignant conditions such as colon cancers, gall bladder carcinoma, hepatoma, various types of lymphomas and leukaemias, lymphomatoid granulomatosis, malignant histiocytosis, pancreatic cancer, renal cell carcinoma, sarcoma and benign condition such as atrial myxoma.

Collagen Vascular Diseases

These include adult Still's disease, Bechet disease, giant cell arteritis/polymyalgia rheumatic,

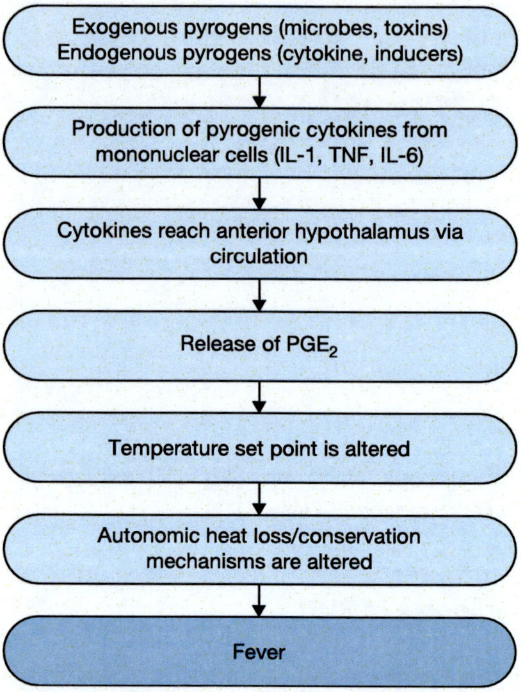

Fig. 71.1 Pathogenesis of pyrexia of unknown origin

hypersensitivity pneumonitis, hypersensitivity vasculitis, mixed connective tissue disease, polyarteritis nodosa, relapsing polychondritis, rheumatic fever, rheumatoid arthritis, systemic lupus erythematosus, Takayasu aortitis and Wegener granulomatosis. Granulomatous diseases include Crohn's disease, granulomatous hepatitis, midline granuloma and sarcoidosis.

Miscellaneous Conditions

These include aortic dissection, drug fever, gout hematomas and haemoglobinopathies.

Laboratory Diagnosis

The correct diagnosis of PUO is essential in dealing successfully with PUO. Because of many possible causes of PUO, the **diagnostic approaches are divided into various stages.**

Stage I

This includes a comprehensive history, analysis of fever and use of routine investigations such as a complete haemogram, chest radiograph, etc.

Stage II

This includes the use of specific diagnostic tests and other non-invasive investigations. The specific tests include (a) direct examination of blood for malaria, filariasis and relapsing fever; (b) blood, urine, sputum, body fluid, pus, etc., for culture and (c) serological tests for diagnosis of brucellosis, tularensis, Salmonella, rickettsial infections, CMV (*cytomegalovirus*) and EBV (Epstein–Barr virus) infections. These also include serological tests for detection of antinuclear antibody, antineutrophil cytoplasmic antibody, rheumatoid factor and serum cryoglobulins for ruling out other collagen vascular diseases and vasculitis.

Table 71.1 Causative agents of nosocomial, neutropenic, HIV-associated and classic pyrexia of unknown origin (PUO)

Feature	Nosocomial	Neutropenic	HIV-associated	Classic (PUO)
Patient's status	Hospitalised, acute care, no infection when admitted	Neutrophil count either <500/μL or expected to reach that level in 1–2 days	Confirmed HIV-positive	All others with fevers for ≥3 weeks
Duration of illness while under investigation	3 days (2 days incubation of microbiology cultures)	3 days (2 days incubation of microbiology cultures)	3 days (or 4 weeks as outpatient)	3 days or three outpatients visits
Examples of infections	Septic thrombophlebitis, sinusitis, *Clostridium difficile* colitis, drug fever	Perianal infection, aspergillosis, Candidemia	MAI (*Mycobacterium avium* intracellularae complex) infection, tuberculosis, non-Hodgkin lymphoma, drug fever	Infections, malignancy, inflammatory disease, drug fever

Table 71.2 Infectious causes of pyrexia of unknown origin (PUO)

Localised pyogenic infections Appendicitis Cat-scratch disease Cholangitis Cholecystitis Liver abscess Mesenteric lymphadenitis Osteomyelitis Pancreatic abscess Pelvic inflammatory disease Perinephric/intrarenal abscess Prostatic abscess Sinusitis Subphrenic abscess Suppurative thrombophlebitis
Intravascular infections Bacterial aortitis Bacterial endocarditis Vascular catheter infection
Systemic bacterial infections Typhoid fever Bartonellosis Cat-scratch disease Gonococcemia Legionnaires disease Leptospirosis Listeriosis Lyme disease Melioidosis Meningococcemia
Rat-bite fever *Campylobacter* infections Relapsing fever Salmonellosis Syphilis Tularemia *Yersinia* infection *Mycobacterial infections* Tuberculosis *Mycobacterium avium/Mycobacterium intracellulare* infections Other atypical mycobacterial infections
Rickettsial infections Ehrlichiosis Q fever Rickettsialpox Rocky mountain spotted fever
Mycoplasmal infections Chlamydial infections Lymphogranuloma venereum Psittacosis TWAR (*Chlamydia pneumoniae*) infection
Fungal infections Aspergillosis Blastomycosis Candidiasis Coccidioidomycosis Cryptococcosis

(continued)

Table 71.2 (continued)

Histoplasmosis
Mucormycosis
Paracoccidioidomycosis
Sporotrichosis
Pneumocystis infection
Viral infections
Epstein–Barr virus infection
Cytomegalovirus infection
Hepatitis A, B, C, D and E
Dengue
Human herpes virus 6 infection
Human immunodeficiency virus infection
Parvovirus B19 infection
Parasitic infections
Amebiasis
Babesiosis
Chagas disease
Leishmaniasis
Malaria
Strongyloidiasis
Toxocariasis
Toxoplasmosis
Trichinosis
Presumed infections, agent undetermined
Kawasaki disease (mucocutaneous lymph node syndrome)

Other laboratory investigations include erythrocyte sedimentation rate, liver enzymes, liver function tests, etc. Elevated serum transaminases are demonstrated in infectious mononucleosis, CMV infection, psittacosis, drug-induced fever, leptospirosis, toxoplasmosis and fever. Non-invasive investigations include abdominal ultrasonography to detect intrahepatic lesions, intrasplenic, intranephric mass lesions, percutaneous or transesophageal echocardiography for detecting vegetations and intracardiac myxomas and CT and MRI in detecting intracranial, intra-abdominal and pelvic diseases.

Stage III

This includes invasive tests such as a liver biopsy to demonstrate miliary tuberculosis and metastatic carcinomas, bone marrow biopsy for demonstration of miliary tuberculosis, or typhoid fever, etc.

Treatment

Treatment includes specific treatment by drugs if the aetiology is known. It also includes therapeutic trials with corticosteroids prostaglandin inhibitors for non-infectious causes of PUO. Empiric treatment with antibiotics or cytotoxic drugs is less warranted; however, antitubercular drugs are given in suspected tuberculosis cases without microbial evidence.

Respiratory Tract Infections

Of several aetiology syndromic infections, respiratory tract infections are among the most common ones prevalent worldwide, with substantial morbidity and mortality, especially in the two extremes of age. Both upper and lower respiratory tracts are prone to infections by different microbes, viz., bacteria, rickettsia, viruses, fungi and protozoa. Some microbes cause both upper and lower respiratory infections.

Table 71.3 Causative agents of common cold

Rhinoviruses
Coronaviruses
Coxsackie virus A
Influenza virus
Parainfluenza virus
Adenoviruses
Echoviruses

Based on the site involved, respiratory tract infections may be classified as upper respiratory tract infection (URI) and lower respiratory tract infection (LRI). The treatment modalities are different for both.

Definitions

Upper Respiratory Tract Infection (URI)

The upper respiratory tract infection can be defined as an infection occurring above the glottis or vocal cords, although some authorities think that any infection above the cricoid cartilage qualifies as URI. Normally the infection involves the nose, sinuses, pharynx and larynx.

Lower Respiratory Tract Infection (LRI)

Lower respiratory tract infection is defined as an infection of the trachea, and the structures beyond it which include the bronchi, bronchioles and lungs proper. The severity of lower tract infections is much more than upper tract infections. It is the leading cause of death among all infectious diseases. The two most common LRIs are bronchitis and pneumonia.

Upper Respiratory Tract Infection (URI)

The upper respiratory tract includes the epiglottis, larynx, pharynx, oropharynx, etc. It is lined with squamous epithelium and continues with mucosal linings of sinuses, Eustachian tubes and middle ear.

The upper respiratory tract diseases are named according to the anatomical site involved. These include common cold, laryngitis, laryngotracheobronchitis, epiglottitis, pharyngitis, tonsillitis and peritonsillar abscesses.

Aetiology

The common cold is the most common clinical condition affecting the upper respiratory tract. The condition is usually self-limiting and is caused mostly by viruses (Table 71.3). Acute sore throat, which is pharyngitis or tonsillitis, is also mostly caused by viruses. Nearly, 30% of the condition is caused by bacteria and other pathogens (Table 71.4). In sore throat, viruses and bacteria produce similar manifestations. These are pain during deglutination, inflamed pharynx and tonsils, cervical lymphadenopathy and fever. The causative agents of laryngitis, laryngotracheobronchitis and epiglottitis are summarised in Table 71.5.

Diphtheria is an important cause of childhood disease, with very high mortality and morbidity. Vaccination has greatly minimised both morbidity and mortality. Despite this, diphtheria is re-emerging, and cases have been reported now in India and other countries.

Laboratory Diagnosis

Specimens

Throat swab from fauces, nasopharynx, etc., is the specimen of choice. Cotton dacron or calcium alginate-tipped swabs are suitable for collecting these specimens. Moist swabs need to be transported and processed within 4 h. However, transport media like Amies and Regan–Lowe transport medium can be used if a delay is anticipated.

Gram Staining

It is useful to identify budding yeast like *Candida*, Gram-negative spirochetes such as *Borrelia vincenti* and *Fusobacterium* spp. Gram staining,

Table 71.4 Causative agents of sore throat

Bacteria
Streptococcus pyogenes
Streptococcus group C and G
Corynebacterium diphtheriae
Haemophilus influenzae
Borrelia vincenti
Viruses
Rhinovirus
Coronavirus
Coxsackie virus A
Influenza viruses
Parainfluenza virus
Adenovirus
Epstein Barr virus
Fungus
Candida albicans

Table 71.5 Causative agents of laryngitis, laryngotracheobronchitis and epiglottitis

Viruses
Parainfluenza viruses
Rhinovirus
Coronavirus
Adenoviruses
Respiratory syncytial virus
Bacteria
Mycoplasma pneumoniae
Haemophilus influenzae type B
Bordetella pertussis

however, is unreliable for the detection of other causative agents of pharyngitis. **Albert staining** is useful for presumptive identification of *Corynebacterium diphtheriae* by demonstrating green-coloured bacilli with bluish-black metachromatic granules arranged in a Chinese letter pattern.

The causative agents can be isolated from clinical specimens by **culture**. The selection of culture media depends on the organism suspected. For example, sheep blood agar is used to demonstrate pinpoint beta-haemolytic colonies of *Streptococcus pyogenes* after 24–48 h of incubation at 37 °C in 5–10% CO_2.

Loffler serum slope and potassium tellurite media are used to isolate *C. diphtheriae.* Freshly prepared Bordet–Gengou medium inoculated directly at the bedside is useful for demonstrating small, shiny, mercury-drops like colonies of *Bordetella pertussis*. chocolate agar and Sabouraud's dextrose agar (SDA) are used for isolation of *Haemophilus influenzae* and *Candida* spp., respectively. *Neisseria gonorrhoeae* on modified Thayer–Martin medium produces small, greyish white, convex and translucent colonies on incubation in the presence of 5–10% CO_2. Identification of the bacteria is based on colony morphology, typical morphology on Gram-stained smears of the bacteria, biochemical tests, toxigenicity tests, etc.

Direct Fluorescent Antibody (DFA) Test

It is a rapid method for detecting *C. diphtheriae* and *B. pertussis* directly in clinical specimens.

Treatment

Treatment of the condition depends on the bacteria causing the disease. For example, penicillin, erythromycin, etc., are used to treat sore throat caused by *S. pyogenes*, and penicillin is also useful for the treatment of diphtheria caused by *C. diphtheriae*.

Lower Respiratory Tract Infection (LRI)

The lower respiratory tract (LRI) infection involves the trachea, bronchi, bronchioles and lungs. It can be acute or chronic. Acute LRI includes acute bronchitis, acute exacerbation of chronic bronchitis, bronchiolitis (in the paediatric age group) and pneumonia. Chronic infections include tuberculosis, aspergillosis, etc.

Types of LRI

The two most common LRIs are bronchitis and pneumonia.

Aetiology

Bronchitis

Bronchitis may be acute or chronic.

Acute Bronchitis

It normally develops from an innocuous infection such as common cold and resolves within 1 or 2 weeks. Chronic bronchitis is a more dangerous condition which develops gradually. Symptoms may wax or wane but never resolves completely.

Acute Exacerbation of Chronic Bronchitis (AECB)

It is a condition which is superimposed on chronic bronchitis. This manifests as deranged lung function with features of airflow obstruction and other signs and symptoms.

Pneumonia

Pneumonia is an infection of the alveoli, distal airways and interstitial tissue of the lung, leading to consolidation. Pneumonia can be broadly categorised as community-acquired pneumonia and hospital-acquired (nosocomial) pneumonia.

Lobar Pneumonia

It is an acute inflammation involving an entire lung lobe relatively homogeneously. *Streptococcus pneumoniae* is the causative agent.

Bronchopneumonia

It is a secondary infection usually followed by a viral infection. This is a patchy consolidation involving one or several lobes of the lungs. *S. pneumoniae* and *H. influenzae* are the common causative agents.

Interstitial Pneumonia

It is due to the inflammation affecting the lung interstitial tissue, along with the alveolar walls and the connective tissue around the bronchovascular tree. The causative agents are viruses, *Mycoplasma pneumoniae*, *Chlamydia pneumoniae*, *Legionella pneumophila* and *Coxiella burnetii*.

Miliary Pneumonia

It is a condition associated with numerous well-defined lesions due to the spread of the pathogen to the lungs via the bloodstream. *Mycobacterium tuberculosis* is a pathogen that causes this condition.

Fever, cough (which may be dry or productive with purulent or rust-coloured expectoration), pleuritic chest pain, chills or rigors and dyspnoea are typical manifestations of pneumonia. One very useful clinical sign of the severity of pneumonia is a respiratory rate more than 30/min in a person having no other underlying lung disease.

Nosocomial Pneumonia

Nosocomial pneumonia or hospital-acquired pneumonia (HAP) is one that develops 48 h or more following hospital admission and which was not in the incubation period at the time of admission. Ventilator-associated pneumonia (VAP) is a subtype of HAP which is seen in intensive care units (ICUs) and is defined as pneumonia which presents itself 48–72 h after tracheal intubation in a mechanically ventilated patient. It has a prevalence of 10–20% in patients receiving mechanical ventilation. Among patients

Table 71.6 Causative agents of pneumonia in immunocompromised patients

Haemophilus influenzae
Streptococcus pneumoniae
Mycobacterium tuberculosis
Rhodococcus equi
Pneumocystis carinii
Histoplasma capsulatum
Coccidioides immitis
Cytomegalovirus

in hospital wards who have no associated risk factors, the pathogens most frequently causing nosocomial pneumonia include *S. pneumoniae*, *H. influenzae* and *Staphylococcus aureus*, enteric Gram-negative bacilli (e.g. *Escherichia coli*, *Klebsiella* spp., *Proteus* spp. and *Serratia marcescens*) the pathogens causing pneumonia in immunocompromised patients are summarised in Table 71.6.

Chronic Infections

Pulmonary tuberculosis is a chronic infection diagnosed by chest radiography and typical clinical presentation like chronic cough, fever, night sweats, loss of weight and haematological picture such as elevated ESR, decreased haemoglobin, etc. **Systemic aspergillosis** may manifest as pulmonary, bronchopulmonary and allergic aspergillosis. Repeated culture positivity is an important confirmatory step.

The pathogens causing LRI are listed in Table 71.7.

Laboratory Diagnosis

Chest radiography is the usual and most common method for diagnosing pneumonia.

Specimens

Sputum, blood, serum, pleural fluid, etc., are the frequently used specimens for laboratory diagnosis of pneumonia. **Sputum** collection is best done in the morning before breakfast. Transtracheal aspiration, protected specimen brush (PSB) and bronchoalveolar lavage are the other useful specimens.

Microscopy

Gram stain is useful in identifying bacteria and fungi. The specimens may also contain normal bacterial flora and fungi like *Candida albicans*. A careful interpretation is needed. Giemsa stain is used to diagnose *Pneumocystis carinii* pneumonia (PCP). Ziehl–Neelsen staining for acid-fast bacilli is the frequently used method to diagnose pulmonary tuberculosis.

Culture

Sheep blood agar, chocolate agar, Lowenstein–Jensen (LJ) media, etc., are used for the culture of sputum specimens. A sputum sample showing less than 25 white blood cells and less than 10 squamous epithelial cells per low-power field by microscopy are suitable for culture. An upper airway colonising organism can be suspected if the organism is isolated from sputum culture but similar organism is not found on Gram staining. The presence of a predominant organism may suggest a presumptive etiologic diagnosis of pneumonia.

All patients who are admitted to the hospital for community-acquired pneumonia should have at least two sets of blood cultures done before initiation of antibiotic therapy. *S. pneumoniae*, *S. aureus* and *E. coli* are the most common isolates.

Serology

Serodiagnosis depends on the demonstration of IgM antibody or the demonstration of a fourfold rise in the titre of IgG antibody between acute- and convalescent-phase serum samples against a particular pathogen. Serodiagnosis is used in the diagnosis of pneumonia caused by various

Table 71.7 Aetiological agents of lower respiratory tract infections

Disease	Bacterial agents	Fungal agents	Viral agents	Parasitic agents
Tracheitis	*Staphylococcus aureus*	*Candida albicans*	Rhinovirus	–
Bronchitis	*Haemophilus influenzae*, *Streptococcus pneumoniae*, *Moraxella catarrhalis*, *Mycoplasma pneumoniae*	–	–	–
Pneumonia	*Mycobacterioum tuberculosis*, *Streptococcus pneumoniae*, *Haemophilus influenzae*, *Klebsiella pneumoniae*, *Chlamydophila pneumoniae*, *Chlamydia trachomatis*, *Chlamydia psittaci*, *Mycoplasma pneumoniae*, *Pseudomonas aeruginosa*, *Escherichia coli*, *Legionella pneumophila*, *Bordetella pertussis*	*Coccidioides immitis*, *Histoplasma capsulatum*, *Cryptococcus neoformans*, *Aspergillus* spp., *Pneumocystis jiroveci*	Influenza and parainfluenza virus, respiratory syncytial virus, mumps, measles, adenovirus, metapneumovirus, and SARS-CoV-2	Malaria parasites, *Toxoplasma gondii*, *Entamoeba histolytica*, *Leishmania donovani*, *Trypanosoma cruzi*, *Toxocara* spp., *Paragonimus* spp., *Dirofilaria* spp., *Schistosoma* spp., *Echinococcus* spp., *Ascaris lumbricoides*
Lung abscess	Anaerobic non-spore forming bacteria (*Bacteroides*, *Fusobacterium*), Actinomyces, *Nocardia asteroides*, *Staphylococcus aureus*, *Streptococcus anginosus* complex	*Aspergillus*, *Cryptococcus*, *Histoplasma*, *Blastomyces*, *Coccidioides*	–	*Entamoeba histolytica*, *Paragominus westermani*

infectious agents. For example, complement fixation test (CFT) is used for the diagnosis of LRI caused by adenovirus, parainfluenza viruses, influenza virus A, *M. pneumoniae* and *C. burnetii*; ELISA is a frequently used test diagnosis for infections caused by *C. pneumoniae* and *M. pneumoniae*. Rapid microagglutination test is used for diagnosis of infections caused by *Legionella* spp.

ELISA is also used to detect *L. pneumophila* serogroup 1 antigen and *S. pneumoniae* urinary antigen in the urine of patients with Legionnaires' disease and with bacteraemic pneumococcal pneumonia, respectively

Biomarkers

Serum procalcitonin (PCT), serum prolactin and CRP are a few biomarkers detected for supplementing the diagnosis of community-acquired pneumonia (CAP). Estimation of serum PCT may help differentiate between bacterial and non-bacterial causes of infection, using a cut-off value of 0.5 ng/mL. Furthermore, serum PCT is useful in predicting bacteraemia and assessing disease severity in CAP patients.

Molecular Detection

Since bacterial, viral, fungal and protozoan cause respiratory tract infections, the most sensitive and specific way of confirming the diagnosis is by PCR/multiplex PCR. Commercial kits available using different panels/combinations of panels are available currently.

Radiology

Chest X-ray (CXR) is considered the gold standard for diagnosing pneumonia. The diagnosis of CAP can be made by the presence of an infiltrate in the lungs on chest imaging in a patient with a clinically compatible syndrome (e.g. fever, dyspnea, cough and sputum production). Posteroanterior and lateral chest radiographs are helpful for most patients with suspected CAP.

Treatment

Macrolides such as clarithromycin, azithromycin or doxycycline are antibiotics recommended for treatment of LRI in an outpatient with no cardiopulmonary disease or risk factors. Cefuroxime/ceftriaxone or cefotaxime or ampicillin/sulbactam in combination with azithromycin or quinolone are recommended for treatment of hospitalised patients with LRI.

Urinary Tract Infections (UTI)

The urinary tract consists of the lower urinary tract encompassing the urethra and bladder and the upper urinary tract encompassing the ureters and kidneys. Infections at various sites of the urinary tract can manifest simultaneously or in individual organ and may be asymptomatic or symptomatic. The female urethral anatomy is of particular importance; hence, UTI is one of the most common bacterial infections in females.

Definitions

Urinary Tract Infection (UTI)

It is defined an infection of the genitourinary tract, extending from the kidney to the urethra.

Bacteriuria

It is the term that denotes the presence of detectable bacteria in urine, whereas **pyuria** is the term that indicates the presence of pus in the urine. Pyuria usually accompanies UTI.

Resident Microbial Flora

The epithelium in the distal portion of the urethra is colonised with resident flora. The urine of a healthy individual is generally sterile. The resident microbial flora of urine consists of coagulase-negative staphylococci, viridans and non-haemolytic streptococci, diphtheroids, *Lactobacillus* and saprobic *Neisseria* spp.

Aetiology

1. **Depending on the anatomical sites of infection**, the UTIs can be classified broadly into two groups: (1) lower UTIs such as urethritis, cystitis and prostatitis and (2) upper UTIs such as acute pyelonephritis and acute pyelitis. The former is an infection of the parenchyma of the kidney, whereas the latter is an infection of the pelvic of the kidney. A single organism mostly causes UTIs. The causative agents of UTIs are listed in Table 71.8.
2. **Depending on the aetiological agents**, UTIs can also be classified as community-acquired UTIs and nosocomial-associated UTIs. The pathogens causing community-acquired and hospital-acquired UTIs are listed in Table 71.9.

UTIs can also be acute, uncomplicated, usually seen in females, or complicated following catheterisation, diagnostic instrumentation like cystoscopy, anatomical or functional abnormalities of the urinary tract, calculi, obstruction, immunosuppression, diseases of the kidneys or diabetes.

Table 71.8 Causative agents of urinary tract infections

Bacteria
Gram-negative bacilli
Escherichia coli *Proteus mirabilis* *Klebsiella pneumoniae* *Pseudomonas aeruginosa* *Enterobacter* spp. *Serratia* spp. Gram-positive cocci
Staphylococcus saprophyticus *Staphylococcus aureus* *Staphylococcus epidermidis* *Enterococcus* spp. Other uncommon bacteria
Salmonella spp. *Mycobacterium tuberculosis* *Citrobacter* spp. *Streptococcus pyogenes* *Streptococcus agalactiae* *Gardnerella vaginalis* *Chlamydia trachomatis*
Fungi
Candida spp. *Histoplasma capsulatum*
Parasites
Trichomonas vaginalis *Schistosoma haematobium*
Viruses: rare
Herpes simplex virus Adenovirus Polyomaviruses JC and BK viruses Cytomegalovirus Hantavirus

Table 71.9 Uropathogens causing community-acquired, nosocomial and miscellaneous urinary tract infections (UTIs)

UTI	Causative agents
Community-acquired UTI	*Escherichia coli, Staphylococcus saprophyticus, Proteus mirabilis, Klebsiella pneumoniae*
Nosocomial UTI	*Escherichia coli, Proteus* spp., *Klebsiella* spp., *Pseudomonas* spp., *Serratia* spp., enterococci, staphylococci, *Candida* spp.
Miscellaneous UTI	*Acinetobacter* spp., *Citrobacter* spp., *Gardnerella vaginalis*, beta-haemolytic streptococci, *Aerococcus urinae, Chlamydia trachomatis, Ureaplasma urealyticum, Campylobacter* spp., *Haemophilus influenzae, Leptospira* spp., *Corynebacterium renale* and adenovirus type 11 and 21 (rare)

Pathogenesis of UTI

UTI is an endogenous infection. It is usually caused by bacteria which constitute the patient's own intestinal flora. These bacteria form the intestinal flora cause an ascending infection by entering the urinary tract in a retrograde fashion via the urethra (Fig. 71.2). *E. coli, Klebsiella* spp. are intrinsic gut organisms that have fimbriae (pili) for attachment to the mucosa of the urinary tract and most of them are motile because of the presence of flagella.

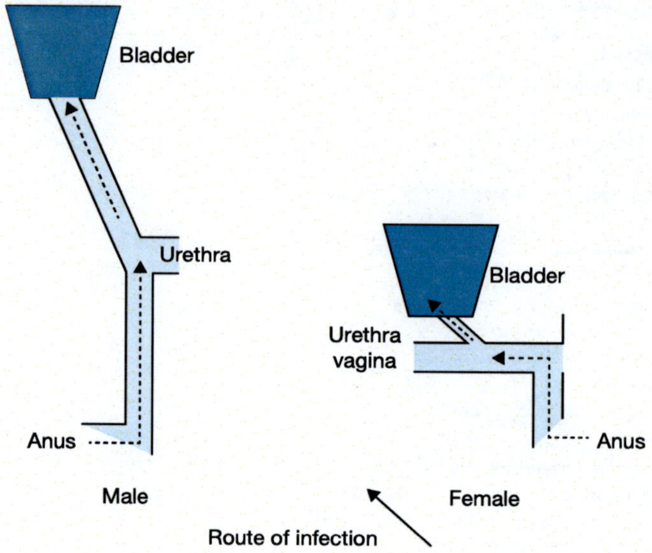

Fig. 71.2 Routes of urinary tract infections

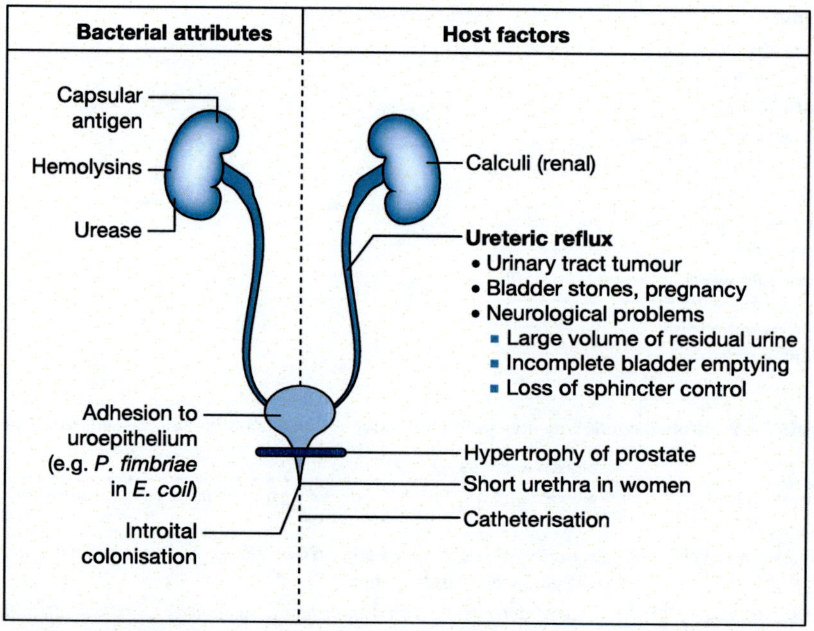

Fig. 71.3 Bacterial attributes and host factors in UTIs

UTI can also occur through the haematogenous route. Haematogenous pyelonephritis is mostly seen in patients who are either chronically ill or suffering from endocarditis, bacteremias, tuberculosis, disseminated infections or receiving immunosuppressive therapy.

Predisposing Factors

These include both host and microbial factors (Fig. 71.3).

Host Factors

These include the followings:

Gender and Sexual Activity

Females are more prone to UTI due to the short urethra (4 cm) which ends below the labia and also proximity to the anal and perianal areas. Bacterial infection occurs in the bladder during sexual intercourse and is linked to the onset of cystitis, and hence is an important factor in the pathogenesis of UTIs in sexually active women. Factors predisposing bacteriuria in men include urethral obstruction, insertive rectal intercourse, lack of circumcision and an HIV infection with CD4+ T cell counts of less than 200 cells/μL.

Pregnancy

In 2–8% of pregnant women, UTI presents in various ways ranging from asymptomatic bacteriuria to pyelonephritis. During pregnancy, temporary incompetence of the vesicoureteral valves and decreased ureteral tone and ureteral peristalsis are associated with increased susceptibility to upper UTIs. Additionally, bladder catheterisation is another important factor for susceptibility during or after delivery.

Obstruction

Obstruction to free urine flow due to tumour, stricture, stone or prostatic hypertrophy may result in hydronephrosis and a greatly increased risk of UTI.

Neurogenic Bladder Dysfunction

Various conditions like spinal cord injury, tabes dorsalis, multiple sclerosis, diabetes and other diseases interfering with bladder enervation can lead to UTI, initiated by catheterisation for bladder drainage. The UTI is facilitated due to the prolonged stasis of urine in the bladder.

Vesicoureteral Reflux

Reflux of urine from the urinary bladder cavity back to the ureters and occasionally to the renal pelvis happens due to the increased pressure in the bladder or voiding. Vesicoureteral reflux is common in children with anatomically normal urinary tracts but is infected.

Genetic Factors

Host genetic factors are associated with an increased risk of UTI. A history of UTI in the mother has been frequently found in women who have recurrent episodes of UTIs compared to the controls. Bacteria attaching to the uroepithelial cells are based on the number and type of receptors that are partly genetically determined. For example, nearly all strains of *E. coli* can attach to the P-positive RBCs due to the presence of *P fimbriae* and are found on almost all strains causing acute pyelonephritis.

Microbial Factors

Various virulence factors of bacteria such as fimbria, haemolysins, aerobactin and ureases increase the uropathogenicity of bacteria like *E. coli, E. coli* and *Klebsiella* spp. These are intrinsic gut organisms that have fimbriae (pili) for attachment to the mucosa of the urinary tract.

Proteus, *Morganella* and *Providencia* have urease enzymes which can split urea and produce ammonia, thus increasing the urinary pH. This results in the formation of crystals leading to the formation of stones. Biofilm produced by certain bacteria such as staphylococci facilitates colonisation of catheters and protects them from host defences and from the action of antibiotics.

Clinical Syndromes

Frequent urination, pain or burning when passing urine, fever, dysuria are the most common symptoms of UTI. Five major types of clinical manifestations of UTI are recognised as follows:

Urethritis

The condition manifests as symptoms of dysuria, increased frequency and urgency, which may be associated with sexually transmitted diseases (STDs).

Asymptomatic Bacteriuria

It denotes the isolation of a specific number of bacteria (colony forming units/ml) from an appropriately collected urine specimen from a person without symptoms or signs of UTI. The presence of more than 10^5 organisms per millilitre from a properly collected midstream *'clean-catch'* urine sample indicates significant bacteriuria and infection (Kass concept).

Cystitis

It is an infection of the urinary bladder. Patients with cystitis present with dysuria increased frequency, urgency and suprapubic pain. The urine often becomes grossly cloudy and malodorous and sometimes bloody.

Acute Urethral Syndrome

Symptoms of dysuria, urgency and frequency unaccompanied by significant bacteriuria are termed acute urethral syndrome. The condition primarily occurs in young, and sexually active women.

Acute Pyelonephritis

The condition refers to inflammation of the kidney parenchyma, calyces and pelvis usually caused by bacterial infection. Symptoms have a rapid onset and include fever, shaking chills, nausea, vomiting and diarrhoea.

Catheter-Associated UTIs

Bacteriuria develops in about 10–15% of hospitalised patient with indwelling urethral catheters. The increased prevalence of UTI in catheterised patients is associated with the following factors: female sex, severe underlying illness, catheterisation for prolonged period, detachment of the catheter and drainage tube, other types of faulty catheter care and lack of systematic antimicrobial therapy.

Infection appears as the bacterium goes into the bladder either (1) by intraluminal route, that is, by migrating through the static urine in the catheter lumen or (2) by a periurethral route that is, by moving up the mucous sheath outside the catheter.

The incidence of catheter-associated UTIs can be minimised by many precautionary methods such as (1) inserting catheters only for right indications and for the right duration, (2) following proper aseptic techniques for urinary catheter insertion, (3) following aseptic insertion of the urinary catheter, and maintaining a closed drainage system and (4) implementing proper standard operating procedures and strategies to enhance appropriate use of indwelling catheters.

Laboratory Diagnosis

Specimen

Determining the number and type of bacteria in the urine specimen is an extremely important diagnostic procedure.

Urine is the specimen used for the diagnosis of UTI. Prevention of urine contamination by resident flora is the most important consideration for urine collection of urine specimens; urine can be collected in three ways: (1) midstream urine, (2) catheterised urine and (3) urine collected by suprapubic aspiration.

Midstream urine is the specimen of choice. It is collected in sterile containers, preferably before the administration of antibiotics. **Catheterised urine** is usually indicated in post-operative cases. Urine from a urobag is always avoided. However, catheterised urine is avoided as far as possible due to the inherent risk of introducing infection in hospitalised patients through the catheters. **Suprapubic aspiration** is usually carried out to collect urine from infants and also from adults when uncontaminated urine specimens cannot be obtained by other methods (Table 71.10).

Urine Microscopy

Microscopy of urine from a symptomatic patient can reveal useful information. The centrifuged deposit of urine may show the presence of pus cells, epithelial cells, RBCs and bacteria. A Gram-stained smear of uncentrifuged urine is very useful for assessing microscopic bacteriuria. This is positive in more than 90% of patients whose urine samples show a colony count of at least 10^5/mL. However, Gram stain is not useful to detect bacteria with fewer colony counts (10^2 to 10^4/mL) in urine specimens. Urine microscopy constitutes strong evidence of infection caused by bacteria; however, the non-existence of bacteria

Table 71.10 Collection of urine for diagnosis of UTIs

Specimen	Container	Procedure	Comments
Clean-catch midstream urine	Sterile screw-capped container	*Females:* Cleaning of the area with soap and water, cleansing with water; labia held apart and midstream urine is collected *Males:* Glans, penis cleaned with soap and water, rinsed with water; foreskin is retracted and midstream urine is collected	Non-invasive; should be processed within 2 h or stored at 4 °C or with boric acid as preservative
Straight catheter	Sterile screw-capped container	Urethral area is cleaned with soap and water. Catheter is introduced into bladder; first 15 mL is allowed to pass, the remainder is collected	Invasive; chances of introducing the urethral organism into the bladder
In-dwelling catheter (Foley)	Sterile screw-capped container	After disinfecting catheter collection port by using 70% alcohol, 5–10 mL urine is aspirated using needle and syringe	The catheter tubing should be clamped off above the port to collect freshly passed urine
Suprapubic aspiration	Sterile screw-capped container or anaerobic transporter	Skin is disinfected. Urine is aspirated by needle above the symphysis pubis through the abdominal wall into the full bladder	The specimen should be transported to laboratory immediately at room temperature; the method prevents contamination of urine; used in infants, small children, etc.; invasive

in microscopy need not essentially exclude the diagnosis.

Microscopical enumeration of pus cells in a Neubauer chamber is very sensitive indicator of UTI in symptomatic patients. Pyuria without bacteriuria **(sterile pyuria)** may be due to UTI caused by unusual bacterial agents like *Ureaplasma urealyticum*, *Chlamydia trachomatis* and *M. tuberculosis* or fungi. Also, sterile pyuria may be explained in non-infectious urological conditions such as anatomic abnormality, calculi, interstitial nephritis, nephrocalcinosis, vesicoureteral reflux or polycystic disease.

Screening Tests

These tests demonstrate the causative bacterial enzymes or breakdown products of these bacterial enzymes in urine specimens. These tests also detect the polymorphonuclear enzyme in the urine to detect bacteriuria or pyuria.

Griess nitrate test is used to detect urinary nitrite, an indicator of UTI. Nitrite is a breakdown product of nitrate due to an enzyme nitrate reductase produced by bacteria present in urine. Similarly, the leucocyte esterase test demonstrates leucocyte esterase, an enzyme produced by inflammatory cells that can be detected by the 'dipstick' method. In this test, the presence of catalase in urine is demonstrated by the formation of bubbles on the addition of hydrogen peroxide, indicating the presence of catalase-producing bacteria in the urine. Triphenyl tetrazolium chloride test (TTC) and glucose test paper are other tests used for indirect evidence of UTI.

Culture

More than 95% of UTIs are caused by a single bacterial species. Hence, a culture that shows mixed bacterial species is contaminated and, therefore, needs to be repeated. The presence of squamous epithelial cells in the urine specimen is a good indicator of contamination. **Polymicrobial infections**, however, may occur in certain clinical conditions that include (i) long-term catheterisation, (ii) incomplete bladder emptying because of neurologic dysfunction and (iii) the presence of a fistula between the urinary tract and the gastrointestinal or genital tract. A urine culture can be carried out by semi-quantitative or quantitative methods.

Semiquantitative Methods

The **calibrated loop method** is a simple, more rapid, less expensive and fairly satisfactory semi-quantitative method of urine culture. In this method, a standardised, calibrated, 4-mm platinum or chrome wire loop is used to deliver 0.005 mL of uncentrifuged urine. The fixed volume of uncentrifuged urine is inoculated onto the blood agar and McConkey agar or CLED agar (cystine-lactose-electrolyte-deficient agar) by this calibrated loop. The media after inoculation are incubated at 37 °C for 24 h, after which the total number of colonies formed on the medium is counted. The total bacterial count per millilitre of urine is obtained by multiplying the number of colonies by 200.

Kass in 1956 defined the criterion for determination of active bacterial infection of the urinary tract by suggesting the concept of **significant bacteriuria** as follows:

1. The urine specimen showing more than 10^5 bacteria/mL is considered significant bacteriuria and suggests an active bacterial infection of the urinary tract. The term *significant bacteriuria*, however, is applicable only for (1) midstream urine specimen, (2) Gram-negative bacteria and (3) for the urine specimen collected prior to antibiotics treatment. Isolation of even a few Gram-positive bacteria such as staphylococci in urine is significant.
2. The urine specimen showing colony count between 10^4 and 10^5 bacteria/mL is considered as of doubtful significance. Hence, the culture of the same urine specimen is repeated.
3. The urine specimen showing colony count less than 10^4 is considered insignificant. A urine specimen showing three or more types of colonies with no predominating organism is considered a contaminant.

Quantitative Method

The quantitative method is performed by **the pour plate method** to count an exact number of bacteria present in 1 mL of urine. This method, however, is not used for routine diagnosis of UTI. Identification of bacteria is carried out by colony character, Gram stain, motility, biochemical reactions and serological tests.

Treatment

Treatment of UTI is based on antimicrobial therapy. Cystitis usually responds within 3 days of treatment, uncomplicated pyelonephritis within 2 weeks of treatment and complicated infections within 6 weeks of therapy. Intravenous antibiotics may be required in seriously ill patients, but oral drugs are mostly effective. Nitrofurantoin, norfloxacin, cephalosporins and aminoglycosides are the commonly used drugs for community-acquired UTIs. Higher antibiotics are used for hospital-acquired and complicated UTIs.

Meningitis

Infection of the central nervous system (CNS) is one of the most life-threatening medical emergencies and hence needs accurate, rapid diagnosis and prompt treatment.

The brain and spinal cord are protected from mechanical pressure and infection externally by the rigid skull and vertebra column and internally by meninges. The meninges consist of three layers: (1) dura mater (outermost covering), (2) arachnoid and (3) pia mater (innermost covering). They enclose the epidural, subdural and subarachnoid spaces. The cerebrospinal fluid (CSF) flows in the brain and spinal cord and fills the subarachnoid space and ventricles.

Definition

Meningitis

Menigitis is the infection and inflammation of the brain and spinal cord within the subarachnoid space or throughout the leptomeninges.

Acute Encephalitis Syndrome (AES)

This is a group of clinically similar neurologic manifestations caused by several pathogens such as viruses, bacteria, fungus, parasites, spirochetes, chemicals/toxins, etc.

Types of Meningitis

Meningitis can be acute, sub-acute or chronic, depending on the clinical presentation.

Acute Meningitis

This is rapid in onset, and according to the host response to the causative organism, meningitis can be pyogenic or aseptic.

Acute pyogenic or purulent meningitis is characterised by the presence of acute inflammatory exudates with an increased count of polymorphonuclear cells in the CSF. The causative agents of acute pyogenic meningitis depend on the host's age and other predisposing factors. This condition is caused mostly by bacteria (Table 71.11). *Neisseria meningitidis*, *S. pneumoniae* and *Pseudomonas aeruginosa* are the usual pathogens responsible for causing meningitis in immunocompromised hosts and *S. pneumoniae*, *H. influenzae* and group A beta-haemolytic streptococci in individuals with a basilar skull fracture. *S. aureus* and coagulase-negative staphylococci, *P. aeruginosa*, *S. marcescens*, *Salmonella* spp., etc., are associated with meningitis in patients with head injury or following post-neurosurgery.

Aseptic meningitis is characterised by the presence of high lymphocyte count and of other mononuclear cells in the CSF without the presence of either bacteria or fungi on culture. It is mostly caused by viruses and occasionally by other pathogens (Table 71.12).

Sub-acute Meningitis

Patients with sub-acute meningitis typically suffer from continuous headache, nuchal rigidity, low-grade fever and restlessness for prolonged periods before being diagnosed in hospital. The causative agents are listed in Table 71.13.

Chronic Meningitis

It is characterised by the presence of symptoms and signs of meningitis for at least 4 weeks. Table 71.14 lists the causative agents of chronic meningitis.

Table 71.11 Causative agents of acute pyogenic meningitis

Age of the infected host	Common causes	Less common causes
Neonates or infants of 2 months or less	Gram-negative bacilli (*Escherichia coli* K strain and *Klebsiella pneumoniae*) Group B streptococcus (*Streptococcus agalactiae*)	*Listeria monocytogenes* and *Chryseobacterium meningosepticum*
Preschool children up to 5 years	*Haemophilus influenzae*, *Neisseria meningitidis* and *Streptococcus pneumoniae*	*Mycobacterium tuberculosis*
Older children and adults	*Haemophilus influenzae*, *Neisseria meningitidis* and *Streptococcus pneumoniae*	*Listeria monocytogenes*, *Haemophilus influenzae* and other aerobic Gram-negative bacilli

Table 71.12 Causative agents of aseptic meningitis

Pathogens	Organisms
Viruses	
Common viral agents	Enteroviruses (polioviruses, echovirus, Coxsackie viruses), arboviruses, HIV and herpes simplex virus 2 (HSV-2)
Less common viral agents	Herpes simplex virus 1 (HSV-1) and varicella zoster virus
Rare viral agents	Adenoviruses, cytomegalovirus, Epstein–Barr virus, influenza A and B, parainfluenza, mumps and rubella viruses
Bacteria	*Mycobacterium tuberculosis*, *Treponema pallidum*, *Leptospira interrogans* serovars, *Leptospira icterohaemorrhagiae* and *Leptospira canicola*
Parasites	*Naegleria* spp., *Acanthamoeba* spp. and *Toxoplasma gondii*
Fungi	*Cryptococcus neoformans*

Table 71.13 Causative agents of sub-acute meningitis

Mycobacterium tuberculosis
Treponema pallidum
Cryptococcus neoformans
Histoplasma capsulatum
Coccidioides immitis

Table 71.14 Causative agents of chronic meningitis

Pathogens	Organisms
Bacteria	
Common bacterial agents	*Mycobacterium tuberculosis, Borrelia burgdorferi* and *Treponema pallidum*
Uncommon bacterial agents	*Actinomyces* spp., and *Tropheryma whipplei*
Rare bacterial agents	*Leptospira* spp.
Fungi	
Common fungal agents	*Cryptococcus neoformans, Coccidioides immitis, Candida* spp., *Histoplasma capsulatum* and *Sporothrix schenckii*
Rare fungal agents	*Cladosporium curvularia, Drechslera mucor* and *Pseudoallesheria boydii*
Parasites	*Toxoplasma gondii, Trypanosoma gambiense, Trypanosoma rhodesiense, Acanthamoeba* spp., *Cysticercuscellulosae, Gnathostoma spinigerum* and *Angiostrongylus cantonensis*
Viruses	Mumps, echoviruses, HIV and herpes simplex virus

Pathogenesis of Meningitis

The blood–brain barrier and blood–CSF barrier are natural barriers to invasion by the organism. Microorganisms causing infection of the CNS overcome these barriers by their virulence factors, such as capsule, pili, lipopolysaccharide, teichoic acid, peptidoglycans, IgA proteases, etc. The organism can invade the CNS by various routes. Haematogenous spread is most common. The main invasion routes are the blood vessels that traverse the skull and vertebral column.

The respiratory tract is the primary portal of entry of most CNS pathogens. Microorganisms initially colonise the nasopharynx and attach to nasopharyngeal epithelial cells. Bacteria traverse across the epithelial cells into the bloodstream. They evade the neutrophils and classic complement-mediated bactericidal activity by their polysaccharide capsule and the IgA proteases.

Blood-borne infection takes place through the blood–brain and blood–CSF barrier. Microorganisms spread directly from infected sites, such as otitis media, sinusitis, mastoiditis, etc., causing meningitis. Anatomical defects such as surgery, trauma and congenital defects lead to easy access of microorganisms to the CNS. Retrograde spread along the peripheral sensory nerves can rarely occur, such as in rabies virus infection.

Microorganisms reach the intraventricular choroid plexus either by infecting the choroid plexus epithelial cells and thus reaching the CSF or else by adherence to the cerebral capillary endothelial cells. Subsequently, they traverse between these cells to reach the CSF. The normal CSF having only a few white blood cells and trace amounts of complement proteins and immunoglobulins provide an ideal niche for the bacteria to multiply rapidly. This is because the

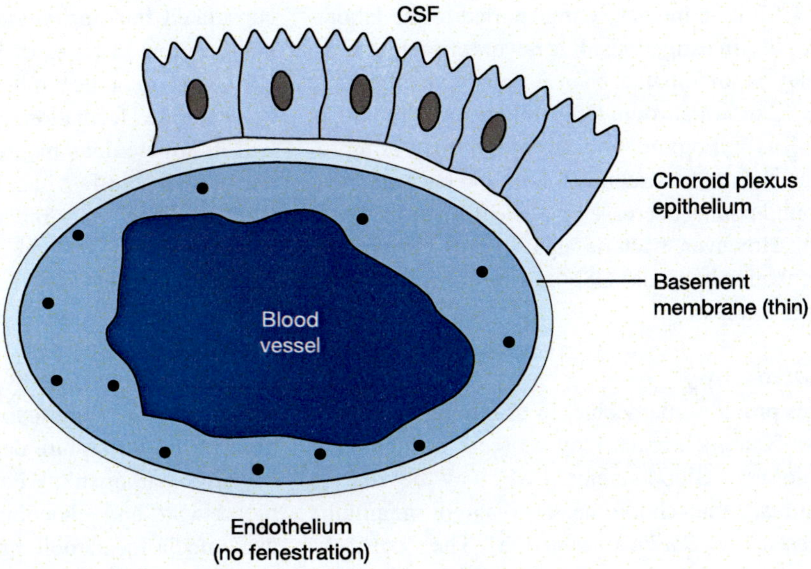

Fig. 71.4 Blood–CSF barrier

CSF does not support phagocytosis to a great extent (Fig. 71.4).

Clinical Syndromes

CNS manifestations are the main manifestations. Fever, stiff neck, headache, nausea, vomiting and neurological abnormalities are frequently associated with changes in mental states. Seizures are more common in children. Cerebral oedema, hydrocephalus and cerebral herniation are some known complications. Permanent deafness is another complication that occurs in some cases of meningitis following *H. influenzae* infection.

Acute Encephalitis Syndrome (AES)

AES is a group of clinically similar neurologic manifestations caused by various pathogens such as viruses, bacteria, fungus, parasites, spirochetes and also by chemicals/toxins. Japanese encephalitis (JE) is an important member of this group. Young children eating unripe litchi fruit on empty stomachs develop serious manifestations. Unripe fruit contains the toxins hypoglycin A and methylene cyclopropyl glycine (MCPG), which are poisonous. This condition was reported recently from India in Odisha.

Laboratory Diagnosis

Specimen

CSF is the specimen of choice. The CSF is obtained by subdural tap, ventricular aspiration or lumbar puncture. In the lumbar puncture procedure, the patient lies in a lateral position with knees flexed and back arched to make the lumbar vertebrae separated and prominent. Surgically draping is done, and the space between L3 and L4 vertebrae is located by palpation with a gloved finger after disinfecting the area. The lumbar puncture needle is carefully aimed between the spinous processes, through the interspinous ligaments into the spinal canal. CSF is allowed to free flow and is collected in a sterile container. A maximum of 5 mL of CSF is collected in sterile and dry containers. The sample is then divided into three aliquots, and one each is sent for pathology, microbiology and biochemistry investigations.

CSF is immediately transported to the laboratory for investigations. It is never refrigerated; if a delay is anticipated, then it can be incubated at 37 °C or left at room temperature for at most 1 h. CSF is hypotonic, and any delay may compromise the demonstration of cells and bacteria; neutrophil counts decrease by 32% in 1 h and 50% in 2 h. However, bacteria also do not survive long incubation periods; CSF can be refrigerated for viral studies.

Macroscopy

This provides many clues in presumptive diagnosis. Xanthochromia may suggest a tubercular aetiology. Blood-stained CSF may be due to trauma, subarachnoid haemorrhage or meningitis caused by *Bacillus anthracis*. The **'cobweb'** coagulum, a fibrin web in the CSF, is seen on standing in case of tubercular meningitis.

Microscopy

The microscopic examination of a direct wet mount of the CSF is useful to demonstrate free-living amoebae. The supernatant obtained after centrifugation of the CSF sample is kept separately in a sterile tube. The smear is made from centrifuged sediment and is stained by Gram stain. Gram staining is one of the simple, rapid, accurate and inexpensive methods for diagnosing meningitis. **Gram staining** is reliable when the bacterial count is more than 10^5 colony-forming units (CFU)/mL of CSF. Gram staining shows varying sensitivity for different pathogens. The sensitivity of the staining is improved by centrifuging the specimen at 1500–2000 *g* or by cytospin (Table 71.15). **Acridine orange staining** is found to be slightly more sensitive than Gram staining. **Modified India ink preparation** of CSF is useful for demonstrating capsulated budding yeast cells of *Cryptococcus neoformans*. In cases of tuberculous meningitis, CSF kept in a test tube for 10–12 h leads to cobweb-like formation. **Acid-fast staining** of CSF in tuberculous meningitis reveals long slender beaded acid-fast bacilli.

Cytological and biochemical parameters in normal individuals and in infected cases with meningitis are listed in Table 71.16.

Culture

Isolation of bacteria by the **culture of CSF confirms the diagnosis of meningitis**. The media used for culture depends on the suspected bacterial pathogen to be isolated. 0.5% sheep blood agar and chocolate agar are commonly used media for aerobic bacteria culture. Cultures should be incubated at 37 °C at least for 72 h in the presence of 5–10% CO_2 before being reported negative. Identification is made by colony morphology, Gram staining, biochemical reaction, etc. Mycobacterial culture is done on an LJ medium. The specimen is cultured on SDA agar for the isolation of fungi. **Blood culture** is useful for diagnosing meningitis caused by *H. influenzae*, *N. meningitidis* and *S. pneumoniae*.

Direct Detection of Antigen

Direct detection of antigen in the CSF, serum or urine is useful for diagnosis of meningitis. Latex agglutination test is used to detect antigens of *S. pneumoniae, N. meningitidis, H. influenzae* type b, group B streptococcus, *E. coli* K1 strains and *C. neoformans* in the CSF. The test is useful for rapid diagnosis of meningitis, particularly in situations where antibiotic treatment has already been given or where bacteriological culture and staining are negative. Major drawbacks of the test are cross-reactions and false-positive reactions

Table 71.15 Sensitivity of Gram stain of CSF smears in meningitis caused by bacterial pathogens

Bacteria causing meningitis	Sensitivity of Gram stain (%)
Streptococcus pneumoniae	90
Haemophilus influenzae	86
Neisseria meningitidis	75
Gram-negative bacilli	50
Listeria monocytogenes	<50

Table 71.16 Cytological and biochemical parameters in normal individuals and in patients with meningitis

Characteristic	Normal individual	Pyogenic meningitis patients	Tuberculous meningitis patients	Viral meningitis patients
Physical characteristic				
Opening pressure (mmHg)	Normal	Highly elevated	Moderately elevated	Slightly elevated
Cytological parameters				
Total leukocyte count (per mm^3)	1–3	1000–20,000	50–500	10–500
Predominant cell type	Lymphocytes	Neutrophils (90–95%)	Lymphocytes (90%)	Lymphocytes
Biochemical parameters				
Glucose (mg%)	40–80	10–20 (decreased or absent)	30–50 (decreased)	Normal
Total proteins (mg%)	30–45	100–600 (highly increased)	80–120 (moderately increased)	60–80 (slightly decreased)
Bacteriological parameters				
Direct microscopy				
Gram staining *Acid-fast staining* *Culture*		Gram-positive or gram-negative bacteria can be seen Specific organism may grow	Acid-fast bacilli may be seen Growth of *Mycobacterium tuberculosis* may be seen on LJ medium	Cell may reveal viruses

due to the presence of rheumatoid factor, blood or high protein concentration in CSF.

Limulus Amoebocyte Lysate

The test is very sensitive for detecting circulating endotoxin in the CSF. Serological tests for detecting serum antibodies are not useful for diagnosing meningitis caused by bacteria or fungi but are useful for viruses.

Molecular Diagnosis

Molecular assays such as PCR offer an alternative in diagnosing central nervous system infections. They are useful for detecting viral agents of meningitis.

Multiplex polymerase chain reaction such as Film Array Meningitis/Encephalitis (FA-M/E) panel is a molecular tool which allows the simultaneous detection of 14 pathogens (bacteria, viruses and one fungus). The entire process is fully automated and has a turn-around time of only 1 h. **GeneXpert assay** is a cartridge-based real-time PCR that can be used to detect *M. tuberculosis* and enteroviral meningitis (Xpert EV). The turnaround time of this test is 2.5 h.

Treatment

Treatment of meningitis is based on antimicrobial therapy, which depends on the microorganism responsible for the condition. For example, cefuroxime/ceftriaxone or cefotaxime or ampicillin/sulbactam in combination with azithromycin or quinolone are recommended to treat meningitis caused by *S. pneumoniae*.

Bloodstream Infections

Invasion of the bloodstream by microorganisms is one of the most serious situations in infectious diseases and if not treated fast, can lead to shock, multi-organ failure, disseminated intravascular coagulation and even death. Invasion of the bloodstream by microorganisms may result in various clinical conditions.

Definitions

Bacteraemia
It is the presence of bacteria in the bloodstream without evidence of multiplication, as evidenced by positive blood cultures.

Septicaemia
It is a clinical syndrome in which bacteria circulate and actively multiply in the bloodstream.

Septic Shock
It denotes sepsis with hypotension (arterial blood pressure less than 90 mmHg systolic or 40 mmHg less than the patient's normal blood pressure) for at least 1 h despite adequate fluid resuscitation.

Multiple-organ Dysfunction Syndrome
It is a condition of dysfunction of more than one organ, requiring intervention to maintain homeostasis.

Endotoxaemia
It is a condition of circulating bacterial endotoxin in the blood.

Types of Blood-Stream Infections

The causative agents of septicaemia are listed in Table 71.17.

Table 71.17 Causative agents of septicaemia

Gram-negative bacilli
Salmonella Typhi; *Salmonella* Paratyphi A, B, C
Haemophilus influenzae
Brucella spp.
Proteus spp.
Escherichia coli
Klebsiella pneumoniae
Pseudomonas spp.
Enterobacter spp.
Bacteroides spp.
Gram-positive cocci
Streptococcus pneumoniae
Staphylococcus aureus
Staphylococcus epidermidis
Streptococcus pyogenes
Gram-negative cocci
Neisseria meningitidis
Gram-positive bacilli
Listeria monocytogenes

There are two major classes of bloodstream infections: intravascular and extravascular. **Intravascular infections** originate within the cardiovascular system. Examples are infective endocarditis, intravenous catheter-associated bacteraemia, etc. **Extravascular infections** are caused by the bacteria reaching the bloodstream via the lymphatics from another focus of infection.

The vascular compartment is normally sterile and usually intact, and the entry of bacteria is facilitated by some damage to the blood vessels adjacent to the skin or mucosal surfaces. Subsequently, the bacteria are transported by phagocytic cells into capillaries or lymphatic systems. The most probable portals of entry are the genitourinary system, respiratory tract, surgical wound infection, etc. The introduction of contaminated material directly into the vascular system also induces blood-borne infections.

Clinical Syndromes

Clinical manifestations of bloodstream infections are highly variable. These include fever, chills and sweats, hyperventilation and resultant respiratory alkalosis. **Septic shock** presents as fever, acute respiratory distress and renal failure. It is mediated by cytokines, such as interleukins and tumour necrosis factors. Gram-negative bacteria produce endotoxin, the lipopolysaccharide, which causes septic shock responses. Lipid A, the core component of the lipopolysaccharide, is responsible for many systemic disturbances like fever, activation of complement and certain clotting factors. Gram-positive bacteria produce a similar response mediated by the exotoxins.

Laboratory Diagnosis

Laboratory diagnosis depends on the demonstration of the causative pathogen in the blood by culture. This is also supplemented by non-blood culture tests, including serological tests in certain microbial infections.

Blood Culture

Isolation of the causative microorganism from blood cultures is critical in diagnosing and treating bloodstream infections. Blood culture can be carried out by either manual or automated methods.

Manual Blood Culture

These systems rely on the use of appropriate blood culture bottles containing enrichment broth BHI (monophasic) or containing both BHI broth and an agar slope (biphasic). These bottles are placed in a conventional static incubator and inspected daily to detect visible signs of bacterial or fungal growth.

In the case of **infective endocarditis**, in the absence of prior antibiotic therapy, a total of 2–3 blood culture sets, 10 mL each, obtained from different venipuncture sites over 24 hours. They are inoculated in blood culture bottles containing glucose broth. Each blood culture set consists of one aerobic and one anaerobic blood culture each. Blood is diluted in the ratio of one: 5 in the enrichment broth for optimum recovery of microorganisms. Cultures are incubated for at least 3 weeks, with subcultures done on blood agar and MacConkey agar after 24 h, 48 h and once a week thereafter. The inoculated media are incubated at 37 °C for 24 h. Isolated organisms are identified by colony morphology, Gram staining, biochemical properties and serological tests.

Automated Blood Culture

In these systems, the carbon dioxide production by microorganisms in the culture bottles is continuously monitored, either by colourimetric, fluorescent methods or else by detection of O_2 consumption and/or CO_2, H_2, and/or N_2 production (e.g. equipment such as BACTEC, BacT/ALERT, Versa TREK System). During incubation, the bottles are continuously shaken in the automated equipment. The current automated systems have a much greater sensitivity compared to the manual systems. These systems have largely replaced manual blood culture systems in healthcare facilities.

Non-blood Culture Tests

These include serological tests to detect bloodstream infections are used to identify certain bacteria, such as *Brucella*, *Bartonella*, *C. burnetii* and *Legionella*, that are hard to be detected by using the traditional blood culture methods. *Tropheryma whipplei* can be detected in vegetation with the help of a microscopic examination using periodic acid–Schiff stain.

Other Tests

Other tests include chest radiography, electrocardiography, complete blood counts and creatinine measurement. These tests also include tests for estimation of erythrocyte sedimentation rate, C-reactive protein level, circulating immune complex titre and rheumatoid factor.

Laboratory Diagnosis of Catheter-Related Blood Stream Infection (CRBSI)

Differential Quantitative Cultures

Two blood cultures—one from a peripheral site and the other from the suspected infected line is performed to diagnose an IV catheter-related bacteremia. A colony count ratio greater than 4 to 10: 1 between the central venous blood and a peripheral blood specimen indicates a probable CRBSI.

Treatment

Treatment by appropriate antibiotic forms the mainstay in treatment of bloodstream infections. In infective endocarditis, it is rather difficult to clear most of the bacteria present in the avascular vegetation, as this site is comparatively unreachable for the host's immune response. Additionally, the bacteria are metabolically inactive at this site and show no growth. Thus, prolonged bactericidal therapy most of the bacteria causing infectiod infected individuals.

Diarrhoeal Diseases and Food Poisoning

Foodborne diseases are caused by food contamination and occur at any stage of the food production, delivery and consumption chain. There are more than 250 known foodborne diseases. A majority of them are due to microorganisms including parasites. Other foodborne diseases are essentially poisonings caused by toxins and chemicals contaminating the food.

Definitions

Diarrhoea

It is the passage of abnormal liquid or unformed stools at an increased frequency. The condition is identified as acute if it persists for less than 2 weeks, persistent if it persists for 2–4 weeks and chronic if present for more than 4 weeks.

Dysentery

Dysentery means the passage of blood and mucous with faeces, often associated with tenesmus.

Gastroenteritis

It denotes a syndrome where inflammation of the gastric mucosa and intestinal mucosa results in nausea, vomiting, diarrhoea and abdominal pain.

Enterocolitis

The condition denotes an inflammatory condition involving the intestinal mucosa, both small and large intestine.

Types

According to CDC, US; there are six important foodborne diseases, known as 'Big 6'—*Salmonella*, *Salmonella* Typhi, *Shigella*, Escherichia coli, Norovirus, and Hepatitis A. Globally, there are nearly 1.7 billion cases of childhood diarrhoeal disease every year. Each year diarrhoea kills around 525 000 children under five. About 250 pathogens can cause diarrhoea. The most common ones are bacteria (e.g. ETEC, *Vibrio cholerae*, *Shigella* spp., *Campylobacter* spp.), viruses (Rotavirus, Norwalk virus, Calicivirus), parasites (*Entamoeba histolytica*, *Giardia intestinalis*, *Balantidium coli*, *Cryptosporidium*, spp. *Strongyloides stercoralis*) and fungi (*Candida* spp.).

Clinical Syndromes

Acute Infective Diarrhoea

Acute infective diarrhoea is most commonly infectious. Risk groups include travellers, consumers of certain foods such as chickens, fried rice, etc., immunodeficient persons, daycare participants and their family members and institutionalised persons. Toxins and invasive pathogens cause infectious acute diarrhoea, whereas non-infectious diarrhoea is caused by antibiotics and many other substances (Table 71.18).

Chronic Infective Diarrhoea

This is caused by *Giardia intestinalis*, *Cryptosporidium parvum*, *Cyclospora cayetanensis*, *Cystoisospora belli*, *Mycobacterium avium-intracellulare* complex, enteroadherent *E. coli*, *Campylobacter* and *Candida* spp.

Table 71.18 Causative agents of acute diarrhoea

Pathogenic mechanism	Pathogen
Infectious causes	
Production of preformed toxins	*Bacillus cereus, Staphylococcus aureus* and *Clostridium perfringens*
Enterotoxins	*Vibrio cholerae*; enterotoxigenic *Escherichia coli*; *Klebsiella pneumoniae*; *Aeromonas* spp.; enteropathogenic and enteroadherent *Escherichia coli*; *Giardia intestinalis*; *Cryptosporidium parvum*; and helminths
Cytotoxins	*Clostridium difficile* and haemorrhagic *Escherichia coli*
Neurotoxins	*Staphylococcus aureus, Bacillus cereus, Clostridium botulinum*
Invasive organisms	*Vibrio parahaemolyticus*, enteroinvasive *Escherichia coli, Salmonella, Campylobacter* spp., *Aeromonas* spp., *Yersinia* spp., *Shigella* spp., *Entamoeba histolytica*
Adherence	Enteropathogenic *Escherichia coli*, Enterohemorrhagic *Escherichia coli*, *Cryptosporidium parvum, Isospora belli*, rotavirus, hepatitis A, B, C, Norwalk virus
Non-infectious causes	Antibiotics, antihypertensives, non-steroidal anti-inflammatory drugs (NSAIDs), certain antidepressants, chemotherapeutic agents, bronchodilators, antacids and laxatives

Traveller's Diarrhoea

Individuals who travel into developing geographic areas with poor sanitation are at particularly high risk for developing diarrhoea. In areas with poor sanitation, enteric pathogens heavily contaminate the water and food. Enterotoxigenic *E. coli* is the most common cause of traveller's diarrhoea.

Gastroenteritis

It is the inflammation of the stomach and intestines, mainly caused by bacterial toxins or viral infection and causing emesis and diarrhoeal disease (Fig. 71.5).

Food Poisoning

Bacterial food poisoning is an acute condition caused by ingestion of food or drinking contaminated with either living bacteria or toxins. It can be (i) infective, (ii) toxic and (iii) intermediate or infective-toxic type (Table 71.19, Fig. 71.5).

Infective Food Poisoning

The condition is caused by invasive bacteria such as *Salmonella* spp (*Salmonella* Typhimurium, *Salmonella* Enteritidis, *Salmonella* Heidelberg, *Salmonella* Indiana, *Salmonella* Newport and *Salmonella* Dublin), *Vibrio parahaemolyticus*

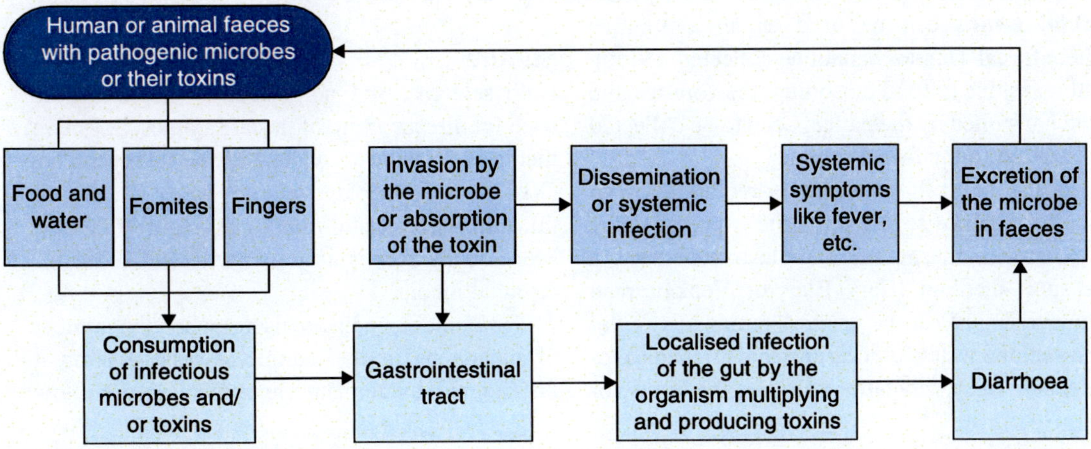

Fig. 71.5 Pathogenesis of diarrhoea

Table 71.19 Bacteria causing food poisoning

Toxic type (Incubation period <6 h)	Intermediate type (Incubation period 8–16 h)	Infective type (Incubation period >16 h)
Staphylococcus aureus	*Clostridium perfringens*	*Escherichia coli*: Enterotoxigenic, enterohaemorrhagic
Bacillus cereus (emetic type)	*Bacillus cereus* (diarrhoeal type)	*Salmonella* spp.: *Salmonella* Enteritidis, *Salmonella* Typhimurium
Clostridium botulinum		*Vibrio parahaemolyticus* *Shigella* spp. *Campylobacter jejuni*

and *Campylobacter jejuni*. The multiplication of these invasive bacteria occurs *in vivo*, followed by an invasion of the intestinal mucosa.

Toxic Food Poisoning

It is caused by preformed toxins of *S. aureus*, *Bacillus cereus* and *Clostridium botulinum*. This type of food poisoning occurs following ingestion of the food contaminated with preformed toxins of these bacteria.

Intermediate or Infective-toxic Food Poisoning

It is caused by *Clostridium perfringens*. In this type of food poisoning, the bacteria secrete toxin in the intestine.

Laboratory Diagnosis

Specimens

A wide-mouthed, sterile screw-capped container is used to collect **faecal sample** for diagnosis. **Rectal swabs** can be used as an alternative where faecal samples cannot be collected. During outbreaks due to food poisoning, the vomitus and food consumed by the patient need to be collected in addition to the faecal sample.

Ideally, a fresh specimen is required for an accurate diagnosis. In case of the expected delay or wherever transport of sample is necessary, a transport medium (Cary–Blair or Venkatraman Ramakrishnan) can be used. Alternatively, alkaline peptone water (used as enrichment media) for detection of *V. cholerae* can also be used for transport. The specimen must be stored in a thermal box maintained at 4 °C or transported in dry ice.

Microscopy

Wet mount preparation, both saline and iodine wet mounts of fresh faecal samples are examined microscopically for demonstrating ova or cysts and trophozoites of parasites, presence of any motile organisms, and pus cells if any. **Hanging drop preparation** is a well-established technique to examine live, motile microbes. For example, darting motility can be observed for *V. cholerae*. It can be further confirmed by adding specific antiserum so that the motility is inhibited.

Gram staining is of little use for faecal samples, but used to detect curved bacilli indicative of *V. cholerae*. The coccidian parasites (*Cryptosporidium* spp., *Cyclospora* spp. and *Cystoisospora* spp.) can be demonstrated by **modified acid-fast staining**. Apart from conventional microscopic techniques, **electron microscopy** is used for the detection of certain viruses.

Culture

Both selective and non-selective media can be used for direct culture of faecal samples. Selective media include xylose-lysine-deoxycholate (XLD), deoxycholate citrate agar (DCA) or thiosulphate citrate bile salt agar (TCBS). MacConkey agar is used as non-selective or differential media.

Enrichment media are useful when the number of pathogens in the sample is less, making it difficult to isolate. The enrichment media allow selective growth of the specific pathogen. The fresh faecal specimen is directly inoculated in a liquid media (alkaline peptone water for 6–8 h; selenite F broth for 12–18 h) and then sub-cultured in a solid medium. The pathogen is then identified by serotyping or classical biochemical tests, followed by antibiotic sensitivity testing.

Cell Culture

It is relatively more helpful in detecting viruses, as tissue culture is extensively used for culturing of viruses. Some bacteria can also be detected by tissue culture. The cytotoxic effect of EHEC verotoxin (VT) on vero cells and penetration of HeLa and Hep-2 cells by ETEC can be demonstrated by tissue culture.

Serology

ELISA is the most widely used serological test for detecting *E. coli* O157:H7 (EHEC), Shiga toxin and *Clostridium difficile* toxins. Faecal antigen detection ELISA has recently gained importance for detecting protozoal parasites. For example, *Entamoeba histolytica* can be detected from the direct faecal specimen by Techlab II stool antigen detection ELISA kit. Certain viruses (Rotavirus) also can be detected.

Immunodiffusion, ELISA, latex agglutination tests and neutralisation are different methods used for detecting enterotoxin.

Molecular Detection

Conventional PCR and real-time PCR-based pathogen detection is gradually gaining importance globally. Molecular methods can detect a wide range of bacteria, parasites and viruses.

Treatment

Management of infective diarrhoea is based mostly on symptomatic management and treatment by antibiotics whenever indicated. The replacement of electrolytes is essential and life-saving in preventing dehydration and hypovolemic shock. Protozoal dysentery may be treated with inidazole/metronidazole. Bacillary dysentery/diarrhoeas require antibiotic therapy. Third-generation cephalosporins, ciprofloxacin and tetracyclines are frequently used. Viral diarrhoeas do not need any antibiotic therapy.

Sexually Transmitted Infections and Reproductive Tract Infections

Sexually transmitted diseases are many of the various communicable diseases that can be transmitted primarily by sexual contact. These include the diseases such as syphilis, gonorrhoea, chlamydiasis and genital herpes, as well as hepatitis B and AIDS that are often contracted by non-sexual means.

Reproductive tract infections as per the CDC (2003) guidelines include three types of infections: (1) sexually transmitted diseases like chlamydial infections, gonorrhoea, chancroid and human immunodeficiency virus (HIV) infections; (2) endogenous infections, due to overgrowth of commensal organisms of the genital tract of healthy women, such as bacterial vaginosis or vulvovaginal candidiasis and (3) iatrogenic infections, arising out of improperly performed medical procedures such as illegal abortion or poor delivery practices.

Definitions

Sexually Transmitted Infection (STI)

The condition denotes an infection with any organism capable of causing STD irrespective of the presence of disease, for example, asymptomatic gonorrhoea in females and early stages of HIV infection.

Sexually Transmissible Disease (STD)

The condition denotes a condition when the pathogen may not qualify as a full-fledged sexually transmitted pathogen because non-sexual routes of infection predominate. The condition includes bacterial vaginosis, candidiasis, trichomoniasis, genital ulcers with pain and lymphadenopathy, chancre, genital herpes, chancroid, inflammation, etc.

Table 71.20 Causative agents of sexually transmissible diseases (STDs)

Transmitted in adults predominantly by sexual intercourse
Bacteria
Neisseria gonorrhoea *Chlamydia trachomatis* *Treponema pallidum* *Haemophilus ducreyi* *Klebsiella granulomatis* *Ureaplasma urealyticum*
Viruses
Human immunodeficiency virus (types 1 and 2) Human T-cell lymphotropic virus type I HSV type 2 Human papillomavirus (multiple genotypes) Hepatitis B virus Molluscum contagiosum virus
Parasites, ectoparasites
Trichomonas vaginalis *Phthirus pubis*
Sexual transmission repeatedly described but not well defined or not the predominant mode
Bacteria
Mycoplasma hominis *Mycoplasma genitalium* *Gardnerella vaginalis* Group B *Streptococcus* *Mobiluncus* spp.
Viruses
Cytomegalovirus Human T-cell lymphotropic virus type II HSV type 1 Kaposi sarcoma-associated herpes virus Transfusion-transmitted virus (TTV)

Reproductive Tract Infections (RTI)

The condition denotes infections of the genital and the reproductive tract caused primarily by sexually transmitted organisms. However, they may also result from endogenous infection due to the overgrowth of organisms normally present in the genital tract of healthy individuals. The condition may also be iatrogenic following the instrumentation or improperly performed medical procedures such as unhygienic delivery practice or unsafe abortion procedures.

Types

The causative agents of STI and RTI are summarised in Table 71.20. STDs can be differentiated into two types: (i) those causing ulcerative lesions and (ii) those causing non-ulcerative lesions.

Infections with Ulcerative Genital Ulcers

These lesions are caused by *Treponema pallidum*, *Haemophilus ducreyi*, *Calymmatobacterium granulomatis*, *C. trachomatis* (LGV, or lymphogranuloma venereum strain), and herpes simplex virus (HSV). The comparative features of these conditions are summarised in Table 71.21.

Infection with Non-Ulcerative Lesions

These can be classified according to anatomical sites involved in men and women.

Men

The infections include urethritis caused by *N. gonorrhoeae*; non-gonococcal urethritis by

Table 71.21 Comparative features of syphilis, herpes, chancroid, lymphogranuloma venereum, donovanosis associated with ulcerative genital ulcers

Feature	Syphilis	Herpes	Chancroid	Lymphogranuloma Venereum	Donovanosis
Causative agent	*Treponema pallidum*	HSV type 2	*Haemophilus ducreyi*	*Chlamydia trachomatis*	*Calymmatobacterium granulomatis*
Incubation period	9–90 days	2–7 days	1–14 days	3 days to 6 weeks	1–4 weeks
Early primary lesions	Papule	Vesicle	Pustule	Papule, pustule or vesicle	Papule
No. of lesions	Usually one	Multiple, may coalesce	Usually multiple, may coalesce	Usually one	Variable
Diameter	5–15 mm	1–2 mm	Variable	2–10 mm	Variable
Edges	Sharply demarcated, elevated, round or oval	Erythematous	Undermined, ragged, irregular	Elevated, round or oval	Elevated, irregular
Depth	Superficial or deep	Superficial	Excavated	Superficial or deep	Elevated
Base	Smooth, non-purulent, relatively non-vascular	Serous, erythematous, non-vascular	Purulent, bleeds easily	Variable, non-vascular	Red and velvety, bleeds readily
Induration	Firm	None	Soft	Occasionally	Firm
Pain	Uncommon	Frequently tender	Usually very tender	Variable	Uncommon
Lymphadenopathy	Firm, non-tender, bilateral	Firm, tender, often bilateral with initial episode	Tender, may suppurate, loculated, usually unilateral	Tender, may suppurate, loculated, usually unilateral	None; pseudobuboes

Table 71.22 Syndromic classification of sexually transmitted diseases (STDs) caused by various pathogens

Syndrome/disease	Microbial aetiology
Painful genital ulcers	
Herpes genitalis	HSV types 1 and 2
Chancroid	*Haemophilus ducreyi*
Painless genital ulcers	
Syphilis	*Treponema pallidum*
Donovanosis	*Calymmatobacterium granulomatis*
Lymphogranuloma venereum	*Chlamydia trachomatis*
Urethritis	
Gonorrhoea	*Neisseria gonorrhoea*
Non-gonococcal urethritis	*Chlamydia trachomatis, Ureaplasma urealyticum, Mycoplasma genitalium, Mycoplasma hominis, Trichomonas vaginalis*
Vaginal discharge	
Vaginitis	*Gardnerella vaginalis, Mobiluncus* spp.
Vulvovaginitis	*Candida albicans*
Trichomoniasis	*Trichomonas vaginalis*
Gonorrhoea	*Neisseria gonorrhoea*
Genital and anal warts	Human papilloma viruses (30 genital types)
Systemic illness	
Hepatitis	Hepatitis B and C viruses
Acquired immunodeficiency syndrome	Human immunodeficiency virus types 1 and 2
Miscellaneous	Scabies, *Pthirus pubis*, Molluscum contagiosum virus *Shigella* spp., *Campylobacter* spp., *Entamoeba histolytica, Giardia* spp.

C. trachomatis, *Mycoplasma genitalium*, *U. urealyticum*, *Trichomonas vaginalis*, and HSV; prostatitis or epididymitis by *C. trachomatis*, *N. gonorrhoeae*; and proctitis by *N. gonorrhoeae, C. trachomatis* and *T. pallidum*.

Women

The infections include cystitis and urethritis caused by *C. trachomatis*, *N. gonorrhoeae*, HSV; vulvitis by *C. albicans*, HSV; vaginitis by *C. albicans*, *T. vaginalis*; cervicitis by *C. trachomatis*, *N. gonorrhoeae*, *M. genitalium*; and bacterial vaginosis by *Gardnerella vaginalis*, *Mobiluncus* spp., and *Mycoplasma hominis*. The pathogens causing STDs are listed in Table 71.22.

Laboratory Diagnosis

Specimens

Vaginal, cervical and urethral discharge can be directly collected in a sterile container.

Swabs are used if the infected area is thick or scanty. Discharge from the base and edge of the infected area is collected with a sterile swab or endocervical curette. Fluid from genital vesicles can be collected using sterile capillary tubes. Blood or serum may be collected in suspected cases of gonococcal bacteraemia.

Immediate transport and analysis need to be carried out as the bacteria may not be viable for longer. Modified Stuart medium is a widely used transport medium for *N. gonorrhoeae*.

Microscopy

Motile trophozoites of *T. vaginalis* in case of trichomoniasis and pseudohyphae of *C. albicans* and yeast cells in case of candidiasis are demonstrated by microscopy of wet mount preparation of vaginal discharge.

Bacterial vaginosis is diagnosed by examining the Gram-stained smear of vaginal discharge. This is frequently helpful for the demonstration of (1) intracellular Gram-negative diplococci

N. gonorrhoeae, along with the presence of other polymorphs in cervical and urethral discharge and rarely in vaginal discharge of women suffering from gonorrhoea; (2) clue cells, the vaginal epithelial cells with bacteria attached to it causing a stippled appearance in the diagnosis of infection caused by *G. vaginalis*; (3) *H. ducreyi*, a Gram-negative coccobacilli in pairs in chancroid, and (4) Gram-positive budding yeast cells in vaginal secretions in the infections caused by *C. albicans*. Giemsa-stained smears are useful to demonstrate the presence of intracytoplasmic cysts organisms (Donovan bodies). These bodies are demonstrated in mononuclear cells filled with deeply staining bodies, in granuloma inguinale caused by *Klebsiella granulomatis*. Tzanc smears may detect the presence of multi-nucleated epithelial cells in the scrapings from Herpes ulcers.

Dark ground microscopy is used to demonstrate spirochetes in specimen suspected cases of primary or secondary syphilis. A direct immunofluorescence test using monoclonal antibody is employed to demonstrate elementary bodies of *C. trachomatis* in endocervical swabs. Direct immunofluorescence is also employed to identify Treponemal antigens and Herpes simplex virus in scrapings from ulcers in vagina, penis or urethra.

Culture

C. trachomatis is cultured in tissue cultures, and the yolk sac of 7 days old chick embryos. Herpesviruses are also isolated in tissue culture and on the chorioallantoic membrane of chick embryos. *N. gonorrhoeae* can be isolated by enriched media such as Thayer-Martin. Chocolate agar is used for isolating other bacteria. Mycoplasma organisms can be isolated in PPLO agar. *T. vaginalis* is cultured in Trichosel broth.

Serology

N. gonorrhoeae antibodies in the serum are demonstrated by the CFT and IFA. Chlamydial and Herpes viral antibodies can be detected by CFT, ELISA and IFA. Mycoplasma antibodies are detected by agglutination test, CFT and IFA.

Standard tests for Syphilis (STS) such as VDRL/RPR and Treponemal such as TPI, TPHA, FTA and FTA-ABS are used for serodiagnosis of syphilis. TPI/FTA-ABS is highly specific and considered gold standard for Syphilis. FTA-ABS is employed for demonstration of serum IgM and IgG antibodies to confirm neonatal syphilis in neonates.

Molecular Detection

Commercial multiplex PCR kits are available for the diagnosis of major sexually transmitted infections (STI-7). The clinical specimen include endocervical swabs from women and urethral swabs from men.

Other Tests

Frei's Test

This is a **skin test** using heat-killed bacterial antigen from the aspirated bubo pus from cases of LGV. This test is frequently used to diagnose lymphogranuloma venereum.

Whiff Test

The vaginal fluid of women with BV often has a foul odour. This is caused by decarboxylation of certain amino acids such as lysine or arginine resulting in the production of the foul-smelling amine compounds. A simple Point of Care test is available for diagnosing this condition in the outpatient clinics, known as the Whiff test. The test is performed by adding a few drops of potassium hydroxide to the vaginal secretions on a glass slide or small test tube. These amines immediately become volatile and produce the typical fishy odour.

Treatment

Treatment of the condition depends upon the causative agent. Acyclovir is recommended for genital Herpes, imidazole or metronidazole for trichomoniasis, tetracycline or macrolides and 3rd generation cephalosporins for *Chlamydia* infections, tetracycline or macrolides and 3rd generation cephalosporins for *Mycoplasma* infection, long-acting penicillin for syphilis; and spectinomycin, ceftriaxone, etc. for gonorrhoea.

Skin and Soft Tissue Infections

The skin is the largest organ covering the human body, composed of three layers: epidermis, dermis and hypodermis. The layers have varying but significant anatomical and physiological functions. The skin provides a large barrier against invading pathogens, chemicals and mechanical injuries. It also regulates temperature of the body.

Skin and soft tissue infections (SSTIs) are associated with varying presentations, and severity. It ranges from mild pyoderma to life-threatening necrotising fasciitis. The presence of inflammation in the skin characterised by redness, swelling, heat and pain indicates SSTIs. The affected area can also become functionless. Co-existing or pre-existing diabetes and HIV also increases the severity of illness. These easily transforms a milder variant of skin infection into an advancing illness leading to mortality status. The SSTIs, therefore, present clinically diverse challenges requiring different management and intervention either medical or surgical.

There is an estimated incidence rate of 24.6/1000 person-years. Most SSTIs tend to resolve within 7–10 days, but the prevalence rate may vary from place to place and remains to be the third common cause next to chest pain and asthma. It is more prevalent among males (60–70% of all cases) and individuals 45 to 64 years of age. Almost three-fourth of these patients can be treated in the outpatient setting.

Types

SSTIs can occur as primary or secondary infections. Most of the primary skin infections like impetigo, folliculitis, boils, etc., are caused by a single pathogen, including *S. aureus*, Beta hemolytic streptococci, coryneform bacteria and others. These bacteria enter through a break in the skin, such as injury, minor abrasions, or insect bites. Many systemic infections such as measles, varicella, staphylococcal scalded skin syndrome and gonococcal infection manifest as skin infections. Secondary infection of the skin also occurs in already diseased skin the skin infected earlier.

Various classification systems have been used for SSTIs. One such classification is based on anatomical location and presence and absence of tissue necrosis.

1. Superficial infection: Impetigo, erysipelas and cellulitis.
2. Deep non-necrotising infection: Pyomyositis.
3. Deep necrotising infection: Necrotising fasciitis, and gas gangrene.
4. Miscellaneous infections: These include
   (a) infections associated with bites and animal contact.
   (b) surgical site infections, and
   (c) infections due to the immunocompromised status of the patient.

Varicella and measles involve the epithelial layer, while dermatophyte infections involve the keratin layer. The epidermis is affected in impetigo caused mostly by streptococcii and staphylococci. Dermis is affected in erysipelas, caused by *S. pyogenes*. Hair follicle infection leads to folliculitis, boils, and carbuncles.

The sebaceous gland may be infected by *Propionibacterium acnes*, and subcutaneous fat lesions leading to cellulitis is precipitated by beta-hemolytic streptococci. Necrotising fasciitis caused by *S. pyogenes* and anaerobes is a life-threatening condition with systemic manifestations.

In 2003, an expert panel headed by Eron classified SSTIs based on the severity and patient's presentation. Eron's classification is as follows:

Class I: SSTIs but no signs/symptoms of systemic toxicity or comorbidities. The patient is healthy and afebrile.

Class II: Either systemically unwell /well with comorbidity that may complicate or delay resolution. The patient is febrile and appears sick.

Class III: Toxic and unwell (fever, tachycardia, hypotension). The patient has at least one unstable comorbidity or a limb-threatening infection.

Class IV: Sepsis syndrome with a life-threatening infection.

Laboratory Diagnosis

Specimens
Pus swab or pus aspirate (in syringe/anaerobic transport medium) collected from abscess, cellulitis ulcer, lymphadenitis, myositis or other related conditions are useful. The surface exudate is cleaned with sterile saline or 70% alcohol for ulcers. It is desirable to spirate by needle from deep tissue. In case, needle aspiration is not feasible, swabs (two) are used to collect the specimen from base of the ulcer. For closed abscess, the surface is disinfected with 70% alcohol and contents are aspirated with a sterile syringe and needle and aseptically transferred to an anaerobic transport medium and transported to the laboratory within a few hours.

Microscopy
Staphylococci, streptococci, diphtheroids, etc., are readily demonstrated by Gram stain. Anaerobic bacteria are identified by their typical morphology.

Culture
S. aureus, β hemolytic *Streptococcus* and Coryneform bacteria grow well on sheep blood agar. RCM broth is used to transport the clinical specimens and isolation of anaerobes. Varicella and measles are isolated in tissue cultures.

Molecular Detection
PCR is used for the molecular diagnosis of viruses such as varicella and measles.

Other Tests
Other tests, such as *Nagler's reaction*, is carried out for the demonstration of alpha-toxin (lecithinase) production by *C. welchii.*

Treatment

Penicillin is the drug of choice for the infection caused by beta-hemolytic Streptococci. Antibacterial susceptibility testing of the bacterial isolate is carried out to treat the infection with the appropriate drugs. Topical acyclovir is prescribed for secondary infection with varicella. There is no specific antiviral therapy for measles.

Hepatobiliary Infections

Hepatobiliary infection includes hepatitis, liver abscess and cholangitis. It is one of the important causes of morbidity among diseases of public health importance. Early identification of such infections helps in reducing short- and long-term complications. The systemic infections involving viral, bacterial, fungal and parasitic pathogens may show primary and secondary hepatobiliary involvement.

Definitions

Hepatitis is inflammation of the liver.
Cholangitis is inflammation of the bile duct.
Cholecystitis is inflammation of the gall bladder.

Clinical Syndromes

Hepatitis
It is a self-limiting condition. Nevertheless, it can progress to fibrosis, cirrhosis, or hepatocellular carcinoma. The pathogens causing hepatitis are listed in the Table 71.23. Liver abscess denotes the localisation of the abscess in the liver. The three major forms of liver abscess are caused by *E. coli, Klebsiella pneumoniae, Streptococcus milleri* group, etc. These infections are polymicrobial and are known to cause a pyogenic abscess. The amoebic liver abscess is the most common one caused by *E. histolytica.* The *candida* species is the common cause of the fungal abscess.

Cholangitis
Cholangitis, which is inflammation of bile duct is caused by *E. coli, Klebsiella, Enterobacter, Enterococcus* and anaerobes like *C. perfringens* and *Bacteroides*. It results in biliary tract obstruction and bile stone formation. The parasites such

Table 71.23 Causative agents of hepatitis

Infectious agents	Non-infectious agents
Virus—Hepatitis virus, yellow fever virus, cytomegalovirus, Epstein–Barr virus, herpes simplex virus, adenovirus, rubella virus, mumps virus and enterovirus **Parasite**—*Echinococcus granulosus, Trypanosoma cruzi, Leishmania, Entamoeba histolytica, Fasciola hepatica, Clonorchis sinensis, Leishmania donovani,* and *Plasmodium* spp. **Bacteria**—*Leptospira interrogans, Neisseria meningitidis, Neisseria gonorrhoeae, Bartonella henselae, Borrelia burgdorferi, Salmonella* spp, *Brucella* spp. and *Campylobacter* spp.	Alcohol Drugs e.g. isoniazid Toxins Autoimmune disease Non-alcoholic fatty liver disease Granulomatous hepatitis Liver abscesses

as *Opisthorchis, Clonorchis, Fasciola hepatica, Fasciola gigantica, Ascaris lumbricoides* and *Echinococcus granulosus* can causes fibrosis and cholangiectasis of bile duct. The hepatitis virus and HIV can also cause cholangitis.

Cholecystitis

Cholecystitis, gall bladder inflammation, can manifest as fever, abdominal pain, nausea and vomiting. The presence of stone in the gallbladder may result in blockage of bile flow aggravated by the infection caused by *E. coli, Klebsiella, Streptococcus, Clostridium* spp., Epstein–Barr virus and HIV.

Laboratory Diagnosis

The evaluation of Hepatobiliary infection involves multiple imaging modalities such as USG, CT, MRI and nuclear imaging. Serum liver enzymes are altered with elevated serum aspartate aminotransferases and alanine aminotransferases. Elevated serum bilirubin level is most commonly seen in acute infection.

Diagnosis of pyogenic abscess is made by microscopy and culture of ultrasound-guided pus aspirate. If an amoebic liver abscess is suspected, microscopy of the wet mount of aspirated pus is carried out for the presence of amoebic trophozoites. Molecular methods such as nested multiplex PCR can be done targeting the 18S rRNA gene of *E. histolytica*. Imaging modalities like USG, CT, MRI, Nuclear scan can be used. For fungal abscesses, Sabouraud's dextrose agar followed by KOH wet mount are used to isolate fungal pathogenes in cases of bile duct infection. Bile is examined by microscopy and culture for detection of a specific bacterial pathogen(s).

ELISA, chemiluminescence assays, immunochromatography tests, etc. are of value in the diagnosis of hepatitis. RT-PCR is frequently used to estimate viral load.

Wound Infections

Wound infections occur primarily because of a break in the skin layer as a result of the complications such as surgery, trauma, bite wounds and other infections that interrupt the mucosal or skin integrity. Old people and patients with diabetes, malignancy and other immune deficiency disorders are more at risk. The condition manifests as redness, warmth and tenderness at the wound infection site. Fever and Pus discharges are frequent. Frequent complications include the spread of infection through the bloodstream to other organs; and septic shock, which may require intensive care and support. Death of the surrounding tissue such as muscle, connective tissue, bone, etc. which may require debridement.

Post-operative wound infection or **Surgical site infections** are one of the most common health-care-associated infection and occurs in 3% of all surgical procedures. The source may be patients' own microbiota or from the environment. *S. aureus* is present in patients' own skin, *E. coli* or anaerobes present in the gastrointestinal mucosa, and Gram-negative bacilli such as *Pseudomonas* and *Acinetobacter* present in the environment are also other sources of infection.

Table 71.24 Pathogens causing wound infection

Organism
Staphylococcus aureus
Coagulase-negative staphylococci
Streptococcus pyogenes
Streptococcus anginosus group
Enterococci
Proteus, Morganella, and *Providencia* spp.
Escherichia coli
Pseudomonas spp.
Bacteroides
Fusobacterium spp.

Wound infection is associated with significant morbidity and mortality.

A bite wound infection results in inoculation of the organism from the saliva of animals and humans into the deeper tissues. The infection due to bite ranges from cellulitis to systemic infection such as bacteraemia or even encephalitis.

Burn injury is another example that results in massive destruction of the skin and its appendages. This can cause the entry of microorganisms from patients' own flora or from the hospital environment. The common organisms causing wound infections is listed in Table 71.24.

Laboratory diagnosis of the condition is made by the culture of debrided tissue collected from an infected wound in a sterile universal container. The tissue or pus sample is inoculated on the bedside into Robertson cooked meat medium and sent for culture for anaerobic bacteria.

Treatment of the condition depends on the nature of the wound, degree of wound infection and the organism causing infection.

Congenital Infections

Congenital infection occurs when vertical transmission occurs from mother to child during pregnancy, delivery or after delivery. The infection becomes severe if the mother gets infected during the first or second trimester of pregnancy. It gets transmitted through the placenta and causes developmental defects, stillbirths and neonatal deaths. Vaccination, prompt detection and treatment can reduce the number of adverse pregnancy outcomes. If there is any suspicion of maternal infection during pregnancy, laboratory testing has to be done for congenital TORCH infections, which include congenital **T**oxoplasmosis, **O**ther infections like syphilis, Varicella-zoster virus, Human Parvovirus, Zika virus, malaria, Trypanosomiasis, **R**ubella, **C**ytomegalovirus, and **H**erpes Simplex Virus.

Congenital Rubella Syndrome

Congenital Rubella syndrome (CRS) is caused by the Rubella virus, a togavirus of the genus *Rubivirus*. It is transmitted through droplets shed from the respiratory secretions of infected persons.

In the infected mothers, it runs a mild course with a low-grade fever (< 39 °C), headache, conjunctivitis and rhinitis. An important feature is post-auricular, occipital and posterior cervical adenopathy, which precedes a red, maculopapular rash by 5–10 days. The rash occurs, begins on the face and neck, and progresses to the lower parts of the body, lasting about 3 days. Arthralgia is also another common symptom of rubella infection.

In the infected infants, the placenta and the foetus are affected if primary rubella infection occurs during pregnancy. It causes a variety of malformations labelled CRS. The classic triad of CRS include hearing impairment, congenital heart defects—in particular, branch pulmonary artery stenosis and patent ductus arteriosus—and eye anomalies such as cataract(s), pigmentary retinopathy (salt and pepper type), chorioretinitis or congenital glaucoma. Additional clinical signs include blueberry muffin skin lesions, splenomegaly, microcephaly, developmental delay, meningoencephalitis, low birth weight, radiolucent bone disease and jaundice within 24 h after birth. The highest risk (90%) of CRS is during the periconception period and early pregnancy (8–10 weeks). If maternal rubella infection is diagnosed beyond 18 weeks of gestation, the foetus might be infected (Congenital rubella infection) but does not typically develop signs and symptoms of CRS.

Serological tests and nucleic acid amplification tests such as reverse transcriptase-polymerase chain reaction (RT-PCR) are the mainstays in diagnosing CRS. The ideal sample is a throat swab; however, nasal swabs, blood, urine or cerebrospinal fluid are acceptable.

In the infants, the CRS is confirmed if (1) Rubella IgM is present in the blood in infants <6 months old, (2) There is sustained elevation of IgG in blood; or is present on at least two occasions between 6 and 12 months of age (in the absence of having received rubella vaccine or being exposed to rubella); or (3) Rubella viral RNA is detected from a clinical sample by Nucleic acid amplification test (NAAT). The optimal sample is a throat swab; however, nasal swabs, blood, urine or cerebrospinal fluid are acceptable.

In pregnant women, CRS is confirmed if IgM for rubella is detected with no history of illness or contact with a rubella case or if NAAT is positive.

Congenital Toxoplasmosis

Toxoplasmosis is caused by infection with the coccidian parasite *T. gondii. The* major routes of transmission of *Toxoplasma* to humans are **oral** and **congenital**. Tachyzoites are the infective form. The incidence of congenital toxoplasmosis depends on the trimester during which mother is infected. The transmission rate is approximately 25% in the first trimester, 54% in the second trimester, and 65% in the third trimester, if untreated.

Clinical manifestation is subclinical in approximately 75% of infected newborns. The severity of the disease in infants is inversely proportional to the gestational age at the time of primary maternal infection—with first-trimester maternal infection leading to the most serious manifestations. It can cause spontaneous abortion, prematurity, or stillbirth. Involvement of the CNS is a hallmark of congenital *Toxoplasma* infection. The presence of chorioretinitis, intracranial calcifications, and hydrocephalus is considered the classic triad of congenital toxoplasmosis. Other occasional findings include hepatosplenomegaly, rash (maculopapular, petechial or both), myocarditis, pneumonitis, hearing defects, an erythroblastosis-like picture, thrombocytopenia, lymphocytosis, monocytosis, and nephrotic syndrome.

Laboratory diagnosis of toxoplasmosis usually is diagnosed based on antibody detection. In acute infection, serum IgG and IgM antibody levels generally rise within one to 2 weeks of infection. The most common laboratory method used worldwide for diagnosing CT in newborns and infants is serological detection of various isotypes of *Toxoplasma* antibodies in peripheral blood. Polymerase chain reaction (PCR) testing of amniotic fluid is performed in the foetus to diagnose congenital toxoplasmosis. The Sabin Fieldman dye test is the reference method for *Toxoplasma* IgG detection and remains the gold standard. Other more commonly used methods rely on enzyme-linked immunosorbent assay (ELISA) and ELISA-like assays, agglutination, and indirect immunofluorescence.

Congenital Cytomegalovirus (CMV) Infection

Congenital Cytomegalovirus (CMV) is transmitted when the virus crosses the placenta during pregnancy and causes infection in the foetus. The highest risk of transmitting the infection is seen among a woman who has primary infection during the first and second trimesters of pregnancy. The infection is most common in pregnant women who are immunocompromised. It causes enlargement of infected host cells and causes the cell-cell spread of the virus. They primarily infect kidney and salivary glands and undergo latency. The most common symptoms include petechiae, hepatosplenomegaly, jaundice, microcephaly. The long-term sequelae involve sensorineural hearing loss, cerebral palsy and visual impairment.

Laboratory diagnosis depends on screening antenatal mothers by CMV IgG serology to rule out the high risk of acquiring the infection during pregnancy. IgM CMV in pregnancy indicates congenital infection. IgG may be present in either past or recent infection and can be differentiated by IgG avidity test. Antigen detection can be done targeting pp65 antigen in peripheral blood leukocytes. Perinuclear cytoplasmic and intranuclear inclusions (Owl's eye appearance) is demonstrated in infected cases' urine.

Other Congenital Infections

These include neonatal Herpes, Varicella-Zoster virus infection, Parvovirus B-19 infection, Zika virus infection, congenital syphilis and malaria in pregnancy.

Bone and Joint Infections

Bone and joint infections are, also known as infections of the musculoskeletal system, can occur both in children and adults. They can be grouped as follows:

1. Osteomyelitis and diabetic foot osteomyelitis.
2. Joint infections that includes acute septic arthritis, prosthetic joint infections, and spondylodiscitis.

Whatever may be the site, the organisms commonly gain entry through the haematogenous route, although exogenous infection due to trauma or extension from an infection at an adjacent site can also introduce the pathogen. *S. aureus* is the most common pathogen involved in causing both bone and joint infections. Group A Streptococcus, aerobic Gram-negative rods bacteria such as *E. coli*, *Klebsiella*, and *N. gonorrhoeae* are other commonly encountered pathogens. *P. aeruginosa*, apart from *S. aureus*, may be involved in intravenous drug users. *S. aureus* also remains the most frequent pathogen in the paediatric population, in addition to other bacteria such as *Kingella kingae*, *H. influenzae*, *Salmonella*, and Group B Streptococcus. Patients commonly present with pain in the affected region, limitation of joint movement, limping, and fever.

Microbiological investigation centres around blood culture and analysis of joint aspiration material. Blood culture must be performed in all suspected cases despite the low yield (up to 50% positivity). Synovial fluid or bone samples should be examined microscopically and by culture. Cultures are often negative, but the sample should be subjected to molecular diagnostics (polymerase chain reaction to detect 16S rDNA). In addition to microbiological investigations, ESR, CRP, complete blood count, and imaging techniques are valuable in diagnosing bone and joint infections.

Eye Infections

Eye infections include conjunctivitis, keratitis or corneal ulcer, and endophthalmitis. Viral conjunctivitis is the most common eye infection caused most commonly by adenovirus and sometimes by Herpes simplex virus or Varicella Zoster virus. Bacterial conjunctivitis is frequently caused by *H. influenza*, *S. pneumoniae* and *S. aureus,* while gonococcal conjunctivitis can be part of sexually transmitted infections. Bacterial infection is the most common cause of keratitis involving *S. aureus* and coagulase-negative staphylococci, *S. pneumoniae* and *P. aeruginosa*. Less commonly, acanthamoeba or some fungi, including microsporidia, can also cause keratitis. The herpes simplex virus remains the main causative organism for keratitis among viruses. Endophthalmitis is caused by coagulase-negative staphylococci, *S. aureus*, streptococci and Gram-negative bacilli. Cataract surgery is the most common reason for developing this condition worldwide. Table 71.25 lists the comprehensive list of organisms causing different types of eye infections.

Conjunctival swab, corneal scrapings or samples from vitreous or aqueous humour are obtained, depending on the type of infection. Routine Gram stain and culture, supplemented with nucleic acid amplification test, if available, are commonly used to detect the causative organism.

Table 71.25 Pathogens causing eye infections

Clinical condition	Pathogens causing infection
Bacterial pathogens	
Conjunctivitis	*Staphylococcus aureus* *Streptococcus pneumoniae* *Streptococcus pyogenes* *Haemophilus* spp. *Moraxella* spp. *Pseudomonas aeruginosa* *Neisseria gonorrhoeae* *Chlamydia trachomatis*
Keratitis	*Staphylococcus aureus* *Pseudomonas* spp. *Enterobacteriaceae* *Neisseria* spp.
Uveitis	*Staphylococcus aureus* *Streptococcus pneumoniae* *Gonococcus* spp. Granulomatous uveitis – *Mycobacterium tuberculosis* *Mycobacterium leprae, Treponema pallidum*
Endophthalmitis	*Staphylococcus aureus* *Staphylococcus epidermidis* *Streptococcus* spp. *Enterococcus* spp. *Pseudomonas* spp. *Streptococcus pneumoniae*
Blepharitis, Hordeolum	*Staphylococcus aureus*
Acute dacryocystitis Chronic dacryocystitis	*Staphylococcus* spp, *streptococcus* spp. Tuberculosis, leprosy, syphilis
Orbital cellulitis	*Staphylococcus aureus* *Streptococcus pneumoniae* *Streptococcus pyogenes* *Haemophilus* spp.
Viral pathogens	
Keratoconjunctivitis	Herpes simplex virus Adenovirus Varicella zoster virus Cytomegalovirus
Acute haemorrhagic conjunctivitis	*Enterovirus 70* *Coxsackie virus*
Retinitis	Cytomegalovirus (in immunocompromised individuals)
Endophthalmitis	Herpes simplex virus *Varicella zoster virus (VZV)* *Cytomegalovirus* Measles viruses
Acute dacryocystitis	Mumps Infectious mononucleosis
Parasitic pathogens	
Keratitis	*Acanthamoeba* (in contact lens wearers)
Chagoma	*Trypanosoma cruzi*
Endophthalmitis	*Toxocara* spp.
Chorioretinitis	*Toxoplasma gondii*
Keratitis	*Onchocerca volvulus*
Chronic dacryocystitis	*Schistosoma* infection
Ocular cysticercosis	*Taenia solium*

(continued)

Table 71.25 (continued)

Clinical condition	Pathogens causing infection
Fungal pathogens	
Keratitis	*Aspergillus* spp. *Candida* spp. *Fusarium* (in contact lens wearers)
Endophthalmitis	*Candida albicans* *Aspergillus* spp. *Cryptococcus* spp. *Coccidioides* spp. *Sporothrix* spp. *Blastomyces* spp.

Ear Infections

Otitis externa is an infection of the external auditory canal commonly caused by *S. aureus, P. aeruginosa* and occasionally by *Aspergillus* or *Candida*. Severe otalgia and fever are the hallmarks of the disease.

Otitis media is an infection of the middle ear and is commonly seen in children between 6 and 12 months of age and declines by the age of 5. It commonly follows an acute upper respiratory tract infection, and it includes in its spectrum acute otitis media, chronic suppurative otitis media, and otitis media with effusion. The most common bacteria implicated in otitis media are *S. pneumoniae*, followed by non-typeable *H. influenzae* and *Moraxella catarrhalis.* Viral pathogens include respiratory syncytial virus, coronaviruses, influenza viruses, adenoviruses, human metapneumovirus, and picornaviruses. Diagnosis is made by physical examination and otoscopy. Laboratory diagnosis is seldom needed and may be done by tympanocentesis to collect the middle ear accumulated fluid for culture and staining purposes.

Quality Management System in Diagnostic Microbiology

72

One of the important aspects of a diagnostic laboratory is to ascertain whether the testing methods and the test results which are generated are trustworthy and reproducible. Towards this end, quality assurance is of utmost importance.

Quality Management System

The accuracy, reliability, and timeliness of the test reports are the three cornerstones of a medical diagnostic facility. The fundamental aim is to prevent or minimise laboratory errors. These errors can adversely affect effective patient management. It includes, among others, inappropriate or improper treatment, unnecessary adverse effects, complications arising out of treatment failure, diagnostic failure, escalation of expenditure and the need to do additional or repeat tastings. From a broader perspective, from the point of view of public health aspect, this can result in a delay in instituting proper public health initiatives, and in the times of outbreaks or epidemics, a breakdown of proper control measures and proper treatment guidelines due to the wrong choice of vaccines or antimicrobial agents.

Definitions

Quality

Quality means meeting the predetermined requirements of the users for a particular substance or service. It can be defined as the accuracy, reliability and timeliness of the test reports.

Quality Assurance

It can be defined as 'all the planned and systematic activities implemented within the quality system that can be demonstrated to provide confidence that a product or service will fulfil requirements for quality'. It is the sum total of all the measures undertaken to ascertain that the test results are not only accurate but also correct.

Quality Control

It is defined as the 'part of quality management focused on fulfilling quality requirements' or the 'operational techniques and activities used to fulfil requirements for quality'. In diagnostic microbiology, it includes those steps which are incorporated in every test run to confirm that the run has been proper and satisfactory.

Thus, quality assurance describes the performance of a procedure while quality control involves the inspection section of quality management.

Components of Quality Management System

All laboratory procedures can be subdivided into pre-analytical, analytical and post-analytical components. The laboratory workflow begins with the patient and ends with the generation of test report; and quality management procedure has to be implemented during the entire process. As per CDC, one such model which is widely used incorporates 12 essential activities which make up the backbone of quality management and includes:

1. Organisation and supervision.
2. Manpower.
3. Instruments.
4. Purchase and inventory.
5. Documentation and record keeping.
6. Process control.
7. Information management.
8. Occurrence management.
9. Analysis.
10. Customer service.
11. Process improvement.
12. Infrastructure and safety requirements.

Quality Assurance

Quality assurance (QA) encompasses carrying out all laboratory activities in a faultless and errorless manner so that the generated results are highly accurate. It ensures that a maximum number of errors is avoided and also pinpoints any error which may creep in. The term Good Laboratory Practice (GLP) is also used for QA. It takes into account all aspects of a diagnostic cycle: Pre-analytical, analytical and Post-analytical phases. Computers and information technology have tremendously helped in improving the quality of laboratories and most laboratories nowadays use some form of the information management system. The components of QA are listed in Table 72.1.

Quality assurance programs have multiple benefits which include the generation of quality products and reliable services, motivation for the staff to work better, building up of the reputation of the laboratory and prevention of legal suits and other problems.

Quality Control

Quality Control (QC) is principally concerned with the analytical phase of diagnostic cycle. This deals with the laboratory activities undertaken to make certain that the test carried out is accurate. Test errors are taken care of and test results are also verified. The sources of error are controlled. As per WHO, analytical quality control depends upon the following factors:

- Equipment reliability.
- Reagent stability, integrity and efficiency.
- Instrument calibration.
- Procedure reliability using standard operating manuals.
- Specificity, precision and accuracy of the test.
- Proficiency of personnel and continuous updating of their knowledge.
- Selection of the right technique of proven efficacy for which reliable reagents are available.
- Good internal quality control.
- Participation in external quality assessment programmes.

Table 72.1 Quality assurance or good laboratory practices

• Appropriate sample collection
• Proper specimen identification
• Immediate dispatch to the laboratory to maintain the correct temperature
• To collect and maintain the sample such that its quality remains unchanged
• Correct test procedure
• Review of the report before issue
• Prompt communication of the report to the correct destination
• Good customer or user relationship

The Standard Operating Procedure (SOP) manual is an integral part of any laboratory. It helps to improve and to maintain GLP and also deals with biosafety in the laboratory. The written manual helps the laboratory personnel to perform the tests to an acceptable standard and also deal with errors which may be encountered in day-to-day practice.

Assessment of Quality Systems

As per WHO, assessment is the 'inspection of some part or the whole of the quality management system to show that the laboratory is meeting all regulatory, accreditation and customer requirements'. Assessment of the laboratory performance is commonly done by quality assessment and audit. The International Organization for Standardization (ISO) uses the term 'audit' instead of 'assessment', and these two terms are synonymous. In quality assessment, the performance of a laboratory is assessed in either of the two ways: Comparison with some predetermined standards or by comparison with some other standard laboratory. It may be done internally by the staff members of the laboratory or it may be performed by some external agency able to accredit the laboratory performance.

Internal Quality Assessment/Control or Internal Audit

Here the laboratory inspects its own processes. It is easy to institute at the departmental level and can be performed frequently and corrective actions can be taken immediately. It is important that it prepares the laboratory for the external audit. Each laboratory should have a clear-cut policy about internal quality control and international- or national-level directives should be followed stringently.

External Quality Assessment/Control or External Audit

Here all or some aspects of the laboratory are assessed in an objective manner by an external agency. Three methods are available for external audit:

Proficiency Testing

Here the external agency sends a set of specimens to the laboratory and asks it to perform some specific tests with the specimens. The results are sent back to the agency and correct results are given a pass while wrong results are designated as fail. In India, such proficiency testing is being done by the Indian Association of Medical Microbiology and the Indian Academy of Tropical Parasitology.

Rechecking or Retesting

Here sets of isolates or slides are sent to a higher or specialised laboratory for confirmation or correct identification.

Site Visit

This is done for inspection, certification and for accreditation of the laboratory. It helps to obtain the ground level actual picture of the laboratory, measures the gaps or deficiencies in the laboratory and thus helps the laboratory to take the necessary corrective actions to remove the deficiencies. Normally, on-site visits are done by central or government agencies, and in India, it is performed by National Accreditation Board for Laboratories (NABL) and National Accreditation Board for Hospitals (NABH). The ISO remains the benchmark for all accreditation agencies.

Glossary

Acid-fastness Ability of bacteria to resist decolourisation by acids during staining procedures.

Adjuvant Substance which enhances the immune system's response to antigens, commonly used to improve effectiveness of vaccines.

Agar A complex gelatinous carbohydrate obtained from seaweed.

Amplifier host A host in which infectious agents multiply to high concentrations.

Anaerobiasis An environment that is free of oxygen.

Anoxomat Automated equipment which evacuates the air from jar and replaces by hydrogen gas from a cylinder.

Antimicrobial resistance Ability of microorganisms to resist the effects of drugs.

Antimicrobial stewardship Group of activities promoting appropriate use of antibiotics and reducing antimicrobial resistance.

Antibody A protein produced mainly by plasma cells that is used by the immune system to neutralise pathogens such as pathogenic bacteria and viruses.

Antigen A toxin or other foreign substance which induces an immune response in the body, especially the production of antibodies.

Antigenicity Refers to the ability to be specifically recognised by the antibodies generated as a result of the immune response to the given substance.

Autoimmunity Immune responses of an organism against its own healthy cells and tissues.

Autoinfection Reinfection by a pathogen that is already in the body.

Bacteraemia Presence of bacteria in the blood.

Bacteriocin Toxins produced by bacteria to inhibit the growth of similar or closely related bacterial strain.

Bacteriophage Type of virus that infects bacteria.

Bioterrorism Use of infectious agents or other harmful biological or biochemical substances as weapons of terrorism.

Biotyping The identification of different types of bacteria based on reaction to biochemical tests.

Capnophilic bacteria Bacteria requiring higher amounts of carbon dioxide for growth.

Carrier A person who harbours the microorganism without showing any signs or symptoms of the disease.

Cellulitis Inflammation of subcutaneous connective tissue.

Chemotherapy Chemicals used to treat disease.

Coagulase Enzymes produced by a bacterium which converts fibrinogen and fibrin.

Commensals They are microbes that live on a living host as normal flora of the host without causing any injury to the host.

Congenital infection Infections transmitted from mother to child during pregnancy (transplacentally) or delivery (peripartum).

Convalescence Gradual recovery of health and strength after illness or injury.

Culture media It is liquid or solid medium designed to support the growth of microorganisms.

Cytokines Large group of proteins, peptides or glycoproteins that are secreted by specific cells of immune system.

S. C. Parija, *Textbook of Microbiology and Immunology*, https://doi.org/10.1007/978-981-19-3315-8

Cytopathic effects Structural changes in host cells due to viral invasion.

Diarrhoea A condition in which faeces are discharged from the bowels frequently and in a liquid form.

Dysentery Presence of blood and mucus in the faeces.

Endemic Characterise diseases that are generally found in a particular area.

Endospores A resistant asexual spore that develops inside some bacteria cells.

Enterotoxin Toxin which acts on the intestines.

Erysipelas Bacterial infection which is characterised by large raised red patches on the skin.

Eukaryotes Organisms made up of cells that possess a membrane-bound nucleus that holds *genetic material* as well as membrane-bound organelles.

Facultative aerobes Anaerobes which can also grow aerobically.

Fastidious microorganism Organism that requires specific nutrients for its growth to be included in the culture medium.

Gastroenteritis Inflammation of the stomach and intestines.

Gene Segment of DNA that carries codons specifying for a particular polypeptide.

Graft A piece of living tissue that is transplanted surgically.

Ground itch Inflammatory reaction resulting from certain helminthic invasions into the skin.

Halophilic Organism that needs high salt concentrations for growth.

Hapten Small molecule which, when combined with a larger carrier such as a protein, can elicit the production of antibodies which bind specifically to it.

Haemolysis The rupture or destruction of red blood cells.

Hypersensitivity Exaggerated or inappropriate immune responses against an antigen or allergen.

Hyphae Branching filaments that make up the mycelium of a fungus.

Immunity Capability of multicellular organisms to resist harmful microorganism.

Immunogenicity Refers to the ability of a substance to induce cellular and humoral immune response.

Immunoprophylaxis Prevention of disease by the production of active or passive immunity.

Inclusion bodies Virus-specific intracellular globular masses which are produced during replication of virus in host cells.

Infection A process when an organism enters the body, multiplies and causes damage to the host.

Meningitis Inflammation of meninges of brain.

Mesophilic bacteria The bacteria grow within a temperature range of 25 °C–40 °C.

Microaerophilic bacteria Grows in the presence of low oxygen tension.

Microscope Instrument used to see objects which are too small to be seen by naked eye.

Mycelium Vegetative part of a fungus consisting of a mass of branching, thread-like hyphae.

Mycetism Poisoning due to mushrooms.

Mycetoma Chronic infection of the skin and the subcutaneous tissue.

Mycoses Disease caused by any fungus that invades the tissues, causing superficial, subcutaneous or systemic disease.

Mycotoxin Toxic compounds that are naturally produced by certain types of moulds.

Necrotising fasciitis An acute disease in which inflammation of the fasciae of muscles or other organs results in rapid destruction of overlying tissues.

Neurotoxins Poison which acts on the nervous system.

Nosocomial infections Infections that are contracted within a hospital environment.

Obligate aerobes Grow only in the presence of oxygen.

Obligate anaerobes Grow only in the absence of oxygen.

Opportunistic infections An infection caused by the microorganisms when the host immune system is depressed.

Pandemic An epidemic occurring worldwide, or over a very wide area, crossing international boundaries and usually affecting a large number of people.

Panencephalitis Encephalitis that affects both the grey and white matter of the brain, resulting in progressive loss of mental and motor functions.

Parasites Microbes that live on a living host, derive nutrition from the host and also causes harm to the host.

Pathogen The microbes capable of causing disease.

Pathogenesis The process of production and development of disease.

Peritonitis Inflammation of the peritoneum.

Pharyngitis Inflammation of the mucous membrane of the pharynx.

Plasmids Extra-chromosomal DNA substances.

Prions A tiny proteinaceous particle devoid of nucleic acid.

Prokaryotes Organisms made up of cells that lack a cell nucleus or any membrane-encased organelles.

Psychrophilic bacteria The bacteria that grow at temperatures below 20 °C.

Pyoderma A bacterial skin inflammation marked by pus-filled lesions.

Recombinant Produced by the combining of genetic material from more than one origin.

Recrudescence Recurrence of an undesirable condition.

Reservoir A place or a person where the infectious agent can live and multiply.

Saprophytes They are the free-living microbes that live on dead or decaying organic manner.

Sensitivity Ability of a test to correctly identify those with the disease.

Septicaemia Invasion of the bloodstream by virulent microorganisms along with their toxins.

Seroconversion Change from seronegative to seropositive.

Specificity Ability of the test to correctly identify those without the disease.

Superantigens Antigens that result in excessive activation of the immune system.

Suppurative Formation of pus.

Surveillance Ongoing collection, collation and analysis of data and the ongoing dissemination of information to those who need to know so that action can be taken.

Swarming Moving in or forming a large or dense group.

Taxonomy Branch of science which deals with classification of organisms.

Thermophilic bacteria The bacteria grow at a high temperature range of 55 °C–80 °C.

Transposons Jumping genes or a type of mobile DNA.

Urethritis Inflammation of the urethra.

Vaccine Substance that helps to prevent/protect against certain diseases.

Viraemia Presence of viruses in the blood.

Virulence Measure of a strain within species to cause the disease.

Yeast Single-celled fungi.

Zoonosis Disease which can be transmitted to humans from animals.

Further Reading

Practical Guidance

Carey, Roberta B. et al. Practical Guidance for Clinical Microbiology Laboratories: Implementing a Quality Management System in the Medical Microbiology Laboratory. *Clinical microbiology reviews* 2018; 31(3): e00062-17. doi:https://doi.org/10.1128/CMR.00062-17

Charlton, C. L., Babady, E., Ginocchio, C. C., Hatchette, T. F., Jerris, R. C., Li, Y., ... & Drews, S. J. (2019). Practical guidance for clinical microbiology laboratories: viruses causing acute respiratory tract infections. Clinical microbiology reviews, 32(1), e00042-18.

Doern GV, Carroll KC, Diekema DJ, Garey KW, Rupp ME, Weinstein MP, Sexton DJ. Practical Guidance for Clinical Microbiology Laboratories: A Comprehensive Update on the Problem of Blood Culture Contamination and a Discussion of Methods for Addressing the Problem. Clin Microbiol Rev. 2019;33(1): e00009-19. doi: https://doi.org/10.1128/CMR.00009-19.

Forbes BA, Hall GS, Miller MB, et al. Practical Guidance for Clinical Microbiology Laboratories: Mycobacteria. *Clin Microbiol Rev*. 2018; 31(2):e00038-17. Published 2018 Jan 31. doi: https://doi.org/10.1128/CMR.00038-17

Humphries RM, Linscott AJ. Practical Guidance for Clinical Microbiology Laboratories: Diagnosis of Bacterial Gastroenteritis. Clin Microbiol Rev. 2015;28(1): 3-31. doi: https://doi.org/10.1128/CMR.00073-14

Leal Jr, S. M., Rodino, K. G., Fowler, W. C., & Gilligan, P. H. (2021). Practical guidance for clinical microbiology laboratories: diagnosis of ocular infections. Clinical microbiology reviews, 34(3), e00070-19.

National and International Guidelines

Department of Biotechnology (2018). Guidelines for Recombinant DNA Research and Biocontainment

Indian Council of Medical Research (2015). ICMR Guidelines on diagnosis and management of Rickettsial diseases in India

Indian Council of Medical Research (2019). Standard Operating Procedures (Bacteriology): Antimicrobial Resistance Surveillance and Research Network

Indian Council of Medical Research (2019). General Guidelines for Establishment of Biosafety Level-3 Laboratories.

National Centre for Disease Control, Directorate General of Health Services, Ministry of Health and Family Welfare, Government of India (2020). National Guidelines for Infection Prevention and Control in Healthcare Facilities

The Food and Agriculture Organization of the United Nations, The World Organisation for Animal Health and The World Health Organization (2019). A Tripartite Guide to Addressing Zoonotic Diseases in Countries.

US Centers for Disease Control and Prevention (2008, updated 2017). Management of Multidrug-Resistant Organisms In Healthcare Settings.

US Centers for Disease Control and Prevention (updated 2019). Guideline for Disinfection and Sterilization in Healthcare Facilities.

World Health Organization (2002). Handbook for good clinical research practice (GCP): Guidance for implementation

World Health Organization. Regional Office for South-East Asia (2014). A brief guide to emerging infectious diseases and zoonoses. WHO Regional Office for South-East Asia.

World Health Organization (2018). Latent TB Infection: Updated and consolidated guidelines for programmatic management

World Health Organization (2022). WHO guideline on control and elimination of human schistosomiasis

World Health Organization (2020). Laboratory biosafety guidance related to coronavirus disease (COVID-19): interim guidance.

World Health Organization (2020). Laboratory biosafety manual, 4th ed.

World Health Organization (2020). WHO guidelines for the prevention of sexual transmission of Zika virus

World Health Organization (2021). Consolidated guidelines on HIV prevention, testing, treatment, service delivery and monitoring: recommendations for a public health approach

World Health Organization (2021). Guidelines for the management of symptomatic sexually transmitted infections

World Health Organization (2021). Guidelines for Preventive Chemotherapy for the Control of Taenia solium Taeniasis

Books

Carroll, K. C., Butel, J., & Morse, S. (2015). Jawetz melnick and adelbergs medical microbiology 27 E. McGraw-Hill Education.

Chaturvedi S, Parija SC (Editors). Immunizing agents for tropics: success, failure and some practical issues (BPKIHS Monograph Series I). p 25–34. BPKoirala Institute of Health Sciences, Dharan, Nepal. 1996.

Cornelissen, C. N., Harvey, R. A., & Fisher, B. D. (2012). Microbiology (Vol. 3). Lippincott Williams & Wilkins.

Cowan, M. K., & Talaro, K. P. (2009). *Microbiology: a systems approach*. McGraw-Hill Higher Education.

Dutta T.K. & Parija, S.C. (2013) Emerging & Re-Emerging Infectious diseases. Jaypee Brothers Medical Publishers (P) LTD, New Delhi. ISBN: 978-93-5025-960-3. 2013 .Pages: 212.

Gladwin M, Trattler B, Mahan CS. Clinical microbiology made ridiculously simple. MedMaster Inc., 2004, p. 400.

Hall G. S. (2013). Bailey & Scott's Diagnostic microbiology, 13th ed.

Levinson W. (2014). *Review of medical microbiology and immunology*. McGraw-Hill Education.

Murray P. R., Rosenthal, K. S., & Pfaller, M. A. (2020). Medical microbiology E-book. Elsevier Health Sciences.

National Academies of Sciences, Engineering, and Medicine. (2015). Improving diagnosis in health care. National Academies Press.

Parija S.C. (2019) Textbook of Practical Microbiology. Ahuja Publishing House, New Delhi. ISBN 978-93-80316-69-7, 2019. 2nd ed. Pages 316.

Procop, G. W., Church, D. L., Hall, G. S., & Janda, W. M. (2020). *Koneman's color atlas and textbook of diagnostic microbiology*. Jones & Bartlett Publishers.

Schaechter M., Engleberg, N. C., DiRita, V. J., & Dermody, T. S. (2012). Schaechter's mechanisms of microbial disease. Lippincott Williams & Wilkins.

Tang Y. W., Stratton, C. W., & Tang, Y. W. (2013). *Advanced techniques in diagnostic microbiology*. New York: Springer.

Scientific Publications

Arumugam P, Habeebullah S, Parija SC (2018). Knowledge and Awareness of Cervical Cancer, Human Papillomavirus (HPV), and HPV Vaccine among Screening Women: A Cross-Sectional Study from a Tertiary Care Hospital in South India. Int. J. Curr. Microbiol. App. Sci.;7(3):2431-4.

Bose AC ,Kate Vikram ,Ananthakrishnan N , Parija SC (2007).*Helicobacter pylori* eradication prevents recurrence after simple closure of perforated duodenal ulcer .Gastroenterol Hepatol. . 22; 345-348.

Brunham, R. C., & Plummer, F. A. (1990). A general model of sexually transmitted disease epidemiology and its implications for control. *The Medical Clinics of North America*, *74*(6), 1339-1352.

Christou, L. (2011). The global burden of bacterial and viral zoonotic infections. *Clinical Microbiology and Infection*, *17*(3), 326-330.

Connor, T. R., Loman, N. J., Thompson, S., Smith, A., Southgate, J., Poplawski, R., & Pallen, M. J. (2016). CLIMB (the Cloud Infrastructure for Microbial Bioinformatics): an online resource for the medical microbiology community. Microbial genomics, 2(9).

Dahiya S, Sharma P, Kumari B, Pandey S, Malik R, Manral N, Veeraraghavan B, Pragasam AK, Ray P, Gautam V, Sistla S, Parija SC, Walia K, Ohri V, Das BK, Sood S, Kapil A (2017). Characterisation of antimicrobial resistance in Salmonellae during 2014-2015 from four centres across India: An ICMR antimicrobial resistance surveillance network report. Indian J Med Microbiol.;35(1):61-68.

De Martel, C., & Franceschi, S. (2009). Infections and cancer: established associations and new hypotheses. *Critical reviews in oncology/hematology*, *70*(3), 183-194.

Despommier, D., Ellis, B. R., & Wilcox, B. A. (2006). The role of ecotones in emerging infectious diseases. *EcoHealth*, *3*(4), 281-289.

Jain V, Bhahini N, Sistla S, Mithun C, Parija SC, Negi V. (2017).Effect of disease modifying antirheumatic drug (DMARD) therapy on immune response to trivalent influenza vaccine in rheumatoid arthritis: APL15-0225.Intl J Rheuma Dis;18:464-470.

Khan S, Sistla S, Roopesh Kumar VR, Srinivas Acharya N, Parija SC.(2012) Recurrent shunt associated ventriculitis caused by methicillin resistant *Staphylococcus aureus*: Case with fatal outcome despite linezolid therapy. Annals Trop Med Pub Hlth, 5:373-375.

Kojima, K., Booth, C. M., Summermatter, K., Bennett, A., Heisz, M., Blacksell, S. D., & McKinney, M. (2018). Risk-based reboot for global lab biosafety. Science, 360(6386), 260-262.

Madhulika U, Harish BN, Parija SC (2004). Current pattern in antimicrobial susceptibility of Salmonella Typhi isolates in Pondicherry. Indian J Med Res.120: 111-114.

Mandal J., Sangeetha V., Nivedithadivya, Das A., Parija S.C. (2013) Characterization of extended-spectrum β-lactamaseproducing clinical isolates of Shigella flexneri. J Hlth Ppln Ntrition 31:405-408.

Maurer, F. P., Christner, M., Hentschke, M., & Rohde, H. (2017). Advances in rapid identification and susceptibility testing of bacteria in the clinical microbiology laboratory: implications for patient care and antimicrobial stewardship programs. Infectious disease reports, 9(1), 18-27.

Menezes GA, Khan MA, Harish BN, Parija SC, Gossens W, Vidyalakshmi K , Baliga S , Hays JP. (2010) Molecular characterization of antimicrobial resistance in non-typhoidal salmonellae associated with systemic manifestations from Pondicherry, Bangalore and Mangalore in India. J Med Microbiol 59:1477-1483.

Miller, R. R., Montoya, V., Gardy, J. L., Patrick, D. M., & Tang, P. (2013). Metagenomics for pathogen detection in public health. *Genome medicine*, *5*(9), 1-14.

Morens, D. M., Folkers, G. K., & Fauci, A. S. (2004). The challenge of emerging and re-emerging infectious diseases. *Nature*, *430*(6996), 242-249.

Padmanabhan, R., Mishra, A. K., Raoult, D., & Fournier, P. E. (2013). Genomics and metagenomics in medical microbiology. Journal of microbiological methods, 95(3), 415-424.

Rajkumar S, Sistla S, Manoharan M, Sugumar M, Nagasundaram N, Parija SC, RayP, Bakthavatchalam YD, Veeraraghavan B, Kapil A, Walia K, Ohri VC. (2017) Prevalence andgenetic mechanisms of antimicrobial resistance in Staphylococcus species: Amulticentre report of the indian council of medical research antimicrobialresistance surveillance network. Indian J Med Microbiol. 35(1):53-60.

Riley, L. W., & Blanton, R. E. (2018). Advances in molecular epidemiology of infectious diseases: definitions, approaches, and scope of the field. *Microbiology spectrum*, *6*(6), 6-6.

Sachs, J. L., Essenberg, C. J., & Turcotte, M. M. (2011). New paradigms for the evolution of beneficial infections. *Trends in ecology & evolution*, *26*(4), 202-209.

Singh G., & Raksha, A. U. (2013). Candidal infection: epidemiology, pathogenesis and recent advances for diagnosis. Bulletin of Pharmaceutical and Medical Sciences (BOPAMS), 1(1), 1-8.

Venkatesh K, & Parija SC.(2010) Electroimmunotransferblot assay for the detection of mycobacterialantigens in the cerebrospinal fluid for diagnosis of tubercular meningitis. Indian J Tuberc 57(3):141-7.

Walsh, T. J., Katragkou, A., Chen, T., Salvatore, C. M., & Roilides, E. (2019). Invasive candidiasis in infants and children: recent advances in epidemiology, diagnosis, and treatment. *Journal of Fungi*, *5*(1), 11.

Index

N

O

Q

R

S

T